T0122066

COMPUTING, CONTROL, INFORMATION AND EDUCATION ENGINEERING

PROCEEDINGS OF THE 2015 SECOND INTERNATIONAL CONFERENCE ON COMPUTER, INTELLIGENT AND EDUCATION TECHNOLOGY (CICET 2015), GUILIN, P.R. CHINA, APRIL 11-12 2015

Computing, Control, Information and Education Engineering

Editors

Hsiang-Chuan Liu
Asia University, Taiwan

Wen-Pei Sung
National Chin-Yi University of Technology, Taiwan

Wenli-Yao
Control Engineering and Information Science Research Association, Hong Kong

CRC Press is an imprint of the
Taylor & Francis Group, an **informa** business
A BALKEMA BOOK

CRC Press/Balkema is an imprint of the Taylor & Francis Group, an informa business

Typeset by diacriTech, Chennai, India

Published by: CRC Press/Balkema
P.O. Box 11320, 2301 EH Leiden, The Netherlands
e-mail: Pub.NL@taylorandfrancis.com
www.crcpress.com – www.taylorandfrancis.com
ISBN: 978-1-138-02800-5 (Hardback)
ISBN: 978-1-315-68589-2 (eBook PDF)

Computing, Control, Information and Education Engineering – Liu, Sung & Yao (eds)
© 2015 Taylor & Francis Group, London, ISBN: 978-1-138-02800-5

Table of contents

Computing, Control, Information and Education Engineering – Liu, Sung & Yao (eds)
© 2015 Taylor & Francis Group, London, ISBN: 978-1-138-02800-5

Preface

This book contains selected Computer, Intelligent Computing, Information and Education Engineering related papers from the 2015 International Conference on Computer, Intelligent Computing and Education Technology (CICET 2015) which was held April 11-12, 2015 in Hong Kong. The aim of the conference was to provide a platform for researchers, engineers and academics as well as industry professionals from all over the world to present their research results and development activities in Computer Science, Intelligent Computing, Information Technology and Education Engineering. This volume is divided into two sections. The first section covers Computer Science and Intelligent Computing. The second section presents the papers on Mechanical, Energy, Information and Education Engineering. I hope the proceedings will promote the development of Computer Science, Information Technology, Intelligent Computing and Education Technology, strengthening international academic cooperation and communications and the exchange of research ideas.

I am very grateful to the conference chairs, organizational staff, the authors and the members of the International Technological Committee for their hard work. We look forward to seeing all of you next year at CICET 2016.

February, 2015

Wen-Pei Sung
National Chin-Yi University of Technology

Computing, Control, Information and Education Engineering – Liu, Sung & Yao (eds)
© 2015 Taylor & Francis Group, London, ISBN: 978-1-138-02800-5

CICET 2015 Committee

Conference Chairmen

Prof. Hsiang-Chuan Liu, *Asia University, Taiwan*
Prof. Wen-Pei Sung, *National Chin-Yi University of Technology, Taiwan*
Prof. Ming-Hsiang Shih, *National Chi Nan University, Taiwan*

Program Committee

Yan Wang, *University of Nottingham, UK*
Darius Bacinskas, *Vilnius Gediminas Technical University, Lithuania*
Viranjay M. Srivastava, *Jaypee University of Information Technology, Solan, H.P. India*
Liu Yunan, *University of Michigan, USA*
Lin Chao, *Chongqing University, China*
Ming-Ju Wu, *Taichung Veterans General Hospital, Taiwan*
Wang Liying, *Institute of Water Conservancy and Hydroelectric Power, China*
Chenggui Zhao, *Yunnan University of Finance and Economics, China*
Li-Xin Guo, *Northeastern University, China*
Mostafa Shokshok, *National University of Malaysia, Malaysia*
Ramezan ali Mahdavinejad, *University of Tehran, Iran*
Wei Fu, *Chongqing University, China*
Anita Kovač Kralj, *University of Maribor, Slovenia*
Tjamme Wiegers, *Delft University of Technology, The Netherlands*
Shyr-Shen Yu, *National Chung Hsing University, Taiwan*
Yen-Chieh Ouyang, *National Chung Hsing University, Taiwan*
Shen-Chuan Tai, *National Cheng Kung University, Taiwan*
Jzau-Sheng Lin, *National Chin-Yi University of Technology, Taiwan*
Chi-Jen Huang, *Kun Shan University, Taiwan*
Yean-Der Kuan, *National Chin-Yi University of Technology, Taiwan*
Jiunn-Min Chang, *National Chin-Yi University of Technology, Taiwan*
Yu-Lieh Wu, *National Chin-Yi University of Technology, Taiwan*
Chi-Wun Lu, *National Chin-Yi University of Technology, Taiwan*
Homer C. Wu, *National Taichung University, Taiwan*
Hua-Zhi Hus, *National Kaohsiung Marine University, Taiwan*
Shih-Heng Tung, *National University of Kaohsiung, Taiwan*
Kuo-Tsang Huang, *National Chiayi University, Taiwan*
Chen-Yi Sun, *National Chengchi University, Taiwan*
Jun-Hong Lin, *Nanhua University, Taiwan*
Lei Wei, *National Chin-Yi University of Technology, Taiwan*

Co-sponsors

International Frontiers of Science and Technology Research Association
Control Engineering and Information Science Research Association

Computer Science and Intelligent Computing

Computing, Control, Information and Education Engineering – Liu, Sung & Yao (eds)
© 2015 Taylor & Francis Group, London, ISBN: 978-1-138-02800-5

Texture image-based 3D entity modeling method

Z.Q. Zhao
Science and Technology on Complex Land Systems Simulation Laboratory, Beijing, China
College of Mechatronics Engineering and Automation, National University of Defense Technology, Changsha Hunan, China

J.X. Hao
Science and Technology on Complex Land Systems Simulation Laboratory, Beijing, China

ABSTRACT: In the visual simulation, drawing modeling method and model transformation method are usually used in three-dimensional entity modeling. But there are such shortcomings as high requirements on material quality, heavy workload, disorder or loss in texture, etc. respectively existing in the two methods. The image-forming principle of photo imaging is the same as that formed by the axonometric projection imaging method used in the mechanical drawing, both belonging to central projection. Therefore, when conducting reverse design, photos or pictures may be regarded as axonometric projection, and the three-dimensional entity model can be earned by reversing the modeling of position relation among the characteristic faces abstracted from the different coordinate plane of real images based on a specific angle shooting. Then by applying texture photos taken as reference when modeling to the plane corresponding to the model, finally realizing the highly unity between texture and model. The texture image-based 3D entity modeling method does not only reduce workload of modeling, and can also achieve a good texture effect.

KEYWORDS: Visual simulation; Texture Image; Three-Dimensional Entity Modeling.

1 INTRODUCTION

A great deal of three-dimensional entity models needs to be created in three-dimensional visual simulation. But such problems as heavy workload, lack of corresponding data, poor overall visual effect of model, etc. challenge this task to be completed in a short time period, furthermore, affecting the progress of projects. Meanwhile, with the improvement of computer performances, especially its graphical features, higher requirements on the reality of the entity model in visual simulation are put forward. To reach the real time simulation effect, less effort made on the details of the three-dimensional entity model in simulation application may improve drawing speed and reduce "lag time" (refer to the time between the corresponding responses made by the user's input and application programs), while the model realism may be achieved by united adjustment on such elements as texture, materials, lighting, etc.

So far the commonly used entity modeling methods include drawing modeling method and model transformation method. In the former method, geometric models of several smaller components are firstly built according to CAD drawings, then assembled and modeling, and finally textures are pasted on each face of the model. This work takes a lot of time and efforts to find information, to do physical mapping, or to collect texture materials. For the latter, it has its own limitation. Although the three-dimensional model format conversion may be made in CAD models or animation software by some tools software, but in the course of transformation and rebuilding, the structural relationship of models disordered. Therefore, even the best CAD model needs a great deal of the arrangements, manual organization and structuring must be done on the polygon data nearly randomly placed so as to turn it into useful polygon groups. Otherwise, after format transformation of many three-dimensional models in animation software, texture disorder or loss may appear. So the model effectiveness established with a great deal of efforts is still not wholly satisfactory, and it has a long development period, unable to meet the requirements on heavy entity modeling works. Therefore, the texture image modeling based on special angle shooting, plus referring to the modeling of CAD techniques in portions is adopted and achieved a sound effect.

2 BASIC PRINCIPLE OF TEXTURE IMAGES-BASED THREE-DIMENSIONAL SOLID MODELING

Known from the analysis of the photo imaging and the axonometric projection imaging method used in the mechanical drawing, they have the same imaging principles, belonging to central imaging, therefore, when doing reverse design, photos and pictures can be regarded as axonometric projection images, as shown in Figures 1 and 2.

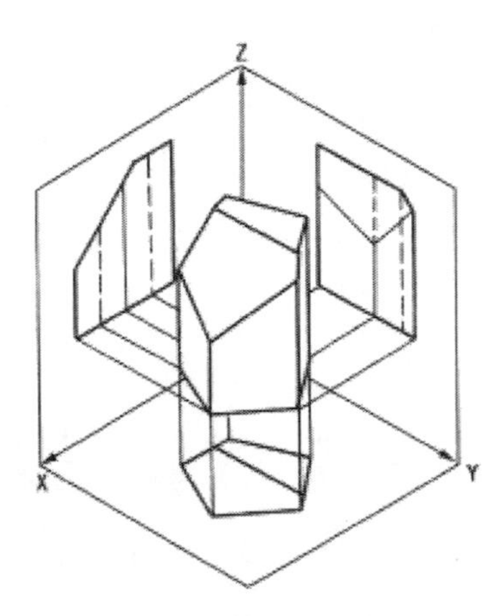

Figure 1. The Axonometric projection imaging method used in the mechanical drawing.

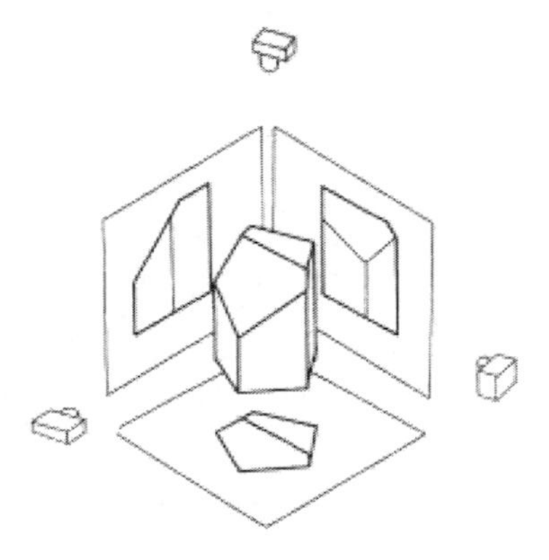

Figure 2. Texture photo shot at specific position.

Specifically, in the axonometric projection imaging method used in the mechanical drawing, objects are projected onto different coordinate planes to produce a vertical view, lateral view and front view, then the appearance characteristics of an object may be plotted based on these views. Similarly, on the specified shooting position (with camera placed in the position perpendicular to the center position of shot plane), the photos similar to the vertical view, lateral view and front view may be taken. After that, by abstracting the solid edges on the texture photos containing the geometric position information of the object and by reversing modeling on the position relation among the characteristic faces abstracted from the different coordinate planes, solid model can be created by reversing. And by applying the texture photos referred while modeling onto the corresponding face of model, eventually the high unity of texture and model can be reached.

3 TEXTURE IMAGE-BASED 3D SOLID MODELING METHOD

The flow of texture image-based 3D solid modeling work is shown as Figure 3.

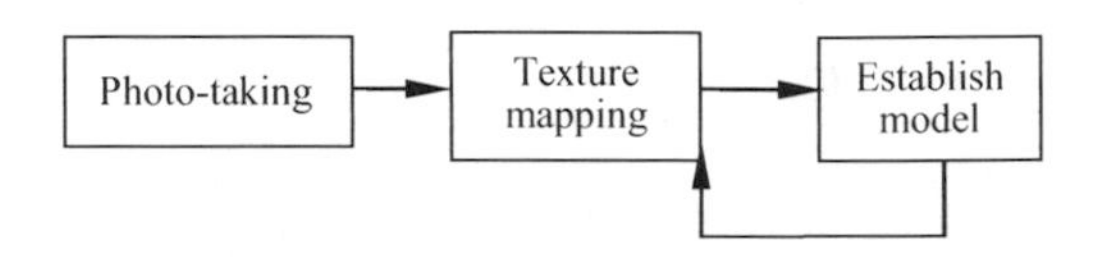

Figure 3. Flow of 3D solid modeling work.

3.1 *Photo-taking*

The shooting of texture photo is a very critical step in the three-dimensional solid modeling. The quality of texture photos will directly affect the reality of the future solid models and the accuracy of proportional relations among each big component, therefore, some special requirements should be observed while taking photos.

3.1.1 *Camera calibration*

Camera calibration is widely used in such fields as photogrammetry, pattern recognition, and computer vision and scene analysis. Whether in getting known of three-dimensional information about object space from images, or in getting known of two-dimensional image coordinates from space three-dimensional information, the space position and direction of camera in the reference coordinate, as well as the geometrical and optical parameters of the camera itself must be determined. These parameters shall be determined through tests and calculations. By completing the calibration of the camera, a certain relation determined between two-dimensional image and three-dimensional space is established accordingly. Based on the research of the texture mapping method of MultiGen Creator and multiple practices, for making the photos taken similar to the vertical view, lateral view and front view in mechanical drawing, when calibrating camera, the camera must be placed perpendicular to the surface to be shot, and for avoiding the forming of angle, the distance between camera and object to be shot should be a moderate one, too close or too far away will cause deformation and distortion.

3.1.2 *Points to note when taking photos*

When shooting photos of some bulky objects, segmentation shooting should be taken.

In order to truly reflect the textures and the detailed features of objects, too strong or too weak light, especially the flashlight should be avoided when shooting, because the strong color contrasts produced by flashlight may affect the appearance effect of models.

Some of the photos shot at any angles and the detailed portion photos should be taken, so as to be used as reciprocal reference in modeling, ensuring the details of the model will not lose shape.

The photos so the shot will not only contain many appearance information about the object, but will also include the information about the complex material of the object surface and the illumination effect, which may meet the demands for any simulation of three-dimensional solid modeling.

3.2 *Texture mapping*

The top priority for texture mapping lies in the strengthening of appearance geometrical information on related models in the texture photos, so as to facilitate the contour extraction when modeling in the future, meanwhile removing overall redundant portions by cutting which may interfere with modeling.

An important feature in the simulation of three-dimensional solid modeling lies in the expression of model details by texturing, which is relatively more realistic and trustful than the texture made by free-hand drawing. To highlight the texture, the sense of volume and a sense of space of models, refining, processing and grammaticalization should be conducted on some portions of texture under the principle of making the photo with different textures reaching a unified vision and strengthening its art effects.

The photographs can be input into the computer as digital images, and will be processed by graphic processing software PHOTOSHOP. The processing of photos mainly includes overall hue adjustment, position adjustment, making of Alpha channel, etc.

3.2.1 *Overall hue adjustment*

Figure 4. Comparison photos before and after color gradation adjustment.

Due to the impact of light, the color contrast between the photos taken from different positions remains a big one. If directly used as texture, the texture variance may occur to the adjacent textures, further affecting the overall effect of the model. The Auto level tool in Photoshop can be adopted in adjustment, making the textures unified. Individual adjustment may be conducted after selecting some portions. Sharpen treatment may be carried out on the detailed textures of some portions to make them clearer.

3.2.2 *Position adjustment*

In the imaging process of camera, slight deformation may occur, in order to overcome this impact imposed on modeling, horizontal and vertical reference lines may be generated to the proper locations of photos, then the horizontal and vertical positions of objects can be adjusted by rotary tools, and the partial correction may be done by utilizing the Distort tool or Perspective tool in the Free Transform. Similarly, proper adjustment should be done on the texture photos taken in segmentation while combining them together.

3.2.3 *Making an Alpha channel*

To demonstrate the effect of special materials such as glass, simply by making the corresponding parts of textures into a transparent (Alpha) channel, can the transparent, or translucent effect be produced. Tactfully using of Alpha channel may reduce the number of polygons of models to a great extent. The effect of antenna appeared in Figure 5 could only be realized by adding a great deal of polygons with the traditional modeling method, whereas, if using Alpha channel, less polygons may produce a transparent effect on the portions to be not shown. Figure 6 is a display of wireframe after closing texture. In addition, the rational using of transparent texture together with the multi-layer texture technique of MultiGen Creator may generate a very complex visual effect.

Figure 5. Effects comparison before texture mapping.

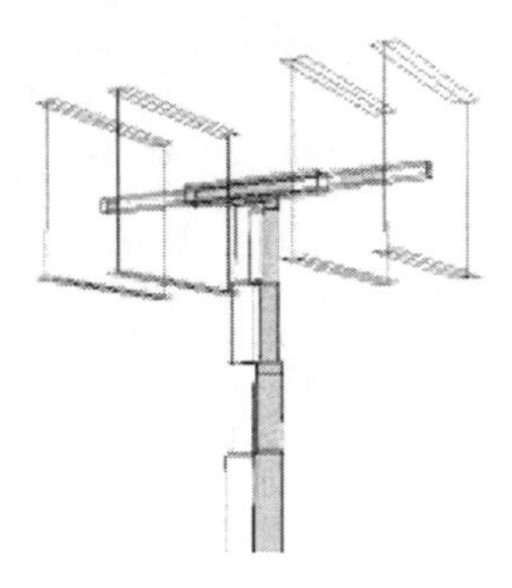

Figure 6. Effects comparison texture mapping.

3.3 *Establishing model*

In MultiGen Creator, a great deal of effective tools is integrated for establishing solid models, the number of layers may be decided by database environment, by which such nodes as the group, object, face, edge, vertex, etc. of the model can be controlled. By adjusting the relations of these nodes, the appearance characteristics of the model and the subordination relation of components may be changed.

3.3.1 *Establishing reference plane and characteristic face*

Enter MultiGen Creator, select appropriate coordinate plane to establish a reference plane with its size corresponding to the width-to-length ratio of texture photo, then map the texture photo onto reference plane. The (P1) in Figure 7 is a reference plane, by clicking each high point of the information about the relevant object position relation, the corresponding polygons may be established, which will become the characteristic face of the model.

Figure 7. Establishing reference plane.

3.3.2 *Geometric modeling*

When conducting the operations such as stretching, rotation and lofting on characteristic faces, the geometric relationship between this characteristic face and other characteristic faces should be put into consideration, the object photos and relevant detailed photos taken from any angles can be referred if necessary, and small solid models (object) are formed one by one by selecting operations on the corresponding coordinate plane. And these small entity models may be placed in respective groups by regrouping and optimization according to the positional relationship of models.

3.3.3 *Texture mapping*

In MultiGen Creator, there are several texture mapping methods: planar three-point mapping, planar four-point mapping, cube mapping, cylindrical mapping, spherical mapping, etc., among them the most frequently used one is planar three-point mapping, namely selecting three mapping points on the texture and mapping them onto the receiving objects, thus creating a corresponding relation between them.

When conducting texture mapping, select overall polygons established according to the same reference plane, and conduct mapping in accordance with the plane vertex positions, thus a perfect fit between mapping and model can be ensured. Refined adjustment may be conducted with the Modify UVs tools embedded in ModTexture. As shown in Figure 6, adjustment may be conducted by clicking the points need to be adjusted and be completed by using the tools on the left side of Figure 8.

Figure 8. Interface by modifying USs tools.

3.3.4 *Production materials*

There are millions of materials in nature. Different materials give us a different visual perception. For instance, placing a plastics product painted black together with a metal product painted black, human may easily distinguish them with their naked eye, just because they are made from different materials, so they have different reflections of light, and the differential absorption of ambient light. Comparing with the animation software such as Maya, 3dmax and others, the materials used in MultiGen Creator are very simple, easy to be adjusted. To improve the quality of entity models, different objects should be fitted with different materials, materials and textures should be combined for using, without any one being neglected.

3.3.5 *Making DOF and LOD*

For completing modeling, the following processes should be followed: making overall zooming on the model according to the actual size of entity object, determining the Degrees of Freedom (DOF) of the moving elements of the model according to concrete applications, manufacturing entity models with different Level of Details (LOD).

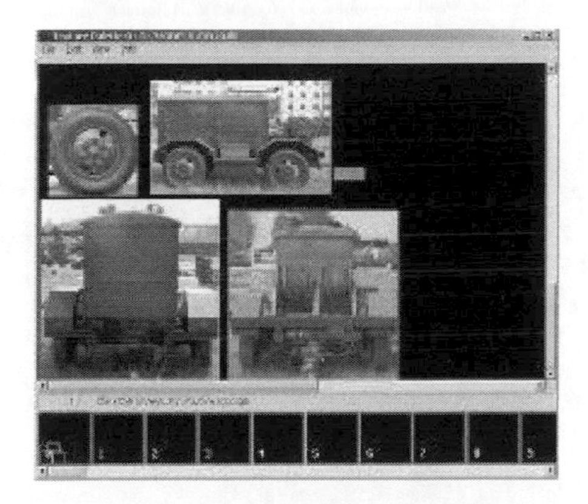

Figure 9. Overall textures used in modeling.

Figure 10. Entity model after texturing.

4 INTERACTION BETWEEN TEXTURES AND MODELS

In order to increase the detailed characteristics of entity models and ensuring it to get better visual effect, refining and adjusting on portions of the texture should sometimes be conducted. By returning to such image processing software such as Photoshop, etc. and calling in the textures to be changed and making corresponding changes, the corresponding changes may occur in the solid models in MultiGen Creator. Similarly, some small changes may be done on the surfaces textured in the models. In traditional modeling method, the texture will be adjusted to stiffly fit models, which one of the important reasons is causing the overall effect of the model so bad. The three-dimensional modeling method based on photo texture may avoid the occurrence of this case, meanwhile model and texture may be amended, model and texture make a cross-reference, and both achieving best results.

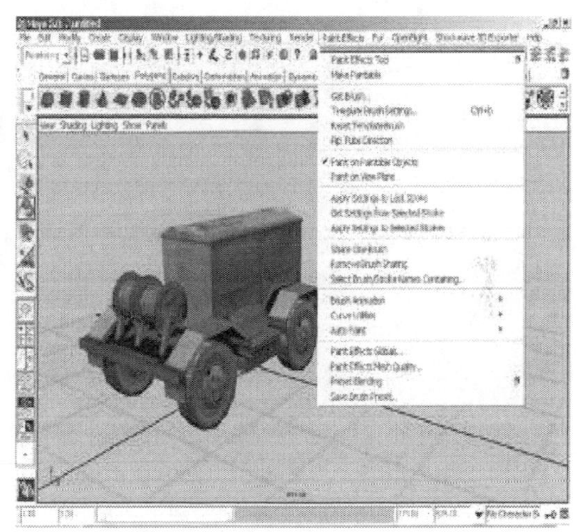

Figure 11. Interfaces of PaintEffects in Maya.

There is another solution, by which, the "*.flt" files built in MultiGen Creator may be called in 3D animation software, Maya, where the Rendering module of Maya may provide PaintEffects tool which may directly paint texture on model's surfaces. There are several brushes for option in PaintEffects, with their sizes, transparency, etc. being able to adjust. After finishing painting and saving, the corresponding texture files may be changed accordingly. The interface of PaintEffects is shown as Figure 8. In Maya 5 or above version, the models may be output in the format of open flight.

5 CONCLUSION

This paper has offered a three-dimensional solid modeling method based on texture images. The three-dimensional solid model can be created by reversing the modeling according to the position relation between the characteristic faces abstracted from the different coordinate plane of real images based on a specific angle shooting, and then the texture photos taken as reference when modeling will be applied the to the plane corresponding to the model, finally the higher unity between texture and model are achieved. The method does not only reduce workload of modeling, but also achieves a good texture effect, having better promotion and practical values.

REFERENCES

M.Z, Yang. & J.Y, Shi. 2000. A method for reconstructing three-dimensional model on the basis of photo's textures. *Software Journal* 11(4): 502–506.

X.W, Zhang. & S.Y, Chen. & H.W, Xiong. 2003. Today and tomorrow of solid modeling method based on photos. *Journal of Mechanical Engineering* 39(11):23–27.

Xiang, Hou. 2006. *Study on three-dimensional modeling method in the virtual reality technology*. Chongqing China: Chongqing University.

L.L, Wang. & Rong, Liu. 2001. Geometric three-dimensional reconstruction method based on image. *Journal of System Complexity* 13(Suppl):77–81.

Computing, Control, Information and Education Engineering – Liu, Sung & Yao (eds)
 ISBN: 978-1-138-02800-5

Design and implementation of web-based research supervisors information management system for postgraduates' education

Ping Cheng, Xue Ping Hao & Chun Xiao
Automation School, Wuhan University of Technology, China

ABSTRACT: With the rapid development of network technology and the popularity of computer applications, using computer technology to manage supervisors' information is imperative. This paper describes a supervisors' information management system using ASP.NET and SQL technology based on WEB. The system can be used to manage supervisors' basic personal information, postgraduate training program and scientific research information. Administrators also can review and assess supervising performances online with user-friendly interface.

KEYWORDS: supervisor; postgraduate; information management system; database; ASP.NET; SQL.

1 INTRODUCTION

Currently, the postgraduate recruitment scale becomes gradually larger and larger, the reforming of postgraduate education develops continuously, the traditional master - apprentice model has been unable to meet the needs of cultivation of postgraduates. As for the management of supervisor team, selection of the new supervisor, assessment of supervising performances and discipline construction have large amount of relevant and crossing information, the workload of the data processing grows exponentially and it makes the data maintenance difficult[1]. Hand or stand-alone processing approach can no longer meet the needs of modern supervisor team managing. For example, in the process of selection of the new supervisors and review of supervising performances, the circulation of paper documents is often used in the past, and the data inconsistencies caused by wrong operation are unable to provide accurate and efficient support for the supervisors' information management [2]. In addition, there are many limitations with the services of supervisor management: (1) It can not be refined to every supervisor's data, and it can not achieve hierarchical management; (2) It can not offer scientific, rational and special services to the varieties of roles of users; (3) the information of supervisors can not update timely, and so postgraduate candidates can not get the latest information [3].

The large-scale data processing services in the management of supervisors team needs a corresponding and efficient processing mechanism. How to realize the identification, the standardization and the informatization of the supervisors' information management has become an important task of education administration [4]. With the development of information technology and computer networks, information resources greatly enriched, the increasing promotion of relevant technologies has facilitated the integration, the sharing and the quick update of information, which provides the technical support and the objective conditions to build the supervisors' information management platform[5]. Meanwhile, the construction of the supervisors' information management system can realize to transmit and share all kinds of information and data online, which can avoid duplication of effort and reduce unnecessary workload. Not only the system can improve the efficiency and level, but also it can make the work more detailed and coherent, which provides good technical reserves for the construction of high-level disciplines. So the supervisors' information management system has a very important practical significance.

2 DESIGN OF OVERALL PROJECT

The overall project design follows up the scientific, standardized and automated design principle. On the basis of the achievement of information management of supervisors' team, making the best of use of this information to take on the statistics, comparison and analysis of supervisors' information in longitudinal, transverse, cross-level and interdisciplinary. The regular selection of supervisors, the reviewing, the real-time tracking and the leaders' decision-making are based on the system.

The design of the system functions has the following characteristics:

1 Web-based architecture and design. Supervisors' basic personal information, academic achievements,

research cases, admissions professional, research and other information are compiled into the database to provide conveniences of classified statistics, information reporting and other managements. Supervisors can update the related information via the Internet information management systems to reflect the latest scientific research projects in time, and let the postgraduate have an accurate understanding of the supervisors' research program and their research interests. The two-way exchange of information makes the selection process of between supervisor and postgraduate more compatible. This design can expand the scope of applications and it is convenient for the management department to do the supervisors' inquiry, the data maintenance and the management whenever and wherever possible.

2 Online review and assessment. Preliminary review of eligibility declaration about the information submitted by new supervisors and selection of the new supervisors can be performed. The managing operations such as the review of the instructors qualification, the assessment of supervisors' academic achievement, the research project, the project funding, the quality of students training, and the execution of supervisors' duties can be conducted. This online review and assessment not only provides users with convenience but also improve the management efficiency and ensure the timeliness, reliability and accuracy of data transfer.

3 Management with classified authorization. The data operation authorizations in deferent levels are configurable and customizable. User licensing mode can be authorized gradational by the initial system administrator account. According to the above features, the frame diagram and the system function module structure of the supervisor information management system are shown in Figure 1 and Figure 2.

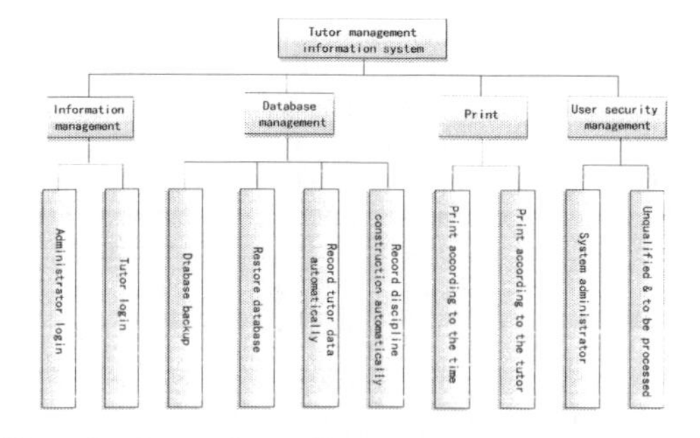

Figure 1. System frame diagram.

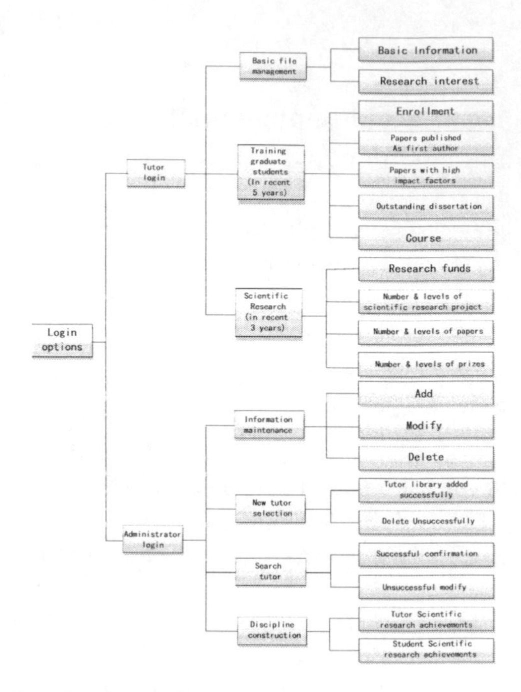

Figure 2. System function module structure.

3 WEB-BASED SOFTWARE DESIGN AND IMPLEMENTATION

3.1 *Framework of system software*

Supervisor information management system is designed in web-based models. The system framework and database management adopt SQL Server, and the connection of dynamic web pages and the database employs the ASP technology [6].

Software entrance including two parts of the supervisor log in and administrators log in. Supervisors or administrators can user name and password to log in the system. Supervisors can view their information, and the administrator can operate on viewing, adding, modifying and deleting supervisors' information. System will do the login authentication based on the user name and password inputting by users, and then make prompt. The interface is shown in Figure 3.

Login authentication parts of the program are as follows:

```
if (this.ddlstatus.SelectedValue == "supervisor")
{
    if
(BaseClass.CheckTeacher(txtNum.Text.Trim(),
txtPwd.Text.Trim()))
    {
```

Figure 3. System login screen.

```
        Session["ID"] = txtNum.Text.Trim();
        Response.Redirect("Daoshi/Info.aspx");
    }
    else
    {
        Response.Write("<script>alert(' You are
not a supervisor or the user name and the password
is wrong ');location='Login.aspx'</script> ");
        txtNum.Focus();
    }
```

Figure 4. Administrator interface.

3.2 *Database design*

Database management includes four parts of the database backup, the database restore, the supervisors and disciplines information automatic data entry. Administrators have the functions of finding, adding, editing, and deleting the supervisors' information. The administrator interface is shown in Fig.4.

Visual Studio's connection with SQLexpress can be achieved with the following fields [7]:

```
public static SqlConnection DBCon()
    {
        return new
SqlConnection("server=.;database=db_Daoshi;
uid=sa;pwd=");
    }
```

GridView control's bingding can be achieved with the following fields:

```
public static void BindDG(GridView dg,string
id,string strSql,string Tname)
    {
        SqlConnection conn = DBCon ();
        SqlDataAdapter sda =new
SqlDataAdapter(strSql,conn);
        DataSet ds=new DataSet();
        sda.Fill(ds,Tname);
        dg.DataSource = ds.Tables[Tname];
        dg.DataKeyNames=new string []{id};
        dg.DataBind();
    }
```

The data can be bound with paging if there are many query data items. The specific method is to add the following fields in the GridView's PageIndexChanging event [8]:

```
    gvTeacherInfo.PageIndex = e.NewPageIndex;
    strsql = "select * from tb_Teacher where
TchNum like '%" + txtKey.Text.Trim() + "%"';
    BaseClass.BindDG(gvTeacherInfo, "ID", str-
sql, "tchinfo");
```

Execute SQL statements can be achieved with the following fields:

```
public static void OperateData(string strsql)
    {
    SqlConnection conn =DBCon ();
    conn .Open ();
    SqlCommand cmd=new SqlCommand(strsql ,
conn );
    cmd .ExecuteNonQuery ();
    conn .Close() ;
    }
```

3.3 *Assistant functions*

Printing of supervisors' information is in a fixed format and users can enter the print query interface by inputting the name or the number of the supervisors. All detailed data records according to customized time period can be browsed and printed in the read-only modes.

To ensure the data security and system operation stability, the system determines user permissions based on user identity which allows the user can use particular features. There are a system administrator (staff in charge of full-time management in school and department) and ordinary users (supervisors and postgraduates). The user authorizations of

Table 1. Authorization of system users.

user	Authorization		
	Information services	Information modification	print
System administrator	√	√	√
Ordinary users	√		√

the supervisor information management system are shown in Table 1.

4 CONCLUSION

In the current environment of information technology, a supervisor information management system based on analysis of the need of postgraduate education activities are designed and implemented. The design framework, some key technologies and ideas of realizing functional modules in the supervisor information management system are presented.

The operation of this system contributes to the promotion of related systems and data acquisition. In the planning and construction process of supervisor management system, it strengthens the data exchange and data sharing with external systems. Supervisors' personal information, qualification information, research projects and the research papers publication information are sharing with the data from school's personnel management system and research management system and so the authority of accumulated supervisors' data and the consistency of each updated data are guaranteed.

Supervisor information management system comprehensively shows supervisors' academic achievements and scientific developments via data integration, and efficient integration of their personal information. On the one hand, the system plays a good role in promoting the management, the inspection and the evaluation of supervisor's roles. On the other hand, the system ensures the accuracy of the information exchange and the reliability and timeliness of the management data processing. The system can improve the work efficiency, which is important to the process of the modernization in the management of postgraduate education. Meanwhile, it also provides an assistant tool for the leadership to make the decision in the education reform.

REFERENCES

[1] Ling Guizhen. Analysis of the Informationization of the Educational Management [J] Computer Knowledge and Technology. 2005, (24): 85–86.

[2] Shou Xiaoping. College Senate NMS [J] Shaanxi Normal University .2005, 34(7): 342–344.

[3] Liu Kainan. Construction of Educational Management Information [J] China Higher Education Research. 2005, (2): 84–85.

[4] Cheng Ping, Chen Jing. The Design and Implementation of Postgraduate Training Funds Management System [J]. Wuhan University of Technology, 2007, 29(1): 87–90.

[5] Qian Yong. The Design and Implementation of ASP. NET Technology-based Educational Management System [J] Computer and IT .2009, 17(4): 69–72.

[6] Du Meiping. The Design and Implementation of .NET-based Educational Administration System [J] . Office Automation, 2009, (12): 53–56.20.

[7] Tang Chaoli, Huang Yourui. Design and Implementation of Web-based Teacher Information Management System [J] .2006 China Science and Technology Information, (8): 267, 270–271.

[8] Mei Hong, Chen Feng, Feng Yaodong. The Software Development Method Based on Architecture and Component-oriented [J]. Journal of Software 2003, 14(4):721–732.

Computing, Control, Information and Education Engineering – Liu, Sung & Yao (eds)
© 2015 Taylor & Francis Group, London, ISBN: 978-1-138-02800-5

The query optimization algorithm based on distributed heterogeneous database

Peng Wang & Hong Yu Li
College of Computer Science and Technology, Changchun University of Science and Technology, Changchun, China

ABSTRACT: Genetic algorithm has good optimizing results in the global query optimization of distributed heterogeneous database. However, this algorithm may have a shortcoming of convergence prematurely, which cannot achieve an optimal query efficiency. This paper puts forward a genetic simulated annealing algorithm towards the shortcoming, by using the local convergence ability of the simulated annealing algorithm to improve global convergence performance of genetic algorithms for ensuring to obtain a global optimal solution. Simulation experiments are designed to verify the improved algorithm has a better optimizing efficiency.

KEYWORDS: Distributed Heterogeneous Database; Query optimization; Genetic simulated annealing algorithm.

1 INTRODUCTION

The distributed heterogeneous database system is united in logical, but separate in physical. Due to the heterogeneous characteristics of distributed heterogeneous database system, the query process becomes more complex, reducing the query efficiency of the database in a large degree. Some scholars introduced the optimization algorithm to reduce the amount of data transmission for solving the problem of inefficiency. Comparing with the optimizing strategies that based on graph theory, dynamic programming, heuristic query [1] and other traditional optimization strategies, genetic algorithm has a better optimization efficiency in query optimization problem of distributed heterogeneous database. But its query optimization is not perfect, there exists the problems of premature convergence and low local search capability. For solving the problem, this paper has improved genetic algorithm, composed a hybrid optimization algorithm by combining the local search of genetic algorithm and the global search simulated annealing algorithm. At last, the designed simulation experiments prove that the improved algorithm can improve the query efficiency of distributed heterogeneous database.

2 THE QUERY OPTIMIZATION MODEL OF HETEROGENEOUS DATABASE

The query processes of heterogeneous database are divided into four steps: grammar analysis, conversion decomposition, global optimization and results merge [2]. Global optimization is the major step of the whole query processing, the query trees may be obtained which is optimized after the operations of the first two steps. Generally speaking, a query statement can have several different query trees, then do an assessment of the execution cost plan in which these query trees correspond, and the execution engine will execute the execution plan of the least cost. A good query execution plan needs to select a suitable replica site to find the optimal execution sequence of database and select the best execution way of join operation.

The cost value is a parameter which chooses the optimal query plan. The query cost includes I/O cost, CPU cost and communication cost[3]. The total cost of distributed heterogeneous database is shown in formula (1):

$$C_{total} = C_{I/O} + C_{CPU} + C_{TC} \tag{1}$$

The communication cost among different sites is shown in formula (2):

$$C_{TC}(X) = C_0 + C_1 * X \tag{2}$$

In formula (2), $C_{TC}(X)$ means the value of the communication cost; C_0 means the time, which the initialization of communication takes between the two sites, it is a constant which the communication system decides, the unit is s; C_1 means the size of data transfer rate, the unit is s/bit; X means the amount of data transmission, the unit is a bit.

In different network environments, the mode which treats the communication cost is different, the communication cost is often higher than the I/O cost and the CPU cost in remote communication network or the system in which data transmission rate is low. But their gap is not too large in the LAN or the system which data transmission rate is high. Typically, the communication cost is the cost of joint operations and merge operations between different sites [4].

3 APPLICATION OF GENETIC ANNEALING ALGORITHM

3.1 *Design ideas of genetic annealing algorithm*

Genetic algorithms GA is an evolutionary algorithm based on the law of natural selection and survival of the fittest in biological evolution theory, that can be well used in the query optimization problem of distribution heterogeneous database [5-6]. The algorithm gets local optimal solution easily because of its poor local search ability. For solving the above problem, this paper introduces the simulated annealing operation to achieve quadratic optimization in the processes of genetic algorithm. The simulated annealing algorithm has a strong local search ability which improves the overall efficiency of the optimization algorithm [7].

The genetic annealing algorithm has the following advantages:

1 Simulated annealing reduces the selection pressure of genetic algorithm in a large scale, it enhances the global convergence of the overall algorithm, accelerates the convergence speed of late optimization by its strong climbing ability, and improves optimal efficiency of the algorithm;
2 Genetic algorithm expands the search area of the simulated annealing algorithm in the solution space, accelerates the search speed and improves the local convergence ability of the simulated annealing algorithm;
3 Due to genetic algorithm has the strongest global search ability which can guide the direction of optimization to improve the search process; simulated annealing algorithm has a good local search ability that can solve the problem of local search to prevent premature convergence.

The basic design concept of the genetic simulated annealing algorithm: genetic algorithm and simulated annealing algorithm have different division of labor in hybrid algorithm. First of all, searching the global optimal solution in the initial population; then generating a new generation of population by executing selection, crossover and variation of genetic operations; at last, the individual new generation of population executes simulated annealing operation for further optimizing individual genes, and the results will be saved the next generation population. The execution process is iterative, the algorithm is completed, when the termination condition is satisfied.

3.2 *Query optimization*

In the optimization process, assuming the data is distributed uniformly and obtaining the information of relational tables between the site and the other site by accessing global data dictionary and dynamic data dictionary. This operation is helpful for the optimal selection of genetic annealing algorithms.

The algorithm is designed as follows:

1 Encoding
This paper uses a tree structure encoding to execute the encoding work in according to the query characteristics of heterogeneous. Generating the query trees by encoding, then searching all nodes in the query tree from top to bottom, non-leaf nodes of the tree mean the root node and the join relation of tables, with a 0 instead. The ID of the table means Serial number of leaf nodes, which is stored in the data dictionary. For example, the chromosome which includes five join relations is (0,0,0,1,2,3,0,4,5). Leaf nodes must attach a value for more clearly indicating the information of the resources which leaf nodes correspond, with a v[i] instead(1<=i<=n), i is Serial number of leaf nodes. Its structure mode: TableNo (number of table), TableName (the name of table), Computer_No (number of computers where table stores), In-Field (the number of fields in table), FieldsNo (number of field), Count_bond (the number of contact fields), InResult (the number of result fields), OperaterType (join type of tables). The value which leaf node j correspond is V [j] {TableNo ... OperaterType}, these information are stored in the dynamic data dictionary.
2 Fitness function
The fitness function employed in this paper is based on the total cost of distributed heterogeneous database. Formula of fitness function is as follows:

$$f(i) = e^{-C_{total}(i)} \quad (3)$$

Above formula shows, f(i) means fitness value, T(i) means the total cost, the total cost is more, fitness is less, the total cost is minimal when fitness is almost 1.
3 Selection strategy
In the study of the genetic algorithm, genetic algorithm converges to the optimal solution when the probability of selection strategy is 1.

Selection probability is Pi, fitness is fi, then formula is as follows:

$$p_i = \frac{f_i}{\sum_{k=1}^{n} f_k}, \quad i = 1,2,3...n \quad (4)$$

4 Crossover operator

The crossover operation is the process in which individual genes of two parent chromosomes exchanged for producing new chromosome. This paper uses the method of crossover which is based on the method of sample hybridization; first, selecting two individual chromosomes randomly; second, obtaining two new individual chromosomes by exchanging the sub-tree gene of the chromosomes whose the number of nodes is same, At this time, two new chromosomes will exist the same genes, they will be replaced by the genes which do not appear; last, producing two new chromosomes correctly. Cross-process is shown in figure 1 and 2:

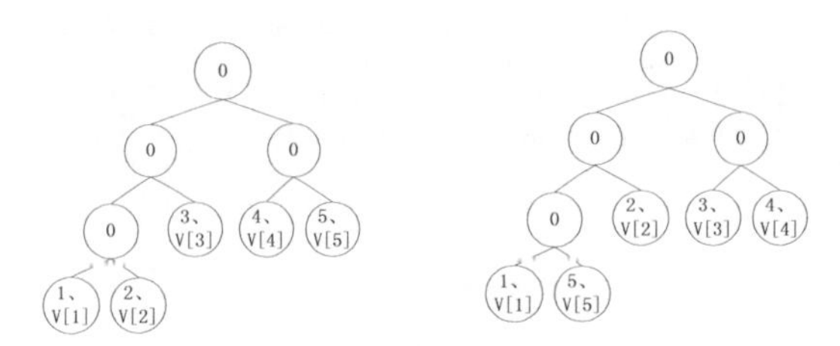

Figure 1. Two coding trees before the crossover.

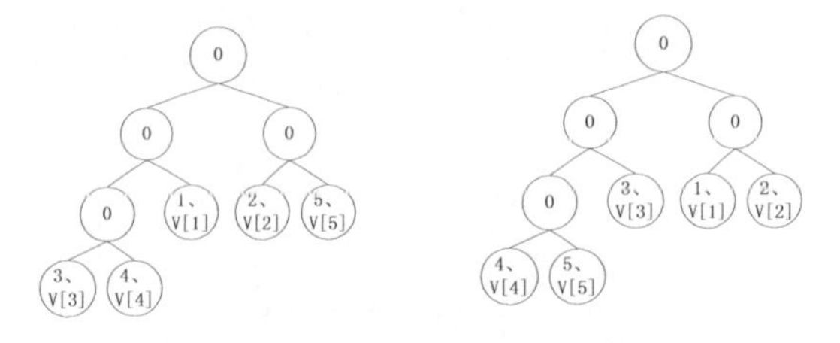

Figure 2. Two coding trees after the crossover.

Exchanging (1,V[1]; 0; 2,V[2]) and (3,V[3]; 0; 4,V[4]) in figure 1, two new coding trees has duplicate genes which include (3,V[3]),(4,V[4]),(1,V[1]) and (2,V[2]), they will be replaced by the genes which do not appear, then obtaining coding trees in figure 2.

5 Mutation operator

The mutation operation is an important technology, which protects the population diversity of chromosomes. The genes will execute small probability of a gene mutation for jumping from the trap of the local convergence when the populations have local convergence phenomenon. In according to the above encoding, the mutation operator includes two aspects that are the mutation of non-leaf gene locus and leaf gene locus. Leaf gene locus exchanges two leaf nodes in a chromosome simply; but non-leaf gene locus is more complex: at first, generating a number of nodes randomly; then finding a sub-tree gene in the chromosome which has the same number of nodes; finally, selecting a gene which is different with the gene in the sub-tree and executing the mutation operation.

6 Simulated annealing operation

By means of genetic manipulation to gain a global search, a new population is obtained as a population of the simulated annealing algorithm after selection, crossover and mutation operations finish, then executing simulated annealing operation, It selects a new generation of individuals based on Metropolis rule. The details are as follows:

1 To retain the best individual to the next generation;
2 To generate new individual j in the area of chromosome i randomly, then individual i and j competed for going into the next generation.

The formula is shown as, if, so x_j goes into the next generation instead of x_i, otherwise x_j must satisfy the condition of $\exp(-\triangle \cos t(x)/T(n)) > random[0,1]$, so it would be accepted by the next generation. $\cos t(x)$ is the energy function of chromosomes, $T(n)$ is the n-th generation temperature parameters.

7 Termination condition

If the algorithm achieves the maximum number of iterations, the algorithm will stop.

The flowchart of the genetic simulated annealing algorithm is shown in Figure 3:

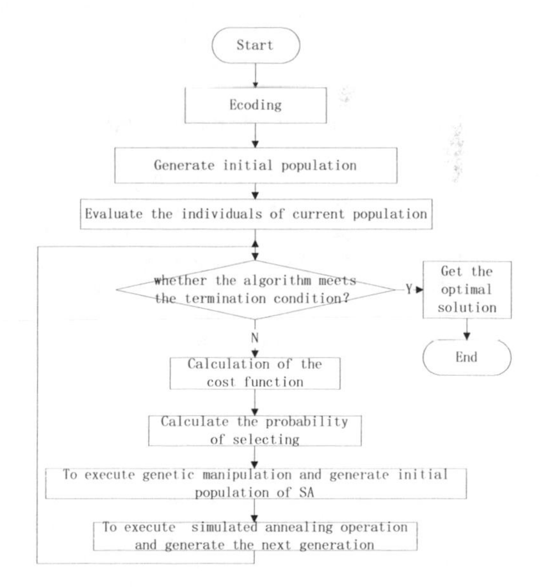

Figure 3. The flow chart of genetic simulated annealing algorithm.

4 EXPERIMENTS AND RESULTS

For complete simulation experiments in a heterogeneous database system for testing execution plans of genetic simulated annealing algorithm and genetic algorithm in the different number of associated tables. In according to the literature[8] which analyses the control parameters of genetic algorithm, the control parameters are set as follows: population size is 20, number of iterations is 30, the mutation probability is 0.06, crossover probability is 0.95. Taking the average query time, often times in the different relational tables. The results are shown in Table 1 and Table 2:

Table 1. The results of genetic algorithm.

Name	Number of relational tables	Response time (s)	Generation time(s)
Genetic algorithm	5	2.12	0.41
	7	2.84	0.61
	9	3.92	0.75
	11	5.75	1.03

Table 2. The results of genetic annealing algorithm.

Name	Number of relational tables	Response time (s)	Generation time (s)
Genetic annealing algorithm	5	1.83	0.59
	7	2.01	0.73
	9	2.51	1.10
	11	3.27	1.52

We can conclude that the query response time of the genetic annealing algorithm is less than genetic algorithm and the plan generation time is longer by comparing experimental results in table1 and table 2. But the genetic simulated annealing algorithm is usually able to generate a better query execution plan to remedy the time, which the plan generated wastes, the number of relational tables is more, its query efficiency is higher. Therefore, the experiment proved that this research achieves the desired achievement in an objective view.

5 CONCLUSION

Through analyzing the query optimization insufficient of genetic algorithm in distributed heterogeneous database, this paper employes genetic annealing algorithm to optimize the query operation of distributed heterogeneous database. On the basis of solving the local convergence problem of genetic algorithm, query time of the database is reduced, query efficiency is improved and a better application value is gained.

REFERENCES

[1] Legato P, Paletta G, Palopoli L. Optimization of join strategies in distributed databases[J]. Information Systems, 1991, 16(4): 363–374.

[2] Yuanyuan F, Xifeng M. Distributed database system query optimization algorithm research[C]//Computer Science and Information Technology (ICCSIT), 2010 3rd IEEE International Conference on. IEEE, 2010, 8: 657–660.

[3] Haraty R A, Fany R C. Query acceleration in distributed database systems[J]. Revista Colombiana de Computación, 2001, 2(1): 19–34.

[4] Bin Liu, Elke A. Rundensteiner: Optimizing Cyclic Join View Maintenance over Distributed Data Sources[J]. IEEE Trans.Knowl.Data Eng, 2006, 18 (2):363–376.

[5] Horng J T, Yeh C C. Applying genetic algorithms to query optimization in document retrieval[J]. Information processing & management, 2000, 36(5): 737–759.

[6] Cao D Y, Cheng J X. A genetic algorithm based on modified selection operator and crossover operator[J]. Computer Technology and Development, 2010, 20(2): 44–47.

[7] Bouleimen K, Lecocq H. A new efficient simulated annealing algorithm for the resource-constrained project scheduling problem and its multiple mode version[J]. European Journal of Operational Research, 2003, 149(2): 268–281.

[8] Elhaddad Y R. Combined Simulated Annealing and Genetic Algorithm to Solve Optimization Problems[J]. World Academy of Science, Engineering and Technology, 2012, 68.

Computing, Control, Information and Education Engineering – Liu, Sung & Yao (eds)
© 2015 Taylor & Francis Group, London, ISBN: 978-1-138-02800-5

Design and implementation of mobile learning system based on cloud computing

G.X. Wang & F.T. Jiang
Information Technology Center of Jiujiang University, Jiujiang, Jiangxi, China

ABSTRACT: Combination of cloud computing and mobile learning provides a new way for the development of education and training industry. On the basis of cloud computing and mobile learning support based on the design of a complete cloud-based mobile learning system , and using LASJ architecture implementation, to the system of education and training in the field of mobile learning help.

KEYWORDS: Cloud Computing; Virtualization Technology; Mobile Learning; Wireless Network Technology.

1 INTRODUCTION

With the development of information technology and wireless communication technology, mobile learning is receiving attention that as a branch of digital learning, and has become the new hot spot in the field of educational technology and related research. Mobile learning as a new form of learning has immeasurable potential applications in the traditional areas of education and training. In recent years, large-scale heterogeneous distributed computing systems and resource sharing technology development, as well as on the development of the popular intelligent terminal and related application systems intelligent terminal Android system is increasingly popular, mobile learning more convenient. However, so far, both theoretical research and practical applications in cloud computing based mobile learning are still in the early stages of development.

2 CLOUD COMPUTING AND MOBILE LEARNING

2.1 *Cloud computing*

Cloud computing is a style of computing in which dynamically scalable and often virtualized resources are provided as a service over the Internet. It includes the distributed computing, parallel computing, utility computing, network storage, virtualization, load balancing, hot standby redundancy and other technology, is a fusion of traditional computer and network technology development. Learner -based learning system in the human brain cloud features in the learning process of the law as a reference, set up situations, motivational, training ability, to cultivate an interest, thereby improving achievement and quality. Cloud computing realized from the application to interpersonal, from the computer to the user, anywhere access to data from the isolated, random shared data transformation, the future will be a cloud computing driven by collaborative computing world. In China, cloud computing applications in the field of education has been highly valued and affirmed. Whether cloud network school of higher education, or primary and secondary cloud computer-assisted instruction, training and IT enterprise cloud- based remote centers, cloud computing applications are preliminary, and showing a good momentum of development. Cloud computing has had a profound impact of human learning is not limited to technical dimension, a new application model will encourage people to rethink the approach to learning, learning resources and learning environment design, thereby improving the human learning environment.

2.2 *Mobile learning*

Mobile learning is the use of wireless mobile communication network technology, wireless mobile communication devices (such as mobile phones, personal digital assistants PDA, Pocket PC, etc.) to obtain educational information, teaching resources and educational services in the form of a new type of digital learning. Currently, mobile learning research focused on the part of foreign developed countries in Europe and North America. There are two research purposes: one is from the current E-Learning provider initiated, and more for corporate training; another sponsored by educational institutions, to improve teaching, learning and management. China's mobile learning research started late, low level of research , a smaller scale , mainly in the planning carried out under the Ministry of Education, focusing on education and research campus LAN and mobile short message based build. For example, the Ministry of Education, "Mobile Education" project. The project uses China Mobile's GPRS platform for teachers and students to provide messaging services, while allowing teachers and students to enjoy more preferential mobile phone business, to establish a " mobile education" service station system for the participating users "mobile education " program offers a variety of services. Compared with the traditional mode of education, "Mobile Education" has made great progress, but does not implement mobile learning in the true sense.There is system instability and poor maneuverability, poor curriculum learning resources, learning mode in real time, flexibility, interactivity poor, and many other problems.

2.3 *Cloud computing support for mobile learning*

Cloud computing is an Internet- based supercomputing model, which bring together a wealth of information stored in the processor resources and personal computers, mobile phones and other devices, providing work together. It distributed computing tasks in the resource pool formed by a large number of computers, making all kinds of applications can get the computing power needed storage space and a variety of software services. The advantages of cloud computing with its powerful and amazing speed development that is affecting the current development of the information industry. With its development and popularization , educational institutions, educators, learners' information will gradually migrate to the "cloud", which is the development of mobile learning is undoubtedly a good opportunity. Cloud computing support for mobile learning mainly in four areas: First, based on the current short message service SMS, WAP browser way to restrict access and interactive learning resources are scarce, real shortcoming relatively poor. Second, the integration of the wealth of learning resources to achieve integration, storage and sharing of educational information , and further maximizes optimization and use of educational resources. Third, the system is able to support the creation of a diversified autonomous virtual learning community, there is conducive to learning and multicultural exchange and cooperation, and integration. Finally, the system reduces the need for mobile learning equipment. Learners only need to use personal smart terminal devices via a browser or application APP access to cloud servers, they can easily achieve self- learning and mobile learning, reducing the requirements for terminal equipment.

3 DESIGN CLOUD COMPUTING BASED MOBILE LEARNING SYSTEM

3.1 *Function mobile learning system design*

Mobile learning and independent study are the same as the emphasis on learner-centered. Learner-centered, personalized learning environment cloud-based service is one of the development trends of cloud computing era. Therefore, under the cloud computing support to build a mobile learning system is an urgent demand for the cloud era of personalized learning, this system has a full-featured, easy to use, real-time, flexible and strong interactive features, and self- learning equipment adapts. In the mobile learning system design, we need to consider, such as learning outcomes, learning resources, learning objects, learning cost, learning time and other factors. Through research and analysis, we conclude that mobile learning system based on cloud computing should have the following features: hierarchical management and provide software support, provides dynamic storage space, provide dynamic data services, provide a wide range of resources and provide a more flexible service interaction.

Provides dynamic storage space. Learners will be able to own all the data stored in the cloud, storage readily accessible at any time, have the capacity of storage space with the number of users and dynamic allocation of resources, unlimited expansion. Because of the cloud of virtual storage pool for data backup and disaster recovery protection mechanism, the user need not worry about the loss of the stored resources. Users do not need to purchase or carry a mass storage device, just to have an intelligent equipment and the ability to access the cloud server, mobile learning can be achieved.

Provide dynamic data services. Cloud server resources through mining and finishing, forming the structure of the data, which can provide large-scale data sharing, management, mining, search, analysis and other intelligence services. Subject Learning Roadmap in the formation of cloud server technology roadmap to facilitate learners selective learning and communication.

Provide a wide range of resources. In the cloud-based mobile learning system, all learners can become perfect by developers and learning resources in order to achieve the diversification of resources to meet the needs of today's diverse society of knowledge. At the same time, the learners' own learning materials stored in the cloud, through sharing, making learning more abundant resources, thus meet the needs of all types of learners.

Diversification of resources makes the learner can no longer confined to a discipline of study, is conducive to the growth and expansion of knowledge in the field of learners.

Provide a more flexible service interaction. In the mobile learning system based on cloud computing, the learner can learn through the intelligent mobile devices curriculum, published an article, submit the job, browse, and even video discussion, anytime, anywhere with the teacher or other learners to interact. When the user is not online, the system will send the text messages, e-mail, QQ and other instant messaging to notify the user to view with their related information.

3.2 *Structural design of mobile learning system*

Clients, wireless network system (mobile communication network or WLAN) and cloud computing network system constitutes a mobile learning system based on cloud computing. The client is running on the terminal device application software, and is intermediary through which the user interact with the system. It can run on smart phones, tablet PCs, PDA and other mobile devices, and offer different interfaces and functions for different categories of users and the stronger adaptive ability of the device. Wireless network system includes mobile communication network and the wireless network, such as 3G, 4G networks and wireless LAN, is the link between the user access to the system. The cloud computing network system is not only the core part of the system, but also the entire mobile learning system functions final execution module. The service requested by the user through the interface is passed to the network cloud computing system which processing, and the results once again by the interface data feedback to the user. The entire cloud computing network system consists of application clusters, data storage, computing and management modules. The application Clusters module provides access to the user's applications. The main calculation module from the user's computing tasks split, and then distributed to the nodes in the cloud having the corresponding functions of parallel distributed computing. After all related calculations performed, and then the final results are collected, the consolidation and feedback to the user. The management module monitors the state of the whole system and load balancing, and to achieve self-management and tuning data for queries, search. Storage module is mainly to save user resources. The system framework model as shown in Figure 1.

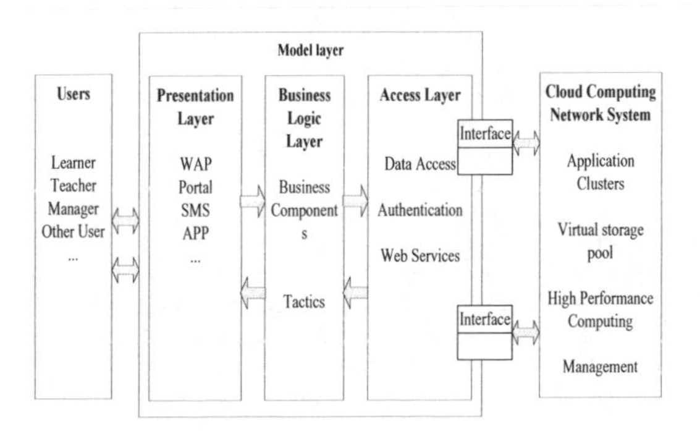

Figure 1. Mobile learning system based on cloud computing framework model.

4 IMPLEMENTATION OF CLOUD-BASED MOBILE LEARNING SYSTEM

4.1 *System architecture*

Cloud adoption LASJ model development, such as Linux, Apache, SQL sever, Jsp. This development model is employed to establish a stable and efficient mobile learning portal. More importantly, the development of the Web system via LASJ can bring better user experience. The intelligent Device terminal uses the popular Android-based development model which mainly used in smart phones, tablet computers and PAD. We mainly introduce cloud services implementations in this section.

In the service side, the server has a physical infrastructure monitoring and management mainly uses a master-slave architecture cluster is composed of five high-performance server clusters, namely cluster master control, application-node cluster, cluster compute nodes, virtual storage pools, monitoring cluster. Architecture as shown in Figure 2.

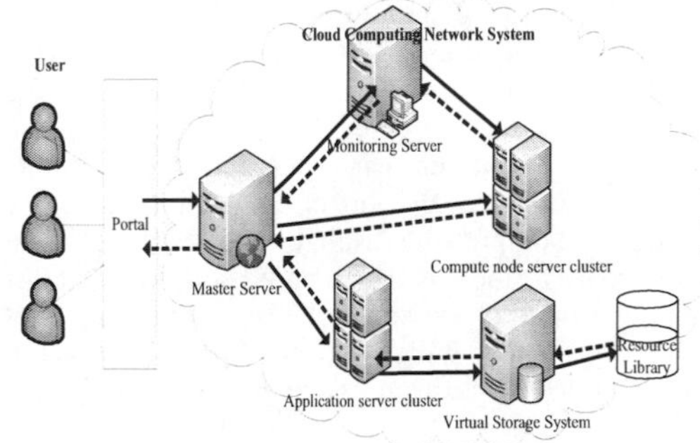

Figure 2. Cloud-based mobile learning system architecture diagram.

The main service control cluster is part of the entire cloud computing control system, which is composed of one or more groups and multiple master boot server group classification Controller cluster system consisting primarily responsible for receiving user requests, authenticating users, and classification and load balancing to achieve application requests. Applications node in the cluster is to provide learners with a variety of application services portal applications usually have different business processing logic cluster system consisting of one or more groups, is responsible for handling a variety of storage applications and sophisticated user application logic. The virtual storage pool is a cloud computing system resource storage unit, mainly by the large disk array system or group has massive storage capacity of the storage system consisting of a cluster system. Compute nodes in the cluster is operational and processor cloud computing system architecture consists of multiple sets of complete high-performance cloud computing clusters, whose main job is to calculate the amount of computation required to handle large. Monitoring Cluster is the manager and scheduler cloud server, primarily responsible for monitoring the working conditions of each server.

4.2 *Implementation of high-performance computing services*

High-performance computing service works as follows: When the user invokes the background of high-performance computing cluster compute nodes, mobile learning system will call the Hadoop Map Reduce API framework offers Map and Reduce processing. In this way, we have implemented the following process: First, we need to be copied to MPI program execution in the Hadoop framework Master and each worker working machine, the worker by the Master choose which machine to perform the Map and Reduce by program. Secondly, all the data blocks allocated to the implementation of Map of worker computer for Map (cut into small blocks of data), and then stored after recording Map worker computer. Finally, the computer that executes Reduce remote gets the results of each Map, mixing, aggregation and sorting, and outputs the result to the user. All calculations are performed by the compute nodes in the cluster background processing, which not only improves efficiency, but also will not affect other learning activities. Data is passed from the client to the server via write servlet code, and other work completed by the cloud server. The servlet code parameter passed to the server core code is as follows:

```
Socket s= new socket(InetAddress.getByName(null),4000);
  //definition connection socket
System.out.println("Connect to the server!"+s); //display
  connection server status
InputStream is=s.getInputStream(); //get the size of
  uploaded resources
byte[] buf=new byte[1024]; //resources for upload into
  blocks
```

5 CONCLUSION

Currently, mobile learning which supported by cloud computing technology is still in the preliminary stage of exploration, but its unique advantages heralded cloud computing has a good potential for application in the field of mobile learning. In this paper, we had put cloud computing and mobile learning together, designed the system from the function and structure, and given the structure model of the system. Finally, we use LASJ framework adopted to achieve the cloud system. Although mobile learning system based on cloud computing is not perfect, but with the rapid development of cloud computing technology has become more sophisticated and applications, cloud computing technology to support mobile learning system will in the field of education and training to play a significant role.

REFERENCES

[1] Zhang zehua.Research on Model and Key Realization Issues of Cloud Computing Federation: The library of Yunnan University. China (2010), p. 77.

[2] Zhao meng.Research and Application on Mobile Learning Based on Cloud Computing: The library of Henan University. China (2012) , p. 30.

[3] Gao hongqing,Qu yanjie. Research on the model of mobile learning based on Hadoop: China Educational Technology. Vol. 1 (2011), p. 124.

[4] OracleCloudComputing http://www.oracle.com/technetwork/cn/architect/cloud/oracle-cloud-computing-final-new-331118-zhs.pdf(2010).

[5] Guangxing Wang,Yanling Gao.Research on Construction of Cloud-Based Learning Platform MOOC: Advanced Materials Research. Forum Vol. 971–973 (2014), p. 1718.

Computing, Control, Information and Education Engineering – Liu, Sung & Yao (eds)

 ISBN 978-1-138-02800-5

Analysis on factors influencing the mental health of rural left-behind women—based on a survey conducted in GT village, Zhaoling district, Luohe city, a central subregion of Henan province, China

Xiao He Hu
Hohai University, Jiangsu, Nanjing,China

ABSTRACT: A special group called "Rural Left-behind Women" is an indirect result of rapid population flow caused by economic and social development. And the mental health of such a group has something to do with changes of social environment, but is more closely linked with the reality of rural communities. Profound analysis of the factors affecting mental health of rural left-behind women so as to work out reasonable and effective solutions and safeguard mechanisms is not only beneficial to improve the mental health of rural left-behind women, safeguard social harmony and rural stability, but also one of the important measures of promoting the construction of new socialist countryside.

KEYWORDS: rural areas; left-behind women; mental health; analysis on factors.

1 INTRODUCTION

"Left-behind women refer to married women in rural areas whose husbands leave their domicile places for working, going into business, etc. for more than half a year or cumulatively half a year in total."[1]. With the deepening of reform and opening up as well as the rapid pace of urbanization, rural young workforce flows into cities, thus numerous couples are separated. Long-term husband absence leaves the left-behind women undertake high labor intensity, heavy mental burden and low sense of security. Thus, their physical and mental health is severely hurt. Therefore, in order to construct a harmonious society and promote economic and social, sustainable development in rural areas, there is an urgent need to cope with such a problem.

2 SURVEY ON MENTAL HEALTH OF RURAL LEFT-BEHIND WOMEN

GT village, an administrative village located in the central subregion of Henan province of northern China, now has a registered population of 3,467, including 1,683 men and 1,784 women with a sex ratio of 94.34%. Sample number of matched left-behind women is 316. According to the questionnaire and the research, main "left-behind" reasons are as follows: children's education, 196 people, accounting for about 62%; poor physical condition, 47 people, about 15%; taking care of elders, 29 people, about 9%; unadaptable to the outside working environment and voluntarily staying, 32 people, about 10%; other reasons, 12 people, about 4%. On physical and mental feelings: those who often feel pleasant, 57 people, about 18%; those who are often accompanied by abnormal psychology, such as anxiety, loneliness and fear, 167 people, about 53%; those who are occasionally seized by mental abnormalities, 66 people, about 21%; those who are unaware of their situation, 24 people, about 7%, those who were diagnosed with anxiety or depression, 2 people (considering that most people are indifferent to mental illness, the data is too conservative), about 1%. On marital problems: those who think there is no need to worry, 148 people, about 47%; those who have a certain degree of concern, 127 people, about 40%, those who suffer from marriage problems, 18 people, about 6%; and those who don't want to talk about their marriage, 23 people, about 7%. On mental adjustment: those who choose to ask relatives for help, 139 people, about 44%; those who choose to seek help from friends and neighbors, 117 people, about 37%; those who choose to bear alone, 47 people, about 15%; those who choose to solicit help from social organizations or professional organizations, 13 people, about 4% (overlapped treatment methods are excluded). On the factors causing mental stress: economic conditions, 133 people, about 42%; family (parents-in-law) relationship, 54 people, about 17%; agricultural labor intensity, 57 people, about 18%; children's education, 63 people, about 20%; violence, 9 people, about 3% (overlapped factors are excluded). On cultural entertainment: those who only watch TV, 183 people, about 58%; those who like dropping around and chatting, 98 people, about 31%; those who read newspapers, listen to the radio and play the computer, 22 people, about 7%; those who watch dramas and dance square dance, 13 people, about 4% (overlapped ways are excluded). On the awareness of *Mental Health Law of the People's Republic of China*: those who said that they had never heard of it, 246 people, about 78%; those who know something about it, 44 people, about 14%; those who do not care about it, 26 people, about 8%.

3 ANALYSIS ON THE INFLUENCING FACTORS OF THE MENTAL HEALTH OF RURAL LEFT-BEHIND WOMEN

3.1 *Compelled physical experiences formed by heavy farm labor*

At present, although the level of agricultural modernization has undergone significant changes and the widespread use of large agricultural machinery and equipment has greatly reduced the physical labor of agricultural producers, part of farm labor, such as handling and drying, top dressing in field management, harvesting crops in hilly regions, management of greenhouse crops, livestock breeding in the tertiary industry, still cannot be done by machine. What's worse, during the busy season, in order to avoid agricultural losses caused by changes in the weather, it is a common phenomenon to harvest and grow crops in a rush. Therefore, in a general sense, farm labor is still heavy. Since the amount of labor per unit time and during a concentrated time period is large, without the support of man, it is difficult for rural left-behind women to organize and complete production activities independently. Without the assistance of sufficient labor support, women can only complete the production activities by raising their endurance. As a result, women's physical pleasure is greatly decreased, causing mental fatigue and anxiety. The combination of physical and mental factors is a great threat for the mental health of rural left-behind women. According to the investigation results, the majority of rural left-behind women are also engaged in some sideline production work, most of which are manual work, to increase the family income. The enduring and boring labor leads to lasting and intense damage to their mental health.

3.2 *Persistent mental stresses formed by trivial daily life*

Compared with the concentrated agricultural production labor, the pressure of daily life is a penetrating and aggressive damage to the mental health of rural left-behind women. The most tedious thing is to take care of children, especially their living and educational conditions. Judging from the investigation results, most women chose to stay in rural areas for this consideration. Investment in education is a major contributing factor to the healthy and sustainable development of the society, and also the most costly investment, especially in terms of mental costs. Subject to the existing conception of "knowledge changes fate" and "a good scholar can become an official", rural left-behind women show much enthusiasm in education, which is not inferior to any urban women and also involves a lot of time and energy. In addition, the relationship with their in-laws and the neighborhood also affects their daily lives. In general, this invisible, gradual and sustained pressure of daily life is a major cause of the anxiety and depression of rural left-behind women. However, the inducing factors are trivial and invisible, thus are easy to be overlooked and ignored.

3.3 *Hidden emotional crises formed by distant space*

Adult males tend to migrate to major cities with high demand of economic factors, which breaks the fated stay-forever marriage pattern and forms a spatial and geographical separation. Although it is offset by advanced communication method and other technological means, but it cannot replace the face-to-face sincere exchanges of couples. On the one hand, the distant spatial separation leads to poor communication between the couple, thus there will be a huge emotional communication gap between husband and wife. The continuing influence of loneliness, depression and yearning forms a huge psychological wave, which is a shock to mental stability and marriage solidarity. On the other hand, this hidden mental barrier is aggravated by the single rural cultural life. The entertainment facilities in the rural areas are insufficient, so women spend most of their time in watching TV; if their children are studying or working in foreign lands, they will have less association with them. Their loneliness will be even more severe and the mental stress will be artificially magnified. Thirdly, due to cultural and skills constraints, most adult males who make a living in urban areas engage in dirty, tired, bitter and dangerous work, which is a threat to their personal safety. Concerns for the husband's personal safety also increase the mental and psychological burden of rural left-behind women.

4 CONCLUSION

In the context of social changes, rural left-behind women, as a special social group, endure much pressure on physical and mental health, which cannot be ignored. The attention paid to the mental health of rural left-behind women and the exploration of reasonable and effective protection mechanism for them to solve practical difficulties is not only conducive to their physical and mental health, but also conducive to the harmony and stability of rural communities. Furthermore, it is also one important measure to promote the stable and healthy development of rural areas.

REFERENCES

[1] Wu Huifang, Rao Jing. Review of Rural Left-Behind Women [J]. Journal of China Agricultural University (Social Sciences), 2009 (2), p. 18–23.

[2] Commentator of this newspaper. Make Service the Striking Theme of the Grass-roots Organization of CPC [N]. People's Daily, May 29, 2014 (01).

[3] Harold J. Berman. Law and Religion [M]. Translated by Liang Zhiping. Shanghai: Joint Publishing, 1990, p. 48.

[4] Lu Fuying. Conflict and Coordination–Game in Rural Governance [M]. Shanghai: Shanghai Jiaotong University Press, 2006, p.63.

[5] Xi Jinping. Further Attaching Great Importance to the Building of the Grass-roots Organization of CPC [EB/OL], February 4, 2013.

Computing, Control, Information and Education Engineering – Liu, Sung & Yao (eds)
 ISBN: 978-1-138-02800-5

Exploration on enlightenment and significance of stratum differentiation in social management—based on investigation on social stratum differentiation in rural areas of China

Xiao He Hu
Hohai University, Jiangsu, Nanjing, China

ABSTRACT: With the transition of the economic system and the promotion of the urbanization process, the pace of social stratum differentiation in rural areas of China is quickened, and this differentiation breaks long-term rural inherent and the relatively steady social formation and causes obvious changes in rural social structure and social order. To deeply analyze the status quo of social stratum differentiation in rural areas and explore its enlightenment and significance in rural social governance can not only help our on-going construction of new socialist countryside, but also avail long-term peace and order in rural society.

KEYWORDS: Rural; Social Stratum; Differentiation; Enlightenment.

1 INTRODUCTION

"The crowds in the same social status demonstrate their consistency in the aspects of interest requirement, political attitude, values, etc., but those with different social status show their difference in the above aspects. This is stratum differentiation." [1] Since the late 1970s, the social stratum differentiation was grown in rural areas. In recent years, with the development of the economy and the speeding up of the urbanization process, the pace of social stratum differentiation in rural areas is enlarged, and the characteristics of all levels of society become more distinctive.

2 MANIFESTATION OF THE SOCIAL STRATUM DIFFERENTIATION IN RURAL AREAS

2.1 *Elites are gradually formed, and dominate the discourse right in rural society*

Elites are situated at the top of the social structure in rural areas, have the ability to influence the rural economic and social development, and even are capable of determining the distribution pattern of various resources in rural society. Assuming the larger social responsibility at the same time, this class makes the highest value contribution, is a benchmark and a model of moral and success in rural society, is respected and admired by rural residents. They control the allocation of resources, so they have the sufficient discourse right. Elites are the focus of interest groups in rural society. In particular, the urbanization process generates agricultural land requisition, resident migration and other economic affairs, strengthening benefit distribution and control right of elites. This class is a higher political literacy, is enthusiastic about political participation, and is capable of carrying out national policies in combination with the actual conditions of rural society. Therefore, at the bottom of state power and the social system, elites play a fundamental supporting role, and have the ability to control state power.

2.2 *The primary class tends to be stable, and becomes the dominant force of social development*

The primary class is an important part of the social structure in rural areas, has the characteristics of the largest population and the most stable structure, and plays a decisive role in the rural economic and social development. There will represent the voice and requirements of rural residents, and is the basis of making a national agricultural policy. They are the main undertaker for policy implementation effect, and represent the "majority" of rural society. Their opinions and suggestions are valued by the upper management. The majority in the primary class has a certain ability to resist risks, including economic strength, technical skills, vision and insight, social relationship, etc., and have reached a mature state. Therefore, even if they encounter some unexpected misfortunes, they will not fall the marginalized class. This class should be preferentially cultivated and developed in rural society.

2.3 *The marginalized class is worrying, and needs to be supported by national policies*

The marginalized class refers to some people who lose their capacity to work and in the survival predicament caused by diseases, misfortunes and other unforeseen circumstances. They are the "minority" in rural society, and it is appropriate to count them in the unit of family. The marginalized class is roughly involved in poor family because of sickness, the family losing their only child, the disabilities congenital or caused by accidental injury, group of divorced women without the land contract right, the family with no sons and all daughters marrying, villagers enjoying the "five guarantees", villagers with social personality disorder (mostly, idlers), etc. As vulnerable groups in rural society, this kind of family or villagers should be given special attention and care. Their living conditions are the measurement scale of social policy and morality.

2.4 *The free class indirectly penetrates, and accelerates the process of the development of rural society*

The free class refers to the crowd who does not directly participate in agricultural production and rural social management, but keeps close ties with rural society for some reason, also called the shadow class. "People with a peasant origin in different regions or on different businesses have the quality of "farmer" to different degrees, and do not cut their connections with the parent population." [2] The crowd with a stable profession externally (such as civil servants, teachers, staff from the state-owned enterprises, etc.), leaders retiring from government departments, public officers, etc., constitutes the principle force of this social class. Most of them have transferred their household registration from their native place, in a legal sense, and are no longer local residents. However, their social relations are widely involved in rural areas, and they have a high social status and influence, so local residents are still willing to regard them as members of the group. Encountering affairs beyond their reach, villagers say "asking someone for help", "making connections" and so on, the objects herein belong to this class. They are a pathway for rural society to obtain external resources, are a model of social status, successful career and morality in rural society, and have a strong demonstration effect and a special influence on the development process of rural society.

3 ENLIGHTENMENT AND SIGNIFICANCE OF SOCIAL STRATUM DIFFERENTIATION IN RURAL SOCIAL GOVERNANCE

3.1 *Looking forward to a major breakthrough in the rural economic system*

As three major economic rights for farmers, the right to land contractual management, the right to use of rural homestead and the property right of rural collective assets are not the only basis on which farmers rely for survival, but also are the footstone to maintain rural economy and society stable and the foundation maintain the existing rural political system. However, judging from the actual situation of the current rural social stratum differentiation, these three rights serve as a kind of immobilized productive relations, and cannot adapt to the development of productive forces in rural society. First of all, the right to rural land contractual management should be expanded as soon as possible. Farmers are entitled to possess, utilize, profit from and in part transfer rural land on the basis of the right to rural land contractual management, but have not right to mortgage and guarantee right, as explicitly prohibited in the property law. In order to satisfy capital demand of the development of rural society, it is necessary to expand those rights. Secondly, farmers' right to use of the rural homestead should be expanded as soon as possible. As a result of social stratum differentiation and fast mobility in rural areas, farmers' transferring their right to use of rural homestead is a common phenomenon, resulting in the actual transfer of the right to use of the rural homestead closely associated with the housing property right. In legal practice, the property law explicitly stipulates that farmers are banned to deal with the right to use of rural homestead and other collective land. There is a discrepancy between reality and law, leading to frequent occurrence of economic disputes in rural areas. In order to put an end to this phenomenon, it is imperative to expand the farmers' right to use of rural homestead. Thirdly, the shareholding system transformation is requested of rural collective assets. Traditionally, rural collective assets are owned by members of the rural collective economic organization, cannot be circulated, mortgaged or guaranteed. With the acceleration of mobility of rural resource factors, this pattern has already restricted appreciation in and returns on rural collective assets. Only when the joint-stock cooperative system is transformed to promote transfer and transaction of collective assets property right, the efficiency and property right of rural collective assets can give full play.

3.2 *Promoting the benign development through social integration*

Social stratum differentiation speeds up the mobility of the population and other resource factors, creating the conditions of the system and mechanism for the integration of rural society while effectively promoting the urbanization process. First of all, relying on the enthusiasm of elites and the primary class participating in political affairs, the village-level power balance mechanism is established. The organization to oversee a third party's economic affairs is established, implementing open-style supervision of the major decisions of the village committee and operation mode of village collective economies, so as to promote the transparency of the village-level power operational mode, to improve the degree of villagers' political participation, to enhance the cohesion of grass-roots organizations, and to prevent expansion of class antagonism and gap. Secondly, the influence of the free class is used to increase the supply of rural public products and services. The free class has similar functions of non-governmental organizations. For participation in rural economic affairs, objectively, they provide economic resources for rural society. Especially, they can make up for the deficiency of public products and services supplied by the government. For example, the free class provides the assistance, contributes money for education, repairs bridges and roads, holds theatrical performances, etc., which is a beneficial addition to the functions of the government. While receiving word-of-mouth buzz and social esteem, they have created the conditions for the benign development of rural society.

4 CONCLUSIONS

Based on two main inducements, i.e. economic development and promotion of the urbanization process, social stratum differentiation in rural society is not only an inexorable trend, but an inevitable result. "Modernity breeds stability, but the modernization process breeds some potential turmoil" [3]. Great importance must be attached to many variables in the process of social stratum differentiation in rural areas. To give full play beneficial effect of social stratum differing in rural areas and strive to avoid contradictions and conflicts between different classes can not only be conducive to the harmony and stability in rural society, but be of an important practical significance of building a new socialist countryside.

REFERENCES

[1] Chen Zhenming, Politics——Concept, Theory and Method [M]. Beijing: China Social Sciences Publishing House, 1998, p. 183.

[2] Li Shoujing. Rural Sociology [M]. Beijing: Higher Education Press, 2000, p. 175.

[3] [USA] Samuel P. Huntington. Political Order in Changing Society [M]. Translated by Wang Guanhua and others. Beijing: Sanlian Bookstore, 1989, p. 38.

Computing, Control, Information and Education Engineering – Liu, Sung & Yao (eds)
© 2015 Taylor & Francis Group, London, ISBN: 978-1-138-02800-5

Evaluation method for e-commerce transaction credit based on cloud model

X.K. Liang & L.M. Tao
Hangzhou Institute of Service Engineering, Hangzhou Normal University, Hangzhou, China

T. Cui
Department of Control Science and Engineering, Zhejiang University, Hangzhou, China

ABSTRACT: Based on the theory of cloud model and combined with the actual situation of e-commerce credit evaluation, a method is discussed for e-commerce transaction credit evaluation. This method is mainly established on the core algorithm of Normal Cloud Generator. In the scheme given below, some important factors corresponding to credit evaluation are taken into account, which includes the distribution of the value of the credit scores, the stability of the credit data, the extent and direction of the credit data deviating from the mean, etc. Since this method may effectively reflect the essential attributes of credit and its evaluation as the fusion of randomness and fuzziness, it can be selected as a new tool for the long term credit evaluation.

KEYWORDS: Cloud model; E-commerce; Transaction credit; Normal Cloud Generator; Algorithm.

1 INTRODUCTION

It is well known that the accuracy and completeness of the credit data have an important impact on e-commerce transaction. Usually, scientific and reasonable credit evaluation scheme for e-commerce can provide the stable and accurate credit data so as to give a technology basis for trading decisions and business administration.

Current transaction credit evaluation method can be summarized as two categories. The first is qualitative method which is represented by the schemes to divide the credit rating [1-6]. The second is quantitative scheme which is represented by the statistical model or fuzzy reasoning method [7-10]. Unfortunately, there are some limitations and shortcomings in the current evaluation projects. In qualitative method, the difference among the evaluation models or scales resulted in the bad comparability and sharing of the credit data or conclusions. In the statistical model, the random factors are fully taken into account but the fuzzy ones are neglected. In the fuzzy method, the fuzzy characteristics have attracted much attention while the random ones are ignored. Both of them are not fully reflect the compatibility of the randomness and fuzziness in e-commerce credit evaluation.

Normally, transaction credit evaluation focuses on the sellers or the shops credit status. The credit scores are mainly reflected in the weighted average of the score of each transaction. In fact, in order to fully reflect the real situation of the sellers' credit, some important factors should be taken into account. Which include the size of the sample, the concentration degree for sample data, the symmetry degree of the sample and the deviation degree or direction for the individual evaluation value from the mean value, etc.

In this paper, an evaluation method for e-commerce transaction credit based on cloud model will be discussed in details. It is hoped that this method can be applied in the practice of credit evaluation.

2 THE BASIC THEORY AND ALGORITHM OF CLOUD MODEL

2.1 *Cloud, cloud drops and the numerical characteristics of cloud*

On the assumption that U is a quantitative universe expressed by an exact numerical value, and a qualitative concept C is defined as a semantic representation in U. Let $x \in U$ and $\mu(x) \in [0,1]$, if x is a random occurrence of C and $\mu(x)$ is a stable random number, then the distribution of the variable x in U is called cloud, and a single x is called a cloud drop [11, 12]. Where x satisfies $\mu : U \to [0,1]$, $\forall x \in U$, $x \to \mu(x)$.

The quantitative characteristics of the qualitative concept C are reflected in the three digital features noted as Ex, En and He. Which are respectively called the expectation, the entropy and the hyper entropy of the cloud. Generally, the expectation is regarded as the most typical sample of the concept quantification. While the entropy is used to indicate the discrete degree of the cloud drops and the range of cloud droplets accepted by the given concept in the universe. Both the randomness and fuzziness of the concept and the relevance between them are revealed in the numerical value of entropy. As for the hyper entropy, it is mainly used for measuring the change of uncertainty status. In other words, the hyper is defined as the entropy of the entropy [13, 14].

2.2 *Visualization of cloud-cloud figure*

Thegeometricalformofcloudiscloudfigure.Generally, the cloud figure can be drawn by the following three different ways: the method of coordinates-grayscale, the pattern of coordinates-circle and the form of joint distribution. In coordinates-grayscale method, the position and the certainty degree of the cloud drops are measured by the coordinates of the points and the magnitude of the grayscale respectively. In the pattern of coordinates-circle, the position and the certainty degree of the cloud drops are represented as the coordinates and the size of the circles respectively. In the form of joint distribution, the cloud figure is shown as the joint distribution graph of the cloud drops and their certainty degrees.

2.3 *Normal cloud model*

Theoretically, the type of cloud model depends on the form of the distribution function $u(x)$. If $x \sim N(\mu, \sigma^2)$, then this cloud is named normal one, where the parameters μ and σ represent the expectation and the standard deviation of normal distribution respectively. In other words, if $x \sim N(E_x, E_n'^2)$ and $E_n' \sim N(E_n, H_e^2)$ and $\mu(x) = e^{(-(x-E_x)^2/(2(E_n')^2))}$, then the distribution of the variable x in the universe U is defined as the normal cloud [15, 16].

Previous research has indicated that the normal cloud model is universal to describe most of the uncertain phenomenon in nature or human society.

As a data processing tool, cloud generator is the most important part of the normal cloud model, which includes the forward normal cloud generator (denoted as CG) and the backward normal cloud generator (denoted as CG^{-1}).

If the expectation, the entropy, the hyper entropy of the cloud and the number of cloud drops are given, then cloud drops x_i and the corresponding certainty degree $u(x_i)$ can be obtained by running the algorithm of CG. On the contrary, given a group of sample data x_i, with or without the values of $u(x_i)$, the three numerical characteristics of the cloud can be calculated by the procedure of CG^{-1}.

Studies indicate that normal cloud model has some important properties as follows. The first, the distribution of the cloud drops about the expectation is symmetrical. This means that the number of cloud drops distributed on each side of the expectation is roughly equal. The second, the certainty degrees corresponding to a pair of symmetric cloud drops are also equivalent. Normally, the greater the distance between the cloud drop and the expectation, the smaller the corresponding certainty degree is. The third, the distribution of all cloud drops in the universe conforms to the rule of '*3En*'. This rule requires that about *99.99* percent of the cloud drops fall in the principal value interval *[Ex-3En, Ex+3En]* [11, 16].

3 CLOUD MODEL METHOD FOR TRANSACTION CREDIT EVALUATION

3.1 *The evaluation indexes*

The evaluation indexes include four second-grade indexes and one first-grade index. The former includes 'the Goods Consistent Degree to the Description on the Web', 'the Delivery Speed of the Seller', 'the Quality of the Seller Service' and 'the Logistics Service' respectively. The later is named as 'the Overall Rating'. The five indexes are denoted by the variables *G, S, D, L* and *I* respectively. They are valued from the continuous interval [*0.5*]. Where, the values of the indexes *G, S, D, L* is gotten form the web evaluation system and the variable *I* is calculated by the following formula.

$$I = 0.5 * G + 0.2 * S + 0.2 * D + 0.1 * L$$

3.2 *The cloud model algorithm of credit evaluation*

Suppose that the credit samples of some shops are given. By running the following cloud model algorithm, a large number of simulation data can be obtained so that the long term credit status of these shops can be achieved.

The cloud model algorithm of credit evaluation is designed as follows.

- Getting the values of the variables G, S, D, L form the original records of the web transaction and calculating the value of the variable I.
- Computing the three cloud numerical characteristics of I by running the algorithm of CG^{-1} according to the samples gotten above.
- Running the algorithm of CG and producing a certain amount of analogy cloud droplets on the basis of the calculated numerical characteristics.
- Calculating the numerical characteristics of the simulation data.
- Drawing the distribution figure and the certain degree graph of the simulated data.
- Carrying on the comprehensive analysis and evaluation by the results of the above two steps.

4 EMPIRICAL ANALYSIS

Suppose that the size of samples is 10, 20 and 30 respectively, the original credit evaluation data and the numerical characteristics of them for the shop A, B and C are now shown in table 1 and 2 respectively. The distribution status is demonstrated in figure 1.

Table 1. The original credit data of the shop A, B and C.

Shop	1	2	3	4	5
A	4.50	3.75	4.25	4.60	4.40
	3.75	4.30	4.70	4.80	4.90
B	4.20	4.90	4.75	4.30	4.10
	4.25	4.40	4.50	4.25	4.00
	4.35	4.85	4.70	4.28	4.09
	4.16	4.33	4.46	4.21	4.05
C	4.75	4.00	3.75	4.60	4.70
	3.50	4.80	4.90	4.10	4.20
	4.50	4.42	4.63	4.35	4.65
	4.92	4.78	4.87	4.15	4.25
	4.67	4.21	4.68	4.45	4.51
	3.88	3.92	4.96	3.97	4.31

Table 2. The numerical characteristics of the original data.

Mean	Variance	Ex	En	He
4.3950	0.3982	4.3950	0.3835	0.1072
4.3650	0.2636	4.3565	0.2587	0.0502
4.4127	0.3823	4.4127	0.3989	0.1140

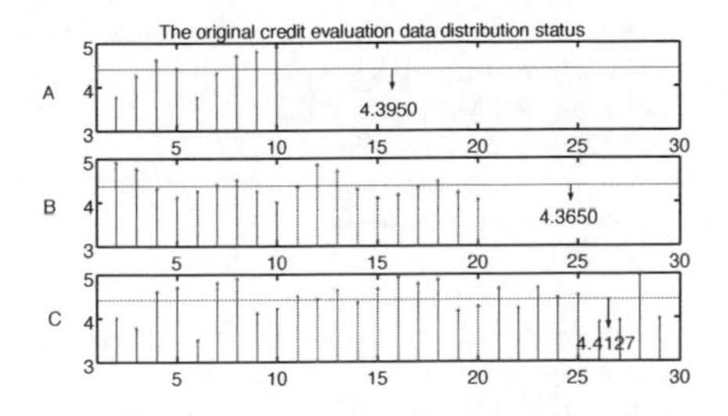

Figure 1. The distribution of original credit data.

From the view of the mean, the sort of the credit is C>A>B. However, this conclusion may not reliable since it is simply drawn from the point of statistics. Actually, because of the difference among the sizes of the samples, the data are not strictly comparable. On the other hand, whether the samples are reliable and stable or not should be considered in evaluation.

The processing and analysis of the credit data for three shops based on the cloud model will be carried out in the following text.

According to the algorithm in 3.2, some simulated data of transaction credit for the shop A, B and C can be easily gotten (the data size is valued as 40). The simulated data are denoted as the vector A, B and Crespectively. And the certain degrees of the data are expressed by the vector A_{cd}, B_{cd} and C_{cd} respectively.

$$A = [3.9129, 4.2783, 4.6984, 4.3059, 4.1347, 4.3350, 4.7715, 4.1600, 3.4456, 4.8785, 4.2358, 5.0000, 4.3338, 4.3516, 4.5731, 4.1005, 4.5801, 3.5529, 3.9905, 4.8868, 4.9800, 4.1553, 4.9900, 4.9700, 4.8916, 4.3492, 4.5336, 4.0070, 4.0523, 4.1476, 4.2260, 4.9500, 4.0149, 4.1800, 4.9139, 4.6952, 4.3665, 4.2464, 4.1902, 4.3012]$$

$$B = [4.8515, 3.6398, 4.1474, 4.4502, 4.1986, 4.0567, 4.4867, 4.0774, 4.5494, 4.5031, 4.7554, 4.3945, 4.2149, 4.1637, 3.8868, 4.0895, 4.6862, 4.4294, 4.3578, 4.6216, 4.7603, 4.4976, 4.4582, 4.4624, 4.5065, 4.2407, 4.3259, 3.8125, 4.1389, 4.3696, 4.3182, 4.5760, 4.3479, 4.5267, 4.6694, 4.1080, 4.4689, 4.3491, 4.6046, 4.1620]$$

$$C = [4.5788, 4.8988, 4.4997, 4.5836, 4.1589, 4.5142, 4.4538, 4.3940, 3.8619, 4.7353, 3.8695, 4.3758, 4.0353, 4.9180, 4.4368, 4.9800, 4.3072, 3.4144, 4.4089, 4.4922, 4.6761, 4.7262, 4.3911, 4.0246, 5.0000, 4.6094, 4.7561, 4.8415, 4.1993, 3.9085, 4.1486, 4.1335, 4.7084, 3.7110, 4.5191, 4.6892, 3.8266, 4.1973, 4.9900, 4.5013]$$

$$A_{cd} = [0.5268, 0.9813, 0.6031, 0.9511, 0.8664, 0.9832, 0.6609, 0.8957, 0.0622, 0.2950, 0.8626, 0.5067, 0.9854, 0.9883, 0.9078, 0.7458, 0.7636, 0.4387, 0.6941, 0.2800, 0.3522, 0.7265, 0.1125, 0.3169, 0.1609, 0.9947, 0.9545, 0.7341, 0.8481, 0.8475, 0.8931, 0.5565, 0.5582, 0.8421, 0.3559, 0.7217, 0.9961, 0.7629, 0.7756, 0.9444]$$

$$B_{cd} = [0.0985, 0.0302, 0.7959, 0.9358, 0.8892, 0.6402, 0.7276, 0.3779, 0.7756, 0.8806, 0.3652, 0.9903, 0.8855, 0.8268, 0.3974, 0.7159, 0.6521, 0.9764, 0.9999, 0.5251, 0.2498, 0.8360, 0.8250, 0.9368, 0.8938, 0.8371, 0.9950, 0.0787, 0.5534, 0.9987, 0.9804, 0.6707, 0.9990, 0.8877, 0.6734, 0.6810, 0.9004, 0.9995, 0.4974, 0.7005]$$

$$C_{cd} = [0.8251, 0.6377, 0.9868, 0.9415, 0.5836, 0.9813, 0.9972, 0.9973, 0.4587, 0.5842, 0.5107, 0.9980, 0.7137, 0.4748, 0.9971, 0.4626, 0.9304, 0.0651, 0.9999, 0.9793, 0.7248, 0.8417, 0.9989, 0.6439, 0.4454, 0.7641, 0.7934, 0.3906, 0.9023, 0.5611, 0.5476, 0.8929, 0.8380, 0.2880, 0.9633, 0.7264, 0.3942, 0.7962, 0.5828, 0.9512]$$

Furthermore, the simulated data and their numerical characteristics are shown in table 3. In order to get a visual impression, the simulated data distribution is displayed in figure 2. Meanwhile, the distribution of the corresponding certain degrees and the expectation curve of certain degrees are shown in figure 3.

Table 3. The numerical characteristics of the simulated data.

The numerical characteristics of simulated data				
Mean	Variance	Ex	En	He
4.3949	0.3982	4.3922	0.3940	0.1045
4.3566	0.2635	4.3566	0.2635	0.0603
4.4126	0.3823	4.4119	0.3812	0.1069

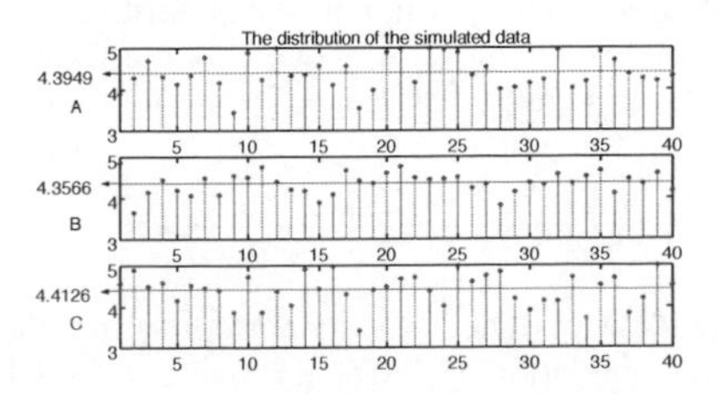

Figure 2. The distribution of the simulated data.

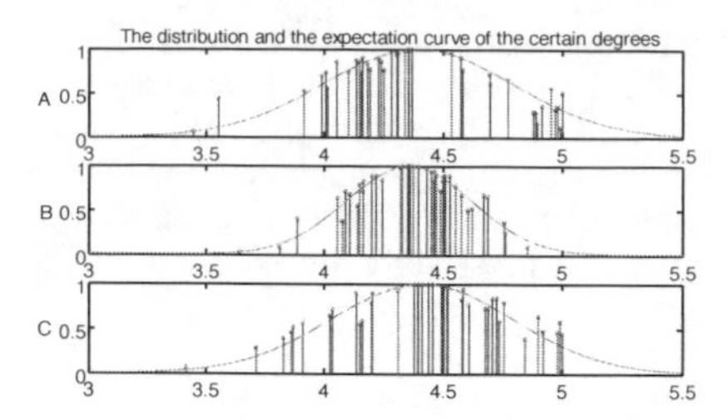

Figure 3. The distribution status and the expectation curve of certain degrees.

From the table and the figures above, some conclusions can be drawn as follows.

- The simulated data of the first shop are mainly located to the left side of the mean. And the ones of the second shop are evenly distributed to the both sides of the mean. Moreover, the reductive data of the third shop are primarily situated to the right of the mean. It is obvious that only the second group of simulated data meet the symmetry of normal cloud model.
- In the three groups of simulated data, both the entropy and the hyper entropy of the second group of data are the smallest. The smallest entropy indicates that the distribution of this group of data is more centralized than that of the others. And then, the smallest hyper entropy hints that the change degree of the uncertainty for the data of the shop B is lower than the one of the other shops. These means that the distribution of the second group of data is the most stable and reliable.
- From the view of certain degree, the certain degree of the second group of data shows a good symmetry and the most of the data are very close to the expectation curve of the certain curve.

5 CONCLUSIONS

Based on cloud model, an effective evaluation method of the credit data for the e-commerce shops are discussed in this paper. The empirical analysis shows that, even in the case of the different sample size, a scientific conclusion can still be drawn by running the cloud model evaluation algorithm. It should be emphasized that a large number of data can be simulated by running the algorithm of CG. So the method mentioned above lays the foundation for long-term credit evaluation.

ACKNOWLEDGEMENT

This work was financially supported by the Natural Science Foundation of Zhejiang Province (Y6110178) and the Science and Technology Planning Project of Zhejiang Province (2012C21031). We would like to express our gratitude to all those who helped us during the writing of this thesis.

REFERENCES

[1] Domenico Bianculli, et al. (2010) Automated dynamic maintenance of composite services based on service reputation. Lecture Notes in Computer Science, 2010(4749), pp. 449–455.

[2] Florian Skopik, et al. (2010) Modeling and mining of dynamic trust in complex service-oriented systems. Information Systems, 35(7), pp. 735–757.

[3] Zaki Malik & Athman Bouguettaya. (2009) RATEWeb: Reputation assessment for trust establishment among web services. The VLDB Journal, 18(4), pp.128–136.

[4] WU Jian-bin & Lü Gang. (2009) Trust and reputation evaluation for Web services based on user experience. Journal of Computer Applications, 2009(8), pp.68–76.

[5] Yuan Haiyan. (2009) Reform of the Credit Evaluation System in the View of Credit Speculation in the C2C Trade of China. Science Mosaic, 2009(10), pp.169–177.

[6] Guan Xiaoyong. (2009) The logics of technology development for the credit evaluation between China and Occident with its comparative research. Science Research Management, 2009(04), pp.63–70.

[7] Wang, Zhi-Bing Li & Chang-Yun. (2010) Research on credit assessment based on trade behavior for e-commerce. Application Research of Computers. 27(3), pp. 945–947.

[8] He R, Niu JW & Zhang GW. (2005) CBTM: A trust model with uncertainty quantification and reasoning for pervasive computing . Berlin: Springer.

[9] X.K. Liang & L.M. Tao. (2013) Fuzzy Reasoning and Calculating Method for the sellers' Credit of C2C Based on the Evaluating Indexes of Taobao Net, Applied Mechanics and Materials, 2013(411), pp.2262–2270.

[10] Liang Xikun & Zhu Zhengzheng. (2010) A Credit Model and Evaluation Method Based on Multivariables Fuzzy Reasoning. The international conference on Management Science and Artificial Intelligence, 2010(1), pp.32–35.

[11] Li Deyi & Du Yi. (2005) Artificial intelligence with uncertainty, National Defense Industry Press, Beijing.

[12] Li Deyi. (2000) Uncertainty in knowledge representation. Chinese Engineering Science, 2(10), pp.75 -81.

[13] Li Deyi et al. (1995) Membership cloud and membership cloud generators. Computer Research and Development, 32(6), pp.15 -20.

[14] Li Deyi & Liu Changyu. (2004) Study on the universality of the Normal Cloud Model. Chinese Engineering Science, 6(8), pp.32 -38.

[15] Lu Huijun et al. (2003) The Application of Backward Cloud in Qualitative Evaluation. Chinese Journal of Computers, 26(8), pp.1009–1014.

[16] Li Deyi et al. (2004) Artificial Intelligence with Uncertainty, Chinese Journal of Software, 15(11), pp.1583–1594.

Computing, Control, Information and Education Engineering – Liu, Sung & Yao (eds)
© 2015 Taylor & Francis Group, London, ISBN: 978-1-138-02800-5

The design and implementation of intelligent home control system

Hong Mei Zhu
College of Electronic Information Engineering, Changchun University of Science and Technology, Changchun, Jilin, China

Hong Wei Yang
College of Computer Science and Technology, Changchun University of Science and Technology, Changchun, Jilin, China

ABSTRACT: In this paper, an intelligent home control system is designed and it can monitor indoor temperature, lights and smog in real time and remote control the home through a pc. This system is composed of an ARM controller which servers as the core controlling chip, a temperature sensor, a smog sensor and a light sensor which function as signal acquisition units, and the peripheral equipments of keyboard, liquid crystal display. What's more, 485 level conversion chip in the system can increase communication distance and realize remote control. With the help of GPRS module, people can control the system via their cellphones and know the real time situation in their house even though they are not there.

KEYWORDS: intelligent home; ARM controller; remote control; GPRS module.

1 INTRODUCTION

With the improvement of living standards and development of the society, people's demand on home environment is rising day by day. Back home, people want to enjoy coolness when it is hot outside and feel warm when it is cold outside[1]. People hope that the temperature indoors can keep constant in accordance to their will. People, especially the ones who have to work with computers for a very long time and the students who work hard, wish the indoor light to maintain constant in order to protect their eyesight. If their house is on fire, people hope to lose no time to know it and take measures in time. To deal with the situations mentioned above, this intelligent home control system is designed. This system can effectively monitor indoor temperature, brightness of lights, smog, etc. via cellphones and conduct automatic control according to real time situations. When something abnormal happens, the users can be informed in the form of text messages. In the meantime, users can use cellphones to control this system in the form of text messages. For instance, they can use cellphones to turn on and off indoor lights and get to know indoor situations.

At present, many intelligent home systems use general single chips as their chief controlling unit. So their function is too simple and their operation is too single. Although cable transmission of distant data and order is safe, it needs too much maintaining cost and has very poor extended performance[2]. What's worse, it is often limited by stringing environment, transmission distance, etc. With people's rising demand on high-quality living standards and the development of consumer electronics technology, this traditional home control pattern has become a stumbling block to the intelligent process of home information. Therefore, the development of a more advanced and intelligent home control system not only meets people's demand for comfortable and safe home lives but also caters for their psychological demands for automation, intelligence and advancement[3].

2 STRUCTURE OF THE INTELLIGENT HOME CONTROL SYSTEM

2.1 *General structural platform of the system*

As is shown in Diagram one, the hardware of the controlling system includes ARM9 processor, system clock module, FLASH memory, GPRS module, LCD module, wireless communication module, various sensor modules, power module, voice alarm set, keyboard controlling module, daylight lamp, etc[4].

The smoke detecting sensor, temperature sensor, light sensitive sensor, etc. are adopted in the users' home to detect whether the house is in danger. Since these sensors assemble different forms of signals, it is necessary to deal with the signals effectively. Then by means of external interrupt, these signals can be transferred to CPU through S3C2440 interrupt interface. To improve the reliability and safety of the system, the assembled alarm signals have to be dealt with correspondingly before they are transferred to the processors.

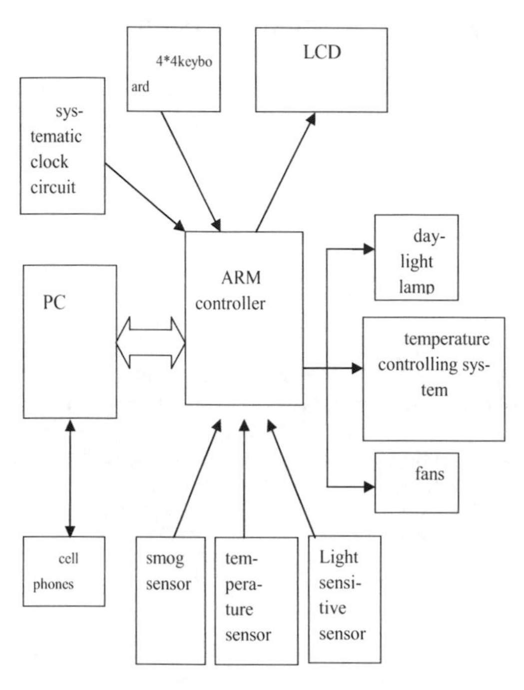

Figure 1. General framework of the intelligent home controller.

2.2 *Software realization of the system*

The original intention of intelligent home is to deal with the automatic control of household equipment–controlling and managing household electrical appliances collectively and making them have intelligent and information network function. The system in this paper offers different solutions to various household electrical appliances which are of different intelligent levels. For traditional hand control household equipments like light switches, etc[5]. , the system adds some assisting controllers. In this way, users can indirectly control these household equipments through Ethernet or text messages. For those electrical appliances that can be controlled by infrared rays, such as TV and air-conditioners, the system can remote control them collectively with a universal controller which has Ethernet interfaces. As to those intelligent electrical appliances that have network access function and support remote control platform, they can be directly hooked up to family Ethernet. The appearance and development of intelligent electrical appliances will have a great impact on traditional electrical appliances, computer and communication industries.

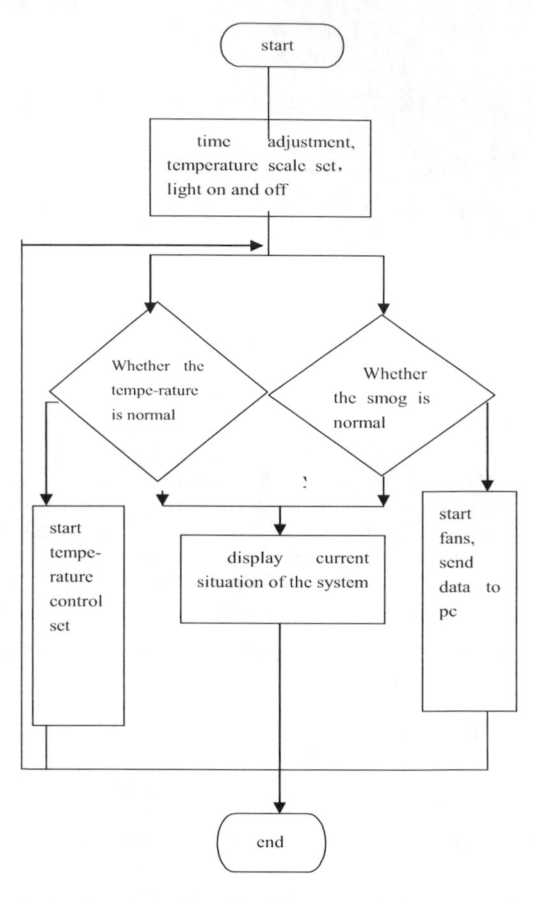

Figure 2. Controlling flow diagram of the systematic sensors.

2.3 *Remote monitoring part*

More and more products in both fields of industry control and consumer electronics start to use Internet technology as their supporting platforms. The intelligent home control system designed in this paper just makes full use of the network image data transmission function of the Internet and makes it possible for users to check their homes at any time via the Internet. The Ethernet chip DM9000 has a multipurpose processor interface, 10/100M self-adaption, 4K double byte sram. DM9000 provides a MII interface to link HPNA equipment or other transceivers which support MII interfaces and also supports 8-bit, 16-bit, and 32-bit interfaces so as to adapt to various processors' internal memory access [3]. DM9000 and RJ45 interface islinked by Network Isolation Transformer HR601627, as is shown in Diagram Four. Network isolation transformer has the functions of signal transmission, impedance matching, waveform restoration, clutter suppression, high voltage isolation, etc. Therefore, this transformer can protect the system[7].

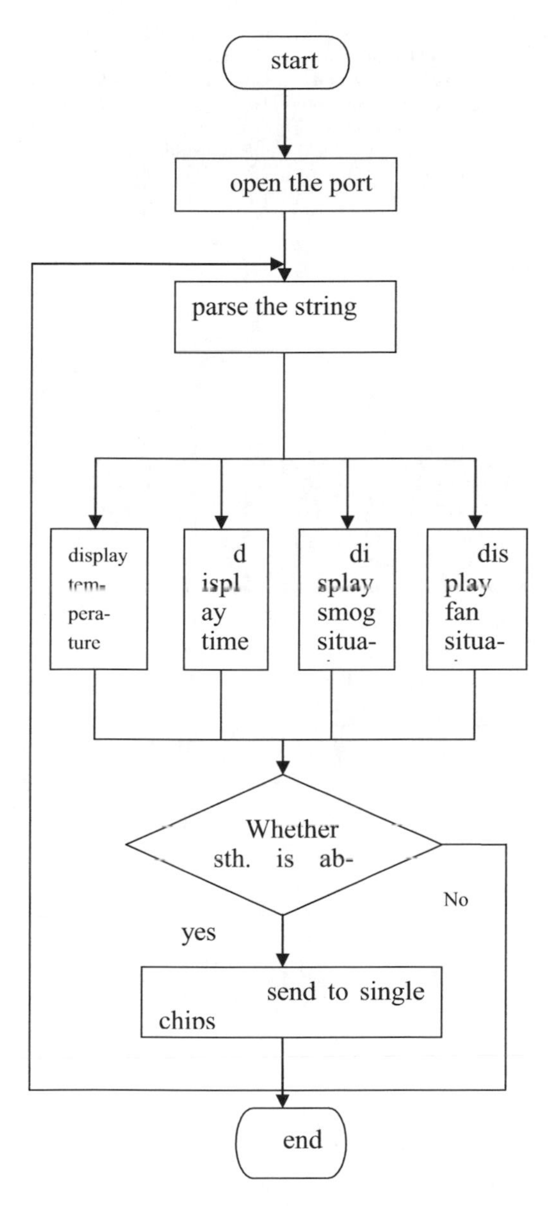

Figure 3. Software flow diagram of a pc.

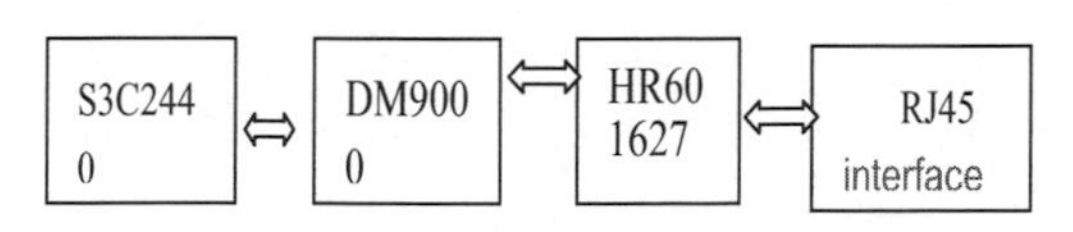

Figure 4. Linking structure of the internet interfaces.

The video signals acquired by the camera are zipped into video data streams and then they are sent to the built-in BOA Web server through the internal bus. Users can look at the video images on the Web server directly by logging onto the network and monitor their home in real time[8].

3 CONCLUSION

The major functions of the system are:

1 Analyzed data transmission form and chose RS485 Level Translator as data communication form.
2 Finished the fieldsampling design of smog, temperature and light information and this design included the hardware and software design of the detecting and controlling parts.
3 Adopted LCD display technology to show temperature, light, smog information on the monitoring terminal.
4 Adopted GPRS module to let the users control the system via cellphones and know real time situation in their homes when they are out.

This system used ARM embedded single chips as the controlling unit and realize the functions of systematic detection of temperature, light and smog, remote control by cellphones and LCD display. In terms of final controlling effect, this design has fairly strong adaptive ability[9].

REFERENCES

[1] The Design Method and Application of One Intelligent Integrator[A]. Proceedings of 2010 International Conference on Computer,Mechatronics, Control and Electronic Engineering (CMCE 2010) Volume 4[C]. 2010.
[2] The Intelligent Real-Time Protection Device for Mesh Optical Network[A]. Microelectronic and Optoelectronic Devices and Integration—Proceedings of SPIE 2008 International Conference on Optical Instruments and Technology[C]. 2008.
[3] Aptitude-Based Methodology for Agent-Oriented Software Engineering[A]. Proceedings 2010 IEEE International Conference on Software Engineering and Service Sciences[C]. 2010.
[4] Xi Yu,Fuquan Sun. Urban Emergency Intelligent Decision and Processing System Based on Decision Table in Rough Set Theory[A]. Proceedings of 2013 IEEE 4th International Conference on Electronics Information and Emergency Communication[C]. 2013.
[5] Suyun Li. Gear Vibration Signal Analysis and Fault Diagnosis[A]. Proceedings of 2013 3rd International Conference on Social Sciences and Society(ICSSS 2013) Volume 42[C]. 2013.
[6] Xue Gao, Le Yang, Li Peng. Enhanced Bilinear Approach for Sensor Network Self-Localization Using Noisy TOF Measurements[A]. Proceedings of 2014 Conference on Sensors and Networks (CSN)[C]. 2014.

[7] Xingchuan Liu, Zhengfeng Wu, Wei Ben. Experimental Analysis and Modeling of CSS Ranging in LOS and NLOS Environments[A]. Proceedings of 2013 22nd Wireless and Optical Communication Conference[C]. 2013.
[8] Feng Geng, Shengjun Xue. A Comparative Study of Mobility Models in the Performance Evaluation of MCL[A]. Proceedings of 2013 22nd Wireless and Optical Communication Conference[C]. 2013.
[9] Ji Huang, Danpu Liu. A High-Reliability Data Gathering Protocol Based on Mobile Sinks for Wireless Sensor Networks[A]. Proceedings of 2013 22nd Wireless and Optical Communication Conference[C]. 2013.

Computing, Control, Information and Education Engineering – Liu, Sung & Yao (eds)
© 2015 Taylor & Francis Group, London, ISBN: 978-1-138-02800-5

Design and analyze a worktable motion control system

Wei Yan
Sichuan Vocational and Technical College of Communication, China

ABSTRACT: Many control systems are designed to obtain high steady-state accuracy, but subjected to extraneous disturbance signals in practice use that cause the system to provide an inaccurate output. The Routh-Hurwitz criterion ascertains the absolute stability of a system by determining whether any of the roots of the characteristic equation lie in the right half of the s-plane. The steady-state accuracy of many feedback systems can be increased by increasing the amplifier gain in the forward channel. This study concludes design and analyze of a worktable motion control system.

KEYWORDS: stability; control system; compensation; transfer function.

1 INTRODUCTION

The performance of a control system can be described in terms of the time-domain performance measures or the frequency-domain performance measures. The performance of a system can be specified by requiring a certain peak time, Tp, maximum overshoot, and settling-time for a step input. Furthermore, it is usually necessary to specify the maximum allowable steady-state error for several test signal inputs and disturbance inputs. These performance specifications can be defined in terms of the desirable location of the poles and zeros of the closed-loop system transfer function, T(s). Thus, the location of the s-plane poles and zeros of T(s) can be specified.

The design of a system is concerned with the alteration of the frequency response or the root locus of the system in order to obtain a suitable system performance. For frequency response methods, we are concerned with altering the system so that the frequency response of the compensated system will satisfy the system specifications. Hence, in the frequency response approach, we use compensation networks to alter and reshape the system characteristics represented on the Bode diagram and Nichols chart.

An unstable closed-loop system is generally of no practical value. Therefore, we seek methods to help us analyze and design stable systems. A stable system should exhibit a bounded output if the corresponding input is bounded. The stability of a feedback system is directly related to the location of the roots of the characteristic equation of the system transfer function. The Routh Hurwitz method is a useful tool for assessing system stability. The technique allows us to compute the number of roots of the characteristic equation in the right half plane without actually computing the values of the roots. Thus, we can determine stability without the added computational burden of determining characteristic root locations. This gives us a design method for determining values of certain system parameters that will lead to closed-loop stability.

2 DESIGN A WORKTABLE MOTION CONTROL SYSTEM

An important positioning system in manufacturing systems is a worktable motion control system. The system controls the motion of a worktable at a certain location. We assume that the table is activated in each axis by a motor and lead screw. We consider the x-axis and examine the motion control for feedback system, as shown in Figure 1. The goal is to obtain a fast response with a rapid rise time and settling time to a step command while not exceeding an overshoot of 5%.

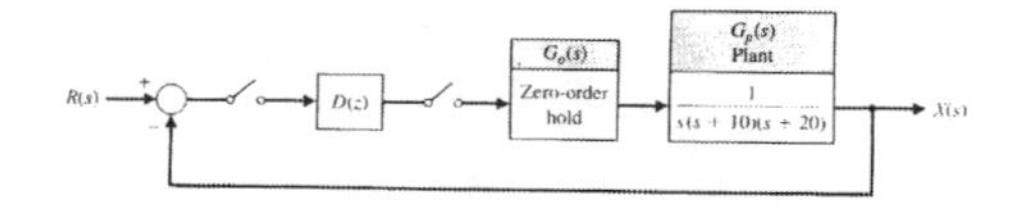

Figure 1. A worktable motion control system.

The specifications are then (1) a percent overshoot equal to 5% and (2) a minimum settling time (with a 2% criterion) and rise time. To configure the system, we choose a power amplifier and motor so that the system is described by Figure 2.

The transfer function of the motor and power amplifier is

$$G_p(s) = \frac{1}{s(s+10)(s+20)} \tag{1}$$

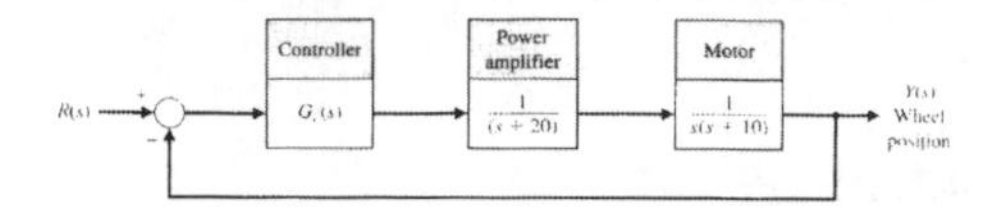

Figure 2. Model of the wheel control for a work table.

We will initially use a continuous system and design Ge(s) .We then obtain D(z) from Gc(s). First, we select the controller as a simple gain, K, in order to determine the response that can be achieved without a compensator. Plotting the root locus, we find that when K = 700, the dominant complex roots have a damping ratio of 0.707, and we expect a 5% overshoot. Then, using a simulation, we find that the overshoot is 5%, the rise time is 0.48 second, and the settling time (with a 2% criterion) is 1.12 seconds. The next step is to introduce a lead compensator, so that

$$G_c(s) = \frac{K(s+a)}{(s+b)} \tag{2}$$

We will select the zero at s = −11 so that the complex roots near the origin dominate. Evaluating the gain at the roots, we find that K = 8000. Then the step response has a rise time of 0.25 second and a settling time (with a 2% criterion) of 0.60 second. This is an improved response, and we finalize this system as acceptable.

It now remains to select the sampling period and then to obtain D(z). The rise time of the compensated continuous system is 0.25 second. Then we require T << TR in order to obtain a system response predicted by the design of the continuous system. Let us select T = 0.01 second.

We have $G_c(s) = \frac{8000(s+11)}{(s+62)}$ (3)

Then $D(z) = C\frac{z-A}{z-B}$ (4)

Where A = e^{-llT}= 0.8958 and B = e^{-62T} = 0.5379.

We now have

$$C = K\frac{a(1-B)}{b(1-A)} = \frac{8000(11)(0.462)}{62(0.1042)} = 6293 \tag{5}$$

Using this D(z), we expect a response very similar to that obtained for the continuous system model.

3 DESCRIBE A FEEDBACK SYSTEM WITH DIGITAL COMPUTER COMPENSATION

A feedback control system with a digital computer used to improve the performance is shown in Figure 3

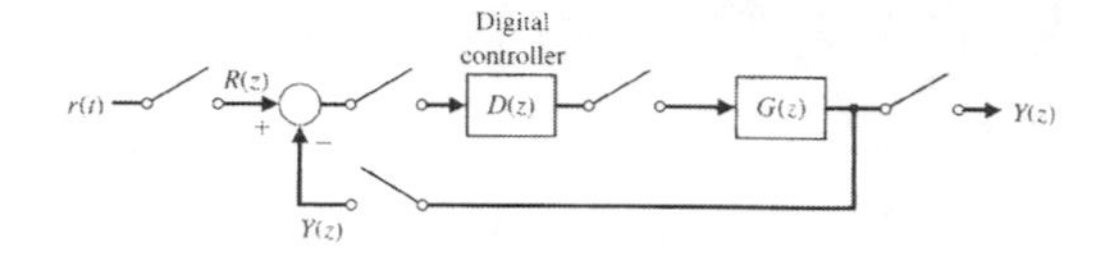

Figure 3. A closed-loop, sampled system.

The closed-loop transfer function is

$$\frac{Y(z)}{R(z)} = T(z) = \frac{G(z)D(z)}{1+G(z)D(z)} \tag{6}$$

The transfer function of the computer is represented by $\frac{U(z)}{E(z)} = D(z)$ (7)

D(z) was represented simply by a gain K. As an illustration of the power of the computer as a compensator, we will consider again the second-order system with a zero-order hold and a plant

$$G_p(s) = \frac{1}{s(s+1)} when T = 1. \tag{8}$$

Then $G(z) = \frac{0.3678(z+0.7189)}{(z-1)(z-0.3678)}$ (9)

If we select $D(z) = \frac{K(z-0.3678)}{(z+r)}$ (10)

We cancel the pole of G(z) at z = 0.3678 and have to set two parameters, r and K. If we select

$$D(z) = \frac{1.359(z-0.3678)}{(z+0.240)} \tag{11}$$

We have $G(z)D(z) = \frac{0.50(z+0.7189)}{(z-1)(Z+0.240)}$ (12)

If we calculate the response of the system to a unit step, we find that the output is equal to the input at the fourth sampling instant and thereafter.

4 ANALYZE THE STABILITY OF A FEEDBACK CONTROL SYSTEM

When considering the design and analysis of feedback control systems, stability is of the utmost importance. We can say that a closed-loop feedback system is either stable or it is not stable. Given that a closed-loop system is stable, we can further characterize the degree of stability. This is referred to as relative stability. One outcome of the relative instability of modem fighter aircraft is high maneuverability. A fighter aircraft is less stable than a commercial transport; hence it can maneuver more quickly. In fact, the motions of a fighter aircraft can be quite violent to the passengers. We can determine that a system is stable (in the absolute sense) by determining that all transfer function poles lie in the left-half s-plane, or equivalently, that all the exigent values of the system matrix A lie in the left-half s-plane. Given that all the poles (or exigent values) are in the left-half s-plane, we investigate relative-stability by examining the relative locations of the poles (or exigent values).

The location in the s-plane of the poles of a system indicates the resulting transient response. The poles in the left-hand portion of the s-plane result in a decreasing response for disturbance inputs, similarly, poles on the $j\omega$-axis and in the right-hand plane result in a neutral and an increasing response, respectively, for a disturbance input. Clearly, the poles of desirable dynamic systems must lie in the left hand portion of the s-plane [1-3].

A common example of the potential destabilizing effect of feedback is that of feedback in audio amplifier and speaker systems used for public address in audio rums. In this case, a loudspeaker produces an audio signal that is an amplified version of the sounds picked up by a microphone. In addition to other audio inputs, the sound coming from the speaker itself may be sensed by the microphone. Due to the finite propagation speed of sound waves, there will also be a time delay between the signal produced by the loudspeaker and the signal sensed by the microphone. In this case, the output from the feedback path is added to the external input. This is an example of positive feedback.

5 CONCLUSIONS

To ascertain the stability of a feedback system represented by a transfer function, we investigate the characteristic equation and utilize the Routh-Hurwitz criterion. We may use the z'-transform of a transfer function to analyze the stability and transient response of a system. Thus, we may readily determine the response of a closed-loop feedback system with a digital computer serving as the compensator (or controller) block.。To improve the transient behavior of a feedback controller, we often choose a better motor for the system. If the system designer is able to specify and alter the design of the process that is represented by the transfer function, G(s), and then the performance of the system may be readily improved. Then the addition of compensation networks becomes useful for improving the performance of the system.

REFERENCES

[1] R. C. Doff, The Encyclopedia of Robotics, John Wiley & Sons, New York, 1988.

[2] G. F. Franklin, et al., Digital Control of Dynamic Systems, 2nd ed., Prentice Hall, Upper Saddle River, N.J., 1998.

[3] R. C. Garcia and B. S. Heck, "Enhancing Classical Controls Education via Interactive Design," IEEE Control Systems, June 1999, pp. 77–82.

[4] H. Kazerooni, "A Controller Design Framework for Tele-robotic Systems," IEEE Transactions on Control Systems Technology, March 1993, pp. 50–62.

[5] F. G. Martin, the Art of Robotics, Prentice Hall, Upper Saddle River, N. J., 1999.

[6] D. Raviv and E.W. Djaja, "Discretized Controllers," IEEE Control Systems, June 1999, pp. 52–58.

[7] R. C. Doff, Electrical Engineering Handbook, 2nd ed., CRC Press, Boca Raton, Fla., 1998.

[8] W. J. Palm, Modeling, Analysis and Control, 2nd ed. John Wiley & Sons, New York, 2000.

Computing, Control, Information and Education Engineering – Liu, Sung & Yao (eds)
© 2015 Taylor & Francis Group, London, ISBN: 978-1-138-02800-5

Application of Newton interpolation in static nonlinear correction for MAF sensor

Yuan Yuan Zhang &Yan Lin Wu
College of Electrical Engineering and Automation, Anhui University, Hefei,China

Yao Hua Xu
School of Electronics and Information Engineering, Anhui University, Hefei,China

ABSTRACT: Piecewise Newton interpolation is used to compensate static nonlinear errors of automobile Mass Air Flow (MAF) sensor. Thirteen pairs of experimental data are divided into two groups according to the characteristic of the MAF sensor curve and static nonlinear correction models of small flow and big flow are separately established. Output signals of MAF sensor can be regarded as input signals of the corrector, the errors between correction results and input signals of MAF sensor are compared and analyzed. Experimental results show that: (1) Linearity of MAF sensor decreases from 22.91% to 1.53%, linear degree has improved significantly. (2) The distributions of errors are concentrated in scope of 0.2g/s and correcticorrecton is good, the correction results of MAF sensor nearly lossless- recthe MAF sensor. pu) signals of MAF sensor. (3) The algorithm has advantages of small sample size data, simplicity and practicability. The method can effectively correct the nonlinear errors of sensors with the same characteristics.

KEYWORDS: Newton Interpolation; Static Nonlinear Correct; Mass Air Flow Sensor.

1 INTRODUCTION

Mass Air Flow sensor is one of the most important sensors of Electronic Control Units for automobile engines. MAF sensor can provide mass flow information of engine intake for Electronic Control Unit. MAF sensor has characteristic of static nonlinear. It's necessary to carry on the correction and correction precision can effect on control accuracy of ECU.

The neural network has good capability of nonlinear mapping and it is used to establish a contrary model of sensor for realizing static nonlinear correction. These kinds of methods had got more research[1][2]. Because of slow astringency and the solution uniqueness of the BP neural network, document[3] made use of the Particle Swarm Optimization algorithm combined with BP neural network to establish the inverse model of static nonlinear characteristic for sensor, then the precision of static nonlinear correction is improved. But neural network has some shortcomings such as more sample data is needed, the overfitting problem appears easily when sample data is limited, the choice of structure and type depends on experience too much, it's difficult to realize in real time. Support Vector Machine (SVM) is proposed based on the structural risk minimization principle, it can preferably solve the problems, including limited sample data, nonlinearity, high-dimensions, local minimal value. Document [4-6] realized static nonlinear correct for the pressure sensor and temperature sensor based on Support Vector Machine, but it was difficult to choose kernel function. Above algorithms have the common characteristics of high complexity, so that real time implementation of microprocessor is effected. Document [7-8] implemented static nonlinear correct for sensor based on the least squares curve fitting. This method obtained the optimal solution based on the gradient theory and it's local search method, then calibration precision for sensors need to be improved. Document [9-10] used interpolation algorithm to establish contrary model of static nonlinear correct for sensor, it's a good attempt. The error of the linear interpolation algorithm was greater. In this article, contrary model of static nonlinear characteristics of MAF sensor was established based on a piecewise Newton interpolation algorithm. This method needed less data and less storage space. It's simple and easy to realize in real time. It's possible to realize static nonlinear correction online. The practicability of algorithm was improved greatly and the availability of the algorithm was proved.

2 STATIC NONLINEAR CORRECTION PRINCIPLE FOR MAF SENSOR

MAF sensor has static nonlinear characteristic and it can be expressed as:

$$y = f(x) \quad (1)$$

where x is the measured flow and unit is g/s; y is the measured voltage and unit is V; $f(.)$ is a nonlinear function. In order to compensate static nonlinear error of MAF sensor, the output signal of the sensor should firstly pass a static nonlinear correction link, and it is illustrated in Figure.1.

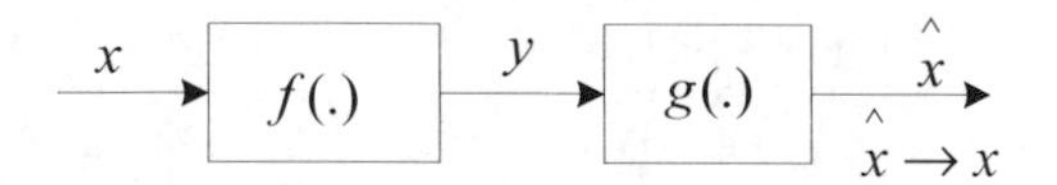

Figure 1. Principle of static nonlinearity correction.

The characteristic of static nonlinear correction link can be expressed as:

$$\hat{x} = g(y) \quad (2)$$

Where $g(.)$ is nonlinear function and it is contrary function of $f(.)$, namely $g(y) = f^{-1}(x)$.

The output signal(y) of MAF sensor is used as the input signal of static nonlinear correction link, the result of nonlinear correction ($\hat{x}$) should be numerically equal to the measured flow so that the sensor after nonlinear correction has linear input-output relationships and it is defined as Formula 3, input flow is reconstructed.

$$\hat{x} = g(y) = g(f(x)) = x \quad (3)$$

3 NEWTON INTERPOLATION ALGORITHM

Polynomial interpolation uses polynomial function to approximate known data and the approximate function needs to go through the data points. The common uses interpolation methods, including Lagrange interpolation, Newton interpolation, Hermite interpolation, cubic spline interpolation. Characteristic of Lagrange interpolation as follows: intuitive meaning, symmetrical form, tidy structure, convenient analysis theory. But polynomial corresponding is needed to be recalculated when interpolation points are changed, so the amount of calculation is large. Hermite interpolation and cubic spline interpolation are effective in overcoming Runge phenomenon. But derivative information of nodes is needed to know.

The advantage of Newton interpolation is not to need derivative information of nodes and only a calculation of the Formula is increased meanwhile the previous calculations are still effective when interpolation nodes increase. Coefficients of Newton interpolation polynomial are difference quotients and higher order difference quotient can be obtained by repeated calculations of first-order difference quotient on the basis of the lower order difference quotient. The calculation has recursive characteristic so that it is convenient to calculate by computer and realize in real time.

Interpolating sequence $(x_i, f(x_i)), i = 0,1,...,n$ is given, Newton interpolation polynomial $N_n(x)$ is constructed and defined as Formula 4:

$$\begin{aligned} N_n(x) = & f(x_0) + f[x_0, x_1](x - x_0)(x - x_1) \\ & + f[x_0, x_1, x_2](x - x_0)(x - x_1)(x - x_2) \\ & + ... \\ & + f[x_0, x_1 ... x_n](x - x_0)(x - x_1)...(x - x_{n-1}) \end{aligned} \quad (4)$$

Where zero-order difference quotient is defined as $f[x_0] = f(x_0)$, first-order difference quotient is defined as $f[x_0, x_1] = \frac{f(x_1) - f(x_0)}{x_1 - x_0}$, second-order difference quotient is defined as $f[x_0, x_1, x_2] = \frac{f(x_1, x_2) - f(x_0, x_1)}{x_2 - x_0}$, N- order difference quotient is defined as $f[x_0, x_1, ..., x_n] = \frac{f(x_1, x_2, ..., x_n) - f(x_0, x_1, ..., x_{n-1})}{x_n - x_0}$, remainder of Newton interpolation polynomial $R_n(x)$ is defined as Formula 5:

$$R_n(x) = f(x) - N_n(x) = f(x_0, x_1 ... x_n, x)(x - x_0)(x - x_1)...(x - x_n) \quad (5)$$

The difference quotient is shown in Table 1.

Table 1. Difference coefficient of Newton interpolation polynomial.

x_i	$f(x_i)$	one-order	second-order	third-order	......
x_0	$f(x_0)$				
x_1	$f(x_1)$	$f[x_0,x_1]$			
x_2	$f(x_2)$	$f[x_1,x_2]$	$f[x_0,x_1,x_2]$		
x_3	$f(x_3)$	$f[x_2,x_3]$	$f[x_1,x_2,x_3]$	$f[x_0,x_1,x_2,x_3]$	
......					

The remainder of interpolation polynomial is smaller and the similarity degree of data is bigger when absolute value of difference quotient is smaller. Good effectiveness can be obtained when an interpolation function is calculated by this group of data.

The effectiveness does not become better accompany with interpolation nodes increasing. The increase of interpolation nodes will bring the increase of interpolation polynomial order. The increment of interpolation polynomial order cannot always bring the improvement of precision. Moreover, higher order polynomial with additional restriction point and the Runge phenomenon will be produced when the distances between added nodes and requested data are very far. In this article, interpolation polynomial is calculated based on piecewise Newton interpolation method and the model of static nonlinear correction for MAF sensor is constructed, so that the above Runge phenomenon can be effectively avoided.

4 APPLICATION AND EXPERIMENT

The above algorithm applies to static nonlinear correction for the MAF sensor. There are 36 pairs of static standardization experiment data[3], the nonlinear characteristic curve is drawn as shown in Figure 2.

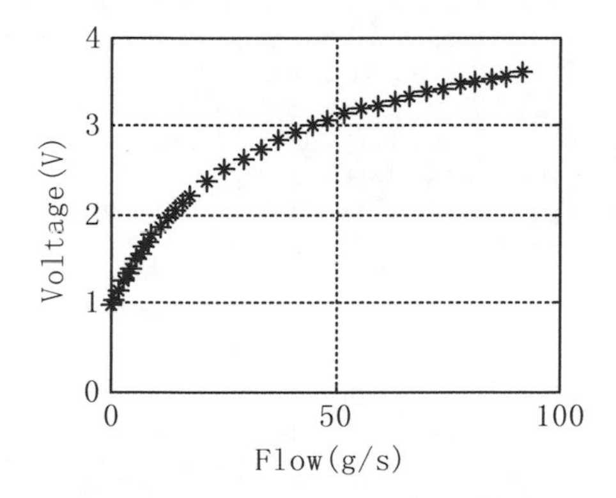

Figure 2. Static characteristic curve of the MAF sensor.

The data are divided into two groups according to the characteristic of the MAF sensor curve. Seven pairs of data (Be illustrated in Table 2 and Table 3) are chosen as interpolation nodes in order to avoid the Runge phenomenon because of the too high order. Note: Data knot (*25.3796g/s)* should be a common node in two groups of data so that the interrupt of flow range can not happen.

Table 2. Interpolation nodes of small flow rate range.

Number	Flow(g/s)	Voltage(V)
1	0.0000	0.9600
2	2.9852	1.2409
3	4.7978	1.3958
4	8.3000	1.6912
5	10.8448	1.8595
6	17.4377	2.1936
7	25.3796	2.4975

Table 3. Interpolation nodes of big flow rate range.

Number	Flow(g/s)	Voltage(V)
1	25.3796	2.4975
2	33.0522	2.7311
3	48.0486	3.0629
4	62.7996	3.2876
5	69.9736	3.3734
6	77.3039	3.4535
7	91.5092	3.5902

The output signal of MAF sensor (voltage value) can be used as interpolation nodes of Newton interpolation polynomial, input signal of MAF sensor (flow rate) can be used as functions of interpolation nodes, then a contrary model of the sensor static nonlinear characteristic based on Newton interpolation polynomial is constructed. Correction part polynomial of the small flow range is obtained by MATLAB programming and it is shown as Formula 6. Correction part polynomial of big flow range is shown as Formula 7.

$$\hat{x} = 16.2179y^6 - 170.799y^5 + 734.796y^4 - 1647.02y^3 + 2026.88y^2 - 1287.3y + 327.481 \quad y \in (0.9600, 2.4975) \tag{6}$$

$$\hat{x} = 297.539y^6 - 5578.76y^5 + 43436.8y^4 - 179730y^3 + 416775y^2 - 513480y + 262572 \quad y \in (2.4975, 3.5902) \tag{7}$$

Where y represents an output voltage signal of sensor, namely an input signal of static nonlinear correct part; Where $\hat{x}$ represents an output flow signal of static nonlinear correct parts.

The characteristic curve and interpolation nodes representing full-scale static nonlinear correction part of the sensor are drawn based on piecewise Newton interpolation and illustrated as Figure 3. Where 'circle' represents full-scale interpolation nodes and there are thirteen circles. Where 'line' represents a nonlinear characteristic of correction link.

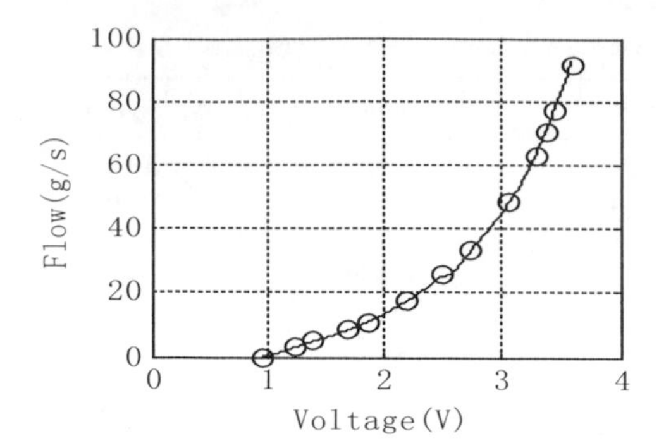

Figure 3. Nonlinear characteristic of sensor calibration part.

The peculiarity of interpolation polynomial is that polynomial function values of interpolation nodes are equal to actual data. Namely, nonlinear correct results ($\hat{x}$) of all thirteen interpolation nodes are equal to actual input flow values of the sensor (x), $\hat{x} = x$. Put other 23 pairs of static calibration experiment data into Formula 6 and Formula 7, the results of static nonlinear correction are calculated. Absolute errors and relative errors between correction results and actual values are calculated based on Formula 8 and Formula 9, they are shown as Table 4. The linearity before and after correction are calculated by Formula 10.

$$\Delta x = \hat{x} - x \tag{8}$$

$$\gamma_x = \frac{\Delta x}{x} \times 100\% \tag{9}$$

$$\gamma_L = \pm \frac{\Delta x_{\max}}{Y_{FS}} \times 100\% \tag{10}$$

Where Δx is absolute error, γ_x is relative error, $\Delta x_{\max}$ is the maximum absolute error, Y_{FS} is maximum scale output ($Y_{FS} = 91.5092 g/s$), γ_L is linearity. The absolute error is maximum ($\Delta x_{\max} = 1.3996 g/s$) when flow achieves *29.2504g/s*. The linearity of MAF sensor before correction is 22.91%; the linearity of MAF sensor after correction is 1.53%. The linear degree is very good and the nonlinearity is greatly reduced.

These 36 pairs of data $(x, \hat{x})$ are drawn in Figure 4. The linearity of MAF sensor output signal after nonlinear correction improves greatly. The result of correction is almost really reflected input signal of the sensor and the linearization static characteristic of MAF sensor is basically achieved. The errors (Δx) between the nonlinear correction results of 36 pairs of data and actual data are illustrated in

Table 4. Static experimental data and correction results of MAF sensor.

Number	Input (g/s)	Output (V)	Results (g/s)	AE (g/s)	RE %
1	1.0737	1.0268	0.4942	-0.5795	-53.98
2	2.0486	1.1403	1.7521	-0.2965	-14.47
3	3.9020	1.3230	3.9640	0.0620	1.59
4	5.6922	1.4796	5.7382	0.0460	0.81
5	6.5646	1.5553	6.6026	0.0380	0.58
6	7.4357	1.6266	7.4617	0.0260	0.35
7	9.1367	1.7551	9.2004	0.0637	0.70
8	12.5213	1.9460	12.3731	-0.1482	-1.18
9	14.1827	2.0384	14.1563	-0.0264	-0.19
10	15.8030	2.1109	15.6483	-0.1547	-0.98
11	21.4397	2.3569	21.3008	-0.1389	-0.65
12	29.2504	2.6139	27.8508	-1.3996	-4.78
13	36.8506	2.8261	37.3091	0.4585	1.24
14	40.6625	2.9153	41.1453	0.4828	1.19
15	44.3720	2.9864	44.2988	-0.0732	-0.17
16	51.8177	3.1257	51.5569	-0.2608	-0.50
17	55.4100	3.1866	55.4069	-0.0031	-0.01
18	59.0761	3.2323	58.5995	-0.4766	-0.81
19	66.3960	3.3324	66.4525	0.0565	-0.09
20	73.7962	3.4138	73.5972	-0.1990	-0.27
21	81.0318	3.4899	80.8382	-0.1936	-0.24
22	84.5575	3.5268	84.5775	0.0200	0.02
23	88.0683	3.5575	87.8391	-0.2292	-0.26

Figure 5(a). According to 36 pairs of static calibration experiment data, six-order polynomial contrary model of a static nonlinear part based on the least square method is established and the errors between correction results and actual values are drawn in Figure 5(b). The errors are relatively scattered in Figure 5(b), meanwhile, the required data of the nonlinear correction part modeling attains 36 pairs. The errors are relatively concentrated in Figure 5(a) and the required data are only 13 pairs, meanwhile, the errors mostly focus on among $\pm 0.2 g/s$. Among the errors of 13 pairs of data are $\pm 0.0 g/s$ and large errors locate two sides of the piecewise interpolation interval.

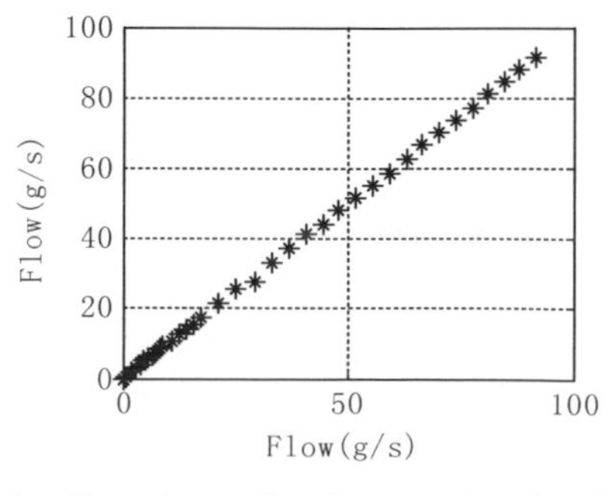

Figure 4. Correction results of sensor output signal.

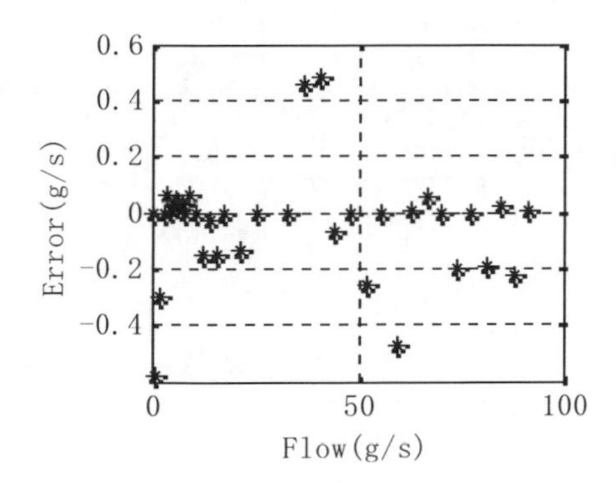

(a) Errors of correction based on Newton interpolation polynomial.

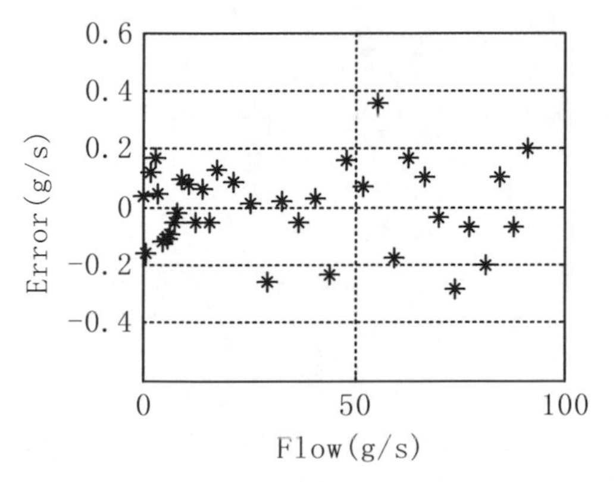

(b) Errors of correction based on the least squares polynomial.

Figure 5. Errors of nonlinear correction results and true data.

5 CONCLUSIONS

The 36 pairs of static calibration data were divided into two groups through the analysis of MAF sensor static nonlinear characteristic and 7 pairs of data were selected from each group. The contrary model of MAF sensor static nonlinear correction part was established by Newton interpolation polynomial. Static characteristic linearization was completely realized in 13 interpolation nodes and these results of nonlinear correction completely restructured the input flow values of the MAF sensor.

The remaining 23 output values of the sensor were used as input values of static nonlinear correction part, the differences between correction results and sensor input flow values were analyzed. The experimental results showed t hat the errors were relatively concentrated and mostly focus on among ±0.2g/s. Linearity of MAF sensor after correction was 1.53% and linearity was improved greatly.

The contrary model of static nonlinear correction for MAF sensor was established by Newton interpolation. This method had advantages of fewer data, occupying less system resources, high calculation efficiency, microprocessor realization convenience and good practicability.

ACKNOWLEDGEMENTS

I would like to express my gratitude to all those who helped me during the writing of this article. This project was supported by Anhui province college funds for outstanding young talent fund, NO: 2012SQRL022 and the new teacher foundation of the education ministry, NO:20123401120013, 2011 3401120007.

REFERENCES

[1] PANG Hongfeng, LUO Shitu, CHEN Dixiang, PAN Mengchun, ZHANG Qi. Temperature Compensation of Fluxgate Magnetometer Based on BP Neural Network*[J]Journal of test and measurement technology, 2011,25(3):278–283.*

[2] XING Hongyan, PENG Jiwei, LU Wenhua, XU Wei, WU Xiangjuan. A Fusion Algorithm for Humidity Sensor Temperature Compensation*[J] Chinese Journal of Sensors and Actuators, 2012,25(12):1711–1716.*

[3] Zhang yuanyuan, Xu kejun,Xu yaohua,Huang shengchu. Application of PSO and BP Neural Network in Static Nonlinear Corrector for Sensor[J]. *Acta Metrologica Sinica, 2009,30(6):526–529.*

[4] GUO Fengyi, GUO Changna,WANG Yangyang. Research on Nonlinear Correction of Pressure Sensor Based on MPSO-SVM*[J].Chinese Journal of Sensors and Actuators, 2012,25(2):188–92.*

[5] Liu tao,Wang hua. Nonlinear correction of sensor using genetic algorithm and support vector machine*[J]. Journal of Electronic Measurement and Instrument, 2011,25(1):56–60.*

[6] Peng Jishen,Yu Jingzhe,Xia Naigin. Approaches to Non-linearity Compensation of Copper Resistor Based on Neural Network in Temperature Measurement*[J]. Computer Measurement and Control, 2011, 19(1):243–245.*

[7] Xie Yu,Yang Sanxu,Li Xiaowei. Nonlinear compensation of capacitance weighing transducer based on inverse fitting*[J].Chinese Journa of Scientific Instrument,2007,28(5):923–927.*

[8] Zhang yuanyuan, Xu kejun. Comparison of Nonlinear Dynamic Correcting Methods Based on Block Oriented Models*[J].Control Engineering of China, 2009,16(2):121–124.*

[9] BAI Shou-jun,LANG Lang,YIN Cheng-zhu. Design of high-precison LVDT transmitter based on ADuC845*[J]Journal of Anhui Polytechnic University, 2013,28(3):47–50.*

[10] Chen Liangzhu,Teng zhaosheng. Nonlinear effect of electronic analytical balance and its compensation *[J].Chinese Journal of Scientific Instrument, 2012,33(3):581–587.*

[11] Xu kejun. Sensor and Detection Technology*[M]. Beijing: Electronic Industry Press,2006:1–33.*

Computing, Control, Information and Education Engineering – Liu, Sung & Yao (eds)

Improved method for the recovery of regular broken paper

Yu Jia Gong
School of Electrical Engineering, North China Electric Power University, Baoding, Hebei, P.R. China

ABSTRACT: This paper studied the stitching and the recovery of the regular broken paper by similarity matching model. As for the low degree of automation of the existing algorithm when the broken paper has a little gray level information, this paper proposes a method increasing the column number of pixels and changing the image binarization threshold to hand it out. It is proven the algorithm has a higher degree of automation and less intervention.

KEYWORDS: similarity matching; recovery of regular broken paper; gray; intervention.

1 INTRODUCTION

The splicing of the broken files is significant in the recovery of judicial evidence, the repairing of historical documents and the obtaining of military intelligence. Traditionally it was completed by men, which has high accuracy but low efficiency. Especially when the amount of the broken files is huge, much more time is needed to complete it by men's work. So it is needed to develop the technology to complete it automatically.

Because there are patterns more or less on the files, they can be extracted to splice the broken files. Literature [1] proposes a method that firstly extracts the contour of the broken files, then judge if the contours of them are matching based on boundary rules and area rules. But it focuses on irregular broken files. Literature [2] proposes a commonly used algorithm which finds the similar matching part by comparing and cut-and try and it works well. Literature [3] proposes an algorithm which focuses on regular broken files and the efficiency is enhanced. But when the gray level of the paper's border is much less, the degree of automation is lower. This paper proposes an improved algorithm aim at the problem.

2 THE PRETREATMENT OF THE BROKEN PAPER

Before splicing the broken paper, it need to be pretreated as follows after it has been scanned into images. The software "Matlab" is used.

1 Pixel[4][5]
A pixel is a unit used to calculate the digital image whose continuous tone is made up of many small square point of similar color. These squares are called pixel.

2 Gray level
Gray is the transition between the white and black color. Set black as the base color. The image can be displayed by different saturation of black. Every pixel's gray level is divided into 256 levels. Black's gray level is 0, accordingly white's gray level is 255. Black-and-white photos contain all kinds of gray level between black and white.

3 Binary image
For the sake of simplicity problem, the images are binarized. The pixel's gray level is set as 0 or 255 through image binaryzation by setting an appropriate threshold, then the image will display obvious effect of white and black. Literature [3] set threshold as 127. When the gray level is bigger than 127, then adjust it to 255. Accordingly adjust the gray level to 0 when it is less than 127.

3 PROVED MODEL FOR THE RECOVERY

The main basis of splicing the broken paper is whether the border pixel's gray level of two images. This paper uses annex 2 of problem B of 2013 Higher Education Press Cup National Mathematical Contest in Modeling to do the test.

For the 19 pieces of broken paper, corresponding gray level matrix $G_0, G_1, G_2 \cdots\cdots G_{18}$ $(i = 0,1,\ldots 18)$ can be got, every matrix is 1980×72. Literature [3] have proposed the steps to splice the image: firstly extracts every matrix's first column and 72th column, then calculate the similarity by using one matrix's 72th column and another's 1st column, lastly match them by using it. But a letter is simpler than a Chinese character, there are less pixel, more columns are needed to match them.

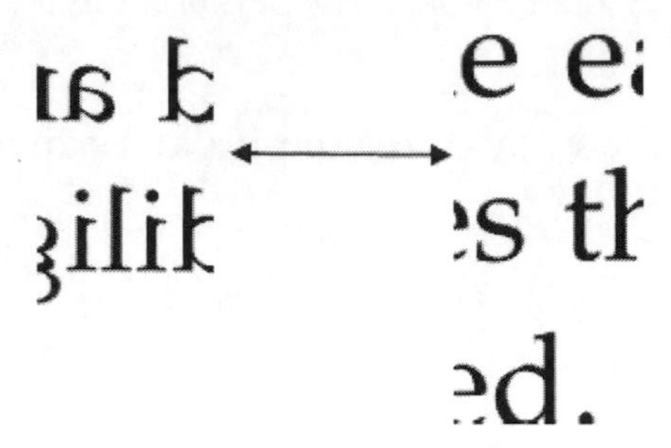

A broken image Another broken image

Figure 1. The comparison of the broken image.

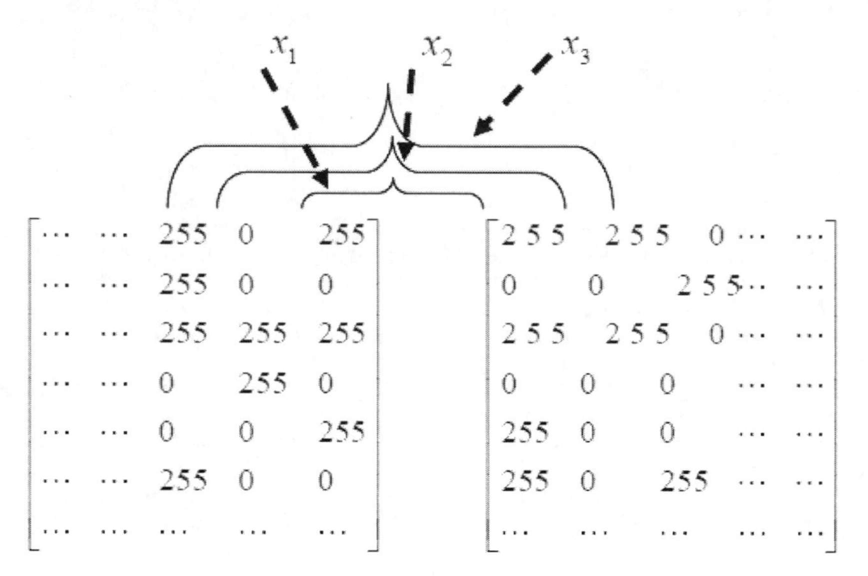

Figure 2. The comparison of grey level matrix.

As shown in Figure 2, extract every matrix's first three columns and last three columns, then calculate similarities to use an image's last 3 columns and another image's first 3 columns according to the following steps.

Compare the symmetrical elements of two matrixes. Then define as the function of the two symmetrical elements.

$$x_m = \begin{cases} 0 & \text{The grey level of the two elements are not equal} \\ 1 & \text{The grey level of the two elements are equal} \end{cases} \tag{1}$$

And the similarity function is modified to formula (3).

$$\sum_{m=1}^{P} x_m = \sum_{m=1}^{P} x_1 + \sum_{m=1}^{P} x_2 + \sum_{m=1}^{P} x_3 \tag{2}$$

$$p(i,j) = \frac{\sum_{m=1}^{P} x_m}{3P} \tag{3}$$

Use the data above and the similarity can be got as following.

$$\sum_{m=1}^{980} x_m = 5+4+4 = 13. P = 980,\ p = 4.42\times10^{-3}.$$

Literature [3] set threshold as 127. But when the gray level information is much less, so many points with high gray level is binarized to white points. So the threshold is increased to 252 to improve the use ratio.

Before starting to splice the image, the image on the far left (or right) is needed to find first. The basis can be that the 1st column's gray level of the image on the far left is 255. Then make it as the first one and splice them in turn. Because there are 1980 pixels in a column, it is not possible that there are two images meeting the condition.

The algorithm can be concluded as below.

STEP1: use Matlab to binarize the image and get a gray level matrix, then extract every matrix's 1st, 2nd, 3rd, 70th, 71st and 72nd column.

STEP2: find the matrix whose 1st or 72nd column's gray level is all 255, the image who has the feature is the first one.

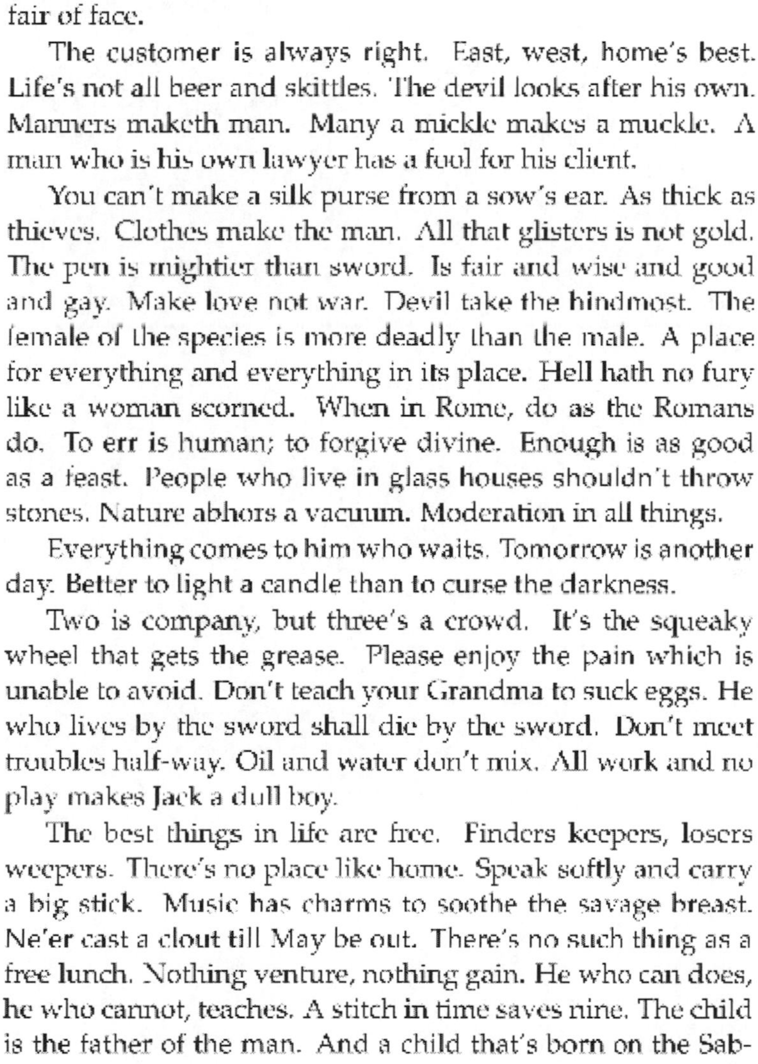

fair of face.

The customer is always right. East, west, home's best. Life's not all beer and skittles. The devil looks after his own. Manners maketh man. Many a mickle makes a muckle. A man who is his own lawyer has a fool for his client.

You can't make a silk purse from a sow's ear. As thick as thieves. Clothes make the man. All that glisters is not gold. The pen is mightier than sword. Is fair and wise and good and gay. Make love not war. Devil take the hindmost. The female of the species is more deadly than the male. A place for everything and everything in its place. Hell hath no fury like a woman scorned. When in Rome, do as the Romans do. To err is human; to forgive divine. Enough is as good as a feast. People who live in glass houses shouldn't throw stones. Nature abhors a vacuum. Moderation in all things.

Everything comes to him who waits. Tomorrow is another day. Better to light a candle than to curse the darkness.

Two is company, but three's a crowd. It's the squeaky wheel that gets the grease. Please enjoy the pain which is unable to avoid. Don't teach your Grandma to suck eggs. He who lives by the sword shall die by the sword. Don't meet troubles half-way. Oil and water don't mix. All work and no play makes Jack a dull boy.

The best things in life are free. Finders keepers, losers weepers. There's no place like home. Speak softly and carry a big stick. Music has charms to soothe the savage breast. Ne'er cast a clout till May be out. There's no such thing as a free lunch. Nothing venture, nothing gain. He who can does, he who cannot, teaches. A stitch in time saves nine. The child is the father of the man. And a child that's born on the Sab-

Figure 3. The result of image splicing.

STEP3: according to formula (1) (2) (3), compute the similarity between the matrix's last three matrix and other matrix's first three matrix, then find the biggest similarity, the corresponding image is the second one.

STEP4: do the same work for the rest of image until 19 images is all spliced.

STEP5: use Matlab to display the whole image.. The result is shown in Figure 3.

4 CONCLUSION

From the image above, the result is very accurate with no erratum. And the work is without intervention. So it can be concluded that the improved algorithm has a high efficiency, it can be used in the practical application.

REFERENCES

[1] JIA HaiYan, ZHU Liangjia, ZHOU Zongtan, etc. An Automatic Stitching Shredding Shape Matching [J] Computer Simulation, 2006, 23 (11): 180–183.

[2] C Papaodysseus, etal. Contour - Shape Based Reconstruction of Fragment, 1600 IEEE Transactions On Signal Processing, June 2002, 50(6) : 1277– 1288.

[3] Wang, Kang Cheng. "A New Simple Automatic Approach for Image Retrieval [J]."*Advanced Materials Research*. Vol. 971. 2014.

[4] XIA Lei. Study On Fabric Image Mosaic Method Based On Cellular Automata. Wuhan Textile University, B. C. W all Paintings [D]. 2011.

[5] Luo Zhizhong. Semi-automatic Text-based Document Shredding Feature Film Splicing [J]. Computer Engineering and Use. 48:5.207–210 2012.

[6] HOU Feng. Debris Automatic Image Stitching Technology Research [D]. Beijing Jiaotong University. 1-33.2007.12.

[7] China Society for Industrial and Applied Mathematics. 2013 Higher Education Press Cup National Mathematical Contest in Modeling tournament title [DB / OL]. [2014-11-02]. Http://www.mcm.edu.cn/problem/2013/ 2013.html.

Computing, Control, Information and Education Engineering – Liu, Sung & Yao (eds)
© 2015 Taylor & Francis Group, London, ISBN: 978-1-138-02800-5

The design of intelligent hydraulic vulcanizing machine control system

Ji Ming Sa
Wuhan University of Technology, Wuhan, Hubei, China

Ai Cheng Sun
Key Laboratory of Broadband Wireless Communications and Sensor Networks, Wuhan, Hubei, China

Shao Yun Wu
Wuhan Tianhe Airport Co., Ltd. Wuhan, China

Yu Jun Gu
Wuhan University of Technology, Wuhan, Hubei, China

ABSTRACT: This article introduces hydraulic vulcanizing control system which is for the object to produce slurry pump accessories, mainly discussing its control principles, programming methods, advantages and system methods of applications.

KEYWORDS: Slurry pump, hydraulic vulcanizing machine, PLC.

1 INTRODUCTION

Hydraulic vulcanizing machines are mainly for rubber production. In vulcanization, temperature, pressure and vulcanizing time are all monitored by Omron PLC. With the alternations of products, vulcanizations are in a different time and numbers, which brings some troubles in the vulcanizing executions. The traditional vulcanizing machine was robotistic controlled by hydraulic pressure, could not meet the requirements of vulcanizate for producing slurry pump accessories. Besides, people had to operate valves by themselves to control vulcanizing time, which were not only in high labor intensity, but cannot execute operation strictly thus having negative effects on productions.

In this paper improves tradition vulcanizing machines, achieving one key automatically accomplish those productions and spontaneously supplement pressure, which reduces labor costs largely and guarantee production qualities.

2 DEFECTS IN TRADITIONAL VULCANIZING MACHINE

Vulcanizing machine controlling system is easily influenced by exterior factors such as vulcanizing pressure and time.

1 Nowadays, vulcanizing machine, universally adopts PLC that only has a digital logic control function, Do module drives Contactors control electromotor to open and close vulcanizing machine, being responsible for the logic programming of controls such as timer, counter and supplementary relay in order to control the vulcanization technology of the machine. However, the machine can not meet the needs of different vulcanizates and one machine is only suitable for one production and as a consequence increasing the cost of productions.
2 Nowadays, vulcanizing machine, universally adopts robotistic pressure gage to reflect pressure points, manual pressurizes products and thus could not control evacuation time accurately. Those executions rely on experiences and largely influence the final quality of products.

3 THE DESIGN OF AUTOMATIC HYDRAUTIC VULCANIZING INTELLIGENT CONTROL SYSTEM

The accurate control of pressure in vulcanizing machine by advanced PLC control technology can not only improve automatic control and manage in enterprises, but also lower the cost of maintaining vulcanizing machines when enter high maintenance periods. Improving control precision of the process parameters can guarantee the control of vulcanization process conditions, improve the qualities of products and decline energy consumption.

1 With the method of molds, and according to different requirements of molds, setting vulcanizing time, exhaust time and exhaust stay in different conditions, which can adjust products and lower the cost.
2 Accomplished some executions, such as inflating by using hydraulic pressure, adopting digital pressure gages and through Digital-to-Analogue Conversion transmit data to PLC to control the pressure products to control pressure accurately.
3 Adopt the emergency stop mode. Take the equipment failures into consideration, emergency stop button have been designed. When press this key, manual mode starts and template forward and back machine rise and fall,vacuum hood rise and fall in vulcanizing machine can be done. Besides, in manual mode, inflating can also be achieved in exam product performance tests.

4 THE IMPLEMENT OF AUTOMATIC HYDRAUTIC VULCANIZING INTELLIGENT CONTROL SYSTEM

Products of vulcanizing machine mainly consist of following parts: vulcanizing machine equipment mainly consists of the following parts: Omron PLC control system, hydraulic control system, touch screen control system, display system, temperature control system. In pic 1, Omron PLC control system completes activities such as template forward and back, machine rise and fall. Temperature control system heats, and display system inputs product information and molds information, fault query system.

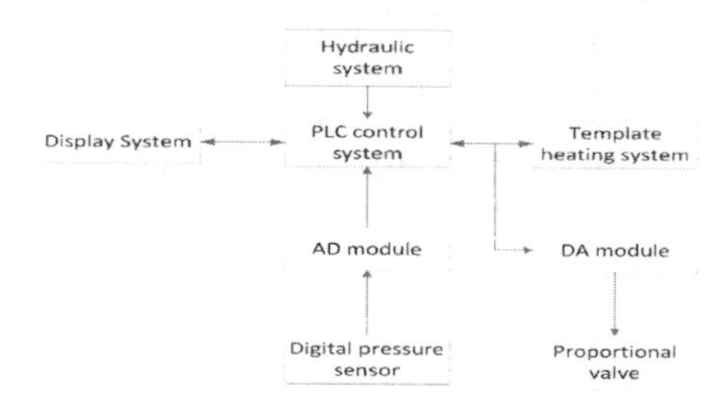

Figure 1. Vulcanizing machine system components.

The operator puts fetal embryoin mold and press automatically start button. After getting input orders and processing information done, the PLC sends modes to a certain place by hydraulic pressure, then mainframe control speed by molds given hydraulic pressure, inflating slowly, and according to the set of open mold clamping, pressure, exhaust and other actions in order to achieve automatic control of vulcanization. Control flow picture of hydraulic pressure products displayed in figure 2.

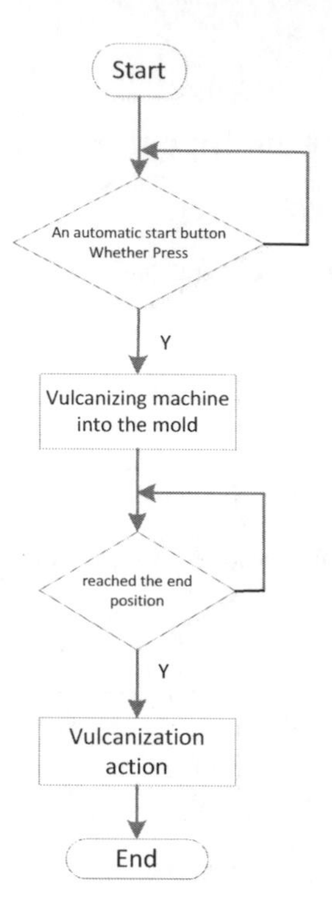

Figure 2. PLC control flow chart.

4.1 *Heat of vulcanizing machine*

in the productive process of products, in order to decrease unnecessary time in the long heat period of template, operators can set the start and stop time to achieve automatically start heating, accelerate production. Timing design screen shown in Figure 3:

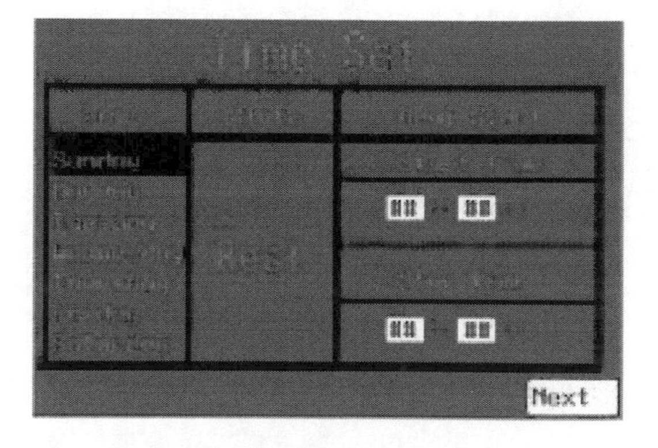

Figure 3. Timer set screen.

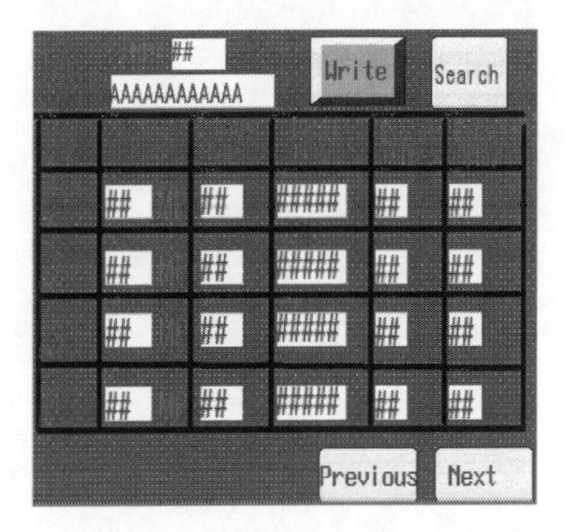

Figure 4. Template settings screen.

4.2 *Mold setting interface*

product process varies in different products, thus the periods in different activities are also different, but all of those can be divided into the following parts: vulcanizing pressure, vulcanizing time, evacuation time, exhaust times. Traditional vulcanizing machines could not adjust these parameters optional, but only add parameters in certain patterns. This product can meet the requirements of ten pressure points, besides, in order to decrease frequent launch of hydraulic system valves, Hysteresis curve of the pressure setting method has been adopted, which enable operator design varies pressure ranges according to different products. The mold setting interface is displayed in figure 4.

5 CONCLUSION

Hydraulic control system decreases labor forces dramatically. Besides, it enables various parameters of vulcanizates to meet requirements of productions to the greatest extents. It has been tested that service time of vulcanizates has been extended largely. Besides, the operation has been simplified and labor forces have also been decreased.

REFERENCES

[1] HUANG Ai-rong,XIANG Zheng-tao,ZHANG Tao,CHEN Yu-feng.Design and Implementation of Vihicle Information Real-timeMonitoring System. COMPUTER ENGINEERING AND DESIGN,Magn. China, vol.31, no.8 , pp.1839–1843, 2010.

[2] Airong Huang, Zhengtao Xiang, Weirong Jiang, Yufeng Chen.Vehicle Auxiliary Anti-collision Warning System Based on Positioning in Electronic Map. The International Conference on Information Engineering and Computer Science(ICIECS 2009), vol. 3, pp.1730–1733, December 2009.

[3] Hu Yi, Yu Dong, Liu Ming-lie. The research status and development trend of industrial control network [J]. Computer Science, 2010, No.1.

[4] Wang Shi-hua. The Development and Application in body Welding of the DeviceNet Bus Technology[J]. Equipment maintenance technology, 2011,12.

[5] Guo, Lei, Wang, Chuansheng; Liu, Qingkun; Su, Dongjian.New-type vacuumizing flat vulcanizing machine structure design and experimental research[J] Key Engineering Materials, v 561, p 218–222, 2013.

[6] Liu, Cai Jun ; Wang, Xu Juan; Xu, Wei.Structural design and experimental study of a new vacuum vulcanizing machine[J].Applied Mechanics and Materials, v 427–429, p 166–169, 2013.

Computing, Control, Information and Education Engineering – Liu, Sung & Yao (eds)
© 2015 Taylor & Francis Group, London, ISBN: 978-1-138-02800-5

Driving habits analysis on vehicle data using error back-propagation neural network algorithm

Wen Jing Zhang, Shu Xi Yu & Yun Feng Peng
A.A.University of Science and Technology Beijing, Beijing, China

Zi Jing Cheng & Chong Wang
Space Star Technology Co., Ltd, Beijing, China

ABSTRACT: In this paper, we designed an Android application to collect real-time vehicle data and analyze the driving habits. Vehicle data, such as speed, and acceleration can be collected from an On-Board Diagnostics module and transferred to a mobile phone through Bluetooth, and then deliver the data from the mobile phone to the backend computing server using HTTP over GPRS or 3G/4G mobile communication networks. An Error Back-Propagation Algorithm (BP neural network algorithm) is developed to classify and analysis the vehicle data to achieve the evaluation on driving habits.

KEYWORDS: Android Platform; OBD; BP Neural Network; Web Service.

1 INTRODUCTION

Vehicles play an increasingly important role in our daily lives. Due to the high cost of fuel, Drivers start to pay attention to eco-driving. Eco-driving[1] can deduce gasoline consumption through good driving habits to avoid rapid deceleration, fierce acceleration, fierce brake. Therefore, the real-time data needs to be collected and analyzed for the drivers to achieve eco-driving.

In recent years, vehicle status sensing system becomes a hot topic. Many researchers conduct in-depth research on various aspects of the vehicle situational awareness system. Literature [2] designed a vehicle status sensing system based on embedded vehicle terminal using LPC2294 microcontroller, to achieve real-time monitoring of the vehicle via communicating with the server through GPRS. It focuses on the design of embedded vehicle terminal and Communicating with the server.

In recent years, vehicle status sensing system becomes a hot topic. Many researchers conduct in-depth research on various aspects of the vehicle situational awareness system. Literature [2] designed a vehicle status sensing system based on embedded vehicle terminal using LPC2294 microcontroller, to achieve real-time monitoring of the vehicle via communicating with the server through GPRS. It focuses on the design of embedded vehicle terminal and Communicating with the server.

The main problem that the eco-driving system facing is to identify the factors which affect fuel consumption. Literatures [3,4] suggests that, avoiding sudden changes in acceleration, and maintaining a low degree of deceleration, as much as possible to use the fourth and fifth gears, and to avoid continuous gear changes will deduce the energy consumption to be good driving behavior.

Following these suggestions, in this paper, we developed a complete system for vehicle data acquisition and driving habits analyzing. An Android application is developed and run on a smart phone to collect vehicle data from the OBD system by Bluetooth communication technology. These vehicle data is uploaded in real time to a central server though GPRS network. An error back-propagation neural network algorithm is developed and run on the server to classify the vehicle data.

This paper is mainly organized as follows. The error back-propagation neural network model and algorithm is presented in Section II. The vehicle data monitoring and driving habits system is presented in detail in Section III.

2 ERROR BACK-PROPAGATION MODEL

Error Back-Propagation Algorithm is an adaptive learning algorithm. It learns the relationship between a given training data set to be network weight, and calculates errors and feed back the errors to the network weights [5]. The differentiable activation functions, i.e., logistic sigmoidal function and hyperbolic tangent, are used in the Error Back-Propagation Algorithm..

A multilayer perceptron neural network based on Error Back-Propagation Algorithm is for nonlinearly multiclass classification tasks. MLP is a parallel computational architecture with fully connected layer's neurons between input and output. It consists of one output layer and one or multiple hidden layer. Weights are used to connect neurons between layers in the network. It usually uses the same activation functions. In our work, sigmoid characteristic function is used in hidden layer and output layer, which can be expressed as,

$$f(x) = 1/(1+e^{-x}) \tag{1}$$

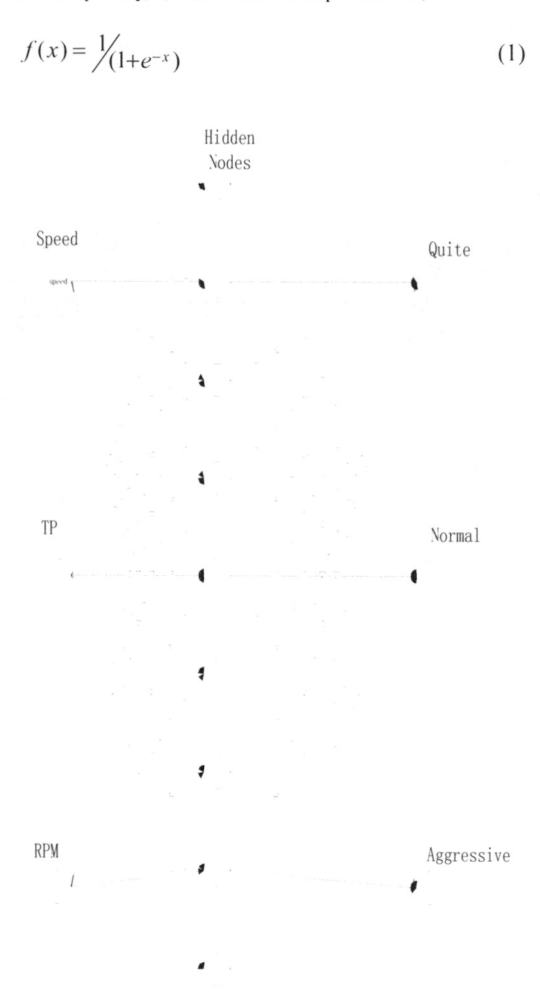

Figure 1. Neural network model.

3 SYSTEM ARCHITECTURE

In the proposed system architecture, the Android smart phone is responsible to collect vehicle data such as speed, revolutions per minute of the engine (rpm), and throttle opening, from the OBD model via Bluetooth, and display these data at the same time. The Android smart phone is also upload these data in real time to a central server though the commercial mobile communication networks. The Server is responsible for storing data and processing data through running BP neural network algorithm. All authorized users can view the car's real-time status on the Internet (Figure 2).

The developed application system consists of three parts: android application, web interface, and BP neural network.

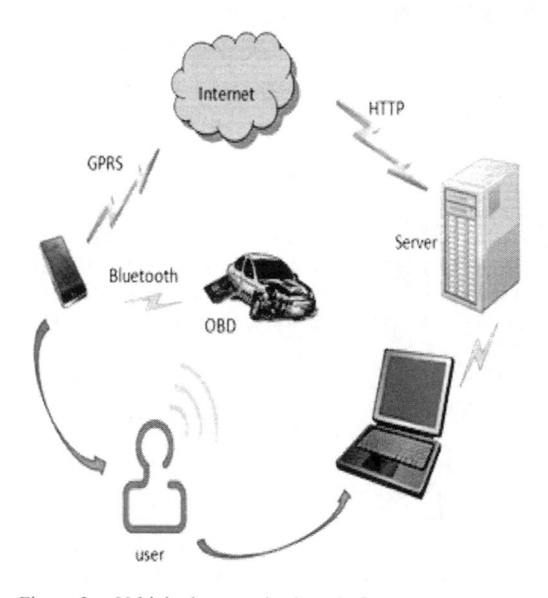

Figure 2. Vehicle data monitoring platform.

3.1 *Android application*

The Android operating system, is an open sourced platform. It has no compatibility issues between different devices, which is convenient for developers to maintenance and upgrades application software. The proposed Android application is a key element, to connect the vehicle and the web server.

The available functionalities of the Android application are described as follows:

OBD Connection: Before collecting the samples, the smart device must be paired with the OBD-II interface. Once the smart device is paired with the OBD, it can receive the data collected.

Data Display: The users can individually select the vehicle information which collected by the sensor being the followings: acceleration, throttle position, engine revolutions per minute (rpm) and put these data show on the screen. It can reflect the driving state intuitively and display diagnostic codes if need (Figure 3).

3.2 *Web interface*

The second main element of the developed application system is the backend information process system involving the database center and the web interface. Two open source softwares, Apache, and Tomcat are selected in this work to build backend system. It can update data in real-time on the web interface. Once the authorized users are logged in, they are able to access all of the vehicle data. The other function of the web server is to convert the data received from the smart phone into the MySQL database, So that the user can view the history of traffic data (see Figure 4).

3.3 *BP neural network*

The BP neural network is a hierarchical neural network whose upper neurons are fully associated with lower neurons. When the training samples are supplied to the network, the transferred value is propagated from the input layer through the hidden layer to the output layer, so we can get the input response from neurons in the output layer[6]. A neural network was created by defining the number of input nodes, along with the number of hidden nodes. In this paper, we have created a neural network, with three input nodes, nine hidden nodes and three output nodes. The neural network allows to classify the type of user's driving habits (quiet, normal or aggressive)[7]. Random weights between 0 and 1 are assigned to links neural networks during initialization.

Figure 3. Android smart phone.

In this work, a problem we faced is to classify the input data, involving the speed, the throttle position, and the revolutions per minute of the engine. The output of the classification process is the type of driving habits. We use Error Back-Propagation Algorithm to implement the classification task. The Neural Network model is trained using a well-defined set of historical dataset and then to classify the driving habits of the individual driver. The input data of each parameter should be normalized between 0 and 1, which can result in a better result of the network training. The equation used to normalize the input data is the following:

$$x'_p = \frac{x_p - x_{min}}{x_{max} - x_{min}} \quad (2)$$

x'_p is the normalized results of x_p, x_{max} and x_{min} represent the maximum and minimum values for the input variables. The training set was generated, after normalizing the input variables. These data set are used to adjust the weights of the neural network. After the neural network implemented successfully, the platform automatically returns the type of driving habits.

Figure 4. Web interface.

3.4 *Level rule*

Driving habits are divided into three types, i.e., quiet, normal, aggressive. In common, if the engine speed is greater than 3000 rpm, the speed covers from 60- to 120km/h, and the throttle position is set at 4-4.9, driving behavior will be sorted as aggressive, which is not the environmentally friendly driving habits. If the engine speed covers from 1900 to 2700 rpm, the speed ranges from 40- to 80km/h, and the throttle position is set to 2.5-3.5, we assume that it is a normal driving behavior which is encouraged. If the engine speed is less than 1700 rpm, the speed covers from 20-60km/h, the throttle position at 2.5-3, we assume that the driving behavior is the level of quiet.

The simulation is carried out on MATLBTM. The neural network model is firstly trained using normalized training data and got to convergence after 17 iterations (Figure 5). Then, normalized testing data, obtained from the OBD, are input to the trained model to get the classification results (Figure 6). We presented the result to the driver in the form of a pie chart intuitively (Figure 7).

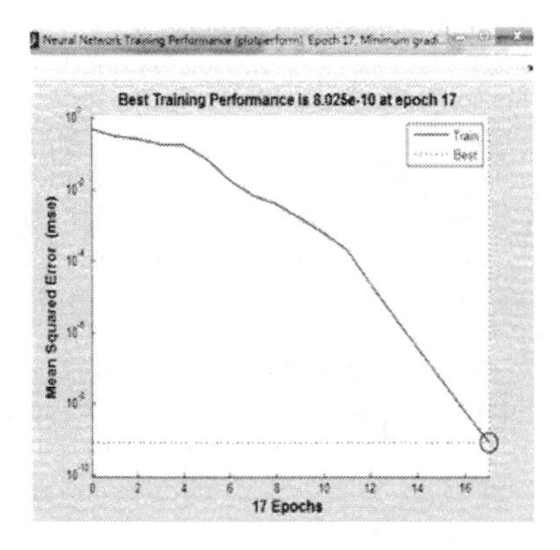

Figure 5. The training result of neural network.

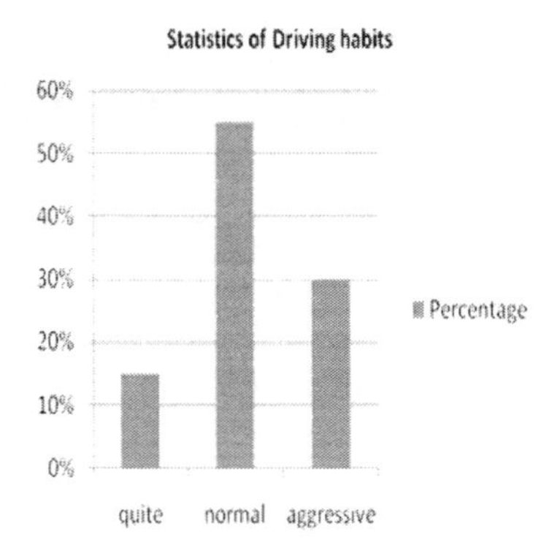

Figure 6. The classification result of test data.

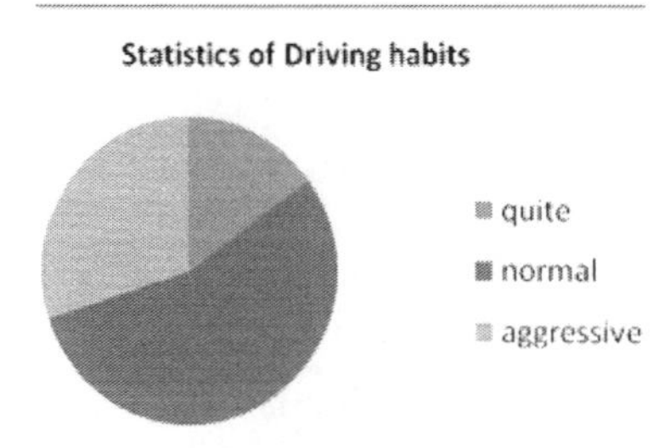

Figure 7. The pie chart of classification result.

4 ACKNOWLEDGEMENT

This work is supported by the Open Research Fund of The Academy of Satellite Application under grant NO. 2014 CXJJ-TX 14.

5 CONCLUSION

In this paper, we demonstrated a vehicle data acquisition and driving habits analysis system, which can deduce the driving habits from vehicle data obtained from the OBD. The main objective is to help drivers improve their driving habit by real-time notification during driving. The driving habits classification is performed using a BP neural network model and simulation results verified its function.

REFERENCES

[1] Doo Seop Yun, Jeong-Woo Lee, Shin-Kyung Lee, "Development of the eco-driving and safe-drivingcoponents using vehicle information". Oh-Cheon Kwon ICT Convergence (ICTC), 2012 International Conference on, 10.1109/ICTC.2012.6387198.

[2] Wang Yanni, "Vehicle Remote Monitoring System Based On LPC2294[D] ". Taiyuan University of Technology, 2010.DOI:10.7666/d.d082692.

[3] Erikcsson, E., "Independent driving pattern factors and their influence on fuel-use and exhaust emission factors" Transportation Research Part D: Transport, Vol. 6(5), pp. 325-345, 200l.

[4] Johansson, H., Gustafsson, P., Henke, M., Rosengren, M., "Impact of EcoDriving on emissions". International Scientific Symposium on Transport and Air Pollution, Avignon, France, 2003.

[5] Albarakati.N, Kecman,V. Southeastcon, " Fast neural network algorithm for solving classification tasks: Batch error back-propagation algorithm" 2013 Proceedings of IEEE DOI: 10.1109/SECON.2013.6567409.

[6] Zhu Li,Qin Lei, Xue Kouying ,Zhang Xinyan, «A novel BP Neural Network Model for Traffic Prediction of Next Generation «. Natural Computation, 2009. ICNC '09.

[7] Meseguer, J.E., Calafate, C.T., Cano, J.C., Manzoni, P., "DrivingStyles: A smartphone application to assess driver behavior"Computers and Communications (ISCC), 2013 IEEE Symposium on, 7-10 July 2013, 10.1109/ISCC.2013.6755001.

Computing, Control, Information and Education Engineering – Liu, Sung & Yao (eds)
© 2015 Taylor & Francis Group, London, ISBN: 978-1-138-02800-5

Use of cloud computing for distributed simulation: A way forward

Xi Hui Fan, Zhi Fan, Qian Xuan & Ming Xi Li
New Star Research Institute of Applied Technology, Hefei, P.R.China

ABSTRACT: Many of the current, in-use simulation systems today are difficult to install, configure, and execute. Cloud Computing has received a significant amount of attention leading to large information technology (IT) initiatives at the corporate and government levels. It is natural to explore the cloud computing with the modeling and simulation (M&S) community. The paper puts forward a distributed simulation framework use of Cloud computing.

KEYWORDS: Cloud Computing; Simulation; Virtualization; SOA; OWL.

1 INTRODUCTION

The formal definition of Cloud Computing as defined by the National Institute of Standards and Technology (NIST) is, "Cloud Computing is a model for enabling ubiquitous, convenient, on-demand network access to a shared pool of configurable computing resources (e.g., networks, servers, storage, applications, and services) that can be rapidly provisioned and released with minimal management effort or service provider interaction. This cloud model is composed of five essential characteristics, three service models, and four deployment models."

Cloud computing is being seen as a means for reducing the cost of IT support by consolidating resources, reducing the hardware and energy footprint, and more efficiently managing resources. Cloud computing is a paradigm which allows the use of outsourced infrastructures in a "pay-as-you-go" basis, thanks to which scalable and customizable infrastructures can be built on demand.

M&S has solved the hard problem of how things actually connect, where cloud promises that things will work without ever defining how that happens. However, distributed simulation has come at a price:

- Each participating M&S site has to maintain its own facilities and equipment in order to participate in exercises.
- Set up for a particular distributed simulation exercise can take months for coordination and weeks on the ground at various sites for installing and integrating participating simulation systems.

Cloud seeks to consolidate resources, providing the ability to co-locate many of the "back-end" components and all within a common hardware infrastructure. Cloud continues to support remote access, utilizing the same network infrastructure already in use with more traditional distributed simulation configurations. Zero, thin or thick client access options are available depending on the display performance needs and location.

2 USE OF CLOUD COMPUTING FOR DISTRIBUTED SIMULATION FRAMEWORK

Cloud Computing is composed of three service models, which is IaaS (Infrastructure as a Service), PaaS (Platform as a Service), and SaaS (Software as a Service). IaaS is the variety of the underlying peer computing and storage resources as a service to users. PaaS is to develop an application and deployment platform as a service to users. SaaS is a Web-based application to provide to the user. To use of Cloud Computing for Distributed simulation, a simulation framework is proposed, which is divided into three layers, shown in Fig.1. The Resource layer use IaaS, the Bus layer use PaaS, and the Application layer use SaaS. The framework is hierarchical and loose coupling in order to achieve the underlying resource, separation and terminal emulation service applications to improve simulation platform scalability, extensibility and maintainability.

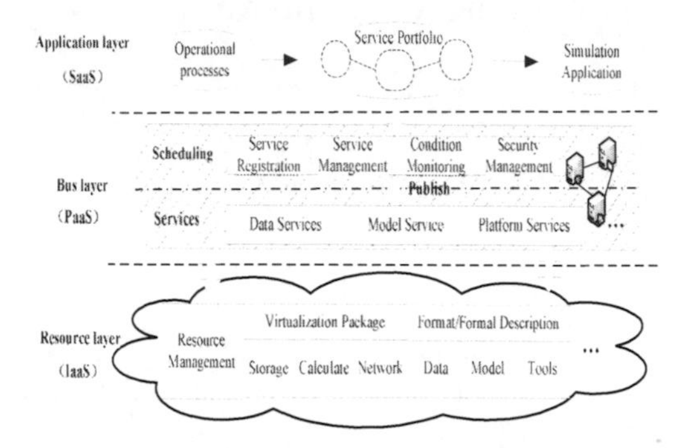

Figure 1. Distributed simulation framework use of cloud computing.

2.1 *Resource layer*

Resource Layer, also known as the resource pool, the purpose is to solve the integration and management of physical resources and simulation software resources needed, including computing, storage, network, hardware, data, models and tools resources. For hardware resources, virtualization technology is adopted. For software resources, using the resource description method based on metadata and ontology. Create a resource pool is conducive to enhancing the reusability of simulation resources, reduce simulation development costs and improve the efficiency of system development.

2.2 *Bus layer*

bus Layer is to support the communication, integration, and management of automated interaction between simulation services, implemented by a variety of services assembled into simulation applications, to provide support for simulation services portfolio, and its function is similar to the traditional run support environment (Run-Time Infrastructure, RTI), meet the requirements of the specification and simulation software management entity, can be the same as the plug is inserted into the bus, which effectively supports simulation system interconnection and interoperability. Not only in the form of bus services layer provides functional simulation run support, while also providing the necessary discovery of existing simulation models of various services. Bus layer is divided into services and scheduling, services are responsible for scheduling to publish simulation-related services, including data services, models services and platform services. Scheduling include service registration, service management, condition monitoring and security management.

2.3 *Application layer*

application layer use SaaS for on-demand self-service. Through needs analysis, service discovery, service matching, service composition and other processes to meet simulation applications. Its technical essence is a member of the Federal Web Services Oriented software design and development, by implementing stateful Web services provide the ability to dynamically configure the federal member, through the introduction of semantic Web service description for simulation services portfolio.

3 IAAS SUPPORT FOR RESOURCE LAYER

Virtualization is one of the core technologies of Cloud. Cloud seeks to consolidate resources, providing the ability to co-locate many of the "back-end" components and all within a common hardware infrastructure. The virtualization component of cloud reduces the need for hardware (and space and power). For example, a 50 PC configuration for running a large simulation exercise would require a large lab space for all the machines and monitors (chairs, etc.). The same exercise using virtualization takes a fraction of the space with 50 PC's virtualized and running on a server. Zero or thin clients are needed only for the participants or pucksters of the exercise. This reduces the lab space footprint of the space necessary to support the operators and eliminates traditional PC workstations from the exercise. The overall hardware, power and space footprint is greatly reduced. Resources can be reallocated for other applications and exercise configurations without the need to wipe and reinstall the operating system and applications every time.

The virtualized architecture used during the exercise is depicted in Figure 2. We used 6 servers, each with 32 GB of RAM and 1 TB of HD in this architecture. One of these servers was for backup. Three of the servers were for the simulation server processes. The other two servers were for virtual desktops. VMWare ESX was used for server virtualization and VMWare View was used for desktop virtualization. We also used 27 thin clients to provide the end users with CAX services. Each virtual machine for end users were dedicated 3 GB of RAM in our server pool. In conventional architecture, i.e., not virtualized, we use 11 powerful servers with 16 GB of RAM and 512 GB of HD in the average, and 27 powerful PCs.

Our tests show that virtualization is practical and more cost effective for the distributed simulation environment. We also investigated the results of the Intercloud approach for distributed simulation. There are important parameters that depend on the location of a service, such as the time required starting a federation and the capability of federation to keep up with real time. These metrics are mostly related to the quality of service (QoS) parameters and especially delay.

4 PAAS SUPPORT FOR BUS LAYER

In the M&S field, the RTI software which implements the High Level Architecture (HLA) interface standard plays an important role in the distribution interaction simulations. In Cloud computing age, the study of distributed interaction simulation is expanding from

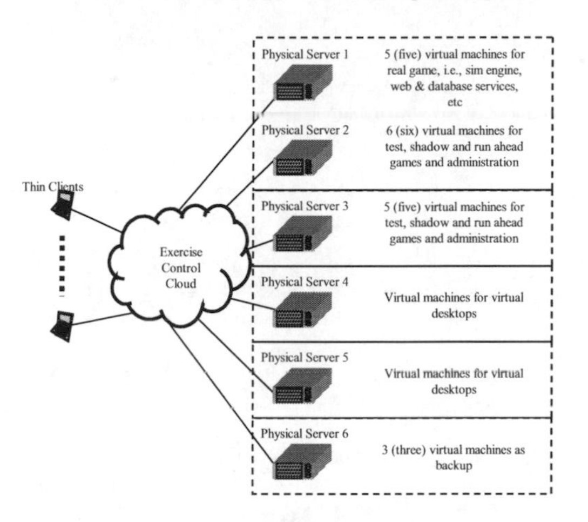

Figure 2. Virtualized simulation services.

LAN to WAN. For the simulation applications on the WAN, the communication on WAN should be paid much more attention. As a type of Cloud computing, Platform as a Service (PaaS) takes revolutionary changes to the software application. It facilitates deployment of applications without the cost and complexity of buying and managing the underlying hardware and software layers. The essence of the PaaS is transferring the computing resources from desktop to Internet, and managing these resources efficiently. The users of PaaS interoperate with the others using Web Services.

4.1 *Web services*

Web Service technology is a way for a software module to call other software modules, usually across a wide area network, in order to make use of their services. The calls are performed using the http protocol, originally developed for transferring web pages. The content that is transmitted is formatted using XML.

Using the Web Service Description Language (WSDL), a set of services can be described to support various application domains. There are a number of software frameworks that make it possible to generate all necessary program code to perform the Web Service calls. Code can be generated in a variety of languages (C++, C#, Java J2SE and J2ME, Visual Basic, ADA, even PL/1, Fortran and Cobol) for deployment in a large number of technical environments.

The HLA Web Services API is described using the WSDL language, which is a precise description of the services, not an actual programming API. Using the WSDL it is possible to generate code that carries out the correct calls. This provides one of the most important benefits of the WSDL API. A wide range of development and deployment platforms are supported. Note that the portion of the federate that communicates with the rest of the federation, the "Local RTI Component", is not necessarily specific to any particular RTI implementation. It is worth noticing that the semantics of what an LRC is has thereby shifted somewhat from a "fatter" to a "thinner" approach. To connect to an RTI you must find somewhere to connect to, which is known as the Web Service Provider in the Web Service world. For this purpose a new RTI component is necessary, known as the Web Service Provider RTI Component (WSPRC), as shown is Figure 3.

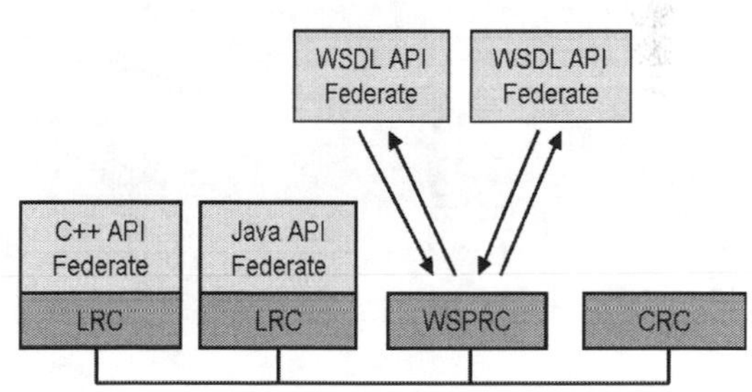

Figure 3. The web service provider RTI component.

4.2 *Paas RTI*

Key technologies of the PaaS RTI include building RTIWS, scheduling and migrating RTIWS (RTI Web Services) instances, and so on. RTI WS is the core of the system. Constructing RTI WS is encapsulating the HLA/RTI API to be Web Services.

The RTI software (Portico RTI for example) includes two parts. The first part is the member methods of the RTIambassador class, and they can be invoked by the federates to request RTI services. The other is callback functions, which is invoked by the

RTI itself to send message to the federation, and they are provided by the FederateAmbassador class.

Figure 4 is the class diagram of the RTI WS. The well known factory/instance patterns are adopted. The factory service creates a resource, and the instance service provides the operations on the resource. The classes of the RTI software RTIambassador and FederateAmbassador_impl are encapsulated into class RTICore which is involved in the class ConceptionalResource. So an instance of the class ConceptionalResource is the real federate in the HLA federation and it is the proxy for the remote user. The remote users (or remote federate) can invoke the member functions of the class CentralRTIFactoryService to create a federate, and then operate through the class CentralRTIService. The class ConceptionalResource also implements the interface RTICoreCallback, it receives the callback information from RTI through class RTICore, then notifies the callbacks to the remote users.

Under this PaaS architecture, the user interacts with others by call the WSRF interfaces of an RTI WS instance. So system keeps an RTI WS instance for each user. Each RTI WS instance also consists of an instance of the class FederateAmbassador and an instance of the class RTIAmbassador, and it can be taken as a "proxy" federate for the remote user. The remote users interoperate through their proxies on the data center.

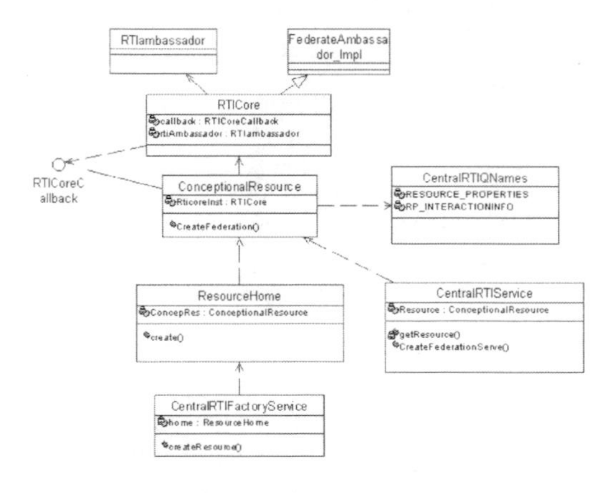

Figure 4. Class diagram of the RTI WS.

5 SAAS SUPPORT FOR APPLICATION LAYER

A lot of researches have been done about Web service discovery, such as UDDI-oriented, P2P-oriented and Semantic-oriented Web service discovery, etc. Among them the Semantic-oriented Web service discovery is more advanced. This paper attempts to apply it in the field of distributed simulation. On the base of WSDL service descriptions and UDDI service discovery, this paper adopts OWL-S as Semantic Web service description ontology to describe its functional requirements, increase the semantic description of the mapping of WSDL to OWL-S and OWL-S to UDDI, and improve the efficiency, security, reliability, stability, and response time of simulation services discovery.

Figure 5 shows the simulation service composition process, first of all is the mapping of WSDL to OWL-S, on the assist of domain ontology adds the semantic annotation for OWL-S service, so that the realization of a unified domain concept and customized combination process, to achieve a combination of Web services. Two important issues are the mapping of WSDL to OWL-S and OWL-S to UDDI.

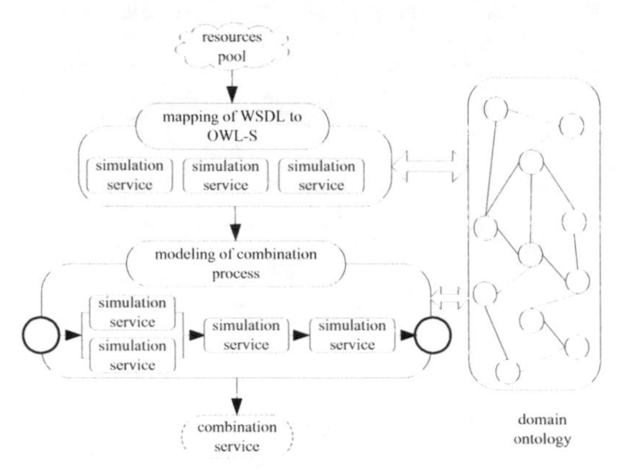

Figure 5. Process of service composition.

The methodology of service discovery and composition was tested by an example of a weather service. Its Web service interface is shown in Figure 6.

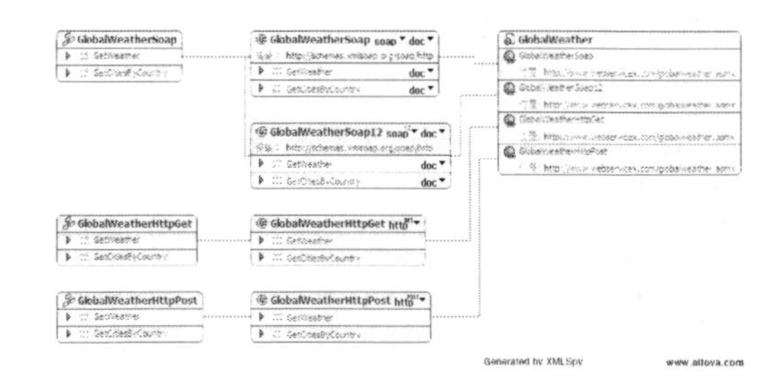

Figure 6. Web service interface.

6 CONCLUSION

Cloud computing is not so much a destination as it is another way point on the journey to applying standards to simulation environments that are growing ever larger and more distributed. Having management strategies in place for providing software as a service in a platform as a service (virtualization) environment, we are now working on the formalization of infrastructure as a service to allow users to completely provision their own resources. Measuring the services provided remains a "nice to have" feature for the near term with research planned for exploring what further metrics make the most sense for future users. These metrics may become more interesting as users begin to make expanded use of self-provisioning capabilities in order to execute large Monte-Carlo type studies.

ACKNOWLEDGMENT

This paper supported by the Anhui Provincial Natural Science Foundation (Grant No. 1408085QF129 and 1408085QF124).

REFERENCES

[1] NIST Special Publication 800–145,The NIST Definition of Cloud Computing,http://csrc.nist.gov/publications/nistpubs/800–145/SP800-145.pdf.

[2] Christina Bouwens,Kurt Hawkcs,Jay Duncan. Modeling and Simulation as a Service: Utilizing Cloud Technologies for Cost Effective Simulation Delivery[C]//Spring Simulation Interoperability Workshop(SIW),Orlando, FL,USA,number 12S-SIW-020, 2012.

[3] Paul Lowe.A Path Taken to Create an M&S Cloud for LVC Simulations[C]//Fall Simulation Interoperability Workshop(SIW), Orlando, FL, SA, number 11F-SIW-009, 2011.

[4] David F.Perme,William H.Crain.Got Simulation?Private Cloud Computing-Remove a Roadblock[C]//Fall Simulation Interoperability Workshop(SIW),Orlando,FL, USA, number 09F-SIW-044,2009.

[5] Bo Hu Li,Xudong Chai,Lin Zhang,Baocun Hou, et al.New Advances of the Research on Cloud Simulation,AsiaSim2011,PICT4,pp.144–163,2011.

Computing, Control, Information and Education Engineering – Liu, Sung & Yao (eds)
© 2015 Taylor & Francis Group, London, ISBN: 978-1-138-02800-5

Research of the color descriptor SIFT

Ye Ran Wang, Tong Bo Zhang, Guang Li Li, Chun Xiao Dong & Hong Wei Zhao
College of Computer Science and Technology, Jilin University, Changchun, China

ABSTRACT: In the SIFT algorithm, the choice of the color model determines the representation of the color description. The suitable color descriptor can improve the accuracy of the matching algorithm. However, the existing researches often do not analyze the color model together with algorithms, nor consider the particularity of the specific algorithm. Focusing on SIFT algorithm, this article compares the color description of the color space RGB with that of color pace HSV. Aiming at the shortcomings of the various types of descriptor, we put forward the researching direction for the color descriptor in the future.

KEYWORDS: Color Model; Color Descriptor; SIFT Algorithm.

1 INTRODUCTION

1.1 *SIFT algorithm*

SIFT (scale invariant feature transform) algorithm was first proposed by Lowe [1] in 1999. This algorithm can extract the stable feature points, and deal with the matching problems of processing translation, rotation, affine transformation and viewpoint transformation between images. It is a kind of registration algorithms based on feature, and it has strong robustness which makes its own the best matching performance. What's more, it is studied most widely currently.

This algorithm uses Gaussian functions to make the images blurred, and establishes the scale space based on Gaussian pyramid. As for the Gaussian pyramid, it uses DOG to normalize the scales and to detect the local minima points. It removes the minima points which have low contrast, and it removes the edge effect. Then it distributes the directions of key points, and makes description of key points to generate feature descriptor which can describe the extracted feature points.

1.2 *The SIFT algorithm based on grayscale*

Classic SIFT algorithm treats the color information as the gray information, and it can be calculated on one channel. Although we can greatly reduce the time complexity with gray information, we ignore the color information and reduce the accuracy of image matching for the descriptor of classical SIFT algorithm is generated by gray images.

1.3 *The classification of noise*

After the image have been mixed up with noise, we should judge all the dimensions extracted. If it does not change, we can make sure that the descriptor has robustness. The SIFT with color information and that without color information have different performances in the conditions of different noise effects.

Table 1. Comparing of two kinds of SIFT.

	Gray-SIFT	Color-SIFT
The same scale change	YES	YES
The same incremental change	YES	YES
Scale change and incremental change	YES	YES
Scale changes in different color components	NO	YES
Linear changes in different color components	NO	YES

In Table 1, the existing grayscale SIFT feature extracting algorithm has the invariance for the first three changes. On the one hand, they both adopt the gradient histogram which eliminates the intense changes, on the other hand, the linear changes of RGB do not affect the directions of the gradient. In addition, they both normalize the descriptor vectors ultimately which have solved the problem of the scale changes of gray level. Gray SIFT algorithm cannot solve the following two dimension changes while SIFT algorithm combined with color information can solve all the problems.

2 THE SIFT ALGORITHM COMBINED WITH COLOR INFORMATION

2.1 *The meaning of the introduction of color descriptor*

Color is important image information. It can describe the information on objects and the scenes in the images more accurately. However, SIFT algorithm based on grayscale ignores the color information so that it will lead to decline in its matching performance to some extent, and cause a certain target identification or classification errors. In conclusion, introducing the color descriptor can improve the accuracy of the algorithm.

2.2 *SIFT color descriptor*

For RGB color space and HSV color space, this article has screened several good color models from the aspects of universality of application and practicability. In this paper, color descriptors for the SIFT algorithm are compared.

RgSIFT [2] is a color model, after normalizing the classic RGB color model. It can show color information by the r/g expression after being normalized. Because the value of each channel of RgSIFT is obtained by the linear-transformed RGB features, it has invariance to the linear changes of all the channels. After normalization, each component has the scale invariance, which defined as follows:

$$\mathrm{r} = \frac{R}{R+G+B} \quad \mathrm{g} = \frac{G}{R+G+B} \quad \mathrm{b} = \frac{B}{R+G+B} \tag{1}$$

m1 m2 m3 model [3] is in RGB color space. It establishes the ratio model by calculating the colors between two adjacent regions. Also, it has certain robustness of color changes which are caused by the change of illumination. When the rotation and scaling of images happen, the pixel points of comparison will also change, its expressions are as follows:

$$m_1 = \frac{R^{x_1} G^{x_2}}{R^{x_2} G^{x_1}} \quad m_2 = \frac{R^{x_1} B^{x_2}}{R^{x_2} B^{x_1}} \quad m_3 = \frac{G^{x_1} B^{x_2}}{G^{x_2} B^{x_1}} \tag{2}$$

Transformed color SIFT[4] is in the RGB space model. The change of light intensity will make the color distribution change irregularly, affect the stability of RGB histogram, and result in the RGB histogram offset further. Subtracting the mean value can offset the offset of each channel, and dividing the variance can offset the value-scale changes of each channel. So after normalization like this, the histogram is invariant to illumination changes, the histogram will not shift. The space model is defined as follows:

$$R' = \frac{R-\mu_R}{\delta_R} \quad G' = \frac{G-\mu_G}{\delta_G} \quad B' = \frac{B-\mu_B}{\delta_B} \tag{3}$$

3 THE DEVELOPMENT TREND OF THE COLOR DESCRIPTOR

3.1 *Reduce the time complexity of SIFT algorithm*

SIFT algorithm based on grayscale only calculates the brightness of one gray channel while the SIFT algorithm based on color descriptor finishes the match of colorful targets through calculating the eigenvector of three channels respectively, and synthetizing a feature vector. Their dimensionality of feature vectors is much higher than that in classical SIFT algorithm so that the time it takes will be greatly increased. The existing means of reducing the time complexity basically can be divided into two kinds, dimension reducing and paralleling.

In terms of dimension, one of the most famous algorithms is PCA - SIFT algorithm put forward by Yanke [7] in 2004. He has applied the principal component analysis (PCA) to the gradient image to obtain PCA - SIFT feature vectors to do image matching. In 2006, Bay [8] has proposed SURF algorithm which has scale invariance and rotation invariance basing on SIFT algorithm and combining integral image with Haar wavelet. In 2010, Liu Xiangzeng [9] has proposed a way to build FKICA - SIFT descriptor by using fast kernel independent component analysis to extract affine invariant independent components of SIFT descriptor focusing on SAR image registration.

In terms of parallel processing, in 2010, Nianhua [10] has designed and implemented the SIFT on CUDA platform d in parallel, and made comparisons with the implementation of the CPU. In 2012, Jiang Guiyuan [11] has adopted data parallel strategy, and put forward a distributed parallel algorithm (DP-SIFT algorithm) that can extract the SIFT features of an image in the PC fleet or COW (cluster work station): according to the feature of characteristic Gaussian scale pyramid, they put forward a data block partition algorithm whose height and width are limited, and they designed the way to distribute data and adjust characteristics. In 2013, Li BF [12] has proposed module parallel SIFT algorithm(pSIFT

- noBE).It automatically ignores some boundary feature, and it is suitable for the processing of large-size images.

Dimension reduction can effectively reduce the time complexity, but in a large number of image information processing, it still needs a lot of time. Therefore, it is needed to further reduce the time complexity. With the researches that many scholars at home and abroad have done in recent 10 years, the research method of dimension reduction has been basically perfect, so that, it is difficult to further reduce the time complexity of algorithm greatly.

Parallel methods can greatly reduce the time complexity of SIFT algorithm, but they actually need a lot of hardware overhead to shorten the time of getting the results while the amount of the work is still the same. In the next process of research, we should consider doing some theory optimization for SIFT algorithm.

3.2 *The selection of color model*

Gevers [2] mentions the saturation S, it shows how much image information has been filled with white light (the information of illumination in an image). Due to the adaptability of light changes in different color models, we can determine what color model we choose according to the difference of illumination information.

In the process of capturing an image, the target needed to be identified is mainly in the center of the image, so when we want to get the illumination information of the image, we can just calculate a part of the image. What's more, getting the color information of a part of the image only need to spend very little time, and obtaining an accurate color model can greatly improve the accuracy of the SIFT algorithm. We can choose the corresponding color model after getting the light degree of an image by using the expression (4).

$$S = 1 - \frac{\min(R, G, B)}{R + G + B} \tag{4}$$

RgSIFT can adapt to the general lighting occasion. The advantage of m1m2m3 model is its strong ability of adapting to color changes. After comparison and analysis by Gao Jian[13], m1 m2 m3 has a lower matching error rate than RgSIFT while its success rate of matching has obviously decreased. At this point, we can make sure whether the image has changed colors according to the calculation of the degree of saturation S, and get the best matching effect. Transformed color SIFT has the invariance in the migration and changes of gray as well as color. The average can offset the value of the deviation of each channel, and the value dimension changes of each channel can be offset by divided by the variance.

However, when the values of each channel in RGB color space are close to zero, the denominator of these color model components will also be close to zero. In this case, the RGB color model could show the instability. In practical applications, the descriptor provided by the RGB color space is not fit to scene with little light. In the dark scene, the HSV color space has obvious advantages on this point relative to RGB color space. But the HSV is a color model which is more intuitive, it is not suitable for being used in illumination model. So many operations of mixed light and light intensity operations cannot be realized by HSV.

4 CONCLUSION

In this paper, we have analyzed several existing color models with high recognition. On the basis of the existing improvements of SIFT algorithm which uses dimensionality reduction and the parallel method, we think that the improvements of SIFT algorithm should not only be finished by those two methods above. Therefore, the problem is that there is the lack of theoretical researches and improvements to the research of SIFT algorithm.

To make choice of color model, we should understand the illumination information about the environment so that it can be more suitable for the environment as far as possible. In view of the improvement of SIFT algorithm and the problem of the selection of color model, the paper shows some views on the development direction of the SIFT in the future, and these conclusions have very important significance to reduce the time complexity of SIFT algorithm and to improve the matching efficiency of SIFT algorithm.

ACKNOWLEDGEMENTS

This paper is supported by Jilin University students' innovation and entrepreneurship training plan (Grant No.2014A53226). The corresponding author of this paper is Hong Wei Zhao.

REFERENCES

[1] Lowe D G. 1999. Object recognition from local scale-invariant features. *The proceedings of the seventh IEEE international conference on. Ieee, 1999, 2*: 1150–1157.

[2] Van De Sande K E A, Gevers T, Snoek C G M. 2010. Evaluating color descriptors for object and scene recognition. *Pattern Analysis and Machine Intelligence, IEEE Transactions on, 2010, 32(9)*: 1582–1596.

[3] Gevers T, Smeulders A W M. 1999. Color-based object recognition. *Pattern recognition, 1999, 32(3)*: 453–464.

[4] Van De Sande K E A, Gevers T, Snoek C G M. 2010. Evaluating color descriptors for object and scene recognition. *Pattern Analysis and Machine Intelligence, IEEE Transactions on, 2010, 32(9):* 1582–1596.

[5] Bosch A, Zisserman A, Muoz X. 2008. Scene classification using a hybrid generative/discriminative approach. *Pattern Analysis and Machine Intelligence, IEEE Transactions on, 2008, 30(4)*: 712–727.

[6] Van De Weijer J, Gevers T, Bagdanov A D. 2006. Boosting color saliency in image feature detection. *Pattern Analysis and Machine Intelligence, IEEE Transactions on, 2006, 28(1)*: 150–156.

[7] Ke Y, Sukthankar R. 2004. PCA-SIFT: A more distinctive representation for local image descriptors. *Proceedings of the 2004 IEEE Computer Society Conference on. IEEE, 2004, 2*: II-506-II-513 Vol. 2.

[8]]Bay H, Tuytelaars T, Van Gool L. 2006. Surf: Speeded up robust features. *Springer Berlin Heidelberg, 2006*: 404–417.

[9] Liu Xiangzeng,Tianzheng,Shi Zhenguang,Chen Zhanshou. 2010. Synthetic aperture image multi-scale registration based on FK1CA - S1FT characteristic. *Optical precision engineering,2010, 19(9)*: 2186–2195.

[10] Nian Hua. 2010. GPGPU and Image Matching Parallel Algorithm Based On SIFT, *Xian electronic science and technology university 2010,1.*

[11] Jiang Guiyuan, Zhang Guiling, Zhang Dakun.A. 2012. Distributed Parallel Algorithm for SIFT Feature Extraction. *Journal of Computer Research and Development, 2012,99(5)*: 1130–1191.

[12] Li, BF, Zheng, ML, Jiang, JP, Zhang, XM ,Tian, BH. 2013. Evaluating the Block-Parallel SIFT Algorithm without Boundary Extension.*2013 6TH INTERNATIONAL CONGRESS ON IMAGE AND SIGNAL PROCESSING (CISP), VOLS 1–3, 2013*: 422–426.

[13] Ganjian, Huang Xinhan, Penggang, Wangmin, Wu Zuyu. 2007. The extraction and matching of feature points based on the SIFT. *Computer Engineering and Applications, 2007,43(34).*

Computing, Control, Information and Education Engineering – Liu, Sung & Yao (eds)
© 2015 Taylor & Francis Group, London, ISBN: 978-1-138-02800-5

Analysis of computer-aided design of the detector appearance

Xin Ting Yu
College of the Arts, Qilu University of Technology, Jinan, China

ABSTRACT: Computers are used more and more widely in industrial design. From inspiration capture and sketching in the pre-design to 3D modeling in the post-design technology, computer-aided virtual design occupies an important position in the whole design process. We can use three-dimensional design software SolidWorks to design the detector's appearance so as to actually reflect the cutting specifications of the detector's housing parts and to show the precise spatial location of each component. Applying computer-aided design to simulating product development and design process can not only shorten the product development cycles, but also reduce the cost of product development and design, enabling businesses to gain more benefits.

KEYWORDS: Industrial design; Computer-aided virtual design; SolidWorks; Detector design.

1 INTRODUCTION

Since China's manufacturing industry is always lagging behind other countries, it's necessary for all companies to try every way to use the least cost in order to ensure maximum benefits in the product design process. The rise of computer-aided virtual design has satisfied the businesses to achieve the highest quality and fastest product update cycle at the lowest cost of product development and design.

Virtual computer-aided design is based on computer three-dimensional software. From the initial concept renderings and engineering drawings to the final product styles and addition assembly relationships in various parts of the product, all are presented one by one in the eyes of designers and producers so that everyone can have an intuitive visual experience. Virtual computer-aided design represents a brand new manufacturing model and design concept. Pieces of software are commonly used in modern computer-aided three-dimensional virtual design: SolidWorks, Unigraphics (UG), Pro / Engineer (Pro / E) and Catia, etc. These pieces of three-dimensional software are integrated with design, correction of motion and finite element analysis in one. Their modeling is fast and intuitive and can show the mutual coordination between all parts of the product. At the same time, it can also achieve a series of processes such as product design renderings, three-dimensional structural design and engineering drawing generation. Among these, SolidWorks, made by American SolidWorks Corporation, established its undefeated market position as soon as it launched in 1995 for its excellent performance, ease of use and innovation. SolidWorks is the first design software of all three-dimensional software to use the Windows operating system. Because of its simple operation and easy data modification, which greatly improves designers' efficiency, SolidWorks was quickly accepted by designers from different countries.

This article is mainly about applying SolidWorks to achieving the design process of detector's appearance renderings, engineering drawings and parts assembly drawings, so that the readers can directly understand the application of computer-aided virtual design in product development process.

2 SOLIDWORKS PRESENTING DETECTOR'S EXTERIOR RENDERINGS

Detector is widely used and can be seen in the plants of many industries such as petrochemical, power, aviation, shipbuilding, paper and textiles industry. Temperature detector is used to check the temperature of a gas or liquid working environment, ensuring the safety of working environment. When the temperature does not meet the requirements in working environment, the instrument will emit a warning sound and the electronic screen will display an error message to give people a warning.

Using SolidWorks to present team pre-drawn renderings in the computer, which requires the use of bottom-up design model, designers put every parts of the detector on the SolidWorks based on the actual size of the hand-painted renderings (Fig.1), and then use the assembly relationships of SolidWorks to unite every parts into products. After assembling a good customization ,designers not only need to make the size parameter perfectly correlated, but also need to get the automatic perfect correlations among geometric shapes or the various components. SolidWorks provides designers with exactly the same interface and commands related to a fully automated design environment (Fig.2).

In this step, there may be some conflicting data, where the advantage of SolidWorks has manifested. Designers can find conflicting data to modify. Once any part is modified, other associated model, size and etc. are automatically update, without human intervention.

As is vividly shown in the detector renderings production: SolidWorks can provide comprehensive assembly characteristics of the product-level functions to create and record a particular assembly design process. In the actual design, according to the design intent, many features generate only in the assembly environment and after the assembly operation, so there is no need to consider about them in the components design. After completing the products'assembly drawings, components are produced with cooperation. Such as the functions of Parts' welding, punching, cutting and interfering check. After assembly check is completed, you can also use SolidWorks in the assembly environment to make out a explosion picture by setting appropriate explosion perspective for the shell of the detector (Fig. 3).

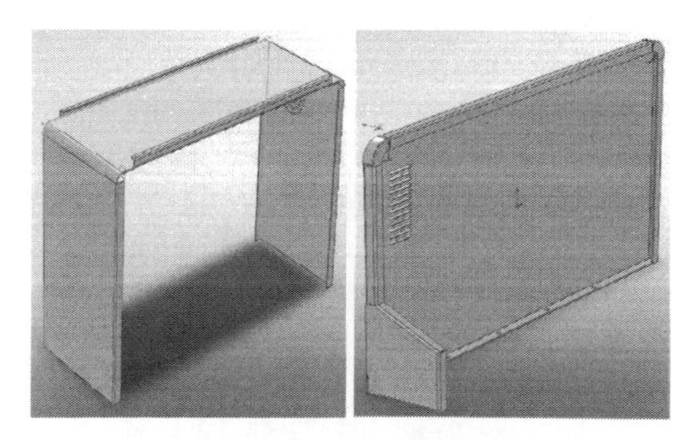

Figure 1. Impression drawing of various parts.

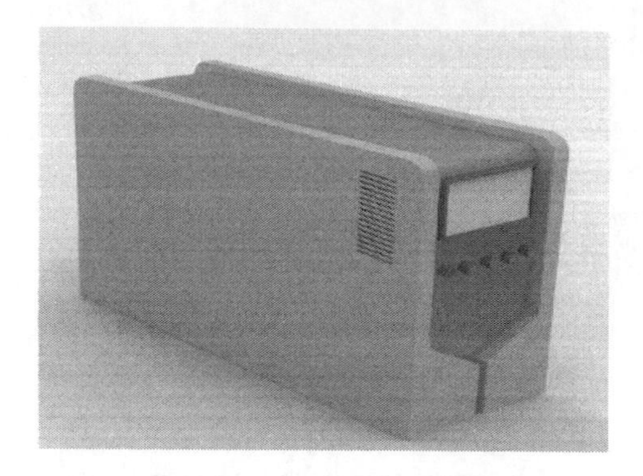

Figure 2. Impression drawing of the detector assembly.

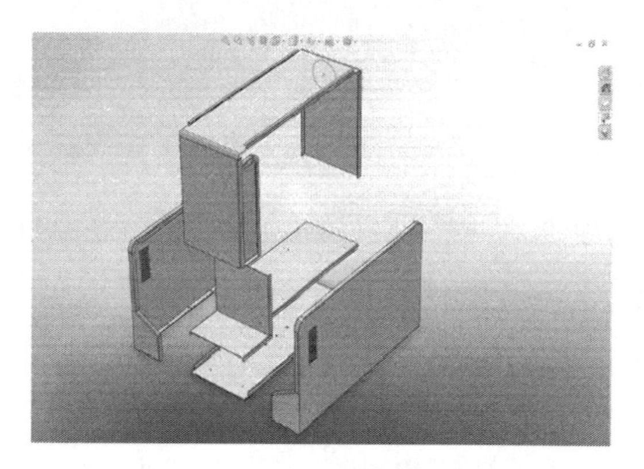

Figure 3. Explosive view of the detector appearance.

3 SOLIDWORKS AUTOMATICALLY CONVERTING DETECTOR ENGINEERING DRAWING

SolidWorks provides full-related class two-dimensional drawings. In the real world, products may consist of thousands of parts, so it is essential to generate engineering drawings, and its speed and efficiency is an issue all pieces of 3D software have to be faced. SolidWorks uses a means of generating quick engineering drawing, making large assembly engineering drawing generation and labeling become very fast.

After finishing making appearance drawings, SolidWorks can easily convert a three-dimensional model into a two-dimensional engineering drawings. The graphics having been converted are not only very accurate, but also featured with a variety of views. For example, cross-sectional views, isometric views, partially enlarged views and so on can automatically generate; the notes of a variety of dimensions, geometric tolerances, welding symbols, etc. can automatically produce, and can freely adjust the size of the type and location (Fig.4).

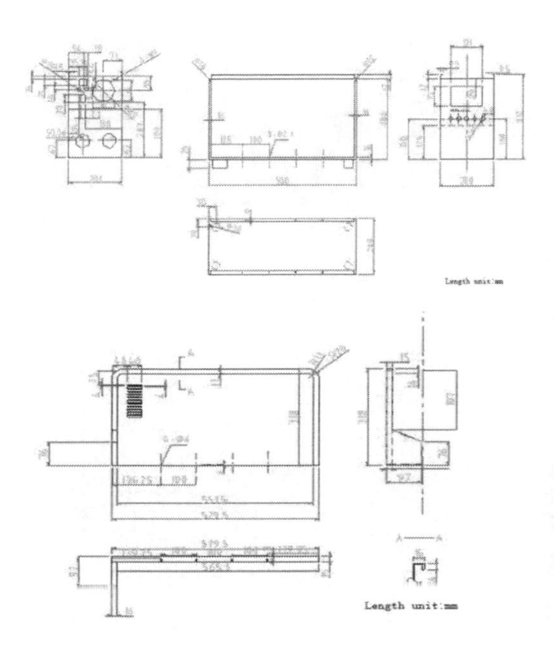

Figure 4. 2D engineering drawing of the detector.

4 SOLIDWORKS DETAILING DETECTOR PARTS WORKING DIAGRAM

SolidWorks provides direct three dimensional model drawings. In the friendly user interface, SolidWorks is not limited in pure 3D or pure 2D view mode. Under the base of three dimensional model drawings, designers can select the complex parts separately, using the way of SolidWorks quickly generating 2D engineering drawing to rapidly form the parts material drawing. In this way, the accurate size of materials used can be displayed and the illustrative diagram in the process can be shown to the producers precisely (Fig.5).

5 CONCLUSION

During the process of using three-dimensional design software SolidWorks to achieve the virtual design of the detector's 3D model, designers can modify the feature data of model to perfect the final productaccording to the need in reality at any time. Thus, when developing new products, this process can shorten the product development cycles, reduce the cost of product development, enabling businesses to gain more opportunities to profit.

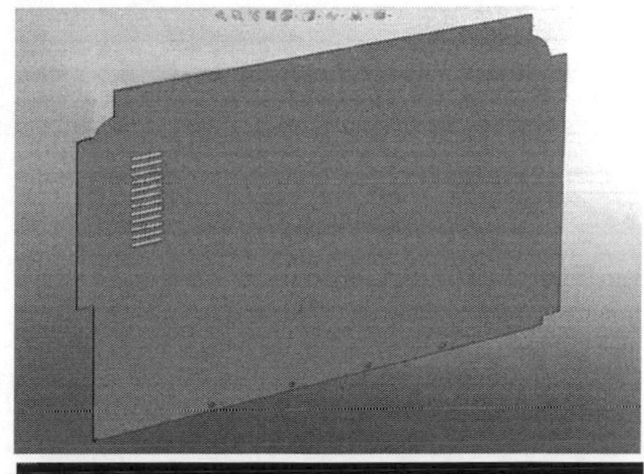

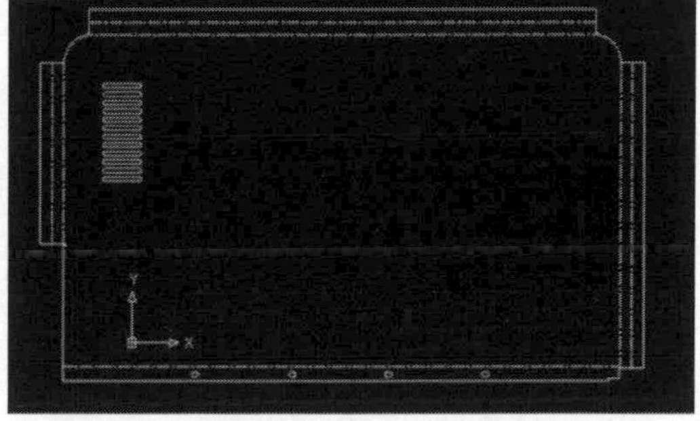

Figure 5. 3D Design drawing and material drawing.

We can see the importance of computer aided virtual design in the whole progress. At the same time, during the progress of making three-dimensional entity model in SolidWorks, the foundation of drawing is also needed so that designers can catch the characteristics of products precisely and present three-dimensional shape quickly and accurately.

REFERENCES

[1] LIU Jin-xia, ZHANG Ming-ming, Loader Work Equipment Design and Simulation Based on SolidWorks[J], Coal Mine Machinery,vol.34(2013), P.21–22.

[2] Zhao Yongjun, Wan Desheng, Application of Solidworks to the Design of Loader Work Attachments[J], Construction Machinery and Equipment,vol.41(2010), p.40–43.

[3] Zhan Diwei, SolidWorks Chanpin Sheji Shili Jiaocheng[M], China Machine Press,BeiJing,2008.

[4] Liu Chunrong, Chanpin Sheji Chuangyi Biaoda-SolidWorks[M], China Machine Press, BeiJing, 2012.

[5] SolidWorks 2007 Training Manuals: Mold Design Using SolidWorks, Dassault Systemes S.A, 2007.

Computing, Control, Information and Education Engineering – Liu, Sung & Yao (eds)

 ISBN: 978-1-138-02800-5

Design of IP core of DPSK modulator with variable bit rate and carrier frequency based on FPGA

Xin Zheng
Department of Physics and Electronic Engineering, Guang Xi Normal University for Nationalities, China

ABSTRACT: DPSK is widely applied in communication because of high-bandwidth utilization and strong capacity of anti-jamming. In connection with the defects of the DPSK digital modulator, The IP core of DPSK modulator with variable bit rate and carrier frequency based on FPGA and DSS is designed, whose bit rate and carrier frequency are programmed and adjusted by controller. The Signal Tap II Logic Analyzer is used to test and verify design.

KEYWORDS: DPSK; Digital modulator; IP core; FPGA.

1 INTRODUCTION

Differential binary phase shift keying that is one of important digital modulations, is widely applied in modern communication for its featuring wide bandwidth utilization, strong anti-interference and eliminating the phase ambiguity. Because of the carrier frequency and bit rate of the current modulator in a market being changeless, there is a high error rate in the process of communication, when burst interference occurs in the communication channel. How to enhance the communication quality in that severe condition is very important. The paper describes a design schedule of FPGA-based based IP core of modulator with changed carrier frequency and bit rate, in which the carrier frequency and bit rate of the modulator are both adjusted to fit different communication environments. The modulator we designed features wide bandwidth utilization and low error rate.

2 THE DESIGN OF BLOCK DIAGRAM OF IP CORE

Figure 1 is a block diagram of the IP core that is consisted of Avalon-MM interface, resister file, task logic unit, D/A convertor and low pass filter. Avalon-MM interface is one of Avalon interfaces, which is an address-based read/write interface typical of master-slave connects. Avalon-MM interface designed in the modulator enhances the interoperation of the IP core. In this design, NiosII processor can easily access the control and status resisters in the resister file using the Avalon –MM interfaces, as shown in Figure1. Resister file is a group of resisters substantially where includes control resister, frequency control resister, status resister, data resister and division factor resister.

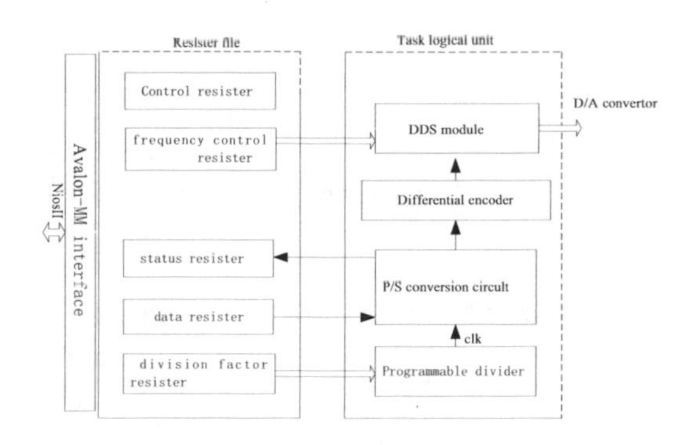

Figure 1. The block digraph of IC core.

Some important information and data, such as transmitted data, control instruction and status, are stored in these resisters. The task logical unit is an important part of modulation that implements the function of IP core. The task logical unit consists of a programmable frequency divider, a parallel-to-serial conversion circuit, a DDS module with the function of phase shift and a differential encoder. The bit rate of modulation is controlled by the programmable frequency divider by setting different data in clock division resister. Parallel-to-serial convert circuit implements the function of converting a parallel digital signal to serial digital signal and set the status resister when conversion is ended. The DDS module can control the frequency and phase of carrier frequency, according to the data in control resister and data transmitted by differential encoder, in which the digit "1" represents phase shift 180 while the digits "0" represents phase shift 0.

Table 1. Registers definition and address assignment.

Resister	address	W/R	bits
Div_Factor_Reg	0	W	32
Data_Reg	1	W	32
State_Reg	2	RW	2
Fre_Control_Reg	3	W	32
Control_Reg	4	W	1

3 FPGA-BASED DESIGN OF HARDWARE

3.1 *The design of resister file*

A group of resisters needs to be defined to store control instruction, status and data, by which the communication between NiosII processor and task logical units is set up. These resisters are listed in the table 1.

3.2 *The design of avalon memory mapped interface*

the Avalon Memory Mapped (Avalon MM) interface is one of the Avalon interface family, which allow system designed to connect component in an FPGA. An Avalon Memory Mapped (Avalon MM) interface is designed in the IP core to read/write date between Nios II and IP core, as shown in Figure 1. The IP core typically includes some Avalon-MM Slave Port Signals, such as readdata, writedata read, and write, required for the task logic unit.

3.3 *The design of DDS module with the function of shift phase*

Figure 2 is block diagraph of the DDS module with the function of shift phase, which consists of a phase accumulator, a phase resister, a shift phase adder and sinusoid storage. 12-bit phase accumulator continuous add the 12-bit frequency control word K and 12-bits data transmitted by later phase resister, and phase resister stores the accumulated results. The shift phase adder do add 2048 compute or do add 0 compute to implement the function shift phase, according to the bit transmitted by parallel-to-serial conversion circuit, in which if the bit is 1 do add 2048 compute. The result, used as address, is transmitted to sinusoid storage to query the data in sinusoid storage. When an overflow in phase accumulator is occurred, it is indicated that the module of DDS outputs a periodical sinusoid. Listing and numbering.

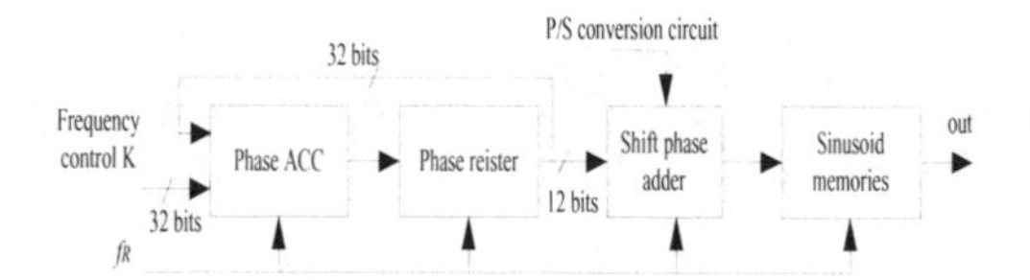

Figure 2. The block digraph of DDS module with the function of shift phase.

3.4 *The design of differential encoder*

A differential encoder is applied before the DDS module to eliminating the phase ambiguity. The Figure 3 is a block diagram of differential encoder. The formula of differential encoder is:

$$b(n)=a(n)\oplus b(n-1)$$

where $\oplus$ is a model 2 adder, b(n-1) is delayed 1 tick of b(n).

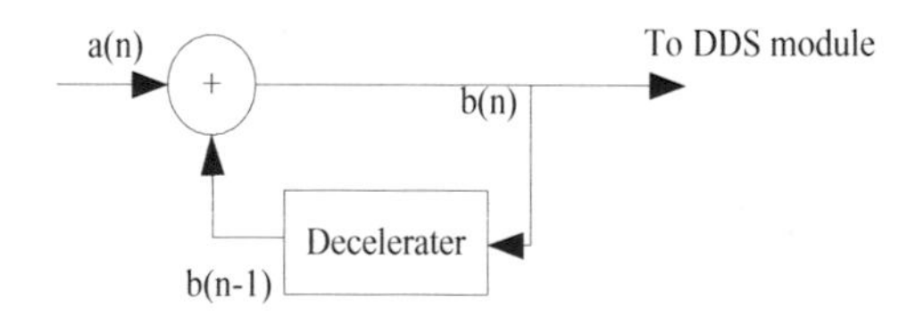

Figure 3. The block diagram of differential encoder.

3.5 *The algorithm of parallel-to-serial conversion*

Figure 4 is the algorithm flow chart of parallel-to-serial conversion. If there is a new data in data resister, the status bit in status resister is set "1", and then the 12-bits data are converted to continuous serial data. When the conversion is over, the status in status resister is set to "0".

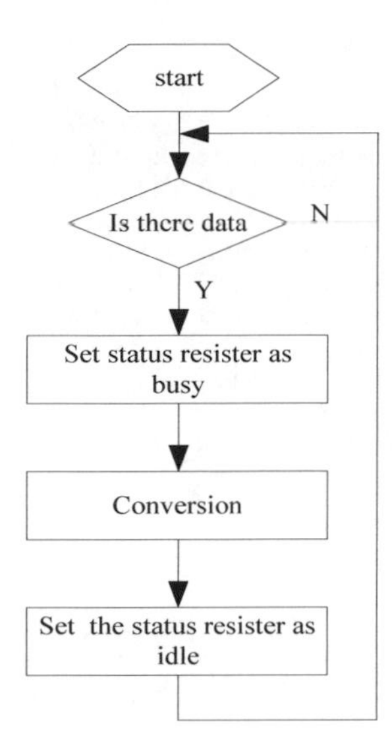

Figure 4. The algorithm of P/S conversion.

4 THE ALGORITHM OF SOFTWARE

Figure 5 is the software algorithm flow chart about NiosII soft CPU how to control the modulator. NiosII write tow certain values to frequency control resister and division factor resister to control carrier frequency and bit rate, and then write "1" to the control resister to start the modulator. Before writing data to data resister, the status bit in the status resister is read. If the modulator isn't busy, NiosII write data to data resister and set the first bit in status resister.

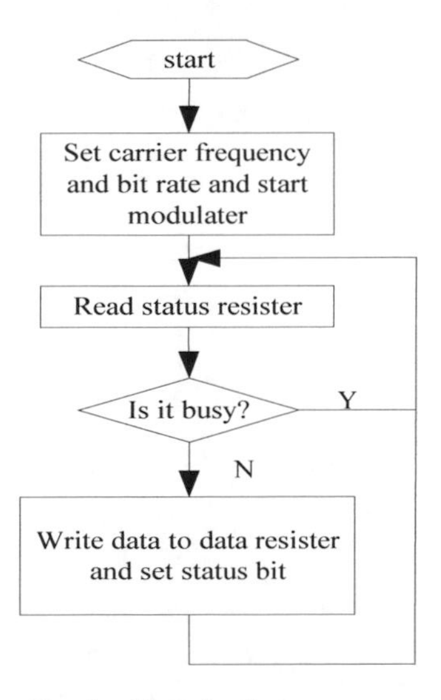

Figure 5. The algorithm of software.

5 VERIFICATION AND ANALYSIS

The signal Tap II logic Analyzer is applied to verify the design. The QuartusII software comes with the signal tap II logic analyzer that allows designer to debug their design in real-time and at high-speed. Before you set up the Signal Tap II Logic

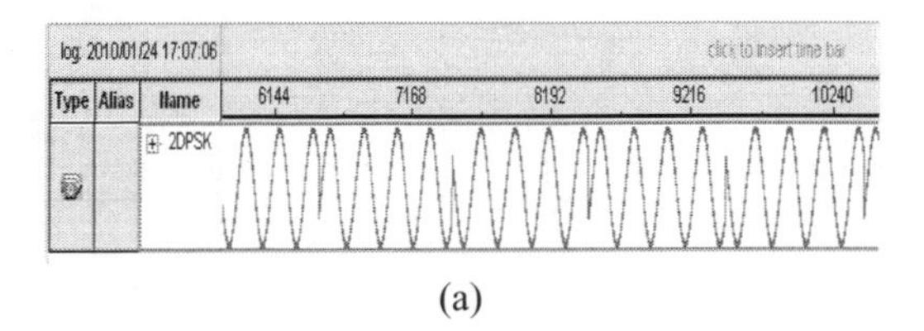

(a)

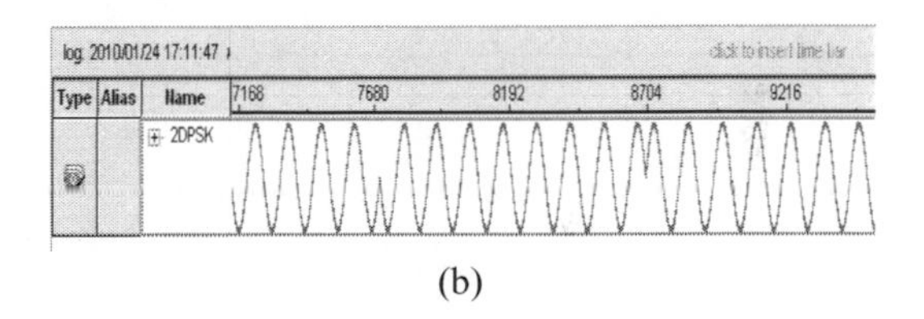

(b)

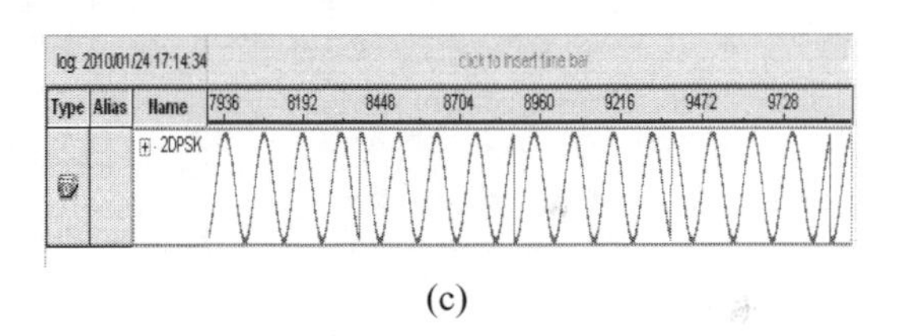

(c)

Figure 6. (a) A 2DPSK modulation signal with 200 kHz and 50kbaud. (b) A 2DPSK modulation signal with 400 kHz and 50kbaud. (c) A 2DPSK modulation signal with 400 kHz and a100kbaud.

Analyzer, a Signal Tap II File that contains the data you capture must be created. The application of Signal Tap II logic Analyzer facilitates verification in FPGA deign. Figure 6 is some modulated 2DPSK signals observed in a Wave Display pane of Signal Tap II Logic Analyzer. Figure 6(a) is 2DPSK signal whose carrier frequency and bit rate are set at 200 KHz and 50k baud. Figure 6(b) is a 2DPSK signal whose carrier frequency and bit rate are set at 400 KHz and 50k baud. Figure 6(c) is a 2DPSK signal whose carrier frequency and bit rate are set at 400 KHz and 100k baud. Figure 6 illustrates that the different phase between adjacent elements is π, and the carrier frequency and bit rate can adjusted-the two bit rates of signal are equal in Figure 6(a) and Figure 6(b), while whose carrier frequencies change 2 times; the two carrier frequencies are equal in Figure 6(b) and Figure 6(c), while whose bit rates change 2 times.

6 CONCLUSIONS

The 2DPKS-based digital communication system with variable parameter features strong anti-interference capability. The 2DPKS-based IP core we design features circuit briefness, low cost.

ACKNOWLEDGEMENTS

This work was financially supported University scientific research project funding projects of the Guangxi (YB2014418).

REFERENCES

[1] LONG Guang-li. Design of 2DPSK Modulation and Demodulation Circuit Based on CPLD. Telecommunication Engineering. Vol.49(2009), 49(4)p.29.
[2] YAN Ying-Jian, REN Fang, FU Xiao-Bing. Design of Voice Encryption Transmission System Based on NiosII. Application of Electronic Technique. Vol.19 (2009), p.61.
[3] He Qian, Hang Chun-lin. Design and Imple-ment of General Digital Modulation Base NiosII. Application of Electronic Technique . Vol.16 (2006), p.120.
[4] YAN G Xiu–zeng . Design of a Signal Gener-ator Base on FPGA and DDS. Electronic Design Engineering. Vol. 17 (11), p.7.

Computing, Control, Information and Education Engineering – Liu, Sung & Yao (eds)
© 2015 Taylor & Francis Group, London, ISBN: 978-1-138-02800-5

Design and implementation of remote control-based autonomous collision avoidance system prototype

Yan Zhang, Xu Ying Zhao, Ke Jun Zhang, Jing Wei Li & Zhen Zhi Yang
Department of Computer Science and Technology, Beijing Electronic Science and Technology Institute, Beijing, P. R. China

ABSTRACT: A remote control-based Autonomous Collision Avoidance System (ACAS) prototype is realized based on the interoperation of Android and Arduino. The prototype comprises a car model and a smart phone. The former collects the distance with an obstacle by a sensor and sends it to the smart phone. The latter computes with the distance by a control program and returns a control command. The car model follows the control command to adjust its movement, which can avoid a collision. All communication between the two parts is based on 3G signal. Arduino integrates the sensor and motors in the car model. Android supports the communication and the control program running. The correctness of the control program is guaranteed by a model checking on hybrid interface automata. An experiment is conducted to test the soundness of the implementation. This prototype helps reduce the cost and improve the quality for developing ACAS.

KEYWORDS: Autonomous collision avoidance system; Android; Arduino; XMPP; Hybrid interface automata.

1 INTRODUCTION

Today, autonomous collision avoidance systems (ACAS) are widely applied in the world, such as driverless cars, Mars exploration rovers, information collection vehicles or robots under dangerous environments. A prototype of ACAS is necessary for finding some latent defects in an early phase of its development. For example, checking or testing the prototype might detect some design flaws in the design phase. This could reduce the cost and improve the quality for the development of ACAS.

Accordingly, we design and implement a remote control-based ACAS prototype by the interoperation of Android (Meier 2012) and Arduino (Warren et al. 2011) and 3G communication. The prototype consists of a car model and a smart phone. A sensor built on the car model collects information about its environment, and send it to the smart phone. A control program deployed on the car model computes with the information, and makes decision. In terms of the decision, a control command is sent to the car model. The car model can react immediately according to the command for evading collision with a moving or static object in the environment. In this paper, we present the design and the implementation of the prototype.

The remainder of this paper is organized as follows. Section 2 gives the architecture of the prototype. Section 3 describes the implementation of the prototype. Section 4 explains an experiment for validating the implementation of the prototype. Section 5 considers related works. Finally, we conclude this paper in Section 6.

2 DESIGN OF PROTOTYPE

The remote control-based ACAS prototype comprises two parts. One is a mobile controller; the other is a controlled moving object. The former controls the movement of the latter without manual intervention. The latter reports the environment information to the former, and accepts the control commands from the former. Additionally, the latter changes its dynamics in terms of the control commands. The architecture of the remote control-based ACAS prototype is shown in Figure 1.

2.1 *Controlled moving object*

The controlled moving object is composed of an information collection module, a dynamic generation module and a communication module. The information collection module gathers the environment information, such as the distance between the controlled moving object and an obstacle, the velocity or acceleration of an obstacle. The dynamic generation module adjusts the behavior of the controlled moving object, such as moving forwards, speed-down or stop, according to the control command sent from the mobile controller.

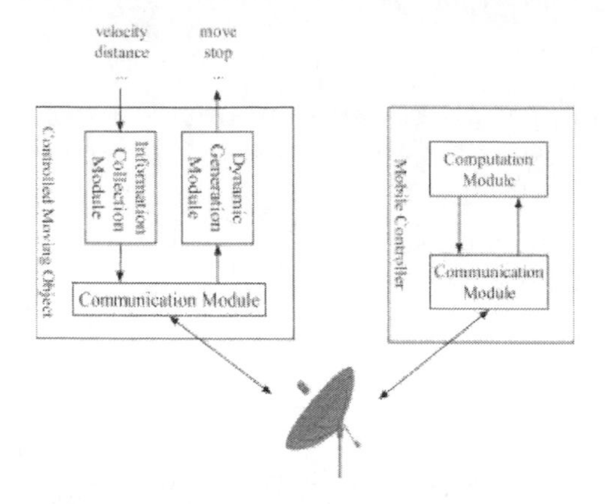

Figure 1. The architecture of the remote control-based ACAS prototype.

An important thing in this part is the interaction among the information collection, the dynamic generation and the communication modules. Especially, the interaction between the communication and the dynamic generation modules involves the integration of discrete and continuous behaviors. Thus, a platform is required to combine the three modules for their cooperation.

2.2 *Mobile controller*

The mobile controller consists of a computation module and a communication module. The communication module takes the task of messages exchange between the mobile controller and the controlled moving object, namely, receiving the environment information from the controlled moving object and sending the control command to it. The computation module processes the environment information, and output control commands, such as acceleration, deceleration or stop, in terms of the computing results.

The kernel of this part is the collision avoidance algorithm. We had made a behavior model for the ACAS by hybrid interface automata (HIA), and had verified the reachability of the behavior model, i.e. HIA, by model checking (Zhang & Zhang, 2012). If an error state that stands for the collision is unreachable, the ACAS is sound. The collision avoidance algorithm can be derived from the behavior model of the ACAS.

3 IMPLEMENTATION OF PROTOTYPE

We adopt 3G signal to remotely control the moving object, since 3G signal can broadcast more widely than WiFi or Bluetooth with the same quality. Moreover, we attempt to deploy the computation module in a smart phone, which is a vogue mode of applications currently. The controlled moving object is realized by a car model. The implementation of the remote control-based ACAS prototype is presented as follows.

3.1 *Communication protocol*

All messages between the car model (i.e. the controlled moving object) and the smart phone (i.e. the mobile controller) are transmitted by 3G signal. Since an access termination only obtain an intranet IP address in Chinese 3G network, two devices in 3G network cannot communicate in point-to-point fashion. As a result, we use Extensible Messaging and Presence Protocol (XMPP) to realize 3G-based communication between the car model and the smart phone. XMPP is an instant messaging (IM) protocol based on Extensible Markup Language (XML), which bases on Client/Server (C/S) architecture. We adopt the server of Google Talk (Gtalk) as the server of XMPP and Android platform the client in our prototype. Fig.2 shows the communication between the car model and the smart phone.

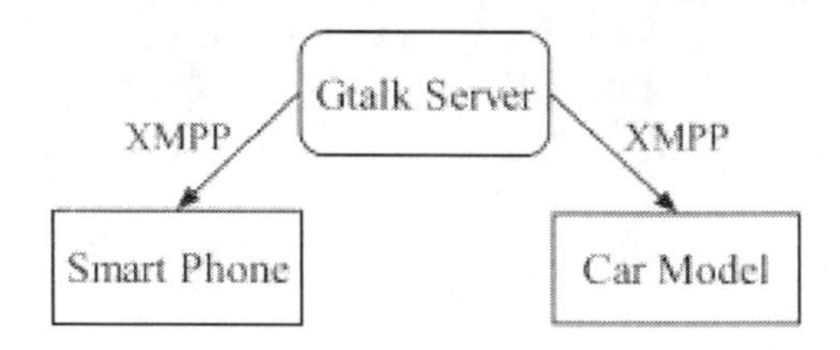

Figure 2. XMPP used in the prototype.

3.2 *Implementation of controlled moving object*

We use a car model to realize the controlled moving object, which is shown in Figure 3.

Figure 3. The car model.

The body of the car model is a 4-wheel-drive electric vehicle (4WD EV), which uses four motors to independently drive the car and have powerful surmounting and climbing capacity.

An ultrasonic sensor HC-SR04 is used to measure the distance to the obstacles. The sensor can detect a range from 2cm to 450cm with a sensitive angle no more than 15°. The precision of the sensor is 3mm.

LFF 130 DC motors are used to control the speed of the car model. The rotation speed of the motor is 10,000 RPM with rated voltage from 4.5V to 6V. The maximum velocity might meet to 68cm/s. The reduction ratio of the reducer is 1:120.

Arduino is an open source hardware platform, which consists of a circuit board with simple I/O capability and a suite of program development environment. Arduino is able to receive digital signal from sensors, and control electric lights, electric motors and other physical devices. We use Arduino to integrate the ultrasonic sensor and the DC motors. The Arduino transforms the control command to Pulse-Width Modulation (PWM) signal, and adjust the rotational speed of the motors for controlling the speed of the car model.

For preventing the Arduino mainboard to be damaged due to the excessive current, we use L293 Motor Shield to derive four DC motors.

The communication module is implemented on an Android platform. In Android, only one Activity is active at the same time. However, in the communication module of the car model, there must be two active Activities, one for real-time monitoring the control commands over XMPP, the other for receiving the environment information from Arduino or sending the control command to Arduino. For this reason, we use an Android Service to implement the former, which run in background, and an Android Activity the latter. The interaction between the Service and the Activity is implemented by Intent and Message. The interaction between the Activity and Arduino is implemented by Android Open Accessory (see Fig. 4).

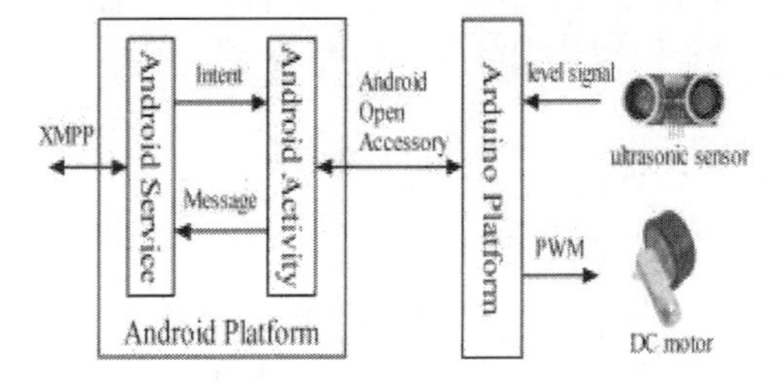

Figure 4. The implementation of the controlled moving object.

3.3 *Implementation of mobile controller*

We use a smart phone based on Android platform to realize the mobile controller, which is shown in Figure 5.

Since the implementation of the communication module on the smart phone is similar with that on the car model, we focus on the implementation of the computation module.

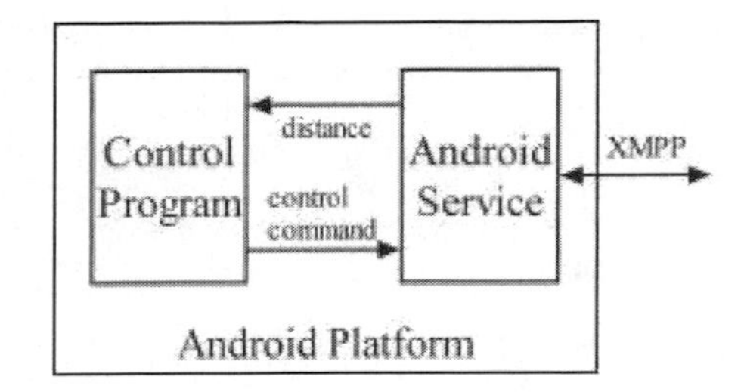

Figure 5. The impltementation of the mobile controller.

We had introduced hybrid interface automata (HIA) to model Cyber-Physical Systems (CPS) and proposed an algorithm for checking the reachability of HIA. CPS is a kind of real-time reactive systems that integrate computation with physical processes. Since the remote control-based ACAS is very similar with CPS, we use HIA to model the behavior of the remote control-based ACAS.

The HIA P of the remote control-based ACAS is shown in Figure 6. x_1 and x_2 are the travel distances (from an origin) of the obstacle and the car model respectively. $\dot{x}_1$ and $\dot{x}_2$ are the velocities of the obstacle and the car model respectively. q_0 is an initial state and q_4 an error state that means a collision between the car model and the obstacle. The initial valuation of x_1 and x_2 are 0 and 150cm respectively. By the algorithm given in Zhang & Zhang (2012), we check the HIA P, and conclude that q_4 is unreachable in HIA P under current setting. And then, we derive the collision avoidance algorithm from HIA P, and implement it on Android platform.

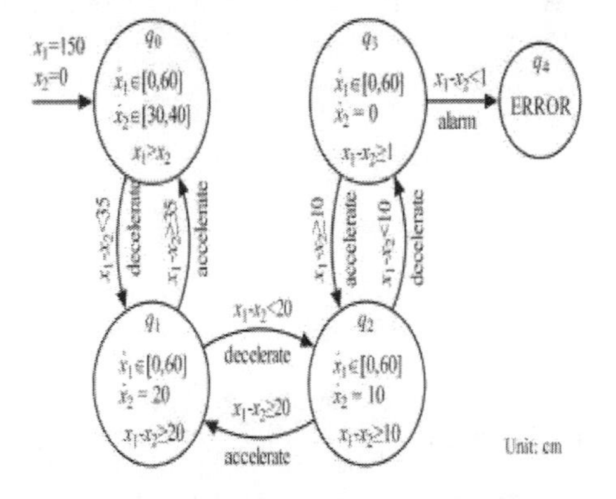

Figure 6. The HIA P of the remote control-based ACAS.

4 EXPERIMENT

We had conducted an experiment in a 100 × 5m² planar space to test the prototype. A remote-controlled toy excavator played the obstacle, and an experimenter manually controlled its movement at random. The setting of parameters of the prototype was consistent with the HIA *P* shown in Figure 6. The experiment was repeated 100 times without any occurrence of a collision.

Figure 7a, b give the scenarios of deceleration and stop of the car model respectively. The right parts of Figure 7a, b are the moving state of the car model displayed on the smart phone.

(a) scenario of deceleration

(b) scenario of stop

Figure 7. The experiment of the prototype.

5 RELATED WORKS

The most relevant work is Xu et al. (2014). They built a remote control system for a mobile object. The system is deployed on Android platform. By the interaction between Android and Arduino, the mobile object can be controlled not to collide with some obstacle. Authors did not show their real system, and not explain the implementation of the mobile object in detail, such as the integration of Android and Arduino on the mobile object and the way of driving motors by Arduino. In our implementation, we adopt pcDuino to integrate Android with Arduino on the car model.

In (Xin 2014), author designed an Arduino intelligent vehicle, which is controlled by Android platform and can measure distance, move under control, and transfer image wirelessly. However, their control is based on WiFi rather than 3G. Thus, the control range of their system is limited.

In (Zhao & Zhu 2013), an autonomous obstacle-avoiding robot car is designed by using Arduino. The robot car is not remote control-based. The control program is embedded in Arduino.

6 CONCLUSION

We design a remote control-based prototype of ACAS. The controlled moving object is implemented by a 4WD car model, on which distance sensors are fixed. The mobile controller is realized by a smart phone with Android OS, on which a control program that is developed in terms of a collision avoidance algorithm is deployed. The communication between the car model and the smart phone is implemented by XMPP based on 3G signal. We use Arduino platform to integrate the distance sensors and the motors, and Android platform the 3G communication module on the car model. By the interoperation of Android and Arduino, the car model can send the environment information collected by the sensors to the smart phone, and accept the control commands from the smart phone to adjust it dynamic behaviors. We use HIA to build a behavior model of the prototype, and verify that a collision between the car model and an obstacle is impossible in the system. Finally, the collision avoidance algorithm is derived from the behavior model of the car model. An experiment demonstrates that the prototype is sound.

The implemented prototype can be used to find the design flaws of ACAS in the early phase of their development. The implemented prototype is useful for reducing the cost and improving the quality of ACAS development.

REFERENCES

Meier, R. 2012. Professional Android 4 application development. UK: Wrox Press.

Warren, J.D., Adams J. & Molle H. 2011. Arduino robotics. New York: Apress Media.

Xin, G. 2014. Arduino intelligent vehicle design based on Andriod. *Computer & Telecommunication* 3(1): 62–64.

Xu, R., Liang, J.C., Yang, H. & Teng, J.W. 2014. Design and implementation of remote control system for mobile platform based on interoperation of Android and Arduino. *International Journal of Smart Home* 8(4): 105–112.

Zhang, Y. & Zhang, T. 2012. Hybrid interface automata. In Karl R.P.H. Leung & Pornsiri Muenchaisri (ed.), *Proceedings of the 19th Asia-Pacific Software Engineering Conference, Hong Kong, 4–7 December 2012*. USA: CPS.

Zhao, J. & Zhu, S.C. 2013. Dsign of obstacle avoidance robot car based on Arduino microcontroiler. *Automation & Instrumentation* 28(5): 1–4.

Computing, Control, Information and Education Engineering – Liu, Sung & Yao (eds)
© 2015 Taylor & Francis Group, London, ISBN: 978-1-138-02800-5

Modeling and analysis of uphill climbing for the electromagnetic-driven spherical robot

Sheng Ju Sang
School of Information Science and Technology, Taishan University, Taian Shandong, China

ABSTRACT: This paper describes a dynamic and analytical studies of an electromagnetic-driven spherical, discusses the influence on uphill climbing with the ratio of the equivalent pendulum mass to the spherical robot mass in detail in order to derive the rolling conditions without sliding on the slope with an inclination angle. In conclusion to the study of the typical motions of spherical robot on the slope, it becomes obvious that the tilting angle of the equivalent pendulum, as a main influence factor, should not be too large for the spherical robot to roll without sliding on the slope.

KEYWORDS: Spherical robot; Dynamic models; Uphill climbing; Electromagnetic driven; Equivalent pendulum.

1 INTRODUCTION

Spherical robot as a new type of mobile robots has made its debut in recent years. It has a ball-shaped outer shell to accommodate all its mechanism, control devices and energy sources. it is characterized as simple, compact, locomotion with minimal friction, walking mainly in a rolling way, constrained spaces, omnidirectional movement without ever overturning and so on[1]. These advantages provide the spherical robots with stronger viability than the traditional mobile robots in many fields, such as military, transportation, surveillance, search and rescue, toys, entertainment etc.

Spherical rolling robots have a long history. As early as in 1893, Tate J.L. invented a spherical toy, and applied for a patent [2]. Many recent studies have been performed on the theoretical and experimental exploration, and have made some significant progress in mathematical modelling, mechanism designing and so on, which was carried out by a lot of scientists from many countries all over the world, such as United States, Finland, Iran, China, Switzerland, and Belgium etc.[3–6].

However, some assumptions has been done when building the dynamics models in some previous studies, which result in some errors even some considerable errors occurring in control. Moreover, little research has been devoted to the uphill climbing Analysis of the spherical robots. The purpose of this paper is to model the dynamic behavior of an electromagnetic-driven spherical on the slope with an inclination angle and analyze the influence on uphill climbing with some factors including the equivalent pendulum mass, the spherical robot mass, and the tilting angle of the equivalent pendulum and so on.

2 MECHANISM OF THE SPHERICAL ROBOT

The present spherical robot is composed of a left hemispherical shell, a right hemispherical shell, a magnetic steel ring and an inner driving mechanism, which is called electromagnetic-driven spherical robot as shown in Fig.1.

The left hemispherical shell and the right hemispherical shell form a spherical shell assembled through the fixing screws coaxially with the magnetic steel ring on which the permanent magnetic steel pieces are uniformly disposed. N-electrodes of two adjacent permanent magnetic steel pieces are opposite in direction.

The inner driving mechanism positioned in the spherical shell mainly comprises an inner driving bracket, a main shaft, a flywheel, a steering motor, a motor support, a flywheel shaft, an electricity power supply, a controller, electromagnetic coils and the like. The main shaft, on which the other components are hang up through the inner driving bracket, is fixedly connected with the spherical shell through a couple of bears respectively fixed on inner surface of the left and right shell in opposite direction. The flywheel is hung on the inner driving bracket through a bearing. The steering motor shaft is connected with the flywheel through a coupling in order to drive the lower part of the motor to rotate together with the flywheel shaft vertical to the main shaft and the flywheel. The electromagnetic coils disposed on the bracket are arranged symmetrically relative to the steering motor shaft.

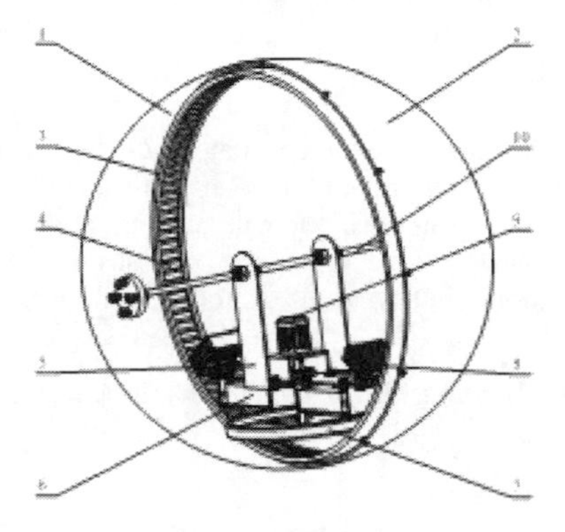

1, left hemispherical shell, 2, right hemispherical shell, 3, magnetic steel ring, 4, permanent magnetic steel, 5, bracket 6, power and controller, 7, flywheel, 8, electromagnetic coils, 9, steering motor, 10, main shaft

Figure 1. Schematic view of the mechanism of the electromagnetic-driven spherical robot.

3 UPHILL CLIMBING

The robot is capable of uphill climbing on a slope in certain inclination as shown in Fig. 2.

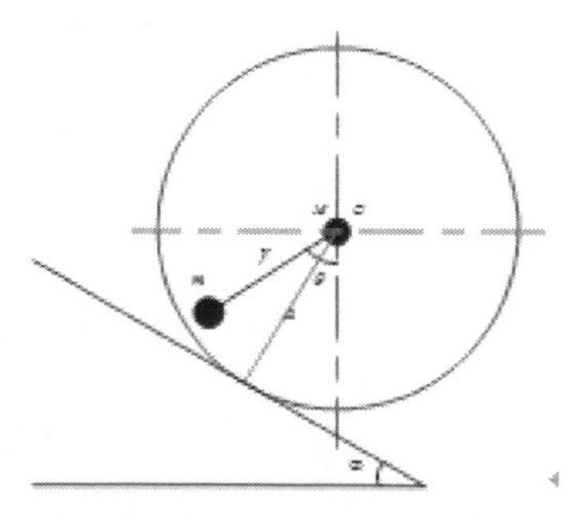

Figure 2. Uphill motion diagram of the spherical robot.

The equation for uphill climbing of the spherical robot is given as follows:

$$mg(r\sin\theta - R\sin\Phi) = MgR\sin\Phi \tag{1}$$

Where, R is the radius of the robot, r is the distance between centers of the robot and the pendulum, g is the gravitational acceleration, M is the mass of the robot, m is the mass of the equivalent pendulum, Φ is the slope angle of the inclination, and θ is the rotation angle of the equivalent pendulum with respect to the robot.

Assuming the robot rolls without slipping, the slope angle of the inclination can be expressed as

$$\begin{aligned}\Phi &= \sin^{-1}((m\times r)/((M+m)\times R)) \\ &= \sin^{-1}(\sin\theta\times m/(M+m)\times(r/R))\end{aligned} \tag{2}$$

As a consequence of the driving angle of the equivalent pendulum increasing, the inclination slope angle which the robot is cable of climbing would be greatened proportionately.

Considering θ can be close to $\pi/2$, that is $\sin\theta=1$, the maximum slope angle of the inclination Φ_{max} can be written as

$$\begin{aligned}\Phi_{max} &= \sin^{-1}(mr/(M+m)R) \\ &= \sin^{-1}(m/(M+m)\bullet r/R)\end{aligned} \tag{3}$$

Hence,

$$\Phi_{max} = \sin^{-1}(\lambda\bullet\kappa) \tag{4}$$

Where, $\lambda = m/(M+m), \kappa = r/R$.

1 The bigger the ratio of r to R, the larger the inclination slope angle for the spherical robot would be. If r can be made close to R, Φ_{max} can be rewritten as

$$\Phi_{max} \le \sin^{-1}(\lambda) = \sin^{-1}(m/(m+M)) \tag{5}$$

2 The bigger the ratio of m to (m+M), the larger the inclination slope angle for the spherical robot. If M can be close to 0 compared with m, Φ_{max} will be only influenced by the ratio of r to R, thus, the equation of the maximum of the inclination slope angle becomes,

$$\Phi_{max} \le \sin^{-1}(\kappa) = \sin^{-1}(r/R) \tag{6}$$

Taking the ratios including r to R and m to (m+M) as the horizontal coordinate axes, h_{max} and Φ_{max} as the vertical coordinate axes respectively, A diagram can be obtained, which as shown in Fig. 3.

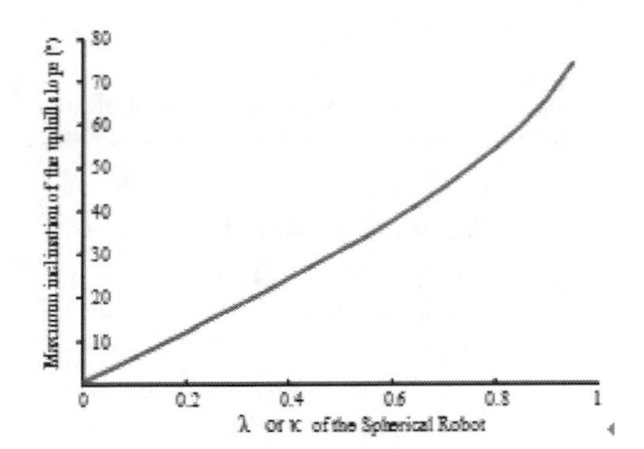

Figure 3. Maximum height of obstacle and inclination angle of slope.

4 SLIDING CONDITIONS

According to the intersection point of the robot's gravity resultant line with the slope surface, the analysis for the sliding conditions of the robot on the slope surface can be divided into three scenarios as shown in Fig. 4. (1)The intersection point of the robot's gravity resultant line with the slope surface is coincident with the contact point; (2) the intersection point is above the contact point; (3) the intersection point is below the contact point.

4.1 *The intersection point coincidence with the contact point*

as shown in Fig. (4a), if the intersection point of the robot's gravity resultant line with the slope surface is coincident with the contact point, the friction force should be big enough to prevent sliding on the slope. That is, the friction force should satisfy the succeeding equation.

$$F_S \geq (M+m) \times g \times \sin\Phi \tag{7}$$

While the friction force can be expressed as,

$$F_S = (M+m) \times g \times \cos\Phi \times \mu \tag{8}$$

Thus,

$$(M+m) \times g \times \cos\Phi \times \mu \geq (M+m) \times g \times \sin\Phi \tag{9}$$

Therefore, we have,

$$\tan\Phi \leq \mu \tag{10}$$

Where, M is the mass of the spherical shell, m is the mass of the equivalent pendulum, g is the acceleration of gravity, μ is the coefficient of sliding friction, and Φ is the inclination angle of the slope.

The fact revealed here is that the friction force should satisfy Eq. (10) to prevent the robot sliding on the slope, when the intersection point of the robot's gravity resultant line with the slope surface is coincident with the contact point.

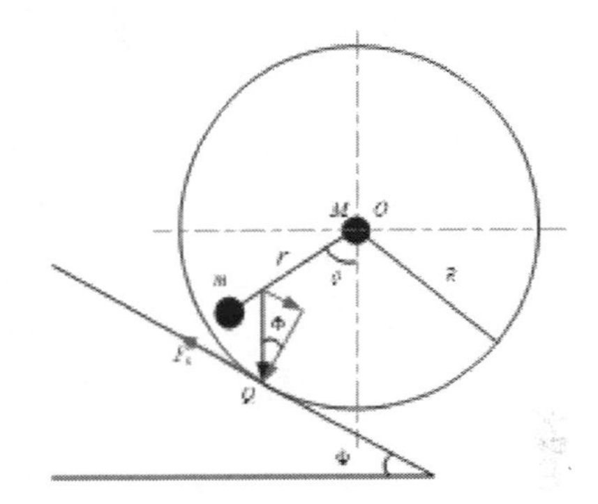

(a) Coincidence

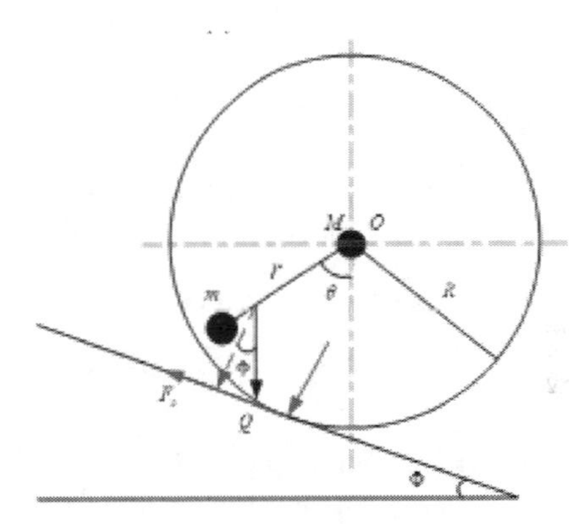

(b) Above the contact point

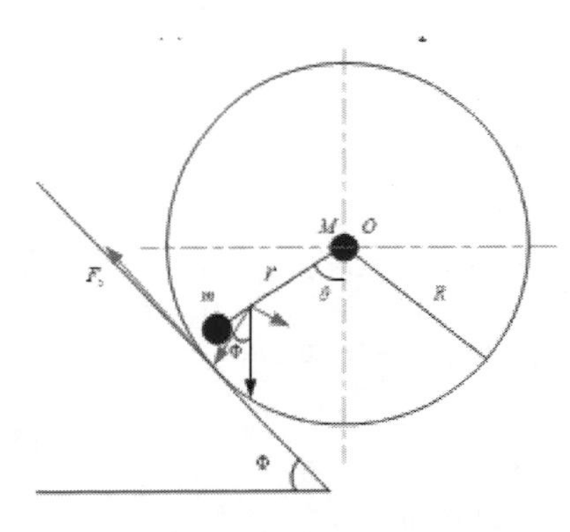

(c) Below the contact point

Figure 4. Force diagram of the spherical robot on slope.

4.2 *The intersection point above the contact point*

in the case when the intersection point of the robot's gravity resultant line with the slope surface is coincident with the contact point, the friction force would be small than the case of coincidence due to counteraction by the component of the gravity as shown in Fig. (4b). Therefore, the friction force should satisfy,

$$F_S - (M+m)\times g\times \sin\Phi \geq (M+m)\times a_C \quad (11)$$

By substituting the Eq. (8) into Eq. (11), we have,

$$a_C \leq (\mu\cos\Phi - \sin\Phi)\times g \quad (12)$$

Thus,

$$\varepsilon_C \leq (\mu\cos\Phi - \sin\Phi)\times g/R \quad (13)$$

From the literature [12], we get,

$$\begin{aligned}\sin\theta &\leq (\frac{2}{3}MR(\mu\cos\Phi - \sin\Phi) + (M+m)\delta)/mr \\ &= \frac{2}{3}\frac{M}{m}\frac{R}{r}(\mu\cos\Phi - \sin\Phi) + \frac{(M+m)}{m}\frac{1}{r}\delta\end{aligned} \quad (14)$$

Eq. (14) indicated the condition of rolling without sliding for the spherical robot on the slope. The fact suggested here is that the maximum of the tilting angle of the equivalent pendulum should be,

$$\theta \leq \sin^{-1}[\frac{2}{3}\frac{M}{m}\frac{R}{r}(\mu\cos\Phi - \sin\Phi) + \frac{(M+m)}{m}\frac{1}{r}\delta] \quad (15)$$

Where, M is the mass of the spherical shell, m is the mass of the equivalent pendulum, g is the acceleration of gravity, μ is the coefficient of sliding friction, δ is the coefficient of rolling friction, Φ is the inclination angle of the slope, and θ is the tilting angle of the equivalent pendulum.

4.3 *The intersection point below the contact point*

Due to superimposition of the component of the gravity as shown in Fig. (4c), the friction force should satisfy the following equation,

$$F_S + (M+m)\times g\times \sin\Phi \geq (M+m)\times a_C \quad (16)$$

On substituting the Eq. (8) into Eq. (16), yields,

$$a_C \leq (\mu\cos\Phi + \sin\Phi)\times g \quad (17)$$

Therefore, the angular velocity should be content with the flowing,

$$\varepsilon_C \leq (\mu\cos\Phi + \sin\Phi)\times g/R \quad (18)$$

Thus, the condition of rolling without sliding for the spherical robot on the slope can be reached as,

$$\begin{aligned}\sin\theta &\leq (\frac{2}{3}MR(\mu\cos\Phi - \sin\Phi) + (M+m)\delta)/mr \\ &= \frac{2}{3}\frac{M}{m}\frac{R}{r}(\mu\cos\Phi - \sin\Phi) + \frac{(M+m)}{m}\frac{1}{r}\delta\end{aligned} \quad (19)$$

As well as the other cases, the condition can be rewritten by the coming equation.

$$\theta \leq \sin^{-1}[\frac{2}{3}\frac{M}{m}\frac{R}{r}(\mu\cos\Phi + \sin\Phi) + \frac{(M+m)}{m}\frac{1}{r}\delta] \quad (20)$$

5 CONCLUSION

Form the above discussion, the conclusions can be reached as following:

1 If the intersection point of the robot's gravity resultant line with the slope surface is coincident with the contact point, the inclination angle of the slope should satisfy $\Phi \geq \arctan(\mu)$ in order to lead the robot roll on the slope without sliding.
2 If the intersection point are inconsistent with the contact point, the tilting angle of the equivalent pendulum should not be too large for the spherical robot to prevent sliding.

Finally, it can be concluded that the tilting angle the pendulum should be big enough to e drive the robot roll, on the contrast; it should not be too large for the robot to roll without sliding. Therefore, the tilting angle should lie into a certain range, which depend on some factors, such as the mass of the spherical robot, the mass of the pendulum, the coefficient of sliding friction, the coefficient of rolling friction, the radius of the spherical, the length of the pendulum and inclination angle of the slope and so on.

ACKNOWLEDGMENTS

The Research was partially supported by the Natural Science Foundation of China (No. 61303022), the Natural Science Foundation of Shandong, China (No. 2009AA01105), the science and technology development fund of Tai'an (No. 20133011), and the Foundation provided by Taishan University (No.Y-01-2013010).

REFERENCES

[1] Shengju SANG. Study on Some Driving Mechanisms and Control Strategies for Spherical Robots [D]. Doctor's thesis, East China University of Science and Technology, 2011.

[2] J.L., Tate. Toy [P]. USPTO: 508558, 1893-11-14.

[3] Aame Halme, Jussi Suomela, etc. Motion control of spherical mobile robot.4th IEEE International Workshop on Advanced Motion Control AMC'96, Mie University, Japan 1996.

[4] Brown H. Benjamin, Jr. and Xu Yangsheng. A single wheel gyroscopically stabilized robot. Proc. IEEE Int. Conf. on Robotics and Automation, 1996(4): 3658–3663.

[5] Bicchi, A., Balluchi, A., Prattichizzo, D., Gorelli,A. Introducing the 'SPHERICLE': An experimental testbed for research and teaching in nonholonomy. Proc. IEEE Int. Conf. on Robotics and Automation, 1997, 3:2620–2625.

[6] Das T., and Mukherjee R. Reconfiguration of a rolling sphere: a problem in evolute-involute geometry. Transactions of the ASME Journal of Applied Mechanics, 2006, 73(44): 590–597.

[7] Bruhn Fredrik C., Kratz Henrik, Warell Johan, et al. A preliminary design for a spherical inflatable microrover for planetary exploration [J], Acta Astronautica, 2008, 63(5–6):618–631.

[8] Kang Hou, Hanxu Sun, Qingxuan Jia and Yanheng Zhang. An Autonomous Positioning and Navigation System for Spherical Mobile Robot, Procedia Engineering, 2012, 29:2556–2561.

[9] Yao Cai, Qiang Zhan and Caixia Yan. Twostate trajectory tracking control of a spherical robot using neurodynamics. Robotica, 2012, 30:195203.

[10] Tuanjie Li,Zuowei Wang,and ZhifeiJi.Dynamic Modeling Simulation of the Internal and externa-driven spherical robot. Joural of Aerospace engineering, 2012, 10:626–641.

[11] Zhao Bo, Wang Pengfei,Hu Haiyan. Design of a spherical robot based on novel double ballast masses principle [J]. High Technology Letter, 2011, 17(2):180–185.

[12] SANG Shengju, ZHAO Jichao, WU Hao, CHEN Shoujun, and AN Qi. Modeling and Simulation of a Spherical Mobile Robot, ComSIS Vol. 7, No. 1, Special Issue, February 2010: 51–62.

Computing, Control, Information and Education Engineering – Liu, Sung & Yao (eds)
© 2015 Taylor & Francis Group, London, ISBN: 978-1-138-02800-5

Starting performance simulation system of the three-phase synchronous and asynchronous motors based on Matlab

Ji Ming Sa & Xiao Shuang Sun
Wuhan University of Technology, Wuhan, Hubei, China

Jin Lin Zhou
School of Art and Design, Wuhan University of Technology, Wuhan China

Yu Jun Gu
Wuhan University of Technology, Wuhan, Hubei, China

ABSTRACT: In order to analyze the performance of motors at start-up stage, a simulation system is designed by using Graphical User Interface (GUI) of Matlab. Based on the motor status equations, the proposed system simulated and analyzed this performance for three types of motors: three-phase asynchronous motor, three-phase synchronous motor and three-phase synchronous generator by applying Simulink models and programing emulation. In addition, capable of dynamically and efficiently adjust the parameters of these motors, we believe our work can benefit the research on motor design.

KEYWORDS: motor; starting performance simulation system; Simulink models; programing emulation.

1 INTRODUCTION

The production and manufacture of the motor are demanding and costly, as well as its complex working characteristic makes a certain degree of difficulty in terms of motor's maintenance and control. Therefore, analyzing the performance of motors at the start up stage (simply called starting performance), operating or something of motors by computer emulation technology contributes to cost reduction and improvement in its design or control.

Therefore, models of the three-phase asynchronous motor and synchronous motor and generator are built in Simulink, which are used to simulate and analyze their starting performance. Finally, a motor starting performance simulation system is designed based on GUI for making simulation and analysis more systematically and efficiently. Excepting for simulating by Simulink, programming simulation is taken in this system to emulate motors in dynamic and real-time way for different parameters, which provides an effective way for acquiring the optimal performance index of motors [1].

2 SIMULATION AND ANALYSIS OF MOTORS

2.1 *Programming simulation and analysis*

Due to the multivariable, nonlinear and strong coupling characteristics[2] of the motor mathematical models in ABC three-phase coordinates, traditional analysis seems much complex. However, the models can become completely decoupled and its analysis variables are reduced largely by taking dq0 coordinate transformation, which greatly simplifying the task of modeling and simulating. Taking the three-phase line-start permanent magnet synchronous motor as an example, in the case of neglecting iron losses and additional losses but considering damper winding, the motor equations are built as follows[3-4]:

Voltage equations:

$$\begin{aligned} u_d &= R_a i_d + p\psi_d - \omega_r \psi_q \\ u_q &= R_a i_q + p\psi_q + \omega_r \psi_d \\ u_D &= R_D i_D + p\psi_D \\ u_Q &= R_Q i_Q + p\psi_Q \end{aligned} \tag{1}$$

Flux equations:

$$\begin{aligned} \psi_d &= L_d i_d + L_{md} i_{2d} + \psi_f \\ \psi_q &= L_q i_q + L_{mq} i_Q \\ \psi_D &= L_D i_D + L_{md} i_d + \psi_f \\ \psi_Q &= L_Q i_Q + L_{mq} i_q \end{aligned} \tag{2}$$

Motion and torque equations:

$$\begin{aligned} J\frac{d\omega_r}{dt} &= N_p\left(T_{em} - T_L\right) - R_\Omega \omega_r \\ T_{em} &= N_p\left(\psi_d i_q - \psi_q i_d\right) \end{aligned} \tag{3}$$

Where ud, uq , uD , uQ , id, iq, iD and iQ express voltage and current in d-q axis; Ψd ,Ψq ,ΨD and ΨQ express linkage flux in d-q/D-Q axis and winding damping; Ψf express the flux generated by the permanent magnets, equal to sqrt(3)*E0/ω; E0 express the back EMF; Ra, RD and RQ express stator and damping resistance; Ld , Lmd, Lq and Lmq express inductance and mutual inductance; ωr , Tem , TL ,NP ,J and RΩ express Rotor angular velocity, electromagnetic torque, load torque, number of pole pairs, moment of inertia and viscosity coefficient.

Based on the equations (1-3) and the motor data in Table 1, the results of programing emulation for the start-up stage are shown in Figure 1.

Table 1. Known parameters of the motor.

Setting	H		others		others
Ld	0.0174	Ra/Ω	0.79	Rn	0.001
Lq	0.0972	RD/Ω	3.35	Np	2
LD	0.0198	RQ/Ω	2.94	E0/V	357.5
LQ	0.1002	UN/V	380	J/kg*m*m	0.0204
Lmd	0.0142	F1/HZ	50		
Lmq	0.0939	TL/N*m	10		

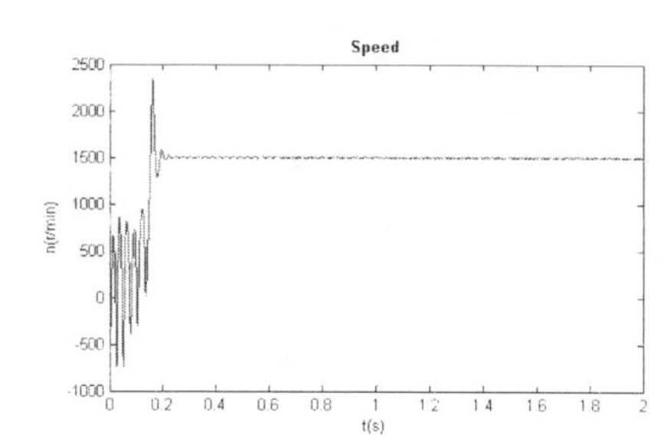

(a) Time(t) –rotational speed(n) waveform.

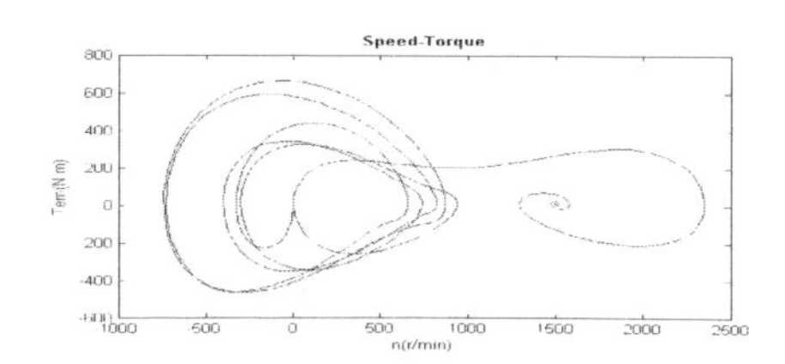

(b) Speed(n)-torque(Tem) waveform

Figure 1. Simulation results of a three-phase synchronous motor.

As depicted in Figure 1(a), the motor rotational speed denoted by n fluctuates dramatically and rises sharply in the first 0.2s after start-up. The reason is a rotating magnetic field is created after three-phase AC being inputted in synchronous motor's stator, which causes the rotor cuts magnetic field and then produces an inductive electromotive force that makes the rotor keep rotating until it becomes synchronous with the stator. Once the motor could not be pulled into the synchronization, the motor won't do work, with n dramatically changing and itself shocking [5]. Therefore, by emulating and analyzing the start-up stage of motors, unreasonable motor design can be effectively avoided as well as irregular motor's operation.

And as shown in Figure 1(b), rotational speed and torque eventually run into a stationary point (1500, 10) in a spiral form after the initial motor fluctuation. It is that electromagnetic torque will finally stabilize at 10 (N*m) to balance the load torque after going through a very brief fluctuation.

2.2 *Modeling simulation on Simulink and analysis*

Taking the three-phase asynchronous motor as an example, functional modules provided by Simulink are used to build sub-models according to the motor state equations[6]. Then these sub-models consist of the whole simulation model, shown in Figure 2, through being encapsulated layer upon layer. One of the simulation results is shown in Figure 3.

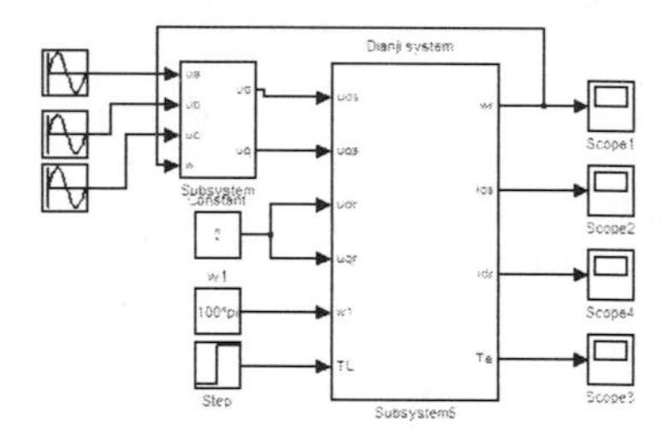

Figure 2. Simulation model on Simulink of the three-phase asynchronous motor.

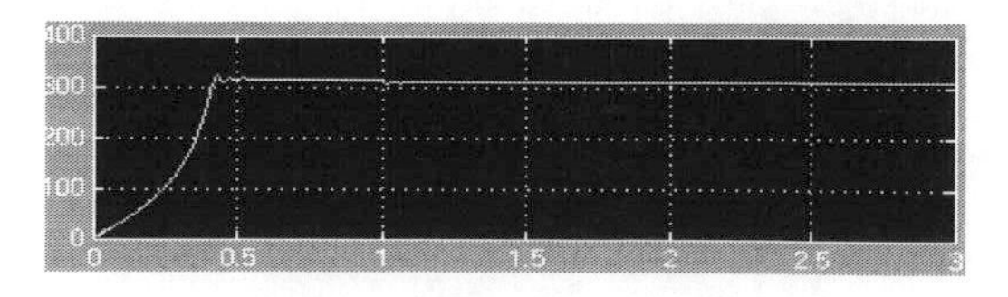

Figure 3. w_r waveform.

As illustrated in Figure 3, a slight decline of the angular velocity of the rotor called ω_r emerged at the first second due to the increment of slip, caused by a suddenly inputted load at this moment, made ω_r decline.

3 DESIGN AND IMPLEMENTATION OF GUI SIMULATION SYSTEM

3.1 *System design*

In order to achieve efficient motor simulation analysis, by using GUI's user-friendly merit, a GUI-based starting simulation system of motors is designed. The related flow diagram is shown as Figure 4.

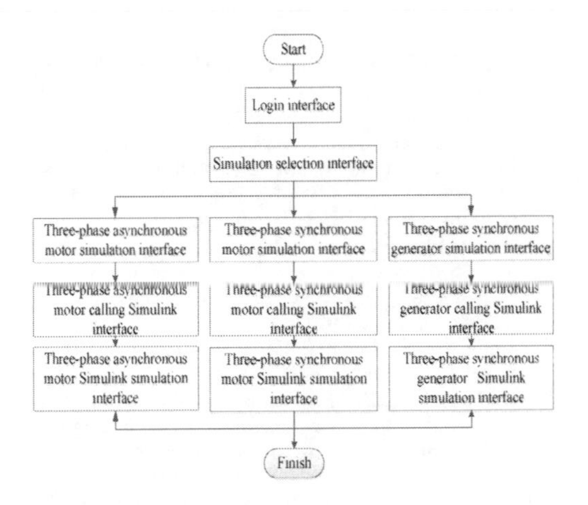

Figure 4. GUI system design flow.

3.2 *Function introduction*

A part of designing interfaces are shown in Figure 5. Motor known parameters are stored in the table. The function buttons and pop-up menus are used to control the corresponding coordinate axes. Here supposing that the user chooses the "three-phase asynchronous motor Simulink simulation interface", the system can jump to the Simulink environment to run the simulation model as shown in Figure 2.

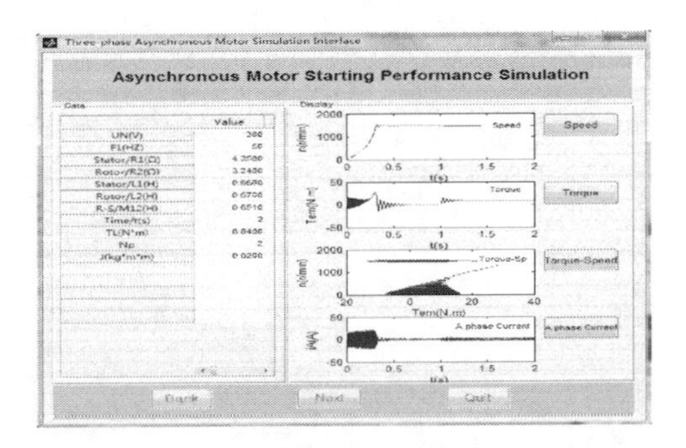

(a) Schematic diagram of "Three-phase asynchronous motor simulation interface."

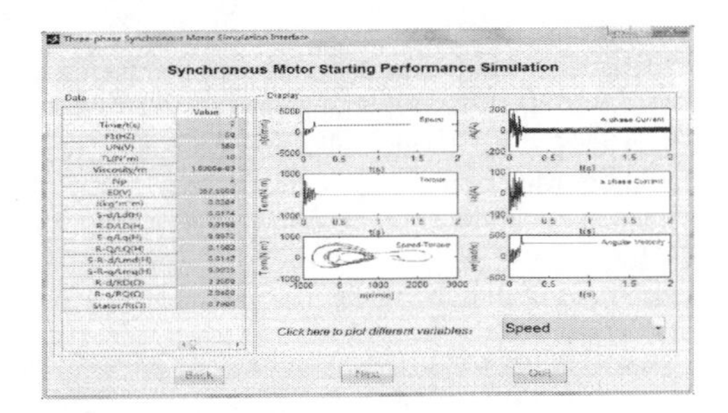

(b) Schematic diagram "Three-phase synchronous motor simulation Interface."

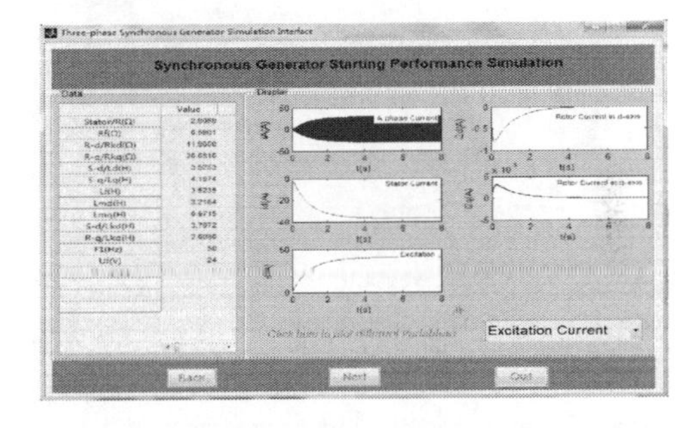

(c) Schematic diagram "Three-phase synchronous generator simulation interface."

Figure 5. Three designed simulation interfaces.

Above interfaces have two advantages: (1) Implement fast simulation for different motor parameters; (2) Zoom in the simulation results to analyze easily. The details are described as follows:

1 The data in table can be freely adjusted to implement different simulation. After modifying and clicking buttons like "Speed", new simulation results will be quickly emerged in the coordinate. Therefore, this function can help users simulate different motors or repeatedly simulate the same motor at different parameters to acquire perfect performance. As shown in Figure 6, the result is from another three-phase asynchronous motor.

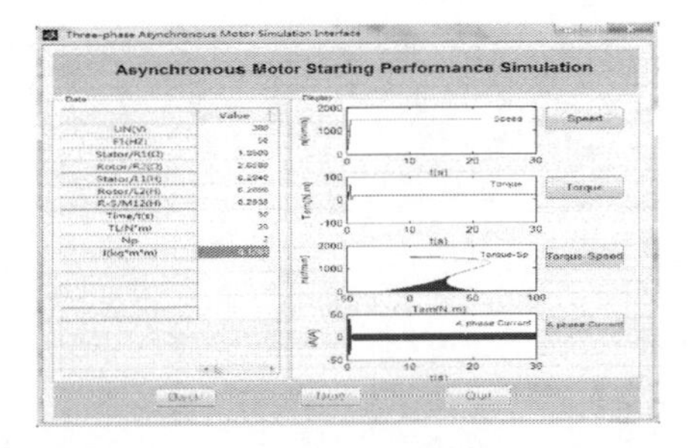

Figure 6. Simulation results of another three-phase asynchronous motor.

2 Due to the waveform region in the coordinate is a little small, a new figure to amplify the waveform to analyze the simulation results carefully can be created by first clicking the blank area of the coordinate and then clicking the 'Open plot in new window' option followed. The amplified waveform of n of the three-phase asynchronous motor is shown in Figure 7.

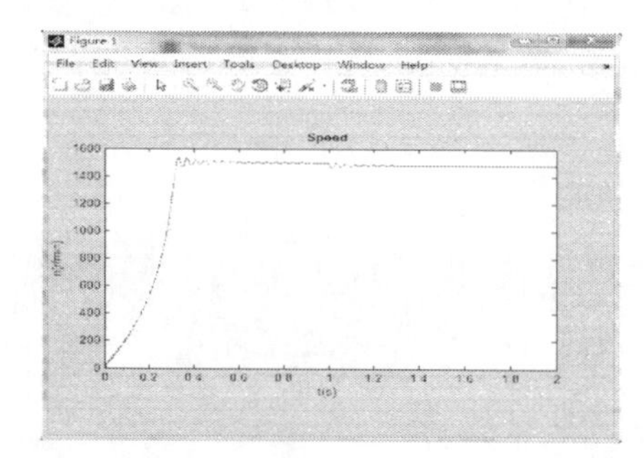

Figure 7. Amplified waveform of n.

4 CONCLUSION

Through emulating and analyzing critical parameters of three-phase asynchronous motor, three-phase asynchronous motor and three-phase synchronous generator in start-up process, it's helpful for users to more profoundly understand the starting character of these motors and the design or research of motors. In addition, designing the motor GUI start-up simulation system makes the simulation and analysis of motors systematic and diversified. It presents a certain priority on achieving fast simulation for different motor parameters and obtaining optimal performance. Supposed the system is enriched in the types of motor and the conditions needing simulating, it would further improve the design and control, even in motor experiment teaching.

REFERENCES

[1] HAN Qinging, XIAO Qianhu, YUAN Qi, etc. GUI-based Function Design of Motor Simulation Test System[J]. Water Resources and Power,2011,01:129–131.

[2] SUN Cai-qin,WANG Bei-bei,ZHANG Ling-jie,etc. Modeling and simulation for synchronous generator in marine power station system[J].Journal of Dalian Maritime University,2013,01:99–102.

[3] WANG Chun-min, JI Yan-ju, LUAN hui, etc. Simulation of PMSM Vector Control System Based on MATLAB/ SIMULINK[J]. Journal of Jilin University(Information Science Edition),2009,01:17–22.

[4] Wang Xiuhe.YONG CI DIAN JI[M].Bei Jing: China Electric Power Press,2007:191–209.

[5] QU Feng-bo,LI Zhi-peng,LI Wei-li.Analysis of starting performance for rotor structures of high-voltage permanent magnet synchronous motors[J].ELECTRIC MACHINES AND CONTROL,2010,09:93–99.

[6] Zhang Hongbao, Wang Xiuhe, Zhong Hui.MATLAB Simulation Research of the Three Phase Asynchronous Motor[J].Electrical Machinery,2009,04:14–18.

Computing, Control, Information and Education Engineering – Liu, Sung & Yao (eds)
© 2015 Taylor & Francis Group, London, ISBN: 978-1-138-02800-5

The application of online examination system of medical science in hospitals

Sheng Li Hu
Information Center, Central Hospital of Cangzhou, Cangzhou, Hebei, China

ABSTRACT: Object: Promote the professional skills of medical staff and improve the work efficiency of hospitals. Method: Expound the advantages and technologies of online examination system of medical science, as well as its functions such as user management, system settings, test bank management, powerful autonomous learning ability, test paper composition, marking, score management, statistic analysis, import and export, anti-cheating. Result: The article analyzes the problems that may occur in the process of operating the system as well as the corresponding solutions. Conclusion: It is of great significance to apply the online examination system of medical science in hospitals.

KEYWORDS: Online; Examination; System.

1 THE ADVANTAGES OF ONLINE EXAMINATION SYSTEM

1.1 *Background*

The online examination system of medical science, as a management platform of efficient examining, learning and training, combines traditional examination models with such advanced technologies as network technology, database warehouse, data mining and statistic analysis. It can make a fast implementation of composing, answering and marking test papers online, which not only shortens examination period and save paper, but also reduce the cost of human and material resources in each examination link.

As the online examination system of medical science is fair, open, random, accurate, secure, etc. it can be widely applied to recruiting, training, assessing, selecting and psychologically testing medical staff, etc. The system can make a quick appraisal of the hospital in-service staff and know their professional skills. Meanwhile, it is beneficial to improve medical quality, ensure medical security and provide a fast and efficient management mode (Zhao Ning et al.2012).

1.2 *The advantage of traditional online examination system*

The examination content involves clinical medicine, preclinical medicine, three basics, medical care, pharmacy, licensing examination and other various professional types. The test bank is capable of storing more than hundreds of thousands of test questions and updated once every six months, which ensures that the questions keep up with latest medical development. Test papers, scores and the like can be preserved for a long time as historical data and referenced at any time.

Medical staff can make full use of spare time to login the system and enter the daily learning module according to their actual situations. They can make a simulation exercise of various questions and reinforce learning, which can not only promote their professional skills and effectively improve pass rate of qualification test, but also enable them to keep abreast of the latest medical development home and abroad.

2 THE ADVANTAGES OF THIS SYSTEM COMPARED TO TRADITIONAL ONLINE EXAMINATION SYSTEM

2.1 *True-time preservation and breakpoint test recovery technologies*

The system adopts the real-time preservation technology, which can preserve in real time the answers in local disk and server's database every certain time when examinees answer the questions. In this way, the examinees' grades can be preserved in local disk if there is something wrong with the network; the grades will be uploaded again to the server to be preserved when the network recovers. Besides, the system has the function of breakpoint test recovery. When examinees login again, the system reads the remaining time from the database and recovers at the fault point, and the examinees can continue the test.

In the process of examination, examinees often run into such unforeseen circumstances as network interrupt, power outage, computer change or computer crash, which will bring the result that the client computer cannot be connected to server's database. The real-time preservation and breakpoint test recovery technologies are capable of preventing data loss and failure of handing in papers.

2.2 *Effective concurrent processing mechanism*

The system optimizes the mechanism of dealing with concurrency conflicts. When examinees are conducting the write operation of the files in the database, the system first carries on the locking operation, and then it conducts the unlocking operation when write operation is finished. The system will close the files when they are read from the database.

When there are a comparatively large number of examinees taking the test at the same time and submitting papers in a short period, the effective concurrent processing mechanism can ensure that all examinees submit papers normally.

2.3 *The enhancement of anti-cheating function*

The template of anti-cheating is further enhanced. The system automatically identifies the login IP. Only one user name can login one computer at the same time, and one examinee can only take the same subject test once. Thus the following situation can be avoided, that is: in the process of the system operation, due to the system bugs, some examinees may answer the papers for those who haven't take the test by using different user names and passwords to login the system on the same computer.

3 THE DATABASE DESIGN AND MAIN TECHNOLOGIES OF THE ONLINE EXAMINATION SYSTEM

3.1 *Database design*

Based on the features of online examination of medical science, the system has designed 8 lists and they connect with each other through primary keys and foreign keys, which can effectively prevent mistakes from occurring during information maintenance. And the primary keys of each list can increase automatically. To deal with the emergencies in the process of operation, the system can set a super account, which can be controlled by codes in procedure and used for maintaining the system when emergency occurs.

3.2 *Main technologies*

By adopting JAVA language to exploit, the system has the function of cross-platform. It employs the asynchronous invocation technology of AJAX (Asynchronous JavaScript and XML) and sends asynchronous request to the server for getting data through the object of XmlHttpRequest. Then the server returns a pure text flow. After acquiring corresponding results, the client processes the results through JavaScript instead of directly showing them on the page. And finally the results will be shown on the page. During the whole process, JavaScript needn't wait for the response of the server, which makes it possible to partially update the page according to actual requirements. Thus the user interface can be optimized and the system performance be improved. When examinees are answering papers, the system will preserve the answers in the server's database in real time through AJAX every certain time, which can greatly improve the reliability of the system.

The system adopts the database access technology of disconnected ADO.NET and fills the object of Dataset with the data by Data Adapter. This function is realized mainly through the Fill (). The object of Dataset can be compared to transcript of the data in the internal storage. And these data can be searched and updated through the object of Dataset. Meanwhile the transcript must be kept consistent with the data source.

The data interactive data of the users in database and data warehouse record the information that examinees access the system. Then the system can figure out the knowledge mastery condition of examines and the difficulty of examination papers by employing Apriority method of the association rule. Thus the difficulty coefficient of the examination paper can be accurately ascertained.

4 MAIN FUNCTIONS OF ONLINE EXAMINATION SYSTEM

4.1 *User management*

The function modules that users see after entering the system vary from one role to another. The following is about the main functions of the system.

The system has set 3 roles: super administrator, administrator and common user. Super administrator can add single user manually and endows this user with the identity (administrator or common user) and with the corresponding authority. It also allows importing personnel from EXCEL according to fixed template. Meanwhile, super administrator can add, delete or examine the personnel and conduct other maintenance operations. The system can record the information of those who haven't submitted papers, haven't taken the test or cheated in the test. And it can also query the records of previous tests according to examinees' ID number.

4.2 *Bank management*

Test bank management mainly refers to the management of the massive examination questions in the database. The administrator can add, modify, delete or search the examination questions. It can add a single question, and also allows bulk importing form EXCEL. The examination questions can be accompanied with various formats of pictures, such as jpg, gif and bmp. At the same time, it supports various complex medical symbols and formulas and allows searching for questions according to key words, question types, difficulty coefficient, etc.

4.3 *Powerful autonomous learning module*

Users can enter the autonomous learning module and randomly choose questions from the test bank to assess their own learning results. They can set that the subjects already mastered do not occur when they make a random choice next time. Thus the learning efficiency can be improved.

4.4 *Powerful anti-cheating module*

The system takes multiple measures to guarantee its security. After examinees submit the papers, the system will automatically lock the computer and they are not allowed to operate the computer again. If something special happens, then the administrator will unlock the computer.

The system adopts the MD5 encryption for such key data as examination questions and answers. Thus the following illegal operations can be prevented: illegal users bypass the login interface, enter the database of the system through other webpage on the address field and then steal examination papers and answers.

The system also adopts certain measures to prevent the multi-point login in case that surrogate examtaker appears. Thus the fairness of the examination can be ensured. To prevent examinees from cheating or misoperation, the system will conduct full screen to stop examinees from seeing the buttons of closing, backspace and refreshing when the examination is conducted. During examination, window-shifting operation is forbidden. Once the pre-established shifting times are surpassed, the system will automatically submit the paper. Meanwhile, the right mouse button is forbidden to use. And some auxiliary keys and functions keys, such as CTRL+C and F5 are shielded to prevent examinees from replicating, refreshing and other cheating methods (Zhang Yanjun et al.2010).

4.5 *Powerful import and export function*

For users who do not have access to network environment, the system supports the export of WORD test paper, reedits and prints it, and then generates the test paper. Thus the examination can be arranged in traditional way. When exporting the test paper, the users can choose from various templates.

4.6 *Test paper composition*

The system provides two patterns: stimulation exercise and online examination. Both of these two patterns have two ways of composing test paper: fixed composition and random composition. The former is employed when fixed test questions are used to conduct the examination. The questions are selected from the test bank in advance and can be replaced according to actual requirements. The latter does not make the question fixed. During examination, the questions are randomly selected from the test bank. These two means of test paper composition need to set the name of the test paper, test time. They also need to set whether the IP block is restricted, whether the grates are shown on the spot, whether question orders are disorganized, etc. And then the range of examinees needs to be added. The markers when composing test paper include the total points of the test paper, the number, point, complexity and percentage of different types of questions.

4.7 *System settings*

The system can set the type of test questions, score of the each type, difficulty coefficient, degree of distinction, etc. It can conduct the processing according to the content set in advance.

4.8 *Test paper marking*

The system configuration, which is flexible, supports the automatic and manual test paper marking. If there are mainly multiple choices, completion, true or false items and the like on the paper, the system will automatically mark the paper based on the answers and give score. If the paper is mainly essay questions and case analysis, after program processing, the system will find key statements in the answer and judge whether the question can get score. As there are different presentations for the same question, the error of the system resulted from the subjective question judgment is relatively big. Thus manual marking is needed.

4.9 *Score management*

Examinees can login the system and check their own scores and papers through the score management module. The mistakes are marked by red ink and the correct answers are shown on the paper. The system can directly print the score online and support the export of EXCEL format. Through the system daemons, administrator can check the scores of all examinees by filling key words such as name and number. It can also export the paper and score of examinees and turn them into encrypted data files for backups and protection.

4.10 *Statistic analysis*

After entering the daemon management procedures, the users, who have the management authority, can make a statistic of the pass rate, highest score, lowest score, average score, the number of examinees and percentage at each score section, which can help the administrator to analyze the effect and complexity of the examination.

5 THE USING EFFECT OF ONLINE EXAMINATION SYSTEM

In practical use, the system is capable of high stability, security and performance, which plays a significant role in the work of the department of personnel, medical service, medical care and science and education. The system also improves the professional quality of medical staff, enables hospitals to show their core competence and truly realizes the integration of assessment, learning and management of hospitals.

REFERENCES

Zhao Ning,Zhang Zhongtao, Hou Yian etc.2012. "Study on the application of online examination system in surgical theory tests" China Medical Education Technology26(5):542–545.

Lu Min, ZhouLin, Wang Guoqing.2012 "Online examination system in the role of the medical three groups examination" Continuing Medical Education(1):57–58.

Zhang Yanjun.2010." Stop cheating Strategies of Asp net-Based Online Exam System " Computer Knowledge and Technology6(33):9660–9662.

Yang Hong,Jia Meijuan,Jie Longmei.2010. "Design of Online Examination System Based on ASP Technology." Computer Study(4):90–91.

DING Nu,ZHANG Hu-yan,YV Juan-juan. "Design and Implementation of the on Line Testing System on Lan" Computer Knowledge And Technology. 6(33):9447–9448.

Computing, Control, Information and Education Engineering – Liu, Sung & Yao (eds)
© 2015 Taylor & Francis Group, London, ISBN: 978-1-138-02800-5

Building image classification methodology using Gaussian kernel combined with Nonparametric kernel

Li Na Ma
Shenyang Urban Construction University, Shenyang, China

ABSTRACT: With the rapid development of machine learning and image processing technology, the combination of these essential techniques is urgently needed for building image analysis. In this research article, we adopt the Gaussian and Nonparametric kernel analysis to conduct an in-depth research on the topic. We assess the empirical performance of the kernel is proposed for simulation and real data sets, and compare their choice based on density estimation or parameter approximation. We show that the kernel implementation performance to match or beat the state of the art in some image classification task. With optimization analysis and experimental simulation, we prove the robustness and correctness of our proposed method. Further potential research areas are also discussed as the appendix material.

KEYWORDS: Image Classification; Gaussian Kernel; Nonparametric Kernel.

1 INTRODUCTION

1.1 *Background introduction*

There are numerous examples in computer vision where images are represented by unordered sets of features [1-2]. For example, the shapes of objects can be represented by sets of local descriptors at edges and corner points. In modern times, every day by millions of digital photos and the problem of how to store, representation and retrieval is proposed. In recent years, the image of the digital management system can effectively deal with the problem are put forward. Many online systems with the development store, manage, and share pictures. For example, Google images can allow the user to the Web search engine image content. Proper use of these services, we can classify the images according to their concept. These images can provide effective information for classification of images, the concept of the theme contains more information. In addition, Flickr can represent a rich set of topics. By comparing image feature set, a simple method is to set seems to contain instance samples from an unknown and likely higher dimensional distribution. Distribution of a deal with these common methods is to use a high dimensional histogram, and compare the histogram by some appropriate target. Based on the Bag-of-words image processing algorithm using a similar approach, but include an additional cluster steps: They treat each image as a bag of visual words, where the words are clustered feature vectors from local regions. Then each image is represented by the empirical, one-dimensional histogram of these words. The collection of these words is called a codebook or a dictionary. Building image classification is becoming more and more important for nowadays smart building development. In this paper, an effective representation is learned for an image by using these Google image class memberships. If two images share a set of similar class memberships, we regard them as a pair of similar images. Thus we use the class memberships to construct the new representation of the image. Based on this new presentation, we could gain an accurate retrieval of images from an image database. A Gaussian kernel classifier combined with Nonparametric is used to measure the image of the image of a special class member.

1.2 *Overview of the research*

We assess the empirical performance of the kernel is proposed for simulation and real data sets, and compare their choice based on density estimation or parameter approximation. We show that the kernel implementation performance to match or beat the state of the art in some image classification task. The experimental results show that the method is an effective and efficient kernel learning classifier. However, the end of the kernel is not a good learning classifier classification, but using the representation of the image classifier. We also show that the kernel classification response vector implementation than the original histogram characteristic. We use the proposed method to retrieve and classify building images from large scale image dataset from Google and Flickr. The experimental result shows that our proposed methodology outperforms other methods. Further work is also discussed in the end.

2 OUR METHODOLOGY

2.1 *The Gaussian kernel*

The overview and framework of the proposed method are shown in figure1. According Wang's [3] research paper, we first download many images of different classes from the internet and then train SVM classifiers to separate the images of each class to other classes using the one-against-all rule. Given a test image, we will support vector machine (SVM) classifier is applied to calculate the response, and then the response vector as a new said. Finally, the different images is to use the measured by Euclidean distance between the response vector.

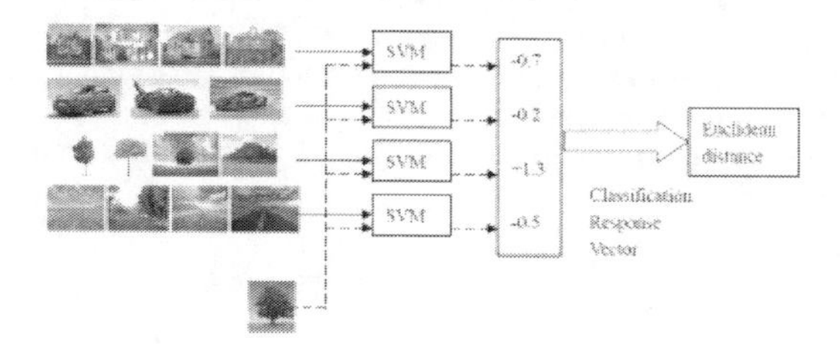

Figure 1. Overview of the proposed Gaussian kernel method.

Class for each of the images, we train a kernel image classifier to distinguish between these and other images. To this end, we first download a lot of such images, and a number of different types of images. Then, we extracted from each image histogram feature vector. The set of image histogram vectors is presented as $\{X_i\}_{i=1}^{N}$ where $X_i = [x_{i1}, \ldots, x_{id}]^{\mathrm{T}}$ denotes the dimension vector of the image. A HIK kernel is used to match a pair of histograms as:

$$K(X_i, X_j) = \sum_k \min(x_{ik}, x_{jk}) \tag{1}$$

Finally, based on this kernel, we design a classifier to predict if an image belongs to this class or not:

$$\begin{aligned} f(X) &= \sum_{i=1}^{N} \alpha_i \phi(X_i)^{\mathrm{T}} \phi(X) \\ &= \sum_{i=1}^{N} \alpha_i K(X_i, X) = \alpha^{\mathrm{T}} K(\varsigma, X) \end{aligned} \tag{2}$$

The definition of parameter for the kernel is an optimization problem and the objective-loss function can be expressed as the formula 3 and 4:

$$\min_\alpha \frac{1}{2}\|\alpha\|_2^2 + C\frac{1}{N}\sum_{i=1}^{N} \Delta(f(X_i), y_i) \tag{3}$$

$$\Delta(f(X_i), y_i) = \max(0, 1 - y_i f(X_i)) \tag{4}$$

We use the following formula to update the α:

$$\begin{aligned} &\alpha^t = \alpha^{t-1} - \tau^t \frac{\partial O(\alpha)}{\partial \alpha}\Big|_{\alpha^t = \alpha^{t-1}} = \\ &\alpha^{t-1} - \tau^t \left(\alpha^{t-1} - C\frac{1}{N}\sum_{i=1}^{N} \beta_i^{t-1} y_i K(\varsigma, X_i) \right) = \\ &(1 - \tau^t)\alpha^{t-1} - \tau^t C \frac{1}{N}\sum_{i=1}^{N} \beta_i^{t-1} y_i K(\varsigma, X_i) \end{aligned} \tag{5}$$

2.2 *Nonparametric kernel*

In literature [4], Liang assumes that the elements of these sets are sample points from unknown distributions that characterize the images. In order to classify the images, we classify these distributions based on their sample set representations. The kernel-based approach is adopted: we introduce and estimate the kernel functions between these distributions. Divisions of the kernel function set definition/distance, like the Euclidean distance is used to define the carrier of Gaussian RBF kernel/to individual. To this end, we need to estimate the difference between distributions. A simple method is to estimate the density of potential and insert it into the or-responding divergence theorem. Histogram and bow methods follow the same pattern. Density estimation, however, the most difficult problems in statistics are due to the curse of dimensionality. Earlier Kondor and Jebara [5] introduced a kernel distribution defined as Bhattacharyya between finite dimensional Hilbert spaces in the affinity between the Gaussian models. This method accord with the characteristics of the Gaussian distribution in the Hilbert space, but it can lead to a huge bias when the data is not Gaussian in Hilbert space. In addition, the method of development is only Bhattacharyya measures. Our method is asymptotically unbiased and can be used for many other differences.

The Pyramid Matching Kernel [5], which also operates over unordered sets, has recently become popular in computer vision. In this approach, each feature set is mapped to a multi-resolution histogram. These histogram pyramids are compared using a so-called "weighted histogram intersection computation." A shortcoming of this approach is that it needs to calculate d-dimensional histograms, which can become very inefficient for large d due to the curse of dimensionality. Selecting appropriate bin sizes is also a difficult problem for which only heuristics are known. We can formulate the ideal into the following formula:

$$D_{\alpha.B}(p\|q)=\int p^{\alpha}(x)\,q^{\beta}(x)\,p(x)d \quad (6)$$

In order to estimate the values of the kernels prior, we could formulate it as:

$$D\alpha,\beta=\frac{B_{k,\alpha,\beta}}{n(n-1)^{\alpha}m^{\beta}}\sum_{i=1}^{n}\rho_{k}^{-d\alpha}(i)v_{k}^{-d\beta}(i) \quad (7)$$

There are RKHS based approaches for defining kernels on unordered sets as well. The method proposed by Smola et al. [6] uses the interaction between pairs in the sample set, and hence its computation time is O(m2). The divergence estimator we propose, by contrast, uses only k-NN distances in the sample set, a well-studied problem with efficient solutions such as k-d trees. Note also that choosing an appropriate kernel function for the RKHS can be a difficult model selection problem, a challenge not faced by our proposed divergence estimator. Sricharan et al. [6] developed k-nearest-neighbor based methods similar to our method for estimating non-linear functional of the density, of which divergences are a special case. In contrast to our approach, however, their method requires k to increase with the sample size m and diverge to infinity. K-NN computations for large k values can be very computationally demanding. In our approach we fix k on a small number.

3 SUPERVISED OPTIMIZATION

3.1 *Overview*

To better analyze out method we plan to use some mathematical optimization techniques to modify the current model. In the particular case of MV analysis, the game theoretic approach to time inconsistency was first studied (in discrete and continuous time) in, where the authors undertake a deep study of the problem within a Wiener driven framework. The case of multiple assets as well as the case of hidden Markov process driving the parameters of the asset price dynamics is also treated. The authors derive an extension of the Hamilton–Jacobi–Bellman equation and manage by a number of very clever ideas, to solve this equation explicitly for the basic problem, and also for the above mentioned extensions. The methodology of [2] is, among other things, to use a "total variance formula", which partially extends the standard iterated expectation formula. This works very nicely in the MV case, but drawback of this particular approach is that it seems quite hard to extend the results to other objective functions than MV.

3.2 *The revised reward function*

From the discussion in the previous section it is clear that we should extend the simple MV problem treated in [11] to the more realistic case when the reward function is given by:

$$J(t,x,u)=E_{t,x}\left[X_T^u\right] \quad (8)$$

Therefore, the function can be revised as:

$$J(t,x,u)=E_{t,x}\left[X_T^u\right]-G\left(x.E_{t,x}\left[X_T^u\right]\right)$$

$$F(x,y)=y-\frac{\gamma(x)}{2}y^2 \quad (9)$$

$$G(x,y)=\frac{\gamma(x)}{2}y^2$$

4 EXPERIMENTAL ANALYSIS

In this section, we show the simulation and experimental analysis of the proposed kernel based methodology in theoretical and real-world applications. The full kernel matrices are projected to be symmetric positive semi-definite and given to a multi-class SVM for classification. Nonparametric divergence kernels are based on the proposed nonparametric R´enyi-divergence estimators (NPR-) and Hellinger distance estimators (NPH). From the figure 2-5, we simulate the proposed methodology.

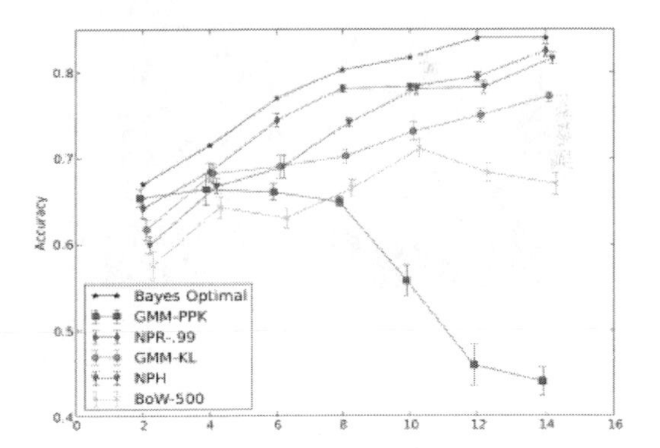

Figure 2. Mean and standard deviation.

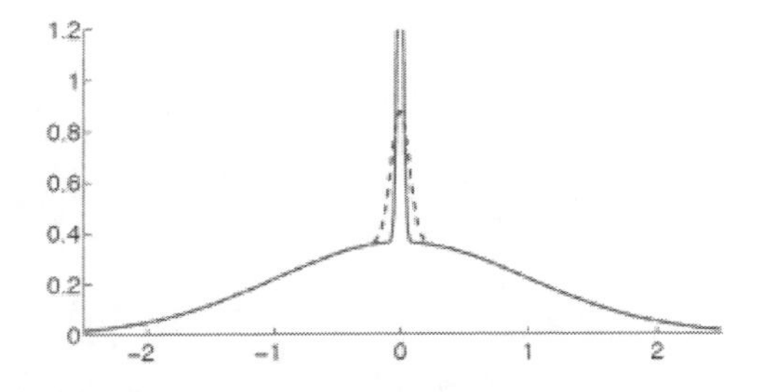

Figure 3. The densities of the matrices.

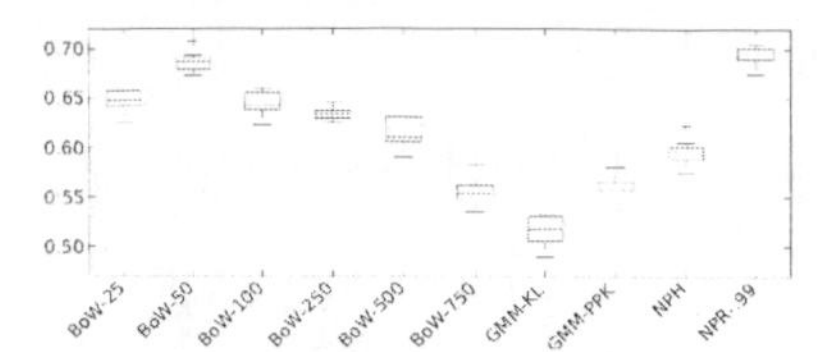

Figure 4. The classification accuracy.

Figure 5. The Test images' dataset.

From the figure 6-7, we compare the accuracy and robustness of the proposed methodology.

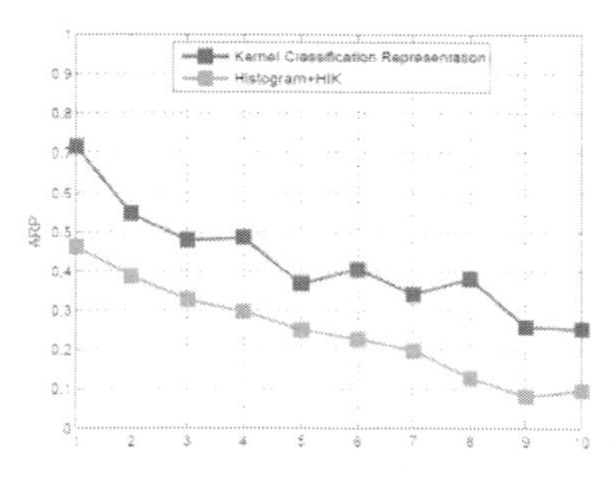

Figure 6. The retrieval ARP of the proposed method and compared method.

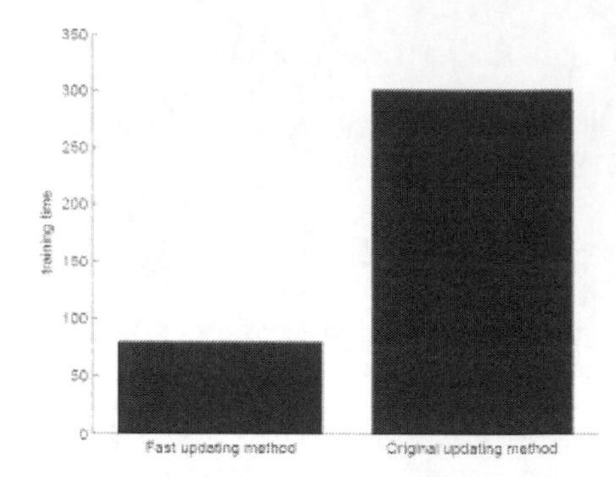

Figure 7. The training time of the updating algorithms.

5 SUMMARY AND CONCLUSION

In this paper, we propose a novel method to represent image using the kernel classifiers. Is that the motivations behind the two images are regarded as similar to if they respond classifier is similar. We developed a rapid and efficient iterative algorithm to train the kernel classifier training image quantity is huge. Parametric methods for divergence estimation are usually biased, since the true distributions may not belong to assumed parametric families. Our nonparametric divergence estimator, however, is asymptotically unbiased. It is also easy to compute, requiring only certain k-NN distances. The experimental analysis illustrates the effectiveness of our method.

REFERENCES

[1] Chen, Zhihua, et al. "Kernel sparse representation for time series classification." Information Sciences 292 (2015): 15–26.

[2] Fernandez-Lozano, Carlos, et al. "Texture classification using feature selection and kernel-based techniques." Soft Computing (2015): 1–12.

[3] Haoxiang Wang. "An Effective Image Representation Method Using Kernel Classification".

[4] Póczos, Barnabás, et al. "Nonparametric kernel estimators for image classification." Computer Vision and Pattern Recognition (CVPR), 2012 IEEE Conference on. IEEE, 2012.

[5] Maldonado, Sebastián, Richard Weber, and Jayanta Basak. "Simultaneous feature selection and classification using kernel-penalized support vector machines." Information Sciences 181.1 (2011): 115–128.

[6] Chen, Yi, Nasser M. Nasrabadi, and Trac D. Tran. "Hyperspectral image classification via kernel sparse representation." Geoscience and Remote Sensing, IEEE Transactions on 51.1 (2013): 217–231.

Computing, Control, Information and Education Engineering – Liu, Sung & Yao (eds)
© 2015 Taylor & Francis Group, London, ISBN: 978-1-138-02800-5

Research on Human Computer Interaction (HCI) and UI pattern design for mobile devices: A review

Xiao Xia Bai
Dalian Art College, Liaoning, China

ABSTRACT: The rapid growth of technological progress and the availability of mobile devices presents opportunities for innovation and feasible context-aware applications and the ability to access information anytime and anywhere. This paper intends to review the latest mobile context-aware application development. Review articles to choose according to their application domain covers the following six categories: smart space, healthcare, advertising, mobile guide, memory aid and disaster alerting applications. We discuss the popular adopted applications based on the design techniques, user experience and areas to illustrate the contribution they have made for Human Computer Interaction (HCI).

KEYWORDS: Human-Computer Interaction; Mobile Device; UI Pattern Design.

1 INTRODUCTION

The two basic elements of pervasive computing (ubiquitous) are environmentally sensitive and mobility. Context is defined as any need knowledge to identify an entity, person, place or object, according to the interaction between the users and applications, and environmentally sensitive is to provide appropriate services to the user or application by understanding their environment and adapt to changes based on their [1]. Therefore, mobile devices such as smart phones, PDAs, portable digital assistant) and tablet is very suitable for the platform realization of context-aware applications due to its availability and technological progress. A combination of context-aware applications and mobile devices provide a novel opportunity for both end users and application developer to obtain context and response to any change in the context consequently [2]. Hence, the main privilege of mobile context-aware applications are to provide an effective, usable, rapid service and reactions by considering the environmental context (such as location, time, weather condition, seasons and other attributes) and adapting their functionality according to the changing situations in the context data without explicit user interaction. Numerous researches have been conducted to show the effectiveness and robustness of the context-aware services. These efforts focus on developing novel technology from: network infrastructure needed for implementation (such as network requirement, network protocol, and sensor); middleware foundation which responsible for sharing and processing required information and application services that can be categorized in six classes [3] as: web service, m-commerce, information systems, communication systems, tour guide and smart space according to their services.

The main purpose of this article is a review of some recent studies context-aware applications to display their important role in human-computer interaction (HCI). Therefore, the principal aim of this work can be summarized as: (1) To review the most relevant and recent context-aware applications of specified categories; (2) To analyze and discuss the techniques and approaches used in each application; (3) To identify the robustness and weakness of each reviewed application; (4) To show how HCI can benefit from mobile context-aware applications [4].

2 THE ROLE OF CONTEXT-AWARE APPLICATIONS

Context definitions and types. A general definition of context was given by Dey and Abowd [1]: any information that can be used to characterize the situation of an entity. An entity is a person, place, or object that is considered relevant to the interaction between a user and an application, including the user and applications themselves. Combination of spatial, temporal, activity and personal contexts makes the primary context to understand the current situation of entities, these types of contexts can response basic question about when, where, what, who. The other types classify as secondary context to share a common attribute of entities and show more detail of entity.

Context-aware definition and categories. According to Dey and Abowd a system is context-aware if "it uses context to provide relevant information and/or services to the user, where relevancy depends on the user's task." They consider three main characteristics for context-aware application as follows: Presenting information and service; Automatic execution of a service; Tagging context information for later retrieval. A human-computer interaction can benefit from context-aware applications through the use of smart devices, can the independent operation, smartly by reducing the recessive requires the user to input and interrupts [5–7].

3 CONTEXT-AWARE APPLICATIONS

General Analysis. Some examples, the existing research on mobile context-aware application will review in this section. However, the scope of this survey does not include those applications focused on the network, based on the middleware infrastructure or the adaptive desktop interface. Application classification according to the characteristics in the following six categories: smart space, medical, advertising, mobile guide, memory aid and disaster warning. Figure1, 2 show these categories and check the application.

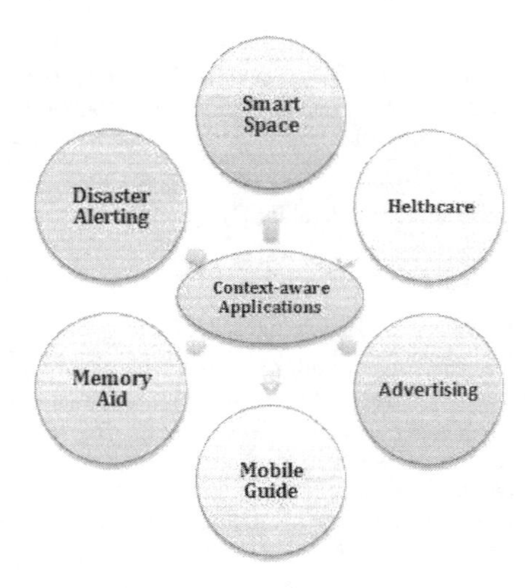

Figure 1. The gesture set.

Category	Application	System Description	Context
Smart Space	Lighting control system [6]	Automated lighting control system for meeting room	Activity, brightness
Healthcare	UCHS - Ubiquitous Context-aware Healthcare Service System [7]	An integrated service platform according to user's life vital signal	Heart rate, respiratory rate, blood sugar, blood pressure
Advertising	iMAS - An Intelligent Mobile Advertising System [8]	A Location-based advertising system	Time, location, personal, user's preferences
Mobile guide	iMuse mobile tour [9]	Tour guide	Location
Memory Aid	Reminder system [10]	Prompting a reminding messages for elders in appropriate time and way	Time, location, activity
Disaster Alerting	intellectual disaster alerts system[11]	Broadcasting alerting information about time and place of disaster	Time, location, temperature, weather condition

Figure 2. The application of context-aware categories.

Smart Space and Healthcare. Smart space is one of the growing fields in context-aware applications that make a space intelligent by employing sensors. This category endeavors to enhance the usability of space to make an easier life for people. Healthcare applications are used for monitoring patients' physical activities and their health condition by employing context-aware computing. These systems can track normal physical activities of patients to make an alarm in certain situations, for instance, provide information about calories used for those actions or other predefined action. Figure 3 illustrates the information about the calorie of user's food, after using a PDA to connect RFID readers in order to sense user's diet content and calories.

Figure 3. The computing BMI value.

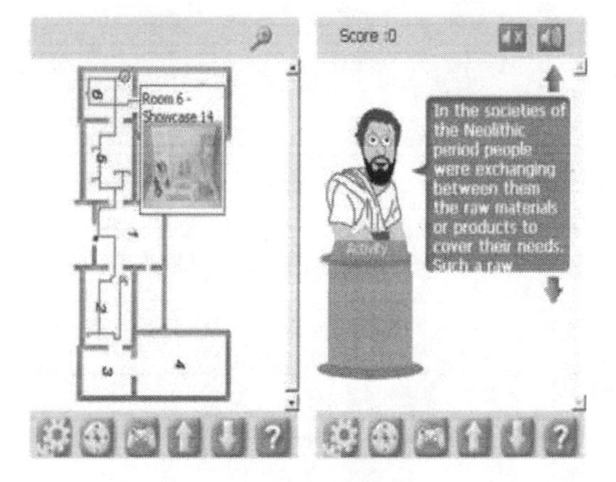

Figure 4. The imuse mobile guide screenshots.

Advertising and Mobile Guide. The main characteristics of a good advertisement are to be shown in a correct way, at the right time, to the proper person, by using multiple modalities. Context-aware advertising systems can tackle the aforementioned requirements and provide a recommendation service according to its users' preferences. The main functionality of mobile guide applications is to provide the specific information to the user via mobile devices such as mobile phone, PDA, and tablet PC. Examples of such applications are museum mobile guides. Museum visitors can be equipped with smart devices and interact actively with the museum environment in order to receive various services. iMuse Mobile Tour is an example of mobile guide applications that provides interactive games, predefined and self-defined tours and enable the visitor to obtain necessary information via their personal mobile phone. iMuse provides context-aware information services by using UHF RFID technology. Guidance services of iMuse can be pushed to the visitor's personal mobile phone by using the group support service and without installing any specific software. Figure 4 shows the screenshots of iMuse mobile guide.

Memory Aid and Disaster Alerting. Memory support and diary applications are developed by tagging special information on specific time or events. The context aware reminder system is proposed by Zhou et.al that prompt reminder message in a proper time and manner for elderly people according to fuzzy linguistic model. To ensure elder receive reminder message in a convenient time with low unpleasant impact two issues should be considered: first, to select an appropriate way to deliver a reminder and second, to determine a reasonable time to present reminder service. Context awareness of mobile platform disaster change system should be knowledge transfer disaster information via mobile devices. Change of disaster system is an effective application in order to reduce disaster influence and ruins the timely transmission of change notification about the disaster. As a result, the system is made up of disaster alert module (receive send alarm information, each user of the equipment), disaster early warning receiver module (disaster change system receives information), disaster early warning database (store all disaster alarm information, rules, reason, and analysis), the risk analysis module (risk) of analysis of the received information, modelling module (coding information to the appropriate type according to the user's equipment and transport) and the project location module provide appropriate information based on the analysis of risk (send).

4 EXPERIMENTAL DISCUSSION

The main idea of context-aware applications is to provide perceptual ability for computer, by means of sensors, in such a way that they can recognize how users interact with their surroundings in specific situations. While, sensors detected situations. They can classify them as contexts and then the system will recognize the appropriate interaction against context, this information will be utilized to trigger, change, and adapt the system or application behavior. Several problems should be considered before developing and wide adoption of the mobile computing technology. Hardware limitation such as small size of mobile devices' monitor, reducing energy consumption, interaction and maintaining privacy are major concerns. Therefore, a feasible and efficient context-aware application should utilize lighter and cheaper sensors, try to gather and inference context information accurately by deploying novel techniques, decrease battery consumption and finally consider the legislations and ethics.

5 SUMMARY AND CONCLUSION

This paper reviewed some recent mobile context-aware applications that were developed based on context-aware computing technology. The basic definition of context was provided and different context categories were listed. Then the scope of review applications was defined by specifying classification categories, and a sample application for each group was briefly discussed. Later context-aware application in the human-computer interaction explanation varies a lot. It can be concluded from the current research and application of the concept of context awareness is widely used in information technology and height will affect the future of human-computer interaction. In the future, we plan to do more research in the review literature.

REFERENCES

[1] Koepfler, Jes A., et al. "Values & design in HCI education." CHI'14 Extended Abstracts on Human Factors in Computing Systems. ACM, 2014.

[2] Giaccardi, Elisa, et al. "Growing traces on objects of daily use: a product design perspective for HCI." Proceedings of the 2014 conference on Designing interactive systems. ACM, 2014.

[3] Dragicevic, Pierre, Fanny Chevalier, and Stéphane Huot. "Running an HCI experiment in multiple parallel universes." CHI'14 Extended Abstracts on Human Factors in Computing Systems. ACM, 2014.

[4] Freitas, Carla MDS, Marcelo S. Pimenta, and Dominique L. Scapin. "User-Centered Evaluation of Information Visualization Techniques: Making the HCI-InfoVis Connection Explicit." Handbook of Human Centric Visualization. Springer New York, 2014. 315–336.

[5] Jackson, Steven J., and Laewoo Kang. "Breakdown, obsolescence and reuse: HCI and the art of repair." Proceedings of the 32nd annual ACM conference on Human factors in computing systems. ACM, 2014.

[6] Nyström, Tobias, and Moyen Mustaquim. "Sustainable Information System Design and the Role of Sustainable HCI." Proceedings of the 18th International Academic MindTrek Conference (MindTrek'14): Media-Business, Content, Managment, and Services. 2014.

[7] Chinthammit, Winyu, Henry Been-Lirn Duh, and Jun Rekimoto. "HCI in food product innovation." Proceedings of the extended abstracts of the 32nd annual ACM conference on Human factors in computing systems. ACM, 2014.

Computing, Control, Information and Education Engineering – Liu, Sung & Yao (eds)
© 2015 Taylor & Francis Group, London, ISBN: 978-1-138-02800-5

Research on the image measurement method of cycloidal gear pitch deviation based on OpenCV

Cui He
Sir Joseph Swan Centre for Energy Research, Newcastle University, Newcastle Upon Tyne, UK
Electrical Engineering Department, Guangxi Electrical Power Institute of Vocational Training, Nanning, Guangxi, China

Zhen Feng Huang & Li Li Cheng
College of Mechanical Engineering, Guangxi University, Nanning, Guangxi, China

ABSTRACT: Aiming at the present problems of the cycloidal gear measurement, a pitch deviation measurement method which is based on OpenCV image processing is put forward. In CCD camera platform, image denoising, background segmentation and edge detection are used to process the image, and the contour coordinates are extracted, and the problem of coordinates' ascription are solved, by using OpenCV functions. Based on the definition of pitch deviation, coordinate system is transformed and the data processing procedure is determined. It is realized the fast and accurate measurement in cycloidal gear pitch deviation.

KEYWORDS: pitch deviation; OpenCV; image processing; contour extraction.

1 INTRODUCTION

Cycloidal gear speed reducer is a new type of gear drive mechanism. It has been applied in the field of petroleum, chemical, metallurgy, mine, pharmacy, food, spinning, hoisting and conveying industry, etc. Cycloidal gear is the main transmission part of cycloidal gear speed reducer. The transmissive performance is determined by the quality and the precision of cycloidal gear. There are some disadvantages of the traditional mechanical testing method, such as low accuracy and low detection speed, so it can't adapt the requirement of modern industry produce. With the development of science, it is a hot topic for scholars to research that substitute computer for human activity. By using a CCD industrial camera as a platform, OpenCV as a tool, the measurement method of pitch deviation is studied. It provides a new train of thought for measuring cycloidal gear pitch deviation.

2 THE COMPOSITION OF THE MEASURING SYSTEM

The measuring system is composed of PC, executing device, imaging device and measurement software, as shown in Figure 1. The image signal collected by a CCD industrial camera is transferred to upper-computer. Then the software system completes the process which contains feedback regulation, camera calibration, image processing and the calculation of cycloidal gear pitch deviation. The measurement results will be displayed on the output device finally.

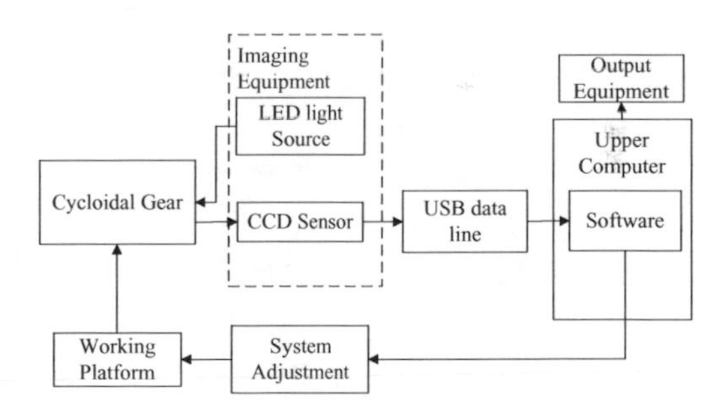

Figure 1. The composition of the measuring system.

3 GEAR IMAGE PROCESSING

For further processing, it is necessary to adjust the clarity of the camera and the verticality of the optical axis to the gear before an actual measurement. By this way, the image which clearly reflects the characteristics of cycloidal gear is obtained, then transmitted to the measurement software system.

As shown in Figure 2(a), the original image of cycloidal gear, which is obtained by computer cannot be directly extracted contour for much noise and low contrast in it. So image processing is needed. Though the image processing methods are various, there is not a common method can adapt to all kinds of occasions. Every method should be chosen based on different environments. If programming and testing various algorithms one by one, it will be difficult to program, at the same time, the speed of the program will be slow. The gear image is processed by using the OpenCV function library. By this way, the time of programming is reduced and the running efficiency of the software is improved. Firstly, the noise points are eliminated by cvSmooth() function. Then mean shift method is applied to segment the image by using cvPyrMeanShiftFiltering(), the result as shown in Figure 2(b). In this function, both input and output dates are 8-bit and three-channel color images, and space radius is set to 2, colors radius is set to 40. Finally, canny edge detector is done by using cvCanny(), and the output image signal is high quality because it is continuous, obvious, and no noise interference, as shown in Figure 2(c).

(a) The original image of gear

(b) Mean shift

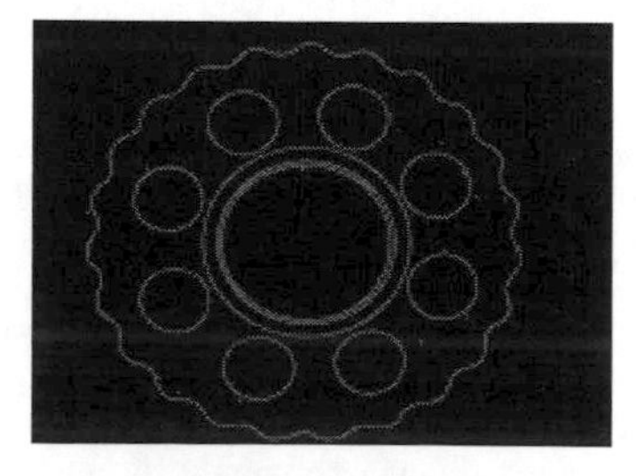

(c) Canny edge detection

Figure 2. Cycloidal gear image processing.

Boundary contour can be detected by Canny edge detection operator, but there are two problems to solve when we want to receive the quantitative result. One is how to get the pixel coordinates of edge points, the other one is how to determine which boundary a point belongs. The traditional method is that scanning the binary image data of edge line-by-line or column-by-column. It is slow, and the problem of attribution is still not solved. In this paper, a method is used to search contour from the binary image based on cvFindContours() of OpenCV. The program statement is as follows:

cvFindContours(contour_image, img_storage, &contours, sizeof(CvContour), CV_RETR_CCOMP, CV_CHAIN_APPROX_NONE, cvPoint(0, 0));

There are six parameters in this function. The input image is called contour_image which is 8-bit, single channel, binary. Output data which exists in the form of CvSeq structure tree sequence is kept in the memory space allocated by cvCreateMemStorage(). The arrangement of contour tree is set to CV_RETR_CtrCOMP[2]. The first floor data are extracted, and every contour is stored in the global dynamic array to calculate. A data flowchart of this process is shown in Figure 3. The default search sequence of Cv Find Contours () function is as shown in Figure 4. These array data can be one to one correspondence with every contour. The problem of edge ownership is solved.

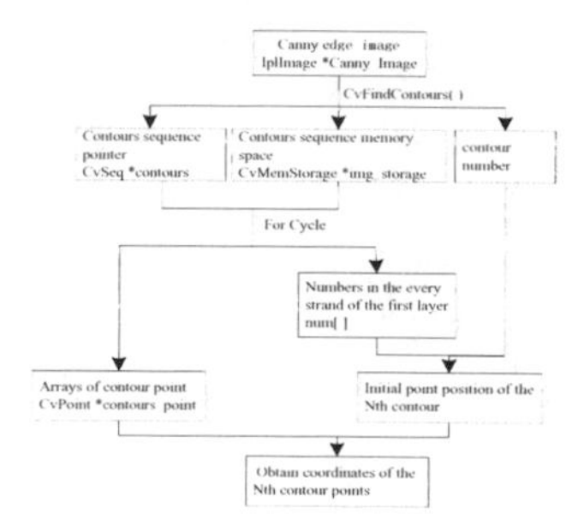

Figure 3. Date flow chart of edge extraction.

Figure 4. The default search sequence of fucntion.

After edge detection, the coordinates of edge points are shown in the shape of full-pixel, which means that the values of the coordinate points are integer type. But it can't meet the high-precision measurement system. In this paper, the sub-pixel edge coordinates are obtained by secondary processing of the edge coordinates, through sub-pixel processing method which is based on cubic spline interpolation.

4 MEASURING THE PROFILE DEVIATION

4.1 *Geometric center of cycloidal gear*

How to obtain the coordinate of the gear center is the premise behind measuring. Known from the processing method and technology of cycloid gear, the rotary axis of the center hole coincide with the geometric center axis of tooth profile. It means that the center of the circle can be approximated as the geometric center of the cycloidal gear. In this article, the coordinates of the gear geometric center are gained by the least square method.

4.2 *Coordinate system conversion*

There are two coordinate system in this measurement system, one is the image coordinate XOY, the other is the part coordinate xoy. The values of XOY are transformed into the xoy after all the required coordinate values are obtained, and the coordinate transform formula is shown in the formula (1). The part coordinate xoy is a rectangular coordinate system which consider the geometrical center of the cycloidal gear as the origin, and the value of this origin in XOY is (X0, Y0), as shown in Figure 5.

$$\begin{cases} p_x = P_Y - Y_0 \\ p_y = X_0 - P_X \end{cases} \tag{1}$$

All the coordinates of the contour are expressed as them in xoy after transforming the coordinate system. While measuring the pitch error, using polar coordinates is more convenient for computing. Therefore, Cartesian coordinates can be converted to polar coordinate values, then polar radius and phase angle are stored into a memory array.

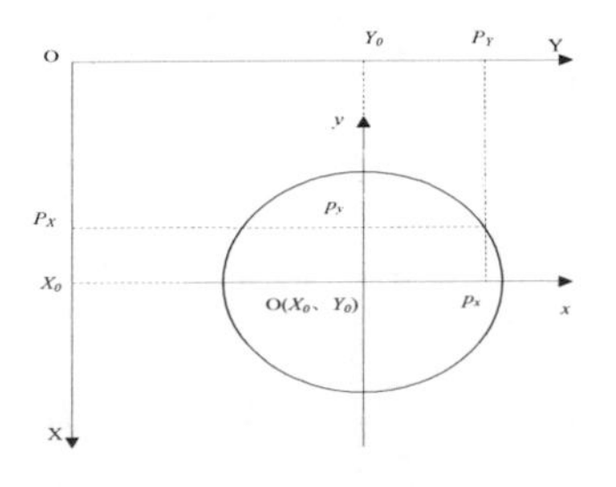

Figure 5. Rectangular coordinate system.

4.3 *Process of software measurement*

The pitch deviation is used to reflect whether the gear teeth in the middle of tooth height are evenly distributed. It affects the stability of the gear directly[5]. And the realization of the measurement process is shown in Figure 6: (1) a middle circle between the addendum circle and root circle is taken as referencing circle to evaluate pitch deviation, and referencing circle ρ_0 is calculated; (2) scanning the polar radius values of the contour polar coordinates array, finding the left point of intersection between contour and the circle which radius is ρ_0;(3) One pitch corresponding central angle will be get, if the adjacent polar angle coordinates are subtracted. Then the results are subtracted from the theoretical central angle of single pitch $\varphi = 2\pi/Z_b$, times ρ_0; (4) the result whose absolute value is the largest is taken as the basis to evaluate whether a single pitch deviation is qualified or not.

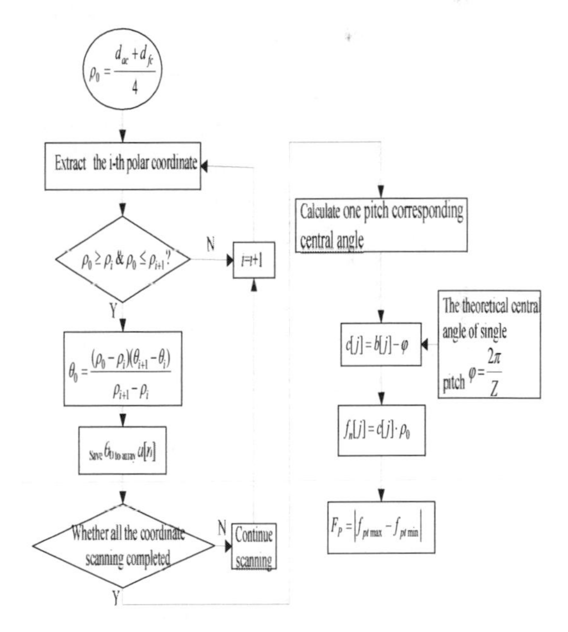

Figure 6. Program flowchart of pitch error.

5 EXPERIMENT RESULTS AND ANALYSIS

The data measured by 19JC digital model universal tool microscope are used to the standard values. And the data measured by this system should be compared with it to obtain measuring error.The Pitch deviation value measured by 19JC digital model universal tool microscope is 0.0779mm.The date measured by this method is recorded in table 1, and the relative error is computed as shown.

Table 1. Measured data of the system (unit: mm).

measurement time	1	2	3	4	5	average	relative error
individual circular pitch error	0.0843	0.0756	0.0788	0.0810	0.0819	0.0845	0.3519%

Relative error reflects the accuracy of measurement level. The result shows that the relative error is less than 0.5%, that is, the measuring accuracy level of this method is 0.5.

6 CONCLUSIONS

In this paper, machine vision technology is introduced to cycloidal gear measuring. An OpenCV vision function is used to process the gear image and extract the contour data of every edge. By this way, the problem the traditional method brings is solved, complexity and low efficiency are avoided. The results show that the accuracy grade of the measurement method which is based on OpenCV is 0.5. This cycloidal gear pitch deviation measurement method is high precision, and it provides a reference for processing and measuring other parameters.

REFERENCES

[1] Jingbin Guo, Xian Wang, Haijun Liu, et al: Measurement of Cycloidal Gear Error and Calculation of Modification, Journal of Tianjin University, vol. 44 no.1 (2011), p. 85–89.

[2] Gary Bradski, Adrian Kaebler, Shiqi Xu, et al: Learning Open CV, Tsinghua University Press, (2009).

[3] Qingyang Li, Nengchao Wang, Dayi Yi: Numerical Analysis, Tsinghua University Press, (2001).

[4] Gang Jiao: Cycloid Gear Process and Special Clamping Apparatus Design Theary, Journal of Shandong Institute of Architecture and Engineering, vol.6 no.1 (1991), p. 70–73.

[5] Shuo Fan: Research on Measuring Technology of Ultraprecision Gear Total Cumulative Pitch Error, Liaoning: Dalian University of Technology, 2006.

Computing, Control, Information and Education Engineering – Liu, Sung & Yao (eds)
© 2015 Taylor & Francis Group, London, ISBN: 978-1-138-02800-5

Research on college major construction of automobile engineering in high engineering education with computer technology education

Y.D. Tian
Shanghai Dian Ji University, Shanghai, China

ABSTRACT: Automobile service engineering is a new undergraduate major authorized by China Education Ministry in 2002, and there are some difficulties in the college major construction and the teaching project. To aim at the problems, the history of automobile engineering is presented in China at first, and then the development, the content and the characteristics of automobile service engineering are analyzed, finally the college major construction of automobile service engineering is explored on the basis of the computer technology education. It carried on the practice in automobile engineering department of the Shanghai Dian Ji University, thus strengthened the connotation construction of automobile engineering.

KEYWORDS: Automobile engineering; college major construction; computer technology education; high engineering education.

1 INTRODUCTION

With the development of the economic society and the car after-markets, in particular the establishment of the uniform automobile service system, the automobile service starts to step into the way of the specialization and professionalism. The high engineering education of automobile engineering is shouldering the historical mission that raises the high quality talented person of the automobile servicing for the car after-markets and the entire society. However, the high quality talented person of the automobile servicing is not only the person grasped the knowledge system of automobile engineering, but also the entity owned the specialized knowledge, the professional accomplishment and the professional skill of the servicing of the automobile servicing.

To achieve this goal, it is necessary that must carry on the college major construction of automobile engineering, namely constructing new idea, new system and new mechanism of undergraduate education of automobile engineering. It has the vital practical significance and the education value. This is the current main problem. In this paper, the college major construction of automobile service engineering is researched on the basis of the computer technology education.

2 CHINESE COLLEGE MAJOR SYSTEM

2.1 *Old college major system*

In China, the college major establishment deferred to the system of European and American university before 1949. As a result of well-known reason, it was imitated in the system of Soviet university after 1949. Since 1978, Chinese college majors catalogue passed through three times of adjustment and the revision, The majors scope was opened up to a certain extent, and some majors were adjusted gradually and combined. At 1993, the number of national ordinary college majors reached to 624 kinds, in which the number of the basic college majors was 504 kinds in the catalogue. In 1998, the number of national ordinary college majors amounted to 635 kinds.

2.2 *New college major system*

The present catalogue of the national college majors was the 2012 edition, and its classification is consistent with the knowledge classification of the graduate student's degree that was promulgated by the State Department Degree Committee and the State Educational Committee in 2010, namely "Knowledge major Catalogue of Awarding Doctor & Master Degree and Raising Graduate students". The study fields have 12 kinds, including philosophy (the code is 01), economics sciences (the code is 02), law (the code is 03), education (the code is 04), literature (the code is 05), history (the code is 06), natural sciences (the code is 07), engineering (the code is 08), etc. The major kinds classification has 92 kinds under the study fields, such as the mechanics (the code: 0801), the machinery (the code: 0802), the instrument (the code: 0803), the materials (the code: 0804), the energy and power (the code: 0805), etc. The major classification is the college major that has been reduced from 535 kinds to 506 kinds in the catalogue that are 354 kinds of basic majors and 154 kinds of special majors [1]. And the college major of automobile service engineering belongs to machinery of the major kinds classification.

3 COLLEGE MAJOR DEVELOPMENT OF AUTOMOBILE ENGINEERING

3.1 *College major about automobile industry*

After 1949, the college major of the automobile and tractor was established the in our country for the first time. In 1988, the college major of the automobile utilization engineering was presented and opened for the first time. In 1993, the college major of the automobile utilization engineering was renamed to the college major of the ship-equipment utilization engineering. In 1998, the college major of the automobile and the tractor was renamed to the college major of the vehicle engineering, and the college major of the ship-equipment utilization engineering was united into the college major of the traffic transportation. Simultaneously, the National Education Ministry established the college major of the car servicing engineering education. And then the college major of the car servicing engineering education was cancelled in 2002. The college major of automobile service engineering was authorized and opened for the first time. At present, there are two college majors about the automobile industry in China, which are the vehicle engineering (the code: 080207) and the automobile service engineering (the code: 080208).

3.2 *University about automobile service engineering*

In 2002, through authorized by the national Ministry of Education, Wuhan University of Science and Technology became the first university in establishing the automobile service engineering in China, and started the first term recruitment of the college students in September 2003.

In 2003, via authorized by the national Ministry of Education, Tong Ji University became the second university to establish the college major of automobile service engineering in the whole country.

In 2004, after the national Ministry of Education authorized, there were seven universities to the establishment the college major of automobile service engineering, such as a Ji Lin University, Chang A University, etc.

At present, about 70 universities have opened the college major of automobile service engineering in the whole nation. And the number of recruitment is approximately 8000 college students in the college major every year.

3.3 *Development about automobile service engineering*

In recent years, our country automobile production presented the spanning type growth situation, the automobile volume of production and marketing climbs fast in the world sequence order [2]. 2000 year Chinese entire automobile production and marketing were 2,070,000 and 2,080,000. In 2014, Chinese automobile production and sales volumes reached 23.7229 million and 23.4919 million, respectively, and had set up six consecutive years to become the world's automobile morality. Chinese automobile sales volume growth and development is as shown in figure 1. Therefore, the servicing of car project specialized has the very big development space.

In our country, Wuhan University of Science and Technology, Ji Lin University, Chang An University, Tong Ji University, Jiang Han University, Changsha University of Science and Technology, Tianjin Technology Normal College, Heilongjiang Technology College, Xi Hua University and so on ten universities occupy the front row in the college major of automobile service engineering.

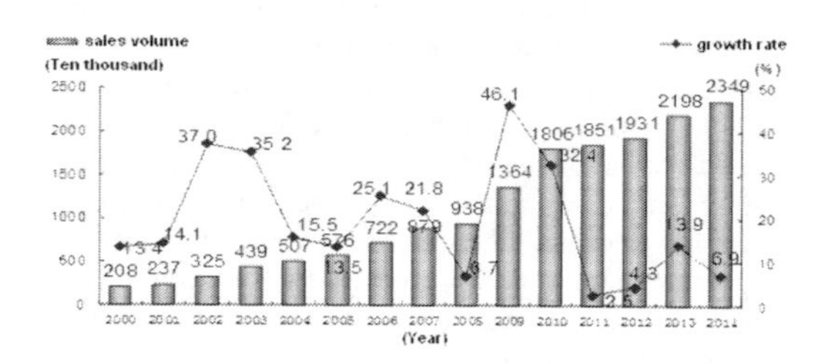

Figure 1. 2000–2014 China automobile sales volume and growth rate.

But has already opened and this similar automobile specialties and so on utilization project and automobile marketing, physical distribution in Shanghai's Tong Ji University and Shanghai University of Technology Science, also the strength is abundant. But because the recent years Shanghai automobile industry development was rapid, and the automobile industry took the pillar industry, also started in 2004 to prepare for construction "the international automobile city", therefore the existing servicing of car specialized graduate has not been able to satisfy the market demand. Therefore, my courtyard start servicing of car specialized conforms to the Shanghai industry development tendency, has the very big development space.

4 COLLEGE MAJOR DIFFICULTY OF AUTOMOBILE ENGINEERING

Along with the rapid increase of the automotive inventory and the unceasing development of the automobile service industry, the automobile engineering more and more becomes a large-scale complex system. The scope and category of automobile service engineering are wider and wider, such as the raw material and the spare part of the automobile industry upstream, the automobile marketing, the automobile service maintenance, the automobile insurance, the automobile replaces, the automobile recycling regeneration, the automobile distribution of the automobile industry downstream, and so on [3]. It brings many difficulties for the college major construction and teaching of automobile service engineering.

4.1 *Course system*

The basic content of automobile service engineering includes four basic aspects that are is the selling, repairing, managing, and using of the automobiles. With the unceasing development of the automobile servicing, the content of automobile service engineering becomes more and more, wider and wider, and it causes the curriculum system of automobile service engineering to raise too large. The automobile electronic commerce, the automobile consumption credit, the automobile decoration beauty, the automobile insurance and compensation, the automobile replacement and appraisal, the automobile recycling and regeneration, the automobile information and communication, the automobile distribution, and the automobile policy are unceasingly expanded in it at all. Of the causes, its theory system becomes more and more, but its course system is limited.

How to establish the curriculum of automobile service engineering suitably, and how to choose the typical course content? It is really a difficult problem.

4.2 *Theory teaching*

Although the automobile service engineering belongs to the engineering major in higher education, it is a union (or intercrossing) of the engineering, the economics and the management science actually. Teaching the theory of automobile service engineering needs not only the automobile specialized knowledge, but also the certain knowledge of the economics, the management, the marketing, the finance and the insurance. Thus the request of the theory teaching is high. With the popularization of computer technology, more and more computer knowledge and skills are applied in the automobile service industry.

How to choose the content of each aspect, and how to unify them reasonably in together? It becomes the second difficult problem.

4.3 *Practice teaching*

As a strong practical major, the experiment of the automobile service engineering needs the many automobile test instruments, equipments and expendables that are very expensive things. Moreover, the wastages are many, the maintaining requests are high, and the maintenances are difficult. So the educational price of automobile service engineering is higher than the other major. Of course, the current computer network communication and electronic commerce is more and more popular.

On the other hand, the practical teaching of automobile service engineering is limited by various aspects of the practice location, the producing arrangement, the location request, the safe operation and the national laws and regulations and so on. Thus, it causes the third difficulty that the practice teaching is developed difficultly.

5 MAJOR CONSTRUCTION OF AUTOMOBILE ENGINEERING WITH COMPUTER TECHNOLOGY EDUCATION

The college major of automotive engineering is a new undergraduate major with the strong practice. In the process of the undergraduate major construction, it needs to explore many kinds of educational patterns in order to find the characteristic of the computer technology education. Since the construction of the undergraduate major starts up, the research has been practiced in the continual four terms of the education and teaching process of automobile service engineering. Passing through the unceasing educational improvement, some educational experiences are accumulated.

5.1 *Research of the training scheme*

It is very important to research the training scheme of automobile service engineering. The automobile service engineering is a new undergraduate major that was authorized to meet the need of the domestic and foreign recent automobile servicing development by the National Education Ministry in 2002, so its standard training scheme is short and each university has a different plan about the major. The Shanghai Dian Ji University is a new undergraduate college that likes many institutes in China, but it's attached importance to the high technical education. In recent years, we have investigated and studied the related teaching material of the other colleges and universities, in particular the key universities and characteristic colleges, and then rise and consummated the training goal and plan of automobile service engineering to conform to our characteristics. The computer theory and technology, communication, network and electronic commerce are added to the curriculum system.

5.2 *Research of the theory and practice teaching plan*

The theory and practice teaching plan of automobile service engineering is an essential matter in the college major construction. At present, about 70 colleges and universities set the undergraduate major of the automobile service engineering in our country, and its development time is not long. The first graduated college student from Wuhan University of Science and Technology less than ten years. Many colleges and universities have various theory and practice contents in theory and practice teaching base construction, and the experience is insufficient. Some colleges and universities stress on the theoretical teaching, and have some difficulties to complete the undergraduate major request. We have designed the reasonable theory and practice teaching plan, and established the union our college and factory with the cooperation background in the theory and practice teaching construction, and then set up our theory and practice teaching base to complete the project, Including computer and electronics business base.

5.3 *Research on the teaching base construction*

Produce-Learn-Research is a very important loop in the high engineering technical education. As an engineering college major, the automobile service engineering must have the production practice, the specialty practice, the graduation practice, and the graduation design (or graduation thesis) according to the Chinese high engineering education pattern. Each practice study period is above two weeks. In our university, more than 80 students need to complete their practices in some practice base, thus only laboratory is insufficient, and most students must go to the factory and company to do it. Actually, it is good for their future working. Therefore, we have established as many practical teaching bases out of our university as possible. In view of the automobile service engineering, it is essential to construct many Produce-Learn-Research bases of car servicing. Relying on the Shanghai automobile industry pillar, the computer and electronic business practice are strengthened.

5.4 *Research on the teaching running pattern*

In the process of the course teaching of automobile service engineering, the relation order of teaching the theory courses and the practice teaching courses is analyzed through the investigation and the study, and the discussion, then the teaching running project of automobile service engineering is established. After the typical teaching content is presented, several key guidebooks are compiled. At last, the corresponding inspection and the appraisal methods of the teaching are established to satisfy the request the undergraduate course. In the past ten years, we have developed six batches of a total of 493 college students.

6 CONCLUSIONS

The college major of automobile service engineering is a new undergraduate major. There are many difficulties in the college major construction and teaching. To aim at the problems, the training plan and the teaching system of automobile engineering is presented, and then the content and the pattern of the theoretical teaching and the practice teaching are established, finally the college major construction of automobile service engineering is explored on the basis of the computer technology education. It carried on the practice in automobile engineering department of the Shanghai Dian Ji University, thus strengthened the connotation construction of automobile service engineering, and made a certain foundation for the next characteristic of computer technology education.

ACKNOWLEDGMENT

This paper was supported by the Discipline Foundation Project of Shanghai Dian Ji University under Grant 12XKJC02.

REFERENCES

[1] High Education Department of Education Ministry of P. R. China, Undergraduate Major Establishment of Chinese University, Beijing: High Education Press, 2012, pp.368–380.

[2] X. Y. Xu, Investigation and Forecast of Automobile Markets, Mechanical Industry Press of China, 2007, pp.13–15.

[3] Y. D. Tian, "Research on college major construction of automobile service engineering with technical undergraduate education", in Journal of Technical Education, 2009, Vol. 3, No. 3, pp.55–58.

Computing, Control, Information and Education Engineering – Liu, Sung & Yao (eds)
 ISBN: 978-1-138-02800-5

Exploration of construction technology of bridge pier column

Shui Ping Huang & Mei Ling Guo
Guangdong University of Science and Technology, Guangdong, China

ABSTRACT: Construction of Pier Column is the key chain for bridge construction, which has a great influence on completion of the bridge engineering mission. Taking Saihu City of Jiujiang County as an example, it is to formulate the documents of bridge pier turnover and sliding formwork's from the viewpoint of construction unit so as to deeply master of essentials of technology for bridge pier construction for the guidance of assurance for quality and quantity of bridge pier mission according to construction progress.

KEYWORDS: Pier Column Construction; Turnover Formwork; Sliding Formwork; Design.

1 INTRODUCTION

Pier Column Construction would be one of an important chain of bridge Construction which determines the progress of bridge Construction, safety and quality and has a great influence on completion of the bridge construction mission. Formulation of practical construction pier column scheme design has an important guiding significance of the deeply master of key technology for construction of bridge pier and timely completion of construction mission. Currently taken the Saihu City of Jiujiang County Bridge Pier column construction as an example, to formulate Bridge Pier Column, Turnover and Sliding Formworks Construction Design Scheme were an important guiding significant of completion of construction mission for actionable guidance of assurance for quality and quantity of bridge pier mission according to construction progress and provides for useful reference of other bridge pier column constructions.

2 OVERVIEW OF PROJECT OF SAICHENGHU BRIDGE

Saichenghu Bridge was located at Saihu City in Jiujiang County, Jiangxi Province, connecting the port area at the western city to its north, a Chengmenxiang area to its south, and was an important key control engineering for communication with Hangrui Expressway and fleet channels of western cities and Cima land plate (Shicheng Avenue). The bridge would be in alignment with connection to the north bank of Saicheng Lake with a full length of 1022m and width of the bridge of 32m. Total investment of Saichenghu Bridge amounts to 0.2 Billion Yuan. It was constructed by the China No.10 Engineering Group. Ltd and planned for construction on June, 28th, 2012 and completed on May 28th, 2014. The upper structure adopts 50m long, 24-hole first-simple beam and next-to-continuous priestesses concrete I-Beam. It would be rectangular as well as thin-wall pier columns at the lower structure. Bank pier adopts column type and ribbed slab type. Bridge foundation adopts drilling, placing pier type. The full bridge was designed as demarcated with 4 horizontal lines. Min. The radius would be 1485m. Max. The longitudinal slope on the bridge surface would be 2.42%.

Main quantities: foundation pier column: 448; Bearing platform: 46; Pier column and ribbed plate: 46; Bent cap: 48; 50 mT beam: 282.

3 PIER CONSTRUCTION DESIGN OF SAICHENGHU BRIDGE

3.1 *Conditions for matching construction team and machinery*

In consideration of scale and quantities for pier column, turnover and sliding formwork of Saichenghu Bridge, the matched construction team and machinery were as follows: 25T crane: 4; QCT4810 tower crane: 10; Turnover formwork: 4; Sliding Formwork:16 sets; HNJ5263GJB concrete tank car: 8; JS1000 agitator: 2; reinforcement-processing equipment: 4 sets; labor force: 150.

3.2 *Construction technology of pier column turnover formwork*

3.2.1 *Inspection for template*

Bridge pier template would be special molded steel designed; each node of the template has a height of 2m; when constructing the turnover formwork, placing height each would be 4m. For ensuring its flatness on the surface of the pier concrete, after placing the template, the template would be pre-sliced and carefully checked up, ensuring that all dimensions of parts, joints and flatness attain up to standard.

3.2.2 *Processing and bundle of steel reinforcement*

Main steel reinforcement on the bridge pier column would be connected with strong rolling, straightened threads, for ensuring the quality of reinforcement, prior to use, the end of steel reinforcement would be grinded to be leveled, for ensuring two reinforcements being seamless butt-welded; for each rolling, straight thread steel reinforcement adopts on-go and no-go gauges for inspection; when constructing the pier column, it must carefully refer to the drawing for design and marking of the bundle points and steel reinforcement joints so as to make them to be controlled within the scope of approval. When initially pre-burying reinforcement, it would be fabricated according to 4m, 6m and 8m. Connecting position would be fabricated according to 4.5m, to ensure the quantities of connecting joints of stressed steel reinforcement being less than 40% of main reinforcements. It would be practically strengthened for quality control of the connecting joints, spacing of joints as well as stirrup.

3.2.3 *Installation for template*

After completion of bundle-up of steel reinforcement and inspection for ensuring quality of bundle-up up to standard, it would then be installed in the template. Prior to splicing of the template, an effort would be firstly made to grinding of the surface of template so as to ensure its cleanliness and tidiness on the surface and then to smear mold discharging agent on the template, during the smearing process, make well light, thinner smearing as well as evenly leveling so as to ensure concrete surface being maintained with the same color. After completion of template, the quality of template would be strictly rechecked with carefully calibration of axle position of template, horizontal elevation, dimensions and perpendicularity of the plate and etc. it was also carefully checked with the seam joints of template, slab staggering quality, to ensure the spacing of seam joints: less than 2mm; in addition, use of steel ruler ensures to measure geometrical dimensions; use of drawing line ensures to rectify template straightness; use of plumb instrument ensures to calibrate template perpendicularity. During installation of template, ensuring to focus on mis-displacement of the axle line being less than 1cm. In case of mis-displacement beyond the standard, electric, manual hoist as well as jack would be used for calibration.

3.2.4 *Control of steel reinforcement protective layer*

It is very important for Control of steel reinforcement Protective Layer, in case of improper control, not only can steel reinforcement be reduced for its endurability, but it can also have an influence on its overall outer appearance. Therefore, during the construction process, it was ensured to stirrup seamless tightly twined up the main reinforcement. A framework of steel reinforcement would be perpendicular with ground level, with its body of pier column set up high-strength concrete cushion block protective layer, to ensure the top of Framework of steel reinforcement appropriately pulled and pushed as well as tightly fastened when connected with template; when placing the concrete, to ensure discharge hopper, tumbling barrel being separated from template so as to eliminate collision of reinforcements occurred as soon as possible.

3.2.5 *Installation for concrete*

Firstly, agitation and transport of the concrete. Concrete would be transported by tank car; crane and hooper would be used for placing the concrete on construction of the turnover template at the position of pier column 1#-5# and 30#-25#. In order to avoid the concrete due to gravity to be separated, free inclination of concrete would be generally less than 2m in the placing course, with the help of two courses of the tumbling barrels on the pier body placed into the turnover formwork.

Secondly, placing the concrete. Prior to placing the concrete, its cleanliness was ensured and no sundries found in the body of pier column and keep the concrete seam joints moisturized. Insert type vibrator would be used to vibrate the concrete, during the vibration process, it was ensured to the spacing between a vibrator and template to be controlled within 5-10cm. When vibrating the upper course of the concrete; single thickness: 30-40cm or so; it would be insisting on the principle of "Quick-insertion and Slow-Extraction" in operation, prevented mis-staggering position of concrete template. When placing the concrete, the special personnel would be assigned for inspection of template stability, in case of loosing, deformation, displacement and other conditions, it would be immediately settled.

3.2.6 *Control of outer appearance of concrete*

For ensuring an artistic outer appearance for engineering, not only can it focus on construction of structural concrete, but it's also for inspection and acceptance for template and can strengthen quality control for the template.

Control of the concrete: Do not randomly exchange cement, sand, broken stones and additives.

Quality Control of the Placing Technique: In general, the following measures would be taken: the mix ratio of concrete would be slightly regulated, i.e. under the condition of the quantity of cement kept unchanged, slightly increase sand content and slight reduction of water content. Properly extend the molding maintenance period. Under the condition of the quantity of cement kept unchanged, increase the quantity of cement, effectively control the degree of collapse.

Agitation and Transport of Concrete must conform to: meteorological precision, agitation in place and stability of degree of collapse.

Focus on the placing and vibration of the concrete, it would be timely changed for the concept of "Ignoring of the Placing Process". During the vibration, please ensure the time for vibration for concrete with reference to the degree of collapse for a comprehensive balance, to avoid the successive vibration.

3.2.7 *Control of the top height of the concrete*

Concrete would be placed one by one on the body of pier column. Therefore, it would be ensured to strengthen the control of the top height of each piece of concrete so as to ensure seamless butt-joints between two adjacent pier column.

3.2.8 *Chiseling*

For ensuring seamless joints of concrete on the upper and lower sections of the concrete, when the strength of concrete would be accumulated up to 2.5 Mpa, the manual chiseling was then commenced. When chiseling, an accumulated cement mortar would be chiseled off, after laying the gravels on the appearance, it would then be chiseled downwards with a depth of 1-2cm, after completion of the chiseling, the residuals on the surface of concrete would be blown off by the blowing gun and then use high-pressurized water gun for spraying and cleaning up the concrete surface, after chiseling, the surface of chiseled concrete would be tidy and cleaned up.

3.2.9 *Dismantle of templates and turnover template*

After completion of chiseling, it would then be entered into the next sequence-bundling process. When concrete would be condensed up to the strength of template dismantlement, the 1st course of pier column template can be manually dismantled under the coordination of the crane and timely for dismantling of accumulated concrete of the template. Smear the mold discharging agent and then start to the next working sequence.

3.2.10 *Control of perpendicularity*

For ensuring the perpendicularity in conformity with the design requirements, during the construction of Pier Column, use the total station for drawing the axle line at the interval of 4m. The ordinate difference would be carefully calibrated, to avoid many low probability events for accumulating up to a substantial risk.

3.3 *Installation of sliding formwork*

3.3.1 *Preparations for construction*

Firstly, preparation Work. Prior to assembling of Sliding Formwork, Sliding Formwork would be firstly spliced and checked up for its dimensions of all the main components and calibrated for its taper of the template, ensuring the Sliding Formwork up to standard, with the principle of "Being small dimension on the upper end while Being larger on the lower end" for assembling the template.

Secondly, assembling of Sliding Formwork. The general working sequence for assembling the Sliding Formwork would be made from the upper part of the lower one and from the inside to the outside; the detailed operating sequence: Lifting frame, inner and outer circle hoops, brackets and templates as well as hoisting frame, Laying of the platform, installation of handrail as well as jack and other lifting equipments.

Thirdly, inspection for Sliding Formwork. Prior to installation of Sliding Formwork, hydraulic jacks would be checked up one by one and increased up to 10 Mpa; the connector hoses would be pressurized up to 12 Mpa as to ensure no leakage within 30min and then for installation; after connecting the oil pipeline, pressure would be tested up to 10 Mpa and completed for 4-5 circulation cycles, after testing the pressure up to standard and inserting the support top poles and a full inspection for the center position of platform of Sliding Formwork, 20kg large hoisting line would be set up to suspend the pier column and plumb line measuring points would be on the top surface of bridge foundation, on the platform being equipped with level closed tube, to avoid the differences appeared on the horizontal and longitudinal directions of pier column.

3.3.2 *Placing the concrete on the body of pier column*

Concrete would be transported by tank car and placed by using the crane and hopper for Sliding Formwork. In order to avoid separation process occurred, its free inclination ensures to be less than 2m; with inner body of pier column set up two courses of the tumbling barrels for placing the concrete into Sliding Formwork.

Sliding Formwork would be constructed with low-flow rate concrete, with concrete placing ensured to be evenly distributed and each layer of thickness of 20-30cm. Attention would be made to keep a spacing of 10-15m between the surface of concrete and the upper border of template. After condensation of the previous course of concrete, next course of concrete would be placed then, with insert type vibrator for compaction of the concrete. The depth of upper concrete inserted by the vibrator ensures to be controlled within the scope of 5m, ensuring that no steel reinforcement, top pole and template can be touched, during the sliding up, do not vibrate the concrete. To avoid collapse and deformation of Sliding Formwork. The concrete would be placed after mold-unloading strength attains up to 0.2-0.4 Mpa.

3.3.3 *Lifting of sliding formwork*

Throughout the whole process of construction of Sliding Formwork, template-sliding processes include the stages of initial-lifting, normal sliding-up and final sliding-up.

Firstly, initial sliding-up. When initial placing the concrete, the height of concrete would be controlled at the scope of 60-70cm and divide into 2-3 placing courses for rime consumption of 3-4h, after that, the template would then be initially lifted up to 5cm, after the bottom of concrete attains up to mold unloading strength of 0.2-0.4 Mpa, it would be lifted up by jack with a height of 3-5cm again.

Secondly, normal sliding-up. After the system of Sliding Formwork would be fully inspected for acceptance and then normally slid up. The template was then lifted up once with each course of placing the concrete, to ensure the sliding height kept synchronized with placing thickness of concrete. During normal sliding-up, it was alternately conducted for placing of concrete, bundle-up of steel reinforcement and sliding-up of the template. In general, placing of concrete and the normal sliding velocity of 20cm/h or so. Normal sliding-up includes more than one time slowly sliding-ups, each normal sliding height ensured to be less than 30cm. During the sliding-up, the working conditions of structural members and equipment ensures to be often inspected, to ensure each type of operation can be intimately coordinated and conducted in sequence.

Thirdly, final sliding-up. After the template was slid up to an elevation of pier column up to 1m or so, it was then entered into the final stage of Sliding Formwork, so the sliding speed ensured to be reduced and simultaneously calibrated for the elevation and position of the most course of concrete.

3.3.4 *Steel reinforcement bundle-up and extension of vertical steel reinforcement*

After each lifting of template, the steel reinforcement was bundled up being simultaneous with the period sliding the formwork and the vertical reinforcement was applicably extended.

3.3.5 *Construction of transverse diaphragm*

In order to ensure its stability and endurability for the body of pier column, Transverse Diaphragm ensured to be set up inside pier column. Prior to set-up of Transverse Diaphragm, inner template and scaffold ensured to be dismantled, after completion of placing of Transverse Diaphragm, they were then reassembled.

3.3.6 *Dismantle of sliding formwork*

After construction of sliding formwork up to the top of the pier, it it time to dismantle the sliding formwork. Working sequence for dismantling of the sliding formwork would just be on the opposite to that of installation, which generally installed on the first and then dismantlement. For ensuring the personnel in safety returned back to the ground level, an external hanging cage method ensures to be used for dismantlement of Sliding Formwork.

3.3.7 *Linear control*

During the construction of the higher pier column, pier column would be strengthened for its perpendicularity, displacement of axle line as well as the control of altitude. Measurement of the axle line adopts 20kg plumb line for measuring the centerline and full station for calibration. During the construction of Sliding Formwork, full station was used to calibrate the displacement position of the axle line at each lifting height of 5m, to ensure the difference of centerline being controlled within the approved scope.

3.3.8 *Correctness of sliding formwork*

During construction of the Sliding Formwork, due to the restrains being subjected to many factors, it was unavoidable for displacement or torque, especially for Sliding Formwork constructed up to a certain height. Therefore, displacement and torque become more apparently observed. Therefore, it was very important for correctness of Sliding Formwork. During the process of Sliding Formwork, three methods were generally employed, such as Deviation Rectification for Unbalanced Loading, Jack Deviation Rectification Method and Wedge Pieces Rectification Method.

3.4 *Safety measures*

This was an overhead operation for construction of bridge pier columns with large risk coefficient, so it ensures always to be in mind for the construction principle of "Safety First". For ensuring stability of the crane, steel plate would be pre-buried at the pier column at an interval of 15m, making it connected to the tower crane; an operating platform adopts for U-bar welded up to the top of the template with its bottom fully covered with wooden plates, surrounding which were installed with horizontal handrail and a safety net for enclosure. It was liable to fatigue at overhead operation, so construction worker ensures to continuously operate for no more than 8 hours. In case of Wind at more than Level 5 or rainstorm, an overhead operation must be halted for the reason of construction worker safety.

REFERENCES

[1] Gu, A.B. Bridge Engineering, Beijing, *People's Communication Press*, 2002.

[2] Liu, X.P. Bridge Engineering, Beijing, *China Science Press*, 2005.

Computing, Control, Information and Education Engineering – Liu, Sung & Yao (eds)
© 2015 Taylor & Francis Group, London, ISBN: 978-1-138-02800-5

Overshoot analysis and suppression technology for damped INS

Gao Wei Zhang
School of Automation, Beijing Institute of Technology, Beijing, China
Unit 91404 of the Chinese PLA, Qinhuangdao, China

Lu Feng
School of Automation, Beijing Institute of Technology, Beijing, China

ABSTRACT: An INS is essentially an unstable system. Outputs of an INS contain oscillating errors due to error sources like gyro and accelerometer biases. The amplitude of the oscillating errors increases with the square root of time when the sensors contain random noise, which, however, is inevitable in a practical system. Therefore a correction network is usually included in an INS to add damping to the system. Nevertheless overshoot arises each time INS switches from undamped mode to damped mode degrading the system accuracy. In this paper the generating mechanism of overshoot is discussed and a suppression technology is proposed. Simulations are presented to better observe the overshoot process and illustrate the performance of the proposed suppression technology.

KEYWORDS: INS; Overshoot; Damping; Suppression technology.

1 INTRODUCTION

The inertial navigation system is essentially an unstable system[1][2]. There are three types of period oscillating errors, i.e. Schuler period, Earth period and Foucault period errors in the navigation information of an INS[3][5]. The oscillating errors can diverge with time due to random noise of inertial sensors[4]. For a long-term INS the oscillating error can become unacceptably large, making the INS lose the navigation function.

An effective solution to the problem is to add a correction network to the system to make the system damped[1-18]. The advantage of this solution is that it does not rely much on external sources of navigation information, e.g. the GPS [11]. The damping network attenuates the oscillating errors by relocating the closed-loop poles of the INS from the imaginary axis at the left half - plane, thus changing the system to a stable one[4][16].

A major defect of damping is the overshoot generated during the mode changing process[6][15]. Despite the long history of research on damping methods, not much literature has made an elaborate investigation to the overshoot generated during the mode changing process. Wei et al. analyze the generating mechanism of the overshoot and propose an automatic compensation method to compensate the overshoot error[17]. Huang et al. propose a horizontal damping network with variable damping coefficients to mitigate the mode changing process, thus reducing the overshoot error [8]. In this paper a parallel resolving mechanism is proposed to reduce the overshoot error. Moreover the impact of introduction of external velocity on overshoot error is discussed.

The rest of the paper is organized as follows. In section 2, the generating mechanism of overshoot during mode changing process is illustrated. In section 3 a parallel resolution algorithm for strap-down INS is proposed to improve the performance of INS during mode changing process. In section 4 the improvement of velocity damping over inner velocity damping in overshoot is illustrated and a platform INS can work under velocity mode to get better dynamic characters. Conclusions are drawn in section 5.

2 HORIZONTAL DAMPING AND OVERSHOOT

The 84.4 min Schuler period error can be damped by adding horizontal damping to the INS. The block diagram of North channel with horizontal damping is shown in Figure 1 where H(s) is the correction network of which a typical one has the form[9]:

$$H_x(s) = \frac{(s+8.80\times10^{-4})(s+1.97\times10^{-2})^2}{(s+4.41\times10^{-3})(s+8.80\times10^{-3})^2} \quad (1)$$

The dynamic balance is destroyed at the moment the system switches to damped mode and severe overshoot can be observed. A simulation is made and the results are shown in Figure 2. The conditions of the simulation are as follows: the vehicle accelerates from static state to 15m/s^2 with a constant acceleration equal to 0.1m/s^2 and then the vehicle moves at a steady speed. The INS switches to horizontal damping mode at 30min. The error sources are assumed to take the values of

$$\nabla_x = 5\times 10^{-4} m/s^2, \phi_{y0} = 1', \varepsilon_y = 0.01 \ /h,$$

where ∇_x is the accelerometer bias, ε_y is the gyro bias and ϕ_{y0} is the initial North misalignment.

The 10h simulation results are shown in Figure 2 where the solid line denotes undamped North misalignment and the dot-dash line denotes North misalign-

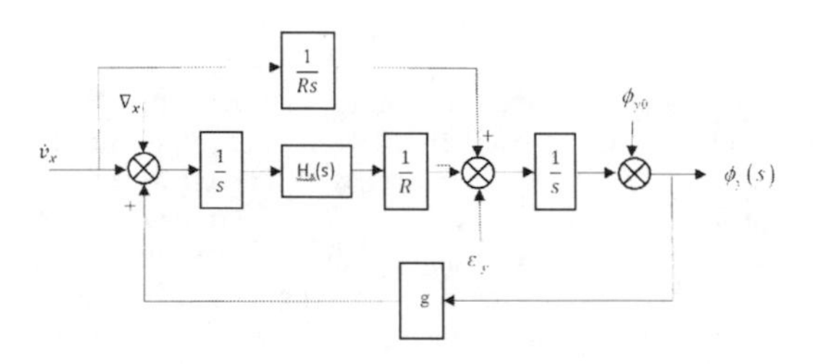

Figure 1. Block diagram of North channel with horizontal damping.

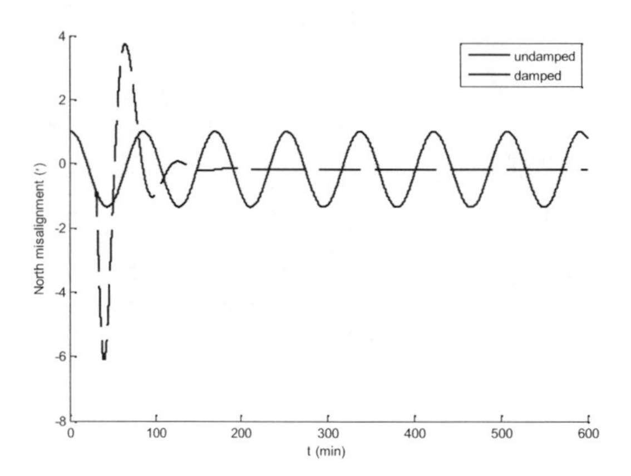

Figure 2. Overshoot during mode changing.

ment with horizontal damping. An obvious overshoot is observed from 30min when damping is added to the system. The North misalignment reaches about -6′ during the mode changing process, about 4 times as large as the amplitude of Schuler period error. The overshoot lasts for about one Schuler period and then approaches to 0. Moreover a higher final speed can cause a more serious overshoot which degrades the system accuracy significantly. Therefore overshoot suppression technology is of great value for a healthy long-term INS[12][13].

3 PARALLEL RESOLUTION METHOD

In Section 2, the overshoot generated from mode switch is displayed. Thus the necessity to eliminate the overshoot is justified. An overshoot compensation method is proposed in [17], however the method can only compensate a major part of the overshoot and

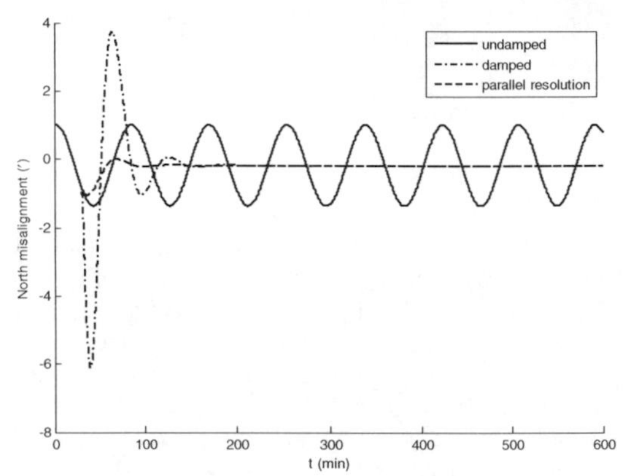

Figure 3. Overshoot with parallel resolution.

the overshoot still exists after compensation. In this section a parallel resolution method will be explained with its performance illustrated by simulations.

Unlike a platform INS, a strapdown INS realizes all the navigation resolution by computer. Therefore a parallel resolution is feasible for a strapdown INS. The parallel resolution method works as follows: The navigation computer solves the navigation solution both in undamped and damped mode simultaneously. In starting phase the INS outputs the undamped resolution results, after the vehicle moves steadily the INS outputs the damped resolution results [14].

A simulation is made with the same conditions as those in section 2. The simulation results are shown in Figure 3 where the dashed line represents the North misalignment with parallel resolution method. It can be obtained that the overshoot is nearly eliminated with parallel resolution method. The damped resolution has gone steadily when the system enters damped mode, i.e. the overshoot has finished before the INS switches to damped mode.

4 VELOCITY DAMPING OVERSHOOT ANALYSIS

In section 3 a parallel resolution method is proposed to control the overshoot during mode changing process of a strapdown INS. However, this method is not applicable to a platform INS due to different resolution pattern. Therefore an alternative method is expected to suppress the overshoot for a platform INS. In fact, the overshoot can be diminished by introducing external velocity to the system, i.e. making the INS operating under velocity damping mode. The overshoot suppressing effect brought by external velocity will be illustrated in this section.

First, we make a variation to Figure 1 in order to analyze the generating mechanism of the overshoot. The modified Figure 1 is shown as Figure 4. It can be observed from the figure that at the moment the system switches to damped mode, an additional path from a to b is connected. We denote the switching time by td. At time td, the value at point a is about 15m/s assuming the conditions in section 2 are satisfied. Therefore a variable of about 15(H(s)-1) is added to the forward path of North channel and a severe overshoot is expected since (H(s)-1) approaches to 0 after a long period.

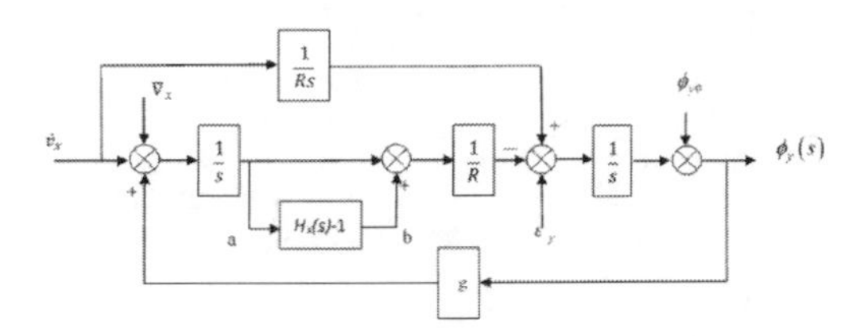

Figure 4. Modified Figure 1.

Next we will analyze the overshoot suppression function of velocity damping. The block diagram of velocity damping is shown in Figure 5. The input at point a is changed to v_x-v_{rx}. If we assume the error of v_{rx} is zero-mean white noise, then we have

$$v_x - v_{rx} = (v + \delta v_x) - (v + \delta v_{rx}) = \delta v_x - \delta v_{rx} = \delta v_x + \upsilon \quad (2)$$

Where v is the real speed, δv_x and δv_{rx} are errors of INS speed measurement and external speed respectively, υ is zero-mean white noise. From (2) it can be figured out that the input at point a is decreased from v_x to $\delta v_x + \upsilon$ which is much smaller, thus weaker overshoot is expected. The overshoot of velocity damping is shown in Figure 6, where the dashed line denotes the North misalignment under velocity damping mode. It can be observed the overshoot under velocity damping mode is much weaker than that under inner damping which is consistent with the above analysis.

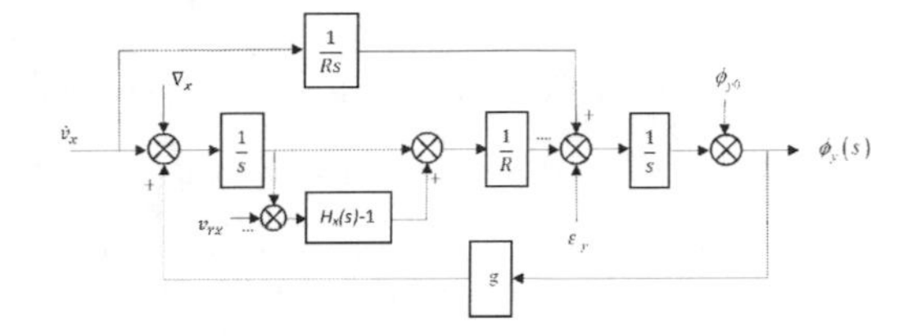

Figure 5. Velocity damping mechanism.

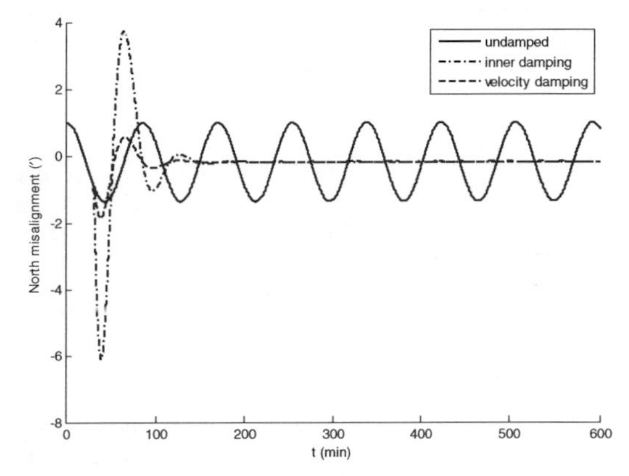

Figure 6. Overshoot under velocity damping mode.

5 CONCLUSION

A parallel resolution algorithm is proposed for strap-down inertial navigation system to improve the overshoot phenomenon during mode changing process in this paper. The proposed algorithm utilizes the advantage in resolution pattern of the strap-down INS, i.e. the navigation resolution is realized all by computer, to make the mode changing phase smoothing by slightly raising the computational burden of the computer. Simulation results demonstrate the effectiveness of the proposed algorithm. However the parallel resolution method is not feasible for a platform INS due to the different resolution pattern. In this paper, we make an analysis of the overshoot suppression function of velocity damping on INS and the analysis is illustrated by simulations as well.

In this paper, we have proposed overshoot suppression technology for both strap-down and platform INS. However, we just investigate the effect of the method for horizontal damping, yet the effectiveness of the method for azimuth damping is not proved. The proposed method should also work well for the azimuth damping intuitively, but a further demonstration by simulations as well as real tests are required to verify the performance of our method.

REFERENCES

[1] W. Gao, Y. Zhang, B. Xu, and Y. Ben. Analyse of damping network effect on SINS. *2009 IEEE International Conference on Mechatronics and Automation, ICMA 2009,* Changchun, China, 2009, pp. 2530–2536.

[2] F. Kasper Jr. Joseph and A. Nash Jr. Raymond. Doppler Radar Error Equations for Damped Inertial Navigation System Analysis. *IEEE Transactions on Aerospace and Electronic Systems*, 1975,vol. 11, pp. 600–607.

[3] Y. Chen and B. Zhong. "Damping of INS" in *Inertial Navigation Theory*, ed: National Defense Industry Press, 2007.

[4] A. Grammatikos, A. R. Schuler, K. A. Fegley. Damping Gimballess Inertial Navigation Systems. *Aerospace and Electronic Systems, IEEE Transactions on,* 1967, vol. AES-3, pp. 481–493.

[5] V. Chueh, L. Te-Chang, R. Grethel. INS/Baro vertical channel performance using improved pressure altitude as a reference. *Position, Location and Navigation Symposium, 2008 IEEE/ION*, 2008, pp. 1199–1202.

[6] R. Cox and S. Wei. Advances in the state of the art for AUV inertial sensors and navigation systems. *Autonomous Underwater Vehicle Technology, 1994. AUV '94, Proceedings of the 1994 Symposium on*, 1994, pp. 360–369.

[7] Y. Hao, Z. Qi, and Q. Wang. Study on dual-system parallel processing technology in SINS. *2012 IEEE International Conference on Mechatronics and Automation, ICMA 2012. 2012 IEEE International Conference on Mechatronics and Automation,* 2012. pp 2334–2339.

[8] W. Huang, Y. Hao, J. Cheng, G. Li, J. Fu, and X. Bu. Research of the inertial navigation system with variable damping coefficients horizontal damping networks. *OCEANS '04, MTTS/IEEE TECHNO-OCEAN '04*, 2004, Vol.3,pp. 1272–1276.

[9] I. A. Mandour and M. M. El-Dakiky. Inertial navigation system synthesis approach and gravity-induced error sensitivity. *Aerospace and Electronic Systems, IEEE Transactions on*, 1988,vol. 24, pp. 40–50.

[10] R. A. Nash, Jr. and C. E. Hutchinson. Altitude damping of space-stable inertial navigation systems. *IEEE Transactions on Aerospace and Electronic Systems*, 1973,vol. AES-9, pp. 18–27.

[11] H. Park. State-Space representation of complementary filter and design of GPS/INS vertical channel damping loop. *Journal of Institute of Control, Robotics and Systems*, 2008,vol. 14, pp. 727–732.

[12] E. Pecht and M. P. Mintchev. Modeling of Observability During In-Drilling Alignment for Horizontal Directional Drilling. Instrumentation and Measurement, 2007,vol. 56, pp. 1946–1954.

[13] E. Pecht and M. P. Mintchev. Observability Analysis for INS Alignment in Horizontal Drilling. *Instrumentation and Measurement, IEEE Transactions on*, 2007,vol. 56, pp. 1935–1945.

[14] F. J. Qin, A. Li, and J. N. Xu. Improved internal damping method for inertial navigation system. *Zhongguo Guanxing Jishu Xuebao/Journal of Chinese Inertial Technology*, 2013,vol. 21, pp. 147–154.

[15] W. S. Ra, I. H. Whang, and H. R. Park. Robust damping loop design for GPS/INS vertical channel. *Electronics Letters*, 2006,vol. 42, pp. 617–618.

[16] R. Stancic and S. Graovac. The integration of strapdown INS and GPS based on adaptive error damping. *Robotics and Autonomous Systems*, 2010,vol. 58, pp. 1117–29.

[17] G. Wei, Z. Ya, B. Yueyang, and S. Qian. An automatic compensation method for states switch of strapdown inertial navigation system. *Position Location and Navigation Symposium (PLANS), IEEE/ION*, 2012, pp. 814–817.

[18] B. Xu and F. Sun. An independent damped algorithm based on SINS for ship. *International Conference on Computer Engineering and Technology. ICCET 2009,* Piscataway, NJ, USA, 2009, pp. 88–92.

Computing, Control, Information and Education Engineering – Liu, Sung & Yao (eds)
© 2015 Taylor & Francis Group, London, ISBN: 978-1-138-02800-5

Summary of the key technologies of robot driver

Ming Ming Sun, Ai Min Du, Zheng Zheng Yuan, Zhong Pan Zhu & Zhi Xiong Ma
School of Automotive Studies, Tongji University, Shanghai, China
Tongji Automotive Design Research Institute, Shanghai, China

ABSTRACT: The application fields of robot driver for vehicle tests were introduced in this paper and the current products of different companies were collected and their performance parameters were compared. At last, the key technologies were summarized based on deep robotic driver study.

KEYWORDS: Automotive Test; Robot Driver; Summary.

1 INTRODUCTION

Robot driver is highly automatic electromechanical integration equipment, which can simulate the behavior and intelligence of human driver. It is a sophisticated product that covers the fields of electronic circuit, signal processing, system control and automobile engineering. In this paper, the application fields of robot driver are introduced. The author compares the parameter of representative products and summarizes the key technologies of robot driver.

2 APPLICATION FIELDS OF ROBOT DRIVER

The robot driver products are applied into two fields: Robot driver for vehicle tests on roller dynamometer in laboratory and robot driver for vehicle tests on the testing road. The robot driver on the roller dynamometer can be applied to the exhaust emission test, durability test and fuel economy test. The robot driver for road tests is attached with the direction control system and the navigation system on the basis of the former.

The exhaust gas from the vehicle, the noise of the test facility, and some severe conditions can be harsh for human test drivers. Besides, high accuracy of the target parameter is required in the comparative tests, while human test driver can not accomplish such tests precisely and repeatedly. The "Human error" is considerable. The road tests cost massive human labor, material resources, capitals and time. In that case, it is the trend of automobile industry to use the robot driver replacing human driver for vehicle tests in the future.

3 COMPARISON OF THE PRODUCTS IN CHINA AND ABROAD

After the first automated robot driver was developed by ONOSOKKI in 1994, only several companies and institutions in the world grasp this technology, such as the SCHENCK,STÄHLE, WITT, LBECO, MIRA, Froude Consine, HORIBA, Nissan Motor, AUTOMAX and the Katholieke University of Belgium. Comparison of some robot products will follow and the key technologies would be summarized.

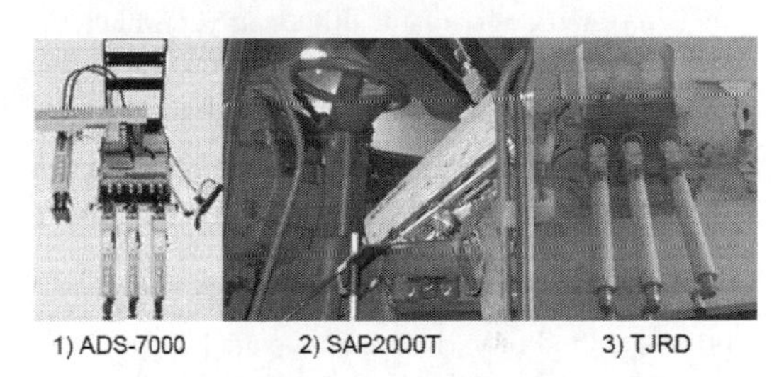

Figure 1. ADS-7000 robot driver is the product of HORIBA.SAP2000T is the product of STÄHLE.TJRD is the product of Tongji Automotive Design Research Institute.

Table1 shows the details of the three products. Through the table we can conclude that the time of self-learning of our product is longer than the products abroad. The products abroad perform better in tracking accuracy.

4 INTRODUCTION OF KEY TECHNOLOGIES OF ROBOTIC DIVER

The key technologies of robot driver can be concluded by the comparison. That is self-learning process, tracking accuracy and multiple manipulator/ leg coordination control.

4.1 *Self-learning process*

Self-learning is the ability of automatically adjusting the control parameters in different conditions of the same vehicle models or different models to complete

Table 1. Details of products at home and abroad.

Products	ADS-7000				SAP2000T				TJRD			
Parameter	Stroke[mm]	F_{max}[N]	V_{max}[m/s]	Power[w]	Stroke [mm]	F_{max}[N]	V_{max} [m/s]	Power[w]	Stroke [mm]	F_{max} [N]	V_{max} [m/s]	Power [w]
Throttle pedal	200	160	-	-	150	100	0.45	45	200	200	0.45	90
Brake pedal	200	220	-	-	150	350	0.3	105	200	350	0.45	157.5
Clutch pedal	200	410	-	-	200	200	0.35	70	200	250	0.45	112.5
Gearshift arm	X250 Y200	220 220	-	-	X250 Y200	250 250	0.6 0.6	150 150	X250 Y250	250 250	0.6 0.6	150 150
Driving mode	electrical				electrical				electrical			
Self learning time	3min				10min				15min			
Tracking accuracy	$\pm 1 km/h$				$\pm 1 km/h$				$\pm 2 km/h$			
Weight	30kg				30kg				40kg			

the vehicle test efficiently. The self-learning ability of robot driver greatly influences the accuracy and precision of the test result. Robot driver which can automatically drive and follow the speed cycle needs to self-study in two aspects, geometric dimensions and performance parameters.

4.2 *The control of speed*

There are many control methods on speed tracking, such as the H_∞ theory proposed by Kai Müler[5]. The control method of variable parameters PID proposed by domestic scholars[6] can also accomplish the speed tracking except the obvious disadvantages. The H_∞ method can restrain disturbances of outside uncertainties and vehicle parameters uncertainties, but its control parameters are difficult to adjust. The variable parameters PID method cannot adjust parameters in the regulator online, and the speed fluctuates is large because of the limit of the regular PID algorithm. The recently proposed Fuzzy Logic theory will be introduced here. The fuzzy control does not need the precise mathematical model. The ability of avoiding disturbances is better than regular PID control. It has good control performance in most test conditions. The specific working process is shown in Figure 5.

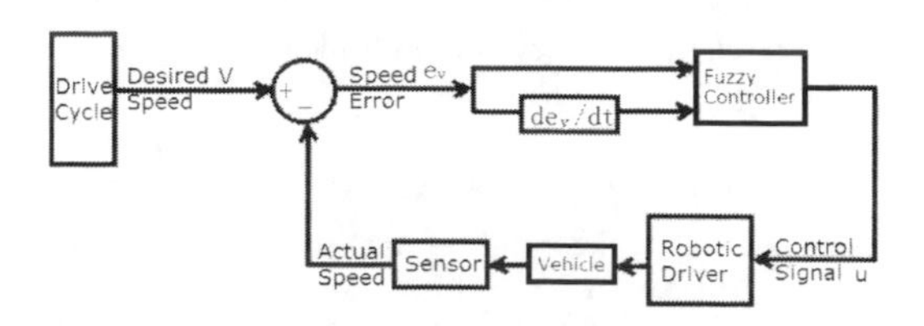

Figure 2. Speed tracking control.

The robot driver calculates the speed error and the change ratio according to the preset cycle speed and real-time data of speed. The difference of each manipulator/ leg steps down or takes back is gained through fuzzy calculation. The accurate track to the preset speed of a robot driver will finally realize with the control of servo motor.

4.3 *Multiple manipulator/ leg coordination control*

The robot driver needs to simulate the real operations of the driver.It should be able to achieve the medium driving level of a skilled driver, coordinately control the actions of ignition or shift manipulator and accelerator, brake or clutch leg, and coordinate the relationship among time, displacement, speed, and force. A good robot diver system should have intelligence similar to human drivers, which can use different control strategies in different conditions. Figure 6 shows the coordination control diagram of manipulator/ leg of robot driver.

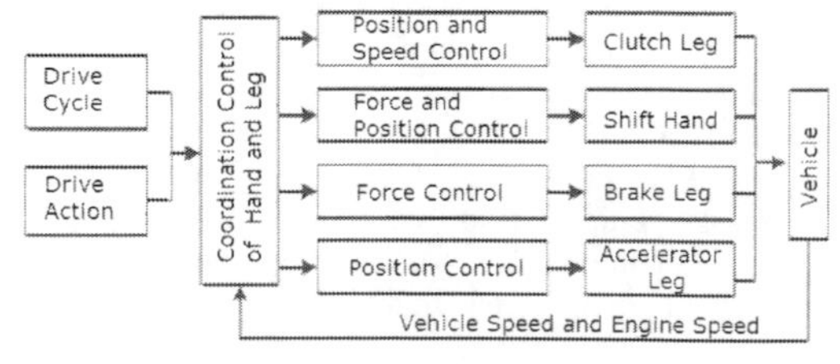

Figure 3. Coordination control of multiple manipulator/ leg.

The robot driver controls the shift manipulator, mechanical legs of accelerator, brake and clutch according to the preset speed cycle table and driving control instruction table to realize the track of vehicle speed.

5 CONCLUSIONS

This paper introduces the concept and application fields of robot driver for vehicle tests, the parameter comparison of typical driver robot productions, and the key technologies of robotic driver system. It can be seen that using robot drivers in vehicle tests is the trend of the automotive industry in the future. The vigorously development from the scientific research institutions all over the world accelerates the process of the application of robot drivers. Due to the limit in basic scientific research level and economic foundation, there is a certain gap between our country and the overseas in robot driver systems. It is not realistic to reach the world advanced level in a short period. However, it is of profound significance to develop high performance robot drivers with independent intellectual property rights no matter for academic development and theory research, or for the development of automotive industry and the international market competition.

ACKNOWLEDGEMENT

Support for this research was provided by STCSM (Science and Technology Commission of Shanghai Municipality), Project NO. 12111101300.

REFERENCES

[1] ADS-7000.http://www.horiba.com/us/en/industry/manufacturing/vehicle/details/ads-7000-973.htm,2014-09-27.

[2] SAP2000.http://www.stahle.com/en/produkte.php?gr=1&prod=1.htm,2014-09-27.

[3] Xue Jinlin,Gong Zongyang,Zhang Weigong.Key Technology and Development of Robot Driver[J]. Robot Technique and Application,2007,(3)_36–40.

[4] Akinobu Moriyama, Isao Murase, Akira Shimozono, and Tohru Takeuchi.A Robotic Driver on Roller Dynamometer with Vehicle Performance Self Learning Algorithm.[J]. SAE paper,No.910036:42–54.

[5] Kai Müller Werner Leonhard.Computer Control of A Robotic Driver for Emission Tests[C].In:Proceedings of the 1992 IEEE International Conference on Robotics and Automati.Anchorage USA pp. 1506–1511.

[6] Chen Xiaobing,Zhang Weigong,Zhang Bingjun. Study on Speed Tracking Control Strategy for Robot Driver on Chassis Dynamometer[J].China Academic Journal Electronic Publishing House,2005,16(18):1669–1673.

[7] M.Masiala,B.Vafakhah,J.Salmon,A.Knight.Fuzzy Self-Tuning Speed Control of an Indirect Field-Oriented Control Induction Motor Drive[J].IEEE Transactions on Industry Applications,2008,44(6):1732–1740.

[8] Tsuneo YOSHIKAWA,Xinzhi ZHENG.Coordinated Dynamic Hybrid Position/Force Control for Multiple Robot Manipulators Handling One Constrained Object[J]. International Journal Robotics Research, 1993,12(3):219–230.

Computing, Control, Information and Education Engineering – Liu, Sung & Yao (eds)
© 2015 Taylor & Francis Group, London, ISBN: 978-1-138-02800-5

Molecular dynamics analysis of PMMA/PVC and PP/PA6 blends

Min Zhang
School of Mechanical Engineering, Shandong University, Jinan, China
Ministry of Education Key Laboratory of High Efficiency and Clean Mechanical Manufacture, Shandong University, Jinan, China

ABSTRACT: In this work, we use Molecular Dynamics (MD) calculations to study the compatibility of polymer blends. The molecular model for two blends, PMMA/PVC and PP/PA6, were set up. Cohesive Energy Density (CED), solubility parameters (δ) and radial distribution function of pure substances and blends were got through MD calculation and were used to analyze the compatibility of blends. The results showed that the difference of the solubility parameter for PP and PA6 is 8.363$J^{1/2}\cdot cm^{-3/2}$, while this vale is 1.921 $J^{1/2}\cdot cm^{-3/2}$ for PMMA and PVC. So PMMA and PVC is miscible, while PP and PA6 is not in the studied system.

KEYWORDS: Molecular dynamic, Compatibility, Polymer blends.

1 INTRODUCTION

Blending of existing polymers is an interesting route towards the development of new polymeric materials. In the past decades only experimental investigation methods have been available: blends were prepared in the laboratory and subsequently analyzed at different temperatures to obtain the phase behavior of the system [1].With the development of hardware and software, molecular simulation has become one of the most important tools to validate structure-property relationship of polymers, blend compatibility, and phase behavior of polymers [1-3]. The Molecular Dynamics (MD), Monte Carlo Simulation (MC) and Mesoscale Simulation (MS) are used the most commonly. In this paper, the molecular dynamics simulations for two kinds of polymer blends were set. One is polymethyl methacrylate(PMMA) and Polyvinylchloride (PVC) blends with the weight fraction 70% of PMMA, the other is polypropylene(PP) and polyamide6(PA6) with the weight fraction 25% of PP. The results were used to analyze the compatibility of two compositions.

2 NUMERICAL MODEL

The Discover molecular mechanics and dynamics simulation module of the Materials Studio software package obtained from Accelrys was used for this study.

Firstly, the molecular model of pure substances and the blends was established in the Visualizer module of the software. The temperature is 298K and the pressure is 1.01 × 10^5Pa. In order to control the amount of calculation and according to the research , the chain length of PP , PVC and PMMA is 50, the chain length of PA6 is 10. The smart minimization method was used to optimize the structure of each model.

The Amorphous module in the software was used to construct the amorphous molecular model of pure substances and the blends. The number of chains and density of pure substances and blends are set as Table 1. The temperature is 298K.

The refined polymer configuration was annealing from temperature 300K to 600K with the step of 50K for five cycles. The total anneal time is 100ps with NPT (Constant Pressure and Temperature) system. After the anneal calculation, the irrationality was eliminated

Table 1. Blends of different compositions considered in MD simulation.

System	Number of chains	Number of atoms	Density [g/cm^3]
1	1 PMMA	752	1.158
2	3 PVC	1128	1.383
3	2PP	904	0.858
4	3PA6	576	1.078
5	2PP2PA6	1288	1.10
6	1PMMA1PVC	1054	1.30

from the model. Fig.1 is the amorphous cells of two blends after generation, minimization and anneal.

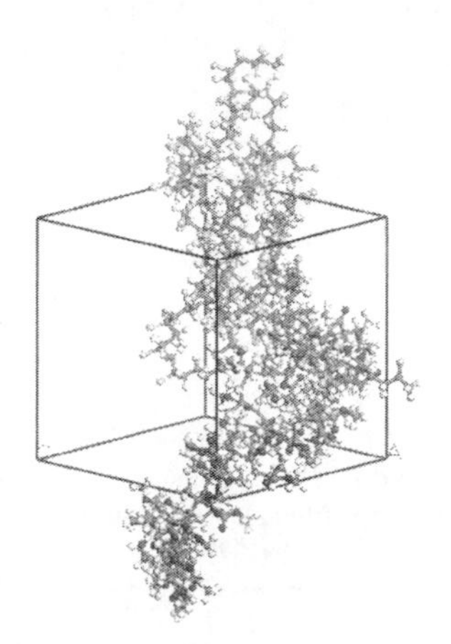

(a)PMMA/PVC blends

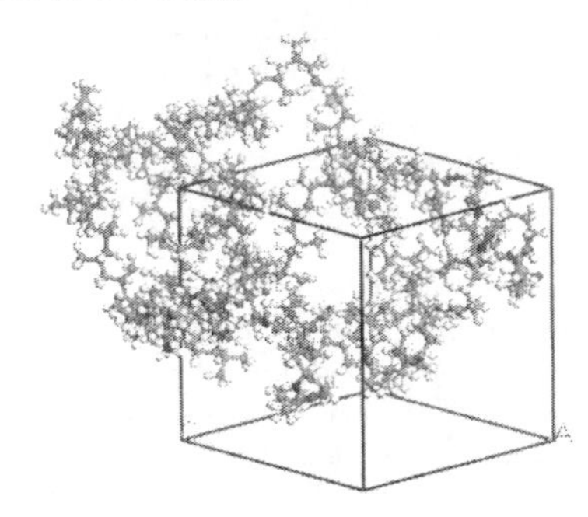

(b) PP/PA6 blends

Figure 1. Amorphous cells of the blends.

MD simulations were performed in the Discover molecular mechanics and dynamics module, and were equilibrated for 350ps under constant temperature (298K) and density(NVT ensemble). The time step of 1fs is used to ensure the stability of simulation. The Berendsen method is used to control the pressure, vdw and coulomb function is tom-based and Ewald method separately, non-bond cut off distance is 0.95nm, spine width is 0.1nm, buffer width is 0.05nm.

3 RESULTS AND DISSCUSSION

3.1 *The pair correlation function*

The pair correlation function (also sometimes referred to as the radial distribution function) gives a measure of the probability that, given the presence of an atom at the origin of an arbitrary reference frame, there will be an atom with its center located in a spherical shell of infinitesimal thickness at a distance, *r*, from the reference atom [4]. Fig. 2 is the pair correlation functions for pure PMMA, PVC and PMMA/PVC blends. Fig.3 is the pair correlation functions for pure PP, PA6 and PP/PA6 blends. Generally, if the peak value of the pair correlation function appears at the distance that is smaller than 0.3nm, it shows that the structure belongs to the amorphous configuration [5]. As we can see from Fig. 2(a) and Fig.3 (a), there isn't peak from 0.3-1.2nm which means the structure we built are amorphous configurations.

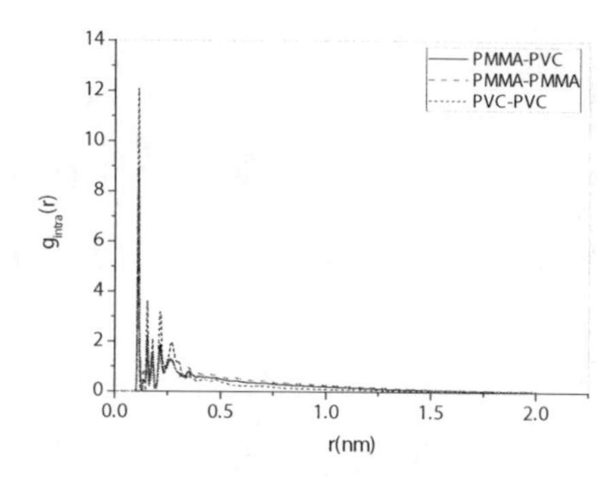

(a) intralmolecular pair correlation functions

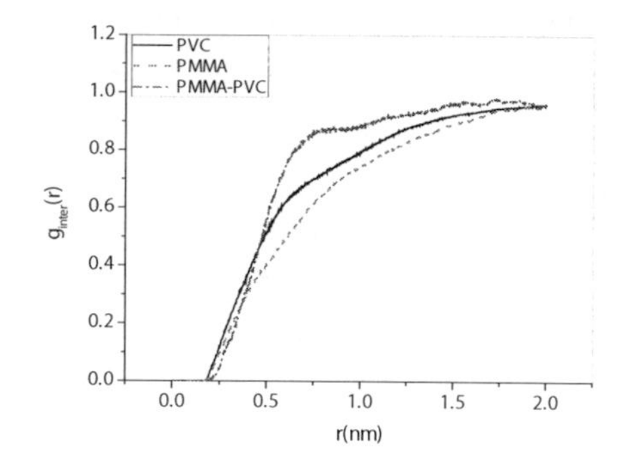

(b) intermolecular pair correlation functions

Figure 2. Pair correlation functions for pure PMMA, PVC and PMMA/PVC blends.

Researchers believe that if the intermolecular pair correlation function value for the same component molecules is lower than that of the two components between, the components in the blends are compatible. The curves in the Fig.2 (b) show that the intermolecular pair correlation function value for PMMA-PVC is greater than that for PMMA-PMMA, PVC-PVC. So PMMA and PVC in this system is incompatible, which is consistent with the previous conclusion. The curves in Fig. 3(b) show that the

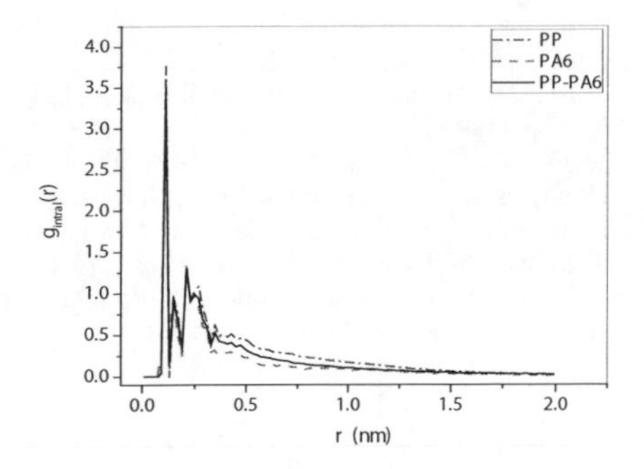

(a) intralmolecular pair corrclation functions

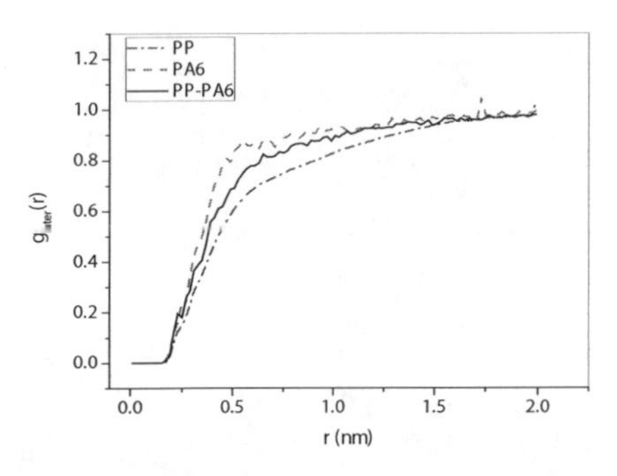

(b) intermolecular pair correlation functions

Figure 3. Pair correlation functions for pure PP, PA6 and PP/PA6 blends.

intermolecular pair correlation function value for PA6 is greater than that for PP-PA6. So PP and PA6 is not incompatible, which is consistent with the previous conclusion.

3.2 *Cohensive energy and solubility parameter*

The concepts of cohesive energy and cohesive energy density were first introduced into the theoretical treatment of mixtures by Hildebrand [7].In the theories, the cohesive energy is used to estimate the energy change on mixing two species. The cohesive energy density, CED, is used in determining the miscibility of polymers. These theories also introduce the solubility parameter, δ, for predict solubility particularly for non-polar materials such as polymers.

$$\delta = \sqrt{CED} = (E_{coh}/V)^{1/2} = [(\Delta H_v - RT)/V]^{1/2} \quad (1)$$

The relationship of the enthalpy difference and the solubility parameters in the blends system is shown in Eq. 2:

$$\Delta H_M / V = (\delta_A - \delta_B)^2 \varphi_A \varphi_B \quad (2)$$

here φ_A and φ_B are the volume fractions of *A* and *B* in the mixed system, δ_A and δ_B are the solubility parameters of A and B.

Table 2 is the CED and solubility parameters of pure substances and blends. The difference of the solubility parameter for PP and PVC is $8.363 J^{1/2}\cdot cm^{-3/2}$, while this vale is $1.921\ J^{1/2}\cdot cm^{-3/2}$ for PMMA and PVC. There is research showed that if the difference of the solubility parameters of two polymers is between 1.3-21 $J^{1/2}\cdot cm^{-3/2}$, they are said to be miscible [8]. So PP/PA6 is hard to miscible and PMMA/PVC is miscible which the same as the experiments [9].

Table 2. CED and solubility parameters of pure substances and blends.

	Cohesive energy density [J/m³]	solubility parameter [J/cm³]$^{1/2}$
PP	2.039×10^8	14.28
PA6	5.127	22.643
PP-PA6	2.976	17.25
PMMA	1.782×10^8	13.349
PVC	2.332×10^8	15.270
PMMA-PVC	2.597×10^8	16.105

4 CONCLUSION

This article use of molecular dynamics with COMPASS field approach for predicting the blend miscibility of PP/PA6 (the weight fraction of PP is 25%) and PMMA/PVC with the weight fraction 70% of PMMA. The amorphous configuration of six systems that include four pure substances and two blends were set. The difference of the solubility parameter for PP and PA6 is $8.363 J^{1/2}\cdot cm^{-3/2}$. So atomistic simulations confirmed blend immiscibility of PP and PA6. The difference of the solubility parameter value is $1.921\ J^{1/2}\cdot cm^{-3/2}$ for PMMA and PVC. These indicate that the compatibility of two blends, PMMA and PVC, is good.

ACKNOWLEDGEMENT

This research was financially supported by the National Science Foundation of China (51103080).

REFERENCES

[1] Inger Martinez de Arenaza, Emilio Meaurio, Borja Coto, *et.al*. Polymer, Vol.51 (2010), p.4431.

[2] Gupta J., Nunes C., Vyas S., *et al*..J. Phys. Chem. B, Vol.115 (2011), p.2014.

[3] Dan Mu, Xu-Ri Huang , Zhong-Yuan Lu, *et.al*. Chem. Phys., Vol. 348 (2008), p.122.

[4] Materials Studio 3.0, Accelys Inc., San Diego, CA, 2004.

[5] Abou-Rachid H., Lussier L.S., Ringuette S. Propellants Explos. Pyrotech., Vol. 4(2008), p.301.

[6] Akten E.D., Mattice W.L. Macromolecules Vol. 34 (2001), p.3389.

[7] Hilderbrand J.H., Scott,P.L. The Solubility of non-electrodytes. Reinhold Publishing Corp., New York, 1950.

[8] Mason J.A., Sperling L.H. Polymer blends and composites. New York Plenum Press, New York, 1976.

[9] Robeson L.M. Polymer blends: a comprehensive review, Carl Hanser Verlag, New York, 2007.

Computing, Control, Information and Education Engineering – Liu, Sung & Yao (eds)
© 2015 Taylor & Francis Group, London, ISBN: 978-1-138-02800-5

Optimization design of flipping table based on dynamic characteristics

Rui Shao, Qiao Chu Liu, Ding Wen Yu & Zhi Jun Wu
Beijing Key Laboratory of Precision/Ultra-precision Manufacturing Equipments and Control, Department of Mechanical Engineering, Tsinghua University, Beijing, PR China

ABSTRACT: This paper presented a simplified 3D model of flap table in the flipping horizontal machining center belonging to APM by SolidWorks. The static and dynamic characteristics of the flapping table were obtained by Ansys Workbench, and the redundancy of the structure was calculated by static characteristics. Based on the topology optimization, the cross beams were removed to reduce the weight. Compared to the previous analysis, the dynamic performance was improved with the modal frequency increasing. The method provides reference value for the subsequent design.

KEYWORDS: Flipping Table Dynamic Optimization.

1 INTRODUCTION

In advanced manufacturing, CNC technology has played a crucial role in industrial production, because of its high speed, high efficiency, high precision, etc.,. More and more applications of CNC machine tools, especially the flap horizontal machining center, are applied in the aerospace sector, In fact, the performance of horizontal machining center flap influenced the development of the aerospace field to some extent. [1]

Static and dynamic characteristics of CNC machine tool has a direct impact on the processing quality, and the performance often depends on large-scale infrastructure. The worktable and the main body of the flipping horizontal machining center can be separated, and the flap table is directly connected to the main body via the tool, so its static and dynamic characteristics will have a great impact on overall performance. [2-4]

2 MODEL OF MACHINE TOOL

The flap table frame is composed flap and frame. The flap is connected by 14 foot frame screws, and it can rotate around a specific shaft driven by in the dorsal motor. The screw moves in a given position after flap movement. The established position of the four hydraulic holders at the top of the flap clamps the table properly. The flap table structure is shown in Figure 1.

Figure 1. Flap table structure.

3 SIMPLIFIED MODEL

The flap table model was simplified by Solidworks to remove a small part of the structure such as the ears and filling welds. The simplified model is shown in Figure 2.

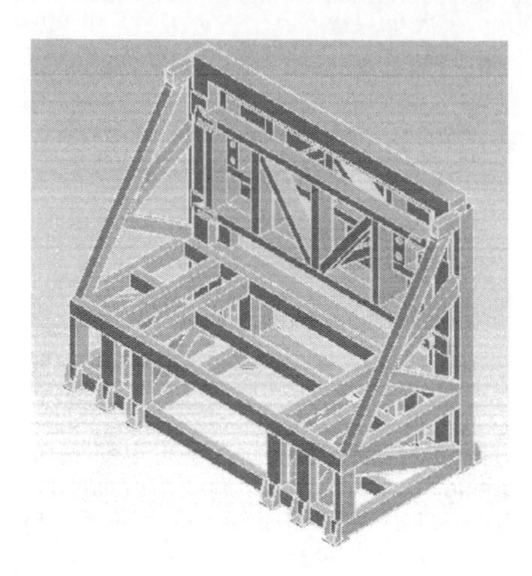

Figure 2. Simplified model.

4 RESULTS AND ANALYSIS

The model was imported into ANSYS through the interface by the two softwares firstly. And then the material properties, boundary conditions, and joint parameters of the spring unit were defined. The deformations in the direction of gravity, and the cutting forces, strain, stress and the modal frequencies were obtained.

According to the early technology research experiences, the maximum cutting forces in X, Y, and Z directions are approximately 1500N, 3000N, 1500N respectively. To improve the reliability, the 5000N cutting force was applied in the table analysis. And the stress, deformation, strain nephograms of the table in +X, −X, −Y, +Z directions were observed under a gravitational acceleration and 5000N forces in several directions.

The table of stress and strain were small, relatively under various loading conditions. The maximum stress under the surface force loading conditions is shown in Table 1.

Table 1. The maximum stress in loading conditions.

Force style	Gravity field	X direction	Y direction	Z direction
Maximum stress	6.15 MPa	0.49 MPa	0.53 MPa	1.14 MPa

The maximum stress of flipping table under a variety of loading conditions were still far less than the yield limit (250 MPa) of structural steels. Therefore, there is no problem about the table in terms of strength, but the strength and stiffness where redundancy largely.

Modal analysis is a common method in the study of dynamic characteristics, and system identification method was applied in the field of vibration engineering [5]. The six natural frequencies were obtained by modal analysis of the flap table in Table 2.

Table 2. Natural frequencies of flap table (Hz).

Modal	1	2	3	4	5	6
Frequency	27.1	37.4	58.5	61.3	71.4	83.4

Using ANSYS topology optimization, the possibility the parts was found, which could be removed. When the material removal rate was 40%, the results were shown in Figure 3.

(a) left parts (b) removed parts

Figure 3. Results of topology optimization.

The results of the analysis indicate that [6], the middle part and the lower part of the main body of the cross beam can be preferentially removed. But only the turning plate location processing state was considered, the lower part supported the frame structure when in a horizontal position. Taking into account the welding process, the removal of lower frame was not considered.

Compared with the lower part of the frame, the removal of top beam concerns will be much smaller. At the same time, the dynamic analysis results showed that the resonance didn't affect the machining precision, but would cause the vibration of the beam in some frequency. Thus the beam was considered for removal.

Compared with the changes of the table mode, which is shown in table 3.

Table 3. The table mode (Hz).

Modal	1	2	3	4	5	6
Before	27.1	37.4	58.5	61.3	71.4	83.4
Now	26.7	34.3	58.6	65.1	89.7	123.7

After modifying, the first four modal frequency model is essentially the same, after a slight increase in fifth and sixth bands. Thus, the dynamic performance has improved.

5 CONCLUSION

1 Generally a three-dimensional model cannot be used directly as the analysis model, we need to simplify the structure of them, one can simplify the model, so that the main structure projecting, on the other hand, the calculation speed can be increased, so that the calculated result converges.
2 If the static analysis was unable to find the weak links of tool,we should analyze the dynamic characteristics, especially modal. We may avoid modal vibration frequency processing, and help to improve structural stability.
3 Through he topology optimization analysis, we may have weight optimized design without any changes of the fundamental basis of the original performance,. By topology optimization, we may remove beams, and improve the dynamic characteristics of the flipping table.

ACKNOWLEDGEMENT

The project is supported by the National Science and Technology Major Projects of the Ministry of Science and Technology of China (Grant No.2013ZX04001-011), The National Basic Research Program of China(Grant No.2013CB035400)

REFERENCES

[1] Shi Jun,Li Gang,APM 2040 Flipping Horizontal Machining Center[J],Aeronautical Manufacturing Technology,2014,04:100–101.
[2] Zhou Xinmin,The Equivalent Structure and The Finite Element Model of The Machine Tool Workbench with "Tee" Slot.
[3] YU Ying-hua, LIANG Bing, WANG Ping-ping,Research on properties of foamed aluminum machine tool table, Machinery[J], 2004,09:4-5+15.
[4] Zhao Ling, Wang Ting, Liang Ming, Li Guomeng Status and Advance of the Lightweight Design of Machine Tool Structures[J],MACHINE TOOL & HYDRAULICS, 2012,15:145–147.
[5] Li Yandong,Statuc & dynamic performance FEA and optimization design for the horizontal machining ceter[D], Shandong Polytechnic University,2012.
[6] Sun Yanfeng,The Structural System of Gantry Based on ANSYS Responds Analyzing Humorously[J], Modern Machinery,2009,05:32-33+56.

Computing, Control, Information and Education Engineering – Liu, Sung & Yao (eds)
© 2015 Taylor & Francis Group, London, ISBN: 978-1-138-02800-5

Fighter digital alarm interface behavioral and eye-tracking experimental evaluation under visual and aural channels

Ya Feng Niu, Cheng Qi Xue, Hai Yan Wang & Jing Li
School of Mechanical Engineering, Southeast University, Nanjing, China

ABSTRACT: In order to achieve objective evaluation on digital interfaces, Eprime software and eye-tracking devices are adopted to conduct experimental analysis on different visual and aural channels for digital interfaces. Accuracy rate, reaction time, fixation point diagram and heap map are obtained according to the experimental measurement for individual channel. Based on these four experimental data, evaluations for each channel of vision and audition are achieved and the optimal channel is identified for further discussion. Here, we report that behavioral data and eye tracking data have important reference values in evaluating both visual and aural channels. The results show that the accuracy rate and reaction time are more susceptible to experimental design and task operation, yet fixation point diagram and heap map has significant differences when presenting on visual and aural channels simultaneously.

KEYWORDS: digital interface. visual channel. aural channel. behavioral experiment. eye-tracking. experiment evaluation.

1 INTRODUCTION

Realistically, vision system assist human acquire more than 80 % information from environment, while for audition about 15 %, which is the most important access to information in addition to visual sense. Visual presentation refers to information presentation and communication through visual stimulation, and voice is way to present and transmit information for auditory presentation. In actual tasks, auditory presentation acts as a significant supplement to visual presentation, which means we can also get information through audition when visual sense is limited, or unworkable, or not suitable.

When fighters are carrying out urgent tasks in the air, with the situation changing rapidly, it can easily lead to fighter fleeting. Pilot's decision is directly related to personal life and national security, and that all channels of information acquisition host on the digital interfaces. In the case of fighter alarm, when single channel of visual stimulation cannot be quickly received, auditory channel is needed for reporting an emergency, so with the visual and aural channels functioning synergistically could help pilots quickly make the right judgment. Under the dual-channel alarm condition of fighter interface, experimental study is conducted through tracking pilot's behavior and eye movement, which could achieve the physiological basis of pilots, making some common multi-channel design recommendations, guiding the interface design and information architecture design, enhancing pilot's ability to acquire information and further reducing pilot's cognitive load.

In the field of eye movement experiments and usability testing research for digital interfaces, multi-channel interaction of fighter cockpit interface design [1], usability evaluation of interface [2], web page [3-4] and digital interface of aircraft avionics systems have been studied and analyzed by both domestic and foreign scholars. Also, eye movement experiments on visual search of different interface layout have already been conducted [6-8]. In addition, several reports provided important contributions to the study on task characteristics of the usability for digital interfaces [9], together with the impact of function area dividing on visual sight [10]. However, synchronous tracking study on behavior and eye movement in visual and aural channels for fighter's digital interface still needed to be further investigated.

Here, we provide evaluation on both the visual and aural channels for fighter interface through behavioral and eye movement experiments. The evaluation indicators include accuracy rate, reaction time, fixation point diagram and heap map. Accuracy rate refers to the ratio of correct number of operations to finish the experimental task and the total number of operations. Reaction time is the response time, timing from the appearance of stimulus to testers reacting. Fixation point diagram consists of fixation location switching within the area of interest, which is described by scanning path and used for investigating visual search path and strategy of testers. Heap map reflects the spatial distribution of fixations, with the depth of color indicating the length of time watching the certain area.

2 EXPERIMENTAL METHOD

Experimental Materials. F18 fighter interface is chosen as experimental materials. For the experimental tasks, the interface needs to be improved to meet the requirements of unified design. As shown in Fig. 1, from left to right, the improved fighter interface is composed of four sub-interfaces, including weapon sub-interface, radar sub-interface, engine sub-interface, and fuel sub-interface. This figure displays the normal state of the digital interface. Each sub-interface has 2 situations, amounting to 8 situations, as displayed in Fig. 2, (a-1) representing Fuel Shortage, (a-2) Plenty of Fuel, (b-1) Find Enemy Aircraft, (b-2) Find Friend Aircraft, (c-1) Engine Anomaly, (c-2) Engine Normality, (d-1) Launch Missiles, and (d-2) Cancel Launching Missiles.

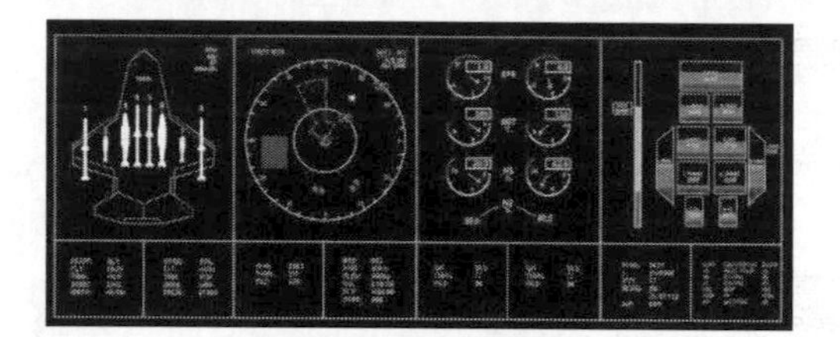

Figure 1. Improved fighter digital interface under normal situation.

Figure 2. Eight situations of improved fighter sub-interface.

Subject. Ten graduate students, with five males and five females, aging from 20 to 25 years old, are chosen as subjects. They all satisfy the following conditions: physical and mental health, normal or rectified vision, and with many years' experience in using graphics device. Subjects are trained to be familiar with the task flow and operation requirements before the formal experiment. During the experiment, subjects are required to wear earphones, with their eyes 550~600 mm away from the screen and both horizontal and vertical perspective controlled in 2.3 degrees.

Experimental equipment and experimental procedures. Behavioral data, including accuracy rate and reaction time, are acquired by Eprime software, while the eye movement data which includes fixation point diagram and heap map, are acquired by Tobii TX300 Eye-tracking Device. Behavioral experiment and eye-tracking experiment were conducted separately as follows.

Behavioral experimental procedure was as follows. Firstly, a white cross appeared in the center of the screen with the background to be black, continuously lasting for 1000ms and then disappeared. Next, fighter interface appeared, presenting for 2000ms and then disappeared. At this stage, subjects were required to judge the warning situation of the present interface. Afterwards, the screen turned to be black, continuously lasting for 1000ms, during which subjects could relax and blink eyes to eliminate visual persistence. Then, warning prompt appeared, which were divided into text warning and sound warning, and these two kinds of stimulus needed to take the ergodicity under 1000ms and 500ms time pressure. Last, blank appeared and lasted indefinitely until subjects had identified whether warning information were consistent with the interface situation in the second stage. What they should do was to press key A if the warning information was in accordant, otherwise press key L.

In the behavioral experiment, eight kinds of different situations appeared randomly and the time interval was 1000 ms for every two trials. Total experiment was divided into 4 blocks according to different channels and different time pressure, such as text alarm prompt under 1000ms, text alarm prompt under 500ms, sound alarm prompt under 1000ms and sound alarm prompt under 500ms. Each block consists of 60 trails and there was a short break between each block. The behavioral experiment process was demonstrated in Fig. 3.

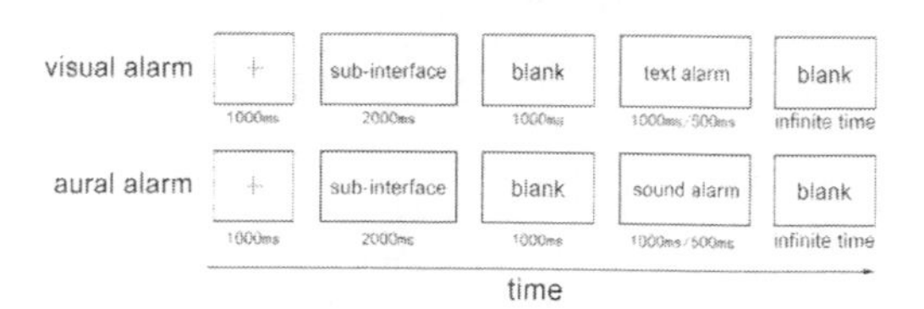

Figure 3. Flow diagram of the behavioral experiment.

The process of Eye-tracking experiment was in consistent with that of the behavioral experiment, while the difference is the removal of time variable. In the eye-tracking experiment, text alarm prompt turned up indefinitely, and sound alarm prompt presented cyclically. Next round experiment procedure began after subjects reacting. By conducting visual search task, visual search strategy of subjects to interface under different channels is investigated in order to evaluate interface reliability for both visual and aural dual channels.

3 EXPERIMENTAL DATA ANALYSIS

Behavioral data analysis. Behavioral data refers to the accuracy rate and reaction time of fighter realizing interface situation recognition under visual and aural channels. As shown in Fig. 4, the change of average accuracy rate under different channels is as follows: sound alarm prompt under 1000ms (0.913) > text alarm prompt under 1000ms (0.909) > text alarm prompt under 500ms (0.900) > sound alarm prompt under 500ms (0.881). Accuracy rate generally shows an increasing trend with the increase of prompt time, which demonstrate that the accuracy rate of sound alarm prompt under 1000ms time pressure is the highest, while the lowest value presents when text alarm prompt working under 500ms time pressure. In Fig. 5, the change of average reaction time is as follows: text alarm prompt under 1000ms (763.428ms) > sound alarm prompt under 500ms (632.366ms) > text alarm prompt under 500ms (620.234ms) > sound alarm prompt under 1000ms (524.028ms). The results demonstrate that the reaction time of text alarm prompt under 1000ms time pressure is the longest while subjects take the shortest time to respond when being sound alarm prompt under 1000ms time pressure.

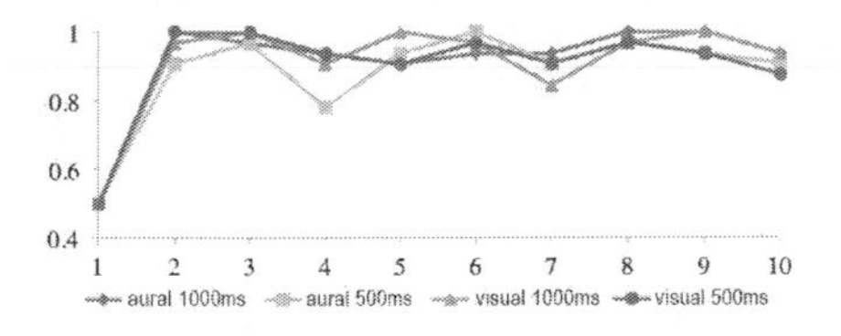

Figure 4. Accuracy rate line chart for fighter interface situation recognition.

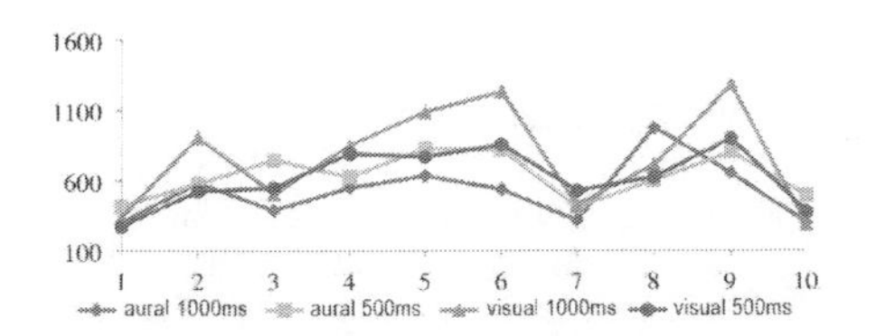

Figure 5. Reaction time line chart for fighter interface situation recognition.

Eye-tracking experiment data analysis. There is nearly no practical significance simply to investigate the eye movement route after listening to the sound under the blank screen situation. Therefore, it is necessary to merge the visual channel into the experiment in the research of eye-tracking data when sound channel occurred. Also, visual and aural channels always emerge simultaneously in a real-world situation, a detailed comparative analysis between visual channel and merging visual-aural channel in the eye-tracking experiment have been conducted. The user cognition law under visual channel and merging visual-aural channel has been investigated by doing the eye-tracking experiment, which could be directly interpreted and described qualitatively based on typical eye-tracking data. Taking Find Enemy Aircraft sub-interface as an example, typical user fixation point diagram and heap map for text alarm prompt are shown in Fig. 6 and Fig. 8 respectively, and for merging text-sound alarm prompt are displayed in Fig. 7 and Fig. 9 respectively.

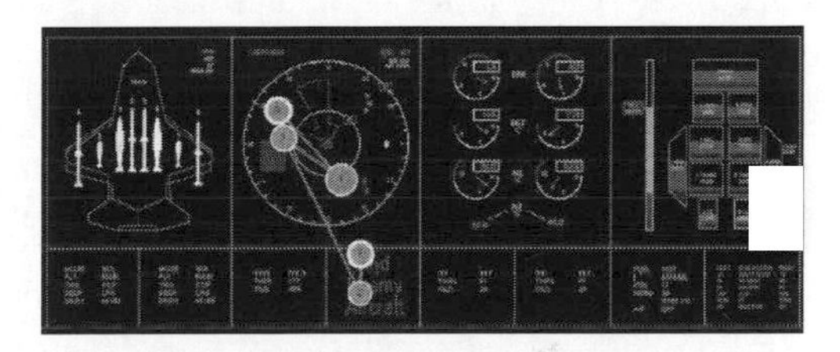

Figure 6. Fixation point diagram for text alarm prompt.

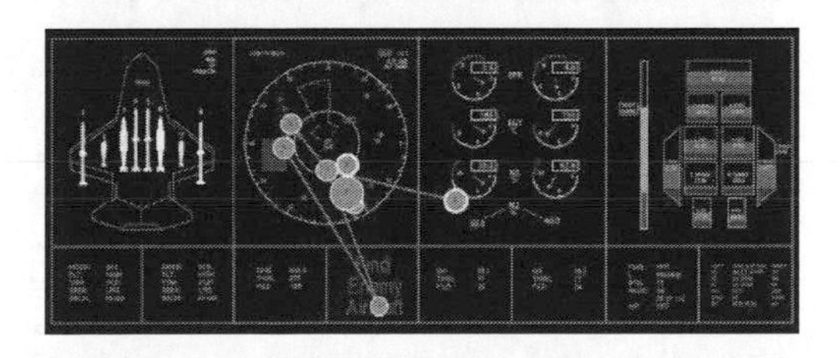

Figure 7. Fixation point diagram for merging text-sound alarm prompt.

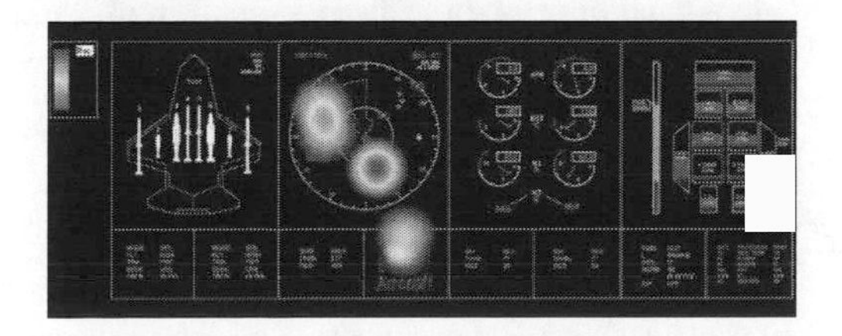

Figure 8. Heap map for text alarm prompt.

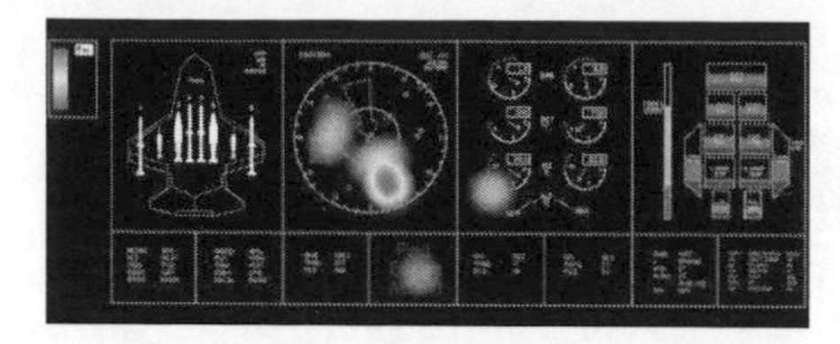

Figure 9. Heap map for merging text-sound alarm prompt.

4 DATA AND OPTIMAL CHANNEL DISCUSSION

Behavioral data discussion. In visual or aural single channel situation, the accuracy rate of fighter interface situation recognition are both higher than 0.8, and the accomplishment of the task is far more than expected. Therefore, considering the reaction time as a focus indicator, results demonstrate that the reaction time of testers is the longest under 1000ms visual channel stimulus (763.428ms) and shortest under 1000ms aural channel stimulus (524.028ms), respectively, while being in an intermediate state under 500ms visual and aural channels. From time pressure variables, neither visual nor aural channels exists significant advantage. This result may be related to the setting of time pressure variables. In this experiment, time pressure variables are set into two kinds only, which means time dimension has a smaller span. The completion and difficulty level of experiment tasks will have impact on the emotion or initiative of testers. The task complexity of this experiment is relatively simple, and there would achieve a better effect by adding task hierarchy and request in the future research.

In addition, sound alarm in the aural channel is reporting verbatim, and there exist no situations of mistaking report or mismatches between interface situation and sound. Testers could generate consciousness before getting the sound stimulus verbatim in the experiment, and especially when key words appear, decision can be made more quickly. However, for text alarm prompt, the information should be presented completely, and it would take more time for testers to make decisions under this stimulus due to the visual search and understanding of words.

Eye-tracking data discussion. From the eye-tracking data of typical users, we can find the number of fixation in Fig. 6 and Fig. 7 is 5 times and 8 times respectively. The reason for this difference is because enemy fighter is more in quantity and task complexity is more difficult in Fig. 7. Previous research has showed that fixation number was closely related to the number interface elements while barely relevant to the processing depth, and the results in our experiment fit very well with this opinion [11]. Scan path in Fig. 7 is longer than which in Fig. 6, and there were intersections and overlaps between scan paths. The reason of this phenomenon is, after adding the aural channel there would be interferences to the comprehensions and decisions of testers, finally leading to longer scan path. Previous studies have also showed that, the longer the scan path was, the lower the search behavior efficiency was, which can be reflected in Fig. 6 and Fig. 7. Fig.8 and Fig. 9 are both heap maps, which can be used to analyze and comprehend the region of interest and visual search strategy qualitatively under visual single channel and visual-aural merging channel for uses. The deeper the color was, the greater the attention rate and interest degree of testers were.

Determination of optimal channel. Behavioral experiment results showed that, phonic decoding process of sound was quicker than semantic decoding process of text in the interface alarm task under a certain range of time pressure. By further eye-tracking experimental verification, we found eye scan path of testers was longer under visual-aural merging channel, which illustrated that the search behavior efficiency was lower. The final results showed that under certain conditions, visual channel had more advantages than visual-aural merging channel in digital interface alarm prompt.

5 SUMMARY

Eprime software and eye-tracking technique were adopted to evaluate visual and aural channels quantitatively for fighter digital alarm interface. Accuracy rate, reaction time, fixation point diagram and heap map were analyzed to evaluate different channels. Behavioral experiment analysis explained that there were no significant advantages on visual or aural channels. In future, behavioral experiment from more kinds of time pressure would be explored and conducted in order to get the relationship between time pressure and behavior reaction under visual and aural channels. Eye-tracking experiment demonstrated that visual channel had more advantages than visual-aural merging channel in certain conditions. Experiment design and result discussion in this paper could provide both scientific analysis and experimental references for future research.

ACKNOWLEDGEMENT

The paper is supported by National Natural Science Foundation of China (No. 71471037, 71271053, 7107-1032, 51405514) and Scientific Innovation Research of College Graduates in Jiangsu Province (No.CXZ-Z12_0093). The corresponding author is Professor Xue Chengqi.

REFERENCES

[1] Y. Zhang: Southeast University Master Dissertation. (2010).

[2] Q. Liu, C. Q. Xue, H. Falk: Journal Of Southeast University (Natural Science Edition) Vol. 40(2) (2010), p. 331–334.

[3] Y. J. Zeng, T. Zhang, X. Chen: Chinese Journal of Ergonomics Vol. 18(3) (2012), p. 83–86.

[4] H. Weinreich, H. Obendorf, E. Herder, M. Mayer: ACM Transactions on the Web (TWEB) Vol. 2(1) (2008), p. 5

[5] Z. B. Wang: Southeast University Master Dissertation. (2010).

[6] H. Y. Wang, T. Bian, C. Q. Xue: Electro-Mechanical Engineering Vol. 6 (2011), p. 50–53.

[7] J. H. Goldberg, M. J. Stimson, M. Lewenstein, et al: Proceedings of the 2002 symposium on Eye tracking research & applications, ACM, (2002), p. 51–58.

[8] Y. Kammerer, P. Gerjets: Proceedings of the 2010 Symposium on Eye-Tracking Research & Applications, ACM, (2010), p. 299–306.

[9] X. Fang, C. W. Holsapple: AMCIS 2000 Proceedings, (2000), p. 411.

[10] Y. Habuchi, H. Takeuchi, M. Kitajima: The 28th Annual International Conference on Cognitive Science, (2006), p. 101–102.

[11] C. Z. Feng: Eye-movement Based Human-Computer Interaction. Suzhou: Suzhou University Press. (2010).

Computing, Control, Information and Education Engineering – Liu, Sung & Yao (eds)
© 2015 Taylor & Francis Group, London, ISBN: 978-1-138-02800-5

Risk assessment of cascading failures in power grid based on fuzzy comprehensive evaluation

Fu Chao Zhang & Jia Dong Huang
School of Electric and Electronic Engineering, North China Electric Power University, Baoding, China

ABSTRACT: The cascading failure is an accident with a low probability but very serious consequences, leading to blackouts or even the whole system crashes. From the perspective of preventing cascading failures, this paper considers the probability and consequences of accidents, using the hidden fault probability and the line fault probability as probability indexes, taking the indexes of lost load, low voltage and overload that reflect static security of system grid as severity indexes. Apply fuzzy comprehensive evaluation to grade the indexes, and determine the level of risk based on the maximum principle membership. Finally, take the IEEE 39-bus system as a numerical example, verifying the rationality and validity of the algorithm.

KEYWORDS: Power system, Cascading failures, Risk assessment, Fuzzy comprehensive evaluation.

1 INTRODUCTION

Modern power system is moving towards "large-grid, large-unit, high-parameter, high-voltage", this trend improves the economic performance of the power system but also poses a threat to the security and stability of the power system. Once a failure occurs, it often diffuses to adjacent areas. And it will lead to cascading failures if not timely and effectively controlled, causing widespread power outages.

Therefore a reliable and effective risk assessment of cascading failures of the power grid is particularly important. Paper [1] presents an assessment of the warning level of the voltage situation. This method can provide an important guarantee for the safe and reliable operation of the power system, but it is a deterministic warning assessment for the single indicator, ignoring the probability of the failures of the power system. Paper [2] analyses the probability and consequence of failures based on the risk theory and the fuzzy reasoning theory. And it presents a risk assessment method for power system transient security, but the method is not able to correctly distinguish a high probability and low severity failure and a low probability and high severity failure.

In view of the above shortcomings, this paper presents a simple and practical risk assessment of the cascading failures. And this method takes the probability and consequence severity of the failures into consideration, using two levels fuzzy comprehensive evaluation to assess the risk level of the occurrence of the cascading failures.

2 RISK THEORY

IEEE standard 100-1992 defines the risk as" the measurement for the probability and severity of the undesired results", and it is often expressed in the form of product of probability and severity. The risk theory takes the probability and consequence severity of the failures into consideration, using logical relations that help the operators quickly acknowledge the consequence of the failures and timely take safety measures.

This paper uses the hidden fault probability and the line fault probability as probability indexes and takes the indexes of lost load, low voltage and overload that reflect static security of system grid as severity indexes, applying risk theory to assess the risk level of the failures.

3 PROBABILITY INDEXES OF CASCADING FAILURES

The development process of the cascading failure can be characterized as follows: the power grid occurs a primary failure that causes large-scale power flow transferring, making the line breaker tripped or line protection installment does not operate correctly and causing further of the failure. When reaches self-organized critical state, it eventually leads to cascading failures. Since the power grid is a complex real-time operating system, making the occurrence of the cascading failures are affected by many factors, but taking all of the factors into account is impossible and unnecessary. Therefore, this paper mainly considers the line fault and hidden fault and takes the line fault and hidden fault as the probability indexes of the failures.

Line Faults Probability. The severe overload of the line plays an important role in the development process of the cascading failures of the power grid. Paper [3] fits a relation curve that the line failures are affected by the change of power flow, which is showed in Figure 1.

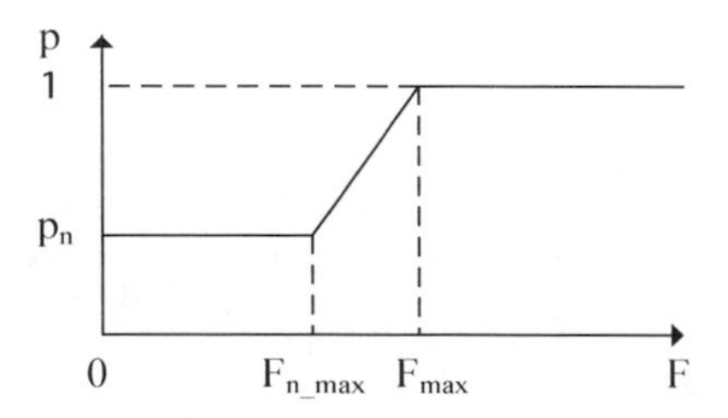

Figure 1. The relation curve of failures probability and power flow.

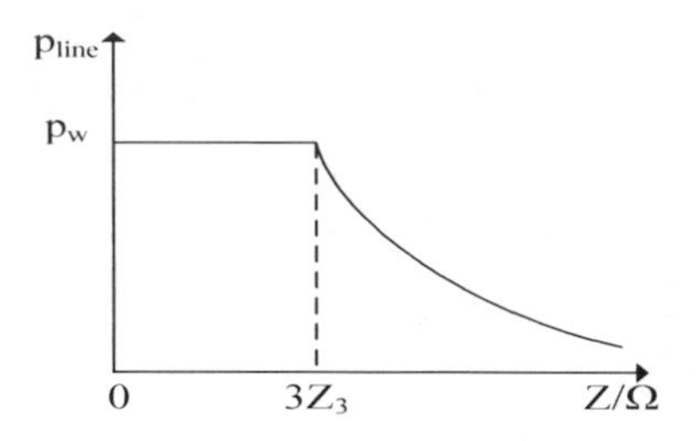

Figure 2. Characteristics of hidden failures of protection devices.

1 If the power flow of the line within the normal range, the probability of line failures objects to Poisson distribution in a given time t:

$$p(t) = 1 - e^{-\lambda t} \tag{1}$$

Where: λ is Poisson constant obtained by statistics.

2 If the power flow of the line is greater than the normal value but less than the limit value, the probability of action of line fuse or protective devices is relevant with the change of the power flow. The relation curve can be approximately fitted by a straight line, and then the outage probability of the line is as follows:

$$p(t) = \frac{1 - p_n}{F_{\max} - F_{n\_\max}} \times F + \frac{p_n \times F_{\max} - F_{n\_\max}}{F_{\max} - F_{n\_\max}} \quad (F_{n\_\max} \le F \le F_{\max}) \tag{2}$$

1 If the power flow of the line is no less than the limit value, overload protective devices will act or the line will fuse because of overheat, and then the outage probability of the line is as follows:

$$p(t) = 1 \qquad (F_{\max} \le F) \tag{3}$$

Hidden Faults Probability. Hidden faults can lead to blackouts. The studies prove that many cascading failures are caused by the hidden faults of protection devices.

Paper [4] proposed a relation curve that the probability of the hidden faults of protection devices changes with measured impedance, which is showed in Figure 2.

It can be drawn from the Figure 2, the relationship equation between the probability of the hidden faults and measured impedance. The equation is as follows:

$$P_{line} = \begin{cases} P_w, & Z \le 3Z_3; \\ P_w \cdot \exp(-Z / Z_3), & Z > 3Z_3. \end{cases} \tag{4}$$

Where: Z_3 is the threshold value of the third stage of the line distance protection; P_W is the probability constant of malfunction of protection devices.

4 SEVERITY INDEXES OF CASCADING FAILURES

This paper takes the indexes of lost load, low voltage and overload that reflect static security of system grid as severity indexes.

Overload Severity Index. The overload severity index is defined as follows:

$$S_{OLi} = \begin{cases} L_i - 0.9, & L_i > 0.9 \\ 0, & 0 \le L_i \le 0.9 \end{cases} \tag{5}$$

Where: L_i is the load ratio of the line i.

The overload severity of the cascading failure is as follows:

$$S_{OL} = \frac{\sum_{i=1}^{t} S_{OLi}}{t} \tag{6}$$

Where: n is the number of overload lines caused by the cascading failure.

Low Voltage Severity Index. The low voltage severity index is defined as follows:

$$S_{Vi} = \begin{cases} 1-U_i, & 0 \le U_i \le 1 \\ 0, & U_i > 1 \end{cases} \quad (7)$$

Where: U_i is pu voltage of bus i.

The low voltage severity of the cascading failure is as follows:

$$S_V = \frac{\sum_{i=1}^{h} S_{Vi}}{h} \quad (8)$$

Where: m is the number of low voltage lines caused by the cascading failure.

Lost Load Severity Index. The lost load ratio after the failure is defined as follows:

$$S_{LL} = \frac{\Delta S}{S} \quad (9)$$

Where: ΔS is the lost load of the cascading, MW; S is the system capacity, MW.

5 MULTI-FUZZY COMPREHENSIVE EVALUATION

Multi-fuzzy comprehensive evaluation is a multifactorial decision for the things that have a variety of attributes or whose the overall merits are affected by many factors. The actual power grid is a real-time system, whose operation is constrained by many internal and external factors. Therefore, the risk assessment of cascading failures in power grid is a complex and comprehensive task, and the selected evaluation indexes have randomness, uncertainty and ambiguity. The comprehensive evaluation is able to take the uncertainties of various indexes into account, making a better evaluation for the risk of cascading failures. Therefore, this paper proposals the fuzzy comprehensive evaluation for the risk assessment of cascading failures in power grid.

This paper establishes a two-level fuzzy comprehensive evaluation model; the specific steps are as follows:

1 Determine the factor set. Divide the above five indexes into two groups based on the risk theory and the characteristics of cascading failures. The two groups are as follows:

$$U=\{U_1, U_2\},\ U_1=\{u_1, u_2\},\ U_2=\{u_3, u_4, u_5\}$$

Where: u_1 is the probability of the line failure; u_2 is the probability of the hidden failure; u_3 is the severity of overload; u_4 is the severity of low voltage; u_5 is the severity of lost load.

2 U_1 and U_2 are belonging to the second-level factor set; U is belonging to the first-level factor set.

3 Determine the evaluation set. The level description of each factor is showed in Table 1.

$$V = \{v_1(\text{I}), v_2(\text{II}), v_3(\text{III}), v_4(\text{IV}), v_5(\text{V})\}$$

4 Determine the weight. The weight vector of the first-level is as follows:

$$W_1 = (w_{11}, w_{12}), W_2 = (w_{21}, w_{22}, w_{23}),$$

$$\sum_{i=1}^{n} w_{ij} = 1 (j = 1,2,\cdots m)。$$

5 Determine evaluation matrix. The evaluation matrix of the first-level factors is obtained by "Expert Judgment" method, which is as follows:

$$R_1 = \begin{bmatrix} r_{11}^{(1)} & r_{12}^{(1)} & r_{13}^{(1)} & r_{14}^{(1)} & r_{15}^{(1)} \\ r_{21}^{(1)} & r_{22}^{(1)} & r_{23}^{(1)} & r_{24}^{(1)} & r_{25}^{(1)} \end{bmatrix}$$

$$R_2 = \begin{bmatrix} r_{11}^{(2)} & r_{12}^{(2)} & r_{13}^{(2)} & r_{14}^{(2)} & r_{15}^{(2)} \\ r_{21}^{(2)} & r_{22}^{(2)} & r_{23}^{(2)} & r_{24}^{(2)} & r_{25}^{(2)} \\ r_{31}^{(2)} & r_{32}^{(2)} & r_{33}^{(2)} & r_{34}^{(2)} & r_{35}^{(2)} \end{bmatrix}$$

6 The evaluation of the first-level factors. Synthetic operation adopts weighted average model, which is as follows:

$$V_1 = W_1 \circ R_1 = (v_{11}, v_{12}, v_{13}, v_{14}, v_{15})$$
$$V_2 = W_2 \circ R_2 = (v_{21}, v_{22}, v_{23}, v_{24}, v_{25})$$

$$v_j = \min[1, \sum_{i=1}^{n}(w_i r_{ij})]$$

7 The second evaluation matrix is obtained from the combination of the first-level evaluation results, which is as follows:

$$R = \begin{bmatrix} V_1 \\ V_2 \end{bmatrix} = \begin{bmatrix} v_{11} & v_{12} & v_{13} & v_{14} & v_{15} \\ v_{21} & v_{22} & v_{23} & v_{24} & v_{25} \end{bmatrix}$$

The distribution of second-level weight is as follows:

$$W = (w_1, w_2)$$

The second evaluation is obtained by fuzzy transformation, which is as follows:

$$V = W \circ R = (v_1, v_2, v_3, v_4, v_5)$$

8 Determine the level of risk based on the maximum principle membership.

6 A NUMERICAL EXAMPLE

Take IEEE 39 bus system for example, the system wiring diagram is shown in Figure 3, and the system parameters refer to paper[5]. Related parameters are as follows: Poisson constant λ is 0.001; the limit value of the line pwer flow is 1.2 times of the normal value; the probability of malfunction delay P_W is 0.02.

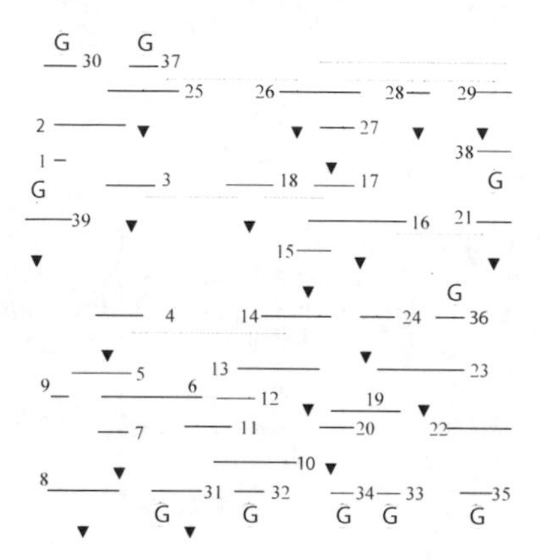

Figure 3. Connection of IEEE 39-bus system.

Table 1. The level description of each factor.

level	Probability level	Severity level	Risk level
I	impossible	neglect	lowest
II	unlikely	consider	lower
III	sometimes	general	general
IV	possible	serious	higher
V	probably	catastrophic	highest

Assess the level of risk of all branches of the system, and the weight of each index is showed in Table 2 and Table 3. The risk levels of part branches are showed in Table 4.

Table 2. The weight distribution of the first-level factor set.

the first-level factor set	weight distribution
line fualts probability	0.2
Hidden fualts probability	0.1
overload severity	0.2
low voltage severity	0.2
lost load severity	0.3

Table 3. The weight distribution of the second-level factor set.

the second-level factor set	weight distribution
failures probability	0.4
failures severity	0.6

Table 4. Risk assessment results of part lines.

lines	Probability grade	severity grade	risk grade
5-6	IV	V	V
21-22	III	V	IV
16-19	III	IV	IV
16-24	III	IV	IV
10-13	II	V	IV
26-27	II	IV	III
15-16	III	IV	IV
28-29	III	IV	IV
6-11	III	II	II
12-13	II	III	II

Seen from Table 4, the probability of some line failures is higher but the severity of the faults is lesser, so the risk level of those lines is lower. While the probability of some line failures are lower but the severity of the faults are more serious, so the risk level of thoes lines are higher. Therfore, the rsk assessment method in this paper can well identify high-probability and low-severity fualts and low-probability and high-severity fualts, and quantitatively consider the probability and severity, comprehensive reflecting the impact that the faults cause to the power system.

7 CONCLUSION

This paper proposes a simple and practical risk assessment of transmission line faults. The method takes factors that reflect uncertainty of transmission lines into account, and identifies the weak lines of power system from a quantitative point of view. Therefore, the higher risk level of the lines should be given focus, curbing the spread of the failures and preventing large blackouts.

REFERENCES

[1] Zhang Shuo, Liu Yongmin, Mei Xiaoli, et al. Risk assessment of cascading failures basing on operating conditions[J]. Central China Electric Power, 2011 (2).

[2] Liu Xindong, Jiang Quanyuan, Cao Yijia, et al. Transient security risk assessment of power system based on risk theory and fuzzy reasoning[J]. Electric Power Automation Equipment, 2009, 29(2):15–20 (in Chinese).

[3] Phadke A G, Thorp J S. Expose hidden failures to prevent cascading outages[J]. IEEE Computer Application in Power, 1996, 9(3):20–23C

[4] Chen Huawei, Jiang Quanyuan, Cao Yijia. Risk assessment of power system cascading failure considering hidden failures of protective relayings[J]. Power Systems Technology, 2006, 30(13):14–19.

[5] PADIYAR K P. Power system dynamics stability and control[M]. Kent, UK: J Wiley Press, 1996.

Computing, Control, Information and Education Engineering – Liu, Sung & Yao (eds)
© 2015 Taylor & Francis Group, London, ISBN: 978-1-138-02800-5

Application of 3D printing technology in product design

Xin Ting Yu
College of the Arts, Qilu University of Technology, Jinan, China

ABSTRACT: With its characteristics of rapid prototyping and vivid display of product model, 3D printing plays a great role in product development stage. Compared with traditional model making, 3D printing can shorten product model development cycle and reduce product development costs; besides, the application of 3D printer in product design can also expand designers' imagination, improve their design capabilities, and enhance their design preciseness.

KEYWORDS: 3D Printer; Model Making; Product Design.

1 INTRODUCTION

Professionally known as "increase material manufacturing" or "additive manufacturing", 3D printing has been expanding its application fields since the first 3D printer was brought into being by an American company named 3D Systems in 1987. Nowadays, 3D printing has been widely used in fields like aerospace, automotive, medical, military, education, architecture, biology, archaeology and so on. With a variety of stack forms and materials, 3D printers vary in processing and molding process. Currently, there are three mainly used types of processing in the domestic market: FDM (Fused Deposition Modeling), 3DP (Three Dimensional Printing), and SLS (Selective Laser Sintering).

FDM printers are typically small in size, with advantages like easy operation and maintenance, low cost, simple system configuration principle and operation, secure system operation, and able to mold parts in any complexity. However, support structures are needed while making three-dimensional computer data model, which will be removed and conduct polish processing when printing is completed. SLS printers can use a variety of materials, including metals, with the simple production process and no need of support structures, while the energy consumption of SLS is relatively large, with lower degree of security and complex operation and maintenance. 3DP printers can produce full-color models, with gypsum powder and glue as its main materials and no need of support structures, and the unused powder can be recycled, thus enhancing its material usage rate, while the price of consuming materials is relatively high.

The wide use of 3D printers in product design enriches the styles of product form, and broadens designers and students' thinking and imagination. Especially in the modeling process, 3D printing modeling is more time-saving, labor-saving, material-saving, and accurate compared with traditional clay modeling and hand modeling.

2 THE INFLUENCE OF 3D PRINTERS ON PRODUCT DESIGN

2.1 *Shorten product development cycle*

Taking 3DP printer with a molding vat of 203*254*203mm as an example, it takes about 6 hours to produce a model with height of 100mm, and the molding process needs no human supervision. Compared with traditional model making, this new technology saves designers more time to conduct preliminary research and scheme scrutiny, thus shortening product development cycle to push new products into market to occupy opportunities.

2.2 *Reduce product development costs*

Compared with traditional processing technology, using 3D printing modeling to trial production can save enterprises a huge sum of open mold trial costs, and reduce product development costs, thus reducing the cost of the product and gaining market share price.

2.3 *Improve product designers' imagination and design capabilities*

3D printers' one-step molding technology makes it possible to manufacture complex items that are difficult or even impossible to process in traditional methods, breaking the limitations of traditional manufacture. As long as designers can conduct design thinking and turn ideas into three-dimensional data model, it will be available to create a three-dimensional object through 3D printer, shortening the distance between ideas and finished product.

2.4 *Enhance product designers' preciseness*

By using a 3D printer to print design schemes directly, designers can conduct analysis and detection on ideas, functions, and structures, identify design issues and modify design schemes accordingly, which forms a process of design, print, modify, and re-print, until the last plan was finalized, thus avoiding production errors in later stages and enhancing designers' preciseness.

3 THE APPLICATION OF 3D PRINTER IN PRODUCT DESIGN-TAKING 3DP PRINTER AS AN EXAMPLE

3.1 *Creative sketches and computer three-dimensional data drawing*

The fact that 3D printers can produce stereoscopic objects lies on the basis that there are qualified computer 3D data models, which on the other hand lies on the basis of good creative ideas and skillful use of computer modeling software. Therefore, the creative design diagram (Figure 1) and computer 3D data model (Figure 2) are the basis for 3D printers to produce stereoscopic objects. While using 3DP printers, avoid tiny sharp structures in the design of products.

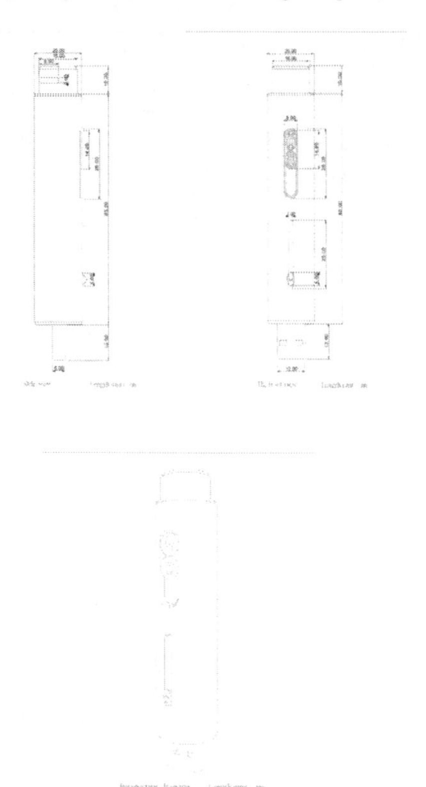

Figure 1. Creative design diagram.

Currently computer 3D data model constructed by 3D modeling software such as SolidWorks, Unigraphics (UG), Pro/Engineer (Pro/E), 3D Studio Max, and Rhino can all be converted into a 3D printer recognizable format. The model constructed must be a solid structure, with body wall thickness controlled within around 3mm, neither too thin nor too thick. Large shell structures should be burrowed at the bottom or other unapparent locations, thus the power inside can be cleaned out to reuse. Currently, all 3D printers can identify *. STL (stereo-lithography) data format, yet the data model in stereo-lithography format are monochrome mold. The color print formats of 3DP printer are *.PLY (polygon file format) and *.WRL format.

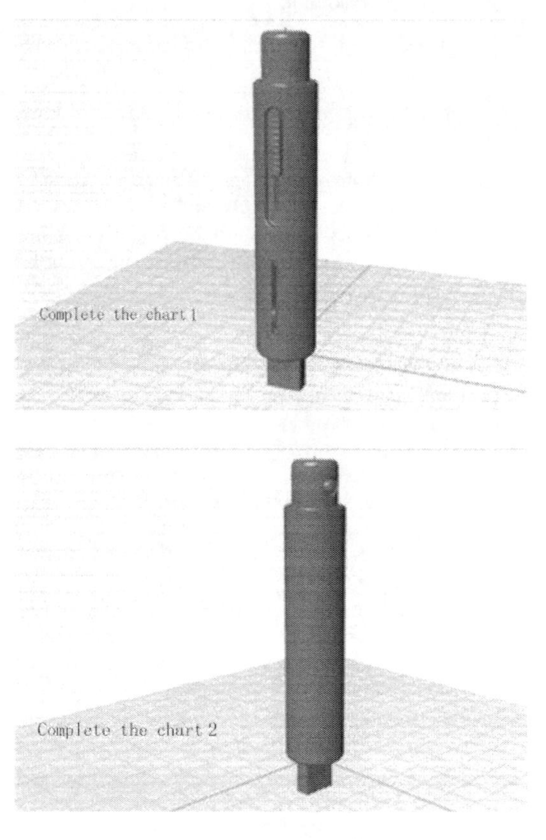

Figure 2. Computer 3D data model constructed by rhino.

3.2 *Detection and reparation 3D data software*

Input the constructed 3D data model into Magics according to the needs of 3D printer to conduct model detection and reparation test (Figure 3). If problems like fracture surface or non-solid appear in the 3D data modeling process, the model should be repaired or re-built.

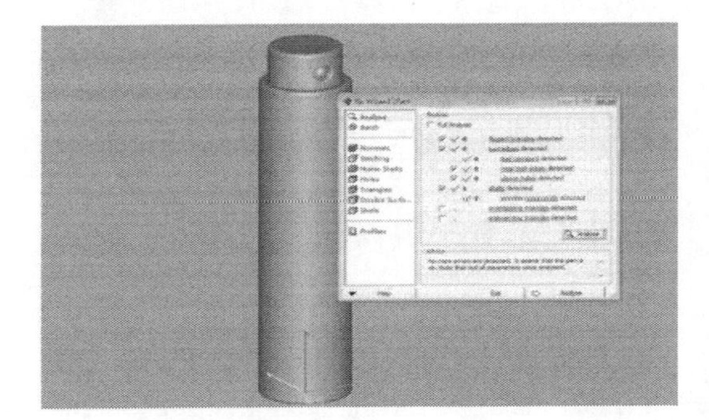

Figure 3. Model detection interface in magics.

3.3 *Detection and maintenance of 3D printer*

Check and clean the printing nozzle before printing, spread powder in the molding vat for 1-2 times, making the powder in the bottom of the molding vat even and flat (Figure 4) to ensure the stability of the model in the printing process. After matching 3DP printing software with 3DP printer, use the Print Time Estimater function in 3DPrint software to check the usage of powder, glue, and three colors as well as the time needed for printing (Figure 5). The software will detect machine status automatically before printing, add powder, replace the glue box, replace print nozzle, or add lubricants according to the guidance.

Figure 4. Finish spreading powder.

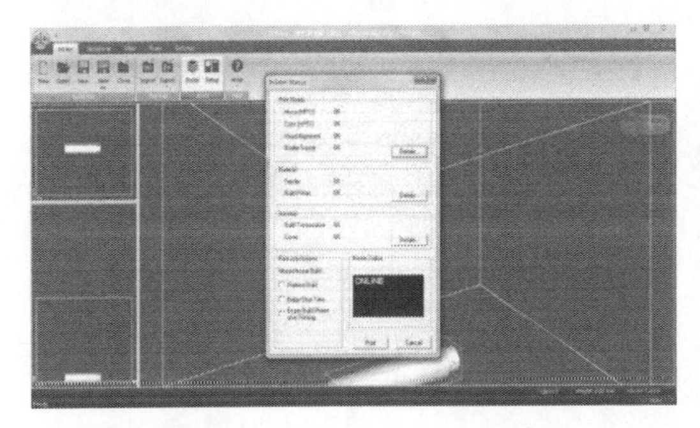

Figure 5. 3D Print material report.

3.4 *3D printer operate*

The printer starts printing model layer by layer (Figure 6) after the computer data was transmitted to it, until the printing is finished. Then the machine will carry on drying operation for an hour or so. Take the model out when drying is finished.

Figure 6. The printer printing layer by layer.

3.5 *Processing the model printed*

When taking the model, predict its position in the molding vat according to the data distribution of the software, and clean the powder with a soft brush (Figure 7). Recover excess powder into the supply vat through powder suction tube in time to reuse the powder. After taking out, the model is put into a drying oven to conduct ventilating thermostat drying (Figure 8) as needed, and then taken into the powder-removing capsule of the printer to conduct final powder cleaning (Figure 9). After the powder is completely removed, carry out curing process under ventilating condition to ensure model's hardness, conduct curing process for 2-3 times, according to the needs, so that the entire model is completed (Figure 10).

Figure 7. Model formed in the molding vat.

Figure 8. Thermostatic blower drying model.

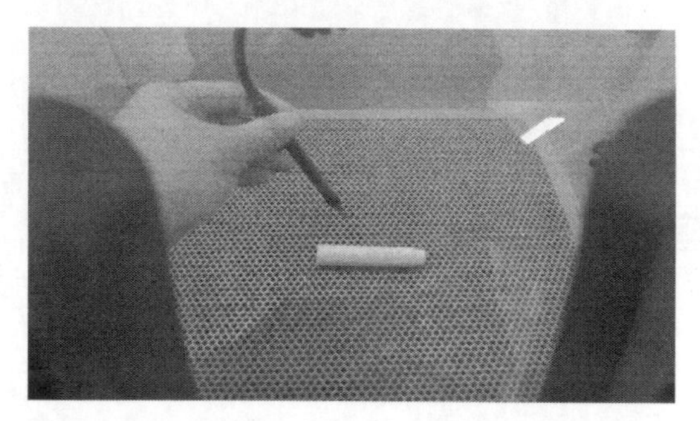

Figure 9. Cleaning the powder from the model.

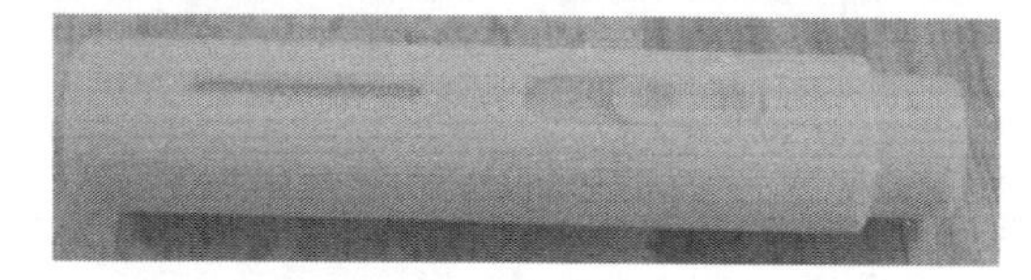

Figure 10. Model after curing process.

In the whole process of model taking, powder removing and curing, operators should wear thin glial no penetration gloves, masks and transparent goggles prevent the powder and volatile pungent curing agent from harming human body.

4 CONCLUSION

In the current or near future time, 3D printer plays an irreplaceable role in product development stage. On the other hand, we should be aware of the fact that 3D printer is merely a tool to demonstrate creative thinking and make preparations for mass production, and without designers' creative thinking and excellent computer data model, 3D printer will lose its unique value.

REFERENCES

[1] XU Jiang-hua,Zhang Min, The Application of Rapid Prototyping in Industry Design [J], Packaging Engineering,vol.06(2004),P. 131–133.

[2] http://www.loverok.com/about.php?id=228

[3] Wang, Bing Zi, The effect of 3D printing technology on the future fashion design and manufacturing[J],Applied Mechanics and Materials, vol496-500(2014), p 2687–2691.

[4] Zhang, Li ,Impact of 3D printing technology on the development of the industrial design[J],Applied Mechanics and Materials, vol437(2013), p 956–960, 2013.

[5] Hod Lipson, Melba Kurman, Fabricated the New World of 3D Printing [M] .BeiJing:China CITIC Press,2013.

Computing, Control, Information and Education Engineering – Liu, Sung & Yao (eds)
© 2015 Taylor & Francis Group, London, ISBN: 978-1-138-02800-5

Carrying capacity analyzing of hydrostatic thrust bearings with linear deformations

Li Gang Cai, Yu Mo Wang, Zhi Feng Liu, Yong Sheng Zhao & Hong Wei Zhao
Beijing University of Technology, Beijing, P. R. China

Xiang Min Dong
Chengde Petroleum College, Chengde P. R. China

ABSTRACT: The carrying capacity of hydrostatic bearing will be influenced by the surface deformation caused by a combined action of the load, flow pressure, manufacture errors, etc. In the present work, a kind of the Reynolds equation based model within surface waviness taken into consideration is established for the analyzing of the load carrying ability of thrust bearings under linear deformations. The superficial deformations, either convex or concave, could be respected linear if it is in a small extent. Then, expression of the fluid film thickness is a regarded as the combination of linear deformations and surface waviness. By adaptive Simpson quadrature, the carrying capacity will be solved as a result of the numerical integration for a partial differential Reynolds equation of second order, which includes flow pressure and fluid film thickness.

KEYWORDS: hydrostatic thrust bearings; Reynolds equation; Adaptive Simpson Quadrature; linear deformations; surface waviness.

1 INTRODUCTION

1.1 *Background*

The hydrostatic thrust bearing is regarded as typical applying of the hydrostatic bearing system, whose carrying capability is based on the hydrostatic effect. Oil pressure is pumped into the oil pocket by external supply system and flow through the edge seal of oil pad, which is essential for the oil pad to keep the pocket pressure. Reynolds equation is widely used for analyzing hydrostatic systems after it is modified to fit the working condition. The pressure distribution on seal edge and carrying performance of the hydrostatic thrust bearing will be influenced by the surface deformation. The surface of hydrostatic thrust bearing could be deformed by various reasons which includes the nonlinear distributed pressure, manufacture errors and etc. However, the deformation of the bearing surface can be respected as linear if it is in a small extent. Moreover, the edge seal of oil pad is always manufactured by milling, which will cause surface waviness and influence on the carrying capability of hydrostatic thrust bearing. Therefore, the design of hydrostatic thrust bearing can be optimized by analyzing the impact of surface linear deformation on the carrying performance.

There are a number of researches, which study the hydrostatic bearings and the resolution of their pressure distribution. The traditional method is conducted based on the flow rate, analyzing which is not always constant that causes a little defect of this sort of method [1-2]. Some studies are carried out by analyzing the N-S Equation, which is able to fit all the fluid questions and difficult to solve [3]. The Reynolds equation is considered as the most widely used methods to analyze the hydrostatic systems [4]. Polar coordinate system with two dimensions is required because it matches the thrust model better [5]. Since the two bearing surfaces are regularly stable and the pressure is mainly distributed on the radius coordinate, the variation of flow, density and viscosity could be ignored [6-7]. Micro texture is found to have influence to the carrying capability of hydrostatic bearings like surface waviness, which should be taken into consideration approximately [4]. With the surface waviness [8-9] taken into consideration, the Reynolds equation will be very hard to solve by analytical integration. So the numerical quadrature can be used to solve the equation with tolerable error [10-11]. The impact of linear deformation of the carrying capability will be got after the pressure distribution is resolved.

The researches about hydrostatic bearings are usually based on the Reynolds equation, but not many of them consider surface deformation. In the present work, the carrying capability of hydrostatic bearing is analyzed with surface linear deformation and waviness taken into consideration. The Reynolds equation is modified to match the working condition after it is converted into dimensionless version. The film thickness is considered as a combination of surface waviness and the linear deformation, which could be divided into two kinds, convex and concave. By adaptive Simpson quadrature, the pressure distribution in the oil pad and the carrying capability will be solved as results of the modified Reynolds equation.

1.2 *Modification of Reynolds equation*

The Reynolds equation for the hydrostatic thrust bearing with the stable bearing surface is written as (1):

$$\frac{\partial}{\partial x}\left(\frac{\rho h^3}{\eta}\cdot\frac{\partial p}{\partial x}\right)+\frac{\partial}{\partial y}\left(\frac{\rho h^3}{\eta}\cdot\frac{\partial p}{\partial y}\right)=0 \tag{1}$$

where p represents pressure, ρ represents density, η represents viscosity, h represents film thickness.

The Reynolds equation can be simplified if the variation of flow density and viscosity could be ignored and written as:

$$\frac{\partial}{\partial x}\left(h^3\cdot\frac{\partial p}{\partial x}\right)+\frac{\partial}{\partial y}\left(h^3\cdot\frac{\partial p}{\partial y}\right)=0 \tag{2}$$

According to the properties of partial derivative, it could be expressed as:

$$3h^2\left(\frac{\partial h}{\partial x}\frac{\partial p}{\partial x}+\frac{\partial h}{\partial y}\frac{\partial p}{\partial y}\right)+h^3\left(\frac{\partial^2 p}{\partial x^2}+\frac{\partial^2 p}{\partial y^2}\right)=0 \tag{3}$$

Thrust bearings are always made in shape of round, so the polar equation will fit the model better. If film thickness h and pressure p is only relative to the polar radius, following statements could be drawn:

$$\begin{cases}\dfrac{\partial h}{\partial \vartheta}=0\\[2mm]\dfrac{\partial p}{\partial \vartheta}=0\end{cases} \tag{4}$$

Based on the properties of the polar coordinate system [5]:

$$\begin{cases}\dfrac{\partial r}{\partial x}=\dfrac{x}{r}\\[2mm]\dfrac{\partial r}{\partial y}=\dfrac{y}{r}\end{cases} \tag{5}$$

According to Equation (5):

$$\begin{cases}\dfrac{\partial h}{\partial x}\cdot\dfrac{\partial p}{\partial x}+\dfrac{\partial h}{\partial y}\cdot\dfrac{\partial p}{\partial y}=\dfrac{\partial h}{\partial r}\cdot\dfrac{\partial p}{\partial r}\\[2mm]\dfrac{\partial^2 p}{\partial x^2}+\dfrac{\partial^2 p}{\partial y^2}=\dfrac{\partial^2 p}{\partial r^2}+\dfrac{1}{r}\cdot\dfrac{\partial p}{\partial r}\end{cases} \tag{6}$$

Based on Equation (4) and (6), Equation (3) could be reorganized as:

$$3h^2\cdot\frac{\partial h}{\partial r}\cdot\frac{\partial p}{\partial r}+h^3\cdot\frac{\partial^2 p}{\partial r^2}+h^3\cdot\frac{1}{r}\cdot\frac{\partial p}{\partial r}=0 \tag{7}$$

After multiplying by r, Equation (7) is rewritten as:

$$3h^2 r\cdot\frac{\partial h}{\partial r}\cdot\frac{\partial p}{\partial r}+h^3 r\cdot\frac{\partial^2 p}{\partial r^2}+h^3\cdot\frac{\partial p}{\partial r}=0 \tag{8}$$

According to the properties of partial derivative, Equation (8) can be reorganized as:

$$\frac{\partial}{\partial r}\left(rh^3\cdot\frac{\partial p}{\partial r}\right)=0 \tag{9}$$

1.3 *Modeling of hydrostatic thrust bearing*

Assuming that r_0 stands for the radius of the oil pocket, R stands for the radius of oil pad, H stands for the film thickness at the border of edge seal.

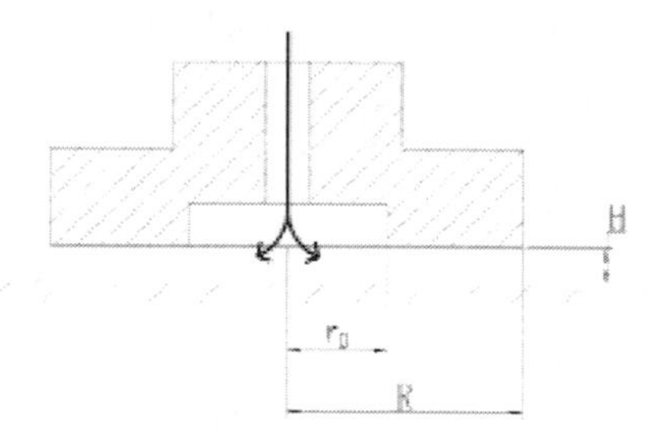

Figure 1. Model of hydrostatic thrust bearing.

To conduct this work, the following dimensionless parameters will be used:

$$\bar{p}=\frac{p}{p_0},\ \overline{p_0}=1,\ \bar{r}=\frac{r}{R},\ \bar{R}=1,\ \bar{r}_0=\frac{r_0}{R},$$
$$\bar{W}=\frac{W}{\pi R^2 p_0},\ \bar{h}=\frac{h}{H},\ \bar{H}=1,\ \bar{\delta}=\frac{\delta R}{H} \tag{10}$$

where p_0 describes the pressure in the oil pocket, W describes the carrying capability, h describes the film thickness, δ describes the linear deformation slope, $|\delta|$ will increase when the slope extend is increasing. $\bar{p}$

stands for the dimensionless pressure, $\bar{r}$ stands for the dimensionless radius, assuming $\bar{r}_0 = 0.2$ [1], $\bar{W}$ stands for the dimensionless carrying capability, $\bar{h}$ stands for the dimensionless film thickness, $\bar{\delta}$ stands for the dimensionless linear deformation slope.

The carrying surfaces of the hydrostatic thrust bearing are considered stable and the dimensionless pressure in the oil pocket is set at 1. The pressure distribution and carrying capability of the ideal model, which ignore the variation of film thickness and surface waviness, is written as [2]:

$$\bar{p} = \begin{cases} 1, 0 \le \bar{r} \le \bar{r_0} \\ \dfrac{\ln(\bar{r})}{\ln(\bar{r}_0)}, \bar{r}_0 < \bar{r} \le 1 \end{cases} \tag{11}$$

$$\bar{W} = \frac{1 - \bar{r}_0^{2}}{2\ln(\frac{1}{\bar{r}_0})} \tag{12}$$

The function of surface waviness $\bar{e}$ on the $\bar{h}$ direction could be expressed as:

$$\bar{h} = \bar{h}(r) + \bar{e}(r) \tag{13}$$

The surface waviness can be respected to be periodic approximately, which is able to be modified by Fourier expansion.

$$\bar{e}(r) \approx \bar{e}_0 + \sum_{i=1}^{n} \left(a_i \sin\sin i\pi\bar{r} + b_i \cos\cos i\pi\bar{r}\right) \tag{14}$$

The value of surface waviness is taken as[9]:

$$\bar{e}(r) \approx 0.05\sin\sin(120\pi r) + 0.02\cos\cos(160\pi r) \tag{15}$$

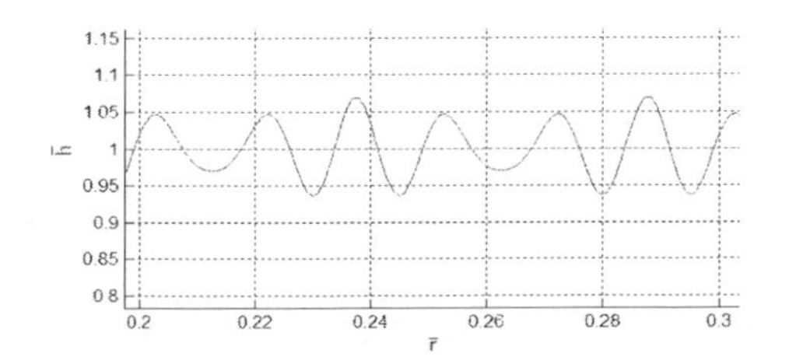

Figure 2. Expression of surface waviness.

The film thickness $\bar{h}(r)$ with linear deformation could be written as:

$$\bar{h}(r) = 1 - \bar{\delta}(1 - \bar{r}) \tag{16}$$

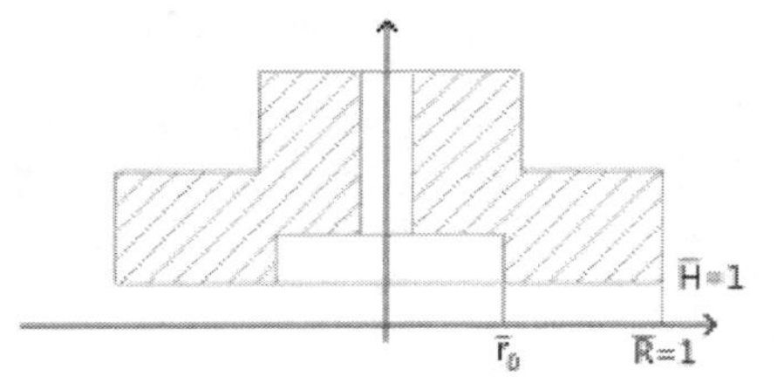

(a) No deformation

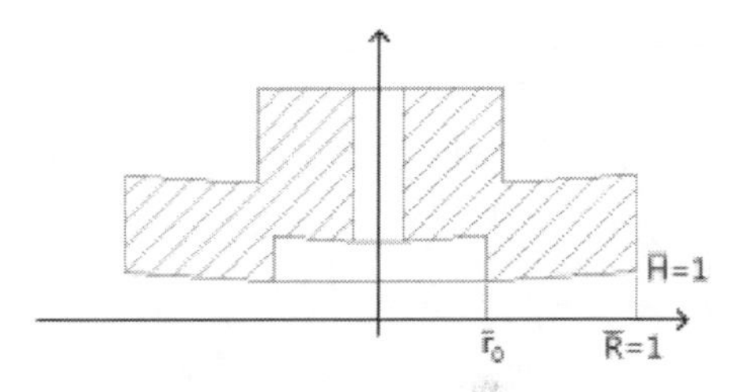

(b) Convex deformation

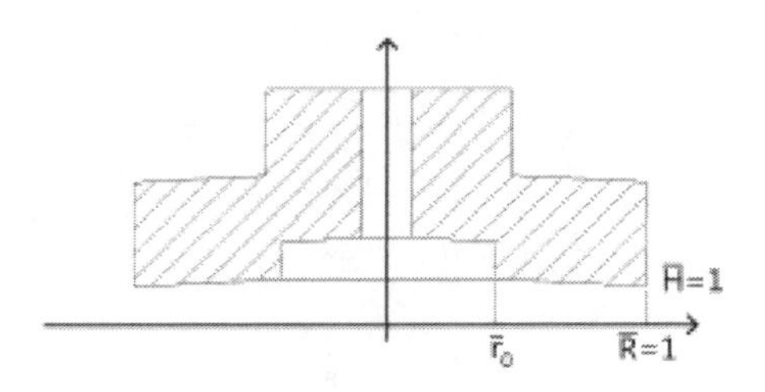

(c) Concave deformation

Figure 3. Linear deformation of bearing surface.

When $\bar{\delta} > 0$, the surface deformation is considered as convex. When $\bar{\delta} < 0$, the surface deformation is considered as concave. When $\bar{\delta} = 0$, the surface is consider as flat. The range of $\bar{\delta}$ is supposed $(-\infty, 1.25)$.

According to Equation (12), (14) and (15), the film thickness is expressed as:

$$\bar{h} = 1 - \bar{\delta}(1 - \bar{r}) + 0.05\sin\sin(120\pi r)$$
$$+0.02\cos\cos\left(160\pi r\right) \tag{17}$$

2 NUMERICAL QUADRATURE

The pressure distribution will be solved as the result of Equation (8) integrating.

$$\bar{p} = C_1 \int \frac{1}{\bar{r}\bar{h}^3} d\bar{r} + C_2 \tag{18}$$

where C_1 and C_2 are undetermined coefficients, which have to settled based on the boundary condition.

The impact of surface deformation on the pressure deformation can be solved analytically by computing software like Maple.

$$\overline{p}=\begin{cases}1,0\le\overline{r}\le\overline{r_0}\\ C_1\left(\begin{array}{c}\dfrac{1}{\left(\overline{\delta}-1\right)^2\left(\delta\overline{r}-\overline{\delta}+1\right)}+\dfrac{\ln\left(\overline{\delta r}-\overline{\delta}+1\right)}{\left(\overline{\delta}-1\right)^3}\\ -\dfrac{\ln\ln\left(\overline{r}\right)}{\left(\overline{\delta}-1\right)^3}-\dfrac{1}{2\left(\overline{\delta}-1\right)\left(\overline{\delta r}-\overline{\delta}+1\right)^2}\end{array}\right)\\ +C_2,\overline{r_0}<\overline{r}\le 1\end{cases}\tag{19}$$

where

$$\begin{cases}1=C1\left(\begin{array}{c}\dfrac{1}{\left(\overline{\delta}-1\right)^2\left(\delta\overline{r_0}-\overline{\delta}+1\right)}+\dfrac{\ln\left(\overline{\delta}\,\overline{r_0}-\overline{\delta}+1\right)}{\left(\overline{\delta}-1\right)^3}\\ -\dfrac{\ln\ln\left(\overline{r_0}\right)}{\left(\overline{\delta}-1\right)^3}-\dfrac{1}{2(\overline{\delta}-1)(\overline{\delta}\,\overline{r_0}-\overline{\delta}+1)^2}\end{array}\right)+C2\\ 0=C1\left(\dfrac{1}{\left(\overline{\delta}-1\right)^2}-\dfrac{1}{2\left(\overline{\delta}-1\right)}\right)+C2\end{cases}\tag{20}$$

However, Equation (18) will be turned into complicated formation, which is very difficult to solve analytically, if waviness equation has been taken in consideration. Therefore, numerical quadrature is needed to solve the equation.

Generally, the numerical quadrature has a computing error that is inevitable but the result is still acceptable as long as the error matched the tolerance. Adaptive Simpson Quadrature is considered as a useful numerical method with relatively higher accuracy. The result of the pressure to a point will be got by each time Adaptive Simpson Quadrature and the entire resolution will be shown as a scatter diagram to demonstrate the pressure distribution. Because C_1 and C_2 only have an influence on the integral range, they could be set as $C_1^{(0)}=1$ and $C_2^{(0)}=0$ to initialize the algorithm. $\overline{p}^{(0)}$will be solved as the result of the first time numerical quadrature and the exact value of C_1 and C_2 will be settled based on the boundary condition.

According to the model of hydrostatic thrust bearing, the boundary condition can be written as:

$$\begin{cases}\overline{p}(\overline{r_0})=1\\ \overline{p}(1)=0\end{cases}\tag{21}$$

Based on Equation (21), C_1 could be express as:

$$C_1=\frac{1}{\overline{p}(\overline{r_0})^{(0)}-\overline{p}(1)^{(0)}}\tag{22}$$

On the basis of C_1 与 $C_2^{(0)}$, $\overline{p}^{(1)}$ could be got and C_2 will be expressed as:

$$C_2=-\overline{p}(1)^{(1)}\tag{23}$$

The final solution of pressure distribution $\overline{p}$ will be resolved after C_1 and C_2 are settled.

The relationship between pressure distribution and carrying capability can be written as:

$$\overline{W}=2\int_{\overline{r_0}}^{1}\overline{p}(\overline{r})\cdot\overline{r}d\overline{r}+\overline{r_0}^{2}\tag{24}$$

The carrying capability could be also solved by numerical quadrature.

3 RESULTS AND DISCUSSION

The pressure distribution $\overline{p}$ is shown in Figure 4 when, which means there is no deformation of the bearing surface.

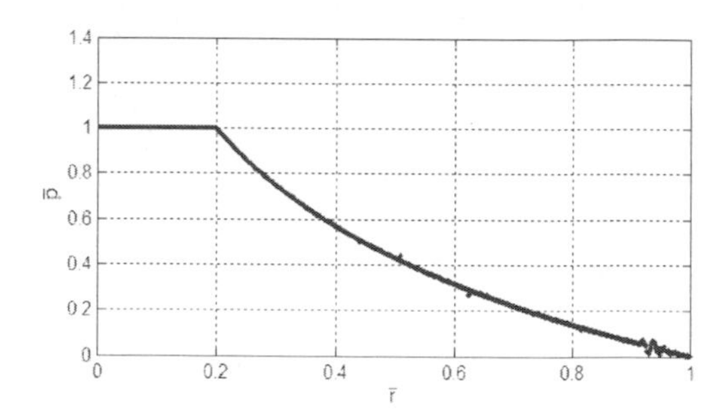

(a) Surface with waviness taken into consideration

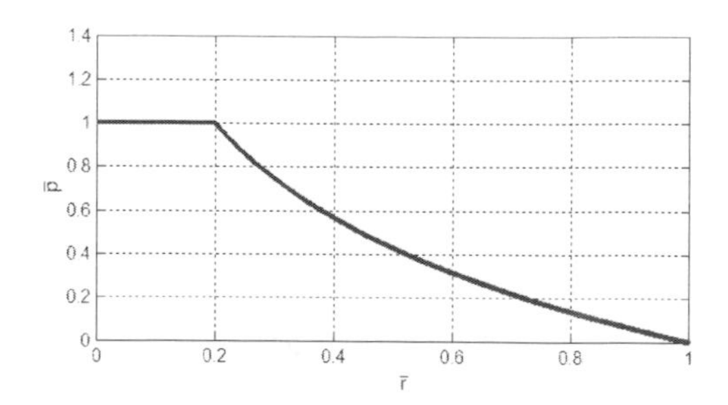

(b) Absolutely smooth surface

Figure 4. Pressure distribution of hydrostatic thrust bearing without deformation.

According to Figure 4, the pressure distribution is seen logarithmic. And several slight fluctuation could be found at the points where $\bar{r} = 0.5$, 0.6, 0.95.

The pressure distribution $\bar{p}$ is shown in Figure 5 when, which means there is convex deformation of the bearing surface.

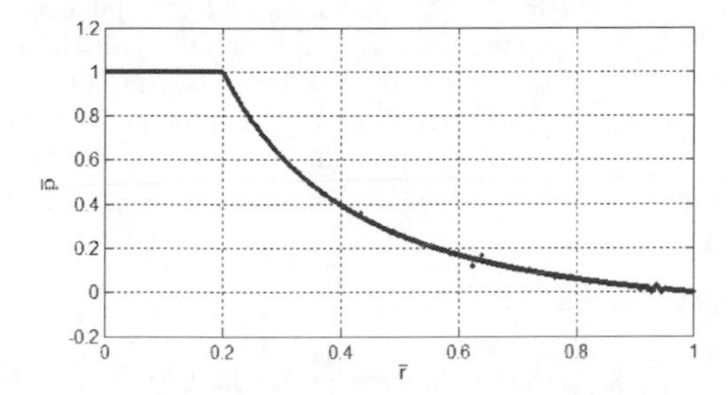

(a) Surface with waviness taken into consideration

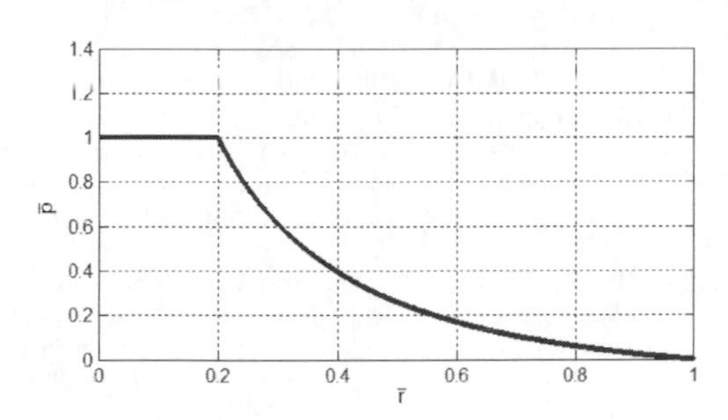

(b) Absolutely smooth surface

Figure 5. Pressure distribution of hydrostatic thrust bearing with convex deformation.

According to Figure 5, several slight fluctuation could also be found at the points where $\bar{r} = 0.5$, 0.6, 0.95. In addition, the pressure distribution seemed to be lower depending on the impact of convex deformation.

The pressure distribution $\bar{p}$ is shown in Figure 6 when, which means there is concave deformation of the bearing surface.

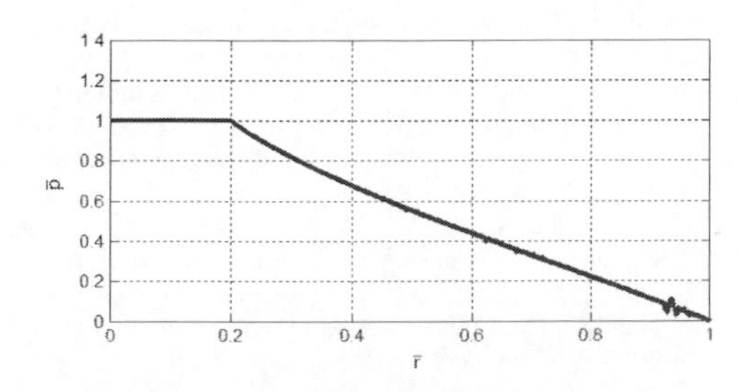

(a) Surface with waviness taken into consideration

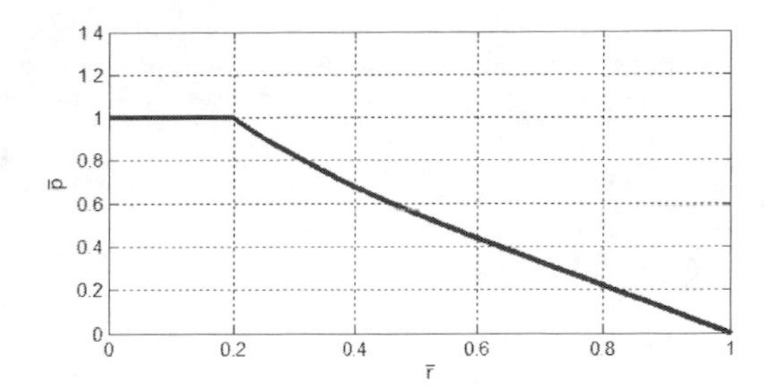

(b) Absolutely smooth surface

Figure 6. Pressure distribution of hydrostatic thrust bearing with concave deformation.

According to Figure 6, less slight fluctuation could be found at the points where $\bar{r} = 0.95$ and the amplitudes are smaller. In addition, the pressure distribution seemed to be greater depending on the impact of concave deformation.

With 11 groups of results synthesized together, the impact of surface deformation and waviness on the carrying capability is shown in Figure 7.

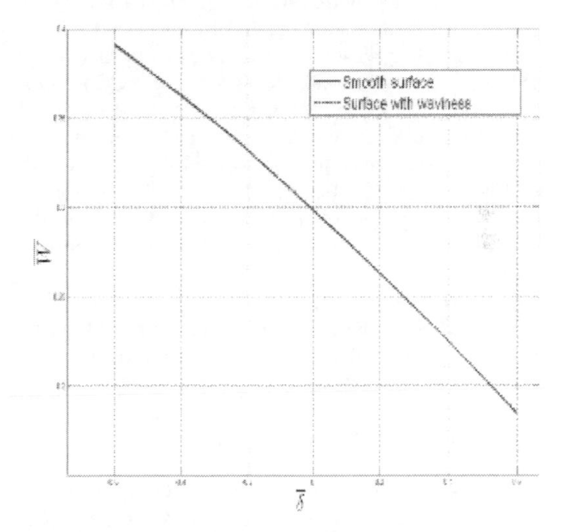

Figure 7. Changes of load carrying capacity in different surface deformations.

According to Figure 7, the carrying capability is slightly improved, by the surface waviness.

From Equation (11), when $\bar{r_0} = 0.2$,the exact carrying capability value of ideal thrust model can be solved as:

$$\overline{W_e} = \frac{1 - 0.2^2}{2\ln(\dfrac{1}{0.2})} \approx 0.2982 \tag{25}$$

In Figure 7, when $\overline{\delta} = 0$ and do not take surface waviness into consideration, the computing value of carrying capability is:

$$\overline{W_*} \approx 0.2985 \tag{26}$$

Computing error is:

$$\Delta = \left| \frac{\overline{W_e} - \overline{W_*}}{\overline{W_e}} \right| \approx 0.1\% \tag{27}$$

It is shown in Equation (27) that, the algorithm in the present work is reliable.

According to figure 7, when $\overline{\delta} > 0$, namely convex deformation, the carrying capability will increase by the rising of $|\overline{\delta}|$. When $\overline{\delta} < 0$, namely concave deformation, the carrying capability will increase by the rising of $|\overline{\delta}|$. The concave deformation is better for the carrying capability improving.

In addition, from Figure 7, the carrying capability of hydrostatic thrust bearing has slightly increased about 0.2% to 0.5% by the surface waviness. The impact of surface waviness is limited.

According to Figures 4(a), 5(4) and 6(a), there several pressure fluctuation caused by surface waviness whose specific reason cannot be drawn in this work. But any kinds of sudden change of the pressure distribution in the edge seal of oil pad could considered to be harmful to the whole bearing system, so the variation of pressure should be avoided. By comparison of Figures 4(a), 5(4) and 6(a), when $\overline{\delta} < 0$, the pressure fluctuation is able to be reduced. Therefore, the concave deformation is better for the carrying performance improving.

Based on the previous analysis, the convex deformation could be thought better in both improving the carrying capability and reducing the pressure fluctuation.

4 CONCLUSIONS

In the present work, an analytic model within surface deformation and waviness taken into consideration of hydrostatic bearing is established to study the impact of deformation and waviness on carrying capability. Based on the previous analysis results, the following conclusion could be drawn:

1. Pressure fluctuations will be caused if the surface waviness at the edge seal of oil pad has been taken into consideration.
2. The pressure fluctuation caused surface will be reduced under the concave deformation.
3. The carrying capability could be improved by the rising of deformation extent under concave deformation. When $\overline{\delta} = -0.6$, the carrying capability increases about 31%.
4. The carrying capability could be deduced by the rising of deformation extent under convex deformation. When $\overline{\delta} = 0.6$, the carrying capability decreases about 38%.

The convex deformation could be thought better in both improving the carrying capability and reducing the pressure fluctuation of the hydrostatic thrust bearing. Concave predeformation can be used in the manufacture of hydrostatic thrust bearing to improve the carrying performance.

REFERENCES

[1] Noah D. Manring, Robert E. Johnson, Harish P. Cherukuri. The Impact of Linear Deformation on Stationary Hydrostatic Thrust Bearing. Journal of Tribology, 2002, 124(8): 874–877.

[2] Chen Yansheng. The liquid static pressure supporting principle and design. 1980.

[3] LI Yun-tang, LIN Ying-xiao, ZHU Hong-xia, SUN Zai. Analysis of the Micro Self-vibration of Aerostatic Thrust Bearing Based on Large Eddy Simulation. Journal of Mechanical Engineering. 2013, 49(13): 56–62.

[4] ZHANG Li-jing, WANG You-qiang, Influence of vibration and shock on transient thermal elastohydrodynamic lubrication of seawater-lubricated plastic bearings, JOURNAL OF VIBRATION AND SHOCK, 2013, 32(15): 203–208.

[5] Nakhle H. Asmar. Partial differential equations (the Second Edition). 2007

[6] Y. Henry, J. Bouyer, M. Fillon, An Experimental Hydrodynamic Thrust Bearing Device and Its Application to the Study of a Tapered-Land Thrust Bearing. Journal of Tribology, 2014, 136(4): 02170301-02170310.

[7] Udaya P. Singh, Ram S. Gupta, Vijay K. Kapur. On the Steady Performance of Annular Hydrostatic Thrust Bearing Rabinowitsch Fluid Model. Journal of Tribology, 2012, 134(10): 04450201-04450205.

[8] REN Zhi-qiang, GUO Feng, WANG Jing. Measurement and Simulation of the Oil Film in a Thrust Ball Bearing Considering the Waviness on Its Raceway. Tribology, 2013, 33(06):586–593.

[9] ZHANG Yao-qiang, CHEN Jian-jun, DENG Si-er , L IN Li-guang. Nonl inear dynamic characteristics of a roll ing bearing2rotor system with surface waviness. Journal of Aerospace Power, 2008, 23(09):1731–1736.

[10] LI Xiao-yang, REN Miao. Numerical Simulation of the Oil Film of Hydrostatic Bearing Under Fluctuation Load[J]. Journal of Beijing University of Technology, 2012, 38(07):992–996.

[11] LIU Zhao-miao, YUE Guang-jie, SHEN Feng. Numerical Simulation of Carrying Capacity in Hydrostatic Oil Cavity Under Different Inlet Reynolds Numbers. Journal of Beijing University of Technology, 2013, 39(7): 965–970.

Computing, Control, Information and Education Engineering – Liu, Sung & Yao (eds)

 ISBN: 978-1-138-02800-5

The heat generation of high-speed spindle bearings under different preload methods

Yong Yang, Qiang Cheng, Li Gang Cai & Qiu Nan Feng
Beijing University of Technology, Beijing, China

ABSTRACT: The bearing temperature of a high-speed spindle increases with its speed. This restricts the rotational velocity and precision of such high-speed spindles. This study numerically simulated the heat generated in a high-speed spindle bearings at different rates of rotation. Moreover, the model distinguished the preload methods, rigid preload and constant preload. The geometrical parameters obtained by improving Harris bearing model and the heat generation are calculated by Palmgren theory. The study shows that the heat generation of constant preload bearing is less than rigid preload bearing in a spindle. This research proposed a method for evaluating the heat generation quantificationally.

KEYWORDS: High-speed spindle, bearings, heat generation.

1 INTRODUCTION

With high-speed and increasingly precise developments in the equipment manufacturing industry, the core functional components high-speed spindles of numerically controlled (NC) machines draw increasing research attention. However, the spindle may be subjected to sudden failure without any warning signals due to the bearing underlying thermal problem. However, the numerous thermal constraint conditions and the extremely complex heat generation and heat dissipation processes in an operational high-speed spindle, the heat transfer model that couples every boundary condition is hard to establish when trying to describe the transient temperature field distribution of the spindle system.

To calculate the heat generation of high-speed spindle bearings, a complete bearing thermal model conforming to the actual preload mechanisms was deemed essential. This study numerically simulated the heat generation of high-speed spindle bearings at different rates of rotation in rigid preload and constant preload. The geometrical parameters obtained by improving Harris bearing model and the heat generation is calculated by Palmgren theory [2]. The study shows that the heat generation of constant preload bearing is less than rigid preload bearing in a spindle.

2 HIGH SPEED SPINDLE- CAT40

The research object is the Siemens CAT40 high-speed spindle as shown in Figure 1. The rigid preload bearing set is installed on the front end of the spindle. The preload is adjustable by adjusting the spacer length between the bearings. A constant pressure preload bearing is assembled on the rear end of the spindle. It applies a near constant preload to the outer ring of the bearing through the spring. It allows the spindle floating backwards and offers thermal compensation space for the spindle.

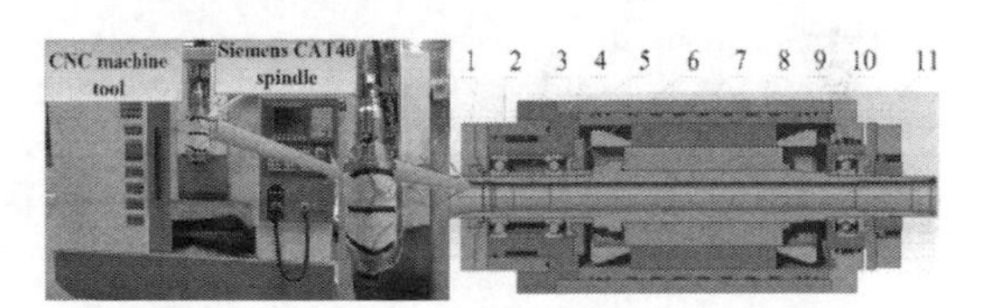

(Key: 1.Bearing 1; 2. Sleeve of inner ring; 3. Sleeve of outer ring; 4. Bearing 2; 5. Spindle; 6. Motor rotor; 7. Motor stator; 8. Sliding sleeve; 9. Bearing 3; 10. Sliding zone of sliding sleeve; 11. Disk spring.)

Figure 1. System structure: the high-speed spindles.

3 THE GEOMETRICAL PARAMETERS OF THE RIGID PRELOADING BEARING

The spindle uses an angular contact ball bearing. The coordinate system of the model is identical to the bearing model established elsewhere [3]. According to Hertzian contact theory, the inertia force of the kth rolling element can be expressed as [4] [5]:

$$F_{ck} = \frac{1}{2} m D_m \Omega^2 \left(\frac{\Omega_E}{\Omega} \right)_k^2 \quad (1)$$

$$M_{gk} = J_b \Omega^2 \left(\frac{\Omega_B}{\Omega}\right)_k \left(\frac{\Omega_E}{\Omega}\right)_k \sin \alpha_k \tag{2}$$

In Eqns (1) and (2), m is the mass of a single rolling element; D_m is the diameter of the pitch circle of the bearing; J_b is the rotational inertia of the single rolling element; Ω is the rotational speed of the spindle (rad/s); Ω_B is the spinning angular velocity of rolling element (rad/s); Ω_E is the angular velocity of the rolling element around the bearing centre; and α_k is the angle between the spinning axis of the rolling element and the axis of the spindle.

The non-linear equation governing the behaviour of the locating bearing can be obtained.

$$\begin{cases} f_1 = Q_{ok}\cos\theta_{ok} - \dfrac{M_{gk}}{Dw}\sin\theta_{ok} - Q_{ik}\cos\theta_{ik} + \dfrac{M_{gk}}{Dw}\sin\theta_{ik} - F_{ck} \\ f_2 = Q_{ok}\sin\theta_{ok} + \dfrac{M_{gk}}{Dw}\cos\theta_{ok} - Q_{ik}\sin\theta_{ik} - \dfrac{M_{gk}}{Dw}\cos\theta_{ik} \\ f_3 = U_{ik} - \Delta_{ok}\sin\theta_{ok} - \Delta_{ik}\sin\theta_{ik} \\ f_4 = V_{ik} - \Delta_{ok}\cos\theta_{ok} - \Delta_{ik}\cos\theta_{ik} \end{cases} \tag{3}$$

Where, $\Delta\delta_x$ is known, while $\Delta\delta'_x$, $\Delta\delta'_y$, $\Delta\delta'_z$, $\Delta\gamma'_y$, and $\Delta\gamma'_z$ are available through data from sensors located on the spindle; the dynamic parameters of the bearing θ_{ik}, θ_{ok}, δ_{ik}, and δ_{ok} are obtained using the Newton-Raphson iterative algorithm.

4 THE GEOMETRICAL PARAMETERS OF THE RIGID PRELOADING BEARING

The mechanism of the constant pressure preload mechanism is indicated as follows: by consolidating the inner ring of the bearing with the spindle, a disc spring applies a near-constant axial pressure on the outer ring. With the thermal extension produced by the spindle, the sliding sleeve connected with the outer ring of the bearing produces an axial displacement. The load-deformation relationship and force balance relationship of the inner and outer rings with the rolling element are identical to those of the locating preload bearing. According to the preload mechanism, a force balance supplement equation is acquired. Where, Fa is the constant pressure preload. The non-linear equation governing the pressuring-fixing bearing can be obtained by integrating as equation(4)

$$\begin{cases} f_5 = Q_{ok}\cos\theta_{ok} - \dfrac{M_{gk}}{Dw}\sin\theta_{ok} - Q_{ik}\cos\theta_{ik} + \dfrac{M_{gk}}{Dw}\sin\theta_{ik} - F_{ck} \\ f_6 = Q_{ok}\sin\theta_{ok} + \dfrac{M_{gk}}{Dw}\cos\theta_{ok} - Q_{ik}\sin\theta_{ik} - \dfrac{M_{gk}}{Dw}\cos\theta_{ik} \\ f_7 = V_{ik} - \Delta_{ok}\cos\theta_{ok} - \Delta_{ik}\cos\theta_{ik} \\ f_8 = Q_{ok} - \dfrac{Fa}{z\times\sin\theta_{ok}} \end{cases} \tag{4}$$

The symbols in Eq. (4) have the same meaning as those in Eq. (3). Similarly, using the Newton-Simpson iterative method, the dynamic parameters of the bearing θ_{ik}, θ_{ok}, δ_{ik}, and δ_{ok} can be obtained.

5 THE TORQUE AND HEAT GENERATION OF THE BEARING

Figure 2 shows the long radius a and short radius b of the elliptical contact spot of the bearing under different rotational speeds and different preload types.

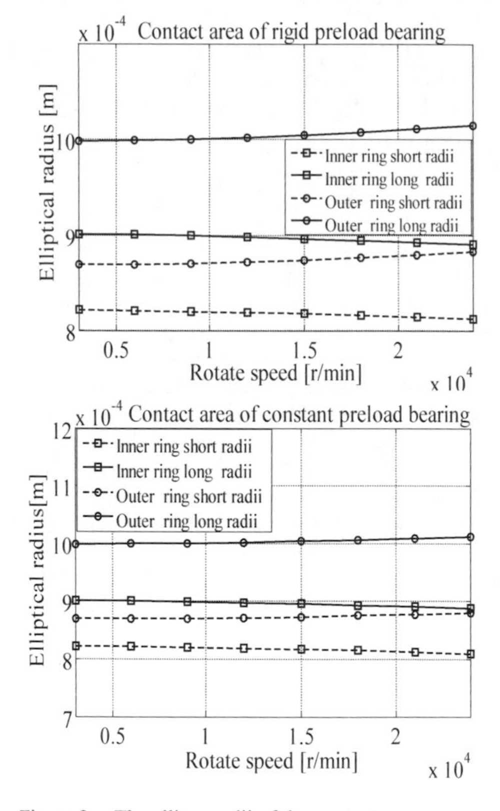

Figure 2. The ellipse radii of the contact area.

As for the kth rolling element, the rotational torques [6,7] of the inner and outer rings is obtained through Eq. (5):

$$\begin{cases} M_{si} = \dfrac{3\mu Q_i a E(e,\pi/2)}{8} \\ M_{so} = \dfrac{3\mu Q_o a E(e,\pi/2)}{8} \end{cases} \tag{5}$$

The heat generated by the contact area of each rolling element is given by:

$$\begin{cases} H_i = \omega_{roll} \times M_i + \omega_{si} \times M_{si} \\ H_o = \omega_{roll} \times M_o + \omega_{so} \times M_{so} \end{cases} \tag{6}$$

In Eq. (6), ω_{roll} is the angular velocity (rotational speed) of the rolling element; ω_{si} and ω_{so} are the rotational speeds of the rolling element with respect to the inner and outer rings respectively; μ is the coefficient of friction of the rolling element; M_i and M_o are the friction torques related to the load and can be calculated by the method proposed elsewhere [2]. Through this analysis, the heat generated by the bearing is calculable as show in Figure 3.

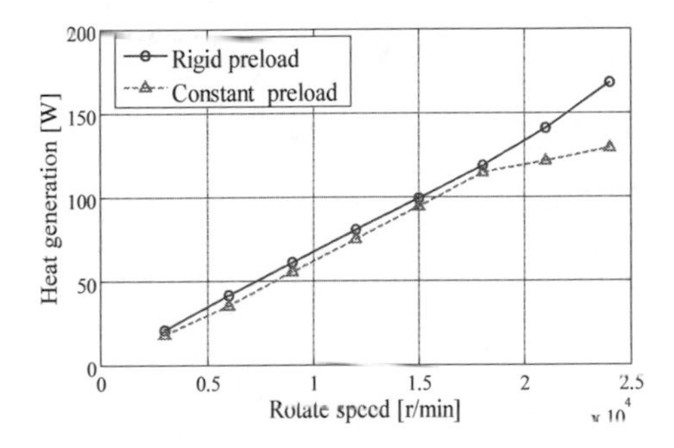

Figure 3. The heat generation of high-speed spindle bearings.

The Figure 3 revealed the relationship between the heart generation and spindle rotate speed. As the rotate speed up the heat generation increases gradually. At the lower speed the heat generation growth approximate linear. However, in the high speed region (up18000rpm), the heat generation of constant preload bearing has a nonlinear trend. The gap between the heat generation lines is no less than 60W, where the speed up to 24000rpm.

6 CONCLUSION

The heat generation of high-speed spindle bearings in different preload mechanism can be calculated by this proposed complete bearing thermal model. Simulation results show that the heat generation of rigid preload bearing is serious than constant preload bearing especially in high speed. The study shows that the heat generation of constant preload bearing is less than rigid preload bearing in a spindle.

ACKNOWLEDGEMENTS

The authors are most grateful to the National Science and Technology Great Special Program (No. 2013ZX04013-011) which support the research presented in this paper.

REFERENCES

[1] Harris T A, in: *Rolling Bearing Analysis, 3rd ed*, John Wiley & Sons,Inc., NY (1991).

[2] Palmgren A, in: *Ball and Roller Bearing Engineering*, Philadelphia, PA (1959).

[3] Cao Y Z, and Altintas Y, in: *A general method for the modeling of spindle-bearing systems*. Journal of Mechanical Design, Vol.126(2004), p. 1089–1104.

[4] Harris T A, in: *Rolling Bearing Analysis, 4th ed*. John Wiley and Sons, NY (2001)

[5] Jones A B, in: *A General Theory for Elastically Constrained Ball and Radial Roller Bearings Under Arbitrary Load and Speed Conditions*. Journal of Basic Engineering, Vol.25(1960), p. 309–320.

[6] Jorgensen B R, and Shin Y C, in: *Dynamics of machine tool spindle/bearing systems under thermal growth*. Journal of Tribology-Transactions of the Asme. Vol.119(1997), p. 875–882.

[7] Li H Q, and Shin Y C, in: *Analysis of bearing configuration effects on high speed spindles using an integrated dynamic thermo-mechanical spindle model*. International Journal of Machine Tools & Manufacture. Vol.44(2004), p. 347–364.

Computing, Control, Information and Education Engineering – Liu, Sung & Yao (eds)
© 2015 Taylor & Francis Group, London, ISBN: 978-1-138-02800-5

Application of adaptive fuzzy theory revision in ATV engine fault diagnosis

Yuan Min Yu & Bo Hu Zhang
Engineering University of Armed Police Force, Xi'an, China

ABSTRACT: All Terrain Vehicle (ATV) is a special vehicle designed for harsh environment and region. The engine is influenced by the environment easily. After a long time of operation, performance parameters of the engine itself will change and these changes lead to a subordinate relationship between fault symptom and fault causes are obscure. Therefore, the fuzzy diagnostic technology is introduced to the ATV engine fault diagnosis, the revised mathematical fuzzy diagnosis improved adaptive model is put forward and set up the fuzzy diagnosis model and use of a model ATV engine specific fault diagnosis examples are analyzed and verified finally. The results show that the use of diagnostic matrix adaptive modified of which diagnosis result is more actually.

KEYWORDS: fuzzy theory, all terrain vehicle, engine, fault diagnosis.

1 INTRODUCTION

The biggest characteristic of the ATV which can easily go through the swamp, snow, mountains, streams, ponds and other harsh terrain is to have low ground pressure and strong traffic ability which also makes the probability of engine failure greatly increased [1]. At the same time, in the process of the ATV engine operation, the operation condition and performance parameters are changing. During that time, it is neither completely "intact", nor entirely "failure", but in a vague state, so is the sign shown. So, the problem of fault diagnosis for ATV engine only depend on qualitative decision can not be solved accurately and quickly, it is necessary to carry out the fault diagnosis by using fuzzy diagnosis technology.

2 MATHEMATICAL DESCRIPTION OF ATV ENGINE FUZZY DIAGNOSIS

Fuzzy diagnosis is a method to solve the various fault cause membership through the symptoms membership. Assumed: there are symptoms that the diagnostic objects may show, express as $x_1, x_2, \dots, x_m$; there are n types of fault causes, express as $y_1, y_2, \dots, y_n$. Fault symptom fuzzy vector $X = (x_1, x_2, \dots, x_m)$, and $x_i (i = 1, 2, \dots, m)$ is the membership of an object with symptom x_i. Fault fuzzy vector is $Y = (y_1, y_2, \dots, y_n)^T$ and $y_j (j = 1, 2, \dots, m)$ is an object with the membership of an object with symptom y_j[2,3]. The fuzzy relation between X and Y is as follows:

$$Y = X \circ R. \quad (1)$$

This is the fuzzy relation equation between the fault causes and symptoms. Where, the symbol "o" is the fuzzy logic operator, and R is the fuzzy diagnosis matrix, which describes the relation between the fault symptoms and fault causes.

$$R = \begin{pmatrix} r_{11} & \cdots & r_{1n} \\ \vdots & \ddots & \vdots \\ r_{m1} & \cdots & r_{mn} \end{pmatrix} = (r_{ij})_{m \times n} \quad (2)$$

Fuzzy diagnosis matrix R is m×n dimensional matrix, the matrix element $r_{ij=}$ i=1,2....,m; j=1,2....n, represents the membership of i-th symptom x_i against the j-th fault cause. It is determined by statistical analysis and expert experience.

Assumed that the symptom domain of ATV model fault after a comprehensive evaluation is X, the diagnosis fault cause domain is Y.

Symptom domain X:

Vibration, shock and noise on the cylinder head surface caused by engine operation contains rich information, it can reflect the real-time engine work state and is an important source of acquiring fault diagnosis information[4]. The fault sign here is based on the spectral data of engine vibration and is divided into nine sections, which are respectively expressed by $P_1, P_2, \dots P_9$.

$X = (P_1, P_2 P_3, P_4, P_5, P_6 P_7, P_8, P_9)$.

Fault cause domain Y:

After investigation, this paper determines that there are 7 fault reasons for ATV engine[5].

Y = {y_1 is the cooling system fault, y_2 the fuel supply system fault y_3, ignition system fault, y_4 boot system fault, y_5 the lubrication system fault, y_6 gas distribution system fault, y_7 curved bar connection system fault}.

3 ADAPTIVE CORRECTION OF FUZZY DIAGNOSIS

The membership degree of the fuzzy diagnosis matrix is obtained by mechanism analysis and expert experience. There are some limitations in practical fault diagnosis, and therefore a membership self-learning automatic correction method for fault diagnosis is needed, in order to improve the practicability and accuracy[6].

If the fault cause is Y_α, but the diagnosis result is Y_β.

That is, $Y_\beta = \max\{Y_j, j = 1,2,...,n\}$ $\beta \neq \alpha$; assumed that $L = Y_\alpha - Y_\beta$, obviously, $L<0$. Given the target value $Y_\alpha = \varphi$, so that $L>0$.

Select a certain step size, correct the r_α row element for correction in the diagnosis matrix, do not correct all elements in the row, and only correct the membership element r_{ij} corresponded to various symptoms in X.

r_{ij} is obtained according to previous diagnostic analysis and expert experience, each r_{ij} has unequal contributions towards the cause determination Y, which can be seen as the association between a symptom and cause. If the element r_{ij} corresponding to in the r-th row directly adding the step size τ, and make Y_a = j, it will undermine the contribution of each element towards the fault cause. For this reason, the weight coefficient K_{ij} is defined to be .

$$K_{ij} = \frac{r_{ij}}{Y_\alpha} \quad (\alpha = 1,2,...,n) \cdot \tag{3}$$

$$K = \frac{R}{Y_\alpha} \ . \tag{4}$$

As a result, despite the increased (or decreased) step size, membership of each symptom will change accordingly, but the membership of each symptom X_i has the same contribution degree towards the cause Y_j, namely, still meet:

$$\sum_{j=1}^{m} r_{ij} = 1 \qquad 0 \le r_{ij} \le 1;\ 1 \le i \le m;\ 1 \le j \le n. \tag{5}$$

After sorting, this paper obtains

$$Y^T = \left(1 + \frac{\tau}{Y_\alpha}\right) RX^T. \tag{6}$$

4 DIAGNOSIS EXAMPLE

The below is the fault diagnosis example for an ATV engine.

This paper uses adaptive correction fuzzy diagnosis method to make fault diagnosis, and the diagnostic process is shown in Figure 1:

According to the specific circumstances and expert experience of ATV engine, its fuzzy diagnosis matrix is:

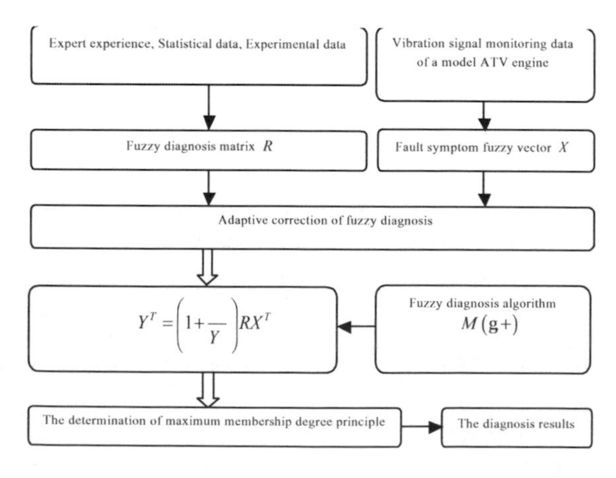

Figure 1. The process of fault diagnosis with adaptive correction.

Table 1. Fuzzy diagnostic matrix of engine.

	x_1	x_2	x_3	x_4	x_5	x_6	x_7	x_8	x_9
y_1	0.00	0.00	0.00	0.76	0.76	0.56	0.56	0.00	0.00
y_2	0.00	0.00	0.90	0.00	0.00	0.00	0.00	0.50	0.00
y_3	0.20	0.10	0.10	0.10	0.00	0.10	0.00	0.10	0.10
y_4	0.10	0.05	0.05	0.10	0.30	0.10	0.20	0.10	0.10
y_5	0.00	0.00	0.00	0.00	0.40	0.50	0.10	0.00	0.00
y_6	0.10	0.80	0.00	0.10	0.00	0.00	0.00	0.00	0.00
y_7	0.90	0.00	0.00	0.00	0.00	0.00	0.00	0.10	0.00

Where, $x_i = (i=1,2 ...,9)$ is an engine fault symptom, and $y_j = (j=1,2 ...,9)$ is the fault cause.

According to a record of ATV engine online monitoring database, this paper gets the symptoms eigenvalues, calculates the membership function, and obtains the symptom membership vector:

X= (0.13 0 0.76 0 0.45 0 0.85 0 0.20)

This paper selects the model $M(g+)$ fuzzy diagnosis algorithm, namely

$$y_{ij} = \sum_{i=1}^{m} x_i r_{ij} \quad (j=1,2 ...,7). \tag{7}$$

Through calculation

$$Y^T = XR^T. \tag{8}$$

Fault membership vector is:

Y=(0.725 0.684 0.122 0.376 0.265 0.013 0.117)

By using the maximum membership principle, this paper can obtain: $y1=max\ \{y_i\ j = 1,2 ...,7\}$, that is, among seven engine fault causes, the cooling system fault is the main cause of the ATV engine fault.

5 CONCLUSION

To get the fuzzy diagnosis matrix which is consistent with the actual situation goes by an example of fault diagnosis analysis and expert experiences and establishes the fuzzy relation between main failure causes and symptoms of engine. The method is simple and does not need to establish accurate mathematical model. Fuzzy diagnosis mathematical model of the engine is adaptively modified by the weight coefficient method that ameliorates the fuzzy relation matrix and improves the applicability and reliability of fault diagnosis. However, the fault symptoms and fault causes arc not the one to one corresponding relationship. Using the maximum membership degree principle in the rules of the diagnosis has certain limitation. The diagnosis result is not accurate enough and need further improvement.

REFERENCES

[1] J. Wu: Science and Technology Innovation Herald No.19,(2012), p.32–35.

[2] H.W. Li, D.S. Yang , Y.L. Sun and J. Han: Computer Engineering and Design No.34,(2013), p.8–10.

[3] J. Yang: Intelligent Diagnosis Technology for Equipment(National Defence Industry Press, Beijing 2004).

[4] Y.X. Xu, W.P. Yang, X. Lv, Z.W. Ma and X.H. Ma: Vibration and Shock Vol.32, No.8,(2013), p.13–16.

[5] X.Z. Wang: Automobile Fault Diagnosis and Detecting Technique(Posts and Telecom Press, Beijing 2005).

[6] L.N. Liu, F.Q. Han, S.Q. Zhang and Z.P. Deng: The Detection of Hydropower Automation Dam Vol.27, No.1,(2003), p.52–55.

Computing, Control, Information and Education Engineering – Liu, Sung & Yao (eds)
 ISBN: 978-1-138-02800-5

An effective tool for optimizing the leaf sequencing algorithms

Jia Jing & Hui Lin
School of Electronic Science and Applied Physics, Hefei University of Technology, Hefei, China

Kai Zhao & Yu Juan Hu
Department of Public Computer Teaching, Hefei Normal University, Hefei, China

ABSTRACT: Objective: To investigate the efficacy of Chen, Bortfeld and Siochi leaf sequencing algorithms, a leaf sequencing optimization tool that compares the results of typical clinical cases based on aforementioned three algorithms of Multileaf Collimator (MLC) leaf sequencing of Intensity-Modulated Radiation Therapy (IMRT) was presented. Methods: The IMRT leaf sequencing calculation program based on Chen, Bortfeld, Siochi algorithms was constructed. The output of this program was the number of segments and the total number of monitor unit. Results: The leaf sequencing program could produce highly conformal dose distributions within a clinically acceptable computation time. Siochi algorithm could yield better result on example 1 while Chen algorithm yielded better results on example 2 though both intensity maps had the same intensity level and maximum value. Conclusions: A unique and effective tool for optimizing the leaf sequencing that can yield shorter delivery time and higher utilization of photons by comparing with the total number of monitor units and number of segments.

KEYWORDS: radiotherapy, monitor unit, segment number.

1 INTRODUCTION

The purpose in the radiation therapy is to irradiate the tumor as efficiently as possible without harming the surrounding organs. To achieve this goal, a special technique, called intensity modulated radiation therapy (IMRT), has been used [1, 2]. Multileaf collimator (MLC) is one of the essential equipments when IMRT is executed. MLC is composed of several pairs of metal leaves, and it can achieve the objective of intensity modulated by series moving of leaves [3]. Although the medical radiation therapy has been studied for a long time, the efficiency of MLC operations can be further improved [4, 5].

Considering the delivery methods of IMRT, the researches about MLC leaf sequencing can also be classified into two categories: for static MLC and for dynamic MLC [6, 7]. For dynamic MLC, an intensity modulate beam is realized by dynamically moving MLC leaves across a treatment field while the radiation beam is on. The advantage of the dynamic technique is its delivery efficiency [8]. For static MLC, always called "step and shoot" technique, MLC leaves move in discrete steps with the beam off from one setup of positions to the next. There are many advantages of this technique, including easy verification, precise dose delivery, and general availability. There are advantages and disadvantages of both two techniques, however, according to Xia's analysis, static MLC is superior to dynamic MLC in IMRT in general [9].

In this paper, the IMRT leaf sequencing calculation tool based on Chen, Bortfeld and Siochi algorithms was constructed. We compared the results of the three aforementioned typical leaf sequencing algorithms so as to choose the suitable one by the clinical demands.

2 METHODS

An MLC consists of metal leaves which can block the radiation. To each row of the intensity matrix there is associated a pair of leaves, a left leaf and a right leaf that can be moved in the direction of the row. The leaf positions of the collimators were modeled by certain 0-1-matrices called segments, where a 1-entry corresponds to a bixel receiving radiation and a 0-entry to a bixel covered by a leaf. This problem could be formulated as decomposing a given m×n integer matrix into a positive linear combination of (0, 1) matrices with the strict consecutive 1's property in rows. The mathematical problem to solve was formulated by the following linear equation:

$$\sum_{i=1} u_i S_i \quad (1)$$

For medical and practical reasons, one aimed at minimizing the total irradiation time and the number of segments, and the segmentation problem could be formulated as follows:

Instance: A nonnegative integer m*n-matrix A.

Problem: Find a segmentation such that:

Total number of monitor unit (TNMU): $\sum_{i=1}^{k} u_i \rightarrow \min$

The number of segments (NS): $k \rightarrow \min$.

However, it was proved that the minimization of the number of homogeneous fields that are needed is an NP–complete problem[1, 13]. That means no single algorithm is the most efficient for all clinical cases or intensity maps [14]. This suggests that it is desirable to have multiple algorithms available in the ARTS which will search through all embedded algorithms automatically and find the most efficient delivery sequence for a given intensity matrix[4, 15].

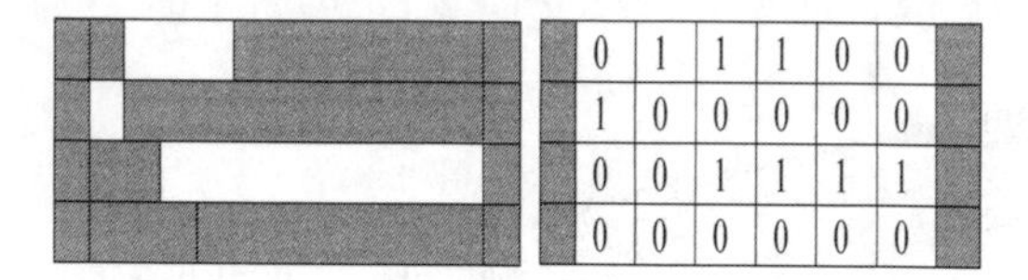

Figure 1. Segment of multi-leaf collimator and corresponding intensity Matrix.

3 THE ALGORITHM OF CHEN

The authors propose a heuristic algorithm which aims at finding a segmentation with a small NS [16]. As several of the algorithms below, it works according to the following general strategy: depending on the given matrix A, a coefficient u > 0 and a number of segments $S_1, S_2, \ldots, S_t$ are determined such that $A' = A - u(S_1 + S_2 + \ldots + S_t)$ is still nonnegative, and then the algorithm is iterated with A' instead of A. It is always clear that this yields a segmentation of A.

4 THE ALGORITHM OF BORTFELD ET AL.

In the method of Bortfeld et al., the intended modulated field was divided into narrow strips each corresponding to a leaf trajectory [17]. It was assumed that prior to the leaf segmentation process the desired modulated fluency profile f (x) along each strip was already known. The optimization process that determined the ideal fluency was based on a finite calculation grid and hence f (x) was spatially discrete.

5 THE ALGORITHM OF SIOCHI

The algorithms presented by Siochi form the basis of the commercial leaf sequencing software IMFAST (Siemens Medical Systems, Oncology Care System, Concord, CA)[18]. The core algorithms are based on two methods, one is termed "rod-pushing" which is analogous to the sliding window method of Bortfeld et al. and the other termed "extraction" which is similar to the method of Galvin et al.. In the methods the intensity map, A, consisted of elements or bixels whose value or intensity was proportional to the amount of time the bixel was open to the beam [19].

6 RESULTS

From the CORVUS treatment planning system, we got two random clinical intensity matrices of oropharyngeal cancer which have slight differences. These two matrices have same maximum value and dimensions. But the results of each algorithm are quite different which are shown in Table I and II.

Intensity matrix 1 and 2 have the same maximum value and dimensions, but after the calculation of Galvin, Bortfeld and Siochi algorithm, we have obviously yielded different results as shown in Table I and II. According to the matrix 1, the user prefers to shorten treatment time, so as to make the shortest switch time of MLC apertures, the Siochi algorithm is the suitable choice. If we choose the smallest TNMU, we can choose Bortfeld algorithm. While on the matrix 2, the trade off between the NS and TNMU is the result of Siochi algorithm.

Table 1. Results of Example 1.

Algorithm	NS	TNMU
Chen	15	76
Bortfeld	48	48
Siochi	25	48

Table 2. Results of Example 2.

Algorithm	NS	TNMU
Chen	26	80
Bortfeld	48	48
Siochi	15	48

Theoretically, narrower strips and more segments can bring finer intensity modulated effect, and make a more homogeneous distribution in the target area and less in the surrounding area. But the static modulated process is unlike the dynamic process, it requires extensive knowledge of MLC leaf positioning accuracy, precision, and long-term reproducibility.

To further study the effectiveness and performance of our compared results, this program is designed to use the intensity maps as input, and the ARTS Treatment System as the target delivery system. We convert the result based on the three algorithms into MLC control files and drive the MLC to work in a static modulated radiation way [20]. The result is in Table III and IV. This column of delivery time in Table III and IV is recorded by the test of MLC whilst the radiation beam is switched off and the beam switched on for variable time with the leaves set to be at each location.

Consider that clinical medical Linac dose rate is usually to be 300 MU/min, the shortest treatment plan is based on Siochi algorithm and the actual time is $135+48\times60/300=144.6$ seconds. The smallest TNMU plan is based on the Bortfeld algorithm, and actually time is $420+48\times60/300+429.6$ seconds.

The similar analysis can be done for example 2, the shortest treatment plan is based on Siochi algorithm and the smallest TNMU plan is based on Bortfeld or Siochi algorithm.

7 DISCUSSION

From the result of matrix 1, the decrease of the number of segments has huge impact on delivery time that can be reduced to 67%. From both the results of matrix 1 and 2, this combined approach will be more efficient than using any single algorithm alone.

Considering that IMRT treatment planning systems such as ARTS generates hundreds of treatment plans before obtaining the optimal plan, this idea of generating many delivery sequences using different algorithms and then picking the best one seems very reasonable, and the additional computing time required is negligible. If the method is implemented clinically, it could result in significant savings in treatment delivery time based on the minimum value number of segments, and also result in significant reduction in the wear-and-tear of MLC mechanics.

Starting point for the calculations is the optimized intensity matrix for all leaf pairs, expressed in monitor units and calculated by inverse treatment planning techniques, taking into account off-axis variations in the primary beam fluence profile of the accelerator. A fluence of 1 MU corresponds to the fluence that results in a dose delivery of 1 cGy at the depth of dose maximum in water for a static 10x10 cm2 field using a source to surface distance of 100 cm. It means that the algorithm which yield the minimum value of MU have higher utilization of photons of linac.

It is assumed that the relative intensity value inside the segment is only 0 or 1. However, the overlaps of bixels that are adjacent along a column receive the smaller one of the doses delivered to the overlapping bixels [21, 22]. The dosimetric characteristics analysis of this issue will be discussed in the following papers.

8 CONCLUSION

An approach on the comparison of three typical leaf sequencing algorithms based on the number of segments and total monitor unit was constructed. The results of this leaf sequencing program can yield shorter delivery time and higher utilization of photons by comparing with the total number of monitor units and number of segments of aforementioned three leaf sequencing algorithms. This program module has been integrated into ARTS treatment planning system [2-4], which takes the intensity map as input so as to pick up the effective strategy to optimize the leaf sequencing process and have a promising future in the field of IMRT.

ACKNOWLEDGEMENT

This work was supported by the Key Foundation for Young Talents in College of Anhui Province No.2013SQRL063ZD, the Key Project of Natural Science in Colleges of Anhui province No.KJ2010A283, Central University research funding No.J2014HGXJ0094, a University Teaching Research project of Anhui Quality Engineering and the SRF for ROCS, SEM.

REFERENCES

[1] T. Bortfeld, "IMRT: a review and preview," Phys. Med. Biol. UK, vol. 51, no. 13, pp. R363–379, Jul, 2006.

[2] Y. C. Wu, G. Song, R. F. Cao, A. D. Wu, M. Y. Cheng *et al.*, "Development of Accurate/Advanced Radiotherapy Treatment Planning and Quality Assurance System (ARTS)," Chinese Physics C, China, vol. 32, pp. 177–182, 2008.

[3] JING Jia, XU Yuan-ying, XU Liang-feng, and LIN Hui, "Multileaf collimator in radiotherapy," Modern Physics, China, vol. 21, no. 03, pp. 28–30, 2009.

[4] J. Jing, R. Cao, Y. Wu, G. Li, H. Lin *et al.*, "Improved Model on Minimizing Static Intensity Modulation Delivery Time." 2nd International Conference on Biomedical Engineering and Informatics, Tianjin, China, Oct.2009, pp. 939–942.

[5] A. Sladowska, "Modern External Beam Radiotherapy Techniques - Intensity Modulated Radiotherapy," Acta Physica Polonica A,USA, vol. 115, no. 2, pp. 586–590, 2009.

[6] D. Carlson, "Intensity modulation using multileaf collimators: current status," Med. Dosim., Elsevier, vol. 26, no. 2, pp. 151–156, 2001.

[7] A. L. Boyer, E. Brian Butler, T. A. DiPetrillo, Mark J. Engler, Benedick Fraass *et al.*, "Intensity-modulated radiotherapy: current status and issues of interest," Int. J. Radiat. Oncol. Biol. Phys., Elsevier, vol. 51, no. 4, pp. 880–914, 2001.

[8] G. A. Ezzell, J. M. Galvin, D. Low, J. R. Palta, I. Rosen *et al.*, "Guidance document on delivery, treatment planning, and clinical implementation of IMRT: Report of the IMRT subcommittee of the AAPM radiation therapy committee," , Med. Phys., Mumbai India, vol. 30, no. 8, pp. 2089–2115, 2003.

[9] P. Xia, J. Y. Ting, and C. G. Orton, "Segmental MLC is superior to dynamic MLC for IMRT delivery," Med. Phys, Mumbai India, vol. 34, no. 7, pp. 2673–2675, 2007.

[10] H. Lin, Y. Wu, and Y. Chen, "'A finite size pencil beam for IMRT dose optimization' - A simpler analytical function for the finite size pencil beam kernel," Phys. Med. Biol., UK, vol. 51, no. 6, pp. L13-L15, 2006.

[11] A. Wu, and Y. Wu, "Effect of CT image-based voxel size on EGSnrc Monte Carlo dose calculation," He Jishu/Nuclear Techniques, Shanghai ,vol. 30, no. 2, pp. 143–146, 2007.

[12] R. Cao, G. Li, and Y. Wu, "A self-adaptive evolutionary algorithm for multi-objective optimization," Lecture Notes in Computer Science (including subseries Lecture Notes in Artificial Intelligence and Lecture Notes in Bioinformatics). pp. 553–564.

[13] T. Kalinowski, "Reducing the number of monitor units in multileaf collimator field segmentation," Phys. Med. Biol., UK, vol. 50, no. 6, pp. 1147–1161, 2005.

[14] G. Wake, N. Boland, and L. Jennings, "Mixed integer programming approaches to exact minimization of total treatment time in cancer radiotherapy using multileaf collimators," Computers and Operations Research, Elsevier, vol. 36, no. 3, pp. 795–810, 2009.

[15] J. Seco, P. Evans, and S. Webb, "Analysis of the effects of the delivery technique on an IMRT plan: comparison for multiple static field, dynamic and NOMOS MIMiC collimation," Phys. Med. Biol., UK,vol. 46, no. 12, pp. 3073–87, Dec, 2001.

[16] Chen DZ, Engel, K and Wang, C, " A New Algorithm for a Field Splitting Problem in Intensity-Modulated Radiation Therapy,"Algorithmica, vol. 21, no. 3, pp. 656–673, Nov,2011.

[17] T. Bortfeld, D. Kahler, T. Waldron, and A. Boyer, "X-ray field compensation with multileaf collimators," Int. J. Radiat. Oncol. Biol. Phys., Elsevier, vol. 28, no. 3, pp. 723–730, Feb, 1994.

[18] R. A. C. Siochi, "Modifications to the IMFAST leaf sequencing optimization algorithm," Med. Phys., Mumbai India, vol. 31, no. 12, pp. 3267–3278, Dec, 2004.

[19] R. A. C. Siochi, "Minimizing static intensity modulation delivery time using an intensity solid paradigm," Int. J. Radiat. Oncol. Biol. Phys.,Elsevier, vol. 43, no. 3, pp. 671–680, 1999.

[20] J. Jing, R. Cao, X. Pei, and Y. Wu, "A program to convert Intensity fluence Map into Optimal MLC machine files." 4th WACBE World Congress on Bioengineering, Hong Kong, China, July 2009, pp. 35–36.

[21] J. Bayouth, D. Wendt, and S. Morrill, "MLC quality assurance techniques for IMRT applications," Med. Phys., Mumbai India, vol. 30, pp. 743–750, 2003.

[22] C. Kwok, G. Lam, and S. El-Sayed, "Suitability of using multileaf collimator (MLC) for photon field matching," Med. Dosim., Elsevier, vol. 29, no. 3, pp. 184–195, 2004.

Computing, Control, Information and Education Engineering – Liu, Sung & Yao (eds)
© 2015 Taylor & Francis Group, London, ISBN: 978-1-138-02800-5

Study on freezing resistance of curing chlorine saline soil

Bo Peng, Guang Kai Yin & Chun Li Cai
School of Chang'an University, Xi'an, China

Wen Ying Li
Xi'an Highway Research Institute, Xi'an, China

ABSTRACT: At present, mass attenuation rate index is a parameter of the research methods of frost resistance to determine the stability of a reinforcement program antifreeze. In the actual project, the strength of the sub grade soil stability under the conditions of freeze-thaw cycles of nature are more considered. This paper puts forward the number of freeze-thaw stability, an antifreeze stability factor as a judgment reinforced soil frost indicator of good or bad performance. The results show that: the number of freeze-thaw stable, antifreeze stability factor can better reflect the variety of reinforced soil frost resistance, which is good or bad and is representative.

KEYWORDS: frost resistance, the number of freeze-thaw stable, antifreeze stability factor.

1 INTRODUCTION

The roadbed of the Shiozawa land area is bound to be subjected to repeated freeze-thaw cycles of action in winter and spring, resulting in the migration of soil moisture, undermining the stability of the roadbed. For this reason, the need for anti-freeze-thaw solidified saline soil properties were studied.

2 TEST METHODS

Inorganic binder stabilized soil frost resistance test is the ability to resist deformation damage under the freeze-thaw cycles.. Because there is sort of standard performance to evaluate solidified soil freeze-thaw method, so we use concrete test specification methods to conduct a variety of solidified soil freezing and thawing tests.

Referencing temperature of Tianjin Binhai New Area, freezing and thawing temperatures of -15°C and -25°C were selected. After 24 h saturated, freeze-thaw tests began. Samples put into -15 °C cryogenic refrigerator12h;and then soaked in a standard curing box (temperature 25 °C, humidity 95%) 12 h was a cycle. The cycle is repeated until the specimen damaged. At the same time, unconfined compressive strength of saturated was measured, to study various solidified saline soil strength variations of freezing and thawing cycles. Finding freeze-thaw stable times of different proportions and different ages and calculating antifreeze stability factor to study frost resistance of various curing program. This paper is better for the future improvement and design saline soil frost resistance test provides a parameter basis.

3 TEST RESULTS

Tianjin Binhai New Area typical chlorine saline soil is used. Four kinds of reinforcement scheme of frost resistance simulation tests were based on lime 6%, cement 5%, lime 6% and fly ash 6%, lime 6% and cement 4%°The test results are shown in Tables 1 through 4.

Table 1. Lime 6% instars intensity relational table with freeze-thaw cycles.

	Freeze-thaw cycles											
Strength	0	1	2	3	4	5	6	7	8	9	10	11
7days compressive strength	0.8	0.241	0.091	0.063	0.058	0.050	0.043	0.033	0.022	0.009	/	/
14days compressive strength	1.0	0.541	0.443	0.241	0.173	0.121	0.091	0.077	0.063	/	/	/
28days compressive strength	1.457	0.650	0.555	0.418	0.377	0.254	0.213	0.159	0.132	0.118	0.107	/
60days compressive strength	1.696	1.042	0.754	0.595	0.432	0.282	0.241	0.213	0.200	0.186	0.159	0.118

Table 2. Cement 5% different ages intensity changes with freeze-thaw cycles of relational tables.

	Freeze-thaw cycles								
Strength	0	1	2	3	4	5	6	7	8
7days compressive strength	0.95	0.404	0.132	0.063	0.050	0.036	0.022	/	/
14days compressive strength	1.05	0.677	0.391	0.213	0.104	0.084	0.053	/	/
28days compressive strength	1.30	0.786	0.664	0.432	0.350	0.268	0.173	0.118	/
60days compressive strength	1.52	0.986	0.853	0.678	0.532	0.396	0.264	0.232	0.176

Table 3. Lime 6% and fly ash 6% change in the relationship between the intensity of different ages table with freeze-thaw cycles.

	Freeze-thaw cycles												
Strength	0	1	2	3	4	5	6	7	8	9	10	11	12
7days compressive strength	1.02	0.527	0.336	0.268	0.152	0.104	0.091	0.057	/	/	/	/	/
14days compressive strength	1.26	0.786	0.555	0.377	0.309	0.268	0.241	0.145	0.118	0.104	/	/	/
28days compressive strength	1.565	1.441	1.182	1.087	0.882	0.623	0.514	0.418	0.295	0.186	0.173	/	/
60days compressive strength	1.825	1.705	1.566	1.250	1.155	0.909	0.609	0.527	0.486	0.350	0.302	0.282	0.241

Table 4. Lime 6% + cement 4% change in the relationship between the intensity of different ages table with freeze-thaw cycles.

	Freeze-thaw cycles												
Strength	0	1	2	3	4	5	6	7	8	9	10	11	12
7days compressive strength	1.325	1.187	0.882	0.445	0.404	0.377	0.323	0.295	0.241	0.213	0.173	/	/
14days compressive strength	1.728	1.500	1.223	0.923	0.827	0.650	0.541	0.473	0.418	0.377	0.364	0.323	0.118
28days compressive strength	2.773	1.782	1.523	1.305	1.114	0.827	0.636	0.582	0.527	0.459	0.418	0.377	0.254
60days compressive strength	3.128	1.942	1.732	1.543	1.372	1.121	0.983	0.763	0.645	0.524	0.432	0.398	0.301

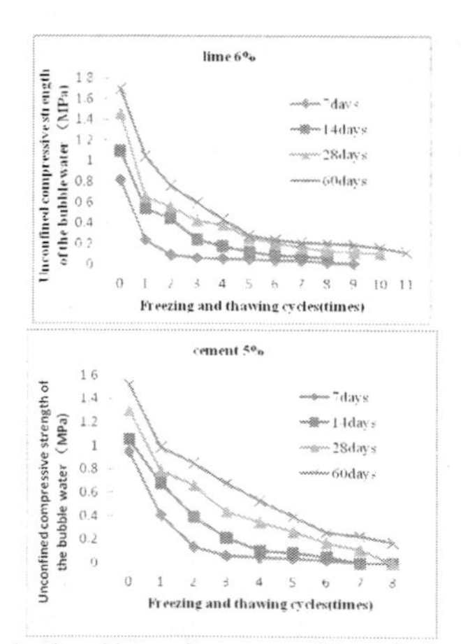

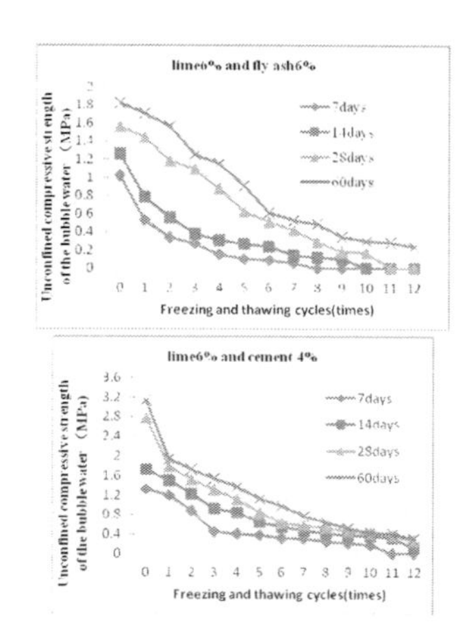

Figure 1. Ratio of intensity maps with each freeze-thaw cycles curve.

We can conclude from the test results, a variety of programs solidified saline soil reinforced with increasing intensity of freeze-thaw cycles were significantly lower. Among them, the lime 6% and 5% cement early strength program fall faster, but the decline rate is relatively moderate of late strength. Lime 6% and 6% fly ash, lime 6% and 4% concrete programs of age strength decline rate are more balanced.

4 FROST RESISTANCE RESEARCH

Currently, the main method of research frost resistance is the quality of the decay rate index as a parameter to measure the stability of some antifreeze reinforcement program. In the actual project, we are more concerned about the strength of the subgrade soil stability under the freeze-thaw cycles in nature. Therefore, this article intends to use the strength of the decay rate method to study the various reinforcement schemes (Equation 1-1).

$$m_i = \frac{P_i - P_0}{P_0} \times 100 \qquad (1\text{-}1)$$

Where: m_i —After the i th freeze-thaw cycles intensity relative to the intensity of the decay rate of the initial value (%);

P_0—Without the freeze-thaw test of the strength of the initial value (MPa);

P_i—Intensity values corresponding to the number of freeze-thaw stable at different ages (MPa).

But how to choose i a unified criterion is the focus of our concern. For example: The strength decay rate of reinforcement scheme A with the freeze-thaw cycles intensity is too fast at first, while the late strength decay rate tends to be higher stable value; The strength decay rate of B with the freeze-thaw cycles intensity is too slow at first, but the late strength decay rates tends to be higher stable value. How to judge two reinforcement frost resistance is good or bad? If you select less freezing and thawing times, ignoring the effects of freeze-thaw cycles of high intensity stabilized soil; while blindly select high freeze-thaw cycles, which will cause unnecessary workload and increase labor time. The reason is that with the increase in freeze-thaw cycles, the intensity decay rate tends to be stable, and the results are not significantly affected. We put forward solidified soil frost resistance index ,namely the number of freeze-thaw stability and antifreeze stability factor.

4.1 *Freeze-thaw stability factor*

When the saline soil in different ages with different curing programs after several freeze-thaw cycles, the strength tending to stabilize, the number is defined as the number of freeze-thaw stability in this age period.

By testing different proportions of different ages, we get freeze-thaw stable times, specifically as shown in Table 5.

Table 5. Different proportions of different ages freeze-thaw cycles.

Ratio	Age	The number of freeze-thaw stability	Ratio	Age	The number of freeze-thaw stability
Lime 6%	7days	3	Cement 5%	7days	3
	14days	5		14days	4
	28days	7		28days	6
	60days	7		60days	6
Lime 6%+Fly 6%	7days	3	Lime 6%+ Cement 4%	7days	3
	14days	4		14days	6
	28days	6		28days	6
	60days	9		60days	10

From Table 5, each mix seven days' the number of freeze-thaw stability is three times; 14 days' the number of financial stability is about five times; 28 days' the number of financial stability is basically six times; 60 days' financial stability is 8 times or so. In order to establish a uniform evaluation criteria, we recommend three times as the number of freeze-thaw stability of seven days using, 14 days using five times, 28 days with 6 times, 60 days using 8 times.

4.2 *Frost stability factor*

Due to the number of freeze-thaw stability does not directly reflect the various curing solutions frost resistance is good or bad, here introducing antifreeze stability factor K for evaluation, to characterize the strength and stability of the attenuation after freezing and thawing conditions, seeing equation 1-2

$$K = \frac{P_i}{P_0} \times 100 \qquad (1\text{-}2)$$

Where: K—Antifreeze stability factor (%);

P_0—The initial value of the intensity without freeze-thaw test (MPa);

P_i—Intensity values corresponding to the number of freeze-thaw stability under different ages (MPa).

Table 6. Different proportions of different ages frost stability factor.

Ratio	Age	The number of freeze-thaw stability	Ratio	Age	The number of freeze-thaw stability
Lime 6%	7days	7.76%	Cement 5%	7days	6.63%
	14days	11.10%		14days	8.00%
	28days	14.62%		28days	13.31%
	60days	12.56%		60days	11.58%
Lime 6% and fly ash 6%	7days	26.27%	Lime6% and cement 4%	7days	33.58%
	14days	21.27%		14days	37.62%
	28days	32.84%		28days	22.94%
	60days	26.63%		60days	20.62%

Stability factor is defined by frost shows that the larger the coefficient values, frost resistance of this kind of curing solution, the better. As can be seen from the table: two reinforced concrete plan of lime 6%,and cement 5% antifreeze stability factor with different age is significantly lower than their peers on two antifreeze stability factor kinds reinforcement program of lime 6% + 6% fly ash and lime 6% + cement 4%. Early frost stability coefficient of the reinforcement scheme fly ash lime 6% + fly 6% relative to the reinforcement scheme cement lime 6% + 4% is lower, but the late strength is higher. This is because the cement improves the early strength of the strengthening scheme of cement + lime, with the consumption of the reaction of cement, the strength tending to increase slowly. Strengthening scheme of lime + fly ash class reinforcement reaction is slow, relatively low early strength. As the reaction proceeds, the intensity of the latter can be improved to enhance the antifreeze stability of fly ash and lime.

5 CONCLUSION

1 The number of freeze-thaw stability is first proposed as a frost resistance evaluation. The results shown that: a variety of mix's freeze-thaw stability with the age of the number 7 days with 3 times, 14 days of age using five times, 28 days of age using six times, 60 days age with 8 times.

2 First proposing frost stability facto as the frost resistance evaluation:

$$K = \frac{P_i}{P_0} \times 100$$

The results showed that: frost resistance index of reinforcement schemes lime 6% + fly ash 6% and lime 6% + cement 4% are better than the reinforcement programs lime 6%,and cement 5%; early frost resistance of lime 6%+fly ash 6% is poorer than the lime 6% + cement 4%, but late is better.

REFERENCES

[1] JTG E30-2005, Concrete testing procedures Highway Engineering cement and cement[S].Beijing: China Communications Press,2005.

[2] Chen Yuan-zhao,Li Zhen-xia. Saline soil engineering properties test[J]. Highway traffic science and technology,2012,29(12):1–6.

[3] Li Qi. Experimental study on coastal saline soils highway embankment filler[J]. Chinese Journal of geotechnical engineering,2005,28(9):1177–1180.

[4] Wang Jian,Ni Wen.Experimental study of chlorine saline soil reinforcement several industrial solid waste[J]. Journal of China & Foreign Highway,2011(4):278–283.

[5] Bao Wei-xing,Yang Xiao-hua. Saline soil under freezing conditions shear strength characteristics of experimental research[J]. Highway,2008(1):5–10.

[6] Wang Jun-sheng. Experimental study on saline soil used as roadbed filler coastal area[J]. Highway traffic science and technology,2006(7):31–33.

[7] Bao Wei-xing,Li Zhi-nong. Testing on transmutation properties of saline soil under freezing and thawing cycles in Kashi, Xinjiang[J].Journal of Changpan University(Natural Science Edition),2008,28(2):26–30.

[8] Li Ying-tao,Yuan Lu. Microscopic study of the mechanism of saline soil strength formation of cement modified[J]. Highway,2013(3):164–168.

Computing, Control, Information and Education Engineering – Liu, Sung & Yao (eds)
© 2015 Taylor & Francis Group, London, ISBN: 978-1-138-02800-5

Research on routing technology of PCE-based multi-layer multi-domain optical network

Xiu Yan Wang, Jian Ling Zhou & Jun Xu
College of Electronical and Information Engineering, North China Institute of Science and Technology, Hebei, China

Xiang Xia Meng
College of Mechanical and Electrical Engineering, North China Institute of Science and Technology, Hebei, China

ABSTRACT: First, make a survey on the developing situation of the optical network, then analyze the characteristics of multi-layer multi-domain optical network. Through the research on the architecture of Path Computation Element (PCE) and method of routing computing, we discuss the existing routing technology based on PCE in the multi-domain network. Finally, based on the research on the mechanism of the PCE, we propose a kind of routing computing scheme combined BRPC with Intermediate domain recursive PCE-based computation (IRPC).

KEYWORDS: multi-domain network; multi-layer network.

1 INTRODUCTION

Optical network has carried more than 80% telecommunication traffic. In order to efficiently manage network resources and provide adequate quality of service (QoS), optical network has developed rapidly in recent years, it has experienced from SDH/SONET to OTN, from OTN to IP over WDM, to MPLS, to GMPLS, to ASON, to WSON. New communication services (e.g., VOIP, IPTV, VOD, cloud computing, and very high speed Internet access) and high-speed bandwidth requirements need more intelligent and higher speed optical network. It has evolved into a multi-layer, multi-domain intelligent optical network. The transmission rate of optical network increases rapidly, it is Mbps in 1985, now it can reach Tbps. With the rapid development of optical network, optical network need to face new challenges: the coexisting of multi-granularity services, the sharing conflicts of network resource by a large number of routing information and signaling, etc. Therefore, there is a need to have corresponding multi-layer multi-domain routing technology so as to achieve rapid intelligent automatic optical switching.

In order to compute a path efficiently and achieve the goal of traffic engineering (TE), the Internet Engineering Task Force (IETF) has promoted the Path Computation Element (PCE) architecture in RFC4655, and proposed a network entity to compute the path. Since then, a variety of different routing computation methods based on PCE have been proposed. A per-domain path computation technique is proposed in and applied in the establishment of label switched paths (LSP) in [2]. A standard backward path computation method is exploited in [3]. A Backward-Recursive PCE-Based Computation (BRPC) Procedure to Compute Shortest Constrained Inter-Domain Traffic Engineering Label Switched Path has been proposed.

2 MULTI-LAYER MULTI-DOMAIN OPTICAL NETWORK

2.1 *Multilayer optical network*

According to the definition of the IETF, multi-layer network (MLN) is a network with one or more kind of switching ability which is controlled by the unified control plane, and can have multiple datum planes, and can support the control of traffic engineering. In practical applications, the multi-layer network can be divided into three layers (as shown in Figure 1 (a)) from the actual physical point of view, and can also be divided into four layers (as shown in Figure 1 (b) [5]) according to the protocol.

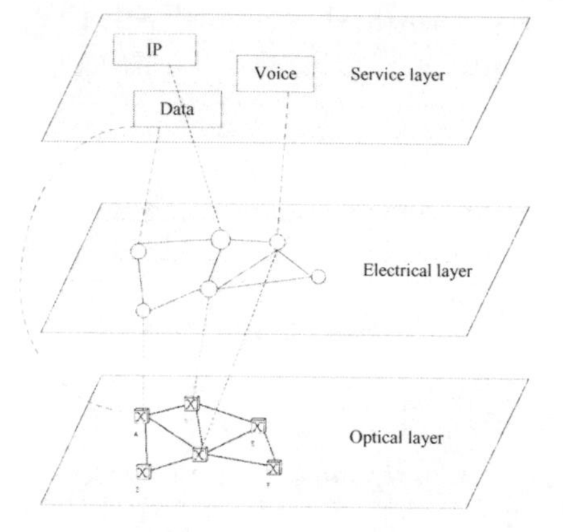

Figure 1. (a) Hierarchical structure of the multilayer network.

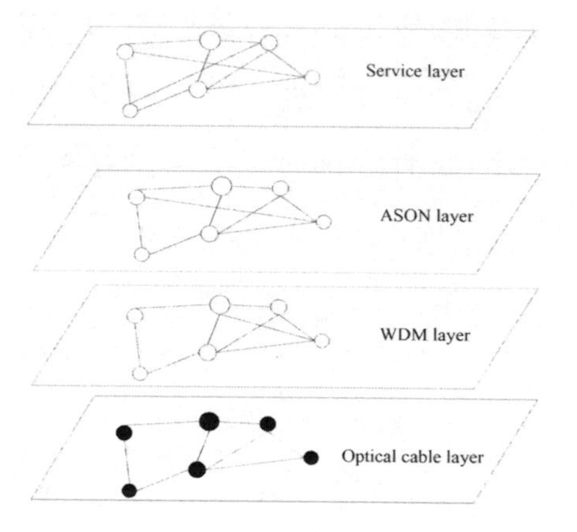

Figure 1. (b) Hierarchical structure of the multilayer network.

2.2 *Multi- domain optical network*

Multi-domain Network (MRN) is composed of two or more than two domains which are controlled by a unified control plane, and each domain is composed of one or more devices. MRN can support the control of traffic engineering and can support a variety of switching types. Each domain is composed of boundary nodes and internal nodes. Boundary node is located in the border, which has link connection with another domain node; Internal node only has link connection with the nodes within the domain. Therefore, a single domain must include the boundary nodes, but may not have internal nodes. In addition, according to the link unrelated protection principle, each two domains at least have more than two boundary nodes connected to other domains. A typical multi-domain optical network is shown in Figure. 2.

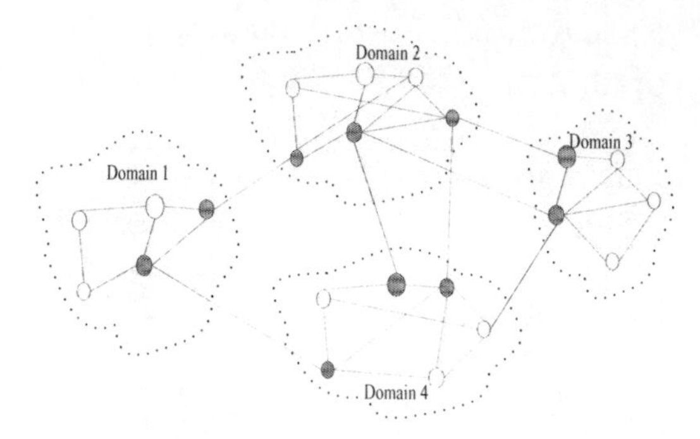

Figure 2. Multi-domain optical network.

3 THE PCE-BASED ARCHITECTURE

PCE has four kinds of working mode, so there are four kinds of PCE-Based Architectures (as shown in Figure 3): Composite PCE, External PCE, Multiple PCE Path Computation, Multiple PCE Path Computation with Inter-PCE Communication.[1]

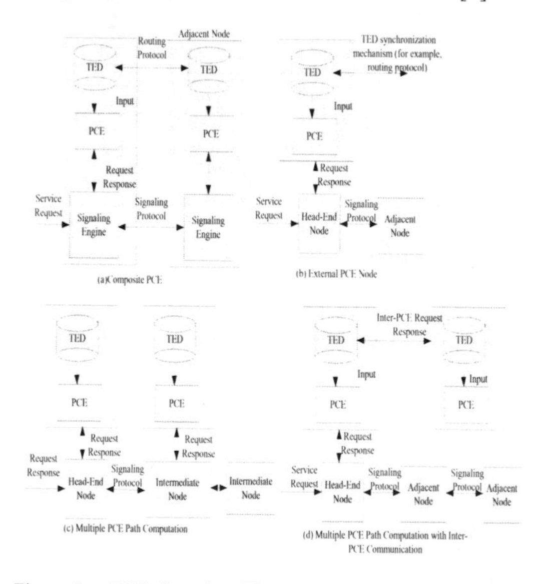

Figure 3. PCE- based architecture.

4 MULTI-DOMAIN PATH COMPUTATION MECHANISM BASED ON PCE

4.1 *Per-domain path computation technique*

In the process of per-domain path computation, each domain is assumed to have a PCE responsible for computing a segment of path exclusively inside that domain. Through this kind of path computing process information from the source node reaches the destination node, and it forms a complete route.

4.2 *Backward Recursive PCE-based Computation (BRPC)*

The BRPC mechanisms obtain the constrained shortest path by the cooperative computing among all the relevant PCEs. Path computation is not starting from the source node domain, but starting from the destination node domain. The BRPC procedure includes two stages according to directions: first, Search the destination node domain using signaling; then, compute the shortest path from the destination domain. [4]

4.3 *An efficient path computation mechanism based on PCE*

Based on the research about the present situation of multi-domain optical network, we propose a combined scheme that uses BRPC combined with intermediate domain recursive PCE-based computation (IRPC) to compute the whole path. This scheme can be used in the condition of known domain sequences in advance. The procedure is shown in Figure 4.

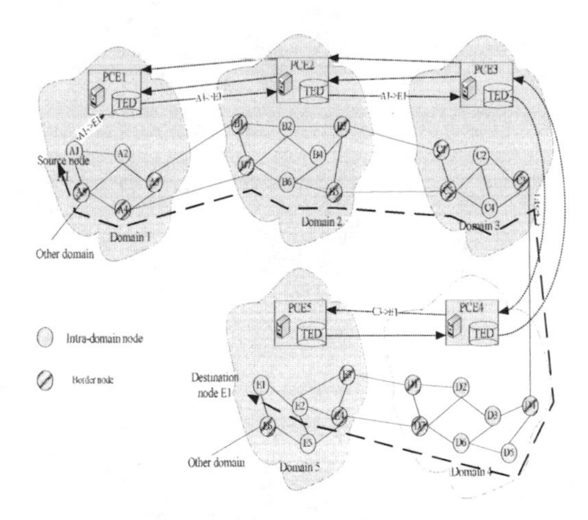

Figure 4. BRPC combine with intemediate domain recursive PCE- based computation (IRPC).

Its working procedure is as follows: first, the source node A1 requests a path to E1 to PCE1 in its domain by sending a PCReq message. PCE1 discover that its destination node E1 is not in its domain, it will forward the request message to the PCEs along the domain path, until it reaches the PCE3. PCE3 discover the C3 border node, though by looking for the traffic engineering databasc (TED), and then, on the one hand PCE3 forward the request message to the next domain PCE4; on the other hand, PCE3 will begin to reverse back from the domain 3 to compute the shortest path from A1 to C3. When the path request message arrives at PCE5, PCE5 will begin to reverse back to the domain3 to compute the shortest path from E1 to C3. After the two computing process, PCE3 return the shortest path from A1 to E1 back to the PCE1 by signaling.

5 CONCLUSIONS

Based on the analysis of the routing technology and the multi-domain optical network, we gave a practical scheme called IRPC combined with BRPC to compute the shortest path, and analyze the working procedure.

ACKNOWLEDGEMENTS

This work was financially supported by the Science Foundation (DX2013B08, JD1202B) of North China Institute of Science and Technology.

REFERENCES

[1] A. Farrel, J. P. Vasseur, and J. Ash: *A Path Computation Element (PCE)-based architecture,* IETF, RFC 4655, Aug 2006.

[2] J. Vasseur, A. Ayyangar, and R. Zhang: *A per-domain path computation method for establishing inter-domain traffic engineering (TE) label switched paths (LSPs), IETF, RFC 5152,* Feb. 2008.

[3] N. Bithar, R. Zhang, and K. Kumaki: *Inter-AS requirements for the Path Computation Element Communication Protocol (PCECP), IETF,RFC 5376,* Nov. 2008

[4] J. Vasseur, R. Zhang, N. Bitar, and J. Le Roux: *A Backward-Recursive PCE-Based Computation (BRPC) Procedure to Compute Shortest Constrained Inter-Domain Traffic Engineering Label Switched Paths, IETF, RFC 5441,* Apr 2009.

[5] Shanguo Huang, Jie Zhang, and Dahai Han: *Optical network planning and optimization,* Published by Posts and Telecom Press, Beijing(2012), in China.

Computing, Control, Information and Education Engineering – Liu, Sung & Yao (eds)
© 2015 Taylor & Francis Group, London, ISBN: 978-1-138-02800-5

Analysis of the effect of RMT parameters on support time

Ting Peng Li, Yue Li , Yan Ling Qian & Xiao Ma
Science and Technology on Integrated Logistics Support Laboratory, National University of Defense Technology, Changsha, China

ABSTRACT: Support time of equipment has great effect on the operational readiness rate and operational availability. Establishing the quantitative relation of RMT parameters and the support time is important to improve the suitability of equipment. Based on the GSPN model of support progress of equipment, the quantitative relation was established and the effect of RMT parameters at support time was discussed detailed. The analysis results can be references for balancing the RMT parameters in the design phase. Furthermore, a case of certain equipment was introduced to validate the method.

KEYWORDS: Integrated Logistics Support; Support Time; RMTS; GSPN.

1 INTRODUCTION

Support time of equipment includes maintenance support time and operational support time, shortening the support time of equipment is the main method to improve the operational availability [1, 2]. RMT parameters are important factors that affect support time of equipment. At the beginning of the product design, establishing the quantitative relation of RMT parameters and the support time not only can conduct a preliminary estimate for support time, but also can analyze the influence of various RMT parameters of support time. What's more, the analysis results can be referenced for the tradeoff of the RMT parameters. Taking the test and maintenance process as the analysis object, which is the typical and important part of support progress of equipment, and based on the GSPN model of support progress, the quantitative relation of RMT parameters and support time was established, and the effect of RMT parameters at support time was discussed detailed.

2 THE GSPN MODEL OF SUPPORT PROGRESS OF EQUIPMENT

RMT parameters of equipment. Equipment Reliability is the ability to complete the required function under specified conditions and within the stipulated time [3]. Reliability consists of basic reliability and mission reliability [1]. Basic reliability refers to the probability of failure-free under specified conditions and within the stipulated time. Mission reliability refers to the ability to complete stipulated function within the prescribed task profile. At the beginning of the equipment design, the basic reliability of the equipment is the designers' main consideration. Therefore, this article chooses failure rate P_F for reliability parameters, which is the main index of basic reliability of equipment. The failure rate refers to the probability of failure under specified conditions and within the prescribed time.

Equipment maintainability refers to the ability to be repaired, maintained or restored to the specified state with specified conditions, prescribed procedures and methods. Maintainability includes inherent maintainability and operational maintainability [3]. Inherent maintainability is decided in the design phase, so we can also call it design maintainability. Operational maintainability is in close with maintenance strategy, using environment, support delay and some other factors [4]. This article chooses accurate maintenance time: T_{AM} and fuzzy maintenance time (T_{FM}) which is the main parameters of inherent maintainability. The accurate maintenance time refers to the time which is spent on maintenance under the premise that the fault isolation is successes. The fuzzy maintenance time refers to the time which is spent by exploratory maintenance on the condition that the fault isolation failed.

Equipment Testability is a design feature that it can timely and accurately determines the state of equipment and effectively isolate its internal fault

[5]. There are many testability parameters, this paper chooses the five most important testability design parameters: Fault detection rate (P_d), Fault isolation rate(P_i), False alarm rate (P_{FA}), Fault detection time (T_d) and fault isolation time (T_d).

$$\begin{cases} P_d = \dfrac{N_d}{N_A} \\ P_i = \dfrac{N_i}{N_d} \\ P_{FA} = \dfrac{N_{fa}}{N_I} \end{cases} \tag{1}$$

Where, N_d is the fault number which is detected correctly by the rule method within the stipulated time, N_A is the total number of the unit fault which is under test, N_i is the fault number which is isolated properly to the ruling replaceable unit with the provision method during the stipulated time, N_d is the number of false alarm which happened during the stipulated time and N_I is the total number of fault indicators.

Fault detection time means the time of detection and fault isolation time means the time of isolation.

The RMT parameters are shown in Table 1.

Table 1. The RMT parameters of equipment.

reliability	maintainability		testability				
failure rate	accurate maintenance time	fuzzy maintenance time	fault detection rate	fault detection time	fault isolation rate	fault isolation time	false alarm rate
P_F	T_{AM}	T_{FM}	P_d	T_d	P_I	T_I	P_{FA}

The GSPN model of equipment. The concept of GSPN is derived from Stochastic Petri Nets (SPN), and it extends timed and immediate transition conceptions [6-8]. Compared with SPN, GSPN not only needs less calculation, but also can model the uncertain process [9-11]. During the progress of equipment support, some process, such as the results of detection and isolation of faults are not sure. Therefore, GSPN is suitable to model the support process of equipment.

The test and maintenance process is the typical support process of equipment, and its workflow is shown as follows: at the beginning of support, the equipment will be tested in accordance with the requirements for testability, if the test result is trouble-free, then directly to the end. Otherwise, it must be brought into fault isolation and diagnosis process. If the fault isolation success, precise maintenance will be conducted, if not, the equipment must be fuzzy maintained.

The GSPN model of test and maintenance process of equipment is shown as Fig. 1.

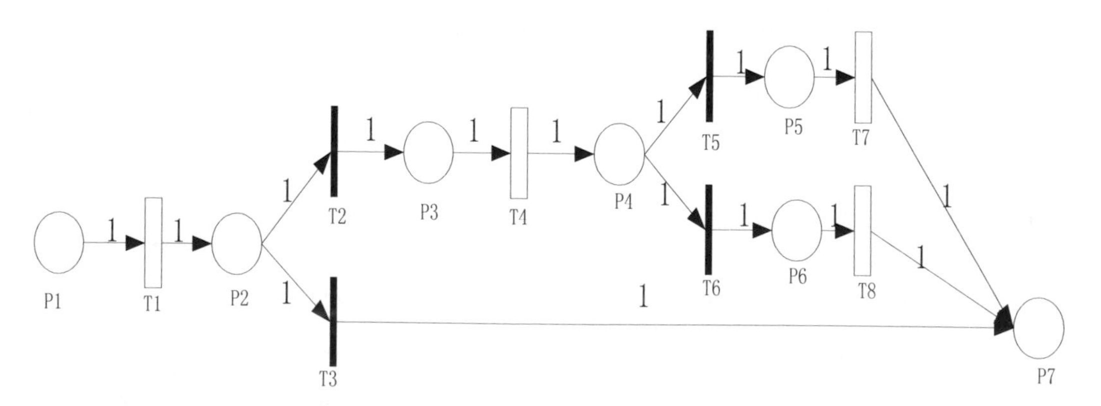

Figure 1. The GSPN model of test and maintenance process of equipment.

There are two kinds of transitions in this model: immediate transition and timed transition. Immediate transition means the transition that has triggered probability and its delay time is zero. Timed transition means the transition that the delay time is not equal to zero while its trigger probability is constant one. The transition type of the model and its practical meanings are shown in Table 2.

The places in the model represent for the state of equipment, and the physical meanings are shown in Table 3.

Table 2. The types and physical meaning of transition of the model.

No.	T1	T2	T3	T4	T5	T6	T7	T8
Type	Timed	Immediate	Immediate	Timed	Immediate	Immediate	Timed	Timed
physical meaning	Fault test	fault	No fault	Fault	Isolation success	Isolation failure	accurate maintenance	fuzzy maintenance

Table 3. The physical meanings of places of the model.

No.	P1	P2	P3	P4
Meanings	Ready for being tested	After tested	Ready for fault diagnose	After fault diagnose
No.	P5	P6	P7	
Meanings	Ready for accurate maintenance	Ready for fuzzy maintenance	Finish testing and maintenance	

In order to highlight the focus of this paper, some assumptions of Fig. 1 model are shown as follows:

1 Ignore the influence of man-made factors to support the process. The number of people and their skills to satisfy the requirements of the supporting scheme.
2 Ignore the influence of spare parts and supporting delay to support time.
3 Ignore the internal time between each work, for example, we neglect the interval time between the fault isolation and maintenance, that is to say, we take maintenance immediately after the fault isolation is done.

3 THE INFLUENCE OF RMT PARAMATERS TO SUPPORT TIME

The parameters of transitions. Time parameter and the probability parameter are the two main parts of the transition parameters in the GSPN model.

The time parameter means the time, which spent in the process of transition. The probability parameter means the probability of transition when the transition trigger conditions are all filled. According to the physical meanings of the timed transition in GSPN model, we can directly calculate the time parameters of timed transition, which are shown in Table 4.

Table 4. The time parameters of the timed transition.

No.	T1	T4	T7	T8
Parameters	T_d	T_I	T_{AM}	T_{FM}

According to the RMT parameters of equipment and the physical meanings, we can directly get the probability parameters of immediate transition T5, T6 are P_I and 1-P_I.

Suppose the probability that the equipment be detected to be fault is P_{TF}; the probability that the equipment have faults and be detected is P_{TF1}; the probability that the equipment have no faults but be detected to false alarm is P_{TF2}. According to the definition of the fault testing rate, obviously:

$$P_{TF1} = P_F \cdot P_d \quad (2)$$

Which P_F is the fault rate and P_d is the testing rate of equipment.

According to the definition of false alarm rate, we can get:

$$P_{FA} = \frac{P_{TF2}}{P_{TF1} + P_{TF2}} \quad (3)$$

Put the Eq. 2 into Eq. 3, then:

$$P_{TF2} = \frac{P_F \cdot P_d \cdot P_{FA}}{1 - P_{FA}} \quad (4)$$

Thus:

$$P_{TF} = P_{TF1} + P_{TF2} = \frac{P_F \cdot P_d}{1 - P_{FA}} \quad (5)$$

The probability parameters of immediate transition of the testing and the maintenance process GSPN model are shown in Table 5.

Table 5. The parameters of immediate transition.

NO.	T2	T3	T5	T6
Parameters	$(P_F*P_d)/(1\text{-}P_{FA})$	$1\text{-}(P_F*P_d)/(1\text{-}P_{FA})$	P_I	$1\text{-}P_I$

The quantitative relation of RMT parameters and support time. Establishing the quantitative relation of RMT parameters and support time not only can make an initial estimate on support time, but also can make an analysis of the influence of the RMT parameters to support time, and the analysis results can be referenced for weighting the RMT parameters.

In the GSPN model of the testing and the maintenance support process of equipment, the mean time ($T_{s\text{-mean}}$) is the time of translating which the state of equipment from P1 to P7:

$$T_{S\_mean} = T_1 + P_{T2} \cdot (T_4 + T_7 \cdot P_{T5} + T_8 \cdot P_{T6}) \quad (6)$$

Taking the data shown in Table 5 into the Eq. 6, we can calculate the average support time:

$$T_{S\_mean} = T_d + \frac{P_F \cdot P_d}{1 - P_{FA}} \cdot (T_I + T_{AM} \cdot P_I + T_{FM} \cdot (1 - P_I)) \quad (7)$$

The influence of RMT parameters to support time. In order to evaluate the influence of the equipment RMT parameters to support time, the transition rate of average support time along with the RMT parameters are calculated with Eq. 7. What's more, the analysis results can be referenced for weighting the RMT parameters in the design phase.

1 The influence of failure rate to mean support time:

$$\frac{\partial_{T_{S\_mean}}}{\partial_{P_F}} = \frac{P_d}{1 - P_{FA}} \cdot (T_I + T_{AM} \cdot P_I + T_{FM} \cdot (1 - P_I)) \quad (8)$$

2 The influence of failure testing rate to mean support time:

$$\frac{\partial_{T_{S\_mean}}}{\partial_{P_d}} = \frac{P_F}{1 - P_{FA}} \cdot (T_I + T_{AM} \cdot P_I + T_{FM} \cdot (1 - P_I)) \quad (9)$$

3 The influence of false alarm rate to mean support time:

$$\frac{\partial_{T_{S\_mean}}}{\partial_{P_{FA}}} = \frac{P_F P_d}{(1 - P_{FA})^2} \cdot (T_I + T_{AM} \cdot P_I + T_{FM} \cdot (1 - P_I)) \quad (10)$$

4 The influence of failure testing time to mean support time:

$$\frac{\partial_{T_{S\_mean}}}{\partial_{T_d}} = 1 \quad (11)$$

5 The influence of failure isolation time to mean support time:

$$\frac{\partial_{T_{S\_mean}}}{\partial_{T_I}} = \frac{P_d P_F}{1 - P_{FA}} \quad (12)$$

6 The influence of accurate maintenance time to mean support time:

$$\frac{\partial_{T_{S\_mean}}}{\partial_{T_{AM}}} = \frac{P_I P_d P_F}{1 - P_{FA}} \quad (13)$$

7 The influence of fuzzy maintenance time to mean support time:

$$\frac{\partial_{T_{S\_mean}}}{\partial_{T_{FM}}} = \frac{(1 - P_I) P_d P_F}{1 - P_{FA}} \quad (14)$$

4 APPLICATION

Determination of RMT parameters. The parameters of reliability, maintainability and testability are determined by the design of equipment, the design RMT parameters of certain equipment are shown in Table 6.

Table 6. The RMT parameters of certain equipment.

P_F(%)	T_{AM}(Min)	T_{FM}(Min)	P_d(%)	T_d(Min)	P_I(%)	T_I(Min)	P_{FA}(%)
10	10	30	95	10	95	5	8

The transition parameters of the GSPN model. Based on the equipment RMT parameters, with Eq. 5, we can calculate the probability that the testing results are failure (P_{TF}) or no failures (P_{TNF}), shown as following:

$$\begin{cases} P_{TF} = \dfrac{P_F \cdot P_d}{1-P_{FA}} = \dfrac{0.1\times 0.95}{1\text{-}0.08} = 0.103 \\ P_{TNF} = 1 - P_{TF} = 0.897 \end{cases} \tag{15}$$

Thus, we can get immediate transition trigger probability of the model and shown in Table 7.

Table 7. The trigger probability of immediate transition.

No.	T2	T3	T5	T6
probability	P_{T2}=0.103	P_{T3}=0.897	P_{T5}=0.095	P_{T2}=0.05

According to the physical meanings of the timed transition and the parameters of RMT, we can directly get the time parameters of timed transition, which are shown in Table 8.

Table 8. The time parameters of timed transition.

No.	T1	T4	T7	T8
Parameters	T_d=10(min)	T_I=5(min)	T_{AM}=10(min)	T_{FM}=30(min)

The average support time of equipment T_s-mean can be calculated with Eq. 7:

$$\begin{aligned} T_{S\_mean} &= T_1 + P_{T2}\cdot(T_4 + T_7\cdot P_{T5} + T_8\cdot P_{T6}) \\ &= 11.65(\text{min}) \end{aligned} \tag{16}$$

The influence of RMT parameters to support time. Taking those data into Eq. 7~ Eq. 10, we can get the influence of the RMT parameters to the support time, which are shown in Table 9.

Table 9. The influence of the RMT parameters to the support time.

failure rate	accurate maintenance time	fuzzy maintenance time	fault detection rate
P_F	T_{AM}	T_{FM}	P_d
16.521	0.098	0.005	1.739
fault detection time	fault isolation rate	fault isolation time	false alarm rate
T_d	P_I	T_I	P_{FA}
1.000	-2.065	0.103	1.796

Making an analysis of the result of Table 9, we can get some conclusions:

1 The influence of the failure rate to the mean support time in the process of testing and maintenance is biggest, for the fault decrease one percent and the mean support time will reduce 16.521 percent.
2 The influence of the accurate maintenance time to mean support time is more obvious than the fuzzy maintenance time. Thus, in order to reduce support time effectively, we should take more attention to shorten the time of accurate maintenance.
3 The increase of the fault isolation rate and the decrease of false alarm rate both will lead to the decrease of the mean support time and the contribution of fault isolation rate will be much more.
4 The influence of the failure testing time to the mean support time is much more than the failure isolation

time. When testability designing, we should devote more resources to shorten the time of fault detection.

5 The decrease of the fault detection rate will lead the mean support time decrease. While, the fault detection rate decrease will lead to the equipment working in the condition of "sick" and this will decrease the operational readiness and operational availability.

The reliability, maintainability and testability parameters equipment should be determined synchronal, so the analysis results shown in this paper can be referenced for weighting and determining the design parameters.

5 CONCLUSION

This paper takes the testing and maintenance process which is the typical processes, the support of equipment as research object. With establishing and solving the GSPN model of the support process, we build the quantitative relation of the RMT parameters and support time. Based on the quantitative relations, this paper analyzed the influence te design parameters of equipment. Finally, an application validated the feasibility and effectiveness of the proposed method.

REFERENCES

[1] Qin Yingxiao. Introduction to The reliability maintainability and supportability [M]. National defense industry press, 2002.

[2] Li Zhishun, Wu Mingxi. Military equipment support Sicence [M]. Beijing: Military science press, 2009.

[3] National military standard of the People's Republic of China (2005). The terms of reliability maintainability and supportability.

[4] Gan Maozhi, Kang jian, gaoqi. Military equipment maintenance engineering [M]. Beijing: National defence industry press, 2005.

[5] Qiu Jing etc. Equipment testability modeling and design technology [M]. Beijing :science press, 2012

[6] Al-Jaar R. Y.. Performance Evaluation of Automated Manufacturing Systems Using Generalized Stochastic Petri Nets[D]. New York:, 1989.

[7] Zhan H. (2006). Study of the Normal Generalized Stochastic Petri nets and its Application in Testing System. In Instrumentation and Measurment Technology Conference (Sorrento, Italy).

[8] Reveliotis S. A Jin Y. C. . A Generalized Stochastic Petri Net Model for Performance Analysis and Control of Capacitated Reentrant Lines[J]. IEEE Transctions on Robotics and Automation, 2003, 19(3):474–480.

[9] H. T. Wang, F. He, H.G. Xiong. Modeling of avionics blueprint architecture based on GSPN and LP [J]. Aerospace Science and Technology. 2013,(26):111–119.

[10] G. Thangamani . Availability Analysis of a Lube Oil System Using Generalized Stochastic Petri Net[C]. Proceedings of the 2011 IEEE ICQR. 2011,186–189.

[11] Elvio Gilberto Amparore. A New GreatSPN GUI for GSPN Editing and CSL^{TA} Model Checking[C]. Proceedings of 11th International Conference, QEST 2014. Florence, Italy. September 8–10. 2014, 170–174.

Computing, Control, Information and Education Engineering – Liu, Sung & Yao (eds)

Parameter correction of ultrasonic horn based on VB platform and finite element method

Hua Wei Ji, Shuang Shuang Zhao & Xiao Ping Hu
School of Mechanical Engineering, Hangzhou Dianzi University, Zhejiang, China

ABSTRACT: With the expanding of ultrasonic applications, the development of ultrasonic machining equipment has become a focus in related fields. As an important component of ultrasonic machining equipment, the ultrasonic horn's properties directly affect machining efficiency, machining quality and the stability of the whole processing system. In order to make the deviation of any order natural frequency and system frequency horn in the allowable range, to ensure the horn's longitudinal vibration effectively, on the basis of theoretical research on ultrasonic horn, the resonance frequency of the ultrasonic horn with the variation of the length of the horn is discussed. The parameter correction of ultrasonic horn was carried out based on the VB platform and the finite element method. The analysis results show that after correction the new resonance frequency of the horn is closer to the working frequency of the system, which proves the validity and reliability of the method.

KEYWORDS: ultrasonic horn; resonance frequency; parameter correction.

1 INTRODUCTION

Ultrasonic machining is a new processing technology. It uses an ultrasonic generator to make the cutting equipment vibrations in ultrasound frequency. The tool's collision and impact energy transfer to the medium, resulting in fatigue fracture of materials to achieve cutting purpose. Ultrasonic horns are an important part of the ultrasonic machining vibration system. Its main role is to enlarge the vibration velocity and displacement and to make the energy concentrate in smaller radiation surface. Its acoustic performance is very important for ultrasonic machining device [1].

The resonance frequency of ultrasonic horn is one of the most important acoustic performances. If the resonant frequency and system frequency deviation is too large, it not only affects the quality of ultrasonic machining, but also leads to ultrasonic vibration system failure, or even damage the ultrasonic generator [2]. The analytical method is a more effective method to study the acoustic performance of the horn. But because of the approximations in wave equation and boundary conditions, the resonant frequency of the horn is often lower than the theoretical value [3]. In order to optimize the performance of the horn, it is necessary to correct parameters of the horn. The influence of structural parameters on the resonance frequency of ultrasonic horn is discussed in this paper. This can provide reference for precise design and site rapid correction. The parameter correction method has practical significance for the horn processing.

2 FREQUENCY CHARACTERISTICS ANALYSIS AND PARAMETER CORRECTION OF ULTRASONIC HORN

The resonant frequency is an important performance parameter of ultrasonic horn.Only when the resonant frequency is equal to the working frequency, it can resonate and can produce greater amplitude at the output. The study found that the resonant frequency is very sensitive about the change of horn length. Figure 1 is the regular relationship between frequency and length of the resonant conical horn when the area coefficient is certain. As can be seen from the figure, in other conditions remain unchanged, with the decrease of horn length, the resonant frequency increases. So, the length of the horn can be changed to adjust the resonant frequency to meet the design requirement.

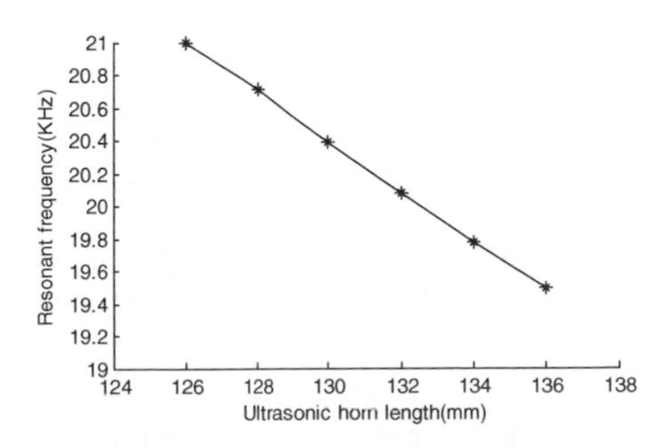

Figure 1. Resonant frequency change with ultrasonic horn length.

By the analytical method, the resonant length of the horn:

$$L = \frac{C}{2\pi f}(Kl)_0 \tag{1}$$

The resonant length correction formula is as follows:

By the formula (1)obtained:

$$L_1 - L_2 = \frac{c}{2\pi f_1}(Kl)_0 - \frac{c}{2\pi f_2}(Kl)_0 \tag{2}$$

Simplification was:

$$\Delta L = \frac{c}{2\pi f_1 f_2}(Kl)_0(f_2 - f_1) \tag{3}$$

In the formula: $K = w/c$ is circular wave number, $c = \sqrt{E/q}$ is the longitudinal wave propagation speed in thin rod, ρ is material density, E is the Young's modulus of the material, f_1 is working frequency, f_2 is longitudinal vibration frequency obtained through finite element analysis or test.

Previous ultrasonic horn correction methods often have long cycle and low design precision. In this study, the powerful computing ability of MATLAB and Visual Basic graphical user interface development was combined. Make full use of the advantages of them to realize the seamless integration of the application systems, effectively shorten the design cycle and improve the design accuracy. MATLAB and Visual Basic integrated in a variety of ways, here is the method with the aid of the ActiveX components. In Microsoft Visual Basic, MATLAB is invoked as an ActiveX component in Visual Basic language through the ActiveX automation interface [4]. The parameter correction interface was as shown in figure 2.

3 PARAMETER CORRECTION EXAMPLE

A parameter correction process based on VB platform and the finite element method is shown in figure 3. A half-wavelength conical horn was designed. Its working frequency was 20kHz and its material was 45#steel. According to product design requirements, big end diameter D_1=50*mm*, small end diameter D_2=30*mm*. According to the traditional wave theory, the resonant length *l*=132.5*mm*.

For a finite element model, when the results of the analysis is not ideal, it is necessary to modify the model size, establish a new finite element model, and then repeat the analysis process. This " design - analysis - modify the design - re-analysis - then modify " process, in finite element analysis, there are a lot of repetitive work, which will directly affect the efficiency of the design. APDL (ANSYS parametric design language) by adjusting the structural design parameters can automatically complete the above cycle, thus greatly reduce time spent on modifying the model and the re-analysis [5]

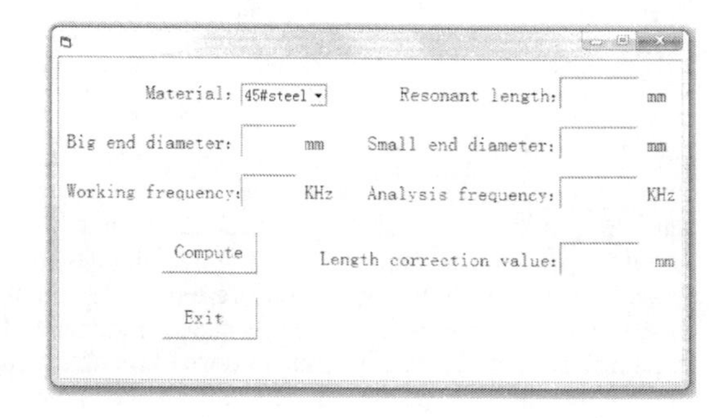

Figure 2. Parameter correction interface.

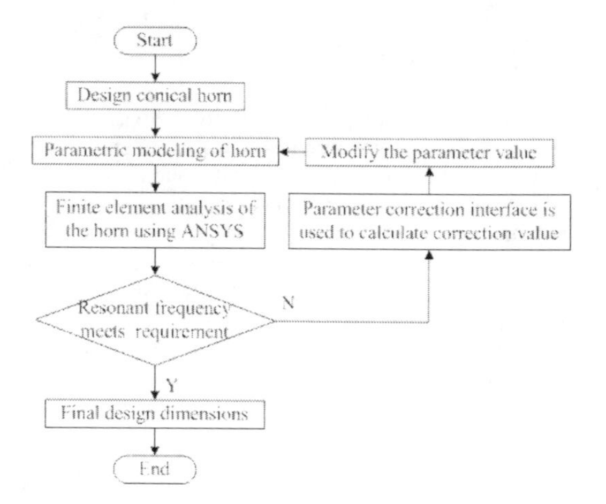

Figure 3. Parameter correction process.

Horn of this paper adopted the bottom-up modeling method. The horn length was parameterized and given the initial value of 132.5mm. Key points, lines, surfaces and body were generated, and then get a conical horn parametric analysis model. The horn was meshed and modal analyzed [6,7].

After modal analysis, 4 modal frequencies were obtained, as shown in figure 4. Read each mode in turn and entered the post processor to analysis the vibration performance. Third mode was the longitudinal vibration. Figure 5 was the vibration diagram at 19.458KHz.

***** INDEX OF DATA SETS ON RESULTS FILE *****

SET	TIME/FREQ	LOAD STEP	SUBSTEP	CUMULATIVE
1	19.002	1	1	1
2	19.003	1	2	2
3	19.458	1	3	3
4	24.179	1	4	4

Figure 4. Each modal frequencies.

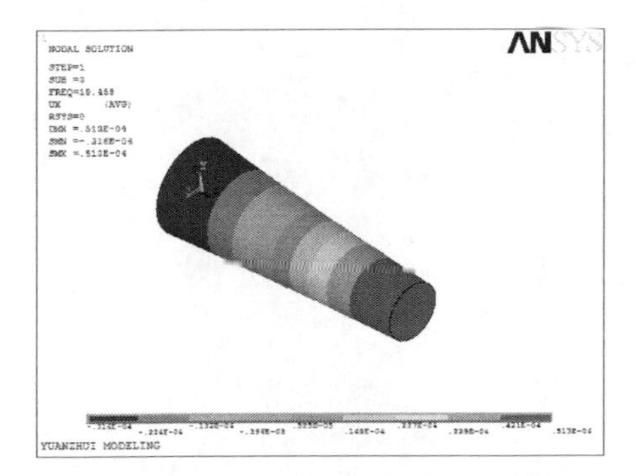

Figure 5. Longitudinal vibration modal diagram.

Modal analysis result show that the longitudinal vibration resonant frequency is 19.459KHz deviating from the design frequency 20KHz.So the developed interface was used to modify the horn length. Run the VB program and input the basic parameters and modal analysis frequency of the horn, then push the compute button, result as shown in figure 6. Among them, the negative length correction value indicates reducing the horn length, otherwise increasing. Therefore,the horn length $L'=132.5\text{-}3.75=128.75mm$. End diameter unchanged, change the horn length and complete modal analysis automatically, the result shown in figure 7. The longitudinal vibration frequency is 20.016KHz after correction. The relative deviation from the target value is only 0.08%. The modified horn meets the requirements of practical application, its longitudinal vibration modal diagram as shown in figure 8.

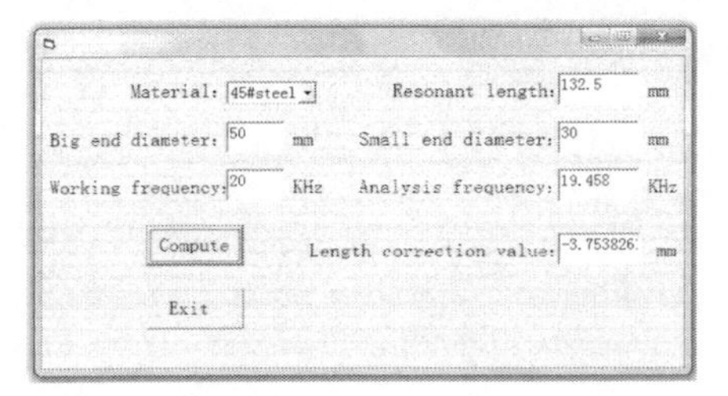

Figure 6. The result of running the program.

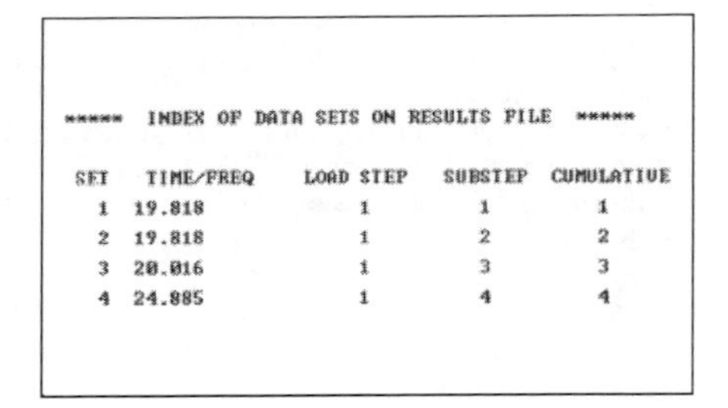

***** INDEX OF DATA SETS ON RESULTS FILE *****

SET	TIME/FREQ	LOAD STEP	SUBSTEP	CUMULATIVE
1	19.818	1	1	1
2	19.818	1	2	2
3	20.016	1	3	3
4	24.885	1	4	4

Figure 7. Each modal frequencies after correction.

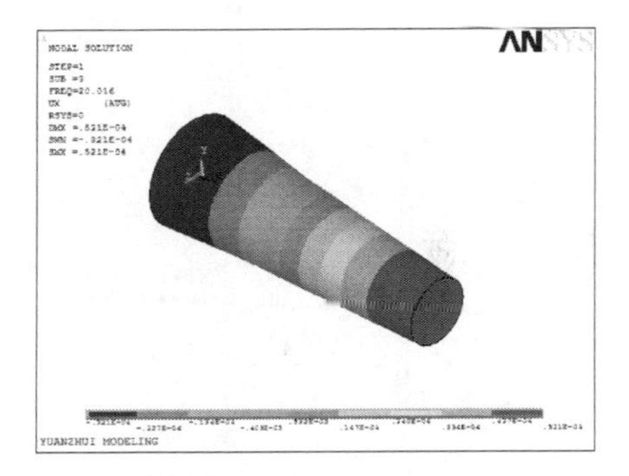

Figure 8. Vibration modal diagram after correction.

4 SUMMARY

The law of changes of the resonant frequency of the horn with its length was studied based on analytical methods. The designed horn was analyzed using the finite element method, and the parameter was corrected using the VB platform in order to make the designed horn meet the requirements of practical application. The method provides design and correction method, and provide references for processing. This program after minor modifications can be used to solve the parameter correction of exponential, linear and catenary horn.

REFERENCES

[1] Zhang Yundian. Ultrasonic machining and its application[M].Beijing: National Defense Industry Press, 1995.

[2] Nad M C L. The effect of the shape parameters on modal properties of ultrasonic horn design for ultrasonic assisted machining[J]. INDUSTRIAL ENGINEERING. 2012: 19–21.

[3] Jia Yang, Shen Jianzhong. Frequency response of stepped ultrasonic horn[J]. Acoustic Technology. 2006(02): 154–159.

[4] Tie Shuxia. Numerical Control Modifying Method with Resonance Frequency of Ultrasonic Transducer and Amplitude Transformer[D]. Hangzhou Dianzi University, 2005.
[5] Zhao Changyong, Zhang Xibin, Zhai Xiaopeng. The structure optimizing design of APDL based on ANSYS parameter language[J]. Shanxi Architecture. 2008(03): 362–363.
[6] M. Roopa Rani R R. Computational modeling and experimental studies of the dynamic performance[J]. Ultrasonics. 2013(53): 763–772.
[7] Gong Shuguang. ANSYS Parametric Programming and Command Reference[M], Beijing: Mechanical Industry Press, 2009

Computing, Control, Information and Education Engineering – Liu, Sung & Yao (eds)
ISBN: 978-1-138-02800-5

The identification of the liquid drop fingerprint combining support vector machine with clustering method

Q. Song, M.Y. Qiao, S.H. Zhang & L. Yang
Automation School, Beijing University of Posts and Telecommunications, BUPT, Beijing, P.R. China

ABSTRACT: In order to effectively reduce the time complexity of the recognition algorithm and improve the recognition accuracy and the generalization ability, a new method combining support vector machine with clustering is put forward. After classifying by iterative dynamic clustering method applied on 38 kinds of liquid samples, all liquid samples can be divided into 8 categories, which are then separately trained with the support vector machine method. Experimental results show that the recognition accuracy can be up to 100% among selected samples, together with the reduced computational complexity of training models and the significantly improved recognition efficiency. Compared with the previous model, the generalization capability has been greatly enhanced, with its estimation of generalization performance been cut exceeded 90%.

KEYWORDS: the liquid drop fingerprint; support vector machine; clustering; pattern identification.

1 INTRODUCTION

The fiber-capacitive drop analyzing technology has been developing rapidly in recent years. Its core is the drop fingerprint. The liquid drop fingerprints are curve images obtained after filtering and normalizing, representing liquid characteristics by collecting the optical fiber capacitance data in the process of droplet formation by the fiber-capacitive drop analyzing technology. By identifying the liquid drop fingerprint, the type of liquid can be identified and the characteristic of liquid can be analyzed [1].

Great quantities of methods aimed at feature extraction and pattern identification have been reported. Support Vector Machine (SVM) is a trendy way of pattern identification in recent years. It performs good identifying effect, but the complexity of a training algorithm largely increased and the identification rate decreased when the type of liquid multiplies, along with the decrease of generalization ability of models. Therefore, other methods are required to reduce the complexity of pattern identification, enhance accuracy of model identification and reinforce the ability to generalize. We report a new method combining SVM and clustering applied to the identification of liquid drop fingerprint in this paper.

2 WAVEFORM ANALYSIS METHOD OF THE LIQUID, DROP FINGERPRINT

Among the liquid drop fingerprint extraction methods, the waveform analysis method is simple, effective and also widely applied. The waveform analysis method takes the liquid drop Fingerprint as a waveform contained peaks and troughs. Calculating waveform characterization parameters, so as to quantitatively express the graphical features of the Liquid Drop Fingerprint. Most of the liquid drop fingerprint mainly present as the shape of "two peak and a Valley".

The minimum capacitance (C_min), the increment of capacitance signal at the secondary peak of liquid drop fingerprint(C_sp), the increment of capacitance signal at the hollow of liquid drop fingerprint (C_h), the increment of capacitance signal at the main peak of liquid drop fingerprint (C_mp), the increment of liquid drop capacitance (C_dif), the minimum optical signal (H_min), the secondary peak height (H_sp), the main peak height (H_mp), Curve length (CL) and Current area (CA) are 10 characteristic values extracted from the liquid drop fingerprint with waveform analysis method [2].

These 10 characteristic values (10-dimensional vector) contain the main features of each liquid drop fingerprint and these values can be considered to represent a kind of liquid. After further pattern recognition to these 10 eigenvalues, the liquid can be identified.

3 APPLICATION OF SVM IN THE IDENTIFICATION OF THE LIQUID DROP FINGERPRINT

3.1 *The support vector machine*

The support vector machine was first proposed by Cortes and Vapnik in 1995, which possesses many unique advantages in resolving the small sample, nonlinear and high dimensional pattern recognition. This method is based on the VC dimension of statistical learning theory and the principle of structural risk minimization. The best compromise between complexity and learning ability of the model is found based on the limited sample of information, in order to obtain the best generalization ability [3,4].

Support vector machine is based on the two types of linear classifier obtaining the best classification effect by maximizing the distance between the decision interface and the data points. Support vectors, which play a role as the real working data points in the process of training classifier, are also the points closest to the decision interface. Support vectors only account for a small part of all data points, which greatly simplifies the operation process and improves classification efficiency. Introducing the kernel function enables the SVM to solve the problem by transforming nonlinear problems to linear ones [3,4].

Support vector machine is established on the structural risk minimization principle of the computational learning theory, whose main idea is aimed at two types of classification problem, finding a hyper plane as the segmentation of two kinds in high dimensional spaces to ensure the minimum classification error rate.

3.2 *Estimation of SVM generalization capability*

The quality of a support vector machine depends on not only the accuracy in recognizing the known samples, but also the generalization and the complexity of machine learning, namely the accuracy of testing unknown data. Leave-one-out and support vectors notations are two widely-used and convenient methods applied to evaluate the SVM generalization ability.

When the error rate is estimated by leave one out method, initially a sample is removed from the training set and then criterion is trained on the remaining samples. And a leave-out-method error happens when classification has errors [5].

This paper uses a support vector notation method, namely counting the number of support vectors in support vector machine [6]. N is the total number of support vectors. l is the total number of training samples. The estimation of generalization ability of support vector machine with support vector notation method is as follows.

According to the principle of support vector machine and the leave-one-out method, only these support vectors trained on all the training samples can cause errors in the leave-one-out method test. The smaller E value is, the less support vectors for training model are used. The model is relatively simpler and the model's generalization ability is better.

In this paper, the recognition accuracy w in the training samples and the estimation of generalization performance E with the support vector notation method are used to characterize the recognition effect of the liquid drop fingerprint.

3.3 *Recognition results of SVM*

This paper selected 3800 samples of 38 kinds of liquid, using the method of exhaustion to optimize parameters in the experiment. Optimal identification results are obtained as shown in Table 1.

Table 1. The recognition results with support vector machine.

Subjects	Accuracy (w)	The estimation of generalization performance (E)
38 kinds of liquid	94.8%	0.7733

From the experimental results it can be seen that when directly using SVM on 38 kinds of liquid samples for training and recognition, the accuracy and the generalization ability need further enhance, because each sample contains 10 dimensional eigenvectors thus the and the training model established is complicated.

4 IMPROVED RECOGNITION ALGORITHM BASED ON CLUSTERING

When analyzing different liquid droplet fingerprint, High similarity and close eigenvalues are found among a part of liquid fingerprints. But some liquid fingerprint curves are found special with great differences in eigenvalues compared with other liquid. In this paper, the liquid shared high similarity are placed in one group by using the clustering algorithm and are identified by suing SVM to improve recognition accuracy and generalization capability.

4.1 *Clustering algorithm*

In the clustering analysis, the number and the structure of sample classes are unknown in advance; the only basis is the character of the samples. Samples with same or similar characteristics are classified as the same group by utilizing a certain similarity measurement thus clustering division is achieved [7].

System clustering is one of the most frequently-used clustering methods. In this paper, union algorithm is used in system clustering, whose basic principles are as follows: at first, considering each sample as a class and setting the distance between samples and the distance between classes, then merging the two nearest classes into one new class until all samples are merged into one class.

The result of hierarchical clustering method cannot be changed once this new class is formed. This requires a classification with high accuracy. A higher requirement of classification method is put forward thus relevant calculation proves to be complex. So, Large sample problems may not be solved by hierarchical clustering method due to limitations of computer memory and computing time. To eliminate the inconvenience in application, the dynamic clustering method comes into being.

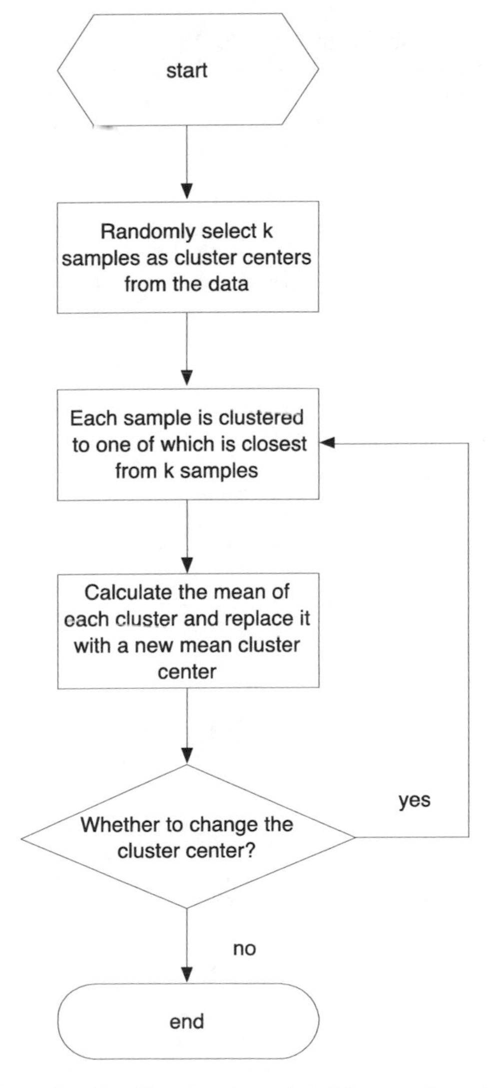

Figure 1. Algorithm flowchart of the K-means dynamic clustering method.

The dynamic clustering method is also called the step-by-step clustering method. The basic idea is to divide classes roughly at first, and then modify unreasonable classification according to a certain optimal principle, until a more reasonable classification is formed thus the final results is obtained. This method reduces the computer memory needed and accelerates the calculation. The Figure 1 shows the algorithm flowchart of the K-means dynamic clustering method.

4.2 *Detailed implementation steps*

In this paper, we use the K-means dynamic clustering method in the R language. A MacQueen dynamic clustering algorithm is used to cluster eigenvectors of optimized 38 groups of liquid fingerprints, and then SVM is used for recognition.

In the process of training, SVM is respectively used for each group after clustering. Then the parameters are optimized to get the best training model.

In the process of identification, test samples are divided into groups first. Training sample closest to the test sample is then obtained. In the following step we incorporate the test sample into the group containing this training sample to narrow the range of identification for test sample. In last SVM training model is applied to all members in the group, thus the final identification result is acquired.

4.3 *Experimental results and analysis*

Results of 38 liquid samples with the dynamic clustering method are shown in Table 2.

Table 2. Results of 38 liquid samples with dynamic clustering method.

Group	Samples
1	Spirit-Beidacang, Spirit-Gujingyuye, Spirit-Jinjiu, Spirit-Quanxing, Spirit-Yililiangchun, Spirit-Beijingerguotou, Spirit-Zhengtongwangpai
2	Wahaha Mineral Spring water, Sea water of Qinhuangdao, River water of Weijin, Wahaha pure water, Kangshifu pure water
3	Grape wine-Nongdahong, Grape wine-Shuangkou
4	Beer-Baotuquan, Beer-Lantian, Beer-Wuxing, Beer-Yanjing
5	Vinegar-Duliu, Vinegar-Shanxi, Vinegar-Shanxibagua, Soy sauce-Hongzhi, Soy sauce-Laochou, Soy sauce-Tesite, Soy sauce-Zhenji
6	Water-Nongfu, Nestle Mineral water, Lake water of JingYe, Lake water of Youth, Wahaha Mineral water
7	White vinegar, Rice wine-Jianlian, Rice wine-Nuomi, Rice wine-Shaoxing
8	Fanta, Coca-Cola, XingMu, Sprite

Table 2 shows that Similar liquid samples can be divided into a group. Eight groups of liquid samples are trained and identified with SVM respectively by using Matlab. Support vector machine is trained with RBF function and parameters optimization processes with the method of exhaustion. The final recognition result is shown in Table 3.

Table 3. The recognition results combining SVM with clustering method.

Group	1	2	3	4
Accuracy(w)	100%	100%	100%	100%
The estimation of generalization performance(E)	0.0400	0.0450	0.0100	0.0500
Group	5	6	7	8
Accuracy(w)	100%	100%	100%	100%
The estimation of generalization performance(E)	0.0167	0.0175	0.0250	0.0375

Comparison of the average recognition accuracy and the estimation of generalization performance between SVM and improved algorithm of classification are in shown in Table 4.

Table 4. Recognition results compared before and after the improvement.

Recognition method	Accuracy (w)	The estimation of generalization performance (E)
SVM	94.8%	0.7733
Improved recognition algorithm based on clustering	100%	0.0302

Table 4 shows the accuracy reaches 100% among these 38 kinds of liquid samples and the estimation of generalization performance is reduced from 0.7733 to 0.0302, the extent of decrease is more than 90%. This shows that not only the complexity of the optimization training model is greatly reduced, but also that its generalization capability has been greatly improved.

5 CONCLUSION

SVM performs well in classification. It has been widely used in pattern recognition and data mining. But the establishment of training model in multi classification problem is excessively complex, which often fail to achieve high recognition accuracy and strong generalization simultaneously.

This paper proposes the support vector machine recognition method based on clustering. For the samples to be classified, we use clustering for preliminary classification of them to ensure the efficiency and accuracy in the recognition process. In further precise identification, high accuracy of classification can be obtained by using the capability to distinguish of SVM.

Experiments show that the support vector machine combining with clustering method is an effective classification method. Preliminary partition of data by clustering the method, the time complexity of the classification model with the support vector machine is greatly reduced, which partly solves the problem of high cost with SVM in training.

The identification of the liquid drop fingerprint combining support vector machine with clustering not only accelerate training process and improve the speed of recognition, but also raise the identification accuracy among 38 kinds of liquid samples to 100%. While the estimation of generalization performance is reduced from 0.7733 to 0.0302, whose extent of decrease is more than 90%. This shows that the complexity of the optimization training model is greatly reduced, and its generalization capability has been greatly enhanced as well.

ACKNOWLEDGMENT

Especial thanks to The Natural Science Foundation of China (NSFC) and Beijing Higher Education Young Elite Teacher Project for the financial support.

REFERENCES

[1] SONG Qing, Research of the identification method based on Drop Analysis Technology and Droplet Fingerprint. Tianjin University. 2005.

[2] Luo Y, Song Q, Li J, Zou C.W. Fuzzy pattern recognition method applied in droplet fingerprint recognition[C], Computer Application and System Modeling (ICCASM), 2010 International Conference on. IEEE, 2010, 14: V14-624-V14-627

[3] Christopher M B. Pattern Recognition and Machine Learning [M]. New York: springer. 2006.

[4] Song Q, Yuan H, Liu X, Qiu C. Support vector machine for the liquid drop fingerprint recognition[C]. Natural Computation (ICNC), 2012 Eighth International Conference on. IEEE, 2012: 280–282.

[5] LUNTS A, BRAILOVSKIY V. Evaluation of attributes obtained in Statistical Decision Rules. Engineering Cybemetics. 1967.

[6] CHAPELL E O, VAPINK V N. Choosing multiple parameters for support vector machines [J]. Machine Learning. 2002.

[7] XUE Yi, CHEN Liping, Statistical Modelin And R Language. Tsinghua University Press. 2007.4.

Computing, Control, Information and Education Engineering – Liu, Sung & Yao (eds)
© 2015 Taylor & Francis Group, London, ISBN: 978-1-138-02800-5

Research on PID leveling system of a rocket launcher based on background trend line prediction model

Zhong Qing Shen & Li Xin Xing
Hefei City, Anhui Province, China

ABSTRACT: For solving the problem of long leveling time and low leveling accuracy generated in the process of artificial leveling, the automatic leveling system based on traditional PID (Proportion Integration Differentiation) is proposed, and by introducing Background Trend Line Prediction Model (BTLPM) into the traditional PID model, the error of information lag caused in the process of filtering and data predicting is deduced greatly. The simulation result of MATLAB show that compared with the accuracy of the traditional PID model, the one of improved PID model which meets actual leveling process better is higher.

KEYWORDS: PID control; Butterworth filtering; Background trend line prediction model.

1 INTRODUCTION

As the eye of rocket launcher, targeting system is very important to the accuracy of rocket launcher. So before firing, the targeting system of every rocket launcher must be checked. After passing the inspection, the rocket launcher can shoot. And because most of the inspection items must be carried on under the condition that the rocket launcher is in the case of leveling, leveling accuracy of the rocket launcher is the key for the inspection results.

So far, the leveling method of the rocket launcher is manual level [1], whose leveling time and leveling accuracy is not ideal due to the involvement of human factors. And most of the leveling system of other military hardware such as missile launching vehicle and radar chooses the traditional PID algorithm as controller model. Because of ignoring the error caused by filtering and information lag, the leveling accuracy of the traditional PID model is bound to be affected.

For solving the problem of information lag existing in the traditional PID model, Butterworth filtering and BTLPM are introduced. And after filtering and predicting, the original lifting height data by supporting leg is put into the traditional PID model to form improved PID model, whose leveling accuracy is better than the traditional PID model.

2 THE COMPOSITION OF THE LEVELING SYSTEM

The leveling system is a hydraulic automatic leveling system of four supporting legs, that is the support structure choose four supporting legs, the servo system is hydraulic, and then automaticity belongs to full automation. The composition of the leveling system is as shown in Fig 1.

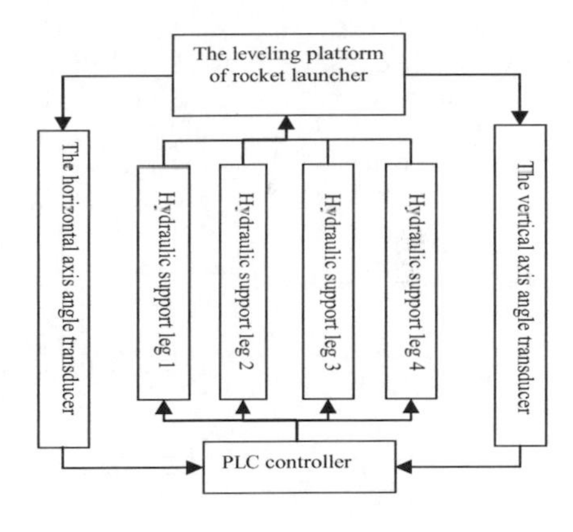

Figure 1. The composition of the leveling system.

3 THE MATHEMATICAL MODEL OF THE HYDRAULIC SUBSYSTEM

As the controlled subject of the leveling system, Hydraulic subsystem has the transfer function, which is only related to the structure of the subsystem, and is not related to the inputs of subsystem [2]. Because the mathematical model and the analytic process of four support legs are similar, it is more convenient for study to take one support leg for example. So any one leg is selected as research object.

3.1 *The flow equation of the slide value*

The flow equation of the slider value describes the relationship between the displacement produced by the spool and the flow into the cylinder.

$$q = C_d \omega (k_x I - \delta)\sqrt{2 \times p / \rho} \ . \quad (1)$$

q is the flow into the cylinder. C_d is flow coefficient. ω is the area gradient of value port. k_x is the proportional coefficient between current and displacement of the spool. δ is the cover volume of the value port. p is the pressure difference between system and load. ρ is liquid density. I is the current delivered to the hydraulic subsystem.

3.2 *The flow continuity equation of plunger tank*

$$q = C_{ep} p_L + \frac{V}{\beta_e} \dot{p}_L + \dot{V} . \quad (2)$$

$$V = A \dot{y} \quad (3)$$

In these equations, C_{ep} is leakage coefficient. p_L is the load pressure of the plunger tank. V is the volume of the plunger tank. β_e is elastic modulus of effective volume. A is area of piston rod of the plunger tank. y is the displacement of the piston rod ,that is, the lifting height of the load.

3.3 *Force balance equation*

$$m \ddot{y} + B_P \dot{y} + ky = p_L A . \quad (4)$$

In the equation, m is a mass of piston that the total mass of the load and piston is converted into. B_p is the viscous damping coefficient of load and piston. k is elastic coefficient of load.

By means of the simultaneous solution of equations (1),(2),(3),(4) and using Laplace transform, the transfer function of controlled subject can be obtained:

$$G(s) = \frac{Y}{I} = \frac{AC_d \omega k_x \sqrt{2 \times p / \rho}}{\frac{mV}{\beta_e} s^3 + (mc_{ep} + \frac{B_p V}{\beta_e})s^2 + (B_p c_{ep} + \frac{kV}{\beta_e} + A^2)s + c_{ep} k} \quad (5)$$

4 TRADITIONAL PID MODEL

As the important theory of the control region, PID control has a history of more than one hundred years. The central point of this control is controlling the controlled subject (hydraulic subsystem) through controlled variable formed by linear combination of Proportion unit, Integration unit, and Differentiation unit of error. And discrete PID model is as follows:

$$I(n) = K_p e_0(n) + K_i \sum_{i=1}^{n} e_0(i)T + K_d [e_0(n) - e_0(n-1)] / T . \quad (6)$$

In the equation, $e_0(n) = Y_d(n) - Y_0(n)$ is the difference between the sampling value of the system and the preset value of the system at the first n moment. T is the sampling period. K_p, K_i, K_d are Proportion coefficient, Integration coefficient, and Differentiation coefficient. $I(n)$ is input current of slide value at the first n moment. So the system schematic diagram of PID under ideal condition is as shown in Fig 2:

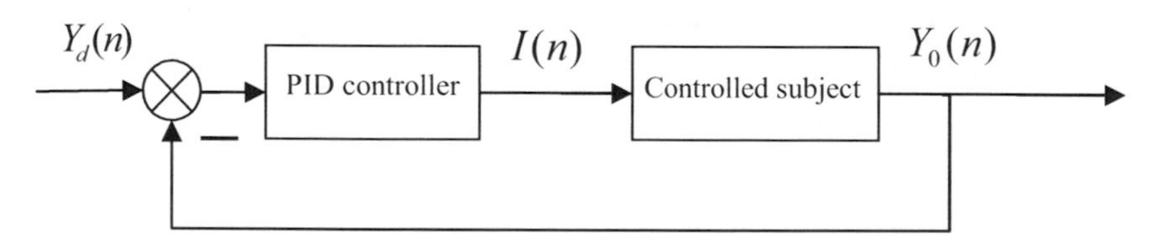

Figure 2. The system schematic diagram of PID under ideal condition.

Take a type of rocket launcher, for example. The sampling period of angular transducer used by the system is 0.1s. The area A of the plunger tank is 0.025m^2. Flow coefficient C_d is 0.7. The area gradient ω of value port is 0.04m. The proportional coefficient between current and displacement of the spool k_x is 0.2. Hydraulic oil density ρ is 870kg/m^3. The actual stroke$L$of plunger tank is 0.19m. So the volume of plunger tank is $V = AL = 0.00475$ m^3. Elastic modulus β_e of system effective volume is 7×108Pa. The leakage coefficient C_{ep} of plunger tank is 1×10-12m^5/N.s. Because of too small, B_p, k are ignored. So equation (5) can be simplified to the equation as follow:

$$G(s) = \frac{Y}{I} = \frac{AC_d \omega k_x \sqrt{2 \times p / \rho}}{\frac{mV}{\beta_e} s^3 + (mc_{ep} + \frac{B_p V}{\beta_e})s^2 + (B_p c_{ep} + \frac{kV}{\beta_e} + A^2)s + c_{ep} k} . \quad (7)$$

$$= \frac{0.01074}{0.00010857s^3 + 0.000016s^2 + s}$$

According to trial-and-error method of PID parameters [3], K_p, K_i, K_d can be come out as 25,0.8 and 10.2. And the MATALB simulation result shows that under ideal condition, the rising height curve of supporting leg is as shown in Fig 3(a). But the actual rising height curve is as shown is as shown in Fig 3(b). There are two reasons lead to the difference between Fig 3(a) and Fig 3(b), that is under actual conditions. The first reason is the shaking of supporting leg in the process of rising, which make an angular transducer swing. So the actual rising height signal consists of low frequency actual height signal and high frequency swinging signal. The second reason is the transducer, which is used to measure angle, not to measure the rising height directly. So there is conversion time between angle and rising height, together with filtering time, which can make the input error of PID be the signal before filtering, that is, information lag signal, not the actual error in current time.

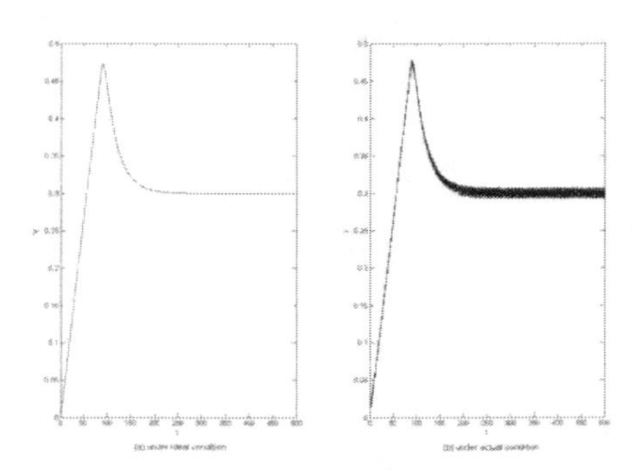

Figure 3. Rising height curve of support leg.

So the traditional PID model of the leveling system is actually:

$$I(n+\Delta t) = K_p e_0(n) + K_i \sum_{i=1}^{n} e_0(i) + K_d[e_0(n) - e_0(n-1)]. \quad (8)$$

5 IMPROVED PID MODEL

Both of aforementioned two circumstances bring the great error to PID control. At first, because of the swing of the transducer, PLC controller may consider the actual rising height signal as the actual height signal. In order to solve the problem, Butterworth filtering is used to filtrate away high frequency swinging signal in actual rising height signal. And at second, the other reason of disturbing PLC controller is information lag. Especially at the moment that the leveling system has already reached leveling condition, hydraulic subsystem continues to be adjusted. In order to solve the problem, the background trend line model is used to predict the actual height of supporting leg real-time.

5.1 *Butterworth filtering*

For filtrating away high frequency swinging signal, the actual rising height signal should be filtered to obtain the actual height signal. So far, because of the satisfactory performance in the time domain and frequency domain, easy to realize, and good stability, Butterworth filtering is used in control and communication region widely [4].

The Butterworth magnitude square function is shown as follows, and The Butterworth amplitude-frequency characteristics at different n is as shown in Fig 4.

$$|H(\omega)|^2 = \frac{1}{1+(\frac{\omega}{\omega_c})^{2n}}. \quad n = 1,2,3\ldots \quad (9)$$

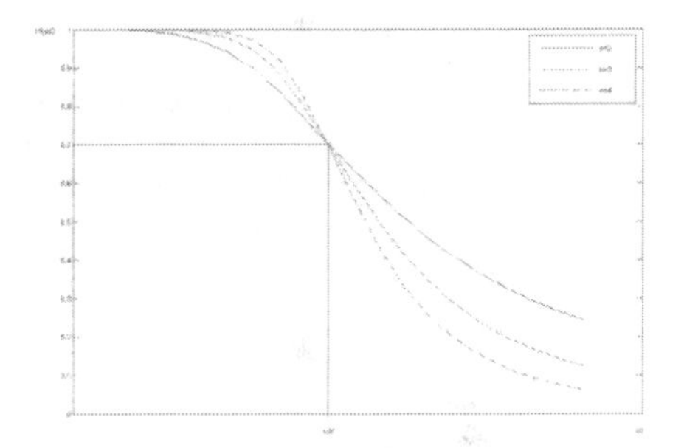

Figure 4. The Butterworth amplitude-frequency characteristics at different n.

In the function, n is the order of filter, and ω_c is the cut-off frequency of filter.

According to the conversion formula $|H(\omega)|^2 = H(s)H(-s)|_{s=j\omega}$ between Laplace transform and Fourier transform, the function (9) can be translated to following function:

$$|H(s)|^2 = \frac{1}{1+(\frac{s}{j\omega_c})^{2n}}. \quad (10)$$

After normalization processing, the function (10) can be translated to following function:

$$H(s)H(-s) = \frac{1}{1+(-s^2)^n}. \quad (11)$$

Let's suppose $1+(-s^2)^n = 0$, so the poles of the above formula are as follows:

$$s_i = e^{j[\frac{\pi}{n}i+\frac{n-1}{2n}\pi]}, i=1,2,\cdots,2n \quad (12)$$

Based on the formula above, the modular and the angle of the first i pole are 1 and $\frac{\pi}{n}i+\frac{n-1}{2n}\pi$. So the following function can be obtained:

$$H(s)H(-s)=\frac{1}{1+(-s^2)^n}=\frac{1}{\prod_{i=1}^{2n}(s-s_i)}. \quad (13)$$

According to function (12), all of the poles are uniform and conjugate; and therefore, through the necessary and sufficient condition of system stability, that is, all of the poles of the system characteristic equation have negative real part, $H(s)$ is as follows:

$$H(s)==\frac{1}{\prod_{i=1}^{n}(s-e^{j[\frac{\pi}{n}i+\frac{n-1}{2n}\pi]})}=\frac{1}{s^n+a_{n-1}s^{n-1}+\cdots+a_1 s+a_0} \quad n=2,3\ldots \quad (14)$$

Through MATLAB calculation, it can be known that the overshoot and the accommodation time of the system will increase as the rising order of Butterworth filtering when inputting step signal, The simulation results are as shown in table 1.

Table 1. The characteristic of Butterworth transfer function.

Order number	2	3	4	5	6	7
Overshoot	4.3	8.3	10.5	12. 0	14.5	15.7
Accommodation time(s)	6.0	6.5	9.8	11.0	14.3	15.3

For the actual height signal and swinging signal, because there is a large difference between their frequencies, the 2 order transfer function can meet filtering demand. So the filtering system of rocket launcher chooses 2 order transfer function $H(s)$

$$H(s)=\frac{1}{s^2+\sqrt{2}s+1}. \quad (15)$$

5.2 *Background trend line prediction model*

The key point of Background trend line prediction model is to fit known time series with three different models, and then calculate three correlation coefficients between original time series and the time series obtained by fitting function. At last, the model that has the largest correlation coefficient is selected as a Background trend line model to predict support leg height real-time.

The commonly used models have the following three forms:

Linear-logic growth composite model: the function is

$$Y=at+b+\frac{K}{1-e^{c-dt}}. \quad (16)$$

Linear-trigonometric function composite model: the function is

$$Y=at+b+c\sin(\omega t+\theta). \quad (17)$$

Exponential- trigonometric function composite model: the function is

$$Y=a\cdot e^{bt}+c\sin(\omega t+\theta). \quad (18)$$

In these functions, a,b,c,d,K,θ,ω are the parameters which can be calculated through putting time series into the above functions.

n monitoring points of transducer are selected to form time group $T=\{t_1,t_2,\cdots,t_n\}$ and original time series $Y_0=\{y_0(1),y_0(2),\cdots,y_0(n)\}$ of support leg height. The three models are used to fit original time series Y_0 to obtain the three time series $Y_i=\{y_i(1),y_i(2),\cdots,y_i(n)\}, i=1,2,3$. i is the type number of time series. The correlation coefficients of three models can be calculated as follows:

$$R_i(Y_0,Y_i)=\sqrt{\sum_{j=1}^{n}(y_i(j)-y_0(j))^2 / n\times\max\{(y_i(j)-y_0(j))^2\}} \quad i=1,2,3 \quad (19)$$

The model which has the largest correlation coefficient is considered as Background trend line prediction model to predict the support leg height $Y_1(n+\Delta t)$ at the time Δt after the first n moment. Δt is the information lag time caused filtering and conversion time between angle and rising height.

5.3 *The PID control based on predicting*

Through the Background trend line prediction model, the support leg height $Y_1(n+\Delta t)$ at the moment of $k+\Delta t$ can be calculated, and at the moment [5], the value of the error is

$$e_1(n+\Delta t) = Y_d(n+\Delta t) - Y_1(n+\Delta t) \quad (20)$$

Through putting the function (20) in the function (6), the traditional PID model can be translated into the following function , which is the improved PID mathematic model. And the system schematic diagram of improved PID control is as shown in Fig 5.

$$I(n+\Delta t) = K_p e_1(n+\Delta t) + K_i\left(\sum_{i=1}^{n} e_0(i)T + e_1(n+\Delta t)\Delta t\right) + K_d[e_1(n+\Delta t) - e_1(n)]/\Delta t \quad (21)$$

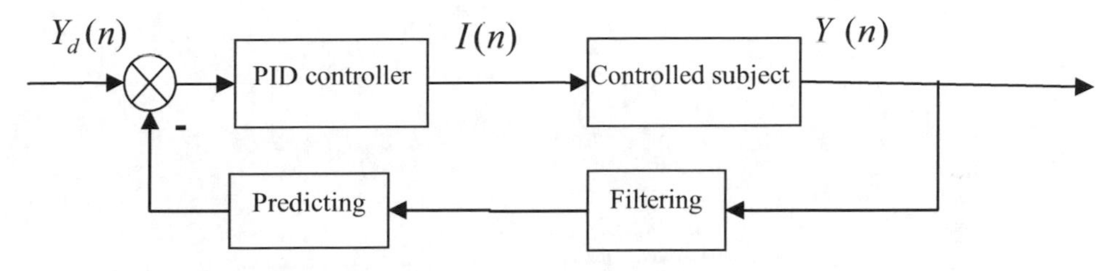

Figure 5. The system schematic diagram of improved PID control.

5.4 *Simulation experiment*

Let's suppose $n=8$. And by using MATLAB calculation, the processing time of filtering and conversion between angel and height is $\Delta t = 0.25s$. So through simulation, the height curve comparing Fig of improved PID control and traditional PID control is as shown in Fig 6. Y_d is the preset height of leveling system. Y_0 is the height curve of the traditional PID model. Y_1 is the height curve of the improved PID model. The Fig 6 shows that the improved PID model has the higher precision.

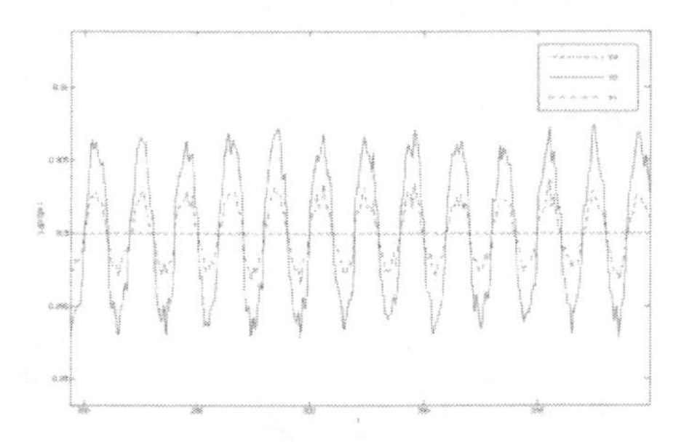

Figure 6. The height curve comparing Fig of improved PID and traditional PID.

6 SUMMARY

Automatic leveling system based traditional PID model not only make the automatic operation of the rocket launcher leveling system come true, but also reduce workload of the operator. And the improved PID model based on Background trend line prediction is proposed to improve the accuracy on the basis of not adding the leveling time. In addition, this model can also be used in the leveling system of other military hardware such as missile launching vehicle and radar. So it has the value of popularization.

REFERENCES

[1] Wu Huaqiang, Zhan Yanmou. Design of a certain type of rocket-artillery's auxiliary leveling system [J]. Ship Electronic Engineering, 2011(12):178–180.

[2] Wang Wei. Research on automatic hydraulic leveling of heavy duty transporter with fuzzy PID control strategy [D]. Wuhan :Wuhan Institute of Technology, 2009:28–29.

[3] Wang Wei, Zhang Jingtao, Chai Tianyou. A survey of advanced PID parameter tuning methods [J]. Acta Automatica Sinica, 2000,26(3):347–355.

[4] Li Zhongshen, Wang Yongchu. Design and analysis of new high order butterworth low pass filter [J]. Electtronic Measurement Technology, 2007,30(12):16–19.

[5] Wang LI, Qian Lin-fang, Gao Qiang, Guo Qi. Research on torque control of servo system load simulator based on grey prediction Fuzzy-PID controller [J]. Acta Armamentarii, 2012(11): 1379–1386.

Computing, Control, Information and Education Engineering – Liu, Sung & Yao (eds)
© 2015 Taylor & Francis Group, London, ISBN: 978-1-138-02800-5

A module-independent solution to localize WLAN access points

Kai Wu, Yong Liu & Xue Rong Gou
School of Network Education, Beijing University of Post and Telecommunications, Beijing, China

Ying Zhe Li
Wireless Network Research Department, Huawei Technologies Co., Ltd., Shanghai, China

ABSTRACT: The location of Wireless Local Area Network (WLAN) Access Point (AP) plays an important factor in indoor localization. But that information is difficult to acquire in most cases because of security. The essay analyzes some traditional methods and introduces a new algorithm that only need take some measured points around the building to locate AP. Simulation claims the algorithm is feasible with high precision.

KEYWORDS: AP Location, Locating area, Module-independent, Measured points, WLAN.

1 INTRODUCTION

With the extensive use of Location-Based Services (LBS), the indoor localization based on WLAN has aroused people's great attention. In most of those positioning systems, a mobile node estimates its position based on the locations of AP, in which is assumed to be known as a priori [1]. However, in many situations, such as shopping malls and offices, the location awareness of AP could not be achieved. It is impractical to be provided with AP location information of all those WLAN networks. Therefore, it is meaningful to localize AP in a reverse way.

A non-monotonic signal propagating model is proposed to locate the AP in an indoor environment [2]. The authors modify the traditional signal propagating model and triangulation algorithm. Although the model has been modified, it's still hardly to find a proper model that is suitable for all indoor situations. So, another model-independent way of localization algorithm is presented [3]. This algorithm leaves out model parameters and simplifies the propagating model between attenuated signal strength and propagating distance. But the authors only test this algorithm on the football field. The interior environment is too complicated to be summarized by the above formula. A new solution based on collecting received signal strength (RSS) data to estimate the location of unknown AP is proposed [4]. This solution depends strongly on the area map, which may be not available in many other scenarios. Some other methods linearly approximate the exponential relationship between RSS and distance by applying the multilateration technique [5]. Because the RSS random process is non-stationary and the physical space distance does not perfectly correspond to signal space distance [6], the linear approximating for exponent will cause a greater error.

In this paper, we innovatively propose a localization algorithm of indoor AP by using the outdoor measured data, in which two steps are implemented. Firstly, the virtual AP that is assumed to be in the open-sided environment is located with the collected data from outdoor by war-driving. Secondly, with the model that reflects the relationship between outdoor/indoor space distance difference and signal strength difference, the real location of AP could be calculated. Simulation results will prove the precision and applicability of the proposed algorithm.

2 VIRTUAL AP LOCATED

The virtual AP is assumed to be in the unobstructed free flat space. So the locating referential area of virtual AP could be achieved by analyzing signal attenuate features and measured data's position that is collected from outdoor by war-driving. In this process, a cross product optimization method is applied. This method is independent of any radio propagating model.

Data Preprocessing. Take each measured point bye per 10 meters to collect WLAN signal around the building and each point data averages by measuring time for every AP. So, the preprocessed dataset is $FS = <FS_1, ..., FS_i, ..., FS_n>$, in which FS_i expressed as Eq. 1:

$$FS_i = < Longitude_i, Latitude_i, AP_1, ..., AP_j, ..., AP_m > . \quad (1)$$

Where $Longitude_i$ and $Latitude_i$ are the i_{th} measured point's longitude and latitude coordinates. $AP_j =< SSID_j, MAC_j, RSS_j >$ represents the j_{th} AP's information. $SSID_i$ is the service set identifier. MAC_i and RSS_i are AP's MAC address and RSS respectively.

Determine Locating Reference Area. Set the target AP to be located as AP_{tag}. Select measured points that include the AP_{tag} from FS as the matched fingerprint dataset FS_{sum}. So there are *sum* measured points in the dataset and *sum* must be not less than 3. In consideration of the WLAN signal propagating distance is very limited compared with a base station signal, AP_{tag} is usually within 100 meters distance from the measured points.

The classic free space propagating model is given by Eq. 2:

$$L(dB) = 32.45 + 20\log d(km) + 20\log f(MHz). \qquad (2)$$

Where L is propagating attenuation and d represents propagating distance with f being the carrier frequency. Considering most of WLAN systems apply 802.11b/g/n protocols, the carrier frequency could be 2.4GHz. And through analyzing indoor and outdoor RSS from the same AP, a fact can be revealed that the outdoor RSS is about 70dB lower than indoor RSS. So, d could be 30 meters. That is to conclude, AP's maximum coverage radius is about 30 meters.

From the matched fingerprint dataset FS_{sum}, the fingerprint geometric center of FS_{sum} can be obtained as $(longitude_{center}, latitude_{center})$. So, under the acquired knowledge, the open-sided and non-blocking virtual AP must be within 30 meters away from the fingerprint center. Therefore, take $(longitude_{center}, latitude_{center})$ as grid center and extend 30 meters all around. Each little grid is a square with 1 meter radium. So, a total 60 * 60 meters flat square area is generated.

Virtual AP Locating Algorithm. Although there are kinds of propagating models, the basic relationship between signal strength and propagating distance is always logarithmic as Eq. 3:

$$s = a\log(d) + b. \qquad (3)$$

Where s and d are srespectively, withtenuation and propagating distance respectively with a and b as parameters. With the help of cross-product optimization method, the location of virtual AP could be obtained as the following steps.

Firstly, construct the signal strength difference matrix R as equation Eq. 4:

$$R = \begin{bmatrix} s_1 - s_2 & s_1 - s_0 & s_0 - s_1 & & \cdots \quad \cdots \\ & & \cdots & & \\ \cdots & s_{i+1} - s_{i+2} & s_{i+2} - s_i & s_i - s_{i+1} & \cdots \\ & & \cdots & & \\ \cdots & \cdots & s_{sum-2} - s_{sum-1} & s_{sum-1} - s_{sum-3} & s_{sum-3} - s_{sum-2} \end{bmatrix}. \qquad (4)$$

Where s_i represents the signal strength of AP_{tag} on the i_{th} measured point.

Secondly, construct the distance vector function matrix as Eq. 5:

$$\vec{L}(x,y) = \begin{pmatrix} L_1 \\ \cdots \\ L_{sum} \end{pmatrix}_{sum*1} = \begin{bmatrix} \frac{1}{2}\log\left[(x-x_1)^2 + (y-y_1)^2\right] \\ \cdots \\ \frac{1}{2}\log\left[(x-x_{sum})^2 + (y-y_{sum})^2\right] \end{bmatrix}_{sum*1}. \qquad (5)$$

Where L_i represents the distance between each one little grid geometric center and the i_{th} measured point. With Eq. 3 and Eq. 5, Eq. 6 and Eq. 7 could be obtained as below:

$$\frac{s_i - s_j}{s_i - s_k} = \frac{L_i - L_j}{L_i - L_k}. \qquad (6)$$

$$(s_j - s_k)L_i + (s_k - s_i)L_j + (s_i - s_j)L_k = 0. \qquad (7)$$

Lastly, the evaluation function is made as Eq. 8. Traverse the 3600 grids to find right grid that makes Eq. 8 minimum. That grid's geometric center is where the virtual AP is.

$$E_1 = \left\| R * \vec{L} \right\|. \qquad (8)$$

3 LOCATING THE REAL AP

Because the real AP is mostly off the ground, the locating candidate area is generated by copying the locating reference area grids in the vertical dimension from 1 to 30 meters. So, a total 60*60*30 meter three-dimensional grids are generated.

The above cross-product optimization method aims to locate virtual AP in the open-sided environment with no obscured property. That is to say, this algorithm does not take into account the attenuation of the signal from indoors to outdoors. This chapter reveals the relationship between the variation of signal strength and the variation of distance to help locate the real AP.

By analyzing some typical indoor and outdoor signal strength difference [3,6] and parameters adapting, the critical relationship that reflects the variation of signal strength and distance is as Eq. 9 shows:

$$\Delta D = -100\log\frac{\Delta P}{100} + 29. \qquad (9)$$

Where ΔP is the variation of indoor and outdoor RSS and ΔD means the distance variation. Calculate signal strength difference from each measured point to the indoor measured signal strength, that is $\Delta P_1,...,\Delta P_{sum}$, and the distance variation $\Delta D_1,...,\Delta D_{sum}$ can be obtained.

Last chapter the virtual AP's location (lon_v, lat_v) is achieved, so the distance variation D_{vi} from each measured point to virtual AP could be calculated as Eq. 10:

$$D_{vi} = \sqrt{(lon_v - lon_i)^2 + (lat_v - lat_i)^2}, i = 1,...,sum. \quad (10)$$

Where (lon_i, lat_i) represent i_{th} measured point's position. Using Eq. 9 and Eq. 10, the distance from each measured point to the real AP could be calculated as Eq. 11:

$$D_i = D_{vi} - \Delta D_i, i = 1,...,sum. \quad (11)$$

So, the real AP should be D_i away from the i_{th} measured point. That means, the real AP should be the intersection of these *sum* three-dimensional sphere surfaces, whose center is the i_{th} measured point and radium is D_i. Due to pratical reasons, these sphere surfaces may not cross at one point, so this algorithm traverses all the three-dimensional grids to get the right grid that makes the sum of distance difference E_2 minimum with Eq. 12:

$$E_2 = \sum_{i=1}^{sum} (L_i' - D_i)^2. \quad (12)$$

Where L_i' is the space distance from the i_{th} sphere's center to each little three-dimensional grid's center. The grid that makes E_2 minimum is where the real AP is.

4 SIMULATION

To evaluate the proposed algorithm, a WLAN signal scanning software is developed in C++ during this research shown in Fig. 3.

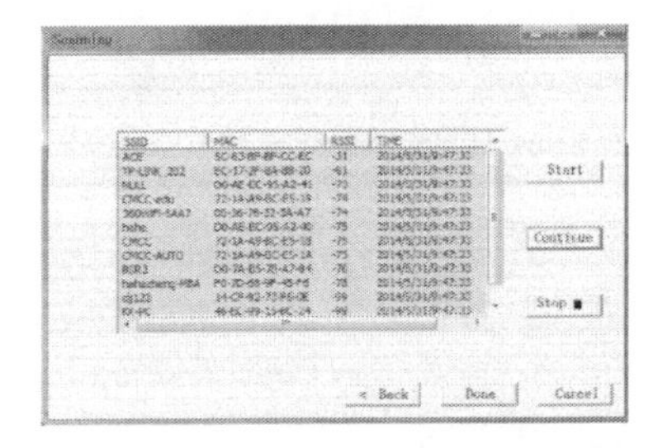

Figure 1. C++-based WLAN signal scanning software.

Figure 2. No.3 teaching building.

Data items in the middle of Fig. 3 are the collected measured data, in which *SSID*, *MAC* and *RSSI* represents an AP's Service Set Identifier, MAC address and RSS respectively, and*TIME* is the local time. Multiple APs' information could be collected during a predetermined period of time, say 1s. In addition, to get the longitude and latitude of each measured point, a GPS device is used with the scanning software together.

The experiment is conducted around No.3 teaching building on the campus of Beijing University of Posts & Telecommunications (BUPT), as shown in Fig. 4. This building is about 30 meters high, totally there are 8 floors. Two APs are chosen in the building, and the locating results of the proposed algorithm are shown as Tables 1 and 2, respectively.

Table 1. Location information of selected APs.

	AP1	AP2
Height(m)	8	12
Building	No.3	No.3

Table 2. Location error of this algorithm.

	AP1	AP2
Horizontal error(m)	6.9	5.2
Vertical error(m)	1.4	2.7
Spatial error (m)	7.0	5.8

From Table 2, the spatial locating error is about 6 meters which is very precise in the reversely locating AP field. In such a case, it can safely confirm that the proposed algorithm is a promising way for high locating precision.

5 CONCLUSION

An indoor AP localization algorithm only using outdoor measured data is proposed with high precision. The algorithm requires neither the indoor propagation model nor the measured data collected in the building and it is evaluated based on the campus of BUPT as the test field. Simulation results show that the proposed algorithm could be used in location estimation of AP without getting into the building.

REFERENCES

[1] H.S. Ahn, W. Yu: Environmental-adaptive RSSI-based indoor localization, IEEE Trans. Automation Science and Engineering, Vol. 6 (2009), p. 626.

[2] S. Varzandian, H. Zakeri and S.Ozgoli: Locating WiFi Access Points in Indoor Environment Using Non-monotonic Signal Propagation Model, 9th Asian Control Conf., Istanbul 2013.

[3] B. Yun, Y. Li and X.C. Sheng: submitted to Journal of Chinese Computer Systems (2013).

[4] M.A.A. Rahman, M. Dashti and J. Zhang: Localization of Unknown Indoor Wireless Transmitter, 2013 Int. Conf. on Localization and GNSS, Turin 2013.

[5] J. Koo and H. Cha: Localizing WiFi Access Points Using Signal Strength, IEEE Communications Letters, Vol. 15 (2011), p. 187.

[6] Information on http://www.sciencedirect.com/science/article/pii/S1574119211001234

Computing, Control, Information and Education Engineering – Liu, Sung & Yao (eds)
 ISBN: 978-1-138-02800-5

A spatial-temporal localization method

Bin Chen
Network and Education Institute, Beijing University of Posts and Telecommunications, Beijing, China

ABSTRACT: The expectations for knowing the outdoor position is increasing along with the demand for Location Based Service (LBS). This paper considers the problems of fingerprinting localization in cellular mobile network base on RSS observations. First, we create a mobile location signal space database as radio map. And then we introduce a new localization method called spatial-temporal localization method to improve the localization accuracy. This mcthod combines the advantage of both the spatial localization method and the temporal localization method.

KEYWORDS: Localization, KNN, L filter, Location Fingerprinting (LF), Received-Signal-Strength (RSS).

1 INTRODUCTION

With the rapid developments of the cellular telephone networks, LBS for personal and commercial applications have been actively researched. For instance, the location of the emergency calls origins, friend finding and photo geotagging, etc.

The prerequisite for implementing the LBS is to be aware of the location of the mobile terminal. Many methods and techniques over the years have tackled the problem of location estimation. One of them is LF technique. It can be easily implemented by exploiting under the existing network infrastructures, without any additional deployment costs and efforts[1]. Generally the fingerprint in the LF technique records the information at some point within a localization area, which includes its GPS positional information, the received Cell Identifiers (CI) and the RSS values.

2 METHODOLOGY

There exists extensive literature on the estimation methods based on the LF technique [2], such as KNN [3], SVM [4] and CS [5], or probabilistic methods.

All the above mentioned estimation methods estimate the user position only from the spatial perspective. However the spatial position information is related with the time variation when a user moves during a call. So by taking advantage of its associated time can get more valuable information of its position estimating.

In this paper, the spatial and temporal localization method are combined together to improve the localization accuracy. In which, the temporal localization method is proposed by L filter. And the spatial localization method is proposed by standard KNN algorithm.

At last, we use the logic and heuristic rules to combine those two localization results together.

The Spatial Localization Method. In this section, we introduce the procedure of the spatial localization

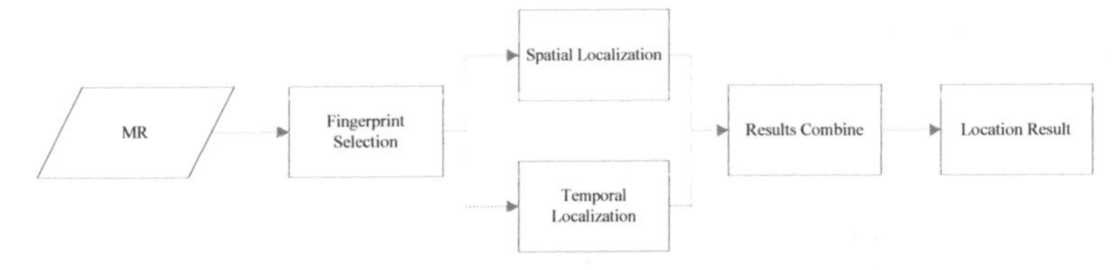

Figure 1. The spatial-temporal localization method implementation block diagram.

method. The spatial localization method is implemented by KNN algorithm.

The principle of KNN algorithm is to find K nearest neighbors of a sample, and then set the average value of those neighbors' property to the sample as its property.

In the location system, first calculate the Euclidean distance between the estimating sample and all selected training samples in signal space. Then sort them order by Euclidean distance and select the nearest K samples. At last, get the predicted sample position by weighting the positions of K samples. We can choose a reasonable K value by experience.

Although the signal strength based KNN algorithm is simple, the accuracy it gets is good enough. This algorithm chose K reference points (RPs) which near the under test RSS. The distance can be calculated by Euclidean distance algorithm.

If we can get the RSS from N APs at testing point (TP):

$$RSS(lat_n, lon_n) = \{RSS_{AP_1}, RSS_{AP_2}, \ldots, RSS_{AP_N}\} \quad (1)$$

Then the Euclidean distance between TP and RPs in database is:

$$d(TP, RP_i) = \frac{1}{N}\sqrt{\sum_{j=1}^{N}(RSS_{TP} - RSS^{i}_{AP_j})^2} \quad (2)$$

In which $RSS^{i}_{AP_j}$ is the average signal strength from AP_j observed at RP_i. RSS_{TP} is the signal strength from AP_j observed at TP. At last, we get the average position of the K nearest RPs.

The Temporal Localization Method. L filter is a smooth method that approximates the discrete points by continuous function. Generally, we consider it must satisfy the following rules when an object moves:

1. The position of the object can't mutate , that is to say the object motion rail must be continuous;
2. The velocity of the object can't mutate too, that is to say the time derivative of its motion rail must be continuous;
3. The acceleration can't be infinite, that is to say the second order derivative of its motion rail must have a certain limit.

Introduce the L filter into location system, due to consider the needs of real-time localization, the current measurement report (MR) filter smoothing work can not consider all MRs before the current. In order to ensure the localization time and hope to get better localization effect, we set a certain window length L (eg L = 5), and use the positioned MRs within the window to calibrate the current positioning MR.

In figure 2, the green dots represent five RPs at the time t_0,t_1,t_2,t_3,t_4. And (a_i, v_i, p_i) is the state of the RP at t_i, where a_i represents the acceleration, v_i represents the speed, p_i represents the position. The red five-pointed star represents the actual value of the current time, and the yellow triangle represents the predicted value of the current time.

According to the above three preconditions of L filter, the object move trail can be seen as a segment connection uniform variable motion. The period of each two RPs can be regarded as a uniform shift. So the acceleration and speed of i-th segment are a_{i-1} and v_{i-1}. Through this assumption, from each discrete RP, we can obtain a continuous function of time, suit the second condition. For any j-th paragraph, we can get the following expression through the uniform movement law:

$$p_j' = 1/2 * a_{j-1}(t_j - t_{j-1})^2 + v_{j-1}(t_j - t_{j-1}) + p_{j-1} \quad (3)$$

$$v_j = a_{j-1}(t_j - t_{j-1}) + v_{j-1} \quad (4)$$

The formula (3) represents the predict position p_j' of next point obtain by the acceleration a_{j-1}, the velocity v_{j-1} and the observed values at time of t_{j-1}. With the above two formulas, it is only need to estimate the acceleration for the new segment, the localization of new position can be competed.

With the five RPs' positions within the window, using the least square method get a straight line to obtain the

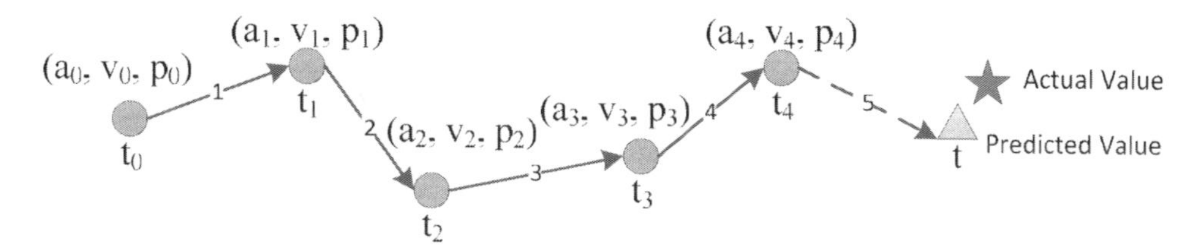

Figure 2. L filter schematic.

preliminary position estimation. With this estimated position and the front five RPs to obtain an estimate curve, and predicted positions of each points (P_0', P_1', P_2', P_3', P_4', P_5')from the curve. To simplify the problem, set the upper and lower limits for the acceleration. The acceleration a_i need satisfy the inequality (5)

$$a_{\min} \le a_i \le a_{\max} \tag{5}$$

The error sum of squares of each RP is a function $S(a_4)$ which related to acceleration a_4. Since the function $S(a_4)$ is derivable, so the minimize problem can convert to the problem of solving the $S(a_4)$'s derivative equals zero. Omit the derivation process, the acceleration values can be expressed as

$$a_j = 2 * \frac{P_{j-1}T_{j-1}^2 + P_j(T_{j-1}+T_j)^2 - v_{j-1}[T_{j-1}^3 - (T_{j-1}+T_j)^3 - p'_{j-1}[T_{j-1}^2 + (T_{j-1}+T_{j-2})^2]}{T_{j-1}^4 + (T_{j-1}+T_j)^4} \tag{6}$$

In which, T_j is the time interval between current time and the previous, that is $T_j=t_j-t_{j-1}$.

After obtain the acceleration which satisfy the condition, and then we can predict the position of the new plus point according to the formula (3) and (4).

In the MR sequence produced by mobile terminal, the adjacent MRs in time dimension have the consecutiveness property and they are coherent in spatial space. So base on the continuity, take count of MR(k) and the former L-1 MRs into localization model, use the dynamic tracking model base on MR(k) and the localization results set of former L-1 MRs to locate MR(k).

The Combine of the Temporal Method and Spatial Method. Combine the spatial localization and temporal localization results to get the final spatial-temporal results. In order to keep combined model comprehensible and as simple as possible. We propose the heuristic rules to combine the above results, the rules are as following:

1. Suppose the longitude and latitude of MR is (lon_i, lat_i), the window length is L=5. So the <(lon_1, lat_1),(lon_2, lat_2),..., (lon_5, lat_5)> store in the window.
2. The temporal localization result which obtained by matching localization algorithm(KNN) is (lon_{61}, lat_{61})
3. The predicted current position (lon_{62}, lat_{62}) is got by motion model
4. Calculate the distance between two adjacent points in the window to get < d_1, d_2, d_3, d_4 >
5. If the adjacent distances are all less than 20 meters, no need to do the L filter dynamic calibration.
6. Calculate the distance between the actual distance between (lon_5, lat_5) and (lon_{62}, lat_{62}) to get d_5
7. Determine the accuracy of the estimate results according to d_5.If d_5 is larger than a certain threshold, the estimated result is considered unreasonable, it is no necessary to calibrate the localization. If d_5 is smaller than that threshold, then calibrate the temporal localization result by setting different weights to the predicted position and temporal localization position. The smaller d_5 is, the temporal result is more accurate, so gives greater weight to the predicted position. In the experiment we set the weights as following:

- If d_5<30m, then lon_6= lon_{61}*1/4+ lon_{62}*3/4, lat_6= lat_{61}*1/4+ lat_{62}*3/4
- If d_5<50m, then lon_6= lon_{61}*1/3+ lon_{62}*2/3, lat_6= lat_{61}*1/3+ lat_{62}*2/3
- If d_5<70m, then lon_6= lon_{61}*2/3+ lon_{62}*1/3, lat_6= lat_{61}*2/3+ lat_{62}*1/3
- If d_5<100m, then lon_6= lon_{61}*3/4+ lon_{62}*1/4, lat_6= lat_{61}*3/4+ lat_{62}*1/4
- If d_5>=100m, then lon_6= lon_{61}, lat_6= lat_{61}

3 EVALUATION

In order to compare the performance of the spatial space localization method and the spatial-temporal localization method, we have conducted experiment in real existing GSM network in urban area. The fingerprints are produced by 3D ray tracing model before the training phase.

The test area is a 1km*1km urban zone. In order to produce the fingerprint we divided the test area into grids of 20m*20m.For the fingerprints, at most 20 cells with strongest RSS are recorded in the radio map. The base station emitted MR which contains RSS and Cell Identity(CI) every 480ms.We take 500 MRs as test samples, they include 8 callids. As the table 1 shows , the result of spatial-temporal localization method improved the accuracy of traditional spatial localization method.

4 SUMMARY

In this paper, we propose a new localization method called spatial-temporal localization method. It takes advantage of the spatial localization method and the temporal localization. There we chose KNN algorithm as the representative of spatial localization

Table 1. Spatial localization and spatial-temporal localization testing result.

Model Samples (Training set/Testing set)	Get the cells set <C1,C2,…,Cn> of every MR according to cell matching flows Extract the RSS of matching cell from FSmatching to get the training samples <S1,S2,…,Sn>FS; Extract the RSS of matching cell from MR to get the test samples <S1,S2,…,Sn>MR	
	spatial localization result	spatial-temporal localization result
Testing results (Localization Accuracy(m))	1. Average error: 76 2. 85.83 @67% 3. Maximum error: 377.93 Minimum error: 1.69	1. Average error: 74.05 2. 82.11 @67% 3. Maximum error: 221.03 Minimum error: 1.46

method and L filter as temporal localization method. Base on the selected fingerprints, the new method works better than temporal localization algorithm in the experiment we conducted.

REFERENCES

[1] Teemu Roos, Petri Myllymaki, Henry Tirri. A Statistical Modeling Approach to Location Estimation. IEEE Transactions on Mobile Computing 1(1), 59–69 (2002).

[2] Qimei Cui, Xuefei Zhang. Research Analysis of Wireless Localization with Insufficient Resources for Next-Generation Mobile Communication Networks. International Journal of Communication Systems, 2012.

[3] C Feng, W Au, S Valaee. Received Signal Strength based Indoor Positioning using Compressive Sensing. IEEE Transactions on Mobile Computing, vol.9, 2011.

[4] Ma Lin, Xu Yubin, Zhou Mu. Accuracy Enhancement for Fingerprint-Based WLAN Indoor Probability Positioning Algorithm. 2010 First International Conference on Pervasive Computing, Signal Processing and Applications, 2010.

[5] Petteri Nurmi, Sourav Bhattacharya, Joonas Kukkonen. A Grid-Based Algorithm for On-Device GSM Positioning. The 12th ACM international conference on Ubiquitous computing ACM New York, NY, USA, 2010.

Computing, Control, Information and Education Engineering – Liu, Sung & Yao (eds)
© 2015 Taylor & Francis Group, London, ISBN: 978-1-138-02800-5

Compatibility study for polymer blends based on MesoDyn calculations

Min Zhang
School of Mechanical Engineering, Shandong University, Jinan, China
Ministry of Education Key Laboratory of High Efficiency and Clean Mechanical Manufacture, Shandong University, Jinan, China

ABSTRACT: Mesoscopic Dynamic (MesoDyn) density function simulation was used to study the compatibility of polymer blends. The mesoscale model for two systems, PMMA/PVC and PP/PA6, were set up. The free energy, enthalpy, entropy, order parameters and density profiles were got from the simulation. The results showed that the free energy density keeps constant negative value for PMMA/PVC system and positive value for PP/PA6 system. The order parameter is much less than 0.1 for PMMA/PVC and about 0.19 for PP/PA6. Phase separation behavior was not observed during the whole simulation time for PMMA/PVC model. But phase separation behavior was emerged in the density fields of PP/ PA6 blends. These showed that PMMA and PVC is miscible, while PP and PA6 is not. The mesoscale simulation results were confirmed with experiments.

KEYWORDS: Mesoscopic dynamic, Compatibility, Polymer blends, Free energy density, Density profile.

1 INTRODUCTION

Blending different polymers has been widely used to obtain products with desirable properties that are not necessarily possessed by an individual component polymer. With the development of hardware and software, molecular simulation has become one of the most important tools to predict or validate structure-property relationship of polymers, blend compatibility, and phase behavior of polymers [1]. The Mesoscale Simulation for the miscibility of polymer blends can be effectively studied using the MesoDyn method [2]–[3]. The phase separation dynamics is described by Langevin equations for polymer diffusion [4]–[5]. In this paper, we report calculations of two different blends, PMMA/PVC and PP/PA6, using the mesoscale simulations based on the Material studio software. Our aim is to explore the compatibility and phase separation behaviors of two kinds of polymer blends.

2 NUMERICAL MODEL

Mesoscopic dynamics simulation is based on the density functional theory in condensed matter physics, so it is suitable for the polymer substance. There are two parameters, beads number n and interaction between beads $v^{-1}\varepsilon$, used in MesoDyn simulations to describe the number of Gaussian beads and the interaction between beads. To map the representative polymer chains onto Gaussian chains, the bead number (n) can be got by Eq.1:

$$n = \frac{N_{mono}}{C_\infty} \tag{1}$$

here n is the number of the beads, N_{mono} is the chain length of monomer and C_∞ is the characteristic ratio of monomer. The chain length is 50 for PMMA, PVC, PP and 10 for PA6. The characteristic ratio is calculated in the Synthia module of the software. We got the number of the beads. The species model is A6B7 for PMMA/PVC and A9B2 for PP/PA6.

The interaction between beads, $v^{-1}\varepsilon$, can be calculated by Eq.2[6]:

$$v^{-1}\varepsilon = xRT \tag{2}$$

here x is the Flory-Huggins parameter, R is the gas constant (8.314J/mol·K) and T is the temperature.

Research showed that the Flory-Huggins parameter could be calculated from Eq.3 [7]:

$$x = \frac{V_m \Delta E_{mix}}{RT} \tag{3}$$

here V_m is the mole volume of the mono, ΔE_{mix} is the energy change of mixing per unit volume. It can be got from Eq.4:

$$\Delta E_{mix} = \varphi_A\left(\frac{E_{coh}}{V}\right)_A + \varphi_B\left(\frac{E_{coh}}{V}\right)_B - \left(\frac{E_{coh}}{V}\right)_{mix} \tag{4}$$

here φ_A and φ_B are the volume fractions of A and B in the mixed system, and $E_{coh}/V = CED$, the Cohesive

Energy Density(CED) can be calculated from Moecular Dynamics(MD) simulation.

We calculated the interaction between beads based on MD simulation. In the PMMA/PVC system (the weight fraction of PMMA is 70%), the interaction is 0.86 KJ/mol. In the PP/PA6 system (the weight fraction of PP is 25%), the interaction is 22.36 KJ/mol.

Solve space is specified to density and potential method which is the traditional method of performing dynamics in MesoDyn. The compression parameter is the proportional factor of local density growth and chemical potential. The higher the value, the more incompressible the system becomes. The Mesodyn simulation recommended value between10-20 [6]. We set the compression parameter to 10.The grid dimensions are 32nm× 32nm×32nm. The time step for the simulation is set to 20ns, and the total simulation time is 200μs. The noise parameter value is 75. The temperature is 298K. The bead diffusion coefficient parameter is default value, 1.0×10^{-7} cm^2/s.

3 RESULTS AND DISSCUSSION

3.1 *Free energy density*

For the mesoscale simulation, it is very important to confirm that the system get into equivalent circumstances before we analyze the reliability of results. As two components mixed, the free energy is a valid parameter to confirm equivalent system. The thermodynamic condition in a miscible system is that the free energy remains negative value.

Fig. 1 shows the free energy density, enthalpy and entropy change with increasing time. As the curves in Fig.1 (a) show, the free energy density keeps constant negative value from the beginning. So the compatibility of two blends, PMMA and PVC, is good in this system. Compared to Fig.1 (a), the curves in Fig.1 (b) are different both in shape and values. The curves drop sharply at the beginning, and then they keep constant values. As the free energy density value is positive, it indicates that the phase in the blends phases would separate as the equivalent occurs.

(a) PMMA/PVC

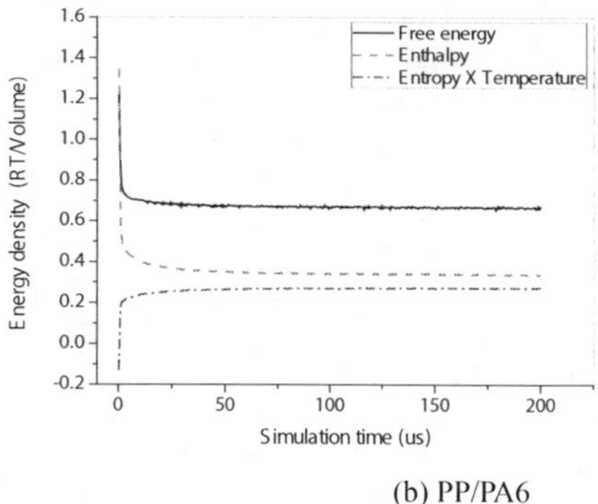

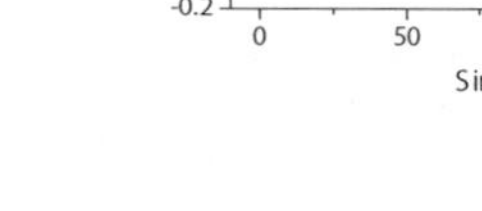

(b) PP/PA6

Figure 1. The free energy, enthalpy and entropy for blends with increasing time.

3.2 *Order parameter*

The order parameter is defined as the volume average of the difference between local density squared and the overall density squared. A decrease in P indicates better miscibility, and the polymer phases mix more randomly. When P values of blends exceed 0.1, macrophase separation will occur [8]. Fig.2 shows the order parameters with increasing simulation time. The curves indicate that the order parameter is less than 0.0014 for PMMA/PVC blends and bigger than 0.13 for PP/PA6 blends. So PMMA/PVC is miscible system and PP/PA6 is not. These are the same as the experiment [9].

3.3 *Isosurface of the density fields*

Fig.3 and Fig.4 are the isosurface of the density fields' evolution with simulation time for two systems. From Fig.3, we can see that there is no phase merge phenomenon for the same phase and no separation behavior for the different phase. It indicates that the PMMA/PVC system is miscible. But there is merge phenomenon for the same phase and the emergence of separation behavior for the different phase in PP/PA6 system (Fig.4). It shows that the phase keeps separation in the equivalent process of PP/PA6. So the phase separation behavior of PP and PA6 can be studied visually by MesoDyn simulation.

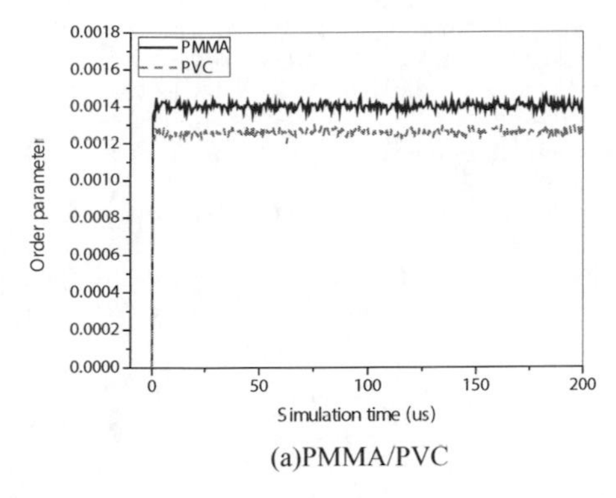

(a)PMMA/PVC

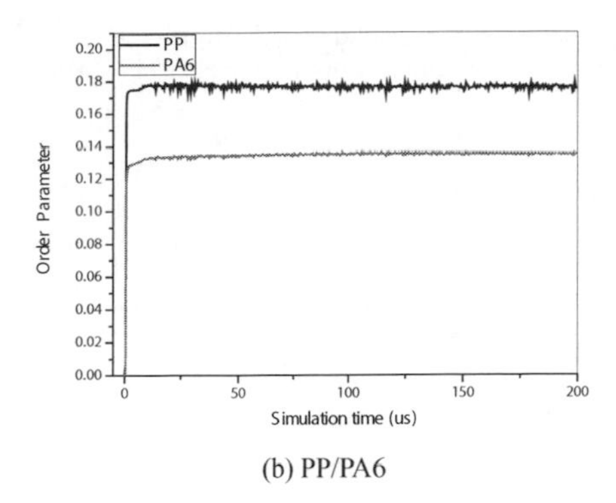

(b) PP/PA6

Figure 2. The order parameter for blends with increasing time.

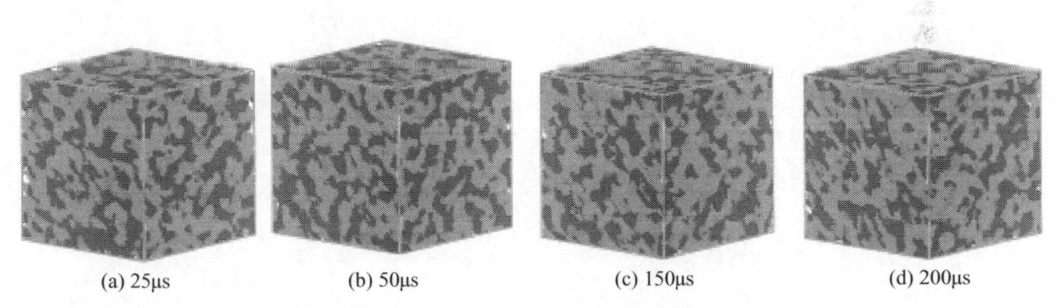

(a) 25μs (b) 50μs (c) 150μs (d) 200μs

Figure 3. The isosurface of the density fields for PMMA/PVC blends with time evolution.

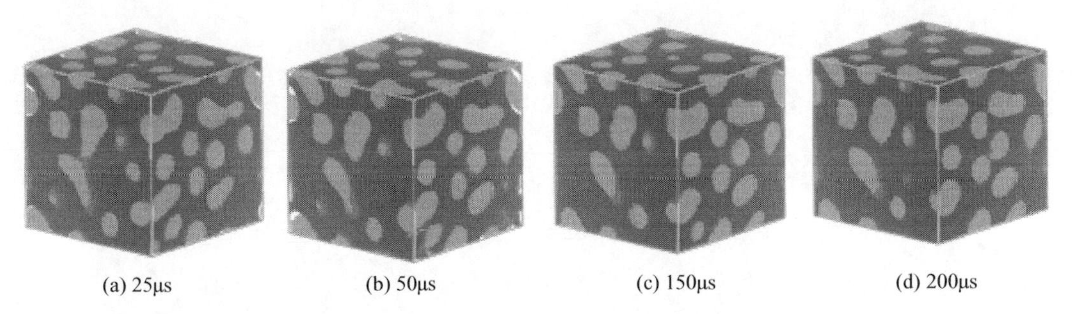

(a) 25μs (b) 50μs (c) 150μs (d) 200μs

Figure 4. The isosurface of the density fields for PP/PA6 blends with time evolution.

4 CONCLUSION

The mesoscale simulations are used to predict the miscibility of PP/PA6 system (the weight fraction of PP is 25%) and PMMA/PVC system (the weight fraction of PMMA is 70%). Phase separation behavior occurs at about 50μs from the beginning for PP/PA6 blends. At the same time, the free energy density attain to a constant value. The order parameters get to constant values also. So mesoscopic simulations confirmed blend immiscibility of PP and PA6. The free energy density keeps constant negative value for PMMA/PVC blends. The value of order parameters is less than 0.01. Phase separation was not observed during the simulation time. So PMMA and PVC is compatible in the system.

ACKNOWLEDGEMENT

This research was financially supported by the National Science Foundation of China (51103080).

REFERENCES

[1] Ahamadi A., Freire J.J. Polymer, 2009, Vol.20 (2009), p. 3871.
[2] Yeng-Ming L., Goldbeck-Wood G. Polymer, Vol. 44 (2003), p.3593.
[3] De Arenaza I. M.,Meaurio E.,Coto B., et al. Polymer, Vol.51(2010), p.4431.
[4] Yin Q., Luo L.H., Zhou G. et al. Molecular Simulation, Vol.36 (2010): p.186.
[5] Gupta J., Nunes C., Vyas S., et al. Journal of Physical Chemistry B, Vol. 115(2011),p.2014.
[6] Materials Studio 3.0, Accelys Inc., San Diego, CA, 2004.
[7] Case F.H., Honeycutt J.D. Trends Polymer Sci., Vol.2 (1994), p.259.
[8] Spyriouni T., Vergelati C. Macromolecules, Vol. 34(2001), p.5306.
[9] Jinghua Yin, Anna Zheng, Jing Sheng. etal. The reactive processes of the polymer. Science Press, Beijing, 2008.

Computing, Control, Information and Education Engineering – Liu, Sung & Yao (eds)
 ISBN: 978-1-138-02800-5

Business intelligence scouting and monitoring

Chen Sheng Wu, Ru Liu, Qiong Wu & Rong Li
Beijing Municipal Institute of Science and Technology Information, Beijing, China

ABSTRACT: In the present under the uncertain and change prevailing environment atmosphere's influence, competitive intelligence without well-grounded methodologies and system for scouting and monitoring dynamic intelligence has become impossible. Dynamic intelligence process refers to the continuous process of gathering, analyzing and reporting information about specified topics to users. The dynamic intelligence process should always be anchored to four key factors (Industry, Technology, Key Person and Customer) within which information will be used. What's more, we also need to know how to set up systems and models to identify and analyze competitors first of all. This project was funded by the Beijing financial fund

KEYWORDS: Dynamic intelligence, Identification of competitors, Scouting and Monitoring.

1 INTRODUCTION

Dynamic intelligence scouting and monitoring as a key success factor of competitive intelligence 21st Century has been marked by increasing uncertainty and rapidly accelerating change in the business environment of many industries. Major change drivers in the Big Data environment, including, continuing economic globalization, rapid technological innovation and moving are well known. But as drivers in the rapid technological innovation environment interact with each industry's own trend and uncertainties, we find that the range of imaginable futures of individual enterprise has grown manifold in comparison to earlier decades. As technological innovation becomes the driving force for competitive advantage, and competitive intelligence becomes the key driver of business innovation, the application value of enterprise-level Dynamic Intelligence Scouting and Monitoring System(DISMS)is reaching new heights.

Dynamic Intelligence Scouting and Monitoring System and methodologies can be used as a stepping-stone in building up strategic intelligence capabilities. Establishing "Dynamic Intelligence Scouting and Monitoring System" has become commonplace in international companies. Behind this development is the idea of the central nervous system in a scenario planning exercise and strategic interactions, it also is in a key position to help the companies deal with uncertainty and change. By using Dynamic Intelligence Scouting and Monitoring System Tools installed in the corporate intranet, people can obtain actionable information about external S&T developments and trends from their competitors which can affect a company's competitive position.

2 GENERAL METHODS OF NEEDS ANALYSIS

Various methods exist for analyzing information needs in an international company, the most of which can be used individually or by combining a number of methods into a cohesive whole. One proven fact is, however, that the analysis has to be conducted in a proactive way, as information users generally do not come and tell about their information needs.

Selecting the analysis method depends on a variety of factors. First, the objective of the data sought must be determined. Standardized techniques are more useful in the collection of objective data than on subjective data. Also the degree of structure sought in the data collection needs to be determined. Unstructured data will require more open-ended questions and more time than structured data.

In general, the mostly used analysis methods are the following:

- Focus Group Discussions & Brainstorming
- Personal Interviews
- Surveys
- Checklists
- Previous Experience, Existing Systems & Common Sense

3 BASIC MODEL OF DISMS

Dynamic intelligence process refers to the continuous process of gathering, analyzing and reporting information about specified topics to users. The dynamic intelligence process should always be anchored to four key factors (Industry, Technology, Key Person and Customer) within which information will be used. In practice, the output delivered by the dynamic intelligence process should find its place as part of the strategic planning process, marketing reviews and technology innovation. So, it is very necessary to construct a new model in order to dynamically track and monitor intelligence information.

The main motivations for dynamic intelligence scouting and monitoring system are future watch, early warning and identification of competitors (Figure 1).

Dynamic industry intelligence scouting and monitoring. Dynamic industry intelligence is the implementation of a new industrial mode, involving significant changes in product design or packaging, product placement, product promotion or pricing.

In the Dynamic intelligence scouting and monitoring system, individual data (main body, time, place, do something, with whom and so on) will be extracted, which describe an intelligence information of certain key person into one timeline.

Dynamic intelligence scouting and monitoring consumer feedback. Production performance and consumer feedback is ultimately one of the major driving factors for any company. Production performance should always be carefully scouted and monitored as it relates to the user experience. By monitoring consumer behavior, companies can improve the production performance to provide the optimal user experience. This will not only attract more customers to those products, but enhance market competitiveness as well.

4 DYNAMIC INTELLIGENCE SCOPE AND SCOUTING & MONITORING

Defining the scope of the dynamic intelligence operation translates as listed out corporate functions that should be using intelligence deliverables, and topics and themes that each of them are more interested in. Additionally, the degree of future orientation needs to be determined; Looking into the rearview mirror is a good starting point, but a mature dynamic intelligence operation also needs to spend a lot of time on outlining possible future scenarios of the operating environment.

The modern business world is affected by global change and turbulence. Companies need to identify and understand new trends, regulatory change, political change, competitor moves and customer perceptions. A well functioning Dynamic Intelligence system is an important capability that will enable companies to identify and understand this change. It depicts the different levels of the operating environment along with the value chain dimension and the perspective of global business through different market regions.

5 IDENTIFICATION OF COMPETITORS

Due to the model we described above, we want to find out about how to scout and monitor dynamic intelligence. But we must know how to set up systems and models to identify and analyze competitors and their products first of all. Enterprises do not merely have to be good at meeting the needs of customers; it has to be better than exact competitors. Competitor identification and analysis play an important role in the planning activities of enterprises. Indeed, enterprises that focus on the dynamic competitor's actions have been found to achieve better business performance than those who pay less attention to their dynamic competitors.

However, competitor identification and analysis that add value to corporate wealth has been a difficult task for many managers. The following figure is to provide an insight from the field of strategic management in general to identify dynamic competitors based on recognizable competitors at present and potential competitors. This has helped to define competitors to be competitors in business domains, competitors in the relevant field, competitors with relevant capacity, new innovative competitors, and unknown competitors.

Furthermore, recognizable competitor analysis and potential competitor analysis have been done by using Dynamic Intelligence Scouting and Monitoring System to address the information needs in competitor analysis. This has helped to create the needed analysis processes and practices to achieve valued competitor analysis. A suitable identification work methodology could be briefly described as follows. The method can be broken down into two aspects along with dynamic change of competitors:

- Identification of recognizable competitors at present
- The impacts on industry, media, government and public
- The competitiveness of the product and service
- Market share and recent increase
- The number of times core services and productions appear
- Patented technology share
- Historic development of product and service competition
- Identification of potential competitors
- The credibility of potential threat of other enterprises

6 EFFICIENCY OF DECISION-MAKING

As an intelligence support function, the key purpose of competitive intelligence is to enable people in the organization to make better informed decisions. It is therefore important not only to investigate whether people consider CI to be beneficial or not, but also whether it actually improves decision-making in organizations

In companies with formal competitive intelligence in place, decision-making is rated significantly more efficient than for companies without a formal function in place. The most noticeable difference in companies that have CI is that information, to a much greater extent, is readily available when needed.

First and foremost, not all relevant information is publicly available and thus becomes more difficult to collect. However, another plausible explanation is that intelligence teams often do not know the exact intelligence needs of their internal customers, and decision-makers' needs in particular. This often results in the CI function, providing intelligence which is either inaccurate, or incomplete with regards to the needs of the decision-makers.

7 THE DEVELOPMENT TREND OF DYNAMIC INTELLIGENCE

In the present under the uncertain and change prevailing environment atmosphere's influence, competitive intelligence without well-grounded methodologies and system for scouting and monitoring dynamic intelligence has become impossible. In the future, additional trends and developments that were raised in the survey through the questions and open comments included:

Increasingly visualized intelligence deliverables: Using graphs, dashboards and scorecards to visualize the analytical output of the dynamic intelligence process as opposed to delivering results in plain text and figure format. This trend again ties in with resourcing the intelligence function adequately: Producing insightful visuals requires time, highly analytical thinking and a solid understanding of the company's business fundamentals.

Adding the Early Warning & Opportunity perspective to existing dynamic intelligence deliverables: Interpreting market signals and analyses from the perspective of both negative and positive risks for the company will increase the strategic value of the intelligence deliverables. The early warning and opportunity perspective will also provide a framework for assessing the relative importance of different developments in the operational environment of the company.

Personalized delivery: While it is not meaningful to the intelligence team to even aim at personally delivering all intelligence output, much of the greatest strategic value is typically created not alone by either decision-makers or intelligence professionals, but in groups of both. Therefore, as the intelligence deliverables develop towards increased sophistication, the survey respondents expect to see more and more of briefings, workshops, and informal discussions as the delivery format of strategic level intelligence output.

Competitive intelligence by every definition is about looking to the future and providing actionable insights. Competition always is in a state of dynamic development; a business competitor is also dynamically changed from potential competitors to recognizable competitor. Innovation has occurred in the context of a rapidly-modernizing and globally integrating business environment, and enterprises are adjusting quickly to the needs of the changing economic environment. They have to monitor their dynamic business environment and update their strategy choices over time. In practice this means repeating the process over and over again.

Dynamic intelligence scouting and monitoring will build around the key success factors of competitive intelligence. The competitive intelligence industry in the future is heading to more sophistication, integration of business processes, impact of decision making.

REFERENCES

[1] Wang Zhijin&Liu Bing, "Enterprises Competitors Tracking Base on Dynamic Environment", Competitive Intelligence, 2008.

[2] GIA (Global Intelligence Alliance) White Paper. "Building Strategic Intelligence Capabilities through Scenario Planning", 2005.

[3] GIA (Global Intelligence Alliance) White Paper. "MI Trends 2015 – The Future of Market Intelligence", 2010.

Computing, Control, Information and Education Engineering – Liu, Sung & Yao (eds)
 ISBN: 978-1-138-02800-5

Energy from waste

Ru Liu, Chen Sheng Wu, Lu Ji Zhang & Qiong Wu
Beijing Municipal Institute of Science and Technology Information, Beijing, China

ABSTRACT: With the growing cost of transporting rubbish, more and more cities, especially some big cities will be forced to build their own recycling plants before long. At that time, everything which goes into dustbin would be made into something useful, everything has its place, and both the dangerous and unpleasant wastes can be put to good use. This article shows us three main ways in which municipal solid waste (MSW) is treated at present. And then we discuss the benefits of these ways. This project was funded by the Beijing financial fund.

KEYWORDS: waste energy, disposal, Municipal Solid Wastes (MSW), biomass.

1 INTRODUCTION

Merriam-Webster defines waste as "refuse from places of human or animal habitation." The World Book Dictionary defines waste as "useless or worthless material; stuff to be thrown away." Unfortunately, both definitions reflect a widespread attitude that does not recognize waste as a resource.

Today, we can burn garbage in special plants and use its heat energy to make steam to heat buildings or to generate electricity. There are lots of electric power companies having already burnt another type of solid material to make electricity all over the world. Energy from waste provides a double environmental benefit: The diversion of waste coming from landfill; The immediate recovery of energy, displacing fossil fuel alternatives and reducing greenhouse gas emissions. And there are three kinds of resources of the waste energy: municipal solid wastes (MSW), industrial/commercial wastes and other wastes.

1 MSW—"basically waste generated in household, schools, markets, gardens."
2 Industrial/commercial wastes— "all types of wastes generated by stores, offices, restaurants, warehouses, industrial processes and manufacturing."
3 Other wastes— "sewage sludge, dredged material and so on". [1]

2 THREE MAIN WAYS TO DISPOSE THE WASTE

As we know, both ordinary wastes and harmful wastes are all utilized as useful sources of energy. What we just need to separate different kinds of waste into the same kind. But the process of conversation is not an easy work. There are different kinds of wastes, such as municipal solid waste, industrial and commercial wastes and so on.. And different wastes can make different effects, for example, the waste from industry can generate some toxic materials, which is harmful to peoples' health. So we need to classify different wastes, and then adopt different methods to dispose them. Generally speaking, the wastes are separated into 1—6 containers, and then be disposed in specific treatment plants. Various systems differ from each other and also set different requirements for further treatment. The waste fractions are forwarded to the most suitable treatment on the basis of fractionation.

In practice, there are three main ways in which municipal solid waste (MSW) is treated at present:

2.1 *Disposal in landfills*

Landfill is the most common means in disposing municipal solid waste, especially in developing countries. Both ordinary wastes and harmful wastes—municipal solid waste, most of household wastes and some harmful wastes from industries—are all utilized as organic materials. Such materials can be decomposed by specialized bacteria under suitable anaerobic digestion conditions during disposal of landfill, which can successfully generate useful landfill gas to benefit human. The condition of landfill is neither warm nor wet, so the process is much slower. And the end product (the landfill gas) is a mixture which is consisted mainly of the gases of CH4 and CH2.

The landfill gas is directly a source of energy to generate electricity and heating. An airtight disposal system for extracting the landfill gas is very necessary during the process of landfill disposal, therefore each area is covered with a layer of impervious materials after the waste is filled, and then the gas generated from waste is collected by an array of specific pipes, finally the end product—landfill gas—will be obtained from gas collection system.

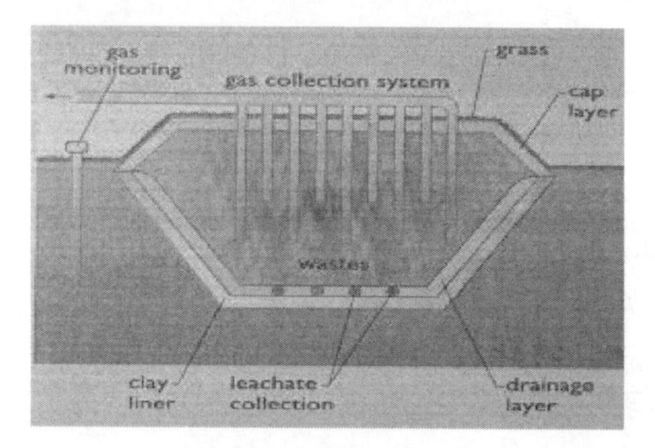

Figure 1. Extraction of landfill gas [2].

2.2 *Combustion*

Combustion is the process of waste to energy. The method is the same as coal for energy. We use the waste plant burning for making electricity as shown in Figure 2.

At first, Burning and releasing heat, and the heat of burning turns water into steam, then, the high-pressure steam turns the turbine generator to produce electricity, at the last, a utility company sends the electricity to homes, schools and offices.

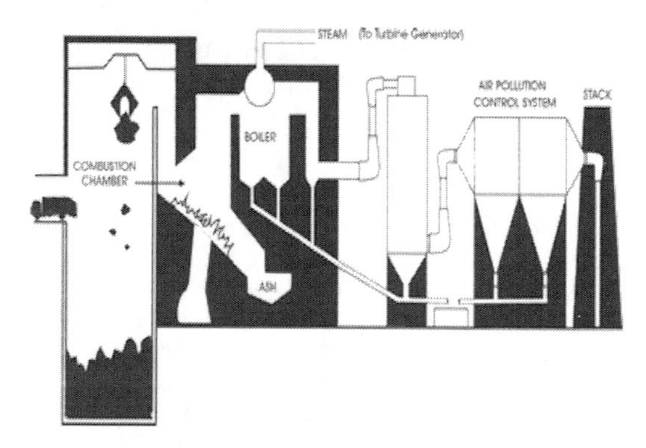

Figure 2. Combustion [3].

The end product is rich energy fuels. We can get the useful garbage from plants. Farmers can breed the special plants. Generally speaking, in 100 pounds of garbage, more than 80 pounds of garbage can be burned. That means the rate of combustion is more than 80%. Besides, a ton of rubbish can generate about 525 kilowatt of electricity, which can heat a typical office building for one day. After high temperature burning, the garbage retune to solid residue, ash. We can use the ash for construction and road building.

2.3 *Disposal in anaerobic digesters*

The organic fraction of domestic waste can base on the landfills. Big digesters will recover of biogas from landfill. This method has several benefits:

a. It will be built over the landfill plants, and it is close to urban. So, transport fee can reduce; b. The methane will release from the digesters. So we can use the combustion heat as electricity; c. The ash after burning can deal with in the way of landfill or can be used construction and road building.

The following diagram shows an integrated waste materials plant. This plant in Florida has facilities for recovery of metals and removal of plastics, followed by anaerobic digestion of the remainder. The solid residue from the digester serves as fuel for power production.

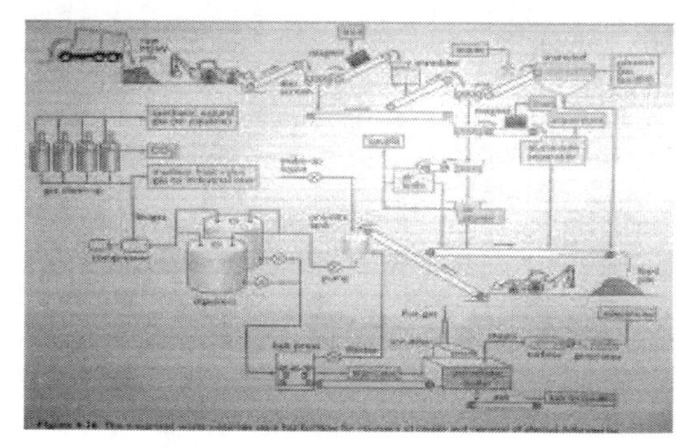

Figure 3. An integrated waste materials plant [4].

3 CONCLUSION

In china, the waste is not used for energy purpose. The significant and critical reason is people have a weak awareness about waste disposal. In addition, invested capital in this aspect is not enough and the advanced technology is deficiency, which is directly factors affecting the use of waste energy in china. People usually do not assort it before they throw them away, which mean all kinds of waste are mixed. The situation brings a giant trouble to the sorting. The factory sorting waste must pay much money, time and energy to separate them before the disposal. Another problem is wastes make the second-pollution when they are mixed, for the wastes can make chemistry-reaction and produce the dangerous materials.

Compare these three disposal ways.

Item	Landfill	Combustion	Anaerobic digester
Locality	Occupy large area and far away from city	Can near the city The distance of transport is less than 10 Km	Need to far away from density area of city for the effect of smell
Production market	Marsh gas can be used to generate electricity	Heat and electricity are easy to be utilized	It is difficult to make production market
Using of resource	Renew the use of land	Electing waste can reclaim the part of matters	As agricultural fertilizer; reclaim the part of matters
The effect to surface water	Need to take some measures to prevent the pollution	It is likely in filling residue, but the degree is low	It is likely in filling the non-compost, but the degree is low
Air pollution	It can be controlled by covering, collecting and transmitting gas	The atmosphere a certain pollution if the smog cannot be disposed well	Generate the slight gas
Soil pollution	The site must be limited in the landfill region	No	Need to control the content of harmful matters

Fossil fuels are limited, it will be used up one day if people use it too much or much quickly. In fact, energy from wastes is a good substitute for fossil fuels. It can diminish the occupied area of the waste, which can beautify our environment to great extent, besides the substitute can also refrain the serious phenomenon of globe warming coming from burning fossil fuels. With the growing cost of transporting rubbish, more and more cities, especially some big cities will be forced to build their own recycling plants before long. At that time, everything which goes into dustbin would be made into something useful, everything has its place, both the dangerous and unpleasant wastes can be put to good use. So we could say: "the rubbish will lose its meaning" That is our final intention.

REFERENCES

[1] guidelines for the storage and collection of residential, and institutional solid waste. Chapter 1, part 243(1995).

[2] Godfrey Boyle, Renewable Energy-power for a sustainable future, PageP121.

[3] EIA, http://www.eia.doe.gov/kids/energyfacts/saving/recycling/images/trasharoundtheworldburn.gif.

[4] Godfrey Boyle, Renewable Energy-power for a sustainable future, Page163.

Computing, Control, Information and Education Engineering – Liu, Sung & Yao (eds)
© 2015 Taylor & Francis Group, London, ISBN: 978-1-138-02800-5

Research of chain-roller type half-feed peanut picking device

Xiao Lian Lü
Key Laboratory of Modern Agricultural Equipment, Ministry of Agriculture, Nanjing, Jiangshu, China
College of Machinery and Automotive Engineering, Chuzhou University, Chuzhou, Anhui, China

Zhi Chao Hu, Bao Liang Peng & Zhao Yang Yu
Nanjing Research Institute for Agricultural Mechanization, Ministry of Agriculture, Nanjing, Jiangshu, China

ABSTRACT: In order to improve the working performance of half-feed picking device, the components' collocation relationship and picking process were studied. The composition and working principle of picking device were introduced. The configuration relationships in between the components were studied. These were analyzed on the motion characteristics and hitting force of the blade, and the influences factors and rules of the effecting picking frequency. The analysis results shown: the configuration mode of the roller and clamping conveyor chain adopt the oblique configuration, and picking roller adopts gradually tight angle and differential phase configuration mode, which can effectively increase picking space, avoid missing picking, reduced the damage rates and power consumption. These factors were comprehensive influences to the picking performance of the roller rotational speed n, picking roller length Lz, the angle between roller δ, roller diameter D and clamping conveyor speed v. The greater the roller rotating speed is, the stronger the blade hitting and comb-brush role is, the greater the influence of the clamping conveyor chain speed is, the higher the picking net rate is, but the greater the damage rate of pod is and the higher the impurity rate is; with the increase of the roller turning angle, the hitting force is reduced, comb-brush role is enhanced, the picking net rate and damage rate are decreased, the influence of the clamping conveyor chain speed is decreased too; the bigger the level deflection angle of the picking roller is, the greater the axial hitting force is, the smaller the radial hitting force is, and the greater the influence of the clamping conveyor chain speed to picking quality is. When the Lz decreased and δ increased, the picking time of the peanut region will shorten, it directly reduces the picking frequency, and then the picking quality has affected. The study provides a theoretical basis and technical reference for research and development of the half-feed peanut picking device.

KEYWORDS: peanut picking device, configuration mode, working parameters, picking space, movement analysis.

1 INTRODUCTION

The peanut picking is an important operation link in peanut mechanization harvest, also is the core technology in the peanut combine harvester. The structure and operation parameters of the picking device determine to the operating performance of the peanut combine harvester. According to the feeing ways of the peanut picking device, it divided into half-feed type and the whole-feed type [1-2]. The whole-feed picking device is mainly used for picking a dry peanut. It has the characteristic of a large amount feeding quantity, highly productive, large power consumption, etc.; The half-feed picking device can be used for picking fresh and dry peanut, It has the characteristic of small feeding quantity, lowly productivity, low power consumption [2-5]. It can better meet the needs of harvesting fresh peanut. It has a large market, especially in peanut production areas in southern China [6]. But compared with the whole-feed picking device, it has relatively complex structure and transmission system and highly technical requirements, and the picking quality is influenced by structure configuration and movement parameters of picking parts, and aspects of the plant. In the paper, the spatial location relationship of components and picking process is theoretical studied for self-development half-fried peanut picking device, the main influence factors and influence rule of the affecting picking quality is analyzed. The study provides a theoretical basis and technical reference for improving the picking performance of picking device.

1.1 *Structure and working principle*

As shown in figure 1, the picking device mainly includes: clamping conveyor chain, picking roller, picking blades, transmission system, etc. [7]. The picking device adopts double roller, the clamping conveyor chain is located above the junction of the picking rollers. The clamping conveyor chain driven by the sprocket, the clamping force can be adjusted by the springs. When the device working, the plant is conveyed by clamping conveyor chain, and one by one enters into the picking range of the double rollers. Under the continuous action of the picking blades, the separation of the peanut pods and seedlings is completed, and the seedling is conveyed to the tail of the device and thrown by clamping conveyor chains.

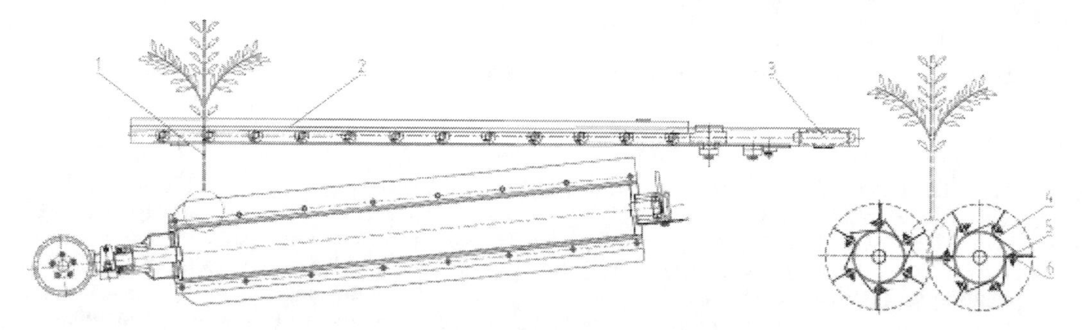

1. Peanut plant 2. Clamping conveyor device 3. Drive sprocket 4. Picking roller 5. Picking blade 6. Mounting bracket

Figure 1. Structural and working principle schematic of peanut picking device.

2 CONFIGURATION MODE OF THE PICKING ROLLER

Configuration mode with clamping conveyor chain. The configuration mode of the picking roller and the clamping conveyor chain has a larger influence to picking performance in the half-feed picking operation. The configuration mode usually has two kinds: the parallel configuration and oblique configuration [7]. The relative position of the pods region and picking roller is fixed in the parallel configuration, as shown in figure 2 (a). In order to avoid missing picking, all pods must be entering into picking range. The parallel configuration mode exist problems as follows: in order to made all peanut enter into the picking range, the size of picking roller should be very big; in actual operation, the distribution range of the pods is bigger, and the space position is difficult to be controlled, so that easy to cause missing picking; in the picking process, picking strength is decreased and power loss and blockage are increasing with picking quantity decreases. The relative position of the peanut region and picking roller is changed with the different clamping peanut seeding height in oblique configuration, as shown in figure 2 (b). The oblique configuration mode exist the characteristic as follows: relative bigger the picking range, good adaptability, and avoiding missing picking; picking pods are in turn completed in the picking range, so that picking strength is even and appropriate, low power loss, and uneasy blockage. Comprehensive analysis, the clamping conveyor chain and picking roller are adopted the oblique configuration mode.

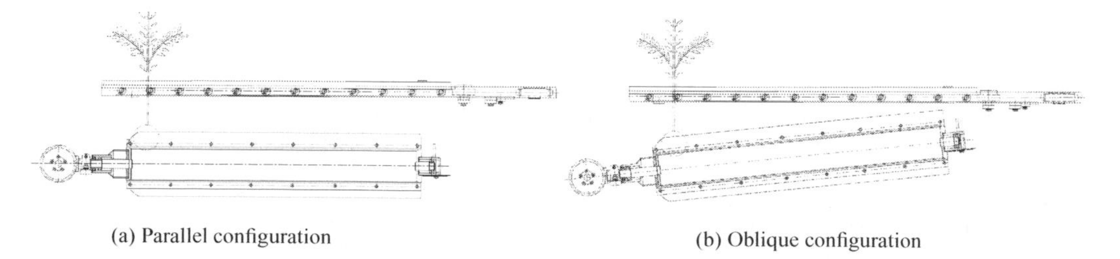

(a) Parallel configuration (b) Oblique configuration

Figure 2. Configuration mode of clamping conveyor chain and picking roller.

The configuration method of the double picking roller. The configuration mode of picking roller has two kinds of the parallel configuration and intersection angle configuration, as shown in figure 3. The shafts of the two picking rollers are parallel in the horizontal plane in parallel configuration. In picking operation, peanut quantity gradually reduces along the axis of the picking roller. In parallel configuration, the pods quantity is relatively more at the inlet of picking roller, easy causes more breakage rate and higher impurity rate. In intersection angle configuration, the center distance between two rollers is gradually decreased along the axis and the picking range is also reduced. By the analysis, considering the evenness of the picking strength and stability of working performance, the picking roller adopts an intersection angle configuration.

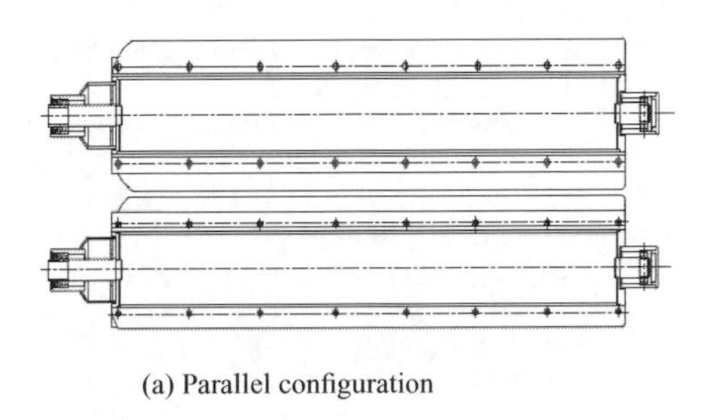

(a) Parallel configuration

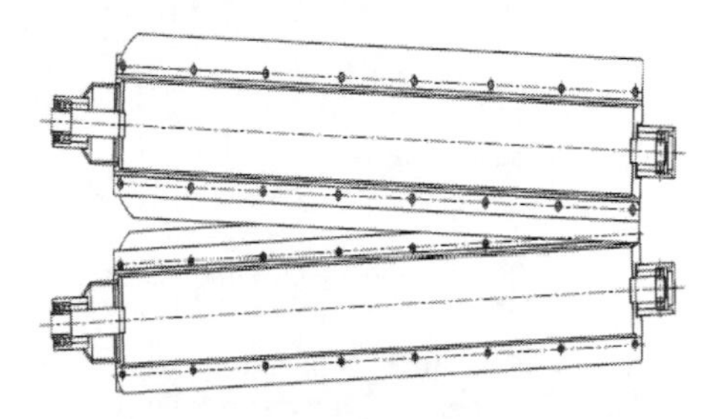

(b) Intersection angle configuration

Figure 3. Configuration mode of the picking rollers.

3 ANALYSIS AND CALCULATION IN THE PICKING PROCESS

Movement analysis in the picking process. In the process of picking, peanut seeding is clamped and conveyor backward at a constant speed, and the picking roller is uniform rotation. When picking device working, the speed V of any point on the picking blade can be divided into the axial velocity Vx, the radial velocity Vy, and vertical downward velocity Vz, as shown in figure 4. The speed V is as follows [8]:

$$V = \sqrt{V_x^2 + V_y^2 + V_z^2} \quad (1)$$

According to the resolution of forces, can be obtained as follows equation:

$$V_x = V_n \sin\beta = V\cos\omega t \sin\beta = \omega R \cos\omega t \sin\beta \quad (2)$$

$$V_y = V_n \cos\beta = V\cos\omega t \cos\beta = \omega R \cos\omega t \cos\beta \quad (3)$$

$$Vz = V\sin\omega t = \omega R \sin\omega t \quad (4)$$

In formula: ω-picking roller angular speed; R-M point on the blade distance from the shaft center of the picking roller; ωt-the angle of the blade turned; β-the horizontal deflection angle of the picking roller.

The analysis result shown that the hitting force of the blade is related to roller rotating speed and the distance from the shaft center. When the rotating speed is constant, the further the point of the blade is the distance from the shaft center of the roller, the greater the speed is. The greater the speed is, the stronger the hitting force is. The speed V of any point on the picking blade can be divided into three speed, the axial velocity Vx is block pods backward movement and play role of the axial hitting pods; the greater the speed of the clamping conveyor chain is, the more significant the role is; the radial velocity Vy play role of the radial hitting pods; the vertical downward velocity Vz play role of the brushing pods. By formula (2), (3), (4), can be known: the three speeds are affected by the angular speed ω of the picking roller and rotation angle ωt of the picking blade; the three speeds increased with the angular speed w increased; when the rotation Angle ωt increased, the Vx, Vy gradually decreases, and Vz gradually increase; it is that the greater the rotating speed of the roller, the stronger the hitting and combing brushing role of the blades are, the greater the influence of the speed of the clamping conveyor chain is, the greater the picking net rate is, but the greater the damage rate of pods are, the higher the impurity rate is; with the roller turning angle is increase, the hitting role is reduced, the combing brushing role is enhanced, the picking net rate and breakage rate of the pods have decreased, the influence of the speed of the clamping conveyor chain is too reduced. In addition, the Vx, Vy are affected by the horizontal deflection angle β of the picking roller. When β *is* increase, Vx is increase, and Vy is decreases. The bigger the β is, the greater axial hitting force is, the smaller the radial hitting force is, and the greater the influence of clamping conveyor speed to picking quality is. The clamping conveyor speed is almost no effect to picking damage rate in parallel configuration, but the influence is still exists to picking net rate.

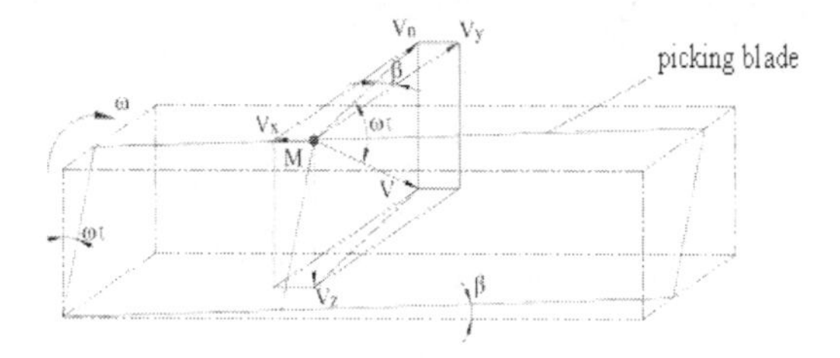

Figure 4. The peanut picking situation.

Analysis of the picking location space. The position state of the peanuts in picking roller is as shown in figure 5. The peanuts distribution area is assumed to sphere, its diameter is d; the shadow zone is picking region; the double section line shadow area is intersection area of the picking roller; the picking frequency and intensity in the area are greater in relative to the single section line shadow zone. In working, the bottom of the peanut seeding began enter to the picking area, and from bottom up gradually into the intersection area of the double rollers. That determined the space position and parameters of the picking roller need to be considered to ensure that all pods are hutted by the blades otherwise will cause missing picking. It is to meet the conditions as follows: (1) the diameter D of the picking roller must be satisfied $D \geq d+\varphi$ (d: the peanuts area; φ: the diameter of roller shaft), on the basis of the D range is 120~150 mm, it requires $D \geq 150+\varphi$ mm. (2) the length Lz of the picking roller, the intersection angle δ of the picking rollers must satisfy with $L_4 = Lz\sin\delta \geq S$ (S: vertical distribution region of the peanuts in picking range); when the clamping height of the peanut seeding is fixing, the S is equal to d, the length Lz of the picking roller, the intersection angle δ of the picking rollers must satisfy with $L_3 \geq d$, namely $Lz\sin\delta \geq d$.

In order to improve the picking quality, two picking rollers are tangent at the entrance and the intersection at the outlet. When the blade contacted with the roller shaft at the outlet, the horizontal deflection angle β of the roller shaft is big. The parameter equation is shown as follows.

$$L_Z tg\beta + L_Z tg\beta \cos 2\beta = (D-\phi)/2 \tag{5}$$

The parameter equation is obtained as follows by formula (5).

$$\sin 2\beta = (D-\phi)/2L_Z \tag{6}$$

The horizontal deflection angle β of the roller shaft is:

$$0 \leq \beta \leq [\arcsin(D-\phi)/2L_Z]/2 \tag{7}$$

The angle of the clamping conveyor chain and picking roller is according to the length of the picking roller. The smaller the length of the picking roller is, the greater the δ is, the compacter the structure of the picking parts is. This is conducive to the whole space layout, but the Lz reduced and the δ increase, the time of peanut picking will be shortened, the picking frequency will directly reduce, so that the quality of the picking has effected. The basis of structure and configuration of Lz=800mm, can get the δ=10.6° and β=10.6°.

The influence of the picking frequency. Picking frequency directly affects the picking quality. Picking frequency is influenced by the machine structure and operation parameters. In order to simplify the problem, the peanuts area is considered as a whole unit (A point in figure 6). The position of the peanut area relative to picking blade is shown in Figure 6.

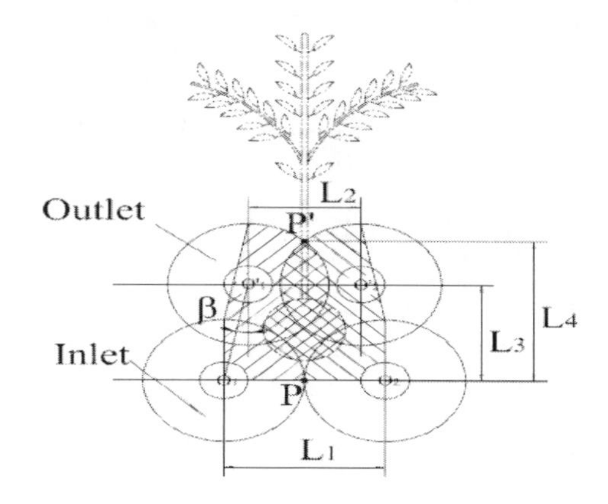

Figure 5. The relative position of the peanut region and picking roller.

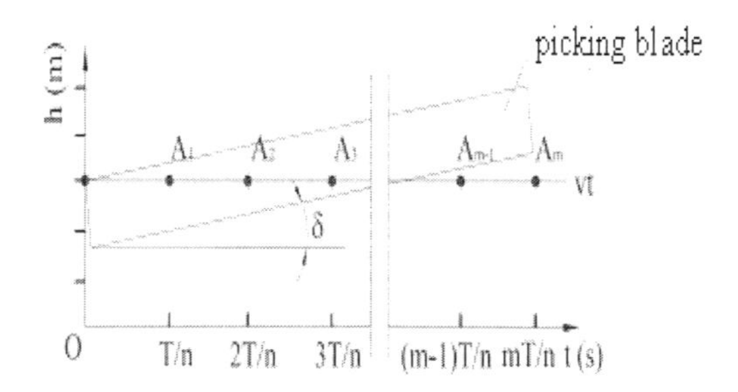

Figure 6. The relative position of the peanut region and picking blade.

The hitting number of the blade in the picking range is calculated as follows, the calculation result is rounded.

$$N_z \le \frac{(D-\phi)m}{vT\sin\delta} = \frac{(D-\phi)mn}{60v\sin\delta} \quad (8)$$

(when $\frac{(D-\phi)}{2\sin\delta} \ge L_Z \cos\delta$, $\frac{(D-\phi)}{2\sin\delta} = L_Z \cos\delta$)

In formula: *N*z-number of the blade hitting pods; δ-angle of the clamping chain and picking roller, °; *v*-clamping conveyor speed, m/s; *n*-rotating speed of the picking roller, circles/min; *T*-the period of the picking roller, min; *M*-number of the blade; *Lz*-length of the picking roller, m.

It can be known By formulas (8), the number of the blade hitting pods is proportional to rotating speed of the picking roller, a number of the blade and length of the picking roller; the number of the blade hitting pods is inversely proportional to clamping conveyor speed and sin value of the angle of the clamping chain and *sinδ*.

4 CONCLUSION

1 On the basis of theoretical analysis, the roller and clamping conveyor chain adopt the oblique configuration mode, and picking roller adopts intersection angle and differential phase configuration mode.
2 The hitting force of the blade is related to roller rotating speed and the distance from the shaft center of picking roller. The farther the hitting point is from the shaft center, the stronger the hitting force is; the greater the rotating speed of the roller, the stronger the hitting and comb-brush role of the blades are, the greater the picking net rate is, but the greater the damage rate of pods are, the higher the impurity rate is; with the roller turning angle is increase, the hitting role is reduced, the comb-brush role is enhanced, the picking net rate and damage rate of the pods have decreased.
3 The bigger the β is, the greater axial hitting force is, the smaller the radial hitting force is, and the greater the influence of clamping conveyor speed to picking quality is. In the parallel configuration of the rollers, the clamping conveyor speed is almost no effect to damage rate, but the influence is still exists to picking net rate.
4 All that Length *Lz* of the picking roller, angle δ of the roller, roller diameter *D*, rotating speed *n* of the picking roller, and the speed *v* of the clamping conveyor chain, have influence to the picking frequency. When the *Lz* reduced and the δ increase, the time of the peanut picking will be shortened, the picking frequency will directly reduce, so that the quality of the picking has effected.

ACKNOWLEDGEMENTS

The study was supported by the Key Laboratory of Modern Agricultural Equipment, Ministry of Agriculture, P.R.China (201303002), the National Natural Science Foundation of China (51375247) and the Natural Science Foundation of Anhui province of china (1408085ME103).

REFERENCES

[1] Weihua Zhong. Jiangsu Agricultural Mechanization, No. 1(2014), pp. 33–34. In Chinese.
[2] Xiangtao Yu, Zhichao Hu, Fengwei Gu. Chinese Agricultural Mechanization, No. 3(2011), pp. 10–12. In Chinese.
[3] Xiaolian Lü, Haiou Wang, Huijuan Zhang, Zhichao Hu. Chinese Agricultural Mechanization, No. 6(2012), pp. 245–248. In Chinese.
[4] Bokai Wang, Lu Wu, Zhichao Hu. Chinese Agricultural Mechanization, No. 4(2011), pp. 6–9. In Chinese.
[5] Dongwei Wang, Shuqi Shang, Kun Han. Transactions of the Chinese Society for Agricultural Engineering, Vol. 29, No. 14(2013), pp. 15–25. In Chinese.
[6] Zhichao Hu, Haiou Wang, Jiannan Wang. Transactions of the Chinese Society for Agricultural Machinery, Vol. 41, No. 4(2010), pp. 79–84. In Chinese.
[7] Zhichao Hu, Haiou Wang, Baoliang Peng. Transactions of the Chinese Society for Agricultural Machinery, Vol. 43, No. S1 (2012), pp. 131–135. In Chinese.
[8] China Agricultural Mechanization Sciences Research Institute. China Agricultural Sciences Press. (2007), In Chinese.

Computing, Control, Information and Education Engineering – Liu, Sung & Yao (eds)
© 2015 Taylor & Francis Group, London, ISBN: 978-1-138-02800-5

Study on the analysis method of the cab vibration transfer path

Yu Tang
Department of Mechanical Engineering, Guangxi University of Science and Technology, Guangxi Liuzhou, China

De Jian Zhou
Department of Mechanical Engineering, Guangxi University of Science and Technology, Guangxi Liuzhou, China
Guangxi Manufacturing Systems and Advanced Manufacturing Technology Laboratory, Guangxi Liuzhou, China

Bing Li
Department of Mechanical Engineering, Guangxi University of Science and Technology, Guangxi Liuzhou, China
Key Laboratory of Sea Machinery and Equipment Design and Manufacturing and Control in Guangxi Province, Guangxi Liuzhou, China

Hai Ten Zheng & Jian Wei Jiang
Department of Mechanical Engineering, Guangxi University of Science and Technology, Guangxi Liuzhou, China

ABSTRACT: Expounds the basic principle of transfer path analysis method. Against a cab model for the study of vibration, stated the basic steps of transfer path analysis methods, And identify the main vibration transmission path, considering the use of the method of the phase and amplitude of vibration contribution analysis, excluding the contribution of the amount of the opposite path. Meanwhile comparative analysis of the Mount transfer characteristics, and TPA results for mutual authentication, improved test effectiveness.

KEYWORDS: Transfer Path Analysis; vibration contribution; cab vibration; Mount transfer characteristics.

1 INTRODUCTION

With the rapid development of China's own-brand cars, people regarded Automotive NVH (Noise, Vibration, Harshness) performance as one of the important indexes of vehicle comfort, which directly affect people's subjective evaluation of the automobile brand, it is very valuable and meaningful to study of automotive NVH issues.

Transfer Path Analysis (TPA) is a diagnosis based on noise and vibration engineering test, mainly used in processing the excitation source - transmission path - in response to system problems, finding out the key path influence. As an approach of comprehensive noise and vibration issues, TPA help setting performance targets for each of the key components, which can help engineers determine the root cause of problems in the early stages of design [1].

2 TRANSFER PATH ANALYSIS METHOD INTRODUCTION

2.1 *Basic principle of transfer path analysis*

Assuming the vehicle is caused by road surface and engine excitation, each excitation force corresponds to many different transmission paths. All transfer functions can be grouped into a transfer function matrix, the response of vibration and sound pressure of all the car noise by vibration point excitation force a pass over and can be expressed as [2]:

$$\{P_k\}=[H]_{MN}\{F_k\}$$

$$[H]_{MN}=\begin{bmatrix} \dfrac{P_1}{F_1} & \cdots & \dfrac{P_1}{F_N} \\ & \vdots & \\ \dfrac{P_M}{F_1} & \cdots & \dfrac{P_M}{F_N} \end{bmatrix}^{-1}$$

$[H]_{MN}$: Transfer function matrix; $\{F_k\}=\{F_1\ F_2\ \ldots\ F_N\}^T$: Each column vector path passing work incentives; $\{P_k\}=\{P_1\ P_2 \ldots P_M\}^T$: Total response of $\{F_k\}$.

2.2 *Load calculation principles*

There are two methods to obtain work excitation which includes direct and indirect methods. This direct method rarely applies because the sensor installation location and vehicle structure complexity constraints. The indirect method is to measure the response of local coupling system, through changes to the inverse transfer function matrix inverse coupling and accurately obtain work excitation [3]. Therefore the inverse matrix method is widely used.

Setting the vehicle body as reference points for j and k, measured FRF and acceleration response signals of two reference points are as follows:

$$\begin{bmatrix} a_j \\ a_k \end{bmatrix} = F_i * \begin{bmatrix} H_{ij} \\ H_{ik} \end{bmatrix}$$

Work motivation can be expressed as:

$$F_i = \begin{bmatrix} H_{ij} \\ H_{ik} \end{bmatrix}^{-1} \begin{bmatrix} a_j \\ a_k \end{bmatrix}$$

3 TEST CASE ANALYSIS

This paper takes a low-speed truck as the experimental object, against the cab seat rail and the instrument panel to TPA, the test process is as follows:

1 Operating condition experiment: in accordance to the relevant standards of experimental TPA, choose a company experiments road, the road conditions are met the GB7031 B class road. The test vehicle selection 40km/h, 50km/h, idle, cobblestone road, washboard road five kinds of conditions in data acquisition. Experimental conditions are shown in Fig.1.

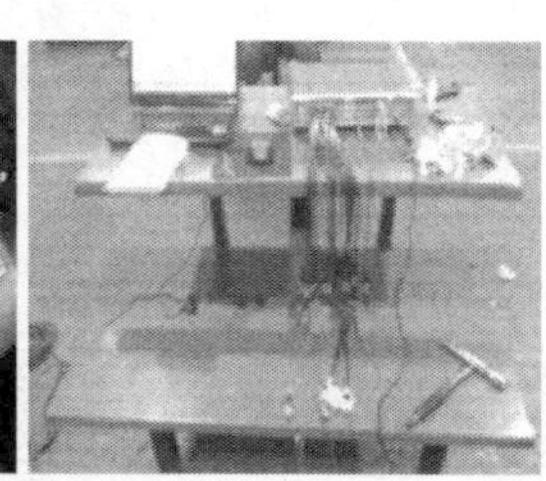

Figure 1. Response point installation location and require equipment.

2 FRF experiment: In order to obtain accurate transfer function, experiments need to disassemble the excitation source. Cab FRF experiments using three-point lifting, contact position using a car tire tube connection, to achieve a flexible connection of the cab.
3 After data acquisition is complete, use the LMS/Transfer Path Analysis function for data processing and transmission path identification, find out the greatest impact on the vibration path to optimization.

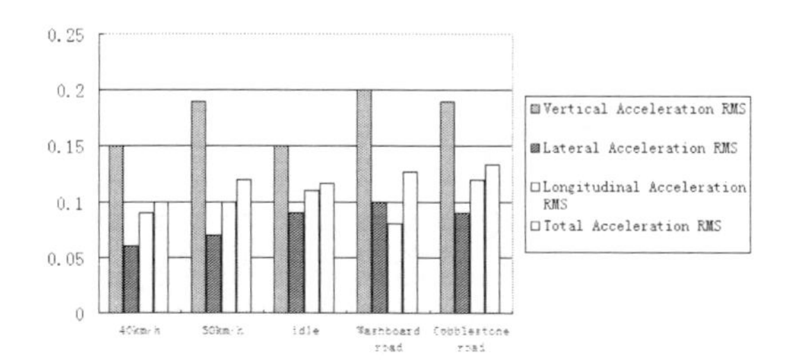

Figure 2. Acceleration of RMS value.

4 ANALYSIS OF TEST RESULTS

4.1 *Ride comfort analysis*

Because each response point has three directions, leading to huge workloads when late optimization, so we can use the amount of contribution under different conditions of the three directions of the summation, and find out the maximum of the influence direction targeted to ride comfort optimization, thus greatly improve work efficiency.

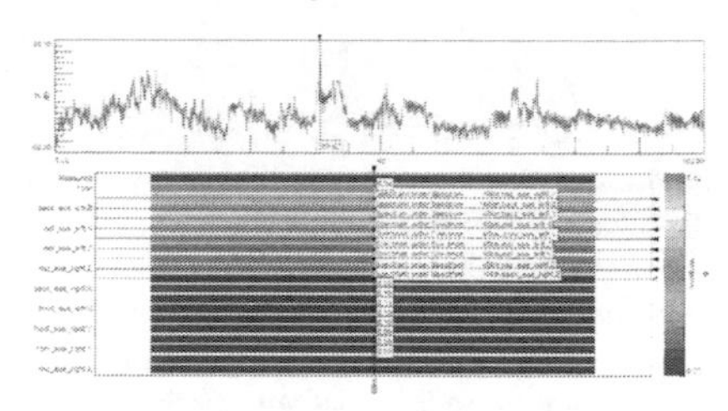

Figure 3. Contribution at frequency.

From the Fig.2, we can see that the RMS value of acceleration in three directions in five different operating conditions, which can draw vertical acceleration RMS is much larger than the other two directions. Due to other two directions contribute little on the cab seat rails and panel ride comfort, in ensuring the accuracy of the premise of the experiment, ignoring the contribution of the lateral and longitudinal values. So the following TPA, mainly consider the response point contribution value of vertical.

4.2 *40km/h operating condition transfers path analysis*

Select 40km/h working conditions for cabin vibration TPA. First, we need a greater impact on the vibration path identification, a large contribution to the value of the path optimization.

Table 1. Vibration spectrum analysis of the main frequency peak.

No.	1	2	3	4	5	6	7
Frequency peak	3.61	14.29	18.72	27.32	40.78	42.72	56.73
Amplitude (g)	0.028	0.01	0.008	0.008	0.03	0.01	0.007

From the table, we can see that the cab seat rails 40.62HZ vibration peak maximum, paper selected for 40.62HZ example peak frequency vibration transfer path analysis.

From Figure 3 we can clearly see that the cab seat rail and the dashboard have a high peak at the 40.62HZ, while according to the chromatogram, we can see the magnitude of the contribution and value of the total contribution of each path in descending order. The following table statistics greater contribution values path:

NO.	Path name	Amplitude (g)
mid_sus_right:Y	Middle mount on the right side in the Y direction	0.24
back_sus_left:Z	Rear mount on the left side in the Z direction	0.22
back_sus_left:Y	Rear mount on the left side in the Y direction	0.13
front_sus_left:Y	Front mount on the left side in the Y direction	0.10
back_sus_right:Z	Rear mount on the right side in the Z direction	0.10

We should not only consider the impact in solving the problem of vibration amplitude of vibration, because the phase also plays a decisive role, often the path of large amplitude, phase may be in the negative direction, so that the path has played a role in offsetting the vibration. If we are blind to reduce the amplitude, without regard to the phase, which will play a negative role, but the vibration strengthened.

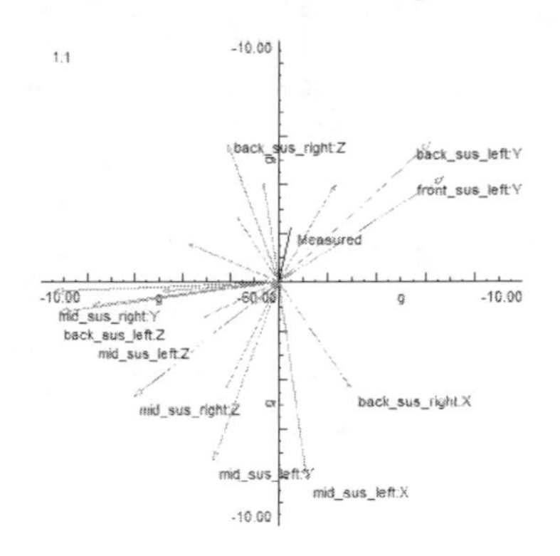

Figure 4. Phase spectrum.

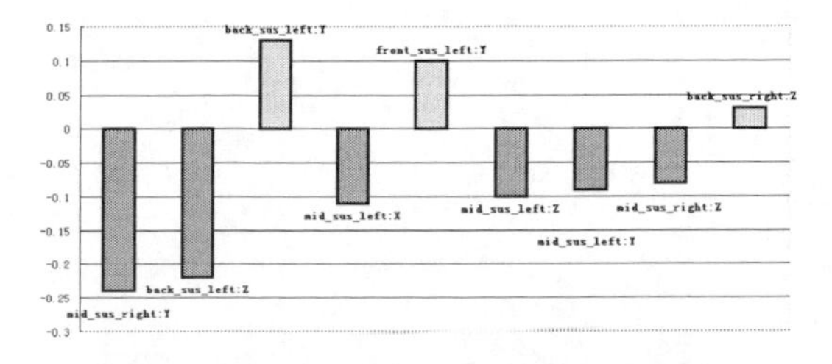

Figure 5. Complex of the phase and amplitude values.

From the Fig.4 and Fig.5 we can see that the path of "back_sus_left:Y" and "front_sus_left:Y", and "back_sus_right:Z" vibration contribution is positive, indicating that the vibration on the driver's seat and dashboard's contribution is a positive effect, play a greater role in vibration, we need to be optimized.

However, the vibration amplitude "mid_sus_right:Y" and "back_sus_left:Z" is relatively large, but the overall contribution of the negative effects, indicating that these paths vibration process plays an active role in offsetting the vibration, so we should keep these paths.

5 MOUNT TRANSFER CHARACTERISTIC ANALYSIS

Mounting transfer rate is the effect of commonly used suspension system evaluation, suspension vibration transmissibility basic concept is as follows:

$$T = A_p / A_a$$

T: mount vibration transmissibility; A_p: mount passive acceleration RMS; A_a: mount active acceleration RMS,

Consider Mount for symmetrical distribution, we consider only the left side of the three Mount for the study. Also from Tab 3 we can see that the transfer rate of the rear mount in the Y direction and front mount in the Y direction to pass rate reached 81.9% and 73.3%. Mount therein is basically not play a role, the results of the above results of TPA coincide, also explains the difference between these transfer rate path is basically caused by the vibration of the main path of the cab, effectively verify the TPA results above.

Tab.3:Transfer rate of the left mount.

Location \ Direction	X	Y	Z
Front mount	0.641	0.733	0.161
Middle mount	0.512	0.638	0.409
Rear mount	0.739	0.819	0.529

6 CONCLUSIONS

1 Use of TPA method can not only identify the contribution of large vibration amplitude path, while also considering the impact of the phase and amplitude of vibration, delete the contribution opposite path, thus providing accurate data for the cab Vibration;

2 By analyzing three directions of cab ride seat rails and dashboard vibration contribution, considering the situation of test efficiency and cost, appropriate choose to ignore the lateral and longitudinal vibration response of impact, and provide a feasible direction for the following experiment complex TPA computing, greatly reduce the workload without affecting the results of the case;

3 Through the analysis of the cab mount vibration isolation effect, that the rear mount in the Y direction and front mount in the Y direction to the vibration isolation rate is very poor, leading to the cab vibration is great, the result is consistent with the TPA results, which verifies the validity of the TPA results.

ACKNOWLEDGEMENTS

This work is supported by a Guangxi Province Education Department of scientific research project (project name: study on vibration control of agricultural transport vehicle based on transfer path analysis method), and the open fund of Guangxi manufacturing systems and advanced manufacturing technology key laboratory (Project: No.1305109011K, name: study on vibration control of agricultural transport vehicle based on transfer path analysis method), and Guangxi universities seaside machinery and equipment design and manufacturing and control Key Laboratory of open funds (project name: Key Technology Research and Application of Hybrid Transfer Path Analysis Method)

REFERENCES

[1] Wellman Thomas, Govindswamy Kiran.Aspects of driveline integration for optimized vehicle NVH characteristics[C]//SAE Technical Paper Series,2007-01-2246.

[2] LMS Inernational Transfer path analysis:The qualification and quantification of vibro-acoustic transfer path[Z].LMS Internation,Application Notes,1995.

[3] Juha Plunt.Finding and fixing vehicle NVH problems with transfer path analysis[J].sound and Vibration,2005(11).

Computing, Control, Information and Education Engineering – Liu, Sung & Yao (eds)
© 2015 Taylor & Francis Group, London, ISBN: 978-1-138-02800-5

Scenario analysis of carbon capture and storage technology in China

Ru Liu, Lu Ji Zhang, Qiong Wu & Wei Zhang
Beijing Municipal Institute of Science and Technology Information, Beijing, China

ABSTRACT: This article seeks to provide an energy planning of CCS (Carbon Capture and Storage) in China to address climate change. This article also provides a non-exhaustive scenario analysis that concerns this issue. CO2 capture and storage technology should be applied in China as well in the next decade. This technique is developed based on human interested in and researching aspect support. This project was funded by the Beijing financial fund.

KEYWORDS: CCS, key stakeholders, scenario analysis, China's market.

1 INTRODUCTION

Along with the development of the economy in China, the energy supply becomes increasingly tense. The coal industry as the conventional energy domain has been put forward. The coal plays an important role in the China's energy system. However, the large amount of carbon dioxide emissions obviously aggravated global warming, along with the utilization of coal resource. One side, so far human daily life and human society cannot continue without the utilization of coal, which is the biggest source of greenhouse gas emissions. On the other side, to reduce carbon dioxide emissions and environmental protection should be processed during the coal utilization.

For many years, exploration and practice of the scientists in the UK has proved that the carbon dioxide from coal combustion can be removed, and the efficiency of coal combustion can be improved significantly. This kind of clean coal technology is called CCS (carbon capture and storage). Fortunately, most of carbon dioxide comes from stationary factories, which is including power stations, petroleum industry, the steel industry and other industrial processes, so it is possible to be captured and stored. Here into, power stations are the most significant. British enterprises started to research a new area of energy saving innovations very early. As one of the world leaders in developing CCS technology and taking it to market, British enterprises have been focusing on new ways to reduce carbon dioxide emissions. Such actions did not interrupt and deter the country's economic growth welfare. What is more, the innovation in CCS technology gains benefits to the country itself, because these new technologies and best experiences have been exported to other countries in the world.

2 THE DEFINITION OF CCS

At present, CCS (carbon capture and storage) is very similar with the collecting carbon dioxide with the condition of high pressure air, it is a technology of separating out carbon dioxide when burning coal, and capture it, then dumping it to underground, or else on or under the sea bed [5]. The collection of carbon dioxide has three ways: post-combustion system, pre-combustion system and the oxyfuel process. Operating conditions define the collection methods.

Carbon dioxide storage also has four ways: ① Carbon dioxide through chemical reaction will be translated into solid inorganic carbonate; ② It can be used directly for industrial applications, or productions of various of carbomateriall compounds as a raw material. ③ To pour the carbon dioxide into the ocean under 1,000 meters. ④ Carbon dioxide is injected into underground formations, this technology is most mature. In many respects, it is the same as the technology of oil and gas industry, and some technologies have begun to use from the late of 1980s. In general, the underground formations include coal seam cracks, deep brine layer, also oil and gas layer. [3]

As matters stand, to achieve a dramatic reduction in carbon dioxide emissions from burning coal the approach must be fussy. CCS requires system integration, large-scale demonstration plants and proof of safety.

3 CCS PROJECT IN CHINA NOW

On November 20th of 2007, The Ministry of Science and Technology of China and the British Government officially launched the agreements in the Near Zero Emissions Coal (NZEC) in Beijing.

The main point of NZEC is about CCS technology, which there will be no emissions of carbon dioxide to the atmosphere in the process of using coal resource. NZEC is also an important aspect of Sino-EU Cooperation on Climate Change. In NZEC program, British agencies include Air Products, Alstom Power Electricity Company, BP, the British Geological Survey (BGS), as well as Doosan Babcock Company; participating institutions from China include China Academy of Sciences, Tsinghua University, and China HuaNeng Enterprise Group. These powerful combinations of companies and organizations will provide CCS programs great technical support and sources of funds. [4]

NZEC project will be divided into three processes: in the first process, NZEC project will try to search for feasibility and blueprint of CCS technology in China in order to achieve near-zero emissions from coal. In the second process, NZEC project will research and design demonstration project. In the last process, NZEC project will establish these demonstration projects.

4 SCENARIO ANALYSIS OF CCS

CCS can be only a matter of time in the future, as long as every country realizes the significant role it plays in our lives and attach importance to it. As presented, the following figure draws the future of CCS. The significant part of all these projects are carbon dioxide capture and separation plant. The carbon dioxide is captured from electricity plants, aluminum plants and plants for gas to liquids. Equally important, CO2 storage can enhance oil recovery and coal bed methane (CBM) gas production, the other carbon dioxide is used by carbonate products plants as well.

Figure 1. The overview of CCS further [2].

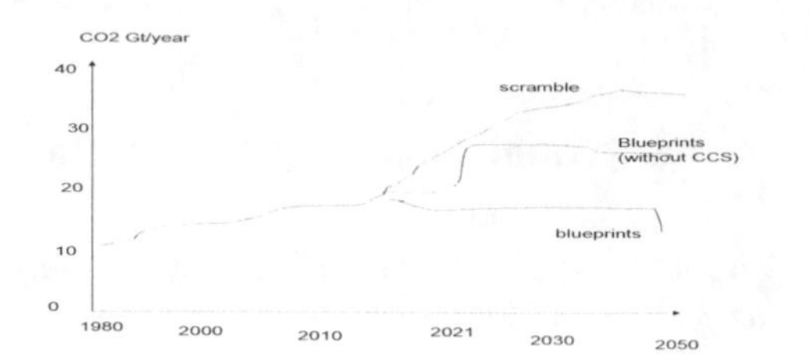

Figure 2. The main assumptions and sources in the scenario definition in 2050.

The development of CCS techniques could allow the vision of the zero-carbon dioxide power plant to be relied on future. There are two kinds of defining policy relevant scenarios for China to 2050: Scramble and Blueprints. Scenario analysis provides a tool for analyzing the consequences of different policies. The "Scramble" means policymakers pay more attention to energy efficient use, and CCS technology is not seriously addressed until there is very grievous climate change. The "Blueprints" means do more sustainable actions to address the economic growing, energy security and environmental pollution. The future is far lower carbon dioxide emissions.

5 DISCUSSION

CO2 capture and storage technology should be applied in China as well in the next decade. This technique is developed based on human interested in and researching aspect support. It is a brand new technique may not apply so quickly in China. So we hope the Chinese coal industry can learn and make use of CCS techniques as soon as possible. To implement CCS technique, China government has responsibilities to establish long-term energy policies, plans for healthy markets, good investment environments. The government also should formulate international cooperation on technology exchanges and mitigation of climate change and enhance public awareness of saving energy. What's more, the Government should give enough financial leg-up and NGO also can pitch in financial support to facilitate the raising of finance. At the same time, Scientists should try to make safer CCS technology as soon as possible. At last, it is not enough for CCS technology development and progress. If we want to spread it out, we need to make more large-scale CCS demonstration projects, not only in China and UK, but also in the whole world. CCS can be only a matter of time in the future, as long as every country realizes the significant role it plays in our lives and attach importance to it. The development of CCS techniques could allow the vision of the zero-carbon dioxide power plant to be relied on future.

REFERENCES

[1] ECOworld, 2008, http://www.ecoworld.com/home/articles2.cfm?tid=379.
[2] John Barry, 2009, Shell International Exploration and Production.
[3] Palmgren C, Granger Morgan M, Bruine de Bruin W, Keith D. 2004, Initial public perceptions of deep geological and oceanic disposal of CO2. Environmental Science and Technology; 38 (24)).
[4] Ministry of Science And Technology Of China, 2007,from: http://www.most.gov.cn/shfzs/sfdtxx/200711/t20071123_57405. htm.
[5] World energy council, January 2007, The World Energy Book Issue 3, page11.

Computing, Control, Information and Education Engineering – Liu, Sung & Yao (eds)
© 2015 Taylor & Francis Group, London, ISBN: 978-1-138-02800-5

Supervision of network bank in Europe and America and its implications

Y. Han
School of Finance, Guizhou University of Finance and Economics, China

ABSTRACT: The rapid development of modern network technology and the continuous improvement of the financial industry, result in the greatest financial innovations of 21st century—network bank. Network bank is a new kind of banking service mode takes internet as the platform, network technology as the means to provide information services and financial transaction services, brings great convenience to the majority of users. However, there are many drawbacks of in the development process of our country's network bank, based on this, learn from and refer the successful experience of network banking supervision model in developed countries, especially the United States and Europe and other countries, wish to have implications for sustained and healthy development of China's network bank.

KEYWORDS: Europe, America, network bank, supervision.

1 INTRODUCTION

As the network finance is developing, the supervision and research on network finance are also in infancy stage. Currently, the Basel Committee has not yet formed a relatively system and complete network banking supervision regime. Regulatory authorities in many countries have adopted quite a cautious attitude on network financial regulation, mainly on account of the coordination issue of innovation, competitiveness and supervision in the country's financial industry. In fact, the network financial risk supervision and control still need to constantly explore in practice. Some countries' regulatory authorities have set up special working organs or groups responsible for timely tracking, monitoring the developments of network financial industry, make some directive suggestions in due course and also develop some new regulatory rules and standards [1]. For example, definite network financial sector, including the network bank in the law. The European Bank Standards Committee defined network bank as those banks using the internet to provide banking products and services for consumers and SMEs, those who use computer, internet TV, set-top box and other personal digital device connect to the internet. The US OCC considers that network bank refers to some banking systems, by using these systems bank customers via personal computers or other intelligent devices can get access to the bank account, and obtain general banking products and services. Due to the strict legal definition of network bank has not yet appeared, at the same time network bank develops rapidly, therefore, the general practice is that according to the setting characteristics of network banking institutions, divide them into branch network banks and pure network banks, then respectively define and manage. Meanwhile, some countries' regulatory agencies, according to the development status of the network finance, revise the existing rules that formulate based on the financial institutions having actual business outlets and cannot extend to the network economy [2]. Such as some financial experts in the US, Germany begin to analyze the applicability of provision: "Basel Agreement" on commercial banks' standard capital adequacy ratio 8% and other regulatory rules in regulating pure network banks, and suggest improvements. Overall, the supervisory mode of network bank in each country is still mainly based on the original division of supervision scope, but increases coordination among regulatory agencies, regulatory agencies with other government departments.

2 NETWORK BANK REGULATORY MODEL IN EUROPE AND UNITED STATES

In terms of network bank supervision, in foreign, mainly formed Europe and United States two modes of regulatory network bank [3].

2.1 *European's regulation on the network bank*

European on regulation network banks takes a relatively new approach. The European central bank requires its members to take the consistency of regulatory principles; domestic regulatory agencies of European Union countries are responsible for monitoring the implementation of uniform standards. The two main objectives of network bank supervision are: First, to provide a clear, transparent legal environment; second, insist on appropriate prudential and consumer protection principles. In accordance with the requirements of harmonization of banking, investment services and insurance services, legal system in EU, European Union implements "single license" ruler on banks registered that business approved to conduct within a country in EU, the same can be carried out in other countries. Specific to the network banking business, the EU requires Member States on the network banking supervision, adhere to a consistent system, undertake the obligations of recognized electronic trading contractual and replace the regulatory rules based on the " register country" and "business happen country" with "original country" rules, to achieve enhanced regulatory cooperation, improve regulatory efficiency, timely monitor the new generational network bank risks. According to these requirements, the monitoring of network banks focused on the following aspects: First, regional issues, including mergers and joint between banks, cross-border transactions and so on; Second, security issues, including the risk generated by errors operations and data processing, network be attacked; Third, service technical capability; Fourth, the increasing reputation and legal risks with the expansion of businesses' number and scope, including the risk may be caused by the different regulatory authorities, different legal systems.

2.2 *United States' regulation on network bank*

By comparison, United States' regulation of Network Bank, basically through the addition of new laws and regulations, makes the original regulatory rules adapt to the network's electronic environment. Thus, in aspects of regulatory policy, license application, consumer protection, etc., the requirement of network banks are very similar to traditional banks. Departments responsible for the supervision mainly conclude the US Comptroller of the Currency (OCC), the Federal Reserve, the Ministry of Finance Thrift Supervision (OTS), the Federal Deposit Insurance Corporation (FDIC), National Credit Union Association, and the Federal Financial Institutions Examination Council (FFIEC). Wherein Comptroller of the Currency and the Federal Reserve are the primary regulators, the Thrift Supervision is responsible for auditing online public savings, such as the business license of the first pure network bank SFNB is issued by OTC. In addition, the network banks in the United States are also under the management of such as liquidation association, any Bank Group and other self-regulatory agencies. However, these managements are only for members, are voluntary, the areas involved in are also mainly technology, standards, etc. The purpose is to create conditions for banks innovation. Like other banking services, internet bank is also constrained by the laws of the federal and state levels. At the federal level, the regulations that the Federal Reserve is proposing to amend for the development of the network bank mainly include: the Federal Reserve Rule B (Fair Credit Opportunity), DD (real savings), E (electronic funds transfer), M (client lease) and Z (real loan). The Federal Reserve has announced new rule DD and rule E. These rules, in case that the customer agrees the banks can use the electronic network means periodically to disclose relevant information, and conclude the legal effect of the spreadsheet. On the state level, the major laws and regulations involving network banking behavior are "Uniform Commercial Code" (UCC) Article three, four, four a term. State implementation of these provisions is varying, but the difference was small.

Most existing financial institutions in the United States when conduct network bank services, do not need prior application and require a statement or record, the regulatory authorities generally through an annual inspection to collect network banking business data. The newly established network banks can either apply for registration in accordance with the standard registration process, can also apply for registration in accordance with the rules of the bank holding company. But an exception for savings institutions, if savings institutions want to carry out network banking services, must accord with OTS requirements, 30 days in advance to make a statement. Overall, the US financial regulatory authorities take a cautious easing policy for the network banks regulatory; on one hand they emphasize networking and transactions security, maintain banks operation stability and protect banks' customers, On the other hand, they consider that network bank is an innovation benefit for financial institutions to reduce costs, improve service, by using the standard Web browser and protocols, this innovation not only can significantly reduce technology maintenance costs and accelerate the development of new systems and software, and also can make banks to achieve resource sharing, cost-sharing, and therefore, they do not substantially interfere with the development of network bank.

3 THE DEVELOPMENT STATUS OF CHINA'S NETWORK BANKING SUPERVISION

Since June 1996 Bank of China took the lead in setting up the website, began to offer online banking services to the community, China Merchants Bank, China Construction Bank, Industrial and Commercial Bank of China, Bank of Communications, CITIC Industrial Bank and some commercial banks one after another have opened online banking services. When 2001 in

China more than 200 branches of 20 banks have the web sites and home pages, wherein branches carrying out substantive network banking amounted to more than 50. Our country's network bank is similar to the development of e-commerce, commercial website, is under the case of relevant laws and regulations are almost empty, then rapidly emerge and constantly evolve, with a strong spontaneity. The administration face the rapid change, have to introduce new management measures with a cautious attitude. This led to that current the rules governing the bank's network are still few; the management system is also unclear. From the view of China's current situation, appropriate network banking supervision is necessary. The necessary norms for basic network banking service behavior facilitate access to consumer confidence, expand the market, and avoid unnecessary transaction friction. Necessary regulatory rules also favor the formation of a relatively fair competitive environment; provide opportunities for the transformation and development of our country's small and medium banks thereby reduce the overall risk of the financial system. If wait until after the relevant agencies take related investment, then regulate, will not only increase the regulation resistance, and also make the early consumers face the loss risk. Overall, China's network banks are managed by two departments: business competent authorities - People's Bank of China and information competent authorities - ministry of information industry. For the network banks providing news & contents, after November 2000, also need to accept the management of public security departments and the press and publication administration. In these sectors, the later three departments are mainly responsible for the information technology and news management, have fewer relationships with the existing bank business, the People's Bank of China is the main management department [4].

From the perspective of regulatory, June 2001 People's Bank of China promulgated and implemented the "Interim Measures for Administration online banking services." The Rule regulated that if banking institutions want to offer online banking services in our country, should apply to the People's Bank prior to carry out, only after review and approval can offer online banking services. People's Bank implements "top level regulation" principle for banking institutions applying for market access to offer online banking services, that all types of banking institutions when offer online banking services for the first time, should apply to the People's Bank, branch or business management department by their head office. People's Bank implements approval system and recording system two systems for Banks applying for offering a new breed of online banking services. Banks use the internet to develop new, and different from the varieties of traditional banking business, online banking services species forming assets or liabilities in the table; use the internet to transact payment and settlement business other than the credit payment; through the internet to offer traditional banking asset classes business varieties within table without the consent of the People's Bank; via the internet to open new variety of businesses directly related to securities and insurance industries, are suitable to approval system. Banks through the internet to increase offering other new varieties of business are suitable to recording system. Banks offering online banking services should comply with the relevant national computer information system security, commercial password management, consumer protection and other aspects of laws, regulations and rules. Banks should develop and implement adequate physical security measures, effectively prevent external or internal unauthorized personnel from illegal access to critical equipment. banks should in an appropriate manner discloses and explain the transaction rules of the various varieties online banking services to customers, when customers apply for the certain online banking services species, should descript the risk of the variety transactions and their rights and obligations in a specific transaction. Without the consent of the People's Bank, the banks will be allowed to close down offering online banking liabilities variety of businesses after censor and approve by the People's Bank [5].

4 THE ENLIGHTENMENT OF US AND EUROPEAN'S NETWORK BANKING SUPERVISION TO CHINA

The degree of China economic informatization is not high, the development of network bank is still in its infancy stage, so our country should take a cautious approach for network banking supervision policy, policy formulation and implementation, do not limit its development also do not give up supervision, through appropriate financial regulation to promote network bank faster and better development. ① improve the existing laws; complement the related legislative provisions for network financial services. First, revise and supplement the not suitable portion of existing laws, and second, predict the future development of the situation, analyze the problems that may arise, take advance legislative protection. ② Combine the characteristics of network banking; improve the existing regulatory approach of business operations. From the aspects of the legitimacy and compliance of business operations, capital adequacy, asset quality, liquidity, profitability, the management level and internal control, etc. according to the network conditions to adjust, supplement, construct a financial supervision indicator system and operating system in line with the survival and development of network bank. ③ Urge the financial institutions carrying out network banking business to strengthen internal management, start from the internal control system to reduce financial risks. ④ Strengthen technical strength of financial regulatory authorities and improve the supervision level. Should gradually achieve to use the advanced electronic network technology for off-site supervision on network banks, establish the operating system such as proportion management of asset and liability, credit ledger management, early warning analysis and intelligent decision-making, through online real-time control to improve the modernization management level of supervision. ⑤ Closely contact with regulatory agencies of other countries, improve regulatory efficiency of the network banks. People's Bank of China and other regulatory agencies should strengthen cooperation with foreign financial authorities; regularly carry out exchanging of regulation situation, discussion on online financial regulatory measures. At the same time strengthen personnel exchanges with the countries that network banks develop relatively fast, increase efforts of training supervisory staff, introduce the advanced regulatory philosophy and technology.

REFERENCES

[1] Jin Chen, Qiang Fu, Network Bank Services [M], Beijing: Tsinghua University Press, 2002.

[2] Zhuoqi Zhang, Mingkun Shi, Online Payment and Online Financial Services [M], Dalian: Dongbei University of Finance and Economics Press, 2002.

[3] Sumei Yu, International experience of network bank supervision laws and its implications [J], Seeker, 2008 (12).

[4] Denghui Shang, Research on china's network bank supervision legal system [J], Southern Finance, 2014 (01).

[5] Kenming Zhuang, Risk regulation of our country's network bank [J], Zhejiang Financial, 2005 (11).

Computing, Control, Information and Education Engineering – Liu, Sung & Yao (eds)
© 2015 Taylor & Francis Group, London, ISBN: 978-1-138-02800-5

A new type of network flow calculation method based on discrete time

Hong Zhang
Key Laboratory of Pattern Recognition and Intelligent Information Processing, Chengdu, Sichuan, China
College of Information Science and Technology, Chengdu University, Chengdu, Sichuan, China
College of Computer Science, Si Chuan University, Chengdu, Sichuan, China

Yun Cheng Shen
College of Information Science and Technology, Zhaotong University, Zhaotong, Yunnan, China

Jun Hu
Chengdu Normal University, Chengdu, Sichuan, China

ABSTRACT: This paper puts forward a new network flow Prediction algorithm (Prediction algorithm based-FARIMA model for Discrete Time, PFDT) in view of the network node congestion or Link disconnected. The algorithm deduces the mathematical expressions of average queue length when the queue exists a failure node with the theory of discrete time and establishes the prediction model by FARIMA. The simulation results show that the algorithm has good adaptability and the standard deviation is 10.28 compared with the original.

KEYWORDS: Prediction; Discrete time; FARIMA model; Average length of queue.

1 INTRODUCTION

With the rapid increase of network scale, the effectiveness of the nodes has produced important influences on the network performance. Therefore, analyzing and computing accurately the nodes effectively depict has important significance for network traffic. So, researchers at abroad and home had done a lot of research work on the flow of network computing. Sahinoglu M and Libby D L [1] evaluate the reliability of the network by using probability expression. In order to consider network traffic factors and connectivity, liu aimin and Liu Youheng[2] put forward the communication network reliability evaluation measure. Wei Juan and Hu Jun[3] put forward the PFB algorithm to depict the network traffic.

At the same time, the queuing theory is applied in the study of node effectiveness and the performance of the network. Such as Ying-hui tang[4], Kim B[5] Dan zhao and Deng Bin[6] and so on.

In this paper, using the knowledge of the queuing theory to compute the node MAC layer packet flow quantitatively based on the related mathematical methods of the discrete time. The mathematical model creates the steady average length of the queue. And give the experiments conducted on the algorithm by simulation software.

2 RELEVANT CONCEPTS DESCRIBED

The assumption in the network, the packets of nodes have the first come first service (FIFO) strategy. The paper describes the packet model of nodes which in the discrete time based on the Geom/Geom/C queuing theory as follow:

A data packet reaches with probability $p(0 < p < 1)$ at head of each time slot (n, n^+), at the same time, no data packet reaches with probability $\bar{p} = 1 - p$. All the packets' arriving constitutes a Bernoulli process with parameters p.

The beginning and ending of all the packets' service occur at the end of time slot, n=1,2.... The service time of packets subject to identically distributed, and

$$P\{S = k\} = \bar{u}^{k-1} u, k = 1,2,... \tag{1}$$

There are c parallel service nodes in the network and the queue of FIFO rules. Arrival interval and service time are independent of each other.

According to relevant thoughts on the queuing theory, using L_n^+ to indicate the number of data packets when $t = n^+$ in instantaneous system. Because the $\{L_n^+, n \geq 0\}$ is a neat MC, there are countable state space $\Omega = \{0, 1, ...\}$, and the transition probability is:

$$p_{ij} = P\{L_n^+ = j \mid L_{n-1}^+ = i\}, n \geq 1, i, j \in \Omega \tag{2}$$

When the number of customers in the system meets $i \le c-1$, the state transition $i \to j(1 \le j \le i \le c-1)$ accords to two kinds of incompatible ways:

$$p_{ij} = \bar{p}\binom{i}{j}u^{i-j}\bar{u}^{j} + p\binom{i}{j-1}u^{i-j+1}\bar{u}^{j-1},\quad 1 \le j \le i \le c-1 \tag{3}$$

The similar analysis is given:

$$p_{i0} = \bar{p}u^{i}, 0 \le i \le c-1 \tag{4}$$

When $i \ge c$, $L_{n-1}^{+} = i$ contains on all nodes in the time slot (n^{-}, n) all busy, the transition probability does not depend on i at this time:

$$p_{i,i-c} = \bar{p}u^{c}, i \ge c, \tag{5}$$

$$p_{i,j} = \bar{p}\binom{c}{j}u^{c-j}\bar{u}^{j} + p\binom{c}{j-1}u^{c-j+1}\bar{u}^{j-1},\quad i \ge c, i-c \le j \le i \tag{6}$$

$$p_{i,i+1} = p\bar{u}^{c}, i \ge c \tag{7}$$

Assume that

$$\binom{c}{-1} = \binom{c}{c+1} = 0 \tag{8}$$

The transition probability in type (6) can be unified represented as:

$$p_{i,j} = \bar{p}\binom{c}{j}u^{c-j}\bar{u}^{j} + p\binom{c}{j-1}u^{c-j+1}\bar{u}^{j-1},\quad i \ge c, i-c \le j \le i+1 \tag{9}$$

In short:

$$a_k = \bar{p}\binom{c}{k-1}u^{k-1}\bar{u}^{c-k+1} + p\binom{c}{k}u^{k}\bar{u}^{c-k},\quad k = 0, 1, \ldots, c+1 \tag{10}$$

3 THE AVERAGE QUEUE LENGTH OF DATA PACKETS

Using L^{+} to indicate the stability limit of L_n^{+}, the distribution can be represented as:

$$\lambda_k = P\{L^{+} = a_k\} = \lim_{n\to\infty} P\{L_n^{+} = a_k\}, k \ge 0 \tag{11}$$

According to the type (11)

$$a(z) = \sum_{k=0}^{c+1} a_k z^k = (z\bar{p} + p)(uz + \bar{u})^c \tag{12}$$

Synthetically type (12), (13) the system average queue length is:

$$\overline{\lambda_k} = \frac{Ka(z)}{(1-a(z))^2} \tag{13}$$

4 ALGORITHM DESCRIPTION

Step 1 Initialization the A nodes state parameter in time slot t;

Step 2 Set the system balance relationship, according to the type (4-7) by reference [7];

$$(\lambda + \mu + w)p_{1,k} = \lambda p_{1,k-1} + \mu p_{1,k+1} + \theta p_{2,k}, k \ge 2 \tag{14}$$

Step 3 Calculate the queue length in steady state with type (11) and use the value to compute the average number packet of node A with type (12);

Step 4 Use the type (15) to pick truncation parameters M and sample N:

$$X(n) = \sum_{m=0}^{M-1} c(m)Z_{\alpha}(n-m) \tag{15}$$

Where n=0, 1, …, N-1;

Step 5 Set t = t + 1, jump to Step 1, until the end of the cycle;

End of the algorithm.

5 THE EXPERIMENTAL SIMULATION

In order to verify the validity of the PFDT algorithm, first of all, set up the relevant network topology in NS2 and produce 1600 relevant data flow based on the FBM (Fractional Brownian Motion) simulation model. Set the average arrival rate is 800 kb/s and the related degree of H = 0.9. At the same time, analyze the simulation results in MATLAB and evaluate the effectiveness of the network nodes. Here, the first 800 data as a priori information and use the results of the PFDT algorithm to predict comparison with data of FBM produced before the last 800 data, as shown in figure 1. From the figure 1, we can get the PFDT algorithm to predict flow with the original flow is relatively close and the standard deviation is 10.28, this shows that the algorithm has any certain prediction accuracy.

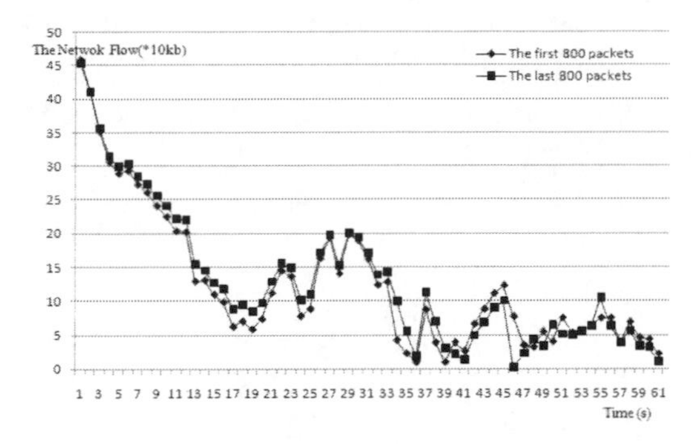

Figure 1. Compared of network flow between PFDT and FBM.

Secondly, consider the depth analysis of the main factors which affect the traffic performance. Assume that w=0.25, θ=0.75 and η=0.6, the figure 2 shows the relationships with the different traffic arrival rate λ between average queue length λ and service rate μ. At the same time, we can see a negative correlation between the average queue length and the service rate. At the small service rate, with the larger λ, the smaller the average queue length and the better the performance. When at the big service rate, the curve mutated. With the smaller λ, the smaller the average queue length and the better the performance.

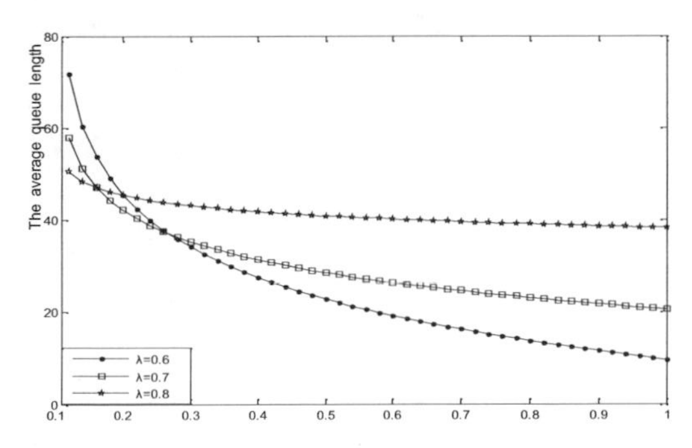

Figure 2. The relationship between average queue length and service rate with different traffic arrival rate λ.

At the same time, assume that w=0.25, θ=0.75 and λ=0.7, the figure 3 shows the relationships with the different traffic arrival rate η between average queue length $\bar{\lambda}$ and service rate μ. We can see that the average queue length is increasing with the service rate increase till the max value, and then decrease to stabilize state finally. This shows that, if one node A failed, the more traffic reaches A, the much greater impact on the system.

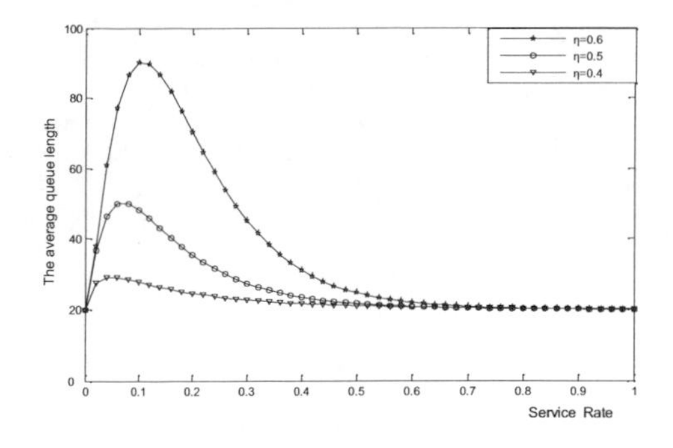

Figure 3. The relationship between average queue length and service rate with different traffic arrival rate η.

6 SUMMARY

This paper puts forward a new prediction algorithm PFDT in view of the network node congestion or Link disconnected. Firstly, the algorithm deduces the mathematical expressions of average queue length when the queue exists a failure node with the theory of discrete time and establishes the prediction model by FARIMA. At the same time, analysis the main factors that influence the properties of flow through the mathematical simulation, and the experimental results also verify the effectiveness of the PFDT algorithm.

ACKNOWLEDGEMENT

The corresponding author of this paper is HU Jun. This paper is supported by Research Project of Sichuan Provincial Department of Education (14ZB0368), School Fund Project of Chengdu University (2012XJZ19).

REFERENCES

[1] Sahinoglu M., Libby D. L.. Measuring availability indexes with small samples for component and network reliability using the Sahinoglu-Libby probability model[J]. IEEE Transactions on Instrumentation and Measurement, 2005, 54(3): 1283–1295.
[2] LIU Aimin, LIU Youheng. Traffic Performance Analysis of Network with Unreliable Components[J]. Chinese Journal of Electronics. 2002, 30(10): 1459–1462.
[3] WEI Juan, YOU Lei, HU Jun. A Novel Depict Method of Actual Traffic Performance[J]. Microelectronics & Computer, 2013,30(3):35–37.
[4] TANG Yinghui, HUANG Shujuan, YUN Xi. Queue-Length Distribution for Discrete Time Geom/G/q Queue with Multiple Vacations and Bulk Arrival. Chinese Journal of Electronics, 2009, 37(7): 1407–1411.

[5] Kim B. Tail asymptotics for the queue size distribution in a discrete-time Geo/G/1 retrial queue [J]. Queueing System, 2009, 61: 243–254.
[6] ZHAO Dan, DENG Bin. The Stationary Queue-length Distribution for the Repaired M/G/1 queue under N-policy with Delayed Startup-Closedown Based on the Maximum Entropy [J]. Mathematics in Practice and Theory, 2010, 40(18): 140–147.
[7] TANG Yinghui, TANG Xiaowo. Queuing theory[M]. Science press, 2006.

Computing, Control, Information and Education Engineering – Liu, Sung & Yao (eds)
© 2015 Taylor & Francis Group, London, ISBN: 978-1-138-02800-5

High resolution actuator using a ballscrew mechanism and the control method

Dong Ya Zhang, Jin Wu Qian, Ya Nan Zhang & Lin Yong Shen
Department of Precision Mechanical Engineering, Shanghai University, Shanghai, China

ABSTRACT: To meet the demand of nano-step actuator which is used in large-diameter grating tiling, a mechanism based on ball screw is designed. It can achieve the requirements of millimeters long range. The theoretical model is built and identification experiment is also conducted to describe the system. The swept and step response simulation of the identification model agrees very well with the experimental result. To achieve nanometer positioning accuracy, an incomplete PID controller is also designed. Using an integral anti-windup method, overshoot value can be very small. The step response experiment is conducted based on laser interferometer feedback. The trajectories in the experiment and the simulation agree very well. Experiments show the actuator can achieve 100nm resolution and keep high stability.

KEYWORDS: Grating tiling; Nano-positioning; Actuator; Ballscrew.

1 INTRODUCTION

With the development of astrophysics and high power laser pulse, a large-scale diffraction grating is more and more important [1]. But, it's difficult to get it by rule grating machine or holographic method [2,3,4]. Grating tiling is through two or more small grating to compose a large grating. By the way of mechanical adjustment, the offset between them can be small enough. So it can be used as a large grating. Fig. 1 shows the tiling process, the static grating in the middle stay a fixed position, then rotate or translate the gratings in both sides. It's need to stay a high adjustment precision during the whole tiling process. So an actuator with nano-positioning precision is needed.

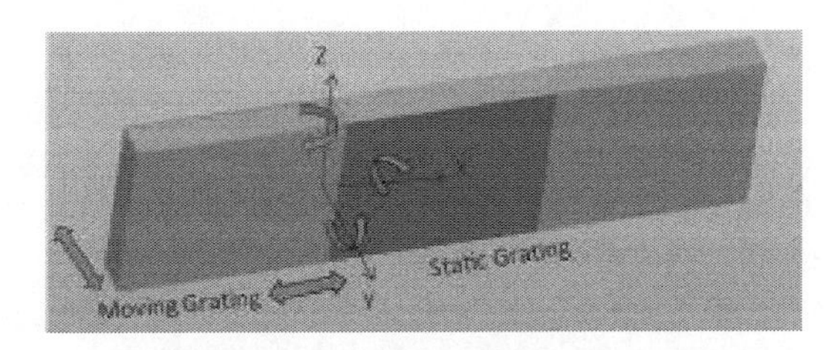

Figure 1. The diagram of grating tiling.

The grating tiling system we designed has five linear actuators in a series and implement adjustment in five-dimensional way. The actuator has a ball-screw–driven structure. A closed-loop control is built based on a sub-micron sensor feedback.

Actuator based on ball-screw has been widely used in large precision machinery. In this paper, an incomplete derivative PID controller is designed, it has high closed-loop gain and the effect brought by friction can be restrained in the micro region [5]. Moreover, by using conditionally freeze antiwidup, the error can be eliminated quickly, and nano-positioning can be achieved.

2 STRUCTURE AND MODELING OF ACTUATOR

2.1 *Structure of actuator*

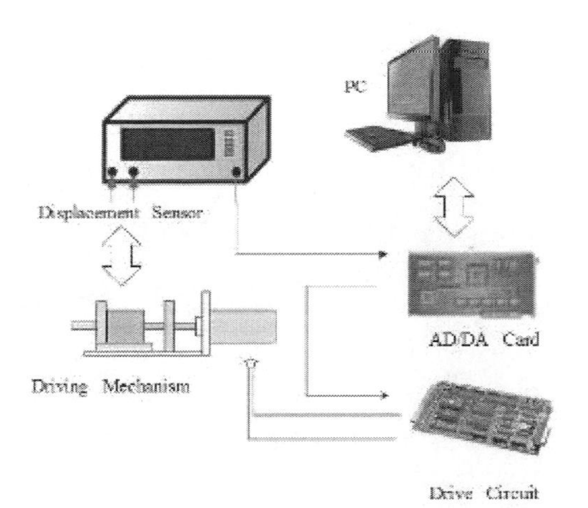

Figure 2. Schematic diagram of actuator's control.

Fig. 2 shows three parts of the actuator's platform, including:

Figure 3. The mechanism diagram of actuator.

1 **Actuating mechanism**. Fig. 3 shows the actuating mechanism, server motor is connected with ball screw by a coupler. Ball screw nut can push a pin-lift into straight-line motion .
2 **Position detection system**. The position of actuator can be measured with a laser interferometer, and the variation of motion can be transferred to the analog voltage output.
3 **PC control system**. The PC control system is composed of PC, DAQ card and control software. DAQ card can finish the sample of position signal and the control of the server-motor. Control algorithm is built by c, it can control the rotation of motion based on motion feedback.

2.2 *Modeling of system*

Fig. 4 shows the mathematic model of actuator, control voltage V of a computer can be transferred to the end motion X of the actuator. The screw nut pair can be equivalent to a spring damping system [6]. The rotational inertia of the motor rotor, coupler, bearing and screw can be converted into J, the mass of the slider can be converted into M.

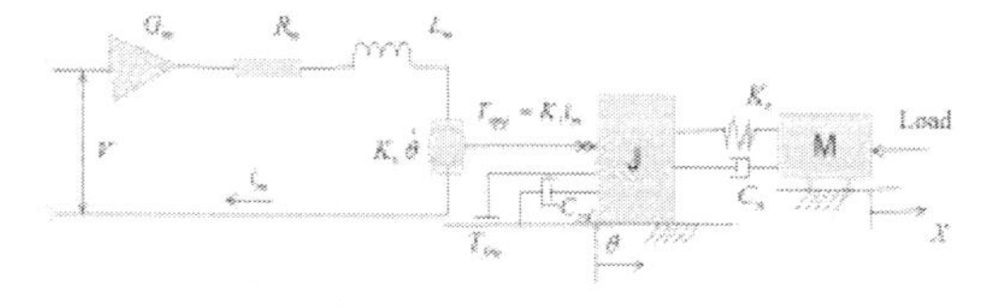

Figure 4. Dynamic model of the actuator.

Based on the model , can list the equations

$$\begin{cases} G_m V_m = R_m i_m + L_m i_m + K_v \theta \\ K_i i_m - T_{fric} - hM x = J \theta + C_{sd} \theta \\ M x = K_n (h\theta - x) + C_n (h\theta - x) \end{cases} \quad (1)$$

Table.1 shoes the parameter value of actuator test platform and its meaning.

Table 1. The test platform parameters.

Parameter	Value	Meaning
M	$0.394 Kg$	Mass of slide table
J	$0.4804 \times 10^{-4} kg\, m^2$	Inertia of rotating parts
R_m	1.1Ω	Motor resistance
K_i	$0.47\, Nm / A$	Torque constant
K_v	$0.47\, Vs / rad$	Back EMF constant
C_{sd}	$3.67 \times 10^{-4}\, Nms / rad$	Rotational damping
K_n	$8.3 \times 10^7\, N / m$	Stiffness of nut-screw
C_n	$7000 Ns / m$	Damping of nut-screw
h	$3.18 \times 10^{-4}\, m / rad$	Transition ration
G_m	1.5	Amplifier gain

In the experiment, the slider position X is the measured, based on the table parameter and simplified calculation, the transfer function from the input voltage to the slider position can be described by

$$G(s) = \frac{X(s)}{U(s)} = \frac{1.046s + 12405.18}{2.08 \times 10^{-5} s^4 + 0.4574 s^3 + 5938.82 s^2 + 18368207.1 s} \quad (2)$$

With the poles, which has little influence to the system, the system can be simplified as a second-order model

$$G(s) = \frac{X(s)}{U(s)} = \frac{a}{s(s+b)} \quad (3)$$

2.3 *System identification*

In the actual experiment, the friction and damping of actuator might not be equal to the theoretical calculating value. In order to be sure the actual model, a system identification test is conducted. A 0-20Hz swept-frequency voltage signal and 1.5V voltage step signal are applied though ADVANTECH PCI-1716 DAQ card. Measure the change of position. Fig. 5 shows the output and input result .

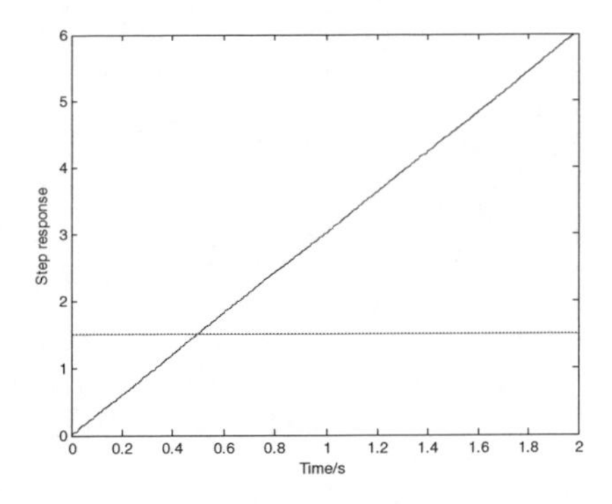

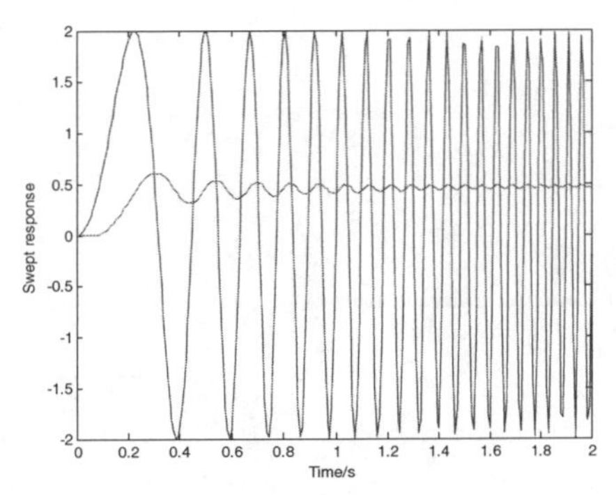

Figure 5. Swept and step signal response.

Using the system identification tool of Matlab,conduct the system identification based on the system response.The identification, calculation result of model is 93.01% and can be used as an approximate mode of the actuator.

There are 15 experimental data results, the transfer function can be described by

$$G(s)=\frac{X(s)}{U(s)}=\frac{247.6}{s(s+120.5)} \tag{4}$$

Fig. 6 shows the comparison of the identification model simulation curve and experiment curve.

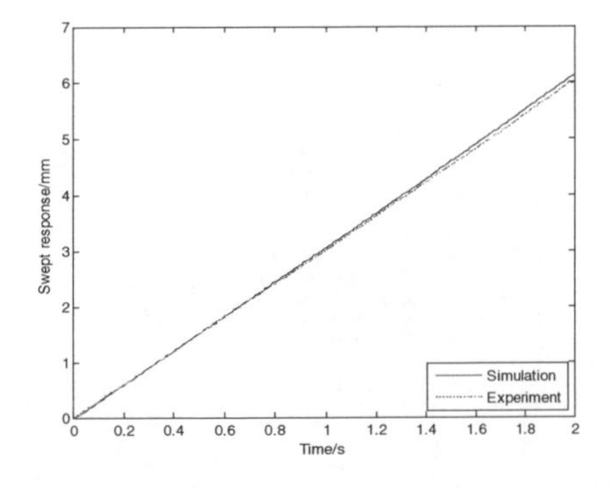

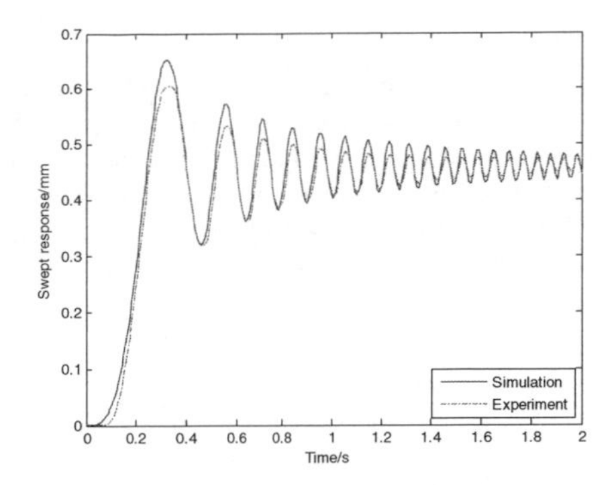

Figure 6. Comparison of identification model curve and experiment curve.

3 CONTROLLER DESIGN

3.1 *Controller structure*

A standard PID controller is composed of proportion, integration, differentiation. In order to deduce the dis turbance of high frequency signal, a first-order inertia link is added to make up an incomplete differentiation PID controller, as shown in Fig.7.

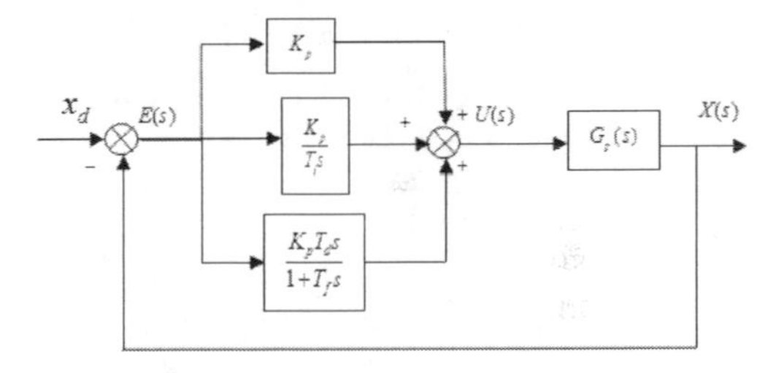

Figure 7. Block diagram of closed-loop system.

Transfer function of PID controller is

$$G_c(s)=K_p\left(1+\frac{1}{T_i s}+\frac{T_d s}{T_f s+1}\right) \tag{5}$$

Closed transfer function of screw-nut is

$$T(s)=\frac{G_c(s)G_p(s)}{1+G_c(s)G_p(s)}= \frac{K_p a[(\frac{T_d}{T_f}+1)s^2+(\frac{1}{T_i}+\frac{1}{T_f})s+\frac{1}{T_iT_f}]}{s^4+(b+\frac{1}{T_iT_f})s3+[K_p a(\frac{T_d}{T_f}+1)+\frac{1}{T_iT_f}b]s^2+K_p a[\frac{1}{T_i}+\frac{1}{T_f}]s+K_p b\frac{1}{T_iT_f}} \tag{6}$$

Let multiple closed-loop poles be placed at p. By emulation and experimental verification, we choose a set of parameters as follows, then the closed-loop system has four same poles and $p=20\pi$.

$$K_p=20,\ T_i=0.5,\ T_d=0.004,\ T_f=0.003$$

3.2 *Integral anti-windup*

Integral gain can cause an undesirable overshoot during step input responses because of the integrator windup effect. To improve the performance , an integral anti-windup method is used, as shown in Fig.8 and with the following rule as Eq(7), where u_m represents the saturation voltage of server motor.

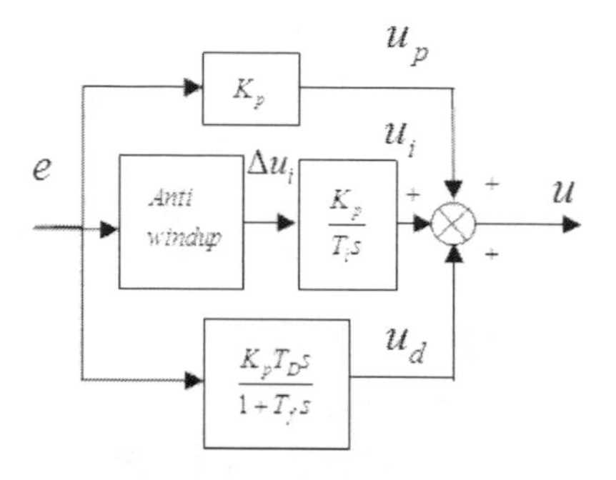

Figure 8. PID controller with the conditionally freeze antiwidup.

$$\Delta u_i = \begin{cases} 0 & |u_p + u_i + u_d| > u_m, and\ e \cdot u_i \geq 0 \\ e & otherwise \end{cases} \tag{7}$$

The control algorithm is compiled by language c in industrial computer. The sampling period is 1ms and saturation voltage u_m is 3V.

4 EXPERIMENTAL RESULTS

Figure 9. Photograph of the test platform.

Grating tiling machine requires that actuator can move ±2mm, and the resolution should be less than 100nm. In the experiment, the integral saturation threshold voltage of the controller is tested with 3V voltage. 1mm, 20um and 100nm stepwise response is respectively given in Fig.10(a), Fig.10(b) and Fig.10(c).

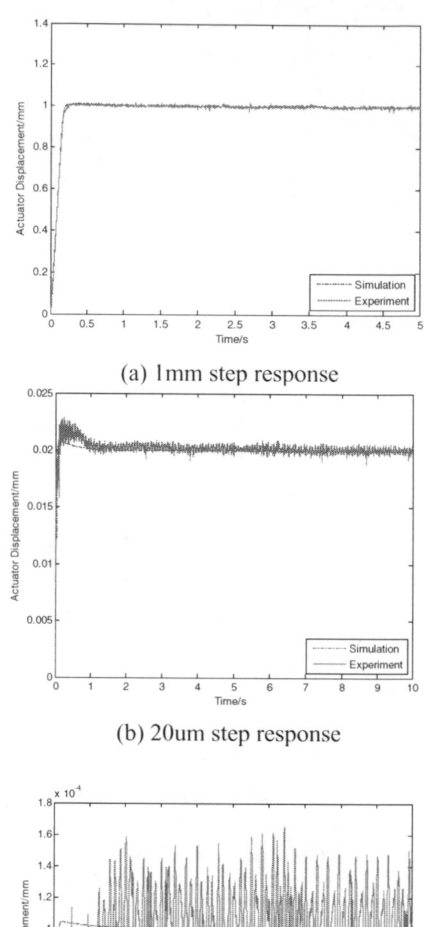

(a) 1mm step response

(b) 20um step response

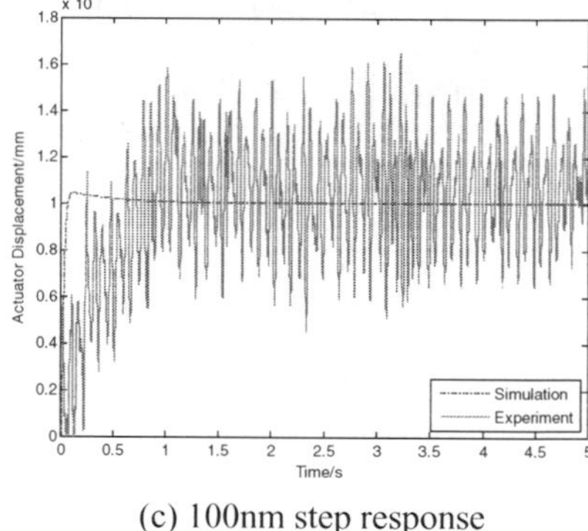

(c) 100nm step response

Figure 10. Simulated and experimental step response.

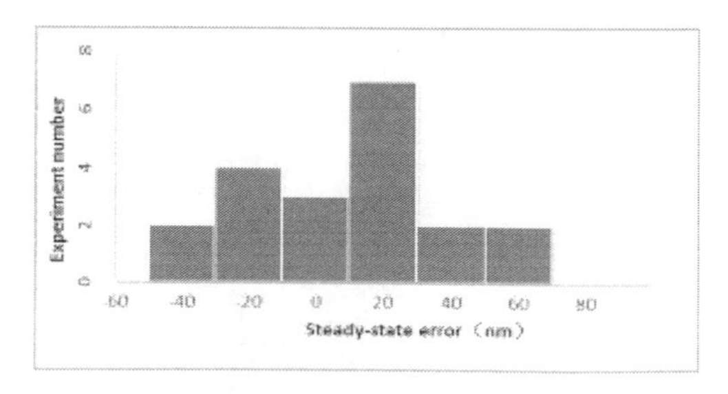

Figure 11. 20um response steady-state error.

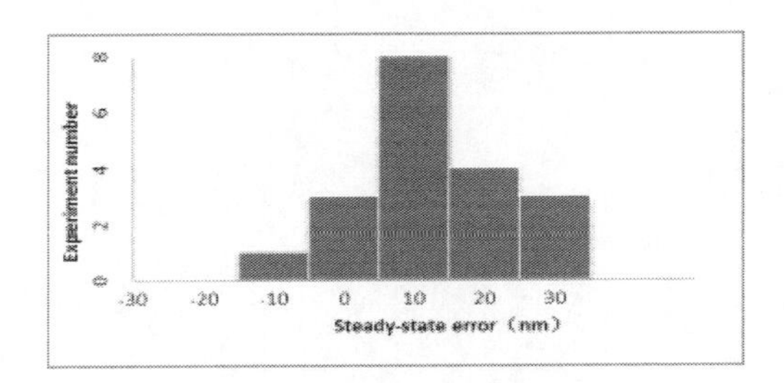

Figure 12. 100nm response steady-state error.

The test is repeated 20 times for 20um and 100nm step response, Fig. 11 and Fig. 12 shows the steady-state error. It can be seen that the steady-state errors in the cases of 100nm and 20um step size are within ±50nm. The trajectories in the experiment and the simulation agree very well for both 1mm and 20 um step sizes. The response speed of 100nm step size is quite slow because of the effect of friction. Still, when the system response goes into steady state, the error is almost the same with the measurement noise.

5 CONCLUSION

The actuator designed in the paper is based on the structure of server motor and screw nut. And the feedback sensor is a laser interferometer with sub-micron resolution. The experiment result shows that the incomplete PID controller can effectively reduce the effect of friction, and the experiment result agrees very well with simulation results. But because of the effect of unstable environment, there are tens of nanometers fluctuation. In the future, improve the stability of the platform and use higher resolution sensor, it's expected to achieve higher positioning accuracy.

REFERENCES

[1] G. PATRICK, H. SERGE and L. PHILIPPE. Mosaiced and High Line Frequency VPH Gratings for Astronomy [J]. Proc SPIE. 2004, 5494(8): 207–215.

[2] Qiao,J. Demonstration of large-aperture tiled-grating compressors for high-energy[J], petawatt-class, chirped-pulse amplification systems. Optics Letters, 2008. 33(15): 1684–1686.

[3] HOOSE, E. LOWEN, R. WILEY and T. BLASIAK. A New Generation of Larger Telescopes Presents an Un-precedented Challenge to Optics Manufactures [J]. Photonics Spectra. 1995, (12):118–120.

[4] WangCong, Zhang Junwei. Technology Progress of Grating Tiling[J]. Laser & Optoelectronics Progress, 2011. 48(8): 1–6.

[5] G.J.Maeda and K. Sato, Practical control method for ultra-precision positioning using a ballscrew mechanism[J]. Precision Engineering, 2008. 32(4): 309–318.

[6] J.Mao, H.Tachikawa, A.Shimokohbe, Precision positioning of a DC-motor-driven aerostatic slide system[J]. Precision Engineering, 2003. 27(1): 32–41.

Computing, Control, Information and Education Engineering – Liu, Sung & Yao (eds)
 ISBN: 978-1-138-02800-5

Anisotropic property of the multiple-dimensional rough surface on moth wing

Gang Sun & Yan Fang
School of Life Science, Changchun Normal University, Changchun, China

ABSTRACT: The micro-morphology, superhydrophobicity, self-cleaning property and chemical composition of the moth wing surface were investigated by a Scanning Electron Microscope (SEM), a contact angle meter and a Fourier transform infrared spectrometer (FT-IR). The wetting mechanism of the moth wing was discussed from the perspective of biological coupling. The moth wing displays multiple-dimensional structural anisotropism. The micrometric scales constitute the primary structure, the submicrometric vertical ribs and horizontal bridges on the scales constitute the secondary structure, and the nano stripes on the vertical ribs and horizontal bridges constitute the tertiary structure. The moth wing surface is of superhydrophobicity (contact angle 150.5~158.4°) and low adhesion (sliding angle 1~3°). In addition, the water droplet exhibits anisotropic sliding behavior on the wing surface. The scale plays a crucial role in determining the self-cleaning property of the moth wing. The coupling effect of material element and structural element contributes to the special wettability of the wing surface. The moth wing can be potentially used as a template for biomimetic design of functional surface with complex wettability. This work may offer insights into the preparation of smart interfacial material and directional self-cleaning coatings.

KEYWORDS: Superhydrophobicity, Adhesion, Self-cleaning, Biomaterial, Moth wing.

1 INTRODUCTION

Anisotropism is one of the most important properties of a patterned solid surface. The anisotropic rough surface results in special wetting and dewetting characteristics in different directions. One example is the anisotropic dewetting phenomenon on the superhydrophobic rice leaf surface. The sliding angle (SA) of a water droplet is greatly influenced by the anisotropic arrangement of the papillae on the rice leaf [1]. The anisotropic wettability has drawn much attention and has been applied for fabrication of self-assembling patterned surfaces [2], submicrometric channel lattices with alternating wettability [3] and rice-like aligned carbon nanotubes (ACNT) film [4]. The anisotropic dewetting property may bring interesting insights into design of lossless liquid transportation channels and novel microfluidic valves, in which liquids can be driven in a preferred direction. Insect wing is an ideal bio-template for artificial fabrication due to some excellent properties, such as attractive iridescence, superhydrophobic characteristics and quick heat dissipation ability [5]. In the present paper we investigated the anisotropism of the rough surface on the moth wings and discussed the wetting mechanism from the perspective of biological coupling. This work can not only promote our understanding of anisotropic wetting phenomenon on bio-surfaces, but provide inspiration for design and preparation of smart fluid-controllable interface and directional self-cleaning material.

2 MATERIALS AND METHODS

Materials. The moth specimens of twelve species were collected in Changchun City and Jilin City, and identified by systematic taxonomy. The moth wings were cleaned, desiccated and flattened, then cut into 5 mm × 5 mm pieces from the discal cell (Fig. 1). The distilled water for the measurements of contact angle (CA) and SA was purchased from Tianjin Pharmaceuticals Group Co. Ltd., China. The volume of water droplet was 5 μl.

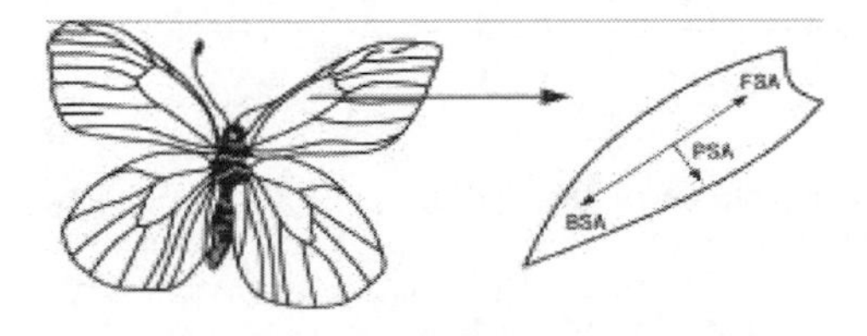

Figure 1. The experimental area and the SAs of water droplet in different directions (FSA: forward SA; BSA: backward SA; PSA: perpendicular SA).

Methods. After gold coating by an ion sputter coater (Hitachi E-1045, Japan), the wing pieces were observed and photographed by a SEM (Hitachi SU8010, Japan). A video-based contact angle measuring system (DataPhysics OCA20, Germany) was used to measure the CA of water droplet on the wing surface by sessile drop method at room temperature of 25±1 °C. The SA was measured in three directions, including forward SA (FSA, the SA of droplet from the wing basal to the wing terminal), backward SA (BSA, the SA of droplet from the wing terminal to the wing basal), and perpendicular SA (PSA, the SA of droplet perpendicular to the major axis of the wing) (Fig. 1). The chemical composition of the wing surface was investigated by means of FT-IR (Nicolet FT-IR200, USA).

3 RESULTS AND DISCUSSION

Anisotropism of the Micro-morphology on the Wing Surface. The moth wing surface exhibits multiple-dimensional rough structures. The micrometric scales constitute the primary structure. The shapes of the scales of various moth species are similar, including latifoliate shape [Fig. 2(A)] and angustifoliate shape [Fig. 2(B)]. The length of the scale is 126~438 μm, the breadth is 44~126 μm, the spacing is 85~155 μm. The submicrometric vertical ribs and horizontal bridges on the scales constitute the secondary structure. The vertical ribs and horizontal bridges are linked as grids, some vertical ribs have branches [Fig. 2(C)]. The height of the vertical rib is 354~1316 nm, the breadth is 376~1182 nm, the spacing is 1543~2241 nm. The nano stripes distributing regularly on the vertical ribs and horizontal bridges constitute the tertiary structure [Fig. 2(D)]. The primary, secondary and tertiary structures all display remarkable anisotropism.

There are extremely significant differences between FSA and BSA ($P<0.01$), between FSA and PSA ($P<0.01$), as well as between BSA and PSA ($P<0.01$). The self-cleaning performance displays remarkable anisotropism on the moth wing surface. The moth wing surface is of low adhesive superhydrophobicity. Such a special complex wettability of the moth wing surface resembles that of the butterfly wing surface [5].

Mechanism of the Anisotropic Wettability on the Wing Surface. The wing surfaces of different moth species have highly similar absorption characteristic of FT-IR spectra. The absorption peaks are at 3292, 2942, 2878, 1650, 1533, 1382, 1227, 1145, 1070 cm^{-1}, respectively. These absorption bands result from stretching vibration, skeletal vibration, deformation vibration or in-plane bending vibration of the bases (e.g. $-CH_3$, $-CH_2$, $-C-CH_3$, C-H, O-H, C=O, C-O, N-H) in chitin, protein or fat. The wing surface is composed mainly of naturally hydrophobic material with an intrinsic CA of 95° [6]. However, much higher hydrophobicity cannot result from the chemical composition alone. Due to the hierarchical rough structures on the wing surface, the water droplet stands on the tips of the vertical ribs. Much air is left under the droplet. The actual contact area between the water droplet and the wing surface is so small that the droplet forms an almost perfect sphere. The solid-liquid-gas triple contact lines (TCL) are expected to be contorted and unstable. The multiple-dimensional microstructure plays a crucial role in the complex wettability of the moth wing surfaces. In a contrast test, the scales were removed from the wing surfaces. The CA decreases by 38.5~55.4° (Table 1), all the SAs (FSA, BSA, PSA) increase above 65° (the maximum inclination angle of the sample table is 65°). The low adhesive superhydrophobicity of the wing surface ascribes to the coupling effect of material element and structural element.

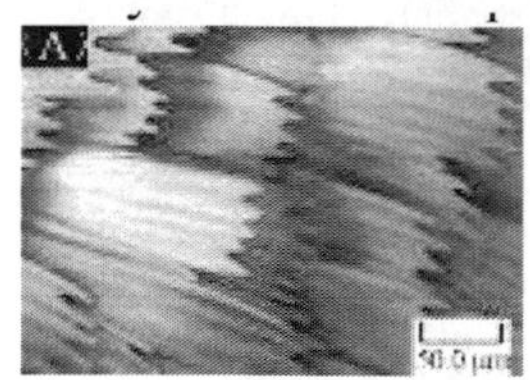

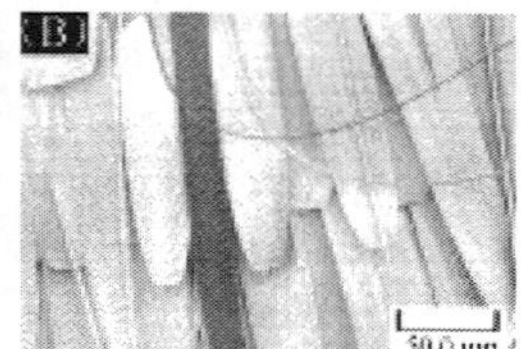

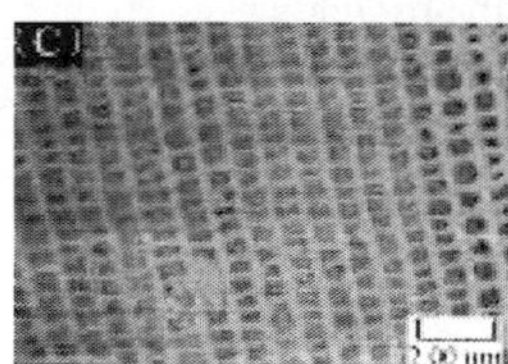

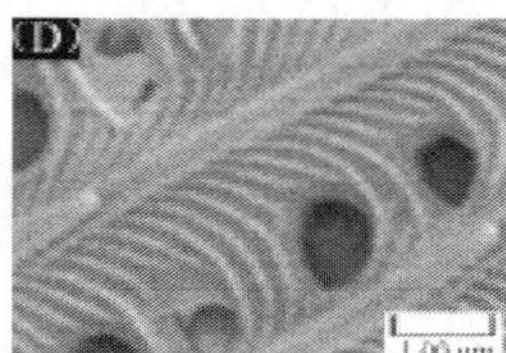

Figure 2. The multiple-dimensional micro/nano microstructures of the moth wing surface. (A), (B) Primary structure (micrometric scales); (C) Secondary structure (submicrometric vertical ribs and horizontal bridges); (D) Tertiary structure (nano stripes).

Anisotropism of the Self-cleaning Property on the Wing Surface. The moth wing surfaces exhibit low adhesion. The range of FSA is 1~3°, BSA 5~9°, PSA 3~9° (Table 1). Meanwhile, the wing surfaces are superhydrophobic (CA>150°). The range of CA is 150.5~158.4°. The large CA and small SA imply outstanding self-cleaning characteristic of the wing surface.

Biological Significance of the Anisotropic Property on the Wing Surface. The anisotropism of SA and self-cleaning property on the moth wing surface is the result of the oriented micro-morphology and the energy barrier difference in various directions. Compared with the sliding of droplet from the wing terminal to the wing basal (BSA), when the

Table 1. CA and SA of water droplet on the moth wing surfaces.

No.	Moth species	CA(°)		SA(°)			
		With scale	Without scale	With scale			Without scale
				FSA	BSA	PSA	FSA/BSA/PSA
1	*Alcis secundaria*	153.2	114.7	3	9	7	>65
2	*Amphipyra erebina*	152.6	98.0	3	8	4	>65
3	*Aspilates geholaria*	155.6	107.5	3	6	5	>65
4	*Autographa nigrisigna*	152.6	97.2	2	6	6	>65
5	*Chlorodontopera mandarinata*	151.8	118.4	2	6	5	>65
6	*Chloromachia gavissima*	153.7	103.2	1	7	4	>65
7	*Dolbina tancrei*	153.2	107.3	2	7	6	>65
8	*Mamestra brassicae*	155.3	106.6	1	7	5	>65
9	*Naxa seriaria*	150.5	103.6	1	5	3	>65
10	*Parum colligate*	152.6	109.8	3	8	6	>65
11	*Scopula pudicaria*	152.3	113.1	1	9	6	>65
12	*Sphinx ligustri*	158.4	112.8	1	6	9	>65
Average		153.5	107.7	1.9	7.0	5.5	>65

droplet slides from the wing basal to the wing terminal (FSA), the anisotropic microstructures exert less influence in hampering droplet sliding, the droplet is highly instable and can roll off more easily, so the SA decreases. This marvelous property results from long-term co-evolution of the moth and the environment, and endows the moth wings with the ability of directional easy-cleaning in a watery environment. Such a self-cleaning function is of essential biological significance for the moth. Even a very slight tilting (1~3°) of the wing is sufficient to cause the water droplet to roll off and take away the contaminating particles effectively. The moth can lighten body burden readily, increase flight stability and efficiency, optimize energy budget. Thus, the moth can get more opportunities to survive and thrive.

4 SUMMARY

The moth wing surface is of low adhesion (SA 1~3°) and superhydrophobicity (CA 150.5~158.4°). The wing surface is composed of naturally hydrophobic material (chitin, protein, fat, etc.), and possesses hierarchical rough structures. The cooperative effect of material element and structural element leads to the complex wettability of the wing surface. The moth wing surface exhibits anisotropism of micro-morphology and self-cleaning property, which is of critical biological significance for the survival of moth. The moth wing surface is an ideal bio-template for development of novel functional materials. This work not only promotes our understanding of wetting mechanism of bio-surfaces, but also brings insights into biomimetic preparation of novel self-cleaning coatings and anisotropic wetting substrate.

ACKNOWLEDGEMENT

This work was financially supported by the National Natural Science Foundation of China (50875108), the Natural Science Foundation of Jilin Province, China (201115162) and the Science and Technology Project of Educational Department of Jilin Province, China (2010373, 2011186).

REFERENCES

[1] T.L. Sun, L. Feng, X.F. Gao, L. Jiang, Bioinspired surfaces with special wettability, Accounts Chem. Res. 38 (2005) 644–652.
[2] A.M. Higgins, R.A.L. Jones, Anisotropic spinodal dewetting as a route to self-assembly of patterned surfaces, Nature 404 (2000) 476–478.
[3] M. Gleiche, L.F. Chi, H. Fuchs, Nanoscopic channel lattices with controlled anisotropic wetting, Nature 403 (2000) 173–175.
[4] H. Liu, J. Zhai, L. Jiang, Wetting and anti-wetting on aligned carbon nanotube films, Soft Matter 2 (2006) 811–821.
[5] Y. Fang, G. Sun, T.Q. Wang, Q. Cong, Hydrophobicity mechanism of non-smooth pattern on surface of butterfly wing, Chin. Sci. Bull. 52 (2007) 711–716.
[6] X.J. Wang, Q. Cong, J.J. Zhang, Y.L. Wan, Multivariate coupling mechanism of NOCTUIDAE moth wings' surface superhydrophobicity, Chin. Sci. Bull. 54 (2009) 569–575.

Computing, Control, Information and Education Engineering – Liu, Sung & Yao (eds)
© 2015 Taylor & Francis Group, London, ISBN: 978-1-138-02800-5

Cost-benefit analysis of Donghai Bridge offshore wind farm

Chen Sheng Wu, Ru Liu, Chen Song & Lu Ji Zhang
Beijing Municipal Institute of Science and Technology Information, Beijing, China

ABSTRACT: Clearly Donghai Bridge offshore wind project is advisable in respect of energy reform and emission mitigation as wind power generation is concerned as "emission free". However, there are several obstacles for wind power development in China, which involves financial issues. We are unable to solve the issues, but we would like to find out if the project is advisable as well regarding the cost and benefit in short and long term. It will be a simplified calculation of cost and benefit. This project was funded by the Beijing financial fund.

KEYWORDS: offshore wind power, CBA (Cost-Benefit Analysis), Donghai Bridge offshore wind farm.

1 INTRODUCTION

As one kind of clean energy, offshore wind power development will be a long-term strategy, but it is not best investment way in the short term. CBA could confirm and compare the project cost and benefit that value through economic market as well as some social cost and benefit. As an investment analysis, CBA focuses on all benefits and all costs. As the outcome of the analysis, the Benefit ratio (Benefits/Costs) could determine whether the project is socially valuable and financially feasible, or if another project should be pursued.

CBA could be analyzed from different points of view. And it could be divided into two aspects: economic CBA and social CBA. From project stakeholders' point of view to consider this project's cost and benefit, economic CBA will use the market price to measure all benefit and cost of the project.

2 SOCIAL BENEFITS OF DONGHAI BRIDGE OFFSHORE WIND FARM

Offshore wind energy development has both positive (benefits) and negative (costs) social impacts. The successful future of offshore wind power will rely on big share of benefits whilst small negative impact.

Offshore wind power could help China deliver its recently agreed target of 15% [9] of all energy must come from renewable sources. Donghai Bridge offshore wind farm in China has several benefits as follow: ①It can help China contribute to CO2 reduction. ② It will contribute to a more supply of electricity beyond the coastal area at east of China. ③It could strengthen China's export: Both onshore and offshore technologies, Project experience and Excess wind energy. ④It could increase more employment opportunities; It also could develop regional economy.

3 SOCIAL COSTS OF DONGHAI BRIDGE OFFSHORE WIND FARM

When compared to other fuel sources power, offshore wind power has the lowest social costs. Offshore wind turbines just have three possible environmental impacts: noise, wildlife and visual impact.

Offshore wind turbine noise has not been observed to disturb birds because of the few birds at Donghai Bridge area. After wind farm construction, mussels will grow on the foundations of the wind turbines, which could improve the variety of flora and fauna in the area. The Danish Environmental Agency has researched the bird population nearby the offshore wind farm in 2002, and the result show that the birds' presence is well correlated with the suitable food, but no big impact on bird population has been detected [7]. Offshore wind farm is far away from land, which produce less noise and visual impacts.

From the Social CBA's standpoint, the Benefit ratio (Benefits/Costs) could determine this project is socially valuable.

4 ECONOMIC COST-BENEFIT ANALYSIS

In fact, the cost of energy power is a very significant method for selecting the way of power generation. It is very important to analyze the local cost of offshore wind power and the related factors to the cost, so that investors could reduce the cost and promote development of offshore wind power in East China Sea in the future.

The basic cost of wind farm includes both upfront and O&M (operation and maintenance) cost. And moreover, there are some variable costs such as tax, salary of employees, depreciation of the offshore wind turbine, institutional arrangements, etc. So the cost of wind energy could fall into initial investment and the annual recurring costs. And there are in general three factors that affect the cost of wind power projects and they are:

1 Capital cost, including purchase wind turbines, infrastructure costs (cables, transport, and foundations), grid connection, installation, licensing procedures, and consultancy. The capital cost of wind farm currently ranges 80% of the total project cost over its whole lifetime [5].
2 Annual recurring costs: Mostly Operation and Maintenance costs (site works) and other variable costs such as tax, salary of employees, depreciation of the offshore wind turbine and institutional arrangements, which varies between different countries.
3 Wind farm lifetime: The lifetime of this case is assumed to be 25 years. Figure 1 shows the cost sensitivity change with the different wind farm lifetime. In this graph, to compare with 25 years project lifetime and 20 years, it makes 9% cost lower. ((0.355-0.325)/0.325DKK/kWh) [1] So, longer project lifetime produces the cheaper cost of electricity.

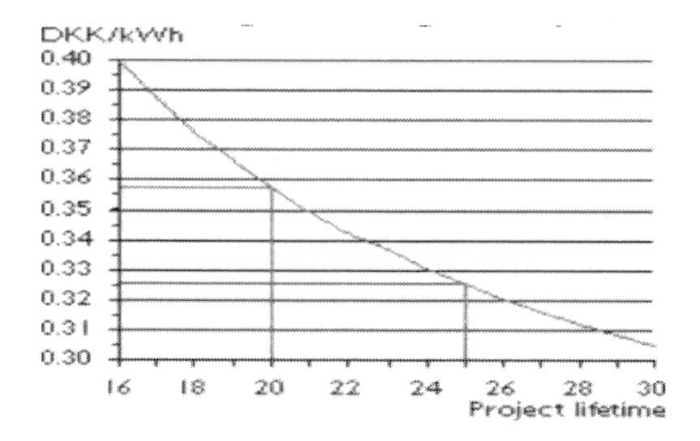

Figure 1. The relationship between cost and project lifetime[1].

The cost per unit of electricity generated have the relationship with the capital cost of wind farm, wind power annual energy output and cost of operating and maintaining the wind farm per unit of energy output. We will calculate some costs of Donghai Bridge offshore wind power so that we can make a comparison between cost and price from the economic aspect.

The known data:
Project life: 25years
Manufacturer: Sinovel Wind Co., Ltd
Model: Sinovel V90-3MW – 90m
Capacity per unit: 3MW
Number of units: 34
Capacity: 102MW [4]
Electricity export rate (Price of the wind energy): 0.64 CNY/kWh = 640 CNY/MWh [2]

The estimated data:
To estimate the annual wind energy output (electricity exported to grid) E (KWh):
E = (hPF)T [3]
h is the number of hours in a year (365days×24hours=8760hours/year)
P is the rate power of each wind turbine in kilowatts (3000kw/each)
T is number of wind turbine (34)
F is the net annual capacity factor of the turbines at the site: 2600hours/8760hours =29.7% (The yearly operation hour of the turbines is 2600 hrs, and there is 8760 hrs per year)
Here, hF = 2600 hours
So, E=(hPF)T =(hF)PT=2600hours×3000KW/each×34=265.2 GWh/year

To estimate the operating and maintenance cost, M, per unit of output: M=KC/E[3]

K is a factor representing the annual operating costs of a wind farm as a small part of the capital cost. EWEA has estimated K to be 0.025, which mean 2.5% of capital cost [3].

C is the capital cost of the offshore wind power which is currently ranges 80% of the total project investment over its whole lifetime. And this case project needs 2.365 billion RMB investments [4].
C = 80% × 2365 million RMB = 1892 million RMB
So, M=KC/E = (0.025×1892 million MB)/265.2(Mkwh/year) = 0.178 CNY/kWh

The operating and maintenance cost per year
M'=KC/N= (0.025×1892 million MB)/25years = 1,892,000 CNY

We use the RETScreen Clean Energy Project Analysis Software to evaluate the offshore wind power and savings, costs, emission reductions, financial viability. This software is one of decision support tools developed with the contribution of numerous experts from government, industry and academia (http://www.retscreen.net). Here, it is used to help us estimate costs associated with the Donghai bridge offshore wind farm. We have calculated some data above, and then put them into the software form, such as power capacity, manufacturer, electricity export rate 0.64CNY/kWh and calculated data (electricity exported to grid 265MWh and capacity factor 29.7%).

The second step, choose the country in emission analysis part, we could get the GHG emission factor 0.74t CO_2/MWh. Multiply GHG emission factor of annual offshore wind power output is GHG emission amount which can be saved by wind farms. And we get the GHG reduction credit rate (239CNY/t CO2) [8] in China, so that this software could calculate the GHG reduction income in 25 years (almost 47 million CNY) as shown in the following figure.

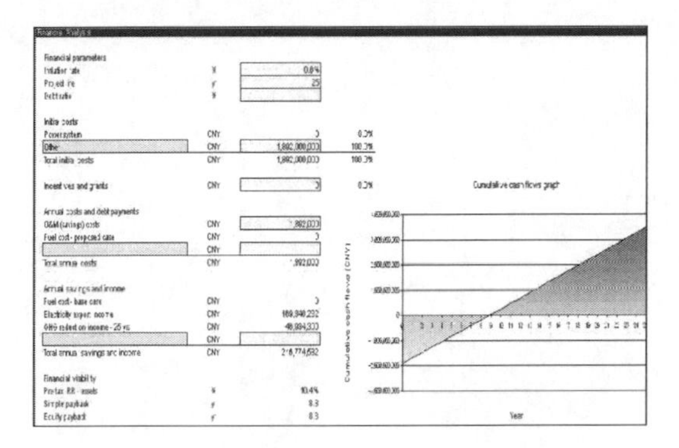

The last step, we type the capital cost and O&M cost in the form, software calculate the rest data and make a cumulative cash flow graph by lifetime. The figure shows the financial viability (the payback time is 8.8 years). So the Donghai offshore wind power is a long-term investment. Although it has large initial cost, the whole project is expected to be profit-make after 9 years.

5 CONCLUSION

Donghai bridge offshore wind power in China is clearly an economically viable energy power in future. Along with the trend with fuel costs on the rise, any renewable power (like natural gas) with free fuel costs becomes economically viable. The wind power is not an attractive investment compare with coal fired power plant and natural gas power plant in the short term. But the coal and natural gas power plant produce GHG emission into the atmosphere from the environmental aspect. They can not to avoid GHG emission cost, because of the energy shortage of the world and the heavy demand for CO_2 sent the price up in future. Thus, Donghai bridge offshore wind power is very close to being competitive with rising fuel and carbon prices, not only comparison with onshore farm, but also in comparison with other traditional power stations.

REFERENCES

[1] Danish Wind Industry Association web site, 2003, http://www.windpower.org/en/tour/econ/offshore.htm.

[2] InfoLib EXPRESS, Wind power development report in 2008, page 17.

[3] Godfrey Boyle, 1996, Renewable Energy-Power for a sustainable future, page 302.

[4] Power International Energy, 2008.

[5] Maria Isable Blanco, the Economics of Wind Energy, page6, 2008.

[6] Power International Energy, 2008, http://www.in-en.com/power/html/power-1137113756151367.html accessed on 2009.

[7] Soren Krohn, Managing Director of Danish Wind Industry Association, Offshore Wind Energy: Full Speed Ahead, page8, 2002.

[8] Soren Krohn, 2009, The Economics of Wind Energy, page13, 49, 69, a report by the European Wind Energy Association.

[9] Vestas Annual report 2008, page 4.

Computing, Control, Information and Education Engineering – Liu, Sung & Yao (eds)
© 2015 Taylor & Francis Group, London, ISBN: 978-1-138-02800-5

Semantic priming effect triggered by sound from different directions: Evidence from N400 effect

Hong Yan Wang, Gao Yan Zhang & Jun Hai Xu
School of Computer Science and Technology, Tianjin University, Tianjin, China

Jian Wu Dang
School of Computer Science and Technology, Tianjin University, Tianjin, China
School of Information Science, Japan Advanced Institute of Science and Technology, Japan

ABSTRACT: *Purpose* This paper is aimed to explore the relationship between the direction of sound and semantic priming effect by Event-Related Potentials (ERPs). *Method* 16 healthy subjects participated in this study. We used the common pictures in daily life as the visual priming stimuli and corresponding sounds as the auditory target stimuli. The sounds were filtered by the head-related transfer function to obtain the different directions. *Results* Our study found that the N400 semantic priming effect was triggered by the incongruent sounds under both conditions, suggesting that it reflected the contextual integration process. Moreover, we also found that there was no significant difference between this two N400 effect. *Conclusion* This study suggested that the direction of sound could not impact on the amplitude size of N400 effect.

KEYWORDS: Semantic priming effect; Direction; N400 effect; ERPs; Head-related transfer functions.

1 INTRODUCTION

Target stimulus is recognized faster when a semantically congruent prime stimulus is presented before, which is named the semantic priming effect [1]. Moreover, many Event-Related Potentials (ERPs) studies have found that the incongruent target induces a larger N400 component (a negative wave evoked around 400ms after target stimulus onset) than the congruent target, which is named N400 effect, suggesting that it reflects the semantic processing [2].

Studies have found that N400 effect can be triggered by different stimulus materials such as words, sentences, natural pictures, sounds, gestures and so on[3]. Researchers also found that the effect can be induced not only in the case of mono-modal, but also cross-modal. Schneider et al.[3] found that the incongruent sounds could triggered N400 effect compared to the congruent sounds by playing the target sound after the appearance of visual priming picture.

Margorie et al. [4]studied the semantic priming effect in the cerebral hemisphere, using the visual words (semantically related or unrelated) as stimulus materials, which appeared randomly in the left or right visual field. They found that words appeared in the right visual field were recognized faster, suggesting that the left hemisphere played a crucial role in the semantic priming process. Although we can obtain spatial orientation information from auditory sound, it is not clear whether the sound from different directions can induce the same semantic priming effect. In this study, we tried to reveal their relationship.

2 PARTICIPANTS

16 graduate students from Tianjin University (9 males, age: 23.9± 2.11 years) participated in this study. Inclusion criteria were as follows: (1) right-handed; (2) physically healthy, and have normal hearing and vision or corrected vision. Written informed consent was obtained from all participants, and they got paid after the experiment.

3 MATERIALS

This experiment chose 30 common pictures in daily life from photographs [5] as the visual priming stimuli and corresponding sounds as the auditory target stimuli. All sounds were transformed into the same format (16-bit, 22kHz sampling rate, 70db sound intensity), and selected 400ms epoch which is the most characteristic part of the sound by Cool Edit Pro 2.1. And then each sound was filtered by the head-related transfer function [6] to obtain two different directions (in a horizontal plane, 45° to the ahead direction on the left and right).

4 PROCEDURE

At first, a fixed point in the center of the screen was lasted 200-300ms to keep the subject focused. Then, a picture was showed in the screen for 200ms and followed by a 400ms target sound with a 400ms inter-stimulus interval. Finally, a circle was presented in the screen to ask subjects to make a response to the target sound. If they heard the sound of an animal, button '1' was pressed, otherwise button '2'. And the next trial started after the response was delivered. The sound and picture in each trial were either semantically congruent or incongruent.

5 DATA RECORDING AND ANALYSIS

32 electron poles through the Neuroscan Synamps system (1000Hz sampling rate, DC filter) were used to record EEG signals referenced to the right mastoid, and HEOG and VEOG were monitored simultaneously. Impedance of each electron pole was kept below 5kΩ. The half value of left mastoid was used as the offline reference.

A band-pass filter (0.1-30Hz) was made to all data. The data polluted by ocular movement and whose amplitude value exceeded a threshold value (± 100μV) were removed. Then the data were divided into epoch, and the interval of each epoch was 100ms before sound onset and 600ms after sound onset. The interval from -100ms to sound onset was treated as the baseline.

This study focused on N400 in the interval of 300-500ms. We selected 15 electrode sites of interest : F3, FZ, F4, FC3, FCZ, FC4, C3, CZ, C4, CP3, CPZ,CP4, P3, PZ, P4. This experiment consisted of three factors: Congruency(Congruent, Incongruent)* Electrode site (fifteen electrode sites of interest) *Direction(Left, Right). We performed repeated measures ANOVA and adopted the Greenhouse-Geisser correction when the degree of freedom exceeded one.

6 RESULTS

Firstly, we respectively conducted two-way (Congruency * Electrode site) repeated measures ANOVA for the average amplitude (300ms-500ms) in each direction condition. The results were shown in Table 1.

Table 1. The results of ANOVA in both conditions.

	Left	Right
Congruency (2)	F(1,15)=48.407 P<0.001	F(1,15)=22.558 P<0.001
Electrode site (15)	F(14,210)=4.405 P=0.015	F(14,210)=5.898 P=0.001
Congruency * Electrode site	F(14,210)=6.458 P<0.001	F(14,210)=0.911 P=0.459

The incongruent sounds triggered a greater N400 effect than the congruent sounds (P<0.001) in both direction conditions (see Fig.1).

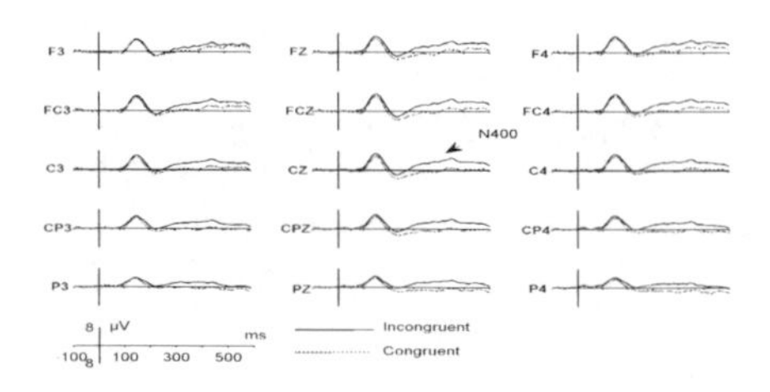

(a) Left

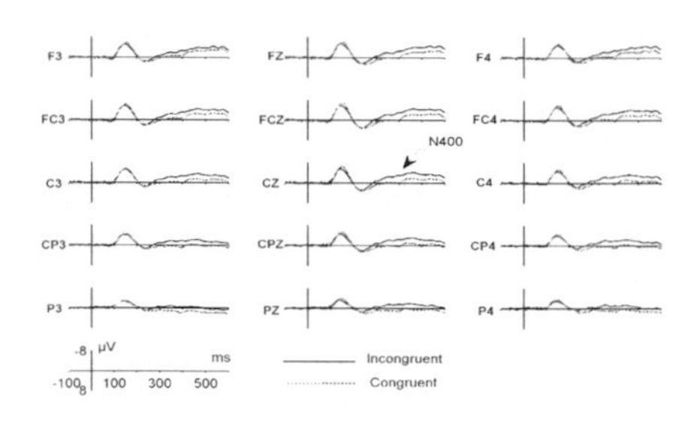

(b) Right

Figure 1. Average ERP triggered by sounds under Left (a) and Right (b) conditions, respectively. The sound onset is at 0ms, and positivity is plotted downward. (a) The incongruent sounds (black line) triggered a greater N400 than the congruent sounds (red line) in the time window of 300ms-500ms . The N400 effect was observed under Left conditions. (b) The N400 effect was also observed under Right conditions.

Secondly, we tested whether there was any significant difference between this two N400 effect (averaged ERPs under the incongruent condition subtract averaged ERPs under the congruent condition). A two-way (Direction * Electrode site) repeated measures ANOVA was conducted for the averaged amplitude of N400 effect. The results were shown in Table 2.

Table 2. The results of ANOVA.

	F value	P value
Direction (2)	F(1,15)=3.482	0.082
Electrode site (15)	F(14,210)=5.245	0.002
Direction * Electrode site	F(14,210)=1.548	0.194

The N400 effect under Left direction condition (Left-N400 effect) was slightly (P=0.082<0.01) greater than the N400 effect under Right direction condition (Right-N400 effect) (see Fig.2), but there was no significant difference.

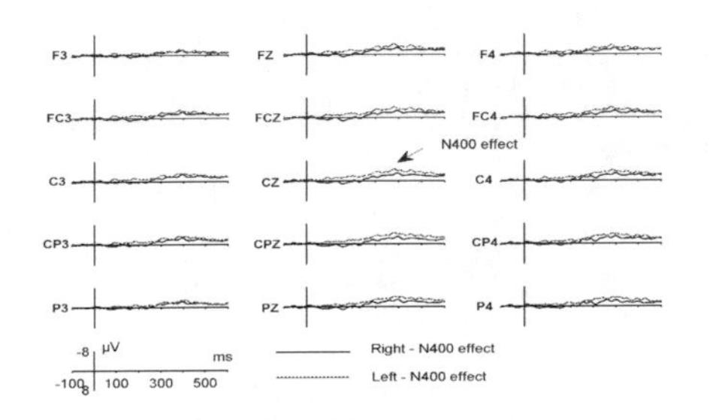

Figure 2. N400 effect under Left (red line)and Right(black line) direction, respectively. The Left-N400 effect was slightly greater than the Right-N400 effect, but there was no significant difference.

7 DISCUSSION

In our study, under both direction conditions, the incongruent sounds triggered a negative-going deflection between 300ms and 500ms compared to the congruent sounds, which was consistent with the previous studies on the cross-modal semantic prime, suggesting that the N400 effect reflected the contextual integration process [3].

Moreover, our results also showed that there was no significant difference between Left–N400 effect and Right -N400 effect, which indicated that the direction of sound has not an influence on the amplitude size of N400 effect.

8 CONCLUSIONS

In this study, we adopted a cross-modal semantic priming paradigm, and used sounds from two different directions by the head-related transfer function, to explore the relationship between the direction of sound and semantic priming effect. The results indicated that sounds from different directions always induced semantic priming effect and there was no significant differences between them. Our study would give us a great help to understand the cognitive mechanism of semantic priming effect.

REFERENCES

[1] Meyer, D. E. & Schvaneveldt, R. W. 1971. Facilitation in recognizing pairs of words: Evidence of a dependence between retrieval operations. J Exp Psych 90 (2) : 227~234.

[2] Kutas, M. & Hillyard, S. A. 1980. Reading senseless sentences: Brain potentials reflect semantic incongruity. Science 207: 203~205.

[3] Schneider, T. R. & Debener, S. & Oostenveld, R. & Engel, A. K. 2008. Enhanced EEG gamma-band activity reflects multisensory semantic matching in visual-to-auditory object priming. Neuroimage 42: 1244~1254.

[4] Koivisto, M. & Revonsuo, A. 2000. Semantic priming by pictures and words in the cerebral hemispheres. Cognitive Brain Research 10 (1- 2): 91~98.

[5] Schneider, T. & Engel, A. K. & Debener, S. 2008. Multisensory identification of natural objects in a two-way crossmodal priming paradigm. Exp Psychol 55: 121~131.

[6] Gao, H. & Ouyang, M. & Zhang, D. 2011. An auditory brain-computer interface using virtual sound field. Engineering in Medicine and Biology Society, EMBC, Annual International Conference of the IEEE. IEEE: 4568–4571.

Computing, Control, Information and Education Engineering – Liu, Sung & Yao (eds)
 ISBN: 978-1-138-02800-5

The research of piezoelectricity ACV and its control technology

Ning Zhuang Liu
School of Electric and Control Engineering, Xi'an University of Science and Technology, Shanxi Xi'an China

ABSTRACT: In this paper, the piezoelectric ACV (Air Compensation Valve) has applied to the system of Electronic controlled carburetor on a motorcycle. The mechanical structure and its basic principle are analyzed according to the piezoelectric equation. After the theoretic analysis of piezoelectric effect, the moving displacement of piezoelectric bimorph and response characteristic are derived and the actual size and shape are designed. Driving 100V&15Hz PWM waveform on the ACV, it not only appears a good stability and consistency, but also completely meets the air compensation for the idle air channel. The stability of design principle and mechanical structure of the piezoelectric ACV have passed the small batch production according to the experimental results.

KEYWORDS: Carburetor; ACV; piezoelectric effect; Bimorph.

1 INTRODUCTION

With the rapid development of the automobile and motorcycle industry, many countries face the very strict exhaust emission regulation about main pollution gas, such as CO, NOx, HC in order to protect the ecological environment and inhibit the global oil shortage. This is a strong shock to the traditional motorcycle industry. In order to obtain a chance of survival, the enterprises must greatly improve their kinds of level, such as management, manufacture and technology level.

The mechanical equipment using inverse piezoelectric effect has been used in many industrial applications, such as warp weaving machine, intelligent valve position. However, the piezoelectric ceramics is involved in motorcycle fields for exhaust emissions has become a new research focus on flexibility, high sensitivity and low power.

2 E_CARBURETOR INTRODUCTION

Electronic Controlled Carburetor is referred to as E_ Carburetor below. The schedule keeps the oil system less changed to avoid too much mechanical modification. The only change is that the ACV is installed on the idle air hole in order to compensate proper mixed air, according to the ECU (Electronic Control Unit). The ECU detects load engine correctly in real time through the acquisition of pressure sensor, oxygen sensor, cylinder temperature, engine speed and other useful information. The ECU controls the idle air compensation by different driving PWM (Pulse Width Module) according to the actual working condition of the engine in order to achieve the best Oil-Air Ratio. The whole schedule aims to reduce the emissions of HC, CO and NOx and to meet the theory Oil-Air Ratio that is 14.7:1 and completely pass the Euro-IV Regulation [10]. In this system schedule, there are two core components just as piezoelectricity ACV and the ECU. They play an important role and decide whether the system is in normal [1].

3 MECHANICAL STRUCTURE OF ACV

The mechanism of ACV is composed of piezoelectric bimorph, package shell and output wire etc. [2]. While the piezoelectric bimorph is composed of two single piezoelectric ceramics, fiber plate and the base body of the ACV, such as the figure 1.For the bimorph has the charging and discharging characteristics of capacitors, the piezoelectric bimorph is forced to vibrate and caused displacement and control the amounts of air into the carburetor. When the proper voltage is driven on the ACV, the polarization direction of upper single will cause to contract. On the contrary direction, the under single chip one will extend. When the ACV blocks the outlet of the carburetor, the amount of air into the carburetor is decreased. On the contrary, when the outlet is open, the amount of air is increased. The Air-Fuel Ratio dynamically adjusted to the Theory Air-Fuel Ratio. (14.7:1)

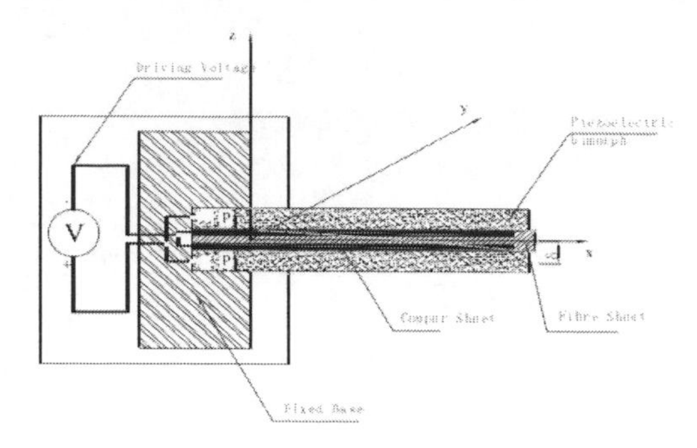

Figure 1. The principle of piezoelectricity ACV.

4 DISPLACEMENT AND RESPONSE OF ACV

According to the switching property of piezoelectric bimorphs, so the PWM pulse is adopted to drive the two piezoelectric ceramics. The piezoelectricity equation of upper bimorph is as follows formula (1)[3]:

$$\begin{cases} S_1^U = S_{11}^E T_1^U - d_{31} E_3 \\ -D_3^U = d_{31} T_1^U - \varepsilon_{33}^T E_3 \end{cases} \tag{1}$$

So as the down bimorph as follows formula (2):

$$\begin{cases} S_1^D = S_{11}^E T_1^D + d_{31} E_3 \\ D_3^D = d_{31} T_1^D + \varepsilon_{33}^T E_3 \end{cases} \tag{2}$$

For the thickness, length, crystal structure of both piezoelectricity bimorph is exactly on same, the deformation under the action of the electric field is equal in size, on the contrary the direction is opposite. According to the above piezoelectricity equation (1) and (2), the deflection of the cantilever beam is as follows formula (3)[4]:

$$y = (3d_{31}ux^2)/(4h^2) \tag{3}$$

When the x in equation (3) is length(x=l), the displacement of free end is as follows as below formula (4)

$$\sigma = (3d_{31}ul^2)/(4h^2) \tag{4}$$

In the formula (4),the d_{31} is piezoelectricity coefficient, u is driving voltage, l is the length of bimorph.

When in the actual design of bimorph, l is equal to 33cm and u is equal to 100V, the thickness (h) is 0.25mm, the width (w) is 0.71mm, piezoelectricity coefficient (d_{31}) is 250pm/V.when the above value into the formula (4), the max displacement (σ_{MAX}) is 0.51mm. According to the above calculation, the conclusion can be drawn that the displacement of free end is decided by the vital parameters, such as outline dimension, the characteristics of material and the driving voltage. The way of improving the displacement is to increase the length and voltage properly or decrease thickness.

For the calculation of bimorph flexibility, the formula as follows (5):

$$S = S_N^E (4l^3 / wh^3) \tag{5}$$

$$F = \sigma / S \tag{6}$$

In the formula (5), the compliance coefficient of bimorph (S_N^E) is $1.8 \times 10^{-12} m^2 / N$. Above the value of the formula (4), the compliance(s) is $0.72m / N$.

According to the above calculation, the max displacement (σ_{MAX}) is 0.51mm, so the force of bimorph (F) is 0.71N as follows formula (5):

$$F = \sigma / S = 0.51 \times 10 / 0.72 \times 10 = 0.71N \tag{7}$$

When the ACV is applied to the actual E_ Carburetor, the voltage on the bimorph is a given PWM pulse, so as Figure 1. The piezoelectricity equation of charge and discharge is as follows (8):

$$\begin{cases} x_1 = Nd_{33}U_0(1 - e^{1/R_1C}) \\ x_2 = -Nd_{33}U_0(1 - e^{1/R_2C}) \end{cases} \tag{8}$$

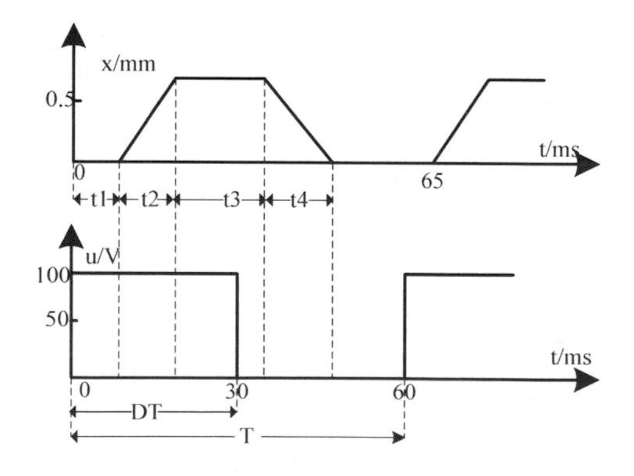

Figure 1. The characteristics of PWM and displacement.

Because of the delay characteristics of the piezoelectric bimorph, the displacement characteristics are expressed by two subsection functions, just as follow (9):

$$u(t) = \begin{cases} U_m & t \in (0, DT) \\ 0 & t \in (DT, T) \end{cases} \quad (9)$$

In the formula (9), Symbol of T is expressed signal period, D is duty, U_m is the amplitude of voltage .

The subsection function is converted into a step function as follows (10):

$$u(t) = U_m\varepsilon(t) - U_m\varepsilon(t - DT) + U_m\varepsilon(t - T) + \cdots \quad (10)$$

Because the piezoelectric bimorph is a capacitive device for double parallel bimorphs and its capacitance is 50nF and its amplitude is 100V,it can be decomposed into pulse function follow as(10),while the response process can be considered as complete response caused by step function[3].

The zero state response expression of $U_m\varepsilon(t)$ is follow as (10):

$$u(t) = D_n U\ (1 - e^{-t/R_1C}) \quad (11)$$

5 THE TEST OF THE FORCE AND DISPLACEMENT

In order to verify the characteristics of piezoelectric ACV, the instruments should be prepared for the test, the driving circuit for ACV, one test computer and some test accessories. With the above conditions, we can draw the following table 1:

Table 1. The test data of force and displacement [N].

Number	Force	Positive	Reverse
1	0.731	0.5050	0.5258
2	0.755	0.5111	0.5236
3	0.725	0.5247	0.4799
4	0.737	0.5274	0.4693
5	0.725	0.5555	0.4867
6	0.717	0.5454	0.4653
7	0.833	0.5759	0.5284
8	0.763	0.5305	0.5124
9	0.79	0.5174	0.5119
10	0.70	0.4873	0.5006

Through the table 1, the conclusion can be drawn that the error range of force is less than 0.07N ($\Delta F < 0.07N$) and the error range of displacement is less than 0.04mm ($\Delta\delta < 0.04mm$). The piezoelectric ACV has good stability and consistency.

6 CONCLUSION

The piezoelectric is just using the inverse piezoelectric effect of ceramic bimorph converting the energy from electrical to mechanical. The ECU outputs some proper duty PWM signal according to the actual working condition of carburetor and causes the piezoelectric deformation displacement for controlling the idle air into the carburetor. Through the adjustment, the emission of CO, NOx, and CH will be greatly reduced, aiming for the theory Oil-Air Rate.

Comparison between the electromagnetic ACV, the piezoelectricity ACV has below advantages that are less power loss and high sensitivity and high reliability. For the piezoelectric ACV has been applied many engine tricycle and proved by test machine, accomplished the emission regulation Euro-4, so the piezoelectric ACV will be applied to many industry fields and has great application prospect.

REFERENCES

[1] Nader Jalili, John Wagner, Mohsen Dadfarnia.A piezoelectric driven ratchet actuator mechanism with application to automotive engine valves. Mechatronics 13 (2003) 933–956.

[2] A. Dolla, M. HeinrichsaAuthor Vitae, F. Goldschmidtboeing etc. A high performance bidirectional micro pump for a novel artificial sphincter system [J]. Sensors and Actuators A 130–131 (2006) 445–453.

[3] Kan Junwu, Yang Zhigang, Peng Taijiang etc.Design and test of a high-performance piezoelectric micropump for drug delivery [J].Sensors and actuators A-physical. 121 (2005) 156–161.

[4] Wang Hai-ning, Cui Da-fu,Geng Zhao-xin etc. Study on Piezoelectric Micro pump Driven by PZT Bimorph [J]. Piezoelectrics & Acoustooptics, 2007, 29(3):302–304.

Computing, Control, Information and Education Engineering – Liu, Sung & Yao (eds)
 ISBN: 978-1-138-02800-5

Study of microstructure of steel samples deformed in a step-wedge die

A.O. Tolkushkin, E.A. Panin, O.N. Krivtsova & Zh.K. Amanzholov
Karaganda State Industrial University, Temirtau, Kazakhstan

ABSTRACT: This work is devoted to investigation of the influence of the deformation process of the blanks of rectangular cross-section in a step-wedge dies on the microstructure of metal. The expediency of the use of new design tools - step-wedge dies, instead of currently used flat dies, since the proposed technique allows to obtain high quality billet with a uniform equiaxed fine-grained structure with less of forging reduction.

KEYWORDS: step-wedge dies, forging, microstructure.

1 INTRODUCTION

The accelerating paces of development of mechanical engineering and metallurgical industries confront in metallurgy the essential tasks of improving the quality of metal and metal products with minimal energy costs. The same goal is also facing forging production, as the forging operation provides not only ingot or billet desired shape and size, but also in conjunction with other treatments significantly improve the mechanical properties of the metal.

Current forging technologies are based on using traditional blacksmith tools and deformation modes, characterized by a low level of mechanical properties and their uneven distribution throughout the cross section of billets. Using the same new forging tools, implementing the alternating deformation can improve the quality of metal forgings [1].

At the chair «Proceeding of metals by pressure» of Karaganda state industrial university developed a number of forging tools allowing to intensify the alternating deformation in the whole volume of the deformed billet. These tools include step dies [2], trapezoidal and locking dies [3-7], dies with elastic elements [8, 9]. All if these tools underwent successful pilot testing at JSC «ArselorMittal Temirtau». The latest development of Karaganda state industrial University scientists are step-wedge dies (Fig. 1). In paper [10] proved, that using step-wedge dies for forging slabs and plates allows to implement conducive strain-stress state for production of metal with a fine grain structure.

This work is devoted to investigation of the influence of the deformation process of billets rectangular cross-section in a step-wedge dies on the microstructure of metal.

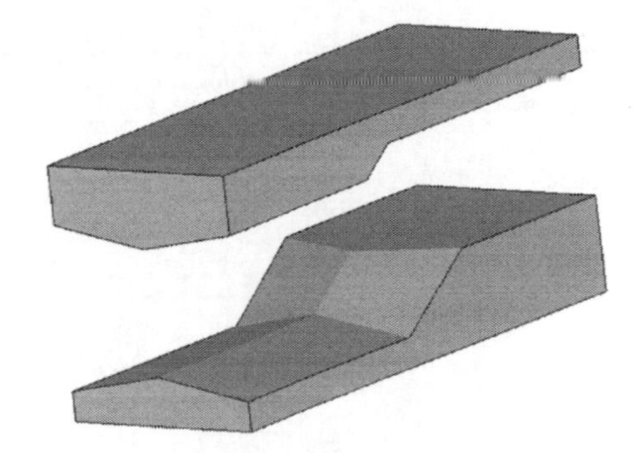

Figure 1. New design of step-wedge dies.

2 RESEARCH METHODOLOGY

Deformation of billets produced under laboratory conditions in a hydraulic press with a force of 1.25 MN. For this experiment billets made of AISI 1020 with proportions $h \times b \times l = 30 \times 60 \times 250$ mm were made. To restore the initial structure of billets before deformation they were subjected to recrystallization annealing at a temperature of 680°C for 4 hours endurance in a resistance chamber furnace. Billets drawing was carried out in a step-wedge dies with angle 30° and the angle of the wedge 160°, step dies with angle 30 and the flat dies.

The deformation of billets carried out as follows. The billet was heated up to the temperature of the beginning of forging 1200°, and then they were applied in a step-wedge die on the first step with a wedge. After the compression of billet at the first stage advancing of billet was carried out on a sloping section and also compression was made. After this, billet advanced on the second flat section, in which straightening of billet was produced. The scheme of deformation of blanks in a step-wedge dies presented in Figure 2. The proportions of deformed billets after their drawing over the entire length constituted $b \times h \times l = 26{,}3 \times 62{,}3 \times 274{,}6$ mm, forging reduction was equal to 1,14. Blanks of the same size were deformed in step and flat dies with the same forging reduction, as in the step-wedge dies. Of all the deformed blanks templets in the longitudinal and transverse directions were cut.

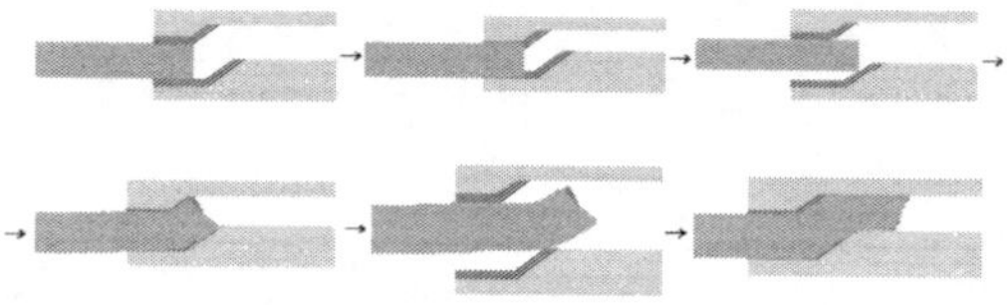

Figure 2. Deformation scheme of billets in the step-wedge dies.

To determine the grain size of the metal forged by the proposed and existing technologies used GOST 5639-82 "Methods of detection and determination of grain size". When determining the amount of grain an optical microscope Leica were used.

3 RESULTS AND DISCUSSION

Data, obtained during the study of microstructure of AISI 1020 are shown in Table 1.

Table 1. The results of determination of the average grain diameter of AISI 1020.

Tool	Study direction	Original grain size, mm	Average grain size after deformation
Step-wedge dies	cross	0,0451	0,0138
	cross		0,0142
Step dies	cross	0,0449	0,0243
	cross		0,0231
Flat dies	cross	0,0449	0,0415
	cross		0,0314

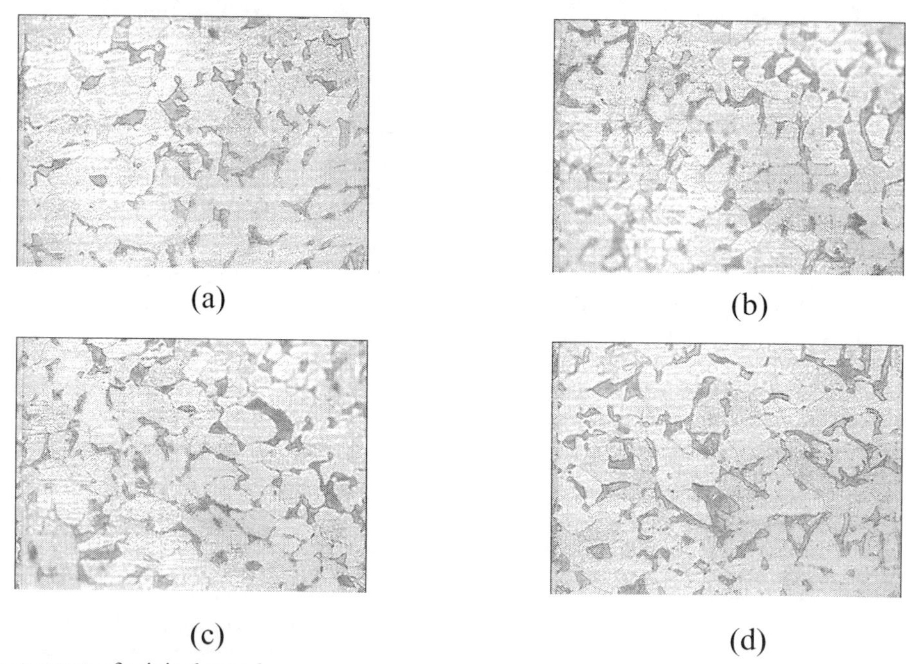

a) microstructure of original sample;
b) microstructure of sample, deformed in step-wedge dies;
c) microstructure of sample, deformed in step dies;
d) microstructure of sample, deformed in flat dies.

Figure 3. Microstructure of sample made of AISI 1020, ×200.

The microstructure analysis of the deformed billet has shown that the microstructure of the metal forgings obtained by deformation of material in a step-wedge dies fine and equiaxial throughout the volume of the billet and smaller by 1-2 points than the microstructure of forgings made in step dies and 2-3 points than the microstructure of forgings made on flat dies. At the same time during the deformation of billet in flat dies, there is a strong anisotropy of the structure in forging volume unlike deformation in forging at step and step-wedge dies.

From the comparison of the results of microstructure during deformation of billets in a step and step-wedge dies, it can be concluded that in both cases there is a uniform structure throughout the body, but when using a step-wedge dies provide a more intensive deforming of the cast structure for one cycle. And this, in turn, allows us to provide the necessary forgings quality with a given level of mechanical properties in fewer passes.

4 SUMMARY

The expediency of the use of new design tools - step-wedge dies, instead of currently used flat dies, since the proposed technique allows to obtain high quality billet with a uniform equiaxed fine-grained structure with less of forging reduction.

REFERENCES

[1] Kolmogorov V.L. Stresses, deformations, damage.- M.: Metallurgy, 1970 .– 229p.

[2] Patent of USSR №1409394. The method of manufacture of forgings and tools for its implementation / V.K. Vorontsov, A.V. Kotelkin, A.B. Naizabekov and others.1988, Bul.№ 26.

[3] Naizabekov A.B., Lezhnev S.N., Bulebaeva A.Zh. Investigation the process of billet deforming in special tool without significant variation of initialsizes. Izvestiya vysshikh uchebnykh zavedenij. Chernaya metallurgiya (6). 2001. – pp.23–25.

[4] Naizabekov A.B., Lezhnev S.N. Investigation of utilization degree of plasticity resource during deforming the billets in trapezoid – shaped block. Izvestiya vysshikh uchebnykh zavedenij. Chernaya metallurgiya (8). 2001. – pp.40–42.

[5] Naizabekov A.B., Lezhnev S.N. Reserves of plasticity in billet deformation by trapezoidal hammers. Steel in Tranlation, 31 (8). 2001. - pp. 68–70.

[6] Naizabekov A.B., Lezhnev S.N. Influence of various methods of deformation on use degree of metal plasticity resource. Izvestiya vysshikh uchebnykh zavedenij. Chernaya metallurgiya (6). 2003. – pp.20–21.

[7] Naizabekov A.B., Lezhnev S.N. Influence of the deformation method on the utilization of the metals potential plasticity. Steel in Tranlation, 33 (6). 2003. - pp. 53–55.

[8] Naizabekov A.B., Lezhnev S.N. Investigation of the billet deformation process in the dies with elastic elements realizing a cross and a longitudinal shear. Materials Science Forum, Stafa-Zurich, Switzerland, 2008, Vol. 575–578.

[9] Naizabekov A.B., Nogaev K.A., Ashkeev Zh.A. Deformation of workpieces with flat dies with the imposition of additional shear strain. Journal «Izvestiya VUZ. Chernaya Metallurgiya. Moscow, 2005 № 2. -P. 16–18.

[10] Development and computer modeling of new deformation technologies of workpieces of the type of slabs and plates of ferrous and non-ferrous metals and alloys/ S.N. Lezhnev, E.A. Panin, A.O. Tolkushkin, D.V. Kuis, G.A. Sivyakova.// Proceedings of VI international congress «Non-ferrous metals and minerals-2014», Krasnoyarsk, 15–19 of September 2014. - P. 1213–1221.

Computing, Control, Information and Education Engineering – Liu, Sung & Yao (eds)
© 2015 Taylor & Francis Group, London, ISBN: 978-1-138-02800-5

Design and development of an omnidirectional mobile robot using fuzzy directional control

Anan Suebsomran
Research Center of Intelligent Machines and Robotics (IMR), Science and Technology Research Institute, King Mongkut's University of Technology North Bangkok Bangsue, Bangkok, Thailand
Department of Teacher Training in Mechanical Engineering, King Mongkut's University of Technology North Bangkok Bangsue, Bangkok, Thailand

Jessadang Thanomsin
Science and Technology Research Institute, King Mongkut's University of Technology North Bangkok Bangsue, Bangkok, Thailand

ABSTRACT: This research aims to design and control of an omnidirectional wheel robot which consists of three elements: robot design and development, C# based GUI and Android based GUI development. It can be divided into two operational modes: manual and autonomous mode. In manual mode, users will be able to control directly position of a robot to their desirable position. While autonomous mode, users will need to enter both distance and angle to command a robot to reach the desired destination. After executing command stage, the robot will start to move according to user input while trying to avoid any obstacle that blocks its path of travel. User can use either C# based GUI or Android based GUI user interface for applying the command and display data during communication with an Arduino microcontroller as a robot controller via wireless LAN. When data is retrieved by an Arduino microcontroller, robot will be controlled according to the specified command from the user. During its motion, two feedback signals are used for design robot control in closed loop control method. The encoder signal applied to control the motion of the robot with the specified distance and direction of robot's movement. Experiments are also conducted to constraint the ground friction conditions with high and low friction of floor material. From the experiment, the robot can move to the target with accurate position with bounded error ±10% of specified distance of robot motion. From the result of robot's development, this development platform can be extended for the robot prototype with a wide range of applications in different places such as industrial factories, hospital or museum with the high performance of the proposed methods.

KEYWORDS: Omnidirectional mobile robot, Position control, Fuzzy directional control, GUI user interface.

1 INTRODUCTION

The conventional wheel mobile robot has applied as the platform for robotic research. Limitation of such a wheel subjects to flexibility of robot control performance, such as robot motion capability, low rotation torque of the wheel, required high friction of ground, etc., To reduce such limitation, a new platform of robot wheel is introduced [1], an omnidirectional mobile robot, to be applied in research and practical use in robot development. The ability of robot has increased function of step-climbing in mechanical design aspect [2]. For application, the omnidirectional mobile robot has extensively to design a robot for hospital environment [3]. They presented a novel wheel design by using anisotropic friction properties. From above literature, the omnidirectional mobile robot was described in the design technique of mechanical system. Moreover, robot control and navigation system were carried out to extensive research. To resolve the problem of plant parameter disturbance, the adaptive PI control system for an omnidirectional mobile robot was proposed by [4]. Control gains were required to readjustment based on error signals. In positioning robot control, linear visual servoing-based control of position and attitude of omnidirectional mobile robot. Position and attitude of target were marked with a binocular vision system. To reach the specified target, robot motion and binocular vision were linearly approximated to control omnidirectional mobile robot. From such the approximate matrix and difference of binocular vision between target marker and robot marker, the translational and rotational velocity of robot ware generated to command the robot to reach the destination. The main problem of the proposed control method has some problem with the distortion of image information. Robot navigation of omnidirectional mobile robots was proposed by [5].

From literatures, an omnidirectional mobile robot was emphasized in the design of the robot mechanism, robot control and navigation system for enhancing robot maneuverable and capability including to the accuracy of the robot positioning system. Thus, this paper proposes a new control robot with user friendly based control by using two modes of control via user: C# GUI based and Android GUI based control for command robot motion. The design of robot controller is based on fuzzy direction control of the robot. Robot positioning control is also designed for feedback control approach. Accuracy of robot positioning control under load condition has to show by experimental methods.

2 DESIGN AND CONTROL OF AN OMNIDIRECTIONAL MOBILE ROBOT

The robot structure is the kind of three wheels omnidirectional mobile robot. Weight of designing robot is totally P. Each wheel takes the average load about 3 kg. Torque of motor is found by using Eq. 1 with specific velocity of motor at 300 rpm. The result of torque is obtained about 28 W. Torque is also obtained about 0.98 Nm.

$$P = F \cdot V \tag{1}$$

The shaft design for power transmission is calculated using Eq. 2.

$$d = \sqrt[3]{\frac{16}{\pi \times 90} \times \left[\left(M \times \propto_b \right)^2 + \left(T \times \propto_b \right)^2 \right]^{\frac{1}{2}}} \tag{2}$$

d is the diameter of shaft for transmission $d = 5.76$ mm from reference standard table, we selected diameter of the shaft is about $d = 6$ mm

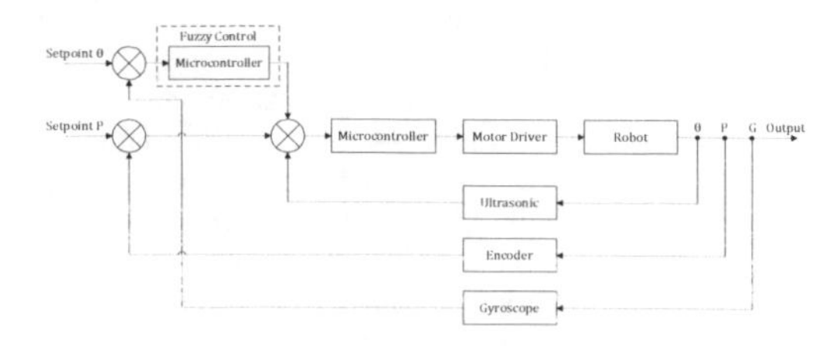

Figure 1. Robot motion controller.

3 POSITION CONTROL OF AN OMNIDIRECTIONAL ROBOT

Position control of robot applies encoder measurement as a position feedback signal. Let's start with PID controller that can be implemented by Eq. 3 formed as:

$$u(t) = K_p e(t) + K_i \sum_{i=0}^{i} e\Delta(t) + K_d \frac{\Delta e}{\Delta t} \tag{3}$$

Next, error of position control can be obtained by Eq. 4:

$$E_p = P_{ref} - P \tag{4}$$

Where E_p is the error of force controller; P_{ref} is a reference signal (mm) ; and P is a position feedback signal.

4 FUZZY DIRECTION CONTROL OF AN OMNIDIRECTIONAL ROBOT

To control the robot, direction and position command are sent to the robot controller. Due to robot configuration, error of robot motion caused by many sources will occur during the robot move with used specified command. Especially in direction control of robot, robot needs more precision direction of the robot. Fuzzy robot control applies as a controller of robot direction. Constructing input-output of fuzzy inference systems, the relationship of fuzzy rules has taken the feedback signal from the Gyroscope sensor to measure the angular acceleration of robot motion.

Table 1. Fuzzy rules construction.

ΔG \ G	N	Z	P
N	N	N	Z
Z	N	Z	P
P	Z	P	P

From Table 1, it shows the design of fuzzy rules constructed with two inputs of three membership function and nine rules. G is defined as a reading previous value from gyroscopic sensor and (ΔG) is reading the current value of gyroscopic sensor. Table 2 is specified, output membership function.

Table 2. Output membership of fuzzy sets with 3 memberships.

Membership function	Shape	Point
N	Trapezoid	–6;–1
Z	Triangle	–1;0;1
P	Trapezoid	1;6

5 EXPERIMENTAL RESULTS

5.1 *Robot prototype*

Development of an omnidirectional mobile robot completely assembled. This platform is three omnidirectional wheels for each wheel separate of each other by 120 degrees. Robot control is a using Arduino microcontroller. User interface to control robot applies wireless communication via TCP/IP protocol.

Accuracy testing of robot motion on the woolen fabric floor is tested on position about 100 cm and angular with each increasing step from 0 degree to 30 degrees to 330 degrees. The result of control robot motion is verified by accuracy of robot motion with specifying position and direction command can be found by Fig. 2 with no load and with load condition. Load condition is varied from 1 kg to 3 kg.

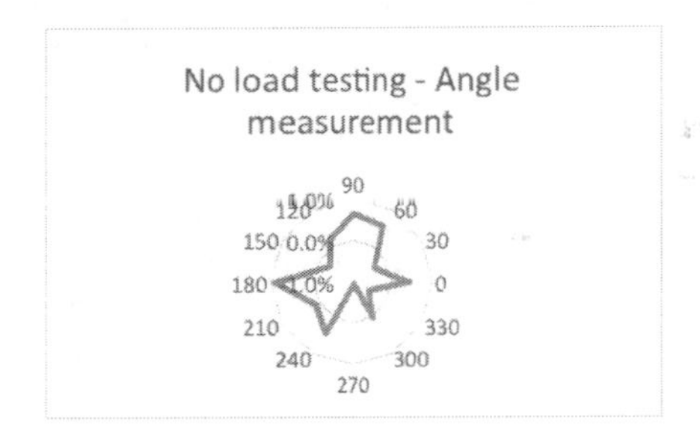

Figure 2. Experiment of robot with no load condition for position and direction control.

6 CONCLUSION

A omnidirectional mobile robot designed and developed for achieving prototype platform was described. Applied load to robot with the ability to carry out is maximum 3 kg. Control of the robot can be controlled by two ways, computer and Android. Displacement and angle command are sent to control robot displacement and orientation. Displacement is controlled by using a simple PID control. Due to a large set error caused robot motion with directional command inaccuracy, thus the orientation control of robot has applied the fuzzy control technique. Fuzzy inference system is based on three input-output membership functions such low, medium, and high. The results from the experiment are found that the error of robot motion with displacement and directional commands is not over than 10 percent. Positioning robot control will have a more precise motion by carry payload, because the source of error is effected mainly with friction and a command signal to generate torque at the omni-wheel.

ACKNOWLEDGEMENTS

This research was funded by King Mongkut's University of Technology North Bangkok. Contract no. KMUTNB-GEN-54-18.

REFERENCES

[1] K. Sato, K. Watanabe, K. Izumi and M. Watanabe: *An Adaptive PI Control System for an Omnidirectional Mobile Robot*, Journal of Robotics and Mechatronics, Vol.11, No.5, pp. 349–355, 1999.

[2] T. Ogino, M. Tomono, T. Akimoto, and A. Matsumoto: *Human Following by an Omnidirectional Mobile Robot Using Maps Built from Laser Range-Finder Measurement*, Journal of Robotics and Mechatronics, Vol.22, No.1, pp. 28–35, 2010.

[3] K. Tadakuma, R. Tadakuma, and S. Hirose: *Mechanical Design of VmaxCarrier2: Omnidirectional Mobile Robot with Function of Step-Climbing*, Journal of Robotics and Mechatronics, Vol.17, No.2, pp. 198–207, 2005.

[4] A. Ozato and N. Maru: *Linear Visual Servoing-Based Control of the Position and Attitude of Omnidirectional Mobile Robots*, International Journal of Automation Technology, Vol.5, No.4, pp. 569–574, 2011.

[5] T. Ogawa and T. Nakamura: *Path Tracking Method for Traveling-Wave-Type Omnidirectional Mobile Robot (TORoIII)*, Journal of Robotics and Mechatronics, Vol.24, No.2 pp. 340–346, 2012.

Computing, Control, Information and Education Engineering – Liu, Sung & Yao (eds)
© 2015 Taylor & Francis Group, London, ISBN: 978-1-138-02800-5

The study of axial forces with the purpose of realization combined process "helical rolling-pressing"

A.B. Naizabekov
Rudny Industrial Institute, Rudny, Kazakhstan

V.A. Talmazan
Karaganda State Industrial University, Temirtau, Kazakhstan

A.S. Arbuz
Kazakh National University Named After Al-Farabi, Almaty, Kazakhstan

ABSTRACT: The key factor for practical realization of the combined process «helical rolling-pressing» is the axial force of the helical rolling, which should provide the continuous pressing in the matrix after rolls. For the measurement of maximum axial rolling force, which is actually the reserve of friction forces, the special strain gauge was made. The work was carried out at the three-roll helical rolling mill "10-30" for the case of hot rolling of steel bars with diameters of 16-25 mm at a reduction of 6 % of diameter and a temperature of 1000 C. Results shown are about 24 tests.

KEYWORDS: helical rolling, axial force, strain gauge, combined process.

1 INTRODUCTION

Usage of ultrafine-grain (UFG) and nanostructured (NS) metals and alloys as constructional and functional materials of the new generation promises big advantages caused by their properties. A number of ways to get such materials with the use of intensive plastic deformation [1-2] is developed. However, the industrial production of UFG and NS materials is still often interlinked with high expenditures of time and energy, restrictions of sizes, and, as a whole, low adaptability to manufacturing that is reflected in their cost price. Therefore the development of new principles for using severe plastic deformation (SPD) for production of volumetric NS metals with perspective properties is an actual target currently.

To remove restrictions on length of blank piece and provide the process continuity with probable combination of continuous process (for example, rolling) and equal channel angular pressing (ECAP) into the one process. The theoretical substantiation and practical realization of the combined process on the basis of longitudinal rolling in smooth rolls and equal channel angular echelon matrix [2] were made in the works [3-6]. The results of these works have revealed the prospectivity of creation of such combined processes.

It is known that during the helical rolling, owing to trajectory and speed features of the metal flow, the formation of UFG structure is also possible, however, owing to the same features the central part of the bar can remain not treated [7]. The combination of helical rolling and ECA pressing can eliminate this lack, thus providing the continuity of the process and potentially better treatment of the structure during the pass in comparison with the above described combined process. This idea is described in details in the work [8], however, its realization is interlinked with the difficulties of the practical nature caused by the complexity of the helical rolling process. The key factor is the value of the maximum axial force (that is, essentially, the reserve of friction forces) during helical rolling which should be sufficient for maintenance of continuous pressing in a matrix standing after the mill. Thus, owing to the complexity of the process, experimental researches of axial force maximum value distribution have great value in designing of the combined installation, with the purpose of the accounting of the total influence of random factors that is the very purpose of this work.

The object in view is reached by the consecutive solution of the following issues: experiment statement, designing and manufacturing of the measuring equipment, carrying out of experimentation and processing of its results.

2 RESEARCH METHODOLOGY

Measurements of axial force during helical rolling were repeatedly made by various authors [9-10] for the case of pipe blank piece piercing into the sleeve and had the target of the optimization of mandrel form and the piercing process as a whole. However, measurements of arising axial force for the case of full compulsory braking of the bar in mill rolls of this type were not carried out.

For reception of the demanded data about maximum axial effort, there was a decision made regarding carrying out of the experiment in 3 sequences with 8 experiments in each, with one influencing factor – ratio between roll diameter (*dev*) and the diameter of a blank piece (D_0), accordingly using different values of the factor for each sequence. Such statement will allow to collect the statistical material for estimation of the maximum axial force along all the product mix of the mill. Besides this, the use of relative indicator (D_v/D_0) as the factor will allow the expansion of the database of experimental data in future at the expense of experiments at other mills with the same type of calibration and adjustment of rolls.

An experiment was carried out in the mill of radial-displacement rolling (RDR) "10-30" designed by National Research Technological University "Moscow Institute of Steel and Alloys" (Russia), as the one providing the demanded structure of the bar after rolling [7, 11]. For experiment carrying out there were initial profiles chosen with diameters: 16 mm, 20 mm and 25 mm, as the most typical for the mill product mix. The ratio between the diameter of conical rolls (71mm) and the diameter of a blank piece (D_v/D_0), as well as initial (D_0) and final dimensions (D_l) of bars by sequences of experiments are given in Table 1.

Table 1. Ratio values by sequences of experiments.

Sequence of experiments	D_v/D_0	D_0, mm	D_l, mm	Number of experiments
I	2,8	25	23,5	8
II	3,6	20	18,8	8
III	4,4	16	15	8

The percentage reduction (ε, %) for all cases is accepted as constant, and was equal to 6% by diameter, as convenient for mill adjustment, as well as due to the fear of possible damaging of rolls and camp structure under high load.

Round hot rolled bars GOST 2590-88 with the length of 300 mm were used as blank pieces for the experiment, with diameters according to Table 1. The steel grade St3 (0,14-0,22 % C) as one of the most globally widespread constructional material was chosen as the material for blank pieces. The temperature of bars heating was defined at the level of 1000°C, and it has expressed the average value of temperature of heat treatment for this class of steel. The distance from the deformation region to the measuring plate corresponded to prospective distance to the matrix during the combined process and was equal to 100 mm.

The scheme of carrying out of the experiment for measurement of the maximum axial force is shown in Fig. 1.

The methodology of experiment carrying out is as follows. Bars for experiments with a length of 300 mm, with diameters according to Table 1, in sets of 2 pieces are loaded into the tubular furnace warmed up to 1000°C with staying of 16-30 minutes, depending on the section and arrangement of blank pieces. After this bar by turns are put in mill rolls. In the process of rolling the bar (1), moving forward, touches the measuring plate (3) with fixed edges and bends it elastically under the influence of axial force (F). Plate deformation is perceived by resistive strain sensors (7) pasted on it and registered by strain sensor station in the form of effort schedule. The peak value of loading brings into the table of results.

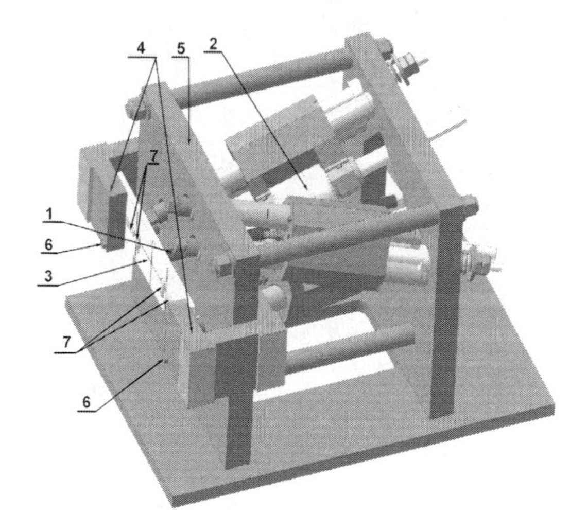

1 – bar; 2 – rolls; 3 – measuring plates; 4 – brackets; 5 – frame; 6 – adjusting bolts; 7 – resistive strain sensors; F – axial force.

Figure 1. Scheme of maximum axial force measurement carrying out.

In addition to the mill of RDR "10-30", the equipment involved in the experiment, include the following: tubular furnace Nabertherm R120/1000/13 (Nabertherm GmbH, Germany); strain sensor station

ZET-017-T8 ("ETMS" CJSC, Russia); measuring plate with resistive strain gauge sensors TKFO1-2-200 ("ETMS" CJSC, Russia); the laptop for control of strain sensor station and signal recording.

Use of plate with fixed edges (beam scheme) as a measurement device is caused by features of mill design not intended for piercing of blank pieces, and therefore is complicated, unlike works [9-11], application of ready decisions (serial load cells). Besides this, such scheme was chosen owing to the greater linearity of measurements, less dependence on point of force application [13], the bigger security of resistive strain sensors, both from temperature and from mechanical damages, simplicity and convenience of realization. The steel 5XB2C (alloyed spring steel) after tempering was chosen as a material by measuring plate, as capable to perceive considerable elastic deformation. The dimensions of the plate were calculated with a condition of achievement of deformation in places of resistive strain gauge sensors pasting, equal to maximum admissible deformation of choosing resistive strain gauge sensors (2%) after the force of 100 kN that is more than 2.5 times more than expected peak force. Plate edges, realising the beam scheme, lean against thick-wall brackets (4) connected to the front frame of the rolling mill adjusting bolts (6).

Thus, the measuring plate with four resistive strain gauge sensors connected into the bridge scheme providing thermal compensation, was developed and manufactured. Resistive strain gauge sensors were pasted using special glue Z70 (Hottinger Baldwin Messtechnik GmbH, Germany). The scheme of pasting and connection of resistive strain gauge sensors is shown in Fig. 2.

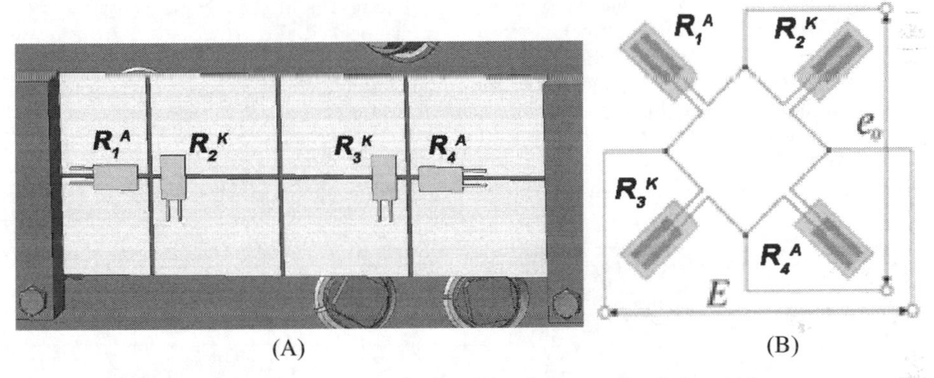

A) – Scheme of resistive strain gauge sensors pasting; B) – Scheme of resistive strain gauge sensor connection (E – feeding of bridge; e_0 – output voltage)

Figure 2. Scheme of resistive strain gauge sensors pasting and connection on the measuring plate.

Resistive strain sensors are pasted symmetrically in the middle of the distance between the center of the plate and support points. Measuring (active) resistive strain sensors (R^A) are pasted along the plate, compensatory sensors (R^K) – across it, thus perceiving only the temperature disturbance. During the connection into the bridge scheme, measuring (R^A) and compensatory (R^K) elements alternate. Such connection provides the increase in sensitivity of the scheme during its protection against temperature distortions [13]. The scheme receives the feed of direct current 5V from strain sensor station. Signal recording is made with a frequency of digitization of 1 kHz.

The measuring plate was calibrated at the torsion and tensile testing machine MI-40KU (Ukraine) in a mode of compression testing, as per methodology minimizing the influence of hysteresis. The essence of calibration is the composition of dependence connecting the electric voltage in the scheme and bending force applied to the plate. Dependence should have the linear character.

For carrying out of calibration the plate was consistently loaded with force with the step of 5 kN in a range from zero to 35 kN. The corresponding values of voltage in the scheme under loading and after its removal were fixed. With a view of hysteresis reduction and the accuracy increase there were 3 passes on the specified range of forces – upwards, downwards and upwards carried out, or 42 measurements in total (including measurements of zero values).

The data received in the process of calibration tests were statistically processed, and there was the regression equation developed on their basis connecting the force applied to the plate (F_i, N) with the voltage (U_i, mV) in the scheme. The equation looks like: $F_i = -3631.2\ U_i + 10122$. The ratio of determination $R^2 = 0.99998$; the standard error of measurements attributed to the value of the working range of plate measurements (40 kN) based on results of 42 tests was less than 0.2%. The received data were uploaded into the program of measurements registration and processing of strain sensor station ZET-017-T8 for the possibility of signal recording in the form of the force trend. After the experiment carrying out there were little random control loadings made, in which the deviation of values has not exceeded the value specified above, that has confirmed the accuracy and the stability of work of measuring plate.

3 RESULTS AND DISCUSSION

All steps of the experiment have passed in the normal mode. At the moment of visible bend achievement of the bar rested against a plate mill drive engines were synchronously stopped to make possible the taking out of the bar from the stand without interfering with mill settings. The general view of the experimental installation and the bar taken out after rolling are shown at Fig. 3. At the Fig 3*A* the bend of measuring plate under the influence of axial force is clearly seen. At the Fig. 3*B* in the bottom part of the bar there are roll marks seen, characterizing the deformation region.

Force trends fixed by the strain sensor station are generally similar and have same distinctive sections. As the illustration of this the Fig. 4 shows trends of tests *II-3* and *II-6* fixed by strain sensor station. At the first section there is sharp (during approximately 0.15 sec) increasing of force with some delay closer to the peak. At this stage there is a bend of the plate under the influence of axial movement of the bar and small deformation of the head end of the bar. The shape of this section of the trend comes close to the parabolic shape. Then there is a bend of the bar, accompanied by the decreasing of force by one third and smooth alignment of force decreasing, probably connected with the start of rolls sliding. Thus, it is important to notice that at the last stage the bar leans not only against the plate, but also against the part of the front frame because at this stage it is generally strongly bent.

The moment of the stoppage of drives is clearly visible at trends in the form of short negative spike to the right of the peak. As it was shown by experiments, the moment of drives stoppage influence neither qualitative nor quantitative picture of force changing at practice.

The results of each sequence of experiences have been checked for presence of gross errors as per Student t-test, and then statistical characteristics have been calculated for each sequence: arithmetic average (F_{ave}), maximum (F_{max}) and minimum (F_{min}) values, standard deviation (F_{st}). The listed characteristics are shown in Table 2. Values of force by all experiments are shown graphically at Fig. 5.

A) – carrying out of the experiment (moment of rolls stoppage);
B) – bar after the test I-7

Figure 3. Carrying out of the experiment.

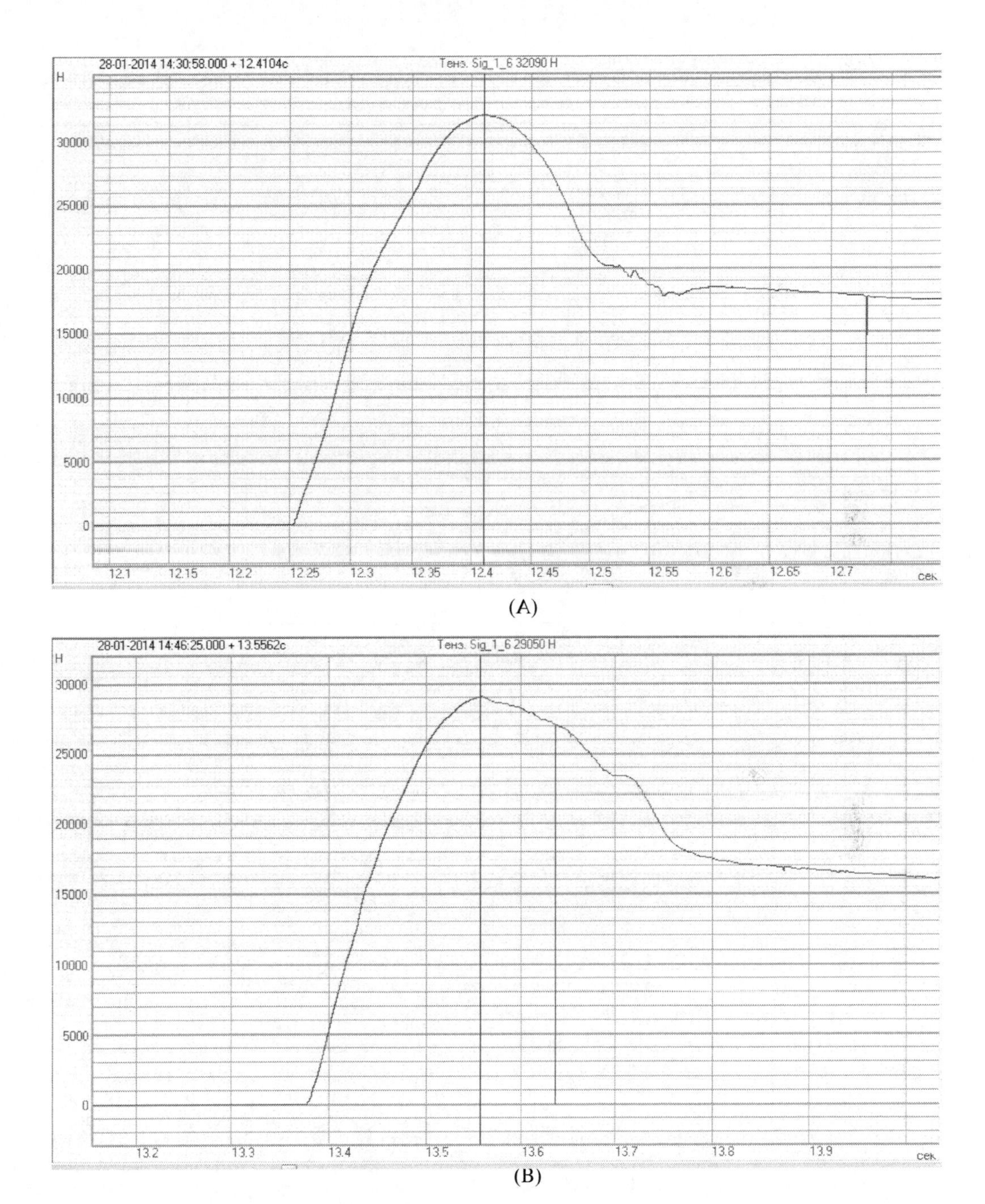

(A)

(B)

Figure 4. Force trends for tests *II-3* (A) and *II-6* (B).

Table 2. Statistical characteristics of experiments.

Statistical parameter	Experiments sequence/ (D_v/D_0)		
	I / 2,8	II / 3,6	III / 4,4
F_{ave}, H	35 541	31 098	20 539
F_{max}, H	39 394	34 484	21 753
F_{min}, H	31 077	27 718	19 512
F_{st}, H	2 402	2 424	787

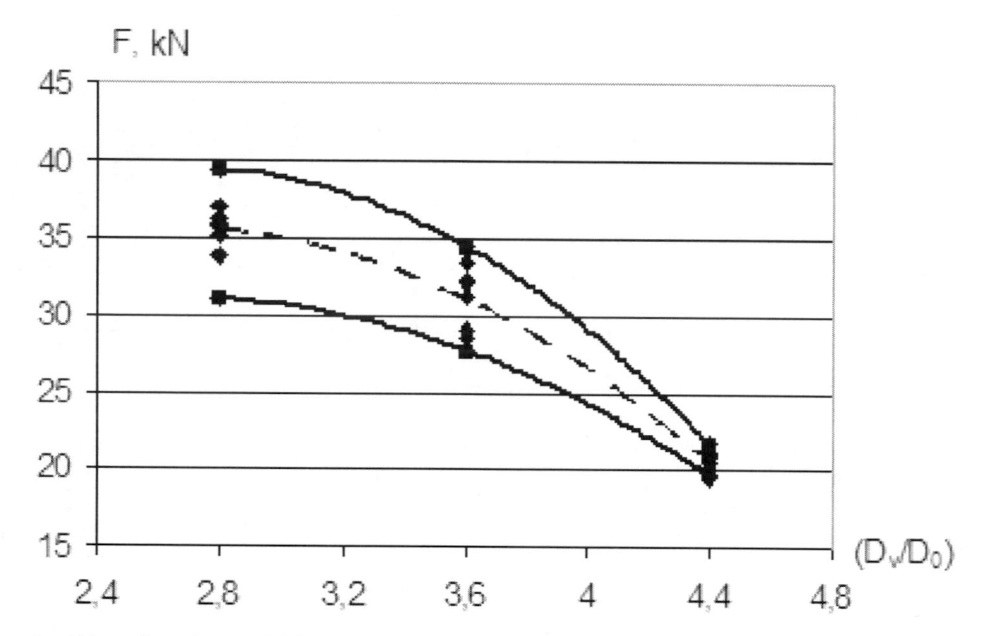

Figure 5. Values of maximum axial force.

4 SUMMARY

There was an equation received $F = -4,78\left(D_v/D_0\right)^2 + 25,05\left(D_v/D_0\right) + 2,89$, characterizing the dependence of maximum axial force at helical rolling from the ratio between the roll diameter and the diameter of a blank piece with a constant reduction $\varepsilon = 6$ % (shown in dotted line). The ratio of determination was equal to $R^2 = 0.92$.

From the resulted data, it is possible to draw the conclusion that during the rolling of thicker profiles essentially bigger disorder of values of forces is observed that can be possibly explained by the increase of influence of following factors – features of internal and contact friction at the blank piece, its rheology, features of the deforming tool calibration.

Measured values of axial force (up to 38 kN) evidence the principal possibility of the combined process of rolling-pressing realization with big values of angles at the joint of channels of the ECA-matrix (140°-150°). It was established that after the reaching of force peak, there is a blank piece bend, thus, during the realization of the combined process the main danger here is not in the lack of force but the possibility of blank piece bend between the rolls and the matrix. Besides this, the value of force can be raised a little by using of notched rolls at maximum reduction. The received results are comparable to results of research of similar profiles piercing force at three-roll mills [9-10].

The results of the research can be used for the optimization of the process of continuous blank pieces, piercing into the sleeve, as results containing the data on the reserve of axial force (friction forces) of the helical rolling.

REFERENCES

[1] Valiev R.Z., Alexandrov I.V. Volumetric nano-structural metal materials: receiving, structure and properties. Monography. - M: Akademkniga, 2007. – 398 pages.

[2] Ashkeev Zh. A, Naizabekov A.B., Lezhnev S.N., Toleuova A.R. Billet deformation in uniform-channel stepped die (2005) *Steel in Translation,* 35 (2), pp. 37–39.

[3] Theoretical studies of the joint " rolling-pressing" process aimed at making sub-ultra fine-grained structure metal. Lezhnev, S. N.; Naizabekov, A. B.; Panin, Ye. A. Proceedings paper of the 20th International Conference on Metallurgy and Materials (METAL) Pages: 272–277. Brno, Czech Republic, 2011.

[4] Panin E.A., Naizabekov A.B., Lezhnev S.N. Simulation of the joint 'rolling-pressing' process using equal-channel step die./17-th International Conference on metallurgy and materials METAL-2008, Ostrava, Czech Republic, 13.-15. 5. 2008. Hradec nad Moravici

[5] Patent of the Republic of Kazakhstan #25863. МПК B21J 5/00. The device for continuous metal pressing. Naizabekov A.B., Lezhnev S.N., Panin E.A. 2013. Applicant and patent owner – Regional state enterprise "Karaganda State Industrial University". - № 2011/0762.1, applied 02.07.2011; published 15.02.2013. Bulletin of inventors, 2013, #7.

[6] Lezhnev, S., Panin, E., Volokitina I. Research of combined process rolling-pressing influence on the microstructure and mechanical properties of aluminium. Advanced Materials Research. 2013 V. 814, pp. 68–75.

[7] Galkin S.P. Trajectory and speed features of radial-displacement and screw rolling. "Modern problems of metallurgy", Dnepropetrovsk. "Sistemni technologii" - 2008, volume 11, pages 26–33.

[8] Combined process of helical rolling with equal-channel angular pressing. Naizabekov Abdrakhman; Lezhnev Sergey; Arbuz Aleksandr. Proceedings paper of the 22nd International Conference on Metallurgy and Materials (METAL) Pages: 422–426. Brno, Czech Republic, 2013.

[9] Teterin P. K. The theory of cross-sectional and screw rolling. Edition 2, redesigned and amended. M: Metallurgy, 1983. – 270 pages

[10] Kovalev D.A. The research and development of the technology for helical rolling process for increase of plasticity hypereutectic silumin alloys. Dissertation of the Candidate of Technology: 05.02.09. protected on 02.03.2011, Kovalev Dmitry Aleksandrovich. - M, 2011. – page 140.

[11] Patent #2009737 of the Russian Federation, MPK[5] B21B19/02. The three-roll mill for screw rolling and the technological tools for screw rolling mill. Romantsev B.A., Mihailov V. K, Galkin S.P., Degtyaryov M.G., Karpov B.V., Tchistova A.P.; - № 5031365/27; applied 13.02.1992, published 30.03.1994.

[12] Patent #2293619 of the Russian Federation, MPK B21B 19/00. The method of screw rolling / Galkin S.P.; Applicant and patent owner - National Research Technological University "Moscow Institute of Steel and Alloys" - #2006110612/02, applied 04.04.2006; published 20.02.2007. Bulletin of inventors, 2007, #5. page 46.

[13] Makarov R.A., Rensky L.B. Tenzometry in mechanical engineering. M: Mechanical engineering, 1975. – 288 pages.

Computing, Control, Information and Education Engineering – Liu, Sung & Yao (eds)
© 2015 Taylor & Francis Group, London, ISBN: 978-1-138-02800-5

Application and study of hard landscape materials in landscape design

Yan Jun Meng
Guilin Institute of Tourism, Guilin, China

ABSTRACT: With the development of age and landscape engineering skill levels more and more advanced, all these contributed to the landscape design of innovation. Reasonable choice and use of landscape materials to fully express the designer's intent and design performance, and therefore the material between the landscape and landscape design are mutually reinforcing. In this paper, the basic types of landscape materials as a starting point, the study design principles for the use of landscape materials, and hard landscape materials selection and application of methods in landscape design, combined with the corresponding design techniques to enrich and highlight the landscape design of the form and content. From the combination of color and texture, detail, lighting, innovative design techniques, such as the angle of the material, study and explore how to properly use landscape materials, and finally summed up the general principles of hard landscape materials and the use of modern landscape design, how to choose a hard material.

KEYWORDS: Hard materials; Landscape design; Garden; Design methods; Color matching.

1 INTRODUCTION

With the improvement of people's aesthetic standards, psychological needs are constantly changing. But not for the physiological requirements of the landscape which has changed since the physiological demands of the landscape is mainly reflected in the use value. Rational use of hard landscape materials can increase the landscape artistry, aesthetics and increase its value.

2 BASIC TYPES OF HARD LANDSCAPE MATERIALS

The first category is the timber, it belongs to the warm material, strong affinity, the appearance of simple, stable performance, with a good touch, space timbered often intimate, relaxed feeling. Wood is widely used in classical gardens, ancient in the Pavilion, Taiwan, House, House, mostly wooden structure [1].

The second category is the stone, which itself is extremely hard at the same time very durable material, through centuries can still be used even after a thousand years. It's beautiful to resist the passage of time, and as glycol, enduring fragrance, the more beautiful the more ancient. Natural stone itself on the diversity of texture, color and appearance of the application makes it in landscape design is very broad.

The third category is the concrete, it can make the best, low cost, its constituent materials of sand, stone and other local materials accounted for more than 80%, in line with the principles of local materials and the economy can be divided into ordinary concrete, decorative concrete. At the same time can be used for floor tiles, blocks, plates and other precast products, can also be cast into a variety of components [2].

The fourth category is the brick and tile, natural clay as the main raw material, mining soil ingredients, blank, bulk material drying, and calcinations process made of clay called. The roasting process can be obtained using different red brick and brick, brick red brick is based on the addition of burn again, and some of the higher strength and gray gives a sense of calm, sense.

The fifth category is metal; the metal is copper, lead, iron, steel, bronze, tin, zinc, aluminum, nickel, etc. in general. It is using a variety of metal materials in the small number of ancient gardens, large garden in modern applications.

3 LANDSCAPE DESIGN PRACTICES

Combinations of Materials. Hard landscape materials and combinations of whether integration with the surrounding environment, the pursuit of the environment as a whole is one of the factors that should be considered. Hard landscape materials in landscape design combinations in different ways, and different designers due to the nature of the material and the design work preferences and personal aesthetic tendencies different, but also in the use of material combinations, each time passing out distinctive design expression [3]. It can be said at any point have a common combination of materials can be coordinated. This commonality can be expressed in a can of color, texture, luster to any one similar, in most cases, you can clear demonstration of this intrinsically linked. Park fence design is shown in Figure 1.

Figure 1. Park fence design.

Color Combinations. In landscape design, the use of hard landscape materials of various color contrast can be achieved in different ways of changing effect. Need to highlight certain parts of the site, such as the turning point or to distinguish by color, you can make it appear out of context from the entire site, designed to increase interest [4]. But also can be different colors with a contrast material, stone paving stock with a hard texture and rich color and tone, you can choose different colors of slate to form a contrast. Different colors of the floor as shown in Figure 2.

Figure 2. Different colors of the floor.

Color and Culture. In the use of hard landscape materials must also be noted that the cultural background color behind the different regions and different ethnic understanding and preferences of color have differences. Some countries, such as people in Italy, France, Spain, Mexico's favorite warm and bright warm colors, such as red, yellow, orange; tropical countries, people are fond of yellow, white; the Americans and Scandinavians prefer blue; Chinese people to as a festive red color [5]. These different regions, different ethnic color aesthetic features of the landscape design by the use of color to produce a certain amount of influence, and thus the color selection of hard landscape materials shall conform to the local aesthetic habits, avoiding some of the taboos on color, reflect certain cultural connotations. It is shown in Figure 3.

Figure 3. Red corridor map.

Texture Combinations. Because different materials have different texture, variety of materials used in a landscape design work, giving the visual impact is greatest in addition to the color texture of the material the material itself. Therefore, the combination of a variety of materials in the hard texture of the landscape design is a prominent landscape feature of the way. In the process used in combination, the degree of similarity between the various materials determines whether it belongs to the combined effect of contrast or millisecond. To highlight the performance of texture without leaving a trace, according to the performance of the resulting beauty of texture is well known. If we try to make it look like the unconscious used, the extent of its beauty but also to highlight. Such as masonry and stone pavement joints seem to be intentionally done some obvious, otherwise the seams too thin, the design intent unclear what became unfunded. The landscape in the rough texture energetic, build stronger joints and powerful use of natural stone and wood such material should be so, especially when [6]. In the design of the external space scales, modulus, interior space than it is appropriate to expand about 10 times, so energetic and rough texture also because there is a good fit. A combination of the texture coordinates leisure seats is shown in Figure 4.

Figure 4. Casual seating.

When several textures quite different materials are combined, the effect will be different spaces, such as the combination of stone and vegetation, it is very easy to match, even if part of the same color, it will not significantly monotonous. As shown in Figure 5.

Figure 5. Combined stones and vegetation.

4 COMBINATION OF TRADITIONAL AND MODERN APPLICATIONS OF HARD LANDSCAPE MATERIALS

Traditional and modern materials, our current landscape design should be a reasonable match. Such as traditional tiles, floor tiles, pebbles and modern stainless steel, glass, metal and other materials together to form a strong contrast of colors, texture and contrast effects.

Mix of Brick and Stone. As traditional brick tiles representatives, both from the color or shape gives the deep, heavy feeling and a sense of a long history, but because of its texture, loose, cannot bear the load of the vehicle, only to pedestrian pavement, hence You can complement each other in combination with rugged and elegant colors of the stone in the paving process, in stark contrast with the contrast. It is shown in Figure 6, to a large area of brick paving, which irregularly dotted light gray granite stone as a separator, the formation of the ever-changing rhythm, dark blue and light gray color contrast, and texture of different materials and texture contrast, reflects the intimacy and sense of history.

Figure 6. Mainly granite brick income side.

As the Main Modern Material Mix. Modern materials are not only able to use as a support structure, you can use it to transfer from traditional materials and transition, and it is the main body to carry out with each other. Figure 7 displays with glass and steel modern light shows, with strong sense of line, texture smooth and metallic luster and other characteristics.

5 SUMMARY

This article is based on the selection and application of hard landscape materials for the study, in-depth analysis of the hard landscape material selection methods and design principles. Design is by understanding the material and the process for making changes in use, so that the landscape design along with the material derived from throughout the history of human civilization. But what kind of material to which the process, with what design techniques to create a perception of what will eventually be what the impact is still worthy of our depth research.

REFERENCES

[1] L.Y. Gan. Building Exterior Skin Material Artistic Research, Chongqing University, (2007).

[2] W.R. Gao. City Hard Landscape Planning and Design Methods of Color , Beijing Forestry University master's degree thesis, (2007).

[3] Y.Fu.Underflow Influence of Modern Western Postmodernism Garden, Nanjing Forestry University master's degree thesis, (2005).

[4] Blanc. Landscape Construction and Detailed Design,China Building Industry Press, (2002).

[5] Li Hong-bin. Local Landscape Design,South Forestry Industry University, (2007).

[6] Hui Yuan.Research of Humanized Landscape Construction of Facilities Space, Northeast Forestry University, (2006).

Computing, Control, Information and Education Engineering – Liu, Sung & Yao (eds)
 ISBN: 978-1-138-02800-5

Secure proxy mobile IP protocol for wireless industrial sensor network based on ECC

Ling Tir & Hong Zhang*
School of Information Science and Technology, Chengdu University, Chengdu, Sichuan, China

Peng Cheng
Shanghai 30wish Information Security Ltd, Shanghai ,China

ABSTRACT: In this paper, proxy mobile ipv6 protocol is adopted to support the network node mobility in the industrial wireless sensor networks. Elliptic curve cryptography is used to achieve security of proxy mobile ipv6 authentication protocol. In the registration and login phase, public key and private key are produced using one-time password method. In the authentication phase, counter and hash chain is used, the authentication process does not need the backend authentication server to be involved into, this procedure reduces the interaction authentication time of authentication protocol. Finally, the authentication test formal analysis method is used to analyze the security of this authentication protocol.

KEYWORDS: industrial wireless sensor network, elliptic curve cryptography, proxy mobile ipv6, authentication protocol, authentication test.

1 INTRODUCTION

1.1 *Elliptic curve cryptosystem*

In this paper, a proxy mobile ipv6 protocol is introduced into the industrial wireless sensor network. It can provide local mobility management. There are two main parts: one is localized mobility anchor (LMA), the other is the mobile access gateway (MAG). When a mobile node enters into the pmipv6 domain, it must authenticate with the MAG.

Elliptic Curve Cryptosystem becomes the most promising public key cryptography. The basic idea is [2]:

p >3 is an odd prime, There is an elliptic curve E, the equation is:

$$E: y^2 = x^3 + ax + b \bmod p \tag{1}$$

Where, $a,b \in F_p$, $4a^3+27b^2 \neq 0$, set E (Fp) include all at this point on the elliptic curve, $(x,y) \in Fp$, there is a point O at infinity. There is a point $G = (x_G, y_G)$. Mobile user can choose a random integer $d_A \in [1,N-1]$ as the private key. According to $Q_A = d_A \times G$, its public key can be calculated. It can be proved that calculation from d_A and G to Q_A is easy, but it is difficult from Q_A to d_A, it is difficulty of discrete logarithm. It can guarantee the security of secret key.

2 ONE TIME PASSWORD AUTHENTICATION PROTOCOL BASED ON ECC

2.1 *Hypothesis*

We assume that there exists the security association among the AAA server, LMA and MAG. They share the pre-shared key.

2.2 *Notations*

All symbols used are listed in Table 1.

Table 1. Notations.

Symbol	Description
d_i	The private key of Object i
Q_i	The public key of object i
ID_s	The ID of sensor node
ID_{MAG}	The ID of MAG
ID_{AAA}	The ID of AAA
PWs	The password of sensor node
PSK	The pre-share key among AAA, LMA and MAG
$SK_{i,j}$	The session key between object i and j
$E_K(M)$	The encryption of message M using key k
$D_K(M)$	The decryption of message M using key.
N_i	Nonce
Counter	The number of sensor nodes to receive service
$OTPS_i$	Sensor's one time password of No i. service
$OTPM_i$	MAG's one time password of No i. service

*Corresponding author: Hong Zhang; tqs64cl@163.com; cdutzh101@163.com; chengp@30wish.net

2.3 *Authentication protocol*

1 Register phase

When the sensor node enters into the domain, it will finish register phase and authenticate with the backbone AAA.

1 sensor→AAA: The sensor node chooses it's ID_s and PW_s, produces N_s, and calculates $PW_S = h(PW_S \oplus N_s)$. It sends PW_S and ID_s to AAA.
2 AAA→sensor, After AAA receives the message, it selects a private key d_A and calculates a public key Q_A according to the ECC, $Q_A=d_A \times G$. AAA calculates $C_1=h(ID_S||d_A) \times G$, $C_2=h(ID_S \oplus PW_S)$, and $C_3= C_2 \oplus C_1$, then, sends Q_A, C_1,C_2 and C_3 to the sensor.
3 AAA→MAG: AAA produces $C_4=E_{PSK}(d_A||ID_{AAA})$ using the key shared between the AAA and the MAG, and sends C_4 to MAG.
4 MAG receives C_4, C_4 and uses PSK to decrypts and have the d_A.

2 The login phase

When a sensor nodes login, it will calculate $PW_S = h(PW_S \oplus N_s)$ and $C_2'=h(ID_S \oplus PW_S)$ and check if $C_2'=C_2$. If they are equal, calculates $C_1=C_3 \oplus C_2$。 Then, sensor chooses a N_A, and calculates N_A C_1 as the private key, then calculates the public key $d_s=(N_A C_1) \times G$.

3 The authentication phase based on one time password

1 sensor→MAG: When a sensor node conducts the first authentication, it produces counter =0, calculates the initial value $OTPS_0=h(ID_S||Counter)$. It constructs a one time password hash chain $OTPS_i=h(OTPS_{i-1}||$ Counter) $OTPS_{i-1}$, $OTPS_0$ is the initial value. The sensor node produces the message $M_1=E_{QA}(IDS||OTPS_i)$, and sends M_1 andQ_s to MAG.
2 MAG→sensor: MAG receives M_1 and decrypts it using d_A.It calculates $OTPS_i'=h(OTPS_{i-1}||counter)$. If the equation is hold, we can authentication sensor, Then calculates $OTPM_0=h(ID_{MAG}||counter)$, and have $OTPM_i=h(OTPM_{i-1}||counter)$. MAG produces $M_2=E_{Qs}(ID_{MAG}||OPTM_i)$ and sends it to sensor.
3 Sensor receives M_2, and decrypts the message with d_s at the same time verifies the equation, $OTPM_i'=h(OTPM_{i-1}||counter)$. If the equation is hold, we can authentication the MAG, at the same time, counter+1.

3 SECURITY PROTOCOL ANALYSIS

We use the authentication test formal analysis method to prove the security of the authentication protocol [6].

3.1 *Authentication for sensor*

Theorem 1: Support C is a bundle in ∑, S∈Sensor strand , and C-height(sensor)=3, then there are regular strands S∈MAG strand, and C-height(MAG)=3.

Proof: $OTPS_i$ is uniquely originating in < Sensor,1>, the edge < Sensor,1> → +< Sensor,2> is output test of the $OTPS_i$, according to the authentication test 1, there exists a regular node m, m′∈C, that make $OTPS_i$ is a test component of m, and m→m′is a transforming edge of the $OTPS_i$. At the same time , m′can only be < MAG,3>. Transforming edge m →m ′must be <MAG,2>→+<MAG,3>, and C-height(Sensor)=3. Because $OTPS_i$ is the only sour ce of Sensor, other components includes the id information of $OTPS_i$, and it is unique. So we can prove that the sensor can authenticate the MAG.

3.2 *Authentications for MAG*

Theorem 2: Support C is a bundle in ∑, S∈MAG strand, and C-height(MAG)=2,there exists S∈sensor strand, and C-height(sensor)=2.

The process of proof is similar as before.

4 SUMMARY

In this paper, we use the elliptic curve cryptosystem to improve the one time password authentication method. The hash chain is adopted in the authentication protocol. Finally, we use authentication test method to prove the security of this protocol.

ACKNOWLEDGEMENT

It is a project supported by the Chengdu University foundation (projectNo.2012XJZ07), Sichuan Province Science and technology support program (No.2013GZ0016), the open foundation of Key Laboratory about pattern recognition and intelligent information processing and the research project about information security situation of industrial control system.

REFERENCES

[1] N. Koblitz. Elliptic curve cryptosystems. Mathematics of Computation, 48(1987) p.203–209.

[2] V.S. Miller. Use of elliptic curve in cryptography. Advances in Cryptology, CRYPTO '85 ProceedingsLecture Notes in Computer Science Vol.218(1986), p.417–426.

[3] An Liu, Peng Ning, TinyECC: A Configurable Library for Elliptic Curve Cryptography in Wireless Sensor

Network. Information Processing in Sensor Networks, 2008.IPSN '08. International Conference on, 22–24 April 2008, P. 245–256.
[4] Jian Li, Yun Li, Jian Ren, Jie Wu: Hop-by-Hop Message authentication and Source Privacy in Wireless Sensor Networks. IEEE Trans. Parallel Distrib.Syst. 25(5) (2014),p.1223–1232.
[5] F J Thayer, J C Herzog, J D Guttman. Strand spaces: proving security protocols correct. Journal of Computer Security, 7(2.3) ,1999, p.191–230.
[6] D.G Joshua, J.T.Fubrega. Authentication test. Proceedings of the 2000 IEEE Symposium on security and Privacy, Berkeley, California, USA, 2000, p.96–109.

Computing, Control, Information and Education Engineering – Liu, Sung & Yao (eds)
 ISBN: 978-1-138-02800-5

Quantitative remodeling of shell structure based on structural mechanics

Kang Qiang Lin & Chao Hao Su
Architecture Design and Research Institute, South China University of Technology, Guangzhou, China

ABSTRACT: From the perspective of structural engineering, with taking the hyperbolic paraboloid shell as the research object , new design ideas and experimental methods called "Prototype structure -Mechanics-Digital remodeling-Structural calculation and optimization" are put forward. Through quantitative test and data analysis of three types of structure textures including circle, triangle, and rhombic shape which all reflect mechanical properties well, the technical route of this article is discussed and proven to be feasible and effective. As a result, a new design pattern, from prototype structure to remodeling of diversity of forms, is created and presented.

KEYWORDS: Synergetic Design; Shell structure; Structural mechanics.

1 INTRODUCTION

In the design circles, parametric design has become more and more popular with the application in SOM, UN Studio, GMP, Foster + Partners and other international well-known design companies, of which Zaha Hadid Architects is the most typical example. As a partner and chief architect of the firm, Patrik Schumacher is also one of the co-teaching directors of the Design Research Lab, which he established and has become the cradle of the Parametricism. [1-2]Arup Advanced Geometry Unit was founded by Cecil Balmond in 2000, and the designers include architects, engineers and scientists. Balmond applied the concept "Structure is architecture" into his multiple design works. In the book named "Informal", Balmond put forward a series of mathematical and physical definitions related to structure and proposed relevant spatial composition and organization[3].

Even though the researches on the combination of complex shapes and structural mechanics are very scarce, they provide valuable clues: structure in complex system context should not be in a subordinate position, however, it can be used as an active factor or "dynamic structure"[3] to guide the entire design process of complex shapes. Therefore, with the start from the structure prototype, new design idea and technical route will be presented on the basis of digital remolding between the force and shape, which will be demonstrated using the example of shell structure.

2 EXPERIMENT METHOD AND TECHNICAL ROUTE

2.1 *Mechanical mechanism of structural prototype*

Two key mechanical logics which are closely related to the form resistance are as follows: First, stress distribution; Second, the stress transfer path, namely "stress flow". [4] In order to make the experiment more focused and interference-free, it is necessary to establish some conditions before it begins.

Boundary conditions. Because different stress distributions and stress flows are generated under different boundary conditions, as for the experiment, the principle of choosing the boundary conditions is to generate more typical stress nephogram and vector plots of stress. After a comparative analysis, symmetric full constrains are set to two nodes nearby the angular points finally.

Material properties. Not only do materials affect the properties of a structure, such as the strength and deformation, but they also affect the theoretical judgment of the structural calculation and analysis in the experiment process. Supposing that the shell is made of reinforced concrete structure (brittle material), the first principal stress is to play a dominant role, According to the first strength theory of Material Mechanics.

Table 1. The result of the static calculation of hyperbolic paraboloid shell.

item	1st principal stress nephogram	3rd principal stress nephogram	Vector plot of stress
Datamap	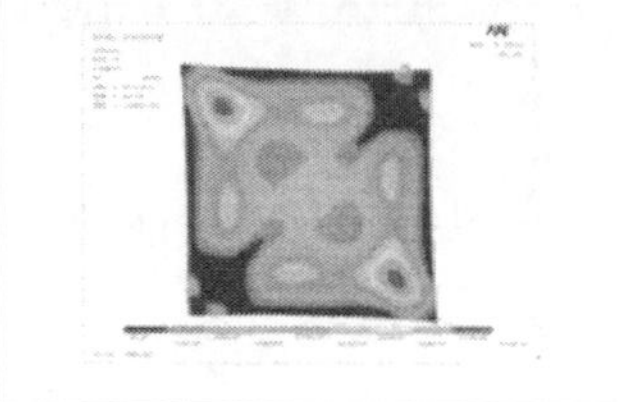	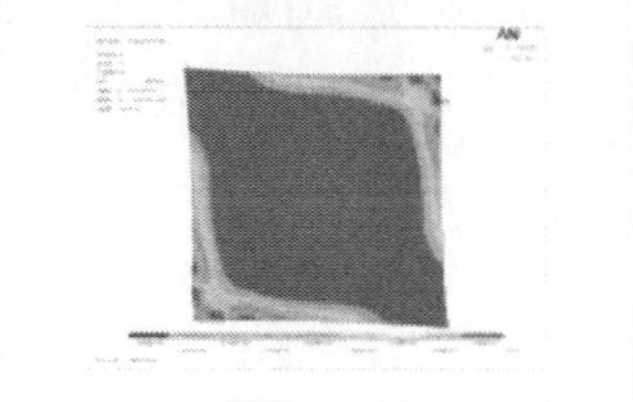	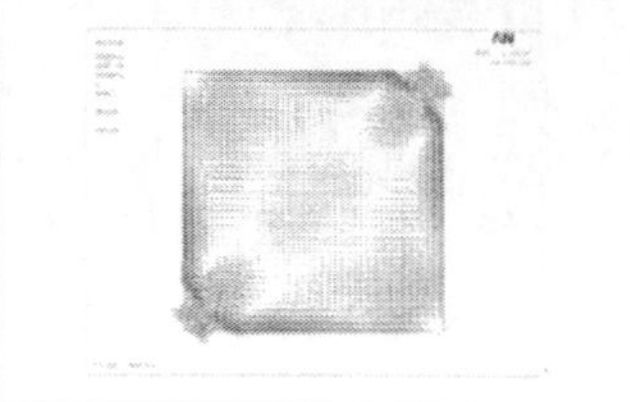

Table 2. The result of the static calculation of two different textures.

Item		Deformation map	1st principal stress nephogram	Vector plot of stress
Square	Data map	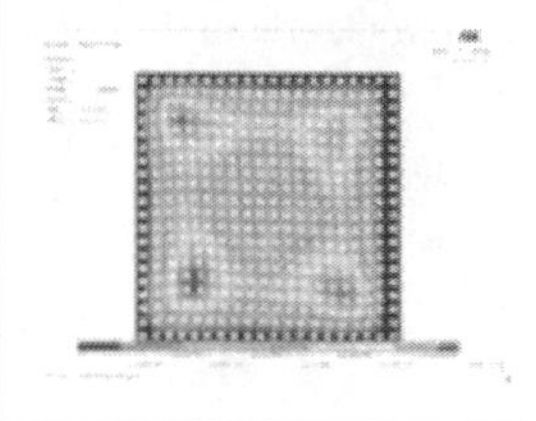	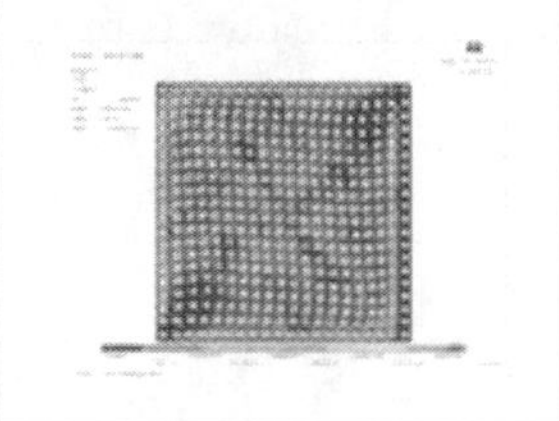	
	Peak value	DMX=0.029m	SMN=15.5MPa	
Rhombus	Data map	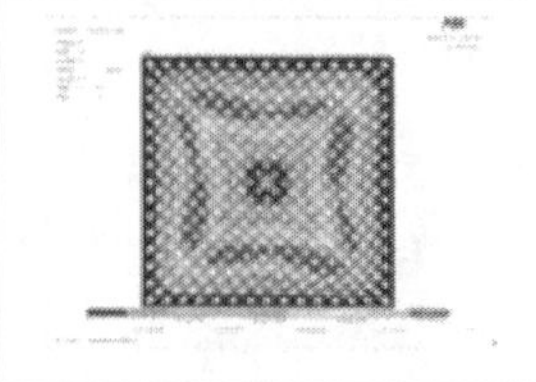	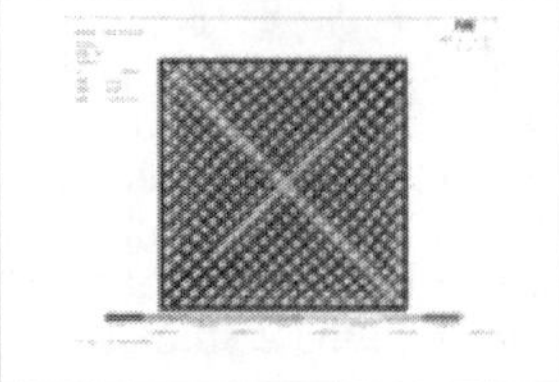	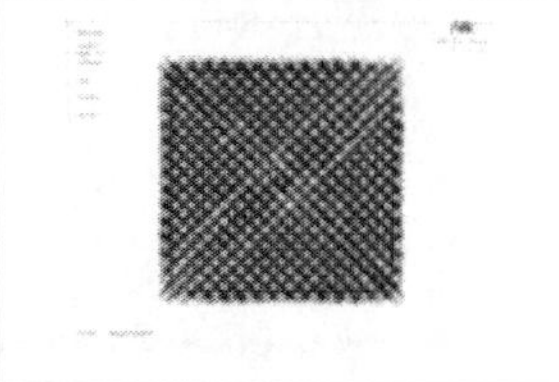
	Peak value	DMX=0.015m	SMN=12.9MPa	

Geometrical conditions. The projected area of hyperbolic is 60m x 60m, the height of the two corner points is 60m, and the thickness is 0.3m.

working conditions. The load of gravity is considered as the single working condition. In order to make the experimental results more comparable, the loads are fairly transformed into double times of deadweight for the different experimental models with different ratios of holes.

2.2 *Digital platform of synergetic design: Reflecting the structural mechanics*

Step 1: Select the area. Considering the 1st and 3rd principal stress nephograms and material properties, the 1st principle stress nephogram of the middle region is selected as the basis for structural morph design, with cutting off the surrounding areas where compressive stress concentration occurs.

Step 2: Decide the direction of the texture. The mechanics of hyperbolic paraboloid shell in the gravity condition turns out to be a mechanism of cable plus arch[6]. The vector plot of stress shows that stress flows regularly along the diagonal direction, rather than along the sides. To guarantee the form resistance effect of opening-hole texture, conforming to the paths of cable and arch is very necessary and important. Therefore, multiple texture forms including circular, triangle, and rhombus shape are designed by GRASSHOPPER program.

Table 3. The result of the static calculation of 13 x 13 matrix.

Item	Circle	Triangle	Rhombus
1st principal stress nephogram	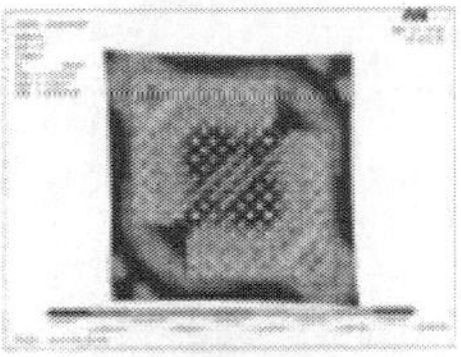	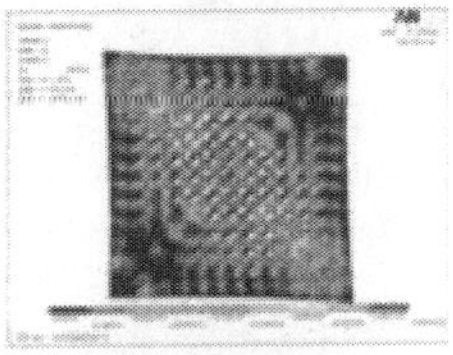	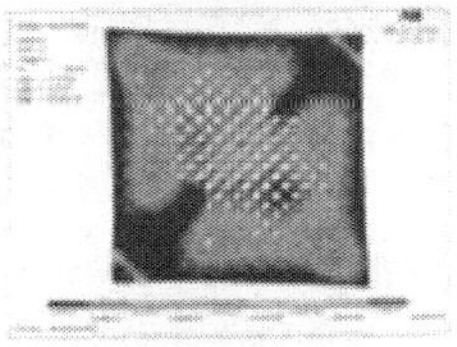
Vector plot of stress	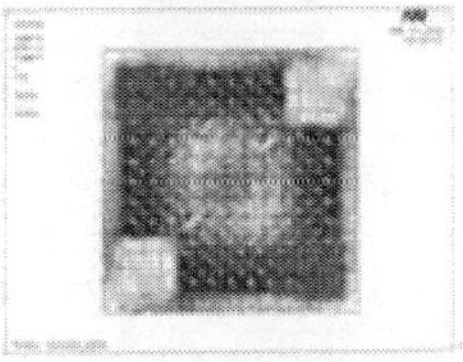	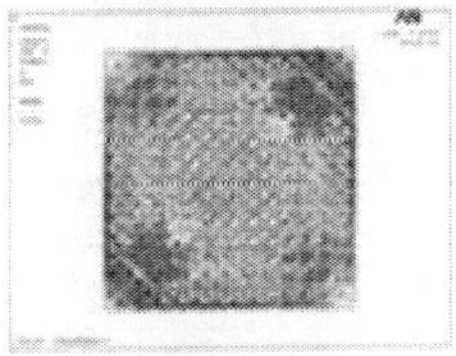	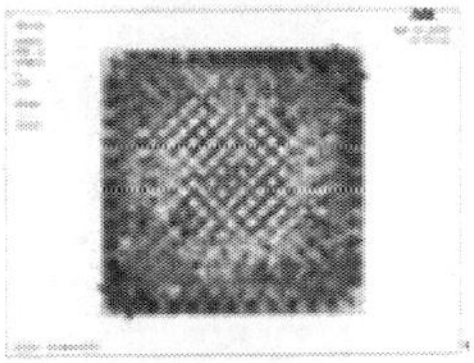

Table 4. The result of the static calculation of 20 x 20 matrix.

Item	Circle	Triangle	Rhombus
1st principal stress nephogram	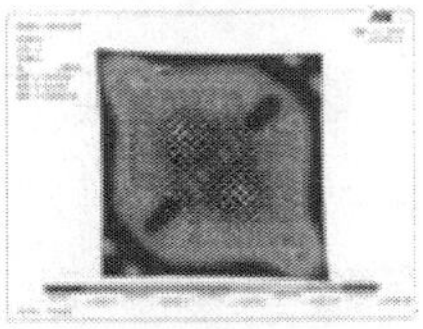	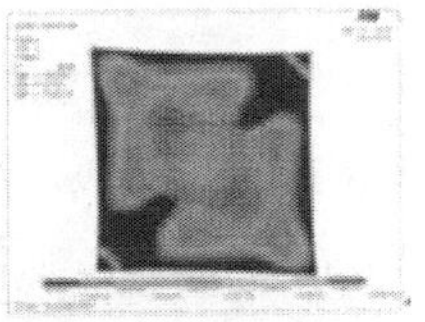	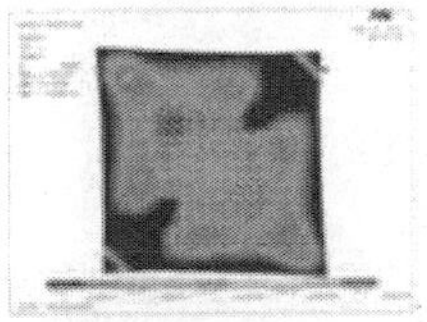
Vector plot of stress	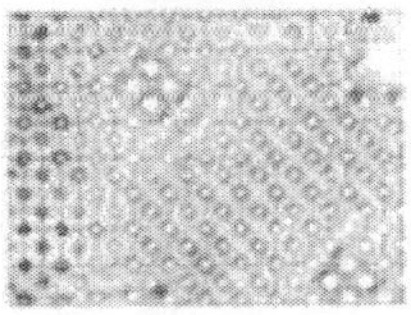	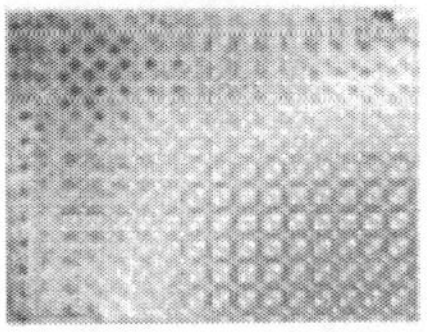	

Step 3: digital modeling oriented to structure software. An important part of synergetic design is the collaboration of different specialty software. The area and volume units controlled by GRASSHOPPER must be designed according to the unit forms defined by ANSYS, such as link, beam, shell, solid, and so on, otherwise they won't be recognized or fail to be meshed in the ANSYS software.

2.3 *Parametric adjustment, structural calculation and evaluation*

Each type of structure forms will be compared by the parameter index: direction, intensity and number of the units, with responding to stress nephogram and vector plots of stress. The structural calculation and evaluation should include: rigidity, strength and interference degree of mechanics.

3 EXPERIMENT RESULT

1 The calculation results of structural prototype are shown in table 1
2 The mechanical properties of two kinds of textures are shown in table 2
3 The calculation results of structural forms in 13 × 13 matrix are shown in table 3
4 The calculation results of structural forms in20 × 20 matrix are shown in table 4

Table 5. Analysis of the static calculation of 13 x 13 matrix units.

Item	Circle	Triangle	Rhombus
Strength	Tensile area is very sensitive to the circular opening, resulting in uneven distribution of tensile stress and significant stress concentration on annular edges.	Tensile stress can't transfers along the diagonal direction, resulting in significant stress concentration exceeding 20.0MPa and hard reinforcement	Tensile stress transfers along the diagonal direction, exhibiting uniformity in the stress range 0.0 to 15.0MPa
Mechanism Interference Degree	The cyclo-hoop effect causes the formation of stress vortex, resulting in significant interference on the force-transfering mechanism of prototype. Therefore, the size and distribution of the opening need to be optimized.	According to the 1st strength theory, the absence of stress flow of the 1st principle stress results in less obvious behavior of prototype mechanics. The direction of units need to be adjusted	obvious force-transfering mechanism of prototype

Table 6. Analysis of the static calculation of 20 x 20 matrix units.

Item	Circle	Triangle	Rhombus
Strength	After adjustment of proportion of the opening in tensile stress sensitive area, the phenomenon of stress concentration is reduced, and the stress distribution becomes more even in the range 0.0 to 15MPa	with the adjustment of cable units along the diagonal direction, stress distribution in the range 0.0 to 15MPa tends to be much more even.	After the adjustment of the number and scale in response to the stress distribution, stress distribution in the range 0.0 to 15MPa becomes more even
Rigidity	The overall deformation is even and small, showing a good integral rigidity.	The max deformation value is reduced to 0.735, and the overall deformation is even, with no mutation.	The overall deformation is even, with no mutation
Mechanism Interference Degree	As the number increases and the scale of the opening becomes smaller, the interference of the cyclo-hoop effect is decreasing, and the force-transfering mechanism of prototype becomes gradually clearer.	Because of the adjustment of cable units along the diagonal direction, and the reducing scale of the opening which makes the stress transfering path shorter, it gradually demonstrates mechanical mechanism of prototype.	It shows quite obvious force-transfering mechanism of prototype

4 ANALYSIS OF THE EXPERIMENTAL RESULTS

1 The mechanical mechanism of prototype

As is shown in table 1, the 1st principal stress nephogram suggests that there are two tensile regions near two cantilevered corner with remarkable stress concentration; the 3rd principal stress nephogram shows compression stress concentration in the periphery area, the closer to the supporting points, the more obvious it is; the vector plots of stress indicate a rule of stress flows along the two diagonal directions: arch + cable force-transfering mechanism.

2 The stress flow

Conclusion can be made from the comparison between two sets of data in table 2: the deformation value and the peak value of stress of rhombus are relatively smaller than those of square, with much clearer stress flows, showing the arch + cable force-transfering mechanism and superiority in forms resistance. So, it also verifies the pre-judgements on the direction of the texture.

3 Analysis of 13 × 13 matrix unit is shown in Table 5

4 Analysis of 20 × 20 matrix units is shown in Table 6

5 Comparative analysis of results of before and after adjustments

First, The relative size of opening causes interference of stress distribution. As the number increases, the stress concentration will be reduced. And as attraction for the stress flows from the opening edges becomes weaker, the mechanics mechanism of shell structure gradually becomes more obvious, with more clear stress flows. Second, Due to the adjustments of the textures in response to the stress distribution, it makes the stress distribution become more and more even.

5 CONCLUSIONS

This paper proposes a synergetic design idea called "Prototype structure -Mechanics-Digital remodeling-Structural calculation and optimization", and has demonstrated the feasibility of the idea and the corresponding technical route through quantitive experiment, as a resulting in the reality of a diversity of complex forms of structure. As for technical route, the key problem is to make clear the mechanical mechanism, especially for the good understanding and uses of stress flows. Because the mechanism determines the distribution and direction of texture units, and therefore determines the characteristics and effectiveness of digital shape remodeling. In this experiment, we have conucted a preliminary study on the typical prototype of the hyperbolic paraboloid shell structure under a single condition. We believe that this is an effective starting point for further studies on shell remodeling and structural stabilities, other or multiple working conditions, on the usage of various prototype of the structure, and other.

ACKNOWLEDGEMENT

The research described in this paper is supported by National Natural Science Foundation of China (51308218), Guangdong Natural Science Foundation (S2012040007467) and "the Fundamental Research Funds for the Central Universities" (2014ZM0012). The corresponding author is Chaohao Su.

REFERENCES

[1] Patrik Schumacher. The Autopoiesis of Architecture: A New Framework for Architecture[M]. Wiley(1 edition), January 18, 2011.

[2] Patrik Schumacher. The Autopoiesis of Architecture, Volume II: A New Agenda for Architecture[M] . Wiley(1 edition), May 7, 2012.

[3] Cecil Balmond. Informal[M]. Prestel USA . April 2007.

[4] Louis L. Bucciarelli. Engineering Mechanics for Structures[M]. Dover Publications, March 26, 2009.

[5] Heino Engel, Ralph Rapson. Structure Systems[M]. Hatje Cantz, October, 2006.

Computing, Control, Information and Education Engineering – Liu, Sung & Yao (eds)
© 2015 Taylor & Francis Group, London, ISBN: 978-1-138-02800-5

Fusion methods of automobile control algorithm with IPG CarMaker

Wen Juan Zhang
Institute of Information, Beijing Union University, Beijing, China

Wen Liang Niu
College of Applied Science and Technology, Beijing Union University, Beijing, China

Tong Jin
Institute of Information, Beijing Union University, Beijing, China

ABSTRACT: The CarMaker software developed by German IPG automobile limited company is a vehicle dynamics simulation software with the integrated virtual driving environment for small passenger cars. The CarMaker software has been widely used in the automotive field to promote the development of the automobile industry. This paper mainly discusses four fusion methods of automobile control algorithm with the CarMaker software. A comprehensive grasp of these four fusion methods will help automobile industry personnel and automobile researcher, research on intelligent vehicle especially in the aspect of automatic driving.

KEYWORDS: CarMaker; Control Algorithm; MATLAB; SIMULINK; FMI; FMU.

1 INTRODUCTION

In recent years, intelligent vehicle has become a research hotspot in the field of the world automobile industry and new growth power of vehicle industry and it has been included in their intelligent traffic system by many developed countries. China also pays a close attention to intelligent vehicle [1~4]. Many colleges and universities carry out research work on intelligent vehicle in succession and do a lot of research, especially in the aspect of automatic driving [5~7]. "The Future Challenge of China Intelligent Vehicle" is held once a year since 2009 in order to facilitate and promote the innovation and development of audio-visual information cognitive computing model, key technology and verification platform research, ensure the realization of the major research plan's overall scientific objectives, promote research and original innovation in intelligent vehicle technology in China.

And now many companies and universities are almost gaining their research results about intelligent vehicle through operating in the real vehicle which leads to several questions like high cost, long research cycle and poor safety. Under these premises, IPG CarMaker [8,9], a vehicle simulation software, will be necessary. Research results should firstly be simulated in the CarMaker software and then be improved according to the simulation results by researchers. Finally researcher can verify their research on the real vehicle. It is a better way than operating in the real vehicle all the time.

2 THE CARMAKER SOFTWARE

CarMaker is a vehicle dynamics simulation software developed by German IPG Automobile Limited company. The CarMaker software is already a mature virtual driving environment and it has been widely used to promote the unmanned vehicle virtual driving test development in the automobile field.

CarMaker kit is composed of two main parts, namely CarMaker interface Toolbox(CIT) and Virtual Vehicle Environment (VVE), which can be seen in figure 1. The virtual environment scene includes roads and traffic while CIT contains a complete set of tools such as simulation control, parameterization, analysis, visualization and file management.

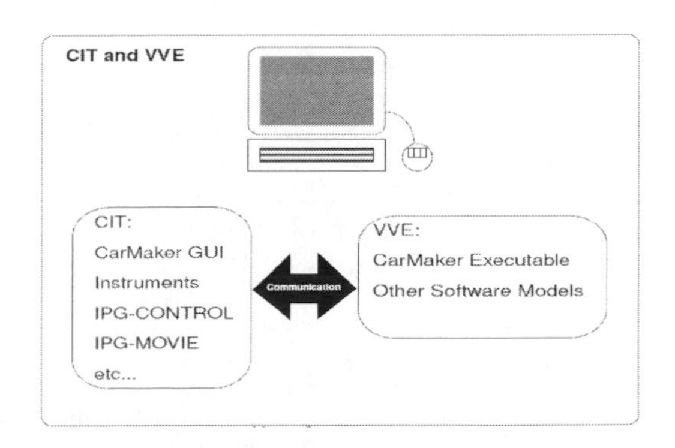

Figure 1. CIT and VVE.

VVE is the integration of Vehicle's computer modeling, including all the components of a vehicle such as power system, frame, tires, road and the driver. It can use the generic model, customization code like various MATLAB/SIMULINK controller model to provide vehicle parts and even the real hardware in the test bench. According to the design task, VVE can not only be operated in the general office computer, but also be operated in real time system. Real time operation can study the deterministic performance. And office operation locks of real-time ability. But because of this, office operation can almost adapt to any host computer and can make simulation carry out more slowly or quickly than real-time operation rely on system performance and complexity of models. In addition, office operation does not require special hardware which can save money greatly.

3 FOUR FUSION METHODS

3.1 *CarMaker for simulink*

CarMaker for Simulink is a complete integration of IPG's vehicle dynamics simulation software, CarMaker, into The MathWorks's modeling and simulation environment, Matlab/Simulink. The highly optimized and robust features of CarMaker software are added to the Simulink environment using an S-Function implementation and the API functions that are provided by Matlab/Simulink. CarMaker for Simulink is not a loosely-coupled co-simulation, but a closely linked combination of these two best-in-class applications, resulting in a simulation environment that has both good performance and stability.

Because of this integration, now it is possible to use the power and functionality of CarMaker in Matlab/Simulink whose functions are powerful. Using or building CarMaker for Simulink is as same as using or building standard S-Function blocks in Matlab/Simulink. It means that the CarMaker blocks can be connected to other Simulink blocks and existing CarMaker blocks can be easily added to CarMaker vehicle models with a few clicks of the mouse.

However, this kind of integration doesn't mean a loss of functionality as all functions of CarMaker software and Matlab/Simulink's rich tool set are all included in CarMaker for Simulink. CarMaker GUI can still be used for not only simulation control and parameter adjustments, but also defining maneuvers and road configurations. IPGControl can still be used for data analysis and plotting. IPGMovie can still be used to display how the vehicle drives in a virtual environment in 3-dimensional space.

Vehicle simulation results can be accessed by the cmread utility which exists in Matlab. The cmread utility can load any CarMaker simulation results into the Matlab workspace so that data about simulation results can be manipulated and viewed using any reliable Matlab tools.

In a word, CarMaker for Simulink is a complete system which can become a part of the Simulink simulation quickly and easily. CarMaker for Simulink is a good method to verify their vehicle control algorithm with S-Function for engineering staff who are familiar with vehicle dynamics and Matlab/Simulink.

3.2 *Integrating C-Code models*

More than one model which is car's subsystem like steering system, brake system can be loaded in the simulation program. Every model has a specific place to be loaded in CarMaker GUI. In the vehicle's parameter file, if Brake.kind = MyModel, then the operation will select user module MyModel instead of the CarMaker internal brake model. All these subsystems are saved in src folder.

This implementation has an advantage that it can be quick and easy to switch models by changing vehicle parameters. It is better to calculate an integrated model at the same time in order to ensure the maximum simulation performance. Another advantage is a modular design that provides the additional possibility to extend the model functionality with user code.

The subdirectory ExtraModels of project directory src folder contains some example modules, such as AccelCtrl.c, MyBrake.c, MyVehicle- Control.c and so on.

Take the integrated MyVehicleControl module as an example to explain how to integrate the C-code into CarMaker simulation environment.

First of all, select the option sources: Extra Models when creating or updating a new project. Copy files which are MyVehicleControl.c and MyModels.h corresponding to MyVehicleControl model into the src folder. Secondly, set up the Makefile. The Makefile needs to know which model needs to be incorporated. It can be done by adding OBJS macro like following sentence.

OBJS = CM_Main.o CM_Vehicle.o User.o MyVehicleControl.o

Thirdly, register the model. Edit the file User.c with an ASCII editor for programming. There are two things to do. One is to add a header file MyModels.h in the include section. The other one is to define a sentence VehicleControl_Register_MyVehicleControl() to call registering function in function User_Register(). Finally, build the simulation program. In the terminal window MSYS-2003 which is installed in the computer when installing the CarMaker software, enter the command make to generate the simulation program in the src folder path.

DoubleclicktheCarMakerIcontostarttheCarMaker GUI and click Application ->Configu->ration/Status to load the executable file CarMaker.win32.exe under the src folder generated before and click Application/ Start & Connect in order to use the generated module in the TestRun related to brake function. Then load MyBrake model by clicking Parameters->Car->Brake ->Model->My- VehicleControl.

Additional parameters can be added using the Additional Parameters field in the Parameters-> Car ->Misc tab. Run the TestRun and monitor the brake torque with IPGControl. Vary the parameters and observe the influence on the vehicle behavior.

Table 1. Subsystems supported by model manager.

Subsystem Type	Model class	Key
Aerodynamics	Aero	Aero.Kind
Brake	Brake	Brake.Kind
Environment	Environment	Env.Kind
Powertrain	PowerTrain	PowerTrain.Kind
Clutch	PTClutch	PowerTrain.Clutch.Kind
Driveline	PTDriveLine	PowerTrain.DriveLine .Kind
GearBox	PTGearBox	PowerTrain. GearBox .Kind
Differential Locks	PTGenCoupling	PowerTrain. DL.FDiff.Cpl.Kind
Engine Torque	PTEngine	PowerTrain.ET .Kind
Steering	Steering	Steering.Kind
Kinematics and Compliance	SuspKnc	Susp<Pos>.<Kin/Com>.<No**>.Kind
Active Suspension Systems	SuspExtFrcs	SuspExtFrcs.Kind
Tire	Tire	*
Vehicle	Vehicle	*
Vehicle Control	VehicleControl	VehicleControl.Kind

3.3 *Functional mockup interface*

Functional Mockup Interface (FMI) is an open standard interface. This interface supports model exchange and co-simulation of dynamic model combined XML file with compiling C-code. The Functional Mock-up Interface offers the possibility to make use of models exported by third party tools in a standardized form as so-called Functional Mock-up Units (FMUs).

The CarMaker supports and only supports functionality of FMI version 1.0. The FMU which is actually a compression package of some specific files with fmu suffix instead of zip suffix must follow the specification of FMI for Co-Simulation Stand Alone instead of FMI for Co-Simulation Tool and implement one of CarMaker model classes. Attention: the FMU must contain binary code of the vehicle control algorithm which is edited in the function DoStep.

We have some work to do before using the FMU. The FMU will display in the dialog box of FMU Plug-Ins after importing it and it will be automatically copied to the Plugins folder under project directory at the same time. Some information of the FMU like name, type, modified time and state will be displayed in the Overview list. There are three possible states which can be seen in Figure 2.

NEW: It means that the FMU is just imported without any more operation.

READY: It means that the FMU is already connected with CarMaker signals and is ready to be used.

ERROR: It means that there is a loading problem or the FMU isn't supported by the CarMaker software. The more information about this error will be found in the lower part of the dialog box.

What's more important for researchers who want to verify if the control algorithm is correct and if the performance of this algorithm is exactly the same as what they thought is to make sure that the FMU's input and output signals should be connected with the CarMaker signals correctly. Figure 3 shows how to set a FMU's input and output parameters.

The FMU could be one of module types as follows: Aerodynamics, Brake, Clutch, DriveLine, Engine, Environment, Gearbox, Steering, Suspension External Forces, UserDriver, Tire and Vehicle Control. There is no doubt that the FMU about vehicle control algorithm should select Vehicle Control module type.

The FMU's input signals could be a constant value, an interface output variable or any DataDict quantity. The FMU's output signal could be an interface variable or any DataDict quantity. In some rare cases, the module type requires additional parameters and these parameters are not available in the FMU. Then these parameters can be edited manually in the Modelclass-specific parameters window.

The FMU with READY state will be used in the same way as the second method.

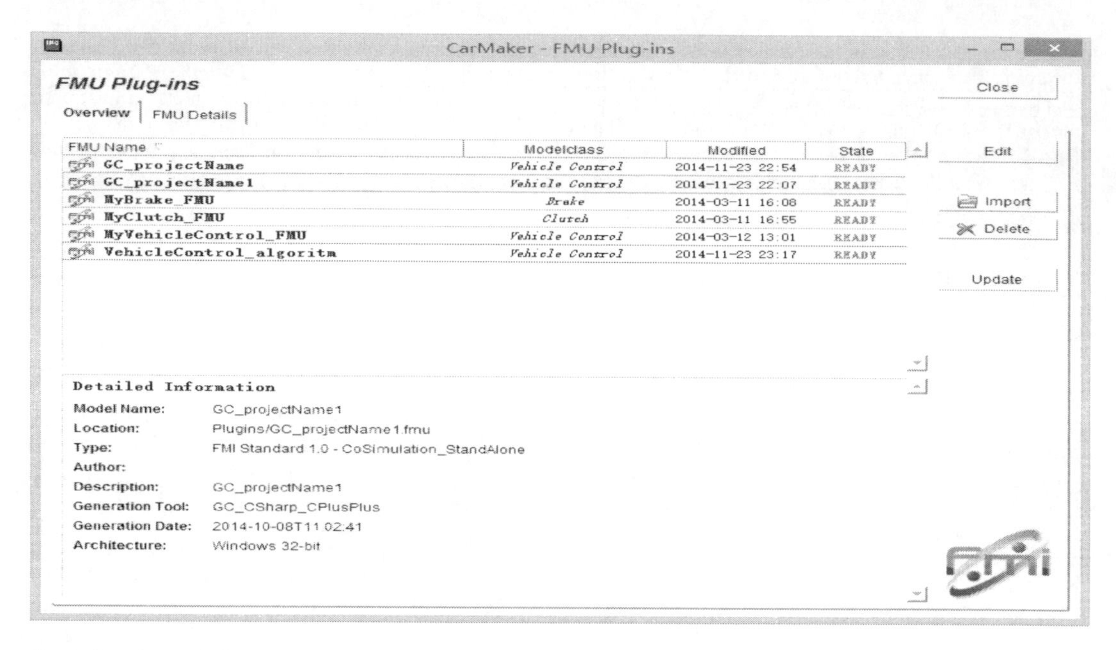

Figure 2. Three possible states of a FMU in CarMaker.

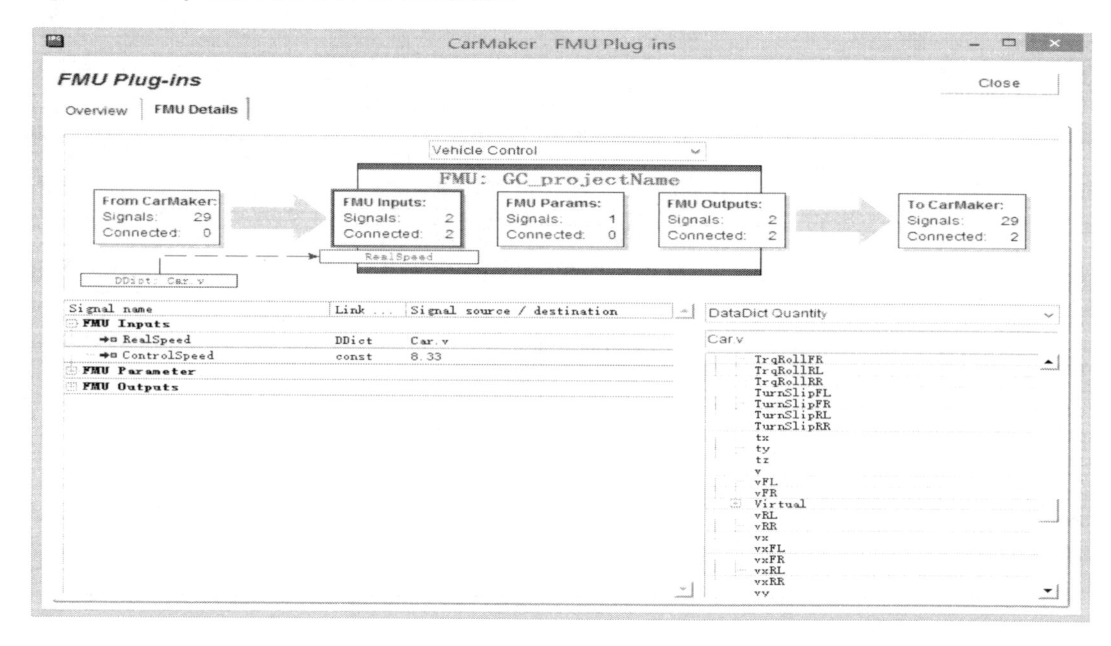

Figure 3. The interface of a FMU's parameters.

3.4 *Model Plug-in*

The fourth fusion method, model Plug-in, is similar to the second fusion method, integrating C-Code method. The model Plug-in method is first built in Matlab/Simulink environment and then called in the CarMaker software. And it is built with Matlab/Simulink environment with S-Function and API function that Matlab/Simulink provides like the first fusion method uses instead of C-Code program. The corresponding model Plug-in will be ready to be used in the CarMaker software in the same way as the second and third fusion methods are used after compiling the model.

Model Plug_in method begins with Create Plug-in Model under CarMaker4SL in Simulink library browser. The VehicleControl model will be defined like Figure 4 shows when researchers want to verify their control algorithm. This VehicleControl model Plug-in could be used in the CarMaker software after compiling the model.

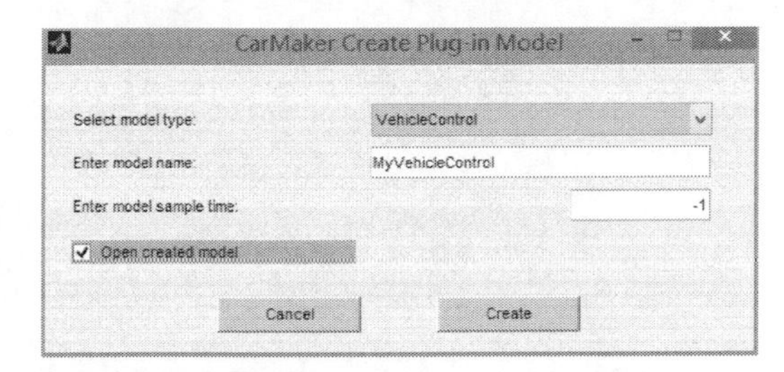

Figure 4. VehicleControl model.

4 CONCLUSION

These four fusion methods make it very easier that integrating researchers' vehicle control algorithm into the CarMaker software to verify the performance and improve the algorithm quickly according to the simulation result. These fusion methods not only shorten the research cycle about intelligent vehicle and reduce cost, but also greatly reduce real vehicle operating frequency resulting in improving researchers' safe. It is a big trend of research intelligent vehicle on autonomous driving combined with vehicle dynamics simulation software such as the CarMaker software within a few years.

ACKNOWLEDGEMENTS

This work was financially supported by"The Project of Construction of Innovative Teams and Teacher vehicular Development for Universities and Colleges Under Beijing Municipality "and "The Importation and Development of High-Caliber Talents Project of Beijing Municipal Institutions" (CIT&TCD201304074 and IDHT20130513). And the corresponding author of this paper is Wenliang Niu.

REFERENCES

Rongben Wang, Bing Li & Shuming Shi. 2001. Overview of intelligent vehicle development worldwide. *Highway Traffic Science and Technology*(18)5.

Haifeng Hu, Zhongke Shi and Dewen Xue. 2004. Research on intelligent vehicle development. *Application Research of Computers*.

Dezhi Gao, Jianmin Duan and Banggui Zheng. 2008. Application statement of intelligent vehicle environment perception sensor. *Modern Electronic Technology*(19).

Shuang Liao, Yong Xu and Shan Chao. 2014. A study on control method for the intelligent vehicles autonomous driving. *Computer Measurement & Control* (22)8.

Qingling Duanmu, Jiewang Ruan and Jun Ma. 2014. Development and advanced technology of driverless car. *Agricultural Equipment & Vehicle Engineering* (52)3.

Xiaoli Ying, Lei Li. 2013. Path tracking control for unmanned ground vehicle in complex road conditions. *Journal of north university of China(natural science edition)* (34)5.

Meixia Zhu. 2013. Research of collision avoidance mechanism for autonomous vehicles at the intersection. *Computer Engineering & Science* (34)5.

Qiang Wang. 2012. Extend IPG CarMaker using the FMI adapter unit. *Journal of Henan Mechanical and Electrical Engineering College* (20)1.

King Tin Leung, James F. Whidborne, David Purdy and Phil Barber. 2011. Road vehicle state estimation using low-cost GPS/INS. *Mechanical Systems and Signal Processing* (25)6:1988–2004.

Computing, Control, Information and Education Engineering – Liu, Sung & Yao (eds)
© 2015 Taylor & Francis Group, London, ISBN: 978-1-138-02800-5

The implication of eco-design for drinking water in polyethylene terephthalate bottle

Taksina Chai-ittipornwong
Faculty of Management Science, Rajabhat Muban Chombueng University, Ratchaburi, Thailand

ABSTRACT: Eco-design within sustainability of bottled water is not new, however, the constructive application has not been largely adopted. The paper was thus aimed to provide the evidence-based application using eco-design for developing the product and process of PET–bottled water, in comparison. Three ways to understand the implications include (i) minimum resource consumption, (ii) maximum resource benefits (iii) waste and impact reduction, with the research method of in-depth interview and field survey. The prominent findings are that the comparative results between the product and the process development somewhat differ, the success goes into the process over the product term, and that the approach of maximum resource benefits is highly valued over the other two indexes. In Thai market, recycled-PET is less considered significant to any new application among the bottlers and the improvement of efficiency in water resource is needed. In conclusion, eco-design can be a primary stage of responsible industry development in the realm of sustainability in which environmental consciousness among all concerns shall be fully developed.

KEYWORDS: Bottled water; Eco-design; Polyethylene terephthalate bottle.

1 INTRODUCTION

This study is particularly timely since waste plastics and water crisis are shifting from one of problem-framing to new agenda that is much more concerned with climate impacts. The world is pushing the planet away from sustainability, according to materially intensive economies. Many disposable products make life possible for a faster and convenient pace of society eventually translate to consumer wastes. Polyethylene terephthalate (PET)-bottled water has become one of the most common plastic products to be consumed and disposed of, on a regular basis, both at and away from home [1, 2]. This actually causes a profound impact on environment and human health as more bottles mean more greenhouse gases from production, transportation, and waste bottles management. The adverse impacts are compounding a framework to justify as to how the businesses respond to the environmental issues at the early stage of creating a product.

PET bottled water. In global context, bottled water has been the top end-used application of PET plastic since its introduction into beverage packaging in the late decade of 1970 [3]. PET-bottled water is expected to maintain annual growth by 15% in total volume terms due to increasing health awareness and hygiene concerns among Thai consumers [4-6].

Bottled-water manufacturing is one of the fastest-growing businesses in Thailand, consuming nearly 40,000 tons of PET plastic and reaching 2.4 billion liters of bottled water a year [7, 8]. There are also large numbers of small players due to the availability of PET plastics and the low cost of water resources, which are mostly tap water. PET bottled water has widely dominated Thai consumers, and cause the decline in returnable glass bottles and cheaper alternative in HDPE bottles[5]. The landscape of drinking water in PET bottles between the leading and small brands exhibit that a 500ml PET bottle highly demanded for on-the-go consumption is the majority pack size focused by the small manufacturers, while larger pack sizes of 1000-1500ml serve the highest sale volume of leading brands [4, 5]. All major brands, as mentioned in alphabetical order: Crystal, Namthip, Purelife and Singha, lead PET bottled water with a 65% in current value and volume terms of the market[6]. The rest are shared among 6,000 small and medium enterprises where most like to operate manufacturing in the provinces near the Chao Phraya River and its tributaries for ease of underground water extraction [8].

Plastics and water are two main raw materials of PET bottled water. PET resin actually constitutes various pollutions to air, water and soil quality as carbon substance is the key element of resin formulation. When producing PET bottles, the resins are melted and polymerized till feasible for a pre-form injection in latter conjunction with a blow molding process. The process consumes both materials and energy while emitting much of chemical substance, including dust and debris of pollutions and acetaldehyde from the polymer degradation [9, 10].

The relevant operations of material extraction, manufacture and distribution throughout the entire processes of PET-bottled water are very close to a reduction of carbon and ecological footprints by a productivity on material and energy consumption, coupled with a recognition to the transformations and released emissions, both at present and in future time [2, 9]. Based on the 100-year time horizon Global Warming Potentials (GWP), PET resins produce higher impact on global warming than do the categories of nonrenewable energy, acidification and eutrophication[11]. It was revealed that the production of 1,000 units of single-serve PET bottle can cause about five kilograms equivalent CO_2 (kg CO_2 eq.) emissions[11–13]. These deal with a responsible industrial development from supply side in sustainability objectives that must be strengthened by all means along with the entire product life cycle [1, 14, 15].

Eco-design. Within characteristics of food container, PET bottle is considered a safe, healthy and convenient packaged food product. A variety of measures on environmental consequences including an eco-design is created in the bottled water industry. Product design is a fundamental factor involving with the beginning stage of product development. Product and process designs enable a less use of material to lower the environmental impact of a product after it reaches the end of its life cycle [16, 17]. They create social value with the essence on production, marketing and distribution aspects. Thus, practices for eco-design involve new products and processes development, manufacturing technology improvement and process planning, with environmental consideration[17-19].

However, changes in the product and process of PET-bottled water of the international brands commercialized in Thai market are highly dependent on the mother brand [20]. It is revealed that product design is particularly used among the leading brands for managing consumer perceptions of corporate image and brand identity. As in Akenji and Bengtsson's research about environmentally-friendly package, the bottles that are designed in packaging harmonization across brands serve both a reduction of waste generation and an increase of resource efficiency [18]. In addition, the common package is advantageous for the processes of logistics and transportation of bottled water to reach resource efficiency in fuel energy consumption and CO_2 emission reduction [1, 13-15, 20, 21]. According to Ramani et al., eco-design is not only potential for taking up life cycle environmental impacts of a product but also enables business opportunities[17].

As a consequence, the research aims to provide the evidence-based application of eco-design for PET-bottled water, with focusing on two development approaches: product and process. The pre-assumption is that the implication of eco-design is best applied for PET-bottled water if product development as well as technological process improvement and process planning are evenly integrated.

2 METHODS

2.1 *Population and sampling*

Sample size derives from two groups of the bottled water business. The first designates the leading brands (LD) of PET-bottled water in Thai market; namely Crystal, Namthip, Purelife and Singha. The other belongs to the small and medium brands (SM) those not only producing Unbranded bottled water, but also providing bottling service for leading brands. An SM for every region (north, northeast, east, south and central part) is purposively randomized for this research. Due to academic and confidentiality reasons, company title of SM brands is not indicated in this analysis. People working in the field of R&D (research and product development job), including policy-makers and engineers, and those concerned about environmental issues are the key informants.

2.2 *Instruments*

A 500ml PET bottle is assigned as a functional unit in this study. The approaches to eco-design are on the basis of product and process development, with three indexes: (i) minimize resource consumption, (ii) maximize resource benefits and (iii) reduce waste and impact generation. A matrix feature of product and process approaches is used to define the possibility of eco-design in PET bottled water. Thus, the achievement in eco-design is calculated with the following equation, including the Rule of Three.

$$ED = \sum_{i=0}^{n}(PDi + PRi) \tag{1}$$

where: ED = represents the achievement in both product and process development

PD_i = represents the product development approached to PET-bottled water

PR_i = represents the process development approached to PET-bottled water

2.3 *Procedure*

First, interviews are arranged in conjunction with plant visits so that observations about the product and process operations could be discussed. The survey transcripts are prepared to record the comparative achievements, in terms of an individual group; LD or SM, and a combined groups, against the indexes of each approach. Then, using content analysis interprets the collective data. A number and the percentage are used to translate the results of eco-design approaches into a matrix feature, with a breakdown between two approaches.

3 RESULTS AND DISCUSSION

According to Table 1, it is evident that the implications of eco-design are applied to the research and development of drinking water in PET bottles commercialized in the Thai market, with 25 approaches to the product development and 30 approaches to the process development. The respondents perform better in the process development (54.55%) comparing to the product term (45.45%).

Table 1. The comparative achievements in product and process approaches to eco-design for a 500ml PET-bottled water.

Items	Product Development (%)	Process Development (%)	Total (%)
Indexes achievements			
(i) Minimize resource consumption	6 (24.00)	4 (13.33)	10 (18.18)
(ii) Maximize resource benefits	15 (60.00)	20 (66.67)	35 (63.64)
(iii) Reduce waste and impact generation	4 (16.00)	6 (20.00)	10 (18.18)
Brand achievements			
- Individual	11 (44.00)	17 (56.67)	28 (50.91)
- Combined brands	14 (56.00)	13 (43.33)	27 (49.09)
Achievement in eco-design	25 (45.45)	30 (54.55)	55 (100%)

The approach of maximizing resource benefits is predominant among three indexes of eco-design achievement (63.64%), while the rest keep the same value of 10%. In identifying the most favorite index against each approach, the results exhibit the same answer of maximum resource benefits, with 60.00% for the product and 66.67% for the process. When comparing group achievements, the performance of individual group is slightly different from that of combined one (50.91% v. 49.09%). In particular, the implications of eco-design for a 500ml-PET-bottled water is more common to the LD than the SM. The reader is encouraged to see the differences in Table 2 and 3 detailed by the groups in later topics, which are drawn with a more discussion on each approach.

Actually, the comparative result of the approaches to the eco-design for PET-bottled water can be understood from the matrix format featured between the product and the process lines. The outcome is consistent with the matrix of process-intensity as shown in Figure 1.

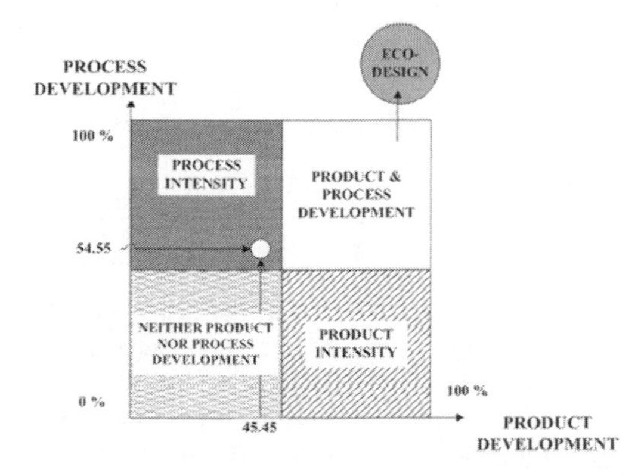

Figure 1. A matrix format featuring the process intensity as the outcome of developing the eco-design for a 500ml-PET- bottled water.

3.1 *Product development*

As the information about product development shown in Table 2, in PET-bottled water business, efficiency in plastics consumption mostly begins with a lighter-weight bottle [20, 22, 23]. A bottle with thinner wall, shorter neck and narrow cap is primary designed for reducing PET-plastic consumption [20]. On the other, the squeezed bottle, which leads to a lighter-weight bottle (12-13 grams) for saving 35% of plastic, is neither attractive nor feasible for any brands, except one of the LD. Additional investment in nitrogen feeding is required to fulfill these special bottles. The eco-design in product development is also advantageous for a reduction of waste generation by handling the bottle problems of bottle-mold quality, uneven thickness, deviations in the spiral neck, bottle clarity, leakage and off-centered bottoms[24].

Additional is the amount of PET plastics consumption in bottled drinking water business. The quantity is rising to 120 tons per month by the SM, while the LD approximates 400 tons, excluding the peak period in the hot season (when consumption reaches 500 tons per month). This is consistent with the information that the most vulnerable impact on the ecological quality derives from waste plastic bottles[13, 23, 25, 26]. Much of the fossil fuel used in the PET bottle is from feedstock energy, which is bound within the product[9, 10, 12, 27]. What should be taken into consideration is the low interest in using the scraps for a new application and new bottles among Thai bottlers. As PET properties, the successes in recovering waste-PET bottles to textile and yarn application, or even new bottles, which are chemically recycled, are in evidence with the advances of recycling technology [2, 10, 28]. According to the results, a legal constraint on using recycled material for food packaging, enforced by the Food and Drug Administration (Thailand) and the threat of knowledge about recycling technology are in excuse to the less account of recycling device.

Table 2. The achievements in eco-design against the indexes of product development.

Indexes		Product Development (500ml PET-bottled water)		Achievement	
				Individual	Combined
1	Minimize resource consumption	1.	Develop shorter-neck bottle (height of 18 mm to 10-12 mm)		✓
		2.	Develop thinner-wall bottle	LD	
		3.	Narrow cap (from 32 to 28 mm)		✓
		4.	Limit bottle weight within 17-17.5 gram		✓
		5.	Initiate 12-13g squeezed the bottle for reduced use of 35% plastic	LD	
		6.	Change from square-shape bottle to drum shape	LD	
2	Maximize resource benefits	1.	Increase toughness of the bottom part for better stiffness in stacking and storage		✓
		2.	Follow common-shape bottle	SM	
		3.	Design in bottle neck for common cap of all shapes and sizes of bottles		✓
		4.	Benefit a reduced use of plastic wrapper for smaller cap		✓
		5.	Recycle in-house PET scrap for new application	SM	
		6.	Develop bottle shape best fit to bundle packing for loading efficiency (up to 16,128 bottles per 6-wheel truck)		✓
		7.	Alternate the emerging market of 350ml bottle used for replacing a 500ml bottled water and providing greater efficiency in PET bottle consumption		✓
		8.	Provide multi-size bottles for a variety of usage	LD	
		9.	Re-circulate lost-filled water to chilling system		✓
		10.	Adjust waste-in-process bottles for refilling	SM	
		11.	Promote the internal use of defective bottled water due to inferior appearance		✓
		12.	Enhance cross-industry value of plastics molding and bottling service	SM	
		13.	Communicate PET qualification as the most recyclable plastic by the numeric symbol 1		✓
		14.	Apply for the Islamic Committee Office of Thailand to facilitate special use	LD	
		15.	Communicate sustainable use for environmental care	LD	
3	Reduce waste and impact generation	1.	Provide cap-seal wrapper to ensure health concern to primary use		✓
		2.	Indicate proper use for health care (keep in cool dry place and away from direct sunlight)		✓
		3.	Communicate appropriate manner of sorting waste bottles (do not litter)		✓
		4.	Promote squeezed bottles to decrease landfill volume	LD	
	Total amount		25 sub-indexes	11	14

3.2 *Process development*

Due to Table 3 detailing the process approach, there exist many attempts on maximum plastics consumption; either by reducing the use of plastics or a limitation of reduced defects. The reuse and recycling of the process-waste plastic are applicable for only some SM those producing not only unbranded bottled water but also providing bottling service for the LD.

It is, however apparent that a huge demand for PET-bottle scraps can distort the interest in renewable PET technology. PET is thus less frequently recycled among Thai bottlers, with remarking on the relevant laws and regulations prohibiting the use of recycled plastics for food packaging. Normal practice among them is to sell them to recycling agents since these flakes are mostly required for export market in China and India [3, 7, 24].

Table 3. The achievements in eco-design against the indexes of process development.

Indexes		Process Development (500ml PET-bottled water)		Achievement	
				Individual	Combined
1	Minimize resource consumption	1.	Invest in bottling technology of closed-loop system for saving 30% lost-filled water from the open loop	LD	
		2.	Use hi-speed blower, instead of drinking water for cleaning the inside of the bottle	LD	
		3.	Use common bottle shape for a decrease of pre-processing materials consumption	SM	
		4.	Optimize capping process by common cap for all sizes and shapes of bottles		✓
2	Maximize resource benefits	1.	Use the proper machine to increase run-hour optimization in production operation	LD	
		2.	Increase productivities of labeling and cap wrapping processes by modifying them apart from bottling system	SM	
		3.	Improve design for mould preparation for less defects	LD	
		4.	Select proper material for defect reduction	LD	
		5.	Use process planning for production optimization	LD	
		6.	Use common bottle-shape for reduced time operation	SM	
		7.	Improve cycle time in filling step down to 0.3-0.4 seconds per bottle		✓
		8.	Re-circulate lost-filled water to chilling system		✓
		9.	Benefit the open-loop bottling system by refilling returnable bottles		✓
		10.	Decrease trip of distribution by a design for multi-layer stacking		✓
		11.	Control cost of fuel energy within US$ 0.01 / bottle		✓
		12.	Minimize distribution cost to US$ 0.02 / bottle		✓
		13.	Control electricity cost within US$ 0.02 / bottle		✓
		14.	Benefit supply-chain management of PET pre-forms and bottles for better process planning	LD	
		15.	Benefit supply-chain management of PET pre-forms for greater productivity		✓
		16.	Benefit supply-chain management of blank bottles for productivity in production and warehouse operation		✓
		17.	Benefit supply chain management of PET pre-forms and bottles for efficiency in stock cost		✓
		18.	Employ multi-plant policy for efficiency in material allocation and product distribution	LD	
		19.	Invest in automatic technology for filler room	LD	
		20.	Optimize production operation between plastics molding and bottling service	SM	
3	Reduce waste and impact generation	1.	Invest in technological improvement of a bottling system for environmental protection	LD	
		2.	Decrease product contamination by setting closed area for filler room	LD	
		3.	Use electrometer and infrared light for dust and debris elimination instead of discharging into the air	LD	
		4.	Reduce GHGs emissions by replacing fuel oil with battery for forklift truck used in stock handling	SM	
		5.	Do sorting collection of in-house waste plastics		✓
		6.	Control wastes-in-process of plastics within 0.2%		✓
Total amount			30 sub-indexes	17	13

Looking at the bottling processes, the closed-loop system is considered possible for a cleaner technology. Its process design enables a reduction of the lost-filled water and a use of electrometers and infrared device for debris and dust elimination, instead of discharging into the air, since the process of bottle blow-molding is actually connected to the filler room. For the open loop, outsourcing of PET-bottle manufacturing is explained for no CO_2 emissions to the air, water and soil generated from bottling system. Nevertheless, the producers using the open loop should be aware of the substantial waste of drinking water, which can reach 30%, which can be reduced through the closed-loop process. The closed loop, which needs an optimized design in space and machine flow, is not operated by most brands those investing in bottled-water business at small and medium scale. In brief, it would be inconceivable for thousands of the SM to switch to the closed-loop system.

It is also found that almost all SM those operate the open loop can get flexible for mobile production of labeling and cap sealing. Moreover, using process planning enables to cut down on unnecessary processes to ensure continuity of the bottling line in case bottle-molding production goes down. A multi-plant policy and truck-loaded management are also used for manufacturers that prioritize cost reduction over impact reduction from reduced consumption of electricity and fuel. As well, logistics and transportation are related to both utility consumption and impact generation [1, 14, 20, 28]. The conclusion comes true that it is potential for both the LD and the SM to take up with the eco-design in the process development of PET-bottled water.

4 CONCLUSION AND RECOMMENDATION

The research highlights that the development of eco-design for a 500ml-PET-bottled water is substantial to a reach of maximum efficiency in resource consumption and minimize waste and impact generation. The matrix result of the product and process approaches is somewhat differ as their successes go to the process intensity. Both the leading and small brands mostly fulfill the ultimate implications of eco-design through a maximization of resource benefits.

This study is by no means exhaustive, but enables a potential for the methods and datasets. The ways of water access and water treatment regarding extraction and acquisition procedures shall be in vision so that sustainable management of land, water and biodiversity can be fulfilled and experienced among the concerns. Thus, the producers using the open-loop bottling system are challenging for improving social responsibility in water resource efficiency. The recycling technology should be truly reconciled with the research and development of PET-bottled water. As eco-design can be a primary stage of responsible industry development towards, or away from, sustainability, the improvement of assessment methodologies from a multi-disciplinary perspective could be set for a more understanding of design for environment.

REFERENCES

Amano, M. and B. Ness, *PET Bottle System in Sweden and Japan: an Integrated Analysis from a Life-Cycle Perspective*, in *LUMES Program*,2004, Lund University: Sweden. p. 1–35.

Takahashi, Y., *Sustainable PET Bottle Recycling System in Japan for Sound Material Cycle*, in *Graduate Program in Sustainable Science (GPSS)*, 2010, The University of Tokyo: Graduate School of Frontier Sciences.

Willett, C. *PET: Is This As Good As IT Gets?* in *World Petrochemical Conference: XVIII Annual. 26–27 March 2003*. 2003. Houston, Texas: Chemical Market Associates Incorporation (CMAI).

Wai Chamornmarn *Bottled Water*. External Environmental Analysis, 2008.

Euromonitor International, *Bottled Water-Thailand*, in *Country Sector Briefing, April 2010*2010, GMID: http://www.euromonitor.com/bottled-water-in-thailand/report.

Ratthanin Sakdamrongrat, *Bottled Water*, in *Marketeer*, October 2010: Bangkok. p. 114–117.

Natheenont Chaisathaporn, *Recycled PET resins (Interview) 18 July 2012*, 2012.

Khokhet Janthalertlak, *Fight Flood*, 2011, Thai PBS: Thailand.

Song, H.S. and J.C. Hyun, *A Study of the Comparison of the Various Waste Management Scenarios for PET Bottles Using the Life-Cycle Assessment (LCA) Methodology*. Resource, Conservation and Recycling, 1999. **27**(3): p. 267–284.

Zhang, B. and F. Kimura Framework Research on the Greenness Evaluation of Polymer Materials. Advances in Life Cycle Engineering for Sustainable Engineering Businesses, 2007. 291–297.

Gironi, F. and V. Piemonte, *LIfe Cycle Assessment of Polylactic Acid and Polyethylene Terephthalate Bottles for Drinking Water*. Environmental Progress & Sustainable Energy, 2010. **30**(3): p. 459–468.

Hauschild, M., J. Jeswiet, and L. Alting, *From Life Cycle Assessment to Sustainable Production: Status and Perspectives*. CIRP Annals-Manufacturing Technology, 2005. **54**(2): p. 1–21.

Huber, M.K., *Bottled Water: The Risks of Our Health, Our Environment, and Our Wallets*, in *School of Public and Environmental Affairs (SPEA)*, 2010, Indiana University. p. 1–39.

Coelho, T.M., R. Castro, and J.A. Gobbo Jr, *PET Containers in Brazil: Opportunities and Challenges of a Logistics Model for Post-Consumer Waste Recycling*. Resource, Conservation and Recycling, 2010. **55**(3): p. 291–299.

Foolmuan, R.K. and T. Ramjeeawan, *Disposal of Post-Consumer Polyethylene Terephthalate (PET) Bottles:*

Comparison of Five Disposal Alternatives in the Small Island State of Mauritius Using a Life Cycle Assessment Tool. Environmental Technology, 2012. **33**(17): p. 2007–2018.

Bare, J. and T. Gloria, *Life Cycle Impact Assessment for the Building Design and Construction Industry.* Building Design & Construction, 2005. **46**(11): p. S22(3).

Ramani, K., et al., *Integrated Sustainable LIfe Cycle Design: A Review.* Mechanical Design, 2010. **132** p. 1–15.

Akenji, L. and M. Bengtsson *Is the Customer Really King? Stakeholder analysis for sustainable consumption and production using the example of the packaging value chain.* Regional 3R (reduce, reuse, recycle) Forum in Asia, Asia Resource Circulation Research Promotion Programme, IGES White paper, 2009.

Sze, T.L. *Spaces for Consumption.* 17 May 2010. 1–12 DOI: http://www.satepub.com/upm-data/35384_01_Miles_CH_01.pdf.

Taksina Chai-ittipornwong, Pomthong Malakul, and Dawan Wiwattnadate, *The Challenges of Implementing Sustainable Production for One-Way Bottled Drinking Water in Thailand.* Advanced Materials Research, 2013. **807–809**(2013): p. 2897–2910.

Feron, V.J., et al., *Polyethylene terephthalate bottles (PRBs): A Health and Safety Assessement.* Food Additives and Contaminants, 1994. **11**(5): p. 571–594.

Nestle Waters North America (NWNA). *Environmental Life Cycle Assessment of Drinking Water Alternatives & Consumer Beverage Consumption in North America.* [webpage] 5 February 2010; Available from: http://beveragelcafootprint.com/Nestlé

Niccolucci, V., et al., *The Real Water Consumption behind Drinking Water: The Case of Italy.* Environmental Management, 2011. **92**(10): p. 2611–2618.

Taksina Chai-ittipornwong, *Application of Life Cycle Management for Sustainable Consumption and Production of Polyethylene Terephthalate (PET) Water Bottle in Thailand*, in *Environment, Development and Sustainability*2013, Chalalongkorn: Bangkok. p. 72.

Barlow, M., *Blue Gold: the Global Water Crisis and the Commodification of the World's Water Supply*, in *The Global Trade in Water,* 2001: The International Forum on Globalization (IFG).

Cellura, M., S. Longo, and M. Mistretta, *The Energy and Environmental Impacts of Italian Households Consumptions: An Input-Output Approach.* Renewable and Sustainable Energy Reviews, 2011. **15**(8): p. 3897–3908.

Hurd, D.J. *Best Practices and Industry Standards in PET Plastic Recycling.* Bronz 2000 Associates, Inc., N.Y. 1–55.

Shen, L., E. Nieuwlaar, and E. Worrell *Life Cycle Energy and GHG Emissions of PET Recycling: Change-Oriented Effects.* LIfe Cycle Assessment, 2011.

Computing, Control, Information and Education Engineering – Liu, Sung & Yao (eds)
© 2015 Taylor & Francis Group, London, ISBN: 978-1-138-02800-5

The optimal parameters of WEDM for material removal rate of hardened tool steel: SKD 61

Komson Jirapattarasilp & Khanchai Kosit
Department of Production Technology Education, Faculty of Industrial Education and Technology, King Mongkut's University of Technology, Bangmod, Thongkru District, Bangkok, Thailand

ABSTRACT: Wire Electrode Discharged Machining (WEDM) has been one of the most machining operations for making tool in the industry. Tool steel was recommended using for making tool for hot working. This research was to study optimal parameters, which were affected on material removal rate (MRR) in WEDM of hardened tool steel JIS SKD 61. A material used in the study was tool steel JIS SKD 61 that was hardening by heat treatment process. Taguchi's method was experimental design and conducted on 5 factors with 3 levels. L27 of the Taguchi standard orthogonal array was chosen for the design of experiments. The parameters consisted of power setting, off time off, wire feeding, wire tension and wire offset. The importance level of the machining parameters was determined by analysis of variance (ANOVA). The optimum parameter combination was obtained by using the analysis of signal to noise (S/N) ratios. Finally, experimentation was carried out an optimal solution of machining parameters for the quality of MRR in WEDM process.

KEYWORDS: WEDM, Hardened Tool steel, SKD 61, Material Removal Rate, Optimal Parameters.

1 INTRODUCTION

Tool steel is the important material to produce the mold and die for industries such as forging die for automotive part industry. SKD 61 is tool steel that recommended using for hot working. High hardness and withstands wear are important properties and also has a relatively high resistance to shock. In order to use for die, this material must be machined by turning or milling operation and hardened by heat treatment process. In order to make molds and die, Wire Electrode Discharged Machining (WEDM) has been most operations. In order to obtain better material removal rate, the suitable setting of WEDM parameters is important before the process takes place [1]. So, there are several factors affected to productivity, such as Power setting, Off time off, Wire feeding, Wire tension and Wire offset. These factors of WEDM could be affected to product based on the machining parameters. Furthermore, there are several researches to study the factors affected to the process of WEDM. Tarng et al. [2] applied a neural network system to determine settings parameter for the estimation of cutting speed and surface finish. Kosit and Jirapattarasilp [3] used a Taguchi's methods to determine factors for surface roughness of WEDM on SKD 61 steel. Mohammadi et al. [4] investigated roundness improvement in wire electrical discharge turning. Scott et al. [5] studied a factorial design method to determine the optimal combination of control parameters in WEDM. Lok and Lee [6] compared the performance of WEDM in terms of material removed rate and surface finish through observations obtained by processing of two advanced ceramics under different cutting conditions.

This study was focused on investigating the material removed rate (MRR) via the WEDM operational condition of tool steel, especially grade JIS: SKD 61 and study optimal solution for its parameters to achieve a better material removal rate by Taguchi's methods. Experiments were employed in this study to consider the effects of power setting, off time off, wire feeding, wire tension and wire offset on the material removal rate. This study was performed on the data obtained from the experiments by using two different analyses. Firstly, analysis of variance (ANOVA) was finding significant of factors. Secondly, analysis of the mean (ANOM) was used for signal-to-noise (S/N) ratio to determine the optimal settings and factor levels.

2 MATERIAL, EQUIPMENT AND EXPERIMENT SETUP

This study was done by use material that was tool steel grade JIS SKD 61. The specimens were prepared by machining in dimension of 50 mm long x 10 mm thick and hardening by heat treatment for

60 HRC. The experiments were performed on CNC WEDM machine '*Mitsubishi model DWC90C*'. The cutting wire was 2.5 mm diameter of brass. The data, average material removal rate (MRR), were evaluated by measuring the cutting time and volume of cut for each section.

3 DESIGN OF EXPERIMENT AND ANALYSIS

Factors were consisting of power setting, off time off, wire feeding, wire tension and wire offset amount. Each factor was set at three levels as shown in Table 1. These factors were tested by a pilot study before running the actual experiment. Five factors and three levels applied experiment design as shown in Table 2. Taguchi's design was used for this experiment. The orthogonal array based on L27 was shown in Table 4. There are consisting of 27 trails.

The actual experiment was run, which run randomly selected. Then, the responses of each trial were measured width of the gap (g), width of specimen (w), thickness of specimen (t) and WEDM time (T) The material removal rate (MRR) was calculated by equation 1. Data were analyzed by statistical methods. Firstly, the finding significant of factors was analyzed by analysis of variance (ANOVA). Secondly, to determine the optimal settings and factor levels, the experimental results were further transformed into a signal-to-noise (S/N) ratio in term of '*larger the better*' that can be calculated by equation 2. There are three S/N ratios available depend on the type of characteristics. The characterization in this study, material removal rate (MRR), that higher value represents better machining performance is called '*larger is better*'. Furthermore, analysis of the mean (ANOM) was used on signal-to-noise (S/N) ratio to determine the optimal level of each factor.

$$MRR = \frac{gwt}{T} \tag{1}$$

$$S/N = -10\log\left[\frac{1}{n}\sum_{i=1}^{n} MRR^2\right] \tag{2}$$

Table 1. The experiment conditions.

Wire Size	2.5 mm.
Wire Material	Brass
Power setting,	4, 7, and 10 Amperes,
Off time off,	8, 12, and 16 millisecond
Wire feeding,	8, 10,and 12 mm per minute
Wire tension	9, 10 and 11 N.
Wire offset	150, 160 and 170 μm
Width of specimen	50 mm
Thickness of specimen	10 mm
Hard ness of specimen	60 HRC

Table 2. Factors and levels of parameter.

Control Factor	Level		
	I	II	III
A: Power setting (Amp)	4	7	10
B: Off time off (ms)	8	12	16
C: Wire feeding (mm/min)	8	10	12
D: Wire tension (N)	9	10	11
E: Wire offset (μm)	150	160	170

4 RESULTS AND DISCUSSION

The experimental results, were measured as shown in and MRRs were calculated by equation 1 as shown in Table 3. The example of surface finish is shown in Fig. 2 Signal to noise ratio (S/N ratio) in term of smaller the better was applied for the analysis of optimal parameters. The loss function (L) for objective of larger is better and S/N ratio can be calculated by equation 2.

4.1 *Effect of factors on Material Removal Rate (MRR)*

The main effect of WEDM parameters of material removal rate (MRR) was analyzed. The finding significant of factors was analyzed by analysis of variance (ANOVA). The ANOVA was shown that the main factors affect to response is only power setting. The effect of power setting on material removal rate (MRR) is shown in Fig 2. This experiment showed that increasing power at high level affected to increase value of the material removal rate (MRR). That means the quality of surface is decreasing while increasing power.

Table 3. Taguchi design (L27 design) and results including S/N ration.

L Trial	*A* Power setting *(Amp)*	*B* Off time off *(ms)*	*C* Wire feeding *(mm/min)*	*D* Wire tension *(N)*	*E* Wire offset (μm)	*g* : width of gap *(mm.)*	*T*: time of WEDM *(min)*	Material removal rate (MRR) *(mm²/min)*	S/N ratio (db)
1	4	8	8	9	150	0.342	118	4.673	3.2214
2	4	8	8	9	160	0.323	76	7.355	6.5472
3	4	8	8	9	170	0.322	110	4.977	3.3108
4	4	12	10	10	150	0.314	105	4.942	3.4928
5	4	12	10	10	160	0.320	105	5.196	3.6597
6	4	12	10	10	170	0.315	95	5.759	4.3917
7	4	16	12	11	150	0.331	115	4.635	3.1612
8	4	16	12	11	160	0.325	125	4.400	2.2789
9	4	16	12	11	170	0.323	125	4.140	2.2253
10	7	8	10	11	150	0.351	30	36.641	15.3431
11	7	8	10	11	160	0.345	35	30.710	13.8552
12	7	8	10	11	170	0.348	90	11.899	5.7246
13	7	12	12	9	150	0.377	41	26.173	13.2514
14	7	12	12	9	160	0.360	43	25.892	12.436
15	7	12	12	9	170	0.345	50	20.500	10.7564
16	7	16	8	10	150	0.345	60	16.833	9.1728
17	7	16	8	10	160	0.357	60	17.593	9.4697
18	7	16	8	10	170	0.351	60	18.170	9.3225
19	10	8	12	10	150	0.378	45	24.105	12.465
20	10	8	12	10	160	0.382	30	43.036	16.0787
21	10	8	12	10	170	0.374	23	55.174	18.2018
22	10	12	8	11	150	0.392	29	46.153	16.5976
23	10	12	8	11	160	0.374	27	50.644	16.8096
24	10	12	8	11	170	0.363	35	38.751	14.2967
25	10	16	10	9	150	0.397	32	43.929	15.852
26	10	16	10	9	160	0.368	32	40.203	15.1934
27	10	16	10	9	170	0.360	31	40.935	15.2775

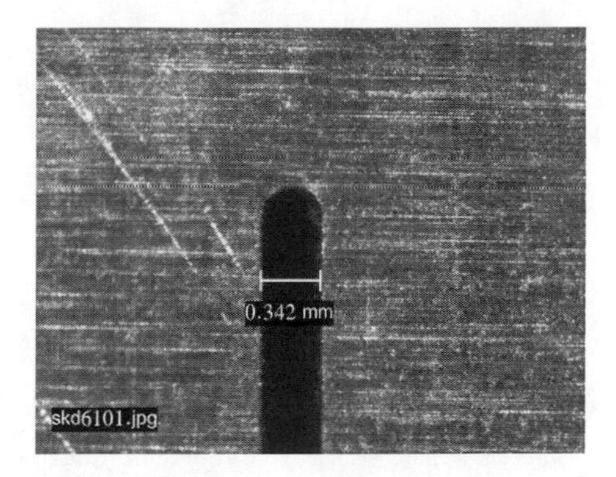

Figure 1. Specimen of trial 1:Width of gap.

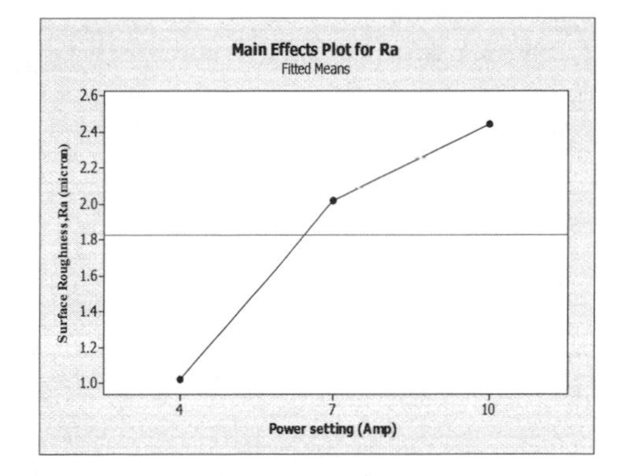

Figure 2. Effect of power setting on MRR.

4.2 *The optimal solution*

The determination of the optimal settings and factor levels was done by the experimental results that were transformed into a signal-to-noise (S/N) ratio. The S/N ratio is shown in Table 4 and overall mean for the S/N ratio was found to be 24.373 db. The mean of S/N ratio in each factor is shown in Table 4. Highest between the maximum and minimum of S/N ratio was factor A: Power setting but the others were closed. It was confirmed that factor A: Power setting was important factors affect to material removal rate (MRR). The optimal settings and parameter levels were determined by analysis of the mean (ANOM). The ANOM result of the S/N ratio response of factors and levels is shown in Table 5. The maximum S/N ratio in each factor was optimal level as shown in Fig. 3. The optimal setting level of each factor for best surface finish is AIII BII CII DI and EII as shown in Table 6.

Table 4. Analysis of mean (ANOM) S/N ratio response of factors and levels.

Control parameters	S/N-Ratio (dB)			Max-Min (dB)
	Level I	Level II	Level III	
A: Power setting	14.065	26.668	32.386	18.321
B: Off time off	24.706	25.044	23.370	1.674
C: Wire feeding	24.269	24.791	24.060	0.731
D: Wire tension	24.967	23.685	24.467	1.281
E: Offset	24.281	25.142	23.696	1.447

Table 5. Effect of parameters on levels shown in dB.

	Control parameters				
Level	A: Power setting	B: Off time off	C: Wire feeding	D: Wire tension	E: Offset
1	−10.308	0.333	−0.104	0.594	−0.092
2	2.295	0.671	0.418	−0.688	0.769
3	8.013	−1.003	−0.313	0.094	−0.677

Table 6. Optimal setting for best surface finish for WEDM of harden tool steel SKD 11.

	A:Power setting	B:Off time off	C:Wire feeding	D:Wire tension	E:Wire offset
Level	III	II	II	I	II
Factor setting	10 Amp	12 ms	10 mm/min	9 N	160 (μm)

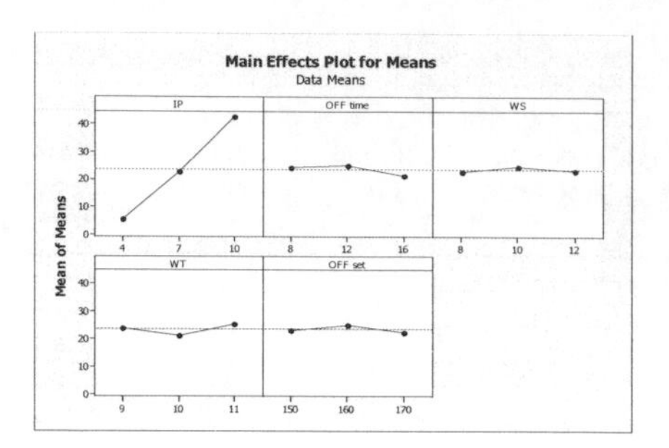

Figure 3. Effect of parameters for S/N ratios.

5 CONCLUSION

Since effects of power setting, off time-off, wire feeding, wire tension, and wire offset on material removal rate (MRR) was experimentally investigated. As the experimental results show, only power setting, has a significant effect on the material removal rate (MRR) of SKD 61 steel in WEDM process. An optimal parameter combination of the minimum material removal rate (MRR) was obtained by using the analysis of S/N ratios. The optimal solution based on ANOM of S/N ratio indicates that it is possible to increase the material removal rate (MRR). The experimental investigation can be confirmed for improving the material removal rate (MRR) and optimizing the machining parameters in WEDM operation for hardened tool steel SKD 61.

REFERENCES

[1] S. S. Mahapatra and A. Patnaik, Optimization of wire electrical discharge machining (WEDM) process parameters using Taguchi method, Int J Adv Manuf Technol Vol. 37 (2006) p.911–925.

[2] Y.S.Tarng, S.C. Ma, L.K. Chung, Determination of optimal cutting parameters in wire electrical discharge machining. Int J Mach Tools Manuf 35:129 (1995) , p.1693–1700.

[3] K.Kosit and K.Jirapattarasilp, The Experimental Investigation of WEDM for Surface Roughness of Hardened Tool Steel: SKD 61, Adv Mater Res, Vol.650 (2013), p.588–592.

[4] A. Mohammadi, A.F. Tehrani, E. Emanian,and D. Karimi, A new approach to surface roughness and roundness improvement in wire electrical discharge turning based on statistical analyse. Int J Adv Manuf Technol Vol. 39 (2008), p. 64–73.

[5] D.Scott, S.Boyina, and K.P. Rajurkar, Analysis and optimization of parameter combination in wire electrical discharge machining. Int J Prod Res 29:11 (1991), p.2189–2207.

[6] Y.K.Lok, and T.C.Lee, Processing of advanced ceramics using the wire-cut EDM process. J Mater Proces Technol 63:1–3(1997), p.839–843.

Computing, Control, Information and Education Engineering – Liu, Sung & Yao (eds)
© 2015 Taylor & Francis Group, London, ISBN: 978-1-138-02800-5

Simulation on microstructure of electrorheological fluid with ultra-fine abrasive particles

X.M. Liu, Y.W. Zhao & D.X. Geng
Engineering Training Center, Beihua University, Jilin Province, People's Republic of China

ABSTRACT: Electrorheological (ER) fluid-assisted polishing process is a precise polishing technology employing the ER effect to gather and stabilize the abrasive particles in the vicinity of the tool tip for the polishing of optical lenses and dies. The ER polishing fluid is composed of ER fluid and abrasive particles with high hardness and good dielectric property. The ER particle and the addition of ultra-fine abrasive particles into ER fluids are dielectrically polarized to form complex microstructure of columns when the electric field is applied. On the basis of dielectric polarization model, the equation of motion of particles is established according to the interacting force between particles in ER polishing fluid. The microstructures of ER polishing fluid under applied external electric field is simulated, and the influence on interaction force between the particles effect to the microstructures of ER polishing fluid is analyzed.

KEYWORDS: Electrorheological fluid; Microstructures; Electrical field strength.

1 INTRODUCTION

ER fluid-assisted polishing is a navel polishing process employing the ultra-fine abrasive particles mixed into ER fluid to complete the material removal by the ER effect. The shear yield stress and the viscosity of ER polishing fluid have undergone tremendous change as the ER particle and the abrasive particles will polarize to aggregate into complex microstructure of columns when the electric field is applied. W.B.Kim et al. [1] investigate the microstructure of ER polishing fluid under electric field and found that the motion of abrasive particles consistent with the ER particles. Kuriyagawa and W.B.Kim et al. [2-3] observe the polishing behavior of ER particles and the abrasive particles by CCD camera in ER fluid-assisted polishing process.

The transform of rheological properties of ER polishing fluid is one of the great concerns in the ER fluid-assisted polishing on surface roughness. Simulations had been conducted for ER fluid when the electric field is applied to reveal the motion of ER particles. This paper simulates the microstructure of ER polishing fluid with diamond particles by the deduced motion equation of polarized particles to analyze the influence on interaction force between the particles.

2 MOTION EQUATION OF POLARIZED PARTICLES MIXED IN ER POLISHING FLUID

Based on Newton's dynamics equation, the motion equation of polarized particles mixed in ER polishing fluid can be described as follows[4]:

$$m_p \frac{d^2 \mathrm{R}_i}{dt^2} = \mathbf{F}_i\left(\left\{\mathrm{R}_j\right\}\right) \tag{1}$$

where $\mathbf{F}_i\left(\left\{\mathrm{R}_j\right\}\right)$ is the sum of the forces exerted on the ith particle in ER polishing fluid, R_j is the distance between other particle and ith particle, mp is the mass of particle.

Ignoring the many-body interference among particles, the forces exerted on the *i*th particle in ER polishing fluid include the interacting force between polarized particles, a short-rang repulsive force among polarized particles, a short-rang repulsive force between particle and electrode, hydrodynamic force, brown force and so on.

2.1 *Interacting force between polarized particles*

The solid particles is made up of ER particle and abrasive particle in ER polishing fluid, so the interacting forces exerted on the ith particle include the action of all polarized particles(j_1 th ER particle and j_2 th abrasive particle) except the ith particle. The interacting forces exerted on the ith particle in ER polishing fluid is given by

$$\mathbf{F}_i^{el}\left(\{\mathbf{R}_j\}\right)=\sum_{i\neq j_1} F_{\mathrm{ep}ij_1}^{el}\left(R_{ij_1},\theta_{ij_1}\right)+\sum_{i\neq j_2} F_{apij_2}^{el}\left(R_{ij_2},\theta_{ij_2}\right) \quad (2)$$

where $F_{\mathrm{ep}ij_1}^{el}$ and $F_{\mathrm{ap}ij_2}^{el}$ are the interacting forces exerted on the ith particle by j_1 th ER particle and j_2 th abrasive particle, R_{ij} is the distance between polarized particles, θ_{ij} is the angle of the electric field direction and the joint line of the two dipoles.

The interacting forces exerted on the ith particle by j th particle using the following equation[1]:

$$\boldsymbol{F}_{ij}^{el}\left(R_{ij},\theta_{ij}\right)=F_0\left(\frac{d_{\mathrm{ep}}}{R_{ij}}\right)^4\left[\left(3\cos^2\theta_{ij}-1\right)\boldsymbol{e}_r+\sin 2\theta_{ij}\boldsymbol{e}_\theta\right] \quad (3)$$

where F_0 is given as

$$F_0=\frac{3\boldsymbol{p}_i\boldsymbol{p}_j}{4\pi\varepsilon_0\varepsilon_f d_{\mathrm{ep}}^4}$$

where $\boldsymbol{p}_i$ and $\boldsymbol{p}_j$ are the dipole moments of two interacting particles, respectively, ε_0 is the permittivity of free space, ε_f is the relative permittivity of dielectric fluid, R_{ij} is the distance between two interacting particles, d_{ep} is the diameter of ER particles.

2.2 *A short-rang repulsive force between particle and electrode*

The short-rang repulsive force on the ith particle by electrode is equal to the interacting force on the ith particle exerted by image of all particles on electrode[5],which can be calculated by

$$F_i^{el.wall}\left(\{\mathbf{R}_j\}\right)=\sum_j F_{ij}^{el}\left(R_{ij}',\theta_{ij}'\right)=\sum_{j_1} F_{\mathrm{ep}ij_1}^{el}\left(R_{ij_1}',\theta_{ij_1}'\right)+\sum_{j_2} F_{apij_2}^{el}\left(R_{ij_2}',\theta_{ij_2}'\right) \quad (4)$$

2.3 *A short-rang repulsive force among polarized particles*

Assuming that the solid particle is similar to hard sphere and can not overlap each other, as a short-rang repulsive force as follows[6]:

$$F_i^{rep}\left(\{\mathbf{R}_j\}\right)=-\sum_{i\neq j} F_0\exp\left[-\frac{(R_{ij}/d_{ep}-1)}{100}\right]\boldsymbol{e}_r \quad (5)$$

2.4 *Hydrodynamic force as the solid particle flowing in ER fluid*

The Hydrodynamic force can be described by the *Stockes* formula[5] as follows:

$$F_i^{hyd}=-3\pi\eta d\frac{d\mathbf{R}_i}{dt} \quad (6)$$

where η is the viscosity of ER polishing fluid.

2.5 *Dielectrophoretic force acting on particle*

The dielectrophoretic force acting on the polarized particle as electric field is applied may be found with the following equation[1].

$$F_{DEF}=\frac{\pi}{4}\varepsilon_0\varepsilon_f\beta_p d_p^3\nabla E^2 \quad (7)$$

where β_p is the particle dipole coefficient, d_p is diameter of particle.

Substituting Eqs. (2)- (7)into Eq. (1) yields, the motion equation of ith particles can be written as:

$$m_p\frac{d^2\mathbf{R}_i}{dt^2}=\sum_{i\neq j} F_{\mathrm{ep}ij}^{el}\left(R_{ij},\theta_{ij}\right)+\sum_j F_{\mathrm{ep}ij}^{el}\left(R_{ij}',\theta_{ij}'\right)+F_i^{hyd}+F_i^{rep}+F_i^{wall} \quad (8)$$

3 SIMULATION ON THE INTERACTING FORCES OF PARTICLES IN ER POLISHING FLUID

The interacting forces among polarized particle play a key role on the distribution of solid particles in ER polishing fluid under electric field from Eq.(8). When the pair ER-ER particle and between the ER-abrasive particle are nearly touching along the electric field ($R_{ij}=r_i+r_j$, $\theta_{ij}=0^0$). The ER fluid used here consists of silicone oil and starch particle mixed with diamond abrasives of a certain percentage. The interacting forces of particles in ER polishing fluid is calculated by Eq.(3) and simulation conditions as shown in Tab.1.

Table 1. Parameters on particles of ER polishing fluid.

Relative permittivity of Al_2O_3 particles	5.7
Relative permittivity of starch particles	20
Relative permittivity of silicone oil	2.7
permittivity of vacuum	8.854×10-12F/m
Diameter of diamond particles	10 μm
Diameter of starch particle	10 μm

The interacting forces of pair particles under the non-uniform electric field are depicted in Figure.1.The attraction force between ER particles is highest, secondary between ER particles and between abrasive particles, the electrophoretic forces on ER particles and abrasive particles is negligible. Figure.2 indicates that the pair interacting forces change with different diameter of abrasive. The interacting forces increases with the increase of the diameter of abrasive particle gradually. The attraction force between ER particles is greater than other pair particles consistently.

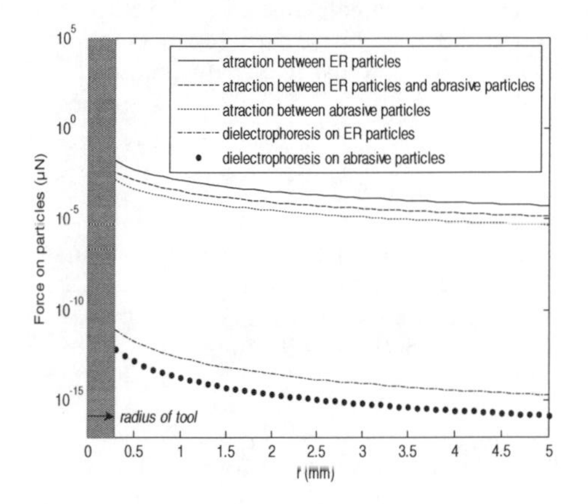

Figure 1. The interacting forces between polarized particle.

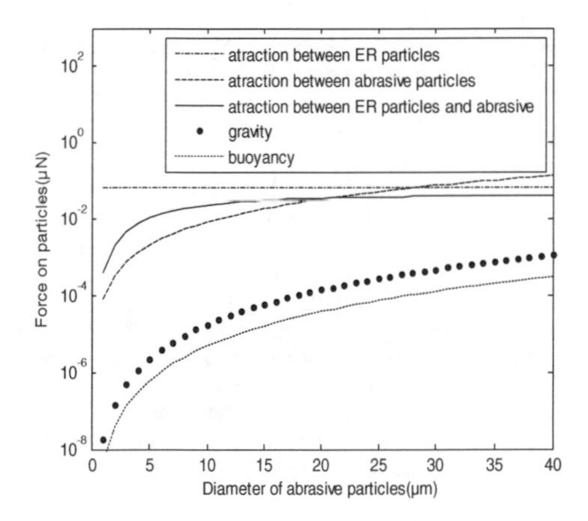

Figure 2. The interacting forces changing with different diameter of abrasive.

4 SIMULATION ON MICROSTRUCTURE OF ER POLISHING FLUID

The solid particles form the microstructure can be given by Eq.(8) when the electric field is applied. The simulation process of the polarized particles is as follow: Firstly, The number of N particles(starch particle and abrasive particle) is randomly placed in the area(L_x=L_y=10). Secondly, the motion equation of polarized particles is calculated by *Euler* formula with the introduction of periodic boundary conditions. The simulation step is $\triangle t^*$=0.001, and the simulation step is 5000, the dimensionless time units is 0.058s. The microstructure of ER polishing fluid with W10 starch particles(white) and W10 diamond particles (black) under the electric field strength (E_0=1KV/mm) is shown in Figure.3.

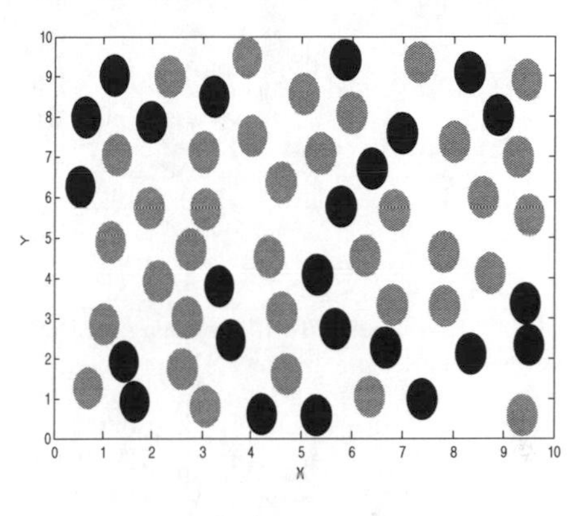

(a) The initial state

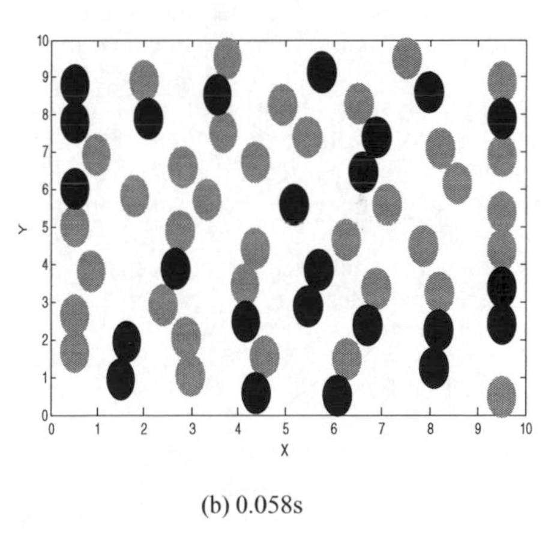

(b) 0.058s

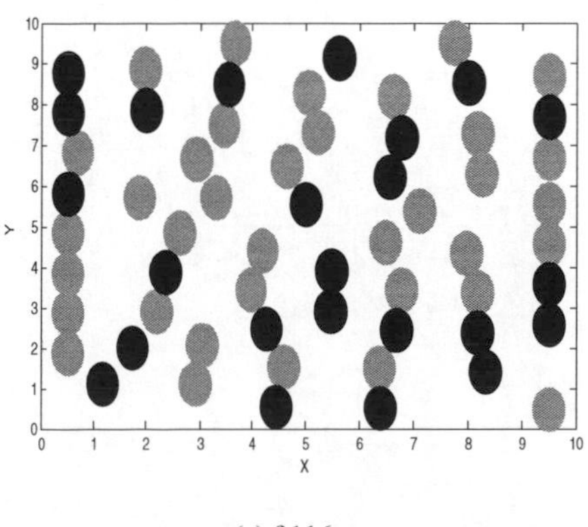

(c) f.116s

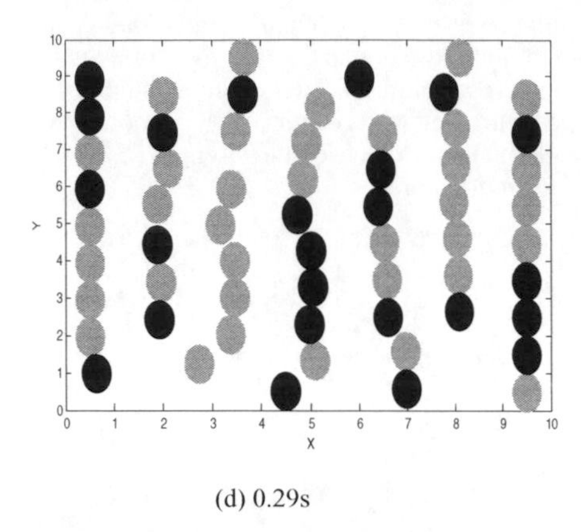

(d) 0.29s

Figure 3. Microstructure of ER polishing fluid mixed of W10 diamond.

The particles are randomly distributed without electric field as shown in Figure.3(a). As the attracting force is the largest among the near particles and particles with stronger dielectric properties, those particles firstly attract each other and gather to preliminary form the chains when the simulation time is 0.058s (see Figure.3 (b)). While the simulation time attains 0.116s, the particles had formed the complete single-chain on both sides as observed from Figure.3(c). As can be seen in Figure.3 (d), the ER polishing fluid had aggregated into the stable structure while the abrasive particles are combined into particle-chain to distribute between ER particles due to the greater attracting force between ER particles and abrasive particles.

5 CONCLUSION

The paper investigates the microstructure of ER polishing fluid mixed of W10 diamond under applied external electric field. A motion equation of polarized particles mixed in ER polishing fluid is derived on the base of *Newton*'s dynamics equation and the distribution of polarized particles is simulated by the interacting forces between particles. The relation of attracting force and the parameters on particles (dielectric properties, grain size) is revealed. According to the simulations, the solid particles of ER polishing fluid aggregated into the stable chain-structure increasing with simulate time when the electric field is applied and the abrasive particles are combined by near ER particle into particle-chain simultaneously.

ACKNOWLEDGEMENTS

The authors would like to acknowledge the financial support for this investigation from program for the Dr. Scientific Research Starting Foundation of Beihua University (199500004), PR China.

REFERENCES

[1] Kim, W. B. et.al. 2003. The electro- mechanical principle of electrorheological fluid-assisted Polishing, *International Journal of Machine Tools and Manufacture* 43:81–88.

[2] Kuriyagawa, T. & Saeki,M. K. Syoji.2002. Electrorheological fluid-assisted ultra-precision polishing for small three-dimensional parts, *Journal of the International Societies for Precision Engineering and Nano technology* 26:370–380.

[3] Kim, W. B. et.al.2004. Development of a padless ultraprecision polishing method using electrorheological fluid, *Journal of Materials Processing Technology* 155–156:1293–1299.

[4] Vicente, J. D. & Ramírez, J. 2007. Effect of friction between particles in the dynamic response of model magnetic structures, *Journal of Colloid and Interface Science* 316:867–876.

[5] Klingenberg,D.J. et.al. 1989. Dynamic simulation of electrorheological suspensions, *The Journal of chemical physics* 91: 7888– 7895.

[6] Tao,R. & Jiang,Q. 1994. Simulation of structure formation in an electrorheological fluid, *Phys. Rev. Lett* 73: 205–208.

Computing, Control, Information and Education Engineering – Liu, Sung & Yao (eds)
© 2015 Taylor & Francis Group, London, ISBN: 978-1-138-02800-5

Waste as a renewable energy source

Ru Liu, Chen Sheng Wu, Lu Ji Zhang & Qiong Wu
Beijing Municipal Institute of Science and Technology Information, Beijing, China

ABSTRACT: This article covers in detail programs and technologies for converting traditionally landfilled solid wastes into energy through waste-to-energy projects. At present, there are lots of electric power companies having already burnt waste material to make electricity all over the world. This article shows us three main ways in which municipal solid waste (MSW) is treated at present. And then we discuss the benefits of these ways. This project was funded by the Beijing financial fund.

KEYWORDS: waste energy, disposal, Municipal Solid Wastes (MSW).

1 INTRODUCTION

With rapid industrialization, the world has seen the growth in the quantum and diversity of waste materials generated by human activity and their potentially harmful effects on the general environment and public health, have led to an increasing awareness about an urgent need to adopt scientific methods for safe disposal of wastes. 'Waste-to-Energy' projects are the latest phenomenon to sweep through the global waste management industry. The technology is more than 60 years old, but it's enjoyed a high degree of popularity in recent years in places like Japan and Scandinavia, thanks to a shortage of landfill space.

At present, there are lots of electric power companies having already burnt waste material to make electricity all over the world. Waste to energy is the process of creating energy in the form of electricity or heat from the incineration of waste source. Energy from waste provides double environmental benefits: The diversion of waste coming from landfill; the immediate recovery of energy, displacing fossil fuel alternatives and reducing greenhouse gas emissions. The following diagram shows us the waste hierarchy from which we can see clearly the waste energy plays so important role in our daily life.

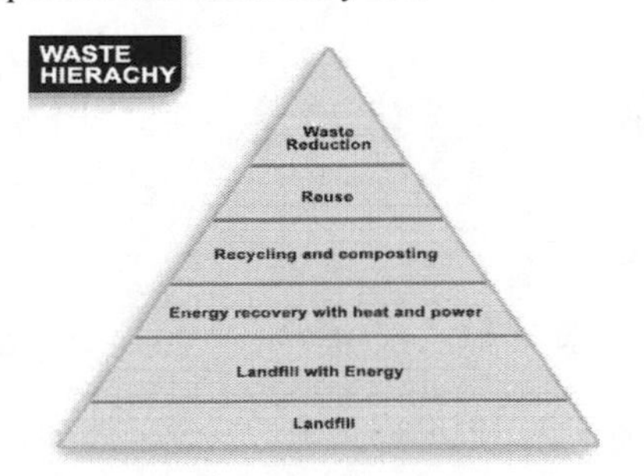

Figure 1. Waste hierarchy [1].

The resources of the waste energy:

- Municipal Solid Wastes (MSW) —"basically waste generated in households, schools, markets, gardens. (e.g., newspapers, books, magazines containers and packaging, food waste.) "
- Industrial/commercial wastes— "all types of wastes generated by stores, offices, restaurants, warehouses, industrial processes and manufacturing. "
- Other wastes— "sewage sludge, dredged material and so on" [2]

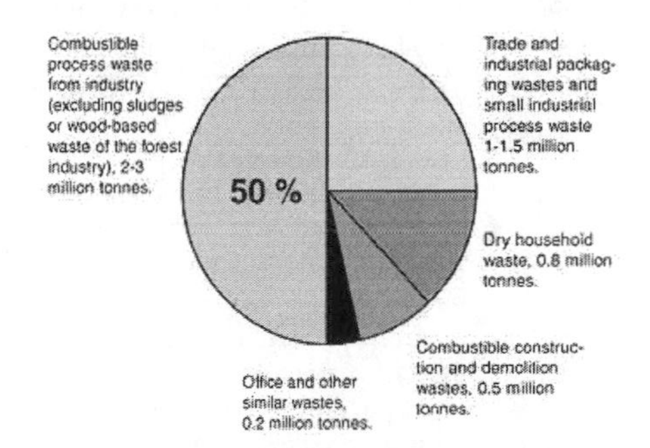

Figure 2. The composition of wastes in the UK [www.vtt.fi/pro/pro2/ pro27eng/indexe.htm].

2 HOW TO DISPOSE THE WASTE

As we know, both ordinary wastes and harmful wastes are all utilized as useful sources of energy. What we just need to separate different kinds of waste into the same kind. But the process of conversation is not an easy work. Just as above saying, there are different kinds of wastes, such as municipal solid waste, industrial and commercial wastes and so on.. And different wastes can make different effects, for example, the waste from industry can generate some toxic materials, which is harmful to peoples' health. So we need to classify different wastes, and then adopt different methods to dispose them.

Generally speaking, the wastes are separated into 1—6 containers, and then be disposed in specific treatment plants. Various systems differ from each other and also set different requirements for further treatment. The waste fractions are forwarded to the most suitable treatment on the basis of fractionation. The following flow diagram is for integrated solid waste management.

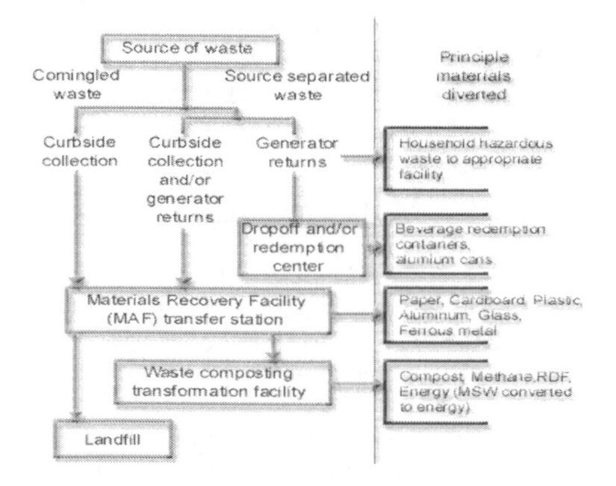

Figure 3. Integrated solid waste management (http://www.ccst.us/ccst/pubs/wm/images/waste.gif).

There are a number of other new and emerging technologies that are able to produce energy from waste. In practice, there are three main ways in which municipal solid waste (MSW) is treated at present:

2.1 *Disposal in landfills*

As is known to all of us, landfill is the most common means in disposing municipal solid waste, especially in developing countries. Both ordinary wastes and harmful wastes—municipal solid waste, most of household wastes and some harmful wastes from industries—are all utilized as organic materials. Such materials can be decomposed by specialized bacteria under suitable anaerobic digestion conditions during disposal of landfill, which can successfully generate useful landfill gas to benefit human. The condition of landfill is neither warm nor wet, so the process is much slower. Generally speaking, the time will last over the years but weeks. And the end product (the landfill gas) is a mixture which is consisted mainly of the gases of CH4 and CH2. In theory, the lifetime yield of a good site should lie in the range 150—300m3 of gas per tonne of wastes, with between 50% and 60% by volume of methane, which suggests that a total energy of 5-6 GJ per tonne of refuse, but in practice yield are much less.

2.2 *Combustion*

It is the process of waste to energy. The method is the same as coal for energy. We use the waste plant burning for making electricity as shown in Figure 6. At first, burning waste and releasing heat, the heat of burning turns water into steam, and then, the high-pressure steam turns the turbine generator to produce electricity, at last, a utility company sends the electricity to homes, schools and offices.

The end product is rich energy fuels. We can get the useful garbage from plants. Farmers can breed the special plants. Generally speaking, in 100 pounds of garbage, more than 80 pounds of garbage can be burned. That means the rate of combustion is more than 80%. Besides, a ton of rubbish can generate about 525 kilowatt of electricity, which can heat a typical office building for one day. After high temperature burning, the garbage retune to solid residue, ash. We can use the ash for construction and road building.

2.3 *Disposal in anaerobic digesters*

The organic fraction of domestic waste can base on the landfills. Big digesters will recover of biogas from landfill. There are four benefits in this way: a. It will be built over the landfill plants. So it is close to urban. The transport fee can reduce; b. The methane will release from the digesters. So we can use the combustion heat as electricity; c. We will get three gases from the gas clean up, one of them is medium heat value gas which can be used industry; d. The ash after burning can deal with in the way of landfill or can be used construction and road building.

The following diagram shows an integrated waste materials plant. This plant in Florida has facilities for recovery of metals and removal of plastics, followed by anaerobic digestion of the remainder. The solid residue from the digester serves as fuel for power production

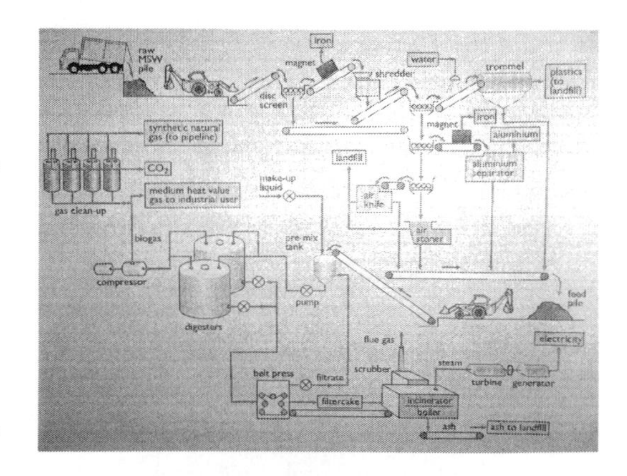

Figure 4. An integrated waste materials plant [4].

3 CONCLUSION

Fossil fuels are limited, it will be used up one day if people use it too much or much quickly. In fact, energy from wastes is a good substitute for fossil fuels. It can diminish the occupied area of the waste, which can beautify our environment to great extent, besides the substitute can also refrain the serious phenomenon of globe warming coming from burning fossil fuels.

Waste-to-energy technology offers two important benefits of environmentally safe waste management and disposal, as well as the generation of clean electric power. Waste-to-energy facilities produce clean, renewable energy through thermochemical, biochemical and physicochemical methods. The growing use of waste-to-energy as a method to dispose of solid and liquid wastes and generate power has greatly reduced environmental impacts of municipal solid waste management, including emissions of greenhouse gases.

REFERENCES

[1] East Riding of Yorkshire Council homepage http://www.eastriding.gov.uk/environment/sustainability/images/waste_hierachy.gif.

[2] Chapter 1, part 243(1995), guidelines for the storage and collection of residential, and institutional solid waste.

[3] http://www.ccst,us/ccst/pubs/wm/images/waste.gif.

[4] Godfrey Boyle, Renewable Energy-power for a sustainable future, Page121–163.

Computing, Control, Information and Education Engineering – Liu, Sung & Yao (eds)
© 2015 Taylor & Francis Group, London, ISBN: 978-1-138-02800-5

Partial discharge pattern recognition in switchgear based on statistical parameters of the support vector machine

Yu Zhou , Wei Guo Zhang, Ji Pan Li, Ke Xu & Xiang Xing Liu
Heze Power Supply Company of SGCC, Shandong, China

ABSTRACT: Online partial discharge pattern recognition can be applied to analyze the possible damages produced by partial discharge. This article based on four typical partial discharge test models and using the TEV sensor to measure the electric signal. Input the statistical parameters which extracted from the electrical signal to the support vector machine for the recognition. The experiment results indicate this method possesses a good recognition accuracy and is proved to be a new method for the partial discharge pattern recognition.

KEYWORDS: partial discharge; pattern recognition; statistical parameters; support vector machine; switchgear.

1 INTRODUCTION

Switchgear is an important equipment in the power system and is widely used in the power system. However, under some certain circumstance, the insulation defects may form the partial discharge (PD) and threaten the stability of the switchgear. PD measurement can effectively detect and remove the switchgear insulation deterioration, and also it has a positive effect on the prediction of the potential accidents .

In the process of the PD pattern recognition, feature extraction and pattern classification are two key aspects. Due to the clear physical meaning and better ability in spectrum distinguishment, the statistical parameter method is applied extensively in the feature extraction. Correspondingly, based on the empirical risk minimization support vector machine (SVM) applies the structural risk minimization, only requiring a small sample to get a better recognition rate and a higher robustness.

This article extracted the statistical parameters from the different types of PD signals from the test models, and finally enter these parameters into the SVM for the pattern recognition. The results indicate the SVM of the statistical parameters possesses a good recognition accuracy.

2 ESTABLISHMENT OF THE TEST PLATFORM AND THE TEST METHOD

According to the characteristics of PD in switchgear, four typical test model were designed: the needle-plate PD model, internal PD model, surface PD model and suspended PD model. The structures of the models are shown in Figure 1 and the test platform is shown in Figure 2.

In Figure 1(a) is for the needle-plate PD model, with a 100mm-in-diameter plate and a 0.2mm-in-diameter needle electrode pinpoint which is 5mm away from the plane. Also, a 1mm thick plexiglass board is placed between the needle and the plate; (b) is for the internal PD model, with two 4mm thick plexiglass board cohesive together, each board crave a 10mm-in-diameter, 4mm-in-thick hole to make a air hole in the sample; (c) is for the surface PD model, to make an obvious discharge, a 20mm-in-diameter 50mm-in-long nail is inserted into the plexiglass block; (d) is for the suspended PD model, the distance between the two plate electrodes is 10mm, a 80mm-in-diameter 5mm-in-thick plexiglass board is placed on the ground electrode and a 10mm high copper is placed on the plexiglass board. All the plate electrodes of the models are 50mm-in-diameter 10mm-in-thick and made of copper. To avoid of the interference from the sharp corners of the electrodes, the electrodes are polished smoothly.

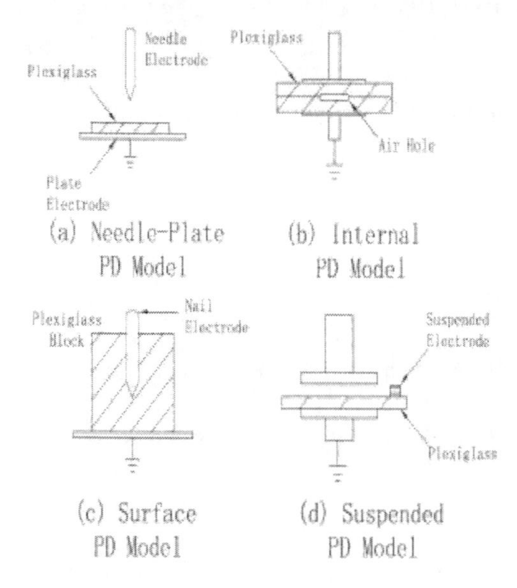

Figure 1. PD test model.

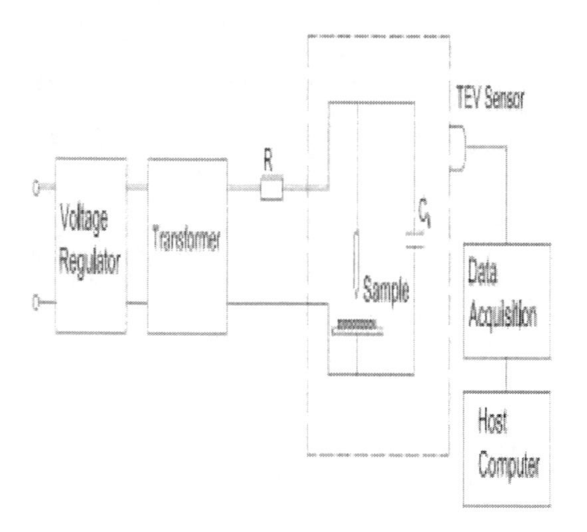

Figure 2. PD test platform.

In Figure 2 R is the protect resistance, C_k is the coupling capacitor, the TEV sensor is the self-made sensor, the signal is collected by the HS5 data acquisition card which is made by Tiepie, each channel sampling rate of the card is 1Mhz, sampling precision is 16bit, the test power is provided by the transformer. The signal collected by the data acquisition card is shown in Figure 3.

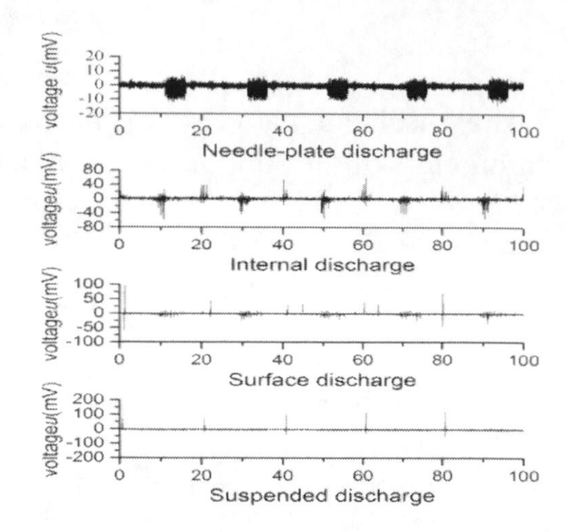

Figure 3. PD signal (From top to bottom: needle-plate discharge, internal discharge, surface discharge, suspended discharge).

3 EXTRACTION OF THE STATISTICAL PARAMETERS

To implement the PD pattern recognition requires to extract the parameters. Due to the randomness of the discharge, this article extracts the statistical parameters which based on the phase-resolved partial discharge (PRPD) pattern. According to the PRPD, the φ can be used as the random variable and the q for the probability density, and then can get the fingerprint characteristics as skewness, steepness, correlation coefficient etc.

3.1 *Definition of the statistical parameters*

Skewness S_k is described the comparison of skewness between the probability distribution of the random variables and normal distribution, which is defined as Eq.1.

$$S_k = \frac{\sum_{i=1}^{W} p_i(\varphi_i - \mu)^3}{\sigma^3} \tag{1}$$

Wherein Eq.1, W represents the number of half-cycle phase windows, p_i represents the probability of x_i in the window i. When $S_k = 0$, the plot is symmetry, $S_k > 0$ means the heavy tail on the right side, the distribution is skewed to the right, $S_k < 0$ means the heavy tail on the left side, the distribution is skewed to the left.

Steepness K_u is described the comparison of steepness between the probability distribution of the random variables and normal distribution, which is defined as Eq. 2.

$$K_u = \frac{\sum_{i=1}^{W} p_i(\varphi_i - \mu)^3}{\sigma^4} - 3 \tag{2}$$

Each volume shares the same definition in Eq.1.

3.2 *Statistical parameters distribution in the test model*

30 samples were got from the test, each sample contains 50 frequency cycles of the PD signal. 20 samples were taken for training the classifier, 10 samples for the testing. Considering φ-q plot and φ-n plot, which are gotten from the PRPD plot, shares some same characteristics, this article only extracted the parameters from the φ-n plot. The results are shown in Figure 4.

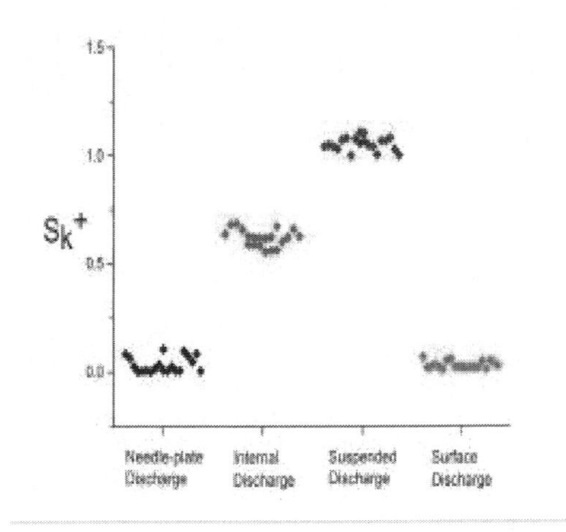

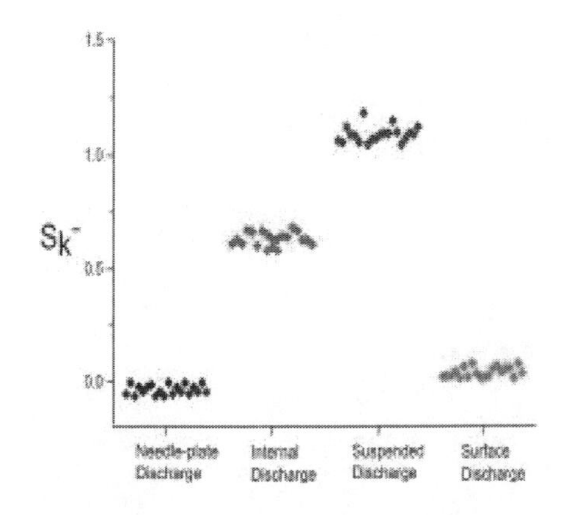

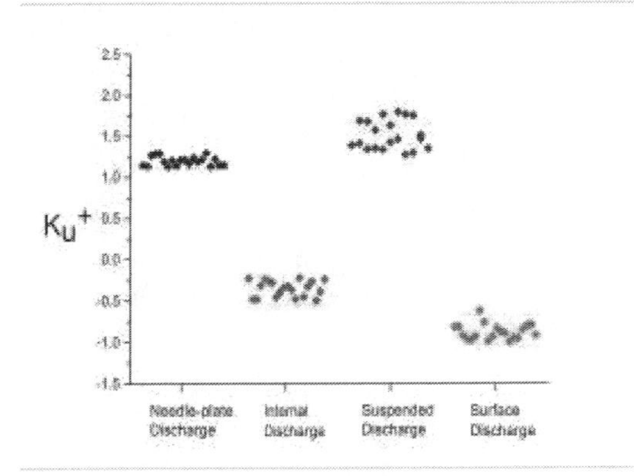

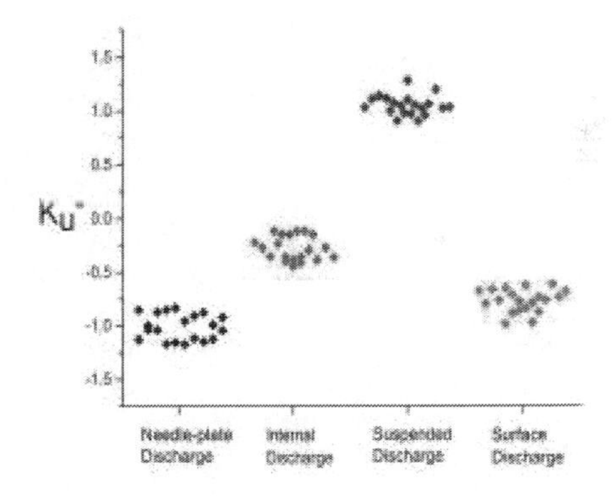

Figure 4. Statistical parameter distribution.
(Each figure from left to right: needle-plate discharge, internal discharge, surface discharge, suspended discharge)

4 PATTERN RECOGNITION BASED ON SVM

4.1 *SVM theory*

Given training samples $T = \{(x_1, y_1), \cdots (x_n, y_n)\} \in (X \times Y)^n$, $x_i \in X = R^n$ for the feature parameters, $y_i \in \{1, -1\}$ for the corresponding expectation. The binary classification problem can be solved by establishing a hyperplane $\mathbf{w}^{\mathrm{T}}\mathbf{x} + b = 0$ where $\mathbf{w}$ for the adjustable weight vector and b for the bias . To meet the hyperplane determined parameters, the Lagrangian function which with the chosen kernel function can be established as Eq. 3

$$Q(x) = \sum_{i=1}^{n} a_i - \frac{1}{2}\sum_{i=1}^{n}\sum_{j=1}^{n} y_i y_j a_i a_j K(x_i, x_j) \tag{3}$$

The goal is to maximize the Eq.3 which satisfy the constraint $\sum_{i=1}^{n} y_i a_i = 0, 0 \le a_i \le C, i = 1, \cdots, n.$ Accordingly the bias can be got as

$b^* = y_i - \sum_{i=1} y_i a_i^* K(x_i - x_j)$, and finally the decision function can be constructed as Eq.4

$$f(x) = sign\left[\sum_{i=1}^{n} y_i a_i^* K(x, x_i) + b^*\right] \quad (4)$$

The widely used kernel function has the three categories: polynomial kernel, RBF kernel and sigmoid kernel.

5 RESULTS AND ANALYSIS OF THE PATTERN RECOGNITION

Based on Matlab, the SVM classify applies RBF kernel and one-against-one algorithm, namely designs an SVM between any two types of samples, thus $k(k-1)/2$ SVMs required for k types. At first each set of data requires internal normalization, i.e., the maximum data of each set as the reference for the normalization, with this method, the recognition rate can be significantly increased. Applied the method above to pattern recognition of the samples, comparing with the BP neural network, the recognition results are shown in Table 1.

Table 1. Recognition results of BP neural network and SVM.

Discharge type	Recognition rate/%	
	BP neural network	SVM
Needle-plate discharge	80	90
Internal discharge	70	80
Surface discharge	70	90
Suspended discharge	80	80
Total	75	85

From Table 1, it can be found the SVM recognition rate is higher than BP neural network. Because of the adoption of the kernel function, SVM can map the feature parameters into a high dimensional space, and make the classification easier than before. And also, the SVM possesses a better generalization ability than a BP neural network, especially for small sample.

6 SUMMARY

a. The statistical parameters in different types of discharge have obvious divergence, these parameters can reflect the PD pattern and be served as the valid input for pattern recognition;
b. Before pattern recognition, the data required preprocessing, such as the normalization, data preprocessing can effectively improve the accuracy of pattern recognition;
c. Compared with the BP neural network, SVM has a better recognition rate, especially in the classification of small samples.

REFERENCES

P. Gill 1998, in Electrical Power Equipment Maintenance and Testing, edited by Marcel Dekker Inc.

J. Tang, J. Y. Lin, Z. Ran & J. G. Tao 2013. Partial Discharge Type Recognition Based on Support Vector Data Description, *High Voltage Engineering Vol. 39, 1046–1053.*

IEC Standard 60207-2003 High-voltage Test Techniques-Partial Discharge Measurements, Geneva, CH, 2003.

L. Niemeyer1995: IEEE Trans Delivery Vol.2 519–528.

B. Fruth and D. Gross1994: ICPDAM, 578–588.

Y. Qian, C.J. Huang, C. Chen, F.N. Huang and X. C. Jiang 2007. Analysis of the Partial Discharge in Generator Based on Sk-Ku Plot *High Voltage Apparatus Vol. 43 , 176–182.*

X.W. Ren, L. Xue, Y. Song, D.D. Guo and Z. Shen 2011. The Pattern Recognition of Partial Disharge Based on Fractal Characteristics Using LS-SVM. *Power System Protection and Control Vol.39 143–147.*

G. H. Feng 2011. Parameter optimizing for Support Vector Machine classification. *Computer Engineering and Applications Vol.47, 123–128.*

Computing, Control, Information and Education Engineering – Liu, Sung & Yao (eds)
 ISBN: 978-1-138-02800-5

Research on evolution model of networked software

Yi Yao, Yu Hong & Hui Li
PLA University of Science and Technology, Nanjing, China

ABSTRACT: Continual evolution is a prominent characteristic of networked software. In this paper, the concept of evolution of networked software and its basic structural model is introduced, a taxonomy model for evolution of networked software which includes the categories and type and fashion of networked software evolution is present.

KEYWORDS: Networked Software; Evolution Model; Software Service.

1 INTRODUCTION

Today's society is increasingly dependent on the networked software providing continuous services [1]. This kind of software systems is deployed in an open network environment, various elements of interaction and cooperation. Compared with traditional software system, the safety, availability and reliability confidence problem becomes more and more difficult to guarantee, especially to provide 7*24 continuous service characteristics, so that it is more and more emphasized in the operation of the dynamic adjustment of structure and function, to ensure that in response to a variety of changes with time still can continue to provide services [2].

The evolution of networked software is a derived concept granularity in software maintenance, evolution, adaptive and other related concepts on the basis of more detailed, more focused, and its purpose is without the services provided by the software interrupt system, enhance the credibility of the software system, improve the ability to adapt to the demand of the software system. The relevant technology mainly solves two core questions: 1) how making the networked software system is the ability to provide service with the runtime can be changed; 2) how to use this ability to realize the evolution of networked software, so as to enhance their ability to adapt to the requirement.

2 BASIC CONCEPTS

Software changes may occur at every stage of software at compile time or run time, loading and other life cycle. In recent years, research on software adaptive, dynamic adaptation and self-organization, direction gives a more profound connotation for networked software evolution. Adaptability refers to software self-evaluate their own behavior, when the evaluation results show that the software can not reach the established goals or the need to increase the function, performance optimization, change their behavior. Dynamic adaptation is a kind of software adaptation processing follows the events at run time: user needs change, system intrusion or failure, running environment and resources change etc.. Self-organization [3] is a dynamic system to adapt to the process, the process in the system without external control of the situation through the interaction of its internal elements to maintain structure. The above concept enriches the intension of networked software evolution from the evolution of morphology, ability.

Evolution of networked software is to improve the networked software system to adapt to the requirement of ability, with no interruption of the services provided under the premise of, or with the help of external dynamic automatic guide software change activities [4].

Here, non-interrupt referring to the users, the software system provides the services continued online; with the help of external dynamic guidance emphasizes the whole evolution process as an indispensable factor involved in; change activities include software function and structure of the change of topology, for example the software architecture and its elements (component, connection the modified activities etc.). Definitions do not explicitly define the non-functional properties evolution, because the general will evolve the trigger function, structure etc.

The evolution of networked software defined above for evolution concept is a kind of operation, but its face is the large scale distributed systems in the network environment. More emphasis on the feature of the whole system level "continuous service online", and try not to demand all system components must continue running. In addition, and software maintenance is often by specialized maintenance personnel to implement different evolution action, not only can be driven by the third party, can also take the initiative to implement by the software system, such as according to some predefined rules.

3 STRUCTURE MODEL

The model is given to participate in Networked software evolution activities of entities and their interactive relationship, the core problem to answer is: how to make the software system has the ability to run time can change and evolve by whom and how to drive the process. This section of the existing research and engineering practice were summarized, and based on the given platform and structure model of engine three based on online evolution, put forward on the basis of the ideal structure model of software services online evolution. These models can be divided into the evolution and the evolution of the two parts to enable decision making.

a. Specific evolution model. The early evolution of technology will be directly related to the evolution of the monitoring module and the execution module is hard coded into the application system, the monitoring module can output, internal state information system application, execution module can implement the online adjustment of the application system through the parameter configuration form. Evolutionary decision is usually done by management and maintenance personnel. This model is shown in Fig. 1, the representative work focused primarily on small scale, single objective and non-system level support for software evolution research.
b. Evolution model based on platform. With the wide application of middleware based software, development and running the platform software has been significantly enhanced to support software evolution. This makes the evolution that can separate mechanism from the application, becomes the public supporting facilities are relatively independent. The separation and improve the software reusability, enhanced the ability of online software system evolution. The model shown in Fig. 2, the representative works mainly focus on the online evolution of enabling platform.
c. Evolution model based on engine. Relative to the platform based on the model, the main characteristic of this kind of model is the increase of evolutionary decision module, trying to achieve evolution engine general software, instead of people to the real-time and automatic make evolutionary decision. This will greatly enhance the autonomy and effectiveness of software evolution, but in practice, due to the evolution engine capacity expectations too idealistic, but to restrict its scope of application. The model shown in Fig. 3, the representative works mainly focus on the Online Evolution in the research of the trusted system.
d. Ideal structure model. The ideal structural model of networked software evolution is (b) model and (c) organic combination model, namely evolutionary decision can advance the rules formulated by the software system to make, also can be in when necessary by the people to drive the evolution of decision making, as shown in Fig. 4. The model integrated (b) and (c) the advantages of the model, answered the software system itself should assume liability for online evolution and the extent of support issues. In view of the fact that a large number of application system deployed in not supporting the software evolution platform, so how to make use of the ideal model to solve a major challenge to evolution of legacy systems is currently facing the transition period.

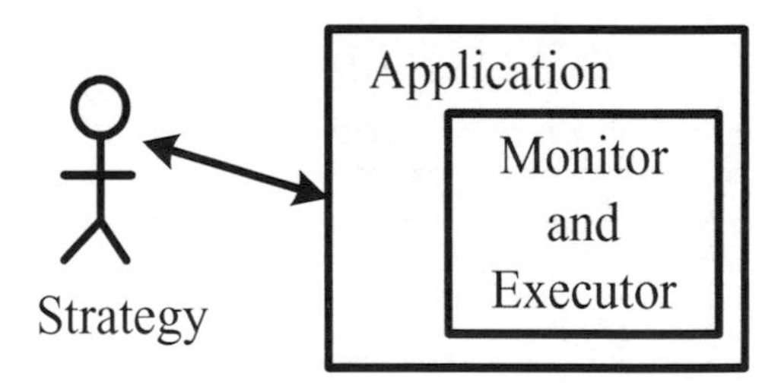

Figure 1. Specific evolution model.

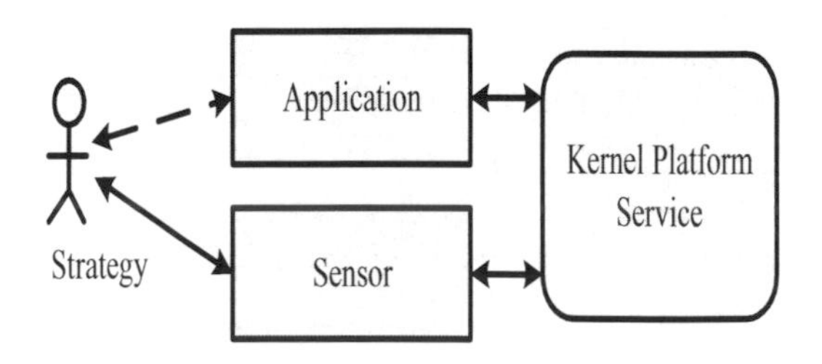

Figure 2. Evolution model based on platform.

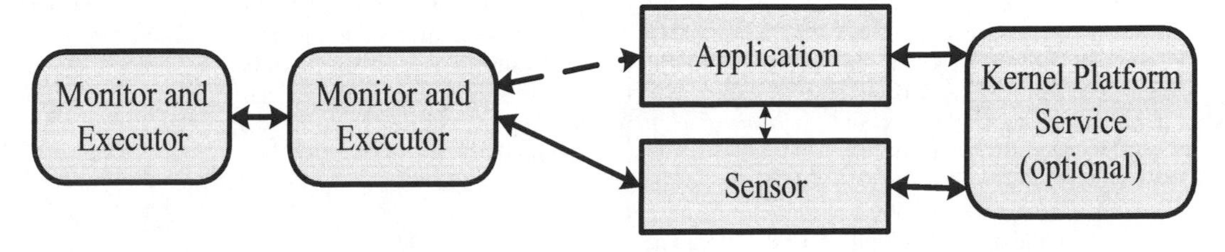

Figure 3. Evolution model based on engine.

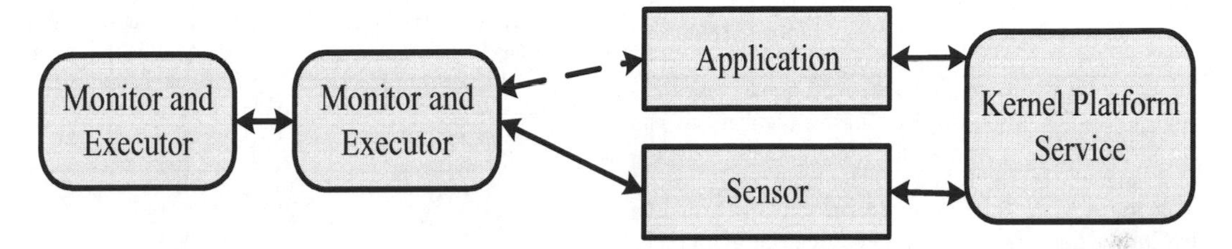

Figure 4. Ideal structure model.

4 CLASSIFICATION EVOLUTION MODEL OF NETWORKED SOFTWARE

Networked software evolution from the initial classification is the classification of software maintenance related reference, the evolution of typical software related classification model arc Swanson, IEEE, Chapin et al., Wang et al. The above classification respectively from the perspective of subjective purpose, objective evidence and reason of evolution for the software evolution, change, adaptation and maintenance carried out classification. Although the classification method is a research hotspot in the field of software evolution, but in view of the current software evolution has no unified definition, evolution and classification method is also difficult to have a uniform standard. Comprehensive studies and in this paper to the definition of software services online evolution, classification model this section presents three aspects from the evolution of a category, type of evolution and evolution mode on the online evolution activity classification [5].

Categories of Evolution. From the perspective of software architecture, the evolution of particle size can be divided into components, connection and configuration evolution. Component evolution refers to the parameter configuration, add, delete, replace the operation of business modules connected; evolution refers to asking the interactive relation adjustment to the business module; configuration evolution involves multiple component and evolution of connection, may be accompanied by the architectural constraints and other element evolution. With local memory, distributed systems are asked to communicate through the new entity known as the node, accordingly, its evolution levels can be divided into single nodes and inter node evolution: single node refers to the evolution of active only in single interior knots; cross node evolution is related to multiple nodes. For example, OSGI mainly focus on the evolution of the internal activity of individual nodes, Rainbow can realize cross node evolution activities through the establishment of a strict hierarchical control mode. Function mainly used to distinguish evolution will change the software system of foreign services provided: if only the code document format, modify, or code modification is only for reconstruction purposes, evolution is known as the function of independent, otherwise known as the related functional evolution.

Types of Evolution. The external characteristics of the evolution of the type from the initiative, predictability and the degree of automation of 3 dimensions characterize online evolution activity.

The evolution of the initiative can be divided into reactive and proactive type two. Reaction type evolution refers to the evolution of active made in external pressure stimulation, such as the necessity of the implementation of the software user demand change after modification; proactive type evolution is in the software system may face risk prediction based on the implementation of the prospective evolutionary activity. Can be expected of evolution is divided into two kinds: the expected type refers to the software development phases can be expected to change activities, in fact, is relatively simple, for example, can through the application of the system in the reserve expansion point and other forms of support; non expected type is online development stages cannot be expected to change activities, generally need to support in the development tools, platform and language, the component model and so on. The degree of automation, automation of the process of evolution is not a required circulation closed people participation, the preset by the software according to the development phase of the evolution strategy to make decision; ginseng and interactive evolution is a need for people to participate in the open cycle, either in whole or in part by humans to make decisions.

Fashions of Evolution. From the general evolution mode, system structure and evolution strategy three dimensions characterize online evolution activities of internal implementation mechanism.

According to whether with the specific architecture style closely coupled evolution mechanism can be divided into two categories: general method is general; specific methods can only be used for a specific system, to solve specific problems. For example, the evolution mechanism in Rainbow can be applied to different styles of architecture, with the general; Arch Studio is mainly aimed at software architectures of C2 and REST, it can be divided into specific method. To the system structure, the method based on architecture refers to the maintenance of explicit architecture model type software running in the process, the implementation of activities under the guidance of evolution in the model; non method based on architecture does not emphasize the significant architecture model type maintenance, such as no dynamic AOP model guided (Aspect-Oriented Programming), JavaHotswap mechanism, etc. Evolution strategies aimed at the system itself has trusted evolution decision ability and how to make decision concerns the automation of the process of evolution, can be divided into two categories based on rules and learning. The former general prediction of the various problems the system may encounter in the development phase, and setting rules; the latter can expand its decision-making capability at run time by the machine learning method.

ACKNOWLEDGEMENT

This work is supported by the Natural Science Foundation of Jiangsu Province, China (Grant No. BK2012060, BK2012059) and the Prep-research Foundation of PLA General Armament Department (No. 9140A05040213JB25068). Resources of the PLA Software Test and Evaluation Centre for Military Training are used in this research.

REFERENCES

[1] H.M. Wang, G. Yin. The trustworthiness-oriented software evolution in the network era. Communications of the CCF, 2010, 6(2): 28–34.

[2] M. Salehie, L. Tahvildari. Self-adaptive software: Landscape and research challenges. ACM Transactions on Autonomous and Adaptive Systems, 2009, 4(2):1–42.

[3] T. Mens, J. Buckley and A. Rashid. Towards a taxonomy ofsoftware evolution. Proceedings of the Workshop on Unanticipated Software Evolution. Warsaw, 2003:50–59.

[4] Q. X. Wang, J. R. Shen and X. P. Wang. A component-based approach to online software evolution. Journal of Software Maintenance and Evolution; Research and Practice, 2006, 18(3):181–205.

[5] H.M. Wang, P.C. Shi and B. Ding. Online Evolution of Software Services, 2003, 34(2):318–328.

Computing, Control, Information and Education Engineering – Liu, Sung & Yao (eds)
© 2015 Taylor & Francis Group, London, ISBN: 978-1-138-02800-5

Experimental study on combustion performance of typical upholstered furniture's fabric

Y. Chu, F. Qu, L.Q. Yang, & Y.L. Yan
College of Safety Engineering, Shenyang Aerospace University, Shenyang, China

ABSTRACT: The cone calorimeter is employed to analyze the combustion performance of three kinds of upholstered furniture's fabrics under different working situations, which are cotton, chemical fiber and terylene. Moreover, the pyrolysis characteristics of seven kinds of upholstered furniture's fabrics, such as cotton fiber blended fabric, pleuche, artificial leather, sheepskin, as well as the above-mentioned materials are also studied by DTA instrument respectively. The experimental results obviously show that peak heat release rate and the risk evaluation coefficient increase with the rise of heat flux. On the contrary, ignition time becomes shorter correspondingly. As for the pyrolysis characteristics, each of the upholstered furniture's fabrics used in contrast experiments has its own outstanding character in different aspects, including mass loss quantity, mass loss rate, exothermic peak, main paralyzing temperature and so on.

KEYWORDS: Upholstered furniture's fabric; Combustion performance; Pyrolysis characteristics; Cone calorimeter; Differential Thermal Analysis (DTA).

1 INTRODUCTION

With the development of social economy, the living standard of people is also on the increase gradually. Due to its beauty and comfort, the upholstered furniture is put into use in modern architecture more and more, such as private house, office building, shopping mall, large venue and entertainment place. However, because most of upholstered furniture is made of flammable material, while not only affording convenience to people, but also bring great risk of fire, especially the covering fabric. Such as cotton fiber blended fabric and leather, typical of combustible materials, even with the fire sources in a non-contact state, are largely vulnerable to thermal radiation to be ignited. Subsequently, the indoor fire will undoubtedly cause property losses, even staff casualties.

As mentioned above, the potential fire hazard and disastrous consequence caused by upholstered furniture's fabric can not be ignored. So, it is high time that we would carry out further study on its combustion performance and pyrolysis characteristics to conduct reasonable fire safety design and reduce fire hazards.

2 OUTLINE OF EXPERIMENTS

The paper firstly selects several common upholstered furniture fabrics to study their combustion properties and relationships under different heat flux through the cone calorimeter and a flue gas analyzer, then the pyrolysis characteristics is explored by DTA instrument separately. Finally, the comparison of fire risk degree is completed on the basis of a comprehensive analysis with data.

3 EXPERIMENTAL RESULTS ANALYSIS AND DISCUSSIONS

3.1 *Influence of heat flux to Peak Heat Release Rate (PHRR), ignition time and risk evaluation factor*

The PHRR, ignition time and risk evaluation factor of cotton and chemical fiber as well as terylene at different heat flux are shown in Table 1.

The risk evaluation factor can be calculated with Equation 1 below:

$$R = \frac{PHRR}{t} \tag{1}$$

Where
R= risk evaluation factor
PHRR = peak heat release rate (kw/m^2)
t= ignition time (s)

Table 1. PHRR, ignition time and risk assessment factor under different heat flux.

Material	Heat flux (kw/m^2)	Peak heat release rate (kw/m^2)	Ignition time (s)	Risk evaluation factor
Cotton	25	184	47	3.914893617
	35	236	38	6.210526316
	45	244	32	7.625
Chemical fiber	25	Not ignited	Not ignited	-
	35	244	98	2.489795918
	45	282	75	3.76
Terylene	25	Not ignited	Not ignited	-
	35	55.9	345	0.162028986
	45	78.7	150	0.524666667

According to Table 1, for the same kind of material, when the heat flux increases, the peak heat release rate also rises steadily after ignition, however, the ignition time is gradually shortened, causing the risk evaluation factor to become bigger, which means that in real fire scenario, the higher energy ignition source has, the greater the risk of fire accident is. Meanwhile, the comparative combustible experiments with different materials under the same heat flux shows that cotton is most dangerous, on the other hand, terylene is much better than the others. To some extent, we may conclude that the selection of raw material with excellent combustion characteristics is helpful to achieve inherent safety and make a significant contribution for fire prevention during the production of upholstered furniture.

3.2 *Influence of heat flux to 180s and 300s average heat release rate*

Table 2 lists the 180s and 300s average heat release rate of cotton, chemical fiber and terylene under three types of heat flux.

As can be seen from Table 2, the impact of heat flux on the 180s and 300s average heat release rate and its impact on the peak heat release rate is exactly the same. At the same time, the 180s average heat release rate of three materials is higher than the 300s average heat release rate, that is, three kinds of materials within180s all have the heat release rate on the rise, nevertheless, within the 300s the heat release rate is lower. Among the three kinds of materials, the terylene can be validated to have better fireproof performance than cotton and chemical fiber.

Table 2. The 180s and 300s average heat release rate under different heat flux.

Material	Heat flux (kw/m^2)	180s average heat release rate (kw/m^2)	300s average heat release rate (kw/m^2)
Cotton	25	39.7	26.1
	35	48.1	24.2
	45	45.5	29.1
Chemical fiber	25	Not ignited	Not ignited
	35	32.5	23.5
	45	34.2	21.9
Terylene	25	Not ignited	Not ignited
	35	1.5	1.3
	45	13.9	8.6

3.3 *Results of DTA*

The pyrolysis characteristics can be obtained through differential thermal analysis, so the ignition temperature and other important parameters of each material can also be estimated; these parameters may be used to provide a reference basis for fire simulation experiments. Figure 1 through 4 separately illustrates the DTA curve of four kinds of materials which are cotton, chemical fiber, terylene and pleuche. Table 3 displays the total pyrolysis parameters of all experimental materials respectively.

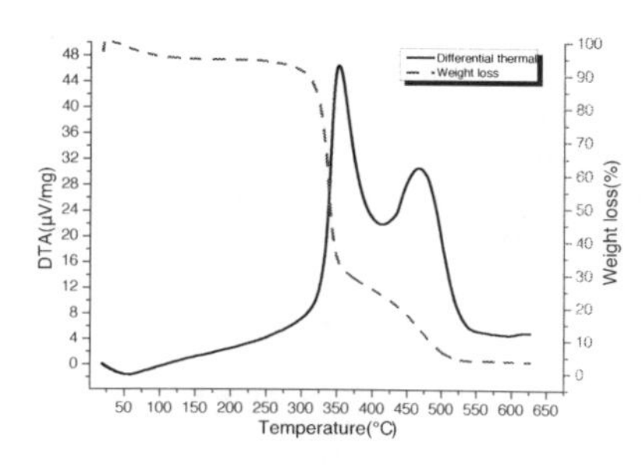

Figure 1. DTA curve of cotton.

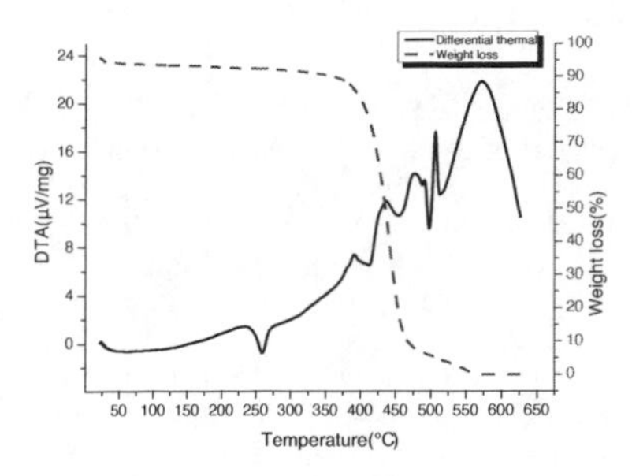

Figure 2. DTA curve of chemical fiber.

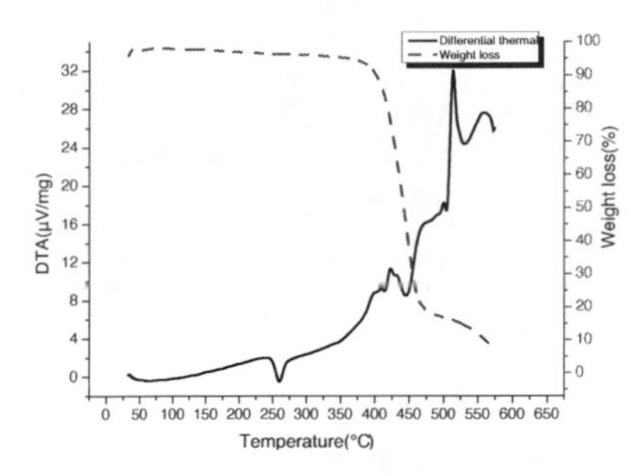

Figure 3. DTA curve of terylene.

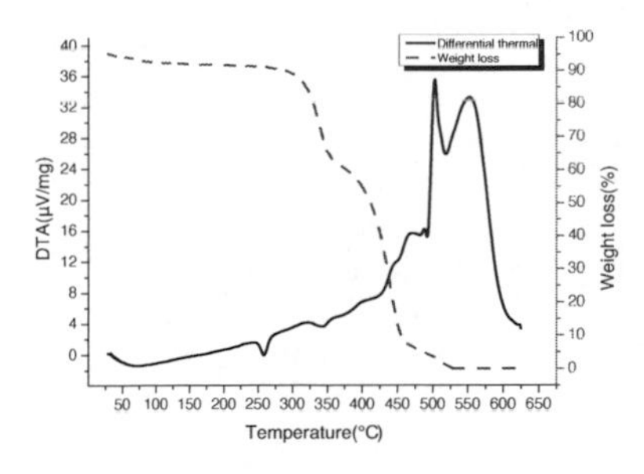

Figure 4. DTA curve of pleuche.

Combining figures with the data in Table 3, we can discover that each material has its own characteristics, and highlighting the risk in a fire accident. As far as the weight loss is concerned, cotton, cotton fiber blended fabric, pleuche, sheepskin and terylene all have two distinct degradation stages, while chemical fiber and artificial leather almost entirely pyrolyses at one-time. Obviously, the total weight loss of sheepskin is the least, whereas, other samples are around 98%.

When the sheepskin began to pyrolyse, its initial temperature is only about 220°C, the lowest of all experimental materials, however terylene's initial paralyzing temperature is 400°C, the most difficult

Table 3. The pyrolysis parameters of all experimental materials.

Sample	Degradation stage	Temperature range(°C)	Weight loss (%)	The total weight loss (%)	Char residue (%)
Cotton	The first	300-360	85.69	99.28	0.72
	The second	360-530	13.59		
Chemical fiber	The first	370-560	99.95	99.95	0.05
Terylene	The first	400-470	77.5	96.25	3.75
	The second	470-580	18.75		
Cotton fiber blended fabric	The first	270-350	30.36	98.98	1.02
	The second	350-580	68.62		
Pleuche	The first	300-360	38.29	99.96	0.04
	The second	360-530	61.67		
Artificial leather	The first	260-550	98.02	98.02	1.08
Sheepskin	The first	55-150	12.04	93.27	6.73
	The second	220-500	81.23		

to be ignited in fire. The curve slope of weight loss represents the weight loss rate of material; we can see that the weight loss rate of the chemical fiber and terylene are much more. The cotton's weight loss rate is very quick in the first stage, but then becomes gentler in the second stage. The including factors all have a key impact on fire, but we need comprehensive consideration when employs fire risk evaluation in order to draw more accurate conclusions.

4 CONCLUSIONS

Through the above analysis and discussions, the following conclusions can be presumably drawn:

1 The peak heat release rate increases with the growth of heat flux, while the ignition time is on the contrary, this also means that the material will be easier to be ignited with high heat flux and conduct more intense combustion reaction; hence, there would be more severe. Thus, when choosing raw materials of upholstered furniture, as for the ignition time and the peak heat release rate, satisfying

the long ignition time and low peak heat release rate at the same time is relatively the best choice.

2 The pyrolysis characteristics of experimental materials are different at mass loss quantity, mass loss rate, exothermic peak, main paralyzing temperature and so on. For example, pleuche's weight loss is the maximum, while sheepskin is the minimum; the main paralyzing temperature of terylene is the highest but the sheepskin's is the lowest.
3 The selection of upholstered furniture's fabrics with good flame retardant, low combustion heat release and high paralyzing temperature is essential to indoor fire prevention and rescue; In fact, it's the aim of inherent safety.

REFERENCES

[1] W.B. Zhu: The Study on Effect of ventilation on Fire Behavior of TyPical UPholstered Furniture. (Doctorate dissertation, in Chinese)

[2] Zh.Ma, Zh.J.Shu,G. X,Y. Jia: *FIRE SAFETY SCIENCE*, Vol.14 (2005), p.132–136. (in Chinese)

[3] ISO: Heat release rate (cone calorimeter method) 5660–1,2002

[4] Ai.M.Li, L.Wang, R.D.Li,L.J.Sun: *China Safety Science Journal*, Vol.14 (2004), p.101–106. (in Chinese)

[5] J.G.Lu, S.L. Liu, X.Q. Peng: *Fire Science and Technology*, Vol.24 (2005), p.414–418. (in Chinese)

[6] J. Zh.Shu, X.N. Xu, Sh.Sh.Yang,Y.Wang: *Chinese Polymer Bulletin*, Vol.5 (2006), p.37–44. (in Chinese)

Computing, Control, Information and Education Engineering – Liu, Sung & Yao (eds)
 ISBN: 978-1-138-02800-5

Remaining charge estimation of $LiFePO_4$ power battery using an Unscented Kalman Filter

Zhaowei Yu, Weimin Li & Taimoor Zahid
Shenzhen Institute of Advanced Technology, Chinese Academy of Sciences, Shenzhen, China
Shenzhen College of Advanced Technology, University of Chinese Academy of Sciences, Shenzhen, China
Jining Institute of Advanced Technology, Chinese Academy of Sciences, Shandong, China

ABSTRACT: $LiFePO_4$ is widely used as the anode materials of power battery for EV but it is difficult to estimate the State of Charge (SOC) of the battery because of material characteristics. In this paper, Unscented Kalman Filter (UKF) method with Thevenin model is used to estimate SOC and Extended Kalman Filter (EKF) is used to compare the results. UKF proves to be an optimal solution for SOC estimation.

KEYWORDS: Electric Vehicle (EV); Lithium-ion battery; State of charge (SOC); Unscented Kalman filter (UKF); Extended Kalman filter (EKF).

1 INTRODUCTION

With government's increasing concerns about the environment, the development of EV technology will rise rapidly and soon electric vehicle will gain popularity among the consumers(Hametner & Jakubek, 2013). However, being a new and under developed technology there still exist a few key problems like short driving range and battery's capacity and cost. Because of its high stability and energy density characteristics $LiFePO_4$ is widely used as the anode materials of power battery used for EV. Battery safety, energy efficiency, memory effect, cycle life and self-discharge rate characteristics gives $LiFePO_4$ battery an advantage over other batteries like NiMH, $LiCoO_2$ and $LiMn_2O_4$(Lu, Han, Li, Hua, & Ouyang, 2013). However, $LiFePO_4$ material structure itself causes small diffusion coefficient and poor electronic conductivity.

To manage a battery package carefully, battery management system (BMS) is used in order to increase the lifetime and improve the safety(Marcicki, Canova, Conlisk, & Rizzoni, 2013). BMS should give accurate information to drivers about the remaining charge that can greatly improve a cars' driving range. BMS monitor key operating parameters of battery package, such as package temperature, current, voltage of package and each cell to make sure a safe operating condition of the package. One of the important functions of a BMS is to estimate the SOC of the battery pack. SOC is defined as the ratio of the current charge $Q(t)$ to the remaining charge Q_{max}:

$$SOC(t) = \frac{Q(t)}{Q_{\max}}. \tag{1}$$

SOC of the electric vehicle's power battery can be directly estimated by calculating the integral of current, though easily realized, but the precision affected by current measurement and cumulative error. Fuzzy logic, neural network or other learning algorithms(Lee, Kuo, & Wang, 2004) shows good estimation results but they need a large amount of data for training and not easily realized in embedded systems. Recently a lot of research has been done on nonlinear Kalman filter that includes EKF and UKF. In this paper, we use UKF to estimate SOC of $LiFePO_4$ battery using OCV functions and compare of two kinds of nonlinear Kalman filter method: EKF and UKF.

2 THEORY OF UKF-BASED STATE OF CHARGE

2.1 *Equivalent battery model*

In the equivalent circuit of Thevenin Model Figure 1, R represents the internal resistance while Rd and Cd are polarization resistance and capacitance. This one step simple *RC* circuit needs less computation to obtain higher accuracy.

The discrete time state equations of the battery model are:

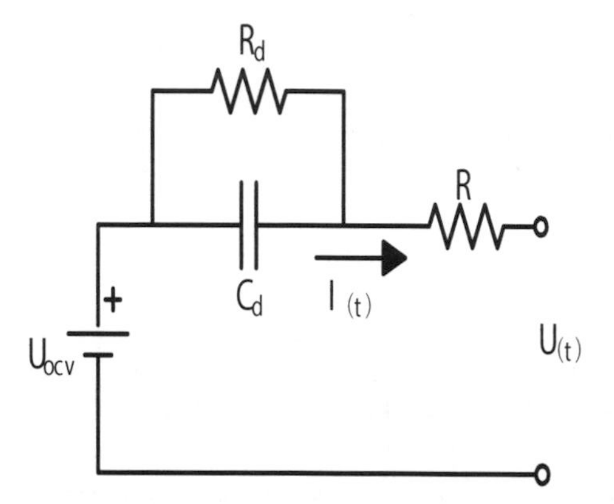

Figure 1. Thenenin equivalent circuit model.

$$\begin{bmatrix} Uc_d(k+1) \\ SOC(k+1) \end{bmatrix} = \begin{bmatrix} -e^{(-T/\tau)} & 0 \\ 0 & 1 \end{bmatrix} \begin{bmatrix} Uc_d(k) \\ SOC(k) \end{bmatrix} + \begin{bmatrix} R_d(1-e^{(-T/\tau)}) \\ -\alpha\Delta t / Q_{\max} \end{bmatrix} i(k) + \begin{bmatrix} r_{uc} \\ r_{soc} \end{bmatrix} \quad (2)$$

$$U_{OCV} = F(SOC(k)) + Uc_d(k) - i(k)R + v \quad (3)$$

Δt is the sampling time and $\tau = R_d C_d$ is the time constant of . The current i and the terminal voltage U(t) can be measured. As SOC is the function of $Uocv$ as a polynomial equation of $Uocv$, so we can use UKF to estimate SOC through U(t) as the measure value. i as input, SOC and voltage of C_d i.e. U_d as the state value. r_{uc}, and r_{soc} are model noise and v is measurement noise respectively.

2.2 *SOC estimation method (Unscented Kalman Filter)*

A nonlinear discrete time function can be described as eq 4. For high order nonlinear systems, UKF approximates the probability density of the nonlinear function rather than the nonlinear function itself. For this purpose UKF uses unscented transform described in 3 sections below:

a. Sigma points select:

The distribution of $y = f(x)$ can be approximately revealed by a set of sigma points s_j (Plett, 2006). State vector has dimension L = 5, mean X and covariance P then sigma points with their weight factors are generated using the following equations; $W_0^{(m)}$ is the weight of s_i to calculate the mean and $W_0^{(c)}$ is the weight to calculate the covariance.(He, Gao, Xu, & Liu, 2009)

$$\overline{X_0} = E(X_0) \quad (4)$$

$$P_0 = E[(X_0 - \overline{X_0})(X_0 - \overline{X_0})^T] \quad (5)$$

$$s = \overline{X} \quad (6)$$

$$W_0^{(m)} = \lambda / (L + \lambda) \quad (7)$$

$$W_0^{(c)} = \lambda / (N + \lambda) + (1 - \alpha^2 + \beta) \quad (8)$$

where β = 2, λ is a composite coefficient.

$$s_j = \overline{X} + (\sqrt{(N+\lambda)P^X})_j ; j = 1,2,3,4,5 \quad (9)$$

$$s_j = \overline{X} - (\sqrt{(N+\lambda)P^X})_{j-5} ; j = 6,7,8,9,10 \quad (10)$$

$$W_j^{(m)} = W_j^{(c)} = 1 / \{2(N+\lambda)\} \quad (11)$$

Where $\lambda = \alpha^2(N + j) - N$.

b. Time update and output state update:

Update the sigma points to get *2N+1* sigma points. Then calculate the mean and covariance of the state vector and the estimated output respectively using below equations:

$$\overline{X}(k/k-1) = \sum_{j=0}^{m} W_j^{(m)} s_j(k/k-1) \quad (12)$$

$$P^X(k) = \sum_{j=0}^{m} W_j^{(c)} [s_j(k/k-1) - \overline{X}(k/k-1)][s_j(k/k-1) - \overline{X}(k/k-1)]^T \quad (13)$$

$$Y(k/k-1) = F(X(k/k-1), X_{j-1}^k) \quad (14)$$

$$\overline{Y} = \sum_{j=0}^{m} W_j^{(m)} Y_j(k/k-1) \quad (15)$$

$$P^Y(k) = \sum_{j=0}^{m} W_j^{(c)} [Y_j(k/k-1) - \overline{Y}(k/k-1)][Y_j(k/k-1) - \overline{Y}(k/k-1)]^T \quad (16)$$

c. Kalman gain and measurement update:

In order to find the Kalman gain we need to calculate the output covariance matrix below:

$$P^{XY}(k) = \sum_{i=0}^{m} W_j^{(c)} [s_j(k/k-1) - \overline{X}(k/k-1)][Y_j(k/k-1) - Y(k/k-1)]^T \quad (17)$$

$$K(k) = P^{XY}(k)P^{Y}(k)^{-1} \quad (18)$$

In the end, we update the state vector:

$$\overline{X}(k) = \overline{X}(k/k-1) + K(k)[Y(k) - \overline{Y}(k/k-1)] \quad (19)$$

$$P^{X}(k) = P^{\overline{X}} - K(k)P^{\overline{Y}}(k)K(k)^{T} \quad (20)$$

3 EXPERIMENTAL RESULTS AND ANALYSIS

In this paper LiFePO4 battery is used for SOC estimation. The results using UKF and EKF method are discussed here using two OCV functions at 25oC shown in equation (21) and equation (22) Photographs and figures:

$$E_{soc} = d_3 soc^3 + d_2 soc^2 + d_1 soc + d_0 \quad (21)$$

$$E_{soc} = d_4 soc^4 + d_3 soc^3 + d_2 soc^2 + d_1 soc + d_0 \quad (22)$$

4 SUMMARY

In this paper, Thevenin model based SOC estimation methods, i.e. UKF and EKF, were used to estimate the SOC of Lithium ion battery. Two different OCV functions were used to help UKF and EKF estimate the SOC. EKF method was used to compare the overall efficiency and accuracy of UKF. As a result, UKF showed outstanding results and due to its better accuracy and estimation properties showed better performance with the help of a proper OCV function. Future research will focus on a finding a better battery model and OCV functions in order to reduce the overall estimation error.

Table 1 presents the overall estimation error between the two methods, i.e. UKF and EKF. It can be seen that UKF shows better results than EKF and is more robust and accurate for estimating SOC of $LiFePO_4$ battery.

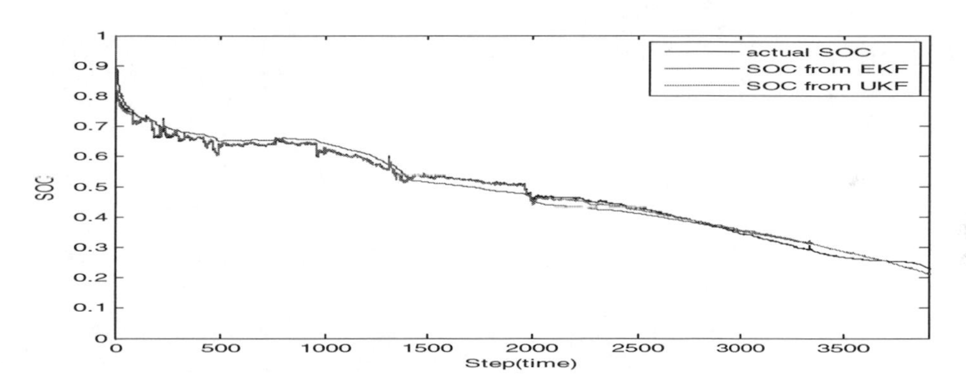

Figure 2. SOC estimation plot using UKF and EKF with Eq.21.

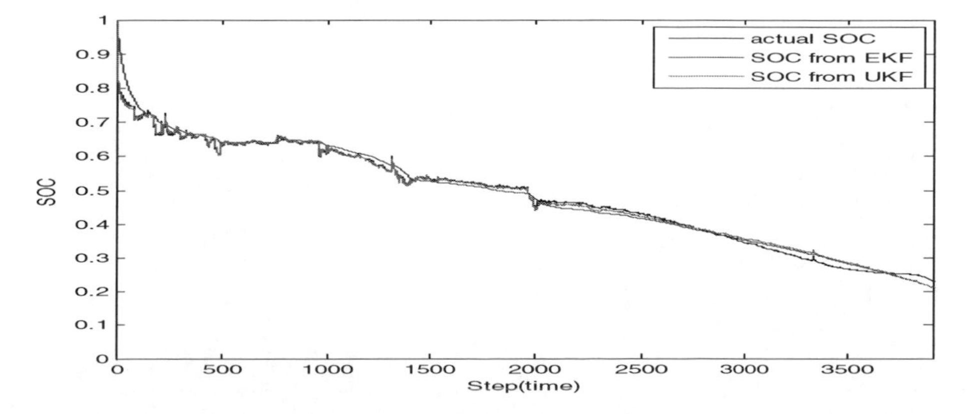

Figure 3. SOC estimation plot using UKF and EKF with Eq.22.

Table 1. Error of SOC estimate use UKF and EKF.

	Unscented Kalman Filter	Extended Kalman Filter
Eq.21	2.24	4.19
Eq.22	2.3	3.23

REFERENCES

Hametner, C., & Jakubek, S. (2013). State of charge estimation for Lithium Ion cells: Design of experiments, nonlinear identification and fuzzy observer design. Journal of Power Sources, 238, 413–421. doi: DOI 10.1016/j.jpowsour.2013.04.040.

He, Z. W., Gao, M. Y., Xu, J., & Liu, Y. Y. (2009). Battery Model Parameters Estimation with the Sigma Point Kalman Filter. 2009 International Conference on Artificial Intelligence and Computational Intelligence, Vol Iii, Proceedings, 303–306. doi: Doi 10.1109/Aici.2009.15.

Lee, Y.-S., Kuo, T.-Y., & Wang, W.-Y. (2004). Fuzzy neural network genetic approach to design the SOC estimator for battery powered electric scooter 2004 IEEE 35th Annual Power Electronics Specialists Conference, PESC04, June 20, 2004 - June 25, 2004 (Vol. 4, pp. 2759–2765). Aachen, Germany: Institute of Electrical and Electronics Engineers Inc.

Lu, L., Han, X., Li, J., Hua, J., & Ouyang, M. (2013). A review on the key issues for lithium-ion battery management in electric vehicles. Journal of Power Sources, 226, 272–288.

Marcicki, J., Canova, M., Conlisk, A. T., & Rizzoni, G. (2013). Design and parametrization analysis of a reduced-order electrochemical model of graphite/LiFePO4 cells for SOC/SOH estimation. Journal of Power Sources, 237, 310–324. doi: DOI 10.1016/j.jpowsour.2012.12.120.

Plett, G. L. (2006). Sigma-point Kalman filtering for battery management systems of LiPB-based HEV battery packs. Part 1: Introduction and state estimation. Journal of Power Sources, 161(2), 1356–1368. doi: 10.1016/j.jpowsour.2006.06.003.

Computing, Control, Information and Education Engineering – Liu, Sung & Yao (eds)
 ISBN: 978-1-138-02800-5

The design of home appliances remote control system based on PSTN and MCU

R. Hao, G.W. Gao & G. He
School of automation BISTU, Beijing, China

ABSTRACT: This paper describes an intelligent home appliance control system which makes it possible to implement remote control of any appliances (air conditioning, water heaters, rice cookers, lights, stereo, DVD recorder) at home through mobile phone or telephone at any time, any places. This paper presents an implementation of remote control system based on telephone network and AT89C51, studies the overall structure, and designs the hardware and software of the remote control system. The system can realize automatic switch intelligent control of objects through telephone or mobile phone based on a voice interactive operation platform. The experiment shows that it is capable of real-time, fast, reliable and stable transmission of data.

KEYWORDS: Remote control system; Dual Tone Multi Frequency; PSTN.

1 INTRODUCTION

Public telephone communication resources nearly cover thousands of households. The connection between home telephone and the outside world is bi-directional and dynamic. Compared with the conventional control method, telephone remote control as a new project shows certain superiority such as no requirements of special wiring and wireless frequency resources which avoid the electromagnetic pollution. With the development of economy and science, people demand a higher-quality life and they hope they can control household electrical appliances at all times and places. In this way, householders can remotely open air conditioner ten minutes early before they get home in hot summer. Besides, they can open the microwave oven and electric rice cooker and so on in advance so that when they get home they can enjoy the delicious food directly. Based on this idea, we design this control system named "home appliances remote phone control system".

2 THE OVERALL DESIGN OF THE TELEPHONE REMOTE CONTROL SYSTEM

The PSTN mean Public Switched Telephone Network.This system uses AT89C51 single chip microcomputer as control core, mainly by the ringing detection circuit, off-hook and on-hook control circuit, DTMF decoding circuit, audio playback circuit, data decoding circuit, data transceiver circuit and driver circuit, etc. If we want to remotely control home appliances we only need to make a phone call, and then the phone will ring. If no one answers, after the phone ringing 4 times it will automatically pick up and put through the telephone and then prompt by voice for a password. We can type the password on the fixed telephone or mobile phone, and the password that the dual tone multiple frequency signals correspond is transmitted to the master control unit AT89C51 after DTMF decoded. If the password is correct, voice will prompt to choose a control channel and then choose the home appliances. Finally the main controller sends out control instructions to drive the appliance achieve on/off action. Therefore the system achieves the goal of the remote control. If the password is wrong, the controller will prompt to enter a password again. When the number is still wrong and reach the required times, the phone will automatically hang up. The overall structure is shown in Figure 1.

3 HARDWARE DESIGN OF TELEPHONE REMOTE CONTROL SYSTEM

3.1 *Ring detection module and simulation off-hook module*

This design uses the resistance to buck, then input to the photoelectric coupler. Waveform signals which are insulated and transformed by photoelectric coupler can directly output to single chip microcomputer interrupt counter input port to complete the whole process of ring detection and counting.

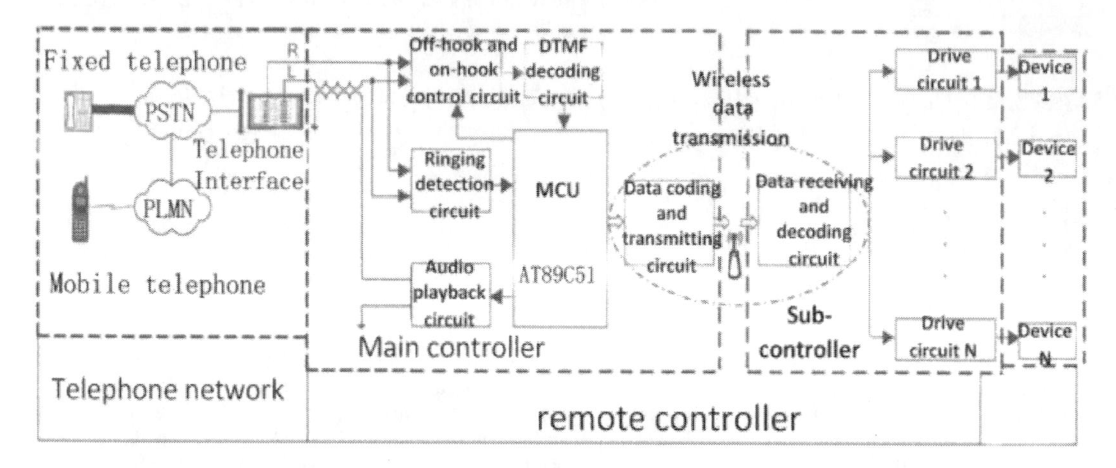

Figure 1. The overall structure figure of remote control system.

The automatic off-hook circuit is done by a relay switch. After MCU sent off-hook signals, the relay make a resistance of 200 ohms put in both ends of the telephone line to make a current of about 30 mA circuit through the whole telephone line. Off-hook and hang-up signal instructions are achieved by the TXD/P3.1 port of the single-chip microcomputer that be changed to a high level. The P3.4 pin of the MCU is used for counting, and when the count value reaches the given value, the P3.1 pin of MCU output high level. Subsequently triode 9012 is connected and relay J21 is acted. Finally, load resistance R21 (200 Ω) is connected in circuit to achieve simulation off-Hook. The design of ring detection and simulation off-hook circuit connection is shown in Figure 2.

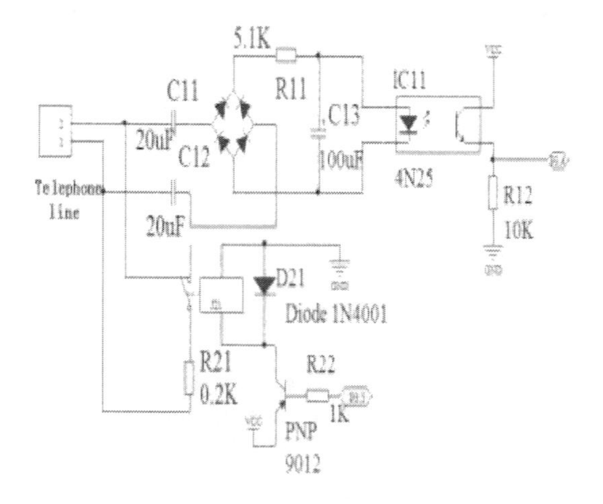

Figure 2. Ring detection circuit and simulation off-hook circuit.

3.2 *The DTMF decoding module*

DTMF is composed of a low frequency group (fb) and high frequency group (fa). Each digit signal is compounded of a low frequency signal or a high frequency signal. We can use the equation (1) to express. According to CCITT recommendations, the definition of DTMF decoding is shown in table 1.

$$f(t) = A\_\{a\}\sin 2f\_\{a\}t + A\_\{b\}\sin 2f\_\{b\}t \quad (1)$$

In order to obtain valid data, the STD pin of MT8870 is connected to the P2.4 pin of AT89C51. When the STD pin changes from low to high, P2 port bus receives effective dial-up the key code after AT89C51 detection. The data output of the MT8870 Q4 ~ Q1 are connected to the P2.0 ~ P2.3 of AT89C51, and then the P2 port of the single-chip microcomputer identify these four codes. In particular, for the "0" number, the 8421 yards that MT8870 output are "1010" but not "0000". In addition, for the "*", "#" word number, the 8421 yards

Table 1. Telephone dialing digits corresponding to the high and low frequency group.

Numeric keyboard		High frequency group/Hz			
		1209	1336	1477	1633
Low frequency group/Hz	697	1	2	3	A
	770	4	5	6	B
	852	7	8	9	C
	941	*	0	#	D

that MT8870 output correspondingly are "1011" and "1100". And invalid double audio signals (phone line noise and people's voice signal, etc.) will not cause the STD pin of MT8870 change. The design of MT8870 circuit connection is shown in Figure 3.

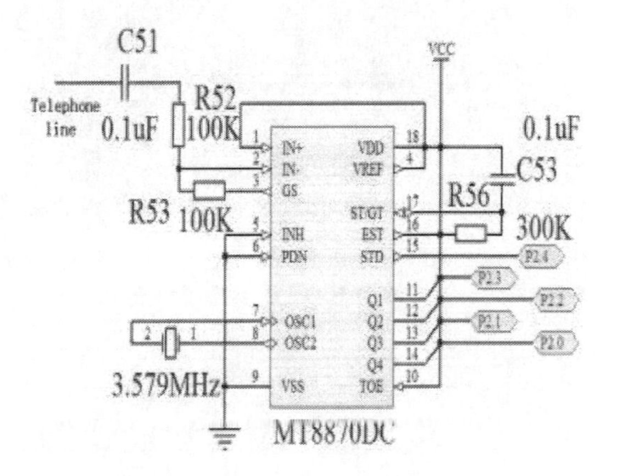

Figure 3. DTMF decoding circuit.

3.3 *Voice prompt module*

To play the voice of section N, we need to give a high level pulse to PD port to make an address pointer reset to 0. All the serial numbers use the memorizer as a benchmark, in addition to the first paragraph, so only need CE port receive 10μs low pulse except for the first paragraph to make the address pointer according to A0 - A7 to address the beginning of section N. The SP+ port is pulled, and then with a low pulse to the CE port can play the first voice messages of section N. This play will continue until the EOM symbol appears. Voice prompt module circuit connection is shown in Figure 4.

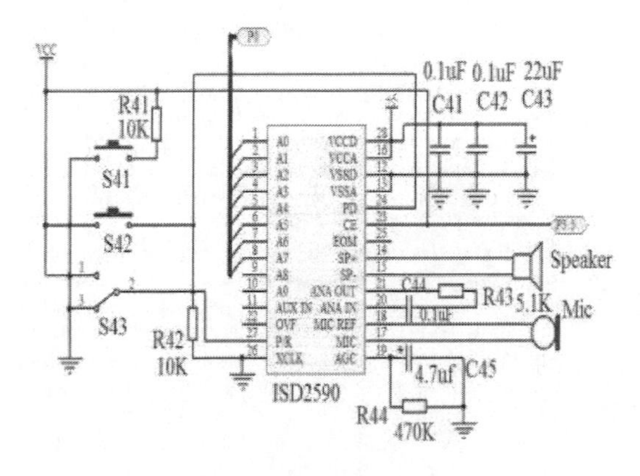

Figure 4. Voice prompt circuit.

3.4 *Electric appliance control module*

This module design is very simple, and its circuit mainly consists of PNP triodes, relays and diodes of anti surge current. And then relays connect controlled household electrical appliances. One control circuit diagram is shown in Figure 5.

The P3.6 and P3.7 pins of single-chip computer AT89C51 are used as the output control feet. They output a high level for triode conduction, and thus drive relay on and off, and at last to realize the multi-channel electrical control. This design adopts controlling power sockets with relays to realize the ultimate control of household electrical appliances, such as rice cooker, water heater, air conditioning, etc. We only need to insert the plug can the master microcontroller through the control of each relay socket to control electrical power on and off. In this device there are two ways can be controlled, and other electric control circuits are same.

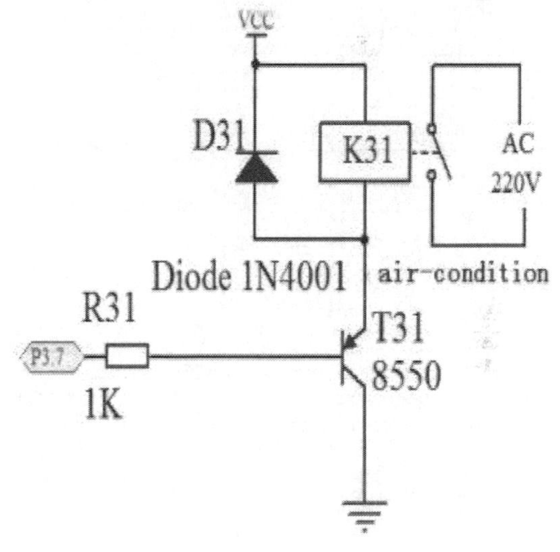

Figure 5. Electric appliance control circuit.

4 THE SOFTWARE DESIGN OF THE SYSTEM

The development tools that this software design selects of are μVision2 IDE integrated development environment, using the Keil C51 compiler and adopting the idea of object-oriented program design. Each small function module as a subroutine and all subroutines are interrelated. Finally the main program calls them successively according to the workflow system. It mainly includes KEY subroutines, paragraphs voice playback subroutine, receive keys subroutine, message/by password control subroutine, voice interaction platform program, timer interrupt subroutine,

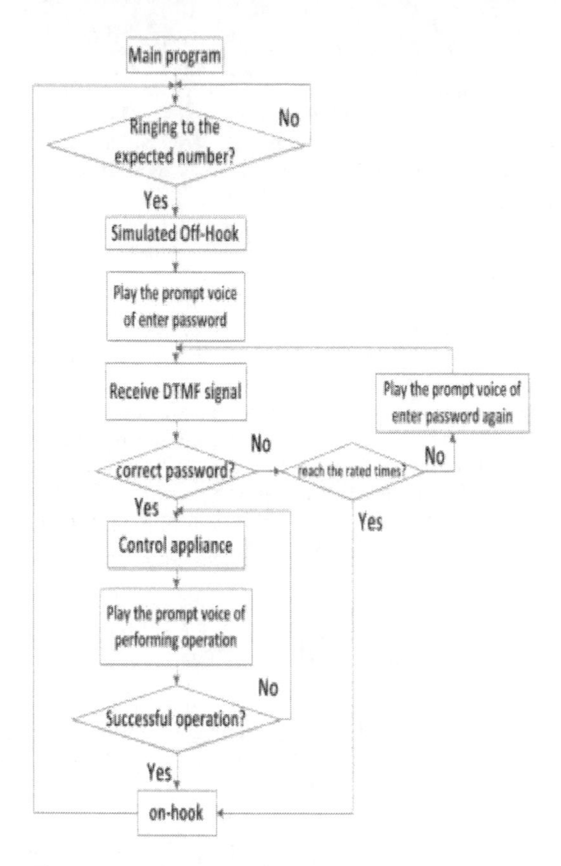

Figure 6. Main program flow chart.

delay subroutine, command/data processing subroutine and so on several modules. The main program flow chart is shown in Figure 6.

5 CONCLUSION

Eventually our achievement can realize remote control of 2 devices at the same time. After the test, detect circuit and DTMF decoding circuit works well. Besides, the voice interaction platform can achieve real person pronunciation and make devices work well by controlling. Furthermore, we can increase control capacity by adding a child controller with double orders of magnitude. It has good expansibility. This system can be embedded into the home appliance equipment as a functional module to achieve the function of remote control. In addition, the system has strong adaptability and security, easy to operate. Users only use fixed telephone or mobile phone can realize remote control of home appliances by entering simple command. Based on this significance, the system has high practical value and broad market prospects. The pictures of my product are shown in Figures 7 and 8.

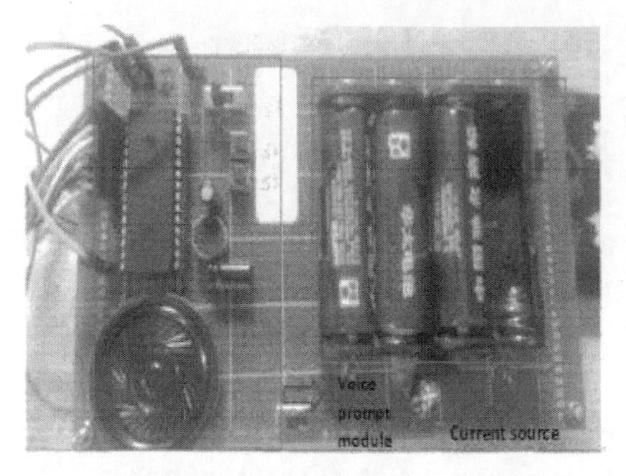

Figure 7. The front of my product.

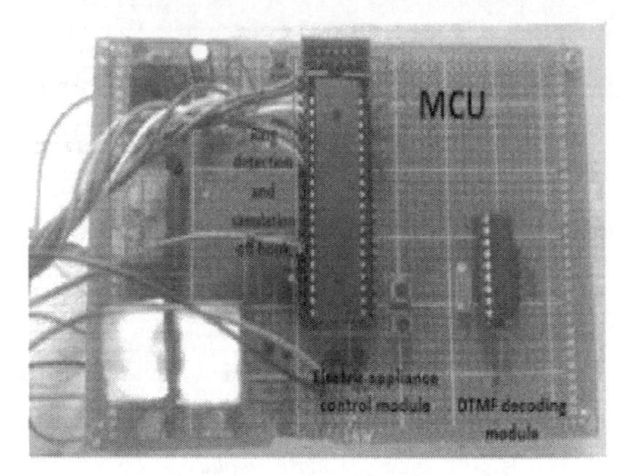

Figure 8. The back of my product.

REFERENCES

[1] Bushby, S.T.*A standard communication infrastructure for intelligent buildings*, Automation in Construction, no.5, pp.529–532, Jun.1997.

[2] Dogan I.2000.*Microcontroller Projects in C for the 8051*, Newnes.

[3] Z.Q Pan. 1998.*Efficient use method of dual tone multi frequency signal receiving chip MT8870*, Telecommunications technology, no.8, pp.49–51.

[4] J.Y Li.2008.*Household appliances remote control system design based on PSTN*, Microcontroller and embedded systems applications, no.12, pp.53–56.

[5] J.H Zhang. *The design of the remote home appliances based on MCU*, *Journal of University of Science and Technology of Suzhou*.vol.19, no.2, pp.79–82, Feb.2006.

[6] H.M Shi, Z.J Yu and D Liu.1998. *The development and application of ISD speech chips*, Modern electronic technology, no.9, pp.1–4.

Computing, Control, Information and Education Engineering – Liu, Sung & Yao (eds)
 ISBN: 978-1-138-02800-5

A new formula of phase contrast imaging for continuous X-ray based on in-line X-ray imaging

Tian Xia, Jun Shan Ma, Jing Hai Cheng & Jun Yang
School of Optical-Electrical and Computer Engineering, Shanghai Medical Instrumentation College, University of Shanghai for Science and Technology, China

ABSTRACT: In clinical applications, the X-ray spectra from tungsten are continuous, and it includes both attenuation effect and phase effect for in-line X-ray imaging. In this paper, a normalized imaging formula (NIF) independent of the light intensity is derived for the polychromatic case. Taking both photoelectric and Compton interactions into account, and according to the NIF, the attenuation and phase effects for different substances could be accurately calculated, which lays the foundation for the further research.

KEYWORDS: phase contrast imaging; attenuation effect; phase effect; electron density.

1 INTRODUCTION

Conventional X-rays have been used in diagnostic imaging since the X-ray was discovered by Roentgen in 1895. In X-ray imaging contrast is generated by the different X-ray attenuation in the object due to its varying composition, thickness or density. For biological tissues containing muscle and bones a sharp image is obtained by conventional X-ray imaging due to the difference in X-ray attenuation between the muscle and the bone. However, soft tissue contrast is limited in X-ray imaging because there are only slight differences in the attenuation of these regions. In contrast, X-ray phase contrast imaging (X-PCI) is a major conceptual and practical advance in the use of X-ray imaging that is based on the phase change effect on the X-ray intensity image[1].

In previous reports, an in-line phase contrast imaging system using a micro-focus X-ray source is comparatively simple and easier to construct and offers important advantages for clinical imaging [2-7].

However, the image directly obtained by this method is a mixed image including both attenuation and phase effect and the X-ray spectrum from tungsten target is usually continuous in clinical applications. Therefore, it is necessary to carry out the theory of phase contrast imaging for continuous X-ray.

In this paper, a new theory of phase contrast imaging independent of the light intensity for polychromatic X-ray is presented for the first time using an in-line phase contrast imaging system.

2 THE THEORY

The degree of phase change is determined by biological tissue dielectric susceptibility, or equivalently, by the refractive index of the tissue. The refractive index n for an X-ray is complex and equal to [8]

$$n = 1 - \delta + i\beta, \tag{1}$$

where δ is the refractive index decrements and is responsible for the X-ray phase shift, and β is responsible for X-ray absorption. For infiltrating ductal carcinoma (IDC) in breast the tumor's δ (10^{-6}–10^{-8}) is about 1000 times greater than that of the β (10^{-9}–10^{-11}) for X-rays in the 10 keV-1,00 keV range[8]. Therefore, phase contrast imaging is much more advantageous than absorption contrast imaging for this application.

For X-rays, the δ is related to the forward scatting from electrons in the sample and in the short-wavelength limit is given by[9]

$$\delta = r_e \rho_e \lambda^2 / 2\pi. \tag{2}$$

The amount of X-ray phase shift from biological tissue is related to δ by

$$\phi(x) = -\frac{2\pi}{\lambda} \int \delta(\eta, z) \mathrm{d}z = -r_e \lambda \rho_e(\eta), \tag{3}$$

where $\rho_e(\eta, z)$ is the electron density at point (η, z), $\rho_e(\eta)$ is the projected electron density, r_e is the classical electron radius with integration over the path of

X-ray beam in tissue. It is assumed that the X-rays are propagating in the z-direction.

In reference [1], Pogany, Gao, and Winkins (PGW) modeled the phase shift and attenuation effects of a piece of biological tissue as a two-dimensional transmission function $T(\eta,\xi)$ at a location (η,ξ) on the object plan perpendicular to the X-ray projection direction (z axis):

$$T(\eta,\xi)=\exp[\mathrm{i}\phi(\eta,\xi)-\frac{\mu(\eta,\xi)}{2}]=A(\eta,\xi)\mathrm{e}^{\mathrm{i}\phi(\eta,\xi)}, \quad (4)$$

where $\phi(\eta,\xi)$ and $\mu(\eta,\xi)$ are the z projection of the object phase and linear attenuation coefficients. In Eq. (4), $A(\eta,\xi)$represents the X-ray amplitude transmission. Concrete expressions of $\phi(\eta,\xi)$ and $\mu(\eta,\xi)$ are

$$\phi(\eta,\xi)=-r_e\lambda\rho_e(\eta,\xi), \quad (5)$$

and

$$\mu(\eta,\xi)=\int\mu(\eta,\xi,z)\mathrm{d}z=\frac{4\pi}{\lambda}\int\beta(\eta,\xi,z)\mathrm{d}z, \quad (6)$$

respectively. Here, $\mu(\eta,\xi,z)$ is the linear attenuation coefficient at point (η,ξ,z) of a given tissue. For calculation convenience but without loss of generality, $T(\eta,\xi)$ is considered a 1-dimension form of $T(\eta)$, which is the same as in reference [1].

According to the relationship between the wave field function and the intensity image, the X-ray intensity image I(x) on the detector plane is[2,5-7]

$$\begin{aligned}I(x)&=E(x,y)\cdot E^*(x,y)\\&=\frac{I_{10}}{\lambda MR_2}\int_{-\infty}^{\infty}\int_{-\infty}^{\infty}\exp[\mathrm{i}\pi M\frac{(\eta_1-\frac{x}{M})^2-(\eta_2-\frac{x}{M})^2}{\lambda R_2}]\mathrm{T}(\eta_1)\mathrm{T}^*(\eta_2)\mathrm{d}\eta_1\mathrm{d}\eta_2,\end{aligned} \quad (7)$$

here I10 denotes the incident X-ray intensity on the object plane and the geometric magnification factor M = (R1+ R2)/ R1,where R1 denotes the distance from the source to the object plane, and R2 denotes the distance from the object plane to the detector plane. The FT of the intensity is written as

$$\tilde{I}(\mathrm{u})=\int_{-\infty}^{\infty}\exp(\mathrm{i}2\pi\frac{x}{M}u)I(x)\mathrm{d}(\frac{x}{M}). \quad (8)$$

Substituting Eq. (7) into Eq. (8) after a tediotus calculation we present Eq. (9) as a generalized imaging formula with respect to spatial resolution u.

$$\begin{aligned}\tilde{I}(u)=&\frac{I_{10}}{M^2}\exp(\frac{\mathrm{i}\pi\lambda R_2u^2}{M})\{\mathrm{FT}[A^2(x)]+\frac{2\pi\lambda R_2u^2}{M}\mathrm{FT}[A^2(x)\phi(x)]\}\\&+\frac{\lambda R_2u}{M}\mathrm{FT}[A(x)\frac{\mathrm{d}A(x)}{\mathrm{d}x}]-2\mathrm{i}\frac{\lambda R_2u}{M}\mathrm{FT}[A(x)\frac{\mathrm{d}A(x)}{\mathrm{d}x}\phi(x)]\}\end{aligned}, \quad (9)$$

It has been established that the amplitude approximation and moderate variation condition for phase can be applied to clinical imaging [5], [6] and [7]. Hence, taking the upper conditions into account, Eq. (9) could be simplified as[6,7]

$$\tilde{I}(u)=\frac{I_{10}}{M^2}\cos(\frac{\pi\lambda R_2u^2}{M})\{\mathrm{FT}[A^2(x)]+\frac{2\pi\lambda R_2u^2}{M}\mathrm{FT}[A^2(x)\phi(x)]\}. \quad (10)$$

In clinical breast X-ray imaging, most X-ray photons range from 10 keV to 100 keV, and the corresponding wavelength varies from 0.124 nm to 0.0124 nm. The distance from object to detector is less than 1 m in general and the maximum spatial resolution needed is 20l p/mm[2]. As a result, πλzu2<<1 and cos(πλR2u2/M)≈1. Thus, Eq. (10) becomes

$$\tilde{I}(u)\approx\frac{I_{10}}{M^2}\{\mathrm{FT}[A^2(x)]+\frac{2\pi\lambda R_2u^2}{M}\mathrm{FT}[A^2(x)\phi(x)]\}s. \quad (11)$$

It should be emphasized that the effect of spatial coherence could be ignored when the X-ray focus is smaller than 5 μm [9]. In this paper, a focus of less than 2 μm in diameter is used.

According to the differential properties of FT, Eq. (11) can be expressed as

$$I(x)\approx\frac{I_{10}}{M^2}\{[A^2(x)]-\frac{\lambda R_2}{2\pi M}[A^2(x)\phi(x)]''\}. \quad (12)$$

Interestingly, $A(x)$ varies very little in most soft tissue, i.e. $A'(x)<<1$ and the upper equation could be simplified as

$$I(x)\approx\frac{I_{10}}{M^2}\{A^2(x)-\frac{\lambda R_2}{2\pi M}[A^2(x)\phi''(x)]\}. \quad (13)$$

Eq. (13) indicates that more desirable contrast can be obtained via phase shift imaging when $A'(x)<<1$. Here, as $R_2=0, I(x)=\frac{I_{10}}{M^2}A^2(x)$, which shows pure attenuation effect.

In clinical X-ray imaging the linear attenuation coefficients μ of a given tissue depends on the X-ray photon energy or on wavelength. In fact, two interactions cause the attenuation for photon energies up to 140 keV (i.e., below the threshold for the pair production): Photoelectric interaction and Compton interaction. Consequently, the linear attenuation coefficients have two contributions. One is from the above mentioned interactions:

$$\mu = \mu_P + \mu_C, \tag{14}$$

Here, the relative mass attenuation coefficient is

$$\frac{\mu}{\rho} = \frac{\mu_P}{\rho} + \frac{\mu_C}{\rho}, \tag{15}$$

where ρ is the density of material.

For the Photoelectric interactions, a scattering cross section of each atom σ_P is proportional to $(h\nu)^{-3}$ and $\bar{Z}^n$[10]. Thus, it may be assumed that

$$\frac{\mu_P}{\rho} = \frac{N_A}{M_A}\sigma_P = k\frac{\rho_e(x,z)}{\rho}\lambda^3\bar{Z}^{n-1}/(hc)^3, \tag{16}$$

where k is the photoelectric interaction parameter, $\bar{Z}$ is effective atom number of the material, M_A is the atomic weight tissue, N_A is Avogadro's number, the n ranges from 4 to 4.8, and $\rho_e = \frac{\rho\bar{Z}}{M_A}N_A$.

For the Compton interaction, the scattering cross section of each electron σ_C is derived from the Klein–Nishina formula[10]:

$$\sigma_C = 2\pi r_e^2\{\frac{1+\alpha}{\alpha^3}[\frac{2\alpha(1+\alpha)}{1+2\alpha} - \ln(1+2\alpha)] + \frac{\ln(1+2\alpha)}{2\alpha} - \frac{1+3\alpha}{(1+2\alpha)^2}\}, \tag{17}$$

and

$$\mu_C = \rho_e(x,z)\sigma_C. \tag{18}$$

In Eq. (17), $\alpha = E_{photon}/m_e c^2$, where E_{photon} denotes the X-ray photon energy, and $m_e c^2$ is the resting electron energy and is equal to 511 keV. Because $E_{photon} = hc/\lambda$, μ_C is also related to λ. Consequently, the linear attenuation coefficients can be expressed as a function with λ and $\mu = \mu(x,z,\lambda)$, where (x,z) denotes the coordinates of any point in the X-ray propagation direction z. Hence, according to Eqs. (4), (14), (16) and (18), we obtain

$$A^2(x,\lambda) = \exp[-\mu(x,\lambda)] = \exp[-\rho_e(x)f(\lambda)], \tag{19}$$

where $\mu(x,\lambda) = \int u(x,z,\lambda)dz$,and $f(\lambda) = \sigma_P + \sigma_C$.

Thus, image contrast will change as long as the wavelength of the incident X-ray changes. Therefore, the applied range of Eq. (13) is very small and is only suitable for the monochromatic case instead of the polychromatic one. To extend the applied field of Eq. (13), it is necessary to consider the effect of polychromatic coherence. We suppose that the source has a number of photons distributed with λ, i.e.$w(\lambda)$. According to the properties of a spherical wave: $I_{10}'(x) = I_{10}R_1^2/(R_1+R_2)^2 = I_{10}/M^2$,

where $I(x) \approx I_{10}'(x)[\langle A^2(x)\rangle - \frac{R_2}{2\pi M}\langle\lambda[A^2(x)\varphi''(x)]\rangle]$ represents the incident light intensity in the image plane without passing through the object. A polychromatic X-ray source then has from Eq. (13)

$$I(x) \approx I_{10}'(x)[\langle A^2(x)\rangle - \frac{R_2}{2\pi M}\langle\lambda[A^2(x)\phi''(x)]\rangle], \tag{20}$$

where

$$\langle A^2(x)\rangle = \frac{\int A^2(x,\lambda)E_{photon}w(\lambda)d\lambda}{\int E_{photon}w(\lambda)d\lambda}, \tag{21}$$

and

$$\langle\lambda A^2(x)\phi''(x)\rangle = \frac{\int\lambda[A^2(x,\lambda)\phi''(x,\lambda)]E_{photon}w(\lambda)d\lambda}{\int E_{photon}w(\lambda)d\lambda}. \tag{22}$$

By using Eqs.(3) ,(19),(20),(21) and (22),we can obtain

$$\frac{I(x)}{I_{10}'(x)} \approx \langle A^2(x)\rangle + \frac{R_2 r_e\rho_e''(x)}{2\pi M}\langle\lambda^2A^2(x)\rangle, \tag{23}$$

where $\langle\lambda^2A^2(x)\rangle = \int\lambda^2A^2(x,\lambda)E_{photon}w(\lambda)d\lambda / \int E_{photon}w(\lambda)d\lambda$, which is termed the normalized X-ray image (NXI).

3 DISCUSSION AND CONCLUSION

The Eq. (23) indicates that the NXI taking $I_{10}'(x)$ as a reference value has no association with the incident light intensity, which is only associated with the spectrum distribution of the incident light. Obviously, a normalized image includes both the attenuation effect $\langle A^2(x)\rangle$ and the phase effect $(R_2r_e\rho_e''(x)/2\pi M)\langle\lambda^2A^2(x)\rangle$.The phase effect is proportional to $\rho''(x,z)$. For example, in the even distribution of the projection electron density $\rho_e(x)$ $(\rho_e''(x) \cong 0)$ the phase effect disappears, so Eq. (23) can be simplified as $\frac{I(x)}{I_{10}'(x)} \approx \langle A^2(x)\rangle$. Consequently, the phase effect may only appear in the projection electron density variation region. As long as we could obtain $w(\lambda)$, the attenuation and phase effects for different substances could be accurately calculated by Eq.(23), which lays the foundation for the further research.

ACKNOWLEDGMENT

Authors Tian Xia, Junshan Ma, Xuelong Zhang, Jinghai Cheng ,Weijun Peng and Xufeng Yao are grateful for the support of this research by Shanghai Education Development Foundation "Morning Plan"(11CGB06). Opinions, interpretations and conclusions in this paper are those of the authors and are not necessarily endorsed by the above funds.

REFERENCES

Pogany A., Gao D, Winkins S. W.1997. Contrast and resolution in imaging with a microfocus X-ray source . *Rev. Sci.Instrum.* 68(7):2774.

Wu X., Liu H.2003 *J. X-ray. Sci. Tech.,* 11: 33.

Gureyev T.E., YakovI.Nesterets,David M.Paganin,2006. A general theoretical formalism for X-ray phase contrast imaging. *J. Opt. Soc. Am. 23*(1)*: 34–42.*

Anna Burvall, Ulf Lundström, Per A. C. Takman, Daniel H. Larsson, and Hans M. Hertz,2011. Phase retrieval in X-ray phase-contrast imaging suitable for tomography *Optical Express* 19 : 10376.

Xia T., Zhang X.L., Zhang G.Y. 2009. Analysis of the absorption and the phase-effect on micro-focus X-PCI *Acta Photonica Sinica,* 38:2516.

Xia T., Zhang X.L., Ma J.S., Cheng J.H., and Huang Y.2011. Effect of Spatial Coherence and in Incident X-ray Photon Energies on Clinical X-ray In-line Phase-contrast Imaging. *Acta Photonica Sinica,* 40:627.

Wu X. Dean A., and Liu H. 200*3.in:X-ray diagnostic techniques in Biomedical photonics Handbook, edited by T. VoDinh* FL,CRC, Tampa.

Davis T.J, Gao D., Gureyev T.E., Stevenson A.W.,and Winkins S.W. 1995. Phase-contrast imaging of weakly absorbing materials using hard X-ray *Nature* 373: 595.

Dyson N.A.1993. *X-ray in atomic and nuclear physics,* New York, Cambridge University(1990).

Computing, Control, Information and Education Engineering – Liu, Sung & Yao (eds)
© 2015 Taylor & Francis Group, London, ISBN: 978-1-138-02800-5

Common design of data communication protocol based on wireless sensor networks

Ping Tang, Xiao Shi Zheng, Lin Wang, Zheng Wei Wang, Guang He Cheng, Ling Yan Han, Feng Qi Hao & Xiang Sun
Guizhou Minzu University, Guiyang, Guizhou, China
Shandong Provincial Key Laboratory of Computer Networks, Shandong Computer Science Center (National Supercomputer Center in Jinan), Jinan, China

ABSTRACT: This paper analyzed shortcomings of existing data communication protocols based on wireless sensor networks, proposed a relatively common data communication protocol. The data frame in the proposed communication protocol was dynamic, and when the terminal ID, the parameter type and the number of parameters were changed, it could adaptively adjust the data frames' length and items with the help of configuration file. The engineering practice showed that the proposed data communication was more common and practical, applying to most wireless sensor networks.

KEYWORDS: Wireless sensor networks; Data communication protocol; Common; Data frame.

1 INTRODUCTION

Data communication protocol also known as data communication control protocol, is a series of agreements to ensure effective and reliable communication between the two sides of communication networks. These agreements include format and sequence of data, data transmission confirmation or rejection, error detection, retransmission control and inquiry operation, and so on. Currently, most wireless sensor networks (WSNs) are developed for specific applications, so the data communication protocols are also specific. Most data protocols of WSNs can adapt to the change of terminal ID by configuration files or dialing. For this kind of communication protocols, when sensor types increase, the type of acquisition terminal will increase correspondingly, usually classify terminal types by dividing all terminal ID numbers; every time when the terminal type is changed, the terminal ID numbers have to be divided, which is troublesome. With the terminal types increasing more and more, this kind of data communication protocols may not be able to assure effective communication of every terminal. As for this question, somebody has designed a data communication protocol with the frame type code to distinguish different types of data frames, which can be used to classify the terminal types. But when how many parameters' data the terminal collect is changed, this protocol cannot achieve effective communication. If by modifying programs to adjust the protocols, then data display and analysis such application programs must also be modified, which greatly reduces the efficiency of software development. Therefore, to design a relatively common data communication protocol is very important, when the demand of terminals changes, including changes of the terminal ID, the parameter type and the number of parameters, the protocol can achieve effective communication with no change of programs. Reasonable and effective data protocol can improve efficiency of network application. This paper designed a data communication protocol for WSNs based on the analysis of existing data communication protocol, which can achieve effective communication without modifying the existing programs when the demand of terminal is changed.

2 COMMON DESIGN OF DATA COMMUNICATION PROTOCOL

2.1 *Terminal type definition and cording rule*

In practical applications, the display equipment received data frames, and then process and display in a certain form. Commonly speaking, there are multiple terminals used for data acquisition in one WSN; when the display equipment receives the data, firstly need to judge this data is sent by which terminal, so terminal ID number is necessary to identify each terminal. Existing data communication protocols in present basically have terminal ID definition, but only having terminal ID can not assuring effective communication in the following two situations, having to change the protocol.

1 Increasing the sensor type, the parameter type changes. For example, there is only a kind of temperature terminal in one WSN, when add a new kind of heat flux terminal, this two kinds of terminal share all the terminal ID numbers. Suppose there are all 256 ID numbers, then 0-127 is the first type, represents temperature terminals; 0-127 is the second type, represents the heat flux terminals. When a new kind of terminals is added, the ID numbers have to be divided again, and all terminals' ID must be reset. With the terminal types increasing, ID number may be not enough, terminal communication failure will come.
2 Increasing the number of collected parameters. As the WSN mentioned above, when add a new kind of terminals which collect "temperature + heat flux" this two parameters data, then original terminal ID numbers and bites of valid data have to be changed, thus programs must be modified to adjust the protocol.

In most applications, data collected by terminals in the same WSN may be different, so terminal type definition is necessary. As analyzed above, when a terminal, its collected parameter type or the number of parameter is different from others, this terminal is a new kind of terminal. Define parameter number according to practical situation and add terminal type through configuration file. Assume there is only one kind of temperature terminal in a WSN, numbered 0, as in Table 1 below. When add a kind of heat flux terminal, for parameter type changing, this is a new terminal type, numbered 1.When add a kind of terminal which collects "indoor temperature + heat flux" two parameters' data, according to the definition, this is a new type, numbered 2. The outdoor terminals collect data of outdoor temperature and heat flux, the meteorological terminals collect data of temperature, humidity and wind speed, and they are different from others, so they are new kinds of terminal, respectively numbered 3 and 4.

In practical applications, in order to observe every terminal' communication directly, all terminals will be labeled. Table 2 shows the useful information of every terminal type, but all information in the label is impossible. Commonly, terminal ID number will be showed in the label, thus personnel can judge which terminal it is by the label. As terminal type definition above, type number and ID number can be duplicate, so terminal type number must be showed in the label too. We mark all terminals in the form of "type number-ID number", for example, 0-0 is the terminal whose type number is 0 and ID number is 0, refer to table 2, we know this terminal is NO.0 node of temperature terminal; 1-0 is the NO.0 node of heat flux terminal. This labeling method is direct and convenient.

Table 1. Terminal type definition.

Type number	Type name	Parameter name	Parameter number
0	Temperature terminal	Temperature	1
1	Heat flux terminal	Heat flux	1
2	Indoor terminal	Indoor temperature Heat flux	2
3	Outdoor terminal	Outdoor temperature Heat flux	2
4	meteorological terminal	Temperature Humidity Wind speed	3

Table 2. Terminal information.

Type number	Parameter name	Parameter symbol	Parameter unit	Parameter order
0	Temperature	T	℃	1
1	Heat flux	Q	W/m^2	1
2	Temperature indoor	Ti	℃	1
2	Heat flux	Q	W/m^2	2
3	Temperature outdoor	To	℃	1
3	Heat flux	Q	W/m^2	2
4	Temperature	T	℃	1
4	Humidity	S	%	2
4	Wind speed	F	m/s	3

2.2 *Node address definition*

Node address is the symbol to distinguish different terminals. In the same WSN, every terminal has its exclusive node address. Node address has useful information of this node. As analysis above, node address should include type number, ID number, and parameter number. Take transmit speed accounted, the length of node address is defined as two bites. Node address definition is shown in Table 3.

Node address takes up two bytes, namely, 16 bits; bit0-bit7 is ID number, can represents 256 different ID numbers from 0 to 255; bit8-bit11 is parameter number, can determine how many parameters the terminals collect; as real parameter number can't be 0, so the parameter number in node address equals real parameter number minus 1. For example, there is one terminal whose parameter number is 0 means that the terminal collect 1 parameter, parameter number is 15 means this terminal's real parameter number is 16, it collect 16 parameters' data.

Table 3. Node address definition.

Bit	Field
15	Type number
14	
13	
12	
11	Parameter number
10	
9	
8	
7	ID number
6	
5	
4	
3	
2	
1	
0	

2.3 *Node address configuration*

We usually set nod address by dialing or configuration files. Currently, most WNSs adopt configuration file to set nod address. Serial communication interface (RS232) is a standard communication interface of the computer. Using serial communication interface for data transmit and acquisition is a important application field of computer. This paper adopts serial communication for node address setting by configuration file shown as follows:

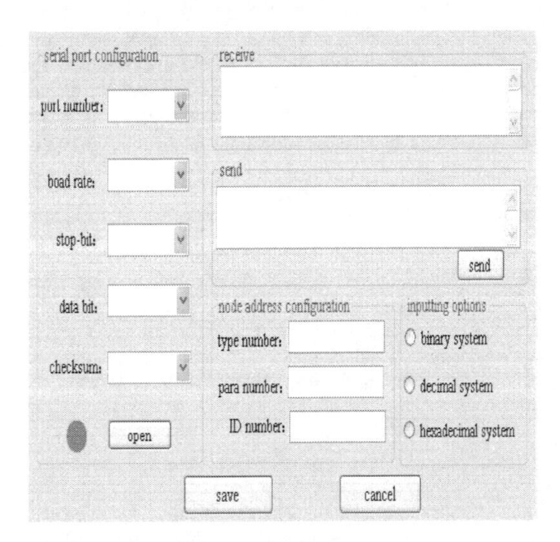

Figure 1. Node address configuration.

Through configuration file above, we can set node address in decimal system, binary system or hexadecimal system, but decode node address just adopt binary system as definition above. The five kinds of terminal in table 5, their NO.10 nodes' node addresses respectively is 0000 0000 00001010, 0001 0000 00001010, 0010 0001 00001010, 0011 0001 00001010, 0100 0010 00001010.

2.4 *Data frame definition*

Data frame definition is an important part of data communication protocol. Terminal collected data and saved as following data frame to transmit.

Table 4. Data frame definition.

Node address	Function code	Data frame length	Valid data	CRC code
2	1	1	2*n	2

The length of data frame above is dynamic, valid data take up 2*n bites, n is the real parameter number, equal to parameter number plus 1; every parameter takes up two bites. Data frame length is determined by the information inputted through configuration file. For example, 0100 0010 0000 1010 is one node address, it means this terminal collects 3 parameters' data, as 0010 in decimal system is 2, real parameter number plus 1 is 3, so in this data frame, the data frame length is 6 bites. Functions of items in above data frame are shown in Table 5.

Table 5. Data frame illustration.

Name	Bite number	Functions
Node address	2	Include type number, ID number and parameter number
Function code	1	Read valid data
Data frame length	1	The length of valid data
Valid data 1	2	Value of NO.1 parameter
…	…	…
Valid data n	2	Value of NO. n parameter
CRC	2	CRC verifying

Terminals collect data and transmit it with above data frame every five minutes. Every parameter takes up two bites, different kinds of terminals have different data frames, the data frame length is determined by the parameter number. Considering transmit speed, packed data in hex plastic; in order to assure the precision of valid data, valid data is real data times 100. For example, a complete data frame sent by NO.18 node of meteorological terminal is shown as follows:42 12 03 06 0B 1D 08 FC 02 58 64 5C.

3 COMMON FRAME DECODING METHOD

When display equipment has received a data frame, programs decode the frame to get valid data. The so-called decoding frame is a data processing, according to the data frame definition to find corresponding items and calculate corresponding values. Figure 2 is the flow chart of common decoding frame method.

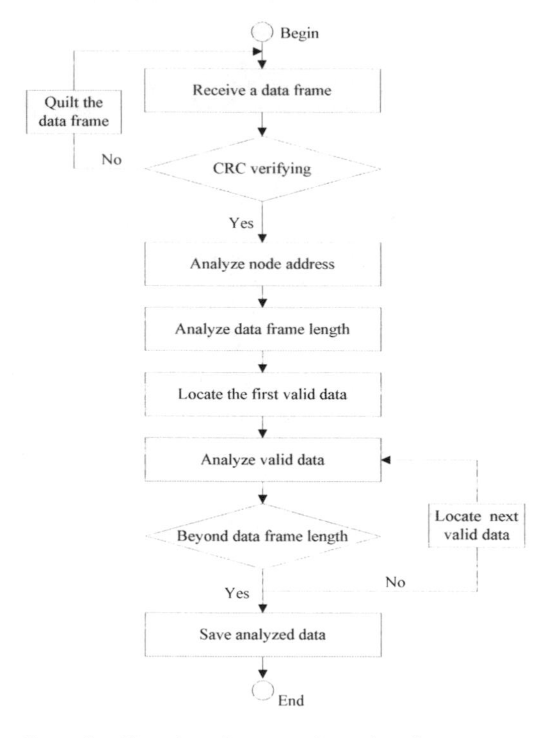

Figure 2. Flow char of common frame decoding.

Node address includes information of terminal type, parameter number and ID number, so the key to frame decoding is decoding node address correctly. Parameter number determines the length of data frame, which is used to judge whether all valid data have been decoded. According to above flow chart, decode following data frame: 42 12 03 06 0B 1D 08 FC 02 58 64 5C. Firstly, verify CRC code "64 5C" with CRC algorithm, the result is right and then decode the remaining bites, or quilt this data frame and receive another data frame. Secondly, decode the node address "42 12", they are numbers in hexadecimal system, 42 equals 01000010 in binary system, 12 equals 00010010, thus the node address is 0100 0010 00010010. By definition, 0100 is type number, 0100 is 4 in decimal system, type number is 4; 0010 is parameter number, so the parameter number is 2, the real parameter number equals the parameter number plus 1, thus this terminal collect 3 parameters' data; 00010010 is 18 in the decimal system, it is the terminal ID number. After decoding node address, read the function code. The function code is 03, which means read valid data from registers. The next step is decoding valid data. The first valid data is 0B 1D, it equals 2545 in decimal system. By definition, this is real valid data times 100, so divide 2545 by 100 to get the first valid data 25.45; similarly, decode the second valid data 08 FC and the third valid data 02 58, get the second real valid data 23.00 and the third real valid data 6.00. Refer to Table 2, we can see the data frame were sent by NO.18 node of meteorological terminal, it collect "temperature + humidity + wind speed" three parameters' data, the first parameter is temperature, temperature was 25.45°C; the second parameter is humidity, humidity was 23%; the third parameter is wind speed, wind speed was 6m/s.

4 SUMMARY

This paper analyzed shortcomings of existing data communication protocols based on wireless sensor networks, summarized two typical situations that existing data protocols don't apply to, designed a relatively common data communication protocol. Data frame in proposed communication protocol is dynamic. When the terminal ID, the parameter type and the number of parameters are changed, it can adaptively adjust the data frames' length and items through the configuration file. And a common frame decoding method was designed, the key lies decoding node address correctly. The proposed data communication protocol is more common and practical, applying to most WSNs.

ACKNOWLEGEMENT

It is a project supported by the international cooperation project (2010DFR10710) and Chinese major science and technology project (2014ZX04015011).

REFERENCES

http://baike.baidu.com/view/1372965.htm?

Huan Huang. The research of wireless sensor network node design and location a masterthesis at the Tongji University. 2007.

Jianfeng Lu . Common design of network data classification [J]. information technology teaching and research. 2010 (23).
Zhidong Zhang. Research of communication protocol based on wireless sensor network. doctoral dissertation of Tianjin University.2007.
Chengjun Wang . Analysis and design of serial port data communication protocol. Computer Engineering. 2004 (30).
Siwei Peng, Qunxiong Zhu. The formal description driven data frame analysis and processing. Computer engineering and application. 2006 (05).

Computing, Control, Information and Education Engineering – Liu, Sung & Yao (eds)
 ISBN: 978-1-138-02800-5

Dynamic adaptive display design based on wireless sensor networks

Zheng Wei Wang, Xiao Shi Zheng, Lin Wang, Ping Tang, Ru Liang Zhang, Guang He Cheng, Qing Long Meng, Rang Yong Zhang & Yang Wan
Guizhou Minzu University, Guiyang, Guizhou, China
Shandong Provincial Key Laboratory of Computer Networks, Shandong Computer Science Center (National Supercomputer Center in Jinan), Jinan, China
China Tobacco Shandong Industrial Co.,Ltd, Jinan, China

ABSTRACT: This paper proposed a dynamic adaptive display method based on wireless sensor networks. Without modifying the original program, the proposed method can flexibly and easily achieve universal dynamic display of data based on the configuration information, and adaptively choose horizontal or vertical mode. It can improve the efficiency of secondary development, reduce development costs and meet the requirements of universal design.

KEYWORDS: wireless sensor networks; dynamic; transmission protocol; adaptive display; display method.

1 INTRODUCTION

The Wireless Sensor Networks (WSN) is formed by a large number of cheap micro-sensor nodes through wireless communication for a multi-hop Ad-Hoc network. Each sensor can perceive, acquire, process and transmit monitoring information of perception objects within geographical area of network coverage, and through a wireless communication network will transmit the information to the user terminal perception, allowing users to monitor the area to fully grasp the situation and react. Therefore, it is essential by the display device in real-time, intuitive and effective display of data collection, and the majority of wireless sensor networks also have a display device so far. Most of the wireless sensor networks were developed for specific applications, so the data also show a particular way. Developers always design a corresponding display screen according to the type of the corresponding sensor, quantity and so on. When increase or decrease the type of collection terminal, you must modify the program to adapt to changes in the front, otherwise it will not achieve an effective display. This is a time-consuming and laborious process, significantly reduces the development efficiency and increases development costs.

Currently, ways of data display based on different sensor networks were also different. For example, the size of the interface, style, visual degree and so on can be described as a wide range, there is no one uniform design standards. Therefore, designing a universal display method based on wireless sensor networks that the display device can adaptively and dynamically adjust the display interface and the data display is very meaningful when data acquisition can be the case at the front end communication protocol is changed.

Under the configuration file, the paper designs a dynamic adaptive display method based on wireless sensor networks. The method focuses on inspection instrument display interface to design, the front-end data protocol changing information is written by pre-designed configuration file, the system can adaptively adjust the display screen, make the display to be universal and friendly.

2 UNIVERSAL TRANSMISSION PROTOCOL

Data transmission protocol refers to a standardized format in the transmission of data between the two devices. Protocol type can determine the error checking methods, data compression method, as well as the end of file and so on. Before universal display, data collection terminals must be universal protocol design. Universal data protocol refers to the type of the parameter is changed when the terminal's type or quantity is changed, the data communication protocol without changing the program's premise, can adaptively adjust the frame format and content to meet the project actual scene requirements.

2.1 *Node address definition*

The node address is defined as follows in Table 1, occupies two bytes, totals 16bits, includes the type number, the number of parameters (number of registers) and node ID; bit0-bit7 represents the node ID (can represent up to 256 different node numbers), bit8-bit11 represents the number of parameters that a terminal can collect those parameters. The number of data collection terminal can't be zero, so the parameter is the actual parameter to subtract one; bit12-bit15 represents the type of number (can represent up to 16), is the type of terminal.

Table 1. Node address definition.

15	14	13	12	11	10	9	8	7	6	5	4	3	2	1	0
Type number				Number of parameters				Node ID							

For example, address of a node is "0011 0010 0000 1010", it means that the terminal is class 3 terminal (0011), collects 3 kinds of data (0010), the node ID of the terminal is "10" (00001010).

2.2 *Data packet format*

In order to achieve adaptive display interface, we have to design a universal transmission protocol. According to the data packet design rules, the transmission protocol is developed as shown in Table 2.

Table 2. Data transmission protocol.

Node address	Function code	Frame length	Valid data	CRC
2	1	1	2n	2

In Table 2, valid data is "2*n", "n" is the number of data in the node address (the number of parameters adds one), "2" is the number of bytes occupied by each of the parameters. The data frame length is variable, node address includes the node ID, module type and the number of parameters. For example, node address is "32 0A", "0011 0010 0000 1010" is its binary form. Among them, "0011" means that the module type, which is type 3; "00001010" means node ID, that ID is 10; "0010" means the number of parameters, which is two parameters; The node address represents that the terminal's type is 3, node ID is 10, collects 3 data, and the valid data occupies 4 bytes.

3 DYNAMIC ADAPTIVE DISPLAY DESIGN

3.1 *Interface design*

Currently, there are many types of data display screen, but when types or quantity of the sensors are changed,

Table 3. Original interface design.

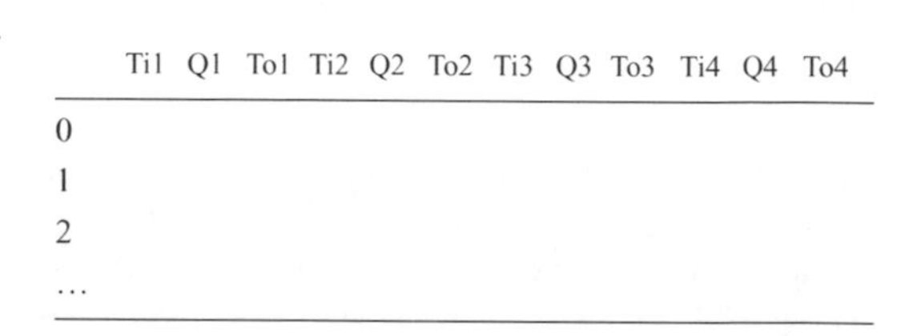

	Ti1	Q1	To1	Ti2	Q2	To2	Ti3	Q3	To3	Ti4	Q4	To4
0												
1												
2												
…												

they can not meet the actual needs, can not achieve the effect of the adaptive display. For example, in a sensor network, it can monitor two kinds of collection terminals, a kind of terminal collects outdoor temperature, another type of terminal collects the indoor temperature and the heat flux, the original design of the display is as shown in Table 3, it has 12 columns, every three columns represent indoor temperature, heat flow and the outdoor temperature.

As shown in Table 3, it can be used as an effective way to display indoor temperature, heat flux, the outdoor temperature. The display mode is fixed size and location when the current wireless sensor networks need a new category of test data that includes temperature, humidity and wind, so inspection instrument can not be effective display new types of data; of course, if reducing type of monitoring parameters, such as only monitor indoor temperature and heat flux, current inspection instrument can display, but the position of display the outdoor temperature can not display any data that will result in more empty space. The interface is not only waste, but also not friendly. the paper proposes a dynamic adaptive display method which can automatically adjust display according to different sensors. The idea is roughly based on sensor data transmission protocol to classify display. You can add or delete terminal type through the configuration file. While changing the rows and columns, it adaptively defines the display interface. Adaptive display frame is as shown in Table 4.

As shown in Table 4 in the design, the parameters in Table 3 are divided into two types, the first category is indoor temperature and heat flux, and the second category is the outdoor temperature. When add a new meteorological terminal which collects

Table 4. Adaptive display frame.

Node type	1			2		....	n	
Node ID	Variable 1	Variable 2	...	Variable 1	...	...	Variable 1	...
0								
1								
2								
...								
n								

Table 5. Adaptive display interface.

ID	Ti0	Q0	To1	T2	H2	W2	...
0							
1							
2							
...							
n							

temperature, humidity and wind, then the adaptive display screen is as shown in Table 5. You can add the type of parameters in turn when add the new terminal.

Table 5 is designed based on adaptive transmission protocol and configuration file by the corresponding configuration information, which makes dynamic adaptive display can flexibly change when quantity or type of sensors is changed.

3.2 *Configuration software*

Make good use of the configuration software can complete complex parameter configuration, which can not only protect control performance of the system, but also simplify the operational aspects and reduce the amount of repetition coding. Thus, configuration software can provide adaptive display interface design to provide effective help. In order to meet the requirements of adaptation, we design an auxiliary software (configuration software), using the configuration software writes the corresponding configuration information to form the desired configuration table. The following shows the configuration software interface shown in Figure 1.

With the help of configuration files above, you can customize columns and rows to determine the size of the display screen based on actual demand, add or delete the terminal types, so you can make the display screen can adaptively change according to the front of adjustment.

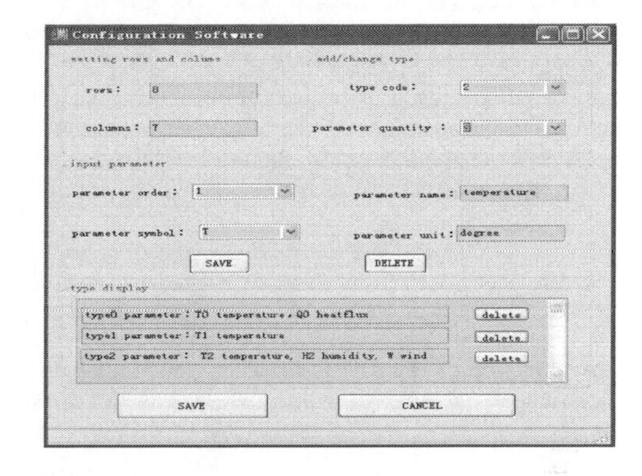

Figure 1. Configuration software interface.

3.3 *Configuration table*

Configuration table is mainly used to help show the adaptive interface design and adaptively display data on the display screen. With the help of configuration information in Figure 1, which can be formed as shown in Table 6, while according to the configuration information in the table to generate the initial display screen is as shown in Table 7 or Table 8.

Table 6. Configuration table.

Type	Parameter name	Parameter symbol	Parameter unit	Parameter order
0	Temperature	T	°C	1
0	Heat flux	Q	W/m^2	2
1	Temperature	T	°C	1
2	Temperature	T	°C	1
2	Humidity	H	%	2
2	Wind speed	W	/	3

3.4 *The initial display interface*

In order to show the beautiful and intuitive interface, the paper automatically designs two different display interfaces based on the number of monitoring parameters. The first is horizontal screen display when the monitored parameter is greater than 6, the row number represents terminal ID, the column represents the parameters being monitored, and when the total number of columns are more than 12. In Table 7, in the patrol inspection instrument display will automatically generate horizontal scroll bar on the screen. In the same way, it will generate the vertical scroll bar when rows exceed 9, so the user is convenient to view the completed parameter information is not displayed; the second is vertical screen display when the number of monitored parameters less, when it usually less than 6, is as shown in Table 8, only three monitoring parameters, includes the number of columns represent the terminal ID, row number represents the monitoring parameters, and horizontal screen display on the contrary, will produce horizontal and vertical scroll bar.

Table 7. Initial display screen - horizontal screen.

ID	T0 (°C)	Q0 (W/m²)	T1 (°C)	T2 (°C)	H2 (%)	W2
0						
1						
2						
3						
4						
5						
6						

Table 8. Initial display vertical screen.

	0	1	2	3	4	5	6
T0(°C)							
Q0(W/m²)							
T1(°C)							

3.5 *Universal display method*

To make the display interface is friendly and easy to operate, in the paper, design of the display interface is divide horizontal or vertical mode. Therefore, this paper designs two display methods based on the universal transmission protocol, as follows.

The first display interface is horizontal mode. After receiving the correct packet, first, according to the corresponding adaptive protocol, the system will analyze data packet in the first two bytes to obtain the device ID, device's type and the number of parameters. Assuming that ID is n, device's type is i, register number (the number of parameters) is m; second, it judges whether i is more than zero. if i is more than zero, in the corresponding configuration table, it gets maximum row where type i-1 is in the table, is set to R_max, so the type is located on the "R_max + j" (j =1,2, ..., m) column, that corresponds to the n-th row and "R_max + j" column on the display , denoted "(n, R_max + j)", so data of the registers are filled to position of "(n, R_max + j)". If i is equal to zero, the type is located in the j-th column(j =1,2, ..., m-1), that corresponds to the n-th row and j-th column on the display , denoted "(n, j)", so data of the registers are filled to position of "(n, j)". Flow chart of horizontal mode is shown in Figure 2.

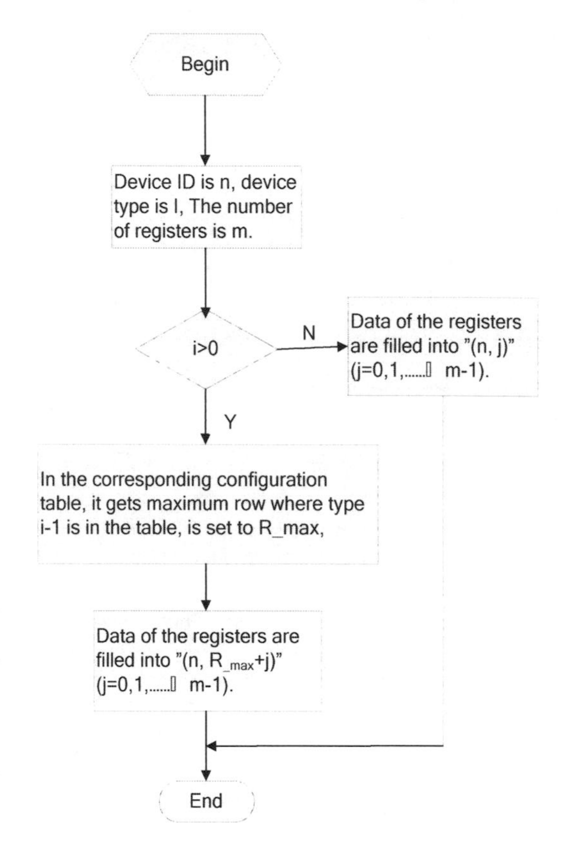

Figure 2. Chart of horizontal mode.

The second display interface is vertical mode. After receiving the correct packet, first, according to the corresponding adaptive protocol, the system will analyze data packet in the first two bytes to obtain the device ID, device's type and the number of parameters. Assuming that ID is n, device's type is i, register number (the number of parameters) is m; second, it judges whether i is more than zero. if i is more than zero, in the corresponding configuration table, it gets maximum row where type i-1 is in the table, is set to R_max, so the type is located on the "R_max + j" (j =1,2, ..., m) row, that corresponds to the "R_max + j" row and n-th column on the display, denoted "(R_max + j, n)", so data of the registers are filled to position of "(R_max + j, n)". If i is equal to zero, the type is located in the j-th row (j =1, 2 , ... , m-1), that corresponds to the j-th row and the n-th column on the display, denoted "(j, n)", so data of the registers are filled to position of "(j, n)". Flow chart of vertical mode is shown in Fig.3.

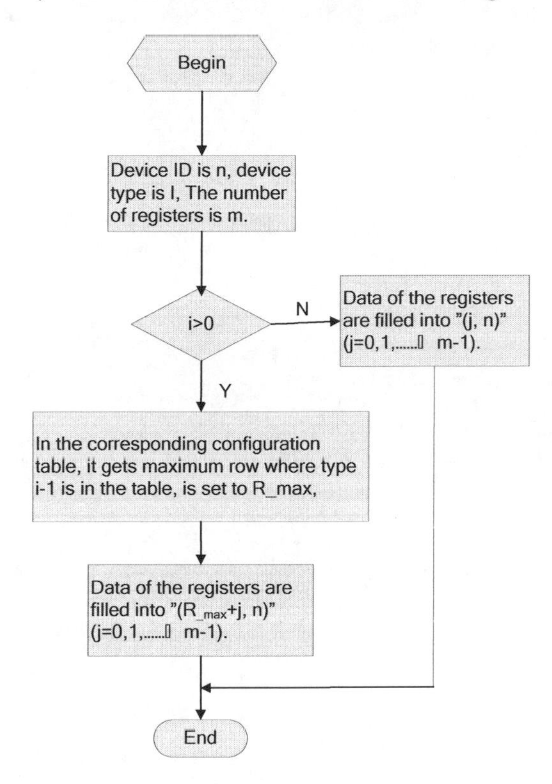

Figure 3. Chart of vertical mode.

4 EXPERIMENTAL VERIFICATION

Entering the configuration information to initialize the display screen is as shown in Table 7, when the received packet is: 22 12 03 06 0B 1D 08 FC 02 58 64 5C, perform data analysis based on adaptive transmission protocols and horizontal mode display protocol. The first two bytes are analyzed, it is node address. 22's binary is 00100010, 12's binary is 00010010, so the node address is 00100010 00100010. According to the definition of node address, 0010 means type number of the terminal, its decimal is 2, so type number is 2, and it is more than 0, the largest row of device type (2-1) is 2. 0010 represents the number of parameters, the number of parameters (number of registers) is 2, indicating that it collects three kind of data (the actual collection data is that number of node address parameters adds one);

00000111 means the node number, the node number of the terminal is 7, namely ID is 7.03 is the function code that reads data of register, 06 means the length of the valid data, so valid data is 6 bytes in the frame structure. "0B 1D" is the first valid data, its decimal is 2545, and then divided by 100 (due to the true value of the transmission protocol of the parameter is enlarged 100 times); in the same way, the second parameter is 23, the third parameter is 6.

In short, experimental procedure and results show that dynamic adaptive display is a good way in this paper, which can display all data of the monitoring parameters, and can intuitively correspond to a specific device's terminal.

5 SUMMARY

This paper proposes a universal display method based on wireless sensor networks. It can add or delete terminal type and detection parameters to change the display screen by the configuration software, achieve adaptive display on the display interface with different data. The system can realize the adaptive dynamically display based on the configuration information, at the same time, adaptively choose horizontal or vertical mode to display based on the detection parameters and the node. Experimental results show that the proposed method meets the objectives. When the parameter type or number of collection terminal is changed, without changing the original program, and with the help of the configuration file, it can adaptively adjust the display interface, reduce development costs and improve the efficiency of secondary development, achieve universal display interface design.

ACKNOWLEDGEMENT

It is a project supported by the international cooperation project (2010DFR10710) and Chinese major science and technology project (2014ZX04015011).

REFERENCES

[1] Wei Fu, Jian Fang. Energy Detector Design Based On Wireless Sensor Networks. Automation Instrumentation, 2013, 07: 49–53.
[2] Zhaohui Jiang, Jun Jiao etc.. Agriculture Universal Wireless Monitoring System Based On ZigBee. Anhui Agricultural Sciences, 2010, 06: 3149–3151.
[3] Jing Li, Jianxin Wang. Research Of Transmission Control Protocol Based On Wireless Networks. Telecom Express, 2010, 07: 37–40.
[4] http://baike.baidu.com/view/175122.htm?fr=aladdin.

Computing, Control, Information and Education Engineering – Liu, Sung & Yao (eds)
© 2015 Taylor & Francis Group, London, ISBN: 978-1-138-02800-5

Recognition and segmentation for fire smoke based HSV

Xue Jun Chen
College of Electrical Engineering and Automation, Tianjin University, Tianjin, China
Putian University, Putian, China

Feng Dong
College of Electrical Engineering and Automation, Tianjin University, Tianjin, China

ABSTRACT: It is important for Video-based fire detector to accurately recognize fire smoke. This paper proposes a HSV-based method for fire smoke segmentation from video image. The kinds of color features including color distribution and histogram are compared. In segmentation based on color distribution, fire color samples are collected and then projected onto S, HV and V planes. In segmentation based on H value of HSV color model, the lower threshold is defined from experience, and the higher threshold value is computed automatically, thus the possible fire smoke regions can be segmented accordingly. Final fire smoke regions are generated from the biggest connected region processing results. Our method is tested on Video-based fire detector with fire and the results are encouraging.

KEYWORDS: Segmentation; Recognition; HSV; Fire; Video.

1 INTRODUCTION

In recent years, with the increasing large space buildings, the suitability of fire detector used in such large space situations has been challenged. Commonly, smoke detector and heat detector are not suitable for high large space or outdoor. For example, even a 5MW fire on the atrium floor would not generate a smoke temperature high enough to activate the detectors on an atrium ceiling higher than 15m [1]. To gain efficient fire detection and protect the lives of many civilians, many fire detectors are developed. There are several methods which are being tested to help improve the quality of detectors throughout the world. They include Visual fire detection [2], multi-spectral sensors [3], robotic fire monitor [1, 4] and other methods.

For the residential fires, being flaming or non-flaming, the general trend is to focus either on the sensor and sensor combinations or detection techniques. Nowadays, visual fire detection is widely studied and used. Existing methods of visual fire detection rely almost exclusively upon spectral analysis using rare and usually costly spectroscopy equipment. Moreover, these methods still produce false alarms in the case of objects whose colors are almost the same as fire. The goal of this paper is to design and test a fire detector which can more reliably segment fire smoke and identify an actual fire. The proposed fire detector technology is based on two cameras. One camera with color video can be monitored as safety monitor when without fire. Thus, when it detected and find the fire it can set as a fire alarm. Another camera with infrared video is used to help the fire detector identify fire. The video-based detector is essential to saving more lives in the future. Therefore, image segmentation for the fire smoke is one of the key works of the video-based detector.

2 VIDEO-BASED FIRE DETECTOR

The fire detector is made up of video acquisition and output module, DSP processor module, communication module and power module. There are two cameras, which one is a color video signal, and the other is an infrared video signal. The two cameras of the video acquisition module will capture color and infrared scene real time and transmitted them to DSP processing center through video encoder. The DSP processing center will process the received videos, including denoise, fire detecting, and output. After the video were detected, they would output through VP0A. Then the monitor will be able to monitor real-time site conditions in the fire control center. The fire alarm signal would be transmitted to the fire control center through the RS485 communication module. Users can monitor the acquisition videos and send the controller signals to the fire detector with interface software or control device. The fire detector diagram is shown in Figure 1.

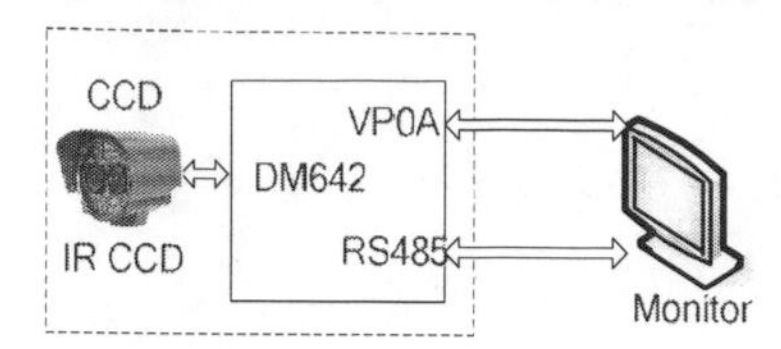

Figure 1. Diagram for video-based fire detector.

3 FIRE SMOKE SEGMENTATION METHOD

Fire smoke segmentation is crucial for the performance of video-based fire detector. Various color spaces are used for processing digital images. For some purposes, one color space may be more appropriate than others. The RGB color space is one of the most widely used color spaces for storing and processing digital image. Aside from raw RGB values, the fire detector uses HSV to aid in the segmentation.

It aims to obtain the region for the fire smoke. Then fire smoke position can be obtained by analyzing segmented image for the fire smoke. The proposed fire smoke segmentation algorithm includes four steps, HSV processing, segmented image for the fire smoke, morphology processing and connectivity and identification. The algorithm flow chart of fire smoke segmentation and processing is shown in Figure 2.

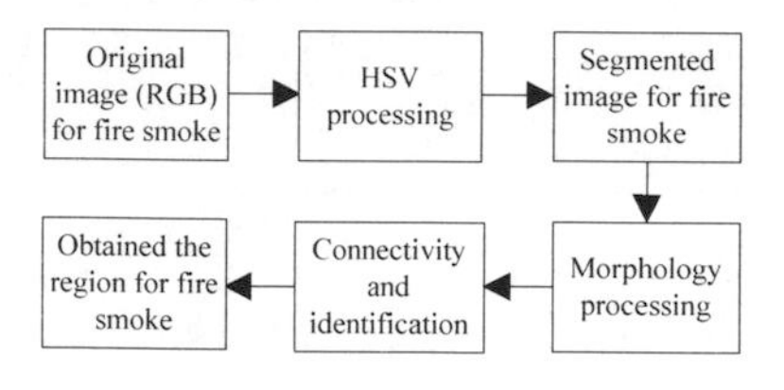

Figure 2. Algorithm flow chart of fire smoke segmentation.

3.1 *HSV processing*

Photographic digital cameras that use a CMOS or CCD image sensor often operate with some variation of the RGB (red, green and blue) model. In the RGB model, an image consists of three independent image planes, one in each of the primary colours: red, green and blue. The color is expressed as an RGB triplet (r, g, b), each component of which can vary from zero to a defined maximum value. Figure 3 (a) shows the original image of the RGB colour model for fire smoke.

One of the first steps is to segment them and find out smoke in them. To find the difference among three independent image planes of the original image, they should be separated. R image plane, G image plane and B image are respectively shown in Figure 3 (b), Figure 3 (c) and Figure 3 (d). To convert them into histograms, the histograms of R image plane, G image plane and B image are respectively shown in Figure 3 (e), Figure 3 (f) and Figure3 (g). From the results, it is difficult to segment them based on RGB model. To separate fire smoke from the background, HSV processing for the original image is used in the paper.

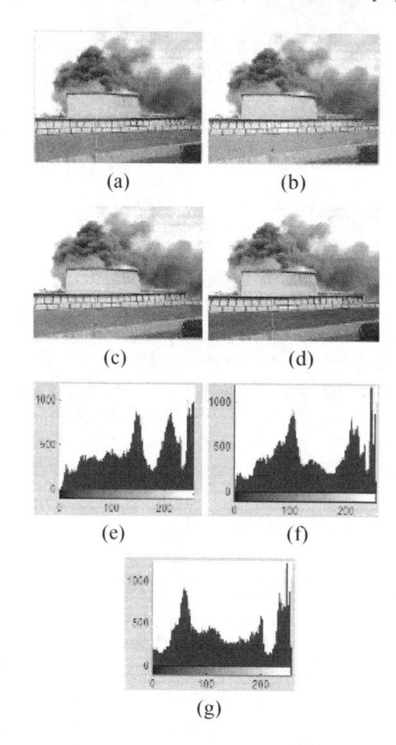

Figure 3. RGB processing of original image for the fire smoke (a) Original image; (b) R image ; (c) G image; (d) B image; (e) the histogram of R image; (f) the histogram of G image; (g) the histogram of B image.

The HSV color space divides the color into three separate components: hue (H), saturation (S) and value (V). The HSV color space describes color with intuitive values. Hue defines the dominant color as described by wavelength, for instance the distinction between red and yellow. Saturation measures the colorfulness of an area in proportion to its brightness such as the distinction between red and pink. Value refers to the color luminance, the distinction between a dark red and a light red. HSV color space is the often used one because of its accordance with human visual feature.

With different illumination, the hue will not change. By means of HSV color space, the segmentation method will be more stable. So, the HSV color space is used in order to separate the brightness from the chromaticity so that stability is enhanced. The segmentation process is performed using appropriately defined ranges of hue and saturation.

For fire smoke detection, the value component is discarded to eliminate the undesirable effect of uneven illumination. The transformation from RGB to HSV is described as follows [5]:

$$H=\begin{cases} undefined, & if \quad M=m \\ 60^{\circ}\times\dfrac{G-B}{M-m}+0^{\circ}, & if \quad M=R\&\&G\geq B \\ 60^{\circ}\times\dfrac{G-B}{M-m}+360^{\circ}, & if \quad M=R\&\&G<B \\ 60^{\circ}\times\dfrac{B-R}{M-m}+120^{\circ}, if & M=G \\ 60^{\circ}\times\dfrac{R-G}{M-m}+240^{\circ}, if & M=B \end{cases} \tag{1}$$

$$S=\begin{cases} 0, & if \quad M=0 \\ \dfrac{M-m}{M}\times 255, & otherwise \end{cases} \tag{2}$$

$$V=M\times 255 \tag{3}$$

Where M=max(R, G, B) and m=min(R, G, B).

Figure 4 shows an example for image content analysis. The original image in Figure 3 (a) is processed by transforming RGB into HSV, getting the HSV image and histogram. The HSV image of Figure 3 (a) is shown in Figure 4 (a). The original histogram is shown in Figure 2(a). H image, S image and V image are respectively shown in Figure (b), Figure 4 (c) and Figure 4 (d). Figure 4 (e) is the histogram of H image. Similarly, Figure 4 (f) and Figure 4 (g) are respectively the histograms of S image and V image. As can be seen from these histograms, they are obviously different distribution. Histogram equalization (HE) and its variations are one of the most commonly used algorithms to perform contrast enhancement due to its simplicity and effectiveness.

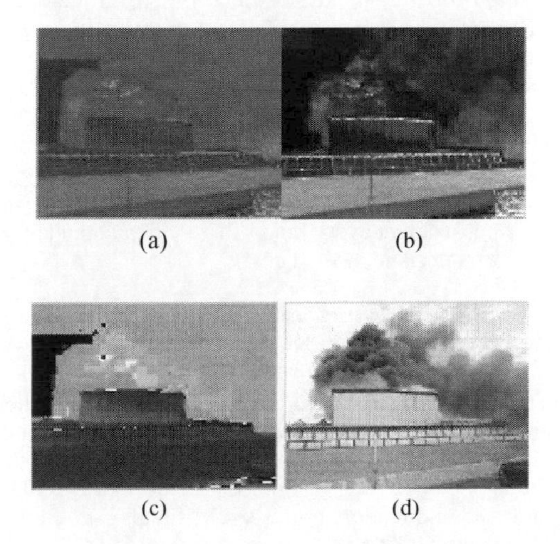

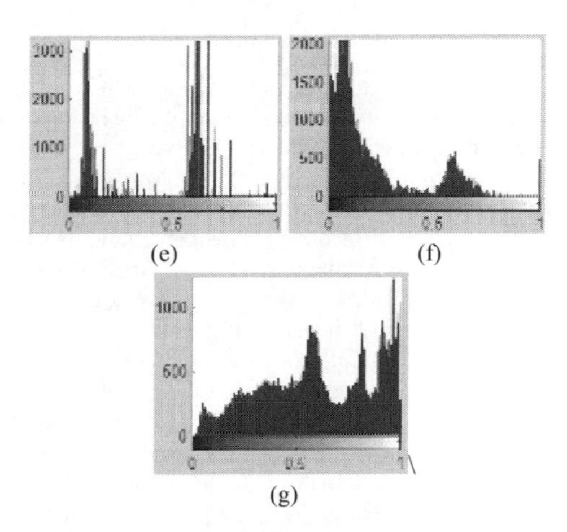

Figure 4. HSV processing of the fire smoke. (a) HSV image of the fire smoke; (b) H image; (c) S image; (d) V image; (e) the histogram of H image; (f) the histogram of S image; (g) the histogram of V image.

It is important, therefore, that the features of interest can be distinguished. HSV is used as a compromise between effectiveness for segmentation and computational complexity. If the goal is object detection, roughly separating hue, saturation, and value is effective. To detect fire smoke, it will compute the threshold and segment hue, saturation or the two parameters of fire smoke from HSV image. The goal for the proposed method is to select a threshold that would segment well the HSV image. The contour of the smoke area is extracted using statistical HSV color space.

Figure 5 (a) was the segmented image for fire smoke from Figure 4 (a) by saturation. Figure 5 (b) shows the segmented image for fire smoke by hue and value. And the segmented image for fire smoke by hue is shown in Figure 5 (c). As it can be seen from Figure 5, Figure 5 (c) has more significant characteristics for the fire smoke. This will be beneficial to segment fire smoke.

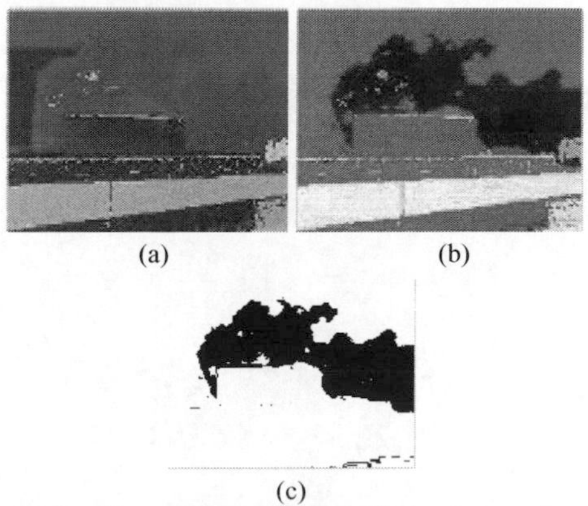

Figure 5. The segmented image of the fire smoke. (a) Segmented image by S; (b) Segmented image by H and V; (c) Segmented image by H.

3.2 *Morphology processing*

After the segmented image of fire smoke by hue, mathematical morphology was used to adaptively process the structuring elements for fire smoke extraction. Among the operation of mathematical morphology, erosion and dilation is the base of other complicated operations' implementation. Erosion is a transformation of shrinking, which decreases the grey-scale value of the image, while dilation is a transformation of expanding, which increases the grey-scale value of the image.

In order to obtain more obvious fire smoke area from segmented image, we use closing and opening operation to denoise the segmented image and detect edge of fire smoke. Let F(x, y) denote a binarizaton segmented image, and A denote structuring element. Opening and closing of F(x, y) by A are denoted respectively by equation (4) and (5) [6-8].

$$F \cdot A = (F \Theta A) \oplus A \tag{4}$$

$$F \cdot A = (F \oplus A) \Theta A \tag{5}$$

Opening is erosion followed by dilation and closing is dilation followed by erosion. Opening can smoothe the contour of the binarizaton image and break narrow gaps. As opposed to opening, closing tends to fuse narrow breaks, eliminates small holes, and fills gaps in the contours.

Figure 5 (c) was dilated shown in Figure 6 (a). Figure6 (a) shows that there are four connected regions, but their sizes are different. In order to gain fire smoke region, connected region processing were used for Figure 6 (a), and Figure 6 (b) show the processing result. Only was one biggest connected region obtained, which is fire smoke. It is shown in Figure 6 (c). Connected region statistics was used for area calculation of fire smoke. After the statistics, the actual fire smoke area is 43591 pixels, and the segmented fire smoke area is 42218 pixels. The segmentation accuracy is 96.85%.

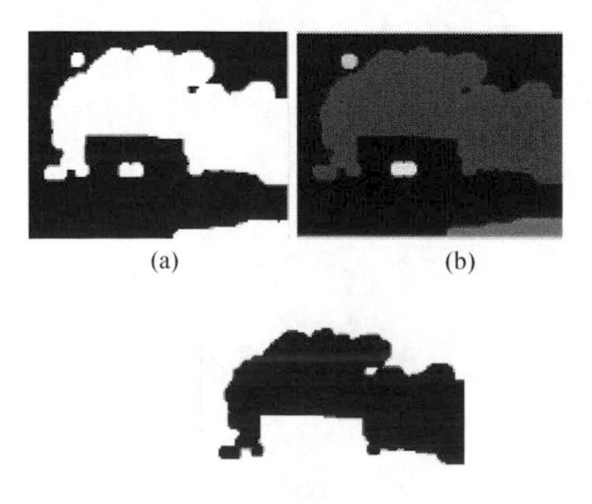

Figure 6. Image of fire smoke preprocessing. (a) The dilated results for Figure5 (c); (b) Connected component labeling; (c) Obtained region.

3.3 *Experiment*

The proposed method can give a threshold value, and then it was transplanted to the fire detector. The lighter was introduced to do the fire detecting test.A sampling rate of 25 frames/s was employed for the video-based fire detector. Figure 7 shows results of using proposed method to segment an image locate the fire Centroid.

In applications such as surveillance the flame of the lighter will be extremely small if a wide field of view is imaged and only a small, distant object has moved. The proposed method can successfully isolate the flame or smoke. In order to see located fire or smoke clearly, a rectangular frame was added to the result, and the coordinates of the flame Centroid of each frame images were shown in monitor.

Figure 7. The segmentation results for fire smoke detecting test.

4 CONCLUSIONS

Authors presented a recognition and segmentation method for fire smoke for video-based fire detector. The method based on HSV and morphology algorithm was proposed and used for the fire smoke segmentation. A lighter was introduced to do fire smoke segmentation test to verify the proposed method for vide-based fire detector. The results show that this method can well locate the fire smoke of the lighter.

The future work, we will use the detector and method to detect every kind of fire video and analyze the characteristic of the fire.

ACKNOWLEDGEMENTS

This work was supported in part by the Fujian Science and Technology Key Project (2013H0038), the Youth Foundation of Putian University (2013052), the Putian Science and Technology Project (2012G03), and the Educational Research Funds of Putian University (JG201309).

REFERENCES

[1] Tao Che, Hongyong Yuan, Guofeng Su, Weicheng Fan. 2004. An automatic fire searching and suppression system for large spaces. *Fire Safety Journal* 39(4):297–307.

[2] Turgay Celik, Hasan Demirel, Huseyin Ozkaramanli, Mustafa Uyguroglu. 2007. Fire detection using statistical color model in video sequences. *Journal of Visual Communication and Image Representation* 18(2):176–185.

[3] S. Briz, A.J. de Castro, J.M. Aranda, J Meléndez, F López. 2003. Reduction of false alarm rate in automatic forest fire infrared surveillance systems. *Remote Sensing of Environment* 86(1):19–29.

[4] A. De Santis, B. Siciliano, L. Villani. 2008. A unified fuzzy logic approach to trajectory planning and inverse kinematics for a fire fighting robot operating in tunnels. *Intelligent Service Robotics* 1(1):41–49.

[5] Chun-Ming Tsai, Zong-Mu Yeh. 2008. Contrast enhancement by automatic and parameter-free piecewise linear transformation for color images. *IEEE Transactions on Consumer Electronics* 54(2):213–219.

[6] Zhao Yu-qian, Gui Wei-hua, Chen Zhen-cheng, Tang Jing-tian, Li Ling-yun. 2005. Medical Images Edge Detection Based on Mathematical Morphology. *Proceedings of the 2005 IEEE Engineering in Medicine and Biology 27th Annual Conference, Shanghai, 1–4 September 2005*. New York: IEEE.

[7] Frédéric Zana and Jean-Claude Klein. 2001. Segmentation of Vessel-Like Patterns Using Mathematical Morphology and Curvature Evaluation. *IEEE Transactions on Image Processing* 10(7):1010–1019.

[8] Lee J.S.J., Haralick R.M., and Shapiro L.G.. 1987. Morphological Edge Detection. *IEEE Transactions on Robotics and Automation* 3:142–156.

Computing, Control, Information and Education Engineering – Liu, Sung & Yao (eds)
© 2015 Taylor & Francis Group, London, ISBN: 978-1-138-02800-5

Excellent resource sharing construction scheme study of automobile theory course—taking Huaxia College, Wuhan University of Technology as an example

Huan Qin Wu & Wan Fu Yang
Huaxia College, Wuhan University of Technology, Wuhan, PR China

ABSTRACT: "Automobile Theory" is a core curriculum of vehicle engineering or related majors. In order to promote its construction, this paper took Huaxia College, Wuhan University of Technology as an example, analyzed the construction target of vehicle engineering professional' "Automobile Theory" resource sharing course, the tactics of teaching staff construction, the reasonable selection of teaching content, the rational use of teaching methods etc.. Finally, this paper colluded that in the process of construction, we need improving levels of faculty through long-term unremitting efforts, strengthen the construction of teaching material, keeping pace with the times, optimize teaching content and curriculum system, promote the reform of teaching methods and teaching means, then improve the overall teaching quality.

KEYWORDS: automobile theory; excellent resource sharing; teaching method.

1 INTRODUCTION

China is a huge country in car sales and car production. In the process of the development of automobile industry, we need a group of technical people, who have a solid foundation of comprehensive and innovative professional knowledge. The knowledge of Automobile Theory is the foundation to the students engaged in the work of the automobile and related professional, which is also basic knowledge for other academics and automobile enterprise technical personnel to understand or application of the vehicle structure system. The curriculum construction affects the teaching effect directly, and then influences the car technical personnel training and the development of the automobile industry. Because of the much content, comprehensive and high requirements on the basis of theoretical knowledge, engaged in vehicle engineering teaching or researching and automobile enterprises technical personnel generally believe learning, understanding or application the automobile theory knowledge is not easy. [1] Therefore, the teaching system structure, basic teaching content and teaching methods must be carefully designed.

"Automobile Theory" is a professional required and core course for vehicle engineering professional's students. "Automobile structure", "engine principle" and other professional courses are the first course, and "automobile dynamics", "car design", are the follow-up courses for "Automobile theory". So, "Automobile theory" is a simulation and a foundation professional course which links car theory, system and design. [1] This course is based on mechanics, and it expounds fuel economy, braking, handling stability, ride comfort, through ability and the dynamic of the automobile, and research how to select automobile reasonable design parameters based on meeting these performance requirements, also introduces how to use modern technology to improve automobile performance. Technical personnel of vehicle engineering must master this course.

2 THE TARGET OF COURSE CONSTRUCTION

Automobile theory can Guide the students to master the basic vehicle knowledge, design and detection principle and means. So, the students' practical and comprehensive solution's ability can be improved greatly. Automobile new technology emerges continuously with the development of technologies. Therefore, in order to keep pace with the changing times, in the teaching process, we must adhere the guidance of a scientific concept of development, taking students as the fundamental, based on knowledge, taking ability as the core, based on market demand, to Integrate of classical and modern knowledge, study and revision the automobile theory syllabus timely. We can reasonable select teaching method, use of modern means of teaching comprehension, and construct this course to be illustrated, rich in content, completely open and sharing excellent network resources. In order to cultivate students' application and innovation ability, we must enforce the construction of laboratory, increase comprehensive and innovative experimental project. In order to guarantee the sustainable development, a team of teachers must be constructed. The teachers must have a high level of theoretical knowledge and reasonable knowledge structure, be able at practice, good at the ethics of teachers, and in different ages.

3 CONSTRUCTION SCHEME

3.1 *The construction of a teaching team*

In order to meet the need of school to transformer towards the direction of "occupation", according to the school location and requirements for students training, it is focused on the team of double-position teachers; they have not only professional theory knowledge and improving the teaching art, but also be paid attention to the cultivation of professional skills and practical experience. In this way, the teachers can not only qualify for teaching and scientific research, but also undertake the responsibility of production and practice.

3.1.1 *Construction measures*

1 Use of laboratory equipment to train young teachers' practical skills

In order to meet the requirements of the development of the automobile industry and to adapt to the transformation of college, it is must adjust the thinking of talents cultivation. So, we can give full play to the school laboratory equipment, in practical skills training for young teachers during the holidays. In order to instruct the students better, we can carry out the teaching activities to practice basic training skills in teaching group at the same time.

2 Held Teaching research activities regularly

We can improve the teachers' level of professional theory and professional skills by holding variety teaching activities, such as speaking a course, dealing a lesson, researching the work of graduates in the enterprise, visiting automobile society annual meeting. At the same time, the teachers are keeping pace with automobile development, and understand the development of the automobile, improve the ability of teaching theory and Practice finally.

3 Strengthening the training of young teachers

According to the characteristics of professional teachers, training them must be in a planned manner and in their direction. Arrange the young teachers to attend the seminar, which takes the curriculum and teaching reform, construction of teaching materials as the theme. Take a refresher course to improve the teaching level. Encourage the young teachers with a master's degree to apply for the doctoral degree. Let the teachers to maximize obtain frontier discipline knowledge, Teaching method and experience to improve the quality and efficiency of training.

4 Encourage young teachers to declare research and education project

Each school has their own research and education projects every year, in order to improve the teaching reform, we can organize teachers to declare or participate, and for practical project, we can combine Enterprise, and then recommend the best to a high level.

5 Broaden the channel of talent introduction

In order to improve the teaching quality, there are three ways in the introduction of talent. The first one, it is introduction of professional teachers from college graduates and emphasize on cultivating. The second one is the introduction of high-level technical personnel from enterprises and institutions or by hiring a higher level Professional and technical personnel to teach professional course. The third one is to hire retired teachers from the first-class university.

6 All professional practice in enterprises

In order to obtain the skills and experience required for practice teaching, we can encourage young teachers to post or credentials practice regularly in cooperative enterprise. Our school arranges teachers to professional counterparts of the enterprise, practicing every year, then the teachers can access advanced equipments, new technology and enterprise culture, Understand operative condition timely, Master of business process system technology, Rich practical experience, Strengthen practical skills, Improved teacher quality. After the practice, new technology and new process can be feedback timely in teaching, and then the effect of class education can be improved.

7 Strengthen the practice teaching

Teachers must undertake practice and Theory Course at the same time. Let the theory teaching into practice Teaching, and then improve teachers' professional skills.

8 Encourage teachers to participate guide the students' innovation work

Most independent colleges, such as Huaxia College, recruited students is the third level. These students are instable ground knowledge, but they are strong at handwork and active thinking. So, according to the characteristics of young teachers, we can plan to arrange them to guide students' innovative work. In the process of innovation, teachers can help students identify and plan of innovation projects, audit drawings and theoretical calculation, guide the students make the product or debugging program, predict the risk which possibly appears in the process of race, at the key time, they must guide students to make important decisions. [2] In the process of innovation guidance, our teachers can understand the trend of the development of science and technology. At the same time, they can enhance their ability innovation and scientific research.

3.1.2 *Achievements in the construction*

There are 9 teachers in our school automobile theory teaching team. Among them, there are 5 theory teaching teachers, 2 teaching guidance teachers, 2 experimental teaching teachers. There is a detailed analysis shown in Table 1.

Table 1. Detail analysis table.

Educational background structure			Age structure		
Dr.	Master	Bachelor	More than 50 years	30~50	Under 30 years
2	3	4	3	4	2
22.2%	33.3%	44.4%	33.3%	44.4%	22.2%
Graduation school structure			Title structure		
Key universities	Others	Professor	Associate Professor	Lecturer	Assist-ant
7	2	2	3	2	2
77.8%	22.2%	22.2%	33.3%	22.2%	22.2%

From the structure of graduate school, most teachers of this team graduated from Wuhan University of Technology, Changan University, which is 985 colleges or 211 colleges, All the teachers have been educated well. From the research direction, there are automobile structural, Automatic transmission, Auto electronic control technology, Automobile performance, Automobile culture etc., which are different. From the staff-student ratio, it is 1:10, which meet the needs of teaching. Integrated view, this course team is reasonable in educational background structure, age structure, graduation school structure, title structure.

3.2 *Establishment of teaching system*

Teaching system construction should be based on the school orientation and training objectives. So, there will be different content in comprehensive and in-depth for different colleges and universities.

Huaxia College, Wuhan University of Technology formulated reasonable personnel training plan according to the training objective application and occupation teaching of the third level college. By examining the curriculum and teaching program carefully, arranged the course content, and the automobile theory course teaching is divided into two parts: theory teaching and practice teaching.

The focus of the theory teaching is automobile basic performances understanding and the analysis methods mastering, such as calculation and analysis of automobile dynamic property, qualitative analytics of its economic and braking performance, two and single degree of freedom control stability model, steady state steering characteristics, ride comfort evaluation standard. About the key content of the course, in order to ensure the students grasp better, we can layout class homework, and reflect in its examine. About the thorny problems, In order to understand for students, we can explain the preparation knowledge in advance. For example, before a vehicle ride comfort explaining, we can supplement spectrum analysis and analysis of multi degree of freedom vibration.

About the practice teaching, there are two parts. One is the basic experiment, another is the extending experiment. Basic experiment is on the basic performance of the automobile test. Its purpose is to understand of the basic concepts and methods. The test environment, process and equipment must be similar with the domestic and international mainstream automobile enterprises which use in the technology research and new product development. So, it can consolidate what the students have learned in class and improve the practical ability. At the same time, the students can understand the technical development level of domestic and foreign automobile performance testing and expand their horizons. In the process of organization and implementation of the test, students must operate test instruments, analyze testing result, and finish the experiment report their own while security is guaranteed. About the extending experiment, its purpose is to cultivate students' ability of using the studied theoretical knowledge. A group of 3-5 students list the experiment name and write the experiment scheme themselves according to what they have learned in the course. Teachers provide assistance as required in the process. The students will burn in hell first and the learning interest will be improved greatly.

3.3 *Improving teaching conditions*

1 The construction of teaching materials

In order to achieve the goal of teaching, we must choose textbooks reasonably or make teaching materials on the basis of the characteristics of school students, product electronic teaching plan and multimedia courseware, write exercises set and experimental guide books, and all displayed on the web with classroom teaching. The teaching group writhed teaching material "automobile theory" of application type on the basis of reviewing the experience of years, maintained the advanced nature of the teaching content. Cooperate with the teaching materials, "automobile culture", and "automotive electrical equipment" etc. has been published at the same time to expand knowledge of students and enhance the enthusiasm of learning.

2 Improvement of experimental equipment

The basic experimental equipment provided by the laboratory. They include experimental car, automobile performance test instrument, speed sensor, fuel consumption instrument, Industrial Personal Computer, gyroscope, three axial seat sensor, human body vibration meter, computer, printer etc.. In order to meet the need of teaching, these instruments should be more advanced.

3 Construction in network

Every classroom, office and dormitory are connected to broadband networks and with free "WIFI" backed by school broadband information network. It provides network support for extracurricular network teaching. At the same time, teachers of the group, especially the old teachers were trained network knowledge for the course resource sharing.

3.4 *Reform in teaching methods*

1 Heuristic method of teaching.

Cancel expository way of teaching and use heuristic method. Then, the students will be on their own initiative instead of studying passively. Because of the limited teaching time, teachers can leave a small part of the teaching content to students' self-study, and let the young teachers answering the questions in spare time. Then, the young teachers get exercise and the students' self-learning ability is trained.

2 Combination of teaching and research

Encourage students to participate in teaching research project of teachers, and enable them to have the opportunity to contact new automotive technology. In this process, students' interest in learning increases slowly.

3 Combined with the MATLAB software teaching

Combined "Automobile Theory" with the MATLAB software course teaching, and assigned car several properties to students to program for a figure at the spare time by young teachers' guidance. [3] So, the students could learn professional knowledge better and time training their ability of computer application be trained at the same.

4 Combination of theory and Practice

In the process of teaching, theory and practice carry out at the same time. Interactive teaching, teaching benefits teachers as well as students, all can improve the teaching effect. In order to get better effect, strengthen practice segments and give students more practical opportunities with the coordination of other relevant courses. By the students' test the performance of automobile themselves, their hand's ability will be stronger and their perceptual knowledge increase.

3.5 *Reasonable application of teaching methods*

Writing on the blackboard and means of computer network teaching will be used mainly in this course teaching in the light of the characteristics of the course.

1 Using the writing on the blackboard of traditional teaching mode in classroom teaching

For the "Automobile Theory" course teaching, if use the computer in classroom, the theoretical formula will be deduced too fast and students will lack of thinking. Then, the teaching effect will be affected.[4] So, using the method of writing on the blackboard in classroom teaching, logical derivation process longer will be visible and students make consecutive thought and take notes of the knowledge possible. In the process of writing, the students will have a strong emotional impact. It will attract the student attention and stimulate the students' interest and have the students take the initiative in learning at the same time.

2 By using the computer at spare learning time

Automobile theory content too much and aspect of knowledge too broad to classroom teaching is only not enough. So, this course must make multimedia courseware. And with the help of the teaching platform for excellent resource sharing, let students learn in spare time. Give full play to the advantages of multi-media teaching, such as image, intuitive, vivid, and informative. By pre-course or review after class, students can know more information and consolidate knowledge learned in the classroom to understand and master. For example, before explaining stabilization of the vehicle, by looking at the video of the car running on the track, students can be more intuitive understanding of the meaning of the stabilization of vehicle, and promote the knowledge of learning in classroom understanding.

3 Part of the homework through the computer to complete and submit

This course includes much curve graph analysis, and it is also an important tool for automobile performance. But in teaching materials, the graphics are given directly. The teacher can choose the appropriate graphics to require students to program as what the teacher said in class, including relevant theories and models. If given the parameters of automobile, students can draw the automobile performance curve and analysis of different factors and parameters on the performance of the automobile. Then the students' interest and positive initiative in learning will be enhanced.

4 CONCLUSION

"Automobile Theory" is a core curriculum of vehicle engineering or related majors. It has referenced value in the actual application and will influence the students directly for their work after graduation or continue their studies. Carry out the "Automobile Theory" boutique resources sharing curriculum construction can broaden students' horizons, improve the students' interest in learning, cultivate students' innovative spirit, and then improve the talent-fostering quality, promote the development of automobile industry at last. However, its construction is not only a long-term process of accumulation, but also a review to improve. [5] In the process of construction, we need improving levels of faculty through long-term unremitting efforts, strengthen the construction of teaching material keeping pace with the times, optimize teaching content and curriculum system, promote the reform of teaching methods and teaching means, then improve the overall teaching quality.

ACKNOWLEDGEMENTS

This work was supported by the Twelfth Five-Year Education Plan of Hubei province (No.2012B433), the Scientific Research and the Teaching Research Project of Huaxia College Wuhan University of Technology (No.13006, No.1423).

REFERENCES

[1] Bao F.B. "Automobile Theory" Course Research on the Practical Teaching Reform[J]. Value Engineering, 2011(33).

[2] Wu H.Q. Study on the Young Teachers' Comprehensive Ability of Vehicle Engineering for Cultivation of Independent College [J], Chinese Science Innovation Herald, 2013(1).

[3] Huang Y.F. Application of MATLAB Software in the "Automobile Theory" Teaching [J]. Science and technology information, 2010(8).

[4] Jang Y.C. the Application Method of the Combination of Theory and Practice in "Automobile Theory" Teaching [J]. Chapter, 2013(5).

[5] Xie C.Q. etc., Exploration and Practice Concerning the Course Construction of "Soil and Fertilizer Science "[J], Journal of Anhui Agricultural Sciences 2012(34).

Computing, Control, Information and Education Engineering – Liu, Sung & Yao (eds)
© 2015 Taylor & Francis Group, London, ISBN: 978-1-138-02800-5

The application of visual simulation in urban rail vehicles driving simulation system

Song Cang
Guidaojiaotong Polytechnic Institute, Shenyang, China

ABSTRACT: Virtual reality is a mimic technology by using computer; it can satisfy the need of immersion and interactivity by building up a 3D surrounding alike to reality. We discussed the key technique of virtual scene constructing, after that the subway driving simulation system was designed and implemented through the visual simulation for urban rail lines, based on MultiGen Creator. The visual scene was integrated into the urban rail vehicle driving simulation platform. The scene we got was lifelike, with a very high running speed. It is not only a 3D digital map, but also a simple GIS system. The research content of this paper has important reference value of the other virtual scene simulation and VR-GIS systems' development.

KEYWORDS: Virtual reality, MultiGen Creator, Simulation system, urban rail vehicle.

1 INTRODUCTION

With the rapid development and increasing sophistication of virtual reality technology, it has been on its closer and closer way to applications in many fields. The virtual reality technology is applied to urban rail train driving simulation system, in this we can use the computer simulation technology to undertake a skill training to drivers. This has numerous advantages: avoiding environmental pollution, reducing energy consumption and the training costs.

The visual system is an important part of urban rail train driving simulation system. To measure whether a simulation system advanced or not, it is a key factor, meanwhile, it is also an important condition to guarantee the training effect.

2 VISUAL SIMULATION TECHNOLOGY AND THE STRUCTURE OF URBAN RAIL TRAIN DRIVING SIMULATION SYSTEM

Visual simulation technology is an important branch of computer simulation technology. Its core technology includes simulation modeling, animation simulation technology and real-time visual generation technique [1]. The visual simulation is based on the purpose of the simulation: building 3D model of simulation object or reproducing the real interactive simulation environment. Its objective is to simulate the real world; users can immerse in it and fully interact with it [2].

The hardware structure of the urban rail train driving simulation system is shown in Figure 1. Environmental simulation generated by the driver room, visual and projection system, meanwhile, we check the performance simulation through the master computer and a variety of computer local area network (LAN). In this process, establishing a mathematical model of train performance helps us to further comprehend the running performance of the train. The train performance simulation is a real-time one. The simulation of the train environmental accomplishes by the visual simulation system, sound system and running simulation system. It is the embodiment of the realistic degree of driving simulation system. Visual simulation part of the system is mainly composed of two parts: the visual scene generation and visual display. Based on the principle and technology of computer graphic, visual generation succeeds, we also get the train running lines and scenes as we expect. Visual display section consists of a single or multi-channel projector and display screens. The paper adopted the related content of visual generation.

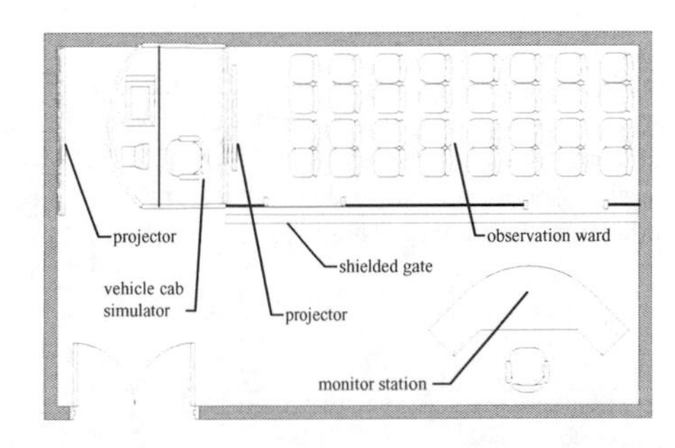

Figure 1. Hardware structure of urban rail train driving simulation system.

3 THE OVERALL DESIGN OF VIRTUAL SCENE

3.1 *The selection of modeling tools*

The entity model of the scene is an important component of the metro train driving system. If the vivid digital model can be established based on the real tracks' condition, it will offer the drivers a real driving experience either.

Most of the solid model in the visual system was created by MultiGen Creator. MultiGen Creator is a new generation software of real-time simulation modeling, developed by MultiGen-paradigm in the United States, which is the world's leading 3 d real-time database generation system. With its superior performance and good stability, it has been widely used in the field of Virtual reality and simulation [3-4].

3.2 *Building digital simulation platform*

In the development of visual system, we chose the encapsulated visual SDK with underlying API to improve the efficiency of development. According to the technical foundation of our laboratory and requirement of the project, we choose the Vega, which is developed by MultiGen-Paradigm, to develop the Visual system in the development environment of Visual C++ integration software.

3.3 *The overall structure of the system*

The design of visual simulation system is a two-step progress: the simulation environment constructing and scene simulation driving. To construct the simulation environment, we need to design the model, scene structure, texture and the special effects, etc.; meanwhile, driving the simulation system mainly includes the scene management and scheduling, distributed interaction and render output action, etc. [5].

In the progress of system development, we build a visual simulation environment by MultiGen Creator, at the same time, the simulation environment is reorganized logically relying on the MC's powerful management of the scene database; it adopts Vega (a three-dimensional real-time visual development software) to drive the simulation scene. Vega encapsulates the multifarious low-level graphics driver function, thus, we can realize the interactive control and schedule the scenes conveniently in the simulation program. The overall structure of the visual system is shown in Figure 2.

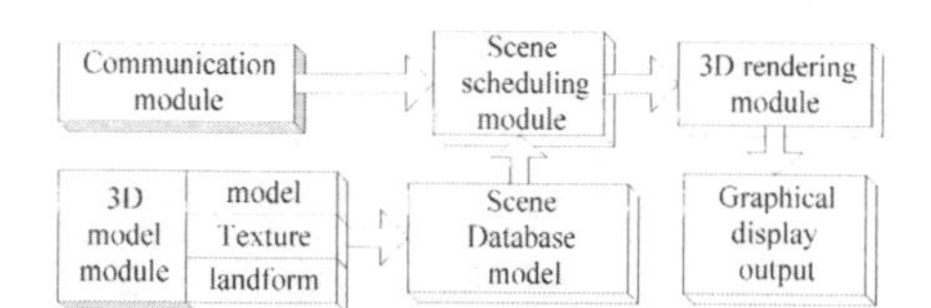

Figure 2. The overall structure of visual system.

The sub-modules' functions of the visual system are as follows:

1 Data communication module: deliver the data to the scene scheduling module
2 3D model-building module: set up scene model by MultiGen Creator
3 Scene database model: manage the attribute data and graphical data
4 Scene scheduling module: According to data communication module's message of the train driving simulation system, we search the visual database and obtain the scene data in the driver's visual field, then, we transit the data to 3D rendering module.
5 3D rendering module: realize the continuous display of moving images of virtual train movement relative to the surrounding environment.

4 3D MODELING BY MULTIGEN

Building environment of virtual is the basis of the visual simulation system, which is the key step for the visual simulation system implementation. And for this we must focus our attention on 3D scene modeling first.

4.1 *The establishment of a subway station model*

Before modeling, we need to explore a field survey of the buildings' information first, such as the length, width and height, the internal situation and other relevant information about the station, accomplishing the sketch on the paper is the next step we should do. In this, 3D modeling of the subway station is established by the applied sketch. Secondly, taking pictures at the station which are needed to reconstruct the three-dimensional entity models, thus, we can modify the pictures using the Photoshop and the texture mapping in the next section. It is an indispensable step to check the 3D graphic database structure. If the database structure is unreasonable, it is not only very difficult to be modified and maintained in the design process, but also influence the simulation effect in the process of the simulation program, which is because that if the data query costs the longer time, the display speed was influenced most greatly.

From the point of architecture, a metro station in the city is composed of platform, stairs, pillars, monitoring room, wall, billboards, road signs, and train operation tracks. For the 3D models of a subway station, we can arrange the various layers of the structure according to the size of its database (from large to small): platform, orbit, ceiling and other independent modules can be placed in the same level. These slave modules can then be linked dynamically, for example, we suppose that the monitoring room is placed on the platform, it is can be lower than the platform level because the monitoring room contains the nodes of the door, window, etc. There, for some common objects, we can establish a set of nodes and give them the unified management. Note that there are some surfaces that are invisible in the simulation in the establishing process of the 3D model: the outer edge of the wall, the station platform base, the bottom and top of pillars as well as the inside of the control room, etc.. If we transit all the built models into the database, there is no doubt increasing the burden of the system greatly, so, deleting the surfaces must be done after all models' completion.

Figure 3. The subway station model 1.

4.2 *The establishment of the rail model*

The establishment of the rail scenario is straightforward. Orbits mainly include the rail, the pad, the sleeper, and the ballast drainage and slope, etc.

To reduce the number of the ballast and improve the strength of the subgrade, the slag, sleepers and roadbed have been replaced by concrete foundation maintenance in modern track construction. So does the orbit of Shenyang metro (line 1).

The main content of the rail modeling is to build a straight body and a variety of curvature of the curve.

With reference to the actual orbit model, the structure of rail scenario contains the following sections: rail, sleeper with fasteners, concrete roadbed and guardrails.

The key work of orbit model building is to build the cross section of rail, sleeper and guardrail model. After that, we can connect several sequential cross sections using loft Tool, the main body of orbit mode formed. The trajectory model is shown as Figure 4.

Figure 4. The curve trajectory model.

4.3 *The integration of virtual subway scene database*

The integrated scene comes from the instance of Creator technology and external references. We can place some recurring models which are in different locations in a certain place through a scenario instance technology, such as the pillars of the platform.

All of the models are introduced into the current database through external reference technology, after that, they are adjusted according to the real scene of Shenyang subway Line one.

External reference technology provides two major benefits: facilitating the scene organization and management, and on the other hand it is easy for model replacement. Nevertheless, the disadvantage is obvious: it is read-only and cannot be edited directly; we can change the location, direction and size proportion only.

4.4 *The scenario model encapsulation*

There is lots of work to be processed when we put the 3D visual system model into Vega.

In general, the running time of system will be unaffected if the amount of the data to be processed is small, or when the data is static. On the opposite, if we are in need of dynamic real-time data, this can result in significant delays. Therefore, reducing the loading time becomes a realistic and urgent demand in situations when the models' visual effect is not affected.

A simple and effective way to improve models' loading speed is to transfer the complex .flt form into a simple .fst.

.fst model is in fast format, which is similar to Vega. Thus, the loading time reduces as well. At the same time it has the effect to the encapsulation of .flt model; we can avoid the chaotic texture when we change paths of the model.

In addition, the model in .fst form cannot be edit directly, which helps the work achievement made by the modelers to be protected.

The .flt form can be transferred into .fst model by the application plug-ins of fstexPort.dll.

4.5 *Key problems in model construction*

1 In Creator, we should choose "m" as the unit because a grid unit is one meter, at the same time it is consistent with the Vega's unit of length.

   So, the geometric modeling should be implemented according to the actual size of the object, that way, the interrelation of each virtual model is ensured.
2 A comprehensive adjustment and optimization to the structure chart is a must during the construction. The regularity of the model facilitates the control in Vega, which avoids too much Default node appearing.

3 Try to reduce the data size: simplify the model on the premise of realistic looking with Vsimplify Function of MultiGen Creator. In this process, we can't follow our inclinations thoroughly.

We should meet the requirements of Virtual reality interaction. While, in order to further strengthen the true feeling of the scene, texture design can be introduced freely.

5 CONCLUSION

As we know, visual simulation is one of the key technologies in urban rail vehicle driving simulation, whose vivid reality sense and running speed decide the performance of the system directly. In this article, we focus on how to establish a model of Multigen Creator software and drive the visual scene simulation by Vega, which provides the implementation means and methods for 3D modeling. Taking establishing the model of a subway station, its rail and surrounding environment as an example, we adopted the work process of scene setting. With the task in the process and the problems we may meet, this paper presented some practical solving methods also. Finally, the paper researches into the encapsulating method of scene model, and sums up the problems that should be paid attention to in the development process. Compared with other simulation system, the visual simulation mentioned in this paper has more advantages, and thus it is playing great promotion role in real-time application.

REFERENCES

[1] MultiGen-Paradigm Inc. MultiGen Creator User's Guide [M].U.S.A: MultiGen-Paradigm Inc,2004.

[2] MultiGen-Paradigm Inc.Veag Programmer's Guide (Version 3.7) [M].U.S.A: MultiGen-Paradigm Inc,2001.

[3] MultiGen-Paradigm Inc.Lynx User's Guide(Version 3.7) [M].U.S.A: MultiGen-Paradigm Inc,2001.

[4] MultiGen-Paradigm Inc.Creating Terrain for Simulations (Version 3.0) [M].U.S.A: MultiGen-Paradigm.

[5] Yang Jianguo, Wang Cheng.Virtual reality technology based on multigen &vega [J].Computer Simulation, 2003, 20(11):75–77.

Computing, Control, Information and Education Engineering – Liu, Sung & Yao (eds)
© 2015 Taylor & Francis Group, London, ISBN: 978-1-138-02800-5

Hadamard like $2\times2\times2$ channel matrix for networking interference alignment

Kai Li Zhang
Institute of Information Technology, Liaodong University, Dandong, China

ABSTRACT: We show that the $2\times2\times2$ interference network, i.e., the multi-hop interference network formed by concatenation of two two-user interference channels achieves the min-cut outer bound value of 2 DoF, for almost all values of channel coefficients, for both time-varying or fixed-channel coefficients. The key to this result is a new idea, called aligned interference neutralization that provides a way to align interference terms over each hop in a manner that allows them to be canceled over the air at the last hop. Like its wireless counterpart, which has 4/3 sum DoF, this channel is shown to have a sum capacity of 4/3 symbols per channel use for most channel realizations. The main insight is that, with a few exceptions that are pointed out, scalar (SISO) finite field channels over F_{p^n} are analogous to $n\times n$ complex vector (MIMO) channels in the wireless setting, so the DoF optimal precoding solutions for wireless networks can be translated into capacity optimal solutions for their finite field counterparts.

KEYWORDS: $2\times2\times2$ Interference Alignment (IA), Zero-forcing, DoF finite field, modular concept.

1 INTRODUCTION

Recently, interference alignment (IA) has been proposed to achieve the maximum degrees of freedom (DOF) for the K user interference channels [1]. It designs the signals transmitted by all users in such a way that the interfering signals at each receiver fall into a reduced-dimensional subspace. The receivers can then extract the projection of the desired signal that lies in the interference-free subspace. It can achieve the optimal multiplexing gain in the interference channels [2], [3]. By extending the desired symbol size in time domain, they can asymptotically achieve the maximum degrees of freedom (DOF) for each a channel. In order to achieve the maximum DOF, relay-aided approaches were investigated to achieve 3/2 DoF for 3-user interference channels [4, 5], where all channels vary every symbol time while both transmitters and receivers need global channel state information. Besides, a lot of feedback information is necessary.

Unlike the interference channel approach which can achieve no more than 1 DoF, Cadambe and Jafar show in [4] that the $2\times2\times2$ IC can achieve DoF almost surely. This is accomplished by a decode and forward approach that treats each hop as an X-channel. Specifically, each transmitter divides its message into two independent parts, one intended for each relay. This creates a total of 4 messages over the first hop, one from each source to each relay node, i.e., the 2×2 X-channel setting. After decoding the messages from each transmitter, each relay has a message for each destination node, which places the second hop into the X-channel setting as well. It is known that the 2×2 X-channel with single antenna nodes has DoF. The result was shown first by Jafar and Shamai [6-8] under the assumption that the channel coefficients are time varying. By using a combination of linear beamforming, symbol extensions and asymmetric complex signaling, Cadambe et al. showed that DoF are achievable on the 2×2 X-channel even if the channels are held constant, for almost all values of channel coefficients. Motahari [9-10] et al. proposed the framework of rational dimensions which allows DoF to be achieved almost surely even if the channels is fixed and restricted to real values. Thus, regardless of whether the channels are time varying or constant and whether they can take complex or only real values, interference alignment through the X-channel approach allows the 2×2 IC to achieve DoF for almost all channel coefficient values.

2 CONVENTIONAL WORK

First, let us deal with trivial cases where some of the channel coefficients are zero.

Theorem 1: If one or more of the channel coefficients $h_{i,j}$ is equal to zero, the capacity is given by:

1. If $h_{1,2}=h_{2,1}=0$ and $h_{1,1}, h_{2,2}\neq0$, then C = 2.
2. If $h_{1,1}=h_{2,2}=0$ and $h_{1,2}, h_{2,1}\neq0$, then C = 2.
3. If $h_{1,1}=h_{1,2}=h_{2,1}=h_{2,2}=0$, then C = 0.
4. In all other cases where at least one channel coefficient is zero, C = 1.

Based on Theorem 1, henceforth we will assume that all channel coefficients are non-zero. Without loss of generality, let us normalize the channel coefficients by invertible operations at the sources and destinations shown in Figure 1.

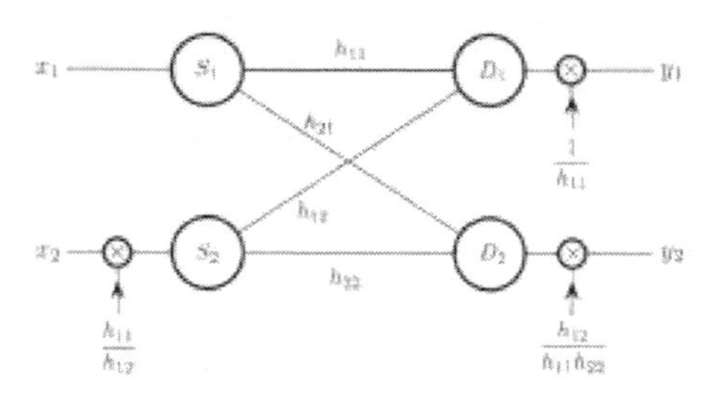

Figure 1. Normalization in X channel.

Destination 1 normalizes symbols by $h_{11}: y_1 = \dfrac{\hat{y}_1}{h_{11}}$

Destination 2 normalizes symbols by $\dfrac{h_{11}h_{22}}{h_{12}}: y_2 = \dfrac{\hat{y}_2 h_{12}}{h_{11}h_{22}}$

Source 2 normalizes symbols by $\dfrac{h_{11}}{h_{12}}: x_2 = \dfrac{\hat{x}_2 h_{12}}{h_{11}}$

Source 1 performs no normalization: $x_1 = \hat{x}_1$
The normalized X channel is represented as

$$\begin{cases} y_1 = x_1 + x_2 \\ y_2 = hx_1 + x_2 \end{cases} \quad (1)$$

wherein we have reduced channel parameters to single channel coefficient h, defined as

$$h = \frac{h_{12}h_{21}}{h_{11}h_{22}} \quad (2)$$

All symbols are still over F.

3 PROPOSED $2\times2\times2$ IA

Recently, $4/3$ is the highest achievable DoF result known so far for the $2\times2\times2$ IC that is applicable to almost all channel coefficient values. Opportunistic interference neutralization is easily understood as follows. For the $2\times2\times2$ IC in Figure 2, consider the setting where the product of the channel $F_{2\times2}\times G_{2\times2}$ matrices is a diagonal matrix. Clearly, in this case, if each relay simply forwards its received signal, the effective end to end channel matrix is a diagonal matrix, i.e., the interference-carrying channel coefficients are reduced to zero, creating a noninterfering channel from each source to its destination. Surprisingly, in this case, even though the channel matrix over each hop may be fully connected the network

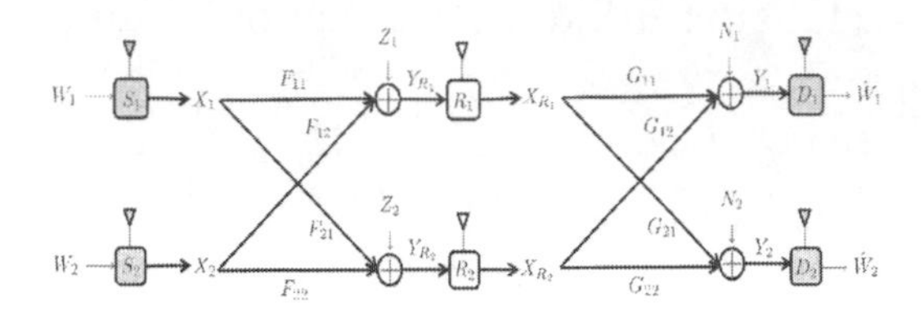

Figure 2. The $2\times2\times2$ IC.

1 Without Symbol Extension

From the source to the relay, it is networking alignment phase, while from the relay to the destination, it is a neutralization decode phase. In this section we should assume that the channel is time varying.

For the first phase, we normalize the channel coefficient similar as system in [4], then the received signals at the relay is

$$\begin{cases} y_1 = x_1 + x_2 \\ y_2 = -Hx_1 + x_2 \end{cases} \quad (3)$$

By this way the parameters of channels have decreased from 4 to 1. Then we forward these two signals to the destinations.

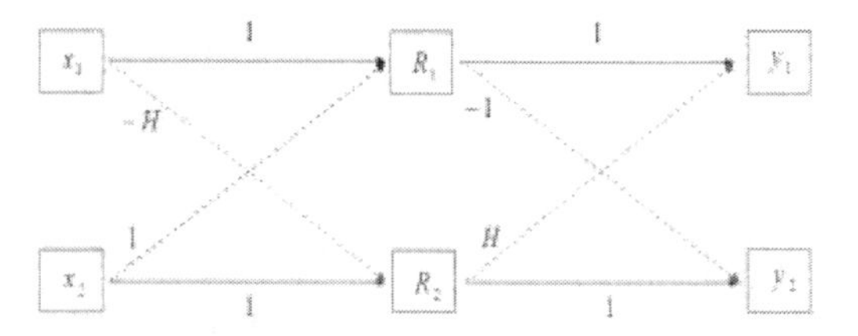

Figure 3. The $2\times2\times2$ IC.

We restructure the channel coefficients from the relay to the destination as shown in Figure 3. For the first phase, we normalize the channel coefficient similar as source to the relay phase, then the received signals at the destination is

For D_1:

$$\begin{aligned} Z_1 &= 1\cdot y_1 + (-1)y_2 \\ &= x_1 + x_2 + Hx_1 - x_2 = (1+H)x_1 \end{aligned}$$

For D_2:

$$\begin{aligned} Z_2 &= H\cdot y_1 + 1\cdot y_2 \\ &= h(x_1 + x_2) + (-Hx_1 + x_2) = (1+H)x_2 \end{aligned}$$

Because of channel state information is perfect known at every nodes, the remained problem is how the relay to the destination phase channel normalization and relates to the source to the relay phase. We should focus on the two Hadamard like matrix

$$\begin{bmatrix} 1 & 1 \\ -H & 1 \end{bmatrix} \text{and} \begin{bmatrix} 1 & -1 \\ H & 1 \end{bmatrix}$$

What is the difference between these two? All of them belong to the matrix

$$\begin{bmatrix} 1 & 1 \\ H & 1 \end{bmatrix}$$

We just change the pre-post-coding + or – at the relay or destination, also we can see the channel from the relay to the destination

$$\hat{H} = \frac{h_{R_1,D_2} h_{R_2,D_1}}{h_{R_1,D_1} h_{R_2,D_2}} \tag{4}$$

Because of our target is to get the same matrix of H, so between the relay and destination there should be exist a post-coder $U_{B,S}$ as

$$U_{B,S} = \frac{h_{1,2} h_{2,1} \cdot h_{R_1,D_2} h_{R_2,D_1}}{h_{1,1} h_{2,2} \cdot h_{R_1,D_1} h_{R_2,D_2}} \tag{5}$$

That means

$$\hat{H} U_{B,S} = H$$

So for the final result, finally, we can see that the two phases can be represented as the equivalent channel

$$\begin{bmatrix} z_1 \\ z_2 \end{bmatrix} = \begin{bmatrix} 1 & 1 \\ -H & 1 \end{bmatrix} \times \begin{bmatrix} 1 & -1 \\ H & 1 \end{bmatrix} \begin{bmatrix} x_1 \\ x_2 \end{bmatrix} = \begin{bmatrix} 1+H & 0 \\ 0 & 1+H \end{bmatrix} \begin{bmatrix} x_1 \\ x_2 \end{bmatrix}$$
$$= (1+H) \begin{bmatrix} 1 & 0 \\ 0 & 1 \end{bmatrix} \begin{bmatrix} x_1 \\ x_2 \end{bmatrix}$$

In the without symbol extension case, the result shows that, we use the network alignment at source to the destination phase to reduce the number of the channel parameters, while at relay to the destination phase structure another Hadamard like channel matrix to make the total system is simply represented.

2 Symbol Extension case

In this section, we consider the case when the channel coefficients are time varying. We will show that over symbol extensions of the original channel. Also, in this case, first we should normalize the channel coefficient at each part. From the source to the relay, we do the networking alignment operation, from the relay to the destination phase, signal neutralization will be introduced.

Consider a $M = 2$ symbol extension of the original network. Then, the channel becomes a 2×2 diagonal matrix with distinct diagonal entries. We will show that W_1 can achieve 2 DoF, while W_2 can achieve 1 DoF for a total of $2M - 1 = 3$ DoF. Source node S_1 sends two independent symbols, $x_{1,1}$ and $x_{1,2}$ along beamforming vectors $v_{1,1}$ and $v_{1,2}$, respectively. Similarly, source node S_2 sends one symbol x_2 along beamforming vector v_2. As shown in Figure 3, we design beamforming vectors such that after going through their respective channels, $v_{1,2}$ and v_2 are along the same direction at relay R_1, while $v_{1,1}$ and v_2 are along the same direction at relay R_2, i.e.,

$$F_{11} v_{1,2} = F_{12} v_2$$
$$F_{21} v_{1,1} = F_{22} v_2$$

Note that v_2 can be chosen randomly and $v_{1,1}$ and $v_{1,2}$ can be solved according to the above equations. Same as the without symbol extension case, the channel coefficient from the source to the relay phase the channel should be normalized as Hadamard like matrix

$$\begin{bmatrix} F_{11} & F_{21} \\ F_{12} & F_{22} \end{bmatrix} = \begin{bmatrix} 1 & 1 \\ -H & 1 \end{bmatrix} \tag{6}$$

After replacing, the alignment conditions can be represented as

$$\begin{cases} v_{12} = v_2 \\ -H \cdot v_{11} = v_2 \end{cases} \tag{7}$$

It is simply that at R_1, x_{11} is in one signal space, x_{11} and x_2 are in one signal space, respectively.

After alignment, $x_{1,1}$ and $x_{1,2} + x_2$ can be isolated through a simple channel matrix inversion operation at R_1 while $x_{1,1} + x_2$ and $x_{1,2}$ can be isolated at R_2. Then, relay R_1 sends $x_{1,1}$ and $x_{1,2} + x_2$ in the presence of noise with beamforming vectors $v_{R_1,1}$ and $v_{R_1,2}$, respectively. Similarly, relay R_2 sends $x_2 + x_{1,1}$ in the presence of noise along beamforming vector v_{R_2}. As shown in Figure 4, to neutralize interference x_2 at destination D_1, similarly, to neutralize interference $x_{1,1}$ at destination D_2, the alignment conditions can be isolated as

$$\begin{cases} H v_{R_{12}} = v_{R_2} \\ v_{R_{11}} = -v_{R_2} \end{cases} \tag{8}$$

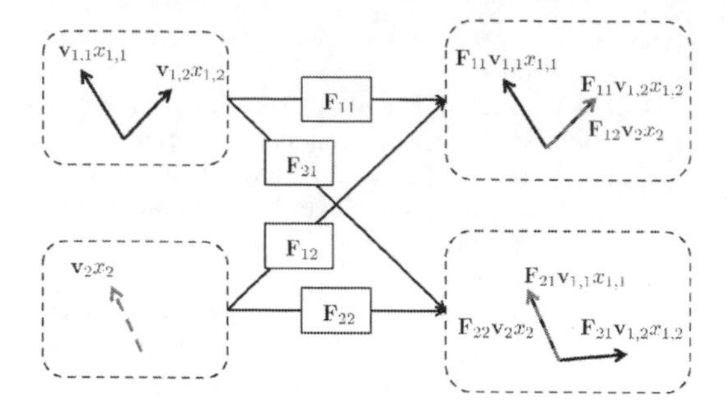

Figure 4. Alignment at relays for $M = 2$.

Similarly from the relay to the destination phase, another Hadamard like matrix will be used

$$\begin{bmatrix} G_{11} & G_{21} \\ G_{12} & G_{22} \end{bmatrix} = \begin{bmatrix} 1 & -1 \\ H & 1 \end{bmatrix} \tag{9}$$

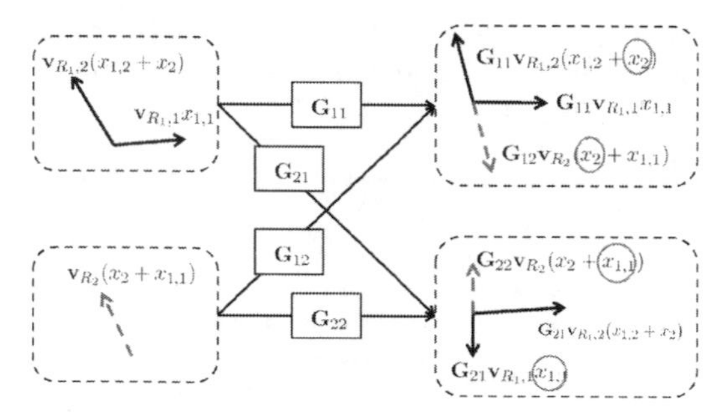

Figure 5. Aligned interference neutralization at destinations.

After alignment,

$$D_1 : \underbrace{v_{R_{12}}(x_{12}+x_2)+v_{R_{12}}x_{11}}_{From\ R_1} + \underbrace{v_{R_2}(x_2 - Hx_{11})}_{From\ R_2} \tag{10}$$

$$D_2 : \underbrace{v_{R_2}(x_2 - Hx_{11})}_{From\ R_2} + \underbrace{v_{R_{12}}(x_2 + x_{11}) + v_{R_{11}}x_{11}}_{From\ R_1} \tag{11}$$

Again, v_{R_2} can be chosen randomly, then $v_{R_1,1}$ and $v_{R_1,2}$ can be calculated. After interference neutralization, D_1 sees $x_{1,2} - x_{1,1}$ along direction $v_{R_1,2}$ and $x_{1,1}$ along $v_{R_1,1}$. Therefore, it can decode $x_{1,1}$ first, and then $x_{1,2}$ to achieve 2 DoF. At D_2, x_2 is received interference free along v_{R_2} and it can be decoded by discarding the dimension $-v_{R_1,2}$ along which interference is received.

4 VECTOR BASED EXAMPLE

First, we introduce the finite field F_{p^n} can be used to generate a n-dimensional vector space as follows [11]. Each element of F_{p^n} can be represented in the form

$$z = x_{n-1}s^{n-1} + x_{n-2}s^{n-2} + \ldots + x_1 s^1 + x_0 \tag{12}$$

wherein $z \in F_{p^n}, x_i \in F_p$.

Next, let us see how multiplication with the channel coefficient $h \in F_{p^3}$ is represented as a multiplication by a 3×3 matrix with elements in F_3. Consider the monic irreducible cubic polynomial s^3+2s+1 which is treated as zero in the field. The field itself consists of all polynomials with coefficients in F_3, modulo s^3+2s+1. Since $s^3+2s+1=0$ in F_{3^3}, it follows that

$$s^3 = -2s-1 = (3-2)s+(3-1) = s-2$$
$$s^4 = s(s^3) = s(s-2) = s^2+2s$$

Since $h, x \in F_{3^3}$ they can be represented as $h = h_2 s^3 + h_1 s + h_0$, $x = x_2 s^3 + x_1 s + x_0$ where $h_i, x_i \in F_3$. The product $y = hx \in F_{3^3}$ can be written as

$$y = Hx = \begin{bmatrix} h_2+h_0 & h_1 & h_2 \\ 2h_2+h_1 & h_2+h_0 & h_1 \\ 2h_1 & 2h_2 & h_0 \end{bmatrix} \begin{bmatrix} x_2 \\ x_1 \\ x_0 \end{bmatrix} \tag{13}$$

wherein x, y are 3×1 vector with entries from F_3 and H is a 3×3 matrix with its 9 entries from F_3.

In the without symbol extension case, the antenna number should be times larger than 3, because the maximize DOF should be $4/3$ to make the total DOF as integer.

So first we should represent some scalar to vector. In the without extension case, there are only ±1 and $\pm H$. We assume that normalized channel coefficient $H=6$, then

$$\begin{cases} 1 = 0\times3^2 + 0\times3 + 1 \\ 6 = 0\times3^2 + 2\times3 + 0 \end{cases} \tag{14}$$

That's meant the channel can be represented as matrix based

$$\pm1 = \begin{bmatrix} \pm1 & 0 & 0 \\ 0 & \pm1 & 0 \\ 0 & 0 & \pm1 \end{bmatrix}, \pm6 = \begin{bmatrix} 0 & \pm2 & 0 \\ \pm2 & 0 & \pm2 \\ \pm1 & 0 & 0 \end{bmatrix}$$

So on the relay part, the received signals can be expressed as

$$R_1 = \begin{bmatrix} 1 & 0 & 0 \\ 0 & 1 & 0 \\ 0 & 0 & 1 \end{bmatrix}(x_1 + x_2)$$

$$R_2 = \begin{bmatrix} 1 & 0 & 0 \\ 0 & 1 & 0 \\ 0 & 0 & 1 \end{bmatrix} x_2 + \begin{bmatrix} 0 & -2 & 0 \\ -2 & 0 & -2 \\ -1 & 0 & 0 \end{bmatrix} x_1$$

Then forward these signals to the destination, the received signals are shown as

$$Z_1 = \begin{bmatrix} 1 & 0 & 0 \\ 0 & 1 & 0 \\ 0 & 0 & 1 \end{bmatrix}(x_1 + x_2) + \left(\begin{bmatrix} -1 & 0 & 0 \\ 0 & -1 & 0 \\ 0 & 0 & -1 \end{bmatrix} x_2 + \begin{bmatrix} 0 & 2 & 0 \\ 2 & 0 & 2 \\ 1 & 0 & 0 \end{bmatrix} x_1 \right)$$

$$= \begin{bmatrix} 1 & 2 & 0 \\ 2 & 1 & 2 \\ 1 & 0 & 1 \end{bmatrix} x_1$$

$$Z_1 = \begin{bmatrix} 0 & 2 & 0 \\ 2 & 0 & 2 \\ 1 & 0 & 0 \end{bmatrix}(x_1 + x_2) + \left(\begin{bmatrix} 1 & 0 & 0 \\ 0 & 1 & 0 \\ 0 & 0 & 1 \end{bmatrix} x_2 + \begin{bmatrix} 0 & -2 & 0 \\ -2 & 0 & -2 \\ -1 & 0 & 0 \end{bmatrix} x_1 \right)$$

$$= \begin{bmatrix} 1 & 2 & 0 \\ 2 & 1 & 2 \\ 1 & 0 & 1 \end{bmatrix} x_2$$

So

$$Z_1 + Z_2 = \begin{bmatrix} 1 & 2 & 0 \\ 2 & 1 & 2 \\ 1 & 0 & 1 \end{bmatrix}(x_1 + x_2) \tag{15}$$

the result is same as scalar's

$$\begin{bmatrix} z_1 \\ z_2 \end{bmatrix} = (1 + H)\begin{bmatrix} 1 & 0 \\ 0 & 1 \end{bmatrix}\begin{bmatrix} x_1 \\ x_2 \end{bmatrix} \tag{16}$$

In the symbol extension case, the pre-post-coding are assumed as $v_{11} = 13, v_{12} = 22, v_{21} = v_{22} = 7$, also we should change them to the vector,

$$\underbrace{v_{11} = \begin{bmatrix} 1 \\ 1 \\ 1 \end{bmatrix}, v_{12} = \begin{bmatrix} 2 \\ 1 \\ 1 \end{bmatrix}, v_{21} = v_{22} = \begin{bmatrix} 0 \\ 2 \\ 1 \end{bmatrix}}_{v_{11}=13, v_{12}=22, v_{21}=v_{22}=7} \tag{17}$$

For the R_1,

$$R_1 = \begin{bmatrix} 1 & 0 & 0 \\ 0 & 1 & 0 \\ 0 & 0 & 1 \end{bmatrix}\begin{bmatrix} 1 \\ 1 \\ 1 \end{bmatrix} x_{11} + \left(\begin{bmatrix} 1 & 0 & 0 \\ 0 & 1 & 0 \\ 0 & 0 & 1 \end{bmatrix}\begin{bmatrix} 2 \\ 1 \\ 1 \end{bmatrix} x_{12} + \begin{bmatrix} 1 & 0 & 0 \\ 0 & 1 & 0 \\ 0 & 0 & 1 \end{bmatrix}\begin{bmatrix} 0 & 2 & 0 \\ 2 & 0 & 2 \\ 1 & 0 & 0 \end{bmatrix}\begin{bmatrix} 1 \\ 1 \\ 1 \end{bmatrix} x_2 \right)$$

$$= \begin{bmatrix} 1 \\ 1 \\ 1 \end{bmatrix} x_{11} + \begin{bmatrix} 2 \\ 1 \\ 1 \end{bmatrix}(x_{12} + x_2)$$

$$R_2 = \begin{bmatrix} 0 & -2 & 0 \\ -2 & 0 & -2 \\ -1 & 0 & 0 \end{bmatrix}\begin{bmatrix} 1 \\ 1 \\ 1 \end{bmatrix} x_{11} + \left(\begin{bmatrix} 0 & 2 & 0 \\ 2 & 0 & 2 \\ 1 & 0 & 0 \end{bmatrix}\begin{bmatrix} 2 \\ 1 \\ 1 \end{bmatrix} x_{12} + \begin{bmatrix} 1 & 0 & 0 \\ 0 & 1 & 0 \\ 0 & 0 & 1 \end{bmatrix}\begin{bmatrix} 2 \\ 1 \\ 1 \end{bmatrix} x_2 \right)$$

$$= \begin{bmatrix} 2 \\ 0 \\ 2 \end{bmatrix} x_{12} + \begin{bmatrix} 2 \\ 1 \\ 1 \end{bmatrix}(x_2 - x_{11})$$

Now moving to the destination to structure the neutralization, we can get

$$Z_1 = \begin{bmatrix} 1 \\ 1 \\ 1 \end{bmatrix}(x_{12} - x_{11}) + \begin{bmatrix} 0 \\ 2 \\ 1 \end{bmatrix} x_{11}$$

$$Z_2 = \begin{bmatrix} 0 \\ -2 \\ -1 \end{bmatrix} x_2 + \begin{bmatrix} 1 \\ 1 \\ 1 \end{bmatrix}(x_{12} + x_2) \tag{18}$$

Using the same reasons described in the proof, it is possible to show that no vector linear scheme for the $2 \times 2 \times 2$ channel over F_{p^n} can achieve rate more than 1 when $h \in F_p$, regardless of the values of p, n. For a definition of vector linear schemes, see [9]. Signal level alignment schemes, such as the recent work in [10] are an interesting research avenue for these channel instances, although we expect that there might be a capacity loss in these cases relative to the 4/3 value.

5 CONCLUSION

In this paper, we explored capacity results for the finite field $2 \times 2 \times 2$ channel, translating precoding based interference alignment schemes from corresponding DoF results for the wireless setting. The main insight is that the finite field$F$can be viewed as analogous to the $n \times n$ MIMO wireless setting, albeit with some structure imposed on the channels. We also described the conditions on channel coefficients under which the linear beamforming scheme would fail. The result for $2 \times 2 \times 2$ IC is also extended to more than two-hop IC with 2 sources and 2 destinations, for which the min-cut outer bound 2 DoF can still be achieved for almost all channels. Thus, regardless of the number of hops, two sources, and two destinations, multi-hop IC has 2 DoF almost surely. Extensions to the setting with more than two sources and two destinations are challenging. This is an interesting question to be pursued in future work.

REFERENCES

[1] C. M. Yetis, T. Gou, S. A. Jafar, and A. H. Kayran, "On feasibility of interference alignment in MIMO interference networks," *IEEE Transactions on Signal Processing*, vol. 58, no. 9, pp. 4771–4782, September 2010.

[2] K. Gomadam, V. Cadambe, and S. Jafar, "Approaching the Capacity of Wireless Networks through Distributed Interference Alignment," *IEEE Global Telecommunications Conference (Globecom 2008)*, pp. 1–6, November 2008.

[3] F. Negro, S. Shenoy, I. Ghauri, and D. Slock, "Interference Alignment Limits for K-user Frequency Flat MIMO Interference Channels," *17th European Signal Processing Conference (Eusipco 2009)*, pp. 1–5, August 2009.

[4] V. Cadambe and S. Jafar, "Sum-capacity and the unique separability of the parallel GaussianMAC-Z-BC network," presented at the presented at the IEEE Int. Symp. Inf. Theory, Austin, TX, Jun. 2010.
[5] S. A. Jafar, "The ergodic capacity of phase-fading interference networks," IEEE Trans. Inf. Theory, vol. 57, no. 12, pp. 7685–7694, Dec. 2011.
[6] S. Borade, L. Zheng, and R.Gallager, "Maximizing degrees of freedom in wireless networks," in Proc. 40th Annu. Allerton Conf. Commun., Control Comput., Oct. 2003, pp. 561–570.
[7] R. Tannious and A. Nosratinia, "The interference channel with MIMO relay: Degrees of freedom," in Proc. IEEE Int. Symp. Inf. Theory, Jul. 2008, pp. 1908–1912.
[8] S. Chen and R. Cheng, "Achieve the degrees of freedom of –User MIMO interference channel with a MIMO relay," in Proc. IEEE GLOBECOM, Dec. 2010, pp. 1–5.
[9] H. Boelcskei, R. Nabar, O. Oyman, and A. Paulraj, "Capacity scaling laws in MIMO relay networks," IEEE Trans. Wireless Commun., vol. 5, no. 6, pp. 1433–1444, Jun. 2006.
[10] V. Morgenshtern and H. Bölcskei, "Crystallization in large wireless networks," IEEE Trans. Inf. Theory, vol. 53, no. 10, pp. 3319–3349, Oct. 2007.
[11] Sundar R. Krishnamurthy and Syed A. Jafar "Precoding Based Network Alignment and the Capacity of a Finite Field X Channel" ISIT, Jul. 2013.

Computing, Control, Information and Education Engineering – Liu, Sung & Yao (eds)
 ISBN: 978-1-138-02800-5

Research on multi-machine nonlinear adaptive excitation control mechanism for power systems: A case study

Ling Chang
Shenyang Urban Construction University, Shenyang, China

ABSTRACT: In this paper, we propose a novel nonlinear disturbance attenuation control scheme based on Hamiltonian theory for the solution of structure preserving multi-machine power systems. The structure remains dissipative Hamiltonian of the power system to realize complete using algebraic equation of the singular perturbation method is regarded as fast dynamics. Moreover, nonlinear excitation controller design pattern without any disturbance attenuation linearization to improve power system transient stability and robustness to unknown exogenous disturbances on the system in the sense of the gain. The experimental result illustrates that the proposed multi-machine scheme can obviously enhance the transient stability of the system regardless of the exogenous disturbance.

KEYWORDS: Multi-machine System; Nonlinear Adaptive Excitation Control; Power System.

1 INTRODUCTION

1.1 *Background research*

With rapid development of modern power system, there become more and more transmission lines with long distance and heavy load, which has a negative influence on damping characteristics. Therefore, the low frequency oscillation happens occasionally, which threaten the system stability and security. In addition to the stability and safety, expected to recover as soon as possible after disturbance frequency, voltage and power balance, that is, the system is expected to have satisfactory dynamic quality. In order to achieve the above goals, better control system is required. Unfortunately, the modern power system is based on a large scale, distributed and highly nonlinear characteristics and complex transient. In view of the characteristics of a traditional controller design based on the linearized model on an operating point, obviously different operating conditions change, nonlinear control theory and method of application is likely to improve power system transient stability. Under certain unforeseen large disturbances the performance of linear controllers might not be satisfactory because of the highly nonlinear nature of real power systems. Beside the selection of base operation points and large disturbances is quite empirical. The nonlinear control theory and applications have got a lot of achievements such as the approach based on differentiall geometric theory [1-4]. Recently, Hamilton's theory of nonlinear control field of electric power system, and many fruitful results have been achieved [5-6].

1.2 *Overview of our research*

In this paper, we will provide a general result for the stabilization of nonlinear differential algebraic systems under input saturation, where the considered system can be transformed to a dissipative Hamiltonian system proposed in [6]. Then, by complete dissipative Hamiltonian realization examination multi - the load of power system, decentralized saturated control scheme is put forward. Central consider linearization technique is to eliminate the internal system of nonlinear feedback equivalent linear system. In addition, the linearization could damage the original structure of the attribute is a useful dynamic performance. Hamiltonian function method is a big advantage is that it can effectively use the internal structural characteristics in the controller design, so it has been widely studied synthesis of nonlinear systems. The simulation and experimental result show the robustness and effectiveness of our proposed method.

2 PRIOR RESEARCH AND PREREQUISITE

Initially, we consider the following system:

$$\begin{cases} \dot{x} = f(x,z) + g(x,z)u \\ 0 = \sigma(x,z) \end{cases} \quad \textbf{(1)}$$

Where, $x \in \mathbb{R}^r, z = \mathbb{R}^s, u \in \mathbb{R}^p$ represent the system state variables, the algebraic variables and the control parameters respectively. Moreover, suppose

$f(\cdot), g(\cdot), \sigma(\cdot)$ is sufficiently smoothed vector functions with proper dimension. Assuming that there is a continuously differentiable function $H(x,z)$ such that the system (1) can be Hamiltonian realized as the following:

$$\begin{cases} \dot{x} = (J(x,z) - R(x,z))\nabla_x H(x,z) + g(x,z)u \\ 0 = \nabla_z H(x,z) \end{cases} \tag{2}$$

Where, $J(x,z)$ is skew-symmetric and $R(x,z)$ is positive semi-definite. We have the following results for the saturated control of nonlinear differential algebraic system:

$$u_i = -S_i sat(K_i\zeta_i / S_i) \tag{3}$$

$$sat(x) = \begin{cases} 1, & if \quad x \geq 1 \\ x, & if \quad -1 < x < 1 \\ -1, & if \quad x \leq -1 \end{cases} \tag{4}$$

We therefore undertake the proof procedure as the follows. Choose the Lyapunov function as the equation $V(x,z) = H(x,z) - H(x_e, z_e)$. The derivative of V along the trajectories of the system satisfies:

$$\begin{aligned} \dot{V} &= -(\nabla_x H)^{\mathrm{T}} R \nabla_x H + (\nabla_x H)^{\mathrm{T}} gu \\ &= -(\nabla_x H)^{\mathrm{T}} R \nabla_x H + \sum_{i=1}^{n} \zeta_i u_i \end{aligned} \tag{5}$$

$$\zeta_i u_i = \begin{cases} S_i\zeta_i < 0 & if \quad -K_i\zeta_i \geq S_i \\ -K_i\zeta_i \leq 0 & if \quad -S_i < -K_i\zeta_i < S_i \\ -S_i\zeta_i < 0 & if \quad -K_i\zeta_i < -S_i \end{cases} \tag{6}$$

According to the La Salle's invariant principle of nonlinear differential algebraic systems, the closed loop system is asymptotically stable. Since power systems are interconnected and distributed over vast areas, the designed controller should have decentralized feature, that is, the variables appeared in the expression of the control law should be locally measurable. In the derived control law for generator I all the variables and parameters are just related to the same generator. Therefore, it is clear that the proposed control law is decentralized and is easy to be expanded to the systems of the realistic size.

3 OUR PROPOSED METHODOLOGY

3.1 *Dissipative Hamiltonian realization*

As you can see from the previous discussion, dissipative Hamiltonian realization of saturated controller plays an important role in the construction of. So, give a paper's main results, we first proposed a dissipative Hamiltonian realization machine more multi- load of the power system. Consider the following pre-feedback control:

$$u_i = \mu_i + \bar{u}_i, \qquad i = 1, 2, \ldots, n \tag{7}$$

$$\bar{u}_i = -\frac{x_{di}}{x_{di}'} E_{qie} - \frac{x_{di} - x_{di}'}{x_{di}'} V_{qie} \cos(\delta_{ie} - \theta_{ie}) \tag{8}$$

It can be directly computing power system that can verify the re-formulated the following dissipative Hamiltonian systems.

$$\begin{cases} \dot{x} = (J - R)\nabla_x H + g\mu \\ 0 = \nabla_z H \end{cases} \tag{9}$$

$$J_i = \begin{bmatrix} 0 & \frac{1}{M_i} & 0 \\ -\frac{1}{M_i} & 0 & 0 \\ 0 & 0 & 0 \end{bmatrix} \tag{10}$$

$$R_i = \begin{bmatrix} 0 & 0 & 0 \\ 0 & \frac{D_i}{M_i^2} & 0 \\ 0 & 0 & \frac{x_{di} - x_{di}'}{T_{d0i}'} \end{bmatrix} \tag{11}$$

The corresponding Hamiltonian function can be formulated as the formula 12 and 13:

$$\begin{aligned} &H(\delta, E_q', \theta, \mathrm{V}, \varphi) = P(\delta, E_q', \theta, \mathrm{V}, \varphi) \\ &+ \sum_{i=1}^{n} \frac{1}{2} M_i w_0 (w_i - 1)^2 \end{aligned} \tag{12}$$

$$
\begin{aligned}
P(\delta, E_q^{'}, \theta, \mathrm{V}, \varphi) = & -\sum_{i=1}^{n} P_{mi}\delta_i - \sum_{k=n+2}^{n+m+1}\left(P_{dj}\varphi_j + Q_{dj}v_j\right) - \\
& \sum_{i=1}^{n}\frac{E_{qi}^{'}e^{v_i}\cos(\delta_i-\theta_i)}{x_{di}^{'}} - \sum_{i=1}^{n}\frac{E_{qi}^{2}x_{di}}{2x_{di}^{'}\left(x_{di}-x_{di}^{'}\right)} + \\
& \sum_{i=1}^{n}\frac{e^{2v_i}}{2}\left(\frac{1}{x_{di}^{'}} - B_{ii}\right) - \sum_{i<j}^{n+1} B_{ij}e^{v_i+v_j}\cos(\theta_i-\theta_j) \\
& -\sum_{i=1}^{n+1}\sum_{k=n+2}^{n+m+1} B_{ik}e^{v_i+v_k}\cos(\theta_i-\varphi_k) - \sum_{i=1}^{n}\frac{E_{qi}^{'}\overline{u}_i}{x_{di}-x_{di}^{'}}
\end{aligned}
\qquad (13)
$$

3.2 *Control of multi-machine multi-load power systems*

In this subsection, we will propose a saturated excitation control for the power system based on the proposed dissipative Hamiltonian realization. Have shown that more machine load of power system and the pre - feedback controller with constant dissipative Hamiltonian realization. In addition, in a proper arrangement, can verify, H (x, z) There is a strict local minimum required operating point. According to the La Salle's invariant principle of nonlinear differential algebraic systems, the closed loop system is asymptotically stable.

4 EXPERIMENT AND SIMULATION

We choose a six-machine eight-load power network system as a paradigm to demonstrate the effectiveness of the proposed control strategy. We also refer to for the data and parameters. The simulation is accomplished by the PSASP package which is professional testing software for power systems designed by the China Electrical Power Research Institute, Beijing, China. Temporary is considered in the simulation of the three-phase short-circuit fault occurs in the middle of the transmission line bus between 11 and 12 in time during 0.1 ~ 0.39 s. The simulation results, as shown in figure 1 and figure 2 in figure 1 shows the generator Angle control configuration. The figure 3-6 shows the additional appendix material.

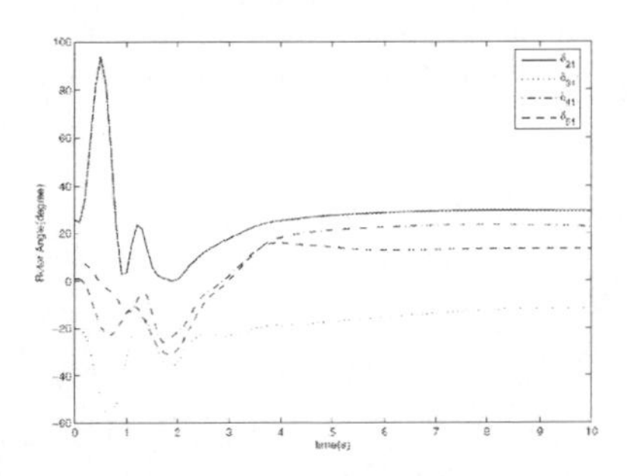

Figure 1. Responses of the generator angles.

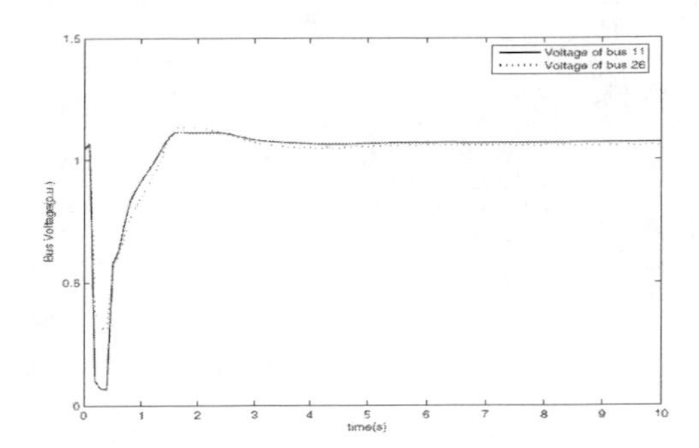

Figure 2. Responses of the bus voltages.

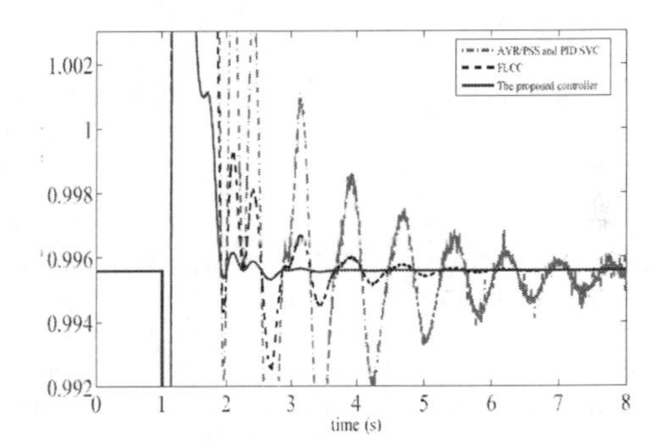

Figure 3. The addition experiment simulation one.

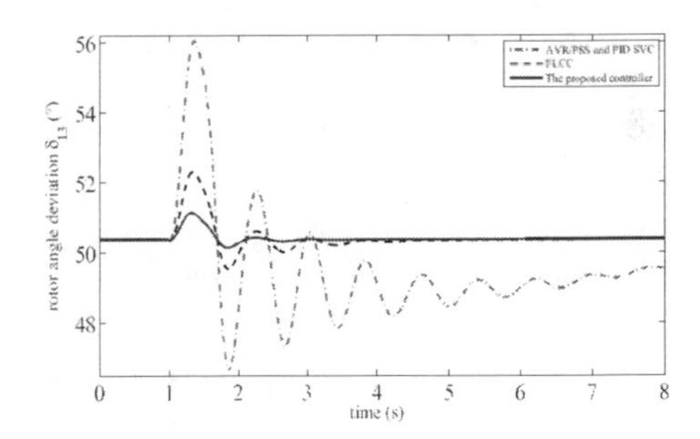

Figure 4. The addition experiment simulation two.

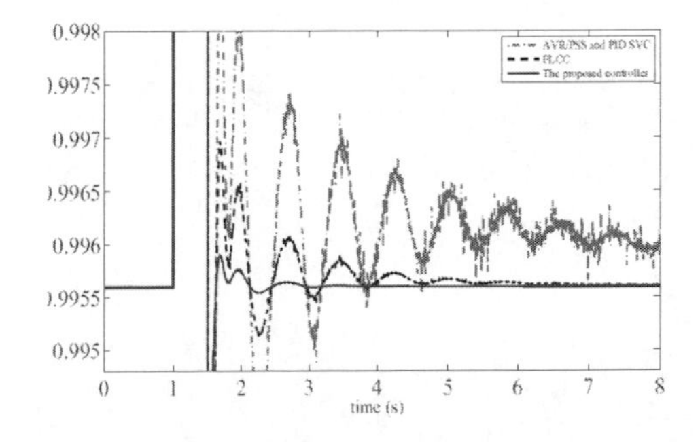

Figure 5. The addition experiment simulation three.

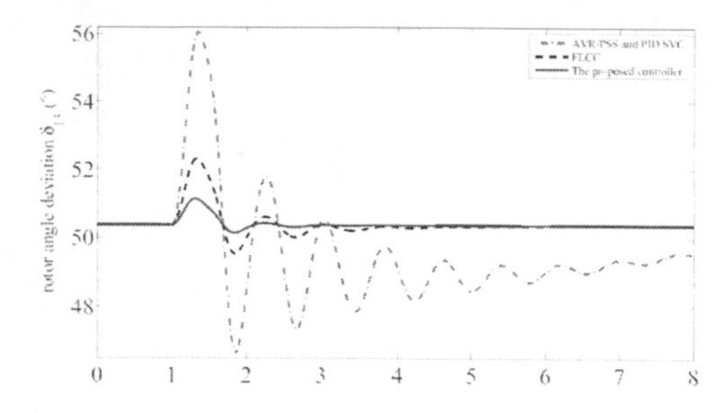

Figure 6. The addition experiment simulation four.

From the simulation results, it can be seen that our nonlinear decentralized saturated excitation scheme can effectively improve the transient stability and dynamic performance of the power system.

5 SUMMARY AND CONCLUSION

Based on the dissipative Hamiltonian realization system, to explore more stable multi - load excitation saturation of the power system. We first proposed a general result of saturated nonlinear differential algebraic system control. The control strategy is dispersed, independent of network parameters. It is easy to implement using microcomputer control law, because all the variables in the expression of this control law are local measurements. Under control is a nonlinear and large disturbance effectively improvement of the transient stability of the system. The results are successfully applied to machine more power to the load with constant active and reactive power system load demand, we get a physical meaning of decentralized with DHR saturated excitation controller of electric power system. Simulation of the nonlinear load power system verifies the effectiveness of the proposed controller. In the future, we plan to use mathematical analysis such as [7-11] to smooth the proposed formulations to get better experimental result.

ACKNOWLEDGEMENT

This paper is supported by The Scientific Research Fund Project by University (XKJ2015003).

REFERENCES

[1] Kamel, Saoudi, Bouchama Ziyad, and Harmas Mohamed Naguib. "An Indirect Adaptive Fuzzy Sliding Mode Power System Stabilizer for Single and Multi-machine Power Systems." Advances and Applications in Sliding Mode Control systems. Springer International Publishing, 2015. 305–326.

[2] Jalili, S., and R. Effatnejad. "Simultaneous Coordinated Design of Power System Stabilizer 3 Band (PSS3B) and SVC by Using Hybrid Big Bang Big Crunch Algorithm in Multi-machine Power System." Indian Journal of Science and Technology 8.1 (2015).

[3] Verdejo, Humberto, Wolfgang Kliemann, and Luis Vargas. "Application of linear stability via Lyapunov exponents in high dimensional electrical power systems." International Journal of Electrical Power & Energy Systems 64 (2015): 1141–1146.

[4] Gholipour, Eskandar, and Seyyed Mostafa Nosratabadi. "A new coordination strategy of SSSC and PSS controllers in power system using SOA algorithm based on Pareto method." International Journal of Electrical Power & Energy Systems 67 (2015): 462–471.

[5] Zhang, Rui, et al. "Post-disturbance transient stability assessment of power systems by a self-adaptive intelligent system." IET Generation, Transmission & Distribution (2015).

[6] Panda, Sidhartha, Sarat Chandra Swain, and Srikanta Mahapatra. "A hybrid BFOA–MOL approach for FACTS-based damping controller design using modified local input signal." International Journal of Electrical Power & Energy Systems 67 (2015): 238–251.

[7] Phelan, Jo C., and Bruce G. Link. "Is Race a Fundamental Cause of Inequalities in Health?." Annual Review of Sociology 41.1 (2015).

[8] Gerdtham, Ulf, et al. "Do Education and Income Really Explain Inequalities in Health? Applying a Twin Design." Scandinavian Journal of Economics (2015).

[9] Agarwal, Ravi, Martin Bohner, and Samir Saker. "Dynamic Littlewood-type inequalities." Proceedings of the American Mathematical Society 143.2 (2015): 667–677.

[10] Xu, Biao, et al. "On reverse Hilbert-type inequalities." Journal of Inequalities and Applications 2014.1 (2014): 198.

[11] John, Fritz. "Extremum problems with inequalities as subsidiary conditions." Traces and Emergence of Nonlinear Programming. Springer Basel, 2014. 197–215.

Computing, Control, Information and Education Engineering – Liu, Sung & Yao (eds)
© 2015 Taylor & Francis Group, London, ISBN: 978-1-138-02800-5

Design and implementation of forest fire monitoring system based on wireless sensor networks

Fei Lao
Information Engineering School, Binzhou Vocational College, Binzhou, China

ABSTRACT: In order to effectively improve forest fire monitoring, the paper presents a new system based on Zigbee and GPRS technology, which consists of some sensor nodes, the coordinator, and monitoring system. This system is able to overcome the traditional drawbacks of forest fire monitoring. This paper discusses in detail the hardware design of the sensor nodes and the coordinator node design. After testing, the system works stability, and is able to effectively complete forest fire monitoring, alarm, and accurately provide the fire alarm according to the node address. It is worthy of promotion and application.

KEYWORDS: wireless sensor; sensor-nodes; GPRS; system design.

Forests are important natural resources for the survival of mankind, which is the important environment for a lot of rare animals and plants. So, how much area and the security of the forest is an important factor in affecting the ecological balance of nature. There are two reasons for forest fires, one is vandalism, and another is spontaneous combustion. A Chinese vast territory, a large gap between the north and south climate, complex and varied terrain, the larger the difference between the type and characteristics of the forest, all these factors cause the complex characteristics in preventing forest fires. In the event of forest fires, which will imperil life and property as well as cause serious damage to the ecological balance. Now, the main way is workers patrol for preventing forest fire. Because of much area in forest, complex terrain, poor visibility and limited factor monitoring of workers, timely and accurate discovering and locating the source of fire is the core of the problem.

Many people do research on the monitoring of forest fires, and have made some encouraging results. For example, the literature [1] proposed a forest fire detection system based on image acquisition, the core idea is that the system analyzes information collected from the forest image through an algorithm to compare a suspected fire analysis, and then gets the location of the fire occurred further. Literature [2] proposed satellite-based early warning system to monitor forest fires, through satellite monitoring to discover, locate the fire position, which focused on solving how to locate the fire position in the blind spot. The drawack of literature[1] is to require further positioning, namely it is not timely warning, and may delay the best time firefighting. Satellite monitoring proposed in the literature [2], because the satellite is too far from the ground prone to error, and if the weather is rainy, it is also easy to receive the impact monitoring results.

Because of the importance of preventing forest fires and the drawbacks of traditional monitoring on forest fire, this paper presents a monitoring system of forest fire based on network technology. The system is capable of real-time monitoring and data collection, and provides accurate positioning for early warning and fighting forest fires.

By analyzing, there are three essential factors in forest fires, which are the source of fire, combustible and supporting combustion environments. First, the air temperature in the forest is an important indicator to determine whether the fire, as well as it is also an important indicator to judge the size of forest fires and the burning time. Therefore, we will set forest air temperature as t one of the parameters in real-time monitoring. Air humidity refers how much moisture in the air. According to the survey, the occurrence of fire and air humidity is inextricably linked. When the air humidity is less than 25%, the probability of occurring fires is large. So, we set air humidity in forest as one of the parameters in real-time monitoring system. In addition, when forest fires happen, the density of smoke produced by burning is also an important factor to determine the size of the fire.

1 FOREST FIRE MONITORING SYSTEM DESIGN

Based on the above analysis, we design terminal nodes to timely perceive and capture the air temperature, humidity, and the smoke density in the combustion. Terminal node collects data information at a predetermined time and then send it via GPRS to the World Wide Web or the user's mobile client. Data that is sent to the World Wide Web will eventually stored in the server's database. The server compares and analyze the received data, once the received data exceeds the set threshold for emergency alarm and identifies as the fire information, and the fire alarm system will be immediately start. According to the number information of sensor node to determine the location the occurrence of forest fires. The system consists of several sensor nodes distributed in the forest, a cluster head node aggregation function, coordinator, GPRS for data transmission, other parts of the World Wide Web, and terminal servers. The components of this specific system shown in Figure 1.

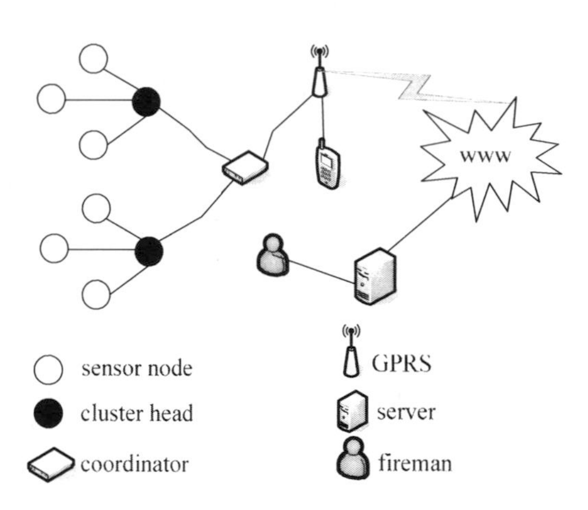

Figure 1. The whole framwork of the forest fire monitoring system.

2 THE HARDWARE DESIGN OF FOREST FIRE MONITORING SYSTEM

2.1 *Sensing node hardware design*

Sensing node is responsible for real-time monitoring and collecting the air humidity, temperature and smoke density in the forest when burning, which plays an important role in the entire system. is used in the design of sensing node, which is an air temperature and humidity sensor. Its ultra-small size is 3.0 * 3.0 * 1.1mm, which communicates with MCU using the I2C interface. Comparing with SHT10 and SHT11, HTU21D has a marked improvement on function, as well as the price is low. Therefore, small size, strong anti-jamming capability, low power consumption, high cost, high sensitivity, all these above factors promote us to select HTU21D. MQ-2 is selected as the sensor of smoke density, which has good stability, high sensitivity, low cost, long life and other characteristics. CC2530 is used as the processing unit, which supports the latest Zigbee protocol. It is an integrated chip, which has full-featured, small size, and low price. The ASE is embedded in CC2530, thus ensures the collected data accuracy, and security in the process of transmission. The components of the sensor shown in Figure 2.

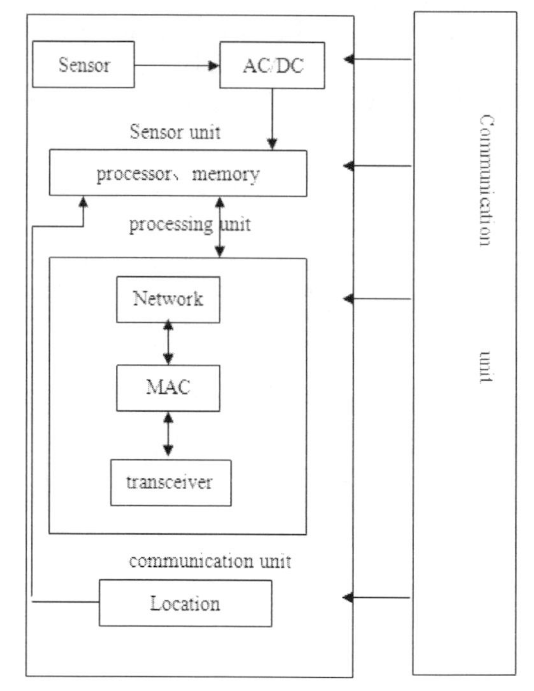

Figure 2. The composition of the sensor.

2.2 *Design of the coordinator node*

A coordinator is responsible for the organization and management of network in the entire system, and for forwarding the collected data. Thus, the coordinator must have a larger memory to save the collected data from all sensing nodes, as well as having a spanking induction ability to timely and effective receiving and processing data. The coordinator must work strongly stability, Once there is a problem because of confounding factors, the entire system may paralyze. In order to improve the communication distance of the coordinator, the amplifier module is used in this system.

2.3 *The design of solar power system*

Generally, forest is far away city, which covers a wide area and has bad weather, and even in some forest, humans cannot survive. Ordinary battery capacity is limited, once the battery is dead, the system will not work. Therefore, replacing regularly battery for the system becomes a major problem faced by the staff. In this system, a solar power supply system is used, which consists of solar systems, batteries and controller. The function of the controller is used to manage the solar battery system for effective and timely charging, and to protect the battery to avoid over-charging and discharging. System battery voltage is 12V, each chip operating voltage varies. Therefore, we must use a switching regulator to convert the voltage of each chip, according to the desired operating voltage.

3 SOFTWARE DESIGN OF FOREST FIRE MONITORING SYSTEM

After power on, sensor nodes and the coordinator will enter the initialization state, and automatically search the channel that is currently not occupied, once finding a channel for their own use, issue the application immediately, and join the network. If the application is unsuccessful, continue to search for the channel. If successfully joined the network, and immediately determine whether the environmental data acquisition time, if not reached the specified time, in order to improve the life of sensor nodes into hibernation. When the collection time reaches, the sensor node has woken up, and gets where the smoke density, air humidity and ambient temperature data, and sends the collected data to the coordinator, and then goes to sleep, waiting again be awakened. The above is the workflows of sensor nodes.

After entering the start state, the coordinator immediately scans the surrounding network. If the network can be used, the coordinator will enter the start state. When the coordinator receives a request sent by sensor nodes, it decides whether to agree the sensor node to join the network based on the system need. After each acquisition cycle, the coordinator is responsible for receiving data transmitted by sensing nodes. The collected data is transmitted to the server through GPRS and WWW. The above is the workflow of the coordinator.

4 SYSTEM STABILITY DESIGN

4.1 *Hardware design system stability*

The stability of the whole system is largely determined by the selection of system hardware and design. In order to ensure that the entire system is stable and reliable, the hardware design is a vital part. In order to ensure timely and effective forest environmental information transmission, and avoid generating an error message, asking that the system must have the ability to identify valid signals and interfering signals. After analyzing the reasons, resulting in fire monitoring system generates an error message mainly from two aspects. An impact is produced within the system, and the other affecting the system externally generated. Mainly due to the impact of internal influence internally generated electrical appliances, such as mutual influence and mutual influence between power lines, transmission lines, underground transmission lines and transmission lines mutual influence, power lines and power lines and so on. External influences mainly from outside the system and external electrical, electromagnetic effects more common, specifically, the impact of lightning on forest monitoring system hardware generated electromagnetic radiation generated by forest monitoring system hardware.

The power supply is susceptible to internal and external from the system, and therefore we pay particular attention to their safety and ability to resist impact in the design. Currently, there are many types of power, which can be selected according to need in the design. For example if you want to ensure the stability of the input power, you can choose to automatically adjust the power of intelligent power supply that also has alerts. In order to avoid overpressure as well as due to an electrical harm undervoltage system does not work due to the occurrence of such phenomena, it is common practice for the power supply system voltage regulator and a voltage regulator and other devices.

Based on the above analysis, in order to avoid overvoltage and undervoltage phenomenon power, supply system of forest fire monitoring system based on wireless sensor, is used by a voltage regulator device. The power supply module, can guarantee the stability of the whole system to work effectively. Due to a power reverse situation caused by misuse occur, the consequences could be disastrous. To address this situation, you can diode in series with the regulator's back, even if the power is reversed, the regulator will not work, will not cause any harm to the power system. Meanwhile, the power supply system effectively avoids voltage overload phenomenon, avoids the individual power supply system failure caused by paralysis of the whole phenomenon. The test proved that the power supply system is worth in terms of dissemination such as coupling, stability, and promote the use of heat.

4.2 *Software design system stability*

Fire detection system is composed of software system and hardware system. From the point of view of stability jamming hardware system settings. In order to avoid the omission of information as well as the serious consequences caused by wrong information reported on the stability and accuracy of the software system has been further optimized.

For example, during the process of monitoring the concentration of smoke, perhaps due to special factors caused by the non-smoke alarm occurs. In order to avoid the occurrence of such phenomena, the ionization smoke detectors connected to a dedicated chip and INT1, can produce delays of about 70ms. In the monitoring process, once found smoke, will immediately start the alarm. After the first 70ms, re-use timer interrupt routine. If the second 70ms not found, then the smoke is not a fire, it will not send the information to the fire alarm. Effectively prevent false alarms caused by fire human, financial and other waste phenomenon.

5 CONCLUSION

With the continuous development of wireless network technology, wire network is replaced by wireless network. Wire network has many shortcoming, for example, space constraints, the age and vulnerability to forest destruction and other shortcomings in the animal gradually. The Zigbee technology for its low cost, low power consumption, easy networking, etc., become the preferred technology for environmental monitoring, biological farming monitoring, real-time traffic monitoring systems, real-time monitoring of the patient. But the technology also has some disadvantages, such as wireless network security are difficult to protect data transmission, data transmission speed is slow and other issues. Next we will focus on resolving these problems, so that the forest fire monitoring system to further research, thereby increasing the level of forest fire monitoring.

ACKNOWLEDGEMENTS

It is a project supported by the Project of Binzhou Science and Technology Development Plans (2011ZC0402).The corresponding author is Feilao.

REFERENCES

[1] Stephenson,M.D.Automatic Fire Detection Systems. Wiring Installations and Supplies,1985,31(3):239~243.
[2] Ma Xinyuan. Research of Microcontroller Development. System Control &Automation, 2004,(6):69–71.
[3] Fieldbus Foundation. Technical Overview. 1996.
[4] M. Weiser . The Computer for the 21'Century. SCLAmer, Sep 1991.
[5] Stanislav Safaric,Kresimir Malaric ZigBee wireless standard 48th International Symposium ELMAR-2006:259–262.
[6] Zhao, JiChun etc. The study and application of the IOT technology in agriculture [J].Computer Science and Information Technology,2010(2):462–465.

Computing, Control, Information and Education Engineering – Liu, Sung & Yao (eds)
© 2015 Taylor & Francis Group, London, ISBN: 978-1-138-02800-5

A creative analysis of business management of the talent training mode based on ERP

T. Li & Y.R. Wang
Langfang Polytechnic Institute, Hebei, China

ABSTRACT: In recent years, ERP technology has developed quickly and ERP system has become more internationalized. Our country gradually pays attention to the application of the ERP system. Therefore, enterprises should attach importance to the innovative establishment of a talent training mode based on the ERP system. This paper has made a brief introduction to the talent training and analyzes innovation and corresponding measures for improvement.

KEYWORDS: ERP; business talent management; training mode; innovative analysis.

With the continuous progress of science and technology, the popularization of network, and the development of software technology, enterprises constantly use relevant systems in their daily management. The application of enterprise resource planning system is more and more popular. In order to keep pace with the new era, the enterprise must carry on innovation and apply advanced ways of cultivating talents. The problem that the enterprise faces is to innovate business management of the talent training mode based on ERP.

1 AN OVERVIEW OF ERP TALENT TRAINING MODE

1.1 *The significance of cultivating talents based on ERP*

To cultivate talents by applying ERP technology can make the enterprise's top leadership have access to relevant information of employees at any time. And ERP sales department can timely upload and feedback accounts receivable, inventory, sales volume. The enterprise's internal management and decision-making leaderships can learn its production and management conditions at any time via the Internet and the system. They can make some changes for their decisions according to actual situations. Besides, the enterprise can strengthen the ties between the controlling company and the subsidiary company in order to receive message feedback from both sides and provide a more convenient and quick information channel. The enterprise can also timely understand customers' feedback on products according to the ERP information feedback and learn their demand in order to grasp market trends and purchase demand and work out marketing strategies and production plans in order to avoid overproduction, establish the balance of supply and demand, and attempt to reach the profit maximization. ERP can get feedback on the sales department's rate of progress and sales plans which will facilitate the enterprises for its further decision-making, strengthen sales control and operability, and improve the efficiency to deal with a series of working plans.

1.2 *Procedures of ERP talent training mode*

The most important precondition of applying ERP to talent training and management is to master basic content of ERP which is the most basic requirement, and to expertly use software systems and a variety of functions. It is necessary to grasp the ERP system for business management and talent training. The enterprise needs to skillfully use the system to maximize its function based on its actual situations and master analogue simulation function of ERP. Basically, the practical operation of ERP is mostly in the computer laboratory and the sand table. The computer application software for simulation in business management has the ERP software. By applying the software or relevant cases, students can experience the enterprise's real business management and its operational efficiency. In the practical training stage of ERP, students or other users have already learned relevant theories of ERP and can skillfully use the knowledge. And the simulation software. When reaching this stage, they should begin fieldwork or combine software with the the enterprise's actual management. They will use the software or system to handle actual business. It is of great help to have a further understanding of ERP by combining theories, practice and operation or by summing up the experience of the application of ERP.

2 THE INNOVATION OF THE TALENT TRAINING MODE OF THE ENTERPRISE BASED ON ERP

2.1 *The framework of ERP talent training mode*

The framework of ERP talent training mode needs to be built which puts concepts into practice, combine scientific research and production management for enterprise resource planning, and build a platform for talent training. It involves the following four steps:

Firstly, the concept of ERP needs to be put into practice which is the first stage of the whole framework. It requires an in-depth understanding of the courses, knowledge of software, and associated market training promotion activities in order to make students and other users learn the application of ERP in enterprises in terms of functions and approaches and help them experience the operation of ERP in practice. The range of its application should be expanded. Professional operators and users have to be trained to be equipped with relevant operational capabilities to use the ERP system to solve practical problems of the enterprise.

Secondly, the simulation stage of ERP relies on corresponding computer software to operate sand table simulation. Practical courses are conducted in computer labs. The combination of the ERP system application with the enterprise's related cases can simulate a real business management environment and make users experience such a real atmosphere during the operation of ERP. Different cases are tested in different operating environment to find out corresponding operational efficiency.

Thirdly, the stage of fieldwork and training ERP requires the user to master related theoretical knowledge and skillfully use simulation software. Students or other users can practice and be trained in enterprises where they can operate relevant software to handle the business. When the ERP management staff operate the software, students or other users can participate in the ERP project via observation and summary. In the real production, operation and management environment, they can be familiar with the use of ERP software and can make full use of a series of its functions.

Fourthly, innovative thinking of ERP should be established which can enhance ERP thinking of the enterprise management staff. After completing the other three basic steps, the last one is to sum up the ERP system, analyze its advantages and disadvantages, and put forward relevant suggestions for improvement. Based on the software's functions and data, innovative methods of ERP can be created.

2.2 *Measures of achieving innovation of the ERP talent training mode*

Firstly, talents training mode and planning need to be recomposed.

It requires to innovate and improve the ERP talent training mode. The most effective way of innovation and change is to make new plans and methods for talent training based on the enterprise's internal ERP system. With the continuous development of enterprises and the expansion of the scale, they need more high-quality and inter-disciplinary talents. So students majoring in management and other practitioners should not only have good ideological and political consciousness and good psychological quality, but also have rich scientific knowledge, especially the knowledge of the ERP system. They need to skillfully operate the software and continuously update their knowledge make themselves in relatively good shape. When training the ERP talents, the enterprise should pay attention to the basic quality and skills of ERP and requires the staff to have certain specialties. ERP talents are more likely to have an innovative development.

Secondly, the exploitation of human resources should be emphasized. The function of developing human resources is a key part in ERP application system. In order to innovate and further improve the the ERP talent training mode, it is necessary to bring in a number of professionals who can master the development and use of the ERP system. Personnel management should be based on the ERP system. Generally speaking, young people are welcomed by enterprises because they know more modern knowledge, accept new things faster, adapt to the changes easily, and advance with the times. Moreover, young people nowadays are equipped with some knowledge of the computer application, and they tend to have more rich knowledge. So during the training process, young practitioners' potential should be exploited and developed and they should be given more funds. Besides, the number of personnel must be controlled strictly. There is no need to leave too many redundant practitioners. The redundancy should enhance the working efficiency. Furthermore, ERP talent training needs to target market to have a more effective allocation and use of human resources. Rewards and punishments should be reasonable. To motivate staff can improve their working efficiency and their enthusiasm and creativity.

2.3 *Suggestions of improving the innovation of the ERP talent training mode*

Firstly, the enterprise needs to invest more money in ERP application. The innovation of talent training should be further explored. In order to meet the needs for ERP talents, some basic ways of innovation have to be used, such as rebuilding the ERP talent training mode and planning. ERP talents are defined as those who have high professional quality and high knowledge system. They tend to have a relatively higher level of ideological and political consciousness, moral cultivation, psychological quality, and advanced knowledge and theories. And they can keep pace with the times to use the ERP system. They are apt to make continuous innovation and can skillfully master related computer technology. The enterprise's ERP talent training is mainly determined by its own conditions for investment. Therefore, the enterprises should treat it from a strategic perspective, and look at it with developmental thinking to make an innovative improvement of the ERP talent training. They should make all-round planning, formulate corresponding steps, and expand the scope and channel of the talent training. Schools, training institutions, and enterprises need to make specific and professional ERP talent training plans.

Secondly, the enterprise's human resources management mechanism should be improved and and the configuration of resources should be optimized. It is essential for related staff to learn the application of the ERP system, so the enterprise's human resources need to be developed appropriately and reasonably and a talent mechanism has to be established which requires to optimize and allocate the existing human resources and pay more attention to the younger generation's potential. Young people grasp more modern knowledge and accept new things faster. They can easily adapt to the changes and advance with the times. Most of them have certain knowledge of the computer application, so they tend to have a relatively rich knowledge, and can use ERP to give better play to its function.

Thirdly, a third party ERP professional training institution should be set up. In addition to universities and enterprises, there should be training institutions for talent training. Colleges and enterprises have certain limitations on the cultivation of ERP talents which can not meet the features of the ERP system's theories and practice. Therefore, it is necessary to have a third party training institution, which can make plans for training professional talents according to the social need for talent and the characteristics of the ERP system, to relieve the pressure of the colleges and enterprises for talent training and increase social talents.

Fourthly, the use of ERP for talent training should adopt diversified teaching and practice methods. Traditional classroom teaching on talent training can't meet the requirement for innovation. In the process of training professional staff, teaching methods, such as citing the enterprise's living examples, scene teaching, sand table simulation, should be used. The content of ERP should be constantly developed, renewed and innovated, and the system and training mode have to be constantly improved. Interactive teaching needs to be used in order to cultivate ERP professionals who have more professional knowledge and innovative spirit.

3 SUMMARY

Enterprise resource planning (ERP), is the enterprise's management planning which has developed quickly, has a higher level of informationization and functions and is a resource-saving one. The application of ERP in business management for talent training has many advantages and favorable factors, so the enterprise should grasp the use of ERP and carry on the innovation to improve its competitiveness and increase economic benefits.

REFERENCES

[1] Hu Xiaoming. An creative analysis of business management of the talent training mode based on ERP [J]. Manager,2013,31:120.

[2] Cheng Duojia. An creative study of business management of the talent training mode based on ERP [J]. Manager,2014,01:83.

[3] Yi Nan. An creative exploration of business management of the talent training mode based on ERP[J]. Knowledge Economy, 2014,02:145.

[4] Guo Wei. An creative study of business management of the talent training mode based on ERP [J]. Enterprise Technological Development, 2013,11:3–4.

Computing, Control, Information and Education Engineering – Liu, Sung & Yao (eds)
 ISBN: 978-1-138-02800-5

Study of servo manipulator based on STM32

G.P. Lu
Beijing Institute of Technology, Zhuhai , Zhuhai, Guangdong, China

Y.H. Zhang
Dongguan Fenggang Sanhesheng Technology Electronics co., ltd, Dongguan, Guangdong, China

ABSTRACT: The manipulator plays an important role in the modern automated production with high flexibility, adaptability and operational characteristics. Most manipulators repeat work according to set a good action. But in some specific locations, such as the dismantling of dangerous explosives, the disaster site, etc., it is needed to complete a series of complex movements through on-site control by staff. Servo manipulator can do it. In this project, the servo motor is adopted as the control of the manipulator, the manipulator has three degrees of freedom, and it can complete a clamping action. The bracket of the manipulator is processed by a 3D printer with light weight and good toughness of ABS material. Hand instant attitude is detected through a detection module implanted in the gloves, and then transmitted to the manipulator control module through the 2.4G wireless communication module. The detection module and the control module are designed by using AltiumDesigner2009 software. All electronic is SMD package form, and the main chip selection is 32 bit microcontroller STM32F series.

KEYWORDS: Manipulator; servo; 2.4G communication; STM32.

1 INTRODUCTION

Servo manipulator is another development direction of industrial manipulator. In some aspects, it has the basic characteristics of the traditional mechanical manipulator, also has its unique superiority. It can do the same actions of people, and equivalent to enlarge the force of hands. This device can help mankind to break through the physiological limit to do a lot of action which hands can not complete. The operator only needs to wear gloves to control the manipulator, because of a piece of attitude sensor is embedded into the gloves. At the same time, the wireless communication mode is adopted to make the operator controlling freely all actions of the manipulator.

2 SYSTEM CONSTRUCTION

2.1 *The mathematical model of the system*

All hand movements can be decomposed into X, Y or Z axis, if the hand is considered as a plane. Then the manipulator instant posture can be obtained by detecting data of each axis as shown in Figure 1.

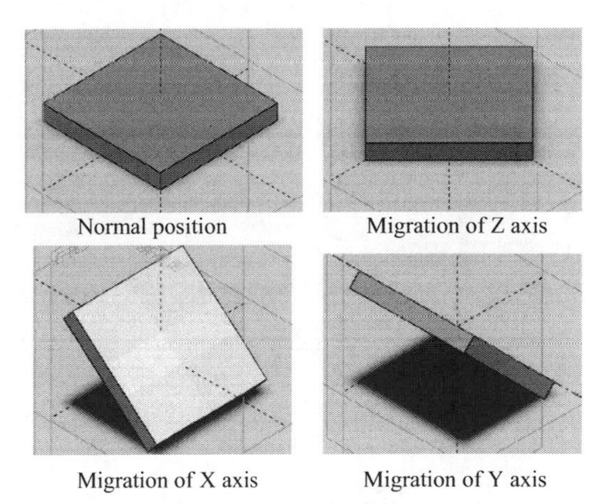

Figure 1. Decomposition of hand gesture.

2.2 *System strategy analysis*

The servo manipulator system is divided into two parts, detection part and a control part. Detection part is mainly responsible for the collection of data, processing data, and sending data. Control part is mainly responsible for receiving data, controlling servo motor. STM32F103 microcontroller is used as the core control chip. Two parts transmit data to each other by the 2.4G wireless transmission module.

3 HARDWARE SYSTEM

3.1 *The frame of hardware*

The system uses STM32F103T8U6 as the core controller, and the diagram of the control circuit of the servo manipulator, as shown in Figure 2. The control circuit is divided into several sub modules, and the following is an analysis and interpretation of each sub module.

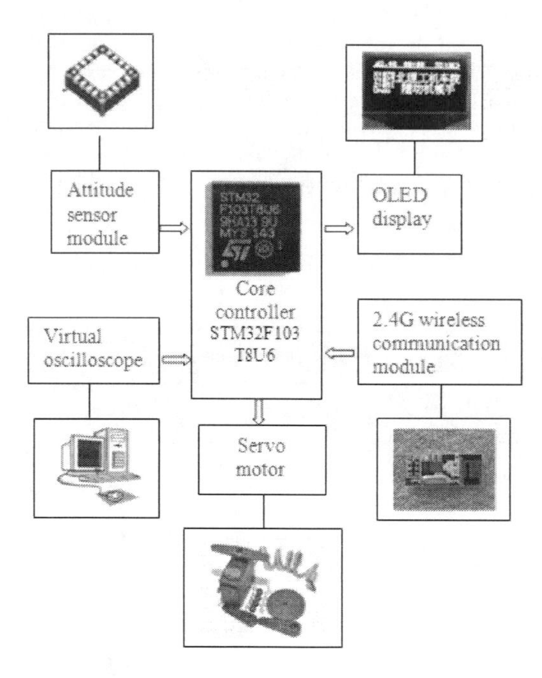

Figure 2. Circuit diagram.

3.2 *Core control module*

The core controller is a 32 bit MCU STM32F103T8U6 from STMicroelectronics with 72MHz maximum operating frequency. This chip owns a lot of on-chip resources such as AD, SPI, IIC, CAN, USB, USART, TIME and so on. Research and development of products is very convenient, because the program can be written using JTAG or serial port.

STM32F103T8U6 contains a number of internal resources, shown as table 1, and each module can be configured to work at different frequencies to satisfy all kinds of work environment.

<table>
<tr><th rowspan="3">Pinout</th><th colspan="2">Low-density devices</th><th colspan="2">Medium-density devices</th><th colspan="3">High-density devices</th></tr>
<tr><th>16 KB Flash</th><th>32 KB Flash(1)</th><th>64 KB Flash</th><th>128 KB Flash</th><th>256 KB Flash</th><th>384 KB Flash</th><th>512 KB Flash</th></tr>
<tr><th>6 KB RAM</th><th>10 KB RAM</th><th>20 KB RAM</th><th>20 KB RAM</th><th>48 KB RAM</th><th>64 KB RAM</th><th>64 KB RAM</th></tr>
<tr><td>144</td><td></td><td></td><td></td><td></td><td colspan="3" rowspan="3">5 × USARTs
4 × 16-bit timers, 2 × basic timers
3 × SPIs, 2 × I²Ss, 2 × I2Cs
USB, CAN, 2 × PWM timers
3 × ADCs, 2 × DACs, 1 × SDIO
FSMC (100 and 144 pins)</td></tr>
<tr><td>100</td><td></td><td></td><td colspan="2" rowspan="3">3 × USARTs
3 × 16-bit timers
2 × SPIs, 2 × I²Cs, USB, CAN, 1 × PWM timer
2 × ADCs</td></tr>
<tr><td>64</td><td colspan="2" rowspan="3">2 × USARTs
2 × 16-bit timers
1 × SPI, 1 × I²C, USB, CAN, 1 × PWM timer
2 × ADCs</td></tr>
<tr><td>48</td><td></td><td></td><td></td></tr>
<tr><td>36</td><td></td><td></td><td></td><td></td><td></td></tr>
</table>

Table 1. STM32F103xx family.

3.3 *Power module*

3.7V and 7.4V lithium battery is adopted for power supply, so the system requires two different sets of power supply system, respectively, the SP6205 regulator circuit to convert 3.7V to 3.3V, as shown in Figure 3; LM1117-5.0 regulator circuit to convert 7.4V to 5.0V, as shown in figure 4.

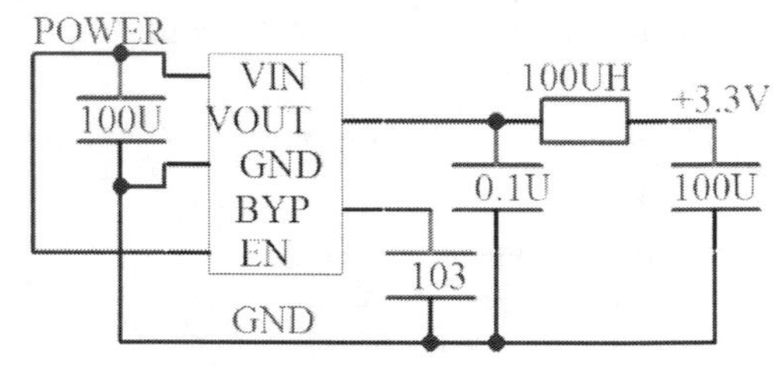

Figure 3. Circuit diagram of SP6205.

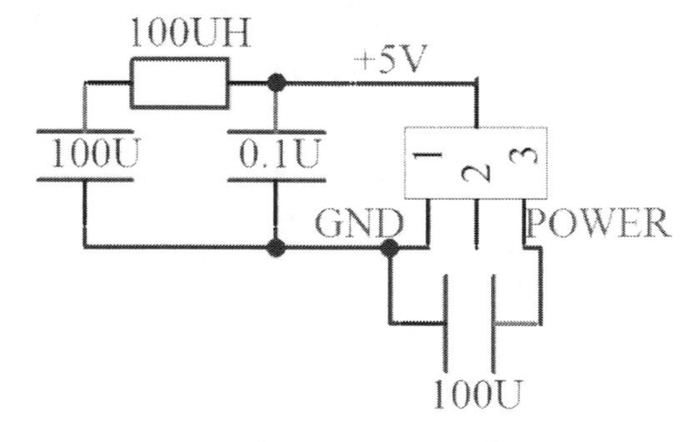

Figure 4. Circuit diagram of LM1117-5.0.

3.4 *Attitude sensor module*

The MPU6050 is a 9 axis motion sensor processing. It is consists of3 axis embedded micro electromechanical systems, acceleration of gravity system and an expanding digital motion processor (DMP). The communication with the third-party sensor is possible, such as electronic compass. When it is connected with an external electronic compass, 9 axis motion processing functions of MPU6050 can achieve complete, and output data through the IIC port. Circuit diagram of MPU6050 is shown in figure 5.

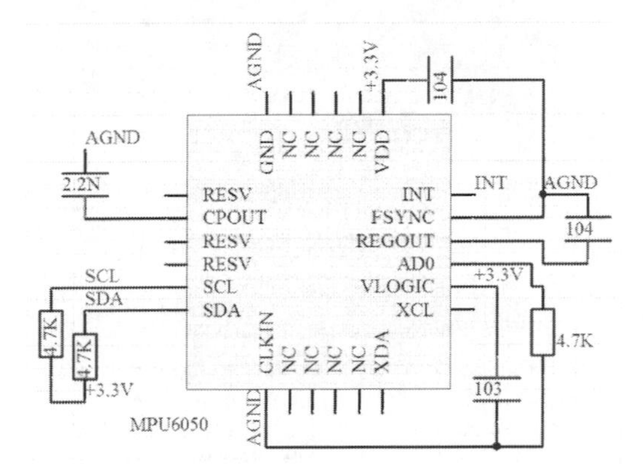

Figure 5. Circuit diagram of MPU6050.

3.5 *Servo motor*

The servo motor is the only power mechanism of the system. According to the different internal structure, servo motor is divided into the digital and analog. Mg995 analog servo motor is adopted in this system, as shown in Figure 6. The structure comprises a DC servo motor, DC servo motor controller, the deceleration gear and group feedback potentiometer. PWM signal is sent the to the controller chip of DC servo motor, then, according to the voltage difference between PWM signal received with the feedback from the mechanical linkage position sensor, it drives the motor to the specified location.

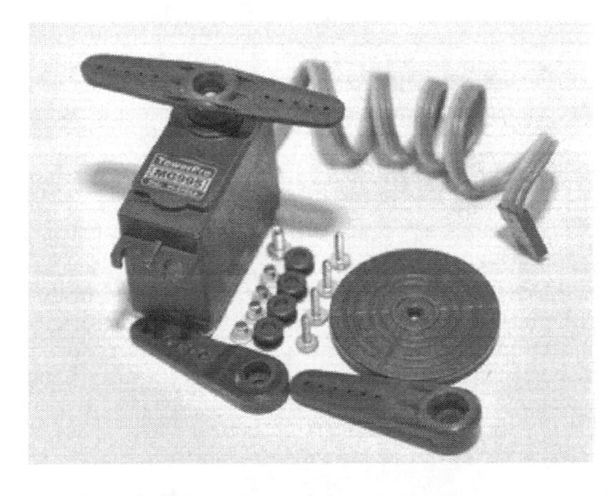

Figure 6. MG995 servo motor.

4 SOFTWARE CONTROL SYSTEMS

4.1 *Virtual oscilloscope*

It adjusts the system parameters based on the MCU internal data monitored, and it is needed communication mode supporting PC in the use of the virtual oscilloscope. The communication mode is similar to the communication format, and only the data stream with the format can be identified by PC, thus it can effectively eliminate data error due to partial loss. The virtual oscilloscope as shown in Figure 7. In the figure, yellow represents the angular velocity of gyroscope; red represents the angle of acceleration sensor; blue represents the angle after complementary filtering.

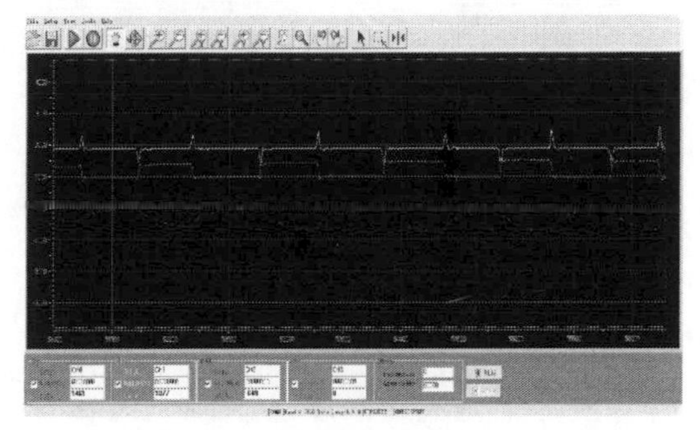

Figure 7. Virtual oscilloscope.

The corresponding program of lower machine is as follows:

```
double OutData[4] = { 0 };
unsigned short CRC_CHECK(unsigned char
*Buf, unsigned char CRC_CNT)
{
unsigned short CRC_Temp;
   unsigned char i,j;
   CRC_Temp = 0xffff;
   for (i=0;i<CRC_CNT; i++)
        {
 CRC_Temp ^= Buf[i];
     for (j=0;j<8;j++)
                 {
 if (CRC_Temp & 0x01)
         CRC_Temp = (CRC_Temp >>1 ) ^ 0xa001;
         else
         CRC_Temp = CRC_Temp >> 1;
      }
   }
   return(CRC_Temp);
}
```

4.2 *Filtering methods of signal*

Measurement of the spatial position of the hand is mainly using the accelerometer and gyro, and tilt angle and tilt angle speed are measured by hand. How to use the algorithm to reduce the noise and error of the sensor itself to obtain fast and accurate information?

The principle of recursion average filtering is the first acquisition of N data as a cohort, then new data is arranged at the end of the queue. At the same time, eliminate the most front-end data of the queue, so the cycle. So that can ensure the flow of data, but also can eliminate the system error. In particular, can suppress the interference to the periodic system, which is suitable for high frequency oscillation. The program is as follows:

```
#define num 20
char text[num];
char j=0;
char first_filter()
{
  char count;
  int sum=0;
  text [j++]=read_mpu6050();
  if (j==num)
j=0;
  for(count=0;count<num,count++)
    sum= text [count];
  return(char)(sum/ num);
}
```

5 THE PHYSICAL SYSTEM

5.1 *Control circuit*

The main function of the control circuit is to control the steering gear, so do not need sensor too much, only stable power supply and signal. In order to facilitate the expansion of the number of actuators, the control circuit can be connected with at most 12 servo motors, as shown in figure 8. Glove with attitude detection circuit is shown in Figure 9.

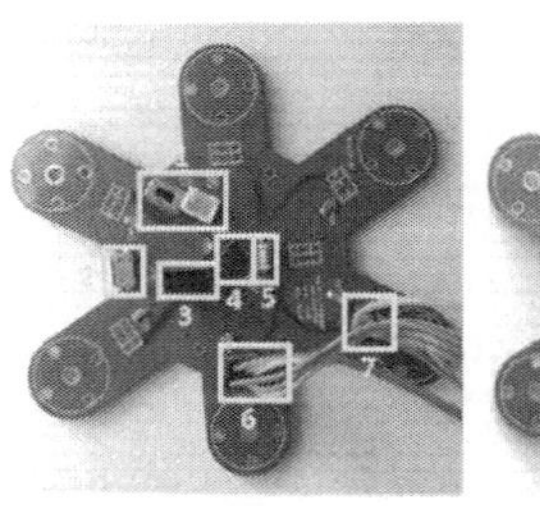

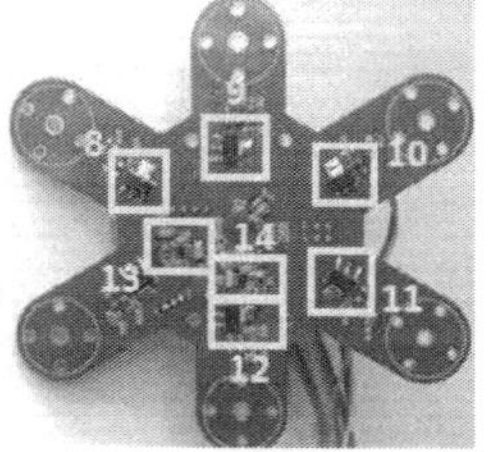

Figure 8. Control circuit.

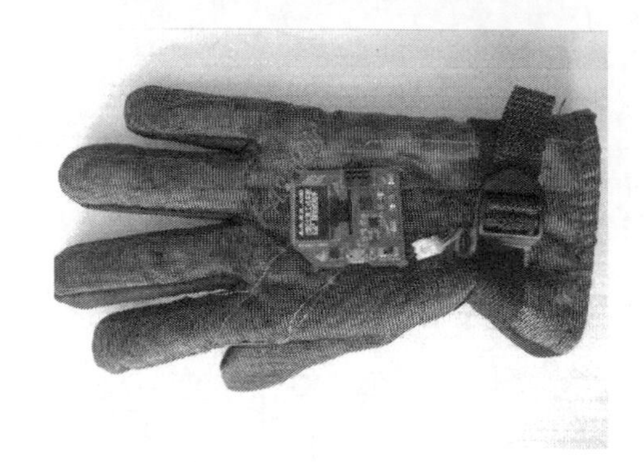

Figure 9. Glove with attitude detection circuit.

5.2 *Manipulator*

The manipulator consists of 4 servo motors, And then stitching with bracket. As shown in Figure 10.

1 X axis servo motor
2 Z axis servo motor
3 Y axis servo motor
4 holding servo motor

Figure 10. Servo manipulator.

6 CONCLUSION

The design is ingenious that a posture sensor is implanted into the gloves, so the operator only needs to wear gloves to control manipulator. Circuit board with the area of only 4.5*5cm integrates single chip, OLED, lithium battery charging IC and 2.4G wireless module and so on. To enable the operator to freely control all movements of the manipulator, the wireless communication mode is adopted, by this way the data transmission operates more easily.

REFERENCES

Yuan Xia,Luo Kelu. 2013. Design of Smart Home Gateway Based on STM32 and CC2520. *Microcontrollers & Embedded Systems*. 13(5): 70–73.

Yuan Xianfeng, Zhou Fengyu. 2013. Design of a Pedometer Based on STM32 and iNEMO Module. *Microcontrollers & Embedded Systems*. 13(9): 42–45.

Zhou Youjie,Xiong Chunhua,Lu Changbo. 2011. Design of hol-mium laser treatment instrument control system based on STM32. *Communication Systems and Information Technology*. 667–672.

Lan Gongjin,He Kai,Xu Duru. 2012. Research on motion con-troller for SCARA robot ased on STM. *Journal of Jiangsu University of Science and Technology (Natural Science Edition)*. 32[A]:417–423.

YANG yuan-zhao. 2005. Study of Web-based integration of pneumatic manipulator and its vision poisoning. *Journal of Zhejiang university SCIENCE*. 5A(5):534–548.

Zhang X D, He J J,Sheng C J. 2005. An approach of micro-step- ping control for the step motors based on FPGA [R]. *IEEE Proceedings of the 10th International Conference on* Industrial Technology. Shanghai, China.

Liu Huiying,FAN Baohan.2010.Study of Control System of Multiple Step Motors Based on STM32. *Measurement & Control Technology*. 29(6)6:54–57.

Huangzhi-cong, SONGBao, TANGXiao-qi. 2012. Dual axis AC servo system based on STM32F103. *Journal of Mechanical&Electrical Engineering*. 29(11):1290–1293.

Computing, Control, Information and Education Engineering – Liu, Sung & Yao (eds)
© 2015 Taylor & Francis Group, London, ISBN: 978-1-138-02800-5

Network media influence on the ideological and political education

Y. Xu
Nanjing College of Chemical Technology, Nanjing, Jiangsu, China

ABSTRACT: With the advent of the internet age, a series of changes in college teaching occurred, with the help of information and network technology, ideological and political education can be carried out smoothly. Ideological and political education should keep pace with the development of the times, use online media to service for education and teaching, when network brings opportunities to ideological and political education, they also bring new challenges. Universities should closely integrate online education with ideological and political work of university students, build a network ideological and political education platform, create a new situation of ideological and political of college students.

KEYWORDS: network media, university students, ideological and political education, influence.

1 INTRODUCTION

With the rapid development of information technology, the internet, as the "fourth media" after newspapers, radio, television's, penetrates into all areas of society, university students as the most concentrated population using the internet, their lifestyles and ideas are deeply influenced [1]. Internet as a new way of transmitting information, there are also two sides, how to strengthen the proper guidance in the educational process of teaching and the effective use of it is an important topic placed in front of university education administrators. Especially the students of higher vocational colleges, with the characteristics of poor educational foundation, lack of ambition, psychological immaturity, etc., and a high proportion of rural students, which put forward higher requirements to us. How to take full advantage of the internet platform to engage in effective ideological and political education work for students, also organically combine the students' daily management and education with online education, to maximize their work efficiency, is the new topic need to be solved urgently in current university ideological and political work.

2 OPPORTUNITIES BROUGHT BY NETWORK TO IDEOLOGICAL AND POLITICAL WORK FOR UNIVERSITY STUDENTS

Since the reform and opening up, China's network technology gets rapid development, the network environment more and more matures, gradually penetrates into the ideological and political education activities. As time goes on, people are fully aware that for the ideological and political education work, introducing high-tech is a rare opportunity for development [2].

2.1 *Provide a wealth of ideological and political education resources*

Network itself has the features of the large amount of dissemination information, rich content and covering a wide range, make ideological and political work information can be greatly expanded. Various databases on the internet are easy to use, all kinds of books, statistics, the party's principles and policies, international and domestic hot news. University students can gain knowledge by browsing the internet, expand the ways to acquire knowledge, share resources, increase students' self-educational opportunities.

2.2 *Expand the channels for ideological and political education*

The development and use of the network has opened up new avenues for the dissemination of information, students either through a network to share ideological and political education resources, but also freedom to carry out exchange of ideas online, it is an interactive equality education between educators and educates. Communication means on the internet are diversified, as a whole with text, sound, images and other communication means, are more likely to attract students' attention, mobilize enthusiasm, initiative and participative to obtain information, broaden channels for the exchange of teachers and students, build a three-dimensional educational space for the university ideological and political education.

2.3 *Improve the efficiency of ideological and political education work*

Internet as a new medium of information dissemination, the information is passed quickly, accurately and efficiently, changes the previous situation of "worn out mouth and leg" in ideological and political work. Today, through a telephone line you can enjoy huge network resources. Through moving the mouse, you can sweep views on learning materials, meeting content, theoretical articles, and other important notices. Facts have proved that the application of network technology greatly improves the efficiency of ideological and political work.

3 CHALLENGES BROUGHT BY NETWORK TO IDEOLOGICAL AND POLITICAL WORK FOR UNIVERSITY STUDENTS

Network technology is a "double-edge-sword", which has brought unprecedented opportunities to university ideological and political work, while also makes university ideological and political work face enormous challenges. University is an important stage for students to establish a correct outlook on the world, life and value, the main purpose of ideological and political education is to help students to establish the correct three outlooks, it is essential to strengthen guidance during the university [3]. ①Due to network information encompassing, broad array, varies greatly, both truth and falsehood, both science and pseudoscience. Teenagers are lack of resistance for adverse lure, and just a fun time, so they are easy to indulge in the network game. ②Some information seriously affects university students' healthy growth, impact their ideological concepts, if not correctly parse and guidance, students can easily be misled cause student health problems, and even cause the network syndrome and mental deformities. ③Long-term facing the computer, serious lack of physical exercise, are not conducive to teenagers' health development. ④ideologies of part network information are too extreme, teenagers are not yet mature, so they are easy to be adversely affected, ideas appeared to change, value orientation to be twisted. Moreover, in the network environment, students can anonymously speak freely, wanton vent discontent. Over time, it is not conducive to the cultivation of students' moral character. ⑤Due to indulging in the online world all the day, ignoring the contact with others, relationships become increasingly tense, personal character becomes more unsociable and eccentric. Coupled with the capitalist ideological concepts poisoned teenagers through the network, make teenagers' ideological concepts fall into confused, these factors are not conducive to the smooth development of university ideological and political education.

4 COUNTERMEASURES OF UNIVERSITY'S IDEOLOGICAL AND POLITICAL EDUCATION WORK UNDER THE NETWORK ENVIRONMENT

Under the network environment, ideological and political education should be kept up the pace of time's development, while introducing high-technology, introduce advanced teaching philosophy, and actively improve the ideological and political education model, meet the needs of university ideological and political education in the new era, actively implement the ideological and political education work [4]. The importance of ideological and political education is self-evident, for further implementation of university ideological and political education under the network environment, we consider that can proceed from the following aspects:

4.1 *By means of network platform to extend classroom teaching content*

Under the network environment, the transfer of information resources is convenient. University ideological and political education should take full advantage of this point to extend classroom teaching, make the form of ideological and political education become more rich and diverse. Through the network platform to extend the classroom teaching content and expand students' knowledge. In the network environment, teaching content will be more rich and varied, in addition to a lot of theoretical knowledge, as well as a wealth of real examples and exercises and other resources. For teachers by means of network platform to enhance communication and interaction with students is necessary. In the traditional teaching mode, the teacher-student relationship is more tense, the students due to the psychology of fearing teachers, don't dare to communicate and interact with teachers. Through the communication platform, teachers can provide counseling to students, solve students' doubts, students can more easily communicate and interact with teachers, is very conducive to improve the teaching results.

4.2 *Construct ideological and political education network platform*

Should explore ideological and political education channels from all the ranges, especially with the aid of the network media, strive to create a better campus network environment for ideological and political education, and enhance effectiveness of ideological and political education work. Ideological and political education workers must combine characteristics of students in higher vocational college, effectively use the platform of the internet, such as in the college website creates party building, group learning, ideological and political online, strive to be advanced and create excellence and other web services section, establish elite ideological and political education network connections, and update website content, focus on education penetration. The content of the website should be enriched and novel, students can expand thinking through the website, strengthen teacher-student communication. Youth League and Students office director, counselors, all grades director can through personal working blog, e-mail, qq group, Fetion and other contact ways, to strengthen communication and exchange with students, to keep abreast of students' ideological dynamics, take full advantage of the network technology, boldly innovate, build a new platform of ideological and political education work [5].

4.3 *Through network technology to strengthen the publicity of ideological and political education*

By means of network technology, post the promotional content of ideological and political education onto university's official website, when students enter the official website, they will get access to the relevant information on ideological and political education at the first time. In each query page, set the flowing window, facilitate students to check all kinds of information, make students influenced unconsciously by ideological and political education. According to my survey found, the information exposure to human in the case of unconscious is usually more likely to translate into the subconscious mind. Human subconscious is the basis for the formation of a variety of ideological concepts; it will have a huge impact on the development of ideas. Actively promote the ideological and political education through network technology, make students in their daily lives imperceptibly influenced by ideological and political education, consolidation the classroom effect of ideological and political teaching, improve teaching quality.

4.4 *Build network ideological and political education working staff*

Team building is the prerequisite and guarantee of good university ideological and political work, and train a number of special combination high-quality political education workforce, with a higher level of business and political literacy, can devote themselves to educational management and skilled of network technology. Online education is through network of this "media" to achieve the ideas exchange with students, is an indirect way of ideological and political education work, for the emergence of some online negative, one-sided or incorrect, untrue opinion tendency or hot issues, consciously carry out correct guidance through the network to make students form a correct ideology, through online forums, chat rooms, qq, etc., make the campus network become the bridge and link for teachers and students to exchange ideas, online counsel, psychological counsel. In addition, hire the ideological and political education experts as part-time education working staffs, participate in students' online Q & A and guide the website construction, invite them regularly to make academic reports or online video lectures for students, all are good choices, not only save manpower, financial resources, but also receive a good education results, further improve the network education system.

4.5 *Strengthen ideological and political education network management*

The internet is a double edged sword, unhealthy information will seriously erode university students' physical and mental health. Therefore, strengthening network management of ideological and political education is essential, we should strive to develop "green network space", and actively create a "green network environment." Increase constructing green campus network; insist standardization and procedure of filtering web content; create a green network space and network carrier for students; strengthen prevention and control of network technology and artificial monitoring; establish strict security system; through the above ways to provide a healthy, positive network environment for students.

4.6 *Create a new model combining students' daily education and online education*

New Period, ideological and political work are diversified development, the ideological and political education as a quick, effective education, gradually develop into the main front of ideological and political work, however, a single form of education has been unable to meet the cultural requirements of contemporary university students, only combining with other kinds of ideological and political education work, can play a greater effectiveness. Especially in the daily management of student education and theoretical classroom teaching, should also penetrate the ideological and political education, if in the ideological and political teaching, in addition to theoretical indoctrination, for changing the single teaching method, can use multimedia to teach and intersperse some cases to improve the students' learning interest, the education effect will be better. Not only base on website, continue to strengthen the effort of online education, but also pay close attention to the daily education and management, constantly improve the offline education. Ideological and political education work is a complex and huge project, only the new model of fully implement both regular education and networking teaching for college students is consistent with the characteristics of the times, meets the educating students' requirements of contemporary colleges.

5 SUMMARY

The Internet is a rapidly developing new media, as well as new carriers, new areas and new topics of college moral education. Ideological and political education is facing unprecedented challenges, but also there is a golden opportunity, We have to keep up with the pace of era development, actively promote the reform of university students ideological and political education, by means of network platform to extend classroom teaching content, build and constantly improve the ideological and political education network platform, through network technology to strengthen the publicity of ideological and political education, strengthen ideological and political education network team and network management, actively create a new model combining students' daily management with online education, adhere to the student-centered, establish service awareness, combine with the characteristics of the higher vocational colleges students, creatively carry out work, and strive to enable students to establish the correct outlook on world, life and value, make the network ideological and political education really play function of educating people, thus create a new situation in university ideological and political education work.

REFERENCES

[1] Zheng Yongting, Methodology of ideological and political education [M], Beijing: Higher Education Press, 2003.

[2] Song Yuanling, Chen Shouping, Network culture and ideological and political education [M], Changsha: Hunan People's Publishing House, 2006.

[3] Cheng Changshou, Luo Donghai, New theory of ideological education network [M], Nanjing: Hohai University Press, 2006.

[4] Zeng Yuwei, Lu Yongzhi, Discussion of network Anarchism's challenge for ideological and political ducation of College Students [J], Journal of Southwest Jiaotong University (Social Science Edition), 2008(05):137–141.

[5] Shen Maofa, Shen Yiming, The architecture and mechanism innovation of higher vocational colleges ideological and political education network [J], China Electric Power Education, 2010(03): 161–162.

Computing, Control, Information and Education Engineering – Liu, Sung & Yao (eds)
 ISBN: 978-1-138-02800-5

Discussion on sports curriculum of multimedia technology application

L. J. Wan
Department of Sports, Guizhou University of Finance and Economics, Guizhou, P.R. China

ABSTRACT: With the rise of science and technology and the rapid development of multimedia technology, Physical Education Teaching methods should keep up with the times, we need to make full use of advanced science and technology, new ideas, bold change and try, thereby improving the quality of teaching. In the field of higher education, appeared new education model educational technology is the representative, in which the use of multimedia technology in teaching is a very prominent part. Multimedia technology can provide teachers and students good teaching conditions, rational use of multimedia technology can better accomplish the task of teaching.

KEYWORDS: Multimedia Technology, University, Physical Education, Application.

1 INTRODUCTION

With the computer information technology, multimedia network technology development, we enter a new era of knowledge and networks. Education Reform and Development and the level of scientific and technological knowledge innovation determine a country's comprehensive national strength and its ability to compete in the world. University Conduct multimedia sports teaching can make better use of various sports teaching resources, make the limited educational resources fully utilized and shared, improve the quality of classroom teaching, enrich students' extracurricular sports activities, break through time and space limitations of traditional physical education, supplement and facilitate the physical education curriculum learning inside and outside the university classroom in a new form, provide a new learning space and time for university students, these will greatly enhance the physical education of university students' learning initiative and enthusiasm, enhance theoretical knowledge of sport, more scientific system to enhance the sports quality of students.

2 THE STATUS OF UNIVERSITY PHYSICAL EDUCATION DEVELOPMENT

1 From the view of university physical education goals. Our goal of Physical Education has gone from skill to physical and then the evolution of health, therefore, physical education suffered [1]. The content and form of Sports teaching depart from the actual situation of university students, make the sports awareness of University Students is poor, a considerable part of the university Students do not know the meaning of exercise on their own and lifelong sports.
2 From the view of university physical education methods and content. Over the years, national physical education course basically uses uniform materials, many content duplicates with high school textbooks. Due to the influence of traditional ideas on sports, there are drawbacks: teaching content narrow, teaching methods monotonous, poor teaching effectiveness. The Physical Education disjointed with social and personal future life, this is the fatal of currently university sports.
3 From the view of o university teachers' objective factors. Affected by the teacher training system and traditional sports teaching objectives, for teachers, there are at least three aspects seriously affected the Physical Education. First, teachers' preferences and specialty impact the materials selection. Second, the teacher's age and gender influence the choice of teaching methods. Third, the technology actions demonstrate difficult affect the teaching process.
4 From the view of university physical education trends.Accordance with the requirements of "National Fitness Program", university sports is not only to promote the full development of university students, but also make students have the correct sports consciousness, after graduation based on objective and subjective circumstances change, can uninterrupted independently engage in physical exercise, so obtain the benefits of lifetime, physical exercise become the main content of throughout life[2].

3 THE ADVANTAGES OF MULTIMEDIA TECHNOLOGY IN THE UNIVERSITY PHYSICAL EDUCATION

1 Multimedia technology is conducive to update university physical teaching concepts.With the development of modern society, center for education has changed from the past teachers "teaching" to the modern university students "learning", emphasis on humane education. The multimedia technology in Physical Education, changed the traditional sports teaching with "teach" centered teaching mode, teachers use modern multimedia teaching methods, meanwhile by means of human-computer interaction to exchange with university students, stimulate students' participate awareness, Stressed university students take the initiative to explore the knowledge, active discovery, reflect the sport of multimedia teaching is "learning" as the center of teaching ideas.

2 Multimedia technology is conducive to strengthening the integration of University sports courses theory.Increasing physical theoretical knowledge teaching, deepening Physical Culture and Education are the consensus reached by the Physical Education Reform, the multimedia technology is widely used in other disciplines have proved it broad applicability in the field of theoretical teaching. Increase the proportion of theoretical lessons, implement including physical health, mental health, physical exercise, and self-help survival, culture and entertainment etc. comprehensive physical literacy education and healthy lifestyle education, construct of sports knowledge system, highlight the deep sports knowledge capacity and cultural implication, are the inevitable trend of Physical Education Reform[3]. But today's university physical education confined to the classroom, teachers impart knowledge is limited, how to improve university sports knowledge base become a problem to be solved, the network features a large capacity can be part of the solution of this problem. In the classroom teachers can make some physical exercise issues related to students in real-life, let students query the answer over the network, meanwhile, university students can online inquiry some issues encountered in their own physical exercise. This process is subtle health education for university students, develop students' spirit of exploration, and improve students' self-training, self-monitoring consciousness. Use multimedia for university students moral education, cannot limit by time, space and the macroscopic, make ideological education content more vivid, contagious. For example, emphasizing the collective spirit of unity and cooperation co-ordination to university student, organizing students watch the Olympics, Asian Games, World Cup and other major international competitions, through the intuitive screen of athletes tenacity to win, five-star red flag raising in the stadium, so that students receive patriotism education in the subtle, to achieve the purpose of teaching, students develop in the compound talents direction [4].

3 Multimedia technologies will help to stimulate University Students' learning interest.Germany educator He Baer noted: teaching should throughout among students' interest, so that students interested in teaching at each stage can be expressed as a coherent attention, wait, study and action[5]. Computer multimedia technology can stimulate the body to vision, hearing and other sensory systems, alternately in different functional areas of the brain activity, make learning content image, intuitive, vivid, strong interesting and easy to understand. Multimedia technology comprehensibly use graphics, animation, music, glitter, color, font, etc. means of expression, enhance the artistic expression and influence of the Physical Education content, truly active classroom atmosphere. Multimedia assisted instruction inspire university students to explore, to create, improve university students' interest in learning, make classroom teaching density, strength, emotional, consistent with university student curiosity, novelty psychological characteristics, also create a better situation and emotional experience, cause and maintain the attention and interest of university students, stimulate the learning enthusiasm of university students, mobilize the enthusiasm of learning.

4 THE APPLICATION OF MULTIMEDIA TECHNOLOGY IN UNIVERSITY PHYSICAL EDUCATION

1 Multimedia images and text's application in university physical education.In the learning process of sports professional concept, theoretical knowledge, stadiums, sports equipment, can make use of text, pictures show, make students understand the sport professional knowledge and concepts, understand space equipment size, shape, and professional terminology and names, the role of each line, the nature, characteristics and use of equipment[6].For example: through football venues pictures and descriptions make it easy for university students to understand the position of football fields in length and width, divided sideline, end line and the center line of the penalty area, goal kick and corner area, the size of the penalty arc and circle, the role and high and wide

of the goal and other basics. We can also download the relevant information picture collection and links through the network, so that students can understand the world famous venue information, the most advanced venues information. such as download Beijing 2008 Olympic Stadium "Bird's Nest" and "Water Cube" design blueprint, make students learn the most advanced design and functionality stadiums.

2 Multimedia animation and video applications in University Physical Education.In sports technology on university physical education teaching, using video animation multimedia, so that students can observe the technical details of technical movements from different angles, quickly establish a correct and complete action representation [7]. As in swimming teaching, we can use animation or video playback of technical movements of water and underwater different angles, allow university students to observe the technical movements and structural features, while accompanied by the teacher to explain and illustrate, make students quickly create action representation. In sports, there are many sports technologies not only complex, but also need to complete a series of complex technical action in an instant, such as athletics jumping events air movement, gymnastics skills tumbling action etc., these technical actions bring great difficulty in teaching. On the one hand, due to the action of teachers demonstrate is limited by their conditions, such as the teachers' understand extent of the action, the teacher's age, physical condition, psychological factors, so many teachers is difficult to complete the action freely, On the other hand, students' viewing angle and timing also subject to certain limitations. Therefore, the students are difficult to see clearly these momentous action, the action cannot form a complete representation, thus bring a certain impact to learn. The use of animation, motion video can show the difficulties and the new technology in the way of continuous slow motion playback, make students carefully observe the details from different angles, and then grasp the technology essentials better, deepen the understanding of the whole movement, establish the correct action representation, improve learning efficiency and shorten the process of teaching.

3 Multimedia courseware's application in university Physical Education.Multimedia teaching has a strong interaction, use multimedia courseware for sports theoretical knowledge teaching can be easily man-machine dialogue, put learning initiative to university[8].The representative picture, beautiful art form, concise text and vivid display plus audio commentary makes university students with great interest to learn sports knowledge, improve the sports culture cultivation. When explaining the various rules of the game and the referee method, can also use courseware targeted for detailed explanation and analysis. University students can test their understanding of the learning content through interactive simulation exercises. Such practice could well stimulate university students' interest in learning, and fully mobilize the enthusiasm of students' learning.

5 CONCLUSION

In summary, the combination of Multimedia Technology and University Physical Education is a novel and effective teaching method in university sports teaching reform, make teacher-student relationship change from teacher-centered teaching to become a university student-centered teaching. With the continuous development of modern technology, multimedia technology as a new form of education and means will bring great impact and influence to traditional education. If multimedia technology can be an important means of university's physical education teaching, the key lies in university sports teachers' information quality level. A university physical education teacher should strive to improve their own quality, especially emphasis on the quality of training information, improve their professional level, to enable reasonable and proper production and use of multimedia technology in physical education. Different teaching media have the characteristics of the respective roles, only fair use, can achieve the optimal effect of teaching, we should take advantage of a variety of media expertise to present different teaching content or reflect different teaching methods. Therefore, multimedia technology in teaching university sports is very deserving of our bold attempt to explore and promote.

REFERENCES

[1] Z F Wang. Discus on the necessity of multimedia technology into the university physical classroom[J]. Journal of Anhui Normal University, 2008(2):197–200.

[2] Q Li. The main task and reform orientation of the 21 century university sports[J]. Journal of Wuhan Institute of Physical Education, 2002(1):58.

[3] Z X Zhong. Constructe the new system of higher physical education curriculum[J].Journal of Xi'an Institute, 2006(4):14.

[4] G M Zhao, K Y Chen. University students' interest in sports and curriculum design[J].Higher Agricultural Education, 2001,(12):15.

[5] H H Zhang, K Y Ji. Experimental study on students' interests[J]. China Sport Science and Technology,2005(5):3.
[6] Z J Zhao. Multimedia technology tutorials[M].Beijing: Machinery Industry Press,2003.
[7] P Wang, ZL Zhao.The multimedia technology application in physical education teaching:advantages, problems and improving strategies[J].Electronic Test, 2014(9):128–129.
[8] J Jia.New ideas for higher physical education reform in multimedia horizons[J].Education and Career, 2014(11):141–142.

Computing, Control, Information and Education Engineering – Liu, Sung & Yao (eds)
 ISBN: 978-1-138-02800-5

Study on model innovation of WeChat network marketing

X. Xia
Guizhou University of Finance and Economics, China

ABSTRACT: Today network develops rapidly, along with the continuous development of network technology revolution, the network nascent product "WeChat" rises to fame quickly in cyberspace, and businesses discover a new way of marketing in the internet "tide player" - WeChat network marketing. WeChat network marketing is an integrated product of today's rapidly developing market economy and network economic; is an important research topic in today's disciplines of economics, management science, etc. Based on this, this paper on the foundation of analyzing the basic connotation and characteristics of WeChat network marketing focuses on analysis of the WeChat network marketing's innovation model, in order to enlighten the relevant personnel.

KEYWORDS: WeChat, network marketing, model innovation.

1 INTRODUCTION

WeChat network marketing is an integrated product of today's rapidly developing market economy and network economics, and is an important research topic in today's disciplines of economics, management science, etc. WeChat is on the network, there is no distance limitation between business sellers and consumers, once the users register WeChat, can form links with other users. Through this connection, consumers can have timely access to business products' information; also can timely communicate feedback quality of relevant products, after-sales service and other issues to management sellers. With the rapid development of e-commerce, WeChat network marketing today has gradually become the most important part of e-commerce, operational sellers through providing product information needed by consumers, in the process of promoting their products, also learn the market and the consumers' needs, then realize the "point to point" service between operational sellers and consumers[1].Therefore, the analysis of WeChat network marketing model also has very important significance. Today, the rapid development of the network, along with the continuous development of network technology revolution, network nascent product "WeChat" rise to fame quickly in cyberspace. Merchants are seeing this point, to seize business opportunities, discover a new way of marketing in this internet "tide player" - WeChat network marketing.

2 CONNOTATION DEFINITION AND CHARACTERISTICS OF WECHAT NETWORK MARKETING

2.1 *Definition of network marketing*

Network marketing is the business activity based on modern marketing theory, through the network, communications and digital media technology to achieve marketing objectives, is contributed by the scientific and technological progress, changes in customer value, market competition and other combined factors, is the inevitable product of the information society [2]. There is broad and narrow network marketing in accordance with its implementation, the broad Network marketing refers to enterprise use all computer networks (including Intranet, EDI and Internet) to carry out marketing activities. The narrow network marketing specifically refers to internet marketing (Internet, the world's largest computer network system). means the whole process of the organizations or individuals base on the convenient internet to make a series of business activities for products, services, so as to achieve to meet the needs of organization or individual, network marketing is an integral part of their overall marketing strategy, is a marketing tool built on the basis of internet via the internet's characteristics to achieve certain marketing objectives[3].

2.2 *Definition of WeChat network marketing*

Currently, there is not a complete and unified definition on WeChat network marketing model. And in today of marketing diverse development, through WeChat network for marketing has gradually become a new marketing model. And micro-blog marketing has the same purpose with WeChat network marketing, is generally known [4]. The CEO of NITC web marketing service center, LIAO Hongfa said: "micro-blog marketing influence is worth, is insightful, is interpersonal——first learn marketing the influence ourselves, product marketing can be more effective." This paper argues that WeChat network marketing and micro-blog marketing are both through the network platform using various resources to undertake a series of marketing activities, through the embedded advertising style of product, those who have purchased feedback so as to achieve the branding publicity of the selling products, as well as through the network to build a new customer service platform.WeChat network marketing is mainly reflected in through mobile clients installed in the mobile phone or tablet with android system, apple system, windowsphone8.1 system to carry out area targeted marketing, business man through WeChat public platform to show micro official website, micro membership, micro push, micro-payment, micro activity, micro CRM, micro statistic, micro stock, micro commission, micro reminder, there has been formed a mainstream online underline WeChat interactive marketing.

2.3 *Characteristics of WeChat network marketing*

WeChat network marketing is a new and unique e-commerce marketing, the breadth of its information network, the shortcut of information dissemination, three-dimension of product promotion, interpersonal interaction, etc., all have their own unique characteristics[5][6]. (1) The extension of information network. Because WeChat registration process is very simple, either real name, or you can also set up a network nickname with product features, product information pass through mutual concern between the WeChat friends circle, and the impact is very extensive. (2) The rapidity of information dissemination.WeChat network marketing is superior to traditional advertising in the information publishing process, eliminates complicated approval procedures, and also eliminates large sums of shooting advertising costs, because the users' feedback is the best advertising. So through WeChat to advertise often is able to reflect the information on selling products in the first time. (3)Three-dimensional of product promotion. Currently, the multimedia technology develops rapidly, merchants in WeChat platform with words and pictures to publicize or with a network connection to a specific online shop, provide for consumers visualize specific varied forms to display the selling goods.(4)Interpersonal interaction. The commodity information released by WeChat network platform can be direct communication with consumers in the form of text or voice in the first time, solve the problem of businessman due to cannot timely communication with consumers then lose customer base; meanwhile, the more timely product information publication, the more resonation or disputation of consumers, through feedback, the more they can cause everyone's attention, the greater the range of information transfer[7] [8].

3 ANALYSIS OF WECHAT NETWORK MARKETING MODEL INNOVATION

3.1 *Grass-roots advertising style-view nearby person*

(1) Product description: Signature column is a major feature of Tencent product, users can update the status in their signature columns, naturally you can make the mandatory advertising, but only the user's contacts or friends can see. The WeChat functionality plug-in " Browse nearby person " based on LBS can make more strangers to see the mandatory advertising. (2) Function mode: After click the "Browse nearby person", the users can find the around WeChat users based on their geographic location. In these vicinities WeChat users, in addition to display the user's name and other basic information, also display the contents of the users' signature. So users can take advantage of this free advertising place to advertise their products. (3) Marketing style: Marketing personnel in the most flow place use WeChat 24 hours, if there are enough users click " Browse nearby person ", the advertising effect will rise with the number of WeChat users rises, it is possible this simple signature column may become a movable "golden advertising" [9].

3.2 *Brand activity type-drift bottle*

(1) Product description: Drift bottle is an application transplanted to QQ e-mail, this computer application is widely acclaimed, and many users prefer this simple and interactive way with strangers. After ported to WeChat, the drift bottle function generally retains the original style simple and easy to use. (2) Function mode: Drift bottle has two simple functions: First is "throw one", users can choose to publish a voice or text and then click "throw into the sea"; second is "pick up one", after "salvage" the drift bottle threw by countless users in the sea, then can dialogue with the thrower, but each user has only 20 chances one day. (3) Marketing style: The WeChat official can change the parameters of drift bottle, and make the number of drift bottles about the cooperation business promotion activities thrown in a period time significantly increase, the frequency salvaged by ordinary users also increase. Plus drift bottles mode itself can send different texts or even a voice game and so on,

if marketing properly, can produce a good marketing result. And this voice mode makes users feel more real. But if only pure advertising slogan, it may cause users' objection [10].

3.3 *O2O discount type-scan the two-dimensional code*

(1) Product description: "Scan QR Code"This function was originally a reference of another foreign social tool "LINE", through scanning to identify another user's identity of two-dimensional code, so as to add friends. But today with the two-dimensional code development, its commercial use is more and more extensive, therefore WeChat follow this trend, combine O2O to undertake business activities. (2) Function mode: put the two-dimensional code pattern into the frame, and then you will be able to get a member discount, business discount or some news information.(3)Marketing style: plug scanning two-dimensional code in mobile application, this O2O style is already popular; WeChat has a million users with active degree high enough, its value is self-evident[11].

3.4 *Interactive marketing type-WeChat public platform*

(1)Product description: For the popular media, celebrity and enterprise, if the open of WeChat open platform + sharing function in friend circle has made WeChat to be a marketing channel cannot be ignored in the mobile internet, then WeChat public platform make this marketing channel more granular and direct.(2)Function mode: through push and attention by one to one, public platform push news, product information, latest events, etc. to fans, and even be able to provide functions including consult, customer service, and so on, become one qualified CRM system. You can say that the on-line of WeChat public platform provides a mobile website based on billions of WeChat users.(3)Marketing style: by publishing two-dimensional code of the public account, make WeChat users can subscribe to the public platform account conveniently, and then after grouping the users and geographic control, public platform can push the precise message directly to the target users. The following is through personal attention page and friends circle to achieve brand's viral spread [10].

3.5 *Social sharing type-an open platform + friends circle*

(1) Product description: WeChat open platform is a new feature introduced in version 4.0 of WeChat, application developers can plug in third-party applications through the WeChat open interface. Applications' LOGO can also be placed in WeChat attachment field, allow WeChat users to invoke third-party applications, to select and share content easily in the session. (2) Function mode: Social sharing has been a hot topic in the e-commercial. In the mobile internet, take the partner before announced by Tencent as an example, the WeChat users spread commodities on the Meitu one by one, to achieve the most direct mouth marketing on social media. (3) Marketing style: In addition to WeChat asynchronous communication features, the open of share function in friends circle, the new feature in version 4.0, offer the best channel for sharing type mouth marketing. WeChat user can directly share mobile phone application, PC client, the compelling content on website in friends circle, and support opening through website links[10].

3.6 *Network operational type-WeChat shop*

This WeChat shop (WeChat Mall) is not Tencent self-operational platform, the upgrade of WeChat selection product channel, but is after merchant apply for permission to pay and then create platform for WeChat shop. By the end of 2013, if the public account want to apply for permission to WeChat payment, need to accord with two conditions: first must be service number; second still need to apply for WeChat certification to obtain WeChat's high-level interface permissions. Only after merchant applied for WeChat payment, can further make use of WeChat's open sources to establish WeChat shop[10].

4 SUMMARY

WeChat marketing is kind of marketing business model in network economic time, is a type of network marketing accompanied by fame WeChat to emerge[12]. In WeChat do not exist distance limitation, after register in WeChat, the user can form a contact with the around "friends" also have registered, subscribe to the information they need, merchants through providing information users need to promote their products, realize point to point marketing. WeChat marketing network, the new type of e-commerce marketing can become the mainstream model of future society, is an effective force and weapons to corporate's brand marketing and personal marketing. Only learn to use the power of WeChat network marketing, make the brand more dynamic, can bring more convenience to people's live and work.

REFERENCES

[1] Wen Danfeng, WeChat Marketing: Weapon on the Fingertip [M], Beijing: People Post Press, 2013.
[2] Feng Yingjian, The Foundation and Practice of Internet Marketing [M], Beijing: Tsinghua University Press, 2002.

[3] Shao Yuxia, The exploration of network marketing communication model in new media era [J], Enterprise Technology Development,2010(06).
[4] Han Yang, How Merchants Bank utilizes WeChat? [J], Business Value, 2013(06).
[5] Zhang Yanhong, Analysis of WeChat application in enterprise network marketing [J], China Trade, 2013(28).
[6] Ou Zhiming, Application of WeChat public platform in enterprise network marketing [J], New Marketing, 2014(06).
[7] Fang Xingdong, etc., Research on WeChat propagation mechanism and governance issue [J], Modern Media, 2013(06).
[8] Dang Haoqi, Deconstruction WeChat information dissemination model from the perspective of Communication Science [J], Southeast Spread, 2012(07).
[9] Tan Kai, Pros and cons of WeChat in enterprise marketing [J], New Marketing, 2012(11).
[10] Shao Xiaotong, Pros and cons of WeChat in enterprise marketing [J], New Marketing, 2012(11).
[11] Yan Xue, Application prospects of O2O business model in the scenic area marketing [J], Tourism Overview, 2013(02).
[12] Wang Xia, The new marketing model and strategy based on WeChat [J], Management Observe, 2013(08).

Computing, Control, Information and Education Engineering – Liu, Sung & Yao (eds)
© 2015 Taylor & Francis Group, London, ISBN: 978-1-138-02800-5

The design and implementation of management information system of hospital of anesthesia operation

Guang Hao Yu & Yang Gao
Medical Imaging Department, Mudanjiang Medical University, Heilongjiang Mudanjiang, China

Li Guo Hao
Medical Technology Department, Qiqihar Medical University, Heilongjiang Qiqiha'er, China

Lian Di Li
Second Affiliated Hospital, Mudanjiang Medical University, Heilongjiang Mudanjiang, China

ABSTRACT: Management Information System of Hospital of Anesthesia Operation was designed in the research. The domestic and foreign clinical medical system and operation anesthesia medicine theory and technology have been analyzed. The present situation of anesthesia operation in clinical medicine has been studied dominantly questions in meeting customer needs. According to the Hospital clinical anesthesia on operation demand and the anesthesia process, the information management system of anesthesia operation in Hospital has been designed to realize the implementation of clinical medicine in scientific, standardized, systematic and anesthesia information automation, convenient management. The computer technology and database technology used in management information system has also been analyzed. The results show the use of these techniques can realize rapidly the system development and system maintenance. Each subsystem modules that are designed in detail and are integrated debugging.

KEYWORDS: Operation anesthesia, Information management, Clinical medicine, system.

1 INTRODUCTION

With the rapid development of modern information technology, the domestic hospitals of the level of more than Grade II have accelerated the pace in information technology, construction, the use of specialized medical software combined with the clinical diagnostics to improve service levels and core competitiveness of the hospital. However, the information technology construction of department of operation anesthesia is in missing or the most backward links in the development of hospital-wide information technology construction[1]. Anesthesia is an important part in the process of clinical practice, does not allow any slack and flawed, but anesthesia involves the enormous amount of information, which is a challenge in the work. Therefore, we urgently need to develop a more advanced information processing method to replace the anesthesiologist hard work, and can be more easily and accurately manage the huge amount of data anesthetic medical information.

At present, most hospital anesthesiology can use the computer artificial record and edit some of the data and print reports, or use some commercial personnel, financial management systems or office system edit and process part of the data or create a database. Anesthesia information system online real sense of integration is almost blank, there is no one system is compatible with all existing databases or completely replace the paper for records[2]. The anesthesia information management system is the use of computers and anesthesia information acquisition equipment instead of doing the artificial recording and analysis about anesthesia information, to complete anesthesia information automation of the collection, storage, analysis and other operations, and make anesthesia information accurate, easy to extract.

The system includes anesthesia information automated collection, integrated management, security, sharing, scientific analysis and other work to achieve clinical anesthesia information scientific, standardized, systematic, accurate, automated management. The language used is C#. It is a new object-oriented programming language, specifically for the .NET running. For the realization of the main functions is to use Visual Studio 2010 to develop server-side, and using PowerBuilder to develop client function.

2 ANESTHESIA INFORMATION MANAGEMENT SYSTEM REQUIREMENTS ANALYSIS

2.1 *Overall requirements analysis*

To anesthesiologists of the hospital, the object of anesthesia data collection, the frequency of anesthesia data collection, the content of anesthesia data collection, anesthesia data statistical methods, the results of anesthesia data analysis, etc., need to collate and analyze these data. Therefore, we should establish an anesthesia data collection convenient, flexible, accurate, easy to use, strict management specification, data security, automated statistical analysis of data clinical anesthesia information management platform[3]; Through hospital Internet or other network realizing patient anesthesia information fast transfer, browse, track between different departments; For anesthesiologists, surgery physicians and other clinical medical personnel it provide a convenient information sharing and information statistics and analysis; establish a strict system of user privilege level, according to the logged-on user privileges difference to provide the appropriate application function module.

2.2 *Functional requirements analysis*

The purpose of the system development is that the function of the developed system is in line with the actual user workflow to help users easily work. The anesthesia management system will focus on operation patients' three stage pre-operation, operation, post-operation, for anesthesia, operation, nurse related personnel to provide timely and accurate patient basic information, diagnostic information, inspection reports, vital signs, etc. comprehensive information for reference, these contribute to help reduce medical risks and improve the operation efficiency.

According to different responsibilities of the staff, the system set different management authority; Managers have all the authority, while other staff can only see the module function of their work need. The function of the system consisting of: basic maintenance module, patient management, surgical application management module, the nurse workforce management, anesthesia record single module, post-operative summary/follow-up (Figure 1).

3 SYSTEM DESIGN

The system, which is operative for the various stages of intelligent management, highly integrated information resource which departments needed to achieve the standardization of medical procedures, so as to effectively improve work efficiency, reduce health care costs and health risks, enhance the research and teaching capacity and improve the hospital overall level of information management.

Anesthesia system mainly is the six modules to handle question: basic data module, patient management, surgical application management module, workforce management, anesthesia record single module, post-operative summary / follow-up.

3.1 *The analysis of system structure*

Entire anesthesia information management system structure is divided into two parts: the server, the client.

The program of the server: Achieve providing the external data sharing interfaces, including HIS interface, anesthesia records interface, anesthesia single interface, equipment management interface, research statistics interface, anesthesia assessment interface, department statistics interface. Achieve some management functions, including department management, system management, and system maintenance.

The program of client: Achieve system the key of performance and higher real-time requirements of business functions, mainly including the system registry, pre-operative management, operative management, PACU management, post-operative management, system maintenance, medical writ management.

The whole system runs on the local area network, the network architecture as shown in Figure 2.

The main work of the client program is completing to automatically collect anesthesia information, manage anesthesia information, compute and print anesthesia information as well as real-time monitoring of anesthesia. The entire program is the management of related client information. By the anesthesia surgical procedure, the main module is divided into pre-operative management, operative management, PACU management, post-operative management, department management.

Preoperative management mainly realizes surgical appointment, surgical applications, surgical arrangements, patients' basic information management, classification management of anesthesia, pre-operative visits and anesthesia pre-operative evaluation. Operative management realizes information management operations during surgery. Its operations have surgical information maintenance and verification, collection of anesthesia information, surgical timely monitoring, anesthesia summary and other operations. PACU management is sustainable to track all kinds of life parameters in patients' recovery stage and instantly record various of measures which the medical staff did to wake up the patient and make them recovery. The postoperative management module is the information management operations after achieving completing the surgery, including the record of analgesic operations, statistical supplies, statistical drugs, post-operative follow-up, anesthesia fees and case submission. Department management module realizes maintenance of departmental information, maintenance of department personnel information and statistics workload of the department.

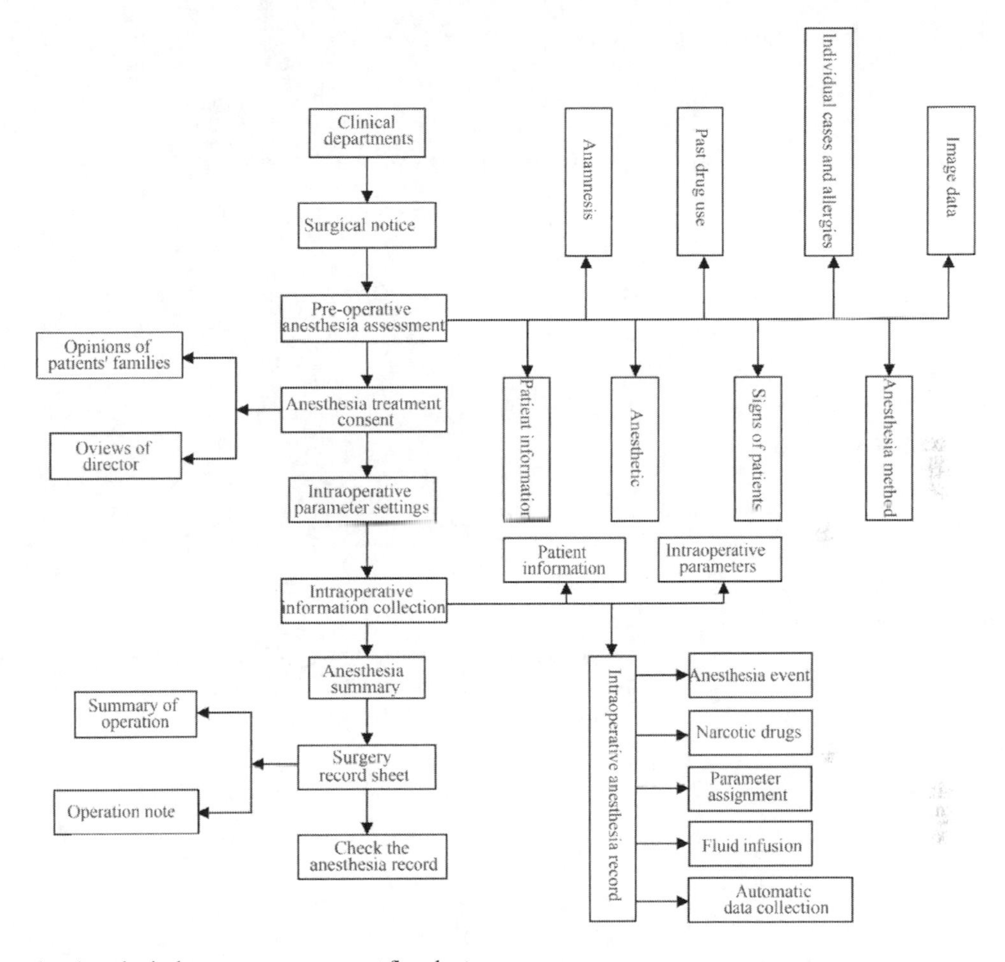

Figure 1. Anesthesia department management flowchart.

3.2 *Database design*

The technology of database development uses a large relational database SQL, the ER diagram and similar diagram design patterns to fully detail systems analysis, remove the unreasonable redundant data structure, before starting the program code designed to carry out a detailed the analysis for system functions, data structures and the plan of data flow to avoid repetitive and invalid work; The theory and technology of system implementation have been fully mature, and therefore the application software that is developed on the basis of these is entirely feasible.

Entity - Relation diagram (Entity-Relation, below referred to as ER) occupy a very important position in the database design. The properties of ER diagram are divided into simple and complex property. Simple property cannot be divided into smaller parts. The ER diagram of surgical anesthesia department management software is below in Figure 3.

Data tables can be converted by the designed ER diagram: It transforms the entity to each of the corresponding table and each entity property to columns of corresponding table; It identifies the primary key column of each table and establishes primary and foreign key between the tables to reflect the mapping relationships between entities.

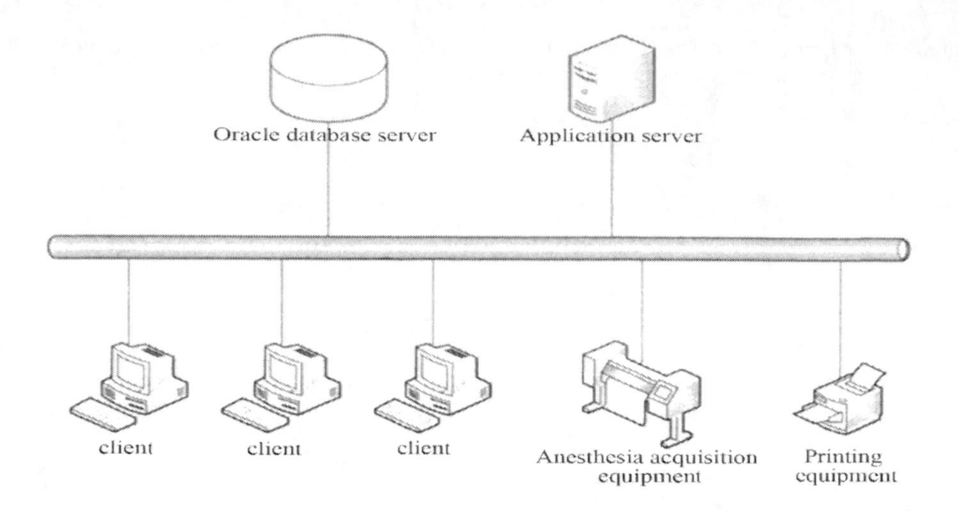

Figure 2. Network architecture diagram.

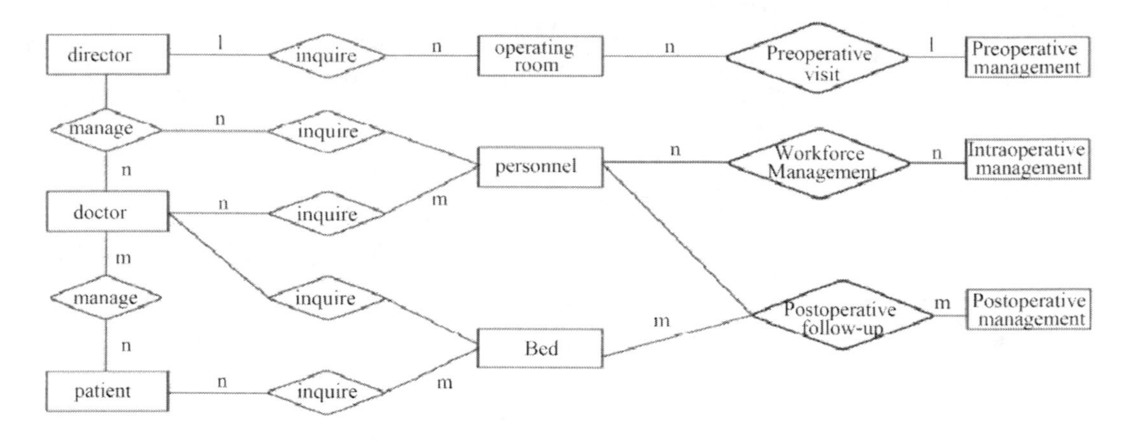

Figure 3. Surgical anesthesia management software E-R diagram.

4 CONCLUSION

The surgical anesthesia information management system is becoming more and more important in the hospital. The paper makes the preliminary design and studies to the surgical anesthesia information management system and mainly consider from the viewpoint of six basic modules: basic data module, patient management, surgical application management module, workforce management, anesthesia record single module and post-operative summary/follow-up. When I design this management system, I design system parameters combined with the actual situation and pay attention to consider its integrated application function to expect widely used in hospitals.

REFERENCES

[1] Yang Shuangying, et al. The guaranteed method of the hospital information management system security operation[J]. Chinese medical equipment, 2007:30–32.

[2] Lai Yan. The written of medical writ[M]. Bei Jing:The Scientific Press, 2008:2–1.

[3] Yu Cheng. Medical equipment management system and clinical surgical anesthesia[J], 2006,12(9):49–52.

Computing, Control, Information and Education Engineering – Liu, Sung & Yao (eds)
© 2015 Taylor & Francis Group, London, ISBN: 978-1-138-02800-5

Design and implementation of cyclic redundancy check algorithm

Sheng Ju Sang
School of Information Science and Technology, Taishan University, Taian Shandong, China

ABSTRACT: This paper focuses on the design and implementation of the Cyclic Redundancy Check (CRC) algorithm. First, the paper provided an overview and principle of CRC, then, given the method of CRC implementation algorithm by hardware as well as using software. Finally, the conclusion can be reached that the design and implementation method here has high practical value.

KEYWORDS: Cyclic Redundancy Check; FCS Frame; Generated polynomial; Galois Field; Modulo-2 Division.

1 INTRODUCTION

In data communication and storage, burst errors are so inescapable in many communication channels, including magnetic and optical storage devices that high reliability of information transmission is required, that is, competitive low bit error rate is required. There are several techniques for checking the burst errors. The cyclic redundancy check (CRC) code is a typical and useful code invented by W. Wesley Peterson in 1961 [1]. It is so called because the CRC codes are a subset of cyclic codes that are also a subset and the algorithm is based on cyclic codes.

The CRC method is a technique for detecting errors in digital data, but not for making corrections when errors are detected. It is used primarily in data communication. In the CRC method, a certain number of check bits, often called a checksum, are appended to the message being sent. The receiver can determine whether or not the check bits agree with the data, to ascertain with a certain degree of probability whether or not an error occurred in transmission. If an error occurred, the receiver sends a message back to the sender, requesting that the message be retransmitted. The technique is also sometimes applied to data storage devices, such as a disk drive. In this situation each block on the disk would have check bits. Blocks of data entering these systems get a short check value attached, based on the remainder of a polynomial division of their contents, on retrieval the calculation is repeated, and corrective action can be taken against presumed data corruption if the check values do not match. CRC implementation can use either hardware methods or software methods. In the hardware implementation, a simple shift register circuit performs the computations by handling the data one bit at a time [2, 3].

The rate of detection error burst errors above 17-bits is 99.9984% by CRC check method [4, 5]. Therefore, CRC can be applied to communication of essential data, such as the detection of running state of slave computers, online reset of running mode or parameters and so on. CRC is not only simple to implement but have the benefit of being particularly well suited for the detection of burst errors, contiguous sequences of erroneous data symbols in messages[6,7]. An n-bit CRC, applied to a data block of arbitrary length, will detect any single burst error not longer than n-bits and will detect a fraction bit of all longer burst errors. For example, the detect method based on the 16-bit CRC standard can detect all the single bit errors, double-bit errors, odd number of bits errors and burst errors which are less than or equal to 16 bits.

2 PRINCIPLE OF THE CYCLIC REDUNDANCY CHECK CODE

CRC code is a subset of cyclic codes and use a binary alphabet, 0 and 1. These arithmetic operations are performed according to the conventions of the algebraic field that has, as its elements, the symbols contained in the alphabet. For binary codes, the field is finite and has 2 elements, 0 and 1, and is called GF (2) (Galois Field).

A binary sequence of bits based on GF (2) can be expressed using polynomials as:

$$M(x)=\sum_{i=1}^{\infty} a_i x^i \quad where \quad a_i \in \{0,1\} \tag{1}$$

For example, the binary 100000111 can be expressed as polynomials as follows,

$$M(X) = x^8 + x^2 + x + 1 \quad (2)$$

According to Eq. (1), for the number of terms in M(x) is called the parity of M(x) or M. We usually interested in even or odd parity which means even or odd terms in M(x) or 1s in M. A polynomial in GF (2) is a polynomial in a single variable x whose coefficients are 0 or 1. Addition and subtraction are done by modulo-2, that is, they are both the same as the exclusive OR operator.

Assume the convention that the leftmost bit represents the highest degree in the polynomial. Suppose M(x) to be the message polynomial, C(x) the code word polynomial and G(x) the generator polynomial, then we have

$$C(x) = M(x)\mathrm{G}(x) \quad (3)$$

We can also rewrite Eq. (3) using the systematic form,

$$C(x) = M(x)\mathrm{x}^{n-k} + \mathrm{R}(x) \quad (4)$$

Where, R(x) is the remainder of the division of $M(x)\mathrm{x}^{n-k}$ by G(x) and R(x) represents the CRC bits.

According to Eq. (4), the transmitted message C(x) contains k-information bits followed by n-k bits of CRC. So, encoding is straightforward as follows:

Step 1: Multiply M(x) with x^{n-k}, that is, Adding n-k bits '0' code to the k-bit information code result in the length of entire code called (n, k) code is n (k + r) bits.

Step 2: For a given (n, k) code, we can prove the existence of a polynomial G (x) whose maximum power is n-k = r.

Step 3: Calculating the CRC bits by dividing $M(x)\mathrm{x}^{n-k}$ by G(x).

The receiving end conducts CRC calculation of the received information and compares it with that transmitted. CRC code of transmitted information is placed at the end of information generally. For the decoding part, the same algorithm can be used. If $C'(x)$ is the received message, then no error or undetectable errors have occurred if $C'(x)$ is a multiple of G (x), which is equivalent to determining that if $C'(x)\mathrm{x}^{n-k}$ is a multiple of G (x), that is, if the remainder of the division is 0.

Sometimes an implementation appends n 0-bits to the message to be checked before the polynomial division occurs. This has the convenience that the remainder of the original bit-stream with the check value appended is exactly zero, so the CRC can be checked simply by performing the polynomial division on the received message and comparing the remainder with zero to determine whether the received information is right.

3 GENERATED POLYNOMIALS

The selection of generator polynomial is the most important part of implementing the CRC algorithm. The polynomial resembles the divisor in a polynomial long division, which takes the message as the divided, and in which the quotient is discarded and the remainder becomes the result, with the important distinction that the polynomial coefficients are calculated according to the carry-less arithmetic of a finite field. The most important attribute of the polynomial is its length (largest degree +1 of any one term in the polynomial), because of its direct influence on the length of the computed check value. The polynomial must be chosen to maximize the error detecting capabilities while minimizing overall collision probabilities. The length of the remainder is always less than the length of the generator polynomial, which therefore determines how long the result can be.

The design of the CRC polynomial depends on the maximum total length of the data block to be protected (information code + CRC), the desired error protection features, and the type of resources for implementing the CRC as well as the desired performance. In fact, all the factors above should enter in the selection of the polynomial. However, choosing a non-irreducible polynomial can result in missed errors due to the ring having zero divisors.

Assuming the polynomial G(x) is given as

$$G(x) = x^8 + x^2 + x + 1 \quad (5)$$

From Eq. (5) the corresponding generated polynomial can be expressed in hexadecimal (HEX) as 107H. For the highest bit of check codes are all 1 in all case, it can be removed for programming convenience. After removing the highest bit, the check code appears as 07H (see Table.1). It can be found whether the highest bit is one or not by comparing with 80H successively.

A common misconception is that the "best" CRC polynomials are derived from either an irreducible polynomial or an irreducible polynomial times the factor (1 + x), which adds to the code the ability to detect all errors affecting an odd number of bits.

The "best" CRC polynomials should satisfy the following conditions:

1 The highest and the lowest exponent of generated polynomial should be one.
2 With error happening in different bits, the remainder should be different.
3 With the generated polynomial divided by mode two, the remainder should not be zero, if any bit error happens.
4 With the reminder divided by mode two circularly, the remainder should be in loop.

The most commonly used generated polynomials are shown in table 1.

Table 1. Common Generated Polynomial

CRC Code	Generated Polynomial
CRC-8-SAE	$x^8+x^4+x^3+x^2+1$
CRC-8-Dallas	$x^8+x^5+x^4+1$
CRC-8-CCITT	x^8+x^2+x+1
CRC-12	$x^{12}+x^{11}+x^3+x^2+x+1$
CRC-16-IBM	$x^{16}+x^{15}+x^2+1$
CRC -16-CCITT	$x^{16}+x^{12}+x^5+1$
CRC-24	$x^{24}+x^{23}+x^{18}+x^{17}+x^{14}$
	$+x^{11}+x^{10}+x^7+x^6+x^5$
	$+x^4+x^3+x+1$
CRC-32	$x^{32}+x^{26}+x^{23}+x^{22}+x^{16}$
	$+x^{12}+x^{11}+x^{10}+x^8+x^7$
	$+x^5+x^4+x^2+x+1$
CRC-40-GSM	$x^{40}+x^{26}+x^{23}+x^{17}+x^3+1$
CRC-64-ISO	$x^{64}+x^4+x^3+x+1$

Numerous varieties of cyclic redundancy check have been incorporated into technical standards. By no means has does one algorithm, or one of each degree, suited every purpose. The numbers of distinct CRCs in use have however led to confusion among developers, which authors have sought to address [8]. Koopman and Chakrabarty recommend selecting a polynomial according to the application requirements and the expected distribution of message lengths [9]. There are three polynomials reported for CRC-12[9], 16 conflicting definitions of CRC-16, and 6 of CRC-32 [9-10].

4 MODULO-2 DIVISION

The main process of implementing the CRC algorithm is modulo-2 division. According to GF (2) (modulo-2 arithmetic), we have:

$$\begin{aligned} &0-0=0+0=0\\ &1-0=1+0=1\\ &0-1=0+1=1\\ &1-1=1+1=0 \end{aligned} \tag{6}$$

From Eq. (6), it is clear that the modulo-2 division is similar to arithmetic division, but the results from division (subtraction) of each bit do not affect the results of other bits, that is, not borrow the higher bit. In fact, the XOR (exclusive OR) operator can implement the division above.

The steps of modulo-2 division are as follows:

1 Subtract the highest bits of dividend data with divider by modulo-2 arithmetic.
2 Shift divider to the right bit: If the highest bit of remainder is one, quotation should be one, on the contrast, if the highest bit of remainder is zero, quotation should be zero.
3 Repeat step (2) until the bits of divider is less than the dividend.

Assume that the row message is 11000101, and the polynomial G(x) is given as

$$G(x)=x^4+x+1 \tag{7}$$

Because of the exponent of generated polynomial above have five bits, the row message should be shifted left by four times to get the appended bits which are called Frame Check Sequence (FCS) bit. The entire message comes into 110001010000. The procedure of modulo-2 division is shown as Fig.1.

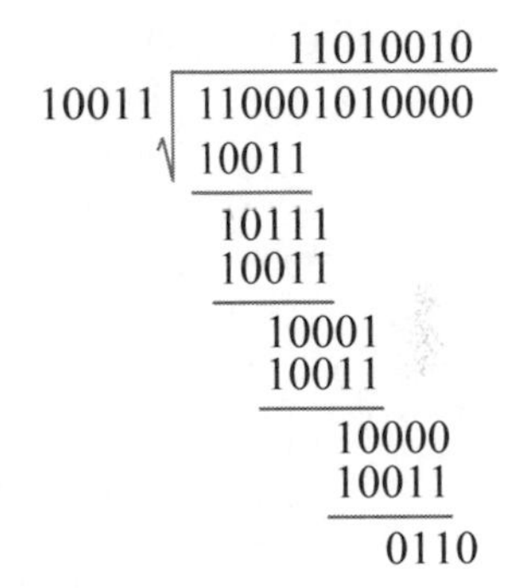

Figure 1. Procedure of modulo-2 division.

The polynomial division can also be thought of as a division of this binary sequence (message bits + 4 appended 0's) by the sequence corresponding to the generator polynomial, here 10011. The FCS bits, or equivalently R(X) is 0110. Therefore, CRC code of the row message 11000101 is 1100010001010110.

5 HARDWARE IMPLEMENTATION

For bit serial sending and receiving, the hardware to generate and check a single parity bit is very simple. It consists of a single exclusive OR gate (XOR gate) together with some other control circuitry. The general circuit generating CRC code for given generated polynomial with R0, R1, ..., Rk-1 as shift registers is shown as Fig.2. The Division by mode 2 is implemented by the exclusive-OR gates (XOR gates) with ki (i=1, 2,…, k-1) switches ON or OFF. If the bit exponent of generated polynomial is 1, the corresponding switch should be ON, otherwise, it should be OFF.

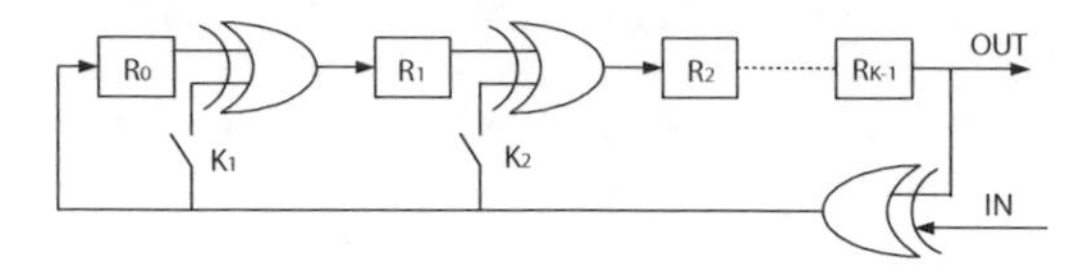

Figure 2. General Circuit of CRC code .

According to Eq. (2), the bit exponent of generated polynomials of R1 and R2 are 1, the corresponding switches, namely k1 and k2, are closed as shown in Fig.3.

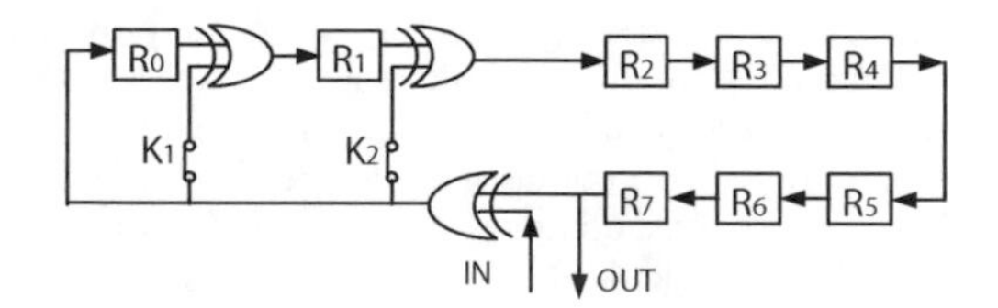

Figure 3. Circuit of sending for CRC-8.

The circuit of receiving for CRC-8 is very similar to the sending circuit shown in Fig. 4. The only difference is that the rightmost XOR is now moved over to just before the leftmost shift register.

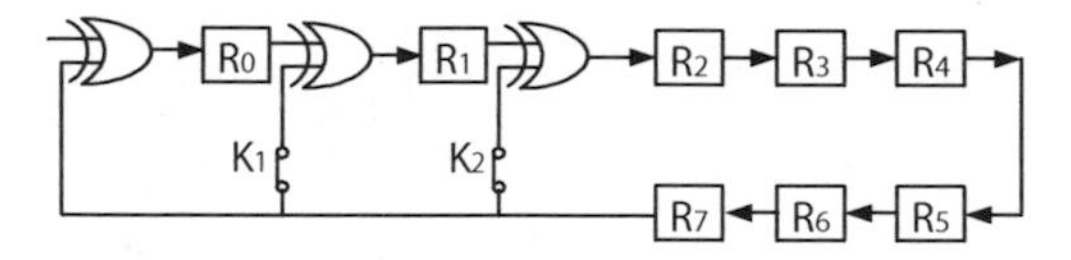

Figure 4. Circuit of receiving for CRC-8.

The input bits in Fig. 4 are clocked in from the left, again with MSB bit first, and with the registers initially containing 0's. After the message and the FCS bits have been all clocked in, the shift registers contain the remainder which must be all 0 if there are no errors. Since this receiver implements division, it could have been used as the transmit encoder also. These simple digital circuits are practical implementations for both producing the FCS and checking a received message, and are possible because of the cyclic structure of the CRC.

6 SOFTWARE IMPLEMENTATION

In order to improve coding efficiency of software implementation, the author proposes a calculation method based on MCS-51 assembly language and implementation program, which are briefly described in this section.

Suppose that the five bytes which are going to be transmitted as Byte0, Byte1, Byte2, Byte3 and Byte1 respectively. The construction process of (48, 40) CRC-8 is as following:

$$\begin{aligned} CRCtmp &= CRC8\ (Byte0)\ XOR\ Byte1 \\ CRCtmp &= CRC8\ (CRC)\ XOR\ Byte2 \\ CRCtmp &= CRC8\ (CRC)\ XOR\ Byte3 \\ CRCtmp &= CRC8\ (CRC)\ XOR\ Byte4 \\ CRCOK &= CRC8\ (CRC)\ XOR\ \&HFF \end{aligned} \tag{8}$$

When the highest bit of temporary storage unit turns into 1, which means 8-bit data have been moved into, then shift temporary storage unit left once more, performs XOR operation with 8-bit check code, which is equivalent to division by mode 2. Therefore, the final remainder in CRCOK (shown in Eq. (8)) is the CRC code needed.

7 CONCLUSION

The CRC code is a subset of cyclic codes and use a binary alphabet, 0 and 1. The paper provided an overview and principle of CRC, also proposed the hardware and software implementation process. It is proven that the CRC algorithm has high practical value in many application fields, such as communication, magnetic and optical storage and so on. The further study will focus on the implementation of the parallel N-bit-wide circuit.

ACKNOWLEDGMENTS

The Research was partially supported by the Natural Science Foundation of China (no. 61303022), the Natural Science Foundation of Shandong, China (no. 2009AA01105), the science and technology development fund of Tai'an (no.20133011), and the Foundation provided by Taishan University (no.Y-01-2013010). The author gratefully acknowledges Professor SHEN Ding for discussion and suggestions, also thank the associate editor and the anonymous reviewers for their valuable comments and kind suggestion to improve the quality of this paper.

REFERENCES

[1] Peterson, W. W. and Brown, D.T. 1961. Cyclic Codes for Error Detection. Proceedings of the IRE, January: 228–235.

[2] Wolf, J.K. and Blackeney, R.D. 1988. An Exact Evaluation of the Probability of Undetected Error for Certain Binary CRC Codes. Proceedings of MILCOM-IEEE: 15.2.1–15.2.6.
[3] Henriksson, T. and Liu, D. 2003. Implementation of fast CRC calculation. Proceedings of the Design Automation Conference, Asia and South Pacific, Jan.2003, 563–564.
[4] Wolf, J.K. and Chun, D. 1994. The single burst error performance of binary cyclic codes. IEEE Transactions on Communications COM-42:11–13.
[5] Sheinwald, D., Satran, J., Thaler,P. and Cavanna,V. 2002. Internet Protocol Small Computer System Interface Cyclic Redundancy Check (CRC)/Checksum Considerations. IETF RFC: 3385.
[6] Gammel,B.M.2011.Crypto-Codes. http://users.physik.tu-muenchen.de/gammel/matpack/html/LibDoc/Crypto/ MpCRC .html . Retrieved 10 February 2011.
[7] Fujiwara, T., Kasami, T., Kitai, A. and Lin, S. 1985. On the undetected error probability for shortened hamming codes. IEEE Trans. on Communications.33(6): 570–57.
[8] Williams, Ross N. 1996. A Painless Guide to CRC Error Detection Algorithms V3.00. http://www.repair-faq.org /filipg/LINK/ F_crc_v3.html. Retrieved 5 June 2010.
[9] Koopman, Philip, Chakravarty, Tridib. 2004. Cyclic Redundancy Code (CRC) Polynomial Selection For Embedded Networks. The International Conference on Dependable Systems and Networks: 145–154.
[10] Cook, G. 2011. Catalogue of parametrised CRC algorithms. http://regregex. bbcmicro.net /crc-catalogue.htm. Retrieved 17 October 2011.
[11] Fiala D., Mueller F., Engelmann C., Ferreira K., Brightwell R., and Riesen R. 2012. Detection and correction of silent data corruption for largescale high-performance computing. Dept. of Computer Science, North Carolina State University, Tech. Rep. TR 2012–5.
[12] SANG Sheng-ju, LU Ying, ZHAO Ji-chao and AN Qi. 2008. The Analysis and Implementation of CRC Code Based on CDT Protocol. Journal of Qufu Normal University (Natural Science), 2008, 34(4): 71–75, (in Chinese).

Computing, Control, Information and Education Engineering – Liu, Sung & Yao (eds)
© 2015 Taylor & Francis Group, London, ISBN: 978-1-138-02800-5

Ensemble global and local features for single-sample face recognition

Zhi Ming Qian, Hai Fei Qin & Xiao Qing Liu
School of Information Science and Technology, Chuxiong Normal University, Chuxiong, China

Mei Ling Shi
School of Physics and Electronic Engineering, Qujing Normal University, Qujing, China

ABSTRACT: This paper proposes a face recognition method integrating both global and local features for single-sample face recognition. First extract global feature of human face using 2DPCA, then extract SIFT feature from the designated area of candidate faces, and finally confirm the candidate face according to feature matching results. The experimental result has verified its effectiveness.

KEYWORDS: Face recognition; Global feature; Local feature.

1 INTRODUCTION

Face recognition is a biometric authentication technology based on human facial features. Compared with other recognition technologies, human face recognition has its natural and imperceptible advantages. The past two decades have seen the wide application of this technology and its more and more social and commercial value. Most existing face recognition methods succeed on the premise of massive training samples within its system. However, the system can only maintain one face image a person in actual practice, due to the difficulty in sample collection and limitation of other requirements, which is then also called single-sample face recognition [1]. As to common face recognition methods, the limited information in single-sample recognition leads to the weakening performance or even failure in recognition. Therefore, the accurate recognition of face in single sample becomes a challenging problem.

Currently face recognition methods are mainly divided into two categories: global feature based and local feature based. Global feature mainly describe the color, contour and distribution of facial organs for rough recognition; while local feature describe characteristics of human facial organs and some other special features for accurate identification. Studies in recent years have shown that relying solely on either feature recognition is not a good solution. Therefore, more and more researchers focus on a combination of global and local features to improve recognition performance.

This paper proposes a single-sample face recognition method based on global and local features, which is to first use global feature for pre-recognition to find the candidate faces and then to confirm the face with local feature to get final result. The method not only retains the quick and efficient performance of global feature, but also takes advantage of the accurateness and robustness of local feature. The experiment shows that the method can achieve effective face recognition in single-sample condition.

2 THE PROPOSED METHOD

Generally an input image goes through two parts of processes. In the detection part, conduct face detection to identify face area and locate two eyes to get model area based on eye position. In the recognition part, use global feature to pre-identify face image to find candidate faces with high similarity, and then extract local feature from the designated area of candidate faces and finally confirm the candidate face according to feature matching results. Figure 1 shows a flow chart of the method.

2.1 *Face and eye detection*

This paper applies cascade-based AdaBoost detection method proposed by Viola [2] to achieve face and eye detection. This method uses AdaBoost algorithm to train classifiers and generates final strong classifiers by combining weaker ones. In the proposed method, we use two strong classifiers, face classifier and eye classifier. First use face classifier to process the input image for face detection and then use eye classifier for eye detection in the detected face image, Figure 2 shows an example of the face and eye detection.

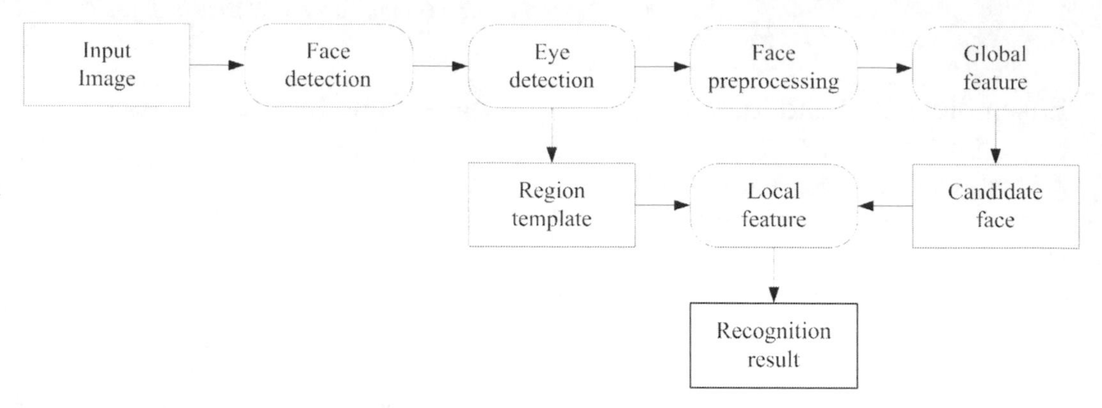

Figure 1. The flow chart of the proposed method.

2.2 *Face pretreatment*

Within the rectangular area of eyes, position accurately on eye pupil with vertical and horizontal projection respectively [3], adjust the horizontal alignment of face image based on the connecting line of two pupils. Do geometry normalization and histogram equalization for the aligned face image, in order to lower the impact of face size and illumination altering on face recognition.

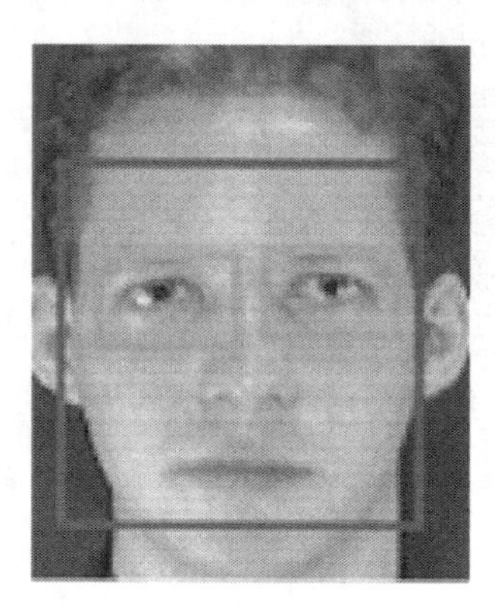

Figure 2. An example of the face and eye detection.

2.3 *Extraction and matching of global feature*

For face image after pretreatment, we first use 2DPCA method proposed by Yang [4] to extract global feature. Much improved from PCA, 2DPCA can construct covariance matrix directly from two-dimensional face images without conversion to one-dimensional vector. Compared with PCA, 2DPCA is more simple and intuitive in image feature extraction and achieves a better extraction performance and faster speed.

First calculate the average image matrix of all the training samples:

$$\overline{A}=\frac{1}{N}\sum_{i=1}^{N}A_i \tag{1}$$

N is the number of training samples; then calculate the covariance matrix of image:

$$G_t=\frac{1}{N}\sum_{j=1}^{N}(A_j-\overline{A})^T(A_j-\overline{A}) \tag{2}$$

Sort the eigenvalue of G_t in a descending order, take the eigenvector corresponding to the first d eigenvalues to form the best projection axis $X_1, X_2, \ldots, X_d$, and the final extracted feature is: $M=[Y_1, Y_2, \ldots, Y_d]$ $Y_i=AX_i$.

After 2DPCA extraction, each image has its feature matrix, which can be classified by nearest neighbor method. Define the distance between any two feature matrix $M_i=[Y_1, Y_2, \ldots, Y_d]$ and $M_j=[Y_1, Y_2, \ldots, Y_d]$ as:

$$d(M_i, M_j)=\sum_{k=1}^{d}\left\|Y_k^i-Y_k^j\right\|_2 \tag{3}$$

Assuming the training sample is $M_1, M_2, \ldots, M_N$, with regard to the input image feature matrix M, the matching result between two images is $d(M, M_k)<\theta_1$.

Find n (n=3) images with biggest similarity among training samples as the candidate face images.

2.4 *Extraction and matching of local feature*

After obtaining the candidate face images, SIFT feature [5] extraction is then conducted. SIFT is a local feature of image, which maintains constant to its rotation, scaling and brightness variation, and keeps relatively steady for angle changes, affine transformations and noises. Due to the difference number of SIFT feature on each face, the simple extraction and matching of SIFT feature on the whole face may lead to large calculation work and inconsistent space in matching results (features from different area match each other). In order to improve the extraction and matching efficiency, we divided face area into models according to eye position (Figure 3), and then conduct SIFT feature extraction in the designated area by model. The reasons are as described below. a: Eyes, nose and mouth are area integrated the most SIFT features on face, with very good discrimination on different faces. b: The limited area saves lots of time in feature extraction and matching to ensure the spatial consistency of matching results.

Extract SIFT feature from the designated area of candidate faces, calculate the feature similarity by the ratio of Euclidean distance of the nearest neighbor and next-nearest neighbor vectors. Assuming the features are $V_i = (v_i^1, v_i^2, \ldots v_i^{128})$ and $R_j = (r_j^1, r_j^2, \ldots r_j^{128})$ in two corresponding area, then the similarity between the features is:

$$d(V_i, R_j) = \sqrt{\sum_{k=1}^{128} (v_i^k - r_j^k)^2} \tag{4}$$

Set the minimum and second smallest values of similarity between two features d and d', if $d/d' < \theta_2(\theta_2 = 0.75)$, then the two features match each other. Finally calculate the total matching numbers between candidate faces and original image, the one with the most matching features is the recognition result.

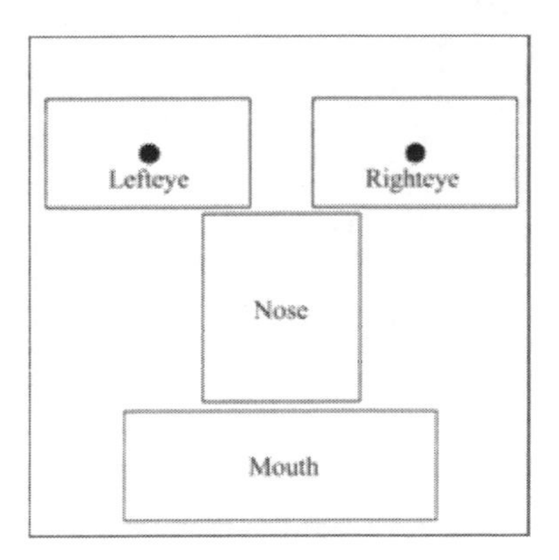

Figure 3. The model of local feature according to eye position.

3 EXPERIMENT

The experiment uses a total of 400 face grayscale images, 40 people, 10 images per person, from ORL face database [6]. All the images are frontal face, 112 × 92 pixels, with certain changes in illumination, facial expressions and tilting and rotation of heads (Figure 4). Since this paper is mainly on single-sample face recognition, the experiment select the first face image as training sample for each person and the rest 360 pieces as testing images.

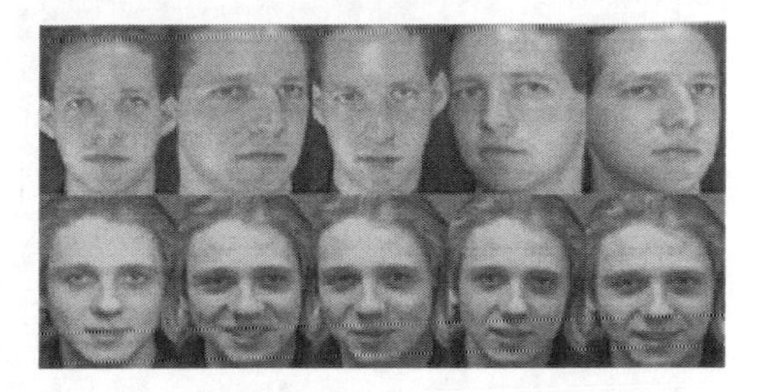

Figure 4. Some images in the experimental datasets.

In order to better compare the performance of this method, we have made a contrast among the testing results of this method and other methods of either global or local feature, see in Table 1.

Table 1. Performance comparisons among different methods on the ORL database.

PCA	2DPCA	LBP	The proposed method
67.78%	71.39%	55.83%	83.89%

The result has shown that the proposed method has a higher recognition rate than using global feature or local feature alone. It proves that the improvement of integrating both global and local features is effective for single-sample face recognition.

4 CONCLUSION

This paper proposes a single-sample face recognition method integrating both global and local features. First use 2DPCA to extract global feature to obtain candidate faces with high similarity, and then use SIFT to extract local feature from the designated area of candidate faces for further matching to achieve the final result. The experiments on ORL face database shows that, the proposed method performs better than the single use of either global feature or local feature recognition.

ACKNOWLEDGEMENTS

This work was supported by the Scientific Research Fund of Yunnan Provincial Education Department (No. 2013Y053), and the Program for Innovative Research Team (in Science and Technology) in University of Yunnan Province.

REFERENCES

[1] Tan X, Chen S, Zhou Z H, et al. Face recognition from a single image per person: A survey. Pattern Recognition, 2006, 39(9): 1725–1745.

[2] Viola P, Jones M. Rapid Object Detection using a Boosted Cascade of Simple Features. Proceeding of the International Conference on Computer Vision and Pattern Recognition. 2001, 511–518.

[3] Zhou Z H, Geng X. Projection functions for eye detection. Pattern Recognition, 2004, 37(5): 1049–1056.

[4] Yang J, Zhang D, Frangi A F, et al. Two-dimensional PCA: a new approach to appearance-based face representation and recognition. Pattern Analysis and Machine Intelligence, IEEE Transactions on, 2004, 26(1): 131–137.

[5] Lowe D G. Distinctive image features from scale-invariant keypoints. International journal of computer vision, 2004, 60(2): 91–110.

[6] Samaria F, Harter A. Parameterisation of a Stochastic Model for Human Face Identification. Proceeding of IEEE Workshop on Applications of Computer Vision. 1994, 138–142.

Computing, Control, Information and Education Engineering – Liu, Sung & Yao (eds)
© 2015 Taylor & Francis Group, London, ISBN: 978-1-138-02800-5

Research and implement based ARINC 659 bus test system

Jin Xu & Zhi Yun Hu
School of Electronic Information, Xi'an Polytechnic University, Xi'an, China

Jie Xia
Xi'an Xiang Teng Microelectronics Technology Co, LTD, Xi'an, China

ABSTRACT: Avionics System is an important part of modern aircraft, constructing a Universal Testing System for ARINC659 Airborne Bus can test the performance of specific devices and detect the fault location, it also provides a debugging platform for the development of the Avionics System. This paper analyzes the ARINC659 Airborne Bus and proposes an implementation scheme for an ARINC659 Testing System. Moreover, it also emphatically analyzes the function of each module in ARINC659 Testing System, the implementation scheme of hardware circuit and the software of the Testing System.

KEYWORDS: ARINC659, Testing System, Bus.

1 OVERVIEW

Avionics system is developing towards integration and modularization, with the arrival of the era that the technologies of electronics and computer are developing at top speed. Recently, especially, the system is required to be more secure and reliable. ARINC659 bus protocol is the backplane data bus specification which was being established by AEEC (Airline Electronic Engineering Commission). It defines the standard of the data transmission among the LRM (local replaceable module) in the rack of IMA (integrated modular avionic), the LRM connected by backplane connectors and the electrical characteristics of the backplane. The ARINC 659 bus is high reliability; high data availability and fault tolerant ability. It supports robust segmentation, controls the sharing of resources reliably, and reduces the complexity to improve the deterministic operation. Abroad, it has been successfully used in the aircraft management system on the Boeing 777 and the VI- A (Virtual Integrated Avionics) system of the Boeing 717N, MD-10, KC-130 and other aircraft, with determination of higher data transmission. The research of ARINC659 in China has just begun as a result of the confidentiality of related technology and limitation of application areas.

In view of previous study of ARINC 659 bus emulation test system, this paper focuses on the analysis of the research and implementation of a ARINC659 bus test system, which will promote the application of ARINC659 bus in China.

2 ANALYSIS OF ARINC659 BUS

ARINC659 bus is a high integrity back-plane bus with high fault tolerance and robust segmentation during the bus transmission time and in storage space. It is a linear multipoint communication bus used for the data-transmission between the local replaceable module in the rack with half-duplex transmission mode and the error correction mechanism of cross- check. The bus media access adopts TD-PA (table-driven proportion access) protocol. The scheduling table defined by F- DL (Frame Description Language) has stipulated bus operation of each node. It has better fault tolerance than the traditional dual redundancy, and less complexity than the traditional redundant. It uses a channel with the data maintenance and the test for loading table command which is separated from the main backplane bus. The ARINC659 bus interface and bus connections are presented in Figure 1.

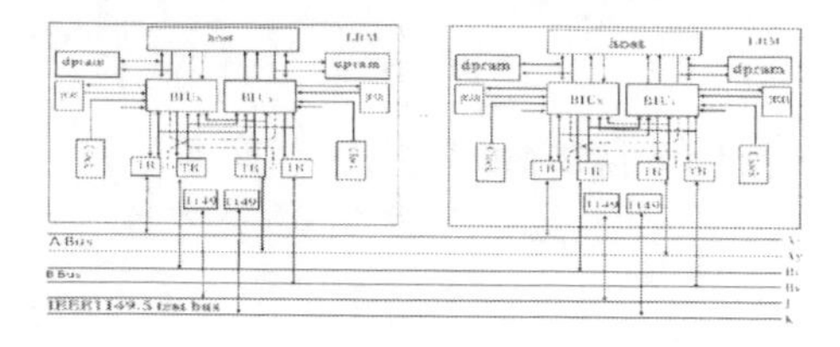

Figure 1. The ARINC659 bus interface and bus connections.

ARINC659 bus is both configurations composed of dual bus pairs (A and B), bus pairs consist of two buses named “x” and “y”. Each bus of the four buses (Ax, Ay, Bx, By) has its own clock line and two data lines, transmit two bits per circle, therefore complete bus consists of 12 lines.Every LRM has two bus interface units (BIUx and BIUy), BIUx sends message with x bus and BIUy with y, each BIU receives all 4 buses. Every bus is driven by the independent transceiver in the LRM that prevents a single failure adversely affects more than one bus. AR-INC659 bus is designed to have the ability to correct all one-bit mistake and detect all dual-bit mistake of the backplane bus.

3 IMPLEMENTATION OF ARINC659 BUS TEST SYSTEM

To completely finish the design of ARI-NC659 test system in accordance with the functional verification specification, we need to build the system according to the actual situation. The test system is mainly to achieve the monitoring of AR- INC659 bus and simulate the actual application of the system according to fault injection. ARINC659 bus communication system supports module-to-module communication and module-to-set communication. Analog the actual operation of ARINC659 bus.

The test system consists of the following modules: fault injection module, CP-U module, bus monitor module, measured module. Test diagram of system hard- ware composition is shown in Figure 2.

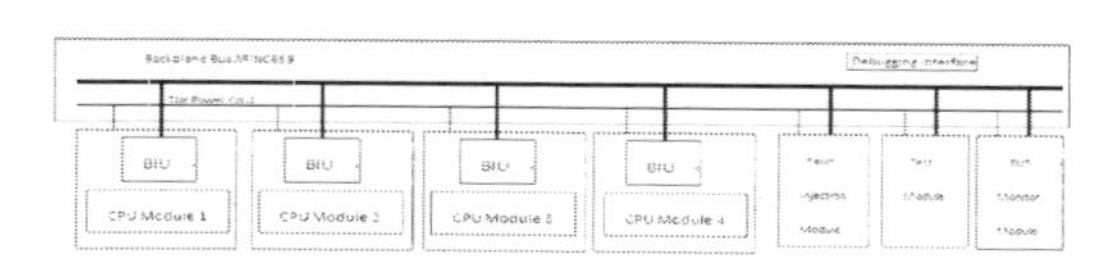

Figure 2. Test diagram of system hardware.

3.1 *CUP module*

According to the requirements of the ARINC659 bus, the bus communication supports one point to point communication and one point to multiple point communication. The type of message is divided into two types: the basic message and primary backup message. The basic message is used for communication between a single source and o- ne or more destination node. The primary back-up message is used for communication between multiple reserved sources and one or more destination node. This communication mode of primary backup messages requires at least four nodes, so the test system includes four CPU nodes, one main equipment, three following equipments, which of them can communicate between them.

The schematic diagram of CPU module is shown in Figure 3.

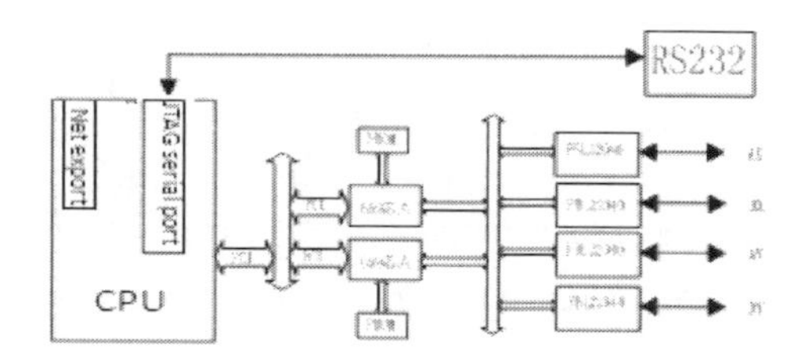

Figure 3. The schematic diagram of CPU module.

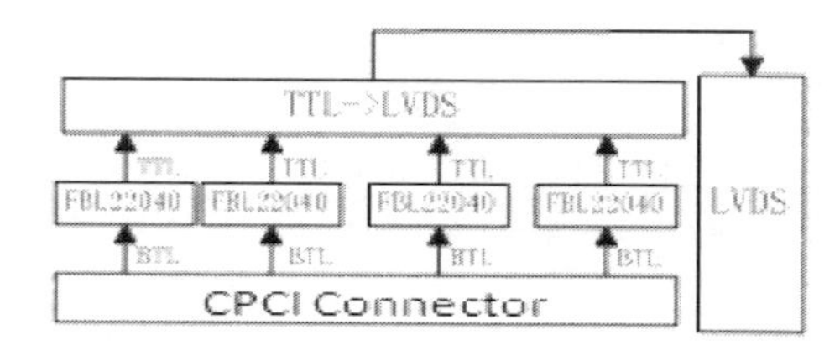

Figure 4. The system diagram of bus monitoring module.

The CPU module consists of ARINC-659 interface board and the processor subcard. The CPU host mainly includes PCM interface board, network interface and JTAG interface used for testing. The interface chip of the ARINC659 bus integrates controller, clock circuit, memory, checking, fault tolerant module. Four road bus transceiver communicates with ARINC659 backplane bus in which Interface chip X sends messages through the X bus and interface chip Y sends messages through the Y bus which bus has an independent transceiver The ARI- NC659 bus interface board contains a ARINC659 chip which completes the communication function in support of the host. Interface board provides a command list about download interface which is used for controlling the bus operation. Interface chip implements the connection between subsystem and subsystem interfaces through a PCI bus controller, completes the exchange of information between the host and the subsystem, achieve the data communication between modules.

3.2 *Bus monitoring module*

The Bus Monitoring Module is made up of CPCI connector, four path transceiver, voltage converting circuits, etc. The CPCI connector can be used to connect the transceiver and the backplane. Using the serial backplane bus of BTL level realizes the Data Acquisition and Data Exchange. The functions of Bus Monitoring Module are described as follows: Identifying the correctness of the waveform from the measured module; Detecting the implement situation of tested module; Monitoring the effectiveness of the fault injection and all kinds of messages.

The system diagram of bus monitoring module is shown in Figure 4.

3.3 *Fault injection module*

Because ARINC659 bus ,a kind of double-double backup communication bus has 4 road serial buses, which has a strong fault tolerance, so the test for the bus fault is also very important. According to the config- uration on each bus, fault injection includes transient fault, fault for a long time, burr and interference etc.Test system needs to design the interface for injecting bus fault, which can make the single way or multiplex communication circuit signal transmission failed, in order to test the tolerance of ARINC659 bus.Fault inj- ection which cannot damage the system is only disrupt the normal transmission of bus signal.The module consisted of FPGA processor, 32 discrete output, 12 road fault type, fault injection serial port, clock and reset circuit, FPGA configuration circuit itself can complete many functions of 659 fault injection, PC communication, and the dispersed quantity signal generation, etc. The fault injection can be imitated on the 12 signal lines of the ARINC659 bus, which respectively injects fault "1", "0" and provides 1 road of 115200 BPS RS232 interface. Master machine will connect with fault injection module through RS232. The fault control program is developed on the master machine.

This module uses the FPGA as the processor. Because of this module is lower demand for logic,only has a serial port to send and receive logic,fault injection logic,and discrete quantity control module,each of these can be controlled by basic FPGA. According to above an- alysis,FPGA is chosen XC3S100E-4TQ-144I of Spartan-3E series. Fault injection has three state outputs in the absence of fault , meanwhile can inject fault "1", "0"when it is fault. From the viewpoint of control, it need two IO at least to control one state. As ARINC659 has 12 roads date, the fault injection module needs at least 2*12=24 IO; The discrete quantity needs 32 road output and 32 IO as well, and a serial port needs 2 IO. In conclusion, the number of IO about this module is 58 at least.

The schematic diagram of fault injection module is shown in Figure 5.

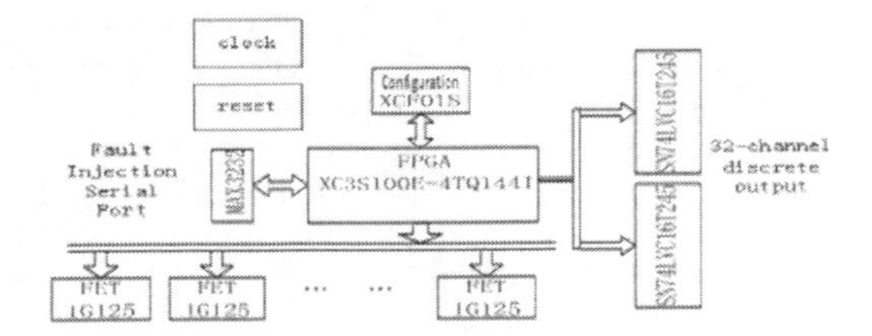

Figure 5. The schematic diagram of fault injection module.

3.4 *Tested module*

The module to be tested is mainly used to verify the protocol conformance of ARINC659 Bus for Physical Layer and Data Link Layer.The verification of Physical Layer mainly in- cludes bus power supply, slot identification,electrical properties,clock accuracy, bus encoding,physical isolation, etc. The verification of Data Link Layer mainly includes detecting whether the time sequence waveform from the tested module bus' output is conform to the protocol, whether the tested module receives and deals with the standard bus'time sequence waveform,whether the tested module performs identifying and dealing with the error existing in the bus' time sequence waveform in consistent with the protocol, etc.

4 IMPLEMENTATION OF VERIFIED SOFTWARE

This authentication system used VxWorks as verification platform operating for system platform. Running VxWorks operating system via a serial port connection to test platform. The test program was operated by PowerPC. The test program of test platform was run by test program which was written based on ARINC659 verify functional specifications.The whole of ARINC659 test system could be tested comprehensively and test results were outputted through a serial port for debugging terminal.

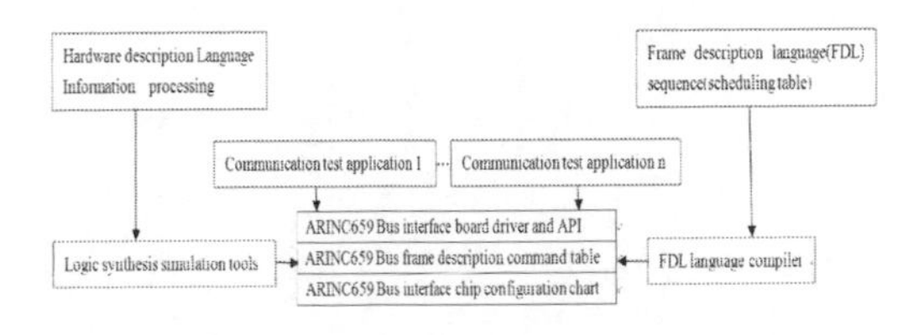

Figure 6. Software system block diagram.

The test function of ARINC659 mainly included:the main/backup message transmission function,-Faulttolerance, etc. In a host machine, PC, host and each module were connected by serial port. Tornado of a host connected with VxWorks for software debugging. System test software Included: VxWorks operating system, bus interface module driver, API package, ARINC659 bus communication test application and The frame description language (FDL) compiler. The software system block diagram is shown in Figure 6.

The compiler was used to translate the specified bus operation scheduling table into the independent design of the interface chip executable machine code. The compiler was a Windows application developed by C language.

Drivers and API package provided the underlying access to Interface chips. Mainly included: interface chips control, Interrupt service, data communication, Status report, etc. It provided the software interface support for Bus communication test application development. It was VxWorks drive or application developed by c language.

5 SUMMARY

With the rapid development of integrated avionics system, the data bus is a key technology. The bus was asked to safer, more effective and better generality. It is important to build a reliable and stable special communication network. ARINC659 bus communication test platform was constructed in this paper proposed the realized method of each module. Test results show that bus communication system's reliability, security and convenience of application have been greatly improved. It is possible to exchange, safe and secure data between each module.

REFERENCES

[1] The Airlines Electronic Engineer ing Committee. ARINC Specification 659BackplaneDataBus[A]. USA:Aeronautical Radio INC.,December 27, 1993.

[2] [2]ARINC.Arinc project paper 664: Aircraft data network, part7-avionics full duplex switched Ethernet(AFDX) network[Z].2005.

[3] ARINC specification 659 backplane data bus[S].1993–12.

[4] Witwer B.systems integration of the 777 Airplane Information Managent system(AIMS) [J].IEEE AES systems magazine, 1996, 11(4):17–21.

Computing, Control, Information and Education Engineering – Liu, Sung & Yao (eds)
© 2015 Taylor & Francis Group, London, ISBN: 978-1-138-02800-5

Study on the data operation models in the android native development kit

Wen Tao Liu
School of Mathematics and Computer Science, Wuhan Polytechnic University, Wuhan, P.R. China

ABSTRACT: The applications often use the native development kit technology to achieve the core functionality in order to obtain higher efficiency and better safety in the Android platform. The data exchange is essential and very frequently used functions which complete the transfer of application information between local environment and Java layer. This process requires special attention to avoid memory overflow due to the small memory and low computing power in the mobile platform device. The different data formats in two environments can easily cause the program to crash if the data cannot be converted correctly. The key methods are introduced and it implements typical data exchange and some matters needing attention are discussed. The problems of memory management and the methods of the resource release are discussed. The thread processing mechanism in the NDK programming is analyzed and it can provide good interaction for the compute-intensive program.

KEYWORDS: Native Development Kit; Android; Data Exchange; Memory Management; Thread Model.

1 INTRODUCTION

NDK programming is an important part of the Android application development and it can provide many advantages. The Android (Android Developers) program runs with the virtual machine Dalvik (ART and Dalvik) and the NDK (Android NDK | Android Developers) is implemented with native code language. It can significantly improve the security of the program to avoid the program is easily cracked. The Android application with Java can be decompiled easily by some tools. Although the program can be transformed into the code which is difficult to read with some code obfuscation tools, it only increases the time to decipher. Some software pirate can modify the Android application and insert the advertisement or virus into the program. The NDK code is written with C or C++ and it will be linked into a library with binary code. The library can be converted into assembly code and it is difficult to understand. Some core and confidential algorithms of application can be implemented by the NDK. Another advantage is that it has a high operating efficiency and it can complete some time-consuming operation. Some third libraries are written by the C or C++ and it can be called in the Android platform using the NDK format and the libraries also can be ported to other embedded platforms. The data exchange is an important component in the NDK programming and it completes the transfer of data between local code and Java code. The data form is different and it needs data conversion and this will affect the efficiency of the entire process. Some invalid and incorrect conversion will affect the operation of the program. The NDK programming requires strict memory management because it can easily lead to memory overflow and the problem of out of memory due to the capacity limitations of memory.

2 DATA EXCHANGE

The NDK uses the JNI (JNI Tips; Java Native Interface Specification—Contents) technology to send or receive the data and the local function is defined in Java with the native key word. The library is loaded with the System.loadLibrary command and the library file is saved in the libs directory with the file extension so. The NDK code is saved in the JNI directory and it can be built with the NDK build command. In the JNI directory, there is one file named with Android.mk which describes the native source files for the NDK compiling tools. The file is defined by the Makefile format and it includes the LOCAL_PATH, LOCAL_MODULE, LOCAL_LDLIBS and so on. Another file Application.mk in the JNI directory describes some properties needed in the Android application. It can include the APP_ABI, APP_PLATFORM and so on. When the so library is loaded, the NDK function named with JNI_OnLoad will be called. The function has the parameter with JavaVM type which describes the Java environment and can be saved in the global variable used in other functions. In the NDK operation there is another important data type with the name JNIEnv which is associated with the thread. Different thread has a different JNIEnv interface and it can be got by the GetEnv function defined in the JavaVM. If it needs a new thread, the JNIEnv must be created by the AttachCurrentThread function defined in the JavaVM.

The data type conversion is defined as the rules that the data type in local code is added to the j letter from the Java code. For example, the int will be converted into the jint, the double will be converted into the jdouble. The type of Java object will be converted into the jobject. The jclass represents the class object. The local code can be written by the C++ language and the method of calling the JNI code is defined as follows.

```
env->JNI_function_name(parameter list);
```

If the code is written with the C code, the code will be defined as follows.

```
(*env)->JNI_function_name(env, parameter list);
```

The C++ code must be included with the extern "C" code (Understanding "extern" keyword in C). The array data can be operated by the functions provided by the JNIEnv which can use the simple type array and the object type array. It uses the GetXXXArrayElements to get the pointer of array and it can be used to operate the total array. Different type has a different function and the integer array uses the GetIntArrayElements. The following method demonstrates the use of an array of integers. The progress of setting the integer array is shown in Figure 1.

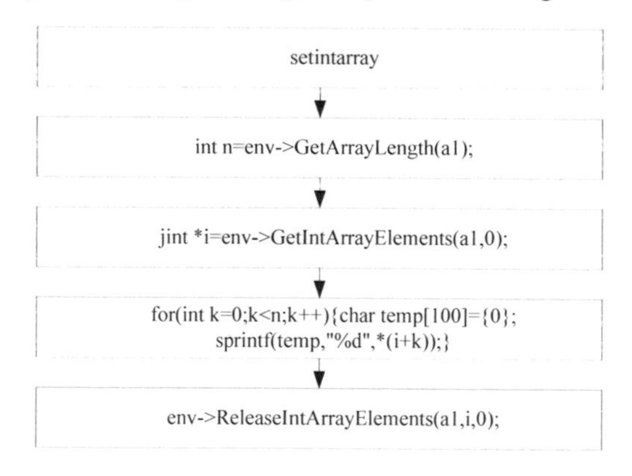

Figure 1. The progress of setting integer array.

When the array access operation is finished, the array must be released by the ReleaseXXXArrayElements function with the array pointer. The function can save the change of array and it will give the feedback to the Java code. The GetArrayLength can get the length of the array and it can be used in the array operation. The Java object array can be operated by the GetObjectArrayElement and SetObjectArrayElement function.

The process of usage of Java object in the local code consists of several steps. The first step is to create a Java object using the JNI function. The next step is to use the object to call the methods of the object. It also can set or read the value of static or instance variables. The NDK uses the type ID to identify the variable or methods. It can use the GetObjectClass to get the Java class object if one object is created by the Java. If the Java class name is known, it can use the FindClass to get the Java class object.

The above operation demonstrates how to pass data from Java to native code. When the native data will be returned to the Java, it can use the related function to send the data to the Java. The simple type data is operated easily. The string type data can be got by the NewStringUTF function.

The Java object can be created in the native code and it also can be returned to the Java from the native code. It uses the FindClass to get the class object with the Java class name which contains the package name. It uses the GetMethodID to get the constructor method id with above class object and the parameter has the function name with <init>. Use the NewObject function to create one object with the above class object. Use the GetFieldID to get the id of the field of class and the returned type is jfieldID. The function needs the name of the field and the data type of field. Use the SetXXXField functions to set the value of the fields and finally return the object to the Java.

The string variable in Java is an object and the returned string array can be used by the jobjectArray type. The core code is defined as follows.

```
JNIEXPORT jobjectArray JNICALL
Java_liuwentao_test1_study_getstringarray
 (JNIEnv *env, jobject obj){
vector<string> sa;
sa.push_back("hello");sa.push_back("test");
int n=sa.size();
jclass objC=env->FindClass("java/lang/String" );
jobjectArray texts= env->NewObjectArray(
(jsize)n, objC, 0);jstring jstr; int i=0;
for(;i<n;i++){
jstr = env->NewStringUTF( sa[i].c_str());
env->SetObjectArrayElement(texts, i, jstr);}
return texts;}
```

If the native code wants to return the object array, it must create one object array variable with the capacity size of the jobjectArray type. It uses the FindClass to get the class object and uses the NewObjectArray to make one object array with the length. It uses the GetFieldID to get the id of field type and makes the GetMethodID to the constructor function id. According to the array size, use the

NewObject function to create all objects and use the SetXXXField to set the value of all fields of each object. Finally, use the SetObjectArrayElement to set the object to the object arrays and return the object array to the Java code.

The basic progress of returning one Java object is shown in figure 2.

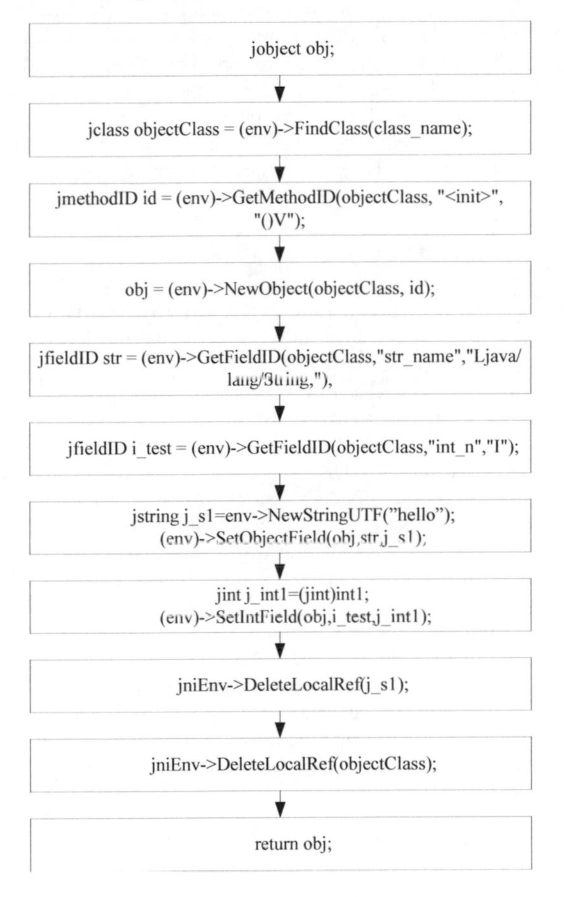

Figure 2. The progress of returning one object.

The process of getting the static member of the class is to get the class object with the FindClass and use the GetStaticFieldID to get the name of static data field. Use the GetStaticXXXField to get the value of one field and use the SetStaticXXXField to set the value of the field. If the native code is to get the instance method of class, it uses the GetObjectClass to get the class of the object and use the GetMethodID to the instance method type. It can use the CallXXXMethod to call the instance method. The coding signature of a function must follow rules. For example, the I code marks the integer type, the symbol C marks the char type, the symbol D marks the double type, and so on.

3 MEMORY MANAGEMENT

The native code and Java code have a different memory management mechanism. If the memory is not used correctly, the application will make the memory leaks and it can lead to the program crash. The memory allocated in the native code should be released in the appropriate place. When the memory reaches the maximum limit, the process will quit unexpectedly. The JVM has its special memory operation methods and incorrect methods can lead to the memory leak of Java heap. If the program is assigned too much Java objects, and there is no timely release, it will cause the JVM to crash. The NDK cannot create a large number of local references, otherwise it will reach the largest number of local reference list and cause the out of memory problem. The local reference is not automatically deleted when the functions exists, unlike the local variables which are stored in the stack. The local variables will be deleted when the function exists. The local object reference is not deleted until the code return to the Java code from the native code unless it is removed by hand. The program will create one local reference list, which will store all local references created in the native method. The local reference will be created when one Java object is used. The native code reads the local reference by the pointer of the local reference list. So the local objects created by the native code should be promptly deleted manually otherwise it will run out of memory because of the maximum capacity limit of the local reference list.

The normal resource release modes are shown in figure 3.

Figure 3. Resource release modes.

The global reference can be created by the NewGlobalRef. The global reference will store the Java object until it is released by the code. So if the global reference is not used, it must be deleted by hand with the DeleteGlobalRef function which will delete the global reference and the related Java object. The object of local and global reference cannot be operated by the garbage collection and the weak global reference can be managed by the garbage collection.

The basic type data does not need to release. The reference type data need to be released. In the NDK, the string type is a reference type and it uses the GetStringUTFChars to get the string and must use the ReleaseStringUTFChars to release the memory. The array data must also be released after finishing operation. The local reference object must be released by the DeleteLocalRef function. The function will find the index of the local reference in the local reference list and delete it.

4 THREAD MODEL

One thread mode is to use the Java thread to call NDK local code. It uses the Thread class and implements the run function. In the run function, the Looper.prepare() (Looper) is called firstly and the local code can be called. The Java thread can use the ProgressDialog (ProgressDialog) to show the thread message. In the NDK local code, the function can be called which is defined in the Java code to show the message in the ProgressDialog view.

The progress of the Java thread operation using NDK code to update the user interface is shown in figure 4. The shaded rectangle represents the Java code and the hollow rectangle represents the local code. The dashed rectangle is used to show the message in the progress dialog and it is called in the local code.

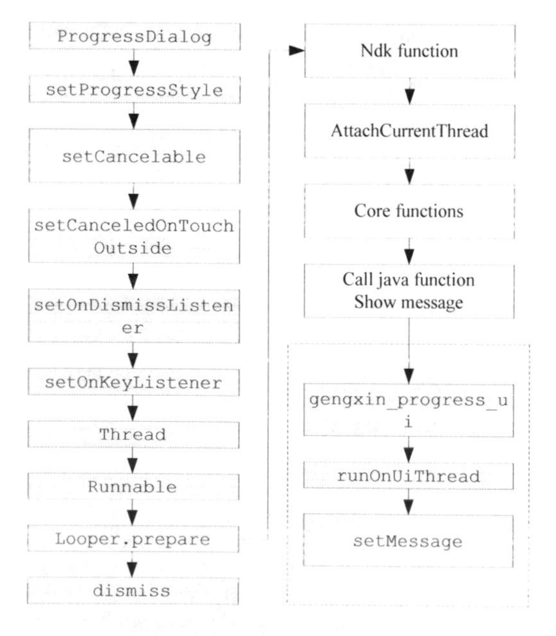

Figure 4. The progress of thread operation.

The thread can be used in both the native code and Java code. In the native thread, the JNIEnv must be related to the thread and cannot be shared with other threads. It can use the function of JavaVM to create one JNIEnv variable. The function is the AttachCurrentThread and its parameter can get the JNIEnv value.

The thread uses the DetachCurrentThread to delete the relationship. The local code uses the pthread_create (POSIX Threads Programming) to create one thread and it must set the tread callback. If one object wants to be used in another thread, it cannot be stored in the global variable directly. It must use the NewGlobalRef function to create one global variable and other thread code can use this global variable to operate this object.

5 SUMMARY

The NDK code can improve the performance of applications and increase the code reuse of local codes. The application implemented by Java can be easily decompiled because of the compilation features. The NDK code can improve the security of the application which can impede the software pirate to crack the program. Although NDK programming can significantly improve operational efficiency, it also can increase the chances of errors. The inappropriate NDK programming may lead to Java heap memory leak and it also may lead to memory leaks of native memory. It may even make the total running program is abnormally terminated. The rules of native code and JNI memory management must all be followed strictly. The thread can improve the interaction and computational efficiency of the program. The thread in the NDK must pay more attention to the message process, especially the data transfer in the different domain.

REFERENCES

Android Developers, http://developer.android.com/index.html

ART and Dalvik, http://source.android.com/devices/tech/dalvik/index.html

Android NDK | Android Developers, https://developer.android.com/tools/sdk/ndk/

JNI Tips, http://developer.android.com/training/articles/perf-jni.html

Java Native Interface Specification—Contents, http://docs.oracle.com/javase/7/docs/technotes/guides/jni/spec/jniTOC.html

Understanding "extern" keyword in C, http://www.geeksforgeeks.org/understanding-extern-keyword-in-c/

POSIX Threads Programming, https://computing.llnl.gov/tutorials/pthreads/

Looper,http://developer.android.com/reference/android/os/Looper.html

ProgressDialog, http://developer.android.com/reference/android/app/ProgressDialog.html

Computing, Control, Information and Education Engineering – Liu, Sung & Yao (eds)

 ISBN: 978-1-138-02800-5

Zernike orthogonal moments for image edge detection

Mai Jiang & Gui Jun Sha
Criminal Investigation Department, National Police University of China, Shenyang, China

ABSTRACT: This paper presents an image sub-pixel edge detection method based on Zernike moment. Firstly, we offered the edge model parameters calculation method through Zernike moment definition and the rotation invariance. Then, the orthogonal complex polynomial from 0 order to 2 order of the Zernike moment was calculated, in order to save time we calculate the image Zernike moment mask values in advance. Meanwhile, to improve the Zernike moment edge detection operator's positioning accuracy and suppressing noise abilities we use 9×9 grid mask. Finally, we applied these Zernike moment edge detection operators for some shoeprint and fingerprint recognition. The experiment results show that this method has higher detection precision compared with traditional methods.

KEYWORDS: Sub-Pixel, Zernike image moment, Orthogonal Complex Polynomial, Edge Detection.

1 INTRODUCTION

According to the current mathematic models the sub-pixel level edge detection methods can be divided into three kinds mainly: fitting method, interpolation method and moment method. The moment is the most widely used sub-pixel level edge detection method. The Zernike image moment is based on the integral operation so it has stronger anti-noise ability compared with the differential operation (Sobel and Robers operator etc.). The moment method mainly including two kinds: spatial moment and Zernike moment. The spatial moment concept was first put forward by Lyvers[1,2] which used six image geometric moments deduced four step edge parameters, this method has redundancy information because lack of the orthogonality. Ghosal and Mehrotal presents an edge detection method based on orthogonal Zernike image moment[3~5]. The domestic scholars also performed related research, paper [6] proposed a height step edge detection which improve the calculation time and sometime lose minimal image detail information.

2 THE PRINCIPLE OF THE ZERNIKE MOMENT EDGE DETECION

2.1 *The zernike moment step model*

The Zernike moment step model is shown as Figure 1. Where b is the unit circle's background gray value, d is the distance between the circle center and edge line($d \in [-1,1]$), θ is the angle between edge line and x axis($\theta \in [-\pi/2, +\pi/2]$).

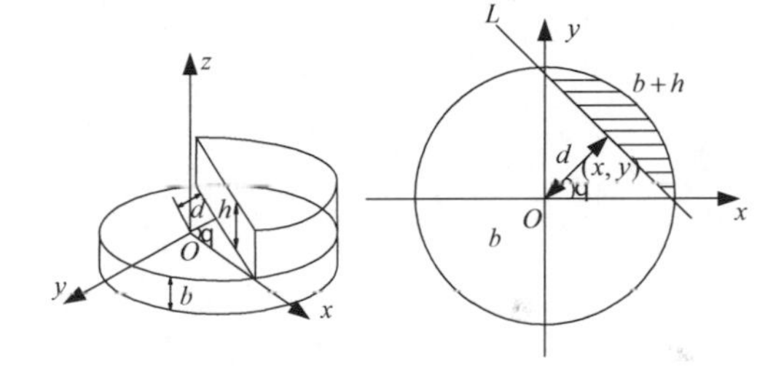

Figure 1. The Zernike step model.

$$\begin{aligned} B_{z,nm} &= \frac{n+1}{\pi}\int_0^1\int_0^{2\pi} f(r,\theta)V^*_{nm}(r,\theta)r\,dr\,d\theta \\ &= \frac{n+1}{\pi}\int_0^1\int_0^{2\pi} f(r,\theta)R_{nm}(r)e^{-jm\theta}r\,dr\,d\theta \end{aligned} \tag{1}$$

Where n is a positive integer or zero, m is a positive or negative integer, meanwhile, $n-|m|$ is even and $|m| \le n$, r(where r = d) is the distance between the unit circle's origin point and the foot point (x,y), $r=\sqrt{x^2+y^2}(-1<x,y<1)$, $V_{nm}(r,\theta)$ is orthogonal complex polynomial and V^*_{nm} is $V_{nm}(r,\theta)$ conjugate complex, $V_{nm}(r,\theta)=R_{nm}(r)e^{jm\theta}$, $R_{nm}(r)$ is foot point (x,y) radial polynomial which is defined as follows[7]

$$R_{nm}(r)=\sum_{s=0}^{(n-|m|)/2}(-1)^s\frac{(n-s)!}{s!(\frac{n+|m|}{2}-s)!(\frac{n-|m|}{2}-s)!}r^{n-2s} \tag{2}$$

2.2 *The rotation invariance of Zernike moment*

If we rotate the $f(r,\theta)$ in clockwise angle α and result after rotation can be expressed as follows

$$f^{\alpha}(r,\theta)=f(r,\theta-\alpha) \tag{3}$$

Change $\theta-\alpha$ to θ' we get Zernike moment after rotation as

$$\begin{aligned} B'_{z,nm} &= \frac{2(n+1)}{\pi}\int_0^1\int_0^{2\pi} f(r,\theta')R_{nm}(r)e^{-jm(\theta'+\alpha)}r\,dr\,d\theta' \\ &= [\frac{2(n+1)}{\pi}\int_0^1\int_0^{2\pi} f(r,\theta')R_{nm}(r)e^{-jm\theta'}r\,dr\,d\theta']\times e^{-jm\alpha} \\ &= B_{z,nm}e^{-jm\alpha} \end{aligned} \tag{4}$$

Form equation (4) we know the rotated Zernike moment remain the module value and only change phase angle which we call it as rotational invariance.

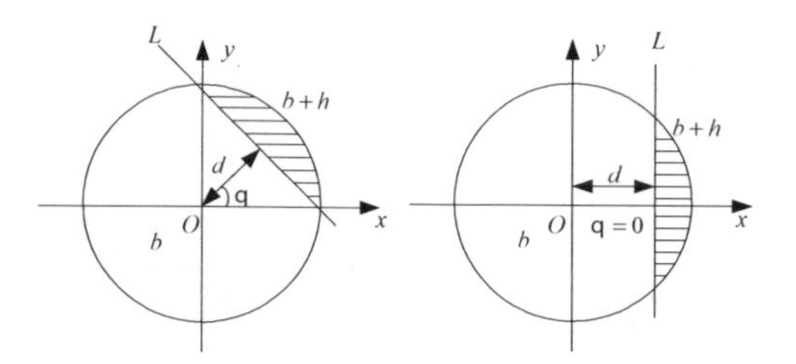

Figure 2. The Zernike rotation model.

As Figure 2 shows if we rotate the Zernike edge model in θ angle and edge line an x axis mutual perpendicular with each other. In this case $f^{\alpha}(r,\theta)$ value is b, b+h in interval $[-1,l]$ and $[l,1]$, respectively. By computing we know the orthogonal complex polynomial V_{11} value is $V_{11}=x+jy$. So the Zernike moment $B'_{z,11}$ value is

$$\begin{aligned} B'_{z,11} &= \frac{n+1}{\pi}\iint_{x^2+y^2\le 1} f^{\alpha}(r,\theta)(x-jy)\,dx\,dy \\ &= \frac{n+1}{\pi}[\iint_{x^2+y^2\le 1} f^{\alpha}(r,\theta)x\,dx\,dy - j\iint_{x^2+y^2\le 1} f^{\alpha}(r,\theta)y\,dx\,dy] \end{aligned} \tag{5}$$

Relative to variable y $f^{\alpha}(r,\theta)y$ is odd function and integral region is symmetrical about x axis so we know

$$\mathrm{Im}(B'_{z,11})=\frac{n+1}{\pi}j\iint_{x^2+y^2\le 1} f^{\alpha}(r,\theta)y\,dx\,dy=0 \tag{6}$$

Through rotational invariance we get

$$\begin{aligned} B'_{z,11} &= B_{z,11}e^{-j\theta_z} = [\mathrm{Re}(B_{z,11})+j\,\mathrm{Im}(B_{z,11})](\cos\theta_z - j\sin\theta_z) \\ &= [\mathrm{Re}(B_{z,11})\cos\theta_z + \mathrm{Im}(B_{z,11})\sin\theta_z] \\ &\quad + j[\mathrm{Im}(B_{z,11})\cos\theta_z - \mathrm{Re}(B_{z,11})\sin\theta_z] \end{aligned} \tag{7}$$

Combined equation (6) and (7) and obtains the result

$$\theta_z = \arctan\frac{\mathrm{Im}[B_{z,11}]}{\mathrm{Re}[B_{z,11}]} \tag{8}$$

Substitution the $f^{\alpha}(r,\theta)$ values b and h into the rotated Zernike moment

$$\begin{aligned} B'_{z,00} &= 2\int_{-1}^{1}\int_{0}^{\sqrt{1-x^2}} b\,dx\,dy + 2\int_{d}^{1}\int_{0}^{\sqrt{1-x^2}} h\,dx\,dy \\ &= b\pi + b\pi/2 - h\sin^{-1}(d) - h\,d\sqrt{1-d^2} \end{aligned} \tag{9}$$

$$B'_{z,11} = \iint_{x^2+y^2\le 1} f'(x,y)(x-jy)\,dx\,dy = \frac{2h\,(1-d^2)^{3/2}}{3} \tag{10}$$

$$B'_{z,20} = \iint_{x^2+y^2\le 1} f'(x,y)(x^2+y^2-1)\,dx\,dy = \frac{2h\,d\,(1-d^2)^{3/2}}{3} \tag{11}$$

By solving the three simultaneous equations (9)–(11) and we finally get the ideal edge two orders model parameters are as follows

$$d = \frac{B'_{z,20}}{B'_{z,11}} = \frac{B_{z,20}}{B_{z,11}}e^{-j\theta_z} \tag{12}$$

$$h = \frac{3B'_{z,11}}{2(1-d^2)^{3/2}} = \frac{3B_{z,11}}{2(1-d^2)^{3/2}}e^{j\theta_z} \tag{13}$$

Where d is the distance parameter, h is the step parameter. Take h and d as the edge criteria for each pixel, we set two threshold value *m1* and *m2* and if $h > m1$ and $d < m2$ we consider the pixel as an edge point.

3 THE ZERNIKE MOMENT EDGE DETECTION OPERATOR MASK CALCULATION

In image the Zernike moment discrete expression is as follows

$$\begin{aligned} B_{z,nm} &= \frac{2(n+1)}{\pi}\int_0^1\int_0^{2\pi} f(r,\theta)V^*_{nm}(r,\theta)r\,dr\,d\theta \\ &= \frac{2(n+1)}{\pi}\sum_x\sum_y f(r,\theta)M_{nm}\Delta x\Delta y \end{aligned} \tag{14}$$

Take convolution operation on the pixel point(x,y) neighborhood and its correspond mask M_{nm}. These edges operator calculation need different orders Zernike moment, to save time we should calculate these different mask values in advance. More grid partition means more calculation and time cost, however, the Zernike moment edge operator's positioning accuracy and suppressing noise ability will corresponding improvement and Figure 3 shows the 3D simulation of the Zernike moment with 9×9 mask(N=9)[8].

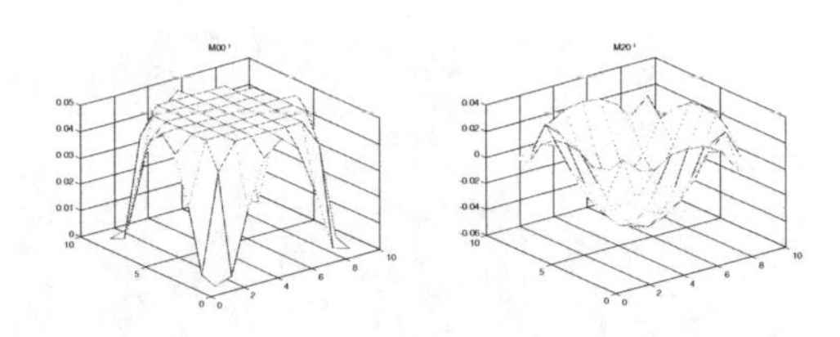

(a)The M00 Mask (b)The M20 Mask

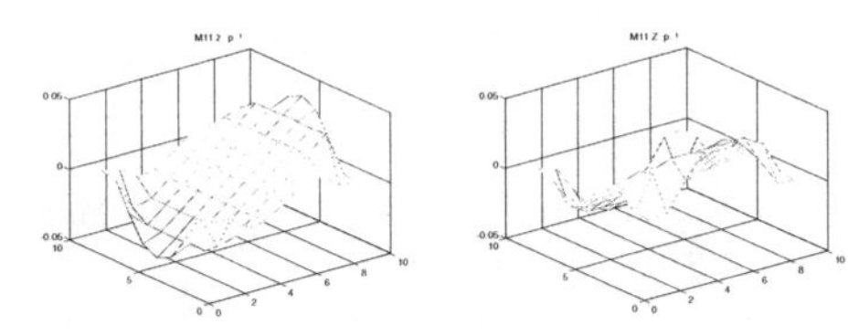

(c)The M11 real mask (d)The M11 imaginary mask

Figure 3. The 3D simulation of the Zernike moment with 9×9 mask(N=9).

The Zernike moment edge detection operator sub-pixel coordinates as follows

$$\begin{bmatrix} x_s \\ y_s \end{bmatrix} = \begin{bmatrix} x \\ y \end{bmatrix} + \frac{N}{2} \times d \begin{bmatrix} \cos\theta_z \\ \sin\theta_z \end{bmatrix} \tag{15}$$

Where (x,y) represents the center coordinates, (x_s,y_s) represents the corresponding sub-pixel coordinates, N is the grid division number, d and θ_z are the Zernike moment edge detection operator parameters.

4 DIFFERENT IMPROVED ZERNIKE MOMENTS STEP MODEL ANALYSIS

4.1 *The zernike image moment transition model*

The four order Zernike image moment was discussed in reference [9]. It assumed that there exists two edges L_1 and L_2 corresponding to different orders as shown in Figure 4.

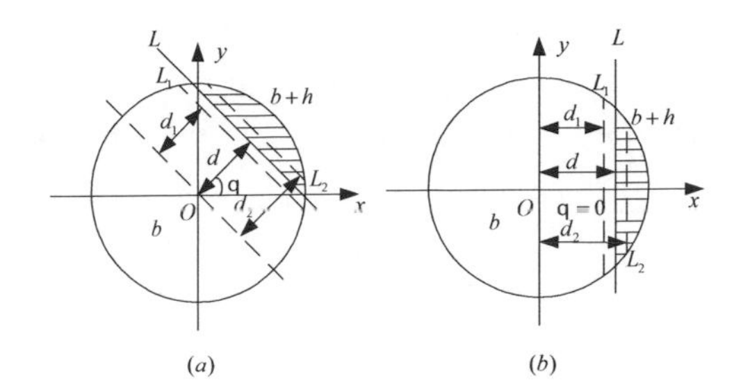

Figure 4. The Zernike image moment transition model.

We can calculate the masks through different orders orthogonal complex polynomial $V_{nm}(r,\theta)$ as follows

$$V_{00} = 1 \tag{16}$$

$$V_{11} = x + jy \tag{17}$$

$$V_{20} = 2x^2 + 2y^2 - 1 \tag{18}$$

$$V_{31} = (3x^3 + 3xy^2 - 2x) + j(3y^3 + 3x^2y - 2y) \tag{19}$$

$$V_{40} = 6x^4 + 6y^4 + 12x^2y^2 - 6x^2 - 6y^2 + 1 \tag{20}$$

Edge distances d_1 and d_2 was calculated as follows[9]

$$\begin{cases} d_1 = \sqrt{\dfrac{5Z'_{40} + 3Z'_{20}}{8Z'_{20}}} \\ d_2 = \sqrt{\dfrac{5Z'_{31} + Z'_{11}}{6Z'_{11}}} \end{cases} \tag{21}$$

Taking the distance value d as $d = |d_1 - d_2|$. Finally, we get the four orders edge model parameters are as shown in equation (22).

$$\begin{cases} d_4 = |d_1 - d_2| \\ h_4 = \dfrac{3B'_{z,11}}{2(1-d_4^2)^{3/2}} = \dfrac{3B_{z,11}}{2(1-d_4^2)^{3/2}} e^{j\theta_z} \end{cases} \tag{22}$$

4.2 *The zernike image moment simplified model*

If change the equation (13) into as follows[6]

$$h = \frac{3B'_{z,11}}{2(1-d^2)^{3/2}} = sll \times B'_{z,11} \tag{23}$$

Where $sll = \dfrac{3}{2(1-d^2)^{3/2}}$ and the relationship between d and sll was shown in Figure 5.

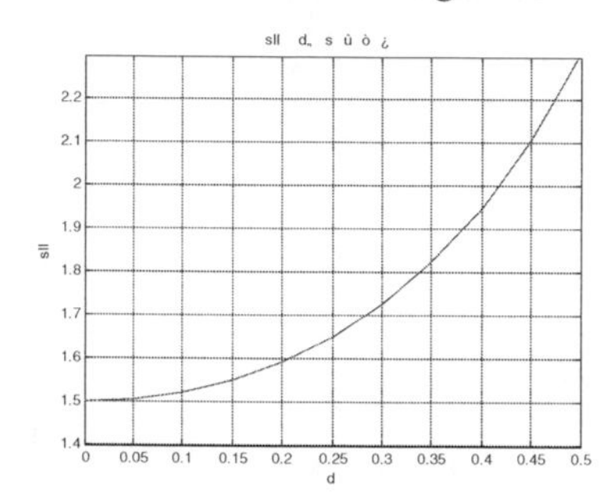

Figure 5. The relationship between sll and d.

Through the simulation curve in Figure 5 we can see with the distance parameter d decreased, there exist a constant 1.5 between them. So we can substitute h into $B'_{z,11}$ if d is small enough and there is no influence to the edge detection results.

5 THE EXPERIMENT RESULTS ANALYSIS

To validate the Zernike moment results of different Zernike image moment models. This experiment used shoeprints as the detection objects. Where the distance parameter d threshold $m1=25$ and the step parameter h threshold $m2=0.15$. The edge detection results were shown in Figure 6.

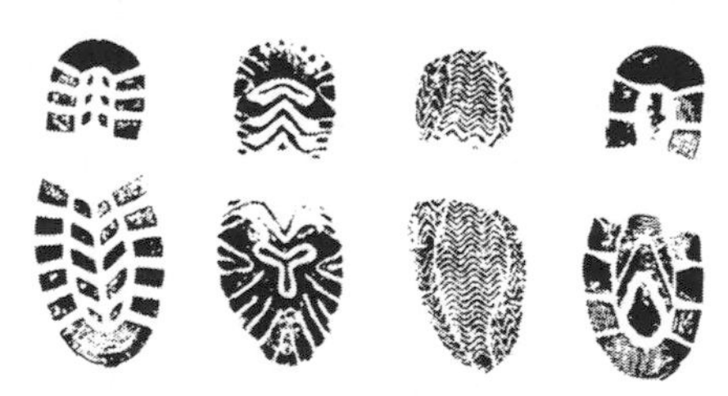

(a) Original shoeprints pictures

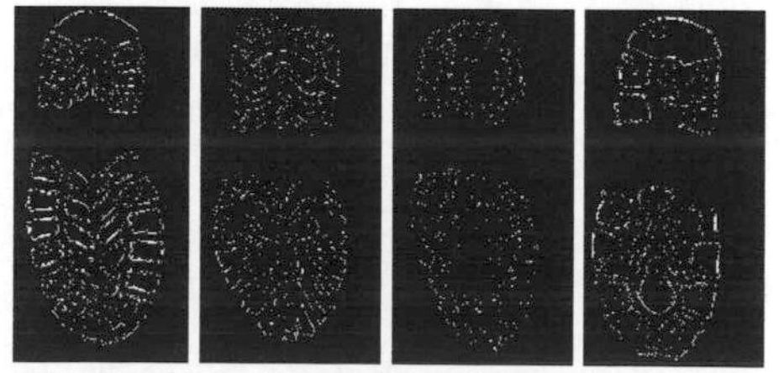

(b) The Zernike Image Moment Transition Model(four orders)

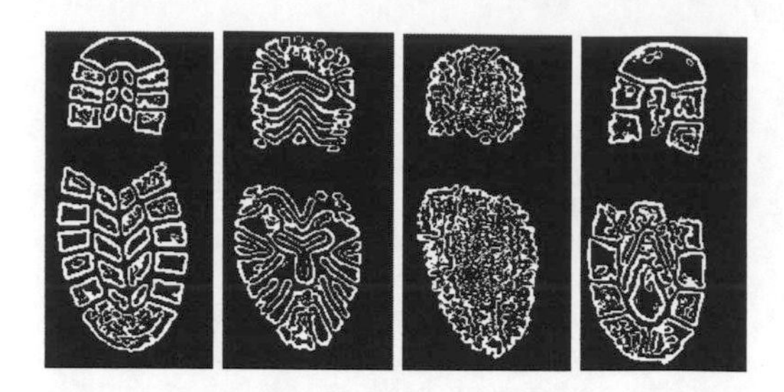

(c) The Zernike Image Moment Step Model(two orders)

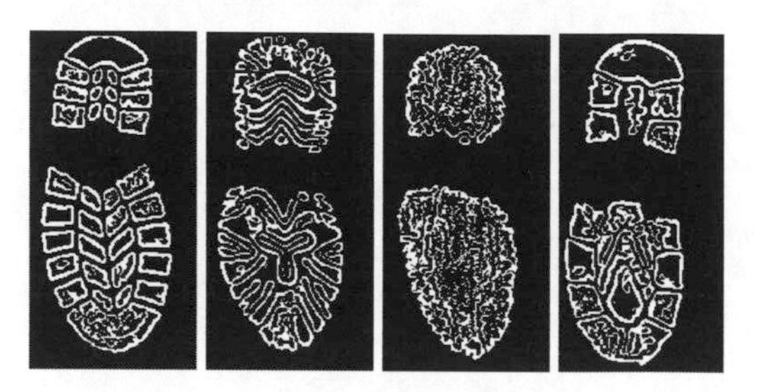

(d) The Zernike Image Moment Simplified Model

Figure 6. Edge detection using different Zernike models.

For further test we still applied the Zernike moment operator (grid partition N=9) to fingerprint edge detection and compared with some regular pixel-level edge detection operators (such as Canny, Laplace). The edge detection results were shown in Figure 7.

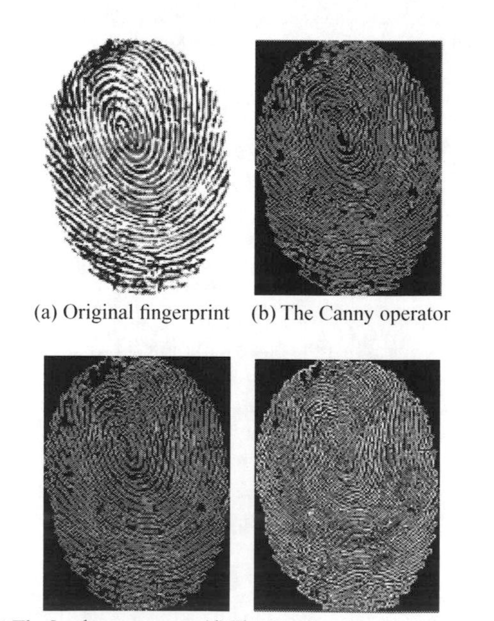

(a) Original fingerprint (b) The Canny operator

(c) The Laplace operator (d) The Zernike simplified operator

Figure 7. Edge detection using different edge detection operators.

In the shoeprint edge detection experiment Figure 6(c) and Figure 6(d) shown an basically identical results and the edge detection results better than Figure 6(c). Meanwhile, by comparing the edge detection results in Figure7 we can see the Canny and Laplace operator generated few edge boundaries and the Zernike simplified moment operator generated more and clear edge boundaries.

REFERENCES

[1] Michael Reed Teague, Image analysis via the general theory of moments, 70(1980): 920–930.

[2] Esward P. Lyvers, Owen Robert Mitchell, Mark L. Akey, Subpixel Measurement Using a Moment-Based Edge Operator[J], IEEE Trans. on Pattern Analysis and Machine Intelligence, 11(1989): 1293–1309.

[3] Sugata Ghosal, Rajiv Mehrotra, A Moment-Based United Approach to Image Feature Detection[J], IEEE Trans. on Image Processing, 6(1997): 781–793.

[4] Sugata Ghosal, An Orthogonal Moment-Based Integrated Approach[J], IEEE Trans. on robotics and automation, 9(1993): 385–399.

[5] Sugata Ghosal, Rajiv Mehrotra, Orthogonal Moment Operators for Subpixel Edge Detection[J], Pattern Recognition, 26(1993): 295–306.

[6] Qu Ying-dong, Li Rong de, A high speed Zernike moments edge operator based on 9×9 Masks, Journal of Optoelectronics.Laser, 21(11): 1683–1687, 2010

[7] Chandan Singh, Ekta Walia, Fast and numerically stable methods for the computation of Zernike moments, Patter Recognition, 43(2010): 2497–2506

[8] Jiang Mai, Liang Ye, Reseach on sub-pixel Zernike moments edge detecting operator model, Computer Measurement and Control, 21(4): 874–876, 2013.

[9] GAO Shi-Yi, ZHAO Ming-Yang, Improved algorithm about subpixel edge detection of image based on Zernike Orthogonal Moments, Acta Automatica Sinica, 34(9): 1163–1168, 2008.

Computing, Control, Information and Education Engineering – Liu, Sung & Yao (eds)
© 2015 Taylor & Francis Group, London, ISBN: 978-1-138-02800-5

Coverage optimization strategy for multilayer mobile sensor network based on artificial bee colony algorithm

Xia Ling Zeng
Jiangxi Science and Technology Normal University, Nanchang, China

ABSTRACT: For the multi-layer structure of mobile sensor network, hierarchical mechanism is proposed to apply in artificial bee colony algorithm, in order to complete the deployment optimization of sensor nodes and to improve the network coverage ratio. Firstly the monitoring area is regionally divided according to equal proportions, and one sink node is deployed in each sub-region. Based on the network coverage optimization model, the sink nodes run an artificial bee colony algorithm to obtain the optimal deployment solution of sensor nodes. Results show that based on the coverage optimization strategy, the coverage ratio gradually increases as the number of mobile nodes increases, and when the iterations of the algorithm are only 300 the network coverage has quickly increased, showing a good coverage optimization performance.

KEYWORDS: Coverage optimization; Artificial bee colony algorithm; Hierarchical; Sensor network.

1 INTRODUCTION

The wireless sensor network (WSN) is widely used in military, environmental monitoring and other fields. The performance of WSN depends on the effective coverage area of sensor nodes after deployment. The mobile node makes the dynamic deployment of sensor network be possible. In the dynamic deployment of sensor network, the coverage optimization strategy is used to adjust the position of mobile nodes after initial randomly deployment, in order to expand the effective coverage area and to improve the monitoring capability of WSN.

In order to improve coverage ratio of monitoring the region, researchers have proposed many effective coverage optimization strategies in recent years [1-4]. In [5,6], they use the artificial bee colony algorithm to optimize the network topology, and achieved good results. The artificial bee colony algorithm is an intelligence swarm algorithm which simulates the behavior of bees collect honey. It has the advantages of less control parameters and the robustness [7]. It can solve continuous and combinatorial optimization problems, which has been successfully applied to the training of neural networks [8], the solving of constrained optimization problem [9] and other fields.

This paper proposes a multilayer mobile sensor network, which using three different nodes. For the multi-layer structure of mobile sensor network, hierarchical mechanism is proposed to apply in the artificial bee colony algorithm. Based on the network coverage optimization model, the sink nodes hierarchically run an artificial bee colony algorithm to obtain the optimal deployment solution of sensor nodes, in order to complete the deployment optimization of sensor nodes and improve the network coverage.

2 THE MULTILAYER STRUCTURE OF MOBILE SENSOR NETWORK

The multilayer structure of mobile sensor network can improve the performance and improve its nodes' poor performance in energy, communication and processing. It is also helpful for lowering the nodes' processing ability, ensuring the stability of transmission and improving the performance of the whole network. According to the different functions of nodes, sensor node can be divided into three categories, including one layer of sensor node (named as S node), two layers of aggregation node (named as Sink node) and three layers of control node (named as C node). The function of these three levels is gradually increasing.

These three level nodes interact with each other in fulfilling the task of wireless sensor network. Among them, the S nodes have the weakest function and its basic function is to complete the data acquisition. They can be fixed in a certain position, and can also move. As a data collector of the network, the Sink nodes have stronger function and transmit the collected data to control node. They are distributed at the beginning and their movements are controlled. The C nodes play the role of the station, and its distribution depends on the monitoring region.

We set the following conditions. The S nodes transmit data to Sink nodes directly in single-hop mode, and there is no mutual data transmission between the S nodes. Without loss of generality, the network topology in certain regions, and the S nodes can move but not beyond topological range.

3 NETWORK COVERAGE OPTIMIZATION MODEL

3.1 *Coverage model*

Assuming the monitoring area is a two-dimensional plane A, and randomly putting N sensor nodes in the region. The parameter of these nodes is same, and there coordinate position is known. To ensure the connectivity of the sensor network, the communication radius R is set to twice of the perception radius r, i.e. R=2r. The set of sensor nodes is expressed as $S\{S_1, S_2, \ldots, S_N\}$, where $S_i = \{x_i, y_i, r\}$ represents the sensor node whose coordinate is (x_i, y_i) and perception radius is r.

The monitoring area A is discretized into m×n grid points. Suppose the coordinate of grid point Q is (x, y), and the distance between Q and sensor node S_i is calculated as in Equation 1

$$d(S_i,Q) = \sqrt{(x_i - x) + (y_i - y)^2} \tag{1}$$

The probability of any one sensor node covering a grid point Q is expressed as $P(S_i, Q)$. Based on the Boolean sensing model, the probability of events is the binary distribution. So the probability of any one sensor node covering grid Q is shown in Equation 2.

$$P(S_i,Q) = \begin{cases} 1, d(S_i,Q) < r \\ 0, \text{Others} \end{cases} \tag{2}$$

In a real environment, the coverage of sensor node is disturbed by the voice sound in the environment. Then the probability model shows a certain probability distribution characteristic, which is as shown in Equation 3.

$$P(S_i,Q) = \begin{cases} 1, d(S_i,Q) \le r - r_e \\ e^{\frac{-\alpha_1\lambda_1\beta_1}{\lambda_2\beta_2+\alpha_2}}, r - r_e < d(S_i,Q) < r + r_e \\ 0, \text{Others} \end{cases} \tag{3}$$

Where r_e is a measurable reliability parameter of sensor node and $0<r_e<r$. α_1, α_2, β_1, β_2 are metrics of coverage probability. They are calculated by Equation 4 and Equation 5 when λ_1, λ_2 are input parameters.

$$\lambda_1 = r_e - r + d(S_i,Q) \tag{4}$$

$$\lambda_2 = r_e + r - d(S_i,Q) \tag{5}$$

3.2 *Regional coverage probability*

In order to improve the coverage probability of a grid point, usually adopting multiple sensor nodes to jointly cover. The coverage probability is calculated as shown in Equation 6.

$$P(S,Q) = 1 - \prod_{1}^{N}(1 - P(S_i,Q)) \tag{6}$$

The evaluation criteria, whether the grid point Q be effectively covered is as shown in Equation 7. Here c_{th} is the threshold of effective coverage.

$$P(S,Q) \ge c_{th} \tag{7}$$

The coverage ratio of nodes sets S to monitoring area A is defined as the ratio between the sums of coverage area of nodes set S and the total area of region A. Since the region A is discretized into m×n grid points and whether a grid point is covered can be measured by Equation 7, the calculation is simplified into the ratio between the number of grid points that are effectively covered and the total number of grid points, as shown in Equation 8.

$$R_{area}(S) = \frac{count}{m \times n} \tag{8}$$

3.3 *The description of coverage problem*

Assuming the monitoring area A is a square with 10×10 side length, and it is discretized into 10×10 grid points. 45 sensor nodes are randomly deployed in the region, as shown in Figure 1. In the figure 1, we use the symbol · to represent the position of th sensor node.

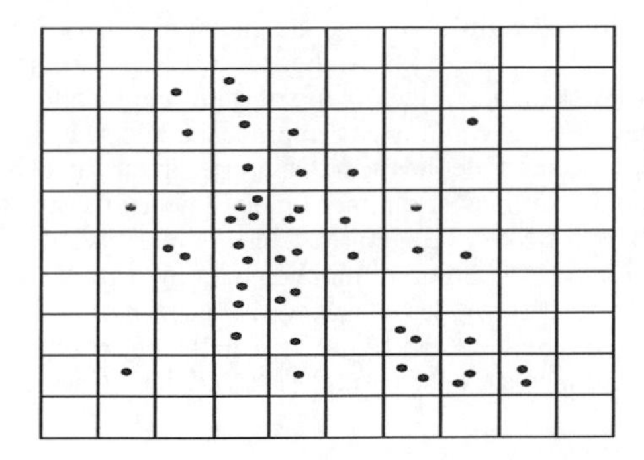

Figure 1. Sensor nodes' distribution map in the monitoring area.

The algorithm to calculate coverage probability is as follows.

Step 1: using the Equation 3, 6 and 7 to evaluate the effective coverage of sensor nodes set to a grid point.

Step 2: repeating step1 to statistic the number of grid points that are effectively covered.

Step 3: using Equation 8 to calculate the regional coverage ratio, and using it as objective function of coverage optimization problem.

4 COVERAGE OPTIMIZATION STRATEGY BASED ON ARTIFICIAL BEE COLONY ALGORITHM

The deployment optimization of sensor nodes in sensor network can be seen as optimization problem which take the solution vector composed by the position of sensor nodes as input and network effective coverage area as a target. For multi-layer structure of mobile sensor network, this paper proposes a hierarchical mechanism to apply in artificial bee colony algorithm.

Firstly the monitoring area is regionally divided according to equal proportions, and one sink node is deployed in each sub-region. The sink node completes the positioning of sensor nodes within its sub-region, and maintains the information table of sensor node position coordinates. Based on the network coverage optimization model, the sink nodes run artificial bee colony algorithm. It uses the regional coverage probability as an optimization objective function to obtain the optimal covering solution of sensor nodes, in order to complete the deployment optimization of sensor nodes and to improve the network coverage ratio.

4.1 *Hierarchical mechanism*

The basic idea of hierarchical mechanism is that for an optimization problem firstly randomly generate N×M samples, and then turn them into N sub-swarms. Each sub-swarm includes M samples. Each sub-swarm independently run artificial bee colony algorithm, which denoted as $ABC_i(i=1,2, \ldots,N)$. Each sink node conducts a search in its corresponding sub-swarm and records the maximum fitness value of search results. The process is repeated until the algorithm terminates, and each sink node obtains an optimal solution. Finally the optimal solutions form each sink node are combined into the optimal solution of the optimization problem.

4.2 *A coverage optimization strategy based on an artificial bee colony algorithm*

According to the network coverage optimization model and proposed hierarchical mechanism, the algorithm based on an artificial bee colony algorithm for solving coverage optimization of multilayer mobile sensor network is shown as follows.

Step 1: Regional division. The monitoring area is divided in to N sub-regions, and each sub-region deploys a sink node. Each sink node completes the positioning of sensor nodes within its sub-region, and maintains the information table of sensor node position coordinates. Assuming the number of sensor nodes in each sub-region is $C_i(i=1,2, \ldots,N)$. The position coordinates of sensor nodes in the i-th sub-region is recorded by a one-dimensional array $R_i[i=1,2, \ldots, C_i]$, where each element is a two-dimensional vector using to represent the coordinates of sensor nodes. Each sink node performs artificial bee colony algorithm beginning at step 2.

Step 2: initializing parameters. Such as the maximum number of iterations for each sub-region, the number of swarm, threshold, monitoring radius, and so on. Similarly, the honey-gathering bees and watching bees are half of each, and a scouting bee is set as 1.

Step 3: generating m initial solution as an initial bee colony.

Step 4: calculating the initial coverage ratio according to the Equations in section 3.3.

Step 5: update the position of the node (nectar) according to the Equation 9 and Equation 10.

$$x_i^j = x_{\min}^j + rand(0,1)(x_{\max}^j - x_{\min}^j) \tag{9}$$

$$v_{ij} = x_{ij} + \varphi(x_{ij} - x_{kj}) \tag{10}$$

x^j_{min} is the lower bound of the j-th component, and x^j_{max} is the upper bound of the j-th component. φ is an randomly generated integer within [-1,1].

Step 6: calculate the new node's coverage ratio (fitness value), and compared with the current value, then retain the better value.

Step 7: memory current nectar as local optima.

Step 8: cycles plus 1.

Step 9: if the termination condition is met (usually set to reach the maximum cycle algebra or target fitness value), go to step 10 and return the best fitness value. Otherwise continue to step 3.

Step 10: repeat from step 3 to step 9 for each sub-region, and merge the optimal solution for all sub-regions to constitute the optimal solution.

5 SIMULATION AND RESULTS

Within the monitoring region of 100m×100m, 50 sensor nodes are randomly deployed. The perception radius of sensor node is set as r=10m, and the communication radius R=2r=20m. The monitoring area is divided into four sub-regions and deployed four sink nodes. In the probability model, r_e =0.5 r=5m, $\alpha_1=1$, $\alpha_2=0$, $\beta_1=1$, $\beta_2=0.5$, the maximum cycle algebra is set to 200. All simulations are completed using Matlab 7.0 in Windows operating system.

Assuming all sensor nodes can move. The network coverage ratio in initial random deployment is about 45%. The bee colony algorithms with iteration from 100 to 500 are respectively run 20 times, and then calculate the average coverage ratio. The relationship between the average coverage ratio and number of iterations is shown in figure 2. In iterations of 300, the network coverage ratio has quickly converged to 81.9%, showing a good coverage optimization performance.

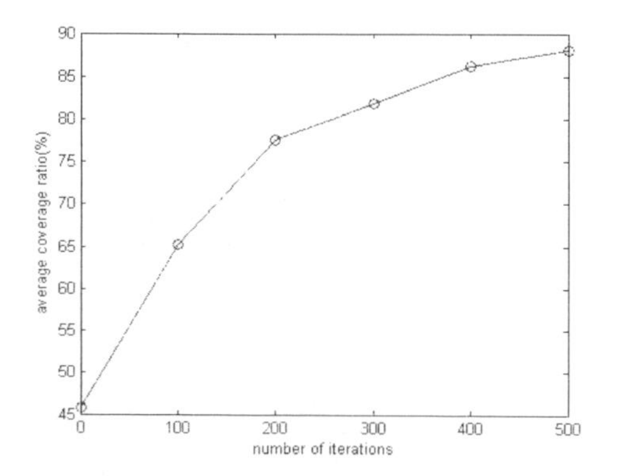

Figure 2. The relationship between the average coverage ratio and the number of iterations.

In the multilayer structure of mobile sensor network, the sensor node can be a static node; also can be a mobile node, or the mixing of static and mobile. The mobile node can move to improve the effective coverage area after deployment. Obviously, in the same total number of nodes, the more mobile nodes the greater the effective coverage area after optimization. As the increase of the ratio of mobile nodes in the total number of sensor nodes is 50, the change of network coverage is as shown in figure 3. With the ratio of mobile nodes increases, the coverage ratio gradually increases.

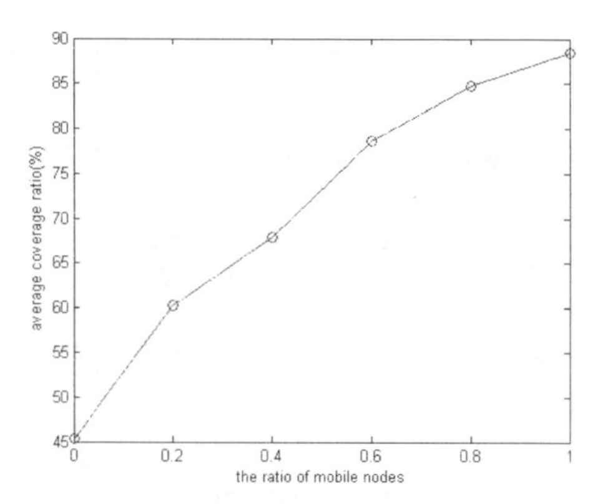

Figure 3. The change of network coverage as the increase of the ratio of mobile nodes.

6 CONCLUSIONS

The deployment optimization of sensor nodes in sensor network can be seen as an optimization problem. For multi-layer structure of mobile sensor network, this paper proposes a hierarchical mechanism to apply in artificial bee colony algorithm. Firstly the monitoring area is divided according to equal proportions, and one sink node is deployed in each sub-region. The sink node completes the positioning of sensor nodes within its sub-region, and maintains the information table of sensor node position coordinates. Based on the network coverage optimization model, the sink nodes run artificial bee colony algorithm. It uses the regional coverage probability as an optimization objective function to obtain the optimal coverage solution, then to complete the deployment optimization of sensor nodes and to improve the network coverage ratio.

Results show that based on the coverage optimization strategy, the coverage ratio gradually increases as the number of mobile nodes increases, and when the iterations of the algorithm are only 300 the network coverage has quickly increased, showing a good coverage optimization performance.

ACKNOWLEDGEMENTS

In this paper, the research was sponsored by the Youth Foundation of Jiangxi provincial education department (GJJ14593).

REFERENCES

[1] Zou Y, Chakrabarty K. 2003. Sensor deployment and target localization based on virtual forces. *Twenty-second annual joint conference of the IEEE computer and communications: 1293–1303.*

[2] Xue Wang, Sheng Wang, Junjie Ma. 2007. Parallel particle swarm optimization of mobile sensor node deployment in wireless sensor networks. *Chinese Journal of Computer* 4:563–568.

[3] Xue Wang, Sheng Wang, Junjie Ma. 2007. Dynamic sensor deployment strategy based on virtual force-directed particle swarm optimization in wireless sensor networks. *Journal of Electronic* 11:2038–2042.

[4] Xue Wang, Sheng Wang, Junjie Ma. 2007. An improved co-evolutionary particle swarm optimization for wireless sensor networks with dynamic deployment. *Sensors* 7(3):354–370.

[5] Ke Hu. 2012. Application research of wireless sensor network coverage optimization strategy based on artificial bee colony algorithm. Chengdu: Electronic Science and Technology.

[6] Ozturk C, Karaboga D, Gorkemli B. 2011. Probabilistic dynamic deployment of wireless sensor networks by artificial bee colony algorithm. *Sensors* 11(6):6056–6065.

[7] Karaboga D, Basturk B. 2008. On the performance of artificial bee colony algorithm. *Applied Soft Computing* 8(1):687–697.

[8] Bin Luo, Peiji Shao, Jinyao Luo, et al. 2011. Customer churn research based on multiple classifier fusing rough sets-neural network-artificial bee colony algorithm. *Journal of Management* 2:265–272.

[9] Xiang Wang, Jianguo Zheng. 2012. A multi-member artificial bee colony algorithm for constrained optimization problems. *Journal of Xi'an Jiaotong University* 2:38–44.

[10] Ozturk C, Karaboga D, Gorkemli B. 2012.Artificial bee colony algorithm for dynamic deployment of wireless sensor network. *Turkish Journal of Electrical Engineering and Computer Sciences* 20(2): 255–262.

Computing, Control, Information and Education Engineering – Liu, Sung & Yao (eds)
© 2015 Taylor & Francis Group, London, ISBN: 978-1-138-02800-5

Pavement crack recognition based on image processing

Z. Gao
Hubei Communications Technical College, Wuhan, China

S.L. Tang
School of Mechanical Engineering, Hubei University of Technology, Wuhan, China

ABSTRACT: Pavement crack detection plays a vital part in highway maintenance and management. As the direction of crack in image is so complex to recognize, and it always has much noise, gray crack image detection algorithm is the most important and difficult point. To achieve the balance between processing efficiency and complexity, this paper presents a new algorithm. Based on defect image segmentation and image edge detection, results show that the proposed detection method can effectively remove the isolated noise point, smooth the edge and improve the segmentation accuracy.

KEYWORDS: pavement crack; edge detection; threshold segmentation.

1 INTRODUCTION

Highway pavement must suffer from various diseases or damages and the service level must be influenced due to repeated wheel wear and influences of such natural factors as icy, rainy or snowy weather. Untimely discovery and improper treatment are bound to shorten the service life and cost a lot of maintenance fees [1-2]. In order to keep the service level and take corresponding measures, we must enhance prevention, maintenance and management of highways and investigate pavement conditions regularly. Undoubtedly, pavement crack is a typical disease. The maintenance fee will be heavily lessened in case that cracks are discovered instantly. So how to investigate the whole pavement fully and figure out the diseases without influences upon normal traffic orders has become a challenge to be solved. While traditional methods based on manual visual detection of cracks cannot adapt to fierce development of highways any more. Prior to 1990s, highway administrations adopted unreasonable pavement maintenance schemes for short of systematic data collection and evaluation, scientific maintenance decision management, limited experience and inadequate quantitative evaluation to pavement damages and damage degree, resulting in deterioration of pavement quality and increase of use cost [3-4].

In order to improve the quality of pavement management and maintenance, we must extract useful information from the pavement crack image and make a detailed description. Based on characteristics of the shape, highway pavement cracks can be divided into regular cracks and irregular cracks; the former contains longitudinal cracks and transverse cracks, and the latter contains massive cracks, fractures and net-shaped cracks. Longitudinal cracks, transverse cracks, fractures, net-shaped cracks and massive cracks are commonly seen in most highway pavement diseases which not only appear earlier but also become increasingly serious.

2 DEFECT IMAGE SEGMENTATION

Since image contains the most information and is the foundation for human vision, vision-based image recognition is of special significance. Image recognition aims at classifying and identifying images on behalf of human through automatic information processing by computer. It is a kind of model recognition and its processes can be divided into image pre-processing, image segmentation, and image feature extraction and image classification which are similar with those of mode recognition [5].

Edge detection method and statistical rules can be adopted to recognize pavement cracks based on significant features of pavement cracks (the image gray level changes suddenly at the cracks). The quality of edge detection algorithm poses great influences upon the effect and accuracy of recognition and detection of pavement cracks. Edge, the fundamental feature of image, is not only the foremost feature for image segmentation but also an important information source of textural features and the foundation for shape features. The classic and simplest edge detection method refers to detect the operator through detecting the edge of an area close to the original image's pixel.

3 EDGE DETECTION

Image edge is quite useful for image recognition and computer analysis [6]. Edge detection can outline the object clearly for the observer; and edge is an important feature in image recognition rich in internal information. Essentially, partial image features will not be shown in the edge continuously but mark the end of an area and the start of another.

Edge detection method can be adopted to recognize pavement cracks due to bigger changes of the gray level. While edge detection algorithm poses influences upon the effect and accuracy of recognition and detection of pavement cracks to a great extent. The simplest edge detection method is to detect the operator through detecting the edge of an area close to the original image's pixel. Presently, edges are often extracted by such methods as template matching method, curve fitting method and edge operator method [7].

Actually, edge detection aims to extract the boundary between the object and the background of the mage through an algorithm in case that the edge is defined as the border of a boundary area whose gray level changes rapidly. Changes of the gray level can be shown by the distribution gradient; so the edge detection operator can be acquired through partial image differentiation technique [8-10]. The typical edge detection method is to detect the operator through detecting the edge of a small area in the original image's pixel.

In case that: $\nabla f(x,y)=\frac{\partial f}{\partial x}i+\frac{\partial f}{\partial y}j$ is the gradient of the image and $\nabla f(x,y)$ contains changes of the gray level, suppose $e(x,y)=\sqrt{f_x^2+f_y^2}$ is the gradient of $\nabla f(x,y)$, e(x,y) can be used as the operator for edge detection. To simplify the computation, we can define e(x,y) as the sum of the absolute value of partial derivative f_x and f_y:

$$e(x,y)=\left|f_x(x,y)+f_y(x,y)\right|$$

Based on above theories, this paper proposes many edge detection algorithms such as Roberts edge detection operator, Sobel edge detection operator, Prewitt edge detection operator, Canny edge detection operator and Laplace edge detection operator. The comparative results of five classic edge detection effects of the pothole image and the net-shaped crack image are shown in the figure below.

It can be seen from the figure that the edge is thicker and not located accurately in Roberts edge detection operator; while the edge is located relatively accurately in Sobel and Prewitt edge detection operator; the result of edge extraction in Laplace edge detection operator is obviously superior; especially, the edge is complete and located accurately. In contrast, the edge extract by Canny edge detection operator is the most complete because of non-maximum suppression and morphological connection.

4 THRESHOLD SEGMENTATION

The histogram of image gray level is changing which has two peaks but unclear valleys or unclear peaks and valleys. Besides, the area ratio of two areas is often hard to be determined; but maximum variance threshold value method can figure out a satisfactory result. Fig. 1 is the histogram of image gray level in two areas; fig. 2 shows the relation between variance σ_B^2 and inter-area threshold valuet.

The ratio and average gray level of area A and area B separated by t in the whole image are shown in the histogram below.

Area A ratio: $\delta_1=\sum_{j=0}^{t}\frac{n_j}{n}$

Area B ratio: $\delta_2=\sum_{j=t+1}^{G-1}\frac{n_j}{n}$

The average gray level of the whole image:

$$\mu=\sum_{j=0}^{G-1}(f_j\times\frac{n_j}{n})$$

The average gray level of area A:

$$\mu_1=\frac{1}{\delta_1}\sum_{j=t+1}^{t}(f_j\times\frac{n_j}{n})$$

The average gray level of area B:

$$\mu_2=\frac{1}{\delta_2}\sum_{j=t+1}^{G-1}(f_j\times\frac{n_j}{n})$$

In the formula, G is the gray level of the image.

Then, the relation between the average gray level of the whole image and area A and B is:

$$\mu=\mu_1\delta_1+\mu_2\delta_2$$

The gray levels in an area are often similar; while those in different areas are different. In case that the gray level of two areas separated by t differs greatly, the average gray level u_1 and u_2 of two areas and the whole image is quite different. The inter-area variance is an effective parameter describing such differences and is expressed as below:

$$\sigma_B^2=\delta_1(\mu_1-\mu_2)^2+\delta_2(t)\mu_2(t)-\mu^2$$

In the formula, σ_B^2 is the inter-area variance separated by t. Obviously, different threshold values t result in different variances; or rather, inter-area variance, mean value of area A, mean value of area B, area ratio of area A and area ratio of area B are all the function of threshold value t as shown in the formula below:

$$\sigma_B^2 = \delta_1(t)\left[\mu_1(t) - \mu\right]^2 + \delta_2(t)\left[\mu_2(t) - \mu\right]^2$$

Based on further derivation, the inter-area variance can be expressed as below:

$$\sigma_B^2 = \delta_1(t)\delta_2(t)\left[\mu_1(t) - \mu_2(t)\right]^2$$

Obviously, it is the best discrete state of two areas in case that the inter-area variance reaches the maximum value; then, T is figured out as below:

$$T = Max\left\{\sigma_B^2(t)\right\}$$

It is a way of selecting the threshold value automatically that other parameters need no setting in case that the threshold value is determined by the maximum variance, which is applicable to select both the single threshold value of two areas and multiple threshold values.

5 CONCLUSION

This paper describes a method of automatic road crack detection, intelligence recognition and feature extraction based digital image processing, and researches modern computer image processing technology in road surface application of automatic detection. At the same time it enhances the road image by using a fast correct method above all, then processes segment threshold and extracts linear feature, and the last is the identification of the crack from the original gray image, thereby it can obtain the quite satisfaction experimental result.

ACKNOWLEDGMENT

In this paper, the research was sponsored by the Department of Transportation of Hubei Province.

REFERENCES

[1] Jifeng N, Lei Z,David Z,Chengke W. Pattern Recognition. 2010.
[2] Yong Gen Huang, Howard Tillotson,Martin Snaith. Massively Parallel Computing TechniquesMight Improve Highway Maintenance. IEEE Concurrency. 2002.
[3] Cheng H.D,Miyojim M.Automatic Pavement Distress Detection System. Journal ofInformation Sciences. 2004.
[4] Cheng H.D,Miyojim M.Novel System for Automatic Pavement Distress Detection. Journalof Computing in Civil Engineering. 2004.
[5] KoutsoPoulos.H.N,Downey.A.B.Primitve-based classification of Pavement cracking images. J.Transp.Engrg. 2003.
[6] Cheng Hengda.Automated real-time Pavement distress detection using fuzzy logic and neuralnetwork. Proceedings of SPIE the International Society for Optical Engineering. 2007.
[7] Velisky S A,Kirschke K R.Design considerations for automated pavement crack sellingmachinery. Proc.2nd Int Conf.On Applications of Advanced Technologies inTransp.Engrg. 2002.
[8] Li L,Chan P,Rao A,et al.Flexible Pavement distress evaluation using image ananysis. J.Transp.Engrg. ASCE. 2003.
[9] Cheng H D,Rong Jim.Novel approach to pavement cracking detection based on fuzzy settheory. J.ASCE. 2007.

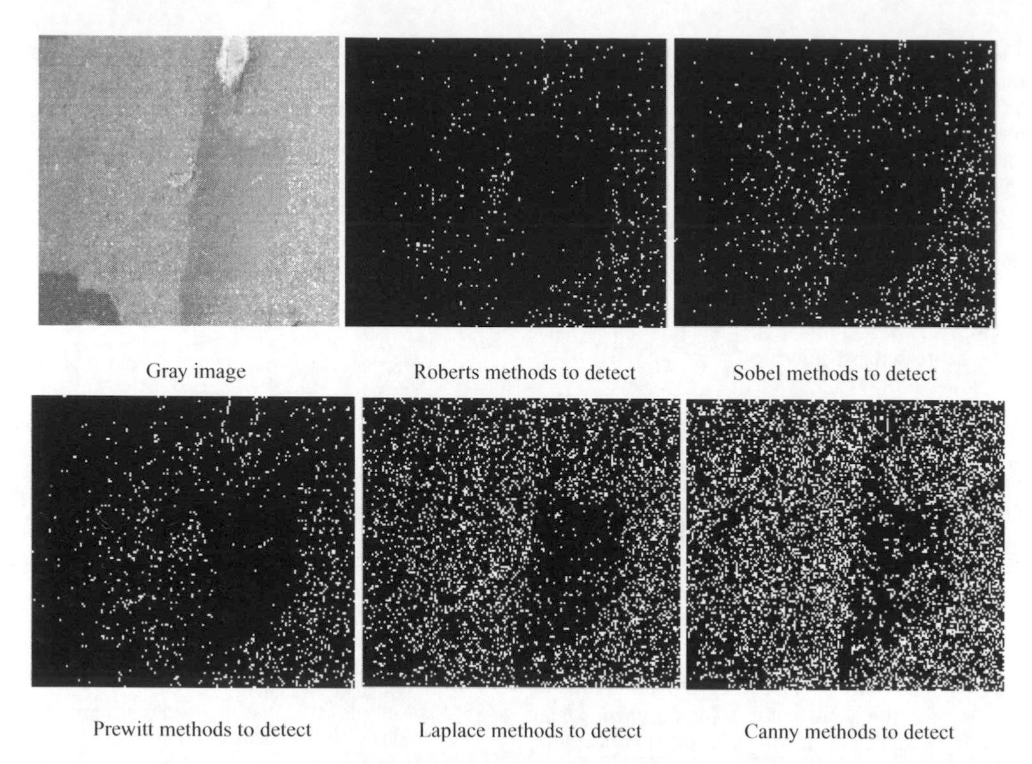

Figure 1. Five classic image edge detection to a pit slot effect comparison results.

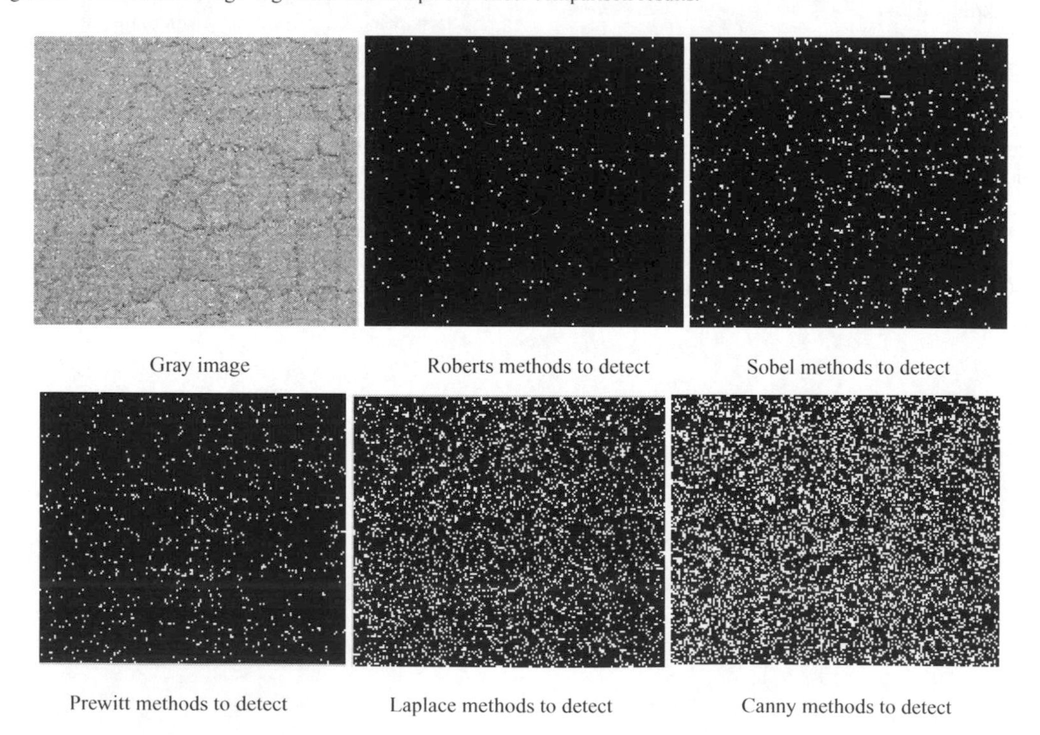

Figure 2. Five classic image edge detection to network to crack effect comparison results.

Computing, Control, Information and Education Engineering – Liu, Sung & Yao (eds)
 ISBN: 978-1-138-02800-5

Research on pavement crack image preprocessing

Z. Gao
Hubei Communications Technical College, Wuhan, China

M.X. Wu
School of Mechanical Engineering, Hubei University of Technology, Wuhan, China

ABSTRACT: Highway pavement crack image will inevitably be disturbed internally and externally in the process of formation, transmission, receiving and processing, resulting in noise to some extent which deteriorates the image quality seriously, fuzzes up the details and makes image processing difficult. It is difficult to detect cracks effectively and accurately through ideal detecting algorithm. This paper improves the pavement crack image processing algorithm from the angle of grey level transformation, median filter and image intensification according to the characteristics of the pavement crack image.

KEYWORDS: Pavement crack; Gray-scale transformation; Image de-noising; Image Enhancement

1 INTRODUCTION

Highway pavement is damaged due to repeated rolling compaction by vehicles so that structural layers are unable to bear the load and various cracks are caused such as fractures, massive cracks, transverse cracks and longitudinal cracks. Crack is the early performance of most pavement crack diseases; so crack is an important index to evaluate the quality of highways [1]. In case cracked roads are not repaired timely, the damage will become increasingly serious on rainy and snowy days and under the action of vehicle load, which can greatly affect the running speed and safety of vehicles. Thus, highway maintenance is of great importance. Presently, manual measurement and analysis are the main methods adopted in our country to detect pavement damages, which can no longer meet the increasingly higher requirements for highway pavement detection [2].

It can be seen through manual detection and analysis of problems existing in pavement cracks recognition that it is extremely urgent to realize automatic detection and recognition of highway pavement cracks by digital picture processing technique because this technique can not only eliminate interference of subjective factors and relieve labor intensity of highway maintainers, but also evaluate pavement behaviors rapidly and accurately. Therefore, this technique has great development prospect, important practical value and significance. Considering characteristics of pavement cracks, this paper studies area segmentation of pavement cracks, feature extraction of pavement and vector machine-supported algorithm, classifies and recognizes the highway pavement crack image, and realize its automatic recognition finally [3-5].

2 DEFECT IMAGE PREPROCESSING

Highway defect recognition mainly consists of image preprocessing, image segmentation, image feature extraction, image edge detection and image recognition shown in Fig.1 below.

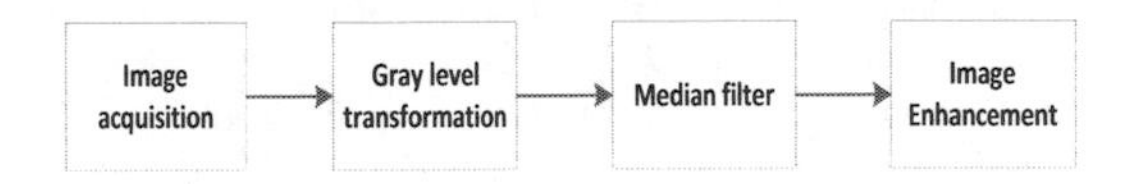

Figure 1. Specific processes of data mining model.

3 GREY LEVEL TRANSFORMATION

The image quality will be worsened in the process of formation, transmission and record due to imperfect imaging system, transmission media and recording equipment, resulting in degraded image, poor visual effect and difficulty in computer processing. Since the image quality is worsened by such factors as imaging system, imaging environment and imaging features, it is hard to show the characteristics by a displayed mathematical expression [6].

Suppose the highway image I(x,y) consists of background of inhomogeneous gray level (illumination) $I_b(x,y)$, pavement crack disease $I_n(x,y)$ and noise caused by stone and pitch, namely

$I(x,y)= I_b(x,y)+ I_n(x,y)+ I_c(x,y)$

Thus, the inhomogeneous gray level can be corrected through finding out the background signal and segmenting the background from the original image. Based on the analysis above, the key is to extract the background image. It is difficult to acquire a flawless background in practical application; if any, the location distance in the timer shaft is large and the illumination distribution is uneven [7].

It is analyzed that the fitted background can be extracted through the bilinear interpolation of the background subset as follows:

1 Acquire the background subset from the original image. Divide the original image into blocks and take a background point from each block. Although illumination is obviously uneven in the image, some parts can be taken as approximately even.
2 Background image acquired through subset interpolation. Since the gray level is changing gradually, 4 adjacent pixels are interpolated through bilinear interpolation for a continuous surface.

4 IMAGE DE-NOISING

Median filter is a kind of nonlinear processing technique which can restrict noises and is easy to use without need for statistical characteristics of the image during actual computation. It is based on an image characteristic; that is, noise often appears as an isolated point corresponding to a few pixel numbers; while the image is made of small blocks with many pixel numbers and a large area.

Although median filter method can eliminate isolated noise points in the image, it damages and fuzzes up edges and details when noise is restricted. So a gradient inverse weighted method is proposed in this model.

The gray level in an area changes less than that between areas in a discrete image and the gradient absolute value on the edge of an area is larger than that between areas [8]. In case the reciprocal of the gradient absolute value between the center pixel and the adjacent points is defined as the weighted value of adjacent points in an nxn window, the weighted value of those inside the area is the maximum; while the weighted value of those near the edge and outside the area is the minimum. Then, the image can be smoothed, and edges and details will not be fuzzed up through averaging the weighted neighborhood partially. In order to control the gray value of the smoothed pixel within the gray unit of the original image, a normalized gradient reciprocal is taken as the weighting coefficient; the implementation algorithm is specific below:

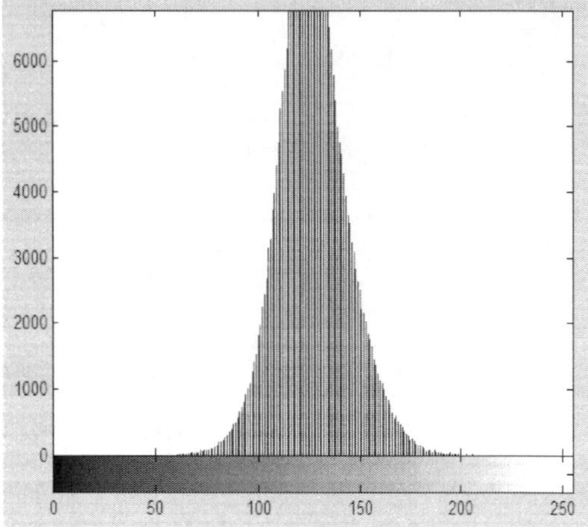

Figure 2. Gray level transformation and its histogram.

Suppose the gray value of point (x,y) is f(x,y), n=3; the gradient reciprocal in 3X3 neighborhood is defined as below:

$$g(x,y,i,j)=\frac{1}{|f(x+i,y+j)-f(x,y)|}$$

In the formula, i,j=-1,0,1, but i and j cannot be 0 simultaneously. In case $f(x+i,y+j)=f(x,y)$ and the gradient is 0 when g(x,y;i,j) value of 8 adjacent points (x,y) is calculated, define the range of g(x,y;i,j) among [0,2] and a normalized weight matrix W as a smooth template.

Thereinto:

$$w = \begin{pmatrix} w(x-1, y-1) & w(x-1, y) & w(x-1, y+1) \\ w(x, y-1) & w(x, y) & w(x, y+1) \\ w(x+1, y-1) & w(x+1, y) & w(x+1, y+1) \end{pmatrix}$$

Provided the central element w(x,y)=0.5 and the sum of other 8 weighted elements is 0.5 so that the total equals to 1. Then,

$$w(x+i, y+j) = \frac{1}{2} \bullet \frac{g(x, y; i, j)}{\sum_i \sum_j g(x, y; i, j)}$$

In the formula, i,j=−1,0,1, but i and j cannot be 0 simultaneously.

Fig. 3 shows the image of pavement defect gray level, the image of Gaussian noise and median filter and the image of weighted average filter of gradient reciprocal.

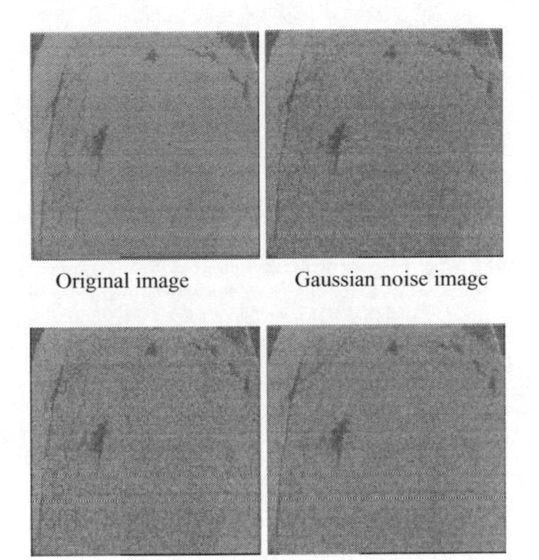

Original image Gaussian noise image

Median filter image Gredient reciprocal weighted average filter image

Figure 3. Image of median filter and gradient reciprocal weighted average filter.

It can be seen from the above figure that the algorithm works on filtering noises; some background noises have been removed, some small noises have been eliminated and loud noises have been weakened to some extent. Besides, marginal information of cracks has been reserved and good effects have been reached to a certain degree, especially cracks and images with quite different backgrounds [9-10].

5 IMAGE INTENSIFICATION

Histogram is the foundation for various spatial domain processing techniques which can be applied to image intensification effectively. Modeling processing of grey level histogram refers to transforming the gray level by the specified gray level distribution; it is applicable to poor contrast and over brightness or over darkness of image, and gray level concentration on the bright and dark side.

Histogram equalization is a kind of effective method; in other words, the gray value of the transformed image is distributed evenly and the overall contrast ratio of the image has been improved. It can be described by a mathematical formula as below:

$$g(x, y) = INT\left(\frac{v(u) - v_{\min}}{1 - v_{\min}} (L-1) \right)$$

In the formula, g(x,y) is the gray value of the transformed image, u is the gray level of the original image, v is the frequency of gray level distribution, V_{min} is the minimum value of the gray level distribution frequency, and INT is the round number.

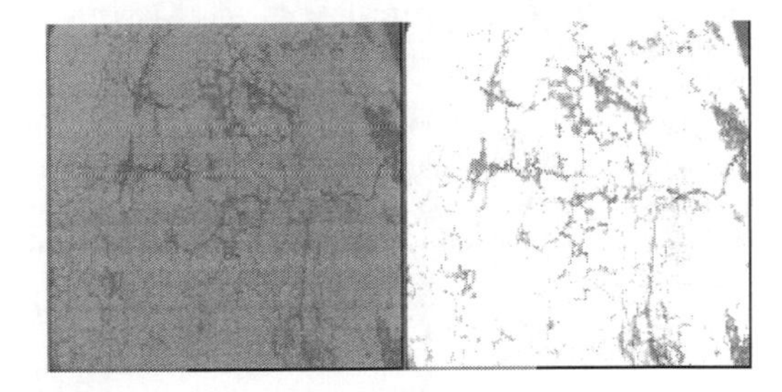

Figure 4. The original image and image enhancement result.

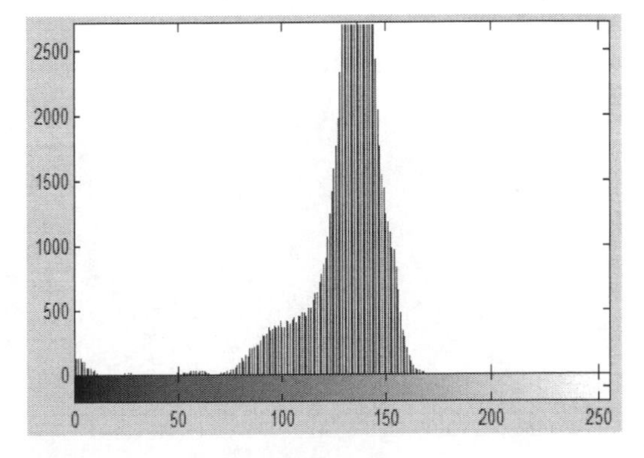

Figure 5. The original image histogram.

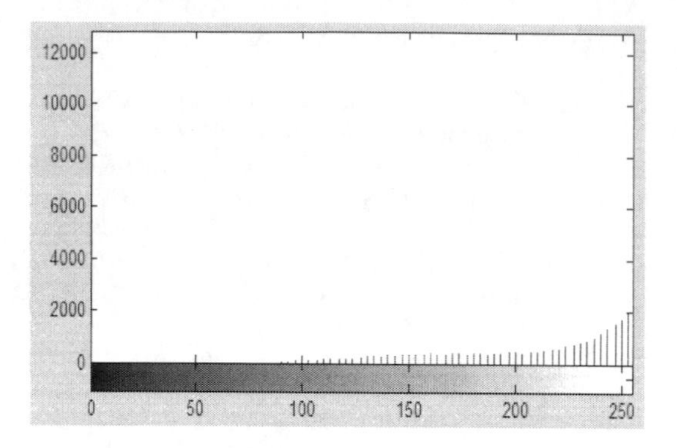

Figure 6. The image after enhancement histogram.

The histogram of the equalized crack image shows that the dynamic range is enlarged, while the "single peak" of the equalized histogram shows unclear features because the edges are exposed unevenly during imaging and the intensified image looks "fuzzy".

6 CONCLUSION

Because of complex crack directivity and loud noises, it is difficult to realize an ideal pavement crack detection algorithm. In order to realize balancing selection between processing effect and process complexity, this paper segments the defective image through edge detection and threshold segmentation now that it is hard and ineffective to process the crack image through current researches and algorithms.

ACKNOWLEDGMENT

In this paper, the research was sponsored by the Department of Transportation of Hubei Province.

REFERENCES

[1] Doyle W. Operations useful for similarity-invariant pattern recognition. Journal of the ACM . 1962.
[2] Otsu N.A threshold selection method from gray-level histogram. IEEE Transactions on Systems Man and Cybernetics . 1979.
[3] Kapur J N,Sahoo P K,Wong A K C.A new method for gray-level picture thresholding using the entropy of the histogram. Computer Vision Graphics and Image Processing . 1985.
[4] Kittler J,Illingworth J.Minimum error thresholding. Pattern Recognition . 1986.
[5] Kroon Aart,Larson Magnus,Moller Iris, et al.Statistical analysis of coastal morphological data sets over seasonal to decadal time scales. Coastal Engineering. 2008.
[6] B. Cadre.Kernel estimation of density level sets. Journal of Multivariate Analysis. 2006.
[7] Siriphan Jitprasithsiri.Development of a new digital pavement image processing algorithm for unified crack index computation. . 1997.
[8] Chou,J. C. O,Neil,W.A.Pavement distress evaluation using fuzzying logic and moment invariants. Transp. Res. Record 1505, Transportation Research Board .
[9] Grivas D A,Bhagvati C,Skolnick M M.Feasibility of automating pavement disease assessmentusing mathematical morphology. Transportation Research . 2004.
[10] Chen T.W,Chen Y.L,Chien S.Y.Fast image segmentation based on K-Meansclustering with histograms in HSV color space. Proc. IEEE International WorkshopMultimedia Signal Process. 2008.

Computing, Control, Information and Education Engineering – Liu, Sung & Yao (eds)

Design and implementation of J2EE Struts architecture based on OA

X. Jia & M. J. Chang
Department of Electronic Information Engineering, Handan Polytechnic College, China

ABSTRACT: The need for government office automation, adopting MVC mode based on Struts, using J2EE platform, analysis and design, and draws a conclusion that the OA system structure, MVC framework and implementation scheme. System in the actual application, to meet the need of high efficiency to deal with daily work, effectively complete the work, has the higher theory and the practical significance.

KEYWORDS: System Structure; MVC; Office Automation System.

1 INTRODUCTION

Along with the information technology, especially the extensive popularization and application of Internet technology, mobile office has become the information to speed up the development of enterprises, improve enterprise efficiency of the office building, an important means to enhance the globalization of business development and international service construction to promote service ability, enterprise management.

The real meaning of mobile information office is not simple to use computers to handle email and other daily office, but through a variety of applications for mobile phone and computer system, so that each of the independent workers can realize the information sharing and cooperative work, remote office, to control the flow of work, enhance the enterprise office efficiency, realize the information sharing and no paper office.

In view of this, this paper mainly studies the design and implementation of the office automation system based on the Struts architecture of J2EE. Office automation system is the realization of office information between enterprise internal departments at all levels and enterprises internal and external collection and processing, call and sharing, and realize scientific decision information system, to increase work efficiency, realize the standardization of office processing, auxiliary leadership decision-making has important significance.

2 TECHNICAL FRAMEWORK

2.1 *MVC model*

MVC English is Model-View-t an application is divided into three layers: model layer, view layer, the control layer. With the rapid increase of network application, MVC model for the development of a Web application is a very advanced design idea, no matter what language you choose, no matter how complicated the application, it can provide the basic analysis method of analysis of model application for you to understand, designed to provide a clear framework for you to construct products, to provide the normative basis for your software engineering.

Using the MVC model can effectively avoid the entanglement of the logical relationship between, so as to simplify the maintenance work. It makes each part of the details are hidden behind the interface, while reducing the code coupling between them. This is the MVC, in the preparation of Java code and maintains HTML and said layer between people, provides a natural boundary.

A successful software needed to have a successful framework, but the establishment of software architecture is a complex process and continuous improvement, successful software developers is not possible for every different item to do different architectures, and always try to reuse the previous architecture, or develop architecture to universal, Struts is one of them, Struts is a scheme of J2EE based architecture of popular, is a relatively good MVC framework, provides the underlying the development of MVC system support.

MVC mode is a kind of the application system is abstracted into Model (model), View (view), Controller (controller) 3 different functional parts, architecture patterns and make them work together. Model is an application object, View is it on the screen of the said Controller user interface definition form of transport people to the user response. Its architecture is shown in fig 1.

The development of this system is using the framework of Struts based on J2EE, the system will use a set of entity bean to represent the data entity, Form

Bean said view layer entity, Model Bean said the data layer entity. The binding Form Bean and JSP pages for presentation to the user data, the user interact with the page, as Form Bean data submitted to the Action Servlet Action Servlet as a Action Controller, the judge, according to the business logic data will be packaged into a Model Bean to the business logic layer, business logic layer is responsible for business processing, completed after treatment if the results need to be returned, then the result is returned as an Model Bean to Action Servlet, Action Servlet Model Bean will be packaged into a Form Bean, and the JSP binding is presented to the user.

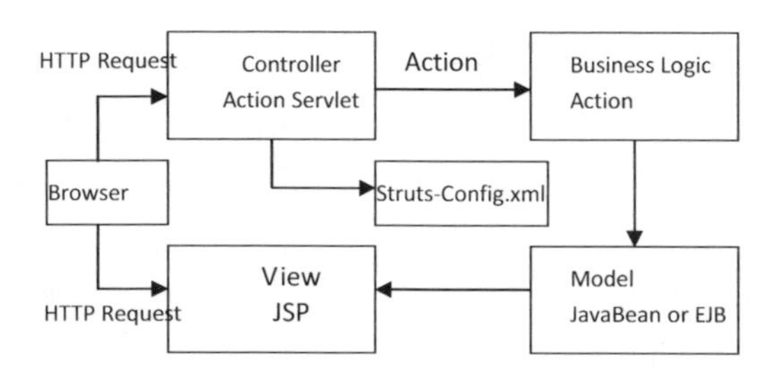

Figure 1. MVC system structure diagram.

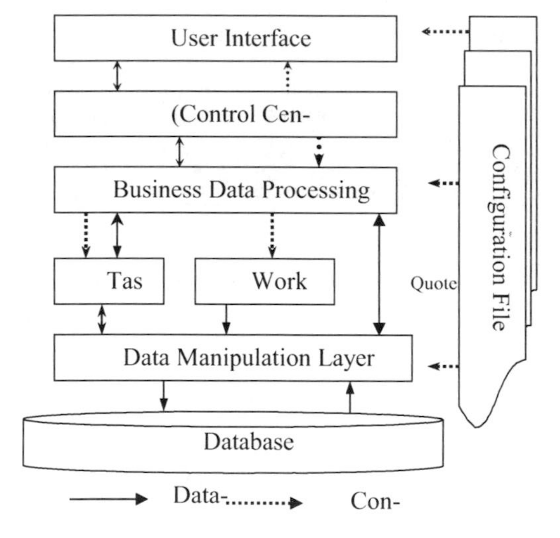

Figure 2. The software system structure diagram.

2.2 *Struts framework*

Struts provide an implementation of the MVC architecture is a highly automated way. Its structure to achieve the MVC, business logic and includes a Servlet controller, a set of JSP pages and applications. The controller user request package, and put them in the other object oriented architecture.

The Struts framework is built around an Action Mapping structure. Controller with Action Mapping to form the HTTP message user requests into application actions. Other information Action Mapping specifies the requested path; plan to deal with the requested object and any service the request. Action Mapping creates a Action object to handle the request. Once the Action object to complete a task, a direct response to a user request it on a JSP page and writes the results to, or it may make an application to other places do respond to flow.

3 SYSTEM STRUCTURE

In the design of the system model, the application layer consists of the user interface; reflect the business interface, business data, configuration data and other functions, this layer is composed of the main work of development. Specific functions are as follows.

4 THE SCHEME OF IMPLEMENTATION

Because Java is an object-oriented, platform independent, network oriented multi thread programming language, so the use of J2EE as a server application development platform is based on a solution of realizing OA system of the Webb. J2EE Struts architecture based on OA, its structure as shown below:

The architecture of the system adopts three layers of the B/W/S model structure, and adopts the interface technology in the underlying data integration unified integration of the underlying data.

By use of three layer B/W/S structure, forming a data layer, business management, business performance level three levels, makes the WEB database access and reduce the burden of the server and improve performance; at the same time as in business management to achieve the business function, make changes to the business of the relevant member only needs to adjust business management, greatly improve the management system; in the aspect of safety system, the three layer B/W/S structure is also compared with the traditional two layer C/S structure has a significant increase, which makes the permission management to control the business function level and data level control is not; in addition, the three layer B/W/S structure is more suitable for the operation in a distributed wide area network environment, you can save the transmission bandwidth more efficiently.

The data interface layer is mainly through between the system and the relevant business system interface and manual data entry, complete raw data acquisition system; business management for data mapping, deformation, summary, analysis, mainly including various business rules and logic rules; business presentation layer is based on customer front-end issued request for the corresponding to the data processing and presentation, is the end user interface, including the function modules of the system, in which all the functions of the authority are controlled by system management function.

The system adopts strict layering design idea, the program processing logic and the processing program of three layer architecture based on Struts separation technology, each logical level provides the basic function of the corresponding module, make the system clear, performance optimization, more scalable and easy to maintain.

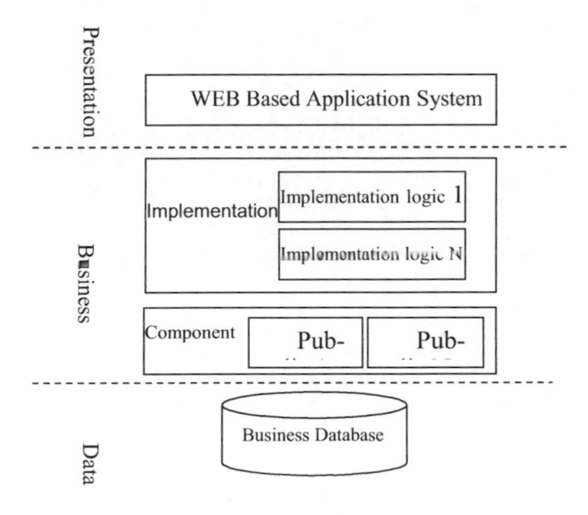

Figure 3. The overall design of B/W/S system.

5 THE FUNCTION STRUCTURE

Information office automation service system mainly includes: system management, integrated office management, report management and other work settings, system interface.

The system includes user management, system maintenance management authority management, basic data and system settings. Integrated office management is mainly responsible for the company documents, Department, company mail notice, the company announcement, meeting management, file management and daily affairs and vehicle management etc.. The main work completed individual settings, set out set and common views set etc.. There is an intuitive statistics for the needs of the user sending and receiving files, a notice of the meeting and other issues, through the way of graphical statistics are mainly to individual user, can only see the sending and receiving text recording statements of their own and for the system administrator can send and receive text record report view of all the people. Also provide other and office application software interface.

6 CONCLUSION

Propose a J2EE based system architecture model of Struts, the realization of the basic framework, the model solution and the functional structure characteristics are presented.

REFERENCES

[1] Xu Renzuo. [M]. software reliability engineering. Beijing: Tsinghua University press (2007).

[2] Li Dong. The management information system theory and its application in [M]. Peking University press, 06(2004).

[3] Chen Ming, Yang Jinsong.PowerBuilder programming database connection method of [J](2006) May micro computing applications.

[4] Bao Yonggang, Wang Degao. The core technology of PowerBuilder8.0 and [M]. Examples of the development of the electronic industry press.06 (2002).

Computing, Control, Information and Education Engineering – Liu, Sung & Yao (eds)
 ISBN: 978-1-138-02800-5

Design of an automatic test system of jiggle switch travel and force

Jing Ping Yuan & Xiao Feng Xue
School of Computer Engineering, Jiangsu University of Technology, Changzhou, China

Fen Fen Yu
Jiangsu University of Technology, Changzhou, China

ABSTRACT: Currently, some small-and-medium sized enterprises adopt manual ways to test the jiggle switch's travel, force and other technical parameters on the assembly line. This leads to problems of labor intensity, low test precision, and likely to get wrong. In order to eliminate these problems, this paper presents a design scheme for an automatic test system of jiggle switch's travel and force based on microprocessors. It analyzes the working principle of the system, and introduces the design of software and hardware of the system in detail. This system can automatically test different types of jiggle switches with high precision, at high speed and effectively reducing the test cost.

KEYWORDS: Jiggle Switch, Automatic Test, Travel and Force, STC Microcontroller.

1 INTRODUCTION

Jiggle switch is a kind of pressure actuation fast switch, also known as sensitive switch. It is the key component of signal transmission and automatic control device, featured with small size and high speed of transforming between closing up and breaking off. Because of the strict parameter requirement of different applications to the inherent action force, release force, motion travel, release travel and movement differential of the jiggle switch, the jiggle switch production enterprises are using different equipment for single detection of jiggle switch's travel and the corresponding action force, which is not only troublesome in operation but also causes low efficiency, low measurement precision and low repetition accuracy. It is the business owners' expectation to improve efficiency, reduce production costs and enhance the competitiveness of enterprises and products by improving technology. Thus we have designed a kind of comprehensive automatic test system which can simultaneously measure all sorts of travel and the corresponding action force of the jiggle switch.

2 WORKING PRINCIPLE OF AUTOMATIC TEST SYSTEM

The automatic test system is an integrated system composed of parameter tests of jiggle switch pre travel (PT), over travel (OT), movement differential (MD), operating force (OF), total travel force (TF) and releasing force (RF), at the same time it can also detect whether there is poor contact on the jiggle switch because of some dirt at the contact point or because of surface concave and convex. It is intended to be used in verification, inspection, selecting jiggle switch for jiggle switch manufacturers and machine manufacturers.

2.1 *Definition of performance parameters of a jiggle switch*

Jiggle switch is a switch contact structure, which has a tiny contact interval and fast mechanism, with specified travel and specified force for the switch action. It has a shell cover and an external driving rod. Jiggle switch usually has three fixed contacts inside: the common contact (COM), the normally closed contacts (NC), and normally open contacts (NO), which are connected to an external circuit. It also has a movable contact, which is connected to the quick-acting mechanism and realizes the opening and closing of electric circuit. Expressions related to jiggle switch action characteristics are shown in Figure 1.

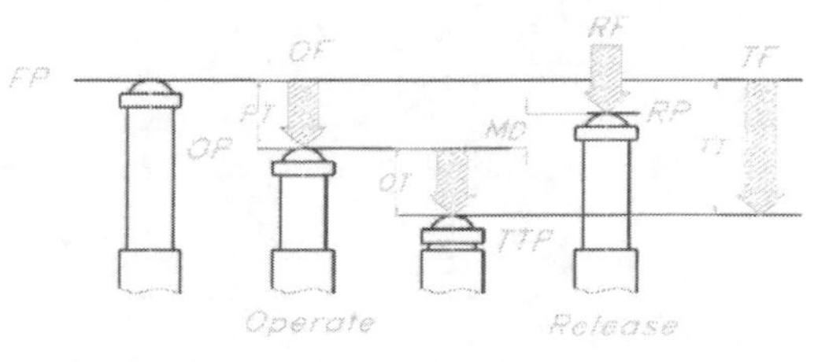

Figure 1. Jiggle switch action characteristics related expressions diagram form.

Position definition of a jiggle switch's driving rod:

1 The position of num any force exerted is called free position (FP);
2 When an external force is exerted to the drive rod, the movable contact just begins to reserve from the free status. At this point, the position of the driving rod is called operating position (OP);
3 The position when the driving rod arrives at the driving-stop position is called total travel position (TTP);
4 When the force on the driving rod is reduced, the movable contact just begins to return from moving state to free status. At this point, the position of the driving rod is called reset position (RP).

Definition of jiggle switch's travel and the corresponding force:

1 The moving distance of the driving rod from the free position to operating position is known as pre travel (PT), the corresponding force exerted onto the driving rod is called operating force (OF);
2 The moving distance of the driving rod from the operating position to the total travel position is called the over travel (OT), the corresponding force exerted onto the driving rod is called total travel force (TF);
3 The moving distance of the driving rod from the operating position to the reset position is called the movement differential (MD), the corresponding force exerted onto the driving rod is called release force (RF) [1].

2.2 *Composition of automatic test system and testing principles of parameters*

Automatic test system of jiggle switch travel and force is composed of the machine workbench and the tester.

Mechanical workbench includes base, clamping components, components of movement, micrometer head, shaft coupling and workbench circuit components, etc., as shown in Figure 2. In the figure, the clamping components are composed of manual clutch, connecting rod, clamping a piece of A and B, and other components such as slide block and slide rod. Clamping piece B is fixed on the base and clamping piece A is connected to the slide block, which can move left or right on the slide rod by operating the clutch. Its function is to clamp the measured jiggle switch during the test , so the shape of the clamping piece A/B must be matched with the shell shape of the tested jiggle switch and form a complete set (different models of jiggle switch usually have different shell shapes); Movement components are connected by a sliding rail, sliding block, mounting plate, push rod and push head; the push rod is exerted force and moves on the slide rail together with mounting plate and sliding block; the connecting mounting plate supports the push rod, causing the push rod to be connected to micrometer head. Horizontal axial pressure from micrometer head is transmitted to jiggle switch's drive rod through the push rod, pressure sensor and push head, making the switch produce corresponding action; Workbench circuit components include 1 ~ 4 probe, pressure sensor, stepper motor, run button and scram button. The pressure sensor is installed between the push head and the push rod, the run button is used for starting the test, when the test process is completed, test data is automatically displayed and a corresponding sound and light alarm will be gave out; if abnormal situation arises during the test, press the scram button, stop the stepping motor and terminate the test.

Components of jiggle switch tester hardware are shown as Figure 3.[2][3] Probe 1 and 2 on the mechanical workbench are respectively connected to the measured jiggle switch's NC and NO, probe 3 is connected to the jiggle switch's COM, and probe 4 connected to the jiggle switch's driving rod. When the stepper motor rotates forward, micrometer head drives push rod and makes horizontal movement to the left, the moving distance from the first contact of the push head and jiggle switch's drive rod to the place where the driving rod reaches the operating position is PT, at this time the driving force on the rod is the OF; moving distance from the drive rod operating position to the total travel position is the OT, at this time the driving force on the rod is the TF; When the stepper motor makes reversal, micrometer head drives push rod and moves horizontally to the right, jiggle switch is reset because of the decrease of the drive force from drive rod, The moving distance from driving rod' operating position to the reset position is MD, at this time the driving force on the rod is release force (RF).

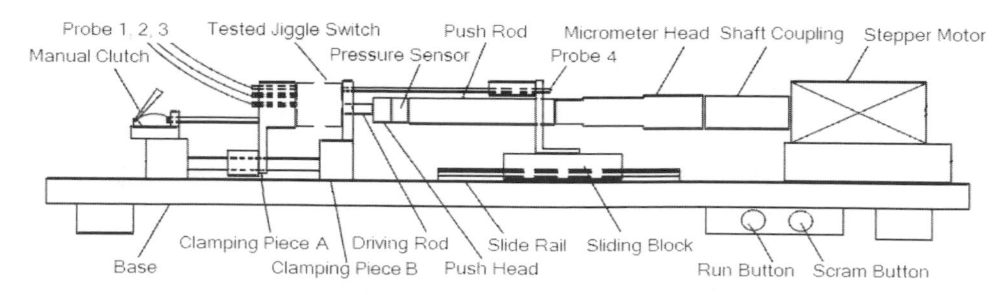

Figure 2. Mechanical workbench schematic diagram.

In Figure 3, two-phase stepper motor is adopted, its step Angle is 1.8 °. It takes 200 steps for the stepper motor to rotate a circle. Through the coupler, the movable casing of the micrometer synchronously rotates a circle. Micrometer screw level drives the rod axial and makes a movement of 0.5 mm. Therefore, for each step the stepper motor moves, the push rod moves horizontally a distance of 0.0025 mm, so jiggle switch's travel parameters can be calculated according to the steps of the stepper motor; Through the pressure sensor, signal conditioning amplifier circuit and A/D converter [4] [5], it is possible to measure the level axial pressure exerted onto the driving rod by the push rod. In Figure 3, filter and comparator are used for hardware filtering unstable and vibration signal caused by contact switch. On the front panel of the jiggle switch tester there are eight-digital tubes, and they form the two groups of display on left and right, which are used for real-time displaying drive rod's travel (unit is mm) and the corresponding action force (unit is N). Before the test, according to the measured jiggle switch model, each travel and force allowance range need to be preset through the button group; At the end of the test, compare the measured values to the set values, whether the parameters of the measured jiggle switch is qualified or not can be indicated by the brightness of each LED lamp, if all the parameters are up to standard, the buzzer will give out warning sound.

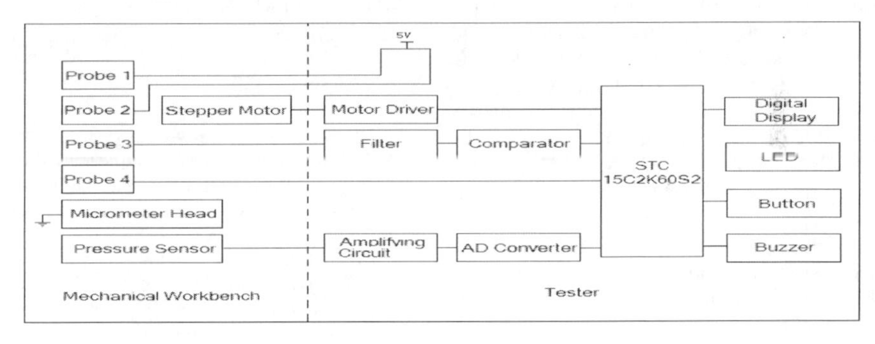

Figure 3. Jiggle switch tester hardware block diagram.

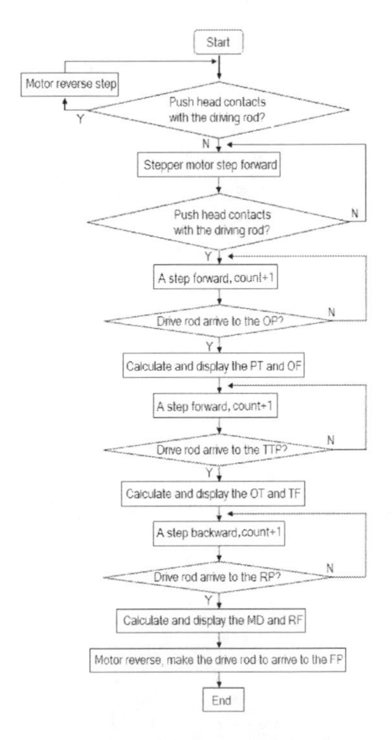

Figure 4. Flow chart of the test module.

Automatic test system of jiggle switch travel and force adopts high-speed microcontroller STC15F2K60S2, a single clock/machine cycle, its instruction code is fully compatible with the traditional 8051, but it's 7-12 times faster. With 60K user program space, integrating 2048 bytes SRAM on-chip and with EEPOM function, it can easily save the value of the test system settings in case of power-fail. The microcontroller has five external interrupt sources, in which INT1/INT2 programmable for interrupt on rising/falling edge, INT2/INT3/INT4 for interrupt of falling edge, so they can well meet the real-time testing requirements of single/double-type jiggle switch [6].

3 SOFTWARE DESIGN OF AUTOMATIC TEST SYSTEM

The following is the whole test process of the automatic test system on jiggle switch travel and force: firstly, preset each parameter value of measured jiggle switch with the aid of the button group on the tester's front panel, and then place the jiggle switch at a regular position on the mechanical workbench, press the run button on the workbench, the stepper motor rotates forward and backward under the control of program, driving the micrometer screw to do horizontal axis movement forward and backward and completing the automatic testing of various parameters of the jiggle switch. In the process of test, a set of digital tubes on the instrument panel's left side display in turn PT, OT, MD; a set of digital tubes on the right side display in turn the travel-corresponding operating force, total travel force and release force. If any unqualified item arises during the test, the corresponding red light is on, if all the parameters are qualified, the green light is on with "beep" sound. If there is poor contact on the jiggle switch because of some dirt at the contact point or because of surface concave and convex, "flashing" red light is lit up. Online workers can fine-tune and repair the unqualified jiggle switch based on the state of the indicator light.

Software function of automated test system includes the main program, parameter setting module, test module, keys and display module. The main program is mainly for completing hardware initialization of the test instrument, reading the allowed parameter value of the tested jiggle switch from the microcontroller's built-in EEPROM, and detecting the presence of key press event. If there is "Settings" button pressed, then it calls the parameters setting module to complete the reset of parameters allowance value and that should be saved into EEPROM, so that they can be read directly when needed; if there is "Run" button, then it calls the data acquisition and processing module and start a complete test process. Test module flow chart is shown in Figure 4.

4 CONCLUSION

Automatic test system of jiggle switch's travel and force is an integrated test system. It makes various parameters of the jiggle switch be automatically tested at one time with high precision, thus improving labor productivity. At the end of a test, online workers can know whether the switch is qualified only by listening to the sound, thus relieving the workers of their labor intensity. For instance, when the test comes to an end, if no qualification-indicating sound is heard, online workers can base on the red light to adjust and repair the jiggle switch. Automatic test system is simple and convenient to operate, and no special training is necessary, so it enjoys good prospects for promotion and application.

REFERENCES

[1] Information on http://www.giemson.com

[2] Wang, Furui. 1999. Monolithic Microcomputer Control System Design Encyclopedia. Beijing: Beijing University of Aeronautics and Astronautics Press.

[3] Yan, Tianfeng. 2005. SCM application system design and simulation debugging. Beijing: Beijing University of Aeronautics and Astronautics Press.

[4] Wang, Yudong. 2008. 400 Cases of sensor application circuit. Beijing: China Power Press.

[5] Information on http://www.analog.com

[6] Information on http://www.stcmcu.com

Computing, Control, Information and Education Engineering – Liu, Sung & Yao (eds)
© 2015 Taylor & Francis Group, London, ISBN: 978-1-138-02800-5

Application of mind mapping to improve the teaching effect of Java program design course

C. L. Li
Research Institute of Higher Education, Changchun University, Changchun, China

L. P. Yang & W. Wang
Computer Science and Technical Institute, Changchun University, Changchun, China

ABSTRACT: Java is an abstract, knowledge points, theoretical and practical course. In order to improve the teaching effect of Java program design course, mobilize the enthusiasm of the students, bring the colorful and high organized mind mapping into the teaching practice of Java program design course will improve the teaching means and strategy based on mind mapping, has obtained the good teaching effect, not only to improve the teaching quality of Java program design course, it also helps students to establish a good cognitive structure, effectively cultivate students' autonomous learning ability and thinking ability, stimulate the students' learning interest.

KEYWORDS: Java program design; Mind mapping; Teaching effect.

1 INTRODUCTION

Java program design is a professional required course of computer specialty in universities, is the necessary courses of learning and application for the Java software engineer. This course is not only to master the basic concepts and the main core technology, especially the need to master the basic concepts of object oriented programming, but also has strong practical, need to train the Java object oriented programming practical skills through the practice. Therefore, it is an object-oriented program design course, students feel difficult to learn it, poor teaching effect. Although colleges and universities have explored various reform measures, put forward many useful teaching models and methods, but in actual teaching, there are many factors leading to the teaching effect is not ideal [1], to solve these problems effectively, it is necessary to pay attention to teachers "teach" to the students' "learn", the two mutually with the essence and learned to do. "Teach" to clear thinking, the method is proper, multiple intelligence and use, to broaden the students thinking, stimulate students' reflection, arouse students' potential. "Learn" to guide and correct method of interest drive, a variety of sensory and use, not only to learn the details of knowledge, grasp the overall curriculum structure, more important to do what the knowledge. I hope that through the introduction of the mind mapping to achieve the effect of teaching and learning.

2 MIND MAPPING OVERVIEW

2.1 *Mind mapping meaning*

Mind mapping is by a British "memory" the father, the famous psychologist education expert Tony Buzan in the 70's according to the brain working principle, has been proved to be a simple revolutionary thinking tool [2]. The mind mapping is the advanced way based on the brain, is the expression of radioactive thinking, is a very useful graphical technique, is to open the brain potential universal key. Mind mapping can easily to make a long list of boring information into color, memory and nonlinear highly organized graphs.

The mind mapping is like the brain map, complete will appear thinking, ideas, content and structure of knowledge expression in the visual image of the way, effective presentation of the thinking process and the relation between knowledge. Its core idea is that the image thinking and logical thinking together, let the left and right brain people operating at the same time in the course of thinking, will eventually structure thinking traces with pictures and lines forming divergent on paper, presents an easy to remember conform to the brain of the divergent thinking of natural expression.

2.2 *Making mind mapping*

The mind mapping follows a set of simple, basic, natural, easy to be brain accept rules, the use of some radioactive lines and graphics to simulate the logical thinking during the brain thinking, and with a different color and special symbols, graphics to mark on the content of the different attention degree. Tony Buzan believes that the mind mapping must have the following four basic characteristics:

The focus of attention clearly concentrated in the central figure

The main theme as a branch from the central figure to the surrounding radiation

The branch is composed of a key graphic or the keywords lines above, not important words in branch form, attached to the branches of high level.

The branches form a connection node structure.

Drawing mind mapping, is to establish a central issue, and then start thinking of the center line to launch around, through the connection and combination of each line is established on the node, and in accordance with the guidance of thinking are connected in sequence, the end branches summarize end node content, gradually construct the hierarchical structure of knowledge [3].

2.3 *The application advantage of mind mapping*

The human brain to remember visual things, easy to get lost in the complex information. The mind mapping can make complicated information to draw on a piece of paper, make complex problems become very simple, it can help us find ideas, links, see clearly the essence, quickly find the key from the complex information. Ideas guiding map adaptation left brain mode of thinking with nature, all kinds of views clearly reflected in the chart, at the same time, the central theme, colors and graphics combined use has been stimulated vision organ, achieved the visual learning, strengthened and deepened memory.

3 APPLICATION OF MIND MAPPING TO "TEACHING"

3.1 *Using mind mapping to prepare lessons*

Scientific and effective method of preparing lessons, not only to the teaching material content are refined, organization, but also to combine the teaching material content and the latest research theory and methods, to guide students to learn actively[4]. Using mind mapping to prepare lessons, it is easy to grasp the framework and structure of knowledge system. It can take many complex content to clear out the form, the relationship between things clear, can greatly shorten the length of the plan. The teaching arrangement will become handy.

Mind mapping is adopted in the teaching design, where necessary, using mind mapping to outline the whole book and each chapter in the book.

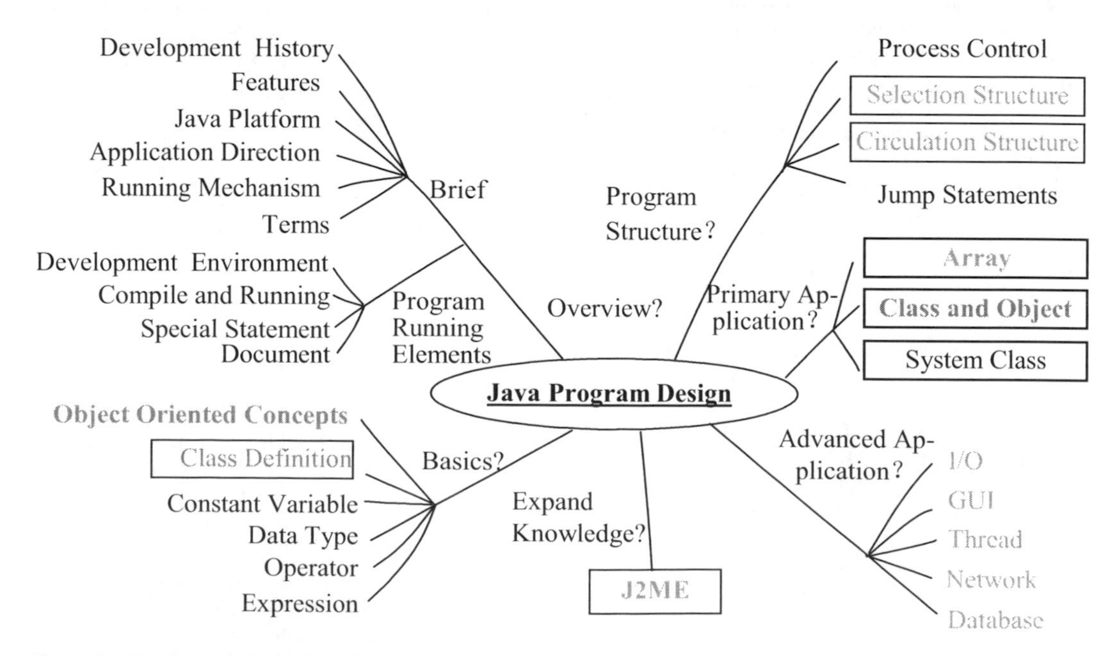

Figure 1. Pre class mind mapping of "Java program design".

By showing the mind mapping of Figure 1, on the first class of Java program design make the students on the content and objectives of this course are introduced, so that students have a general understanding of the course, clear learning goals and tasks. After each class is put on this map content in expansions. So the organization of teaching, teaching of clear thinking, and it is easy to add new content.

3.2 *Using mind mapping to teach*

"Hierarchy of mind mapping", urging the teachers to fully grasp the classroom context, is to find the breakthrough point to communicate with the students. Its unique branch structure is conducive to the students to get to the bottom of the problems. So teachers can naturally explore the content to the depth introduction, mine the individual potential of students. The relationship between teachers and students is no longer confined to the simple teaching and learning, but students think close to the problems under the teacher's guidance [5]. Mind mapping example of "Class and Object" chapter is shown in Figure 2.

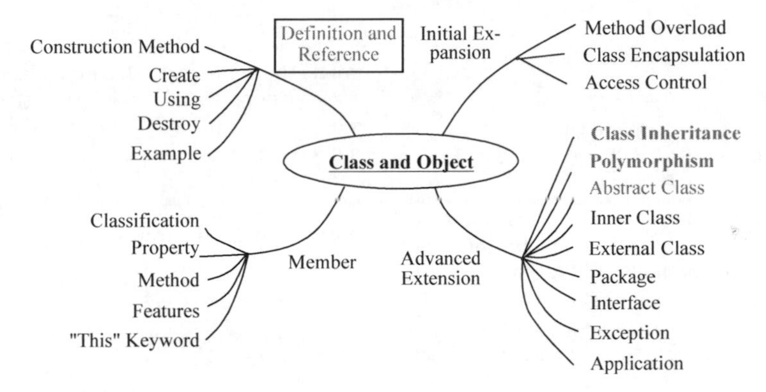

Figure 2. Mind mapping of "Class and Object" chapter.

3.3 *Using mind mapping to summarize*

The traditional teacher use blackboard to write summary, simply for the students to the classroom knowledge point. Most of the students cannot keep up the teachers' ideas, resulting in the effect of lowering. The mind mapping of this excellent memory roadmap, it is the way organized fact and thinking together, that consistent with the brain's natural working way, which means that the brain can more easily remember, after also can more easily remember. Teachers use the "mind mapping" review, is not only a simple review the teaching of knowledge, also the intrinsic link between the sort of knowledge points, in a short period of time to allow students with the teachers review "a lesson". The mind mapping example of "Class and Object" chapter summarizes is shown in Figure 3.

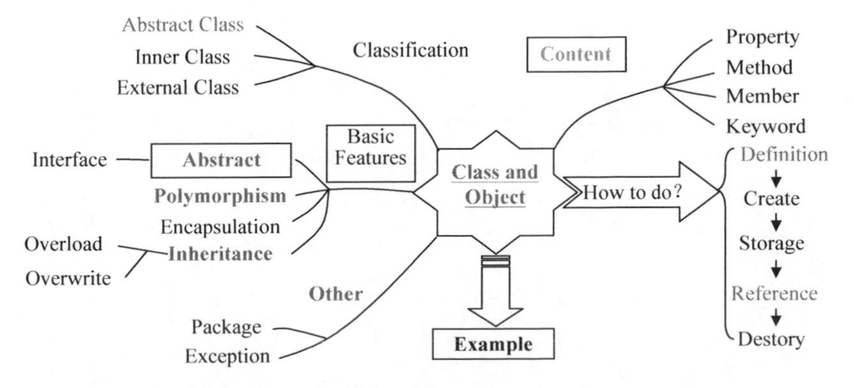

Figure 3. Mind mapping of "Class and Object" chapter summary.

4 THE ADVANTAGE OF USING MIND MAPPING IN JAVA PROGRAM DESIGN TEACHING

4.1 *Mind mapping as a teaching and learning tool, help teachers organize teaching contents*

The mind mapping is to support learning and teaching tools. The mind mapping can put linear dull information into the color easy memory, the highly organized map and highly generalization of keywords, symbols, codes, and use it for teaching design more convenient and flexible, more elastic, easy to modify and perfect. The master mind mapping tool, not only can improve the efficiency of teachers preparing, reduce workload in computer teaching design, improve the process of teaching, also can reach good teaching effect.

4.2 *Mind mapping as a teaching skills and learning, promote the development of teachers and students' own ability*

Divergence features of mind mapping can stimulate the students' imagination and creativity, can make teachers more divergent thinking. Students and teachers' construction process of mind map, is also the process that constantly understanding, digestion and absorption of knowledge. Between teachers and students learn from each other's mind mapping, could change the perspective of thinking after comparative, optimize their thinking mode, can make their own knowledge system more perfect, to think more comprehensively and more active thinking.

4.3 *Mind mapping as a teaching and learning strategy, embodies the "student-centered" teaching philosophy*

Constructivism emphasizes the main role of the learner, emphasis on learning initiative, sociality and situation, advocate learner centered learning under the guidance of a teacher. Students are the main body of information processing, are the active constructors of meaning, rather than passive recipients of external stimuli and was taught the object. In the teaching process based on mind mapping, the whole process from preview to review are to actively participate in the students, students are the main body, teachers act as guides, organizers and facilitators of student learning. Students have completed self construction of knowledge using the mind mapping under teachers' instruction and guidance, fully embodies the "student-centered" teaching philosophy, to maximize the student's main body status.

5 CONCLUSION

The teaching mode of using mind mapping for Java program design courses not only helps students to establish a good cognitive structure, but also can effectively cultivate students' autonomous learning ability and thinking ability, to stimulate their interest in learning, the relationship between the knowledge points to deeply understand the inner factors and the construction of a knowledge system. Using mind mapping aided teaching in Java programming teaching has injected new ideas, to provide new tools and strategies, is a kind of effective and positive way of teaching, not only can promote the teachers' teaching, it can promote the learning of students, is worthy of promotion.

ACKNOWLEDGEMENTS

This work was financially supported by the key project of Jilin Province Education Science "Twelfth Five Year " Plan (ZD14077).

REFERENCES

[1] K.P.Zhang & J.J.Mao. 2011. Thinking on the teaching reform of 'Java language program design' course. *Journal of Chifeng University: Science &Education* 3(10):213–214.

[2] Tony Buzan, translated by D.K.Zhang & K.R.Xu. 2005. Mind map – the brain Manual. *Beijing: foreign language teaching and Research Press.*

[3] Tony Buzan & Barry Buzan. 1994. The Mind Map Book—Radiant Thinking. *EP Dutton.*

[4] L.Wang. 2013. Improving Teaching Effect of Software Engineering by Mind Mapping. *Higher Education Forum* 3:73–75.

[5] C.M.Yu. 2014. Research of classroom teaching mode based on mind map. *Computer Era* 1 :66–68.

Computing, Control, Information and Education Engineering – Liu, Sung & Yao (eds)
© 2015 Taylor & Francis Group, London, ISBN: 978-1-138-02800-5

Computer simulation and numerical calculation of unbalanced rotor in dust blower

Peng Zeng
School of Mathematics and Computer Science, Jianghan University, Wuhan, China

Li Zhao
Dongfeng Honda Automobile Co. Ltd., Wuhan, China

ABSTRACT: Industrial dust blower is the key equipment for gas and dust discharge in industrial enterprises. This paper established the dynamics model of industrial dust blower rotor system by using the finite element method and modal analysis theory. The factors causing the mildly unbalanced rotor were analyzed, and the mathematical model of the rotor system under complex conditions was presented. Based on it, the computer simulation model of mildly unbalanced rotor system was established, which provides the reliable basis for sensor arrangement and rotor structure optimization.

KEYWORDS: Numerical Calculation; Computer Simulation; Unbalanced Rotor; System Modeling.

1 INTRODUCTION

Industrial gas and dust discharge is one of the most important causes for air pollution. In order to deal with the industrial gas and dust, industrial dust blowers are widely used for fluid transport in the fields of chemistry, metallurgy, petroleum, mining and other industry. However, due to the complex working conditions of high temperature, high pressure, high dust, and long-term high load operations, industrial dust blower is easy to failure. In all kinds of faults of industrial dust blower, rotor unbalance is one of the most common faults. According to statistics, the failure number of rotor unbalance occupies more than half of the total faults. If there is no timely and accurate fault detection method, it may lead to abnormal production, or even cause serious air pollution when the rotor unbalance problem arises. Therefore, it is important and valuable to study rotor unbalance fault of industrial dust blower under the complex conditions deeply.

In view of the difficulties that the early vibration characteristics of the unbalanced rotor are not obvious, many new measuring sensors and new measuring approaches are used [1]. Some data fusion algorithms of rotors unbalance fault diagnosis were proposed considering the uncertainty in sensor arrangement and sensor measurement. Zhou Linren et. al. presented simulation models in finite element analysis to study the parameters selection and parameters adjustment of rotor structure[2]. To analyze the influence factors under complex conditions such as high pressure and high temperature, Mersinligil Mehmet et. al. researched the fast response phenomenon in multistage axial compressor and blower impeller [3]. Khlaief Amor et. al. estimated the unbalanced initial rotor position by using numerical analysis, and established adaptive model of rigid rotor [4]. Considering the variable speed of rotor working requirement, Key Nicole L. et. al. researched the nonlinear problems of magnetic bearings-rotor system, and proposed the method of detecting the faults from rotor rotating signals [5]-[6]. From the searched information, current researches put emphasis on the structure design, later faults diagnosis and data fusion control of rotor, and few papers studied the relationship between modal shapes and vibration characteristics of early unbalanced rotor in industrial dust blower.

2 SYSTEM MODELING

Because it is difficult to solve the sensors monitoring problem of unbalanced rotor of industrial dust blower under complex conditions with conventional measurement approach, in this paper, we propose the method of combining modal analysis and finite element simulation, which is used for nonlinear phenomena analysis of industrial dust blower rotor in variable speed mode. Through the vibration response analysis of industrial dust blower rotor modals, the maximum vibration occurrence of the rotor system is determined, which make guidance to the adjustment of sensor arrangement and working parameters of industrial dust blower under complex conditions.

During the working process of industrial dust blower, due to the interaction of large amount of dust in high speed fan impeller, which causes local wear or fouling, the rotor system of industrial dust blower is always unbalanced, and the rotor unbalanced position is always changed. With the accumulation of time, this kind of unbalance may be increasing, eventually leading to the system fault. Therefore, detecting the early unbalance of rotor as soon as possible can effectively carry out device maintenance timely, reduce noise and vibration, and prolong equipment life. To study the relationship between the mildly unbalanced rotor variation and the vibration mode, the rotor system is simplified as a mechanical model of rotor-bearing-foundation, as shown in Figure 1.

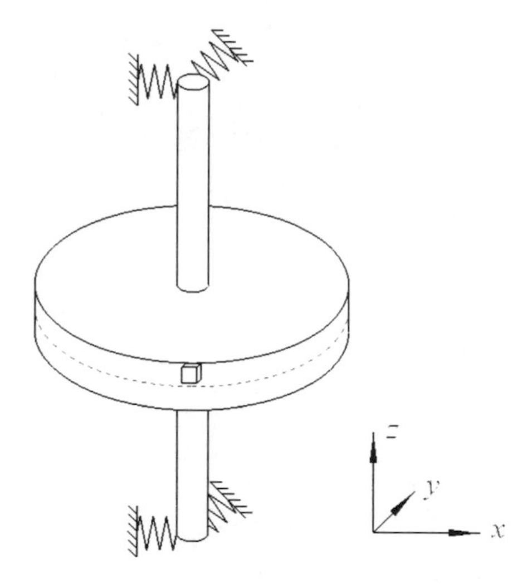

Figure 1. System frame.

Define u_x and u_y as the axial displacements, θ_x and θ_y as the rotor angles around the X axis and the Y axis, m_r as the total rotor mass, m_u as the rotor unbalanced mass, J_d as the rotor inertia around the rotating shaft, J_p as the rotor transverse inertia, z_u as the unbalance axial position, r_u as the unbalance radial position, Ω as the rotor rotating speed, and k as bearing stiffness, the dynamics model can be expressed as:

$$\begin{bmatrix} m_r & & & \\ & m_r & & \\ & & J_d & \\ & & & J_d \end{bmatrix}\begin{bmatrix} \ddot{u}_x \\ \ddot{u}_y \\ \ddot{\theta}_x \\ \ddot{\theta}_y \end{bmatrix} + \begin{bmatrix} & & & \\ & & & \\ & & & J_p\Omega \\ & & -J_p\Omega & \end{bmatrix}\begin{bmatrix} \dot{u}_x \\ \dot{u}_y \\ \dot{\theta}_x \\ \dot{\theta}_y \end{bmatrix}$$

$$+\begin{bmatrix} 2k & & & \\ & 2k & & \\ & & kl & \\ & & & kl \end{bmatrix}\begin{bmatrix} u_x \\ u_y \\ \theta_x \\ \theta_y \end{bmatrix} = \begin{bmatrix} F_x \\ F_y \\ M_x \\ M_y \end{bmatrix} \tag{1}$$

$$\begin{bmatrix} F_x \\ F_y \end{bmatrix} = \begin{bmatrix} m_u r_u \left(\Omega^2 \cos\theta + \dot{\Omega}\sin\theta\right) \\ m_u r_u \left(\Omega^2 \sin\theta - \dot{\Omega}\cos\theta\right) \end{bmatrix} \tag{2}$$

$$\begin{bmatrix} M_x \\ M_y \end{bmatrix} = z_u \begin{bmatrix} 0 & -1 \\ 1 & 0 \end{bmatrix}\begin{bmatrix} F_x \\ F_y \end{bmatrix} \tag{3}$$

In order to simulate the mildly unbalanced rotor, considering various damping, we used the MSC Nastran finite element software to solve the rotor dynamics module. From Eqs. (1)-(3), the Nastran rotor dynamics module is described as follow:

$$\left(B_s + \left(\frac{g}{W3}\right)K_s + \left(\frac{1}{W4}\right)K4_s + B_r + \left(\frac{g_r}{WR3}\right)K_r + \left(\frac{1}{WR4}\right)K4_r + \Omega B^G \right)$$
$$\dot{u}(t) + M\ddot{u}(t) + \left(K_s + K_r + \Omega\left(K^{Cv} + \left(\frac{g_r}{WR3}\right)K^{Cgr} + \left(\frac{1}{WR4}\right)K^{Cge} \right)\right)u(t)$$
$$= F(t) \tag{4}$$

Where M is the general mass matrix, B_s is thesupporting viscous damping matrix, $\left(\frac{g}{W3}\right)K_s$ is the equivalent structural damping supporting viscous damping, $\left(\frac{1}{W4}\right)K4_s$ is the equivalent material damping supporting viscous damping, B_r is the rotor viscous damping matrix, $\left(\frac{g_r}{WR3}\right)K_r$ is the equivalent structural damping of rotor viscous damping, $\left(\frac{1}{WR4}\right)K4_r$ is the equivalent material damping of rotor viscous damping, B^G is the gyroscopic force matrix, K_s is the rotor stiffness matrix, $K4_s$ is the supporting material damping matrix, $K4_r$ is the rotor material damping matrix, Ω is the rotor rotating speed, K^{Cv} is the cyclic matrix lead by B_r, $g_r K_r^{Cgr}$ is the cyclic matrix lead by $g_r K_r$, and K_r^{Cgr} is the cyclic matrix lead by $K4_r$.

3 COMPUTER SIMULATION AND ANALYSIS

According to Eqs. (1)-(4), the finite element rotor dynamics simulation model is established by NASTRAN software to analyze the rotor modal shapes under normal conditions and abnormal conditions of mildly unbalanced rotor, as shown in Figure 2 and Figure 3. From the figures, it indicates that the rotor modal distribution range of mode shapes and natural frequencies under the rotating frequency excitation can be determined, and the location of maximum vibration occurred can be measured. In order to simulate vibration characteristics of mildly unbalance rotor, the local thickness of different blades in the finite element model is adjusted. Assume the overall wear quantity of rotor blades is 100kg and the wear distribution is random, then it can be found that the blades local stiffness are reduced and local vibration frequency is decreased because of the blades wear. However, the rigidity of the shaft is not changed, and the overall vibration frequency is increased slightly because the overall quality of rotor is decreased. When the rotor is mildly unbalanced, compared to the normal rotor, the vibration frequency is declined slightly. Under the condition that rotor unbalance force is constant, increasing the stiffness of the rotor bearing load and non load side may decrease rotor vibration. At the same time, the vibration frequency characteristics of the unbalanced rotor system are nonlinearly changed with system rotating speed.

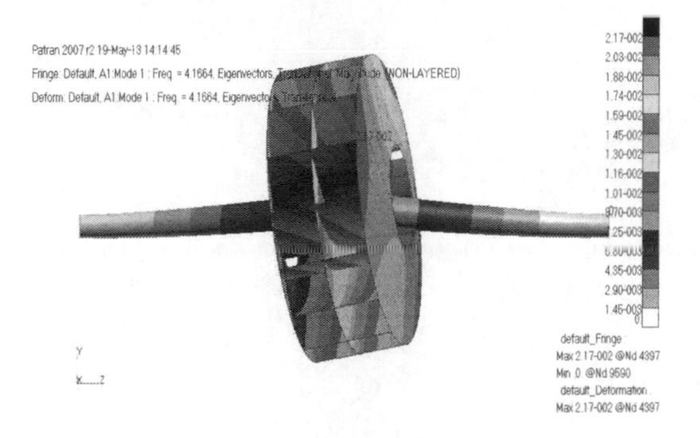

Figure 2. Under normal conditions.

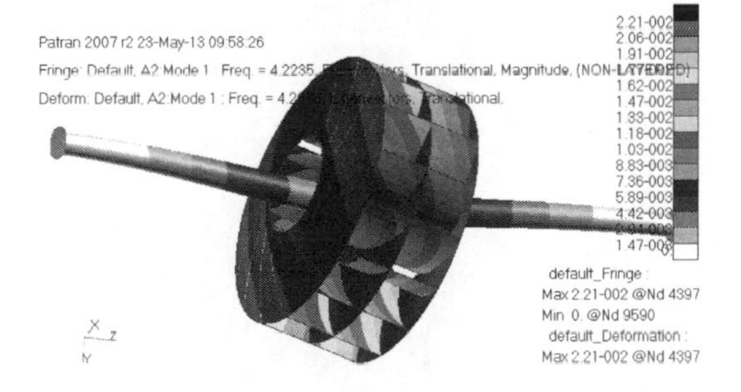

Figure 3. Under abnormal conditions.

4 CONCLUSIONS

The unbalanced fault of industrial dust blower rotor is a difficult problem for industrial equipment repair and management. In this paper, we proposed a finite element simulation method based on modal analysis, established the rotor dynamics model of industrial dust blower under complicated conditions, and revealed the nonlinear relationship among rotor vibration, modal shape and rotor speed. The research results can be used not only for vibration monitoring and faults diagnosis, but also can provide technical support for sensor arrangement and optimization design of industrial dust blower, which is greatly helpful for reducing energy consumption and improving equipment reliability.

ACKNOWLEDGEMENTS

This work is supported by the Industry-university-research project of Wuhan Education Bureau (Granted No: cxy02) and the NSF of Hubei Province (Granted No: 2011CDB172), China. The authors also gratefully acknowledge the helpful comments and suggestions of the reviewers, which have improved the presentation.

REFERENCES

[1] Yi Jiangang. 2015. Modelling and Analysis of Step Response Test for Hydraulic Automatic Gauge Control. *Strojniški vestnik - Journal of Mechanical Engineering*, 61(2): 115–122.

[2] Zhou Linren, Ou Jinping. 2010. Study of parameters selection in finite element model updating based on parameter correction. *ICIC Express Letters*, 4(5): 1831–1837.

[3] Mersinligil Mehmet, Brouckaert Jean-Franois, Courtiade Nicolas, Ottavy Xavier. 2012. A high temperature high bandwidth fast response total pressure probe for measurements in a multistage axial compressor. *Journal of Engineering for Gas Turbines and Power*, 134(6): 117–131.

[4] Khlaief Amor, Boussak Mohamed, Gossa Moncef. 2013. Model reference adaptive system based adaptive speed estimation for sensorless vector control with initial rotor position estimation for interior permanent magnet synchronous motor drive. *Electric Power Components and Systems*, 41(1): 47–74.

[5] Key Nicole L. 2014. Compressor vane clocking effects on embedded rotor performance. *Journal of Propulsion and Power*, 30(1): 246–248.

[6] Jin Fang, Li Qing-Quan, Hao Yong-Ping. Quantized control for parameter uncertain systems with limited information. 2011. *ICIC Express Letters*, 5(3): 4337–4344.

Computing, Control, Information and Education Engineering – Liu, Sung & Yao (eds)
© 2015 Taylor & Francis Group, London, ISBN: 978-1-138-02800-5

Software development of molten iron weight measurement control system

Xiao Yan Zhu & Peng Zeng
School of Mathematics and Computer Science, Jianghan University, Wuhan, China

Li Zhao
Dongfeng Honda Automobile Co. Ltd., Wuhan, China

ABSTRACT: The control of molten iron weight measurement is an important work of production processes in the iron and steel enterprises. This paper discusses and establishes the weight control system for molten iron measurement. Through the construction of communication network based on field bus, the key parameters of molten iron tank and the bottom working parameters of the control system are accurately monitored. With the designed software system, the precise measurement of molten iron weight and the accurate discovery of the hidden equipment troubles is achieved, which improve the management level of enterprise modernization of equipment, and prolong the service life of the control system.

KEYWORDS: Molten Iron; Monitoring System; Automatic Measurement; Industrial Ethernet.

1 INTRODUCTION

In the iron and steel enterprises, the control of molten iron weight measurement is always a difficult problem [1]. First of all, during the molten iron weight measurement process, the working parameters of iron melting number, weighing tank number, gross weight and other important data of tin car are manual copy. The heavy repetition work inevitably brings to mistakes. Secondly, because of high temperature environment where the molten iron tank works, the faults of the control system often occurs [2]. If there is no effective monitoring and control system, once an accident happens, it will influence normal production and even lead to security incidents. Finally, according to the requirements of production, it is better that there are no operators in the working field, which requires the field devices and the control room to be connected into a whole system [3]. In the control room, one can not only start and stop the equipment, but also can monitor the running status of equipment and weighing parameters. In this case, the abnormal condition during production can be detected and treated timely by the operators so as to avoid accident [4].

2 SOFTWARE SYSTEM DEVELOPMENT

The basic method of the molten iron weight measurement is as follows. Firstly, the empty tank is weighed on the weighing scale and numbered. Secondly, the same tank with molten iron loaded is weighed. Finally, the difference of the same numbered tank is the weight of the molten iron [5]. Therefore, in the weighing control system of molten iron, the following functions should be implemented.

1. Two or more weighing scales are able to be operated synchronously, so that they can be scheduled along the tracks as needed.
2. When a weighing scale is on fault, the remains can still finish the weighing task.
3. When the molten iron is weighed, power outage is not allowed for power supply system. Therefore, standby power supply or UPS are necessary.
4. The weight measuring system shall be capable of memorizing past measuring parameters, such as tank number, gross weight, total weight of iron. When measuring, the current parameters should be compared with the last measured parameters to discover whether there exists disoperation, and the measuring data should be saved automatically.
5. The fault alarm lights should be set at the screen of the operating computer and the CRT display to show if the control system is normal. When the alarm occurs, the operators should check the devices immediately.
6. A database and its communication interface based on computer network should be established to connect the operating rooms in different locations.

According to the above discussion, we use RS485 communication interface to connect 4 weighing scales with the control room LAN so as to form a whole network. Based on it, an automatic weight measuring system is developed to monitor the condition of molten iron tanks. The parameters of tank number, gross weight and total weight of molten iron distributed in different locations were sampled and analyzed, and the running state of the equipment is on-line judged. Once the system is abnormal, the system alarms immediately and shows the fault position [6]. Furthermore, it provides maintenance advices for operators to ensure the normal production and avoid catastrophic accidents. At the same time, the working log is recorded and saved in the computer database to provide the important technical basis for the planned maintenance.

3 SOFTWARE SYSTEM IMPLEMENTATION

Figure 1 shows the implementation framework of the control system. The man-machine interface is programmed by Microsoft Visual C++, which is used by operators to monitor the field equipment. The signals of tank number, temperature, gross weight and total weight of molten iron are quickly sent to S7-300 PLC through PROFIBUS - DP network. PLC combines the working parameters and the equipment condition signals to control the actions of the field devices through logic calculation. Meanwhile, PLC delivers the equipment condition signals to the man-machine interface through PROFIBUS - DP network so that the operators are able to check the working state of all equipment and all process parameters. In case that the abnormal phenomenon happens, the operators can quickly response and deal with the faults timely.

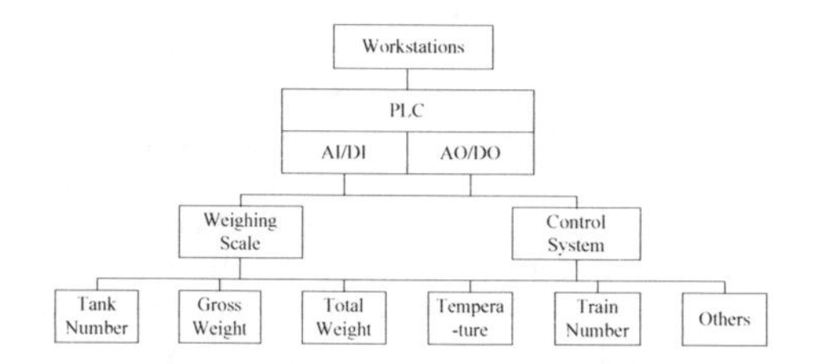

Figure 1. System frame.

Figure 2. The weighing control system.

Figure 2 shows the developed control system platform. In the panel, the "Weight" monitor screen is used to show the real time weight of the tank. The "Out Time" monitor screen is used to show the out time of molten iron. The "End Time" monitor screen is used to show the time when the tank is carried out of the weighing scale. The "Net Weight" monitor screen is used to show the weight of molten iron. When the weight is beyond the pre-warning value, the pre-warning light is alarmed. When the weight is beyond the warning value, the warning light is alarmed, and the audible and visual alarm is worked.

According to the development requirement mentioned above, we established an industrial Ethernet communication system including weighing IPC, PLC and database server, as shown in Figure 3. The key field signals are sent to the PLC through AI or DI modules, and then sent to the weighing IPC through field bus. The weighing IPC deals with the acquired data and send the results to the main computer in the center control room. Meanwhile, the results are saved into the database server. The center control room is connected with the monitor workstations by using PLC communication cards. Considering the bad working environment, the workstation uses the IPC to ensure the reliability and anti-interference of the system.

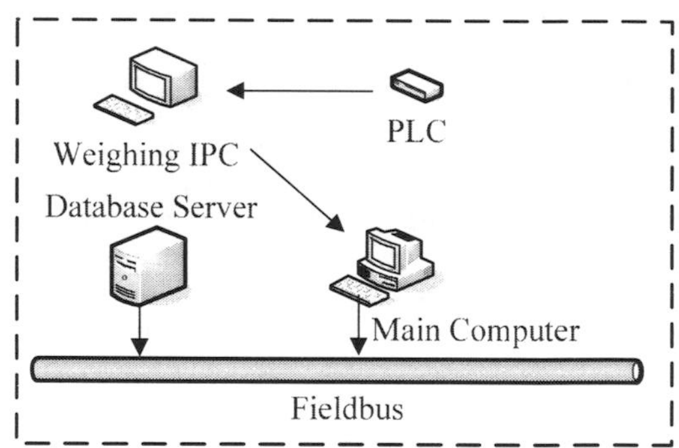

Figure 3. The communication system.

The database system is designed by using SQL SERVER in the developed weighing system. The main functions are as follows.

1. Establishing various data tables for inquiry with different combined conditions, and generating the relative sheets.
2. Connecting with the main control room.
3. Preventing illegal tampering with the database, which is an important means of preventing cheating.
4. Backup and recovery the database in time.

4 CONCLUSIONS

The weight control system for molten iron measurement has significant advantages in the aspects of improving reliability, speed and accuracy of weight control equipment, and saves manpower and energy. Aiming to the weight control problem of molten iron, the weighing control system is developed by using the principle of automatic control, hydraulic transmission, computer signal test and fluid mechanics theory. The practical results shows that the system is stable and works well, which improves the operation effects of the equipment, reduces the labor intensity of operators, improves the reliability of production line, and has good application value.

ACKNOWLEDGEMENTS

This work is supported by the Industry-university-research project of Wuhan Education Bureau (Granted No: cxy02), China. The authors also gratefully acknowledge the helpful comments and suggestions of the reviewers, which have improved the presentation.

REFERENCES

[1] Mersinligil Mehmet, Brouckaert Jean-Franois, Courtiade Nicolas, Ottavy Xavier. 2012. A high temperature high bandwidth fast response total pressure probe for measurements in a multistage axial compressor. *Journal of Engineering for Gas Turbines and Power*, 134(6): 117–131.

[2] Yi Jiangang. 2015. Modelling and Analysis of Step Response Test for Hydraulic Automatic Gauge Control. *Strojniški vestnik - Journal of Mechanical Engineering*, 61(2): 115–122.

[3] Khlaief Amor, Boussak Mohamed, Gossa Moncef. 2013. Model reference adaptive system based adaptive speed estimation for sensorless vector control with initial rotor position estimation for interior permanent magnet synchronous motor drive. *Electric Power Components and Systems*, 41(1): 47–74.

[4] Heung Suk Hwang, Suk-Tae Bae and Gyu-Sung Cho. 2007. Performance Model for Manufacturing Facility Planning Based on System Configuration, Ram and Lcc. *International Journal of Innovative Computing, Information and Control*, 3(6):199–209.

[5] Key Nicole L. 2014. Compressor vane clocking effects on embedded rotor performance. *Journal of Propulsion and Power*, 30(1): 246–248.

[6] Yi Jiangang, Zeng Peng and Sheng Quanchen. 2010. Research on Monitoring and Diagnosis System of Hydraulic Components for Ladle Fining Furnace. *ICIC Express Letters*, 4(1): 219–224.

Computing, Control, Information and Education Engineering – Liu, Sung & Yao (eds)
© 2015 Taylor & Francis Group, London, ISBN: 978-1-138-02800-5

Based on the FPGA serial interface under the control of VGA display system design

Yun Luo & Jian Zheng Cheng
School of Electronic and Electrical Engineering, Wuhan Textile University, Wuhan, China

ABSTRACT: This article presents pictures based on the serial control VGA display system, it is combined with a VGA display principle and serial communication protocol to Altera's DE2 development platform, HDL description language as a design method to complete the system hardware and software design. After testing that the system has good portability, accurate test features, has been successful in railway wagons carrying saddle parts detection system has been very good application.

KEYWORDS: VGA display, FPGA, HDL, Serial port.

1 INTRODUCTION

VGA (Video Graphics Array) is a kind of video transmission standards developed by IBM, it has high resolution, fast speed display, rich color and other advantages, and it has been widely used in the field of color display [1]. But traditional VGA chips are unable to apply a variety of display models, VGA display realized by Altera Corporation CycloneII series of EP2C35F672C6 can overcome this disadvantage. In practical applications, VGA display's startup needs to control by human's operation. Through the feedback information, whether the VGA display ends or not can easily get and what's more the communication between computer and FPGA can be achieved through serial protocol. This paper studies VGA display system under the FPGA serial control. This system can automate control the VGA display to improve the convenience of daily use, the flexibility of application in human-machine interface, testing and other aspects.

2 SYSTEM DESIGN

The hardware design of the system mainly consisted of the FPGA development board, PC machine and computer display. PC will download a program to the FPGA development board, and then it sends command to the EP2C35F672C6 by the serial port aide. After FPGA development receives the instruction, it will instruct the computer screen to display pictures, and through the serial port assistant to send information to the PC machine[2] . The design of the system block diagram is shown in Figure 1, the FPGA development board work mainly includes six modules: the clock multiplier module, serial port receiver module, serial transmission module, image data storage control module, VGA interface, VGA display module timing generator selector. The clock multiplier is the use of PLL IP core inside the ring to implement, through frequency doubling method to get the precise value of the clock frequency of 68.25MHZ. And by using the programming method of state machine to implement the design of an asynchronous FIFO based on serial transmitter, ensure the correct data writing and reading, this avoids the happening of the overflow phenomenon. VGA timing generator display module, data memory and a read and write control uses single ROM core to implement, it solves the problem of display data sources and data storage. The image data stored in the core of the ROM, for there is no color converter in the internal of ADV7123 so when the data displays as an RGB signal in memory, it can be transmitted directly to the ADV7123.

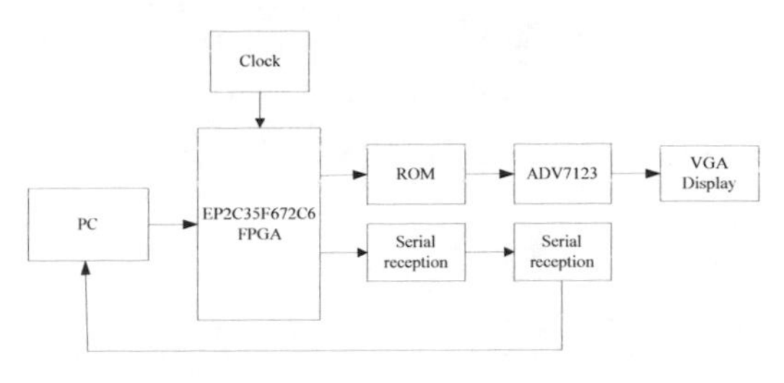

Figure 1

3 CORE MODULE

3.1 *Serial transmission and reception*

The baud rate of sending and receiving was designated at 9600 bit per min and the data bit occupies a bit, the stop bit occupies a bit too, when all this take the Asynchronous serial communication protocol [3]. The sampling baud rate of the system clock is 16 bit per min and the intermediate sample value is needed. The receive timing of UART is show in Figure 2. The timer starts to count when it detects the falling edge of the data, and when the timer counts to 24, the sampled value is a zero data bit. When the timer counts to 40, the sampled value is one data bit. And so on, when the data counts to 168, the sampled value is one, it represents the stop bit and a frame receiving ends.

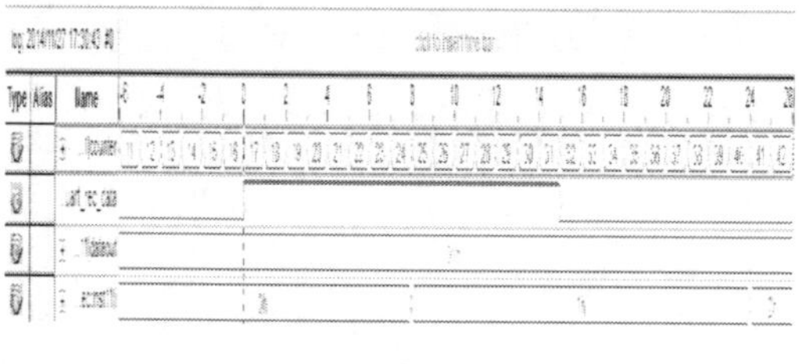

Figure 2

3.2 *VGA display timing generator*

The images data was stored in the 16 bit ROM, and select a VGA interface through the two choose one module. The timing signal generator of the VGA interface includes line point counter, field point counter, point line sync signal, the field synchronization signal, and the effective display area and so on. Among them, the line point counter is 1280 binary counter, field point counter is 768 binary counters. And according to the industry standard of VGA, there are 4 states that are synchronizing pulse signal (Sync), display along the signal (Back Porch),visual display area (Visible area), display (Front Porch) [4]. The timing law of these 4 states is very clear, and the transform of these 4 states can be realized by the finite state machine. As shown in Figure 3.

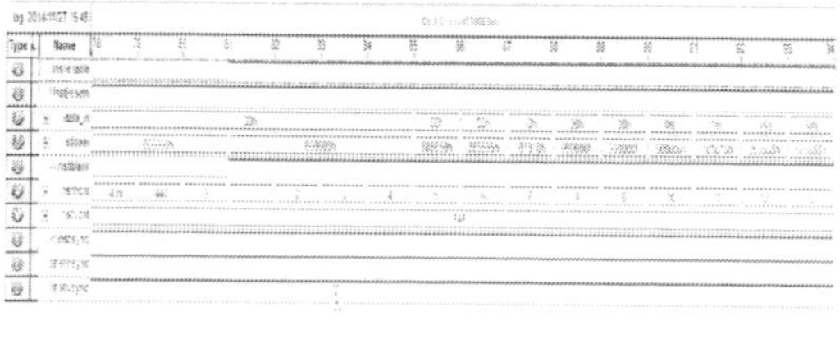

Figure 3

4 SYSTEM EXPERIMENT

The experiment results mainly include the sending and receiving data and computer show pictures. As shown in Figure 4.

By sending sixteen hexadecimal data 01, the system can correctly get the sixteen hexadecimal data 02 and 03. And the end mark of the sixteen hexadecimal data is 04. As shown in Figure 5, black and white stripes appear on computer screen, the image data are no wear of the bearing saddle multiple angles shoot good pictures, and as reference data, these pictures are stored in the ROM in the form of sixteen hexadecimal, so the display is essentially a picture of the left movement, a time interval of only for a few seconds. The serial port sends 01 to the driver board, and then VGA will automatically display pictures sequentially. Finally return data 04, the computer will display a white screen as the end symbol.

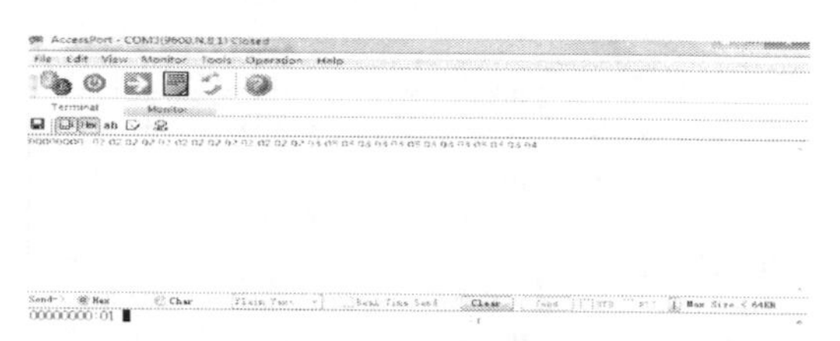

Figure 4. VGA image display.

Figure 5. VGA display end.

5 CONCLUSION

After the serial data is received, it will immediately start the camera in vertical angles to shoot the saddle, and then project the photos. Through the VGA interface, the photos were shown. After the data about the surface of the object was achieved, it was send to compare with the baseline data to finish the calculation of the surface damage. This method can discover the lossy components very easily and can be applied to the railway field. It can effectively improve the efficiency and precision of the truck carrying saddle repair for the railway maintenance work. At the same time, it also shows that this testing system has the character of testing accurate, stable and reliable, meet the design requirements.

ACKNOWLEDGEMENTS

(Fund Projects)

1 National Natural Science Foundation of China (No. 11174230)

REFERENCES

[1] HANSIRTEL, Color-vision demonstrations on an IBM PC/AT with VGA, *Behavior Research Methods, Instruments, &: Computers* 1992, 24 (1), 88–89.

[2] Carlos Allberto Ramos-Arreguín,Juan Carlos Moya morales,Juan Manuel
Ramos-Arreguín,Jesus Carlos Pedraza-Ortega, Saúl Tovar-Arriaga,Marco Antonio Aceves
Fernandez, Jose de Jesus Rangel-Magdaleno, FPGA Open Architecture Design for a VGA
Driver,The 2012 Iberoamerican Conference on Electronics Engineering and Computer
Science Procedia Technology 3(2012)324_333

[3] Hu Zhe, Zhang Jun, Xiling Luo, A Novel Design of Efficient Multi-channel UART Controller
Based on FPGA, Chinese Journal of Aeronautics 20(2007)66–74.

[4] Information on http://tinyvga.com/vga-timing,last consult 12/08/11.

Computing, Control, Information and Education Engineering – Liu, Sung & Yao (eds)
© 2015 Taylor & Francis Group, London, ISBN: 978-1-138-02800-5

The method of classifying requirement documentation into different levels based on COSMIC

T. Liu & H.J. He
National University of Defense Technology, Changsha, HuNan, China

P. Lu
Urumqi Armed Police Command College, Urumqi, XinJiang, China
National University of Defense Technology, Changsha, HuNan, China

ABSTRACT: The international organization Standish Group found that [] the cost of correct requirements specification in the early stage of the project is the lowest, the cost in maintenance stage is up to more than ten times than in the design stage. Therefore, improving the quality of software requirements specification is of great value. In order to guarantee the quality of software, based on COSMIC, we proposed a new method of classifying requirement documentation into different levels in early stage of a project to help improve the quality of requirements documentation. We used the method in a real project, it shows the method is generally because the results of different members measured the same documentation are consistent. We can also get that when documentation quality levels rank from high to low, the accuracy of measurement results ranges from 97.58% to 32.9%, it shows that the method can accurately describe and promote to improve the quality of requirement documentations.

KEYWORDS: COSMIC; Documentation Quality; Functional Size Measurement.

1 INTRODUCTION

The required documentation of low quality can[1] directly affect the project schedule. Measurement activities can classify requirement documentations into different levels, determine the quality of requirements documentation, to find the problems existing in the required documentations and correct them without delay, to improve the quality of requirements documentation. Besides, through classifying requirements documentation into different levels of quality, it can also reflect the accuracy of measurement results. The higher the level of the quality of requirements documentation is, the more accuracy of the measurement, and vice versa.

The method of classifying requirement documentations into different levels of quality is mainly based on COSMIC functional size measurement[2]and relies on recognizing function[3] and data transfer. In recognizing data transfer[4], it needs to recognize function, sub-function and data type, which are closely related to the documentation quality. Through this method, each functional process can include not only the size of functional point, but also the level of documentation quality. Analyzing the measurement results will help to improve the documentation quality and describe the reliability of measurement results. If the level of the documentation is quite low, it means that the deviations of the measurement results from this functional process is too large, therefore, the reliability of this functional process is low, and vice versa. In the mapping stage, classifying requirement documentations into different levels of quality and providing feedback to the organizer of requirement documentations to constraint quality control play an important role through the whole measurement process. In the measurement process, firstly, we classify requirement documentations into different levels of quality of functional process. After the measurement of all the functions process is completed, we calculate the weight coefficient of each functional process in the whole process, convert the level of documentation quality into value, and apply formulas to calculate the mean of the result, and then convert the result value into the level of documentation quality, which is the level of quality of the required documentation.

2 METHOD DESCRIPTION

2.1 *Classify functional process*

The first thing to classify requirement documentations into different levels is to classify all the functional processes into different levels. According to the requirement documentations, if the description of a functional process is complete, data transfer and type are clear, we will define the level of functional process as A;

According to the requirements documentation, if the description of a functional process is complete, data transfer is not clear, in which the input, output, read and write are all described, but not sufficiently clear to recognize data transfer, we will define the level of functional process as B;

According to the requirements documentation, if the functional process can be recognized, but the data transfer cannot be judged, we will define the level of functional process as C;

According to the requirements documentation, if the functional process is described but not in detail, we will define the level of functional process as D;

According to the requirements documentation, if the functional process is not described, but has subtle hints, we will define the level of functional process as E;

According to the requirements documentation, if the functional process cannot be recognized, we will define the level of functional process as F.

The measurement of functional processes into different levels, mainly depends on the some facts of requirement documentations, those facts are:

If there are file descriptors for recognizing every functional process;

If there is information to recognize data transfer (E, X, R, W).

All the functional processes and their corresponding level of documentation quality are shown in Table 1.

Table 1. Quality rating values.

Quality Level	A	B	C	D	E	F
Value Range	90-100	80-90	70-80	60-70	40-50	0-40
Formula Value	90	80	70	60	50	40

2.2 *Determine the level of requirement documentation*

After all the functional processes of the required documentation are classified, we determine the level of requirements document. The formula of determining the level of requirement is documented as follows:

Quality rating values of requirement documentation=(n1*A+n2*B+n3*C+n4*D+n5*E+n6*F)* The weight coefficient

The weight coefficient = 1/ The total number of functional processes

n1, n2, n3, n4, n5, n6 are the level constants of different functional processes.

A, B, C, D, E, F are the formula value of different functional process quality level.

The weight coefficient is given as the scale factor, represents the degree of importance of a number of the functional processes from the whole functional processes.

The sum of the level of all the functional process level is the cumulative sum of quality levels during measurement.

The level of requirement documentation is decided as follows:

a. Apply the formula to calculate the numerical value of the quality level of the requirement document;
b. Find which range of quality level the numerical value is in; c) determine the quality level of requirement document.

3 VERIFY THE METHOD OF CLASSIFYING

3.1 *Experiments*

For example: A project 1P-07 is an information management software, it contains five modules: message protocol processing, data application processing, platform interface processing, end machine interface processing and acoustic processing code. Its function is mainly to connect platform, complete the initialization of end machine, work management, the message processing, system maintenance and etc.

The first measure from three members(M1,M2, M3) is shown in Table 2.

Table 2. List of the quality level of functional processes.

1P-07 List of Functions Process						
Functions Process	CFP			Quality Level		
	M1	M2	M3	M1	M2	M3
Load Initial Parameters	4	4	3	B	B	B
Work Control	18	18	17	A	A	A
Channel Management	6	8	8	B	B	B
Management	10	10	10	A	A	A
Filter Management	5	7	6	C	C	B
Message Reception Processing	8	8	8	A	A	A
Message Response Processing	33	26	28	C	C	C
Message Sending Processing	2	4	2	B	B	B
Self-Test	15	16	16	B	B	B
Platform Condition	4	4	4	A	A	A
Signal Process	2	2	2	A	A	A
Fleet Platform Maintaining	15	16	16	A	A	A
Total Functional Points	122	124	120	-	-	-

According to Table 2, we can get the quality level of the documentation from the three members(M1,M2,M3), shown in Table 3.

Table 3. Classify quality level of documentation.

	Total Quality Level	Function Process Number	Weight Coefficient	Quality Level Value	Quality Level
M1	6A+4B+2C	12	0.08	80	B
M2	6A+4B+2C	12	0.08	80	B
M3	6A+5B+1C	12	0.08	80.8	B

We can compare the measurement results of the three members (M1,M2,M3) with the real project, shown in Table 4.

Table 4. Measurement results.

	Quality Level	CFP	Lines of Code	Convert Rate	Standard Point	Accuracy Rate of Measure
M1	B	122	17837	148	121	99.18%
M2	B	124				97.58%
M3	B	120				99.17%

Note: The Conversion Rate refers from {Software Measurement and Estimation}, which specifies the number of lines of C code into the mean of standard functional points, that 1CFP = 148 lines (lines of code).

As shown in Table 4, for the same required documentation, to measure members (M-1 and M-3) judge that the quality level is B, and the measurement accuracy is separately 99.18%, 97.58%and 99.17%. Since the measurement accuracy of the three measure members is above 97%, it shows that the results of different measure members measured the same requirement documentation are the same. And the higher the quality level is, the more accuracy the measurement. Meanwhile, when determining the quality level of function process, the lower the level is, the less clear from the description of the documentation requirement of the functional process. To modify these unclear descriptions will improve the quality of requirement documentations.

3.2 *Differences analysis*

Classifying the quality level of requirement documentations can help users understand and use them better. How much difference between the different quality levels? If the difference is small, it is little help to improve the quality level of requirement document and we do not need to classify documentations into different levels. Therefore, we continue to demonstrate the differences between measuring documentations of different quality levels.

In order to make sure the differences between different quality levels of documentations, we conduct an experiment. In the experiment, a measurement member respectively measure different requirement documentations and classify the requirement documentations into different quality levels. After the measurement, we put the same quality level of requirement documentations together, calculate the average accuracy of the same quality level and analyze the differences between the average accuracy rate of different quality levels. In order to calculate the accuracy, we convert the software scale of the requirement documentation which means lines of code into functional points, ,then make the functional points as a standard functional point and compare it with the points from measure members, from which we analyze the differences of different quality levels of documentations.

The basic conditions of requirement documentations and measurement results are shown in Table 5.

Table 5. Measurement results of different documentations.

Project Code	Quality Level	Programming Language	Lines of Code	Convert Rate	Standard Point	Measure Results
1P-1	E	C	1785	148	12	40
1P-2	D	C/ Assembly Language	1729/ 75	148/172	11+1	20
1P-3	C	Assembly Language	5640	172	32	27
1P-4	C	Assembly Language	5985	172	34	27
1P-5	D	Assembly Language	6054	172	35	27
1P-6	B	C	15654	148	106	91
1P-7	A	C	17837	148	121	129
1P-8	E	C	12396	148	84	30
2P-1	B	C	5707	148	39	44
2P-2	B	C	25027	148	169	150

We classify the same quality level of the measurement documentation into a class, and compute the accuracy shown in Table 6.

From Table 6, we can get that the documentation quality level ranks from A to E. The average of accuracy of measure members ranges from 97.58% to 32.9% in a descend way. The average accuracy of the measurement on the documentations of quality level of C is more than 80%, which we believe that they achieve the basic requirements of measurement. The average accuracy of measurement on the documentations of quality level below D declines in a geometric trend and such accuracy of measurement is too coarse-grained, which cannot achieve the measurement target. Besides, it can also illustrate that the quality level of documentations below D, whose description is not clear and cannot meet the requirements of measurement.

Table 6. Accuracy of project metric.

Quality Level	Project Code	Accuracy Rate	Average Accuracy Rate
A	1P-7	97.58%	97.58%
B	2P-2	88.8%	87.8%
	1P-6	85.9%	
	2P-1	88.6%	
C	1P-4	79.4%	81.9%
	1P-3	84.4%	
D	1P-2	60%	68.6%
	1P-5	77.1%	
E	1P-1	30%	32.9%
	1P-8	35.7%	

4 CONCLUSION

Using the COSMIC functional size measurement to classify requirement documentations into different levels can directly reflect documentation quality, promote documentation, writes to correct documentation quality, in the early stage of project guarantee the accuracy of requirements of the lowest cost, avoid the big cost of correct documentation in the late stage of the project. The quality level of measurement can improve documentation quality, analysis the reliability of measurement, from Table 6, with the improvement of the quality level, the reliability of measurement increases.

The quality level of documentations below D, which the average accuracy rate below 68.6% can not guarantee the reliability of measurement, if not in time to improve quality of requirement documentations, it will increase the cost of correct requirements defect in the project life cycle. At the same time, from Table 2, classifying the functional processes into different levels can be associated with the content of requirement documentations, help documentation, writes quickly locate the documentation description that needs to be correct, improve the efficiency of correcting documentations.

REFERENCES

[1] "The Common Software Measurement International Consortium (COSMIC): Guideline for Assuring the Accuracy of Measurements Version 1.0.," 2011.

[2] ISO/IEC, IS 19761:2003,Software Engineering – COSMIC-FPP – A functional Size Measurement Method, International Organization for Standardization, March 2003.

[3] Guideline for 'Measurement Strategy Patterns' Ensuring that COSMIC size measurements may be compared VERSION 1.0 March 2013.

[4] Linda M. Laird and M. Carol Brennan. Software measurement and estimation: a practical approach[M]. Hoboken,N.J.: Wiley-Interscience : IEEE Computer Society, c2006.

Computing, Control, Information and Education Engineering – Liu, Sung & Yao (eds)
© 2015 Taylor & Francis Group, London, ISBN: 978-1-138-02800-5

ESHARP: Energy-saving hybrid ad-hoc routing protocol for wireless sensor network

Da Lei Qiao
Nanjing Marine Radar Institute, Nanjing, China
Software Institute, Nanjing University, Nanjing, China

Feng Xue
Nanjing Marine Radar Institute, Nanjing, China

Yong Yang
School of Computer, China University of Mining and Technology, Xuzhou, China

ABSTRACT: For a large number of clustering routing protocols are under the condition that every cluster head can communicate with sink node directly which will not happen in practice, this paper proposed an energy-saving hybrid ad-hoc routing protocol (ESHARP) for wireless sensor network. The protocol enhances the function of Heed and Direct Diffusion by combining them together. During the hierarchical characteristic, the protocol can support fast routing set-up, date aggregation; for the multi-hop communication between cluster heads and sink node, the protocol can reduce the energy consumption of the cluster heads which can make the network service time become longer consequently.

KEYWORDS: wireless sensor network; energy-saving; hybrid routing protocol; trigger clustering.

1 INTRODUCTION

Recently, wireless sensor network (WSN) enjoys a fast development in many fields. For example, WSN can be applied in climate monitoring, military communication, space detection, intelligent agriculture and so on [1]. But WSN is different from the traditional network because there are some limitations. On one hand, the sensor nodes are usually powered by batteries, the energy of each node is limited. When the energy of the node is depleted, the node will lose its working ability. When the deleted nodes ratio increases up to a certain value, the whole network will broke down. So the design of WSN need take the energy consumption as the first target, including deployment strategy, routing protocols, etc. On the other hand, the node transmission range is limited, so the node needs to transmit data through multi-hop communication which needs the routing protocols to point the transmit path for the node. For these reasons, the energy efficient routing protocols has become a hot research filed in wireless sensor network.

2 RELATED WORKS

The routing protocols in WSN can be divided into two types, flat routing and hierarchical routing. In the flat routing protocols, the typical protocols are Flooding, Gossiping, SPIN, SAR and Directed Diffusion [2]. Estrin proposes a routing protocol for wireless sensor network called direct diffusion [3]. In hierarchical routing protocols, the typical algorithms are LEACH [4], HEED [5], and PEAGSIS [6]. Among them HEED is an improved protocol derived from LEACH, it proposes a cluster head competition strategy called AMRP which denotes the average energy consumption level if the node turn into the cluster head. AMRP value is a good assessment of communication cost of the cluster head. HEED can select a set of cluster heads after several iterations, and then cluster heads send the cluster_head_msg to other nodes and notify them to join the corresponding cluster.

Hierarchical routing protocols can decrease the energy dissipation among nodes by make the node become cluster heads in turn. Because the normal nodes communicate with other nodes forwarded by its cluster head, the data aggregation can be easily realized on the cluster head. But these routing protocols are on the hypothesis that all the cluster heads can communicate with sink node which are not feasible in practice. So the inter-cluster routing is the crucial problem needed to be solved. References [7, 8, 9] are hybrid routing protocols based on cluster routing.

3 SYSTEM MODEL

3.1 *Ad-hoc network model*

We assume the following properties about the ad-hoc sensor network:

1 The nodes are densely deployed in the monitored area and the location distribution follows the uniform distribution.
2 The initial energy of each node is the same.
3 There is only one sink node whose energy is not limited.
4 Nodes are powered by batteries, and battery re-charge is not possible, so the routing protocols are required for energy conservation.

3.2 *Energy dissipation model*

In this manuscript, we adopt the energy dissipation model proposed by Heinzelman [10], the receiving and sending energy consumption is:

$$E_{recv} = lE_{elec} \tag{1}$$

$$\begin{cases} E_{send} = lE_{elec} + l\varepsilon_{fs}d^2, d < d_0 \\ E_{send} = lE_{elec} + l\varepsilon_{amp}d^4, d > d_0 \end{cases} \tag{2}$$

Where E_{elec} denotes the energy consumption of transmit circuit; l denotes number of transmitted packages; d denotes the communication distance. If the distance is shorter than a given value d_0, the model is free space model; otherwise, the model is multi-path fading model. ε_{fs} and ε_{amp} are the power amplifier factors for these two different models.

4 ENERGY-SAVING HYBRID AD-HOC ROUTING PROTOCOL

Although hierarchical routing decreases the energy consumption, the disadvantage is obvious. These protocols do not give a routing strategy between cluster heads and the sink node. This paper proposed a hybrid protocol which uses cluster routing in intra-cluster transmission and uses direct diffusion to setup the communication path between cluster heads and the sink node. The hybrid ad-hoc routing protocol has two steps before data transmission: cluster building and inter-cluster routing setup.

As shown in figure 1, the ESHARP started as selecting cluster heads, and we adopt the method proposed in HEED which selects cluster head by calculating the AMPR value of each node. Once the node becomes the cluster heads, it will send announcement message to other non-cluster-head nodes. Normal

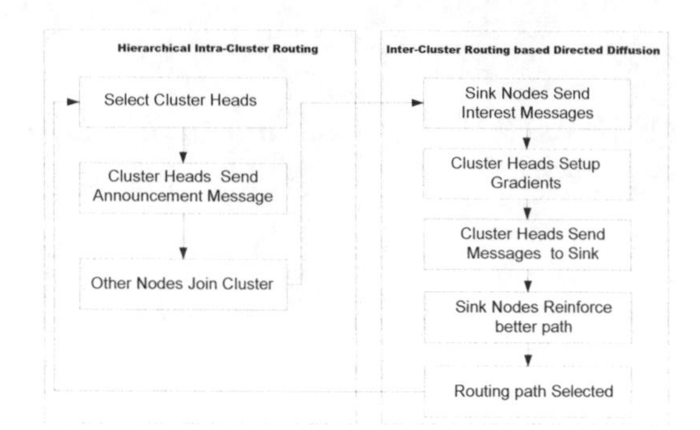

Figure 1. Architecture of ESHARP.

nodes will select the node as its cluster head which has smaller communication cost and stronger receiving signal. After the clusters are formed, the sink node will broadcast interest messages into the network. Then the cluster heads setup gradients at the reverse direction of the receiving message. The cluster heads send messages to the sink node along the gradients. When sink node receives these messages, it will reinforce the path which reaches earlier, and it will send the interest messages at a higher speed. At last, the routing paths between cluster heads and the sink node will be determined.

4.1 *ESHARP details*

During the whole process, the ESHARP goes through three phases: cluster forming, inter-cluster routing build up and data transmission.

Suppose S represents the set of cluster heads, CH_{prob} represents the probability of being cluster heads. Each node calculates the communication cost from itself to the nodes in the S and selects the nodes as its cluster heads which has the lowest cost. If the node id equals the id of itself and $CH_{prob}<1$, it will announce tentative cluster head message; if the node id equals the id of itself and $CH_{prob}=1$, it will announce final cluster head message. If the node id does not equal the id of it selves', the node will compare the CH_{prob} with a random value range from (0, 1). If Random (0, 1) < CH_{prob}, it will announce tentative cluster head message. After each iteration, the node will double its CH_{prob} value until it reaches 1. When the iteration ends, the cluster heads will announce final cluster head messages and other nodes will choose their cluster heads to join the cluster.

In the inter-cluster routing build up phase, the sink node firstly broadcasts the interest package at a low data rate. The nodes received the messages will set up a gradient along the reverse direction which is a tuple likes [direction, hops]. And then the nodes will forward the message to their neighbor nodes. Only the cluster heads will send the messages along the gradients to sink node. When the sink node receives this type of message, it will send packages along the biggest gradient that we call it path reinforcement. At last, each cluster heads will select a shortest path to the sink node.

In the data transmission phase, the common nodes send the monitored message to its cluster head. When the data reaches the cluster head, it will be forwarded along the shortest path to the sink node.

4.2 *Triggered clustering*

The traditional hierarchical routing protocols such as LEACH, HEED and PEGASIS adopt rounds selection. In a designate time interval, the network will perform re-clustering. The value of the interval plays an important role in the routing selection. If the interval is set too large, it will cause the cluster heads consuming too much energy and died before next re-clustering. In contrast, the small interval will bring too much communication cost.

The core idea of triggered clustering is substituting the round selection for the parameters triggered clustering. In this paper we compare the residual energy of the cluster head with a given threshold, if residual energy is lower than the threshold, it will send re-clustering message in its cluster range. Suppose the current residual energy of the cluster head is E_{HR}, the average residual energy is E_{CR}, the condition of the re-clustering is:

$$E_{HR} \le \rho E_{CR} \tag{3}$$

The node will encapsulate its residual energy in the data packages while it sends them to the cluster head node. When the packages reach the head node, it will recalculate the average residual energy of the cluster. If this value satisfies formula 4, the cluster head will send re-clustering message in its cluster range which will cause the intra-cluster re-clustering. Suppose there are n clusters in the network, and the residual energy of the i_{th} cluster head is E_h^i, the average residual energy of the i_{th} cluster is E_c^i, the node number of the i_{th} cluster is num_i, so the average residual energy of the network is

$$E_N = \frac{\sum_{i=1}^{n} E_c^i}{\sum_{i=1}^{n} num_i} \tag{4}$$

The average residual energy of the cluster heads is

$$E_H = \frac{\sum_{i=1}^{n} E_h^i}{n} \tag{5}$$

The condition of the re-clustering is

$$E_H \le \rho' E_C \tag{6}$$

The sink node will recalculate the average residual energy of the network when it receives the packages from cluster heads. If the energy of the cluster heads is lower than the average residual energy of all the nodes, the sink node will send global re-clustering message which will cost the whole network perform re-clustering.

4.3 *Backup cluster head*

For cluster head selection will cost a lot of energy, the network service time will be prolonged if the selection times decrease. We add double cluster strategy into ESHARP. At each cluster head election round, a main cluster head (CH) and a backup cluster head (BCH) are selected according to the AMRP value. When CH runs out a proportion of energy, it will send routing table to BCH, and announce to the other nodes that the BCH will become the new CH. Then the new cluster head will select another BCH according to its residual energy.

4.4 *Triggered clustering with backup cluster head*

Under triggered clustering, the re-clustering condition is $E_{HR} \le \rho E_{CR}$ which requires the backup cluster head also satisfying $E_{BHR} > \rho E_{CR}$ or the backup cluster head need to send re-clustering message once it transfer to new cluster head. But the backup cluster head will lose its energy in the succeeding monitoring time. So there are two situations under triggered clustering with backup cluster head.

1. When the re-clustering condition is satisfied and the backup cluster head does not satisfy the energy threshold, the cluster head just need to flood the message to announce the backup cluster head to be the new cluster head.
2. When the re-clustering condition is satisfied and the backup cluster head also satisfies the energy threshold, the cluster head flood the re-clustering message to start a new process of cluster head election.

5 EXPERIMENTAL SIMULATION

In order to evaluate the performance of ESHARP, we do several experiments based on the simulation framework OMNet++ [11]. The simulation environment CPU is AMD2600+ with 1G memory, operation system is Windows XP.

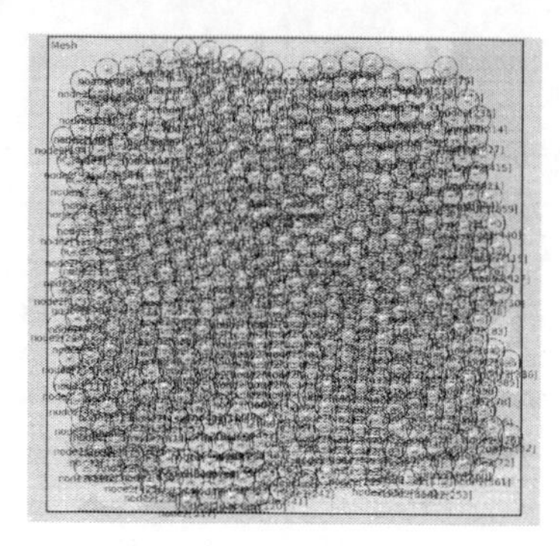

Figure 2. Nodes distribution in monitor area.

As shown in figure 2 there are 300 nodes in the monitor area follow the uniform distribution. The monitor area is a 1000 * 1000 square area, and the location of the sink point is (300,300); the initial energy of each node is 5J; the max communication radius is 10.

Figure 3 shows the network life time under HEED and ESHARP.According to the experiment result, the network lifetime under ESHARP is longer than HEED. When the initial energy of each node is 5J, the first node dead time is 1685s under ESHARP and 1470s under HEED.

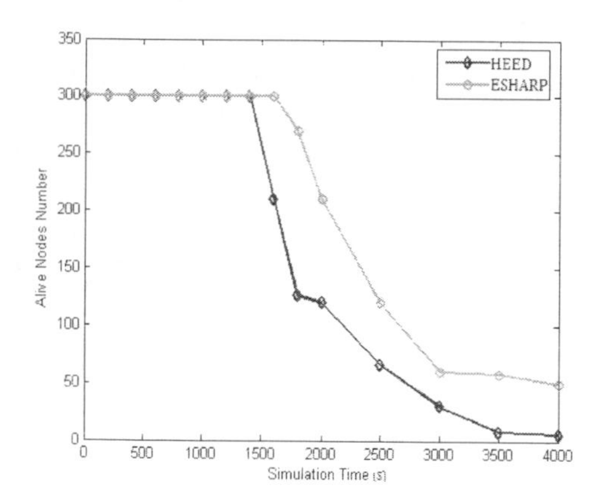

Figure 3. Network life time under HEED & ESHARP.

Table 1 shows the network lifetime in different initial energy levels, from this table we can find that not only the first node dead time but also the last node dead time under ESHARP is bigger than that under HEED. There is an obviously performance enhancement when using ESHARP instead of HEED. Through multi-hop routing between cluster heads and sink node, the network consumes less energy.

The network life time under ESHARP and ESHARP with triggered clustering is shown in fig. 4. The first node dead time under ESHARP with triggered clustering is later than ESHARP while the last node dead time under ESHARP with triggered clustering is earlier than ESHARP. Adding Trigger clustering to ESHARP can decrease the energy consumption speed which needs less re-clustering time. The experiment result shown in figure 5 is similar to figure 4. When applying backup cluster head mechanism in ESHARP, the network lifetime will be prolonged.

Table 1. Network lifetime of HEED & ESHARP in different initial energy.

Initial Energy	Model	First Node die time(s)	Last Node die time(s)
1J	HEED	257	765
	ESHARP	431	1044
2J	HEED	695	1865
	ESHARP	785	2013
5J	HEED	1430	4096
	ESHARP	1732	4532

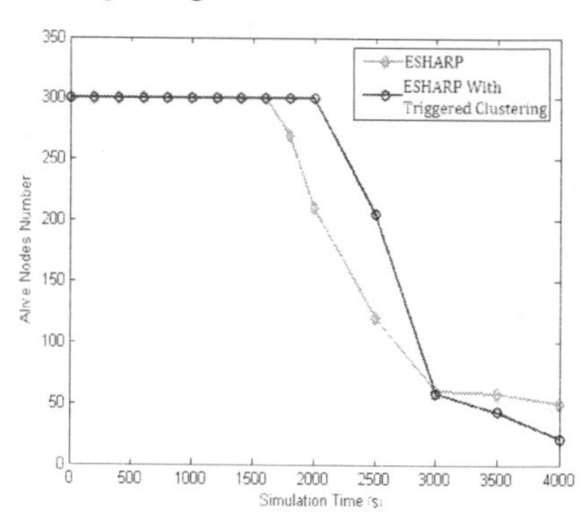

Figure 4. Network life time under ESHARP& ESHARP with triggered clustering.

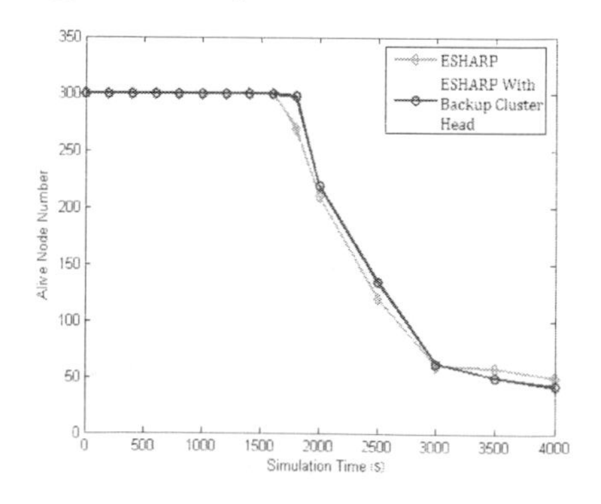

Figure 5. Network life time under ESHARP&ESHARP with backup cluster head.

6 CONCLUSION

This paper proposed a hybrid routing protocol for wireless sensor network. This protocol adopts the core idea of hierarchical routing protocols and setup a multi-hop routing path between cluster head and the sink node. Simulations show that ESHARP can prolong the network life time compared with traditional hierarchical routing protocols. But in this manuscript, we suppose the initial energy level of all nodes is the same. So we need to take into account of heterogeneous node type and different distribution in future research.

REFERENCES

[1] F.Y. Ren, H.N. Huang and C. Lin, Wireless sensor networks. Journal of Software, 2003, 14(7):1282~1291.

[2] C. Nagonwiwat, R. Govindan and D. Estrin, Directed diffusion: a scalable and robust communication paradigm for sensor networks. MobiCom 2000. Proceedings of the Sixth Annual International Conference on Mobile Computing and Networking, 2000, 56~67.

[3] C. Intanagonwiwat, R. Govindan, D. Estrin, J. Heidemann and F. Silva, Directed diffusion for wireless sensor networking. IEEE/ACM Transactions on Networking, 2003, 11(1):2~16.

[4] W. Heinzelman, A. Chandrakasan, and H. Balakrishnan, Energy-Efficient Communication Protocol for Wireless Micro sensor Networks. In Proceedings of the Hawaii Conference on System Sciences, Jan. 2000.

[5] O. Younis, S. Fahmy, HEED: a hybrid, energy-efficient, distributed clustering approach for ad hoc sensor networks. IEEE Transactions on Mobile Computing, 2004, 3(4): 366~379.

[6] S. Lindsey, C.S. Raghavendra, PEGASIS: power-efficient gathering in sensor information systems. 2002 IEEE Aerospace Conference Proceedings. 3: 1125~1130.

[7] L.S. Tan, G. Fei and J. Li, HCEP: a hybrid cluster-based energy-efficient protocol for wireless sensor networks. International Journal of Sensor Networks, 2009, 5(2): 67~78.

[8] H. Zhou, D.L. Qing , X.M. Zhang, H.L. Yuan and C. Xu, A Multiple-Dimensional Tree Routing Protocol for Multisink Wireless Sensor Networks Based on Ant Colony Optimization. International journal of distributed sensor networks, 2012, 1~10.

[9] L. J. García Villalba, D. Rupérez Cañas, A. L. Sandoval Orozco and T.-H. Kim, Restrictive disjoint-link-based bioinspired routing protocol for mobile Ad Hoc networks. International Journal of Distributed Sensor Networks, 2012, 1~5.

[10] W.B. Heinzelman, A.P. Chandrakasan and H. Balakrishnan, An application - specific protocol architecture for wireless sensor networks. IEEE Transactions on Wireless Communications, 2002,1 (4).

[11] F. Bause, P. Buchholz, J. Kriege and S. Vastag, A Simulation Environment for Hierarchical Process Chains Based on OMNeT++, Simulation,86(5–6):291~309.

Computing, Control, Information and Education Engineering – Liu, Sung & Yao (eds)
 ISBN: 978-1-138-02800-5

A new method of file type identification based on two level 2DPCA

Zong Da Han & Bing Long Li
Information Engineering University, Zhengzhou, Henan, China
State Key Laboratory of Mathematical Engineering and Advanced Computing, Zhengzhou, Henan, China

ABSTRACT: The current methods of file type identification are based on random file fragments, but they didn't solve the problem that the time consumed during the process of acquiring fragments and false positive rate when detecting large-scale files. In this paper, a random file fragment achieving architecture based on a pre-cluster index is designed firstly. On this basis, a file type identification algorithm based on 2-Dimension Double Principal Component Analysis (2DDPCA) is proposed. This algorithm used 2D code of file to compute the optimal feature set of files and furthest compressed the set of features which were used to support vector machine (SVM) classifier. It reduces the time overhead and improves the accuracy. Experimental results show that the average accuracy on six types of file is 83%, and it is effective, for large numbers of files and reduces the overall cost of the task of file type identification.

KEYWORDS: Computer forensic; file type identification; byte frequency; file fingerprint; pre-cluster index.

1 INTRODUCTION

Forgery of files is an important part of anti-forensic techniques, the malicious third party utilizes the technologies of forgery and steganography to forge files (Jain et al., 2014), and they are aiming at stealing, faking, hiding important information and spreading viruses, etc. At the same time, the increasing of the compound file types, file with less features and long cycle of detection are challenging the existing method of file identification (Cao et al., 2010). So it is necessary to improve the method of detection to correct the false or unknown file type and protect the security of information in the host.

The methods of type identification are based on suffix and magic number or file content. The former is not safe (Patel et al., 2010) and the detection of the file content is more safe (Roussev et al., 2013). According to a fragment which is analyzed, it also can be divided into two parts:

1 Analysis of the file header. Using the characteristic of the byte frequency distribution (BFD) (McDaniel et al., 2003) to match with the target set, it has a high rate of success (Moody et al., 2005, Karresand et al. 2006, Feinerer et al., 2013). This kind of methods has a good effect with the obvious characteristics in the file's hard, but they are depends on a fixed position excessively, the misclassification rate increases obviously when it is being forged.
2 Analysis of file fragments or data packet which have no relationship with file header. Veenman (2007) used Kolmogorov complexity to identify file type combined with Shannon entropy based on BDF, but it only achieves a high rate on html and JPG. Erbacher (2003) and Mulholland (2007) proposed a method of statistical measure to identify file type, but they only divided files into several categories. Irfan (2011) established an n-gram detection model to identify the file type, but it's over reliance on the fixed gram. In addition, Irfan discussed the method to read file fragment in the random location, the classified accuracy was more credible, but the efficiency and time consume were increased too. Amirani (2013) proposed to use PCA (Principle Component Analysis) and neural network to accomplish fast file type identification, but if where are massive files, MLP learns slowly because "trial-and-error" exists the problem of time consuming.

Ellen (2013), Gregory et al (2010) and Conti (2010) are mapping the different type of files into n-dimensional graph by reverse engineer, and completing the conversion from binary data to a visualize form. It is convenient to distinguish large-scale data, especially just "look" though the disk. This kind of methods divides the features of different file type of n-dimensional graph, but this method depends on manual work primarily and it restricts the efficiency obviously.

For the purpose of detecting large-scale data, this paper proposes the idea as follows: we build a cluster index by side of forensic image and it will provide a rapidly acquirement of data through a random location in a file. When we want to analyze a forensic image, the file system should be reconstructed by forensic program firstly. The process of file type detection will reconstruct and read the file system over and over, because a file is read and a file system will be reconstructed once, it will take a lot of times. At the same time, the cluster index is used to check the accuracy of forensic result by repeating the progress of detection, and it improves the efficiency of the investigation.

The main contribution of the paper as follows: 1) proposes a new mechanism of file access through cluster index, it improves the efficiency of random data acquiring; 2) proposes a detection method based on 2D (two-dimensional) data. File type detection based on file content confirms the maximum accuracy, then we use 2DDPCA algorithm to reduce the time and the number of features during feature selection, and use SVM classifier to distinguish different types of file. 3) File type detection based on six kinds of file. Experiment results show that the time of detection is reduced by our method and accuracy achieves 83% finally.

The rest of this paper is organized as follows. Section II describes the strategy of file type identification. Section III discusses the method of 2D encoding in file fragment. Section IV explains the algorithm of feature extraction. Section V presents the experimental results, followed by the conclusions and future work in Section VI.

2 FILE TYPE IDENTIFICATION STRATEGY

During computer forensic, file type detection is always working on image instead of original storage medium and file content analysis is a safe method comparatively (Ahmed et al., 2009). Specific location can be used by the investigator, but it can also be used by attacker too, so the data in a random location are safer and reduces the probability of being cheated (Roussev et al., 2009, Gopal et al., 2011). In this paper, two important targets in our design scheme are given: 1) Identifying files. One file is complete, the proposed method can correctly identify falsified file and figure out the real part of a file. Another one is file fragments, it will be identified and figured out the type it belong to; 2) Completing the mission of detecting large-scale files and adjusting to the environment with TB-level files.

We use 2D code to convert file data to graph by visualization technology, and take advantage of 2DDPCA algorithm for feature extraction, each 2D node is represented as a feature vector, so a file can be expressed as a set of feature vectors. While each character of two bytes reflects the local characteristic of the set of bytes, but it cannot reflect the overall characteristic of the expressed file content. In order to solve this problem, this paper introduced SVM (Support Vector Machine) (Weston et al., 2005) as the file type classifier, then by using the way of cascaded to accelerate the speed of detection.

3 TWO-DIMENSIONAL ENCODING

File visualization analysis is a concept put forward in recent years, and it has been extended from one dimension to n-dimension. Two-dimensional code has a good result on sampling analysis and adopted in this paper, namely the two consecutive bytes treat as a node, then using frequency similarity between the same type of files, and it can visualize the entire file system or file by mapping binary stream into image. The technology of document visualization makes up the shortage of BFD in the files which have no significant byte frequency characteristics, such as txt, encrypted file and compound document and so on, and it improves the accuracy of detection further (Conti et al., 2011, Ellen, 2013).

The visual characteristics of 2D code are very obvious, but how to identify code automatically there is scarcely mentioned. In this paper, 2D code is abstracted as graph and the size is 256*256, the coordinate of any point in the graph is equal to 2D code, the initial value of each coordinate is 0 and if there is a 2D code equal to this coordinate, the value of this coordinate plus with 1. After the statistic of value of each coordinate in a fragment, the data is mapped to the graph and the value of each point represents the "grayscale" of this point, the value was used to classify different type. This way of encoding can be accepted by 2DDPCA and achieve the purpose of detection by machine. Due to the limit of length of paper, we don't show the graph here.

4 FEATURE EXTRACTION ALGORITHM

Amirani (2013) reduces the set of byte features which represent the original dataset through the algorithm of PCA (Principal components analysis). Due to the high dimension of the sample vector space when there is a large sample space, the consume of calculation is huge during the process of feature extraction, and the result of the size of the sample is higher comparatively, we cannot ignore the problem of high computational complexity and the difficulty of covariance matrix evaluation. In this paper, we introduce 2DDPCA algorithm to improve the speed of feature selection, and leave the main two-dimensional feature perfectly. The calculation of distance function is more exact and compresses the count of features with the precondition of keeping accuracy.

4.1 *Feature extraction based on 2DDPCA*

Supposing that the set of training sample after encoding is $X = \{x_j^i \in R^{m\times n} \mid i = 1,\cdots,N, j = 1,2,\cdots,k\}$, i represents i-th type of file, j represents j-th file fragment in i-th type, N represents the count of types in training sample, K represents there are K fragments in each type, M represents the total number of training samples and $M = N \times K$.

The average graph of the entire training sample as follows:

$$\bar{X} = \frac{1}{M}\sum_{i=1}^{N}\sum_{j=1}^{K} X_j^i \tag{1}$$

The covariance matrix of the training sample as follows:

$$G = \frac{1}{M}\sum_{i=1}^{N}(X_j^i - \bar{X})^T (X_j^i - \bar{X}) \tag{2}$$

φ_i represents the eigenvector of the covariance matrix, eigenvalue is λ_i , $\{\lambda_i \mid i = 1,2,\ldots,N\}$, the relationship as follows:

$$G \cdot \varphi_i = \lambda_i \cdot \varphi_i \tag{3}$$

Eigenvalues are sorted as $\lambda_1 \geq \lambda_2 \geq \cdots \geq \lambda_n \geq 0$, and the group of the optimal projection vectors $\varphi_1, \varphi_2, \cdots, \varphi_d$ can be selected as the uniting orthogonal vectors which are corresponding to the d maximum eigenvalues, the value of d is changed with the actual requirement. If a file type should be searched in a few times, d can be smaller to reduce the time of comparing.

$$P = (\varphi_1, \varphi_2, \cdots, \varphi_d)^T \tag{4}$$

P is the optimal projection matrix.

4.2 *Feature extraction based on two-level 2DPCA*

The above-mentioned 2DPCA removes the relevance between different rows, but the relevance between different columns is existed too, so the dimension of features can be reduced again.

These features are not classified when we got first time, feature matrix are treated as the new training sample after transposition $I_i = P_i^T \in R^{d\times m}, i = 1,2,\cdots,M$, and repeating the process of 2DPCA feature selection, the projection matrix of columns is $P_c \in R^{m\times d_2}(d_2 < m)$. $Y_j^i = A_k \in R^{m\times d}$, the new covariance matrix as follows:

$$G = \frac{1}{M}\sum_{1}^{M}(A_i - \bar{A})(A_i - \bar{A})^T \tag{5}$$

In formula 5, $\bar{A} = \frac{1}{M}\sum_{i=1}^{M} A_i$.

Similarly: the uniting orthogonal vectors $B_1, B_2, \ldots, B_{d_2}$ are calculated which are corresponding to the d_2 maximum cigenvalues in covariance matrix G, and they are treated as the new projection space (feature space). The extracted feature matrix p_2 after the second time as follows:

$$\begin{aligned} P_2 &= A^T[B_1, B_2, \cdots, B_{d_2}] = P^T X^T [B_1, B_2, \cdots, B_{d_2}] \\ &= [\varphi_1, \varphi_2, \cdots, \varphi_d]^T X^T [B_1, B_2, \cdots, B_{d_2}] \end{aligned} \tag{6}$$

In the first time, the dimensions of the extracted feature matrix is $m\times d$, the dimensions of the second extracted feature matrix is $d\times d_2$, it reduced the dimension of the extracted feature matrix again, the time of classify is shorter and faster.

4.3 *Feature extraction by two-level 2DPCA*

In order to balance the speed and the accuracy, we use both 2DDPCA and SVM. The 2DDPCA extracts d_2 features from 256*256 basic features which were taken from 2D code, and these d_2 features will be used by the SVM classifier to distinguish different types. Figure 1 consists of two phases, the left side is training phase and the right side is detection phase. During both the training and detection phase, after 2D encoding of the sample data the dataset must be normalized, because 256*256 is too bigger to raise the speed.

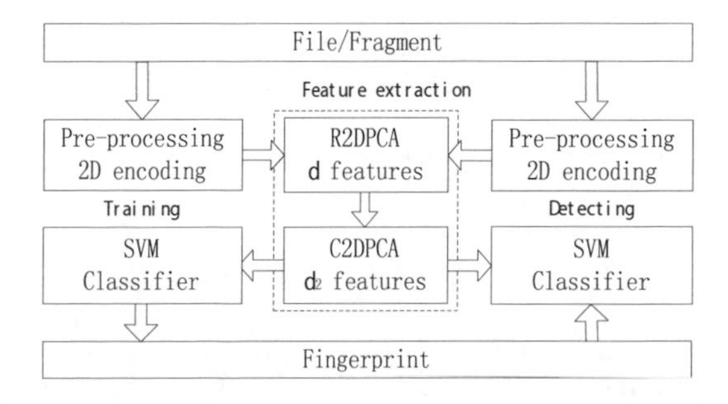

Figure 1. Training and detection phase of the proposed method.

The function is accomplished by Matlab 2010b and LibSVM (CHANG et al) program. LibSVM provides four kinds of Kernel function as linear, polynomial, RBF (Radial Basis Function) and Sigmoid, we apply RBF and it has good accuracy with little additional computational effort. When we are training the sample dataset, the accuracy can be improved by raising the count of training samples. The more samples there are, the generalization of 2DDPCA will be better. We don't concern about the sample, because there are too many files could be chosen, and there will be a lot of time left before computer forensic.

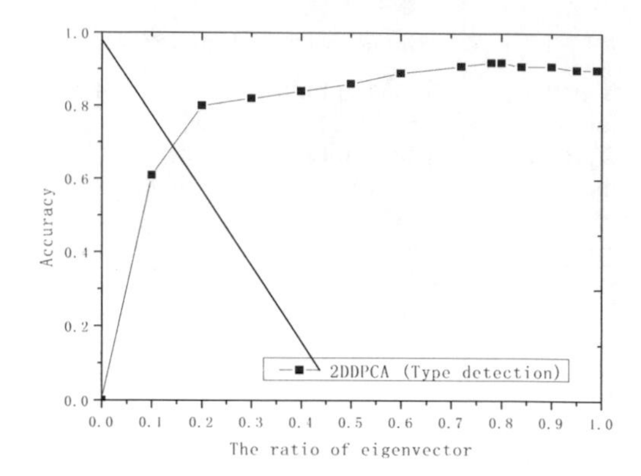

Figure 2. The accuracy with the change of the feature dimension.

5 EXPERIMENT AND RESULTS

5.1 *Dataset*

In the experiment, there are six popular data types (doc, pdf, wav, jpg, gif, txt). The documents such as doc, pdf, txt are collected from Google SE (search engine). The image files JPG are collected from photo sharing websites such as Baidu and Renren. The medium files wav are collected from the databases on the internet. Table 1 shows the details of each file.

There are 2400 files totally and 400 files of each type. 3/4 and 1/4 of the dataset was used for training and testing. There are a large number of files used for training, it is because of the randomizations and we must make sure the training phase has done enough work to cover all the places. Random locations are coming from the order of "rand" in VS2010, and file fragments are as long as 1500 bytes.

Table 1. The details of each file type.

File type	Total number	minimum (Bytes)	maximum (Bytes)	average (Bytes)
doc	400	2,949	7,412,810	296,817
pdf	400	43,008	9,199,616	871,961
wav	400	1,683	18,983,104	7,954,387
jpg	400	9,113	11,744,051	587,219
gif	400	4370	924,759	37,684
txt	400	560	5,691,476	13,250

5.2 *Training and feature selection*

After the progress of 2D encoding, dataset will be normalized as 32*32. The GRBF function of the SVM classifier keeps the default value of $\sigma = 1$, and $d_1 = d_2$. In order to figure out the value of d_2, we make a test as Figure 2 shows the result.

In figure 2, when the ratio is 0.2 would be suitable in our test, the accuracy rises drastically and the number of features is small, it can make a good balance between accuracy and speed. By the way, if we want a better accuracy, the ratio can be increased too.

Table 2 shows the accuracy and timeliness consume of each type of 100 files, txt gets the best result and jpg only reaches 78. Time consumes between different types are small, because the processes of detection are all in the same way.

Table 2. The identified files and the time overhead.

Fragment type	Identified number	Error number	Time sec/hundred
doc	76	24	0.57
pdf	71	29	0.59
wav	90	10	0.58
jpg	84	16	0.60
gif	75	15	0.61
txt	99	1	0.58

5.3 *Comparison results*

In table 3, the average accuracy of the proposed method reaches 83%, which compare with the method of Ellen (2013) by six common types. Ellen used 2D code as the initial input and grain divided different types of grain.

Combining with cluster index, we test the influence of index at the time of detection. In Figure 3, the total number of files is 12000 and each type contains 2000 equally. The number of fragments in every file exponential growth based on 2, the relation between accuracy and the expectation of time shows that the accuracy is growing dramatically when the number of fragments under 512. The highest value of accuracy is closed to 1, because every part in a file has been detected include the obvious features in file header.

The test result shows that the proposed method has a positive effect on these six file types. Some files are easy to figure out like txt and others are not, because inside a file there are other objects. It would confuse the program to judge which type it belongs to, and txt contains words only which make it have a really good performance. Additionally, there also has a lot of file types can be detected by this method and if there have a lot of time left to raise the dimensions of the feature or the count of samples, the accuracy will be higher.

Table 3. The comparison of accuracy between the different methods.

size	classifier	doc	pdf	wav	jpg	gif	txt	CCR
1500Bytes	Grain n	74%	71%	88%	81%	72%	99%	81%
1500Bytes	2DDPCA+SVM	76%	72%	90%	84%	75%	99%	83%

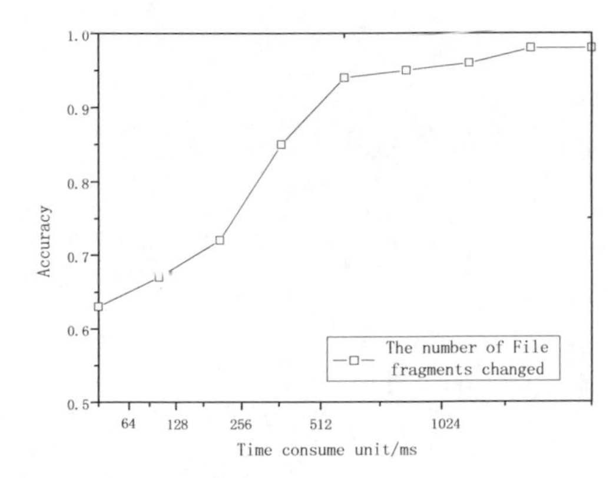

Figure 3. The time over head with cluster index.

6 CONCLUSION AND FUTURE WORK

This paper presents a type detection framework through pre-cluster index and encoding binary data in file fragment, and then exacting features based on 2DDPCA, the file type will be identified by SVM finally. The computing complexity of this algorithm is low, and we can enforce the result by adding file header detection. The proposed method satisfies of large-scale file type detection.

The larger of the data size, the easier to hide information, if the way of encoding objects in a compound file can be connected and limited in a common forensic framework, it will decrease the rate of error and pre-know the content of a file without opening it. If the structure of the compound file can be reconstructed by some tools, the easier of figuring out hidden or falsified files.

REFERENCES

Jain A, Chhabra G S. Anti-forensics techniques: An analytical review[C]//Contemporary Computing (IC3), 2014 Seventh International Conference on. IEEE, 2014: 412–418.

Cao D, J. Luo, M. Yin, and H. Yang, "Feature selection based file type identification algorithm," in 2010 IEEE International Conference on Intelligent Computing and Intelligent Systems (ICIS), Xiamen, China, 2010, pp. 58–62.

Patel P, Mishra S. A Survey On Various Methods To Detect Forgery And Computer Crime In Transaction Database[J]. International Journal of Scientific & Technology Research, 2013, 2(11).

Roussev V, Quates C. File fragment encoding classification—An empirical approach[J]. Digital Investigation, 2013, 10: S69-S77.

McDaniel M, Heydari M H. Content based file type detection algorithms[C]//System Sciences, 2003. Proceedings of the 36th Annual Hawaii International Conference on. IEEE, 2003: 10 pp.

Moody S J, Erbacher R F. SADI-statistical analysis for data type identification[C]//Systematic Approaches to Digital Forensic Engineering, 2008. SADFE'08. Third International Workshop on. IEEE, 2008: 41–54.

Li W J, Wang K, Stolfo S J, et al. Fileprints: Identifying file types by n-gram analysis[C]//Information Assurance Workshop, 2005. IAW'05. Proceedings from the Sixth Annual IEEE SMC. IEEE, 2005: 64–71.

Karresand M, Shahmehri N. Oscar—file type identification of binary data in disk clusters and ram pages[M]//Security and privacy in dynamic environments. Springer US, 2006: 413–424.

Feinerer I, Buchta C, Geiger W, et al. The textcat package for n-gram based text categorization in R[J]. Journal of Statistical Software, 2013, 52(6): 1–17.

Veenman C J. Statistical disk cluster classification for file carving[C]//Information Assurance and Security, 2007. IAS 2007. Third International Symposium on. IEEE, 2007: 393–398.

Erbacher R F, Mulholland J. Identification and localization of data types within large-scale file systems[C]//Systematic Approaches to Digital Forensic Engineering, 2007. SADFE 2007. Second International Workshop on. IEEE, 2007: 55–70.

Ahmed I, Lhee K S, Shin H J, et al. Fast content-based file type identification[M]//Advances in Digital Forensics VII. Springer Berlin Heidelberg, 2011: 65–75.

Amirani M C, Toorani M, Mihandoost S. Feature-based Type Identification of File Fragments[J]. Security and Communication Networks, 2013, 6(1): 115–128.

Ellen J. VMIFF - Visualization metrics for the identification of file fragments[J]. Graduate Theses and Dissertations. 2013, 13131.

Gregory C, Sergey B, Anna S, et al. Automated mapping of large binary objects using primitive fragment type classification[J]. digital investigation, 2010, 7: S3-S12.

Conti G, Bratus S, Shubina A, et al. A Visual Study of Primitive Binary Fragment Types[J]. 2010.

Ahmed I, Lhee K, Shin H, et al. On improving the accuracy and performance of content-based file type identification[C]//Information Security and Privacy. Springer Berlin Heidelberg, 2009: 44–59.

Gopal S, Yang Y, Salomatin K, et al. Statistical learning for file-type identification[C]//Machine Learning and Applications and Workshops (ICMLA), 2011 10th International Conference on. IEEE, 2011, 1: 68–73.

Roussev V, Garfinkel S L. File fragment classification-the case for specialized approaches[C]//Systematic Approaches to Digital Forensic Engineering, 2009. SADFE'09. Fourth International IEEE Workshop on. IEEE, 2009: 3–14.

Weston J, Watkins C. Support vector machines for multi-class pattern recognition[EB/OL]. [2014-6-12]. http://www.cse.yorku.ca/course_archive/200506/F/6002B/Readings /weston99.pdf.

CHANG C C, LIN C J. LIBSVM: Alibrary for support vector machines [EB/OL]. [2014-6-12]. http://www.csie.ntu.edu.tw/~cjlin /papers/libsvm.pdf

Computing, Control, Information and Education Engineering – Liu, Sung & Yao (eds)
 ISBN: 978-1-138-02800-5

A monitor system for big data analytics

Ming Ruo Shi
Beijing Wuzi University, China

ABSTRACT: An application over big data analytics usually needs to build a pipeline on the top of workflow engine which connects these relevant periodic workflow jobs. It's crucial to timely alert pipeline issues, provide an issue diagnosis subsystem to find out root cause from a variety of sources, and make a precise measurement for the entire business performance. In this paper, we identify three indispensable qualities monitor systems must fulfill namely timeliness, accuracy and flexibility. We find that the conventional monitoring tools lack at least one of three qualities, and introduce a monitoring system for big data analytics to keep data freshness, collect measurement metrics and meet SLA.

KEYWORDS: Monitor system (Monitoring Alerting and Diagnosis); Hadoop; Oozie; SLA.

1 INTRODUCTION

A monitoring system becomes increasingly important in analytics over big data falling in a wide family of application scenarios: from online advertising to financial securities exchange, from social networks to medical information systems. The system contains three subsystems i.e. measurement, alerting and diagnosis. Measurement is used to measure if Service Level of Agreement (SLA) is achieved; if Key Performance Indicator (KPI) is met; if system resource is within budget; and other internal measurement indicators such as usage / adoption / coverage / precision / recall of individual components and prediction models. Alerting is targeted to alert about system abnormal situations such as pipeline/service error/over SLA. The purpose of alerting is to shorten MTTD (mean time to detection). Diagnosis provides a tool to better understand the whole system and find the root cause of a system fault more quickly. The target of diagnosis is to shorten the MMTR (mean time to repair). Without monitor, system faults are difficult to detect and hard to track KPI, hence more efforts and time are required to fix a business system issue. In 2006, Khanna [1] developed an external monitor by analyzing external message exchanges. In 2007, Khanna [2] proposed a rule based diagnosis for distributed IT infrastructures. In 2010, Haifeng [3] proposed an invariants based failure diagnosis method for distributed computing systems. In 2010, Joshi [4] proposed a probabilistic model-driven recovery for distributed systems. Some individual software packages [5, 6, 7] e.g. Ganglia, Nagios and Splunk are provided some functionalities for monitor, alerting and diagnosis for distributed systems such as Hadoop [8]. Due to technical complexity, none of these system is general purpose for analytics applications over big data.

MapReduce is the hot distributed and parallel programing paradigm for processing over big data in IT industry. In 2004, Jeffery [9] proposed MapReduce to simplify data processing on large clusters and widely used in Google. In 2006, Hadoop [10] is an open source implementation of MapReduce which is a subproject of Apach Lucene. Hadoop jobs could be aggregated by Pig and Hive. They are widely used in Search engines, data analytics and so forth. It's vital important to provide a general monitoring system for jobs running on both these MapReduce platforms and other online systems.

Workflow engine orchestrates the running of pipeline jobs. These jobs consist of MapReduce job, DB load job, timer job, stream monitoring job or other jobs. A job could be run in hourly, daily, weekly or monthly. Workflow engine is the key components to make job done in the expected way. Oozie [11] is a scalable open source workflow engine for Hadoop.

In this paper, we proposed a general purpose monitor system for analytics over big data in order to keep data freshness, to shorten SLA and to provide metrics to corresponding product system. Current implementation is based on Hadoop and Oozie. It's easy to extend to support Google MapReduce by implementing the predefined interfaces.

The paper is organized as follows: in section 2, we describe in detail the architectural attributes of monitor system that enable timeliness, accuracy and flexibility. Section 3 describes the components of monitor service. In section 4 introduces measurement pipelines. Section 5 introduces deployment details.

Section 6 discusses using the experimental setup to measure the performance. We conclude and summarize our ongoing work in section 7.

2 ARCHITECTURE

Our monitor system is defined on top of big data platform and workflow engine. The current implementation is based on Hadoop and Oozie respectively. It aims at measuring, alerting and diagnosing big data applications. There concerns problems in the architecture collaborated by monitoring service, measurement pipeline, measurement DB, cube and reporting service and Web view. Fig. 1 shows the architecture of the proposed monitor system.

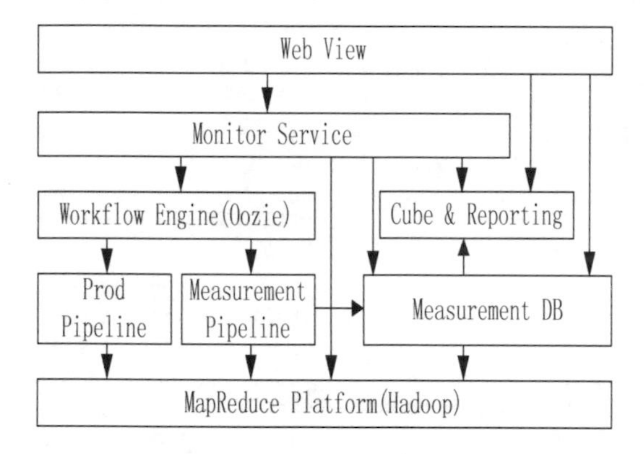

Figure 1. Architecture of propotsed monitor system.

Monitor service is responsible for pipeline monitoring, alerting and diagnosis. Measurement pipeline is responsive for measurement metrics calculation. Measurement DB stores measurement metrics from measurement pipeline and monitor service for dashboard and online retrieval. Based on metrics data, cube is processed. Finally users get pipeline status and measurement metrics by Web view or reporting service. Prod pipeline, i.e. production pipeline for analytics, is monitored by the proposed monitor system.

3 MONITOR SERVICE

Monitor service collects and caches pipeline latest status with which pipeline owners could diagnose pipelines and alert pipeline abnormal behaviors. Current implementation is based on Web Service. The major components show in Fig. 2.

Latest Workflow Status Cache contains latest information of workflow jobs for diagnosis, alerting and pipeline metrics. Job information in the cache is arranged by day. Cache contains last 15 day jobs (configurable). Any unsolved alerts are stored in the cache for tracking purpose.

Pipeline Refresher is the key components to periodically refresh the workflow status. Our period is set to 15 minutes (configurable). During refreshing, the latest job information is refreshed and updated to cache. Logs of running jobs are mined for any workflow errors and added to cache for the newly found workflow errors. Remove any alerts for the newly completed jobs.

Alerting emails are sent by Alert Sender to workflow owner and other desired audience for the newly detected failed jobs. These alerts can be retrieved by Web UI as well since they are maintained in cache.

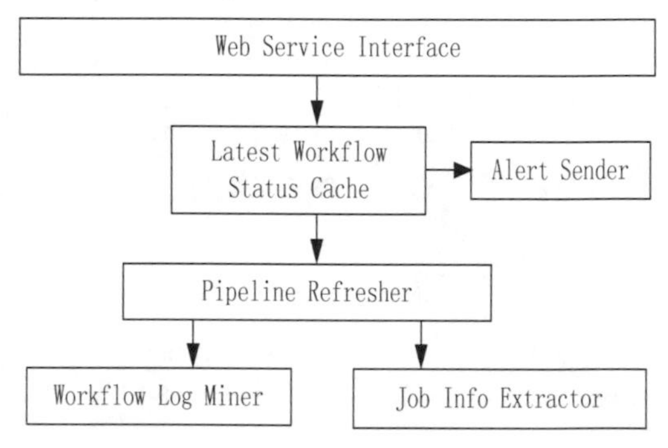

Figure 2. Monitor service major components.

Workflow Log Miner is responsible for mining workflow logs to find workflow errors for alerting and diagnosis. Any individual workflow will write log during its execution. We defines some application specific rules to judge if there is an error.

Job Info Extractor is in charge of parsing Hadoop job metadata including job name, job URL, input streams, output streams, job start running time, PN hour and running time. Multiple threads are used to extract job info since job information extraction is time consuming when there are many jobs.

Monitor's freshness is achieved by the design of monitor service. Monitor will send alert email if workflow meets errors within 15 minutes. Monitor also allows users to set SLA by hours, workflow owners will receive alter email once workflow is over SLA.

By analyzing Oozie log files, the alerting accuracy of workflow errors achieves 100%.

The workflow log minor and job info extractor define interfaces so it can support other big data platform such as Google MapReduce.

4 MEASUREMENT PIPELINE

Measurement pipeline is a key component to make the business system measurable to meet business goal. The Fig. 3 shows the general workflows and dependency for a typical big data applications. For a specific measurement system, measurement workflows vary. Generally, raw log monitor, common log cooking, core measurement, and DbLoad workflows are required.

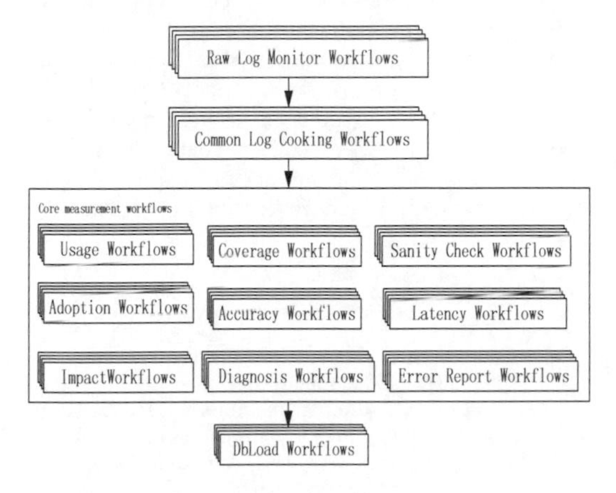

Figure 3. Measurement pipeline for a typical big data application.

Raw log monitor workflow checks if raw log upload is complete. If done, Monitor system will generate a signal file or post an event to indicate its completion. Common log cooking workflow does ETL for later use. Core measurement workflow calculates a variety of measurement metrics which includes and not limited to usage, coverage, sanity check, adoption, accuracy, latency, impact, diagnosis, and error report workflows. Some of these metrics are defined as KPI. DbLoad workflow loads metrics from HDFS to Database for later use by cube and UI.

Measurement pipeline workflows are also monitored by Monitor service which follow the predefined SLA. KPI is very important so KPI related workflows should take higher priority to ensure the freshness and correctness. It's hard to ensure the correctness of metrics in big data analytics since data is from heterogeneous data sources.

Measurement pipeline provides flexibility for data quality check i.e. sanity check. There are two major issues about data quality. One is that measurement result need do sanity check. The other is that streams generated by product environment need also do sanity check so that we provide a general sanity check framework. Feature owner just need write sanity check workflow, the framework to check if the result stream is empty or not. Sanity check is passed if the result stream is empty. Otherwise framework will send alert email for detailed check results to feature owners for follow up.

5 DEPLOYMENT

The Monitor system deployment is shown in Fig. 4 which consists of five machines to host monitor service, measurement pipelines, measurement DBs, cube, and Web UI respectively. Note that Monitor system need access the log and workflow metadata of Prod pipelines. But actually Prod pipelines don't belong to Monitor system. DB, Cube and reporting service are based on SQL Server. Web UI is implemented by Apache Tomcat Java Server Pages.

In practice, each component could be hosted on multiple machines in master slave or always on modes to enhance Monitor system stability. Cube is used for Web UI/Reporting service quick access. Web UI provides four views: (1) Workflow view shows workflow dependency and workflows running status and also provide diagnosis tools; (2) Metrics view shows metrics provided by Monitor system; (3) Dashboard view shows key metrics created by feature owners; (4) Tools view provides other diagnosis tools.

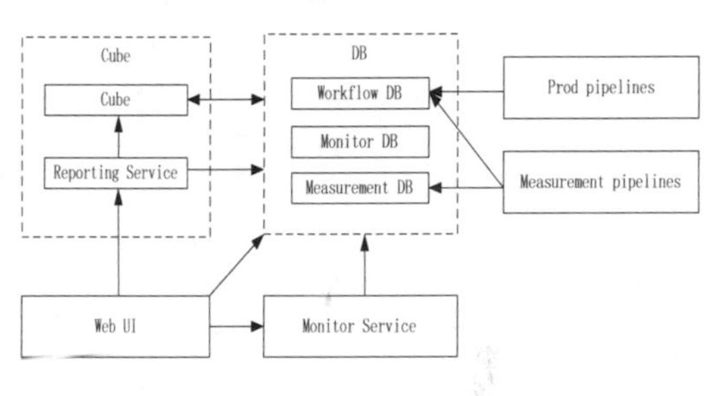

Figure 4. System deployment.

We apply Monitor system to our online Ads test platform. The color on workflow job shows the running status. White, green, blue and red mean "Not started", "Running", "Finished" and "Error" respectively.

6 PERFORMANCE

Accuracy, timeliness and flexibility are the most important identified qualities of Monitor system. For evaluation purpose, we apply our Monitor system to our online Ads test platform. We setup 1000 product workflow jobs per day which is divided into 4 feature areas. Twenty workflows are hourly scheduled. Five hundred and twenty workflows are daily scheduled. These workflows has some predefined dependency. So the start time of these workflows are potentially different. Ten hourly workflow jobs has issue. Ninety daily workflow jobs has issues. The root causes of these issues are: (1) map or reduce function bug; (2) lack dependent jar package; (3) lack required input HDFS file; and (4) workflow configuration error. We running these workflows 2 days.

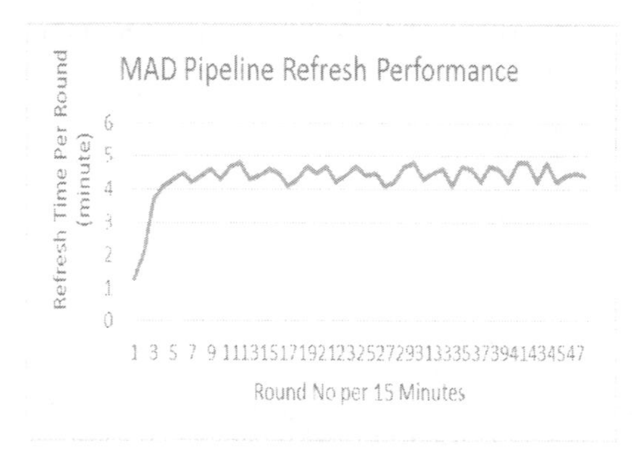

Figure 5. Monitor system pipeline experiment refresh performance.

We found that 101 hundred alerts received of which 1 alerts is false-alarms i.e. job resubmitting message. So recall achieves 100% and precision is 98%.

For timeliness, we refresh the pipeline status every 15 minutes. The actual refresh time per round is shown in Fig 5.

We found the refresh time per round is increased in initial stage then keep stable. This is due to the number of running workflows is increasing in initial rounds due to workflow dependency. Then the number of running workflows keep stable then. Note that we will analyze log files of failed jobs as well as log files of running jobs. The average refresh time is less than 5 minutes.

As discussed in previous section, we could verify sanity check result with a simple and common rule. So alerting for sanity check failures is general by leveraging our sanity check framework. This means the flexibility of data correctness verification.

7 CONCLUSION

A general-purpose Monitor system system is proposed to maintain an online/offline big data application in more efficient and easier way. The Monitor system system can alert timely with 100% recall rate and very low false alarm (less than 1%) and provide fine-grained diagnosis capacity such that the data freshness is kept and maintenance cost is greatly reduced. A general metrics framework is provided for KPI slice-dice by measurement pipelines and UI tools, which provides the indicator for continuous improvement. However, our Monitor system system doesn't resist on single point of failure. ZooKeeper [12] is a good candidate to conquer this issue. Another future work is to support other MapReduce implementations such as SCOPE and other workflow engine although the design itself provide corresponding interfaces.

REFERENCES

[1] Khanna, G.; Padma Varadharajan; Bagchi, S. "Automated online monitoring of distributed applications through external monitors", Dependable and Secure Computing, IEEE Transactions on, On page(s): 115 – 129 Volume: 3, Issue: 2, April-June 2006.

[2] Khanna, G.; Mike Yu Cheng; Padma Varadharajan; Bagchi, S.; Correia, M.P.; Verissimo, P.J., "Automated Rule-Based Diagnosis Through a Distributed Monitor System," Dependable and Secure Computing, IEEE Transactions on , vol.4, no.4, pp.266,279, Oct.-Dec. 2007.

[3] Haifeng Chen; Guofei Jiang; Yoshihira, K.; Saxena, A. "Invariants Based Failure Diagnosis in Distributed Computing Systems", Reliable Distributed Systems, 2010 29th IEEE Symposium on, On page(s): 160 – 166.

[4] Joshi, K.R.; Hiltunen, M.A.; Sanders, W.H.; Schlichting, R.D. "Probabilistic Model-Driven Recovery in Distributed Systems", Dependable and Secure Computing, IEEE Transactions on, On page(s): 913 - 928 Volume: 8, Issue: 6, Nov.-Dec. 2011.

[5] Ganglia. http://ganglia.sourceforge.net/.

[6] Nagios. http://www.nagios.org/.

[7] Splunk. http://www.splunk.com/.

[8] Apache Hadoop. http://wiki.apache.org/hadoop.

[9] Jeffery Dean, Sanjay Ghemawat. MapReduce: Simplified Data Processing on Large Clusters. http://labs.google,com/papers/mapreduce.html. 2004.

[10] Yahoo! Lauches world's Largest Hadoop Production Applications. http://developer.yahoo.com/blogs/hadoop/posts/2008/02/yahoo-worlds-largest-product-hadoop/.

[11] Mohammad Islam, Angelo K. Huang et al. Oozie: Torwards a Scalable Workflow Management System for Hadoop. SWEET 2012, May 20, 2012.

[12] Flavio Junqueira, Benjamin Reed. ZooKeeper: Distributed Process Coordination. O'Reilly, 2013.

Computing, Control, Information and Education Engineering – Liu, Sung & Yao (eds)
© 2015 Taylor & Francis Group, London, ISBN: 978-1-138-02800-5

Improving the performance of AODV routing protocol in VANET

Hong Wei Zou
Huawei Industrial Base, Bantian Longgang, Shenzhen, P.R. China

Xiao Kang Lin
Shenzhen Graduate School of Tsinghua University, Shenzhen, P.R. China

ABSTRACT: Ad hoc On-demand Distance Vector (AODV) Routing is an efficient routing protocol in VANET. AODV is a reactive routing protocol and it works only when the data is transmitted. It communicates with all of its neighbor nodes to exchange the link information. When a node moves with high speed, the route failure happens frequently. In this paper, AODV-PLET (Preferred Link Expiration Time) protocol is proposed to improve the route survival time of AODV. AODV-PLET divides the neighbors to different forwarding priority base on link expiration time. Different priority node forwards the route request packets (RREQ) with a different delay time. During the delay time, node detects the RREQ from neighbors and determines whether RREQ can recover all of their neighbors. If that's so, the node does not rebroadcast the RREQ. Simulation result shows that the AODV-PLET has higher packet delivery rate and lower overhead than AODV. AODV-PLET is more suitable for VANET.

KEYWORDS: Link expiration time, AODV, routing protocol, VANET.

1 INTRODUCTION

Vehicle ad hoc network (VANET) is the most important part of Intelligent Transport System[1]. In VANET, rapid movement leads to network topology change frequently. A high-reliability and high-performance routing protocol is required. Many routing protocols are used in VANET[2]. One of the most effective routing protocols is AODV[3]. AODV works on demand when the node wants to send packets. There are two phases of AODV: route discovery phase, route maintenance phase. When the node wants to send packets and there is no route to destination, route discovery phase works. The source node sends a broadcast routing packet called Route Request Packet (RREQ) to all of the neighbors. If the RREQ receiver node is the destination node or if the node has the route to the destination node, it sends a unicast packet called the Route Reply (RREP) to the RREQ sender. If the node that receives a RREQ does not have the route to the destination then it forwards the RREQ to all of its neighbors. The node receives RREP forward the RREP to the source. The route maintenance phase works when a route is established. Each node sends HELLO packet to its neighbors to keep alive. When a node detects not receiving a HELLO in a long time that means its link was broken, the node unicasts a Route Error (RERR) to the source node. Simulation of existing MANET routing protocols (AODV, DSDV, DSR, TORA) shows that AODV is the best one[4]. However, there are two disadvantages of AODV in VANET. First, each node rebroadcasts the rout request packets (RREQ). This will reduce the network throughput. Second, the algorithm does not consider the stability of a link. When a node moves with high speed, the route failure happens frequently.

In this paper, AODV-PLET (preferred link expiration time) protocol is proposed to improve the route survival time of AODV. AODV-PLET divides the neighbors to different forwarding priority base on link expire time. AODV-PLET selects the stable node as forwarding node, thus reduce the link broken and reduce the route discovery.

Rest of paper organized as below: Section 2 describes AODV-PLET protocol. Section 3 evaluates AODV-PLET with ns2 simulator. Finally, we make a conclusion in section 4.

2 AODV-PLET ROUTING PROTOCOL

2.1 *Concept of LET*

LET(Link Expiration Time) is the link expiration time that two nodes can keep communicating. LET is affected by the relative speed between the two nodes. As photograph 2-1 shown, X(x_1,y_1) is the coordinate of NODE_X. Y(x_2,y_2) is the coordinate of NODE_Y.

Vector v_1 is the speed of NODE_X. v_1 is modulus of vector v_1. θ_1 is the direction of the vector v_1. Vector v_2 is the speed of NODE_Y. v_2 is modulus of vector v_2. θ_2 is the direction of the vector v_2. r is the communication radius.

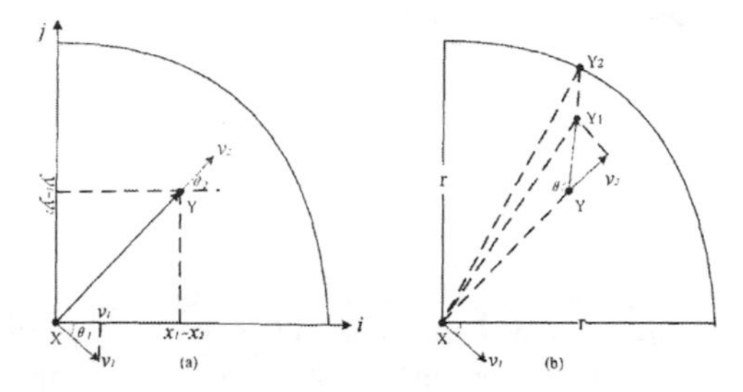

Figure 1. Coordinate of NODE_X and NODE_Y.

Vector YY_1 is the relative speed of NODE_Y moving away from NODE_X. Then $YY_1 = v_2 - v_1$. Assume that NODE_X is standstill and Node_Y moves away from NODE_X. When the distance between NODE_Y and NODE_X get to communication radius r, $|YY_2|$ is the distance between NODE_Y moving away from NODE_X. θ is the moving direction. $\theta, \theta_1, \theta_2 \in (0, 2\pi)$. LET_{XY} is the LET(Link Expiration Time) of NODE_Y moving away from NODE_X.

1) Case $v_2 - v_1 \equiv 0$: $LET_{XY} = \infty$.
2) Case $v_2 - v_1 \neq 0$: $LET_{XY} = |YY_2| \Big/ |\vec{v_2} - \vec{v_1}|$

According cosine of ΔXYY_1:

$$COS\theta = |XY|^2 + |YY_1|^2 - |XY_1|^2 \Big/ 2 \bullet |XY| \bullet |YY_1|$$

According cosine of ΔXYY_2:

$$COS\theta = |XY|^2 + |YY_2|^2 - |XY_2|^2 \Big/ 2 \bullet |XY| \bullet |YY_2|$$

$$|XY_2| = r$$

$$v_1 = v_1 \bullet \cos\theta_1 \bullet i + v_1 \bullet \sin\theta_1 \bullet j \ v_2 = v_2 \bullet \cos\theta_2 \bullet i + v_2 \bullet \sin\theta_2 \bullet j \ v_2 - v_1 = (v_2 \bullet \cos\theta_2 - v_1 \bullet \cos\theta_1) \bullet i + (v_2 \bullet \sin\theta_2 - v_1 \bullet \sin\theta_1) \bullet j$$

$$= v_i \bullet i + v_j \bullet j$$

According to these:

$$LET_{XY} = |YY_2| \Big/ |\vec{v_2} - \vec{v_1}|$$

$$= -(v_i \bullet x_i + v_j \bullet y_j) + \sqrt{(v_i^2 + v_j^2) \bullet r^2 - (v_i \bullet y_j - x_i \bullet v_j)^2} \Big/ v_i^2 + v_j^2$$

$(x_i = x_2 - x_1\ y_i = y_2 - y_1\ v_i = v_2 \bullet \cos\theta_2 - v_1 \bullet \cos\theta_1\ v_j = v_2 \bullet \sin\theta_2 - v_1 \bullet \sin\theta_1)$

2.2 *AODV-PLET*

AODV-PLET (Preferred Link Expiration Time) protocol aims at reducing the link broken. So it selects the node with stable link to forward packets. The AODV-PLET process as follows:

1 Each node broadcasts a HELLO packet to all of its neighbors periodically. The packet header of HELLO is extended by adding the node position, node speed, moving direction information.
2 The node received HELLO packet is getting the position of its neighbor and calculates the distance D between sender and receiver. The receiver marks itself as Preferred-Area (case $\lambda_1 R < D < \lambda_2 R$) or Best-Effect-Area(case $D <= \lambda_1 R$ or $D >= \lambda_2 R$). The λ_1 or λ_2 is a coefficient of range 0 to 1. Default value of λ_1 is $\frac{\sqrt{3}}{3}$. Default value of λ_2 is $\frac{\sqrt{6}}{3}$. Parameter R is the communicate radius.
3 The packet header of Route Request Packet (RREQ) is extended by adding the node position, node speed, and LET(Link expiration time) value .
4 When a node in Preferred-Area receives RREQ, it forwards the RREQ after a $\mu\overline{T_i}$ time. $\overline{T_i}$ is the average one-way delay of Node_i. μ is a coefficient of range 0 to 1.

$$\mu = \frac{LET_i}{n-1}$$

n is a count of its neighbors. LET_i is the order of LET value that Node_i compares with its neighbors. If the LET value of Node_i is the maximal value, then $LET_i = 0$. If the LET value of Node_i is the second max value, then $LET_i = 1$. If the LET value of Node_i is the minimal value, then $LET_i = n-1$.

When a node in Best-Effect-Area receives RREQ, it will wait for a T_{listen} time not forwarding RREQ.

$$T_{listen} = 2\overline{T_i} \quad (2\text{-}10)$$

During the T_{listen} time, the node listens RREQ packets. Then the node judge if all of its neighbors have received the RREQ packets or not. If all of its neighbors have received the RREQ packets, the node does not forward the RREQ anymore. If any of its neighbors does not receive RREQ, the node will forward the RREQ after T_{listen} time.

5 When a node needs to forward the RREQ, it compares its LET value with LET value in the

receiving RREQ packet. Then it puts the minor LET value in RREQ packet.

6 When a node needs to update the backward route table, it compares the local LET value with LET value in the receiving RREQ packet. If its LET value is minor than the receiving, then update else do not update.
7 When the destination node receives the RREQ packet, it reply the first arriving RREQ packet.

Take Photograph 2-2 for example:

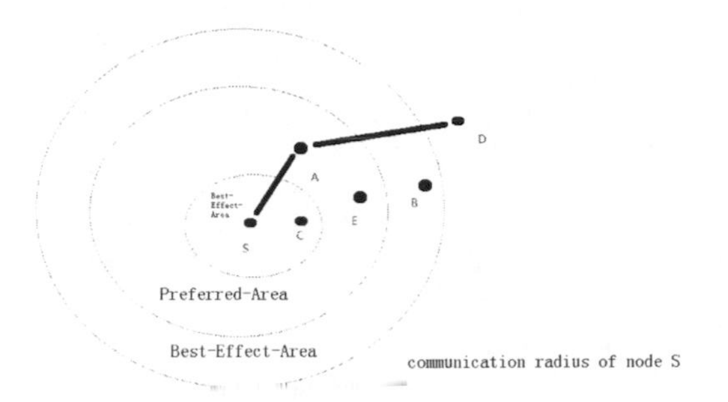

Figure 2. AODV-PLET procedure.

According to the node position, Node A and node E is marked as Preferrred-Area. Node C and node B is marked as a Best-Effect-Area. When Node S sends packet to Node D. S broadcasts a RREQ packet. Node A, B, C, and E will receive the RREQ. Node A and Node E will forward the RREQ in $\mu\overline{T_i}$ time because they are Preferred-Area. Node B and C listens RREQ for $2\overline{T_i}$ time, then judge whether to forward RREQ. In node A, the $\mu\overline{T_i}$ value is set to 0 because A has the maximal value of LET value among all of its neighbors (B, C, E). So node A will forward the RREQ immediately. In node E, the $\mu\overline{T_i}$ value is set to $\overline{T_i}/5$ because E has 5 neighbors and the LET value order is second. So node E forwards the RREQ after $\overline{T_i}/5$ delay time. The destination node D first receives a RREQ from A. Then it sends RREP to A. So the route(S->A->D) is set up. After receiving RREQ, Node B and C waits for $2\overline{T_i}$ time. During the $2\overline{T_i}$ time, Node B and C receive same RREQ from A and E. Node C judge that all of its neighbors (A, S, B, E) have received the RREQ. Then node C gets the result that no need to forward the RREQ. Similarly, node B gets the result that no need to forward the RREQ. So the AODV-PLET reduces the broadcasting of RREQ packets from 4 to 2. AODV-PLET greatly reduces the Routing Overhead.

3 SIMULATION OF AODV-PLET PROTOCOL

Simulate the AODV-PLET in city realistic traffic model with NS2(network simulator). In the 1200*1200 meters range, create 60 nodes. Select 802.11 as MAC protocol. The node communication radius is set to 250 meters, bandwidth is set to 2Mbits/s, application traffic is set to CBR, packet size is set to 200 BYTES, simulation time is set to 200 seconds. Randomly select 40 nodes from 60 nodes to send CBR packets. CBR rate is set to 16Kbits/s. Node moves randomly, with maximal speed 72KM, moving pause time is set to 3 seconds, 6 seconds, 9 seconds , 12 seconds , 15 seconds , 18 seconds , 21 seconds , 27 seconds , 30 seconds. Compare the PDR(packet delivery ratio), NRO(Normalized Routing Overhead), AED(Average End-to-End Delay) between AODV-PLET and AODV

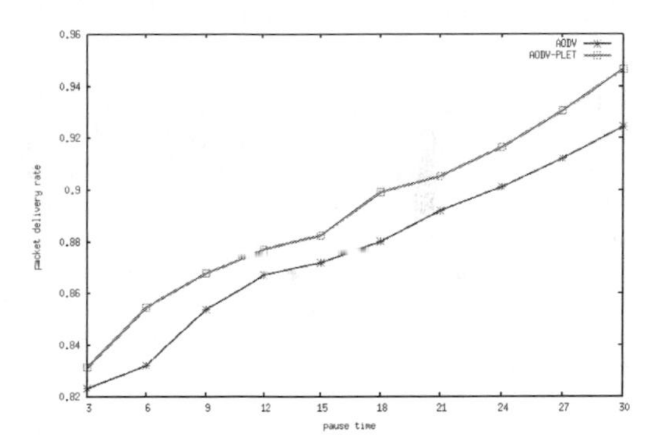

Figure 3. Compare PDR between AODV-PLET and AODV.

Photograph 3-1 shows the PDR(packet delivery ratio) difference between AODV-PLET and AODV in different moving pause time. As the moving pause time increase, PDR of both two protocols is increased. Because, as the pause time increase, the network becomes stable and the link broken reduces and the route discovery reduces. PDR of AODV-PLET is better than a PDR of AODV. Because AODV-PLET first select the stable node as forwarding node, this reduces the link broken.

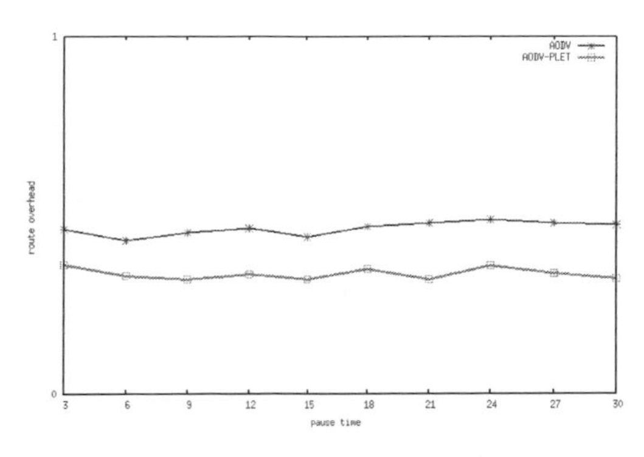

Figure 4. Compare NRO between AODV-PLET and AODV.

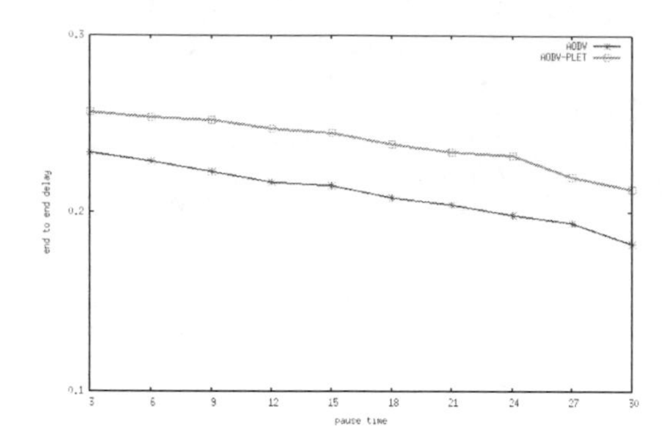

Figure 5. Compare AED between AODV-PLET and AODV.

Photograph 3-2 shows the NRO(Normalized Routing Overhead) difference between AODV-PLET and AODV in different moving pause time. NRO of AODV-PLET is better than NRO of AODV because AODV-PLET drops many redundancy RREQ packets. Another reason is that AODV-PLET first select the stable node as forwarding node, this reduces the link broken and reduce the route recovery.

Photograph 3-3 shows the AED(Average End-to-End Delay) difference between AODV-PLET and AODV in different moving pause time. AODV-PLET has a more AED than AODV because AODV-PLET delay some time to forward the RREQ. Another reason is AODV-PLET select the stable node to forward and the stable node is not the minimal hop node. Consider the delay increase less than 10%, the disadvantage can be ignored

To summarize, AODV-PLET increases little AED delay, highly increases PDR, highly reduces the NRO. As the explosive growth of VANET traffic, PDR and NRO become more and more important. AODV-PLET is more suitable for VANET.

4 SUMMARY

AODV-PLET divides the neighbors to different forwarding priority base on link expire time. Different priority node forwards the route request packets (RRQ) with a different delay time. During the delay time, node detects the RRQ from neighbors and determines whether RRQ can recover all of their neighbors. If that is so, the node does not rebroadcast the RRQ. The result of city realistic traffic model simulation shows that the AODV-PLET have higher packet delivery rate and lower overhead than AODV. AODV-PLET is more suitable for VANET.

REFERENCES

[1] Internet ITS Consortium, http://www.internetits.org.

[2] Royer et al, "A review of current routing protocols for ad hoc mobile wireless networks", IEEE Personal Communications, Apr 99.

[3] C. E. Perkins and E. M. Royer, "Ad-hoc On-Demand Distance Vector Routing". In Proceedings of the 2nd IEEE Workshop on Mobile Computing Systems and Applications, pages 90–100, New Orleans, LA, 1999.

[4] S. Das, C. Perkins and E. Royer, "Performance comparison of two on-demand routing protocols for ad hoc networks," in Proc. of 19th IEEE Conf. on Computer, Communications, pp. 3–12, Mar. 26–30, 2000. Article (Cross Ref Link).

Computing, Control, Information and Education Engineering – Liu, Sung & Yao (eds)
© 2015 Taylor & Francis Group, London, ISBN: 978-1-138-02800-5

Research on digital micro-teaching platform-based novel pattern for art and design education: An innovation methodology

Xin Yue Zhang
Lijiang College of Guangxi Normal University, Guangxi, China

ABSTRACT: With the rapid development and progress in computer science and information science technology, the combination of computer and traditional art and design teaching is needed. In order to build up a better system for micro-teaching, we implement it by adding multimedia editing, film video production, multimedia storage, video on demand, digital broadcast live to form a much powerful digital micro system. As a special kind of applied computer science based application, through adding audio visual technology on traditional educational theory, the micro-teaching method plays a unique role in modern school. The result of the experiment is acceptable.

KEYWORDS: Digital Technique; Novel Pattern; Art and Design Education; Innovative.

1 INTRODUCTION

With the rapid development of information technology, especially in the growing popularity of internet and digital storage which are represented by networking and digitizing, it is necessary to produce a reform that combines Micro teaching with new video recording and storage technology, and it is also necessary to take advantage of the existing network resources to create a digital micro system that combines with Micro teaching, multimedia editor, film production, storage, video on demand, digital live broadcasting. Micro teaching is an applied educational technology that combines with audio-visual technology with educational theories; it can train students' teaching skills effectively [1-3]. Microteaching is a classroom teaching skills training methods for students in school and in-service teachers, which is defined as a Controllable practical system makes normal students and teachers could focus on solving a specific behavior or study under a controlled conditions, built on educational theory, visual theory and technology. Micro-teaching represents a new idea and way of resources development. With the development of society, the progress of time, the wide usage of modern information technology, micro-teaching is a kind of rapidly developed form on innovation education which has many characters such as various methods. It is a small but integrated teaching activity which centers with the resource of micro video, and match with other educational resources. It is a systematic teaching skill training method for teachers and normal students. The rapid development of multimedia technology and information technology, not only provides a large number of digital learning resources, but the digital training environment replaces of simulation, which make great changes in information store. However, micro resource management and teaching skills evaluation methods are not synchronized to follow up. Hence, the study of teaching resource management and evaluation in micro teaching is imperative. Therefore, Micro teaching can be implemented in the relevant teaching theories and psychological theories. However, how to infiltrate Micro teaching into specific curriculum instruction is a permanent problem. This article will focus on the verification of the practical Micro teaching from the perspective of martial arts teaching, thus to further improve and perfect Micro teaching pattern. Teaching resource management (TSM), the module includes resource information import, maintenance, inquiry and micro video on demand. The system classifies recourses by student id, classroom location, training time automatic. The student can also retrieve the error information resources by resource claim. The system offers online play and download play to facilitate users, and several ways to quickly search related resources. According to resources clicks, numbers of evaluation, and assessment grading, the system will selection some outstanding resources for recommending and long-term preservation in order to facilitate students. A subject and a curriculum can be specially designed for students and meet their requirement throng micro-teaching development in which approach is more in line with the needs of curricula. And then, the cognitive sense of student is concerned with the teaching progress, therefore, micro-teaching should also emphases on helping students develop a proactive learning attitude.

2 OUR PLATFORM

2.1 *Structure of the system*

The digital Micro - System is a digital network system, which adopts advanced digital transmitting techniques, digital storing and network applying a plan. Students participate micro training is changing every year, so there is a familiar process for each of them. This process will certainly have some valid resources. Due to a large number of students and large capacity of video the resources can achieve more than 1000G every year. Therefore, preserve all these resources for long-term is unrealistic. Filter the teaching resources to ensure those left behind which are the best is needed. It is a collection of Micro-teaching, multi-media editing, audio and video making, multi-media storing, video on demand, digital living, the publishing of Micro-teaching resources, Micro-teaching experience exchanging, and Micro-teaching evaluating [4]. They are the new Micro-teaching systems on the base of Web2.0 standard, P2P techniques, digital video collecting and processing techniques, digital image storing and indexing techniques, and digital video on demand techniques, which make digital micro-teaching, digital micro-teaching course learning and the releasing of Digital Micro-Teaching resources come true through Blog, BBS, Wiki [5], RSS, video instant-publishing and NOD&MOD [6]. Multimedia digital micro-system has many practical functions in teaching with window type and imaging Chinese software interface. This kind of multi-media digital micro-system is easy to manipulate and often called micro-classroom [7]. As to the structure, function and movement of the digital micro system, please refer to Figure. 1.

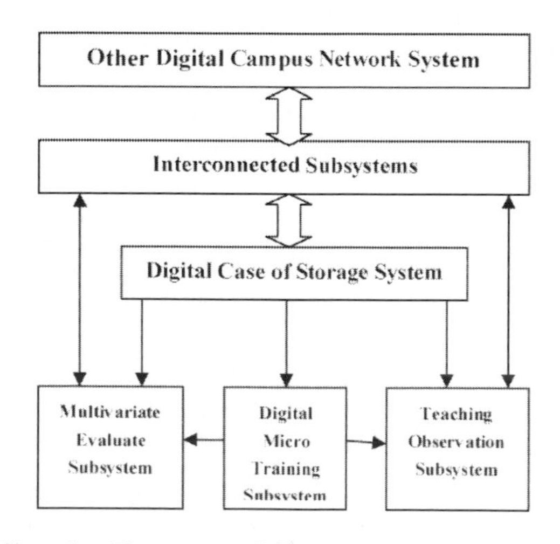

Figure 1. The structure of the system.

2.2 *Design idea of digital micro-system*

integrating training theory with skill practice, the digital micro system is based on digital micro equipment and computer network system, including: micro classroom teaching system, IC card reading system, the control room monitoring system, the classification of the network broadcast, automated micro video categorization, live web casting, Micro teaching and management system, editor and the special effect of nonlinear systems. Function modules which include digital micro training and examination subsystem, teaching subsystem number and case, digital case or network storage subsystem based on multiple evaluation subsystem of a computer. In addition, it also includes other digital connection and sharing resource subsystem.

2.3 *Realization method of digital micro-system*

training and storage subsystem are not only the skeletons of the digital micro system, it is also the core of the whole system and application models. At present the related digital micro's research and the practice mainly concern this. In fact, this has already decided the operating way of the digital micro system. The figure. 2 shows the detailed information.

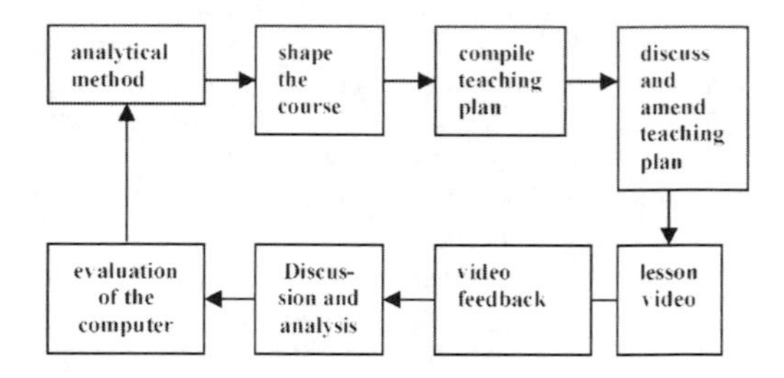

Figure 2. The digital micro-teaching diagram.

This module is responsible for user information management operation, such as importing user information, information changes, rights management and user authentication operation, such as user login, rights distribution. Information both teachers and students are imported by the system to make sure evaluation effectiveness. According to the roles in the process of microteaching and class information, they are divided into guidance teacher, regular teacher, the same professional student and general students. The different user owns different rights and operations. Environment management module responsible for the overall environment settings, including professional management, class information management, micro classroom management, and bulk import combination with individually set by the administrator at the beginning of the practical. The teaching recourse bulk import through background and automatically updated daily.

3 EXPERIMENT AND ANALYSIS

3.1 *General analysis*

martial arts, mostly appear in routines and there are a lot of changeable action and transformation routes. It is difficult for beginners, and thus the high demand for the use of teaching methods will be proposed. For complex action, more difficult and high demanding of martial art teaching technique and some abstract conception exercises, it is difficult for students to understand and master [5]. The research does experiment and control teaching combined the above digital micro system design with realization of the operation mode, through the mathematical statistics to prove the feasibility and teaching effectiveness of Digital micro-teaching mode. Experimental group and Control group are set up to compare experimental, process and analyze the statistical data to make a conclusion. According to status analysis, we designed a digital microteaching management system for the web. The system mainly consists of four functional modules specific structure is shown in Figure 3.

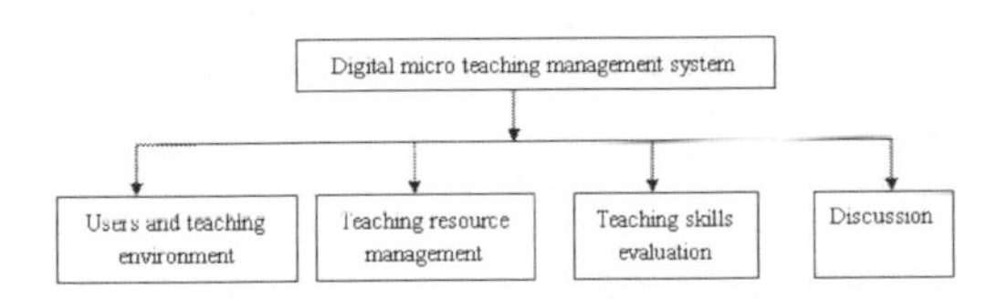

Figure 3. The system function module.

3.2 *Experimental analysis*

We show the result of the experiment and give a detailed explanation of the related result. (1) A Comparison of technology total score. This paper analyzes the total result which examines the student before and after the experiment. Respectively viewed from the horizontal and vertical standpoints, we may discover that there are highly significant differences between the control group and experimental group (See Fig.3). It shows that the students' technique of the two groups has improved. In view of the differential value (T=6.450, P<O.Ol), the act technique of the experimental group is improved more significantly than the control group, indicating that making full use of digital micro-teaching is very beneficial for the students to enhance their overall technical skills. (2) A Comparison of movement accuracy Because martial arts teaching emphasizes the standards of the movement, the students' sensation ability on spatial movement should be paid attention to in actual teaching. The principal means to raise the sensation level is to detect the movement using the visual contrast by watching the video so that students can get timely and accurate act technique. In view of the act accuracy of the two groups (see Fig.4), there is a significant difference for the experimental group before and after the measure (P<0.05). As to the control group, there isn't a notable difference (P>0.05), indicating that the micro-technology can help the experimental group to understand the act accuracy better by video feedback. (3) A Comparison of Movement Range Movement range is a very important target to weigh the martial arts skill. A diversity of physical quality causes different understanding of the movement range in actual teaching, especially for the flexibility quality and the strength quality. Different feeling, perception and visual affect the teaching effect in the actual teaching to certain extent, but the digital micro-teaching technology can effectively eliminate these differences. Firstly, compare with the video material to form correct representation; secondly, get students to understand the condition of them clearly through computer's objective evaluation: Finally, make prompt and effective improvement. (4) A Comparison of Movement Dynamics Action force is the key point of martial arts teaching, because act dynamics directly decide the students' level. Martial arts act dynamics become the essential part in teaching and training. We can specifically target training tools and methods on every student, to effectively improve the students experience different action intensity, leading to raising the drilling level of the martial arts routines. By analysis, we can see that (in Fig.3) there's no significant difference from the control group before and after the experiment (p>0.05). There is a significant difference for the experimental group before and after the experiment, and between the control and experimental groups. This indicates the experimental group students have a higher regulative ability in movement dynamics than the control group, obviously. (5) A Comparison of Movement Mastery Martial arts act skill refers to the memory level of martial arts movement. Because of complicated and diversified act, strong rhythm, the students have to remember the order and the speed of the movement skillfully. Undoubtedly, this will greatly increase the difficulty for the students to learn. Martial arts will not be mastered in one day. One will not have much gain if one works by fits and starts. To avoid forming wrong dynamic stereotype, the students need unceasing improvement and correction. By digital Micro teaching system, students can understand the act technology better and enhance the skill of the martial arts effectively. Form Fig.3 we can see that, before and after the experimental measure there is no significant difference from the control group (p>0.05). There is a distinguished difference for the experimental group before and after the experiment, and between the control and experimental groups ($p < 0.01$). This shows that the control group has not improved the proficiency obviously, but the experimental group is at a higher level of proficiency. (6) Movement Expressive Force Comparison Practicing martial arts, internally "Jing, qi and shen" are refined; externally muscles, bones and skin are strengthened. When the form breaks, the intent connects; when the momentum stops, the "qi" links. The charm of martial arts must be displayed through concrete body movement, which requests the students to understand the martial arts movement deeply. The main solution is to observe, emulate and learn from outstanding students or athletes frequently. Only the digital video technology can solve this difficult problem. It can be seen from Fig.4 that there is a significant disparity before and after the experiment for the experimental group ($p < 0.05$). The differential value ($p < 0.01$) will explain the movement expressive force in the experimental group is better than the control one, which is useful for students to enhance confidence, highlight the expressive action. Therefore, making full use of digital technology may promote effectively the performance ability. The following figures 4-7 are the simulation result.

Groups	Index before experiment	Index after experiment	Differentials	Tested by t
Experimental group	75.7 ± 6.5	84.1 ± 5.4	8.4 ± 1.1	8.017**
control group	79.7 ± 5.9	80.5 ± 4.9	0.7 ± 0.9	4.061**
Groups t check	0.013	3.014**	6.450**	—

Figure 4. The experiment result 1.

Groups	Index before experiment	Index after experiment	Differentials	Tested by t
Experimental group	16.4 ± 1.4	17.0 ± 1.1	0.6 ± 0.3	2.276*
control group	16.4 ± 1.0	16.5 ± 0.9	0.1 ± .01	0.823
groups t check	0.133	1.554	1.408	—

Figure 5. The experiment result 2.

Groups	Index before experiment	Index after experiment	Differentials	Tested by t
Experimental group	15.9 ± 1.4	17.4 ± 1.9	1.4 ± 0.2	6.644**
control group	15.7 ± 1.7	16.1 ± 1.3	0.3 ± 0.2	1.798
Groups t check	0.389	3.101**	3.556**	—

Figure 6. The experiment result 3.

Groups	Index before experiment	Index after experiment	Differentials	Tested by t
Expermental group	15.3 ± 1.9	17.0 ± 1.1	1.7 ± 0.3	5.773**
control group	15.5 ± 1.7	15.7 ± 1.4	0.2 ± 0.1	2.168*
Groups t check	0.304	3.445**	4.532**	—

Figure 7. The experiment result 4.

4 CONCLUSION

Integrating all resources to comprehensive quality control, the digital Micro teaching system has diverse methods of teaching and evaluation and high efficiency. The use of digital micro-system can improve informational attainment of students, teachers and administrative staff, can optimize the procedure of and improve the environment of educating and learning which can expand learning resource and can form student independence and innovation. This kind of educational model contains three systems, that is, e-learning, classroom teaching and extra-curricular activities. In sports teaching, the application of the digital micro educational model will be an important symbol of sports teaching modernization. Applying digital micro-system is an important way of re-allocation and optimization of teaching resources, enhancement of the information management and promotion of academic development.

REFERENCES

Li, M. (2015). Instructional Model Oriented Towards Improving Teaching Ability of Preservice Teachers. In Exploring Learning & Teaching in Higher Education (pp. 3–41). Springer Berlin Heidelberg.

Kesicioğlu, O. S. (2015). The effects of an undergraduate programme of preschool teaching on preservice teachers' attitudes towards early mathematics education in Turkey: a longitudinal study. Early Child Development and Care, 185(1), 84–99.

Rutherford, V., Conway, P. F., & Murphy, R. (2015). Looking like a teacher: fashioning an embodied identity through dressage. Teaching Education, (ahead-of-print), 1–15.

Martin, A. J. (2015). Are These Testing Times, or Is It a Time to Test? Considering the Place of Tests in Students' Academic Development. In Controversies in Education (pp. 55–62). Springer International Publishing.

Al-Humaidi, S. H., & Abu-Rahmah, M. I. (2015). Enhancing Microteaching at Sultan Qaboos University. Studies in English Language Teaching, 3(1), p28.

Eleftheriou, M., Reuver, M., Bostock, J., Sorgeloos, P., & Dhont, J. (2015). AQUA-TNET thematic network: an 18-year chronicle of development and achievement in European aquaculture education. Aquaculture International, 1–11.

Computing, Control, Information and Education Engineering – Liu, Sung & Yao (eds)
© 2015 Taylor & Francis Group, London, ISBN: 978-1-138-02800-5

Brain waves intelligent control and real-time monitoring equipment

Yi Zhang
School of Electronic Engineering, Beijing University of Posts and Telecommunication, Beijing, China

Si Bo Hao
School of Automation, Harbin Institute of Technology, Harbin, China

Tian Yi Qiao
Tsinghua University, Beijing, China

Gao Feng Cui
School of Electronic Engineering, Beijing University of Posts and Telecommunication, Beijing, China

ABSTRACT: The existing intelligent system mainly uses manual operation completes human-computer interaction, so the traditional manual control system still has its limitations, can not get rid of manual control, resulting in some people with disabilities cannot be used. Put the brain wave elements into intelligent manipulation can liberate controller's hands, bring greater convenience. This article will use brain wave extraction equipment as a non-implantable device system, using induced brain waves to form specific EEG brain waves, and for intelligent control. Currently, we have implemented the use of brain waves switch songs and play different music in real-time based on people's mood. This paper has achieved the short-term monitoring of negative emotions, of human emotions and make improvements. People with disabilities can easily use. Meanwhile, spontaneous EEG can also be used in monitoring human states, for example, health and mood. Eventually, the device can be made into portable equipment.

KEYWORDS: intelligent control system; brain wave; portable equipment.

1 INTRODUCTION

1.1 *Introduction*

Electroencephalogram (EEG) is the result of the cerebral cortex activity, and is related to brain's cognition and perception to the surroundings. Neuroelectricity has a high level of temporal resolution, which enabled it to be analyzed in terms of frequency, spectrum, high frequency spectrum and etc. EEG place particular emphasis on the information transmission and processing of time, which can reflect the subjects' information-process condition while dealing with cognizing tasks. Utilizing the high level of temporal resolution, EEG can be real-time processed and transformed into instructions. Thereby touch-less control is realized. It is easy to acquire EEG with the present technology, but in contrast, it's hard to analyze its exact physical and physiological meaning. In the process with certain algorithm, exact physical and physiological meaning is obtained as the degree of concentration and relaxation. These two indexes are subjectively controllable, which can be called induced brain electricity, which can be induced by simple imagination. So it is feasible to realize the control of things subjectively controlling the brain to relax and concentrate—which is proved by the ideational games in the current market. In addition to real-time control, we can also build portable devices with Bluetooth and Wi-Fi technology. The devices can be used to monitor the individual's mental and health index at real-time. Now Beijing university of Posts and Telecommunication project team has realized the functions of playing different music according to the mood of the subjects. Mainstream ways of playing music-direct control and voice control on computers or mobile phones or the most convenient Home button control designed by apple-has no exception of the dependence on manual control. Besides, it is not practical to play different music by monitoring an individual's mood, so as to improve his/her condition. Besides, switching songs backward and forward according to concentration values provides great convenience to people with disabilities and those who need this function. The second function- playing music by examining a person's mood in order to alleviate his/her anxiety or fatigue condition-enjoys high health value too.

1.2 *The character of EEG*

Brain wave frequency range between 1 to 30 times per second is divided into four bands-δ(1–3Hz),θ(4–7Hz),α(8–13Hz),β(14–30Hz). When a person is cheerful or meditating, the beta, delta and theta waves which kept active weak down, while the alpha wave is reinforced. Theta wave is extremely apparent while frustrated and depressed. Beta wave is produced by mental tension and high emotion. So beta wave value will rise while anxiety. Modern research discovered that the state of the participants can be observed by the value of theta and beta wave. Eighty percent of the normal rhythm has given priority to the alpha brain waves. The alpha rhythm among normal and neuropahic anxiety patients declines or disappears. Brain-Computer Interface is a system, taking advantage of EEG. The system can realize the communication and control between the brain and other electronic equipment. BCI is a kind of through the brain electrical signal to realize the human brain and computer or other electronic communication equipment and control system. Brain-Computer Interface is divided into two kinds, direct BCI and indirect BCI.

Indirect BCI is advisable for external system control. Non-implanting BCI is divided into two kinds, evoked EEG and spontaneous EEG. In this application, evoked EEG is adopted, which can produce electrical activity generated by external stimuli. Seeing that Mu/Beta rhythm is related to human activities and sensory function, it can be adopted as a good ideal feature of EEG signals.

Table 1. Peak and mean value of β and θ in contrast group.

High beta max	High beta average	Theta average	High Alpha max	High Alpha Average
<40000	15000	70000	40000	22000

2 INTELLECTUAL CONTROL

2.1 *Intellectual control solution based on the EEGG*

The users wear EEG module.The receiving terminal receives the users' brain signal and the program automatically analyze their brain waves,in order to estimate their mental state.Subsequently,open the music library to search and display the corresponding music.To substantiate, the system will play music suitable for thinking when user's alpha brain wave is stronger—the user is in a state of calm relaxed ; the system will play slow and soft music when user's beta wave is stronger—the user is in a state of anxiety; and

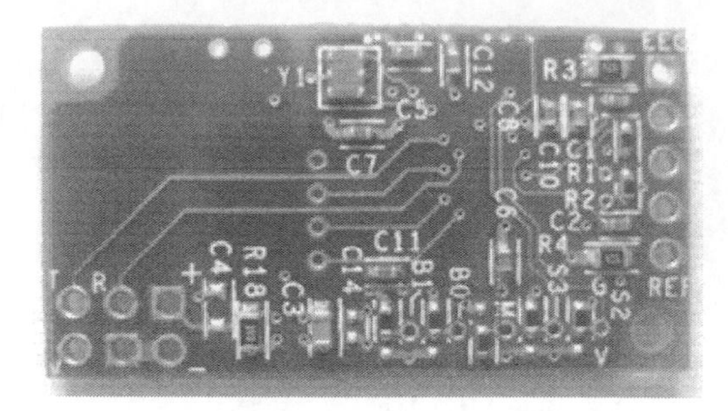

Figure 1. TGAM chip module.

same for theta and delta waves—the system will also play corresponding music.

2.2 *Experiments and data acquisition*

The experimental group: Working at his desk for a long time, mental pressure large student population.

The control group: Mental state is relatively healthy students.

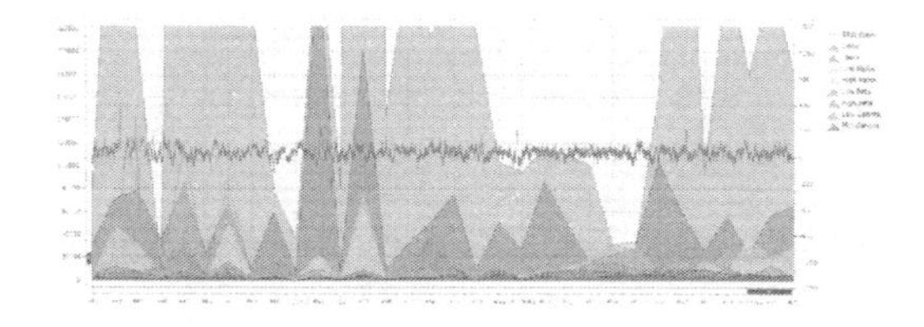

Figure 2. The contrast group.

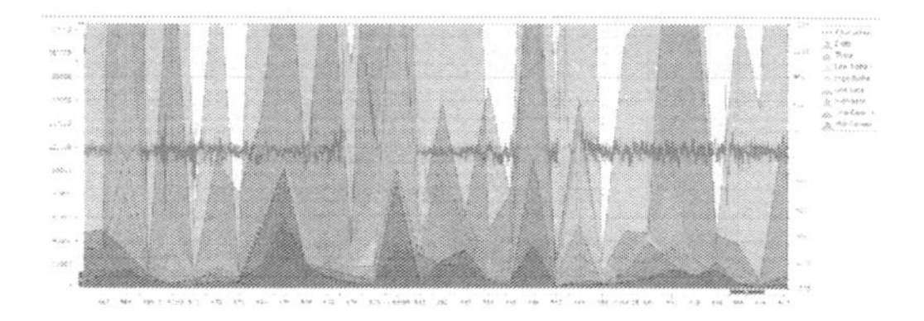

Figure 3. The anxiety group.

Table 2. Peak and mean value of β and θ in anxiety group (Before the experiment).

High beta max	High beta average	Theta average	High Alpha max	High Alpha average
>220000	140000	200000	30000	<20000

Table 3. Peak and mean value of β and θ in anxiety group (After the experiment).

High beta max	High beta average	Theta average	High Alpha max	High Alpha average
<40000	15000	70000	40000	22000

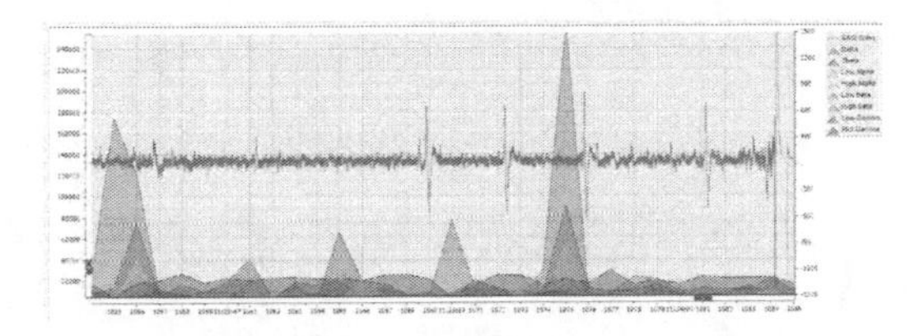

Figure 4. The anxiety group (After the experiment).

Measured with instrument, observation data are displayed on the computer, and sampling record. In anxiety group, is still in the quiet environment, were measured with instrument brainwave state, sampling record. To compare the two groups of data images.

3 DATA ANALYSIS

3.1 *The relationship between data and voltage*

Real-time voltage of forehead skin:

V_{EGG}=[rawdata*(1.8/4096)]/2000

The enlarged voltage:

Vm=rawdata*(1.8/4096)

the value of alpha, beta and theta in the experiment has no unit, and is valid only compared with rawdata.

3.2 *Data analysis before experiment*

Before the experiment, the maximum of anxiety group's Beta value is higher than 220000 while the value of the control group is less than 20000. Obviously the value of the anxiety group is higher than the control group.The average of anxiety group's beta value is about 140000 and the theta value is about 70000.On contrast, the indexes of the control group are 10000 and 15000,much smaller than those of the anxiety group.The beta and theta value of excitement is higher than normal. When people feel good or meditation retreat, α wave relative has been strengthened.

3.3 *Data analysis after experiment*

After the experiment,the maximum of anxiety group's beta value is around 40000 and the average is around 15000.These values declined after the experiment, but are still higher than those of the control group. The values of alpha increased after the experiment.So the alpha wave is reinforced when a person is cheerful or meditating.

4 CONCLUSIONS

According to the analysis of experimental results, playing the corresponding state of music, determined by brain waves' real-time analysis of people's mood, has a positive impact on their mental state. Through the further research using such equipment, a practical monitoring and improving system of mental state has been developed, providing it with a broad prospect. And this new system will bring great improvements human life.

REFERENCES

R.Benzi.A.Sutera, and, A.Vulpiani, J, Phys.A 14,1453(1981)

C.Nicols, Tellus 34,1(1982).

F.Moss, in *Frontiers in Applied Mathmatics*, edited by G.Weiss (SIAM, Philadelphia, 1992).

L.Gammaition, P, Hanggi, P, Jung, and F.Moss, Phys.Rev.A 46,R1709(1991).

R.Benzi, A.Sutera, and A.Vulpaini, J.Phys.Rev.A 46, R1709(1991).

K.Wiesenfeld, D.Pierson, EPantazeloi, C, Dames, and F.Moss, Phys.Rev.Lett.72,2125.

A.S.Mikhailov, *Foundtions of Synergetics I* (Springer-Verlag,Berlin,1994).

M.Locher and E.Hunt, Phys.Rev.Lett.77,4698.

S.J.Simth, *Progress in Brain Research*,edited by L.Hertz,M.D.Norenberg,E.Sykova,and S.Waxman (Elsevier, Amsterdam, 1992), Vol, 94, p.119.

P.Jung, A, H.Cornell-Bell, K.Madden, and, and F.Moss, J.Neurophys.70,1098.

P.Junf and G.Mayer-Kress, Chaos 5,458.

S.Kadar, J.Wang, and K.Showalter, Nature(London)391,770.

Computing, Control, Information and Education Engineering – Liu, Sung & Yao (eds)
 ISBN: 978-1-138-02800-5

Design of radar antenna control system test device

Wei Ming Du, Guo Hong Liu & Li Juan Gu
Wuhan Mechanical Technology School, Wuhan, China

ABSTRACT: Following the test and repair work is difficult and complex in modern radar maintenance support, the automatic testing of radar is necessary for radar maintenance man. Radar antenna control system test device is designed to test work state and performance parameter of the radar antenna control system off-line or on-line. And means while it has expert default diagnosis system to analyze key signals and locate typical faults. The test device is accordingly seen as the best means to radar maintenance support. The design of radar antenna control system test device is given, its hardware solution and software solution is explained. And finally the practicality and reliability of the test device are discussed at least.

KEYWORDS: radar antenna control system; test device; industrial control computer; radar maintenance support.

1 INTRODUCTION

1.1 *The problem of radar maintenance*

Withthe developmentt of radar technology, especially Large Scale Integrated circuit (LSI), Radar becomes more and more complicated, the automation and modularization is higher and higher. In order to test its performance, the number of test instruments rapidly increases. The radar performance test is more and more inconvenient and difficult. Under this condition, the automatic test technology is blossoming [1]. How to fast and automatic test radar is a problem for maintenance support staff. And the antinomy restricts the radar maintenance ability, which is between the high requests of users and their skill actuality in the current radar maintenance system.

1.2 *The use of test device*

The radar antenna control system test device is a kind of comprehensive utilization of modern computer integrated test technology, virtual instrument technology, computer control technology, the fault diagnosis expert system technology and touch screen technology. The use of standardized structure, make test and repair in one of portable device. The test device offers a working environment for the radar antenna control system, mainly used to test work state, performance parameter and locate faults faster and automatically [2]. The device can reduce the work difficulty of maintenance support staff and the maintain time of the antenna control system.

2 THE HARDWARE DESIGN

The hardware of radar antenna control system test device is mainly composed of industrial control computer, DA output module, AD sampling module, controllable silicon rectifier and analog feedback module. The hardware make-up and internal relation block diagram is shown in figure 1.

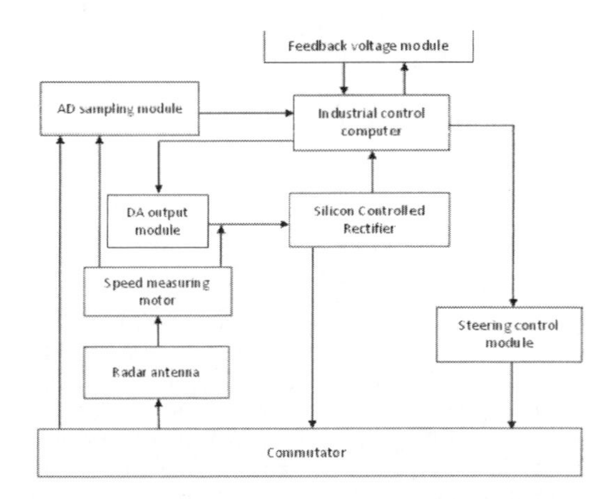

Figure 1. The hardware make-up and internal relation block diagram.

3 PRINCIPLE OF OPERATION

3.1 *How to test antenna*

When the test device is used to control and test antenna, user just needs to set the antenna rotational speed and rotation direction on the touch screen of industrial control computer. The antenna rotation speed is converted to control signal of the DA output module. The DA module outputs the given voltage. The speed measuring motor outputs the feedback voltage. The feedback voltage value is compared with the given voltage value, and then their voltage difference is amplified to control three-phase thyristor conduction angle of the controllable silicon rectifier. And then the value of output voltage is changed and sent into the commutator, which control the rotation speed of the antenna. At the same time, industrial control computer sends the rotation direction information to the steering control module, the polarity of the output voltage of controlled silicon is converted to control the positive rotation or the reverse rotation of the antenna.

3.2 *How to test antenna control system*

When the test device is used to control and test antenna control system, it can provide load to make it work offline. AD module samples the control voltage in load and outputs digital data to industrial control computer. After calculation, the industrial control computer outputs the voltage value of the feedback voltage module. The feedback voltage is simulated and then sent to industrial control computer. With the work of the simulation feedback voltage, the function of the radar antenna control system is checked.

4 INDUSTRIAL CONTROL COMPUTER

Industrial control computer is mainly used to control working mode of test device on user interface. The parameters are displayed on the touch screen in real time, such as input voltage and current, output voltage and current and feedback voltage. Industrial control computer can control the rotated speed and direction of the antenna.

Industrial control computer is composed of CPU module, the input module, output module, memorizer and insulation unit, as normal micro-computer control system. The user program must read-in the memorizer beforehand and is executed in scan mode. When the user operates the switch or knob on the control panel, one trigger signal is produced and sent to input module. When CPU scans the trigger signal, as its response, the different user program is executed. The output signals are used to indicate faults, trigger the time counter, make the switch on or off, then the programmable power supply produces the work voltage. Then the radar antenna control system is working, its electric current and voltage is indicated on the screen. If the current is over range, the protect circuit is made out trigger signal to a PLC, then PLC will cut off the power supply to protect Radar.

5 AD SAMPLING MODULE

5.1 *The composition*

The AD sampling module is composed of AD sampling chip ADC0809, microcontroller chip STC89C52 and communication chip MAX232. The AD sampling module block diagram is shown in figure 2.

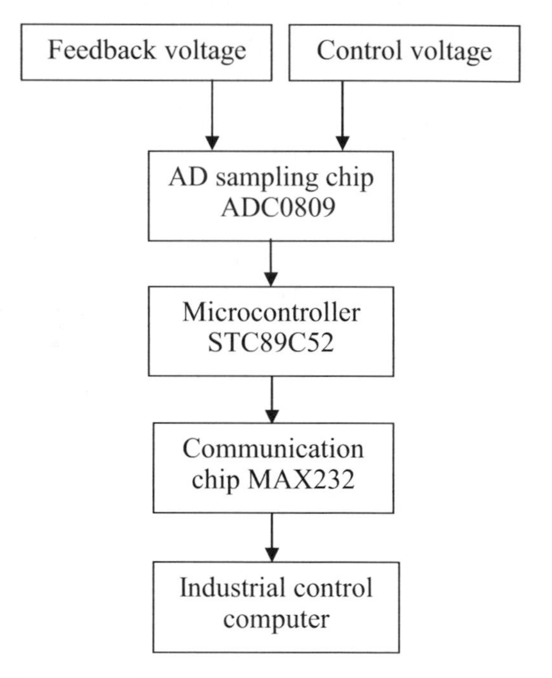

Figure 2. The AD sampling module block diagram.

5.2 *The work principle*

The control voltage of rotational speed of the antenna and the feedback voltage is acquired by AD sampling chip ADC0809 and converted into 8 bit digital signal, transmitted to the microcontroller chip STC89C52. The results are calculated in the computer program and then transmitted to computer through communication chip MAX232, and finally displayed on touch screen.

6 DA OUTPUT MODULE

6.1 *The composition*

The DA output module is composed of DA conversion chip DAC0832, microcontroller chip STC89C52, communication chip MAX232 and amplifier LM324. The DA output module block diagram is shown in figure 3.

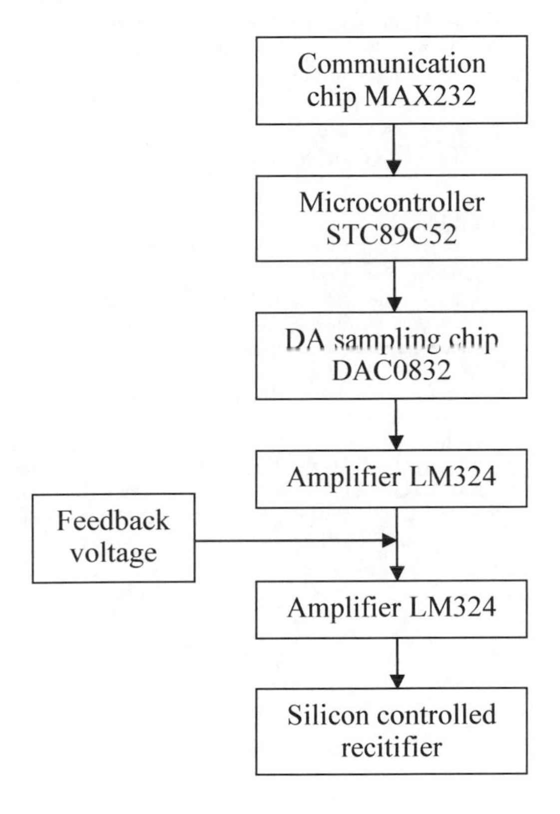

Figure 3. The DA output module block diagram.

6.2 *The work principle*

When the user sets the antenna rotational speed and rotation direction on the touch screen of the industrial control computer, the value of antenna rotation speed is converted to digital instruction. And then the digital instruction is transmitted to microcomputer chip STC89C52 through communication chip MAX232. The digital results is sent to the DA conversion chip DAC0832, which is calculated in microcomputer chip. DA conversion chip produces the corresponding simulation given voltage in certain clock. The voltage is amplified through the operational amplifier LM324, and then added to the feedback voltage from motor speed. Their sum through the operational amplifier LM324 and then sent to silicon controlled rectifier, to realize the real-time control of antenna rotating speed.

7 SILICON CONTROLLED RECTIFIER

7.1 *The composition*

The silicon controlled rectifier is composed of DC amplifier, synchronous pulse generator and the controlled silicon. The block diagram of the silicon controlled rectifier module is shown in figure 4.

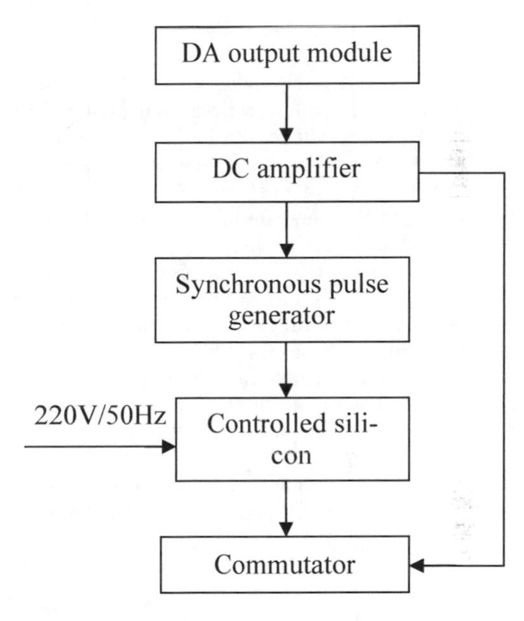

Figure 4. The silicon controlled rectifier block diagram.

7.2 *The work principle*

The sum of the DA module output given voltage and speed feedback voltage, sent to the DC amplifier, the output voltage of amplifier controls the oscillation frequency of three synchronous pulse generator, also controls the appear time of trigger pulses. Trigger pulse is applied to the control pole of controlled silicon, make controlled silicon conducts, converts AC into DC. The size of output DC voltage is determined by trigger pulse arrival time sooner or later. As the given voltage can be continuously controlled, so synchronous pulse generator can be continuously controlled. In this way, the value of output voltage of silicon controlled rectifier can be continuously controlled. Under the action of voltage, drive motor rotates at the certain speed.

8 ANALOG FEEDBACK MODULE

The digital instruction is transmitted to the microcontroller chip STC89C52 through communication chip MAX232. According to the polarity of feedback voltage, electric relay is turned on or off. At the same time, the feedback voltage controls the resistance value of digital potentiometer MCP41010 through the Optocoupler isolation chip TLP521-4. The 50V switch voltage is divided by resistance, and then the required simulation feedback voltage is generated.

9 THE SOFTWARE DESIGN

The windows system is installed in the industrial control computer, and user program is mainly written in Microsoft Visual Basic, meanwhile some DLLs to drive hardware module is written in C language. When the user turns on the power switch, user program runs and initializes equipment automatically, then enter to user interface. User need choose test item, then the user program enters to subprogram. The automatic test system can automatically generate a different supply signal and synchronous trigger signal for different test items. The process is automatically controlled by the user program, as data acquisition, data possess, analyze of the signal and the output of outcome. Finally result and some important parameters are displayed on the screen. The expert default diagnosis system also installed to analyze key signals and locate faults.

10 CONCLUSIONS

The radar antenna control system test device has been successfully used in certain model radar. Through the practical application test, it meets design requirements and functions well. The device uses modularize design to ensure expansibility of system, and use automatic test technique and default diagnosis system to extricate the radar maintenance man form intricate analysis [3]. In summary, the device can provide off-line working environment for the radar antenna control system, and can help radar maintenance man to test performance parameter of the radar antenna control system fast and automatic, so it provides the best means for radar maintenance.

REFERENCES

[1] Weiming Du, Caibin Liu. Research on Automatic Test System of radar Board Based on Virtual Instrument Technology. Applied Mechanics and Materials,Vols.121–126(2012), P.2002–2005.

[2] Xiaohuai Chen, Fangning Tian, Guoqiang Sun. Automatic test system design used in radar antenna control system. Informatization Research: 2010, p40–42, In Chinese.

[3] Wenhua Hu, Limin Liu. The Condition Monitoring and Failure Isolation System for Electronic Equipment. 7th International Symposium on Test and Measurement (2007): p.3181–3183.

Computing, Control, Information and Education Engineering – Liu, Sung & Yao (eds)
© 2015 Taylor & Francis Group, London, ISBN: 978-1-138-02800-5

A search-efficient neighboring optimization algorithm in P2P-VoD networks

X.F. Meng, Q.X. Huang & K.Y. Ding
School of Computer Science and Technology, Dalian University of Technology, Dalian, China

ABSTRACT: A theoretic model for the P2P-VoD search problem which takes both the peers' accessibility and the search's cost into account has been presented, and the algorithm for choosing and adjusting a peer's neighbor peers, under which a peer's neighbor peers always have the similar interests to that of the peer has also been proposed in this paper. Simulations show that this algorithm can largely improve the search efficiency of the peer's requested movies and the performance of P2P VoD systems.

KEYWORDS: P2P; video on demand; Replication; Scheduling; Neighbor selecting.

1 INTRODUCTION

Peer-to-peer (P2P) systems have received much attention and become one of the most popular distributed applications in recent years. With the boosting of multimedia contents over the Internet, more and more researchers have focused their efforts on the using of P2P network to deliver video-on-demand (VoD) services, which is the so-called P2P-VoD system. A P2P-VoD system has many advantages, such as high scalability, low cost and robustness to churn.

A critical problem which should be solved in P2P VoD system is how to mitigate the workload on the server. To this end, a lot of replication algorithms have been proposed, however, most algorithms are based on the assumption that the issued request for a movie could arrive at all the peers which hold the movie or the replicas of the movie. Apparently, this is an obstacle to applying the algorithm to the real P2P VoD systems since such assumption does not always hold due to the existence of TTL, a threshold used to limit the forwarding hops of a request.

In this paper, we take two issues into consideration to improve the performance of P2P VoD systems. The first is how to choose a peer's neighbor peers according to its interests so as to be able for the peer to obtain its requested movies in a given hops. The second is how to adjust a peer's neighbor peers when a change of the peer's interests took place.

2 STARTED SYSTEM ARCHITECTURE AND SCHEDULING POLICY

2.1 *System architecture*

We first describe the architecture of a P2P-VoD system to which our proposed algorithms can be applied. There are two kinds of resources in P2P VoD system: nodes (including servers, trackers, and user nodes) and contents. The P2P VoD system architecture includes: 1) content severs 2) trackers and 3) peers. The content severs store all of the movies and can upload data to requesting peers. After each peer downloads data from system, it reports to the tracker the data it caches in the local storage so as to provide upload service to other peers. The main function of a tracker is helping peers find the data they demanded. As for each peer in P2P-Vod system, it needs to contribute its local storage to cache video data and upload bandwidth to provide upload service to other peers, respectively.

2.2 *Scheduling policy*

We denote peers which are in TTL hop range as the reachable peers. In this case, suppose a peer wants to watch movieM_k, its watching request can be supported by two different sources: 1) the server S, 2) reachable peers. Under this kind of situation, the peer scheduling strategy is that a peer first requests data from reachable peers, and only when reachable peers cannot supply sufficient bandwidth, then this peer requests data from server S.

Here we focus on how to increase the proportion of reachable peers in concurrent and replication peers so as to increase the utilization of replicas and decrease the server's workload.

3 MODEL

3.1 *Relationship between movies and peers*

As we can see, in most P2P-VoD systems such as PPive, PPstream, movies are divided into different categories according to their contents. In this paper, we divide movies into m classes: $K_1, K_2, \ldots, K_m$. The information about this categories are opened to all the peers in the system. As a peer may have different possibilities to choose movies based on its preference. We use node-lists to connect peers' preference and movies. Each movie class has a corresponding node-list, so there are m node-lists $P_1, P_2, \ldots, P_m$ in the system. A node-list P_i is a set of series peers which have high possibility to watch movies in the movie class K_i.

Node-lists are stored in trackers. For a peer i, trackers will collect the statistical data about its history movie viewing records, analyze the data, and then put peer i into node-list according to the analysis result. Among all the movies in the records, check which class they belong to. If the ratio of movie class K_i to records exceeds the threshold, put this peer in to node-list P_i. This means that peer i is interested in the movies belong to movie class K_i and will watch movies in movie class K_i with high probability in the future. Notice that one peer may also join several node-lists, because it may have several different interests.

Node-lists are maintained by both trackers and peers. Trackers communicate with peers regularly, get the latest data about peers' history movie viewing records, analyze the data, and then update node-lists according to the analysis result.

3.2 *Model of server's workload*

After analyzing the relationship between movies and peers, we continue to explore the bandwidth workload of the server. Assuming a P2P-VoD system has N peers and K movies. The playback rate time of a particular movie is r, and every movie has the same playback rate. For, peer i, according to the scheduling strategy, let d_i^{TTL} be the download supported by reachable peers.

Therefore, if we want to minimize the server load, we need to maximize the download bandwidth supported by reachable concurrent peers and reachable replication peers. In this case, the final optimization problem can be described as:

$$\max \sum_{j=1}^{N_{TTL}} d_{ij} \times f_{ij} \tag{1}$$

s.t.

$$\sum d_{ij} < U_j \tag{2}$$

$$N_{TTL} \le \left(\sum_{i=1}^{T} L^i\right) \times N_{percent} \times T_{percent} \tag{3}$$

Where d_{ij} is the upload bandwidth that peer j provides to peer i; f_{ij} is the impact of the distance (i.e., hops) between peer i and peer j, and the more hops between peer i and peer j, the greater impact it has to the system performance; U_j is the upload capacity of peer i; The possible largest number of peers in TTL hops is $\sum_{i=1}^{T} L^i$, due to each peer has the same number L of neighbors, and the TTL hop value is T; $T_{percent}$ is the proportion of peers that can provide upload bandwidth to requesting peer; and N_{TTL} is the number of peers that provide upload bandwidth during the viewing process.

According to this objective function and constraints. To decrease the server's workload, what we need to do is to maximize $T_{percent}$ so as to reach the largest number of peers that can provide bandwidth during the viewing process. Meanwhile, minimizing the delay caused by too many hops between request peer and service peer.

4 NEIGHBOR SELECTING ALGORITHM

4.1 *Connection degree*

In order to maximize the $T_{percent}$, we analyze it first. As far as we are concerned, $T_{percent}$ Can be expressed as:

$$T_{percent} = \frac{N_{TTL}}{\sum_{i=1}^{T} L^i} \tag{4}$$

If we want to maximize $T_{percent}$, what we need to do is increasing the number of peers that can provide bandwidth N_{TTL}. That is, a peer gives priority to other peers within the same interests to be its neighbors, and according to the amount of data transmitted between this peer and other peers to adjust neighbors dynamically so as to make this peer and its neighbors closely linked and frequently transmit data with each other. Here we propose the concept of connection degree.

The connection degree C_{ij} represents the tightness between peer i and peer j. The larger the value is, the more amount and frequency the data transmission between these two peers. In a period of time t, the connection degree between peer i and peer j can be expressed as:

$$C_{ij} = D_{ij}^{\alpha} \times \alpha + D_{ij}^{\beta} \times \beta \quad (5)$$

In which, α and β are weighting factors range from 0-1 and $\alpha + \beta = 1$. D_{ij}^{α} represents the amount data transferred between peer i and j within the first $t/2$ time, and D_{ij}^{β} represents the amount data transferred between peer i and j within the last $t/2$ time. By adjusting the value α and β, we can adjust the inflv.

4.2 *Search-efficient Neighbor Static Selection and Dynamic Adjustment Algorithm*

According to the above analysis, we propose the search-efficient neighboring static selection and dynamic adjustment algorithm. The process for this algorithm can be described as follows:

Static selection algorithm

Parameter specification:

L_i: the number of peer i's neighbors

```
For each peer i:
  If L_i < pre-set value
      Statistic the history viewing record in a period of
time;
      Calculate its node-lists;
      Send request to trackers, then trackers give the
corresponding node-lists;
      Select peers in node-lists randomly to be its new
neighbors;
  If L_i > pre-set value
      For each neighbor j:
         Calculate the connection degree C_ij;
      End for;
      Select the peer n which has the minimum
connection degree;
      Disconnect with peer n;
  Else do dynamic adjustment algorithms
End for.
```

With the growth of time, peer's interest may migrate. Under such circumstance, the neighbors of this peer will change correspondingly. We use dynamic adjustment algorithm to adjust peers' neighbor dynamically. The process for this algorithm can be described as follows:

Dynamic adjustment algorithms

```
When a peer i finished watching a movie, do
For each neighbor j:
Calculate the connection degree C_ij;
End for;
Select the peer n which has the minimum connection C_in
degree;
For each peer j which has exchanged data with this peer:
   If the peer j is in its neighbor list
      No change;
   If the peer j is not in its neighbor list
      Calculate the connection degree C_ij between peer i
and peer j;
      If C_in < C_ij
         Peer i disconnect with peer n;
         Peer i connect with peer j;
      End if
   End if
End for
```

---

5 SIMULATION AND RESULT ANALYSIS

5.1 *Measurement of system performance*

Two important parameters have been selected as the measurement of the system performance, to measure the performance of neighbor selecting algorithms in the P2P-VoD system.

The server's workload: represent by the upload bandwidth provided by the server. This index refers to the total data provided by the server. The decrease in the bandwidth cost of the server represents the improvement in the performance of neighbor selecting strategy. Consequently, the lower the bandwidth cost of server is, the better the effect of the neighbor selecting strategy will be.

The utilization of each replica: represent by the data transmitted between peers. The higher the utilization of each replica is, the more frequently data transmission between peers, which result in higher bandwidth and storage utilization on P2P-VoD system.

5.2 *Simulation environment and result*

We validate the mathematical models and evaluate the performance of Search-efficient neighboring static selection and dynamic adjustment algorithms through simulation. Focusing on the problem of selecting neighbors, the simplest strategy is the random selection strategy (RS). That is, select any online peer as its neighbor without considering the interests and performance of this peer. Another strategy is the random class selection (RCS). Select peers in the same node-list randomly. This strategy takes peers' interests into consideration, but it failed to analysis the information about history viewing record. We compare our algorithm (SS-DA) with these two strategies.

We use Peersim as the simulation tool to evaluate the performance of each neighbor selecting strategies. We use the followings as inputs to our simulator: (1) the total number of peers in the system is N=10000. The average peer's upload capacity is 512 kbps. Each peer can cache 40 movies and its history viewing records contain 20 movies. Each peer has 5 neighbors, and TTL hop is 4. (2) The system provides K=20 kinds of movies, and each kind includes 100 different movies. Since the purpose is to evaluate the utilization of replica and the performance of neighbor selecting strategies, therefore, the length of each movie is equal, which will not influence the analysis of the algorithms performance.

In the experiment, at every time round, the peer requests one movie. If other peers in TTL hop range cannot provide enough upload bandwidth, the peer will search server for help.

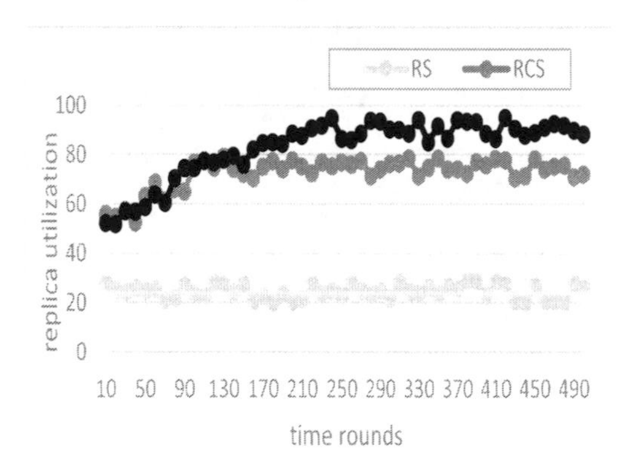

Figure 1. The relationship between the average utilization number of replicas and simulation time.

It can be seen from the figure that, with accumulating of time round, the average replica utilization percent from the RCS strategy and SS-DA strategy is increasing, while the RS strategy had not changed significantly. However, the performance of RCS is limited after about 150 round times. That is because, despite the consideration of peers' preference, there still exist some limitation for RCS to deal with neighbor selecting. On the other hand, SS-DA takes peers' preference and history viewing records into consideration, increases the data transmission between peers, thus the average replica utilization is higher than other two strategies.

As to the server's workload. After each time round, we make each peer selects a movie with a certain probability according to its preference and interests migration. Then statistic the number of peers which failed to get enough bandwidth from other peers. These peers' demands will be supported by the server. Therefore, we use the number of peers which ask the server to provide upload bandwidth to measure the server's workload.

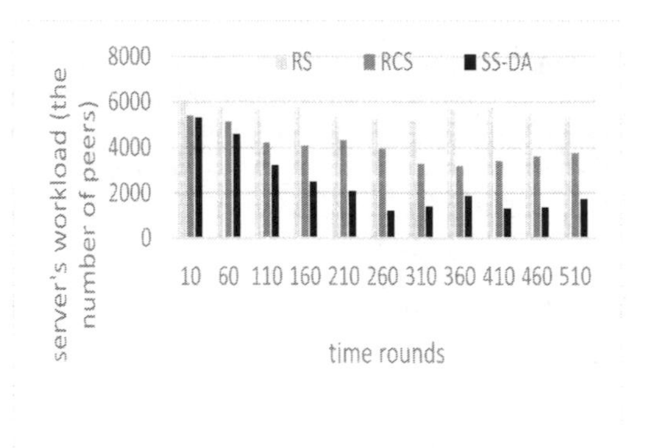

Figure 2. The relationship between the server's workload and simulation time rounds.

The relationship between the server's workload and simulation time rounds is shown in figure 2, from which it can be seen that the server's workload in the SS-DA strategy is lower than other two strategies, showing that when compared with the RS and RCS strategies, the SS-DA strategy can reduce the server's workload and improve the performance of the system.

6 CONCLUSION

Movie replication is an effective technique in the design of P2P-VoD systems. However, it remains an open problem to maximize the utilization of each replica so as to minimize the server's workload. In this paper, we present mathematical models and formulate an optimization framework to understand the impact of replicas' utilization on server's workload, and reveal a search-efficient neighboring static selection and dynamic adjustment algorithms to increase the utilization of the replicas so as to improve the sharing rate of each replica and reduce the load of the server. Through the calculation of each peer's interests and the connection degree between peers, peer selects neighbors which can provide upload bandwidth to it.

We compared our neighbor selecting strategy with other two selecting strategies. Finally, we validate via simulations to show that our algorithms can achieve high quality of performance and reduce the server's workload.

REFERENCES

Ramzan, N. & Prak, H. & Izquierdo, E. 2012. Video streaming over P2P networks: Challenges and opportunities. *Signal Processing: Image Communication* 27(5):401–411.

Yipeng, Z. & Tom, Z, J, F. & Dah, M, C. 2013. On replication algorithm in P2P VoD. *Transactions on Networking* 21(1):233–243.

Hao, Z. & Minghua, C. & Abhay, P. 2011. A distributed multichannel demand-adaptive P2P VoD system with optimized caching and neighbor-selection. *International Society for Optics and Photonics*: 81350X–81350X19.

Carta, A. & Mellia, M. & Meo, M. 2010. Efficient uplink bandwidth utilization in p2p-tv streaming system. *Global Telecommunications Conference*: 1–6.

Jiajun, W. & Cheng, H. & Jin, L. 2008. On ISP-friendly rate allocation for peer-assisted VoD. *Proceeding of the 16th ACM International Conference on Multimedia*: 279–288.

Yifeng, H. & Ling, G. Improving the streaming capacity in p2p vod systems with helpers. *Multimedia and Expo, 2009. ICME 2009. IEEE International Conference on* 27(5):790–793.

Hao, Z. & Jiajun, W. & Minghua, C. & Ramchandran, K. 2009. Scaling peer-to-peer video-on-demand systems using helpers. *Image Processing (ICIP), 2009 16th IEEE International Conference on* 27(5):3053–3056.

Shaowei, S. & Jinjin, W. & Jiali, Y. 2011. Bandwidth Adaptive Data Schedule Strategy in P2P VoD. *Computer Engineering* 37(1):13–15.

Haiyang, W. & jiangchuan, L. & Ke, X. 2010. Measurement and enhancement of BitTorrent-based video file swarming. *Peer-to-Peer Networking and Applications* 3(3):237–253.

Hsu, C, H. & Hefeeda, M. 2010. Quality-aware segment transmission scheduling in peer-to-peer streaming systems. *Proceedings of the first annual ACM SIGMM conference on Multimedia systems*: 169–180.

Computing, Control, Information and Education Engineering – Liu, Sung & Yao (eds)
© 2015 Taylor & Francis Group, London, ISBN: 978-1-138-02800-5

Architecture design of two associated softwares for a MFL inspection device

S. Kong, Z. W. Ling, X. L. Guo, M. L. Zheng & W. C. Guo
Zhejiang Provincial Special Equipment Inspection and Research Institute, Hangzhou, China

ABSTRACT: Depending on the establishment of mechanical structure, the architecture of two softwares was designed. Firstly, the bridge communication the mechanical structure, the data acquisition software and the data analysis software were ascertained. Secondly, the architecture of data acquisition software and some key points were designed. Thirdly, the architecture of data analysis software and some key points were designed. At last, the two associated softwares were developed, experiments showed that the two softwares can satisfy the MFL inspection application.

KEYWORDS: architecture design; MFL; software development.

1 INTRODUCTION

Magnetic flux leakage testing (MFL) technology is the most important or only available inspection method in the field of huge tank bottom plates, underground or undersea pipes, factory tapes and wire ropes. Professional device is necessary for MFL inspection. Correspondingly, associated data acquisition software and data analysis software is needed. In detail, the data acquisition software is integrated in the inspection device, which is used to acquire and save the inspection data, the data analysis software is installed on any other personal computer, which is used to do the statistics, analysis and report job. The data acquisition software should build on the mechanical structure and finish efficient communication with the data acquisition card. The data analysis software should build on the data acquisition software and load the data files saved by the data acquisition software. Therefore, the mechanical structure, the data acquisition software and the data analysis software are three progressive job levels.

In this paper, depending on the establishment of the mechanical structure, the architecture of two softwares was designed. Firstly, the bridge communication the mechanical structure, the data acquisition software and the data analysis software were ascertained. Secondly, the architecture of data acquisition software and some key points were designed. Thirdly, the architecture of data analysis software and some key points were designed. At last, the two associated softwares were developed, experiments showed that the two softwares can satisfy the MFL inspection application.

2 BRIDGE COMMUNICATION

As shown in Fig. 1, the three progressive job levels indicate the program of a typical MFL inspection project. The data acquisition software is installed on the touchscreen computer in job level 2, while the data analysis software is installed on the personal computer in job level 3. The bridge communication between level 1 and level 2 is a data acquisition card, the bridge communication between level 2 and level 3 is a data copy method, such as USB, Email, Wifi, etc.

The data acquisition card was selected in consideration of sampling frequency and required channels. In order to get evenly sampled series along its walking displacement, an external clock was assigned by another encoder. The encoder was fixed to a wheel of the device, it could provide 1000 pulses while the wheel rotated a circle. Then the data acquisition card kept acquiring samples according to the encoder pulses. When the device stopped walking, a data acquisition process was done. The data acquisition card sent these data to the touchscreen computer. However, the signal data acquired contained fake and disturb signals. Digital filtering progress was quite necessary. Because the digital filtering program is professional and complex, it is feasible to handle the digital filtering program with a professional toolkit, such as Matlab software. The bridge communication between Matlab and main programming tool, selected as Visual Studio, could be finished by a DLL (dynamic link library) file, as shown in Fig. 2.

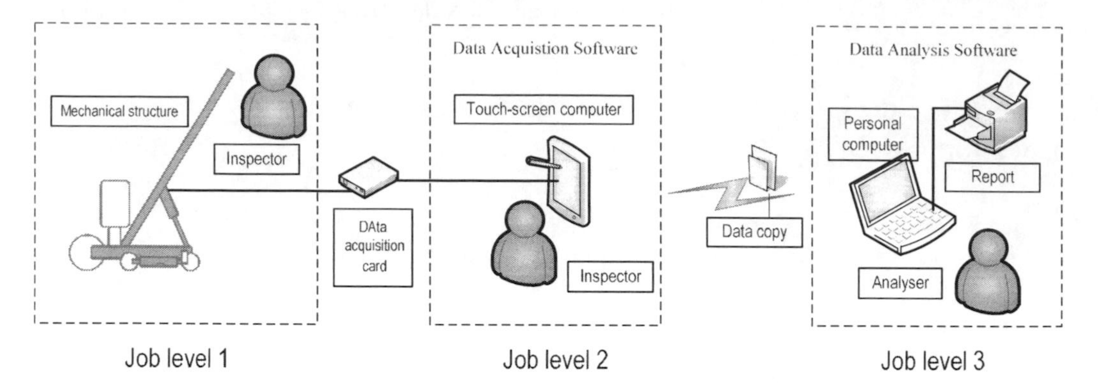

Figure 1. The three job levels of MFL inspection.

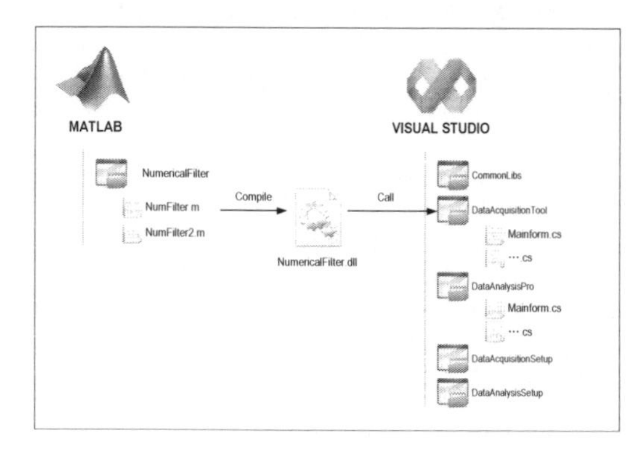

Figure 2. The bridge communication of Matlab and VS.

3 DATA ACQUISITION SOFTWARE

The functions of data acquisition software mainly include calibration, acquisition and history viewing. Calibration is essential by MFL inspection technology, history viewing is convenient for temporary check while inspection progress. Acquisition is the main function, including the setting of scanning type, reference point and plate information. Pause and resume functions are also needed because the device need to overcome some blocks while walking.

3.1 *Channel independent calibration method*

In technical structure design, all 15 channel sensors were placed in three groups, and the interval between groups much bigger than the interval between sensors of the same group. These uneven intervals caused mistakes for calibrated. The experiment showed the middle group sensors got about twice bigger signal amplitude than the left or right groups. While the sensors of every same group got about equivalent signal amplitude. Therefore, a channel independent calibration method was adopted to increase accuracy.

3.2 *Touchscreen oriented GUI design*

As noted before, the data acquisition software is installed on a touchscreen computer, and it is used while inspection. Because the surrounding condition was not suitable for mouse and keyboard usage, the GUI composition was analyzed. A typical form of the software contained three kinds of areas, respectively named display area, touch area and input area. Display area mainly includes labels, that was designed with small font and soft color. Touch area mainly includes buttons, that was designed with big enough size and bright color. Input area mainly includes text boxes, that was designed to be able to call out specially designed virtual keyboard. According to the possible information of one textbox, whether a full keyboard or a number-only keyboard was determined to be called out.

3.3 *Inspection data saving*

There are three kinds of data should be saved to files, respectively called calibration data, inspection data and scanning data. The calibration data file contained a bitmap and a numerical array. The bitmap indicated the numerical array in graphics, in order to represent clearly whether the calibration was good enough for inspection. The inspection data file contains general information about an MFL inspection project, for example, date, time, plant name, inspector etc. These information was essential for an inspection report. The scanning data files contained the signal information of scanned plates. Through these files, the inspector could determine whether the plates contain defects and where the defects locate.

4 DATA ANALYSIS SOFTWARE

The functions of data analysis software mainly include task view, plate view, statistics view, revisions view and report generation. In task view, all plates scanning data files could be loaded and be pieced together. In plate view, any plate could be checked in detail with ruler measurement. In statistics view, all defects could be summed up, classified into different degrees, then be displayed in graphics. In revisions, view, some correction could be applied. At last, an inspection report could be generated.

4.1 *Object-oriented class design*

In different with data acquisition software, data analysis software should deal with lots of plate data in the same time, Object-oriented class was designed for programming. As shown in Fig. 3, five classes were designed to present the objects, including RectShape, Plate, Track, Plates and Tracks. The plates are the assemble of Plate, Tracks is the assembled of the truck. Then, a general method was programmed to operate all tracks with different parameters, such as plate orientation, start corner, reference corner etc.

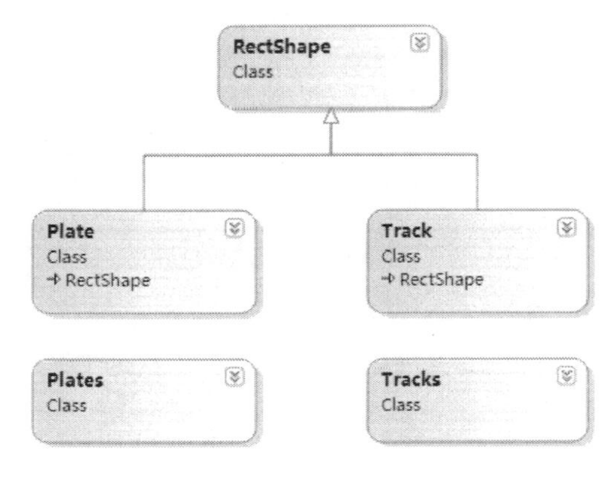

Figure 3. Calibration results of different sensors.

4.2 *Regular express reading*

The data analysis software needed to load lots of plate scanning data, every plate data were saved in a plain text file. Because every plate contained several thousands of lines and every line should be separated into 15 channels, it was a quite heavy work. Several trials showed that normally reading needed lots of time to finish. Therefore, the regular express reading method was adopted to substitute normal reading due to its high efficiency in plain text processing. Experience proved that regular express reading saved much loading time.

5 CONCLUSION

Depending on the establishment of the mechanical structure, the architecture of two softwares was designed. After function design and key point analysis, the two associated softwares were developed, experiments showed that the two softwares can satisfy the MFL inspection application, the main conclusions include the below.

1 A magnetization structure contains three subparts was designed for curvature plates, then a motor drive walking device was integrated, as well as a separated control box was designed, finally the device was established for underground oil tank inspection.
2 A new channel independent calibration method was raised. Through comparison of channel independent and identical calibration methods, the experiment results using channels independent calibration method has better precision of about 1/3 discovery rate.

ACKNOWLEDGMENT

This work was financially supported by the National Department Public Benefit Research Foundation (No. 201210019) & (No. 201410027).

Computing, Control, Information and Education Engineering – Liu, Sung & Yao (eds)
 ISBN: 978-1-138-02800-5

The re-establishment of curriculum system for computer software engineering specialty under the background of undergraduate education restructuring

Xin Hua Xu , Xing Ben Yang, Wen Shan Cheng & Hong Ying Gao
College of Educational Information and Technology, Hubei Normal University, Hubei Huangshi, China

Guo An Ning
Office of Academic Affairs, Hubei Polytechnic Institute, Hubei Xiaogan,China

ABSTRACT: In order to adapt to the needs of the development of undergraduate education restructuring , the re-establishment of talent cultivation scheme is an important link for the restructuring practice, the difficulty lies in the construction of professional curriculum system; Our method is that professional curriculum system revision must follow three guiding principles: adhere to the "optimization of general courses, strengthen the practice ability, guide the personality development, encouraging innovation" principle; to clarify the relationship between curriculum and talents training target; highlight the principles of practice education; The result is that the new courses system of the major is characterized by three new features, such as the curriculum setting with the focus of practice and expertise, docking and mutual recognition for professional education and professional qualification , credit system and modular teaching. This can be concluded that the new curriculum system can be competent for the need for talents cultivation of the restructuring.

KEYWORDS: Undergraduate Education Restructuring; Talent Cultivation; The professional curriculum system; practice education.

1 INTRODUCTION

1.1 *"The Zhumadian consensus"*

The International Forum For Industry & Education had been held in Zhumadian on April 25-26, 2014。178 colleges and universities jointly issued "Zhumadian Consensus" at the closing ceremony of the forum. According to the consensus, technological applied undergraduate colleges and universities emerge at a historic moment, the restructuring development of some local undergraduate colleges is imperative."

1.2 *"The decision of the state council about accelerating the development of modern vocational education decision"*

In May 2014, the release of "the decision of the state council about accelerate the development of modern vocational education "[1] means the completion of top-level design which aims to promote the development of vocational education reform in the new period. In June, the Ministry of Education and six ministries jointly issued "the modern vocational education system construction plan (2014-2020) "[2], and further proposed to encourage the establishment of application technological university. The revision (and even re-establishment) of talent cultivation scheme consequently is an important part for the restructuring practice.

2 THE EARLY EXPLORATION OF THE REVISION OF TALENT CULTIVATION SCHEME AND THE POSITIONING OF TALENT CULTIVATION.

2.1 *The Hubei Normal University had made full preparation for the restructuring*

The Hubei Normal University is located in the industrial city of Huangshi, is a provincial local undergraduate college. On August 28, 2014, Hubei province identified 10 provincial colleges pilot transition, 5 provincial undergraduate colleges and 5 private colleges, including Hubei normal university. The Hubei Normal University had made full preparation for the restructuring, including exploring the industry enterprise research, curriculum development and amalgamation of industry and education, having released "Hubei Normal University principled

Suggestions about revising the talent cultivation scheme". It laid a solid foundation for the construction of the new curriculum system and revision of the talent cultivation scheme. For the talent cultivation scheme for Computer Software Engineering revision, Computer Software Engineering Specialty comply with the basic requirement of undergraduate course of restructuring, according to practical situation of specialty, regional and industry, having made a preliminary exploration in amalgamation of industry and education, university-enterprise cooperation, innovation, entrepreneurship practice bases, "double type" teachers team construction, such as creating a "project team" , "studio", "science and technology business incubator", "student company" and so on.

2.2 *The revised talent cultivation target*

Previously, the cultivation goal is: to cultivate excellent engineers with moral, intellectual, physical all-round development, in the field of electronic information system, with a solid theoretical foundation, reasonable knowledge structure, innovation ability.

The goal of the revision of talent cultivation is: to cultivate talents with moral, intellectual, physical all-round development, with innovative spirit and practical ability, good ideological and moral quality and cultural quality of practical science and applied engineering and technical personnel.

Highlight positioning of the talent cultivation type of applied talents, technological talents.

With clear positioning, the restructuring will follow the direction and avoid the deviation.

3 PRINCIPLE OF THE COMPUTER SOFTWARE ENGINEERING SPECIALTY CURRICULUM SYSTEM RE-ESTABLISHING

3.1 *The new curriculum system*

The new curriculum system of draw lessons from the practice of German university of applied technology, such as "dual system", by reference to higher vocational educational reform is successful "work process-oriented curriculum development theory", in accordance with the "industry enterprise research-determine the training goal and professional personnel training specification, determine the corresponding post group, according to the requirement of post group of characteristics and the ability to ability to structure, module according to ability build a detailed teaching module - module descriptions, define the teaching goal, content, method and evaluation method"[3] of course development process, through the efforts of the before and after nearly 1 year, established a new system of professional courses.[4]

3.2 *Principle 1: Adhere to the "optimization of general courses, strengthen the practice ability, guide the personality development, encouraging innovation" principle*

1 Increase practice courses: specialized course of experiment, practice, course design, project practice, graduation design, accounted for 48%;
2 Optimization theory courses: eliminate or reduce part of the theory courses;
3 Increase elective courses, elective courses in total accounted for 31% of the total credit.
4 Application-oriented, technical characteristics: advocating "double certificates" (diploma and qualification certificate) personnel quality consciousness, on the basis of Hubei Normal University extracurricular credits "specialty" project (including professional ability certificate, all kinds of competition award, all kinds of corporate action classes, innovation, invention, papers and works, other classes, a total of six categories) management approach, set the extracurricular special credit project, takes five credit, the purpose is to promote students' personality and career development.

3.3 *Principle 2: To clarify the relationship between curriculum and talents training target*

1 Fully considering the integrity and continuity of students knowledge structure; optimization theory and does not destroy discipline knowledge structure
2 Both students' individual character development and employment intention: set the direction of the professional courses(professional electives courses to 22%), professional direction of shunt(provided the direction of "Java", ".Net" and "Mobile App Development ")
3 Combined with professional learning, started the "entrepreneurial foundation" course: Numbers TS990103, 8 semester classes, 2 credits, required courses
4 Course examination evaluation system:

Part of the course using "process" evaluation, reduces the weight of the final exam scores (below 50%), Increase the weights of practice and grades(more than 50%);Cancel the request for some courses, replace it with an open-book exam, small paper of science, technology, experiment examination a variety of forms; Strengthen the awareness of professional qualifications, for through the RED HAT series certification (RHCE/RHCT/RHCA/RHCSS/RHCVA), Huawei series certification, The technology of the computer and Software specialized technology qualifications test, national computer rank examination certificate(NCRE), certified

Software Professional (c/c++ or Java), Oracle series certification(OCJP/OCA/OCP/OCM), Tarena series certification (Java+4G/4G-Android/C++/.NET+4G/4G-iPhone/PHP/WEB3.0, embedded system certification engineer (AAE), Microsoft series certification, automatic identification certificate of special technology, electronic information technology certification (EITP), MATLAB certification, students may exempt the Linux system, computer network, Java program design, ORACLE database management system, Advanced programming language design, C# programming language design, .NET programming design, PHP website development, Android applications development, IOS applications development, HTML5, ARM embedded systems, SQL server database management and development, Website design and development, MySQL database and application of MATLAB language courses.

3.4 *Principle 3: Highlight the principles of practical education*

1. Increases the practice teaching content, design, comprehensive and innovative projects increased 20%;
2. Transforming the form and content of practice teaching, setting up the entrance education, social activities in summer, voluntary labor for public welfare, mental health education, professional practice, four comprehensive curriculum, professional and comprehensive practice, innovation and entrepreneurship training and practice, internship before graduation and thesis (design);
3. Strengthen the construction of practice bases, strengthen the guidance of graduation practice and process management;
4. Set up the innovative practice, promote innovation, entrepreneurship training plan the project, through the discipline competition, science and technology creatively, professional development and other activities, enriching the content of practical education, and strengthen the practice innovation ability training.

4 FEATURES OF THE NEW CURRICULUM SYSTEM OF COMPUTER SOFTWARE ENGINEERING

4.1 *Features 1: The new course type outstanding practice and expertise, including general class, professional class, practice class and specialty class*

All kinds of curriculum requirements are as follows:

General courses further optimization theory teaching by adding computer, employment, entrepreneurship, college Chinese courses.

Specialty courses are divided into compulsory courses and professional elective courses, compulsory courses include professional basic courses and professional core courses. The professional basic course of the same subject needs implementation of collaborative teaching.

Practice courses including scattered in various practical content in the teaching activities, concentrated practice teaching links and the student extracurricular autonomous practice of three parts. The theory course with a strong practical should be arranged a certain amount of course, practice, practice or course design. All kinds of experiment, training courses should reduce the verification experiment, strengthen the comprehensive experiments; Additional design and researching experiment, improve the students' practice ability and innovation ability.

Specialty courses in order to improve the students employment competition ability, we encourage students to take an active part in all kinds of skills at all levels certificate test, subject contest (ITAT contest, "Lanqiao cup" national software and information technology professional talents competition, Android app development contest), innovation and entrepreneurship("Challenge Cup" National Science and Technology College academic competition, College students' innovative entrepreneurship competition), social practice, community activities or published literary works or academic papers. Students through such activities get special credits which is closely related to the courses can be offset or exempt the course credits. Qualified identification of specialty credits is according to Hubei Normal University extracurricular special credit management method".

4.2 *Feature 2: The interconnection of professional education and qualifications*

A feature of the new scheme is to encourage the cultivation of students' professional ability, encouraging students to get professional qualification certificate according to their own career planning. By getting this kind of certificate students can get advanced standing. It achieved the interconnection of professional education and qualifications.

4.3 *Feature 3: The credit system and modular teaching*

A feature of the new scheme is to encourage the cultivation of students' professional ability, encouraging students to get professional qualification certificate according to their own career planning. By getting this kind of certificate students can get advanced standing. It achieved the interconnection of professional education and qualifications.

Hubei Normal University introduced "The measures for the implementation of the school year credit system of Hubei Normal University "(Education of Hubei Normal University [2013].5), which stipulated in article 4, in the training scheme, the curriculum is divided into two categories: compulsory courses and elective courses, all kinds of courses for credit standard is:

1 Theory courses: 16 learning time is 1 credit, various curricula designed by multiples of 8 hours, 0.5 credits minimum design course;
2 Practice courses: 32 learning time is 1 credit (practices including experiments, physical education, arts of professional technology, etc.);
3 None-classroom teaching form of concentration or dispersion of practice: for 1 week (or 8 x 5 = 40 hours) 1 credit

According to article 7, the college implements the elective system. General public courses elective classes can be choosen according to the measures for the administration of the public elective course of Hubei Normal University. Upon personal interests, and the actual demands of knowledge structure development, the students can choose the corresponding courses. We encourage the capable students to take more major courses or participate in the minor professional course; the inspection qualified course corresponding points can be obtained.

In the new curriculum scheme, elective courses account 20% of the total 166 credits, for providing diverse courses to meet the different needs of students, and comprehensively promote the credit system and the modular teaching, cultivation programs for different students to develop diversified talents.

Martin Trow think the course characteristic of popularization stage is "flexible modular courses "[5], that is, breaking the highly structured and specialized traditional academic courses, and according to the logic relation between courses, course combined with the corresponding relationship of professions into a different class process module, offering students more options.

The new scheme set three modules of "Java direction", ".NET electronic direction" and "Mobile app development direction" in order to meet the students' vocational interest and planning.

5 CONCLUSIONS

Since the implementation of the new curriculum system, students in this major have achieved good results in national and provincial skills certification, exam, subject competition, "Challenge Cup" National Science and Technology College academic competition, college students' innovative entrepreneurship competition, social investigation report, and publishing academic papers have achieved good results. CCTV also reported that Hubei Normal University gets good performance in the 2014 national youth science innovation experiment competition. The new curriculum system's grasp the key of the student's practical ability, cultivating the ability of technical skills, is suitable for enterprise requirements; the new curriculum system can be competent for the need for talents cultivation of the restructuring.

ACKNOWLEDGEMENTS

The research in this paper was supported by the Hubei Province Education Science "Twelfth Five Year Plan" project (Grant No. 2012B329); Hubei Normal University Talents Project (Grant No. 2014F0330).

AUTHOR'S BRIEF INTRODUCTION

Xinhua Xu (1968-), male, a professor of Hubei Normal University ,master tutor. Mainly engaged in education technology and higher education research. Email:xinhuaxu@163.com.

REFERENCES

[1] http://www.scio.gov.cn/xwfbh/xwbfbh/yg/2/Document/1373500/1373500.htm, 2014-06-24.
Chinese government website. "The state council about accelerate the development of modern vocational education decision".
http://www.scio.gov.cn/xwfbh/xwbfbh/yg/2/Document/1373500/1373500.htm, 2014-06-24.
[2] http://www.moe.edu.cn/publicfiles/business/htmlfiles/moe/moe_630/201406/170737.html, 2014-06-16.
The ministry of education website. "The notice of six departments about print and distribute The modern vocational education system construction plan (2014–2020)".
[3] Zhoufei, chuzhaosheng. Hefei colleage: :Successful transformation on the road of building applied university [N]. China education newspaper, version 1, On April 16, 2014.
[4] Zhangzhengwen. Entrepreneurship education mode should be for the purpose of innovation consciousness[N].china education newspaper. version1. On June 3, 2010.
[5] Martin Trow. The problems in the transition from elite to mass higher education.[j]. Foreign institutions of higher education information,1999.(1)

Computing, Control, Information and Education Engineering – Liu, Sung & Yao (eds)
© 2015 Taylor & Francis Group, London, ISBN: 978-1-138-02800-5

A measurement method based on trinocular stereo vision

Lu Huang, Jing Bo Gao & Jian Yan
School of Astronautics Engineering, Harbin Institute of Technology, Harbin, Heilongjiang province, China

ABSTRACT: Based on the domestic and foreign research and its trend in the field of stereo vision, this article will introduce a measuring method in Trinocular stereo vision and make an experiment to verify the feasibility. Through the experiment, it can be seen that the results are accurate and it can be used in measurement.

KEYWORDS: Trinocular stereo vision; 3D reconstruction: Halcon.

Stereo vision is a kind of technology using the image as information acquisition, transmission and analysis, which applies image processing, pattern recognition and artificial intelligence to reconstructing the 2D or 3D geometry characteristic information and obtain the outline of the analyte or its dynamic parameters. With the development of science and technology, modern visual measurement system is widely used. For example, Olson [1] had researched in binocular robots and made it successful in Martian probe. Gao Qingji[2], Harbin Institute of Technology, used heterogeneous binocular vision in autonomous soccer robot to realizing its navigation. It is monocular or binocular vision that has rapidly developed, but with the development of the research, the visual measurement system is towards on multi-vision. Kleiner Bernhard[3] succeeded in navigating by a mobile 3-D handheld scanner.

Recently, multi-vision is still a direction of doing a research or exploring in the field of Machine vision. This article will introduce a measuring method with Trinocular stereo vision. There are several characteristics below: Firstly reduces the difficulty of matching and enhance the precision of matching. Secondly expand the measurement domain. Lastly improves the precision of result of redundant data.

1 THE PRINCIPLE OF EXPERIMENT

Trinocular stereo vision can be seen as three different groups of binocular stereo vision. Based on binocular stereo vision, we use three cameras to shoot the analyte from different directions at the same time to collect images which are used in processing the image recognition, then calculating the cameras parameters, finally coming to 3D reconstruction. So the first step is to calibrate the camera parameters. Four kinds of coordinates will be used in it: world coordinate system $O_\omega - X_\omega Y_\omega Z_\omega$, image coordinate system $O - XY$, pixel coordinate system $O - UV$ and light coordinate system$O_c - X_c Y_c Z_c$, shown in Figure 1.

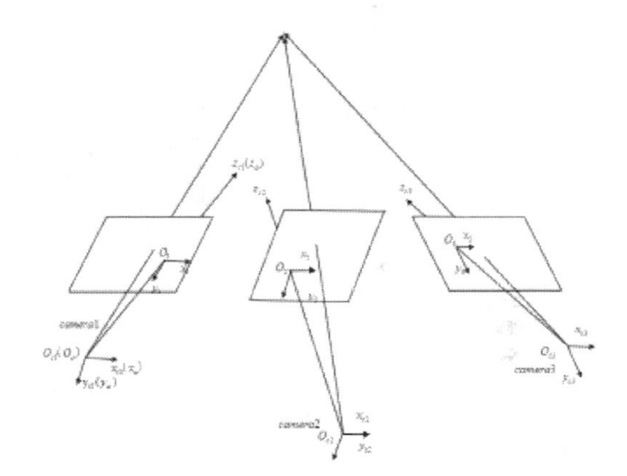

Figure 1. All coordinates.

Through the transformation of coordinates, the relationship between image co-ordinates and world coordinates can be shown in the formula (3):

On the basis of the principle of small hole imaging and without considering the distortion of the cameras, the relation between light coordinate and pixel coordinate:

$$z_c \begin{bmatrix} u \\ v \\ 1 \end{bmatrix} = \begin{bmatrix} f/dx & 0 & u_0 & 0 \\ 0 & f/dy & v_0 & 0 \\ 0 & 0 & 1 & 0 \end{bmatrix} \begin{bmatrix} x_c \\ y_c \\ z_c \\ 1 \end{bmatrix} \quad (1)$$

Where f is focal length; dx, dy are unit pixel size; u_0 v_0 are the center pixel coordinate.

According to the transformation of the world coordinates, finally obtained the relation between the pixel coordinate system and the world coordinate system:

$$\begin{bmatrix} x_c \\ y_c \\ z_c \\ 1 \end{bmatrix} = \begin{bmatrix} R & T \\ 0^T & 1 \end{bmatrix} \begin{bmatrix} x_\omega \\ y_\omega \\ z_\omega \\ 1 \end{bmatrix} \quad (2)$$

$$z_c \begin{bmatrix} u \\ v \\ 1 \end{bmatrix} = \begin{bmatrix} f/dx & 0 & u_0 & 0 \\ 0 & f/dy & v_0 & 0 \\ 0 & 0 & 1 & 0 \end{bmatrix} \begin{bmatrix} R & T \\ 0^T & 1 \end{bmatrix} \begin{bmatrix} x_\omega \\ y_\omega \\ z_\omega \\ 1 \end{bmatrix} = MN \begin{bmatrix} x_\omega \\ y_\omega \\ z_\omega \\ 1 \end{bmatrix} \quad (3)$$

Where R is rotation matrix; T is translation matrix. M is the internal parameter matrix of the camera. N is the outside parameter matrix of the camera. The parameter of cameras will be calibrated next .

The traditional method of calibration costs much; The results ofthe autonomous calibration methodd is not precise; Althoughthe camera calibration methodd based on active vision is simple and its robustness is high, this method requiresa high precision visual platformm for camera calibration and experimental equipment is expensive. On the basis of Zhang Zhengyou calibration[4], the edge detection is not so good. So the Halcon calibration Method [5] is used to calibrate the inner and outer parameter matrix of the cameras in this paper.

2 THE PROCESS OF EXPERIMENT

This paper will do an experiment on the flexible sheet based on Trinocular stereo vision. The scene of the experiment is shown as Figure 2.

Figure 2. The scene of experiment.

In this paper, the experiment adopts three Germany cameras, MG419, shown in Figure 3. Because the measurement domain of three cameras is equal to the measurement domain of one camera, the distance between three cameras and the analyte is about 1.85 meters long.

Light source and landmark are important factors that affect vision measurement. According to a recent study, this experiment will choose photoflood lamp and landmark [6] shown in Figure 4.

Figure 3. Cameras.

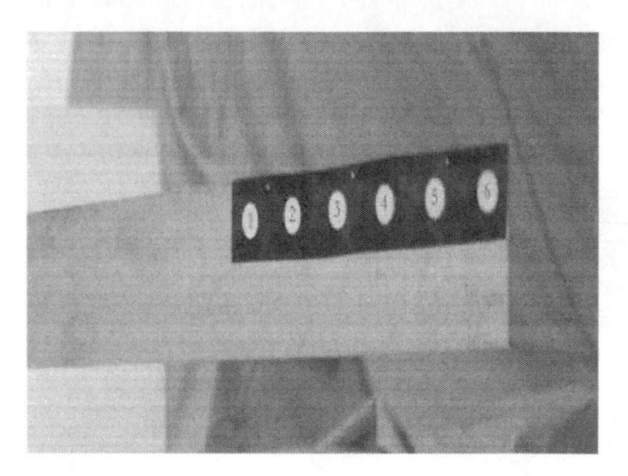

Figure 4. Landmarks.

After shooting, then calibrating the inner and outer parameter matrix of the cameras, the results shown in Table 1

Table 1. Inner and outer parameters of three cameras.

Inner parameters				
	Initial	Camera1	Camera2	Camera3
$f(mm)$	12.5	12.49	12.54	12.50
$dx(\mu m)$	5.5	5.47	5.48	5.49
$dy(\mu m)$	5.5	5.5	5.5	5.5
u_0	300	311.41	309.34	315.62
v_0	225	212.58	220.625	230.98

	Outer parameters		
	Camera1	Camera2	Camera3
$\alpha(^\circ)$	0	1.23	1.15
$\beta(^\circ)$	0	353.89	347.57
$\gamma(^\circ)$	0	359.07	358.44
$t_x(m)$	0	0.22	0.40
$t_y(m)$	0	0.06	-0.003
$t_z(m)$	0	1.23	0.12

Through 3D reconstruction, the coordinates of landmarks are shown in Table 2.

Table 2. The coordinates of landmarks in world coordinate.

	$X(mm)$	$Y(mm)$	$Z(mm)$
Landmark1	-85.01	-16.44	1.841
Landmark2	-45.91	-17.77	1.849
Landmark3	-6.35	-18.69	1.851
Landmark4	32.82	-19.73	1.861
Landmark5	72.14	-20.98	1.867
Landmark6	111.55	-22.04	1.874

In fact, the distance between two landmark is 40mm. Through 3D reconstruction, the distance between them shown in Table 3.

Table 3. The distance between each two landmarks.

The theoretical value(mm)	40
Landmark1and 2(mm)	40.1273
Landmark2 and 3(mm	40.0212
Landmark3 and 4(mm)	39.8974
Landmark4 and 5(mm)	39.9339
Landmark5 and 6(mm)	39.9824

3 CONCLUSION

In Trinocular stereo vision, this paper optimizes the feasibility of Trinocular stereo vision algorithm. From the experimental data, as shown in Table 1, the result of camera calibration is accurate. Seen from table 3, this shows Trinocular stereo vision can use redundant data to improve precision of measurement, it can be used in the measurement system.

REFERENCES

[1] Olson. 1991. Real-time vergence control for binocular robots. *International Journal of Computer* Vision, ISSN 0920–5691.

[2] Gao Qingji. 2003. The autonomous soccer robot navigates based on heterogeneous binocular vision. *Journal of Harbin Institute of Technology.*

[3] Kleiner Bernhard. 2014. Handheld 3D Scanning with Automatic Multi-View Registration Based on Visual-Inertial Navigation. *International Journal of Optomechatronics.*

[4] Yang Lu. 2009. Research on environment per ception technology for menology environments based on binocular vision. *IEEE Computer Society.*

[5] Luo Zhenxi. 2010. Camera calibration based on HALCON. *Video Engineering.*

[6] Gai Shaoyang. 2010. A new dot detection algorithm for optical measurement. *Journal of Southeast University(Natural Science Edition).*

Computing, Control, Information and Education Engineering – Liu, Sung & Yao (eds)
© 2015 Taylor & Francis Group, London, ISBN: 978-1-138-02800-5

Security risk analysis and security mechanism study of virtual machine

Ming Fu Tuo, Jun Zhang, Hong Mei Zhang & Yong Mei Zhao
Air Force Engineering University, Xi'An, China

ABSTRACT: Virtual machine is not only faced with the traditional network security threats, but also introduce new security problem itself. This paper analyzes the security risk of virtual machine, and points out that the virtual machine image file protection is the key of the virtual machine risk prevention. We discuss the portable image security control method and the Hyper - V security mechanism. We put forward a new security mechanism to enhanced the image file security.

KEYWORDS: virtual machine, security, image file, encryption.

1 INTRODUCTION

Virtualization is one of the core technologies used in cloud computing [1]. Virtualization refers to technologies designed to provide a layer of abstraction between computer hardware systems and the software running on them. By providing a logical rather than a physical view of computing resources, virtualization solutions make several very useful functions possible. Most fundamentally, they in essence make an operating system recognize a group of servers is a single pool of computing resources [2-4]. They can allow running multiple operating systems simultaneously on a single machine.

Server virtualization, storage virtualization, network virtualization, application virtualization and data center virtualization can improve the level of concentration and share resources of enterprises, so as to realize the purpose of reducing the cost, optimize the utilization [5].

2 THE VIRTUAL MACHINE SECURITY RISK ANALYSIS

The application of the virtual machine technology is more and more popular, also more and more powerful, but research on the safety of the virtual system has not caught up with the pace of its application development. At present, some research on virtual system security threats is analyzed on the basis of experience, speculation or black box testing for a virtual machine system [6-8]. We will analyze the security risk of virtual machine in this part.

Secure isolation, confining a program to a virtualized environment should guarantee that any action performed inside the VM cannot interfere with the system that hosts it, is basic to virtualization. Consequently, VMs have seen rapid adoption in situations in which separation from a hostile or hazardous program is critical. If the physical host server's security becomes compromised, however, all of the VMs and applications residing on that particular host server are impacted. And a compromised virtual machine might also wreak havoc on the physical server, which may then have an adverse effect on all of the other VMs running on that same machine.

The virtualization tools that allocate host computing resources such as processors, memory, storage, and networking reside on a host with an operating system or console operating system (COS) that assists with the management and administration of those allocations. Any weakness in the COS could result in confidentiality or availability attacks that could negatively impact the virtualization tools, thus potentially providing an attack vector to harm the VMs.

Configuration standards for the COS based on the sources cited above and adjusted for the particular organization's security risk appetite need to be developed first to establish, build procedures, monitoring procedures, and a baseline to which any assessment metrics can be compared. Included in these configuration standards are topics such as secure root access, network port limitations, running service limitations, and many others found in a typical operating system hardening or configuration guide.

Powerful administration access to the COS, particularly the remote variations, including management software, client software, and browser-based and traditional (such as SSH) remote access tools, should be appropriately configured with strong authentication and encryption where appropriate. Virtualization tools, like many other software products, include built-in logical access controls to control access to the features of their software. These default roles, ranging from read only to all-powerful, should be assigned to users based on the least privilege principle, and any new roles created should be traceable to approved business needs and the organization's change control documentation [9].

Another security-related risk is "hyperjacking," in which an attacker crafts and then run a very thin hypervisor that takes complete control of the underlying operating system. A good example of how this risk might present itself is the Blue Pill rootkit developed by security researcher Joanna Rutkowska. A rootkit is a Trojan program designed to hide all evidence of its existence from system administrators and others who look for anomalies and security breaches in systems. The Blue Pill rootkit bypasses the Vista integrity-checking process for loading unsigned code into the Vista operating system's kernel. This code uses AMD's secure VM, designed to boost security, to masquerade itself from detection, and becomes a hypervisor, taking control of the operating system without system administrators and others detecting its presence [10].

In most of the virtual machine technology, the image file is located in the host operating system file system. As a result, the malicious code from the host operating system image file for reading and writing became a destruction of the virtual machine or virtual machine information from a possible way is also a very efficient, and very direct way. Because the virtual machine to the operation of the virtual memory can eventually convert to the host operating system to the operation of the image file, cause the image file is difficult to completely isolate from the host operating system file system. And detection of the host operating system in the virtual machine has been installed and their storage file location is easy. So exposed to the host operating system image files of the file system, easy to malicious code to read and write, in which confidential information may therefore leak or damaged, it has become the image file at the virtual machine system security a major weakness. And prevent the image file to be malicious code to read and write to the host operating system becomes an important link of safety.

Therefore, for the image file stored in the host operating system in a virtual machine, the safety of the guest operating system and isolation are impossible. Image files can be read and write a virtual memory and virtual machine system security. A virtual machine technology can give the guest operating system independent logical partitions on your hard drive, which can be isolated from the host operating system effectively, but this way on the allocation of storage resources is inefficient, because each set up a virtual machine will be assigned a fixed size of hard drive space for it, and don't care about this virtual machine, in fact, how much hard disk space to store the image file, in fact most of the virtual machine technology support for dynamic image file size, can according to the actual demand to increase or reduce the size of the image file. Additionally set up independent hard disk logical partition itself is a very troublesome thing, just think if you want to run at the same time on a physical server hundreds or thousands of virtual machine, build so many hard disk logical partition is very painful thing. Set up independent hard disk logical partitions cannot like build image file can according to need to create or delete at any time, in the management is very troublesome. Finally it is hard to realize virtual machines in a number of physical servers efficient migration, while the virtual machine on a physical server migration is one of important application in the field of virtual machine technology in the server. So most of the virtual machine technology does not support this technology, it is necessary to design a security scheme of image to enhance the security of the image file.

3 VIRTUAL MACHINE SECURITY MECHANISM

3.1 *A portable mirror safety control*

Virtual machine technology is regarded as the spread of the server-side resource sharing and application service foundation. The virtual machine by mirror materialization, the image contains the initial state of file system and virtual machine software. A mirror a client in the virtual machine may be instantiated, and the hardware of this image is provided by a service resource provider. Virtual machine images are spread "virtual application components" convenient tools. These virtual application components by prepackaged good software component distribution, when the client request, the equipment to instantiate the application components. Consider the following one of the most simple example, the virtual machine in the user directly on interface of equipment used to instantiate. Customer controller is a Web page, handle customer request form and according to the demand for the virtual desktop instance into a special component image. At that time, the safety control need is a simple context form, this form can be Key transmission, makes the owner can use transport layer security protocol (for example, TLS/SSH/SSL, etc.) and the owner and the client node mutual authentication, to secure login via a connection. Ideally, the login don't even need a password.

The premise is a binding agreement to establish a safe verified the control channel, the channel connecting the customer controller and the mirror. The safety of the control channel is the key. If the channel is not safe, then an attacker can control a new instance of the virtual machine, imitation or block access right owner. A more experienced attackers may be inserted between the controller and node by the client, to monitor or tamper with the client. So the binding agreement must provide mutual authentication and key exchange. Because the controller is likely to be separated by a wide area network, the agreement needs to be in and above the transport layer provides a complete defense, rather than relying only on an isolated network segment.

An alternative solution is the specific use of a particular user build image. Now it is commonly used on the image to add dependencies, but this kind of practice and image sharing conflict, and equipment provider for mirror providers increased the unwelcome burden, they want a mirror image of the management environment related variables. For example, Amazon's EC2 provides a virtual computing architecture management the AMIs (Amazon Machine Images). Once instantiated, an AMI through a mirror with the HT TP query URL to get the owner's public key. The URL to a domain, the domain name management by the Web server, accept the request, and the corresponding Key returned to the user. So the mirror can only run on EC2.

Another popular solution is for security token (such as encryption, etc.) embedded in the image. This lead to mirror is a different security system Shared, it is not safe. Furthermore, when a portable mirror is a mirror provider permission to install the security token will disable this permission. Mirror providers can provide special customer in accordance with the requirements with mirror. But this method still bans the Shared image, and need device to manage the same image on a variety of variables. Due to the size of the image will be big to the level of GB, a mirror image of the conflict and caching mechanism and multiple instances may affect performance.

The third option is to modify the installation process of portable mirror, installation key or other rely on the information. Software device, for instance, may write key when the mirror after verification. This compromise the portability, because the software must pay attention to the image structure, operating system, file system status, key storage location and authentication service. This choice will limit the operating system and the development of new components.

Ionut Constandache puts forward a simple and practical and flexible binding agreement, simple image providers can safely support. This agreement does not rely on the mirror with safety status. However, it needs the KeyMaster on the mirror in the initial stage. The running performance of this agreement Binding protocol implementation in the smallest on the KeyMaster python scripting language, this protocol exchanges and installs the public key that can set up a root SSH connection.

3.2 *Hyper-V safe mode*

In 2003, Microsoft began to develop a hardware assist virtualization system, and integrate it into the operating system. This project recently released, named the Hyper - V, is a part of the Windows server 2008. The Hyper - V is a platform for investment and will act as a basis of future technology. So from the beginning of the architecture is correct. Through the design of the Hyper - V has added the principle of trusted computing. The Hyper - V has three main parts: the Hypervisor, virtualization stack, the virtual machine.

The Hypervisor is to partition the kernel, no virtualization functions, virtualization are provided by virtualization stack and is a very thin layer of software, no device drivers, but for other operating systems support provides a good interface.

Virtual stack will run on the Root partition, it is the traditional Hypervisor partition on the upper as a miniature Hypervisor, manage customer partitions, resolve conflicts.

The establishment of the safe mode has the following safe assumption that the user is not trusted, the Root must be the Hypervisor trust, the parent node must be quilt node trust. The call interface document is very detailed, and the attacker widely visible. So will try to call interface of the users, that can detect whether the user run on a Hypervisor. Security goals include: isolation between partitions, protect the user data integrity and credibility and so on.

The Hypervisor safe mode includes the following aspects:

Memory: Hy per - V need to maintain physical addresses to the partition map. Memory on the maintenance model of the parent-child relationships. Can be read on the client page tables get right.

1 CPU hardware management caching and registration of isolation, instruction;
2 I/O: user to the actual hardware without permission;
3 the Hypervisor interface: partition privacy model, user mode call interface.

The Hyper - V safe mode includes the following aspects: the use of authentication manager, defines the specific function for the individual, such as start, stop, create, add hardware, changing the driving mirror, etc. Virtual machine administrators do not have to be system administrator.

4 CONCLUSION

Virtualization has brought the change of logic to the machine. Many security risks existing in traditional computer also threaten the virtual machine. Even more complicated, because there are too many different types of virtualization technology on the market, each of which has its own advantages and disadvantages, and virtualization deployment is often not the same. These factors make it is more difficult to overcome the virtual machine security risk.

This paper presents an in-depth analysis of the main security risks existing virtual machine and introduces the typical virtual machine software used in the main security mechanism. In order to enhance the security of virtual machine image files, a method for image file protection is put forward.

REFERENCES

[1] Jianxin Li, Bo Li.A virtualization security assurance architecture for green cloud computing. Future Generation Computer Systems 28, 2012.

[2] S. Subashini, V.Kavitha.A survey on security issues in service delivery models of cloud computing. Journal of Network and Computer Applications34,2011.

[3] William Arbaugh, Brandon Baker, Hyper-VSecurity, talking on VMSec workshop of CCS'08.

[4] http://www.vmware.com/cn/technology/virtual-infrastructure.html.

[5] Yih Huang, Angelos Stavrou, Anup K.Ghosh, Sushil Jajodia, Efficiently Tracking.

[6] Edward Ray, Eugene Schultz. Virtualization Security. CSIIRW '09 Proceedings of the 5th Annual Workshop on Cyber Security and Information Intelligence Research,2009.

[7] Flavio Lombardi, RobertoDiPietro. Secure virtualization for cloudcomputing. Journal of Network and Computer Applications34,2011.

[8] Krešimir Popović, Željko Hocenski. Cloud computing security issues and challenges. MIPRO 2010, May, 2010.

[9] Luis M. Vaquero, Luis Rodero-Merino. Locking the sky-a survey on IaaS cloud security. Computing ,2011.

[10] Ulf Mattsson.Real security for virtual machines. Network Security, April 2009.

Computing, Control, Information and Education Engineering – Liu, Sung & Yao (eds)
© 2015 Taylor & Francis Group, London, ISBN: 978-1-138-02800-5

Research on safety verification technology of cyber-physical systems

Ming Fu Tuo
Northwestern Polytechnical University, Xi'An, China
Air Force Engineering University, Xi'An, China

Xing She Zhou
Northwestern Polytechnical University, Xi'An, China

Li An & Rui Zhu
Air Force Engineering University, Xi'An, China

ABSTRACT: This paper analyzes the challenges for safety verification of CPS, introduces the related works, discusses the key technologies of safety modeling and safety verification, and puts forward a framework for integration of safety modeling and safety verification.

KEYWORDS: CPS; safety verification; safety modeling ; integration.

1 INTRODUCTION

CPS are a new type of hybrid systems which characterized by deep integrations of computation with physical processes [1]. There are usually computed entity, physical entity and communication entity in CPS, which is shown in Figure 1. Application fields of CPS are very wide, such as intelligent transportation, telemedicine, the smart-grid, aeronautics and astronautics, and so on [2]. CPS has become one of the focuses on the study of computer science.

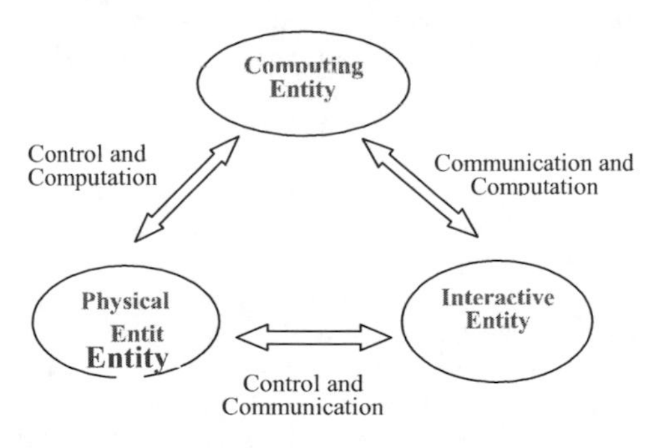

Figure 1. Entities and 3C in CPS.

2 THE SAFETY OF CPS

Safety is a key property for CPS to be applied in critical application fields [3,4]. Whether the properties of CPS can satisfy the requirements can be analyzed in the system design stage by model verification technology. This helps to find the defects of the deign as early as possible, so it can reduce the risk of system development effectiveness.

2.1 *Safety issues in CPS*

CPS consists of embedded computing units, which tightly interact with their physical environment to provide critical functionalities such as early detection of health problems, securing sensitive data, and enabling long term uninterrupted operation [5]. The computing units of a CPS can be characterized by a set of quantitative properties, C. These properties are related to the computing operation and are functions of the type of application executed. For example, members C can be the utilization of a server in a data center, the initial concentration input to a drug infusion control algorithm, the duty cycle of a sensor, or a 128 bit key for encryption during communication. The physical environment in a CPS can be similarly characterized by a set of quantitative properties, P. Examples of physical properties include time varying physiological and environmental signals such as, temperature, humidity, and amount of sunlight.

In a CPS, the properties in C are closely related to those in P through physical processes that cause a variation of the properties in the physical environment. Such physical processes can be characterized by a set of interaction parameters, I. The interaction parameters can be associated with both the computing and physical properties in a CPS. Typical examples of interaction parameters include heat transferred from the servers in the data center to the ambient air, the amount of energy harvested from the environment, or frequency domain features of physiological signals. Both the physical and computing properties affect the interaction parameters. The computing properties are time varying. Hence the mapping between the sets C and I can be represented by $G : C \times t \rightarrow I$, where t is time.

2.2 *The challenges for the safety verification of CPS*

The characteristics of the CPS include the coexistence of computational process and physical process, mass heterogeneity, uncertainty of communication, and so on. All these factors put forward new challenges for the traditional system safety modeling and verification methods.

From the point view of safety modeling, there are many heterogeneous models in CPS, including discrete computational model, continuous physical model and communication models with indeterminate delay. The safety of the system is not only related to the computing time, but also related to the factors such as physical space and network transmission. However, these factors cannot be depicted by one model, so we should use more safety models to verify the safety of CPS, each model verifies some safety attributes of CPS.

3 STUDIES OF CPS ANALYSIS AND VERIFICATION

With the widely used by CPS, studies of CPS analysis and verification become more and more deeply [6]. To extend the classical verification methods is one of the ways, such as extension of FSM, optimization of fault tree, etc. [7]. There are also researchers who use colored petri nets (CPN) to model and verify CPS.

According to the verification principle, safety verification technologies can be divided formal verification and non formal verification.

3.1 *Formal verification method*

Model checking and theorem proving are typical formal safety verification methods. The idea of model checking is that the authenticity of a proposition is determined by checking every state of the system. Its advantage is a high degree of automation, and counterexamples can be generated. Its disadvantage is there is a state space explosion. Cyber-physical systems usually are hybrid systems. They include both discrete states transitions and dynamic continuous variation processes, that is to say, the sates of most CPS are infinite.

The typical model checking tools for real time system include UPPAAL, Hytech, Kronos, etc[8]. SMV developed by CMU and SPIN developed by Bells LABS are widely used in the model checking of concurrent systems. They can find system defects effectively. SMV can detect whether or not the finite state system satisfies the given properties depicted by CTL [9]. It uses OBDD symbolic model checking techniques and is suitable for describing complex finite state systems.

CPS are often a mixture of both continuous and discrete dynamic systems, so the hybrid control theory has become more widely used in the CPS. To verify the safety of CPS is mainly to check whether there is a trace that make the performance of the system can't be satisfied. Model check tools for hybrid system verification include HyTech, HyperTech and MOCHA.

With the development of model driven technology, there appear some auxiliary tools that can automatically generate code from the model. In addition to checking the model of the system model, the code of the system is directly verified to verify the system. For example, SMT is a verification tool which can analyze the bounds of C code, and verify the satisfiability of a program.

3.2 *Non formal verification method*

The non formal verification techniques usually simulate the system dynamically. It monitors the behavior of running a system on-line and test the change of key attributes when the system is running. This method is different from the formal verification method. It can monitor the change of key attributes when the system is running in simulation. According to the requirement of checking the AADL model, a simulation test tool called ADeS(architecture description simulation) is developed based on Eclipse by Cantabria university of Spain. It can parse system description of the system and simulate the global behavior of a system. It can both be used to simulate the global behavior of a system in the early stage of development of system and can also be used to evaluate the accurate behavior of system in the late development.

Stanley Bak and Lui Sha, researchers of UIUC, think that even the system is verified by formal method, the system is still often influenced by realtime operating system, middleware and errors of the microprocessor. It's need to establish a Simplex Architecture in system level to ensure the performance of the system when there is a fault in logical application layer, using hardware and software collaborative design. They suggest that add a sandbox which has been verified to CPS which has not been verified, see figure 2.

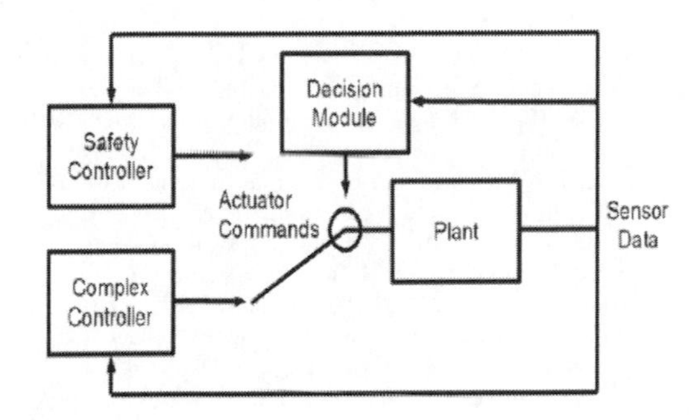

Figure 2. A simplex architecture for CPS safety verification.

CC is a complex controller not verified. SC is a safety controller. DM is a decision making unit. When CC affects the safety of the system, DM enables SC and suspended CC at the same time. One of the most important challenges is to determine the transformation logic of DM. But for all validation method based on the model, the result is only as accurate as the model itself. It requires run-time monitoring to ensure that the behavior of the system complies with the model validation. The algorithm can only be used in simpler hybrid systems.

4 THE KEY TECHNOLOGY FOR SAFETY VERIFICATION OF CPS

4.1 *Multiple model collaborative verification technology*

CPS are usually more complex distributed systems. Their behavior cannot be depicted by only one modeling. Each model describes part properties of the CPS. The models for computing entity include hybrid automata, data flow, etc. The models for physical entity include differential equation, state equation, etc. In the same way, other models are needed for communications. So we should use more than one modeling method and models collectively to verify the safety of CPS.

4.2 *Integrated system modeling and the safety verification*

The modeling tools for CPS are usually graphic platform, such as AADL, UML. They are simple and intuitive. On the other hand, we usually use formal models to verify the safety of CPS. These models are complex and abstract. If we model the safety of CPS in graphic platform first, then transform these models to formal models automatically on the condition that semantic consistency of models is ensured. It's a way to combine the advantages of both.

4.3 *The hierarchical safety verification method*

To verify the safety of CPS, we should not only to check the accessibility of the various states in system, but also to examine whether the safety related physical attributes are in the given scope, such as time, centigrade, location, velocity, and so on. The process of verification is more complex. Usually, verification suitable for simple system is difficult to apply to more complex systems. We can study the verification method for one component or subsystem first, and then study the verification method for the whole system based on this.

5 A FRAMEWORK FOR SAFETY MODELING AND VERIFICATION OF CPS

In this part, We put forward a concrete framework for safety modeling and verification. We extend HYSDEL, a traditional hybrid system description language, and name it E-HYSDEL. Furthermore, we use it to model the behavior of CPS. Based on this, the safety of CPS is verified by means of KeYmaera, a theorem prover. The framework of this new method for CPS behavior modeling and safety verification can be depicted as figure 3.

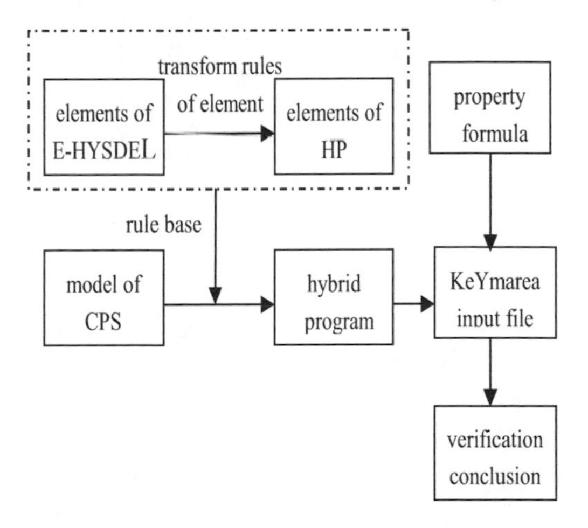

Figure 3. A framework for safety modeling and verification of CPS.

According to this framework, the process for CPS behavior modeling and safety verification can be divided into four steps. In the first, the transformation rules between meta-model of E-HYSDEL and the meta-model of HP are established. Then, the behavior model of a specific CPS is described by E-HYSDEL code. Next, the behavior model is transformed to the corresponding HP based on the transformation rules. Finally, the formula that describes the constraints of variables affecting safety of system and the HP are inputted into KeYmarea to verify whether the safety related constraints are met.

6 CONCLUSION

Safety of CPS is the premise for CPS to be used in many applications, such as intelligent transportation, medical robot, and so on. The safety analysis and verification of CPS based on the CPS model has become one of the hotspots in the field of CPS research. This paper analyzes the challenges of safety verification of CPS and related works. CPS are usually complex hybrid systems. This paper discusses the key technologies of safety verification. The paper puts forward a framework for safety modeling and verification finally. It helps to integrate the safety modeling and the safety verification.

REFERENCES

[1] Lee, E. CPS foundations. Proceedings of the 47th ACM/IEEE Design Automation Conference., 2010. 737–742.

[2] He, J. F. Cyber-physical systems. Communications of the China Computer Federation, 2010, 6(1): 25–29.

[3] Rajkumar, R., et al. Cyber-physical systems: The next computing revolution. in Design Automation Conference (DAC), 2010 47th ACM/IEEE. 2010.

[4] Sha, L., et al., Cyber-physical systems: A new frontier. Machine Learning in Cyber Trust, 2009: p. 3–13.

[5] Lin, J., S. Sedigh, and A. Miller. Towards integrated simulation of cyber-physical systems: a case study on intelligent water distribution. in Dependable, Autonomic and Secure Computing, 2009. DASC'09. Eighth IEEE International Conference on. 2009. IEEE.

[6] Papavassiliou, S., et al., Guest Editors' Introduction: Special Issue on Cyber-Physical Systems (CPS). Parallel and Distributed Systems, IEEE Transactions on, 2012. 23(9): p. 1569–1571.

[7] Broy, M., Engineering Cyber-Physical Systems: Challenges and Foundations. Complex Systems Design & Management, 2013: p. 1–13.

[8] A. Colombo and D. Del Vecchio. Enforcing Safety of Cyber-Physical Systems Using Flatness and Abstraction. IEEE/ACM Second International Conference on Cyber-Physical Systems, Chicago, June 2011.

[9] Jerry Ding, Maryam Kamgarpour, Sean Summers, Alessandro Abate, John Lygeros and Claire Tomlin. A Dynamic Game Framework for Verification and Control of Stochastic Hybrid Systems. EECS Department, University of California, Berkeley, Technical Report No. UCB/EECS-2011-101, September 7, 2011.

Computing, Control, Information and Education Engineering – Liu, Sung & Yao (eds)
© 2015 Taylor & Francis Group, London, ISBN: 978-1-138-02800-5

Current research on situation of DCT starting control technology

Jun Qian Wang, Yuan Yao, Xiao Jing Li, Shan Li & Tao Wu
School of Transportation and Automotive Engineering, Xihua University, Chengdu, China

ABSTRACT: DCT could accomplish the continuity of shifting action, enhance the shifting quality as well as improve the dynamic performance and fuel economy. Two different ways can settle the problem of its starting control in engagement. Jerk and friction work evaluate the quality of starting process, the critical technology is to reduce jerk and friction work during the engagement. Four aspects will describe the current research situation.

KEYWORDS: DCT; starting control; control of clutch; engagement; disengagement.

1 INTRODUCTION

Compared with AMT, DCT not only has the same advantage of simple structure and high transmission efficiency, but also overcome the disadvantage of the interruption in shifting phrase. It can also share the same riding comfort as AT by the coordinated control of the engine and DCT. The essence of DCT starting control is the optimal control of the automatic engagement of the dual clutch. Besides the application of the traditional starting control strategy, both of the two clutches are equally driven in starting phrase. Thus the advantage of the structure of DCT is highlighted, what's more, the heat generated in the engagement would be dispersed and the wearing of the single clutch will also be reduced, which can prolong the lifespan of it.

2 STARTING PROCESS AND EVALUATION

2.1 *Engagement process of the clutch of starting*

The single clutch has 4 phrases shown in figure 1. Tr is the resisting torque of the vehicle, Tc is the transmission torque of the clutch, ω_e is the rotate speed of the drive disc and ω_c is the rotate speed of the driven disc.

Phrase1 0-t_1 the gap between the two discs of is eliminated. No torque is transmitted and the engagement should be fast from the perspective of quick starting.

Phraes2 t_1-t_2 the two discs begin rubbing (Tc=0 now) until Tc could overcome Tr, the vehicle remains still during this time. In order to reduce jerk, the engagement should also be fast.

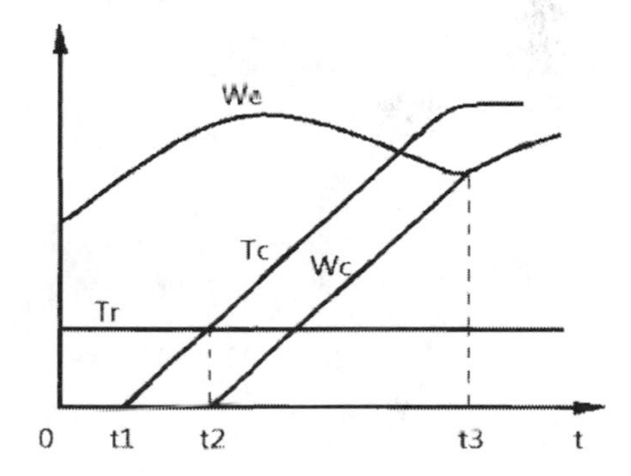

Figure 1. Curve of the process of the clutch.

Phrase3 t_2-t_3 the torque transmitted is equal to Tr in t_2, and then the driven disc begins to rotate. Now the vehicle begins to move and t_2 is defined as the semi-engagement point. In order to improve riding comfort, the engagement speed at this time must be slow, but the speed should not be too slow, which will increase the time of rubbing and threaten the lifespan of the discs.

Phrase4 t_3- in t_3 the two discs would share the same rotate speed for the first time, and the value of the speed is determined by the engine.

The dual clutch starting[1] means the two clutches participate in the rubbing and transmit torque together. When the difference value of the two discs of one clutch is less than a threshold value, the other clutch will stop engaging and begin to separate, while the former would continue to engage. In this way the inconsistency of the wear degree of the two discs will be reduced and the starting time can be shortened.

2.2 *Evaluation of the starting quality*

The starting quality means to ensure the durability, dynamic property and the extent of quick starting. Durability means the lifespan of the various components of the driving system, which requires small jerk, while dynamic property stresses the speediness and high efficiency of starting. Ride comfort is the feeling of passengers, which requires small impact and shakes in starting phrase. Finally starting time, impact allowance and jerk should be considered when evaluating. Thus the process would be simple as the three indexes could quantitatively evaluate the starting quality.

2.3 *Critical technologies of starting control*

The starting control of DCT is the control of the clutch engagement and the torque of the engine at the domination of TCU and the starting intention of the driver, by which the impact allowance and jerk will less than some value. The traffic report determines the tap position of starting, for example, we will select the 2nd gear when the resistance is small. But in general condition we prefer to select the 1st gear which can give more driving force. The starting intention of the driver means the throttle percentage and its rate of change, which are recognized by TCU to control the engine and clutch.

The clutch control is the critical technology of the accomplishment of the automatic starting of vehicles, the complication and difficulty are reflected in four aspects as follows [2]:

1. The targets of the control process are complicated, the evaluation indexes are contradictory and all of the indexes could not be fine at the same time;
2. The stability of the engine cannot be ensured during this time;
3. The counting process of the impact allowance and jerk are complicated, and the time-varying characteristics, non linearity and difficulty in modeling limit the application of the traditional control theory.
4. The intention of starting could not be recognized comprehensively.

3 CURRENT RESEARCH

3.1 *Research on control system*

The control system is the critical and core part of the normal operation of DCT, which would control starting and shifting effect. Liu Huaxin [3] in Chongqing University developed the software of the DCT control system. Lv Jiming [4], the doctor of Jilin University did some research on control strategy of starting on the ramp, and proposed a method to recognizing the total weight of the vehicle and the gradient of the ramp. He also gave an adaptive control method to avoid sliding on the ramp and took the advantage of fuzzy self-adoption method to control the torque to resist sliding. Finally, he proved the control efficiency of his method by simulation.

On abroad, Goetz, Levesley and Crolla [5] designed a closed-loop controller to make the torque of the engine more gently, what's more, it has robustness when controlling the clutch. Kulkarni Manish, Shim Taehyun, Zhang Yi [6] in UMich established a control logic of starting and shifting, and then proved their method by simulating. Borg Warner [7] applied a patent for a series of methods to DCT control in 2004, which included the method of timing and control of transmission torque in each clutch during shifting.

3.2 *Research on dynamic model of starting*

The engagement of clutch is a nonlinear process, an accurate dynamic model is difficult to be built, so researchers have to build an approximate model by supposing and simplifying. The model of DCT consists of engagement model and torque transmission model. Yao Xiaotao [8] simplified the dynamic model of clutch, conformed the basic parameters of the dual clutch according to the design criterion of the diaphragm spring clutch. He also proposed that the slippage rate could be taken as the threshold of starting in the dual clutch starting. Lei Yulong [9] built a model of the clutch actuators according to Bond Graph theory. Then he analyzed the torque transmission model of engagement by the nonlinear relation of the pressure between the two discs and the axial type variables on this basis he built the dynamic model.

3.3 *Research on single clutch starting*

DOLCINI [10] separated the engagement in two phases: the first was applying open-loop control to reduce the frictional time, the second was the last engagement in which the ride comfort was ensured by the optimization of the engagement. Glielmo [11] got the optimal pressure curve of the discs by applying the optimal control on the basis of the starting model of DCT. But it does not tally with the actual situation because the author sets the torque of the engine and loads as constants. SUN [12] considered the position of the throttle, the rotate speed of the engine, transmission ratio, the acceleration of the vehicle and the traffic report comprehensively, and then applied the optimal control by building the objective function of jerk and impact allowance.

Given the difficulty in building accurate model, LUCAS [13] analyzed the operating data of 40 drivers, and concluded the central regulation of the excellent drivers, which could provide reliable data for the application of fuzzy control in starting phrase. TANAKA [14] built a database of the principle of fuzzy control on the basis of the excellent drivers' experience. Then he reasoned out the intention of the drivers by fuzzy control, according to the drivers' operation on the pedal, thus the fuzzy control of the engagement could be realized. Ge Shun [15] optimized the engagement of clutch in starting phase by means of multiple objective functions on the basis of which the combination of quantitative factor fuzzy control and variable coefficient proportional control was applied in the clutch control of AMT. In this way a vehicle can not only start with excellent ride comfort, but also prolong the lifespan of the clutch.

3.4 *Research on dual clutch starting*

Compared with the traditional single type dual clutch starting is more complicated, there are only some universities doing the research which is now in its baby time. GOETZ M [16] is the first who proposed the type of dual clutch starting on the basis of the dynamic model of DCT. Sun Dongye [17] applied the principle of constant speed control of traditional AMT in the control process of DCT and realized the dual clutch starting. They also defined the relevant type of starting according to different throttle opening, analyzed the influence of the separation threshold under the condition of different road loads and throttle percentages. But the threshold is lack of the analysis and support of the theory, its optimal value could not be accurately confirmed.

4 SUMMARY

Two types of starting a pattern of DCT have been analyzed, the difficulty in the control technology of starting phrase is pointed, and the current research on the control technology of starting is summarized in four aspects. The research mainly focuses on the combination of modern control theory and the data of experiments which include fuzzy control, optimal control, adaptive control and the comprehensive of them. Then the simulation of computer is used to prove the correctness of the methods. Although some problems have been solved the research is still beyond the ideal outcome. Because the process of starting is a complicated working condition, what's worse, the indexes of its evolution are contradictory.

REFERENCES

[1] Sun Wei, Yang Yongli, Ma Jin Experimental Study on Dry DCT Starting Using Two Clutches [J]. Agricultural Equipment & Vehicle Engineering, 2012, 50(8) 29~32.

[2] Zhang Dongfang. The Starting Control Strategy of DCT [D]. Chongqing: Chongqing University of Technology, 2011.

[3] Liu Huaxin. Development of DCT Control Software [D]. Chongqing: Chongqing University, 2008.

[4] Lv Jiming. Simulation Study on Vehicle Hill-start Control of Dual Clutch Transmission [D]. Changchun: Jilin University, 2009.

[5] Goetz, M C Levesley, D A Crolla. Dynamics and control of gearshifts on twin-clutch transmissions [J]. Journal of Automobile Engineering, 2005, 219(8): 951~963.

[6] Kulkarni Manish, Shim Taehyun, Zhang Yi. Modeling and control for launch and shift of dual-clutch transmission vehicles [A]. 17th International Conference on Design Theory and Methodology[C]. US:2005. 597~604.

[7] Jing Congbo, Yuan Shihua, Guo Xiaolin. Application and Prospect Analysis of DCT [J]. Journal of Mechanical Transmission, 2005, 29(3): 56~58.

[8] Yao Xiaotao, Qin Datong, Liu Zhenjun. Simulation Analysis Of DCT Starting [J]. Journal of Chongqing University(Natural Science Edition), 2007, 30(1):13~17.

[9] Lei Yulong. Research on Improvement of AMT Property [D]. Changchun: Jilin University, 1998.

[10] Dolcini P J' Canudas C, Bdchart H. Observer-based optimal control of dry clutch engagement. IFP international conference. 2007, 4(62): 615~621.

[11] Glielmo L, Vasca F .Optimal Control of Dry Clutch Engagement [C] SAE World Congress 2000. Transmission and Driveline Symposium 2000. Detroit: SAE, 200-01-0837.

[12] Sun C. Zhang J. Optimal control applied in automatic clutch engagements of vehicles. Chinese Journal of Mechanical Engineering, 2004, 1 7(2):280~283.

[13] LUCAS G, MIZON R. Clutch Manipulation during Engagement [J].Automotive Engineer (London), 1978, 3(2): 81~85.

[14] TANAKA Hirohisa, WADA Hideyuki. Fuzzy control of clutch engagement for automated manual transmission [J]. Vehicle System Dynamics, 1995, 24(4): 365~376.

[15] Ge Shun. Fuzzy Control of Automatic Transmission [J]. Automobile Technology, 1996(6):7~11.

[16] GOETZ M, Levesley, M.C and Crolla, D. A. Integrated power train control of gearshifts on twin clutch transmissions [J]. SAE paper 2004-01-1637, 2004.

[17] Sun Dongye, Qin Datong. Clutch Starting Control with a Constant Engine Speed in Part Process for a Car [J]. Journal of Mechanical Engineering, 2003.11, 108~112.

Computing, Control, Information and Education Engineering – Liu, Sung & Yao (eds)
© 2015 Taylor & Francis Group, London, ISBN: 978-1-138-02800-5

Design of digital herbarium for Hubei medicinal plant specimens based on ASP and SQL SERVER management system

Yang Bo Sun, Lin Bi Zhang & Ke Li Chen
Hubei University of traditional Chinese Medicine, Wuhan, Hubei, China

ABSTRACT: Based on the herbarium specimens of medicinal plants collected in Hubei Province and stored in the Herbarium of Hubei University of Traditional Chinese Medicine, we aim to establish a digital herbarium for medicinal plant of Hubei Province by using ASP technology and SQL SERVER database management system, through designing in three aspects including the need analysis, data storage and functional realization, to realize digitization and networking of herbarium specimen.

KEYWORDS: Medicinal plant; Digital Herbarium; ASP; SQL SERVER; Management system.

With the development of information technology, the databases of Chinese medicine resources are under construction all over the country. Hubei Province, based on the third national Chinese medicine resources survey, after many years of accumulation, acquisition and exchange, now in the herbarium of Hubei University of traditional Chinese Medicine collects a total of 189 families and 2804 species, 8314 specimens of the medicinal plant herbarium specimens, along with the development of the fourth national Chinese medicine resources survey work, some new precious herbarium specimens will newly introduce. These specimens are the basis of the naming and entities for the resources of medicinal plants in Hubei province, and they are the most basic data and information study the medicinal plants and also the scientific basis of the archives and the dynamic change of the medicinal plants.

Medicinal plant specimen collection and production have highest professional and technical requirements, which is time-consuming. Specimen preservation time is limited, viewing specimens are limited geographically. How to make these herbarium specimen open to more people, realize the sharing of resources, so that information can be reused and longer preserved is the very problem that should be solved by information technology.

In order to rescue protection and research these precious medicinal plant specimens and realize the sharing of resources, we use ASP+SQL SERVER 2005 database management system based on the B/S mode in order to establish Hubei medicinal plant herbarium management system and fulfill digitization and networking of the specimen information.

1 REQUIREMENT ANALYSIS

1.1 *System's goal*

Based on the investigation of medicinal plant resources in Hubei Province and Herbarium specimen collection, production, identification, classification, we established Hubei Province medicinal plant herbarium management system to achieve systematic management of specimen information and constantly update and add new data of herbarium specimens. This system is convenient for the collection of specimens for dynamic management, and at any time to understand the whole picture. Meanwhile, it's useful for the pharmaceutical industry in our province in the fields of teaching, scientific research, production and development to provide morphological characteristics, medicinal plants growing environment, regional drug, Latin naming of scientific information and reduce the loss caused by direct reading of specimens.

1.2 *System functional demand*

The system is to realize the information storage of medicinal plant herbarium specimens, and in accordance with the user's permissions, and then respectively achieve the herbarium information query, modification, deletion and user management, and finally achieve the goal of remote access to this system through a network.

1.3 *System development environment*

We use Dreamweaver, Flash, Photoshop and other software to develop the web. The server database use SQL SERVER 2005. A web site set up to use JDK Version 1.6 and above, WEB server use apache-tomcat-6.0.37and IIS. The client only needs to install IE 6 and above browsers.

2 DATABASE DESIGN

2.1 *Tables structure*

Based on the above analysis, the database contains three tables: userlb (user information table), herbinf (The specimen information table), herb picture (The picture information table).

1 userlb: userid, username, password, user type, true name, email, Register_time.
2 herbinf: her bed, herb name, Latin name, common name, collects personally, collect them, collect place, elevation, longitude, latitude, distributionResourcetype, height, breast diameter, base diameter, root, stem, leaf, flower, fruit, seed, Charactersgenus, medicinalparts, drug, name, local medicinal, note, species name, specieslatinname, family name, familieslatinname, genus name, genuslatiname, identify a person, identify them, verifyperson, verify them, store.
3 Herb picture: her bed, PID, Panama, paddr, type, description.

2.2 *Relationship between the tables*

SQL SERVER 2005 database management system is used for the establishment of the three tables, the relationship between the tables in the database is as shown in figure 1.

2.3 *The operation class of the database*

The operation class is mainly set up for adding, deletion, modification and query of the new specimens and user information, and operating the database by calling the ASP class.

2.4 *Database connecting*

You can connect to the database through three practical methods in ASP script: by using ADO components directly, establishing a connection by means of ODBC DSN and OLEDB. The first method is applied in this system, i.e., by using ADO components directly.

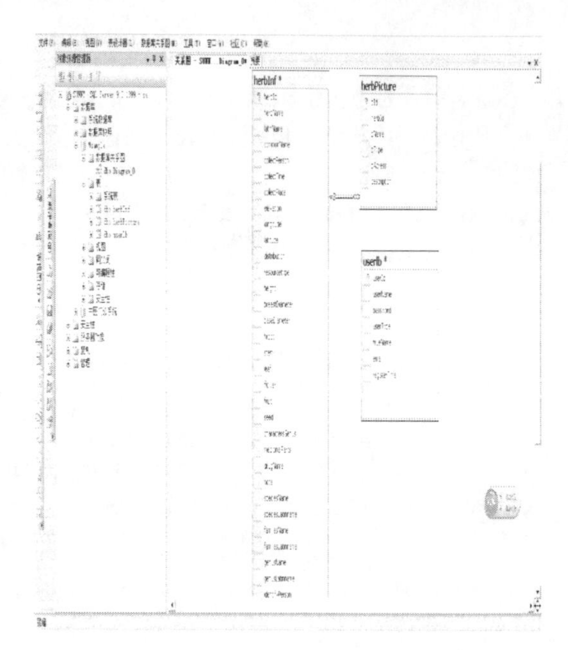

Figure 1. Relationship between database tables.

3 SYSTEM DETAILED DESIGN AND IMPLEMENTATION

3.1 *Foreground inquiry and paging display*

This system uses the gradual entry method to display information specimens. When Users access the medicinal plant herbarium website, they can (1) click medicinal plant type specimen collection in the homepage and the information will be displayed according to the names of the plant family; (2) click the family, the name list of the specimens can be opened; (3) click a specimen name, the information will be displayed in accordance with the basic information, medical information and identification information by separate columns with the corresponding images. In order to facilitate the search, you can conduct a fast search by specimen number or scientific name.

Multiple records will return when a SQL statement in ASP is executed. If all records are displayed in the same page, the efficiency is low and it will be difficult to read. To solve this problem, information has to be paged.

3.2 *Background data management*

After logging in, Users will enter "specimen database entry system background" to operate the main interface, the users' interaction with the database mainly includes: (1) specimen input. (2) The specimen management. If users need to edit the medicinal plant herbarium information, they just need to click "delete/modify/image information" options on the right to delete, modify or upload a picture. (3) Personnel management. User information can be added, modified or deleted. (4) Upload the picture settings.

3.3 *Storage and processing of images*

Medicinal plant herbarium specimen images are categorized into three classes: specimen images, identification of signatures, collecting information of the collectors, which are characterized by a large amount of information, images with high pixels, slow speed of access. So it is necessary to pre-process all the images.

Images can be stored in the SERVER2005 database in two ways: (1) the pictures stored as binary data format (firstly, convert images to binary format data) and (2) the picture stored in the storage path. This system adopts the second method. Firstly, the pictures in the directory structure are stored in the physical device, then in the database the corresponding storage path strings are stored. Image files are stored separately, and direct contact will be established with this system through the field value of stored path in the database.

Image processing uses the popular JPG format to process and store images which can't compress with loss and should retain the original archival data. At the same time in order to reduce the amount of data storage and save the transmission bandwidth, data should be compressed. We select the irreversible compression method to make the compressed image meet the visual requirements and the data amount be greatly reduced so that costs will be eventually reduced.

4 CONCLUSION

The design of Hubei medicinal plant herbarium management system along with the sharing of resources of medicinal plant herbarium specimens is the best choice for teaching, scientific research and social needs. Taking Hubei University of traditional Chinese Medicine Herbarium collection herbarium data for the construction of the Hubei medicinal plant herbarium resources can effectively solve the problem of database remote query difficulty. The query service is open to all the Internet users, so the professionals in different fields can obtain data according to their own needs, which will greatly improve the utilization rate of the medicinal plant specimens. At the same time, it provides the basis for further development of professional database of medicinal plant specimen data.

REFERENCES

[1] The State Administration of traditional Chinese medicine. Traditional Chinese medicine informatization" The 12th Five Year" plan.2012.

[2] Di Guangzhi. The Development of Plant Specimens of Database System[J]. Computer Programming Skills & Maintenance, 2011, (2): 36–37.

[3] Du Xinying. StructureInformationalizedManagement System of Medicinal Plants Specimen [J].Resource Development & Market, 2010, (9): 824–825.

[4] Lin Yuting. Design and Implementation of Digital Museum Specimens of Trees Based on Roaming Technology [J].Forestry Education inChina, 2013,32 (2): 25–27.

[5] Wang GuoPing et al.The Construction of MedicinalPlants Database in Xinjiang[J]. Xinjiang Journal of Traditional Chinese medicine, 2010,28 (2): 86–88.

[6] Li Runmei et al. The Construction ofMedicinal PlantsHerbarium Database [J]. Journal of Clinical Rational Drug Use, 2012, (5): 138:139.

Computing, Control, Information and Education Engineering – Liu, Sung & Yao (eds)
© 2015 Taylor & Francis Group, London, ISBN: 978-1-138-02800-5

A new supervised learning hierarchy clustering classification method

Lu Ping Pu
College of Information Science and Engineering, Guilin University of Technology, Guilin, China

ABSTRACT: This paper presents a novel supervised hierarchical clustering method, SHC (Supervised Hierarchy Clustering), which can divide the class-known samples to bintree of clusters with hierarchy clustering by the rule of maximizing clusters of same class samples and can make up SHCC (Supervised Hierarchy Clustering Classifier) combining with varied classifier. The rule of maximizing clusters of same class samples is to get a partition of fewest clusters of same class samples, i.e. to maximize local same class clusters. The principle, algorithm and computing complexity of SHC are introduced. SHC has been implemented in MATLAB language. SHCC classifier is made of the combination of SHC and the linear classifier of statistics toolbox in MATLAB. The SHCC has been tested with four groups of UCI datasets, its 10-fold cross-validation classification error, test, which show that SHCC has a higher classification accuracy than the other 7 kinds of classifier. The new method the paper proposed automatic can automatically find the number of clusters of same class samples with the rule of maximizing clusters of same class samples.

KEYWORDS: Data mining, Machine learning, Supervised clustering.

1 INTRODUCTION

Cluster refers to the classic unsupervised learning method, whose mission is to group the samples as similar as possible within the cluster and between-cluster differences as large as possible. The cluster is mainly divided hierarchical clustering, model-based clustering based clustering and density grid-based clustering and other types [2]. Due to the lack of sample class labels these teachers supervise signal, the parameters such as the number of clusters should be pre-specified in the cluster when the sample density was not significantly different, with low efficiency and unknown actual classes of clusters.

Classification is a supervised learning method which needs samples with class-known as teachers supervise signal to train and can classify the samples of class-unknown.

Supervised clustering due to using the sample's class label as a teacher's training sign to supervise learning, which greatly reduces the uncertainty of information, raises the higher the efficiency, the result is clearly the true classification class reflect the structure of the sub-sample distribution and so on. Supervised clustering of existing forms, such as learning vector quantization network [3,4], based on dynamic partitioning and incremental clustering method [5,6], support vector machines [6]. Learning vector quantization networks by competitive learning network classification results were wrong incentive to adjust the weights to learn. Based on dynamic clustering method and incremental division, within the class commonly clustered index does not penalize purity minimization method. Support vector machines, constraint information, sample classes through the nuclear non-linear mapping function to a high dimensional Hilbert space, so that in the new space come together in the same sample, heterogeneous sample separation increased, can be divided hyperplane realize supervised clustering. These methods require you to specify the number of clustered, learning and classification efficiency and provide explicit information on the structure of the subclass distribution of the length of each.

Hierarchical clustering analysis is one of major clustering method [4,6,7], with the defined distance of clusters and samples, clusters are merged (or spitted), the samples from each sample as a cluster to all the samples as a cluster- there are multi-level hierarchy consisting of clustered binary [5], but to rely on artificially divided into several sub-clustered, you need to specify various parameters [1], such as the number of clustering, the distance between the clusters threshold or the maximum depth of the subtree. A supervised hierarchical clustering analysis can automatically learn the optimal parameters by using category information.

This paper presents that the target of supervised clustering is the separation of fewest same class clustering, that is dividing the samples into local maxima same class clustering, thus propose a new supervised clustering method: hierarchical clustering method (supervised hierarchy clustering, referred to SHC method) to local maxima of same class clustering principle divided clustered hierarchical clustering, can automatically learn the local maxima of same class clustering, specify the number of clusters free of human-designated inter-cluster distance threshold. SHCC classifier is combined SHC with a variety of classifiers. A SHCC of SHC and linear classification is completed in Matlab language, which is run with artificial simulated data for the explanatory principle experiment and with four groups UCI standard data for classified experiment. The results prove that this method a more high classification accuracy.

2 PRINCIPLE AND METHOD

2.1 *The principle of maximizing local clustering of same class*

Supervised clustering Optimal target evaluation criteria should be different due different viewpoint. The data preprocessing of multi-subclass classification is an important use of supervised clustering, so its optimal target should be evaluated from the perspective of classification.

Similar situation appears when used with a category label sample set training classifiers. Classifier training sample set is a mapping function from the sample's attributes to the mode category, which is actually a mapping function to determine what area of the sample falls into the classification in the training process of the attribute space sample to sample category is divided into different zones when classified as unknown samples using the mapping function to determine the properties fall into a certain area of decision-making in the region belong to the category of training samples. Sample set using the area calculation classifier constructed and application classifiers is proportional to the number of regions, and therefore the number of regions the fewer , the better.

Nearest neighbor classification is an extreme example, in which the test sample is classified in the class of the nearest training sample. It designated an area for each training sample: each sample point adjacent to the sample point perpendicular bisector hyperplane polyhedron composed of super-regional, regional category in which the sample is determined by the type of test samples fall into the category in each area was marked for the region, which is the area of the sample categories [1]. Due to the training sample for each region, the amount of calculation proportional to the product of the number of training samples and the number of test samples, so if the number of training samples and the predicted samples is big, the amount of calculation should become very large.

To reduce the number of regions, the regions of adjacent same class samples should be merged into a new bigger one. This kind of merging continue until it encounters different types of sample area stop, this is the time to get area of Local maxima same class clustering and the number of areas is less, which is the best classification supervised clustering solutions.

If all same class samples are in the same area-cluster, the number of clustered with the same number of classes, which is the minimum of number of areas and can be directly used for general classifier training to classify. But many times, the same class sample is often located in different regions of the other classes are split, forming areas of sub-classes. The general classification usually uses clustering as data preprocessing (such as support vector machine classification method is mapped to a method of using kernel functions to solve high-dimensional space) to get sub-classes regions for training classes.

In summary, from the viewpoint of the efficiency of classification, clustering a similar minimum number of divisions is optimal. The fewer the number of clustered, the more the number of samples within a clustered, clustered number less equivalent to the number of samples within each clustered more, so from the point of view classification supervised clustering optimization goal is to find such a clustered division scheme: the same composition of each sample contains as much as possible clustered samples. Namely supervised clustering maximizing local clustering of similar principles.

2.2 *Supervised Hierarchies Clustering (SHC)*

Supervised hierarchy clustering is a find local maxima of same class clusters supervised clustering method with a hierarchical clustering binary tree by each node clustering being annotated category. Its principles and methods are described in detail below.

2.2.1 *Principle*

Hierarchical merging clustering analysis based on distance equivalent to The merge process from nearest neighbor classification began to sample cluster region described previously. As in the nearest neighbor classification, each sample can be regarded as an object occupies an area of a single sample point clustering, if two recent samples are the same class, you can merge them into a new cluster. If two adjacent clusters are same class, then they should be merged into a new larger cluster.

Every cluster continues to merge like this until the recent cluster is different class so far. At this point is the optimal partitioning of least number of same class clusters which are local maxima same class cluster composed by as much as possible same class samples.

In the hierarchical clustering binary tree generation of hierarchical merger clustering analysis, you can use these ideas to find the largest and the most of the same samples cluster in the local region of attribute space, that is a local maxima of same-class clusters.

The method is: looking for the leaf nodes of the hierarchical clustering binary tree up to their parent nodes, if a same-class sibling node cluster is a cluster of same-class samples of same-class composition, composed of the two parent nodes clustered cluster is same-class, so look for continued up until a sibling node cluster nodes same-class is not the same as same-class clusters, the parent node is composed of two clustered of different class clusters.

Therefore, local maximum clustering binary tree clusters of similar characteristics are the parent node of multi-class clusters.

2.2.2 *SHC method*

The work program of SHC method is composed of three steps of hierarchical clustering, Marking clustered categories and identifying local maxima of same class clusters:

Step 1 Hierarchical clustering: cluster a set of training samples in a binary tree of cluster which is from each sample belongs a cluster to all samples belong a cluster with hierarchical clustering method.

Step 2 Markers clustered categories: from the leaf nodes start clustering binary tree up, according to each sub-category clustering to identify and mark its category by the order of clustering clustered. The principle of marking category of cluster node is that if both it's child nodes are same class, then the class of a cluster node is marked as same as the class of its child nodes, otherwise marked as multi-class cluster, multi-class cluster marked mark should be able to distinguish the category number with similar clusters as -1.

Step 3 Find local maxima of same categories clusters: Find all local maxima of same categories clusters of the binary tree according to the features of local maxima of the same categories cluster is sub-node same class cluster of multi-class clustering. All samples of local maxima of same categories clusters are marked with the no. of the cluster.

2.3 *Supervised Hierarchy Clustering Classifier (SHCC)*

2.3.1 *Principles*

Supervised hierarchy clustering to divide the training samples into the clusters of same class sample. But same class samples may be clustered into different clusters which are a sub - class and can be used for classification.

2.3.2 *SHCC classification method*

Supervised hierarchical clustering classifier (SHCC) first Supervised hierarchical clustering (SHC) is applied to the training sample set, then obtained for each sample clustered category classification categories as training a classifier, the last available this classifier classification of unknown samples, get clustered category to which they belong, to identify the real mode of this subclass clustered category as the sample real mode category. General Classifier such as linear classifier can be used The Proceed as follows:

Step 1: Train samples with supervised hierarchical clustering (SHC) to get the sample set with class of patterns and No. of the cluster.

Step 2: Train a kind of classifier with the samples' attributes and No. of cluster as the class attribute.

Step 3: Apply this classifier to classify the class-unknown sample to obtain it's No. of the cluster.

Step 4: Find out the class of patterns of this No. of cluster as the predicted class of patterns of the sample by search Chart of class of patterns and No. of the cluster.

2.4 *Computational complexity of SHC and SHCC*

Because the first step SHC of SHCC is hierarchical clustering, when the classifier is linear classifier, the subsequent step is linear, so the computational complexity and hierarchical clustering same as O (N2).

3 EXPERIMENTS

3.1 *Experimental environments*

SHC and SHCC implemented in MATLAB programming. SHCC is combined by SHC and the classify function of Statistics Toolbox of MATLAB, and only the linear classifier function of the classify function [9] to be used.

3.2 *Experiment simulated data sets*

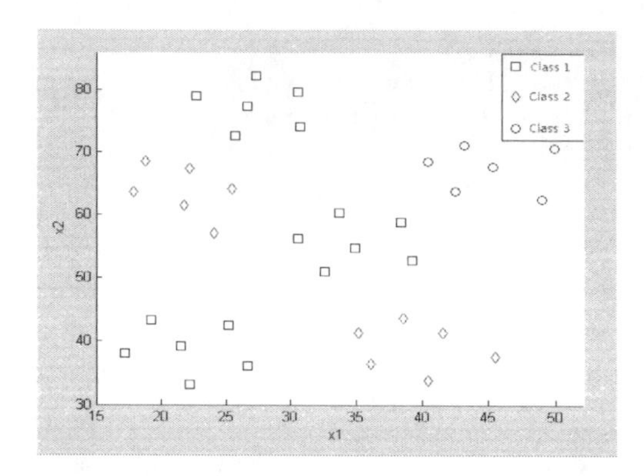

Figure 1. The distribution of samples.

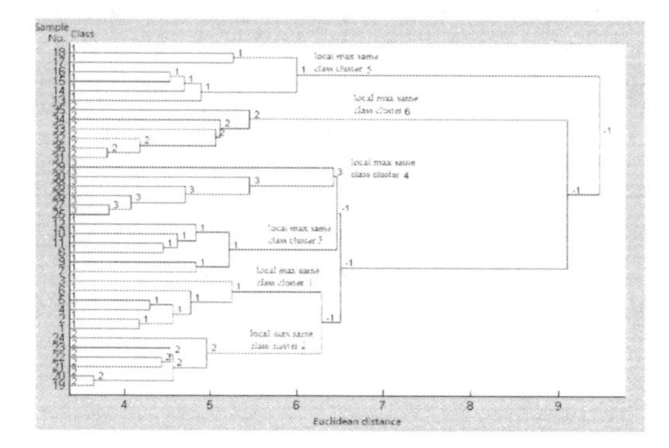

Figure 2. Finding out all max same class clusters by marking each cluster node class from letaf to root node of Binary Tree.

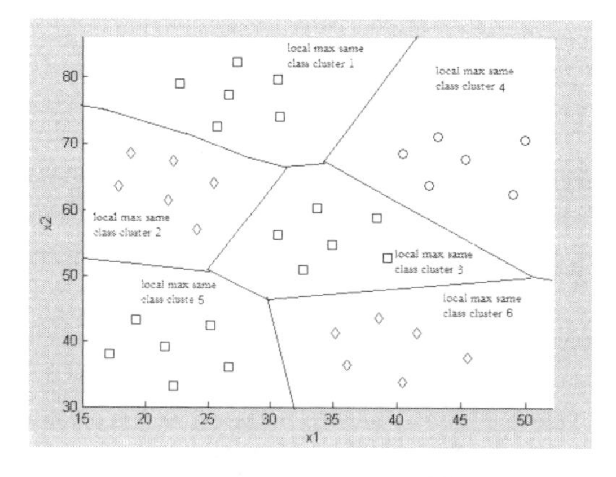

Figure 3. The result of SHCC: linear classification Basing the clusters from SHC.

Designed a set of experiments a total of 36 samples of three classes, each sample attribute data with two features and a class attribute (Figure 1). The similarity between the samples is measured by their Euclidean distance, and the similarity between the clusters is measured with their average distance. The clustering binary tree is marked with a class label by looking for max same class clusters and divided into six clusters (Figure 2). Finally, trained linear classifier with the No. of cluster of samples as their class (Figure 3), each cluster occupies a area. The samples of class 1 are divided into the three adjacent clusters 1,3,5 which form an area of concave profile, the samples of class 2 are divided into nonadjacent two clusters: cluster 2 and 6, and all samples of class 3 are in the cluster 4. With max same class clusters SHC, the distributing information of different class samples instructs cluster and classify.

3.3 *Standard test data set*

3.3.1 *Test dataset*

Experimental test data to select the four UCI data sets cited from http://finalfantasyxi.inf.cs.cmu.edu/MATLABArsenal/MATLABArsenal.htm: Iris, Glass, wine and Ecoli [10], all of the attributes are continuous numeric type, each parameter as shown in Table 1. Clustering the first data before normalization.

3.3.2 *Test method*

The cross-validation method is commonly used test methods of classifying in machine learning, which can ensure that all the samples are treated as test samples independently. In this experiment, 10 sets of cross-validation approach, the sample material was randomly divided into 10 groups, each in turn when the test samples and the training sample, after so perfectly executed 10 times, got 10 group classification error rate, the average rate of 10 groups of classification error after that the average error rate of data collection for this purpose.

3.3.3 *Analysis of experimental results*

Table 1. Datasets used in the test.

Dataset	Iris	Glass	Wine	Ecoli
number of features	4	9	13	7
number of classes	3	6	3	8
number of samples	150	214	178	336

Table 2. The 10-fold cross validation classifying error rate.

Classifier	EColi	Glass	Iris	Wine
libSVM-Linear	0.134	0.070	0.040	0.045
libSVM-Poly	0.172	0.687	0.467	0.117
libSVM-RBF	0.152	0.088	0.047	***0.017***
Naive Bayes	0.140	0.159	0.053	0.028
C4.5	0.193	***0.023***	0.047	0.062
BP	0.143	0.047	0.027	0.039
Perceptron	0.238	0.389	0.327	0.140
SHCC	***0.112***	**0.027**	***0.013***	0.042

4 UCI datasets of EColi,Glass,Iris, Wine are cited from http://finalfantasyxi.inf.cs.cmu.edu/MATLABArsenal/MATLABArsenal.htm

3.3.4 *Analysis of experimental results*

compared the error of classification results applied the current representative seven kinds of classifiers to the 4 datasets with 10-fold cross-validation. In the eight kinds of classifiers on four datasets of *E. coli*, Glass, Iris and Wine classification error SHCC have two best, a second, a 4th (Table 2, Figure 2), which say SHCC are the best.

SHCC on the four datasets classification error is less than the total average 0.0485, indicating high accuracy.

4 CONCLUSION

The paper presents a principle of maximization of the same class cluster as the Optimal target of supervised clustering, that is minimization of number of clusters and maximization of same class samples in a cluster. SHC (supervised hierarchical clustering) is the product of the application of the principle to hierarchical clustering analysis. SHC cluster data by marking each node class label from leaf nodes up to root node to find local max same class clusters to divide clusters, can automatically learn to best Optimal cluster division, the min number of clustered and maximum of same class samples in a cluster., SHC Need not human designated cluster distance threshold and can automatic acquire variety cluster-dividing parameters such as the number of clusters, division designated poly maximum depth of the tree, etc. SHC can be combined with a variety of classifiers into a supervised hierarchical clustering classifier SHCC, 10 group cross four groups UCI standard data validation classification results and compare classifier with seven kinds of other representative to prove SHCC a higher classification accuracy.

Due to the hierarchical clustering computational complexity is O (n2), so the SHC and SHCC for small samples task.

ACKNOWLEDGEMENT

This study was funded by Guangxi Natural Science Foundation (project task No. 2013GXNSFAA019276) and Guilin University of Technology doctoral project start-up capital fund.

REFERENCES

[1] side Hajime Regards, Zhang workers and other pattern recognition (second edition) [M]. Beijing, Tsinghua University Press, 2000.
[2] Richard O.Duda, Peter E. Hart, David G.Stork. Pattern classification [M]. Beijing, China Machine Press, 2003.442 – 447.
[3] Kohonen T. The self-learning map [A] Proc IEEE, 1990, 78: 1464–1480.
[4] Cheng Jianfeng, based on the EM algorithm PROJECTILES supervised LVQ neural network and its application [J], Systems Engineering and Electronics Technology [J], 2005,27 (1): 121–123.
[5] Song Tong, Songbao Jiang learn a new supervised clustering method and its application in fault diagnosis [J], Computer Engineering and Science [J] 2001.23 (5): 63–69.
[6] Dettling M, B¨uhlmann P: Supervised Clustering of Genes [J] Genome Biology 2002,3: Research 0069.1–0069.15.
[7] Thomas Finley, Thorsten Joachims. Supervised Clustering with Support Vector Machines Proceedings of the 22 nd International Conference on Machine Learning, Bonn [A], Germany, 2005.
[8] Zhao big Huwang Liang, Li Zijin. Deposits statistical prediction [M]. Beijing, Geological Publishing House, 1983.157 – 161.
[9] The Workshop, Inc. Classify.m in matlab. [Z], 2002.
[10] Newman, DJ & Hettich, S. & Blake, CL & Merz, CJ (1998). UCI Repository of machine learning databases [EB / OL] [http://www.ics.uci.edu/~mlearn/.MLRepository.html] Irvine, CA: University of California, Department of Information and Computer Science.

Computing, Control, Information and Education Engineering – Liu, Sung & Yao (eds)
 ISBN: 978-1-138-02800-5

Research on intelligent computer system based automobile manufacture and assembly

Cheng Li Pang
Dalian vocational and technical college, Dalian, China

ABSTRACT: RFID technology is already widely applied in manufacturing industry. However, it could be applied more flexibly in production management. This study proposes the use of RFID in production management, not to alter the production system but to enhance it with RFID technology so that it is precise, automated, and gives readings in real time. Although passive RFID is rather similar to barcode reading, capturing data directly with RFID reader can save human resource operation time and also the electronic tag can be reused and is not easily damaged. Our proposed system combines the traditional automobile manufacture with intelligent computer based techniques to achieve the initial goal of designing and implementing the intelligent automobile manufacture and assembly system.

KEYWORDS: Automobile Manufacture; Automobile Assembly; Intelligent Computer System.

1 INTRODUCTION

During the period between the late 1990s and the early 2000s, IT had penetrated deep into society at large due to the spread of the Internet, broadband communication networks, development of equipment using IT, etc. Accordingly, the world economy including advanced countries had changed into a knowledge-based economy due to the development of IT since 1990s [1, 2]. Furthermore, the U.S. National Science Foundation (NSF) expects that the period since the Industrial Revolution and the Computer & Communication Revolution will become the "Fusion Technology Age". This development and spread of IT has brought about sweeping changes in lifestyles by innovating paradigms of the entire society. Recently, IT has raised a new issue called convergence with other technologies or industries, and which is located in the center of changes and innovation [3-5]. Technology continues to advancing manufacturing has shifted their attention from the mass production of lean production in the past, to meet the actual demand. How to reduce waste and improve quality and reduce cost has become one of the biggest challenges. Recently, manufacturing has been use barcode system to transmit information with some goods, make the production process more smoothly and accurately. With the aid of bar code system, production process information can be passed to the backend system; Bar code, however, still needs human intervention, so neglect of operators may lead to abnormal data. Meanwhile, since bar codes can be easily damaged, if any stain or blemish affects the code during the production process, the production information will be hard to collect. The applications of RFID technology in recent years include security management, goods tracing system management, and public transportation system management, such as is used in the Easy Card system. Applications of RFID are becoming more and more popular. Not only the retail industries but the medical industries and manufacturing industries have also implemented RFID systems. For the general public, the Easy Card is the most well-known and frequently experienced RFID application. Other than in the mass transportation system, people can use an Easy Card in supermarkets, to manage student school attendance, and even in cell phone systems that are integrated with the Easy Card in advance. The applications of RFID in other industries and sectors include the development of patient management systems through RFID and a dove tracking system to aid in dove group management and identification. In recent years, the application of RFID and research focuses more on the production car manufacturing process, tracking and stocks, identification, security protection, part of automobile safety system. However, the actual effect of application and related parts inventory levels automobile assembly line rarely discussed. Through the RFID technology, this research is mainly an ID to products, and setting along the lines of the assembly line. Besides, if any omission of production order occurs, the system can send out warning message to avoid the errors that could happen in the assembly line and therefore the actual production situation and production resume in real time can be also controlled. We will discuss these in detail in the later sections.

2 OUR PROPOSED SYSTEM

2.1 *Analysis of as-is and to-be in assembly lines*

Before RFID technology was implemented, only information entering work station can be controlled. State assembly and parts using each line point cannot be determined on the assembly line. After calculating the number of parts the product is the assembly line in time. As the product line inventory levels cannot be determined accurately, there may be unnecessary waste. Of this study was to develop a real-time, accurate and efficient systems will be instant (JIT) concept is applied to the production or transportation quantity only when they need to completely eliminate waste by implementing RFID technology, the production data sent to the back-end system, and monitor part number used in the production, production scheduling, production line, a part of the existing shares. When stock on-hand drops below safety stocks level, stock replenishment staffs will be alerted to replenish the stock so that the assembly line stocks are promptly maintained. When product passes through line point, it is examined in real time to ensure that product sequence is complied. An alarm will be sound by the control system in case of non-compliance. The accumulated production volume, list of products, products on assembly line, and production resume at each line point in each work day can also be reviewed in real time.

2.2 *Structure of the system*

Our system is divided into front-end data collection system and back-end management system as shown in the figure one.

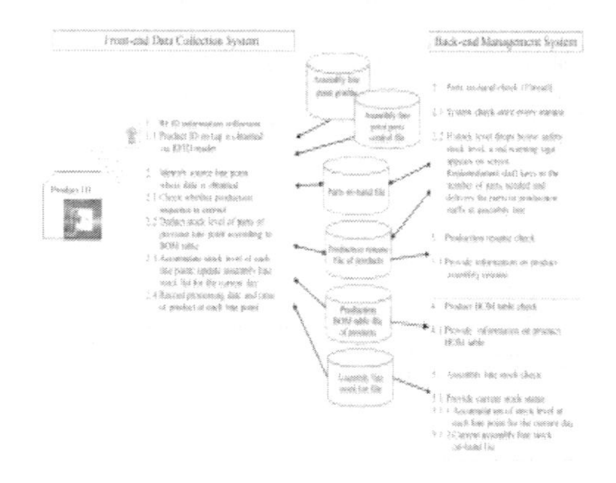

Figure 1. The structure of our system.

Front-end data collection system aims to collect RFID data and review line point sequencing in real time. Using immediately if the data is correct and record the number of parts in the production of the resume. The backend problem is mainly management information. In addition to the current stock information added staff, an assembly line system for production management to provide real-time data about the sample quantity and the status quo of the assembly parts to the assembly line. In the event of abnormal parts usage, production management staffs are promptly notified to deduct the abnormal quantity so that the actual assembly line stock level can be determined.

2.3 *System flow and installation*

Our system flow simulates the actual production flow, and determines when to place RFID tag, installation place of RFID reader, data collection method, and feedback of data to back-end system shown in the figure 2.

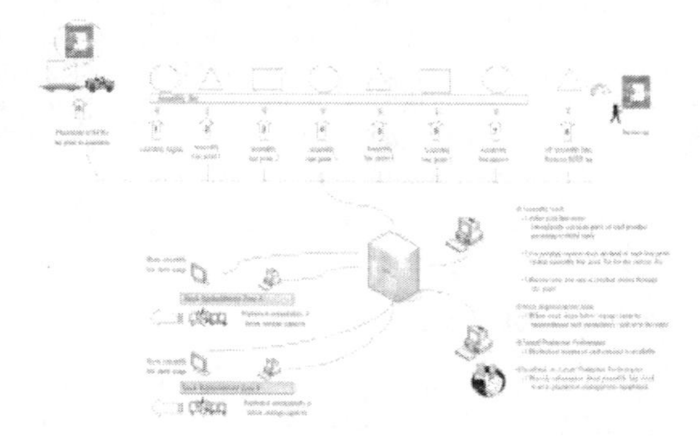

Figure 2. The system flow chart.

Daily production order is fixed, once the product ID system production, can be said as the only product identification number. Product assembly, product ID is written to the electronic tag and label on the tray assembly. The tag is the most important reference in the entire assembly flow shown in figure 3.

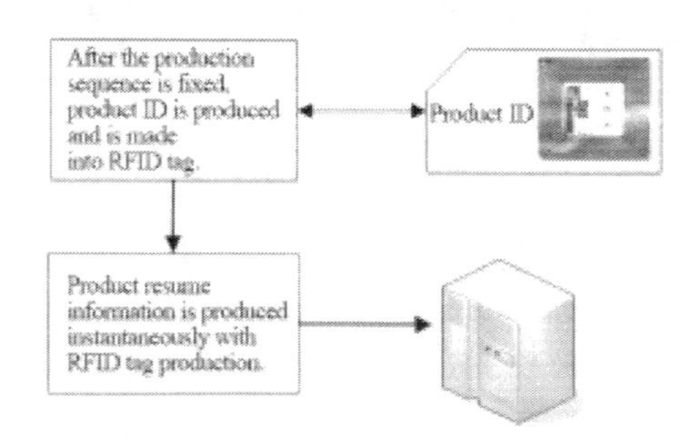

Figure 3. The Product ID: Flow chart of electronic tag production procedure.

2.4 *Assembly line status and off-assembly line product quantity check*

Product assembly is completed once it passes through the gate out line point. Production management staffs may obtain the latest information about number of finished products gated out and assembly line production status from the system at all times. They may also check whether the number of finished products gated out is equivalent to the estimated number of productions. Production management staffs may immediately calculate whether production of the day exceeds or falls behind the projected production quantity. Productions ahead of or behind schedule are considered abnormal. The figure 4 shows the procedure.

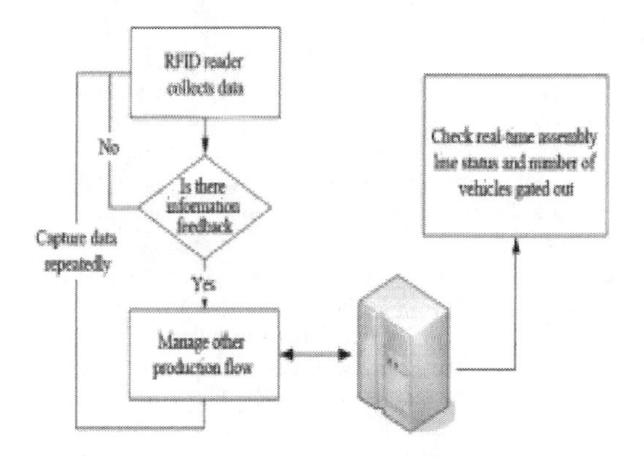

Figure 4. The flow chart of check on Real-time assembly line status and number of vehicles gated out.

When the product passes through line point, parts assembly will begin at the line point. The conference content and production in line with the instructions in the original e - kanban kanban system. System number of production parts will be deducted according to the use of a predetermined part of the BOM list, in each line. There will be a warning signal, when the part below the safety stock levels to inform employees collection and supplement. Products after time points and lines of code record Numbers of components used in product recovery file. After collecting parts according to kanban, parts collection staffs increase stocks of parts and deliver the parts to line point. The figure 5 shows the procedure.

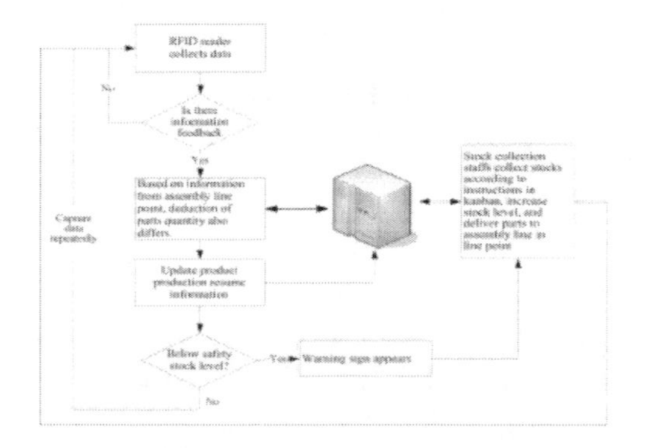

Figure 5. The flow chart of response to product parts usage.

3 SIMULATION AND EXPERIMENT

3.1 *Front-end data collecting*

Once the front-end data collection system is activated, it will begin collecting data via RFID reader. Product ID on electronic tag will be captured and stored in database. Data stored in RFID database is then analyzed via real-time data collection operation to determine the source assembly line and line point of data. The related product information is then sent to front-end "product assembly" screen, the figure 6 shows the system.

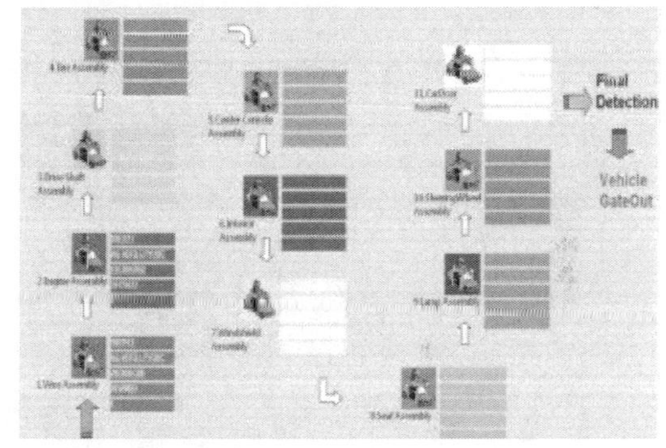

Figure 6. The flow chart for producing product resume and parts usage feedback.

3.2 *Back-end management system*

When the product passes through line point, product ID will be captured via the RFID reader. Product BOM table may then be produced via product ID. Parts in the BOM table and parts in line point may be compared. If both data does not tally, it means there may be product which has entered incorrect line point. A warning sign will appear to alert production management staffs so that they can notify the assembly staffs, find out reasons for abnormalities, and immediately re-transfer the product to its correct line point for assembly (shown in the figure 7).

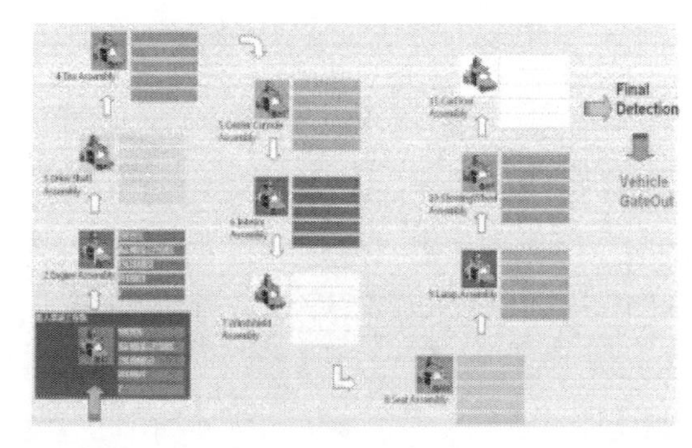

Figure 7. The line point inspection.

When the product passes through line point, parts assembly will begin at the line point. System number of production parts will be deducted according to the use of a predetermined part of the BOM list, in each line. There will be a warning signal, when the part below the safety stock levels to inform employee collection and supplement. Products after time points and lines of code record Numbers of components used in product recovery file. To collect parts according to the kanban, collect employees to stock more parts and delivery of the production line in line. Upon completion of stock replenishment, the staff may add the number of replenished stocks to the current status system to increase total assembly line stocks, see figure 8.

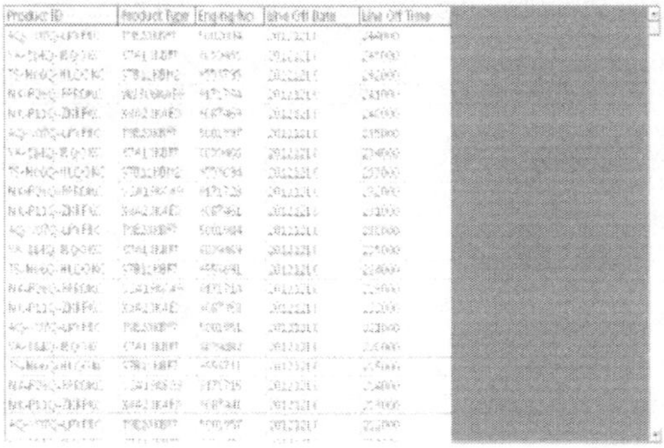

Figure 8. The check on number of vehicles gated out in real time scale.

This system enables check on real-time production status and provides current assembly line status and list of vehicles gated out. Users may find information from the system at any times to check the latest assembly line production status. The number of vehicles gated out may also be checked at the end of the day to determine whether it tallies with the projected number of production for the day. This system also enables other systems to manage abnormalities such as productions ahead of or behind schedule. See figure 9.

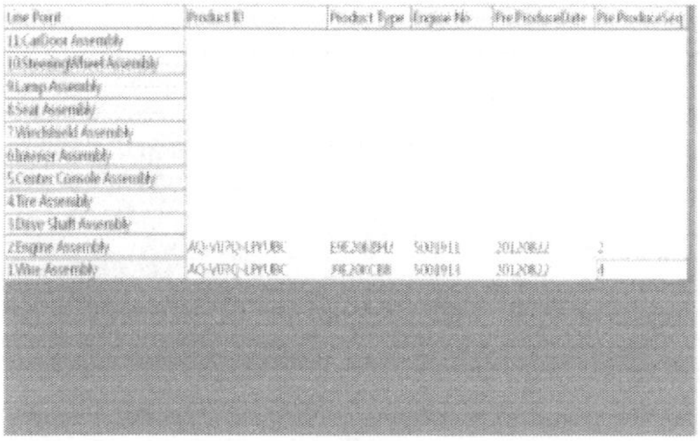

Figure 9. The Screen showing assembly line status.

4 SUMMARY

RFID technology is already widely applied in manufacturing industry. However, it could be applied more flexibly in production management. This study proposes the use of RFID in production management, not to alter the production system but to enhance it with RFID technology so that it is precise, automated, and gives readings in real time. Although passive RFID is rather similar to barcode reading, capturing data directly with RFID reader can save human resource operation time and also the electronic tag can be reused and is not easily damaged. In the entire production process, the RFID electronic tag is the product ID. Other than assembly process, it can also be used in post-processing after the product is off assembly line during packaging and when it leaves the factor. With this system, production management staffs can instantaneously follow the assembly progress and compare it with daily production plan. In the event production is ahead of or behind schedule, production plan adjustments can be made immediately so that lean production can be achieved. Our future research will focus more on the RFID part, to level up the correctness of the system.

REFERENCES

[1] Sengar, Surabhi, and S. B. Singh. "Reliability Analysis of an Engine Assembly Process of Automobiles with Inspection Facility." Mathematical Theory and Modeling 4.6 (2014): 153–164.

[2] Canuto da Silva, Guilherme, and Paulo Carlos Kaminski. "Application of digital factory concepts to optimise and integrate inventories in automotive pre-assembly areas." International Journal of Computer Integrated Manufacturing ahead-of-print (2014): 1–9.

[3] Kobayashi, Hideo. "Current State and Issues of the Automobile and Auto Parts Industries in ASEAN." AUTOMOBILE AND AUTO COMPONENTS INDUSTRIES IN ASEAN (2014): 1.

[4] Wilson, Shellyanne, and Nazma Ali. "Product wheels to achieve mix flexibility in process industries." Journal of Manufacturing Technology Management 25.3 (2014): 371–392.

[5] Otto, Alena. "Minimizing Risks for Health at Assembly Lines." Operations Research Proceedings 2013. Springer International Publishing, 2014. 341–346.

Computing, Control, Information and Education Engineering – Liu, Sung & Yao (eds)
 ISBN: 978-1-138-02800-5

The launch vehicle cowling environment monitor system based on wireless sensor networks

Zhi Yi Fang, Yong Jie Zhang & Jian Jin
Shanghai, China

ABSTRACT: This paper discussed the Cowling Environment Monitor System in Launch Vehicle (CEMSLV), and proposed a new system design based on wireless sensor networks, which had achieved initial experimental verification. The wiring itself is cumbersome, damageable, and vulnerable to a variety of distortion and interference due to the long distance transmission. Because the wiring is often built into the infrastructure, it is difficult for sensor installation in hard to reach locations and changing configuration is almost impossible, such as expansion of the number of sensors or sensing modalities. This paper explored the feasibility of Wireless sensor networks (WSN) in the launch vehicle, and establish a useful base to WSN application in other spacecraft.

KEYWORDS: Wireless Sensor Networks, Cowling Environment Monitor System, Launch Vehicle.

1 INTRODUCTION

The cowling environment Monitor System in launch vehicle (CEMSLV), completed the real-time monitoring of temperature and humidity condition of the satellite and the rocket, transmitted the environment information to the ground. Ground crew controlled the air conditioning system according to the environmental data, to ensure the environmental compliance. CEMSLV starts working from the technical area, through horizontal transport to emitter region, erection and docking to the launch pad, until 45 minutes before launch. CEMSLV plays an important role in satellite-rocket training, horizontal state test, vertical lifting test and transition region. The schematic diagram of CEMSLV is shown in Figure 1.

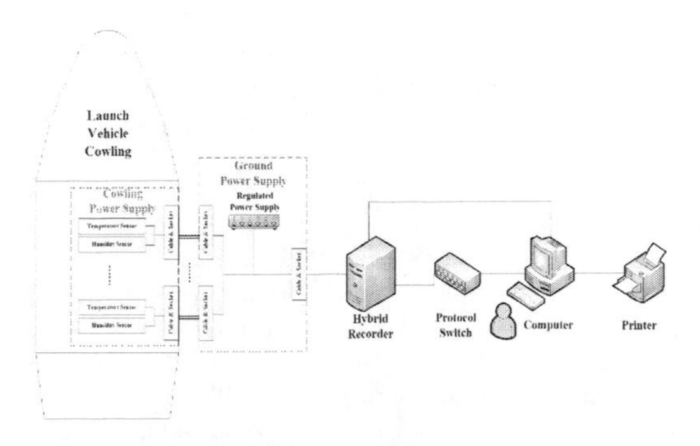

Figure 1. Schematic of active duty launch vehicle cowling environment monitor system.

CEMSLV is composed of cowling grid, ground power grids, hybrid recorder, protocol switch, computer and printer. The cowling grid is composed of temperature sensors, humidity sensors, cable and sockets. Sensor installation and selection should satisfy the needs the satellite measurements demand. Temperature sensors distributed throughout the cowling, such as the inner thin-gauge skin and insulated enclosure surface. Humidity sensor distributed near the temperature sensor on the reference surface. The cable socket located on the cowling surface for connecting cowling grid and ground grid. Regulated power supply in ground grid is powered by the 220V AC voltage, outputs 24V DC voltage, to cover the sensor supply. Hybrid recorder is responsible for receiving, transmitting, processing, displaying the sensor parameters, sends the real time data to a computer. Protocol switch is used to convert the data protocol between hybrid recorder and computer, so as to realize the accurate transmission.

When CEMSLV works, temperature and humidity sensors get the environmental parameters. Afterwards, the data are teleported to hybrid recorder through the cowling and ground grid. Hybrid recorder converts the data to temperature and humidity parameters. After processing by CEMSLV software, the real-time display and storage is completed. Someone in measurement system will be put in charge of environment data observation to provide the basis for the environmentally Controlling system.

2 WIRELESS SENSOR NETWORK SYSTEM DESIGN

2.1 *System design*

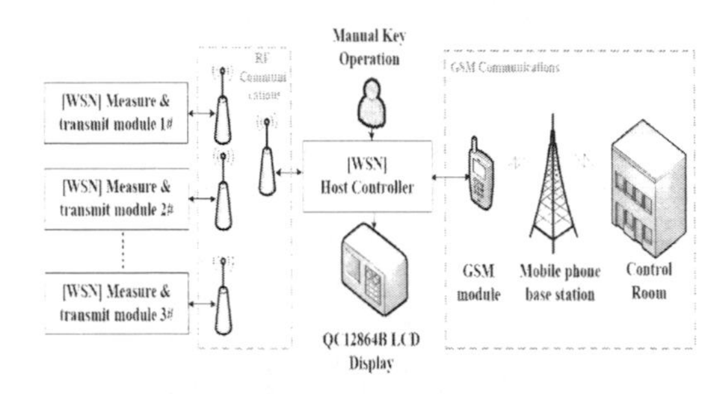

Figure 2. Schematic of launch vehicle cowling environment monitor system based on WSN.

In Figure 2, the cowling environment monitor system in the launch vehicle, based on wireless sensor network, is mainly composed of host controller and measure & transmits modules. Measure & transmit module is used for acquiring the environment parameters, and uploads data to the host controller. Host controller gets data of all modules through RF wireless communication. After data processing, host controller sends the environment information GSM receiving node or a phone through GSM module and mobile phone base. Monitor staff also can call or send text messages to CEMSLV based on WSN to query the current environment parameters or system status. Host controller also has a manual control mode. The operator could complete all system functions through the keys and LCD screen.

When the wireless sensor network is adopted, there will be much less cable and sockets in CEMSLV based on WSN. The long-term interference, joint damage and installation limit problems will be avoided in new system. WSN improves system reliability and flexibility.

2.2 *Hardware design*

2.2.1 *Main control module*

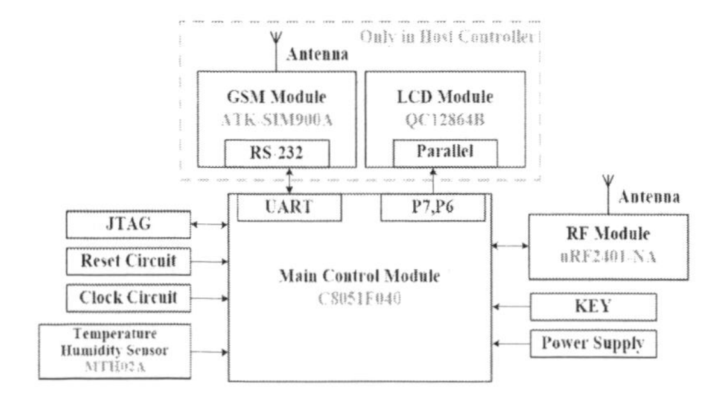

Figure 3. Main control module.

In CEMSLV based on WSN, we use mixed signal microcontroller C8051F040 as a main control module in measure & transmit module and host controller, as shown in figure 3. In measure & transmit module, MCU is responsible for communication with a sensor and nRF2401, and uploading sensor data to the host controller. In host controller, MCU increased GSM module, LCD module and keyboard module function, sends the environment parameters to control center, and completes the HMI function.

2.2.2 *Temperature and humidity sensor*

Temperature and humidity module MTH02A with single line digital interface performs high-speed accuracy temperature measurement. The temperature measurement precision is 0.5°C, and humidity measurement accuracy is 2%, which satisfied the system requirements.

2.2.3 *RF communication module*

The RF communication module is the main part of system design. NRF24L01, which uses FSK modulation, has high speed, small volume, stable performance and other advantages. Module maximum power can reach about 20DBM. The best transmitting distance is 200 meters (open environment), which fully meets the needs of the system design.

2.2.4 *GSM communication module*

GSM module ATK-SIM900A (900/1800MHz) can transmit SMS, data and other information with low power dissipation. GSM module and the main control module are connected to RS232, based on AT command set. The module in SLEEP status consumes less than 2mA, satisfied the node power consumption requirement in wireless sensor networks.

2.2.5 *LCD module*

LCD module QC12864B can display Chinese characters and graphics, 8192 Chinese characters (16X16 matrix) built-in, DC3.3V power supply, controlled by parallel interface.

2.3 *Software design*

The software flow of host controller is shown in figure 4. The system can be triggered by manual operation, GSM call and GSM text. We can get the temperature and humidity of all the sensor and running state of CEMSLV only by a mobile phone. Meanwhile CEMSLV based on WSN also can complete pre-judgment and alarm function, which improves the level of automation of CEMSLV.

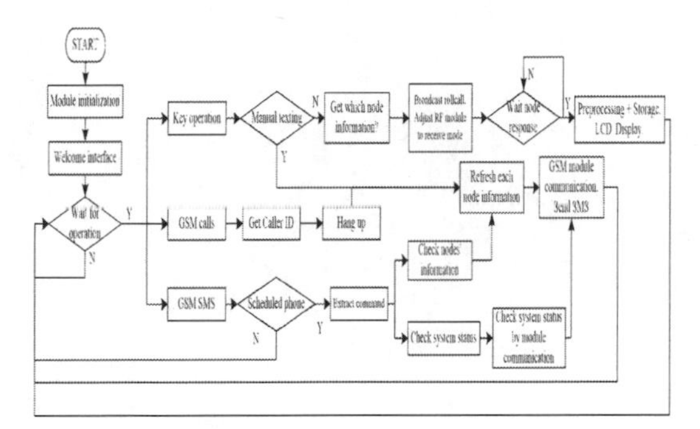

Figure 4. Software flow of host controller.

3 EXPERIMENTAL RESULTS AND ANALYSIS

CEMSLV generally need to continuously work more than 60 hours. The temperature measurement range requires for the -30°C to +50°C, 0.5°C accuracy. The humidity measurement is less than 85%, the accuracy is 3%.

The experimental platform for CEMSLV based on WSN is shown in Figure 5. The platform can achieve the measurement of temperature and humidity requirement. The experimental result shows that the communication distance between RF modules is above 50 meters (no shading condition). The measured accuracy reached 0.5°C, humidity 2%, which satisfies the system design requirement.

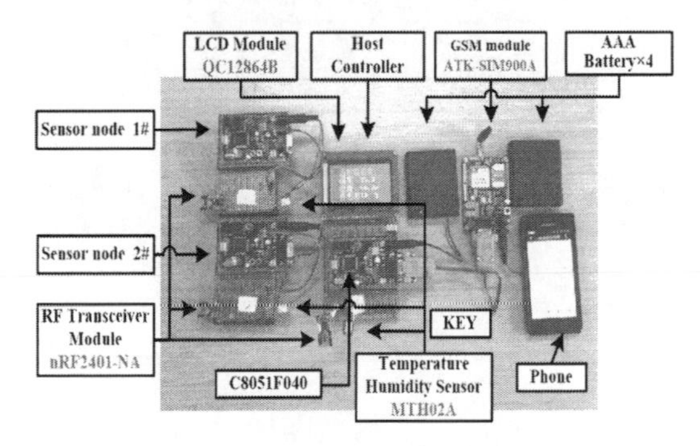

Figure 5. Experimental platform for wireless sensor networks.

4 CONCLUSION

This paper presents a new cowling environment Monitor System in launch vehicle based on wireless sensor network technology, which could effectively reduce the weight of system, and improves the performance. The new system has strong engineering practicability and value.

REFERENCES

[1] W. S. I.F. Akyildiz, Y. Sankarasubramaniam, E. Cayirci, "Wireless sensor networks: a survey," Computer Networks, vol. 38, pp. 393–422, 2002.

[2] B. M. Jennifer Yick, Dipak Ghosal, "Wireless sensor network survey," Computer Networks, vol. 52, pp. 2292–2330, 2008.

[3] H. K. Qureshi, et al., "Poly: A reliable and energy efficient topology control protocol for wireless sensor networks," Computer Communications, vol. 34, pp. 1235–1242, 2011.

[4] T.-W. Sung, et al., "Reliable data broadcast for zigbee wireless sensor networks," International Journal on Smart Sensing and Intelligent Systems, vol. 3, pp. 504–520, 2010.

[5] Z. Rosberg, et al., "Statistical reliability for energy efficient data transport in wireless sensor networks," Wireless Networks, vol. 16, pp. 1913–1927, 2010.

[6] M. T. Penella and M. Gasulla, "Runtime extension of low-power wireless sensor nodes using hybrid-storage units," IEEE Transactions on Instrumentation and Measurement, vol. 59, pp. 857–865, 2010.

[7] A. Koubaa, et al., "Improving quality-of-service in wireless sensor networks by mitigating hidden-node collisions," IEEE Transactions on Industrial Informatics, vol. 5, pp. 299–313, 2009.

[8] S.-J. Park, et al., "GARUDA: Achieving effective reliability for downstream communication in wireless sensor networks," IEEE Transactions on Mobile Computing, vol. 7, pp. 214–230, 2008.

[9] G. A. William Wilson, "Wireless Sensing Opportunities for Aerospace Applications," Sensors & Transducers Journal, vol. 94, pp. 83–90, 2008.

[10] A. B. John Schmalzel, Stephen Rawls, Jon Morris, Mark Turowski,Richard Franzl, and Fernando Figueroa, "Smart sensor demonstration payload," IEEE Instrumentation & Measurement Magazine, vol. 10, pp. 8–15, 2010.

Computing, Control, Information and Education Engineering – Liu, Sung & Yao (eds)
© 2015 Taylor & Francis Group, London, ISBN: 978-1-138-02800-5

Research on interactive E-learning platform based on data mining and cloud computing: A review

Jin Mei Liu
Weifang University of Science and Technology, Shandong, China

ABSTRACT: With the fast development of computer science and technology, a combination of a web service and E-learning platform is a hot research topic. When the learning process becomes digitized, educational data mining employs the information generated from the electronic sources to enrich the learning model for academic purposes. To provide support for e-learning systems, cloud computing is set as a natural platform, as it can be dynamically adapted by presenting a scalable system for the changing necessities of the computer resources over time. We give an overview of the current state of the structure of the cloud computing, we provide details of the most common infrastructure, has been to develop such a system. We also present some examples of e-learning approaches for cloud computing, and finally, we discuss the suitability of this environment for educational data mining, suggesting the migration of this approach to this computational scenario. As the final analysis, we present our future potential research area.

KEYWORDS: Data Mining; Cloud Computing; Big Data; Hadoop; Interactive E-Learning.

1 INTRODUCTION

1.1 *Background research*

electronic learning, better known as e-learning, refers to the issues related to virtual distance education through electronic communication mechanisms, specifically the internet [1-5]. It is based on the use of approaches with several formats and functionalities that may support the teaching-learning process, such as emails, web pages, forums, various learning platforms, and so on. The main advantages defined by studying through online tools include flexibility, convenience, ease of access, consistency and repeatability of the proposed tasks. In addition, the demand for teaching resources is usually dynamic and rapid way each are not identical, to activity heights. At that time no other service request system weakened, it will be necessary to prepare a superior infrastructure than regular work needed for institutions of learning. Another kind is according to the demand for services, and only paying the actual use of resources. The answer to these requirements is a cloud computing environment. A computational grid is more static, from the point of view of hardware resources, and is designed mostly with the aim of obtaining the best computer performance. Cloud computing is intended to allow the user to obtain various services without being aware of the underlying architecture while offering a transparent scalable. It is therefore not so restrictive and can offer many different services, from web hosting, to word processing.

The advantages of this new computational paradigm with respect to other competing technologies are clear. First, cloud application providers strive to provide the same or better services and performance as if the software programs were installed locally on end-user computers, so the users do not need to spend money buying complete hardware equipment for the software to be used. Second, this type of environment for data storage and computing schemes allows companies to get their applications up and running faster, with a lower requirement of maintenance of the IT department since it automatically manages the business demand by dynamically assigning IT resources.

1.2 *Cloud computing introduction*

the philosophy of cloud computing, mainly implies a change in the way the problems are solved by using computers. The design of the applications is based upon the use and combination of services. In contrast to what happens in more traditional approaches, i.e. grid computing, the provision of the functionality relies on this use and the combination of services rather than on the concept of process or on algorithms). There are different categories in which the service-oriented systems can be clustered. One of the most widely used criteria to group these systems is the abstraction level that they offer to the system user. In this way, three different levels are often distinguished, as we can observe in Figure 1. In the remainder of this section, we will first describe each of these three levels, providing the features that define each of them, and some examples of the most well-known systems of each type. Then, we will present some technological challenges that must be taken into account for the development of a cloud computing system.

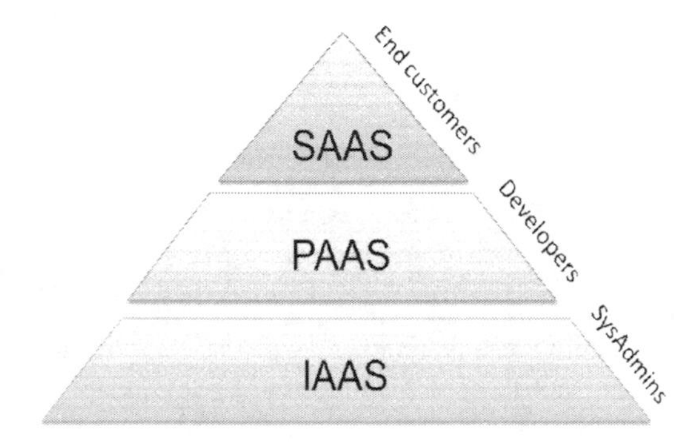

Figure 1. The Illustration of the layers.

Traditionally, in order to process a high amount of data in a short period of time, a grid computing environment was the most suitable solution in order to reduce computational costs and to increase the flexibility of the system. The cloud computing platform contains characteristics of both clusters and grids, since both of them are composed of loosely coupled, heterogeneous nodes, and grids are also geographically dispersed. However, there are significant differences among them, which mainly refer to the target of their application, and the implementation technologies used.

2 OUR METHODOLOGY

2.1 *Pre-discussion for the platform*

as we stated in the introduction of this work, with the huge growth in the number of students, educational content, services that can be offered and resources made available, e-learning system dimensions grow at an exponential rate. The challenges with regard to optimizing resource computation, storage and communication requirements, and dealing with dynamic concurrency requests highlight the necessity of the use of a platform that meets scalable demands and cost control. This environment is cloud computing. Defining the potential of SaaS solutions for efficient and sustainable online learning platforms in contrast to the former 'classic' online learning platforms, may lead us to understand the benefits of cloud computing on both a technological and educational level.

We need to introduce the 'path' for promoting a migration to such a model in order to obtain a valuable system for web tools and collaborative services, such as lesson plans, videos, curriculum resources, student interactions, and so on. By referring to a wide number of current developments, we will measure the real impact of this novel computational scheme and we will describe some examples on how to start using a system with these characteristics.

2.2 *Drawback of the current method*

among learning technologies, web-based learning offers several advantages over conventional classroom-based learning. Its most notable advantages reduce costs, as a physical environment is no longer required and it can therefore be used at any time and place at the convenience of the student. Moreover the number of students that can follow the class is not limited by the site of a physical classroom. Additionally, the learning material is easy to keep updated and the teacher may also incorporate multimedia content to provide a friendly framework and to facilitate the understanding of the concepts. Finally, it can be viewed as a learner-centered approach which may address the differences among teachers, so that all of them may compare their material in order to evaluate and re-utilize common areas of knowledge. However, there are some disadvantages that must be addressed prior to the full integration of e-learning into the academic framework. Currently, e-learning systems are weak on the scalability at the infrastructure level. Several resources can only be deployed and assigned to specific tasks so that when receiving high workloads, the system needs to add and configure new resources of the same type, making resource acquisition and management very expensive.

2.3 *Organization of the cloud computing*

above we have seen the potential goodness of the implementation of a cloud computing environment in the educational sector. A further consideration is the type of cloud that such institutions need, i.e., the option of a private, public or hybrid cloud. Under these circumstances, a private cloud is mostly preferred for three reasons: First, educational centers have the option of utilizing their existing infrastructure which allows them to have a better control over the resources. Regarding this previous point, this implies a low cost process for setting up the cloud. Finally, the security issue is a major factor in the preference for this type of cloud. Certain activities such as the admission process for new students and the policies thereof, examinations/tests conducted for existing students and all research activities require confidentiality and breaches in security cannot be tolerated. Motivated by all the aforementioned benefits of integrating cloud computing in educational institutions, and especially those related to private clouds, throughout this section we will describe the characteristics of building a private cloud inside an educational institution, from both the architectural and functional point of view.

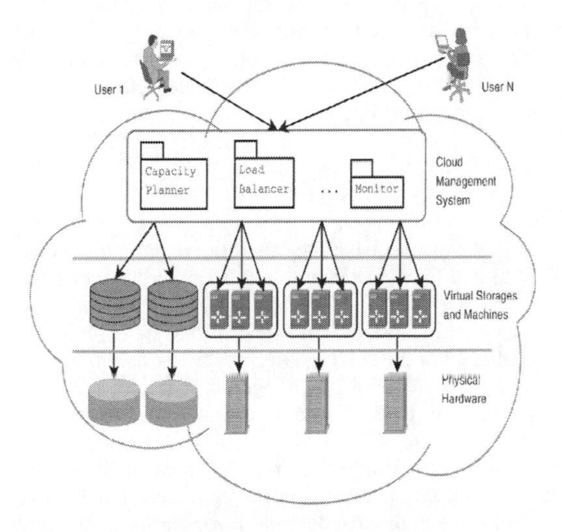

Figure 2. The Overview of a cloud architecture.

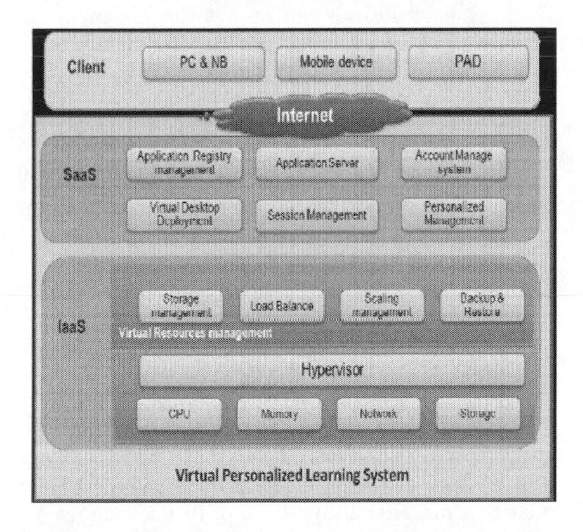

Figure 3. The architecture of the virtual personalized learning environment.

Nowadays, the combination of cloud technologies and e-learning has been insufficiently explored. The pedagogical possibilities of the collaborative aspect of the cloud are studied in Stevenson and Hedberg. In the aforementioned work, the authors refer to the lack of research that might provide a theoretical framework with which pedagogy could be developed. However, the mobility inherent in the cloud approach could be stressed as a major advantage in developing conceptual frameworks and establishing effective teaching and learning strategies. Regarding the use of cloud computing for educational purposes, we can distinguish between the public cloud and the private cloud solutions. We will review these two different approaches in the following sections. It is a framework which provides on-demand creation and configuration of VM images so that students are able to have their own Java servlet environment for experimentation, containing MySQL, Tomcat, PHP, and Apache web server. With this approach, students can focus more on developing, deploying and testing their applications in a servlet container. Another example of an application that can be found in the specialized literature is the BlueSky cloud framework developed by Xi'an Jiaotong University (China). Its architecture has several components aimed at the efficient provision and management of e-learning services, and is able to preschedule resources for the hot contents and applications before they are needed, to safeguard the performance in concurrent accesses, although no details have been found with regard to how this is achieved. This architecture may provide an effective and flexible way of matching resources with the current economic condition through the utilization of unused resources and the abstraction of third party involvements, as well as providing a more flexible environment so that the client can now configure his own security policy.

3 EDM AND CLOUD COMPUTING

3.1 *Introduction to EDM*

The EDM process follows the same steps as general data mining (see Figure 4). First, it starts with the preprocessing of the data in order to prepare it for the learning stage, i.e., using cleaning techniques, reducing the number of instances and/or input variables to facilitate the working of the algorithms, and so on. Then, the learning algorithms are applied to extract useful information from the data, whose type will depend on the final aim of the user, such as classification, clustering, association rule mining, sequential mining or text mining. Finally, a post-processing step can be carried out to enhance the obtained results or to provide a more interpretable representation of the system. However, it is worth pointing out that EDM has a broader purpose than just applying traditional data mining techniques. It is used in a wider sense by including other approaches, such as regression, correlation, or visualization. All these approaches are necessary to carry out a better treatment of the educational data for all kinds of purposes and to generate summary reports that facilitate the understanding of the extracted related knowledge.

From a general perspective, EDM allows us to discover new knowledge based on students' usage data. Its aims include the validation and evaluation of educational systems, to potentially improve some aspects of the quality of education and to lay the groundwork for a more effective learning process. The specific objectives that are addressed in the framework of EDM are applied research to improve the learning process and guide students' learning and pure research to achieve a deeper understanding of educational phenomena. The great advantage of the EDM methodology is its use in research and in building models in various areas that can influence e-learning systems. One of the main areas is user modelling, covering what the student knows and the behavior and motivation, how the user experience is, and how different user requirements are satisfied by e-learning.

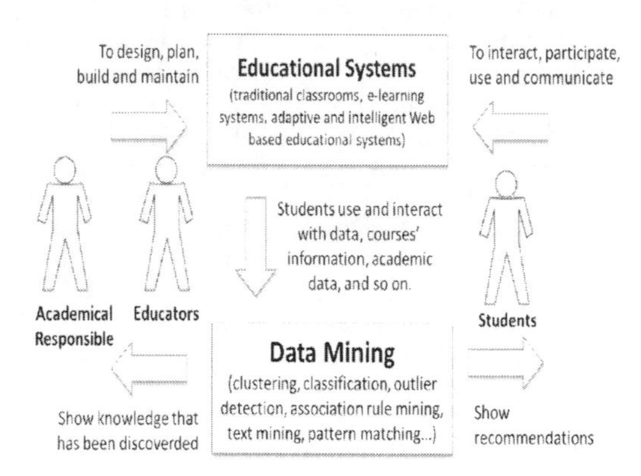

Figure 4. The cycle of applying data mining in educational systems.

3.2 *Cloud combined EDM*

We have noted an increase in the instrumental educational software in recent years. This has made the gathering of more and more information for teaching purposes possible. As we have commented above, instructors may use different frameworks that apply data mining algorithms to the representation and extraction of dynamic learning processes and learning patterns to support students' deep learning, efficient tutoring and collaboration in web-based learning environments. It will be noted that this system has significant similarities to that introduced for e-learning purposes in the prior section, specifically at the lower layers of the infrastructure. In this way, all the guidelines introduced in the former section remain valid for that which we are currently describing. Both systems may coexist, but it is also possible to include a cloud computing framework for EDM to process data from a standard e-learning approach. Additionally, the concept for the development of a cloud computing system for EDM is practically identical as in the case of e-learning tasks. As stated above, the advantages of being able to process more information in a shorter time are critical in allowing a quick response to improve learning quality in accordance with the students' progress with the current pedagogical tools. On the other hand, for a major deployment of the cloud environment, the personal and technological drawbacks that this new paradigm implies, remain necessary to be overcome.

4 SUMMARY AND CONCLUSION

In this work we have identified the main components of e-learning, focusing on the flexibility, convenience, ease of access, consistency and repeatability of this kind of system. An e-learning system faces the challenge of optimizing large-scale resource management and provisioning, in accordance with the huge growth of users, services, education contents and media resources. We have established the goodness of a cloud computing solution in this area of education. We have enumerated several approaches that have already been proposed to address e-learning on cloud computing, describing these models and how they take advantage of this environment to enhance the features of the educational system. The significance of the application of a cloud computing platform for EDM has also been stressed. We have emphasized the goodness of EDM for enhancing the quality of current education from a pedagogical point of view, but it has the disadvantage of requiring the management of a large amount of information, usually generated within e-learning courses. In order to extract useful knowledge from it by running data mining algorithms, and to efficiently present the reports obtained from them to the instructors, the computational advantages of this new paradigm of cloud computing are evident. For the full establishment of this paradigm, further research should include the simplification of the deployment of a private cloud, considering the hardware infrastructure and management issues of the system, as well as stressing the impact cloud delivery may have on greater advancements in pedagogical effectiveness.

REFERENCES

[1] González-Martínez, José A., et al. "Cloud computing and education: A state-of-the-art survey." Computers & Education 80 (2015): 132–151.

[2] Stantchev, Vladimir, Lisardo Prieto-González, and Gerrit Tamm. "Cloud computing service for knowledge assessment and studies recommendation in

crowdsourcing and collaborative learning environments based on social network analysis." Computers in Human Behavior (2015).
[3] Yu, Shenquan, and Xianmin Yang. "A Resource Organization Model for Ubiquitous Learning in a Seamless Learning Space." Seamless Learning in the Age of Mobile Connectivity. Springer Singapore, 2015. 141–158.
[4] Oliveira, Jéssica, et al. "Model Driven Testing for Cloud Computing." Innovations and Advances in Computing, Informatics, Systems Sciences, Networking and Engineering. Springer International Publishing, 2015. 297–304.
[5] Schweighofer, Patrick, Stefan Grünwald, and Martin Ebner. "Technology Enhanced Learning and the Digital Economy: A Literature Review." International Journal of Innovation in the Digital Economy (IJIDE) 6.1 (2015): 50–62.

Computing, Control, Information and Education Engineering – Liu, Sung & Yao (eds)
© 2015 Taylor & Francis Group, London, ISBN: 978-1-138-02800-5

Design of intelligent lighting-control and temperature-measure system oriented to integrated home system

Chen Liang Zhang, Wen Wang & Xue Jun Su
Department of Basic Experiment, Navy University of Aeronautics and Astronautics, China

ABSTRACT: To implement intelligent lighting-control and temperature-measure, a system oriented to integrate home system was proposed. The system took use of master-slave structure, in which slave stations received commands from master station to accomplish lighting-control, temperature-measure and returned data frames. The master received the data and displayed it on the LCD screen. The system is composed of a microcontroller, PLC modules, relays and temperature transducer, which avoids rewiring and wireless communication interference. The proposed system works stably and operates conveniently by testing, which can satisfy application requests.

KEYWORDS: integrated home system; intelligent lighting; temperature measure; master station, slave station.

1 INRODUCTION

Logistics Network and intelligent technology develop greatly in an intelligent household field, and traditional lighting technology is heavily attacked. People want to not only know different regions' temperatures, but also control their lights. Based on many technologies such as computer, tensor, communication, and auto control, intelligent control system booms quickly.

The existing control systems of integrated home system are mainly based on wire and wireless systems, in which wire systems, for example, RS485 bus, require an additional bus to transmit data, while wireless ones use ZigBee wireless technology to transmit data with low power and automatic network. Although ZigBee leaves out wiring question, the wall, maybe cause ZigBee's signal reduction because of low transmitting power. In addition, ZigBee maybe affects system's reliability for mutual interference between wireless systems.

The control system based on PLC (Power Line Carrier) communication could effectively avoid the questions above by use of power-line to transmit control signals. Because PLC can carry signals from low-voltage power-line which no longer requires special control line or wireless network, it has good applications for low cost and operational difficulties. The paper studied an intelligent lighting-control and temperature-measurement system based on PLC and gave a design scheme and implemented the system.

2 SYSTEM DESIGN

The system includes a master station and several slave stations, whose outline is shown in Fig.1. The master station is Sunplus SCM SPCE061A, which communicates with every slave station by PLC modules SENS-09, and slave stations are MCS-51 AT89C2051. The master not only transmits commands of light power on or off, light state or temperature inquiry and so on, but also displays the light state and temperature value. The slave receives commands to control the corresponding light on or off by a relay or measure temperature by DS18B20 or return correspond values.

The intelligent lighting control system mainly takes use of DMX (Digital Multiplex) and DALI (Digital Addressable Lighting Interface) bus communication protocol. DMX512 data package includes initial code and 512 data frames, and each frame includes one low-level start bit, eight data bits, and two high-level end bits. One data frame represents a control channel, so the protocol supports 512 control channels. DALI control bus uses master-slave structure, and one interface can at most connect 64 addressable control equipment and 16 addressable groups, which can connect several interfaces to control more interfaces and lamps.

According to the protocols above, a special protocol for intelligent lighting-control and temperature-measurement system is designed. Based on RS-232, the protocol takes three bytes as a frame including control command, slave station's address, state information and temperature data, the format of Protocol frame is shown in Table 1.

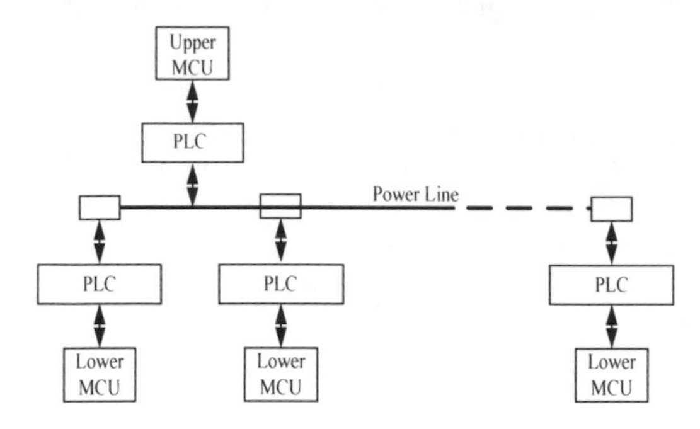

Figure 1. Structure diagram of an intelligent lighting-control and temperature-measure system.

Table1. Format of protocol frame.

Initial field	Address field	Information field
One byte	Two byte	One byte

In table 1, the protocol frame can be light power's control command, state inquiry command, temperature value and light state returned. The initial field of control frame for light power is 0xfa, and the information filed is 0xfe and 0xff which indicate light power off or on respectively. The initial field of state inquiry is 0xfb, whose information field is 0xfe and 0xff, where 0xfe indicates temperature inquiry, 0xff is light state inquiry. The initial field of temperature data returned is 0xfc, and the information filed is temperature value. The information field of state frame returned is 0xfe or 0xff, which respectively indicates light power on or off. The initial one is 0xfd. The address fields above have the same range from 0x00 to 0xff.

3 DESIGN OF SLAVE STATION

3.1 *Hardware design*

Place the cursor on the T of Title at the top of your newly named file and type the title of the paper in lower case (no caps except for proper names). The title should not be longer than 75 characters). Delete the word Title (do not delete the paragraph end). Place the cursor on the A of A.B.Author(s) and type the name of the first author (first the initials and then the last name). If any of the co-authors have the same affiliation as the first author, add his name after an & (or a comma if more names follow). Delete the words A.B. Author etc. and place the cursor on the A of Affiliation. Type the correct affiliation (Name of the institute, City, State/Province, Country). Now delete the word Affiliation. If there are authors linked to other institutes, place the cursor at the end of the affiliation line just typed and give a return. Now type the name(s) of the author(s) and after a return the affiliation. Repeat this procedure until all affiliations have been typed.

The slave station is the final to directly control light on or off and temperature measure. Each is set unique address and has the same circuit and control program. The slave is composed of a smaller MCU system, PLC module, and dial switch for setting up an address, relay, and temperature transducer and so on, as is shown in Figure2, in which the smallest MCU system is the control center of slave station.

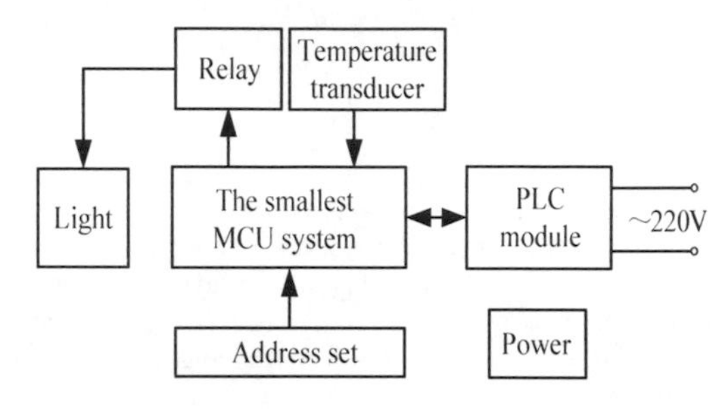

Figure 2. Hardware structure of slave station.

The PLC module is SENS-09, which can accomplish half-duplex communication between the master and the slave by Power-Line with 600, 1200, 2400, 4800 and 9600bps. Every slave MCU has a unique physical address which is set by an 8-bit dial switch, which can be checked up by AT89C2051. The relay controls, power supply on or off. To prolong the lives of relays and decrease interference to MCU, a decreasing spark circuit is parallel with each relay. Because MCU can't directly drive a relay, the power driver is implemented by S9013. The state detection is used to monitor the state of a slave station's power output. To protect AT89C2051, a photo coupler is used to separate 220-voltage detection from I/O pins. The circuit of a slave station is shown in Figure 3.

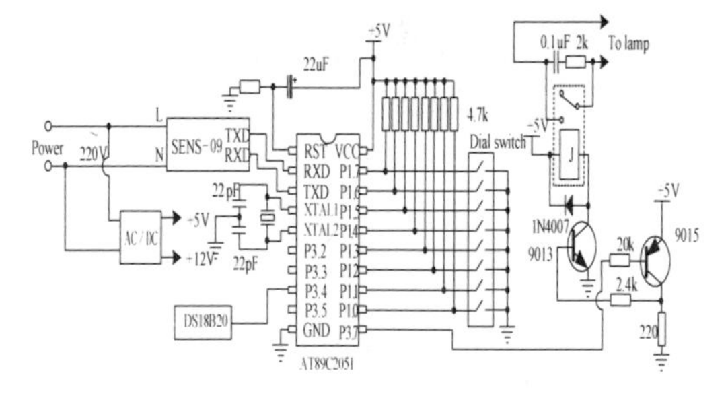

Figure 3. The circuit of lower MCU.

3.2 *Software design.*

After receiving control frames, address field and information field sent by the master, the slave extracts address field and compares with home address. If it is equal, AT89C2051 executes information field corresponding to command for power supply or checking up.

When receiving state or temperature inquiry command, AT89C2051 transmits light state or temperature value frame back to the master, the latter display light state or temperature value on LCD.

The programs of a slave station are composed of main program, subprogram of UART interrupts for receiving and sending out data frame, relay control subprogram and state detection subprogram. The main program implements calling subprogram to complete receiving and transmitting data frame, temperature measure, state check, power on or off and etc. The flowchart of the program is shown in Figure 4.

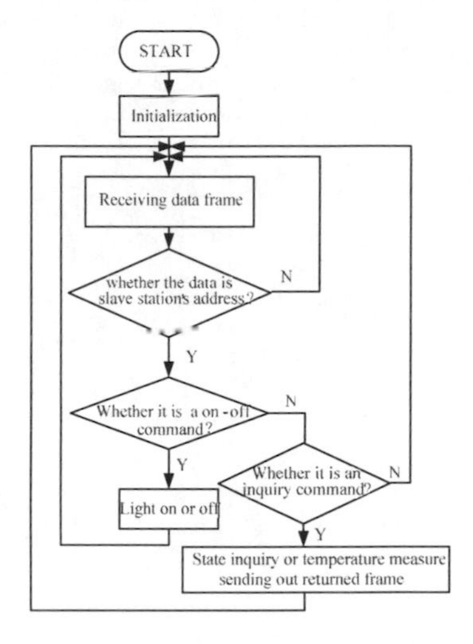

Figure 4. The flowchart of a slave's program.

4 HUMAN-MACHINE INTERFACE DESIGN

The human-machine interface includes keyboard and LCD, in which keyboard is mainly used to set the master and input control commands,the LCDD is OCMJ5*10B that has word stock and can display 50 Chinese characters or ASIC II codes of 8*8 or 8*16. The display interface is implemented by Sunplus SCM SPCE061A, which can show the salve station's address, lamp state and temperature, as is shown in Figure.5.

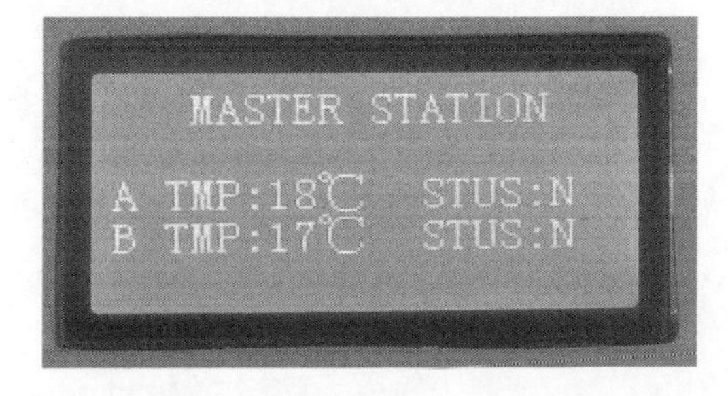

Figure 5. The interactive interface of the master station.

5 SYSTEM TEST

After hardware and software are designed, we must test the system performance.The tested system includes a master station, two slave ones called A and B, and corresponding lamps, all equipment are supported by the same power. Firstly, the two slave stations' addresses are set as 00000001B and 00000010B respectively in ascending order, which are equal to addresses in the master's program. And then the system is powered up, the temperature value and state of A and B can be seen on the LCD, as is shown in Fig.6 (a). When the temperature transducer of A is heated by hand, we can see the value higher on the LCD. At last, light control performance is tested. When the key of A is pressed down, the command to control A-light is sent out, A's light is on and the state Y is shown on the master station. When the key is pressed down again, A's light is power off, which is also shown on the LCD, as is shown in Fig.6 (b). Likewise, B's test is normal. The tests above show that the proposed system works stably and operates conveniently which can satisfy many application requests.

(a)

(b)

Figure 6. The result of test.

6 CONCLUSIONS

The intelligent lighting-control and temperature-measurement system based on PLC is designed by use of master-slave stations, PLC modules, and temperature transducer. The system takes use of existing power-line to transmit data, which not only avoids rewiring, but also eliminates interference in wireless communication. At the same time, the system has the characters of simple settlement, reliable work and low-lost.

The system's performance test indicates that design requirements are satisfied. The system has better generality and can be used in other fields of power control and temperature measure.

REFERENCES

Zhao Yong. Design of Remote Monitoring System of Smart Home Based on ARM and ZigBee [J]. ,Measurement & Control Technology, 2012, 31(11): 52–54.

Wu Yi-juan, Qin Cai-yun, Wan Mi-yang. A Design of Smart Home Environment Monitoring System Based on ZigBee Technology [J]. Journal of Beijing Institute of Petro-Chemical Technology, 2013, 21(1): 46–50.

Wu Yi, Bao Chun-Lan. Design of smart home system based on GSM and zigbee technology[J]. ,Journal of Hebei University of Technology, 2013, 43(1): 15–18.

ZENG Peng, Wang Xu, ,Wang Yang, et al. Multi-channel interference avoidance scheme for ZigBee network in smart grid application [J]. Journal of University of Science and Technology of China, 2012, 42(8): 609–616.

Chen Da-hua, Zhang He-zhen. Development of LED Intelligent Lighting Industry Based on the Internet of Things [J]. China Light & Lighting, 2013, (1): 11–15.

Jiang Xing-ming, Huang Wei. Design of Tire Pressure Wireless Monitoring Automatic Control System Based on ARM[J]. Computer Measurement & Control, 2010, 18(4): 752–754.

Tang Kai-jie, Li Can, Wang Di, et al. Design of digital temperature collection and alarm system based on DS18B20 [J]. Transducer and Microsystem Technologies, 2014, 33(3): 99–102.

Computing, Control, Information and Education Engineering – Liu, Sung & Yao (eds)
© 2015 Taylor & Francis Group, London, ISBN: 978-1-138-02800-5

Research on three-dimensional reconstruction of heart using serial sections based on computer 3D printing

Hai Tao He
Affiliated Hospital, Beihua University, Jilin City, China

ABSTRACT: The visible virtual heart was established by using the three-dimensional of the serial section reconstruction program. The serial section images were obtained by using Motic software and saved as the format of JPEG. Then the three-dimensional model of heart was reconstructed in computer 3D printing by the medical three-dimensional reconstruction program from the two-dimensional images. The virtual heart was succeeded reconstruction by using the medical three-dimensional reconstruction program. And the reconstructed heart can be freely rotated and incised and restored. The method of 3D models reconstructs from serial histological sections of heart is feasible.

KEYWORDS: Three-dimensional reconstruction; Heart; Serial sections; Assemble.

1 INTRODUCTION

"Virtual human" is the modern computer 3D printing information technology and medicine and other disciplines combine results, which not only enables people to see the spatial relationship to each other thousands of human anatomy size, shape and organs between the three-dimensional forms, but also applied to research, teaching, diagnosis, treatment planning, virtual reality and other fields. Currently Visible Human images are based on tomography on. With the "Visible Human project," "visual Korean", "virtual Chinese people," the emergence researcher's three-dimensional reconstruction of the human body has accumulated a lot of experience, but for three-dimensional reconstruction of the heart is still little reported. This study considers the use of three-dimensional computer 3D printing reconstruction of serial sections of biological tissue to rebuild a "virtual heart".

2 MATERIALS AND METHODS

Selection of Sprague-Daley rats, pregnant 9-10d. Pregnant rats were sacrificed by cervical dislocation, cut the skin, muscle and peritoneum along the midline to expose the uterus. Separated out the rat hearts were placed in vials containing penicillin Bouin's fixative fixed 24 h. Under the rat tail with a scalpel cut the number of rat tail collagen fibers extracted root, first into penicillin vials containing Bouin's fixative fixed 24 h. 2~3 min and then stained with eosin, and make it red. According to the requirements of the paraffin-embedded tissue dehydration, transparent, dipping wax, pour the melted paraffin-embedded slot, and then head up the vertical placement of rat heart groove surrounding the heart is inserted vertically slightly longer than three hearts the longitudinal axis of rat tail collagen fiber bundles as a positioning line, so that rat tail collagen fiber bundles in the heart outside evenly distributed and parallel to the longitudinal axis of the heart.

Selection of 5 mm is serial sections of a sheep heart (HE staining, 500). Sliced shot taken using a digital microscope image, due to the larger slices, each slice will be divided into several parts were taken and saved as JPEG format. Then run Motic images assembly1.0, the number of rows and columns load pictures selected, and sequentially load images, select the appropriate consolidation method, the optimal scanning step and scan mode, mosaic image, and save it as JPEG format .

Using Adobe photoshop7.01, and the use of hand-positioning and computer 3D printing positioning method combines two-dimensional image acquisition to be corrected positioning. Get image method in Java then the captured image from the DICOM format into JPEG format, select 256 colors (16-bit depth) image, to obtain black and white DICOM images while the image capture, compression processing. Finally, all converted to DICOM format two-dimensional image into a specific directory, run the three-dimensional reconstruction of medical software, complete three-dimensional reconstruction of a two-dimensional image.

3 THE RESULTS

By AO slicer on rat hearts were embedded in paraffin serial sectioning and staining with HE staining were obtained 293 7 μm paraffin sections of rat hearts continuously. Each slice of the cross-section of the middle of the heart, surrounded clearly visible three break points, namely buried in this experiment set punctuation. Since the rat tail collagen fibers, it is red after HE staining, irregular borders (Figure 1).

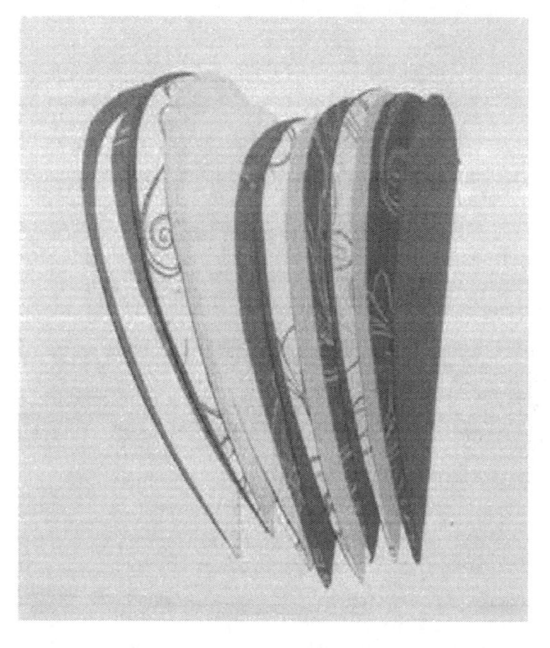

Figure 1. Heart serial sections, given the surrounding tissue visible punctuation, a: HE stain × 20 a Section 014 slices; b: Section 124 slices; c-section 197 slices; d:Section 198 slices; e: first 219 slices; f:section 266 slices.

Heartlogy (referred Heartlogy) is a form of discipline, the contents of the study were mainly fine structure of the body organs and tissues under a microscope. The basic purpose of the experiment is to verify the theoretical knowledge of teaching, students' observation skills, independent thinking, analytical problem-solving skills, ability and practical skills. Successful experimental teaching not directly affect the quality of teaching is good or bad, and teaching equipment experimental teaching directly affects the experimental teaching. However, due to the limitations of optical microscopes, between teachers and students is difficult to communicate quickly and effectively, affecting the quality of teaching experiments and teaching efficiency. With the development of modern science and technology and education reform, the Minister of Education stressed the need to "increase education funding to ensure that teaching needed for the operation." Traditional "box sliced a mirror, a teacher turn stops" requires an experimental model can not meet the current medical education.

The digital microscope mutual system is a new teaching tool set of digital microscope system, language communication, video communication and application software system developed in recent years as a whole, has been widely used in anatomy, histology and heartlogy, pathology and medicinal botany and identification of traditional Chinese medicine and other morphological experiment teaching. The system of teachers and students with a microscope with built-in digital imaging lens, all students under the microscope, synchronous display in the teacher computer 3D printing and screen, also can be selectively displayed the image of any a microscope. The system also has the image adjustment, photographing and video recording. In addition the system for teachers and students with a pair of Headset, microphone, two-way communication can realize the teachers and students, students can put questions to the teacher through the experiment desk questions button, teachers can choose to call mode to communicate with the students, including the call mode, students, teachers and students talk mode, demonstration model discuss in groups and other 4 kinds of mode. In recent years, we put the experiment teaching of digital interactive teaching mode widely introduced in this course, to change the traditional teaching of abstract, boring, boring time, created a new situation of efficient teaching interaction, image sharing, has significantly improved the quality of teaching and the teaching efficiency advantage.

4 THE DIGITAL MICROSCOPE MUTUAL SYSTEM

Because of the same laboratory students the sections of the same, always have students through cheating way to get high scores, deceived clearance. Now use the digital microscope mutual system, teachers can real-time browsing and monitor students, students have to find the characteristic structure of microscopic stick out a mile; at the end of the experimental examination although each student test sections are not the same, but still there are students to test their own to the neighbor table classmates to help see the cheating. And now the teachers not only can monitor, can also ask the students to the characteristics of structure under microscope camera as the judgment basis of his diagnosis of organs, such as students' critical sections of the stomach, he must at least 4 layer structure take a low magnification of the stomach and a high magnification of the funded gland, the chief cells and parietal cells of large glands. The teacher, according to the students' examination results and photographed the comprehensive score. This form of examination method, make students have not been lax, can consciously in earnest good each lab, and actively do the preview before class and review, the teaching quality can be improved. At the same time the student independent thinking, independent analysis and problem solving ability get exercise and improve.

The use of digital microscope camera system on heartnic serial sections were taken, due to the larger slices, each slice will be divided into two parts were taken, received a total of 500 × 2 images, pictures shown clear organizational structure, rich information (Figure 2a,b). By Motic image assembly 1.0 will restore with a slice of two images together to form a complete picture of the heart to get 500 full slice image, the resulting image formation, stitching accurate anatomical structures continuity between the images of the good (Figure 2c).

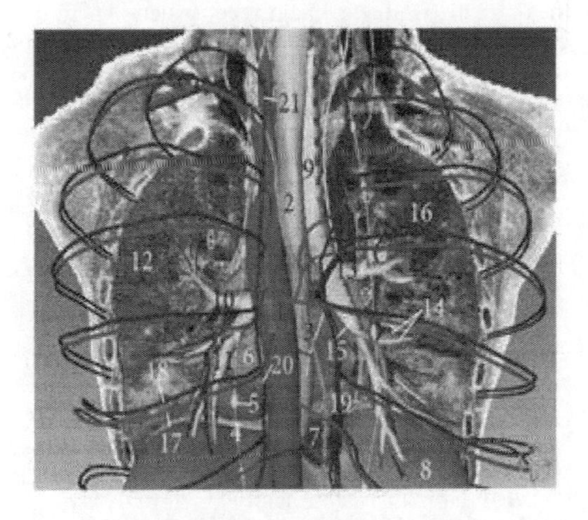

Figure 2. After stitching digital cameras capture images and computer 3D printing, HE staining × 20a, b for digital cameras capture images; c: image after stitching.

In Adobe Photoshop 7.01 is the acquisition of image positioning to ensure the accuracy of the subsequent three-dimensional reconstruction (Figure 3). JPEG image format and then converted to the DICOM format, and the interception, the compression (Figure 4). Finally, under the three-dimensional reconstruction of medical software is to obtain a three-dimensional image of the heart. The resulting heart reconstruction outline clear, can truly reflect the spatial structure of the original heart. Can be arbitrary rotation, observed from different angles; may be cut, and the reduction, re-cutting, and can display three-dimensional tissue structure resection; also arbitrarily zoom and other operations. Therefore, the reconstructed three-dimensional image can be easily obtained the desired section, without damaging the original three-dimensional image (Figure 5).

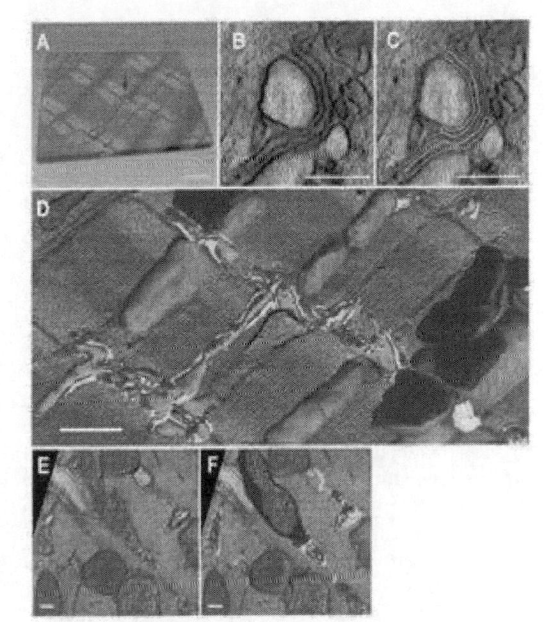

Figure 3. Image positioning, HE stain × 20.

5 DISCUSSION

Whether the original image data is reconstructed to provide sufficient effect depends on that unit length depending on the number of slices taken, the thinner the slice, the more abundant the original data, the degree of distortion of the reconstructed image is smaller. Historical data from the human body in general three-dimensional reconstruction of frozen sections, CT and MRI topographic image, and we want to make a three-dimensional reconstruction of the heart, the data from the continuous heartnic tissue sections, which can be paraffin, sliced collusion, frozen sections, as compared with the three-dimensional reconstruction of human data sources, with fault spacing is small, rich in information and so on. In this experiment, a paraffin sectioning technique, paraffin sections of up to 5 ~ 8μm, compared with collodion slices, frozen sections, greatly reducing the thickness of the slice, and therefore is the preferred method for three-dimensional reconstruction of the organization. To accurately display the three-dimensional reconstruction to rebuild the structure of spatial form, must make consecutive slice images and original objects have the same relationship corresponding three-dimensional space, namely the image positioning, if inaccurate positioning, image reconstruction will be distorted. However, in a continuous paraffin tissue sections, stained piece of the process will inevitably produce a certain bias, objectivity and accuracy of its location are subject to certain restrictions, so we need to set up a method by reference to ensure slice of bit is always accurate. The positioning method currently used method and the outer position location method. External positioning method is to use mechanical or manual methods to make the organization, are the following:

(1) Using a fine needle, laser or micro-electrode perforation; (2) based on a priori knowledge of the biological tissue anatomy and identification; (3) the specimen block face to be cut by trimming machine repair flat or cut with a razor blade flat, Xiu side of the plane.

In this experiment, Motic digital microscope camera system to capture two-dimensional image data, the system of the microscope, a digital camera and Motic image advanced 3.0 software to link up directly capture images stored in the computer 3D printing, and can rotate the image, segmentation, measuring and so a variety of image processing, is simple, fast, real-time observation, process images and so on. Tissue sections of continuous is three-dimensional computer 3D printing reconstruction of the data collection. Because the object reconstruction a large slice of sheep hearts, and therefore it needs to be divided into upper and lower parts were taken, and then spliced into a slice reduction, which involves stitching the image problem. Previous picture stitching generally used are hand-aligned method, as in Photoshop by layer overlay, after a transparent process, the position of the manual adjustment of the picture, so to reach stitching several images stitched neatly in the mosaic's vision. Manual splicing complex operation is time-consuming, and the resulting poor stitching pictures accuracy. The experiment uses Motic images assembly1.0 software for stitching images to a computer 3D printing automatically stitching. The software is looking through the same pixel points and the same two pixels overlap splicing algorithm to the image.

6 CONCLUSION

Experiments show Motic image assembly software is a simple, time-saving, precise stitching and other characteristics. Since the selection of hearts reconstructed slices serial sections are reserved early years before the department prepared, without any alignment mark on the slice, and the continuity of the poor, so we chose a better continuity of which more than 300 slices, using the computer 3D printing or manual positioning and positioning method of combining continuous slice positioning. Tissue sections currently we made on the basis of paraffin, has an initial progress on each slice to add an anchor tag method to study the future of this location marked with a three-dimensional reconstruction can greatly shorten the positioning step process, and can ensure the accuracy of positioning. Experiments show that the three-dimensional medical image reconstruction studio computer 3D printing software used in the continuous three-dimensional reconstruction of tissue sections is entirely feasible. The resulting heart reconstruction outline clear, can truly reflect the spatial structure of the original heart. In the three-dimensional image reconstruction, multiple hearts can be of any shape and depth of cut, the cut can be displayed, and the structure of the tissue to be excised, and these operations can return, so that the reconstructed three-dimensional image can be easily obtained the desired cross-section, without damaging the original 3D image. Image reconstruction may also be any three-dimensional space around the axis of rotation at any angle, hearts from different angles, and can timely measurement of the angle of all the structures in the warp and in any direction. Any hearts can also zoom in or out reconstruction and other treatment. Basically meets the construction of "virtual heart" requirements.

REFERENCES

Ackerman MJ The visible human project Proceeding on IEEE, 1998, 86:.. 504–511.

NLM NLM long range plan: electronic images Report of the board of regents, Bethesda; MD US Department of Health and Human Services, Public Health Service, National Institute of Health, 1990, 4: 2190–2197.

Luoshu Qian virtual human research and medical image processing methodology of medical device information, 2002, 5: 1–3.

ChungMS, Kim SY Three-dimensional image and virtual dissection program of the brain made of Korean cadaver Yonsei Med J, 2000, 41:. 299–303.

The original Lin, Huang Wenhua, Tang Lei virtual human visual overview of Chinese Journal of Clinical Anatomy Research, 2002, 20: 341–342.

Merickel M. 3D-reconstruction: The registration problem Computer 3D printing Vision, Graphics Image Processing, 1988, 42:. 206–208.

Otten E. A 3-D reconstruction package from serial sections images Eur J Cell Biol, 1989, 48: 73–75.

Zhang Jingjiang, Ye Qiao true, Ryder Jin, et reconstruct the three-dimensional structure of mouse kidney mitochondria Sun Yat-sen University (Natural Science Edition), 1990, 129: 73–74.

Zhang Xian Quan, LI Zi double, Mr Young and other three-dimensional computer 3D printing reconstruction of serial sections of liver sinusoid Journal of Biomedical Engineering, 1997, 14: 195.

Computing, Control, Information and Education Engineering – Liu, Sung & Yao (eds)
© 2015 Taylor & Francis Group, London, ISBN: 978-1-138-02800-5

The method of pre-extracting support vectors for SVM based on sample interaction distance

Zhong Min Li & Ping Ling
College of Computer Science and Technology, Jiangsu Normal University, Xuzhou,China

ABSTRACT: SVM's training process often contains the quadratic optimization of large amounts of non-support vectors, which leads to long training time under large-scale data and seriously restricts the application of SVM. Pre-extracting support vector is the most commonly used ways to solve the problem. In this paper, a method based on sample interaction distance for pre-extracting support vectors is proposed. The method calculates the interactive distance between samples. According to the size of the distance complete on the pre-extraction of support vector candidates. It does not need to prejudge whether the training samples are linearly separable and is easy to implement. Experiments of large data sets demonstrate the effectiveness and feasibility of the method.

KEYWORDS: Support vector; Interaction distance; Pre-extracting.

1 INTRODUCTION

Statistical learning theory was the small sample statistical learning rules in the 1960s. Based on this theory, in the mid-1990s, Vapnik, et al proposed a new learning algorithm—support vector machine (SVM) [1]. SVM can be seen as the polynomial neural network based on structural risk minimization, or the classifier based on radial basis function. It has the very strong generalization capacity and shows good classification ability in many practical problems [2-5] such as: handwritten character recognition, face recognition, text classification, intrusion detection, voice recognition, etc. However, the training of SVM corresponds to a quadratic optimization process; its time complexity is higher, which seriously influence the application of SVM in large scale data environment.

The purpose of training SVM is to find support vectors. Hence, if we can remove the non-support vector samples from the initial large-scale training set before SVM training and pre-extract the data points which include support vectors, then the SVM training time and space complexity can be largely reduced. Domestic and foreign scholars have done remarkable research on pre-extracting support vector method [6-12].

Jiao Licheng, Li Qing, et al [6-7] proposed pre-extracting support vector methods which are based on center distance ratio and vector projection. In the papers, the authors defined the Euclidean distance between sample points and the center of the homogeneous samples as the self-center distance; the Euclidean distance from the sample points to another class center point as cross-center distance. It calculates the self-center and cross-center distance of each sample point. According to the ratio r_i of the above distance, all the sample points satisfying $r_i \geq r_0$ will be pre-extracted.

Hu Zhijun, et al [9] proposed a pre-extracting support vector method which is based on random center distance sorting. Firstly, this method randomly selects a certain number of samples in the same class and regards the mean values of all dimensions as the sample center. Then, it calculates the distance from each sample to another random class center point, and sorts the distance. Samples with small distance are the pre-extracted boundary sample set. For nonlinear separable case, the method maps the original samples into high dimensional feature space with the aid of kernel functions and makes the original samples be linearly separable in high dimensional feature space. Then the pre-extracting problem can be solved like linear separable case.

Yang Xiaomin, et al proposed a pre-extracting support vector method based on the projection center distance which is improved from Jiao Licheng's method. That paper also gives two definitions of distance. Self-projection center distance is the projection of self-center distance in the direction of the two classes of sample center. Cross-projection center distance is the projection of cross-center distance in the direction of the two classes of sample center. In that method, the ratio of self-projection and cross-projection center distance determines participate samples. Since the boundary vectors are generally located in the boundary region of two classes, the ratio of samples' two distances in this area is relatively large [12]. Although the above methods can select the boundary sample vectors as the training sample better, it needs to determine whether the sample is linear separable or not. Furthermore, using the kernel functions to solve the problem lacks the consideration of isolated points' interference.

Based on the above problems and considering that the support vectors are generally located near the border of two classes, this paper proposes a pre-extracting support vector method based on sample interaction distance, called Neighbor Boundary Vectors-SVM (NBVSVM). This method considers the distance of heterogeneous samples as the difference between samples, then sorts the distance matrix and extracts two classes of samples which are near each other as pre-extracting support vectors. The experimental results prove that this method can effectively reduce the training time of SVM as well as remain the performance of SVM.

2 THE SUPPORT VECTOR MACHINE THEORY

The main idea of SVM is to construct an optimal separating hyperplane based on structural risk minimization principle and use the optimal separating hypeplane to achieve the best classification results of unlabeled samples. According to the samples is linearly separable or not, SVM can be categorized into linear SVM and nonlinear SVM.

For linear separable samples, set mode sample as $Z=\{(x_i, y_i)|x_i\in R_n, y_i\in\{+1,-1\}, i=1,2,\ldots N\}$, where x_i is an n-dimensional space vector, y_i is the label of x_i. Solving problems of maximal margin plane and separating the two classes of samples can get the optimal separating hyperplane. The original optimization object function of SVM is:

$$\begin{aligned} &\min_{\omega,b,\xi} \quad \frac{1}{2}\|\omega\|^2 + C\sum_{i=1}^{N}\xi_i \\ &s.t. \quad y_i\left(\left(\omega\bullet x_i\right)+b\right)\geq 1-\xi_i, i=1,2,\cdots N \\ &\qquad \xi_i \geq 0, i=1,2,\cdots N \end{aligned} \tag{1}$$

where: $y_i\left(\left(\omega\bullet x_i\right)+b\right)\geq 0$ is the separating hyperplane to be solved, ω is the normal vector of separating hyperplane, $C>0$ is called as the penalty parameter which is generally determined by the practical application problems. When C is relatively large, the punishment for misclassification is large. However, a smaller value of C may cause the system to under-fitting. b is the offset of separating hyperplane, ξ_i are slack variables.

The dual problem of the original optimization problem is:

$$\begin{aligned} &\max_{a} \quad -\frac{1}{2}\sum_{i=1}^{N}\sum_{j=1}^{N}a_i a_j y_i y_j\left(x_i\bullet x_j\right)+\sum_{i=1}^{N}a_i \\ &s.t. \quad \sum_{i=1}^{N}a_i y_i = 0, i=1,2,\cdots N \\ &\qquad 0\leq a_i\leq C, i=1,2,\cdots N \end{aligned} \tag{2}$$

a_i is the Lagrange multiplier which corresponds to the training sample x_i, where the sample with nonzero a_i is called as support vector. Solving the dual problem can get the classification decision function for linear SVM:

$$f(x)=sign\left(\sum_{i=1}^{N}a_i^* y_i\left(x\bullet x_i\right)+b^*\right) \tag{3}$$

For nonlinear separable samples, introduce kernel function so as to map it from the input space to feature space. Using kernel function $K(x_i, x_j)$ to replace the dot product of SVM, the above optimization problem can be written as:

$$\begin{aligned} &\min_{a} \quad \frac{1}{2}\sum_{i=1}^{N}\sum_{j=1}^{N}a_i a_j y_i y_j K\left(x_i\bullet x_j\right)-\sum_{i=1}^{N}a_i \\ &s.t. \quad \sum_{i=1}^{N}a_i y_i = 0, i=1,2,\cdots N \\ &\qquad 0\leq a_i\leq C, i=1,2,\cdots N \end{aligned} \tag{4}$$

The corresponding nonlinear SVM decision function is:

$$f(x)=sign\left(\sum_{i=1}^{N}a_i^* y_i K\left(x\bullet x_i\right)+b^*\right) \tag{5}$$

By the decision functions (3) and (5), one can see that support vectors and support values a_i play a key role in the calculation of the decision function. Therefore, if the candidates of support vectors can be drafted out before training then we can decrease the time and space complexity of SVM and reduce the training time of SVM.

3 THE PRE-EXTRACTING METHOD FOR SVM BASED ON SAMPLE INTERACTION DISTANCE

The study finds that support vectors are generally located in the boundary region of two classes, namely the boundary area of samples. The samples of this area have similar coordinates, but belong to different categories. In this paper, through calculating the distance between the heterogeneous samples and according to the size of the distance, we pre-extract the samples in the boundary region. These samples are well expected to contain the real support vectors of the training samples.

In this method, we give the following definitions.

*Definition*1. Boundary Vector (BV): Boundary vectors refer to the sample points located near the boundary region among different classes.

*Definition*2. Boundary support vector (B-Sv): Boundary support vectors refer to the support vectors of boundary vectors.

*Definition*3. Sample distance: the characteristic difference between two samples is called as the sample distance. For example: giving two known n-dimensional vector samples U and V, the sample distance of them is:

$$d = \|U - V\|_2 = \sqrt{\sum_{i=1}^{n}(U_i - V_i)^2} \tag{6}$$

Where U_i and V_i represent the i-th dimensional data of vector U and V, respectively.

*Definition*4. Sample interaction distance: Assume that there are m samples of class P, n samples of class Q. Then the sample interaction distance between P and Q denotes the distance matrix generated from each sample distance in P and Q.

Specific algorithm steps are:

Step 1. For each sample in P, calculate the sample distance from P_i to each sample of Q and put the sample distance in the distance matrix D. D_{ij} represents the sample distance between the i-th sample of P class and the j-th sample of Q class;

Step 2. According to the first step, we can get the interaction distance matrix D between P and Q, whose size is $m*n$, where m is the number of samples in P, n is the number of samples in Q; then sort the distance matrix D and gets a new matrix f with $m*n$ and 1 column;

Step 3. Set extraction ratio β according to the actual need and extract distances from the smallest value of distances between the first $m*n*p$ samples in matrix f. Based on the position of the value in the matrix D, we can find out the corresponding initial training samples in P and Q, and put these two pre-extracted samples into a pre-extracted set M;

Step 4. Take a unique treatment for the initial pre-extracting M which is obtained in the third step, remove the same sample vectors, and ultimately get the boundary vectors set which contain most support vectors. Finally, replace the original training samples by this boundary vector set to experiment.

4 EXPERIMENTS AND CONCLUSIONS

In this paper, the experiments are done under Matlab R2009b. The computer system is Windows XP-SP3. The CPU is Intel Pentium (R) Dual-Core CPU E5400@2. 70GHz (2700 MHz), and memory is 2GB DDRII. The results are the average results of 10 independent experiments. In the experiment, select the SVM penalty parameter C=1, and Gaussian kernel as the kernel function

$$K(x, y) = \exp\left(-\frac{\|x - z\|^2}{2\sigma^2}\right) \tag{7}$$

where scale parameterσ=3. Experimental data are the artificial data sets and UCI data sets; including: linear separable datasets and nonlinear separable datasets.

4.1 *The artificial data sets*

4.1.1 *The linear separable data set*

Experiments randomly generate two even distributions of samples. Two classes of samples are U_1 ([0, 2] × [0.7, 1.2]) and V_1 ([0, 2] × [0, 0.5]). The number of two classes of samples is 2800, from which select 600 samples as training data and the remaining 2200 samples as the test data. In this paper, experiments are conducted under standard SVM algorithm and NBVSVM pre-extract algorithm. Figure 1 is the result of linear separable data sets, where '.' denotes the initial training set, '+' and '△' denote boundary vectors which are selected from the initial training set and 178 support vectors in the initial training set respectively. From Figure 1, one can see: the boundary vectors of the pre-extracted vector set contain the original support vectors well and remove most common samples. That is, this method can largely reduce the number of training samples at the same time.

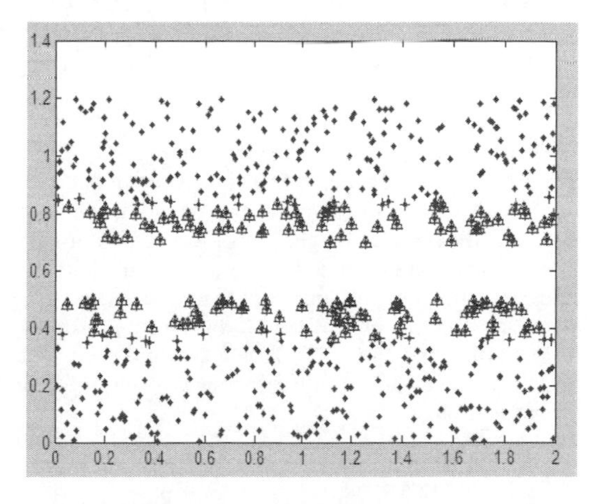

Figure 1. The distribution of BV and support vector in linear separable case.

Table 1. Performance comparison in linear separable case

Algorithm	Training set	BV	SV	Convergence time	Test set	Recognition Rate
SVM	600	-	178	34.2	2200	100
NBVSVM	600	142	142	2.1	2200	100

As can be seen from experimental results in Table 1, 142 boundary vectors selected from the initial training samples are all support vectors and the number of support vectors in initial training set is 178. We can see NBVSVM pre-extracts most of the support vectors from the original training samples. In terms of recognition rate, NBVSVM has reached 100% as same as the standard SVM, but obviously shorten the training time.

4.1.2 *The linear inseparable data sets*

Randomly generate two uniform distributions of non-linear samples, and the generating function of the first class of samples is:

$$\begin{cases} x_1 = r_1 \sin\theta \\ y_1 = r_1 \cos\theta \end{cases} \quad \theta \in [0, 2\pi], r_1 \in [8, 13] \tag{8}$$

The generating function of the second class of samples is:

$$\begin{cases} x_2 = r_2 \sin\theta \\ y_2 = r_2 \cos\theta \end{cases} \quad \theta \in [0, 2\pi], r_2 \in [15, 19] \tag{9}$$

The number of two classes of samples is 1200, from which select 600 samples as training data, the remaining 600 samples as the test data, and use standard SVM and NBVSVM to experiment. The results are shown in Figure 2, where '.' represent the initial training set, '+' and '△' represent the support vectors of initial training set and pre-extracted boundary vectors.

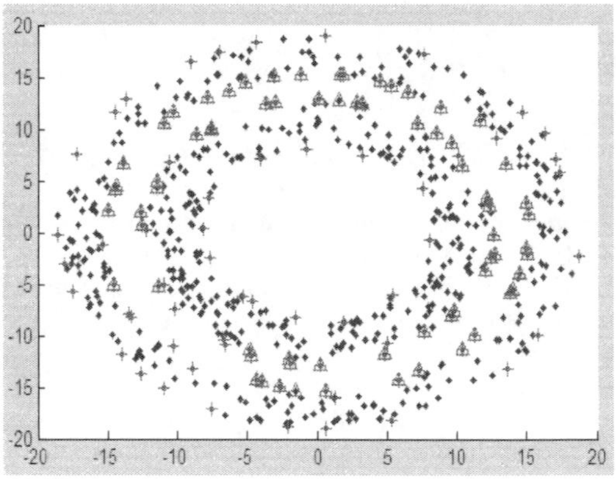

Figure 2. The distribution of BV and support vector in linear inseparable case.

Table 2. Performance comparison in inseparability separable case

Algorithm	Training set	BV	SV	Convergence time	Test set	Recognition Rate
SVM	600	-	115	11.45	600	100
NBVSVM	600	97	69	1.49	600	100

It can be seen from the experimental results, there are 69 of 97 pre-extracted boundary vectors become support vectors, which means the boundary vectors selected by NBVSVM contain most of the support vectors, and indicate the effectiveness of this method. Moreover the recognition rate of the same test sample does not decrease in the case of the shorten of training samples, indicating NBVSVM shortens the SVM training time clearly, while ensures the result of SVM classification.

4.2 *UCI public standard data sets*

Experiments use the UCI standard data sets (diabetes, ionosphere, letter, waveform, wine and poker-hand data sets) to investigate the performance of this method in this paper [13]. We select classes1-4 in letter data set, classes 0 and 2 in waveform data set, classes 1 and 2 in wine data set to implement experiment.

Table 3 shows the number of initial training set, support vectors, test set, boundary vectors and boundary support vectors. The number of BV and B-Sv has been largely reduced with the aid of this method. Table 4 shows the time of standard SVM and NBVSVM. One can see that the convergence time is much shorter than standard SVM. But the recognition rates of test sets have not really gotten lower, especially in ionosphere, waform02 and waform12 data sets.

Table 3. Comparison of the number of BV and B-Sv

datasets	Ionosphere	letter12	waform02	wine12	letter34	waform12
Training set	154	611	943	59	500	1553
SV	73	257	534	59	252	653
BV	81	202	142	34	219	332
B-Sv	37	121	134	34	141	254
Test set	197	944	2410	71	1041	1790

Table 4. Comparison of Convergence time (ms)

datasets	Ionosphere	letter12	waform02	wine12	letter34	waform12
SVM	3.25	31.74	82.53	1.49	22.31	161.92
NBVSVM	1.07	5.39	31.59	0.59	6.83	13.5

Table 5. Comparison of Recognition rate (%)

datasets	Ionosphere	letter12	waform02	wine12	letter34	waform12
SVM	100	99.79	91.62	63.28	99.81	95.36
NBVSVM	99.49	97.14	91	61.97	97.41	95.14

The experimental results show that NBVSVM algorithm can effectively reduce the redundancy samples involved in the training samples and extract boundary vectors containing a higher proportion of support vectors. This method can apparently decrease the training time and ensure the classification accuracy of SVM, and create conditions for the application of SVM in large-scale data.

5 CONCLUSIONS

This paper proposes a method of pre-extracting support vectors based on sample interaction distance. According to calculate the interaction distance between samples, this method can complete on the pre-extraction of support vector candidates while does not need to determine whether the data set is linearly separable or not in advance. The results show that NBVSVM can apparently reduce the size of the training set. Especially when the support vectors are of smaller proportion in the training set (larger sample redundancy rate), the algorithm is more evident. The follow-up research will aim at the calculation and selection of parameters to further improve the efficiency of the algorithm.

REFERENCES

[1] V.Vapnik. The nature of statistical learning theory[M]. New York: Spring-Verlag, 1995.

[2] Deng Naiyang, Tian Yingjie. A new method in data mining——support vector machine [M]. Beijing: Science Press, 2004

[3] Suykens JAK, Van Gestel T, Vandewalle J, et al. A support vector machine formulation to PCA analysis and its kernel version [J]. IEEE Trans on Neural Networks, 2003. 14(2):447–450.

[4] Vamvakas G, Gatos B, Perantonis S J. Handwritten character recognition through two-stage foreground sub-sampling [J]. Pattern Recognition, 2010. 43(8): 2807–2816.

[5] Assheton P, Hunter A. A shape-based voting algorithm for pedestrian detection and tracking [J]. Pattern Recognition, 2011. 44(5):1106–1120.

[6] Luo Yu, Yi Wende, He Dake and so on, fast reduction for large-scale training set [J]. Journal of Southwest Jiao Tong University, 2007.42(04):468–472.

[7] Jiao Licheng, Zhang Li, Zhou Weida. Pre-extracting Support Vectors for Support Vector Machine [J]. Journal of Electronics, 2001.29(3):383–386.

[8] Li Qing, Hu Hanying. Pre-extracting support vectors for Support Vector Machine using K-nearest bound neighbor method. Journal of Circuits and Systems, 2013. 18(2):91–96.

[9] Hu Zhijun, Wang Hongbin, Li Rong. A Random Center Distance Sorting-based Support Vector Pre-extracted

Method [J]. Journal of Microelectronics and Computer, 2013. 30(8): 36–39.
[10] Ding Ailing, Liu Fang, Li Ying. Pre-extracting support vector by adaptive projective algorithm[C]//. Proceedings of the 6th International Conference on Signal Processing. Beijing, China: IEEE, 2002 (1):21–24.
[11] Ding Ailing, Liu Fang, Cao Wei. Adaptive Projective Algorithm for Selecting Support Vector Beforehand [J], Computer engineering and Application, 2002(19): 116–118.
[12] Yang Xiaomin, Wu Wei, Chenmo, He Xiaohai. Pre-extract support vectors based on projection center distance [J]. Journal of Sichuan University (Natural Science Edition), 2010(1):85–90.
[13] http://www.ics.uci.edu/mlearn [OL]

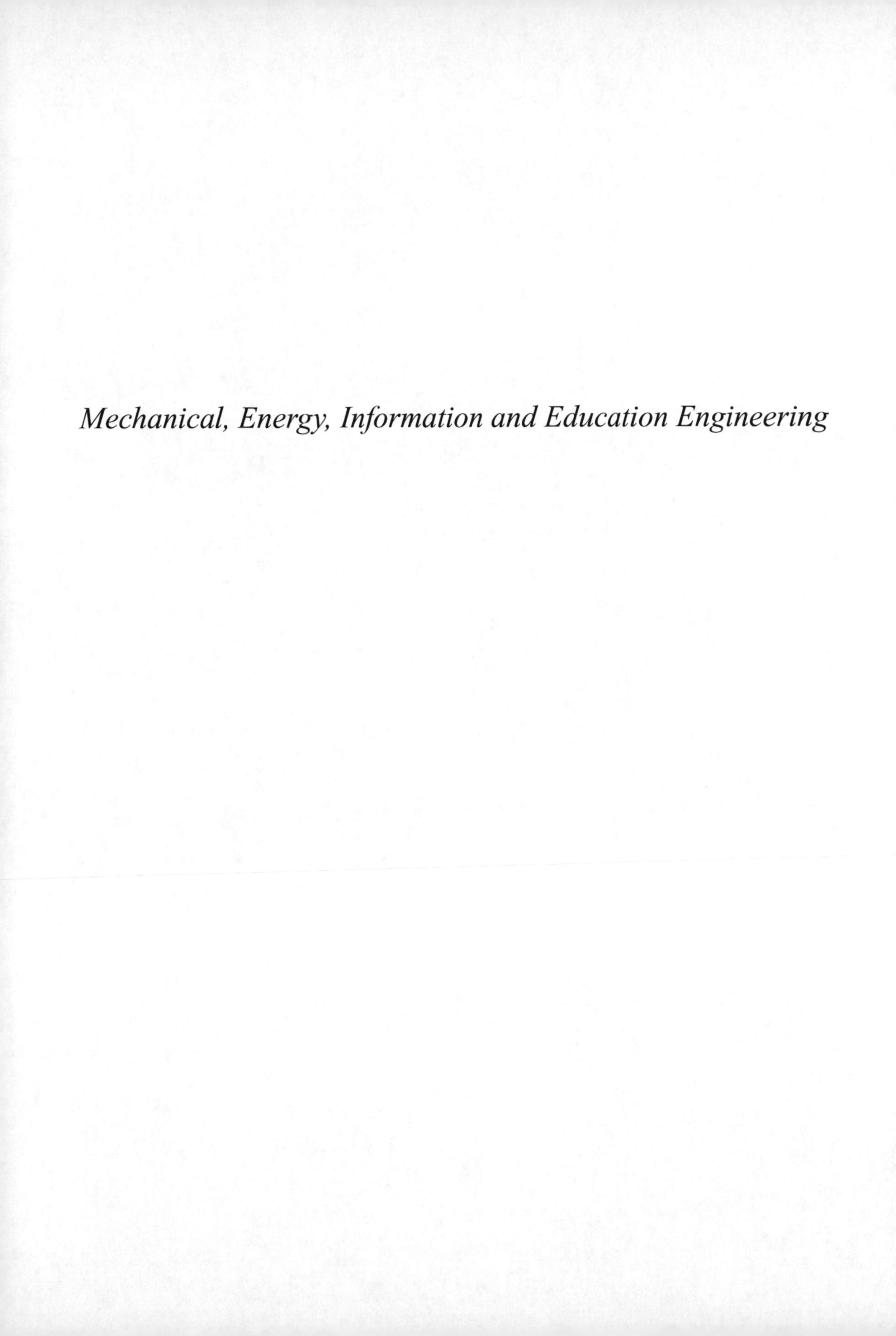

Mechanical, Energy, Information and Education Engineering

Computing, Control, Information and Education Engineering – Liu, Sung & Yao (eds)
© 2015 Taylor & Francis Group, London, ISBN: 978-1-138-02800-5

Use teaching aids to build efficiency "Student-Centered" class in vocational colleges

Mei Ying Fan
Beijing Information Technology College, China

ABSTRACT: In order to explain how to build an efficient "Student-Centered" class in vocational colleges, which emphasizes collaboration between teachers and students, this paper analyses the reasons of using teaching aids from the point of the current situation of vocational college is, after that it gets the first conclusion that teaching aids are necessary in vocational colleges. Then it discusses different kinds of teaching aids in various ways, at last demonstrates if a teacher in vocational colleges can make the interesting, visual teaching aids be involved in the learning process, it could be effective to stimulate students' curiosity in learning, highlighting the focus of teaching, breaking through the difficulties of teaching, optimizing the structure of teaching and learning, developing innovative thinking ability for students.

KEYWORDS: Student-Centered; Teaching Aids; Learning Experience; Vocational Colleges.

1 INTRODUCTION

"Student-Centered" came from "child-centered" view, which was put forward by an American child psychologist and educator John Dewey. He advocated the liberation of children's thinking, organizing learning activities as children focused and making children play subjective role in learning. He also promoted the "learning by doing" beliefs on science education. After the "child-centered" thinking applied to secondary schools and universities, "student-centered" philosophy of education is recommended gradually.

"Student-Centered" has many characteristics, for example, students can create knowledge with ideas which they explore, students construct their own meaning through social interactions, students should be involved in learning activities, etc. Teaching aids could be very useful to build a high efficiency "student-centered" learning environment.

2 REASONS OF VOCATIONAL COLLEGES USING TEACHING AIDS

2.1 *Current situation of vocational colleges*

Higher vocational education is an important part of higher education system in China. As its scale is expanding, the quality of the students in Higher Vocational Colleges has gradually declined. During the teaching practice, it's discovered that logical thinking and abstract thinking of a lot of higher vocational students need to be improved; they have weak knowledge of basic course, such as Chinese, Mathematics. In addition, their behavior and study habits are not fine for vocational education. All of these phenomena requires teacher to update and reform teaching approaches constantly, build a new teaching style based on the "Student-Centered" theory, so that the whole class could change from teaching to learning, and all of the learning activities are established on the students' own experiences, their own activities, and their own exploration. At the end, it could provide a maximum space for students' reflection, exploration, discovery and innovation.

Some researches of education psychology have shown that students who are good at visualizing always start learning new things with the help of feeling. They need to convert the observation and touch the result of specific things into perceptual knowledge which has nothing to do with the specific things, and then make the perceptual knowledge as abstract and rational knowledge. However, in order to help students complete these transformations, teachers need to use some activated methods to stimulate the students' five senses: visual, auditory, touch, taste and smell. In this process, teachers need to use some specific, intuitive demonstration teaching aid to show them out.

2.2 *Stimulate intellectual curiosity for students*

"Curiosity is the best teacher", if students are interested in what they need to learn, they can pay a long attention, keep a clear awareness on it, besides that, they can stimulate their own imagination, generate positive thinking and enjoyable learning experience. With the help of appropriate teaching aids, static knowledge could be active, abstract content could be visualized; students' attention could be attracted.

Learning begins from thinking, while thinking originates from doubt. Teachers can also use teaching aids design some practical activities and put forward problems from them, so that make students have chances to reflect. In this way, teachers can promote students have a successful learning experience, forming a virtuous circle for students' learning activities.

2.3 *Highlight the focus and break through the difficulties of teaching*

In order to focus on the important things during learning, teachers can use teaching aids to design various rich learning activities around the central problem, so as to make the learning content more specific, much deeper, much more clearly.

While breaking through the difficult problems, teachers can use teaching aids to make the abstract things into concrete, complex things to simple and unfamiliar to familiar, so that students can develop their perceptual understanding skills, make the first signal system and the second signal system work at the same time, teachers can break through the difficulties better.

2.4 *Optimize the structure of teaching and learning*

In order to achieve established learning objectives, teachers should design a series of stable, simplified procedures and activities, based on a certain educational ideology for those influential factors including time, space and others things. These procedures and activities pose the structure of teaching and learning. It has a close relationship with whether the curriculum learning objectives could be achieved successfully and whether students are positive during learning.

With the guidance of "Student Centered" view, optimizing the structure of teaching and learning means when teachers make preparation for the curriculum, they should change from the traditional way of how to teach, to a new way of how to make students learn. In this process, study and master the curriculum guideline carefully, position the teaching goal accurately is still necessary, however, there are other things that are more important, for example, understand the students' situation fully; respect students' individual differences and learning mode; study, analyze, optimize teaching resources carefully; design each learning activities and each problems rationally; lead students to launch thinking enough by choosing appropriate teaching aids and approaches, and so on.

2.5 *Develop innovative thinking ability for students*

The core target of higher vocational colleges is to develop the higher technical applied people. In order to realize this goal, various forms of practical learning activities are needed. Vocational college could offers some real or simulated environments for students, vocational teachers could create learning situations for students, all of these could make students have opportunities to face concrete and comprehensive problems, so that students can apply professional knowledge to solve some practical problems, furthermore, they can enhance their awareness of innovation, improve their practical ability.

3 THE CATEGORIES OF TEACHING AIDS

3.1 *What is teaching aids?*

Teaching aids are tools that can help teachers complete teaching activities in the class. From the general point of view, real and simulated environment also belong to teaching aids. However, if according to what sense the teaching aids stimulate, teaching aids can be generally divided into visual aids, auditory aids, tactile aids, taste aids, olfactory aids, etc.

3.2 *Category of teaching aids*

It has been indicated that there are three kinds of learning styles, visual, auditory and tactile styles. So for different students, teachers should use different teaching aids during teaching as much as possible.

For example, pictures, charts, slides, photographs, video, color-board, graphics can be used to show image, they belong to visual aids. Students with visual learning styles can catch useful information under their help. Music, recording, pronunciation tools can attract students' with auditory learning styles and make students learn effectively.

All kinds of objects, models, samples and game cards can make students have the opportunity to touch and use, so that the tactile learning style students could have a positive learning experience.

During teaching activities, teachers can't distinguish the students' learning style strictly, so different kinds of necessary teaching aids should be chosen according to the teaching goal in order to make each student get a vivid perception impression, carry out learning observation and thinking actively. At the same time, through teachers' observation, listening and guidance when necessary, make students understand knowledge and analyze the essence of some phenomenon through the use of teaching aids, make them get perceptual knowledge and rational ideas, forming a scientific concept.

4 TYPICAL CASES OF USING TEACHING AIDS

4.1 *Video VS mathematics*

It is difficult to imagine, at the beginning of an economics class, the teacher played a piece of music video, which spent about 3 minutes carrying the content of the calculating, made all students absorbed in the class without any distraction, and opened the learning process well.

It must be admitted that dynamic music, beautiful images, nervous calculation really made all students exciting; the teacher helped students entered into the lesson unconsciously.

4.2 *Cards VS terminologies*

Terminologies in the professional field for teachers and students are difficulties to teach and understand. However, students also can master 24 professional terms in 30 minutes with the help of teaching aids.

At the beginning of class, the teacher sent each student one card with various printed fruits and its corresponding text, then asked the students to group according to their respective card. It took only about 2 minutes to make group and meanwhile students took this opportunity to master the fruit expressions. After that the teacher took out a stack of envelope mysteriously, students were full of curiosity and expectation, the teacher handed out an envelope for each group. Afterward, students couldn't wait to open the envelope, and then they found that there were 48 cards including 24 terminologies and 24 explanations of them. Within 30 minutes, cards made students understand and master 24 terminologies, which was a myth before! The flow of this process is shown as Figure 1.

4.3 *Poster VS AIDA theories*

AIDA is an acronym used in marketing and advertising that describes a common list of events that may occur when a consumer engages with an advertisement.

A – Attention: Attract the attention of the customer.

I – Interest: Raise customer interest by focusing on and demonstrating advantages and benefits (instead of focusing on features, as in traditional advertising).

D – Desire: Convince customers that they want and desire the product or service and that it will satisfy their needs.

A – Action: Lead customers towards taking action and/or purchasing.

Using a system like this gives one a general understanding of how to target a market effectively. Moving from step to step, one loses some percent of prospects.

Just in such a lecture, the teacher handed out a poster for each student, which demonstrated "ENTER UNIVERSITY IN ONLY ONE YEAR", and explained how AIDA was used in this poster. Then students were asked to do a poster for a hotel in Las Vegas.

In this class, with illustrations of the "poster", the marketing theory "AIDA" is lively and concise. From the point of the students, designing poster task put all students' exited, the whole teaching process was designed remarkable.

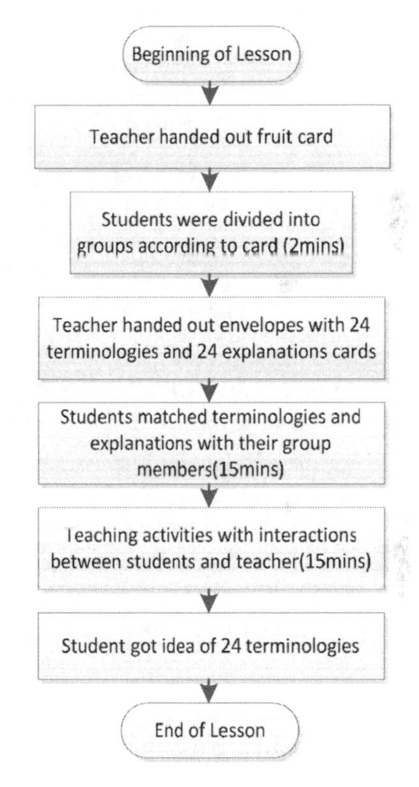

Figure 1. Flow chart of case in using card.

4.4 *Games VS cooperation*

The game is always favored by people, no matter young or old, whenever the game is referred, everyone will feel refreshed. In the class, using teaching aids properly, such as jigsaw puzzles, building blocks game, digital filled game, painting game and folding paper game can effectively improve the communication between teachers and students, promote understanding between teachers and students sell and enhance the partnership between them.

With the bell, the teacher went into the classroom with a digital filled game, the students paid attention to how to win the game attentively. In the game, the students are very dedicated, suddenly there was a "Hooray", a student won, and he did not only get the admiration of another player, also received a gift from a teacher. In the game, students no longer had a psychological barrier to digital, harmonious relationship between teachers and students also quietly grew…

4.5 *Scenarios vs. programming*

In the programming classes, teachers can use teaching aids too. For example, using simulated scenarios will promote students generate power to solve the problem.

It's believed when explaining the branch statement in programming, if students are just told to use if-else to complete some exercise problems simply without any description of the situation and why it should be used, how many students will have responded? In another way, with the help of some real problems in life as the content, students are asked to help solve them, the result may be different.

In this situation, the class would be like this way, the teacher told a story to students, what had happened in a middle school, then students got a task that they were required to help the middle school to develop a program in order to test the first entered the students' level of computer general knowledge. 10 questions had been designed well by the middle school, and what students needed to do was to calculate how many questions were answered right, and what grades each tester got for the total testing. Of course, if the student's achievement was good enough, they would get some payment. Nearly all of the students were eager to have a try.

Using "scenario" as a teaching aid, the teacher should not only need to find out what students are delighted to see and hear, then construct the outline of the scenario, but also need to design each parameter and detail for it. The more details are clear, the more students are able to feel the truth and value of the problem.

5 CONCLUSIONS

In summary, using various teaching aids flexibility in teaching and learning can make the class interesting and spectacular. Rational using teaching aids can make mathematic have sound, terminology have dynamic, abstract become concrete, theory no longer boring and the learning more efficient, make students become the real master of learning.

REFERENCES

Hattie J (2010) Visible learning: what's good for the goose… Department for Education and Early Childhood Development State Government Victoria.

O'Neill G and McMahon T (2005) Student-centred learning: What does it mean for students and lecturers?

Jiang Xuhua (2013), Management & Technology of SME, Simple Teaching Aids Play An Important Role in Vocational Teaching.

Tang Bo (2014), ShangQing, Make Simple Teaching Aids to Solve Difficulties in Teaching.

Computing, Control, Information and Education Engineering – Liu, Sung & Yao (eds)
© 2015 Taylor & Francis Group, London, ISBN: 978-1-138-02800-5

The research platform construction of mobile digital library from the public perspective

Bin Xu
Library, Handan Polytechnic College, China

ABSTRACT: Under the condition of the popularization of the 3G network, intelligent mobile phone, relying on the existing software and hardware facilities rich electronic resources, personnel and technology advantage, professional characteristics, on university library, it is a new service model to set up the mobile digital library platform by WAP\W-IFI technology and to meet demand access to information resources whenever and wherever possible. By this, the university go out of the ivory tower and provide services for the demanders of society, playing its due role in the transformation of service mode and popularization of public culture and make a positive contribution to improving the quality of the whole people and promoting entrepreneurship to employment.

KEYWORDS: the public view mobile digital library platform.

1 INTRODUCTION

Now, "the human being is entering into a digital society and the influence of digital information on human life have been or are being more than traditional media. The library which makes the collection of records of human civilization as the responsibility should have enough preparation."As the library of Higher Vocational College, in the social raging demand like a storm today, we should serve the community, use our own advantages, combined with the local district public libraries and other institutions to construct a public cultural service platform of adapting to the digital age, in order to achieve a better public services and serve society. The application and popularization of mobile network technology in wireless network, represented by 3G technology make the environment more accessible. Its positive, open, fresh ways of service will eventually become normalized, universal handheld tool, therefore, it is imperative to construct a mobile digital library.

1.1 *Overview*

In recent years, with the rapid development of mobile communication technology, and other increased mobile phone mobile terminal users, the public demand increasingly diverse for the culture. In this context, the main goal of this project is to construct a mobile digital library services platform on the base of existing digital library, communicating mobile network, providing the public customized information service, including digital books, newspapers and other general information and features of abstracts of audio and video, reference consultation,etc.

1.2 *The overall platform design*

Making use of the ASP language conforms to the page access to the mobile terminal, combined with Access or SQL as the backstage database; organizing audio and video resources characteristic and digital, in accordance with the use of mobile devices.

2 PLATFORM DESIGN

2.1 *The language choice of platform design*

The ASP Active Server is the abbreviation of Page, which means "active server pages". ASP is an application developed by Microsoft Corp CGI script instead, which can interact with databases and other programs and is a simple convenient programming tool. ASP is a server side scripting environment, which can be used to create and run dynamic Web Page or Web applications. The ASP Webpage can contain HTML tags, plain text, script commands and COM components. The use of ASP can be added to the interactive content(such as Webpage in online forms), can also use HTML to create Webpage as the user interface of the web application program. The ASP page has the following characteristics: 1) ASP can achieve a breakthrough of some functional limitations of static pages, to realize the dynamic web page technology; 2)ASP interpret on the server can execute ASP program on the server, and transfer the results to the client browser in HTML format so compatible with a variety of mobile phone browser; 3)ASP server can use ActiveX components to perform a variety of tasks, such as access to the database, send Email or access the file system and other convenient connection of ACCESS and SQL database.

Due to the above advantages, choosing to use the ASP language platform web is easier to extend and maintain, and suitable for the condition of mobile terminal many types, the many kinds of browser.

2.2 *Function design of platform*

The purpose of the study is to build a platform With a mobile digital library as the main body, combined with the resources and academic advantage of itself, providing multimedia resource services or feature sent mainly via SMS and WIFI mode.

2.2.1 *The platform releasing information*

Mobile library platform, integrated two functions of data management and information release. The following descriptive information release function, its specific access as follows: 1) the WIFI connection access. Because the WAP will generate traffic cost, the use of WIFI access mode is a better choice in reality. And can be provided higher bandwidth in the use of WIFI access, access to multimedia resources better. 2) WAP access WAP access can be access platform page anywhere, Without being limited by the environment If there is a WIFI hotspot, and whose defect is to produce flow, increase the cost of access platform.

2.2.2 *Platform page design*

Resolution to design a page using 240*320 is the mainstream page shows the effect of mobile phone. On the main interface all the content of the module can be shown. The whole page consisting, multimedia audio and video module and electronic books and reference module are arranged in parallel, and the search function module set, can realize the use of keywords retrieval way retrieval to meet the requirements of the content of all the modules, and is shown below.

3 A PLATFORM OF DATA RESOURCES

An important part of the platform construction research is organization and utilization of information resources in the public culture perspective, acquisition of public interest or useful for some industry resources, and digital processing, making it accord with the characteristic of communication, mobile devices such as resolution, the file size etc.. It can be divided into two parts of data sorting and data resources and processing resources.

3.1 *Data collection*

3.1.1 *Data collection principles*

For part of the resources of their data resources and collection, it can be sorted according to three principles:1)the principle of standardization Considering the diversity of smart mobile devices, it should be performed according to both lower the performance of mobile devices for resource conversion: Standard for video resources, in accordance with the 240*320 resolution,MP4 audio format conversion; unified standards in accordance with the MP3 bit rate and format text; unified use TXT text format. Because of the limit of the display screen size of mobile devices, for indexing metadata description and data should be as concise as possible.

2) The order principle. The working process is divided into: the collection, classification, cataloguing, indexing steps. The realization of information carrier, orderly and reached a search engine optimization, make the page more concise. 3) usability principles. In the organization of information resources, the principle should be followed, information resources concise, clear, easy to understand, easy to use by users.

3.1.2 *Data resources composition*

The platform is based on the traditional literature and digital resources in the library of our institute, which can provide to the public, such as: superstar e-books, CNKI, VIP, Wan Fang data, global English, French law, the starting examination of data exchange, widely seen journals, test questions library, New Oriental online and so on ordinary digital resources and the characteristics of audio and video resources of which Zhao Culture Database, HeBei Province Higher Vocational database, experimental training video is our school library characteristic resources. Experimental training video close to the factory workshop, the actual operation, is the learning platform which the enterprise workers can be directly observed, with the stronger learning and reference for primary workers and migrant workers; Zhao culture information database, collects including Handan idioms and allusions, the various schools of thought, cultural relics and photos of Zhao Culture related research results and other information, have higher the reference value for Zhao Culture enthusiasts and researchers; higher vocational colleges database collection of lectures, papers and national and provincial excellent courses courseware of some teachers, these are the crystallization of the theory and practice of teachers in higher vocational colleges, to understand the needs of enterprises, enhance cooperation and exchanges with the benefit of the enterprise, is the precious data resources in Higher Vocational colleges.

3.2 *Conversion of data resources*

3.2.1 *Audio and video resources conversion*

We use the converter to the multimedia resources required for conversion, can be on the video and audio conversion, converse the ordinary video or audio a resource can be watched on the mobile equipment.

Taking an example of video, on the conversion of the video the video resolution should be set, so as that users can get the best viewing results in access. The platform of the resolution of the video resources should be conversed using 320*240.

3.2.2 *Text format conversion*

The majority of mobile equipment with respect to the hardware specifications, due to uneven, so the rate of choice should be suitable for the majority of equipment, selection here is 384KB/seconds. The format type can be chosen mp4.For audio resources, the use of MP3 conversion, the other audio formats such as WAV, wma etc. the unified format is converted to MP3 format, fixed 64Kbps sampling rate to balance the format size and quality.

Because of the support of mobile phones and other devices for PDF or doc format not good for the text part, it uses TXT text format, with better compatibility and ease of use.

4 CONCLUSION

In view of the lack of current mobile devices in some aspects, limits of network technology (Internet) and mobile network (mobile web) and troubled by copyright issues of the digital resources, the user cannot freely enter into the 4ANY (anywhere\ anyone\ anytime\ anything) state to obtain culture resources. Showing: on the one hand, the limitations of the mobile digital platform, providing resources, on the other hand, the society need to pay fees in user access to information. But with the development of information technology, smart phones, PDAs, tablet PCs and other handheld terminal function more powerful, to further canonical m digital copyright management and gradually mature Tripler even four nets fusion, the platform function of the mobile digital library can be extended further, services can be more perfect, many short comings can be changed. By then, the service content of the mobile phone library will be more abundant, service links will be more smooth, the construction of the mobile library platform also will enter a new stage.

REFERENCES

[1] YE Shasha, Ni Xiaojian. Research on leaf inside and outside the school library open service from the perspective of public cultural services. The library(J). 2012.05.

[2] Zeng Li Design and implementat-ion of a mobile service system of Un-iversity Library library work study. [J]. 2010.10.

[3] Ye Aifang.the development status and Prospect of mobile library in China Library and information.[J]. 2011.04.

[4] Song Enmei YuanLin. Mobile bo-ok sea: current status and developme-nt trend of domestic mobile library.[J]. 2010.09.

[5] Guo Xichuan Practice and innovative application based on 3G network mobile digital library at home and abroad.[J]. 2011.05.

Computing, Control, Information and Education Engineering – Liu, Sung & Yao (eds)
© 2015 Taylor & Francis Group, London, ISBN: 978-1-138-02800-5

Enlightenment of ancient comic books of Tsai Chih Chung to Chinese traditional cultural transmission

Jun Fang An & Rui Bo Hu
Guizhou Normal University ,Guiyang, P.R.China

ABSTRACT: Tsai Chih Chung's ancient comic books take the lead in interpreting Chinese classics. He easily interprets unintelligible studies of Chinese ancient civilization through combining pictures and texts, making the readers learn the Chinese traditional culture in a relaxing and burdenless way. A "Tsai Cyclone"has swept across every corner of the world, triggering a "Chinese Classics Craze" that breaks the actuality of the minority privilege of Chinese classic ancient books and making ancient comic books became a new window to understand the Chinese traditional culture. Tsai Chih Chung's ancient comic books urge us to reflect on Chinese traditional culture (figure 1) and explore effective means to alleviate the spiritual crisis and value confusion of contemporary citizens under the current historical background.

KEYWORDS: Tsai Chih Chung's ancient comic books, Chinese traditional culture, transmission.

1 INTRODUCTION

Tsai Chih Chung says, "Comic books are a good way of inheriting civilization, arousing conscience, reshaping morality and justice as well as a sense of responsibility." Tsai Chih Chung has published dozens of Chinese classic comic books which interpret the studies of Chinese ancient civilization through an interesting way loved by the masses, illustrating thousands of years of Chinese ancient culture and taking the lead in interpreting Chinese classics. With a history of several thousand years, Chinese traditional culture is rich in content and complex in composition, and its social function is gradually decreasing after experiencing the baptism of industrialization and urbanization. From the current situation of transmission, the transmission forms and ways are single, the transmission contents are unintelligible and boring, and creative transformation and modernization transformation have not been realized. The ancient comic books are an effective way to spread traditional Chinese culture, reduce the gap between the masses and traditional culture, and relieve the masses' fear mentality of keeping at a respectful distance from traditional culture.

2 EASY INTERPRETATION OF UNINTELLIGIBLE CHINESE ANCIENT BOOKS THROUGH COMBINING PICTURES AND TEXTS

2.1 *Turning abstract into concrete to realize "image conversion" of ancient books*

Words can give people infinite imagination space, while comic books can help people turn the abstract into concrete, so as to cure the "Era Disease of Words Fainting" of the public in the picture-reading area. Images have played a more and more dominant role in people's life. "Contemporary culture is facing the impact of visual phenomena. A discourse system with language as the core is turning into a visual discourse pattern, reading is turning into to picture-reading and screen-reading, language competence is gradually turning into visual literacy, and the prosperity of text publishing is gradually turning into the flourish of the image industry. Soft power competition is more embodied in the image display, and this trend is accelerated by the rapid development of the media industry."[1] People are gradually giving up the language thinking habit and tend to the thinking mode of visual culture. The development of the times puts forward new requirements for the transmission forms of Chinese traditional culture. In modern society, people's ideological position and accepting perspective are different from those of the ancients. There is no doubt that the inheritance of the cultural spirit of ancient classics must go through a reshaping process. Combining fresh and elegant pictures with concise and vivid texts, Tsai Chih Chung's Chinese classic comic books are full of reading interestingness and collection value, which enables people to comprehend the thoughts and life philosophy of sages racily and naturally in a cheerful time. The practical significance of traditional cultural transmission is just like old wine in a new bottle through new ways loved by the masses instead of indiscriminately imitating traditional classics. The quintessence of traditional classics is subtly integrated into the development of modern society, becoming a kind of invisible spiritual power to guide people's life and forming the mainstream of world view.

2.2 *From "unintelligibility"to"humor": dredge readers' reading disorder*

"Traditional culture is visible like bells, Qin bricks and Han tiles, the Great Wall and other historical relics, such as poetry, Ci(a type of classical Chinese poetry, originating in the Tang Dynasty and fully developed in the Song Dynasty), prose, Confucian classics and other cultural classics, all of which are the manifestations of traditional culture."[2] Due to its age, classical Chinese disjointing from oral Chinese becomes difficult to understand or articulate, and uncommon, archaic and abstruse words become elusive "unearthly language"in the eyes of readers, so learning and using classical Chinese has become the "patent"of the minority. The unintelligibility of classical Chinese goes against the transmission and popularization of traditional culture. Through the humorous images with smooth, elegant and vivid lines and straightaway texts created by Tsai Chih Chung, the readers can learn Chinese classics in a relaxing and burdenless way and receive the nurture of traditional culture. Tsai Chih Chung spreads the elusive ancient Chinese classics through transforming them into comic books, which is of great significance to traditional Chinese culture. Ancient comic books, as a effective way of spreading traditional culture, has great development space and potential to grow up; in the future, the combination of comic books and cartoons which are similar in nature and expression techniques will be the main media to spread traditional culture, which has a powerful transmission efficiency. The expression of Chinese traditional culture through the comics, this universal language, is advantageous to the popularization and transmission of Chinese culture in the world.

3 SAGACITY IN HUMOR, PROVIDING MENTAL NOURISHMENT

3.1 *Original versions are cornerstone and comic works are destination: emphasizing the effectiveness of spreading traditional classics*

The original versions are the cornerstone of comic creation, and comic works are the destination through which cartoonists present ancient classics. "Chinese Classics Craze" is an important cultural phenomenon, not to be neglected in contemporary China. Ancient comic books have made a great contribution to the transmission of Chinese traditional culture. It is not a fresh topic to adapt traditional classics into comics, but not all cartoonists have serious and responsible attitude towards the original works. There is no lack of such people who treat adapting traditional classics as merely playing games and interpret traditional classics out of context. Different cartoonists' cultural quality and receptivity will also result in a tremendous difference in the degrees of creation. Comic art is a kind of contemporary art, so the viewers' tastes are bound to be influenced by the dominant public opinions, ideological trends, tastes, etc. of the contemporary era. To be faithful to the original, the cartoonists must respect the thoughts and the essence of the original. Classics are classics after all, so they cannot be deconstructed at will. Among the tremendous books, only a few books become classics passed on from generation to generation. Because of their gravity, greatness and magnificence, they won't fade away with the passage of time. The ancient classics are not only the legacy of history, but also the wealth modern people should cherish. Adapting the classics into comics does not completely exclude commercial interests, but cultural conscience should never be forgotten. Comic art is also a kind of epochal art. On the basis of respecting the original works and history, cartoonists make some creation and elaboration towards the original works to a certain extent through their own interpretation. The benevolent see the benevolence and the wise see the wisdom. The creation and elaboration can add entertainment elements into ancient books, which enables readers to have a new understanding of the classics and masterworks under the new historical background, making classics irradiate eternal light.

3.2 *Sagacity in humor, tasting connotation in humor: surpassing "laugh-it-off"*

The cultural connotation of Chinese entertainment culture has been replaced by "vulgar" entertainment, and spirit cultivation has been replaced by the materialism carnival. "Where are we going? Dad", "The voice of China", "China's Got Talent", "Empresses in the Palace" and other programs catch the eyes of people in the form of entertainment, achieving very high audience ratings and becoming the hottest topics at people's leisure. Excellent entertainment programs should be connotative, high-quality, conscientious, interesting, entertaining and cultivating spirit instead of making the audience laugh it off or remember nothing after reading. The ancient comic books must seek a balance between tradition, culture and entertainment, and entertainment should not become the primary purpose, but should serve noble objectives, which can attract and touch the audiences in a real sense. The entertainment with only humor and consummate painting techniques, but no traditional humanistic connotation will be a flash in the pan. Nietzsche also mentioned that "if we want a lifetime of happiness, there is no way other than promoting and developing national culture."[3]Traditional culture can be a driving force for mental consumption, achieving the purpose of transmission through association with entertainment and providing the public with mental nourishment. Creators need a devout attitude to excavate and explore in the classic cultural tradition, follow the market demand, lead a positive and healthy cultural atmosphere, and make ancient comic books become the window to perceive the elegant demeanor of Chinese culture.

3.3 *Mass media transmission with diversified forms, forging cultural industry chain*

The progress of network technology provides a new platform for the transmission of traditional culture. Comics integrated with classical elements of Chinese traditional culture can be transformed into "nutritional comics" and "nutritional animation" which can be spread through the mass media. Network media, as a new platform for "nutritional animation", has begun to take shape. In the future, the Internet should also try to connect microblog, search results, Taobao, e-commerce websites, TV, movies and other systems through a set of management systems, seamlessly integrating image management, contents, product commerce, business cooperation so that the masses can come into contact with traditional culture through different channels. In the phase of technology research and development, a certain amount of manpower and money should be invested so as to break through the existing form and approach towards to compatibility of new media network and TV platforms. Mobile phones are one of the most popular, convenient and practical new media in this age. "Mobile phone animation" is not only a kind of service, but also a fashionable neologism. Mobile phone "nutritional animation" can be compressed into a small format for convenient storage, and mobile phone classic comics also can be used as a screensaver or boot animation so as to help users achieve personalized settings. Therefore, mobile phone animation has increasingly become a new favorite of the personalized group of people. Moreover, "nutritional comics"can be applied to Chinese classic classes of schools at all levels, which can not only animate the classes with enjoyment when teaching five thousand years of Chinese culture, but also turn learning into a pleasure.

At the same time, "nutritional comics" industry also can form an industry chain and the peripheral products are a large gold mine. Successful comic works won't merely depend on publication amounts and box office of animation, but various subsequent products should be generated, such as audiovisual products, toys, electronic games, costume images, wallpapers, screensavers, etc., so as to achieve multi-level extension of cultural value. "Nutritional comics" industry should also reverse the main profit point from traditional production marketing to copyright benefits and industrial development benefits, create comic images and brands through multiform transmission modes of the mass media, and achieve image authorized operation and derivative product development. In this highly competitive cultural market, culture itself can be a kind of consumption goods bringing huge profits and play a leading role in a market economy.

4 CONCLUSION

Tsai Chih Chung draws the outline of Chinese traditional classics with natural and smooth lines and humanistic care towards Chinese traditional culture. His works have swept around the world for his unique painting style, profound connotation and popular humor, making 1.3 billion Chinese people re-examine their own culture. The lack of culture will inevitably lead to the fault of humanistic spirit. Corruption and immoral events have emerged in an endless stream in the Chinese society and the masses feel pity that public morality is not what it used to be. Shrewd politicians and sociologists often explain these phenomena as unavoidable "negative factors" in the process of Chinese social transformation, but these factors, perhaps will not disappear automatically along with economic growth and better-off life. Such a large China cannot merely rely on foreign culture to construct its own cultural framework, and the cohesion of national spirit can only be achieved by restoring and developing the essence of Chinese traditional culture. In the social environment of "multimedia text", the transmission of Chinese traditional culture should be reformed or optimized in forms, contents or media with era development, breaking through the single mode of transmission and forging a traditional cultural industry chain, so as to penetrate the essence of

Chinese traditional culture into all aspects of the society. Chinese traditional culture not only provides survival value support for individuals to settle down and get on with their pursuit, but also contains universal value and ultimate humanistic care that is the cultural capital by the aid of which Chinese people integrate into the world and win the right of speech.

REFERENCES

[1] Dang Ximin, *Right Operation of Visual Culture*, page 1, People's Press, 2015.

[2] Cao Liping, *Traditional Culture and Modernization*, page 34, National Library Press, 2010.

[3] Marcuse, *Nietzsche Werke*, page 46, Guangxi Normal University Press, 2001.

[4] Chen Long, Chen Yi, *An Introduction to Visual Culture Transmission*, Shanghai Sanlian Bookstore, 2006.

[5] Neil Postman, Zhang Yanze, *Amusing Ourselves to Death*, Guangxi Normal University Press, 2004.

[6] Liu Zhenyu, Apollo, *Current Status Research of Chinese New Media Animation*, Jinghua Press, 2008.

[7] Tsai Chih Chung, *Tsai Chih Chung's Chinese Classic Ancient Comic Series*, SDX Joint Publishing Company, 2005.

[8] Ruibo Hu, Jingyuan Li, Liyan Chen, Renping Xu, Kunqian Wang, Xingran Mao. Application Study of Conceptual Design in the Product Design[J]. Applied Mechanics and Materials, 2012, Vols. 121–126: 730–734. (EI: 20114714533309).

[9] Ruibo Hu, Renping Xu, Kunqian Wang, Jingyuan Li. The Research of Virtual Tactile Design Application in Dongba Sculpture[J]. Advanced Materials Research, 2011, Vols. 228–229: 185–190. (EI: 20112113994664).

[10] Ruibo Hu, Renping Xu, Kunqian Wang, Xingran Mao. The Research of Developing and Constructing Tourist Attractions in Dong Jiahe[J]. Advanced Materials Research, 2011, Vols. 250–253: 3884–3888. (EI: 20112314040542).

[11] Ruibo Hu, Xinyu Suoand Zhen Gao. Application study of the Virtual freeform in product development design [C]. Applied Mechanics and Materials Vols. 496–500 (EI: 20140817357748).

[12] Zhen Gao, Xinyu Suo, Ruibo Hu. A Study of Application of Upholstered Furniture Design[C]. Applied Mechanics and Materials Vols. 496–500 (EI: 20140817357749).

[13] Xinyu Suo, Ruibo Hu, Renping Xu. A Study on the Future Development Trend of Virtual Clay Technology[C]. Applied Mechanics and Materials Vols. 496–500 (EI: 20140817357784).

[14] [`14] Ruibo Hu, Xiaosong Zhang, Kunqian Wang. Application Research of Virtual Sludge Technology FreeForm in Character Design [C]. Applied Mechanics and Materials Vols,556–562 (EI:20142417825966).

[15] Ruibo Hu, Xiaosong Zhang, Kunqian Wang. Application Research on transmission of information in Design Semiotics [C]. Applied Mechanics and Materials Vols,556–562 (EI:20142417826153).

[16] Ruibo Hu, Xiaosong Zhang, Kunqian Wang. Research on Application of Symbol Typesin Design and Semiotics [C]. Applied Mechanics and Materials Vols (EI:20142917961722).

[17] Qiaoling Gui Xinyu Suo, Feng Su1,Ruibo Hu 2. A Probe into Utilization of New mater in Naxi Residential Door Trim of Lijiang Yunnan Province[M] .Advances in Intelligent Sytems Research(ALSR) Vol .23 (EI: 20125215828782).

[18] Xinyu Suo, Ruibo Hu ,Renping xu. Characteristic and its development tendency of virtual product development technology[M] . Advanced Materials Research Vol s.605–607,part 1 (EI:2013045941756).

Computing, Control, Information and Education Engineering – Liu, Sung & Yao (eds)
 ISBN: 978-1-138-02800-5

Application of computer technology in landscape design

Hong Lei Wang, Jun Gang Liu, Ke Jian Li, Xi Lin Zhan & Jin Hua Wang
Jinan Landscape Flower and Plantlet Breeding Center, Shandong, China

ABSTRACT: Landscape design is a comprehensive work, with the popularity of computers, the computer is playing an increasingly important role in landscape design, how to be better applied computer technology landscape design work has become a new topic. This article describes several common landscape design computer software, explore the relationship between landscape design and computer technologies.

KEYWORDS: computer technology; landscape design; auxiliary.

1 INTRODUCTION

With economic development, the improvement of the ecological environment gradually is paying attention. The numbers of green space planning and urban landscape projects are more and more. Landscape design is a highly technical, informative work. In the landscape design development process, it can be finished with artificial drawing early. With the development and widespread use of computers, this work can be done with computer-aided design drawing. Every aspect of landscape design can be by computer-aided design, computer-aided design is populared in landscape landscape designer and it is more and more important.

2 LANDSCAPE DESIGN AND COMPUTER TECHNOLOGY

《Yuan Ye》 "Xing creationism" has said "Three builders, seven masters". It is said the goodness or badness which depends on the spatial awareness of landscape designers, aesthetic ideas, fun and artistic accomplishment in life [1]. The designer's idea, the design concept is the computer can not be replaced, but the computer drawing landscape design can help designers to express design concepts better, but also it is easy to modify the design draft.

Computer design software has powerful editing and drawing features that can help designers to fully and completely express their design intent, and enable designers to be freed from tedious manual drafting work. So the effort can be devoted to the design ideas and concepts. It can solve difficult problems that the manual drawing can not deal with. At the same time, this technology improves the quality, efficiency and speed of drawing draft, and make it easy to modify and save the design draft. It also facilitates communication between designers. Design software's appearance not only breaks the limited nature of the static space, and expands the designers' horizons and improves the design efficiency, but also with artificial intelligence model, it can build space and brilliant color transform, allow designers to switch between computer models and physical models arbitrarily. It can inspire the ideas of designers.

3 LANDSCAPE DESIGN SOFTWARE

Currently in the design process it is commonly used computer software such as AutoCAD, PhotoShop, Coreldraw, 3D MAX, Piranesi, etc. The use of these software combinations can draw designs vividly and complete landscape design in many design elements, such as mountains, water, stone, plants, buildings and so on. AutoCAD can draw lines based plans, construction plans, on this basis, then PhotoShop, Coreldraw and other computer softwares do post-processing, render, and perfect design. Virtual reality and virtual space technology development in recent years, but also to promote the great development of design tools.

3.1 *AutoCAD drawing software*

The software makes use of computer-aided design software in terms of landscape design engineering. The software has a powerful drawing, editing, friendly interactive interface and it is welcomed by many landscape design staff. Meanwhile, the software provides a scalable interface and editing tools. Designers can complete secondary development based on the software. And the software enhances the three-dimensional modeling, image processing functions and a very useful additional tools to support programming interface ActiveX Automation, and it provides developing scripts, macros, and third-party tools Automation applications, but also it has a network graphics capabilities to support a variety of graphics formats accessible, enables graphics to transfer between different CAD systems [2].

The final product is a landscape design drawing using CAD technology. Plotted drawings are directly reflected in the landscape engineering computer applications. Landscape design using computer-aided design includes a drawing topographic map, location map, functional zoning map, and the overall design plan, building layout, planting design and so on.

3.2 *3D MAX software*

In landscape design, the software 3D MAX plays an important role in the three-dimensional modeling. This software is very comprehensive. It can set modeling, lighting, rendering, animation in one. And its simplicity, quick operation quickly laid in the landscape design industry can not shake the leading position.

The landscape design process requires having a good grasp the ratio of space. You can use this software to create a virtual scene to help designers aided design. This can intuitively control the proportion of the surrounding environment and landscape, and can control the spatial scale reasonably accurately . In the course of using the software, to better grasp the size ratio between the terrain, the drawing buildings, to analyze the rationality of design, setting the exact size, proportion and scale on the requirements more in line with the design requirements, At the same time, the effect can be convenient and intuitive design of the show, you can also dynamically display screen.

3.3 *PhotoShop software*

In the landscape design process, using modeling software such as 3D Max drawn images cannot be printed directly onto paper, you need to use PhotoShop software rendering image post-processing, such as color, saturation increases, re-set or modify the image background, adding with King (such as people, animals, etc.), to achieve the synthesis and processing of images, make the design more attractive. After PhotoShop rendered landscapes reflect the true figure when more appropriate, to better reflect the intent of the designer.

Every computer technology has brought change and development landscape design, art design and process design on the subject of change and development, design work performance means changing the original hand-drawn, the computer technology has brought a highly efficient, real, easy, three-dimensional design tools.

4 REMOTE COLLABORATIVE DESIGN, LANDSCAPE DESIGN AND VIRTUAL REALITY SIMULATION TECHNOLOGY

The development of computer technology and networks promotes the emergence of remote collaborative work. Collaborative technology gradually changes designers lackless of communication. It changes repeated design work because without communication. It makes remote collaborative design possible. Because of the exchange of ideas between design members, the research design effect is improved.

Computer simulation technology and landscape simulation are challenged to the traditional methods and concepts of landscape planning and design. It brings an important development direction of modern landscape design, and it is the forefront of the development of information technology to adapt to the modern computer revolution[3]. The key of the simulation is the development of computer graphics technology in the core of the digital landscape. Computer simulation technology and some special equipments as a tool to model are envisaged by the system dynamically test system to achieve outstanding visual effects. Realistic simulation results on two-dimensional images and three-dimensional image rendering are realer, fine and accurate with the development of digital technology. Virtual reality technology is interactive, participatory, vision characteristics[4]. It allows designers to become participants and become part of the environment. It can make designers have relationships with virtual environments in a variety of objects as if in a real environment. This can deepen knowledge and understanding of the environment and make designers generate new ideas. By interacting with their environment, it can enable designers to obtain all the information in all directions.

5 SUMMARY

The design of the current situation that the applications of computer technology in landscape design has been formed from the whole landscape complete computer software industry. Landscape engineering and technical personnel in a variety of computer software should be carried out to develop a deeper and more widespread use, in order to promote the standard of landscape design to a new level.

REFERENCES

[1] Weiquan Zhou. Chinese classical landscape history [M]. Beijing: Tsinghua University Press, 1990.

[2] Liu Hao, Wu Xun. prospects in computer-aided design technology landscape design [J], Technology Square, 2005 (8): 102–103.

[3] Songmao Mao. Applied Research "Digital Landscape" in a residential environment design [D], Hunan Normal University, 2012.

[4] Xiaochuan Fu.computer design applications leaning research in the performance of the landscape[D], Northwest Agriculture and Forestry University, 2010.

Computing, Control, Information and Education Engineering – Liu, Sung & Yao (eds)
© 2015 Taylor & Francis Group, London, ISBN: 978-1-138-02800-5

Reference and influence of Japanese furniture on Chinese national furniture

Zhen Gao
Tibet University, Lhasa, P.R. China

Rui Bo Hu
Guizhou Normal University, Guiyang, P.R. China

Xin Yu Suo
School of Vocational Technical Education, Yunnan University of Nationalities, Kunming, P.R. China

Jun Fang An
Guizhou Normal University, Guiyang, P.R. China

ABSTRACT: During the formation of Eastern Asian national furniture, Japanese furniture is the earliest, the most unique and the most successful. Though comments on the influence of Japanese furniture vary from person to person, a complete system is formed by the development of Japan's modernization. Japanese culture is influenced by China, but has already finished the integration and transformation. It is closely combined with modern industry and constantly tries to integrate art and science and uses modern business ideas, forming national furniture which not only shows the traditional beauty but also with modern style. It is different from Korean furniture, which is diverse, all-embracing and flexible. However, the Japanese ones are mature and complete, originate from nature and have a fresh and noble style. The new Chinese research includes theoretical research, technical research and production practice. This paper analyses the Japanese style from many aspects and also from different perspectives and explores the future direction of the Chinese style.

KEYWORDS: Japanese style, National furniture, Reference points, Innovation, Differences and similarities, Education.

1 INTRODUCTION

When talking about Japanese furniture, Japanese culture comes first. Different from South Korea, its close geographical location with China determines the fact that Japan has an ambiguous relationship with Chinese culture since ancient times. When Chinese culture is strong, Japanese learn from it and admire it heartily. However, when Chinese culture is weak, Japanese take every opportunity to invade. Because Japanese culture has always absorbed from Chinese culture independently and automatically, Chinese culture is even more preserved in Japan. For instance, the Japanese kimono takes example from Han Chinese clothing and its architecture preserves a lot of the best essence of Chinese culture, especially that of Tang Dynasty. However, all these are very rare in China. In the Eastern Han Dynasty, the ironware and clothing of Central Plain were introduced to Japan and thus greatly increased the productivity in Japan. After the Chinese convention of sitting on the ground was brought to Japan, some coarse, shabby and low furniture appeared in Japan. In Tang Dynasty, Japan reached an all-round sinicization through the Great Reform, integrating the Chinese architecture, art and system with its own. Because Japan did not go through the time of "Wu Hu Uprising", high-type furniture failed to spread to Japan, which was also due to the great influence of Buddhism. At the beginning of the Edo period, Japanese furniture formally formed. However the real turning point of Japanese furniture is Meiji Restoration, after whose appeal of "Civilization and Enlightenment" the machine produced-furniture appeared. After the World War II, the company philosophy in Japan changed from plutocracy to the modern joint-stock system. Japan started to pay more attention to quality and environmental protection. Till today, an environmentally friendly and creditworthy sustainable model with many types and small batches has been formed. A comparison between the furniture of China, Japan and South Korea: Chinese furniture values the material and has a very complex structure, sophisticated embellishment, and is majestic and decent and is also jubilant and harmonious.

2 CULTURAL CONNOTATION

Without a rereading of the cultural connotation, it is not easy to supplement "artistic conception". The watershed between China and Japan is in the Tang Dynasty, after which China formed the tradition of sitting on a chair through "ethnic fusion", while Japan developed on its own under a closed environment. The Japanese furniture is greatly influenced by "Buddhism" while it is also the same with the Chinese style, for example, the round and harmonious square tables, round stools and stands; the only difference is the height. There seems to be a big difference between the Chinese and Japanese styles, but they share the same "artistic conception". The complicated and huge Chinese style shows the aggressiveness of "setting examples for other countries", while the simple and delicate Japanese style shows the contemplation of life. In modern society, people run after money and fame, lacking a soul harbor in their heart. The holy atmosphere of the "quiet and aloof" Japanese furniture offers soul comfort and makes people return to innocence and start back on the road. The rich vanity from the Chinese style is only the reward of the successful men and keeps reminding people of the classes between men, Therefore, the cultural connotation of the Chinese furniture needs to be supplemented by the Japanese style's milk of human kindness. The Japanese furniture values Buddhism and the lines of the chairs vaguely show the holy figures of the temple doors and have a strong atmosphere of "Zen". Its straight models and lines and simple decorations emphasize the practical functions. Compared to the Chinese furniture which shows class everywhere, even at the cost of function, the Japanese style fits the public aestheticism more. The aesthetic conception of the Japanese style does not require perfection and never demands the best, but the better and also has the room for improvement. Less is more: they simplify the decorations and structures and decrease the nobility of the materials and even take off the brand and simplify the packages only to preserve the functions. All these show the Buddhist imagery of "nothing is everything". Having no artistic conception is in itself an artistic conception. Therefore, the Japanese style is better compared to the arrogant Chinese furniture. The Japanese furniture is good at learning, assimilating, promoting and improving itself through technological methods. However, the Chinese style is not so good at assimilating and thus cannot always keep the vitality. Reference points: the Chinese style should attach less importance to class and more importance to "Zen" and incorporate modern elements. Based on modern fashion elements, some primitive elements should be added properly.

3 STRUCTURAL INNOVATION

The Japanese models use tender straight lines and wavy lines and have a simple and solid structure, mainly using squares and rectangle. There are no decorations on the surface and the materials are light. In most cases, the joining structures are shown and the color collocation is bright and lively. Its size is very suitable for their short stature and thus suits their life style greatly. The Japanese furniture perfectly integrates modernity and tradition and admires nature and is human-oriented and puts functions in the first place and has delicate workmanship and values every detail. For example, the traditional bracket is square and the practice of sticking out the two ends of the top makes it lighter and more modern. The Japanese chair uses plywood and is formed once and is very easy to move. The body of the chair is hollow and is more air permeable and is not slipping and is not easy to be left with some indentations. The asymmetric structure shows the easiness of letting nature take its course. It is very humid in Japan and the softwood with a high degree of swell-shrinking has a low requirement of mortise and tenon. Therefore, its joints are simpler than the Chinese ones. Chinese furniture likes curves and has more complicated structures. The Japanese furniture has an affinity which comes with its Japanese styles and the "Zen" culture. Even for the most complex chairs, wood is connected in a flat way, and the chair legs are inserted. This method not only saves materials, but also makes it easy to assemble and also uses materials cautiously, catering for the modern people's aesthetic pursuit of simplicity. The Chinese furniture, especially those in the Qing Dynasty, is bigger in size and rougher in materials, giving people a feeling of fear. Even the most delicate Ming styles are too exquisite and would be too complicated for practical use as modern products. The structure can be simplified on the ground by keeping its essence and uses the exact structure and plain texture to attract the public attention.

Reference points: the Japanese design is modularized and uses space better and is close to life. The asymmetrical model can be adopted by the Chinese furniture and therefore stresses the flexibility and the sense of relaxation. Its shape and structure should be simplified and the thickness and weight should be decreased, making it easy for the procession. The firewood furniture in the humid regions of china can try to simplify the mortise and tenon to reduce the complexity of the work. More efforts should be made to change curves into straight lines. And the soft intensity (for example, polish, burnish and soften) of the surface should be increased.

4 FUNCTIONAL INNOVATION

The functions should be specified and is human-oriented and is loyal for life. The Japanese furniture pays more attention to storage for example, storage boxes, storage bags, trunks, cabinets and shelves. Because Japan is a small land with many people and the rate of house-owning per person is low, the storage design is strengthened in order to avoid the sense of the crowd. Reference points: because China has a lot of space, the storage design of the Chinese furniture is relatively weak. If the storage design is improved for example, the treasure chest, then a mess in the room can be avoided and the room will be clean and comfortable. The furniture in the Ming Dynasty is human-oriented but the functional designs are not in detail and they should be more specific.

5 DECORATIVE PATTERNS

The Japanese decorations and images adopt natural materials. The Japanese furniture uses plants as decorations and is very peculiar with the usage of colors and mainly uses unsaturated colors and therefore does not have a very strong eye shock. The Japanese style works towards a more specific direction and is excellent in details. The Chinese style should learn this. However, because the Chinese style is very huge, the details where eyesight fixes should be more delicate and specific while stressing the whole. The Japanese furniture adopts light and elegant colors, but is more harmonious in matching with green plants. The Japanese tradition of substance sadness integrates the love for nature with the intrinsic spirituality of human and nature and inspires the aesthetic beauty of expressing emotions through scenery and combining scenery with emotions and uses emotions to handle the real essence. Reference points: the Chinese furniture should increase the rate of unsaturated colors and decrease the visual shock. More plant decorations can be used, but red flowers should be used to set off the rich atmosphere.

6 MATERIAL CHOOSING

Anti-insect paper with thick texture is placed at the bottom of the Tatami, and when placed in the middle it feels natural and can deaden noise and prevent heat. For the cabinet boxes, paper made from camphor trees replaces glass and it is very light and thin and is also water proof and is very ductile. The Japanese furniture uses its domestic Whitewood and keeps its original color and seldom uses dye to dye the surfaces. The commonly used materials are wood, bamboo, vine and grass. The wooden part would be taken out and the wood would be reused and then be gilded and be decorated by bronze wares. Natural paint and wood oil are used instead of chemical paint and poisoned paint to keep it purely natural. Ma Weidu says "Chinese people have always been peculiar about materials and think that only the heaviest materials are good." ② The Chinese style is intended for the high end users and uses expensive materials to show the status of the users and thus believes that the more expensive the material is the better. The designers dare not to innovate on materials, which makes it difficult for the Chinese furniture to meet the requirements of the public. The choice of the Japanese materials can be learnt by the Chinese style. The ancient poor Japan was unable to import rosewood and had to use their national soft wood. The surface decorated with clear varnish and together with the texture expresses a special beauty. The Chinese style emphasizes too much on the polish of the surface, but if we use hard materials and thin the paint and highlight the decorative function of the texture, then we can decrease the cost while preserving the ancient beauty. In today's society where rosewood is very rare, this is worth a try. Reference points: The Chinese furniture should make the surface in a cleaner and quicker way and avoid methods which break structures such as engraving. Thin the paint and use colorless phenolic resin paint instead of lacquer. The materials should be dewatered, dried, insects killed and sterilized so that they are clean and durable. Use a thin layer of paint and no polish and emphasize the texture. The texture goes with the paint, making it not only primitive and elegant but also modern and fashionable.

7 FAAN ANALYSIS OF THE MARKET

The Japanese furniture has introduced the American business model and is people-oriented and has humanized design and is marketed for its delicateness and preciseness. The domestic power of consumption in Japan is very strong, but there is a lack of resources and labor. Therefore, the overseas product lines are greatly developed. Their consumption situation is different from that in China. The Chinese market is far from being saturated but is in great need of rosewood. Overseas product lines should be developed so that it is easy to use local materials. In addition, the designs should have more alien features in order to open up overseas markets. The retail business in Japan adopts direct selling and brand authorizing and the factories humanly cooperate with the retailers. The business volume of the wholesalers is lower than that of the retailers. However, the former are better at the development and circulation of the products and even change into retailers. The retailers change from selling products to promoting some life styles to the customers and even become decoration and furniture-consulting companies. This tendency fits the modern furniture's direction of "small batches and various types" and is also the future road for the Chinese furniture. Now broadband becomes common in every household and the IT furniture which suits this change appears. The Japanese customers have a difference of being high, middle and low. Different furniture has different production lines which have serial services of design, material choosing, production, package, promotion, after-sales service and market analysis. The Industrial Wood Law also greatly supports the production and wipes out false frauds in order to ensure the safety of the furniture. While looking at the Chinese furniture, the domestic law construction is not complete. This is mainly because of the unbalanced regional development and the weakness and unbalanced distribution of the buying ability. All these make it difficult to form a robust and flexible law protection system. If the Chinese furniture wants to realize the industrial upgrading, it must rely on the developmental advantages of the creative industry and let innovative designs play their role.

Reference points: The Chinese style should develop delicate and detailed furniture and develop small furniture like little square tables and little stools and win by details. According to the special Chinese national situation, a center radiation road should be taken. Big cities like Beijing, Shanghai and Chongqing should make efforts to pull the development of other regions. The Chinese furniture industry should take advantage of regional resources and strengthen the risk management and form close cooperative relations. With the increase of e-families in China, internet furniture should be developed and form high, middle and low consumption groups. The regional advantages should be specified and different measures should be adopted according to the local features. The law protection should be improved and more attention should be paid to the promotion and after-sales services. Its market credit should be improved and international market should be opened up and rational methods to regulating industrial clusters should be put forward. (9) The core competence should be improved.

8 SPATIAL FEATURES

The space of the Japanese furniture is loose and elaborate. The setting of the furniture and the regulation of the building space are precisely calculated and the wooden structures are simple and do not have too much decoration. It has a strong sense of space. Its model is "small, delicate and exquisite" and uses the space of the cabinets to create shadows and soft shadows. Its lines are clear and the paintings on the wall are pure. The inside of the room has lights hanging over and the light is nicely used to go with colors to create imagery. Human and nature become one. Importance is attached to corridors and cornices. The corridors are spatial, bright and free. The multiple uses of the rooms address the house owner's trouble of the low usage of the main rooms and extra rooms, and are the best design for people with small houses. The Chinese furniture is usually in the middle and close to windows and emphasizes symmetry and has inflexible and odd spatial designs with little changes. Reference points: the Japanese style has a precise calculation of the space and has a strong sense of space. The Chinese furniture should learn its advantages of using the lights to create atmosphere. The Chinese light is red and yellow and has a big space. For modern small unit houses, it would be warmer if the height of the furniture is decreased in order to gain a wider view.

REFERENCES

[1] Dang Ximin, *Right Operation of Visual Culture*, page 1, People's Press, 2015.

[2] Cao Liping, *Traditional Culture and Modernization*, page 34, National Library Press, 2010.

[3] Marcuse, *Nietzsche Werke*, page 46, Guangxi Normal University Press, 2001.

[4] Chen Long, Chen Yi, *An Introduction to Visual Culture Transmission*, Shanghai Sanlian Bookstore, 2006.

[5] Neil Postman, Zhang Yanze, *Amusing Ourselves to Death*, Guangxi Normal University Press, 2004.

[6] Liu Zhenyu, Apollo, *Current Status Research of Chinese New Media Animation*, Jinghua Press, 2008.

[7] Tsai Chih Chung, Tsai Chih Chung's Chinese Classic Ancient Comic Series, SDX Joint Publishing Company, 2005.

[8] Ruibo Hu, Jingyuan Li, Liyan Chen, Renping Xu, Kunqian Wang, Xingran Mao. Application Study of Conceptual Design in the Product Design[J]. Applied Mechanics and Materials, 2012, Vols. 121–126: 730–734. (EI: 20114714533309).

[9] Ruibo Hu, Renping Xu, Kunqian Wang, Jingyuan Li. The Research of Virtual Tactile Design Application in Dongba Sculpture[J]. Advanced Materials Research, 2011, Vols. 228–229: 185–190. (EI: 20112113994664).

[10] Ruibo Hu, Renping Xu, Kunqian Wang, Xingran Mao. The Research of Developing and Constructing Tourist Attractions in Dong Jiahe[J]. Advanced Materials Research, 2011, Vols. 250–253: 3884–3888. (EI: 20112314040542).

[11] Ruibo Hu, Xinyu Suoand Zhen Gao. Application study of the Virtual freeform in product development design [C]. Applied Mechanics and Materials Vols. 496–500 (EI: 20140817357748).

Computing, Control, Information and Education Engineering – Liu, Sung & Yao (eds)

Software process measurement model based on generalization quality system

Xin Chen & Yu Ping Zhang
School of Computer Science and Technology, Nanjing University of Aeronautics and Astronautics, Nanjing, China

ABSTRACT: The development of software engineering has greatly enhanced awareness and importance of software process quality. Software process with high quality not only improves software quality, but also reduces unnecessary manpower and resource depletion. As a kind of knowledge product, the definition of software process quality is ambiguous. Hence, the generalization quality system is introduced. Combined with CMM software process model, a generalized software process measurement model is proposed. Further studies on processes of analyzing and selecting metrics have been conducted.

KEYWORDS: software process; generalization quality; process measurement; quality model.

1 INTRODUCTION

Since 1980s, studies on software quality model have conducted worldwide, among which Boehm model, McCall model and ISO/IEC9126 model are the most famous [1]. However, these models are mostly quality-oriented, which are designed to analyze factors affecting software quality, but incapable of evaluating software process capability. Meantime, quality management evaluation systems such as CMM/CMMI and SPICE don't present specific quality models. Today, software process evaluation mainly focuses on primary process of software development while weakening evaluation of organizational process and supporting processes. For example, SW-CMM and CMMI both have OPF (Organization Process Focus) process field, but in execution and evaluation, OPF may be diluted [2], which focuses on the primary process. Hence, software process quality management is important.

However, researches on the software process quality model are different from the traditional quality model as follows: 1. As a kind of product knowledge, metrics of the software product are explicit and easily measured compared with tangible products. But metrics of software process involve some subjective factors which are ambiguous [3], which makes it hard to evaluate software process. 2. Process measurement has larger time and space span compared with traditional product measurement. Hence, when measuring, current software process quality management system will ignore some factors, which makes the results of software process different from actual process quality to some extent [4]. To treat these issues, generalization quality system first designed for industries and finance is introduced. Combined with current software process quality management system, a generalization quality system for software process measurement is established and applied to the software process measurement, which composes a software process measurement model.

2 GENERALIZATION QUALITY SYSTEM AND SOFTWARE PROCESS QUALITY

2.1 *Researches on software process quality management*

There are some different opinions about the management and evaluation of process worldwide. The main management and evaluation system of software process consists of: 1) ISO9000 quality management and international quality standard [5]. ISO9000 family of enterprise quality management includes important quality standard ISO9001. 2) SEI SW-CMM/CMMI system [6]. The idea of improving the quality of software produce by improving the quality of software development process derives from product quality management. 3) ISO/IEC12207 and ISO/IEC 15504 international standards [7]. *ISO/IEC12207: 2008 (Second edition) information technology — software lifecycle processes* standard aims to establish a common framework for software life cycle processes, which provides a procedure to identify, control and improve software lifecycle processes. ISO/IEC 15504 initially was evolved from SPICE (Software Process Improvement and Capability Determination). The standard attaches importance to not only software process, but also staff, technology, management, quality, user support, software development and other issues. The three quality management systems provide a standard for evaluating software process to some extent. However, in practice, problems exist in these standard systems. For example, the cost factor is usually neglected in CMM model. Lots of enterprises cannot afford the high cost [8]. Today, many small businesses are certificated with lower levels of CMM/CMMI. CMM model ignores the evaluation of customer satisfaction [9], which leads to the difference between CMM/CMMI level assessment and practical capability of the software process. Besides, ISO9000 quality management system only generally describe quality management of software process and provide the minimum standard of the capability of software process, which is not applicable to software process. Meantime, the three quality management systems for software process don't present specific quality models, which partly constrains self-evaluation of enterprises.

2.2 *Generalization quality system*

Generalization quality is the extension and development of traditional quality. Definitions of quality are different internationally. One given in ISO9000:2005 is "degree to which a set of inherent characteristics fulfills requirements"[10]. This definition generally interprets the essence of quality, and the quality standard can be applied to the measurement of products, processes and systems. Hence, the generalization quality of the software process is derived from the definition of quality given in ISO9000:2005 that includes three parts: 1. "Fitness for use" from Doctor Juran; 2."Zero Defects"[11] from Crosby; 3. "Total quality control" from Feigenbaum and Ishikawa Kaori. The three viewpoints define quality based on basic quality [12], user satisfaction and organization, which promotes the determination of quality characteristic.

3 SOFTWARE PROCESS QUALITY IN THE GENERALIZATION QUALITY SYSTEM

Not only all quality characteristics in quality management system but also other, non-quality factors should be considered to model software process. Such as user satisfaction and value realization. In this section, based on current software process quality management system, a software process measurement model combined with the generalization quality system is constructed.

To easily measure software process, some rules need to be followed:

1 Based on GQM (Goal-Question-Metric) model;
2 According to process, field-common feature-key activity in CMM, divide software process measurement characteristics layer by layer until the sub characteristic can be easily measured. When sub characteristics of the lowest layer have been measured, adopts bottom-up strategy to calculate comprehensive measurement of the software process.
3 To handle high implementation cost of software process measurement, reduce irrelevant measurement data while not affecting ultimate results. The bottom measurement should be intuitional and simple.

According to characters, software process can be divided into three kinds in current software process quality management systems:

1 Basic process: these are major parts of the software lifecycle, mainly including the processes of getting, supplying, developing, operating and maintaining.
2 Supporting process: these are processes of supporting basic processes, including documentation, configuration management, quality assurance, verifying, determination, joint review, audit, problem solving.
3 Organization process: these are processes of establishing, implementing and keeping improving one kind of elementary structure, including processes of management, basis, improvement and training.

Following two points also needed to be taken into consideration: (1) ISO9000 family pays attention to "Serve customers, satisfy customers"[15]. It develops a customer satisfaction strategy that is interpreted as customer value quality in software process. Hence, software process quality considers not only quality characteristics inside software process. Today in China, most software organizations are commercial institutions or many organizations develop software on commercial purpose. Like generalization quality systems in industries, description of customer value, such as customer satisfaction and value realization, needs to be involved. (2) Software process evaluation differs when software engineering is applied in different fields [17]. Since various fields have different additional requirements for software process quality, measurement of field quality characteristic needs to be involved. In conclusion, quality sub characteristics of software process should consider quality characteristics of basic processes, supporting processes,

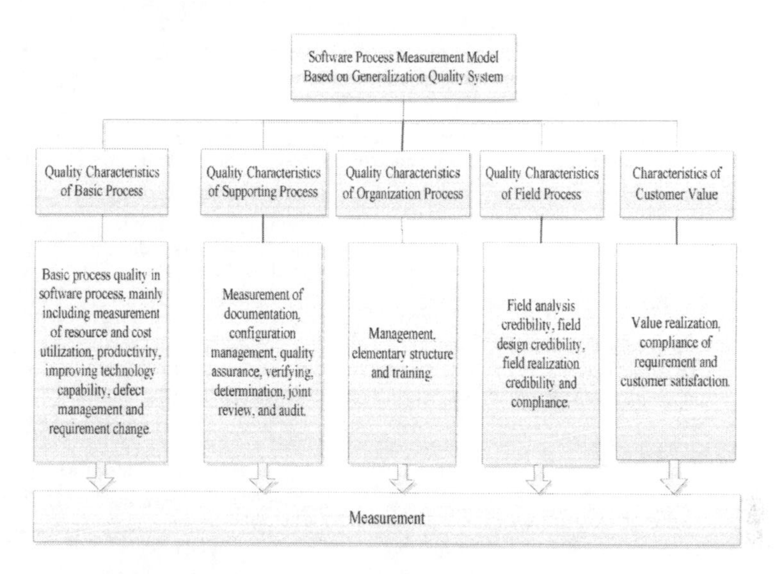

Figure 1. Software process measurement model based on generalization quality system.

organization process, field process and customer value.

Figure 1 is a software process measurement model based on the generalization quality system

Quality characteristics of basic process: basic quality characteristics are description of basic process quality in software process, mainly including measurement of resource and cost utilization, productivity, improving technology capability, defect management and requirement change.

Quality characteristics of supporting process: quality characteristics of supporting process are descriptive of supporting process quality in software process, mainly including measurement of documentation, configuration management, quality assurance, verifying, determination, joint review, and audit.

Quality characteristics of organization process: quality characteristics of organization process are measurement of organizational process, mainly including evaluation of management, elementary structure and training.

Quality characteristics of field process: quality characteristics of field process are evaluation of different software processes applied in various fields, mainly including field analysis credibility, field design credibility, field realization credibility and compliance.

Customer value: customer value is the measurement of customer value, mainly including the evaluation of value realization, compliance of requirement and customer satisfaction.

4 MEASUREMENT OF SUB CHARACTERISTICS

After constructing characteristics of software process based on generalization quality, their sub characteristics will be measured. Measurement of sub characteristic adopts the bottom-up method. When all characteristics of bottom layer have been measured, then measure the upper layers. This is a typical multi-attribute decision-making that needs to consider every attribute to obtain an objective comprehensive evaluation. A lot of attributes can be measured using current measurement methods in software process quality management system. However, other attributes need to construct their measurement methods in a certain order.

Take customer value as an example. Its sub characteristics and measurement items are shown in Table 1.

In sub characteristics, customer value points can be obtained through four ratios: ratio between realization value points and total of customer requirement value points; ratio between number of document types and total of document types related to customers; ratio between number of documents and total of documents in the system; ratio between number of running processes and total of processes. The measured value of compliance of the requirement can be calculated by equation 1:

$$f = \sum_{i=1}^{4} \frac{rv_i}{pv_i} \times w_i \tag{1}$$

Table 1. Customer value, sub characteristics and measurement items.

Characteristics	Description	Sub Characteristics	Example
Compliance of requirement	A group of attributes related to defined requirements	Compliance of requirement	①number of function points/Number of required function points
		Other	①actual progress/require progress ②actual progress/require progress ③following measurement items of value realization
Value realization	A group of attributes complying with value points defined by a group of users.	Customer value point	① realization value points / total of customer requirement value points
			② number of document types / total of document types ③ number of documents / total of documents ④ number of running processes / total of processes
		Supporting status and result	①number of help/scale
			②number of successful help/number of helpers ③time of supporting customer/scale
Customer satisfaction	Preparation for using software and evaluation of a group of users.	Objective satisfaction	Based on function realization, value points and supporting data analysis.
		Subjective satisfaction	Based on customer satisfaction survey.

where rv_i denotes realization value points, number of document types, number of documensts, number of running processes, pv_i denotes total of customer requirement value points, total of document types, total of documents, total of processes, w_i represents weight of i that satisfying $\sum_{i=1}^{4} w_i = 1$. Then calculate consumer satisfaction and value realization as equation 2.

$$\text{VRC} = \sum_{i=1}^{2} w_i \times m_i \tag{2}$$

Where w_i denotes weight of i, m_i represents customer value points and supporting status and results, w_i represents weight of i satisfying $\sum_{i=1}^{2} w_i = 1$.

Similarly, after calculating every sub characteristic using bottom-up method, measurement of customer value characteristic can be obtained by equation 3.

$$CV = \sum_{i=1}^{n} C_i \times w_i \tag{3}$$

where n represents number of sub characteristics C_i represents compliance of requirement, value realization and customer satisfaction.

There are many ways to handle multi-attribute decision-making. In this paper, the weighted average method is adopted. Besides, entropy method, TOPSIS and some other methods are also usually used. It depends on the specific measurement item to select proper decision-making methods.

5 CONCLUSIONS

According to current research on software process quality models, a software process measurement model based on the generalization quality system has been proposed in this study. The model constructs a group of measurable indexes for software process, which provides the basis for future quantitative evaluation of the software process. Following aspects require further study: 1) In order to facilitate measurement, further researches on the measurement indexes of every sub characteristic needs to be conducted. 2) According to different features of sub characteristics, various computing methods should be constructed.

REFERENCES

[1] Lan YQ, Zhao T, Gao J, Jie H, Jin MZ. Quality Evaluation of Foundational Software Platform. Journal of Software, 2009, 20(3):567–582.

[2] Qian H B, Zhu L J, Cao H M. Research and Design of Software Process Measurement System Based on CMM. Application Research of Computers, 2004, 21(6): 49–52.

[3] [3]Stoica A J, Pelckmans K, Rowe W. System components of a general theory of software engineering. Science of Computer Programming, 2014.

[4] [4]Pino F J, García F, Piattini M. Software process improvement in small and medium software enterprises: a systematic review. Software Quality Journal, 2008, 16(2): 237–261.

[5] Wang Q. SQA Model Based on ISO9000. Journal of Software, 2001, 12(12): 1837–1842.

[6] [6]Chrissis M B, Konrad M, Shrum S. CMMI for development: guidelines for process integration and product improvement[M]. Pearson Education, 2011.

[7] Ye S Y, Zhou P, Gu Q. A Software Process Model Based on CMM. Computer Science, 2002, 29(6): 123–127.

[8] Liu D Z, Liang G Q. Research on Relation between Customer Satisfaction and Loyalty. Modern Management Science, 2006, 2: 007.

[9] Han C Y. Discussion on the Principle of Quality Economics Basing on the General Quality Concept. World Standardization & Quality Management, 2010 (3): 18–24.

[10] Philip B C. Quality Is Free: The Art of Making Quality Certain, 2006.

[11] [11]Moen R M. New quality cost model used as a top management tool[J]. The TQM Magazine, 1998, 10(5): 334–341.

[12] Li J. Research of Software Quality Control and Measurement Technology. Beijing: Beihang University, 2000.

[13] [13]Fine C H. Quality improvement and learning in productive systems[J]. Management Science, 1986, 32(10): 1301–1315.

[14] Jiang H, Wang Q. A ISO9000-Based Implement Model on Software Organization. Computer Engineering and Design, 2002, 23(3): 26–29.

[15] Chen J B, Zhang Z. An Economic Analysis on Product Quality Based on Customer Satisfaction. Journal of Nanjing University of Science and Technology, 2003 (z1): 38–42.

[16] Li K Q, Chen Z L. An Outline of Domain Engineering. Computer Science, 1999, 26(5): 21–25.

[17] Guo Y J, Zeng Y, Cheng Q L. Method of Software Process Quality Metric. Computer Engineering and Applications, 2010, 46(9): 227–230.

Computing, Control, Information and Education Engineering – Liu, Sung & Yao (eds)
© 2015 Taylor & Francis Group, London, ISBN: 978-1-138-02800-5

Structural analysis of 800GS80 single-stage double-suction centrifugal pump

Xing Rong Chu, Jia Bin Wang, Wang Yao, Jun Gao & Lei Wang
School of Mechanical, Electrical and Information Engineering, Shandong University, Weihai, China
Shandong Shuanglun Co., Ltd., Weihai, China
Research Institute, CSIC, Harbin, China

ABSTRACT: To ensure the pump design reliability, the structural analysis of 800GS80 single-stage double-suction centrifugal pump was carried out by finite element analysis in this work. Firstly, the pump 3D model was built in Solidworks. Then the model was implemented into ANSYS to compute the statics analysis and dynamics analysis of the designed pump. The deformation, stress and modal characteristics of the 800GS80 pump were obtained. The simulation results prove the reliability of the product design.

KEYWORDS: centrifugal pump; finite element analysis; statics analysis; dynamics analysis.

1 INTRODUCTION

Centrifugal pumps are widely used in the industry production. The traditional pump design is mainly based on the model conversion method and velocity coefficient method [1], for which the design period is long and the prototype validation step is complex. Today, due to the flexible demands of the market, the traditional design method cannot meet the factory design requirement. Because of the complex structures of the centrifugal pump, the Computer Aided Design (CAD) and the Finite Element Analysis methods can greatly improve the product design efficiency and the product reliability [2-3].

In this work, the 3D geometry model of the centrifugal pump is carried out with Solidworks. And the model is implemented into ANSYS for the statics analysis and dynamics analysis. The deformation, stress and modal analysis were obtained to provide theoretical guidance for the centrifugal pump design.

2 3D MODEL AND MESH MODEL OF 800GS80 PUMP

Based on the blueprint provided by Shuanglun Company, the 3D model of the centrifugal pump 800GS80 was built. The pump body and the pump cover are shown in Fig. 1.

After the necessary model simplification for finite element analysis, the assembly model of the centrifugal pump was obtained. Due to the symmetrical structure, only 1/2 model is imported to ANSYS in order to save the computational cost. The space structure of the centrifugal pump is complex. Hence in this work, hexahedral meshes are applied for the part with regular shape and tetrahedral meshes are applied for the part with irregular shape. The whole meshed pump model is shown in Fig. 2.

Table 1. Centrifugal pump parameters for ANASYS analysis.

Material	Density	Young's modulus	Poisson's ratio	Designed Pressure
HT250	7200 kg/m^3	1.1E11	0.28	1.0 MPa

3 FINITE ELEMENT ANALYSIS

3.1 *Statics analysis*

The selected material and the material parameters for the pump body as well as the centrifugal pump designed pressure are listed in Table 1.

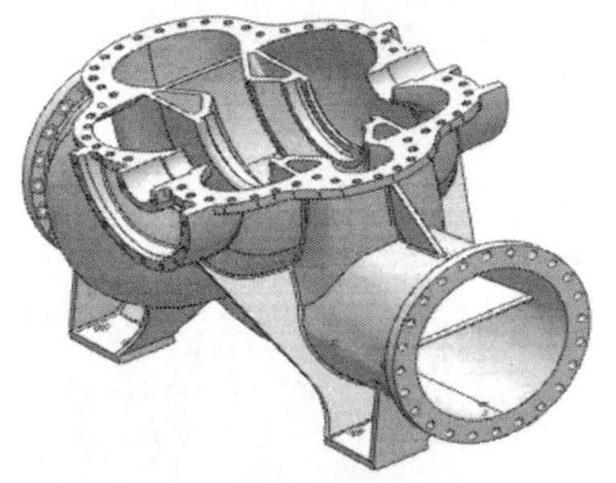

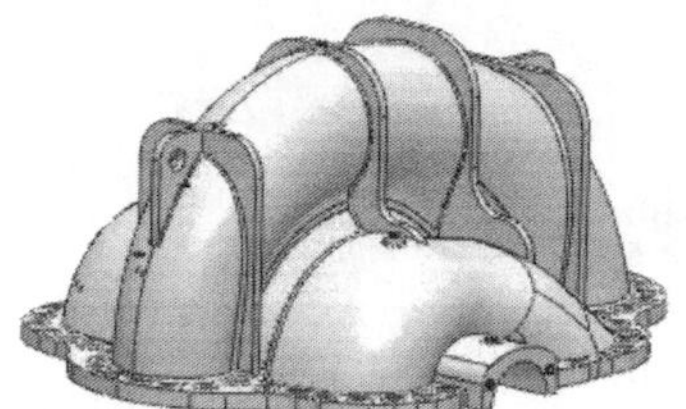

Figure 1. 3D model of the pump body and pump cover.

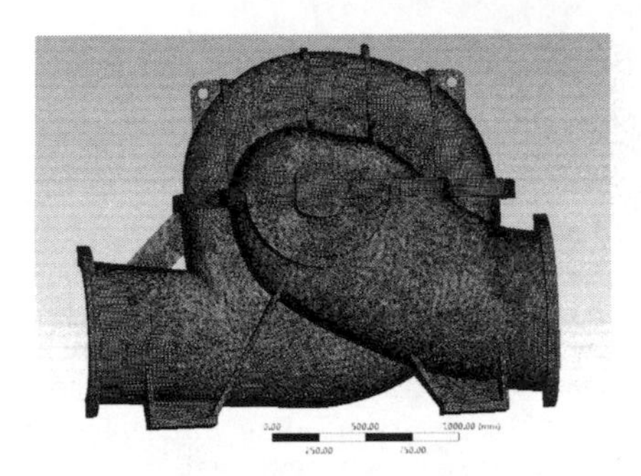

Figure 2. Pump mesh model.

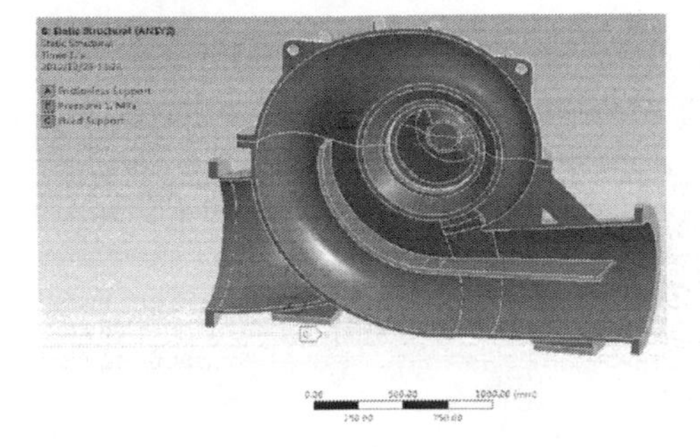

Figure 3. Load and constraints of the model.

The selected parameters are implemented into the centrifugal pump finite element model. For the working condition, the pump body is connected to the base with the bolts. Hence a fixed constraint was applied. Frictionless support in the symmetry plane of the pump was applied. According to the pump design condition, a pressure of 1.0MPa was applied to the pump internal surface. The details of the constraints and the load are shown in Fig.3.

The statics analysis of the pump model was obtained with ANSYS. The pump cover equivalent stress distribution, the pump body equivalent stress distribution, the assembly pump body stress distribution and the deformation of the assembly pump body are shown in Fig. 4. It can be seen from the statics analysis results that the stress, strain and pump body deformation are mainly concentrated in the impeller rotation work area, which coincides with the centrifugal pump real wok condition.

As shown in Fig. 4, the deformation, concentrated strain and stress mainly appear in the stiffer of the pump cover. The maximum total deformation value of the pump structure body is about 1.42mm, the maximum equivalent strain is about 0.0027. The result also reveals that the deformation of the stiffer is smooth and no obvious mutation was found, which proves that the stress concentration is mainly caused by the structure discontinuities. So the position design of the stiffener is reasonable. Besides, the strain, stress and deformation distribution in other parts of the pump is small and uniform. Most of the structure stress is in the range of 70-90MPa. Hence, the simulation analysis proves the stability and safety of the centrifugal pump structure design for normal operating conditions.

3.2 *Dynamics analysis*

The modal analysis is adopted in this part to identify the modal parameters of the designed structure and get the vibration characteristics in order to provide theoretical basis for the product design optimization and vibration fault diagnosis.

The Block Lanczos modal analysis method can calculate the inherent frequency values in the given range of the features. In this part the Block Lanczos modal analysis method is adopted. The parameters in Table 1 are used for the modal analysis. The fixed boundary condition is applied to the pump body and the ground.

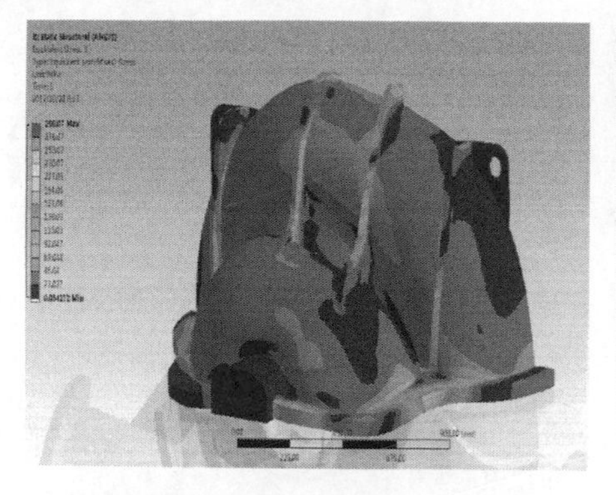

(a) Pump cover equivalent stress

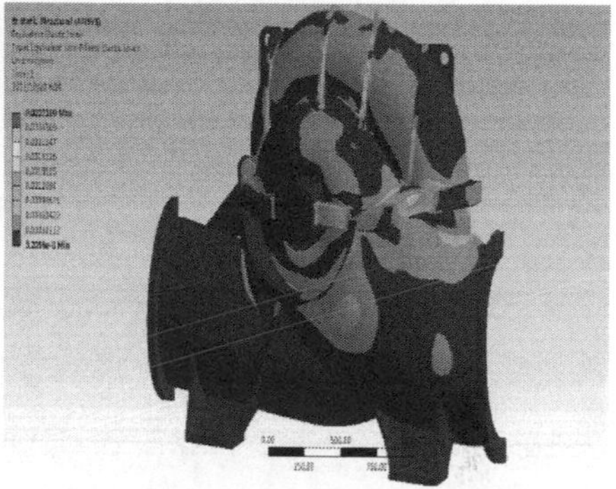

(b) Pump body equivalent stress

(c) Pump assembly structure strain

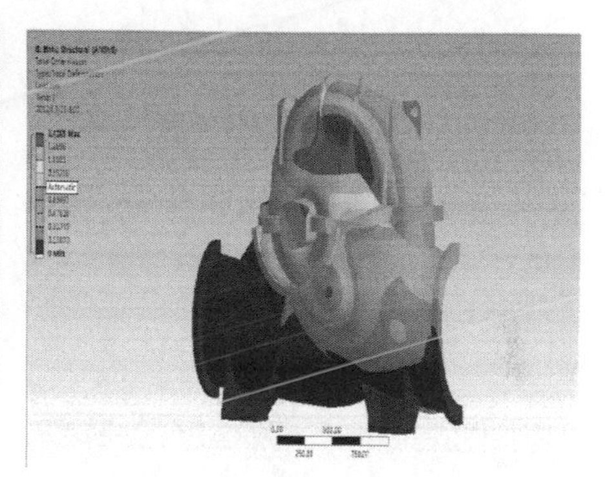

(d) Pump assembly deformation

Figure 4. Statics analysis results of the centrifugal pump.

In this work, only the first three order modal natural frequencies are analyzed. The obtained values are listed in Table 2. The first three order vibration modes of the pump body are shown in Fig. 5.

As shown in Fig. 5(a), the frequency of mode 1 is 47.8Hz and it represents for the swinging of the whole pump. The maximum value of the deformation appears in the upper pump cover and the vibration in this region is the strongest. For mode 2, a swaying back and forth of the pump body vibration is observed. The part where the maximum deformation and the strongest vibration occur is almost the same like the result for mode 1. For mode 3, it shows a twisting up and down mode at the inlet port and the outlet port. The maximum deformation appears at the outlet port where the vibration is the strongest.

Table 2. Natural frequency of the pump body structure.

Mode	Frequency
1	47.8Hz
2	98.9Hz
3	110.3Hz

According to the modal analysis result, the corresponding rotor shaft speed is 47.8×60=2868 r/min at the first order natural frequency. It means that when the rotor shaft speed reaches to this threshold value, the resonance occurs. While for 800GS80 centrifugal pump, the designed rotor rated speed is 960r/min, which is far less than 2867r/min. So it can be concluded that the pump design is reasonable. The resonance can be avoided under normal working condition.

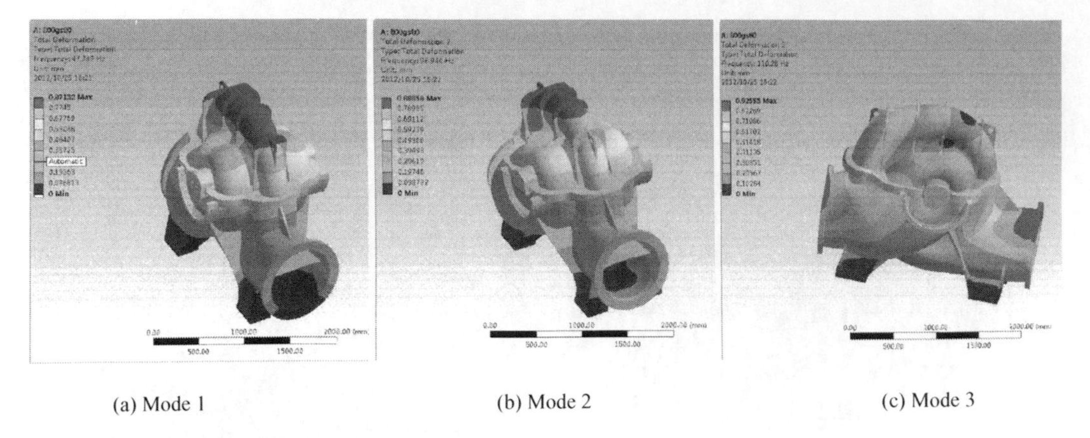

(a) Mode 1 (b) Mode 2 (c) Mode 3

Figure 5. Three different mode shapes of the pump body.

4 CONCLUSION

In this work, the 3D geometry model of 800GS80 centrifugal pump is generated in Solidworks. The statics analysis and dynamics analysis of the designed pump body were carried out by ANSYS. The deformation, strain and stress distribution and the modal characteristics of the pump body were obtained. The reasonable design of the pump was confirmed. The finite element analysis can provide a convenience, fast and reliable method for the centrifugal pump design.

ACKNOWLEDGEMENT

Thanks for the technical support provided by Shuanglun Company. The corresponding author is Jun Gao.

REFERENCES

[1] Labaniff. Centrifugal Pumps: Company Design & Application [M]. Houston: Gulf publication, 1992.

[2] Fujun Wang. Computational fluid dynamics analysis [M]. Beijing: Tsinghua university press, 2004.

[3] Xi Shen. Discussion on ANSYS and Solidworks data exchange. Equipment manufacturing technology, 2006(5): 50 – 57.

Computing, Control, Information and Education Engineering – Liu, Sung & Yao (eds)
© 2015 Taylor & Francis Group, London, ISBN: 978-1-138-02800-5

Challenges and coping methods for the grass-roots organization of our party strengthening social management—based on investigation and survey on GT village in central Henan Province of North China

Xiao He Hu
Hohai University, Nanjing, Jiangsu,China

ABSTRACT: The grass-roots organization of our party is the base of the overall work and battle effectiveness, a bastion where our party leads the masses to carry out the Party's theories, line, principles, policies, and tasks. With the transition of rural economic system and the advancement of urbanization, the grass-roots organization of our party is facing severe challenges while strengthening social management. The cause of the problems is analyzed in terms of the external environment and internal mechanism. From three aspects, i.e., innovation of management mode, improvement of the supervision mechanism and strengthening self-construction, the way to improve the grass-roots organization of our party is explored. Management ability and service level of the grass-roots organization of our party are continuously improved, to better consolidate the Party's ruling foundation in rural areas. It is conductive not only to the ongoing construction of a new socialist countryside, but to the stability of the rural society.

KEYWORDS: Village; the Grass-roots Organization of Our Party; Management; Challenges; Methods.

1 MAJOR PROBLEMS IN THE BUILDING OF THE RURAL GRASS-ROOTS ORGANIZATION OF OUR PARTY

General information about GT Village: The village, located in the east of Luohe in Northern China, is comprised of 9 groups of villagers, village party branch and villager committee, with a population of 3467 people, including 1683 men, 1784 women, 37 party members, inclusive of 7 female party members. The party branch was established in 2012, and consists of a Party branch secretary (holding a concurrent post of political instructor in the militia company), an organization sectary and a publicity secretary. The village committee was also founded in 2012, and consists of a director of villager committee (holding a concurrent post of militia company commander), an accountant, a security chief and a women's director. Based on the investigation of GT Village, there are three challenges in the building of the grass-roots organization of our party, as below:

1.1 *One is that some members of the grass-roots organization of our party pursue profits*

The grass-roots organization of our party is the base of the overall work and battle effectiveness, and its mission is to serve the people wholeheartedly, to act as role models in driving development, serving the masses, uniting people together and promoting harmony. In real terms, however, some members of grass-roots organization of our party are falling spokesmen to interest groups, even have established mutually beneficial relations of cooperation with interest groups, in exchange of personal economic benefits with political resources. On the one hand, some members of the grass-roots organization of our party provide additional political protection for market participants, in exchange of unjustified economic returns. In particular, in villages and towns rich in production factors such as mineral products, water resources and land, some party members provide unreasonable political support for some enterprises, and indulge them to illegally use or plunder its economic resources, to make profits. Some party members, on the other hand, directly get involved in the operation of market participants, and become a beneficial union. Relying on favorable political resources, in different ways, such as joint venture, holding performance shares and sharing out bonus, they form a co-existing relationship with interest groups, and it is justifiable for them to share economic interests. Thirdly, some party members improperly interfere with business activities of market participants, to achieve the purpose of participation in profit distribution, not only seriously undermining the Party's image and cause separation between the Party and the masses, but also shaking the Party's ruling foundation.

At present, such behaviors are the biggest problem in the building of the rural organization of our party.

1.2 *The second is that there are overlapping functions between the grass-roots organization of our party and villagers' autonomous organization*

Village party branch and villager committee of the peasant autonomous organization, generally called the village "two committees", constituting the political basis of rural society, shall be jointly responsible for political, economic, and social life in rural areas, balance the ecological environment of rural society, and maintain the stability of rural social order. Surveys, however, show the division of labor between village "two committees" is not clear, and there is a serious overlapping parts, not only affecting work efficiency, but also inducing conflicts. There is a subtle relation between Secretary of village party branch and village head, and their division of labor and cooperation is not completed fully. On the one hand, Secretary of village party branch often makes an excuse of "Party leads everything" to infringe upon the powers of the office and rights of property of the village head. Village head, on the other hand, with an excuse of "Self-governance of village affairs by villagers", occasionally deprives of decision-making power, the right to know and participation rights of Secretary of village party branch. Third, both of them are likely to shirk responsibilities or compete for a job not specified. If a job is not explicitly stipulated in the system and is lack of actual economic benefits, namely, the "hard but thankless job", including propaganda and implementation of safety production policy and popularization of agricultural insurance measures, etc., Secretary of village party branch and village head will dispute over trifles, and are reluctant to take responsibilities. If those are the profitable affairs, both of them will compete for and even repeatedly administrate it.

1.3 *The third is that self-construction of the grass-roots organization of our party is badly lagged behind*

Self-construction is the premise to guarantee the grass-roots organization of our party can give play to bastion function and ruling foundation. Our party has paid more attention to self-construction, however, based on the practical situation in rural areas, sluggish construction and not in place still exists. The management of party members is a lack of new ideas. In response to an older population, low level of knowledge and high-consciousness of party members in rural areas, the grass-roots organization of our party is the lack of effective quality enhancement measures, so the problem that some party members have no Party's progressiveness cannot be solved. Second, there are defects in the structural mode of the grass-roots organization of our party. In the election of members of village party branch, most candidates are nominated by the Party Committee at the township level, and party members in GT Village are voted in accordance with the requirements of the superior Party Committee. Due to insufficient inner-party democracy, party members' motivation is low. Third, the working mode is old-fashioned. Limited by its own conditions and the actual environment, the grass-roots organization of our party is unable to implement management innovation, so that the current working methods cannot adapt to the needs of economic and social development. Especially, the methods and measures adopted in the ideological and political work give priority to preaching. Due to no obvious educational effect and low acceptance by the masses, they are in need of innovation.

2 MAIN WAYS TO STRENGTHEN AND IMPROVE MANAGEMENT ABILITY OF THE GRASS-ROOTS ORGANIZATION OF OUR PARTY

2.1 *One is the innovation of management style*

In 2014, the General Office of the CPC Central Committee promulgated on *Opinions on Strengthening the Building of the Service-oriented Grass-roots Organization of Communist Party of China*, requesting all regions and departments must earnestly implement it based their actual conditions, and proposing as below: Build the service-oriented grass-roots organization of our party, to achieve the goal of "Six Haves", namely, having powerful leading group, having backbone team with strong abilities, having service places with practical functions, having various service carriers, having the perfect institutional system, having service performance to the masses' satisfaction [1]. In order to build the service-oriented party organization and really serve the mass peasants, first of all, accurately grasp new demands and expectations of the mass peasants, firmly set up the philosophy of "Concerning for rural residents, serving rural residents". Worry what rural residents are worried, and think what rural residents think. By sticking closely to rural residents' thinking, sincerely do a good thing and get things done for the mass peasants. The second is to broaden channels and patterns of the village-level organization of our party and party members serving the mass peasants. Different service modes are used, namely, "Village Party Branch +Peasant Households", "Village Party Branch+Rural Cooperative Economic Organizations", "Village Party Branch +Peasant Association", "Party

Members +Peasant Households", "Rural Cooperative Economic Organizations +Party Organization + Party Members +Peasant Households". Which mode is used must depend on the actual situation. The third is to do more things contribute to the protection of vital interests of the mass peasants, to solve the prominent issues the mass peasants concern and have strong opinions on, and to provide material, spiritual and cultural products and services conducive to safeguard the rights and interests of the mass peasants. The village-level organization of our party must always take it as the important obligation to serve the mass peasants, keep the safety and interests of the mass peasants in the mind, and put it into action. It is guaranteed that a grass-roots organization of our party is a bastion and a party member is one banner.

2.2 *The second is to perfect the supervision mechanism*

First, establish and improve the disclosure system of vital information, and improve the transparency of power operation. "No disclosure, no justice" [2]. Township-level organizations of our party to implement the information disclosure system of economic and social affairs within their own jurisdictions, timely disclosing or offering limited query of major issues concerning production and living of rural residents. Content of disclosing information covers the basis for decision-making, implementing standards, voting procedures, authenticity of information, etc., specifying query conditions, and focusing on improvement of the procedures. This information disclosure system matches with the disclosure system of village affairs prescribed in the *Organic Law on Villagers Committee*, to guarantee the right to know of rural residents from the perspective of establishing rules and regulations. The village-level organization of our party should actively encourage villagers to participate in the management of local economic and social affairs, enhance their political consciousness and right consciousness, provide the subject of activities for the implementation of the information disclosure system, and promote political pluralism and social progress. The second is to set an independent commissary in charge of discipline inspection in the grass-roots organization of our party, strictly regulating internal supervision. "An ideal, complete system should include two parts, i.e. normative institutional arrangements and disciplinary, institutional arrangements." [3] The internal supervision system is an important defense to guarantee the administration of the grass-roots organization of our party in accordance with laws and rules. As required by the *Implementation Plan for System Reform of CPC Discipline Inspection*, the position of a commissary in charge of discipline inspection should be set in the village-level organization of our party, to impose discipline, supervision or monitoring of the activities of party members, to improve the system of punishment and prevention, to prevent violating the law and discipline and the members of the grass-roots organization of our party from pursuing profits. Third, the grass-roots organization of our party should strengthen the supervision of economic and social affairs, and promote the execution of the decision. In addition to fulfilling its duties and responsibilities, the grass-roots organization of our party also takes the initiative to listen attentively to the voice of the people, and focuses on the village-level economic affairs together with the organization of village affairs supervision. By adhering to the principle of "the villager committee and other organizations play a role from the perspective of the executors, and the rural grass-roots organizations of our party from the angle of the supervisors", improve transparency in the process of decision-making and implementation. Especially, take a serious look at hot issues the people concern about, such as identification of the qualification for receiving subsistence allowances, operation of the village collective economy and management of cooperative medical services, to avoid a black-box operation, and to promote the realization of fairness and justice in the name of a spokesman of the people.

3 CONCLUSION

"The grass-roots work is important. Without a solid foundation, everything will shake up." [4] The building of the rural grass-roots organization of our party is a major problem concerning whether the Party's ruling status is stable, and is of great political significance to promote rural economic and social development. Facing with the drastic change of the current rural social environment, only when executing innovation from the aspects such as system improvement, construction of mechanical and self-construction, the grass-roots organization of our party perfectly agrees with the rural social reality, to better consolidate the Party's ruling status in the rural areas. In a word, the building of the grass-roots organization of our party must be strengthened, not weakened, only step forwards not backwards. Only when an innovation path suitable for the development of the grass-roots organization of our party is found in the rural societal movement, the Party's ruling role can give full play.

REFERENCES

[1] Commentator of this newspaper. Make Service the Striking Theme of the Grass-roots Organization of CPC [N]. People's Daily, May 29, 2014 (01).

[2] Harold J. Berman. Law and Religion [M]. Translated by Liang Zhiping. Shanghai: Joint Publishing, 1990, p.48.

[3] Lu Fuying. Conflict and Coordination—Game in Rural Governance [M]. Shanghai: Shanghai Jiaotong University Press, 2006, p.63.

[4] Xi Jinping. Further Attaching Great Importance to the Building of the Grass-roots Organization of CPC [EB/OL], February 4, 2013.

Computing, Control, Information and Education Engineering – Liu, Sung & Yao (eds)
© 2015 Taylor & Francis Group, London, ISBN: 978-1-138-02800-5

Research on the strategy of discipline construction for Chinese university

Li Ping Wang
Office of Development Planning and Disciplinary Constructions, Wuhan University, Wuhan, China

ABSTRACT: Considering the condition regarding discipline construction of Chinese university, this paper proposes some measures to enhance the level of discipline construction for Chinese universities, which contains how to refine characteristics and improve the advantages of disciplines; construction of a first-class basic disciplines; strengthen cross-disciplinary; promote the discipline of international construction; innovate discipline construction management and operation mechanism, etc.

KEYWORDS: Chinese university; discipline construction; Strategy.

1 INTRODUCTION

At current stage, most of the Chinese universities focused on building an international, comprehensive and research based university, while discipline construction became the carrier to achieve the overall objective.

Li Yang[1] studied the experience on disciplines building of world-class university, for example, disciplines to be diversified, focusing on cross-disciplinary integration, development of specialized disciplines, building outstanding academic groups. She suggests that we should establish a rational and well-found discipline system, condense disciplinary construction, adhere to some things, and organize outstanding academic communities. Jiajia Wu [2] discussed the definitions and standard condition of research disciplines, and raised relevant thinking: ideological understanding must be unified for creating a research discipline, meantime, it is also essential to build talent plateau, strengthen fierce competition, and strengthen theoretical guidance. Shaohua Guan[3] thinks the development trend of university discipline construction mainly present in three aspects: firstly, discipline direction will still focus on dominant and special disciplines; secondly, academic team construction emerges in the network, matrix state, and the core is to output innovative talents, so that the university will become a hotbed of social think tank and source of innovation; thirdly, construction of discipline base highlights the development trends of connotation, which will become an important platform of innovation for the development of the country. Yonghong Liu[4] proposed discipline construction of colleges and universities should strengthen the teaching staff, create a high level multidisciplinary team and condense discipline direction so as to build a more comprehensive discipline system; in addition, strengthen reform and innovation of teaching, promote cultivation of innovative talents; strengthen scientific and technological innovation to enhance the influence of discipline at home and abroad; Improve the construction of scientific research platform, strengthen technological and scientific innovation conditions of discipline and to enhance the level of discipline.

This article aims to explore discipline construction and development strategies of Chinese university at this stage through reviewing development history of discipline and analyze present situation of the Chinese university.

2 DISCIPLINE CONSTRUCTION HAS ENTERED A NEW STAGE OF DEVELOPMENT

2.1 *World discipline development trend*

The development path of university disciplines was accompanied by the development of higher education. At the first stage, to the 16th century from the medieval universities, subject structure of University is mainly characterized in a single and hierarchical way, which mainly covered the disciplines of theology, law and medicine; During the second phase, from the 16th century to the end of the 18th century, the generalized humanities rose up and gradually replaced the primacy of theology, natural sciences, humanities and social sciences were slowly specialized in universities; At the third stage, from the beginning of the 19th century to the mid-20th century, differentiation of university disciplines was intensified, along with confrontation between humanities and social science and natural science disciplines, meanwhile the discipline type was diversified; In the fourth phase, from the mid-20th century till now, disciplines highly integrated on the basis of highly differentiated, whose structures presents a comprehensive system of trend[5].

2.2 *The construction of a new stage of domestic discipline*

The highly integrated trend of the world-famous university discipline development gives domestic colleges and universities a great revelation. Since the 1990s, 708 colleges and universities were merged into 302 multi-disciplinary or comprehensive colleges and universities, according to the principle of "Building, adjustment, cooperation, and merger". It was considered that the first reason of merging is to improve the benefit scale of running higher education, and the second reason is to create a new pattern of higher education management, which will improve the higher education resources to get a more reasonable configuration and usage [6]. In fact, the deep-seated reason is that discipline construction with kernel of research, education and development of qualified personnel should adapt to the objective laws of development of science and technology and social. At present, the discipline development is not only highly differentiated, but also highly integrated and majored in high integration[7]. The aim of the reform was to change the situation that the discipline development was not adaptable to social progress in science and technology, through the method of changing the irrational discipline distribution in schools, as well as characters of the industry and singularity.

Meanwhile, with key support of "211 Project" and "985 Project", discipline construction of some universities developed rapidly, and in the position of development, building a world-class, internationally renowned high-level research university colleges and universities is the consensus. At the new stage of development in the discipline building, we have visions of building a world-class university; meantime, we must see discipline construction through the international environment, converging with international standards on concepts and methods in order to build competitive and influential disciplines.

3 THE CONSTRUCTION STATUS OF CHINESE UNIVERSITY DISCIPLINES

3.1 *The core competitiveness of disciplines improved, but the gap is still significant*

The "211 Project" is constructed for the development of key disciplines which have a certain foundation, while in the construction of the "985 Project" discipline platforms, it reflects the focusing on the country's relatively scarce educational resources to support part of the construction of the key disciplines in some key universities in order to achieve a breakthrough. It is also focused in the construction process in these schools. However, the highlights of disciplines are still insufficient in the current situation of Chinese university's construction of key disciplines. Taking the ESI subjects ranking as an example, in 2014, Peking University leads the China's universities, which have 19 disciplines ranking into the top 1% of the total of 22 subjects. But there was a great gap compared to foreign prestigious universities on the subject number and the rank order of precedence. The number of disciplines which have strong influences in the international arena is still small in domestic universities.

3.2 *Disciplinary construction internationally yet startling*

Approaching to international development is the trend of China's discipline construction. The elements of an international university should at least include: 1) sponsoring the concept of internationalization, having an international development concept, to establish the internationalization of the world sent training objectives; 2) the internationalization of teachers and management team, one aspect including their own teachers having international communication skills, international impact and international experience, and the other aspect is keeping a rational foreign-teacher ratio; 3) the internationalization of students, including the number of international students, the good international experiences and perspectives of domestic students; 4) the internationalization of research, including having broad international exchanges and research activities and international influential research achievements[8].

At present, being aware of the importance and urgency of the internationalization, many domestic colleges and universities have been taking effective activities, such as changing the concept, establishing development goals, etc., which is still in its infancy. For team building, some schools constantly absorbing talents with overseas prestigious background, which increasing the percentage of foreign teachers, but the number is still small; For student training, the schools enrolled international students who are mainly studying in the areas of social sciences and humanities, while rarely in other disciplines, which reflect the gap of these disciplines in the international arena; For scientific research, there are rarely substantive cooperation between the colleges and universities with the foreign elite and well-known research institutions, along with the shortage of serious research achievements. It was displayed in the 2009 Academic Ranking of World Universities, produced by the higher education Institute of Shanghai Jiaotong University, there were total 18 universities shortlisted in the top 500 for Chinese mainland universities, with the rankings all out of 200. The situation is still very grim [9].

3.3 *The soft environment needs to be improved*

Most Universities have too much departments in China, which hindered the cross-disciplinary integration. For example, there are over 30 departments in Peking University; Zhejiang University has 7 faculties consists of 20 colleges (including 63 lines), 20 college-level lines and a teaching and research department; Shanghai Jiaotong University has 24 colleges (lines); there are 29 departments in Fudan University. Tsinghua University does well at this aspect, whose colleges are less than 15. On the other hand, using the foreign famous universities as examples, Harvard University has 15 Colleges, Stanford University is located only seven colleges, which is in sharp contrast with the domestic universities. Separated from the original close contact with subjects, with the development bottleneck, there are a variety of resistances when toward the new fusion. Due to the limitations of existing disciplines, it is hard to establish efficient operation mechanism and the interdisciplinary research progress has been slow, with the reaction lagged in strategic disciplines construction. In addition, compared to the foreign tradition of academic freedom, the good disciplinary cultural environment has not yet formed, which is not conducive to the development of the discipline innovation.

4 THE STRATEGY OF DISCIPLINE CONSTRUCTION FOR CHINESE UNIVERSITY

4.1 *Refining characteristics and improving the advantages of disciplines*

The famous universities of the world adhere to the guiding ideology of keeping focused and formatting special disciplines, so come with lively scenes of standing out a lot of personality and having their own strengths. For examples, University of Oxford is known to theology, classical literature; Cambridge University, prominent in physics and biology; Harvard University is famous for its business management, political science; while Stanford University is known for psychology, electrical engineering; the economics and linguistics of Massachusetts Institute of Technology are best; the California Institute' Aeronautics and Applied Physics are strongest; Cornell University dominates in agricultural science, hotel management [10].

Therefore, a university must have an elite minority discipline firstly. They should focus on the development of the advantages of discipline in schools, highlight the characteristics, give great importance to the construction of key disciplines from a strategic perspective and give policy supports. For example, the University of California at Berkeley, with original 14 colleges, whose development slogan is "for each of these areas to maintain the nation's top three", had to adjust the development strategy to concentrate and focus on the development of biological atomic engineering and other disciplines linked with the biological atomic research from different angles when it found the early strategy is not suitable finally. Ultimately, there came a total of 17 Nobel Prize winners in the field of the University of California at Berkeley, and the discipline has become one of the world's most famous subjects. Domestic colleges and universities can also focus on planning, highlighting advantages of features disciplines, support a small amount of these disciplines to convergent teams, highlight the degree of achievements and expand the international influence through policy guidance, special project supporting and so on.

4.2 *Construction of a first-class basic discipline*

The setting of colleges varies during famous foreign universities. Maybe there is not an institute of Technology, or some universities don't set up a business school or medical school, however, they all set up ARTS. The basic discipline is the basis for the development of various disciplines, a lot of significant original scientific research achievements come from the basic discipline area. Without the strong support of the basic disciplines, applied sciences will be A river without water, a forest without trees. However, due to the long investment cycle for the basic disciplines and construction efficiency is not obvious in the short term, lading a substantial market effect like the applied sciences in the social development created, the basic disciplines funding is obviously insufficient and treatment of teachers is bleak in many schools, so comes the bad condition of the development of the basic disciplines.

Domestic colleges and universities should attach great importance to this phenomenon, increase investment in basic disciplines, rational handle on the relationship between basic disciplines and other disciplines, and stabilize and improve the strength and level of it. Here is a good example. Attaching great importance to the construction of basic disciplines, Wuhan University started to take "revitalization plan of action to develop the basic disciplines" in 2005, which is focusing on improving the disciplines of literature, history, philosophy, mathematics, physics and chemistry. When the second round of disciplinary assessment (2007-2009) results announced by Academic Degrees and Graduate development Center, Ministry of Education of China, the ranking of the basic disciplines of Wuhan University significantly increased, thanks to the effectiveness of the building.

4.3 *Strengthen the cross-disciplinary*

To strengthen the cross-disciplinary construction is not only the nature of the development of the discipline demands, but also the real needs of the community. Considering a major national and regional strategic needs and aiming at the forefront of economic, social and technological development, we can plan some platforms for a number of subjects in key research areas, which will promote cross-integration of Humanities & Social Sciences and the Natural Science, coming with the inter-new discipline growth point. It will help to format advantages of integration, and the distinctive characteristics of discipline groups.

4.4 *Promoting the discipline of international construction*

Internationalization of teams. Firstly, it is necessary to Improve the proportion of overseas elite, Ph.D., the world's top 100 Ph.D., the overseas prestigious tenured teachers, and the foreign teachers in the whole school teacher; Secondly, we should select outstanding and potential young teachers abroad for training, to do postdoctoral research, and to develop an international perspective learn from foreign advanced research methods; Lastly, it will substantially improve the international level of the teacher through training their ability of team-oriented international academic exchange and cooperation.

Internationalization of talent cultivation. The quality and level of Talent Cultivation is an important aspect to reflect the effectiveness of discipline construction. The first-class discipline should train the talents of world- oriented and international competitive. It will lay the foundation for talent cultivation by expanding the proportion of international students, setting international curriculum standards, further improving the teachers' level of international channels and so on. Undergraduate Training should strengthen general education, advocate of innovative teaching, guide the research study and implement of open education. Graduate education need to update the concepts, promote in training methods with international standards and establish sound-school graduate training system; Furthermore, we should focus on strengthening the innovation capacity, independent research ability of the graduate students to enhance their level of international competitiveness.

Internationalization of Scientific research. To improve the Internationalization of Scientific research, we can take the following measures. Firstly, Universities should strongly support the disciplinary unit to organize, contract or participate in high-level international academic conferences; Secondly, to establish foreign research basing on recruiting high-level persons is very important; Thirdly, we can cooperate with relevant world-class disciplines in talent cultivation and researching on cutting-edge issues, establishing a joint research team and sharing research progress and results. At the same time, under the guidance of "going out" strategy, it is necessary to strengthen promotion of Chinese culture and release the introduction of scientific research with Chinese characteristics.

4.5 *Innovating discipline construction management and operation mechanism*

We need to develop disciplines construction assessment system. Firstly, It is necessary to take scientific assessment of the construction of various disciplines through selecting reasonable, representative and operability assessment indicators and determining the difference of weights and algorithms for the differences in disciplinary culture and disciplinary paradigm; Secondly, we can enhance the assessment to the performance of the person who is in charge of the disciplines; Lastly, to establish investment mechanisms of special investment, matching investment, investment associated with the performances and so on.

It is important to keep the concept of "Respecting for the scholar and advocating academic", which will drive us to construct the disciplines people-oriented and create a harmonious discipline culture.

REFERENCES

[1] Li Yang. Experience in building world-class university discipline and Enlightenment[J], LITERATURE LIFE,2011(9):242–242,248 (In Chinese).

[2] Jiajia Wu. The construction and thinking of research disciplines[J]. Chines Hospitals, 2011, 15(8):8–9 (In Chinese).

[3] Shaohua Guan. The development trend of discipline construction in our universities[J]. Jiangsu Higher Education,2011(5):36–38 (In Chinese).

[4] Yonghong Liu. The idea and practice for the enhancing of the level of discipline construction[J]. China Electric Power Education,2011(26):16–17 (In Chinese).

[5] Qingshan Pang.The discussion of the university disciplines[M]. Guangdong Education Publishing House, 1st ed., 2006:25–26 (In Chinese).
[6] Deguang Yang. Higher Education Management[M]. Shanghai Education Press,2006:337–337 (In Chinese).
[7] Yongbo Cheng, Yun Luo. Enlightenment and Reference: the research on discipline building practice of foreign famous university [J]. Heilongjiang Researches on Higher Education,2007(3):35–37 (In Chinese).
[8] Zhong Tang. A little understanding of the international and the international influence. http://hbgxyz.yangtzeu.edu.cn/article.asp?articleid=56.
[9] Academic Ranking of World Universities – 2009. http://www.arwu.org/ARWU2009.jsp.
[10] Yun Luo. China's key universities and discipline construction[M]. China Social Sciences Press, 1st ed.,2005:60–65 (In Chinese).

Computing, Control, Information and Education Engineering – Liu, Sung & Yao (eds)
© 2015 Taylor & Francis Group, London, ISBN: 978-1-138-02800-5

Analysis of market discipline in banking supervision

N. Ye
Department of Finance and Economics, Xinyang College of Agriculture and Forestry, Xinyang, China

ABSTRACT: Under the background of international financial crisis, financial supervision will become the popular topic on the basis of financial stability. Because there has continuous and invisible pressure of supervision, more and more countries will focus on the issue of market discipline. This text focuses on a series of issues in market discipline. It described the concept, meaning and role of market discipline, especially paid attention to the limitations of market discipline. Finally, it analyzed the effective market discipline in China's banking supervision, showed that it should strengthen the awareness of recognition and precaution risks, improve the content of information disclosure, and adjust the system of banking supervision in order to promote the development of our banking.

KEYWORDS: market discipline; banking supervision; the New Basel Capital Accord.

1 INTRODUCTION

In recent two decades years all over the world, there occurs a series of banking crises, which has significantly and systematically impact on the financial system and the real economy as a whole. In order to prevent the banking crises occur again and reduce the negative influence on the real economy, market discipline is playing an important role in the banking supervision.

The outbreak of recent financial and economic crisis has brought the serious condition of bank regulation and supervision cannot successfully manage bank behavior, so market discipline views as the "pillar" in the prudential regulatory architecture, is attracted by policy-makers and expected to supplement bank regulations (Basel Committee on Banking Supervision, 1999). Basel II Capital Accord at that time prevents the failure in capital adequacy, but it only could handle with partial problems.

Furthermore, from the Pillar 3 of the proposed, it revises and improves the capital framework (Basel II in 2004) for credit risk-capital requirements, bank supervision, and market discipline, which supports the greater bank disclosure requirement to enhance market discipline. This structure, as shown below in Figure 1, it reflects the main content of the New Basel Capital Accord consists of three pillars: minimum capital requirements, supervisory review process and market discipline. (Benink, H. & Wihlborg, C. 2002)

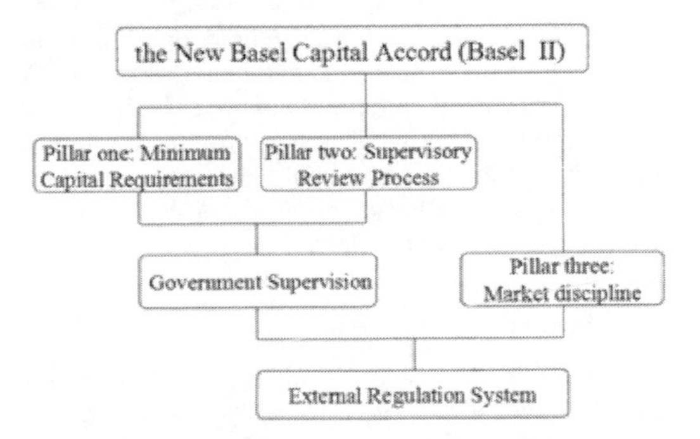

Figure 1. The structure of Basel II framework consists of three main pillars.

2 THE CONCEPT OF MARKET DISCIPLINE

Market discipline has a wide term of supervising and monitoring banks, many articles on bank regulation comments "market discipline" with respect to bank behavior. As discussed in detail by Covitz, D.M. et al. (2004), subordinated debt investors become more sensitive to default risks without protecting large bank creditors. Thus, market discipline becomes an important role to allocate funds reasonably and control risks.

Market discipline can be described that the behavior of bank creditors to deal with risk and then the response of bank to the behavior of their creditors. Market discipline in the banking focuses on the behavior of financial institutions, which is including depositors, creditors, and stockholders face costs that are disciplining excessive risk-taking behavior undertaken by banks. It means that market discipline works with regulatory systems by market participants to make sure the banking sector operate the business in the sound and a safe financial market.

3 THE MEANING OF MARKET DISCIPLINE IN BANKING

The effective market discipline depends on sufficient information by market participants, moderate rewards and punishment of management, and arrangement with investors understood relative decisions. Market discipline should have two major conditions about the meaning of the market discipline with the behavior of risk-taking in banking.

3.1 *Full consideration of the funding cost*

Market discipline requires the behavior of risk-taking financial institutions need to face the cost of funding. In other words, market discipline can be revealed the pricing of default risk. For shareholders of corporations, they provide incentives to transfer risks to bank creditors. For bank creditors, they face the bank risk with incentives, but creditors cannot be fully monitored the default risks. Market discipline operates the non-insured liabilities, so that it is able to ensure banks provide the non-insured liabilities.

3.2 *Healthy governance structure*

Market discipline should consider issues with governance. Financial institutions must provide the correct and sufficient information, which come from different channels of the costs of funding. Under the situation of the market economy operation, the financial management selects the asset and liability positions to make sure shareholder to achieve the maximal value. Moreover, market discipline framework need consider fully legitimate rights of the stakeholders, and gain greater understanding and cooperation by creating opportunity and maintaining financial management system.

4 THE ROLE OF MARKET DISCIPLINE

4.1 *Expand the bank's risk assessment and management*

Under the sufficient and effective information disclosure and transparency of banks, market discipline can expand the power controlling the bank's risk assessment and management.

A lot of people confuse the issues with too many complicated risks, which require information and knowledge, efficiency of financial markets with each asset or costly information. On the contrary, it is not necessary to evaluate the risks. As Mayes, D. G. & Llewellyn, D. T. (2003) emphasize, there may exist sufficient incentives in the bank to be monitored by financial markets. Sufficient incentives provide the relevant and available information for financial institutions to reduce the cost of capital by revealing positive information.

Under the proposed New Accord, market discipline is as a financial instrument for banks to hold sufficient capital. Market discipline reflects the bank provides the incentives to market participants to allocate the cost of funding to the bank creditors with the bank credit risk evaluation from the shareholders and debt holders. That is to say, the central bank and regulators build up the market discipline, in order to make strong concern many incentives for the action of excessive risk-taking. For example, there occurs the financial crisis because of underestimation of risk in Scandinavian countries. It shows that there is no efficient market discipline to monitor this kind of problems. Market discipline will evaluate the credit risk from the monitoring of a larger number of observers in banks.

4.2 *Reduce losses caused by closing down insolvent banks*

In the financial institutions, debt instruments have less sensitivity to insolvency risk than in other industries when creditors without protection against losses. According to this, the implement of market discipline such as the bank resolution regime becomes the significant role of predictable rules for closing down insolvent banks.

The UK banking authority designs a new comprehensive framework rule of bank resolution for a restructured and insolvent banks. Because insolvent banks will cause to the whole financial crisis, the bank resolution regime requires the authority regulators and system to adjust the bank restructure and operation when its failure threatens the stability of the financial system.

It is important to set up bank resolution that balances the rights of shareholders rather than crisis management base on the credit crisis from 2007 to 2009. Bank resolution regime protects rights of shareholders and creditors; it also provides the effective operation and regulation in the economy.

4.3 *Regulate the bank behaviors to keep the safety and soundness of banks*

All of market participants have the well-knowledge and expertise to conduct the monitoring in banks that it eliminates the inherent danger, which conducts by the behavior of supervisors to make the mistake results with ineffective and incomplete information.

Increasing the powerful role of market discipline can be achieved the results of the safety and soundness of banks through restricting the danger of official forbearance. It turns out that it would ensure the most effective and sufficient control over the bank's behavior by market participants than monopolist's judgment.

5 THE WEAKNESS OF MARKET DISCIPLINE

5.1 *The policy of "too big to fail" banks*

Market discipline is weak for the policy "too big to fail" banks to evaluate how explicit and implicit insurance of banks' creditors with risk-taking action.

Most large banks are rescued by the government to compete with the role of market discipline. Similarly, these large banks can be judged to be "too big to fail", it reduces the incentives for effective market disciplining. Supervisors have to concentrate on the cost of smaller banks to against the runs and failures of banking. The results from Sironi, A. (2003) uses a large amount of evidence to indicate that subordinated debt yields of "too big to fail" banks have, the less sensitive to risk proxies than the yields of other banks. Otherwise, different interest rates on the loan level between the large and small banks demonstrate that the weakness of market discipline must be causing the financial crisis.

5.2 *Without reflecting available information*

Another reason of market discipline in weakness could be caused by no reflecting available information about the value of assets and resource of funding. Supervisors and managers cannot take action in time to consider the market price relates to the cost of funding can provide early warnings about coming distress of the individual. As noted by Angkinand, A. et al. (2011), the main indicators of individual banks in distress must have the timeliness of equity prices, subordinated debt yields and credit default. In fact, because the market prices do not reflect timely and correct information, market discipline can't take measures to deal with large financial institutions in distress. Obviously, market discipline does not imply well foresight in advance the financial crisis.

5.3 *Lack of disclosure information and transparency*

The discipline of interbank market is weak in the crisis countries, because the lack of accurate and timely disclosure information and transparency. And then the quality information about accountancy data from bank accounts will not be believed. In fact, many procedures of accounting and auditing data could not reflect the truth; some cases can be deliberately supplied with wrong information. When the transparency and sufficient information is at the low level, the greater the risks and uncertainty cannot be exactly judge by market participants and government. In a word, disclosure requirements need to be an essential part of the market discipline.

5.4 *Less power for small regional banks*

Market discipline only concentrates on publicly traded securities and debt issued by banks, which it offers only works with the available market prices. Therefore, market discipline has less power for small regional banks without securities and debts. If debt issues are not significant, rating agencies are unable to conduct a full credit rating on a bank's subordinated debt. It will be occurred the potential consequence to undermine the effective market stability.

6 THE EFFECT OF MARKET DISCIPLINE IN CHINA'S BANKING SUPERVISION

In China, no matter regulation theory or practice with this principle, it still emphasizes the analysis of government supervision in external regulation system for a long time. Obviously, there is little research on the market discipline, which need full study and deserve extra attention.

Along with China joining WTO and the deepening of financial liberalization, the domestic banking sector is facing further fiercer market competition than before. It's fortunate that Chinese Vice Premier of the State Council Wang Qishan attended the Lujiazui Forum in 2008, said that the stability and safety of the financial system will be the top consideration, strengthens the social financial knowledge and risk education, focuses on the area of system of financial regulation. With the idea of the New Basel Capital Accord, the significance and limitation of market discipline and the present development situation of the Chinese banking sector, there are three broad aspects with market discipline to pay more attention.

6.1 *Strengthen the awareness of recognition and precaution risks*

As described by Morgan, D. P. & Stiroh K. J. (2001), effective market discipline requires investors to recognize some relative risks and watch for financial risks in time. It means that all kinds of risks are the important parts for market discipline. Market investors can improve the capability of sense in the variation of banks through increasing transparency. Moreover, another approach to improve recognition risks is establishing a new series of incentive systems for investors. For the precaution, investors can take some actions such as rising financing cost and reducing investment to decrease financial risks.

6.2 *Improve the content of information disclosure*

Information disclosure is absolutely central to well informed investment decisions. However, one of the biggest problems of market discipline has been the lack of good information disclosure system even though most people understand the meaning of it. For improved information disclosure in banks, there has three parts need to be considered: risk information, governance information and financial information. Commercial banks need to analyze risk information to establish comprehensive risk management. It must clear the duty of directors and the board of directors, and not only that, it also puts a high value on accounting statement and information out of the form, so that it can supplement financial report information with them.

6.3 *Adjust system of banking supervision*

At the situation of overemphasis on official supervision but little on market discipline, it must be changed by some measures. In China, the central bank as a leader in banking sector with much more power and responsibilities than others. Under the market discipline, it makes the role of the central bank evolve and change through time. The central bank has full control of the information disclosure and approval process of the banks for a long time. Nowadays, it must develop the diversified subjects of financial supervision to break the central bank's information monopoly, to support the good management environment and to help investors to regulate the market by achieving timely and accurate information.

7 CONCLUSION

Even there are some problems in market discipline, it is still meaningful to monitor the whole market.

When the pricing of bank securities reflects the bank's true potential risks, market discipline takes measures to limit bank risk-taking behavior directly. Market and supervisory discipline act as the substitution and complement, which shows that appropriate regulations to improve the development of market discipline and enhance the effective supervisory discipline to supervise the financial transactions.

With the new change in the economic situation, bank regulations and supervisions require market participants to provide the effective market discipline for the banking behavior. Obviously, in order to become the effective market discipline, investors should achieve the timely and accurate information from the highest level of disclosure information and transparency in the organization. Disclosure information and transparency requirements are as the key factor to keep the financial market efficiency, which it strengthens market integrity and economic performance.

REFERENCES

Angkinand, A. et al. 2011. "Market discipline for financial institutions and markets for information", in Barth, J., Chen, L. and Wihlborg, C. (Eds), *Handbook on Research in Banking and Governance*, Edward Elgar, Cheltenham.

Basel Committee on Banking Supervision, 1999. "New Capital Adequacy Framework." Basel,Switzerland, June.

Benink, H. & Wihlborg, C. 2002. The new basel capital accord : making it effective with stronger market discipline, *European Financial Management*, 8(1): 103–115.

Covitz, D.M. et al. 2004. Market discipline in banking reconsidered: The roles of funding managers decisions and deposit insurance reform. *Board of Governors of the Federal Reserve System Financial and Economics Discussion Series*: No. 53.

Mayes, D. G., & Llewellyn, D. T. 2003. The role of market discipline in handling problem banks. *Bank of Finland Discussion* Paper No. 21.

Morgan, D. P. & Stiroh, K. J. 2001. "Bond Market Discipline of Bank: Is the Market Tough Enough?" *Federal Reserve Bank of New York*, Working Paper.

Sironi, A. 2003. "Testing for Market Discipline in the European Banking Industry: Evidence from Subordinated Debt Issues." *Journal of Money, Credit, and Banking*: 35, 443–472.

Computing, Control, Information and Education Engineering – Liu, Sung & Yao (eds)
© 2015 Taylor & Francis Group, London, ISBN: 978-1-138-02800-5

Review of transmission line project cost based on probability analysis

Dong Xiao Niu & Bing Jie Li
School of Business and Management, North China Electric Power University, Beijing, China

ABSTRACT: Characteristics of transmission line projects result in a higher rate of its cost over budget rate, so it's important to review its cost. Firstly, the cost of transmission line project should be reconstructed. Then, use the method of probability analysis to establish the risk index system of the review of the representative project. Finally, according to the distribution of the risk index value of the target project to review the target project, and do a case analysis of the Ningbo 110kV long-distance overhead line project.

KEYWORDS: Probability Analysis; Transmission Line Project; Cost Review.

1 INTRODUCTION

The transmission line project has the features of large investment, long duration, geographical span, complex construction environment, unpredictable factors and so on, which can easily lead to greatly enhancing of the transmission line project cost, thus affecting the financial plan, raising investment cost shares, and leading to over budget situation. Usually conventional transmission line project cost review is not conducive to the analysis of the project cost, and to control project cost risks, and therefore it needs to be further refined.

2 ANALYSIS OF RISK FACTORS OF TRANSMISSION LINE PROJECT COST

2.1 *Re-constitution of transmission line project cost*

According to the cost structure of cost analysis data table, the risk factors of transmission line project cost will be re-constituted, and the reconstruction result is as follows: body costs, ancillary facilities costs, preparation of the spread, other expenses (other costs excluding the following costs), construction site acquisition and compensation costs, project preparatory work costs, environmental monitoring and inspection costs, water and soil conservation project acceptance and compensation, pile testing costs, ancillary construction costs, road construction costs, production preparation costs and basic reserve fund.

2.2 *Risk index system of transmission line project cost intelligent review*

A different type of construction and cost constitution, and the corresponding risk review index system is also different. So we should establish a category of risk review index system based on different project types. It's shown here the intelligence review risk index of classic project category –110kV long-distance overhead line construction.

Through the costs analysis of 110kV long-distance overhead line projects in Zhejiang Province between the years of 2006-2012, the contribution of the subjects constituted the project cost and the possibility of the occurrence of over budget can be drawn, as shown in Table 1.

As can be seen from Table 1, the degree of influence of various factors on the 110kV long-distance overhead line projects is also different. Among them, the environmental monitoring and inspection costs, water and soil conservation project acceptance and compensation, pile testing costs and basic reserve fund are not only the contribution rate of the cost is low, and the probability of the occurrence of the over budget cost is nearly zero, so it's sure they are not the main risk factors of line project cost and they will be removed in the below analysis. While the other seven factors are identified as the main factors of the cost risk index system, and compare their corresponding data of the proposed budget and final accounts. As a result, over budget risks faced by the construction management units can be reflected. A risk index system of 110kV long-distance overhead line project is shown in Figure 1.

Table 1. The cost analysis of 110kV long-distance overhead line projects.

Name of pro cost	The contribution rate of cost estimates/%	The contribution rate of final accounts/%	Over budget rate/%
body costs	77.32	78.72	25.00
AFC	0.17	0.12	5.00
other costs	11.38	10.50	22.50
CSACC	6.76	9.20	45.00
PPWC	0.53	0.38	2.50
EMIC	0.01	0.00	0.00
WSCPAC	0.01	0.01	0.00
pile testing costs	0.00	0.00	0.00
ACC	1.62	1.04	7.50
RCC	0.01	0.03	5.00
basic reserve fund	2.62	0.00	0.00
total	100	100	

*AFC: ancillary facilities costs; CSACC: construction site acquisition and compensation costs; PPWC: project preparatory work costs; EMIC: environmental monitoring and inspection costs; WSCPAC: water and soil conservation project acceptance and compensation; ACC: ancillary construction costs; RCC: road construction costs

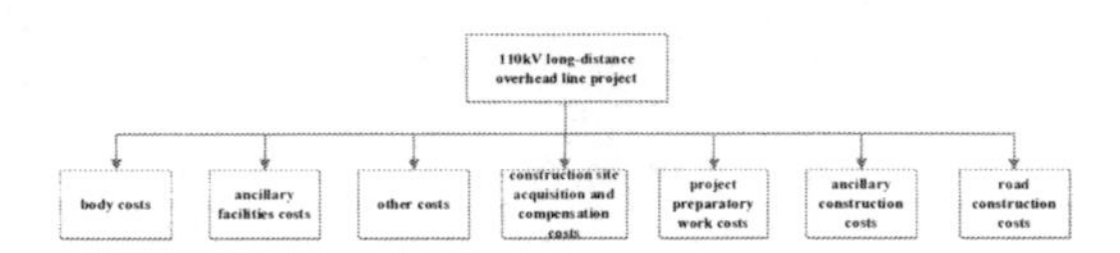

Figure 1. Risk index system of 110kV long-distance overhead line project.

3 COST REVIEW TECHNIQUE OF TRANSMISSION LINE PROJECT

3.1 *The roadmap of cost review technique of transmission line project*

The roadmap of cost review technique of transmission line project studied in this paper is as shown in Figure 2.

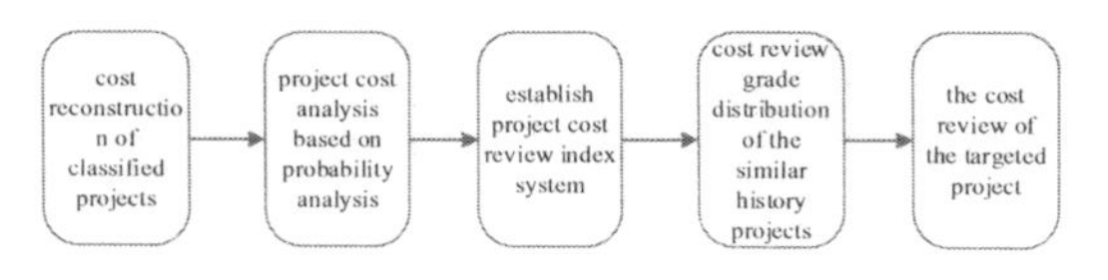

Figure 2. The roadmap of cost review technique of transmission line project.

3.2 *The cost review sample of 110kV line project of XiFeng-LanJiang in Ningbo*

First, according to the cost analysis data of Zhejiang Province between the years of 2006-2012, screened a total of 39 projects which are 110kV long-distance overhead lines. Then, selected the corresponding data with risk indicators from the estimates data of the 39 similar projects. Last, draw the cost review grade distribution figure as shown in Figure 3.

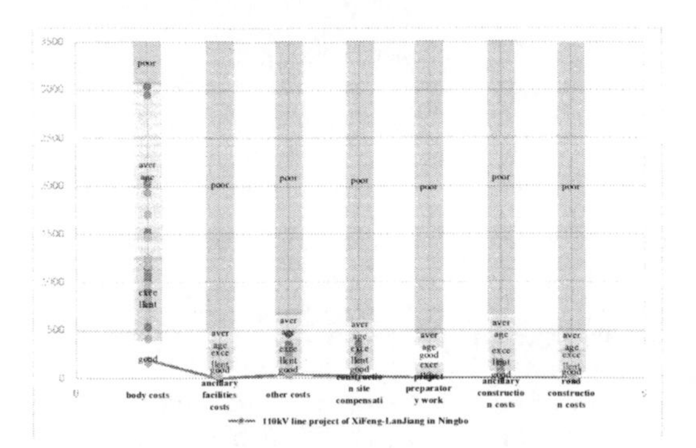

Figure 3. The cost review grade distribution figure of 110kV line project of XiFeng-LanJiang in Ningbo.

Similarly, according to the scatter distribution of the seven risk indexes of 39 projects, the level and of risk costs of the historical projects can be seen clearly. Paint the seven risk index value of the grade distribution figure, the cost review results can be got. The review of the project cost is good, but the review of project preparatory work cost is excellent, and the other individual cost review results are good.

4 SUMMARY

Firstly, this paper teased the cost of transmission line projects and reconstructed their cost structure. Then, using the method of probability analysis, established the cost review risk index system 110kV long-distance overhead line project. Last, according to the distribution of the risk index value of the target project, review the target works. Not only the review results of overall cost level can be obtained, but also you can get the grade evaluation of individual risk costs.

REFERENCES

[1] Wang Mianbin,Huang Yichen,Zhang Jie,Geng Pengyun. Cost Risk Evaluating Model of Power Transmission Project Based on Probability Analysis Method[J].Power System Technology,2011,11(35):26–31.

[2] Wang Mianbin, Zhang Jie. Cost Risk Evaluation Model of Power Transmission & Transformation.
[3] Project Based on Bill-of-Quantity Model[J].Power Construction, 2012,12(33):91–96.
[4] Yu Zheng,Wang Yingying. Construction of Audit Model of the Risk-Based Heat Map[J].Practise,2011,08:48–51.
[5] Yin Tao. Research On The Method About Project Budgetary Estimation And Examination [D]. College of Electrical Engineering Chongqing University,2006.

Computing, Control, Information and Education Engineering – Liu, Sung & Yao (eds)
 ISBN: 978-1-138-02800-5

Mobile context-aware recommender system

Jia Ke
Huaxia College, Wuhan University Of Technology, Wuhan, China

Zhong Wu
Physical Education and Equestrian School, Wuhan Business University, Wuhan, China

ABSTRACT: Mobile context-aware recommender system is a useful tool to solve information overload problem. Improvement of accuracy and user satisfaction in personalized service by utilizing contextual information has become one of the hottest researches in the field. This article presents an overview of the field of mobile context-aware recommender system, introduce a system framework. The future prospect and development are also discussed.

KEYWORDS: mobile Internet, Context-aware Recommender System.

1 INTRODUCTION

With the development of mobile devices, sensors, the Internet of Things, wearable devices, cloud computing, more and more people are accustomed to use smart portable devices in the mobile Internet when they have questions or in free time. Users want to get the useful information (service) quickly at any time, any place through a wireless network. If the product (service) can provide users with accurate content, predict user's preferences, Internet company can increase consumers, improve the consumer satisfaction.

2 RECOMMENDER SYSTEMS

Recommender system is a software tool and technique providing useful suggestions and items (information, commodity, service) for users. It solves the "information overload" effectively some extent. Recommender systems emerged as an independent research area in the mid-1990s[1]. The research on recommender systems is dramatically increasing, it has been widely used in web site, e-commerce, seeking film/music, ordering food, tourism, e-health and so on, especially location-based service (LBS).

Recommender system and search engine have implemented in common. When a user needs specific and clear, he search; when a user demand is not clear or it is difficult to express, he needs a recommendation. On the other hand, when the user need to find popular (hot) content, he search, when a user need to find personalized content, he needs a recommendation. Many scenarios, the user's individualized demands are difficult to express short explicit query, for example 'today at noon I want to find a nearby, taste is conformable, consumption is not expensive restaurants', this demand is common but it is hard to use the query words to express clearly. Recommender system can fill the gap, according to user's behavior history dig out personalized demand, realize it real-time.

3 MOBILE CONTEXT-AWARE RECOMMENDER SYSTEMS

Mobile information system has three characters: user's mobility, device portability and wireless connectivity. [2] E.g. If a person wants to travel, the first thing is determined where to go, whether he has a companion (family or alone), the destination. After he arrives, he may use a mobile device to find optimal route, attractions, nearby hotels, restaurants, how far the attractions, how to get there, whether it will rain next several days, he hopes mobile device can provide relevant recommendations base on specific contextual situation.

Mobile context-aware recommender system has some characters different from the traditional recommender system:

The limitation of hardware: the limits of mobile device (limited screen size, limited storage, limited battery consumption, interactive way, user interface), the limitation of wireless networks.

User's behavior and preference: young users like to get information (service) as long as they have free time or piece time, they use the mobile Internet every day many times, they like new and fun things. If the product (service, APPS) can't attract users in a moment, it means lose the user.

The mobile recommender system also has two peculiar characteristics: The first exclusive property is "location-aware"(the user's physical position at a particular time). The second exclusive property is "ubiquity"(the ability to deliver the information and service to mobile users wherever they are, and whatever they need). In a mobile environment, location, target, time, mood and other contextual factors influence users' decision, so many experts provide a notion of "context-aware"[2].

3.1 *Definitions of context*

Bazire and Brezillon [3] examined and compared 150 different definitions of context , finded it difficult to unify definition.

Dourish[4] distinguished between two different views of context: the representational view and the interactional view. The representational view means context can be described with observable attributes, the structure of these contextual attributes does not change over time. The interactional view means that contextual features are dynamic, this view assumes a cyclical relationship between context and activity, where the activity gives rise to context and the context influences activity.

Gediminas Adomavicius [5,6] sorted four types of context that in a mobile environment: physical context, social context, interaction media context, and modal context.

Physical context: representing the time, position, and activity of the user, but also the weather, light, and temperature when the recommendation is supposed to be used.

Social context: representing the presence and role of other people (either using or not using the application) around the user, and whether the user is alone or in a group when using the application.

Interaction media context: describing the device used to access the system (for example, a mobile phone) as well as the type of media that are browsed and personalized. The latter can be ordinary text, music, images, movies, or queries made to the recommender system.

Modal context: representing the current state of mind of the user, the user's goals, mood, experience, and cognitive capabilities.

3.2 *Context-aware data resource*

Mobile context information can be obtained through a variety of ways, such as user location by GPS, mobile base station location. Time information can be provided by the portable terminal; the weather, the temperature information available in the database via the Internet; user behavior information by the user's session or log.

3.3 *The applications of context information*

Location context

Mobile users in different environments, demand will vary. There are a large number of location-based service in the business.

"Uber" (a APP of mobile phone) locate user position, compute the distance between the consumer and taxi driver, recommend a several nearest driver to the consumer, it also provide price, performance of the vehicle, gender of the driver to consumer, user can choose which taxi they like.

Girardello [7] used the consumer's current location, analysis the location near the usage of mobile applications, recommend the most frequent applications to the user, only consider the location context and the usage of the application, not consider mobile user preferences affect applications choice. Yang [8] analyzed the statistics of the merchants of mobile users visited web pages, get the mobile user preference characteristic vector, using a cosine similarity measure the similarity of user description file with merchants page, at the same time, consider the location of the distance between mobile users and merchants, far from the merchants, mobile user preferences will reduce. Setten[9] recommendation system for mobile users could input the rules (for example based on location rather than the price, recommend the cafe), users could see the recommended context factors.

Table 1. The applications of mobile context-aware recommender system.

Users scenario	location	time	User preference	Ranking(Top-n)
Tourism	√	√	real-time, accuracy	
News	√	√	real-time	√
Electronic commerce	√	√	accuracy ,diversity	√
music		√	diversity	√
movie		√	diversity	√
advertisement			accuracy	
SNS			real-time	

Time context

Zheng [10] used naive Bayesian classification methods to classify context time dividing into working days and weekends, computed items (APPS) belong to different categories of probability, considered mobile user preference in different time. Generally, we set a time attenuation factor in the popular(hot) items.

4 THE PROCESS OF MOBILE RECOMMENDER SYSTEM

We first collect user's data, parse them, set up the model. Figure 1 demonstrates the process of the mobile recommender system.

Item profile: Item profile includes some attributes (features) of the item: for example, a film attributes (features) include the director, actors, release time and type.

User profile: A user profile is a representation of information about an individual user that is essential for the (intelligent) application we are considering. Generally it includes:

User demographics: user's age, gender, profession, educational background, residence, etc. The information obtained from the user registration information commonly;

User interests and preferences: sport, technology, read, movie, music, game, tourism;

User behavior: it includes explicit feedback and implicit feedback. Explicit feedback includes user rating, buying, comment, download, explicit feedback includes viewing, click, staying time, closing, user's session, user's log.

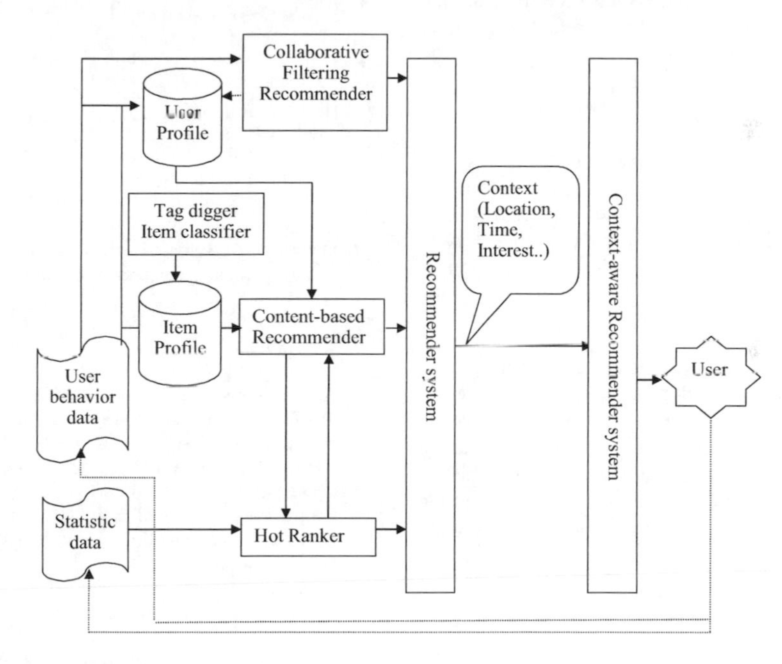

Figure 1. The process of mobile recommender system.

5 RECOMMENDATION APPROACH

Recommendation approach (algorithm) has been classified several categories: collaborative filtering, content-based, knowledge-based and hybrid approach.

A collaborative filtering method based on a set of items in the same users, it is according to the user has a similar preference with neighbor users. The principle of this method is a two-dimensional matrix. This method influenced by TOP-N (Hot Ranker) easily. A new user or low active users will encounter a cold start problem. This method first used by Amazon, which is based on the statistical data, the result can't see the link between the items. So now Amazon use content -based recommendation.

Content-based recommendation is the most intuitive recommendation algorithm, this method is simple, has no cold start problems. The algorithm depends on the depth of content analysis, often limited to text, images, or audio and video.

The knowledge-based recommendation can be seen as a kind of reasoning (inference) technology in a way, it is not recommended based on user needs and preferences. The method based on knowledge due to their different classification item.

The hybrid approach combines content-based, collaborative filtering, and knowledge-based technique in many different ways.

6 FUTURE WORK

6.1 *Data collection*

The research on how to collect user's data of psychological cognition is limited. Existing research mainly access the user's external data (such as time, place, weather, etc.). User's cognition to choose a personalized service is an important influencing factor [11]. Different customer has different interests, cultural value. Get the user cognitive context data can improve customer satisfaction. The existing literatures are not enough to express the cognitive context data, more work needs to be carried out.

6.2 *Data parsing*

Basic data sets have multiple format, types and expression, mobile data acquisition, abstract, parsing need to be finished in a short time. How to obtain the largest amount of historical data storage, how to explore the user's preference are also need to research.

6.3 *User interface*

Recommendation system demand real-time and accuracy, particularly dependent on human-computer interaction, the traditional recommender system users are concerned about the price, service quality, the accuracy of the content and diversity, but mobile recommendation system, user care about the mobile user interface(UI), how to design more better UI is a interesting research.

6.4 *Evaluation of recommender system*

Evaluation of recommender system is necessary in a mobile environment, the needs of users timeliness and easily influenced by the surrounding environment, the existing evaluation index system is relatively single, definition of mobile situation, approval, evaluation and analysis did not perfect. Diversity, real-time and robust characteristics considered inadequate. It is necessary to set up a suitable evaluation system for the mobile recommender system.

7 CONCLUSIONS

The mobile recommender system is the human decision-making process in essence. Accurate data sets are the most important thing to recommender system. Data includes user's positive feedback and negative feedback, negative feedback data also useful to the algorithm. A good mobile recommender system covers data mining, machine learning, algorithm, psychology, sociology and so on. Choose which algorithm depend on scene and business objectives.

ACKNOWLEDGEMENT

Supported by the Humanity and Social Science Youth foundation of the Ministry of Education of China.

REFERENCES

[1] Francesco Ricci and Lior Rokach and Bracha Shapira, Introduction to Recommender Systems Handbook, Recommender Systems Handbook, Springer, 2011, pp. 1–35.

[2] Francesco Ricci," Mobile Recommender Systems", P.3,January 11, 2010.

[3] Bazire, M., and Brezillon, P. "Understanding Context Before Using It". In Proceedings of the 5th International Conference on Modeling and Using Context, Lecture Notes in Artiicial Intelligence, ed. A. Dey, B. Kokinov, D. Leake,and R. Turner, 113–192. Berlin: Springer,2005.

[4] Dourish, P. "What We Talk About When We Talk About Context". Personal and Ubiquitous Computing 8(1):19–30,2004.

[5] [5]Gediminas Adomavicius, Bamshad Mobasher," Context-Aware Recommender Systems", Artificial ntelligence, p.74,fall 2011.

[6] Adomavicius G, Sankaranarayanan R, Sen S, Tuzhilin A. Incorporating Contextual Information in Recommender Systems Using a Multidimensional Approach[J]. ACM Trans. on Information Systems (TOIS), 2005, 23(1): 103–145.

[7] Girardello A, Michahelles F. AppAware: Which mobile applications are hot? In: Proc. of the 12th Int'l Conf. on Human Computer.

[8] Yang WS, Cheng HC, Dia JB. A location-aware recommender system for mobile shopping environments. Expert Systems with Applications, 2008,34(1):437–455. Interaction with Mobile Devices and Services. Lisboa: ACM Press, 2010. 431–434.

[9] Setten MV, Pokraev S, Koolwaaij J. Context-Aware recommendations in the mobile tourist application COMPASS. In: Proc. of the 3rd Int'l Conf. on Adaptive Hypermedia and Adaptive Web-Based Systems. Berlin, Heidelberg: Springer-Verlag, 2004. 515–524.

[10] Zheng VW, Cao B, Zheng Y, Xie X, Yang Q. Collaborative filtering meets mobile recommendation: A user-centered approach. In:Proc. of the AAAI 2010. Atlanta: AAAI, 2010. 236–241.

[11] Rich E. User Modeling Via Stereotypes [J].Cognitive Science, 1979, 3 (4) : 329 – 354.

Computing, Control, Information and Education Engineering – Liu, Sung & Yao (eds)
© 2015 Taylor & Francis Group, London, ISBN: 978-1-138-02800-5

Increase the classroom teaching effectiveness of "mechanical principles" based on the students

Yang Fang Wu, Shu Zhen Yang, Chun Lin Xia & Zhao Feng Zhou
Zhejiang University City College, Hangzhou, Zhejiang Province,China

ABSTRACT: Mechanical principles as a backbone technology pilot course, with the features of various mechanism, more theoretical points and strongly engineering contact, it is more difficult for students to understand and master. In this paper, exploration from a comprehensive understanding of the students, the teaching mode to combination teaching aids, writing on the blackboard and electronic courseware, case guidance system to enhance the engineering awareness and system concepts, promote independent learning, proposed an effective way to improve classroom teaching effectiveness.

KEYWORDS: Student -centered; Mechanical principle; Classroom teaching effectiveness.

1 INTRODUCTION

Mechanical principles as a backbone technology foundation course of the mechanical engineering. It is based on the compulsory curriculum, such as calculus, physics, engineering graphics and theoretical mechanics, and also lay the necessary foundation for the study of mechanical design and related professional courses. It is an important transition course for the students to learn from the basic courses to professional courses, to enlighten the mechanical design thinking, and is the important part to cultivate innovative design capabilities. Task of this course is to enable students to master the basic theory, basic knowledge and basic skills of mechanism and machine dynamics, to learn the analysis and synthesis methods commonly used in mechanism design and have the preliminary ability to design the mechanical motion plan of the system. The course has the features of mechanism as the main line, more theoretical points and strongly engineering contact, it is more difficult for students to understand and master. In the teaching process, teachers should be student-centered, using various means to improve the teaching effectiveness, stimulate students' interest in learning. This paper presents some recommendations based on years of teaching experience.

2 TO TEACH PURPOSEFULLY BASED ON STUDENTS-CENTERED

The students are not only the object of teaching, but also the understanding subject of the teaching process. The ultimate teaching goal of this course is the students' ability to use knowledge to solve practical problems. In the teaching process, teachers should fully understand the students, including students' knowledge base and capabilities, awareness and the level of enthusiasm for the profession, mastery of the relevant curriculum knowledge, in the learning process feedback problems, learning habits and learning methods. Only a comprehensive understanding of the students can be targeted to improve teaching effectiveness. In this course, depending on the characteristics of students continue to adjust the teaching content, take a variety of teaching methods such as aids presentations, courseware presentations, writing on the blackboard, classroom discussions, online learning, engineering practice. It is obtained the student's identity, improved the teaching effectiveness.

3 COMBINING THE AIDS, WRITING ON THE BLACKBOARD AND COURSEWARE TO STIMULATE THE STUDENTS' LEARNING INTEREST

Aids as a traditional teaching tool have some advantages, such as intuitive, strong inspiring. Mechanical principles have the characteristic of mechanism as the main line, more theoretical points, the advantage of aids is particularly prominent. Writing on the blackboard with the feature of coherent, focused on the emphasis, can provide clear knowledge systems, and provide students with a buffer of time to think and digest. The multimedia courseware can provide a large amount of information in the pictures, videos and other information to make up the less information of blackboard and the monotony of teaching aids. In the teaching process should combine these three methods to improve teaching effectiveness, to cultivate the image thinking, design thinking and project awareness.First lines of paragraphs are indented 5 mm (0.2") except for paragraphs after a heading or a blank line (First paragraph tag).

Table 1. The application of "aids, blackboard and courseware."

Name	Application	Target
aids	four-bar hinge mechanism, jaw crusher	Know the significance and drawing process of mechanism motion perceptual.
blackboard	Define the kinematic motion sketch and its step.	Focus on the Emphasis and difficulty.
courseware	Common representation of the motion mechanism and component; the video of the jaw crusher video.	Increase the amount of information, stimulate interest in learning.

As an example, the knowledge of kinematic diagram drawing to demonstrate the effectiveness of the trinity teaching mode-teaching aids, writing on the blackboard and courseware. Kinematic diagram is a motion transmission route for the mechanism drawing proportionally with simple lines, and is an important means of analysis and design of machinery. "Aids, writing on the blackboard, courseware" used in the present knowledge of teaching as shown in Table 1. Specific instructional design is as follows:

1. Show aids four-bar hinge mechanism: ask the question, "Why do I need to draw kinematic diagram? Which is a motion size?" Guide students to recognize what is kinematic diagram deeply.
2. Writing on the blackboard: the definition of kinematic diagram.
3. Courseware gives "representation of commonly used motion mechanism and components": how to master the expression of complex parts with simple lines.
4. Writing on the blackboard: the draw methods of kinematic diagram.
5. Show aids jaw crusher as in Figure1: slowly rotating machine, observe each part of the mechanism.
6. Play the video of jaw crusher: to explain the eccentric structure; observe the motion transmission route, looking for component and motion links.
7. Writing on the blackboard: draw the kinematic diagram of jaw crusher as show in Figure 2.
8. Play the other machinery video, such as internal combustion engine, shaper: guide the students to think how to draw kinematic diagram.

This teaching method made the students to understand the concept deeply, theory with practice, to stimulate student interest in learning, improve the teaching effectiveness.

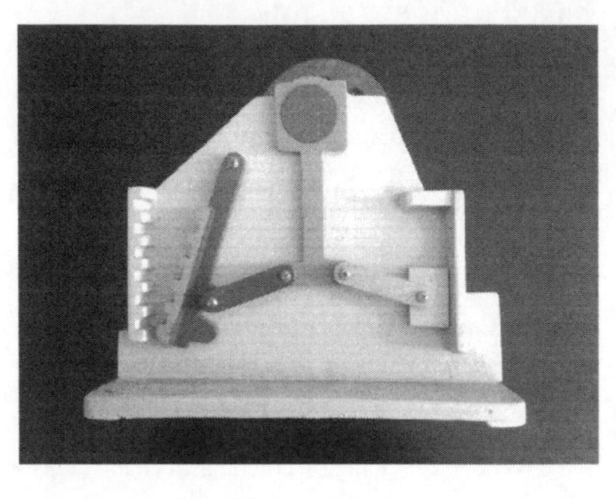

Figure 1. Teaching aids-jaw crusher.

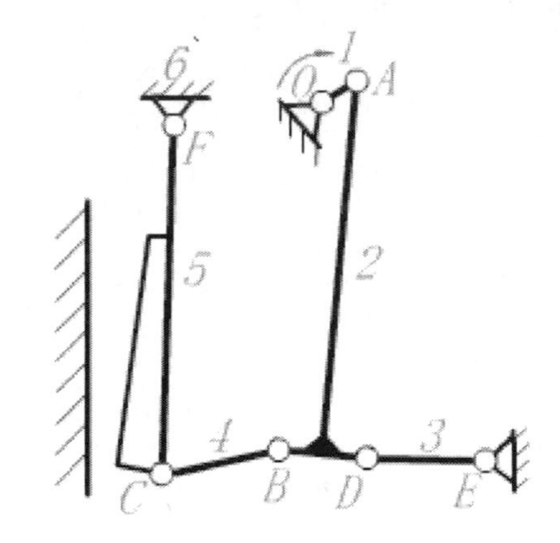

Figure 2. The kinematic diagram of jaw crusher.

4 CASE GUIDE, STRENGTHEN TO CULTIVATE ENGINEERING AWARENESS AND SYSTEMS CONCEPTS

Mechanical principle is the first course to closely relate with mechanical product design, the analysis and design method of different mechanism is relatively independent. And as the future engineering and technical personnel, the students must have a strong engineering awareness and systems concepts. The engineering awareness is not developing immediately, it is the potential which sum from the set of engineering knowledge, skills, abilities, positive and negative lessons, etc., and will be dynamic growth with the development of engineering technology and own experienced dynamic [4].Enlightening the engineering awareness is to enable students to learn to use theoretical knowledge to understand the engineering phenomenon, analysis of engineering problems and guide engineering practice.

This course uses "Animation Project - heuristic teaching - discussion expand- Systematic Summary - engineering problem solving" teaching model to develop the engineering consciousness and systems concept. Second-year students is less of perceptual knowledge of machinery and mechanism, before learning the design approach of various mechanism, through playing a lot of animation of various typical mechanical system, so that students understand the role of various mechanism in the mechanical system, thus the single, isolated mechanical mechanism design becomes part of the system, students learn to think with system concept, helping to train engineering consciousness. For example, the car is one of the typical mechanical products, play the video to explain the work principle of the single-cylinder internal combustion engine, understanding the four stroke internal combustion engine (intake - compression - acting - exhausting), analysis the function of the link mechanism, the cam mechanism and the gear mechanism, then leads to the different design methods and roles of the various mechanism. Under the guidance of heuristic teaching methods, students learn by thinking, with discussions and exercises and other means to acquire knowledge. Based on grasping the basis of knowledge solidly, some questions, which have engineering significance is put forward (such as door opening and closing mechanism, clamping mechanism, etc.). To train the students have analyze problems and problem-solving skills, as well as the ability of using theoretical knowledge to analyze the engineering problems, to formed engineering consciousness and systems concepts consciously.

5 STRENGTHEN THE EXTRACURRICULAR ACTIVITIES DESIGN AND PROMOTE STUDENTS' SELF-LEARNING

University education should focus on the students' extracurricular self-learning ability. Teaching should be student-center, give full play to the initiative of students, through a variety of ways for students to deepen their understanding of course content and the courage to practice. The course takes the mode of additional courses practical projects and optimizing homework to promote student learning by self as shown in Figure 3.

Homework has a very important role for students to acquire the knowledge learned in the classroom to deepen understanding of the content of the curriculum. Homework from the current situation, it takes not much extracurricular time for students to complete the homework, and focus on evaluating the knowledge situation from per class. The tightness link theory and engineering practice there can be improvement. The curriculum reform from the following aspects:

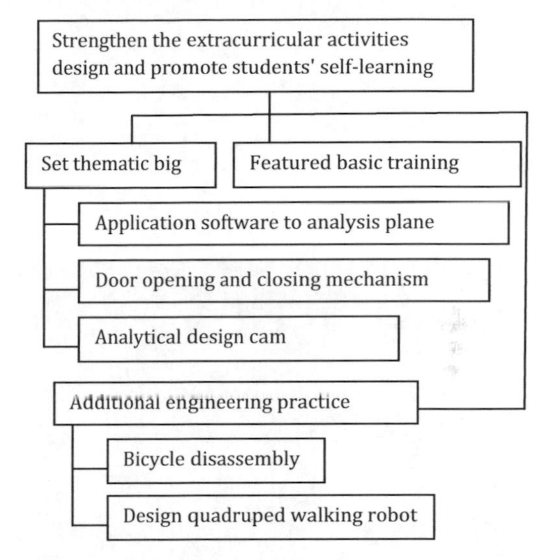

Figure 3. Design Extracurricular Activities content.

Featured basic training topics: Through a small amount of basic training topics to mastery test student knowledge of each class. Improve students' ability to analyze and solve problems.

Set thematic big job: Combined with the feature of mechanical principle to design thematic big job throughout the chapters, specifically shown in Fig. 3. To culture the students systematic application of knowledge and innovation design capabilities.

Additional engineering practice project: Project-driven, stimulate student interest in learning, improve learning initiative, training engineering application capabilities. Engineering Practice project launched in the form of teamwork, when the packet fully consider the advantages of ability within the group of students, according to the mechanical product design features, according to C (Conceive, idea innovation capacity), D (Design, with good design and calculation capabilities), I (Implement, strong hands), O (Operate, communication skills) each one consisting of four-person team, so that students learn from each other in the course of project implementation, and improve together. Bicycle disassembly project have been implemented in the sessions of the students, and achieved good results, Figure 4 is a bicycle disassembly event, Figure 5 is a quadruped walking robot works instance.

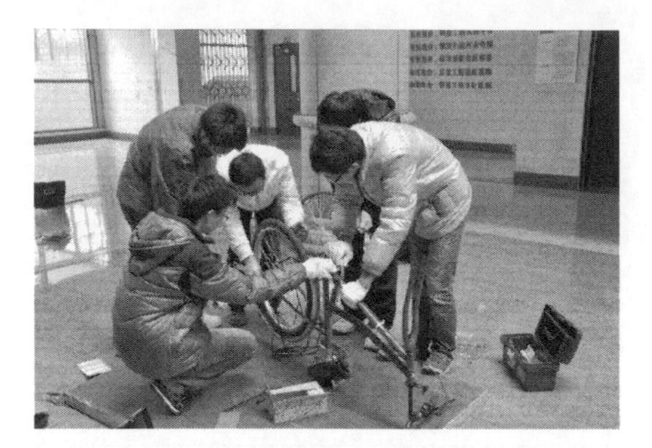

Figure 4. Bicycle disassembly event.

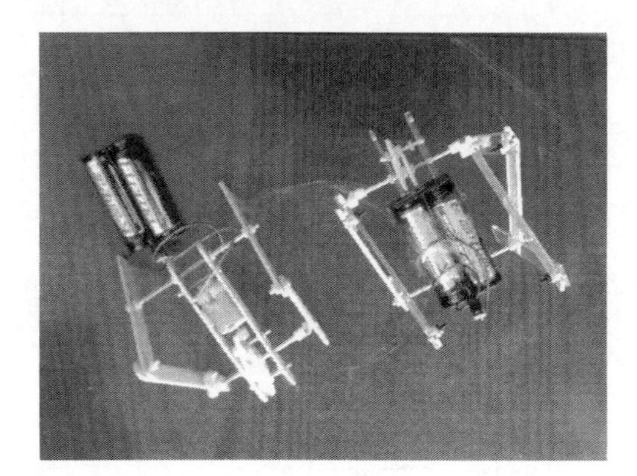

Figure 5. Quadruped walking robot.

6 CONCLUSIONS

Mechanical principle is the core foundation of Mechanical Engineering, teaching effect of this course directly affects students' professional interests and emotions. In this paper, student-centered, comprehensive understanding of the students from the start, taking the combination of "teaching aids, writing on the blackboard, courseware" teaching mode, case guided expand the teaching and extra-curricular activities through carefully designed to guide students' independent learning. After the sessions the students to try to obtain some success, praised by students. How to improve the teaching effect of the course is to explore and practice require long-term problem, we just sum up the experience of many years of teaching, and for reference.

ACKNOWLEDGEMENTS

This work was financially supported by 2013 Annual Classroom Teaching Reform of Higher Education in Zhejiang Province Research Project, 2014 Annual Educational Technology Research and planning issues (JB066), Zhejiang University City College Core Curriculum Group Project (HX1204) and Zhejiang University City College nine reform project (JG1327).

REFERENCES

Wei Chang-wu. 2012.Experience in improving the teaching effect of mechanical principles[J].Journal of the Socialism Institute of Yunnan, 6. (in Chinese).

Shen Yi,Wang Ming-qiang. 2012.Exploration and Practice of Trinity Teaching Mode for Mechanical Principles Course[J].Equipment Manufacture Technology, 2:184–186. (in Chinese).

Zhang Wei,Li Ying-ping. 2013.Exploration and practice-oriented design of mechanical principles Teaching Reform[J]. Education Forum, 36: 213–215. (in Chinese).

Feng Xuan. 2006.The importance of engineering Awareness Training[J]. Mechanical Vocational Education, 1: 44–45. (in Chinese).

Sun Huan,Chen Zuo-mo,Ge Wen-jie.2006.Theory of Machines and Mechanism(seventh edition).Beijing:Higher Education Press. (in Chinese).

P.H.GU,M.F.SHEN & X.H.LU. Rethinking Engineering Education-The CDIO Approach. Beijing:Higher Education Press.2009. (in Chinese).

Yangfang Wu.2014. "Theory of Machines and Mechanism" teaching reform to inspire mechanical design thinking[J] Applied Mechanics and Materials, 678:627–631.

Computing, Control, Information and Education Engineering – Liu, Sung & Yao (eds)
© 2015 Taylor & Francis Group, London, ISBN: 978-1-138-02800-5

Commercial strategies chain suppliers pharmaceutical products to decrease costs logistics

Amelec Jesus Viloria Silva
University Sergio Arboleda, Bogotá, Colombia

ABSTRACT: This article aim to raise Provider Business Strategies of Pharmaceutical Products Chains for its Decreased Costs Logistics. The sample is a multinational production and marketing of pharmaceutical products based in France. The first proposal is to increase the value of MOV, forcing customers to pass orders greater value and volume generating more stock in their stores and therefore make fewer orders. This positively impacts the logistics department budget, given the savings generated in administrative overhead and transportation. The second option is a total restructuring of the business strategy, simplifying trading conditions, normalizing discount rates, establishing new promotions by creating a bonus system that offers the customer different discount rates depending on the total value of the order, the which influence the client and leads to ordering more volume, therefore less requested annually. This proposed strategy seeks to simplify logistics and sales operations, reducing operating costs and the pursuit of incremental business.

KEYWORDS: Supply chain, Business strategy and Minimum order value.

1 INTRODUCTION

At present, the high competitiveness, changes in markets and the financial crisis, generate a great pressure on corporate managers. These are continually demanding their managers and employees perform simplification or process reengineering, seeking to reduce costs and achieve profit maximization [1].

Within the pharmacy channel in France, there is a big difference between all customers, from large pharmacies in major cities such as medium in residential areas or small farming villages. This generates a range of volumes, fairly wide orders and the only existing restriction is the Minimum Order Value ("Minimum Order Value", MOV). Thus arises the need to question the model and current procedures, the main approach is to find a method to induce customers to spend less purchase orders, without generating a financier impact [2] and [3].

Business strategies and business conditions emprersa study are clearly defined. These are based on contracts signed between the company and pharmacies, which offer discounts and benefits specific category. At the beginning of the contract year in which the pharmacy is committed to making a certain volume of sales is established, which leads to different levels of discount promotions and other conditions that benefit the customer.

However, constantly changing conditions occur, new categories are added to the portfolio of new products or promotions are set. This leads to continuous changes in information systems, greater complexity in the monitoring of logistics operations, increased chances of error in billing, among other issues, generating an increase in operating costs.

Therefore, in the present investigation was designed and implemented trading strategies that allow the company to reduce costs studio, specifically those related to logistics and so continue to drive the growth of trade.

2 METHODOLOGY

The proposed methodology is defined by the following steps [4]:

- Diagnose the current management of the supply chain in the circuit pharmacies, their business models and their overall impact on business management.
- Analyze the current business strategy of the chain of pharmaceutical products supplier.
- Propose business strategies for pharmaceutical chain provider.
- Implement and evaluate the proposed strategies.

3 RESULTS AND DISCUSSION

Phase 1: Diagnosis of supply chain and business model

Is presented below in Table 1 is an array SWOT supply chain provider pharmaceuticals.

Table 1. SWOT matrix for supply chain provider.

STRENGTHS	OPPORTUNITIES
-Partnerships With customers Structured supply -chain reputable company -Trademarks With market positioning	-Simplification of trading conditions better integration of information systems -Improvements Logistics
WEAKNESSES	**THREATS**
- Complexity of business strategy and logistics operations - Faulty internal communication	- Competition with better trading conditions and simpler - Lack of motivation of staff

Phase 2: Analysis of business strategy

Table 2 shows the SWOT matrix for the commercial strategy currently pursued by the company under study is identified.

Table 2. SWOT matrix for the current marketing strategy supplier.

STRENGTHS	OPPORTUNITIES
- Known by customers Strategy - Configured in information systems	- Possibility to change the MOV and development of new discounts
WEAKNESSES	**THREATS**
- Arrival of new categories - MOV low - Dependent Information System	-Risk Financial changes -Resistance To change by the sales force

Phase 3: Proposed business strategies

The proposals of changes in MOV indicator and improvements in business strategy are developed.

Strategy 1: Changes in MOV indicator

It is planned to observe the possible scenarios that can be obtained by increasing or decreasing this value. The main objective of this proposal is to seek savings in total billing costs by reducing the number of annual orders and alignment with business growth objectives. The two basic scenarios are decreasing or increasing this value.

In the case of a decrease, this allows the search for new small customers who have no ability to place large orders. While this generates sales growth, a direct increase in costs is because the number of orders, both new customers and old who prefer to make small orders to reduce their inventories in pharmacies increases.

The scenario of increasing MOV, would force customers to spend more orders volume and value. In this case, we are making a push strategy so inventories will be acrecentados customers. Therefore, it is assumed that a reduction occurs in purchase orders. Similarly, one should consider the potential loss of business profits or impact, customers are not willing or able to adapt to increased MOV.

Strategy 2: Improvements in business strategy

The proposal to develop is the simplification of the current business strategy, by eliminating the markets and conditions [5] and [6]. It is proposed to establish a basis for category discount, which will be offered to all customers. To complement customers, discounts and additional promotions by category or subcategory, depending on the volume of units engaged in the order is made.

Aggregation of the order value is then performed and to encourage customers to place orders higher value, an additional discount on the total turnover is offered. Orders will be classified in ranges of value, as the range increases to increase the discount offered. This tool can also be considered as a push strategy that will allow the reduction in orders requiring the customer during the year and therefore a reduction in operating costs.

Phase 4: Evaluation and Validation of proposals

By varying the parameters of the models may have different simulation scenarios using Excel for the chain of distribution of pharmaceutical products under study. Optimum results are only shown in this phase. Table 3 shows the results of calculating the impact on revenue due to increased MOV is presented. It was considered that ranges between 0 to 200, 200 to 300, 300 to 400 euros, an impact of thirty percent (30%), ten percent (10%) and five percent (5%) is generated respectively.

Table 3. Calculation of the impact on admission to establish a MOV 400 €.

Rank	No. bill	Income (euros)	Impact	Impact (euros)
0-200	8849	1.531.811,54	30	459.543,46
200-300	8826	2.256.67859	10	225.667,86
300-400	6127	2.198.909,18	5	109.945,46
Total impact				795.156,78

Table 4 shows the results obtained with defined discount rates are presented. Estimates for determining who can gain approximately four thousand five hundred (4,500) purchase orders, representing a saving locate a quote about seven hundred forty thousand euros (€ 740,000). This can be considered as a great impact, but we must consider that this discount is associated with streamlining business strategies.

Table 4. Evaluation of invoice discount depending on the value range.

Ranges	Discount rate%	% Reduction orders	Order less	Income (Euros)	Order required	# Order	Income undiscounted (Euros)	Income Discount (Euros)
150-500	0	20	4978	1.452.914,35	0	19910	5.811.657,40	5.811.657,40
500-1000	1	15	2411	1.709.770,42	2049	15711	11.141.613,42	11.030.197,29
1000-2500	1,5	5	570	861.868,53	1131	11963	18.085.272,49	17.813.993,40
2500-5000	2	2	52	174.249,57	258	2817	9.400.097,52	9.212.095,57
5000-máx	2,5	0	0	-	20	780	6.717.485,71	6.549.548,57
						51.181	51.156.126,54	50.417.492,23
				Orders won		4553	**Total discount**	738.634,31
				Cost / Order		4,45		
				Budget cattle		20.259,17		

of some twenty thousand euros (€ 20,000). The sales and marketing department must locate a quote about seven hundred forty thousand euros (€ 740,000). Therefore the marketing and sales department must perform the calculations necessary to adapt this discount along with discounts and promotions basis.

4 CONCLUSIONS

In Table 3, it can be seen that increasing the MOV four hundred euros (€ 400) a reduction of about forty-five percent (45%) of the purchase orders is performed. Which allows the supplier to make a saving of around forty eight thousand euros (€ 48,000) in billing expenses, it is for them the most critical budget. Equally significant savings are made in transport costs. Although care must be taken to impact the income of about eight hundred thousand euros (€ 800,000) representing thirteen percent (13%) of lower revenues to MOV.

Analyzing Table 4 it can be concluded that the marketing and sales department of the supplier must

REFERENCES

[1] Metzger, M., & Donaire, V. (2007). Gerencia Estratégica de Mercadeo. México D.F., México: International Thomson Editores.

[2] P&G Pharmaceuticals. (2011). Condiciones comerciales de la Go To Market 2012. Paris, France: P&G.

[3] P&G. (2012). The Power of Purpose. Disponible en: http://www.pg.com/en_US/company/purpose_people/index.shtml. Consultado en: Febrero de 2012

[4] Sainz, J. M. (2001). La Distribución Comercial: Opciones Estratégicas. Madrid, España: ESIC Editoria.

[5] Schultz, D. E., & Robinson, W. A. (1995). Como Dirigir la Promoción de sus Ventas. Barcelona, España: Ediciones Granica S.A.

[6] Soret, I. (2006). Logística y Marketing para la distribución comercial. Madrid, España: ESIC Editorial.

Computing, Control, Information and Education Engineering – Liu, Sung & Yao (eds)
 ISBN: 978-1-138-02800-5

A survey on dimension reduction techniques in text classification

Zhi Juan Wang
College of Information Engineering, Minzu University of China, Beijing, China
Minority Languages Branch, National Language Resource Monitoring and Research Center, Beijing, China

Ruo Song Zhou
College of Information Engineering, Minzu University of China, Beijing, China

ABSTRACT: Dimension reduction is one of the key points for text classification. Feature selection and feature extraction are the two common methods of dimension reduction. In this paper, we mainly discussed some dimension reduction techniques from two aspects including traditional methods (Information Gain, Mutual Information, Document Frequency, Correlation Coefficient) and new methods (Optimization Mutual Information Based on Word Frequency, CDF (Concentration, Dispersion and Frequency), Semantic Relatedness). Then analyzed the principle of these methods and illustrated their advantages as well as disadvantages.

KEYWORDS: Dimension reduction; Text classification; Feature selection; Feature extraction; CDF.

1 INTRODUCTION

Text classification is an important part of text mining, which makes great meaning in improving the speed of information retrieval and the accuracy of the rate. We usually select Token (word appearing in a text) to denote the text's features space $d=(T_1,T_2,\ldots,T_n)$. Therefore, the dimension of features space ranging from thousands to ten thousands is so high that the traditional methods of classification can't process those features. Under the circumstance, Dimension reduction is the primary task as well as the key question to be solved for text classification using the method of machine learning. Dimension reduction includes two methods: 1). Feature selection (including Document Frequency (DF), Mutual Information (MI), Correlation Coefficient (CC), χ^2-statistics (CHI) [1-4]). 2). Feature –extraction (including Random Projection (RP), Latent Semantic Analysis (LSA), Concept Indexing (CI) [5-7]). Feature selection selects parts of meaningful features from the original features to form a new low-dimensional space without changing the property of original features space. However, Feature extraction projects the original features space onto a new feature space by structuring an evaluating function of features. The procedure of dimension reduction is shown in Figure.1.

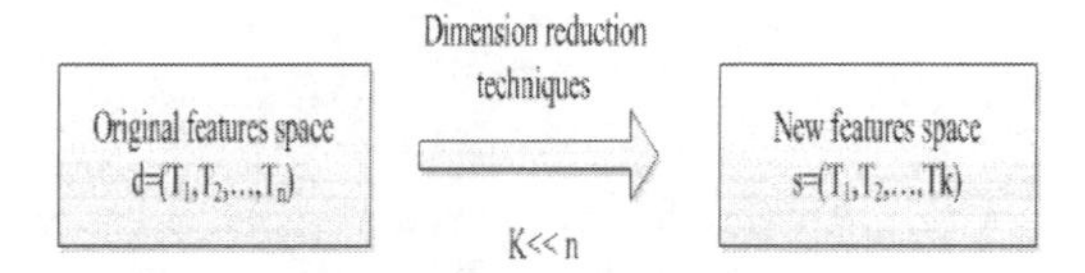

Figure 1. The procedure of dimension reduction.

2 TRADITIONAL METHODS ON DIMENSION REDUCTION

2.1 *Document Frequency (DF)*

DF denotes the occurrence frequency of a feature in the training corpus. When the value of a feature's DF is lower than an appointed value, it means this feature maintaining a little sense for text classification. Therefore, this feature has to be removed out to reduce dimension. This method is very easy, but ignores the dependence between a feature and a category. We usually combine this method with others to improve precision for dimension reduction.

2.2 *Mutual Information (MI)*

MI is used to evaluate the degree of correlation between a feature T and a category C_i. The higher the value of MI is, the greater the degree of correlation is. The computing formula for MI is shown in Eq.1.

$$\mathrm{MI}(\mathrm{T},C_i)=\log\frac{P(\mathrm{T},C_i)}{P(T)*P(C_i)}=\log P\left(T|C_i\right)-\log P(T) \quad (1)$$

Where $P\left(T|C_i\right)$ is the frequency of texts for T in C_i. $P(T)$ is the frequency of texts for T in the training corpus. In this method, if a feature gets higher MI, it will be selected as corresponding categories' feature.

2.3 *Information Gain (IG)*

IG is used to evaluate the degree of contribution made by a feature T to a category C_i. The computing formula for IG is shown in Eq.2.

$$\begin{aligned}&\mathrm{IG}\left(\mathrm{T},\mathrm{C_i}\right)=-\sum_{i=1}^{n}\mathrm{P}\left(\mathrm{C_i}\right)\log\mathrm{P}\left(\mathrm{C_i}\right)+\\&\mathrm{P(T)}\sum_{i=1}^{n}\mathrm{P}\left(\mathrm{C_i}|\mathrm{T}\right)+\\&\log\mathrm{P}\left(\mathrm{C_i}|\mathrm{T}\right)+\\&P(\overline{T})\sum_{i=1}^{n}P(C_i\,|\,\overline{T})\log P(C_i\,|\,\overline{T})+\mathrm{P}(\overline{T})\end{aligned} \quad (2)$$

Where P (C_i) is the frequency of texts for category C_i existing in the training corpus. P(T) is the frequency of texts for T existing in the training corpus. $P(\overline{T})$ is the frequency of texts for T without existing in the training corpus. $P\left(C_i|T\right)$ is the frequency of texts for T existing in C_i. $P(C_i\,|\,\overline{T})$ is the frequency of texts for T without existing in C_i. N is the amount of categories.

The biggest difference between MI and IG is the latter considering if a feature T appears or not, while the former just considers one condition that T appears. According to the paper [8], considering the condition that T doesn't appear makes it more accurate to classify texts than ignoring that.

2.4 *χ2-statistics (CHI)*

The principle of CHI is similar with MI, which can be used to reduce the dimension according to the degree of correlation. The computing formula for CHI is shown in Eq.3.

$$2\left(\mathrm{T},\mathrm{C_i}\right)=\frac{\mathrm{N}*(\mathrm{AD}-\mathrm{CB})^2}{(\mathrm{A}+\mathrm{C})*(\mathrm{B}+\mathrm{D})*(\mathrm{A}+\mathrm{B})*(\mathrm{C}+\mathrm{D})} \quad (3)$$

Where A is the frequency of texts belonging to C_i and including T. B is the frequency of texts not belonging to C_i but including T. C is the frequency of texts belonging to C_i not including T. D is the frequency of texts not belonging to C_i and not including T. N is the amount of texts in the training corpus.

CHI is a procedure of normalization which can reduce more features than MI. But the expenditure of time for CHI is also more than MI.

The above four methods have good stability, but bad robustness in dimension reduction. The following three methods will excavate relation between a feature T and dirent categories deeply, including two methods based on statistics and the other one based on semantic relatedness to improve the robustness.

3 NEW METHODS ON DIMENSION REDUCTION

3.1 *Optimization mutual information based on word frequency*

This method is proposed by HaiFeng L and ZeQing Y in 2014 [9]. As depicted above, the traditional MI just considers the frequency of texts, but ignores the frequency of words between different categories as well as in the same category, which will lead to some deviations. Therefore, this new method adds two weight factors λ, β to original formula (2). The formula of λ, β is shown in Eq.4.

$$\lambda_{jk}=\sum_{i=1}^{n_i}\frac{[\mathrm{tf_{jik}}\left(\mathrm{t_k}\right)-\min((\mathrm{tf_{jiq}}\left(\mathrm{t_q}\right))]}{\max((\mathrm{tf_{jiq}}\left(\mathrm{t_q}\right))} \quad (4)$$

$$\lambda_k=\sum_{i=1}^{n}\frac{\lambda_1}{\sqrt{\sum_{k=1}^{m}\lambda_{jk}^2}} \quad (5)$$

$$\beta_{k-\varepsilon}=\sqrt{\frac{1}{n-1}\sum_{j=1}^{n}(tf_{jk}\left(t_k\right)-\frac{1}{n}\sum_{i=1}^{n}tf_{jk}\left(t_k\right))^2} \quad (6)$$

$$\beta_k=\frac{\beta_{k-\varepsilon}}{\sqrt{\sum_{l=1}^{m}\beta_{l-\varepsilon}^2}} \quad (7)$$

Based on λ, β, the improved method of MI is:

$$\mathrm{MI}\left(t_k\right)=\lambda_k*\lambda_k*\log P\left(t_k|C_i\right)-\log P(t_k) \quad (8)$$

Where $tf_{jik}(t_k)$ is the frequency of words for t_k existing in document D_j of category C_i. N_i is the amount of texts in category C_i. M is the amount of features. N is the amount of categories.

3.2 *CDF*

CDF (Concentration, Dispersion and Frequency) is proposed by Zhongyang X and Jian J in 2009 [10].

Concentration, dispersion and frequency are three important indicators for dimension reduction, which is discussed in document [9]. Based on above indicators, CDF is used to evaluate the degree of correlation between a feature T and a category C_i as a new method. The formula for the CDF is shown in Eq.5.

$$\mathrm{CDF}\left(\mathrm{T}, C_i\right)=\frac{df_i(T)}{\sum_{i=1}^{m} df_i(T)} * \frac{df_i(T)}{N_i} * \frac{tf_i(T)}{df_i(T)} \tag{9}$$

Where $df_i(T)$ is the frequency of texts for T existing in C_i. $tf_i(T)$ is the frequency of words for T existing in C_i. N_i is the amount of texts in category C_i. M is the amount of categories. The first section $\frac{df_i(T)}{\sum_{i=1}^{m} df_i(T)}$ is to evaluate the concentration for T between different categories. The second section $\frac{df_i(T)}{N_i}$ is to evaluate the dispersion for T in C_i. The last section $\frac{tf_i(T)}{df_i(T)}$ is to evaluate the average frequency for T in C_i.

3.3 *Semantic relatedness*

This method is proposed by Yang L in 2013 [11]. Based on the synonymous word forest which includes not only a word's synonymous term, but also its similar-class words [12], it can remove some features by computing different feature's semantic correlation. The computing methods are always based on two categories: Ontology or Taxonomy. The synonymous forest belongs to Ontology, which has greater authority. We can see that the experience's result of text classification is much better than traditional methods based on statistics in paper [13].

4 CONCLUSION

Considering the higher dimensions will have a great effect on the speed and accuracy of text classification. Therefore, dimension reduction is indispensable before text classification. In this paper, we mainly discussed four traditional, but commonly used methods of dimension reduction. Then three new methods were analyzed. All these methods have reduced the feature's dimension in varying degrees. While how to combine with different methods to improve the efficiency for text classification is still a hard way to go.

ACKNOWLEDGEMENTS

In this paper, the research was sponsored by the Key Program of National Natural Science Foundation of China (No. 61331013) and National Language Committee of China (No.WT125-46, No. WT125-11).

REFERENCES

[1] Salton G et al. 1975. A Theory Term Importance in Automatic Text Analysis [J]. Journal of the American Society for Information Science, 26(1):33–44.

[2] R Nattuti. 1994. Using mutual information for selecting features in supervised neural net learning [J]. IEEE Trans. Neural Networks, 5(4):537–550.

[3] Luigi G & Fabrizio S. 2003.Feature Selection and Negative Evidence in Automated Text Categorization [C]. Proceedings of the ACM KDD Workshop on Text Mining.

[4] Zheng Z H & Srihari S H. 2002. Text categorization using modified-CHI feature selection and document term frequencies [C]. ICMLA: 252–263.

[5] Kaski S. 1998. Dimensionality Reduction by Random Mapping: Fast Similarity Computation for Clustering [C]. Proceedings of International Joint Conference on Neural Networks: 413–418.

[6] S Dumais et al. 1988. Using Latent Semantic Analysis to improve Access to Textual Information [C]. Proceedings of the Conference on Human Factors in Computing Systems.

[7] George Karypis & Eui-Hong Han. 2000. Concept Indexing: A Fast Dimensionality Reduction Algorithm with Applications to Document Retrieval & Categorization [C]. ACM CIKM Conference.

[8] Yang Y & Pedersen J O. 1997. A comparative study on feature selection in text categorization [C]. ICML. 97:412–420.

[9] Haifeng L et al. 2014. Optimization Mutual Information Text Feature Selection Method Based on Word Frequency [J]. Computer Engineering, 40(7):179–182.

[10] Zhongyang X et al. 2009. New feature selection approach (CDF) for text categorization [J]. Journal of Computer Applications, 29(7):1755–1757.

[11] Yang L.2013.Feature Selection Method based on Semantic Relatedness [J]. Network Security Technology & Application, 4: 68–70.

[12] Jiaju M et al. 1996. The Synonymous Word Forest [M]. Shanghai: Shanghai Dictionary Press.

[13] Hongzhe L & De X. 2012. Ontology based Semantic Similarity and Relatedness Measures review [J]. Computer Science, 39(2):8–13.

Computing, Control, Information and Education Engineering – Liu, Sung & Yao (eds)
© 2015 Taylor & Francis Group, London, ISBN: 978-1-138-02800-5

An empirical study on the effect of Synergistic Learning instructional design

Zhi Yan Ding
ZheJiang Technical Institute of Economics, Hang Zhou, Zhe Jiang, China

Xiang Bin Zhu
ZheJiang Normal University, JinHua, Zhe Jiang, China

ABSTRACT: According to the new learning system framework of knowledge age—Synergistic Learning model, take the college student poetry teaching in Zhejiang province for instance. The poetry generation system designed based on genetic algorithm. Experiments prove that this is a good method.

KEYWORDS: Synergistic; Poetry; Learning; Construct.

1 STUDY BACKGROUND

The concept of Synergistic Learning is first proposed by Professor Zhiting Zhu, doctoral tutor, east China normal university in December 2009, in the domestic. Synergistic is a word comes from ancient Greek. It refers to a large number of interactions among the parts of the open system and the effect of the whole, collective or Cooperative. Synergistic Learning in essence is different from the usually what we call "Collaborative Learning" and "Cooperative Learning".

Synergistic Learning is more emphasis on "learning technology system elements, including cognitive subject and cognitive object and its interaction to form the Synergistic relationship and structure between the learning fields, the goal is to get teaching synergies" [1]. Sum up, the biggest difference between Synergistic Learning and Collaborative Learning lies in: Synergistic Learning study group members not of primary and secondary points, team learning members advance and retreat together, learning system has a Synergistic Learning strategies (such as Synergistic Learning script and enabling technology) [2].

We take the poetry teaching as an example, for the design of Synergistic Learning. The goal is to make machines learn to write the poems of Tang Dynasty. In this process, only the learners themselves to understand and familiar with the Tang Poems, is likely to teach a machine to write poems.

2 RESEARCH SYSTEM

In carefully studied the synergistic learning theory, the researchers introduce a genetic algorithm in Synergistic Learning instructional design. Poetry generation is essentially the optimization of process in a solution space. And the advantage of a genetic algorithm is solving such problems. Our goal is to build an automatic five-character-quatrain generation system, it can according to the user's input subject word and rhyme categories to generate quatrain. We have completed.

This system includes three basic modules: segmentation corpus and phonological database of the poems, the norms of grammar and semantic measurement, a genetic algorithm for poetry generation.

The statistical method will extract the stronger bond strength, the stable and the metaphor of two-character words, to establish a better word table. And another way is according to the metrical pattern to corpus poetry, we combined with the advantages of the above two methods to establish the Tang Poetry segmentation corpus. On the part of speech tagging of the corpus, parts of speech are simplified to the following six kinds: N: noun; VT: transitive verbs; VI: intransitive verbs; EN: verb (left to do); ADJ: adjectives; ADV: adverbs.

Through the analysis of a large number of sentences of poetry, we find that the number of effective models of sentences is limited, and presented the hierarchical structure. And this kind of hierarchical pattern recognition, with the DFA is a good choice. After producing a large number of alternative individuals, let them one by one through the test the DFA, good stay, bad leave.

Restriction on the collocation of words is not only syntax, but also semantic constraints. Calculating lexical relevancy is to establish a link between words, excavate of the words possibility of co-occurrence and collocation, to ensure that coherent of poetry style and theme. This topic, the meaning of relevance computation is mainly used in:

Lexical similarity is to measure the alternative of words. The purpose of lexical similarity calculation lies in the guarantee under the premise of the subject word on track, make more rich and changeful words in generation poetry. Strict metrical pattern, making poetry creation is different from the general natural language text generated, it embodies is especially apparent in the creation of words. So-called "write lyrics", means not only to find an accurately express the meaning, to convey the feelings of words, also requires the word strictly abides by the rules of level and oblique tones and rhyme, etc. Under this requirement, the vocabulary is rich, the ability of synonymous substitution, metonymy, symbol, and even the use of rhetorical devices such as metonymy, to a large extent determines the quality of poetry creation.

For modern Chinese natural language processing, there are two kinds of common lexical similarity calculation methods, one is using a large-scale corpus statistics, the other is calculated according to some Ontology.

This generated module based on the genetic algorithm is mainly composed of five main steps: Initial population generation, fitness calculation, selection, crossover and variation. Among these steps, code choice is a difficult point. We use "0, 1," to represent "level tones, oblique tones" and use wildcard "*" to represent "can be level can oblique ".

Genetic algorithm includes the following main steps:

1 Coding: Before the GA to search the solution space, express the solution data as a genotype string structure of the space. The different combinations of these string structure data make up the different points. Coding is a kind of means of mapping in solution space.
2 Initial population generation: randomly generated N initial list data structures, each structure is called an individual, N individuals form a group. The genetic algorithm takes the N series structure data as the initial point to iterate. The size of the initial population should have enough scale and randomness.
3 Fitness function assessment test: fitness function is used to show the advantages and disadvantages of a solution. A different problem, the definition of fitness function mode is also different. Fitness combines the genetic algorithm and the original optimization problem.
4 Selection operator: the purpose is to select excellent individuals from the current group. So they have a chance to sons as the parent to the next generation of breeding. By selecting process embodies the idea of genetic algorithm to select principle is an adaptable individual probability of contributing to the next generation of one or more of the offspring. Select operation reference to the evolution of the corresponding point of view: the higher the individual fitness, the more it is a choice, the so-called principle of survival of the fittest.
5 Crossover operator: crossover operation is the main genetic operators in genetic algorithms. Through the crossover operation, two individual chromosome gene swap, resulting in a new generation of individuals. Characteristics of new individual combine the features of their parents. Cross reflects the information exchange of ideas.
6 Mutation operator: in the biological individual variation is an important step in the process of reproduction. By mutations in certain genes on the chromosome location makes new individuals are different from other individuals. Variation selected individuals randomly in the group, for the selected individuals randomly alter string structure data at a certain probability in the value of a string. Like biology, incidence of the mutation in GA is very low, but it provides the opportunity for new individuals.

Selection of the design of fitness function and genetic algorithms is directly related to the operation. In addition, it also affects the genetic algorithm iteration stop conditions. The experiment of fitness evaluation is mainly based on the following four indexes: G: (1) grammar legitimacy: through the test of DFA score is 1, otherwise 0. (2) Subject correlation: equal to the sum of relevancy of each word and subject. (3) Words collocation appropriateness: equal to the sum of relevance of every two consecutive words. (4) Emotion and style of unity: Adaptive value function F design for the weighted sum of the above four quantities.

In addition, we also provide some public tools in the teaching system, such as QQ discussion groups, BBS, E-mail, etc., as teachers and learners to various channels of communication, such as temporary release instruction, teachers will be in the QQ group or sent via email.

3 EXPERIMENTAL ANALYSIS

The system requires the user to input a subject word, and then let the user choose rhymes. Here the user enters the "plum blossom" theme and select the "Xian" as rhymes. According to the meaning of relevance and the calculation of similarity, the lookup in the tang dynasty segmentation corpus, find relevant words, then, according to the requirements of the rhyme categories and five greats formats, through a series of calculations, generate the original work.

The work abides by the metrical pattern. The subject is about plum blossom. The lexical relevancy is high.

4 CONCLUSIONS

The five-character-quatrain generation system has a lot of experience can be summarized. Poetry describes, for example, is a kind of artistic conception. The sentences are logic. The machine's generated poet can abide by the metrical pattern well, but lack integrity. It feels like an associated words pile. The appreciation of poetry has always been an individual process, and is mostly limited in the field of literature. It is difficult to form a standard evaluation system.

Learners also deeply felt: because it is the first time to use the system, it need time to be familiar with it and its tools. In addition, the teacher and group leader's concern for members also significantly affects the learners' motivation.

Synergistic Learning, this new learning mode requires the participation of more teachers and learners. And how to establish a scientific evaluation system of community members, to more objectively evaluate the learners, to reduce their dependence on teachers and the leader of the group, also is our next focus.

ACKNOWLEDGEMENT

This work was supported by the scientific research project of the Department of Education of ZheJiang Province (Y201328703), a teacher professional development program of colleges and universities visiting scholar in ZheJiang Province (FX2012131), the national education information technology research "twelfth five-year"plan youth project (No.136241221), college students of science and technology innovation project in ZheJiang Province (xinmiao talent plan) (No.2014R446 002).

REFERENCES

Zhiting Zhu,Youmei Wang,Hongwei Luo.2007.Synergistic Learning for Knowledge Age:Theoretical Model, Enabling Technology and Analytical Framework. *Proceedings of ICW2007 Lecture Notes in Computer Science.*/Heidelberg:Springer.

Dongmin Qian.2011.The application research of Synergistic Learning.Shanghai.

Xu Bi.2006.Research on imago retrieval based on corpus of Tang poetry.Dalin.

He Jing,Ming Zhou,Long Jiang.2010.Generating Chinese Metrical Poetry by a Statistical MT Approach.*Journal of Chinese Information Processing*24(2):96–103.

Jinsong Su,Changle Zhou,Yihong Li.2007.The Establishment of the Annotated Corpus of Song Dynasty Poetry Based on the Stastistical Word Extraction and Rules and Forms. *Journal of Chinese Information Processing*21(2):52–57.

Junfeng Hu,Shiwen Yu.2002.Word Meaning Similarity Analysis in Chinese Ancient Poetry an its Applicatins. *Journal of Chinese Information Processing*16(4):39–44.

Zhihe Yang, Xiaoqing Gu, Ziting Zhu.2009.The New Development of CSCL Technology.*China Educational Technology* (275):110–115.

Computing, Control, Information and Education Engineering – Liu, Sung & Yao (eds)
 ISBN: 978-1-138-02800-5

Design of remote monitoring station for sewage treatment based on embedded platform

Pei Zhi Wen, Chen Jiao Xu, Wen Ming Huang & Zhen Rong Deng
Guilin University of Electronic and Technology, Guilin, Guangxi, China

ABSTRACT: Aim at problems of sewage treatment stations in the domestic rural areas such as difficult to manage because of geographically dispersed, prior to process data and collect water quality in real-time, less-automated of treatment equipment and so on, wc proposed a design of the remote monitoring station for sewage treatment based on embedded technology. Monitoring station uses S3C2440 microprocessor as the core of local embedded platform, controls the operation of sewage treatment equipment with programmable logic controller (PLC), collects the real-time water quality data with wireless sensor network (WSN) which is built with ZigBee technology, realizes the data remote transmission between monitoring station and monitoring center by GPRS wireless communication technology. Test results show that the design has dependable performance, good real-time, high automatization, Low operation cost and excellent application prospects.

KEYWORDS: Embedded system; Remote monitoring; Data acquisition; Sewage treatment; Rural areas.

1 INTRODUCTION

Along with the rapid development of economy and society in china, urbanization process accelerated, the water pollution problem is becoming more and more prominent in rural areas. In recent years, a lot of small to medium sized sewage treatment stations were built in rural areas, greatly improved the ability of controlling water pollution, but there are serious problems of stations such as low automatic degree of equipment, complicated environment, difficult to collect data, inconvenient to manage for geographically fragmented and so on [1-4]. Therefore, this paper presents a design of the remote monitoring station for sewage treatment, with embedded technology as core, combined PLC technology, Zigbee technology and GPRS wireless communication technology, it realizes automation of sewage treatment, networking of data transmission and centralization of management.

2 HARDWARE DESIGN SCHEME

In this paper, the hardware of monitoring station consists of four parts: embedded processor module, data acquisition module, control module of sewage treatment equipment and wireless communication module, as shown in figure 1. Embedded processor module as the core part of local embedded monitoring system, achieves normal operation of monitoring stations. The data acquisition module is used for three kinds of water quality parameters collection, temperature, PH and dissolved oxygen, then sends to the processor module by wireless sensor network (WSN). Control module of sewage treatment equipment is responsible for the normal running of equipment. Wireless communication module realizes data remote transmission between monitoring station and monitoring center.

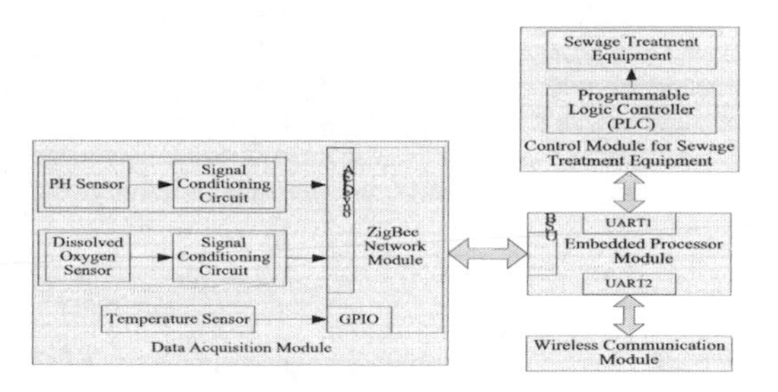

Figure 1. Hardware structure.

The data acquisition module is composed of ZigBee wireless communication and water quality sensors, modular structure as shown in figure 1. Water quality sensors are waterproof temperature sensor (DS18B20), electrode assembly pH (E-201-9) and dissolved oxygen electrode covered (JYD-2)[5][6]. Generally, the impedance of chemically electrode is large and the output signal is weak, so process signal by conditioning and amplifying circuit is needed, then convert analog signals to digital signals by A/D converter. This paper uses addition circuit to amplify the pH electrode's output voltage signal. Figure 2 is the addition circuit schematic diagram. The circuit uses four road operational amplifier LM324, adjusts output voltage by changing the value of resistor R1. The output current of dissolved oxygen electrode is weak, the signal need to be processed by an I-V conversion circuit. So we use AD795 chip which has low drift and low noise performance to make an I-V conversion circuit, as shown in figure 3. We choose CC2530 chip of TI as the rf chip of the ZigBee network module to realize the network transmission of water quality data[7]. CC2530 chip embeds lowest-power 8051 microcontroller core which has the code prefetching function, includes a 2.4GHz high-performance rf transceiver with high performance direct sequence spread spectrum, an A/D converter with 12bit and 8-ch, the chip can process data of sensor by different channel of the A/D conversion.

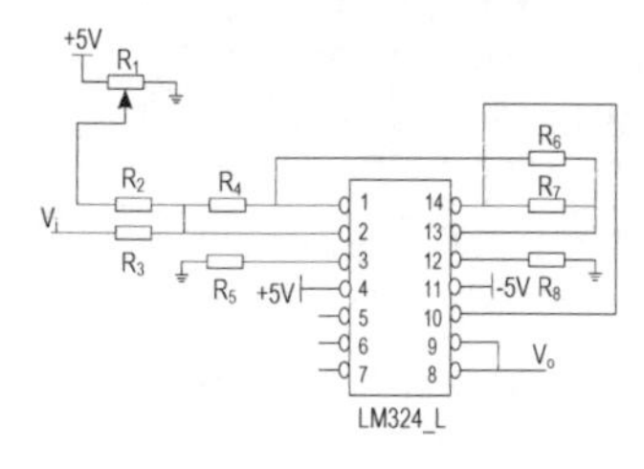

Figure 2. Addition circuit.

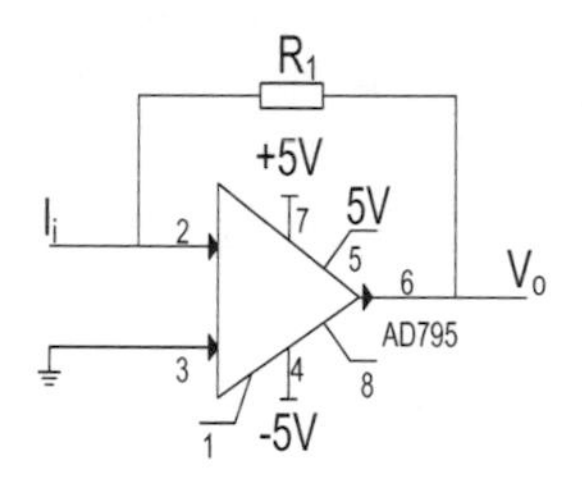

Figure 3. I-V Conversion circuit.

A control module for sewage treatment equipment, we choose a programmable logic controller (PLC) FX1N-60MR-001 made by Mitsubishi to control the operation of sewage treatment equipments. PLC takes a kind of programmable memory which is used for an internally stored program to execute a series of user-oriented instruction and controls various types of machinery or production process by the digital or analog input/output. FX1N-60MR-001 PLC built-in 36 DI and 24 DO, adopts advanced encapsulation method, pin spacing distribution is reasonable, suitable for control in small or medium environment.

Wireless communication module, we use the WG-8010 industrial GPRS wireless data transmission module. This module provides standard RS232/485 data interface, supports TCP and UDP protocols. The module's initialization configuration is more convenient, user device can establish connection with data center via GPRS wireless network by one-time configuration, and realizes transparent transmission of data.

The embedded processor module is a core component of monitoring stations, in charge of normal operation of various functions for entire system includes user operations, device control, resource management, data transmission and so on. The embedded processor module of this design mainly consists of main processor, power management circuit, USB interface, RS232 serial interface, JTAG debug interface, LCD touch-sensitive screen, memory circuit, etc., as shown in figure 4. We use S3C2440 made by Samsung as the main processor [8]. S3C2440 microprocessor is developed upon the 16/32-bit ARM920T kernel that designed by ARM. The chip includes 130-bit general purpose I/O ports,24-ch external interrupt source, LCD Controller (STN&TFT), MMU to handle virtual memory management, 3-ch UART, 2-ch USB Host/ 1-ch USB Device, etc. It is a system level chip with low-cost and high-performance.

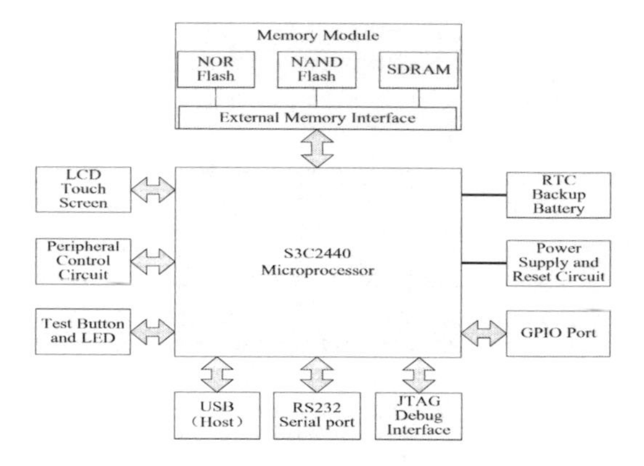

Figure 4. Embedded system hardware structure.

3 SOFTWARE DESIGN SCHEME

The paper adopts the modularized program design method to realize the monitoring station's software system design. It includes establishing WSN and data acquisition, making communication protocols of GPRS, developing application programs of the embedded monitoring system.

The data acquisition module includes establishment of WSN and water quality data acquisition. Wireless sensor network is made up of multiple sensor nodes, We adopt the tree structure to organize WSN and divides nodes into three categories: data acquisition node, routing node and coordinator node. Data acquisition nodes are terminal nodes of the network, collect water information from different sensors. Routing nodes forwards data packets of child nodes to coordinator node. Coordinate node connects to the embedded monitoring system via serial port, realizes data transmission between WSN and monitoring system. In general, the coordinator node begins to establish network and collect data after receiving the data acquisition command from the monitoring system. The program design process of data acquisition part as shown in figure 5.

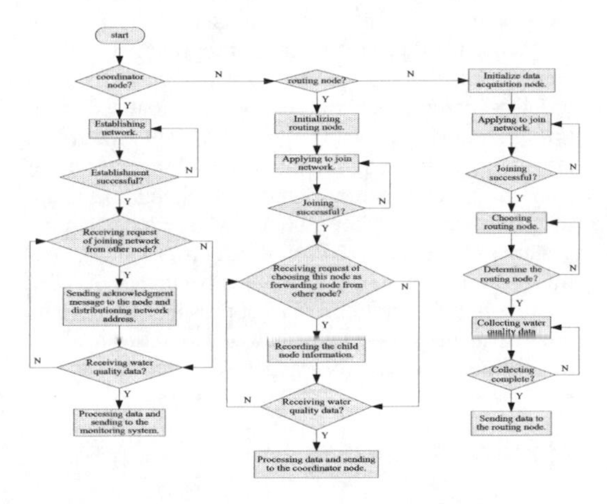

Figure 5. Program flow of data acquisition.

The wireless communication protocol of GPRS part achieves the monitor station uploads local data, receives remote control commands. We set the GPRS state machine variable[4] to display the status of wireless communication, the value is shown in table 1. In addition, we make some rules for the packet frame format[9] to ensure that the monitoring center can read correct data. The packet frame format is shown in table 2.

Table 1. Status value.

status value	operation executive
GS_NET_SEARCH	search the GPRS wireless network signal
GS_NET_STATUS	query wireless network status
GS_NET_LAND	log in wireless network
GS_NET_CONNECT	establish a wireless network connection
GS_NET_LINKCONFIRM	query whether the link successfully established
GS_NET_CLOSE	close wireless network
GS_NET_DISCONNECT	close wireless connection
GS_DATA_SEND	send data

Table 2. Packet frame format.

Frame start identifier	PH start tags	PH value	PH end tags	DO start tags	DO value	DO end tags
MSG	PH	X	PH	DO	Y	DO

TEMP start tags	TEMP value	TEMP end tags	PLC status start tags	PLC Status value	PLC Status end tags	Frame end identifier
TT	Z	TT	PLC	W	PLC	END

Monitoring system application mainly contains the following aspects: control mode, wireless sensor network, sewage treatment equipment, wireless communication and system setup. We choose Linux as the operating system and develop applications in Qt development framework [10] [11]. Figure 6 is the structure diagram of an application.

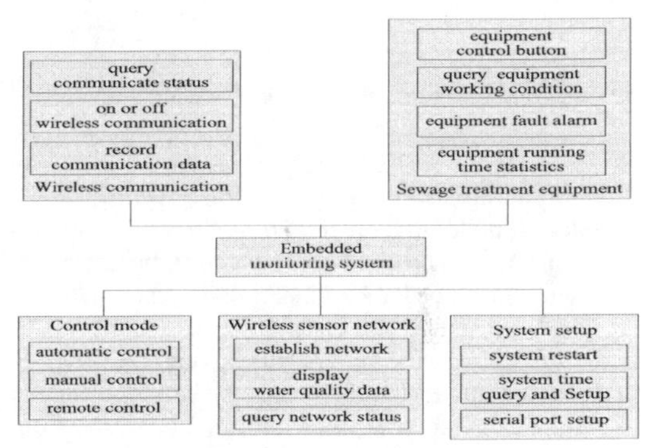

Figure 6. Application structure.

Control mode program determines control mode of monitoring system for sewage treatment equipment, it divide into three conditions: automatic control, manual control and remote control. In automatic control mode , the operation of monitoring system according to the preset running time of equipment. Manual control mode is monitoring personnel operate devices switch on touch screen to control the running of equipment. Remote control mode is the monitoring center according to monitoring data directly send control commands to adjust the running of sewage treatment equipment.

Wireless sensor network application programs are divided into organization and status query of WSN and water quality data display. Organization and status query of WSN program realizes the function of building WSN and the monitoring system can view the current network status. Water quality data display program displays data collected by sensor nodes in real-time on the screen.

Sewage treatment equipment consists of equipment control, fault alarm, equipment working status query and total equipment running time statistics. Equipment control program realizes system directly control the running of equipments by control buttons on touch screen. Working status query, fault alarm and running time statistics of equipment program are monitoring system send related inquiry command to the PLC and display the feedback information on the screen.

Wireless communication program realizes the system connect to mobile networks, query network connection status , sending data and receiving commands of monitoring station, meanwhile, the program could record communication data.

The system setup program contains system restart, system time setting and serial port control function. The process of system time setting will set the Linux system time and reset the PLC internal system time. A serial port control program of monitoring system operates serial port 2, serial port 3 and USB port. We adopt the way of creating a child thread to read and write serial data.

4 REALIZATION OF SYSTEM FUNCTION

We chose the school's sewage treatment station to verify the monitoring station design. We placed three sensor monitoring nodes in a different position in the monitoring area, and connected PLC of sewage treatment equipment to the embedded monitoring system platform. After system started, the data of wireless sensor nodes and the working status of equipment could be displayed on the main interface in real-time. We used a computer to simulate a monitoring center, simulation of the monitoring center could normally communicate with the embedded monitoring station. Table 3 is the monitoring data of one sensor monitoring node in 10 hours. It could be seen from table 3 that data collection of WSN was complete, stable and reliable. As seen from the implementation result of system function, the design of monitoring station in this paper had stable performance, high automatization, good reliability and real-time.

Table 3. Record of water quality data.

Time	TEMP (^{o}C)	PH	DO (mg/L)
08:00	10.8	6.9	6.84
09:00	11.3	7.1	7.45
10:00	11.8	7.0	7.23
11:00	12.2	7.3	6.33
12:00	12.2	6.8	6.81
13:00	12.7	6.9	7.21
14:00	12.6	7.2	7.41
15:00	12.6	7.3	7.12
16:00	12.1	6.9	6.92
17:00	11.8	6.7	7.35
18:00	11.7	6.9	7.12

5 CONCLUSION

In this paper, we did some research and design of the sewage treatment monitoring station. The design adopted embedded technology, programmable logic controller technology, Zigbee technology and GPRS wireless communication technology, realized real-time monitoring of water data, automatic control of sewage treatment equipment, remote transmission of data and other functions. The implementation result of system function showed that the design had simple system structure, good feasibility, convenient for centralized management and other advantages. Finally, the design also had good market prospect, it could be applied to the other environmental monitoring field by changing the type of sensor.

ACKNOWLEDGEMENT

The research work reported in this paper is supported by "Natural Science Foundation of Guangxi Province, China (GKZ No.0991240)", "Science and Technology Plan of Guangxi (GKG No.11107006-10)", "Science and Technology Plan of Guilin (No.20130105-7)".

REFERENCES

[1] Yu Wu. 2013.Research on small mobile water quality monitoring system. Hangzhou: Zhejiang University.

[2] Juwei Peng, Chengwen Wang, Hongxiang Fu, et al. 2010. Analysis for scattered rural sewage treatment pattern. *Environment and Sustainable Development* 1:28–30.

[3] Yufeng Lv.2014.Design and Research of monitoring control system and optimization of operation for sewage treatment automation. Jinan: Shandong University.

[4] Zhihui Tang, Hongchao Li. 2011. Design of intelligent monitoring system of water quality based on embedded processor technology. *Instrument Technique and Sensor* 2:99–101.

[5] Ying Zhao, Deqing Mo, Jian Han. 2013.Research on a remote monitoring system for water quality. *Journal of Guilin University of Electronic Technology* 33(2):118–121.

[6] Xiguang Du. 2010. Design of the weak optical signal detection circuit. *Electronic Component & Device Applications* 12(1): 52–54.

[7] Weicong Zhang, Xinwu Yu,Zhongcheng Li. 2011. Wireless network sensor node design based on CC2530

and ZigBee protocol stack. *Computer Systems & Applications* 20(7):184–120.
[8] S3C2440A 32 - Bit CMOS Microcontroller User' s Manual, Revision 1. 2004.*Samsung Electronic*.
[9] Deqing Mo, Jian Han, Ying Zhao,Ying Liang.2014. Sewage disposal remote monitoring system based on GSM and LabVIEW. *Measurement & Control Technology* 33(4):62–65.
[10] Meixia Duan, Shuxia Yao. 2010.Design and realization of instrument of detecting water quality based on wince remote-data capturing. *Computer Measurement & Control* 18(4):968–970.
[11] Ming Guo. Design of remote wireless monitoring system for sewage based on GPRS. 2013. Nanchang: Nanchang University.

Computing, Control, Information and Education Engineering – Liu, Sung & Yao (eds)
© 2015 Taylor & Francis Group, London, ISBN: 978-1-138-02800-5

An improvement of Chinese text hierarchical clustering algorithm

X.L. Liu, Z.G. Chen, B. Zeng, Q.X. Zhu & T. Chen
Secure Communication Key Laboratory, University of Electronic Science and Technology of China, Chengdu, China

ABSTRACT: Hierarchical clustering algorithm has low computational efficiency and error accumulation problem in iterative clustering process. To deal with the problems, we propose an improvement of hierarchical clustering algorithm based on GAAC (Group-average Agglomerative Clustering), and the improved algorithm is applied to Chinese text clustering. The results of our experimentation show that the improved algorithm have been improved greatly in computational efficiency and the quality of clustering results.

KEYWORDS: Hierarchical clustering algorithm; Text clustering; Chinese text.

1 INTRODUCTION

Text clustering is to divide the series of documents into several text clusters, so that the topic text within the same text cluster is as close as possible, and the text between the different clusters of text have a smaller similarity. Common clustering algorithms include division of clustering[1][2], hierarchical clustering[3], density-based clustering[4], grid-based clustering[5][6] and model-based clustering[7], etc.

Since many people think hierarchical clustering algorithm has better clustering results than the flat clustering algorithm[9], hierarchical clustering algorithm has been widely used in practice. The hierarchical clustering includes agglomerative clustering and divisive clustering. Depending on the similarity calculation method, agglomerative hierarchical clustering algorithm can be divided into single-link, complete-link, group-average and centroid similarity methods.

However, hierarchical clustering algorithm has problems with low efficiency of arithmetic operations, arithmetic errors and iterative error accumulation. In order to solve the problems, we propose to set the threshold of clustering, once merger several clusters and optimize matrix operation, based on the traditional hierarchical clustering algorithm optimized processing.

In this paper, we describe the NLPIR Chinese word system and the traditional GAAC in Section 3. And then in Section 4 we illustrate an improved algorithm based on GAAC. The analysis of corresponding results are given in the Section 5. Finally, our conclusion to this study is given in section 6.

2 RELATED WORK

Classification algorithm is also a common text classification method, such as KNN (K-nearest-neighbor). Shengyi Jiang[10] and others have made some improvements on the KNN algorithm. By combining clustering algorithm and KNN algorithm, they improved the efficiency of the algorithm. However, as all the same classification algorithm, KNN need the sample data set and need to be marked on the sample, this is not suitable for handing with unknown type of massive text data. In addition, Viet Ha-Thuc[11] proposed a large-scale hierarchical text classification algorithm without labelled data. They are clustered by determining the relationship between the characteristics of each text and the relationship between the text clusters and cluster classification, but this algorithm can not choose a different cluster size depending on the text data.

3 TRADITIONAL ALGORITHM

3.1 *NLPIR Chinese word segmentation system*

As the improved algorithm proposed in the experimental part is processing the Chinese text data, we use NLPIR[12] Chinese word system in the data preprocessing stage. The system includes Chinese word segmentation, keyword extraction, and other functions. By using the Chinese word segmentation system, we convert the Chinese document data into a list of keywords to apply it in our algorithm.

First lines of paragraphs are indented 5 mm (0.2") except for paragraphs after a heading or a blank line (First paragraph tag).

3.2 *Group-average agglomerative clustering*

1 Calculating the TF-IDF value obtained in Part A

TF (Term Frequency) which means frequency, represents the frequency of word t_i in the document d_j.

$$tf_{i,j} = \frac{n_{i,j}}{\sum_k n_{k,j}} \tag{1}$$

IDF(Inverse Document Frequency) which means inverse document frequency, is used to measure the general importance of a word.

$$idf_i = \log \frac{D}{|\{j : t_i \in d_j\}|} \tag{2}$$

D: The total number of documents in the corpus

The number of documents containing the word t_i

Finally, we can calculate the TF-IDF value of this word by following formula.

$$tfidf_{i,j} = tf_{i,j} \times idf_i \tag{3}$$

2 Assuming the number of articles which contained in set D is N, we can calculate the TF-IDF value resulting in a N*N similarity matrix. N_{ij} represents cosine similarity between text D_i and text D_j.

$$sim(X,Y) = \frac{\vec{x} \bullet \vec{y}}{\|x\| \bullet \|y\|} \tag{4}$$

3 Find the most similar to the two vectors.

4 Combine the two vectors and obtain a new vector combined; recalculate new (N-i) *(N-i) of the similarity matrix.

5 Repeat step 3) and step 4) N-1 times and obtain the final binary tree.

4 ALGORITHM IMPROVEMENT

From section 3 we know that we have to recalculated the step 3 and step 4 and traverse the entire (Ni)*(Ni) of the similarity matrix every time. Since N can be a great value, then it will be a very large computational overhead for this N - i times calculation and it will have a great impact on the efficiency of the clustering algorithm.

Further, when the step2 calculate the cosine distance, the traditional method does not take into account the cosine function is more distinction between the direction of the vectors, while the absolute value is not very sensitive, could cause the actual similarity of two vectors which are small calculated high similarity, and ultimately affect the clustering results.

Meanwhile, in the traditional algorithm, we find the closest two vectors in one time, and then later in the clustering process use the new vector instead of the original two vectors. Due to the combined vector is not exactly the same as the previous two vectors, and we can only select the most similar to the merger of two. If this minimum is error value, it will result in the accumulation of error iterations, greatly affect the final clustering results.

To solve these problems, we propose a hierarchical clustering algorithm improved. Specific algorithm described as follows:

1 Suppose the text set D contains N articles text data, calculating the TF-IDF value of the feature word ti and obtaining the N eigenvectors;

2 To fix the cosine similarity of values insensitivity, we subtract a mean from all dimensions to avoid because of the similarity of the numerical calculation is insensitive to errors caused when calculating text D_i and D_j;

3 After calculation of TF-IDF values, we obtain a N*N matrix characteristics, where N_{ij} represents the cosine text between Di and D_j text similarity;

4 Because N_{ij} = N_{ji}, and we set the text Di self-similarity is 1, so the algorithm in dealing with N*N similarity matrix will do (N-1)*(N-1) times repetitive work, so the matrix is stored in a linked list containing the (N-1)*(N-1) values, and after that we only deal with this list;

5 Given a minimum similarity threshold *argmin*, the multiple vectors of which similarity is greater than *argmin* are disposable combined. On the one hand, it can avoid the accumulation of errors iterative data, on the other hand, it can reduce the number of iterations of the algorithm;

6 We only update vector associated with the new similarity value when updating the list, without recalculating all the (N_i-1)*(N_i-1) similarity value;

7 Repeat step 5 and step 6 several times until the list is no longer included vector group which is more than *argmin*, then in accordance with the principle of maximum similarity merger, agglomerate the rest of the vector group.

5 CONCLUSION

Our algorithm implementation uses JAVA programming language and runs on Windows 8 operating system with Intel Core i3-2100 processor, 4GB of memory. The experimental data is crawled from Sina news reports by a python crawler and each one is saved as a text xml format.

When we take the different values of the algorithm threshold, we will get the result with different size text clustering. As we can see from Figure 1, with the increase of the number of clustering text, the growth rate of algorithm computation time is increasing, however, the threshold value have little effect on the running time.

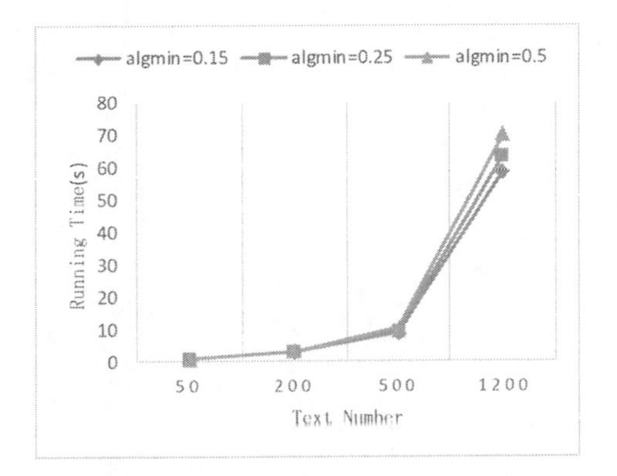

Figure 1. The running time of the improved algorithm with the increase of the threshold.

On the other hand, it can be seen from Figure 2, which compared with the traditional hierarchical clustering algorithm, the improved algorithm increases the number of text processing with a slower increasement of computing time, while the traditional clustering algorithm has become very slow when the text reaches 500 or more. So, the improved algorithm has a great advantage in computing efficiency.

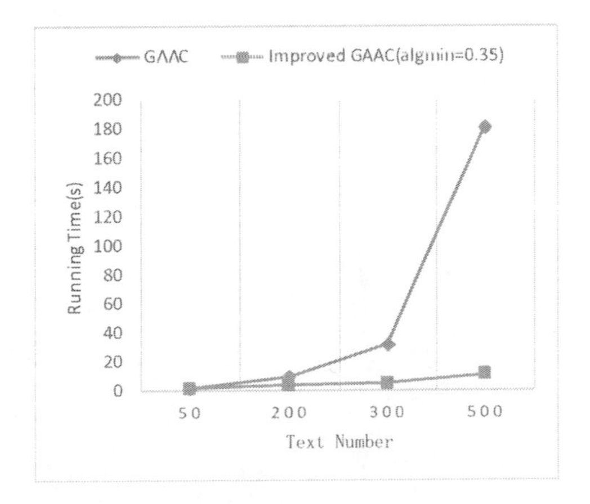

Figure 2. The running time comparison between the traditional algorithm and our improved algorithm.

6 SUMMARY

The operation efficiency of traditional agglomerative hierarchical clustering is low, and it does double counting the same time for most similarity values. If there is unreasonable merge because of the error values, the error will continue to accumulate as the iteration.

To solve this problem, we present an improved hierarchical clustering algorithm based on Chinese documents. The algorithm for similarity calculation method has been improved and only recalculate the vectors which are updated. So our operational efficiency is improved greatly. At the same time, we increase the judgment of threshold and merge multiple vectors once, avoiding the errors caused by the accumulation of pairwise merge. However, how to select the appropriate threshold to get the best clustering effect will be our main future research studies.

ACKNOWLEDGEMENT

This work was supported in part by the National Natural Science Foundation of China, No.61402080 and China Postdoctoral Science Foundation funded project, No.2014M562307.

REFERENCES

[1] MacQueen J. Some methods for classification and analysis of multivariate observations[C]//Proceedings of the fifth Berkeley symposium on mathematical statistics and probability. 1967, 1(14): 281–297.
[2] Huang Z. Extensions to the k-means algorithm for clustering large data sets with categorical values[J]. Data mining and knowledge discovery, 1998, 2(3): 283–304.
[3] `Cilibrasi R L, Vitányi P. A fast quartet tree heuristic for hierarchical clustering[J]. Pattern recognition, 2011, 44(3): 662–677.
[4] Ester M, Kriegel H P, Sander J, et al. A density-based algorithm for discovering clusters in large spatial databases with noise[C]//Kdd. 1996, 96: 226–231.
[5] Su M C, Chou C H. A modified version of the K-means algorithm with a distance based on cluster symmetry[J]. IEEE Transactions on pattern analysis and machine intelligence, 2001, 23(6): 674–680.
[6] Agrawal R, Gehrke J, Gunopulos D, et al. Automatic subspace clustering of high dimensional data for data mining applications[M]. ACM, 1998.
[7] Kohonen T. Self-organized formation of topologically correct feature maps[J]. Biological cybernetics, 1982, 43(1): 59–69.
[8] Jain A K, Dubes R C. Algorithms for clustering data[M]. Prentice-Hall, Inc., 1988.

[9] Larsen B, Aone C. Fast and effective text mining using linear-time document clustering[C]//Proceedings of the fifth ACM SIGKDD international conference on Knowledge discovery and data mining. ACM, 1999: 16–22.

[10] Jiang S, Pang G, Wu M, et al. An improved< i> K</ i>-nearest-neighbor algorithm for text categorization[J]. Expert Systems with Applications, 2012, 39(1): 1503–1509.

[11] Ha-Thuc V, Renders J M. Large-scale hierarchical text classification without labelled data[C]//Proceedings of the fourth ACM international conference on Web search and data mining. ACM, 2011: 685–694.

[12] NLPIR Chinese word segmentation system http://ictclas.nlpir.org/.

Computing, Control, Information and Education Engineering – Liu, Sung & Yao (eds)
© 2015 Taylor & Francis Group, London, ISBN: 978-1-138-02800-5

An empirical analysis of housing price based on hedonic pricing theory: Evidence from Chengdu city

K. He, M.S. Zhang & H. Wang
School of Management, Southwest University for Nationalities, Chengdu, China

ABSTRACT: Taking 231 residential real estates of Chengdu in March, 2014 as the research objects, a hedonic price model is built to quantitatively analyze the factors of urban housing price. Our results indicate that CBD distance, metro station, decoration quality, property fee, and greening rate have a significant impact on the re-al estate prices in Chengdu, whereas the bus route, plat ratio, education facilities and life facilities are not significant. The findings reveal that the buyers and investors in Chengdu are willing to pay an additional price for location characteristics and building characteristics.

KEYWORDS: Housing Price; Chengdu; Hedonic Pricing Models.

1 INTRODUCTION

Despite the Hedonic Pricing Model is widely used for housing valuation. However, the applications to Chinese cities are limited because of data availability, which constrains housing research in China. With the rapid development of the economy, the data are becoming accessible and more and more scholars are starting to devote themselves to hedonic housing price research in China. Housing studies of Chinese cities can offer useful policy implications, because China is in a unique transitional stage of economic development.

The literature on hedonics is voluminous. Lancaster (1966) and Rosen (1974) lay the foundation for Hedonic Pricing Theory. After that, many scholars make use of hedonic models for housing valuation. Ozanne (1985) finds that there are many characteristics of a house. [1] Therefore, variable selection is crucial to hedonic model specification. Haurin (1996) summarizes a number of case studies, and selects out six principal components.[2] Adair (2000) examines the influence of house characteristics, house accessibility and social-economic factors in the Belfast Urban Area.[3] Kestens (2006) uses household-level data to measure the heterogeneity of implicit prices regarding house type, income, education facility and house age.[4]

In 1990s, the literature on hedonic research begins to appear in China. Jiang Yijun and Gong Jianghui (1996) introduce the hedonic theory to the housing price valuation in China at first, and they summarize the way to build hedonic models according to the foreign acquirable literature. [5] Wang De and Huang Wanshu (2005) classify the foreign literature and demonstrate the feasibility to apply in China.[6] He dan and Jin Fengjun (2013) take the houses nearby Beijing Metro Line 4 for sample to analyze the effect of traffic condition.[7]

Although research of Hedonic Price Theory is relatively mature in the west, it is not in China. Most research objects are big cities such as Beijing, Shanghai and Guangzhou. In this paper, we make use of publicly accessible data to build a hedonic pricing model about Chengdu, a leisure Chinese city, to examine the implicit prices of housing characteristics, such as location and building characteristics.

2 THEORY BASIS AND MODEL FORM

2.1 *Theory basis*

Housing price is closely associated with its characteristics such as size, quality and location, so it is a typical heterogeneous commodity. Lancaster (1966) and Rosen (1974) lay the foundation for the hedonic pricing theory. Lancaster proposes the theory of consumer preference, and believes that commodity consists of various characteristics. [8] It is sold as a collection of intrinsic characteristics which form a package that can affect utility. The demand for a good is based on characteristics the good contains inside rather than the good itself. Rosen proposes the market supply-demand equilibrium model. [9] He analyzes the short-term and the long-term equilibrium of the heterogeneous commodity market in the frame of the full competition market. When the Hedonic Price Curve is tangent to the supply curve and demand curve, the maximum utility to the consumer and the maximum profit to the producer are both achieved and the equilibrium is realized.

2.2 *Selection of model form*

The general hedonic model is as follows:

$$p = F c_1, c_2, \ldots, c_n \tag{1}$$

The $c_1, c_2, \ldots, c_n$ respectively indicate the housing characteristics. P stands for the market house price. With the same conditions in all other aspects, working out the partial derivatives of each housing characteristics in the equation will come to the implicit price of each characteristic. So the hedonic price is:

$$p_{c_i} = \frac{\partial p}{\partial c_i} = \alpha_i, (i = 1, 2, \ldots, n) \tag{2}$$

Therefore, the coefficient α_i in the equation stands for the implicit price of each characteristic and the algebraic sum will make the total price of the house. Linear, logarithmic, and log-linear models are mainly three forms of function. Through the trial of the model, it is found that log-linear function has a good explain ability. Therefore, log-linear function is adopted to express the relationship between housing characteristics and housing price:

$$LnP_i = \alpha + \beta_1 C_1 + \beta_2 C_2 + \ldots\ldots + \beta_n C_n + \varepsilon_i \tag{3}$$

3 DATA AND VARIABLES

Five districts of Chengdu are selected as the research area, and they are Wuhou District, Qingyang District, Jingjiang District, Chenghua District and Jinniu District. The sample data is collected from the website of http://www.cd.fang.com in March 2014. The total number of collected samples is 231. The information about second-hand housing, including the listing prices, type of building, degree of decoration, property management fee, administrative regions, as well as external information such as environmental characteristics of the neighborhood and so on.

There are two type variables: independent variables and the dependent variable. The listing price is dependent variable, and the housing characteristics are the independent variables. Considering the real situation, 9 variables are selected as dependent variables, and they are categorized into three dimensions, namely location characteristics, building characteristics and neighborhood characteristics. Table 1 is the total of housing characteristics variables as follows:

Table 1. Total of housing characteristics variables.

Characteristics	Variables	Definition
Location characteristics	CBD distance	Straight line distance to Chengdu CBD(in km)
	Bus route	Number of bus routes within 1 km
	Metro station	Dummy variable:1 if there is a metro station within 1 km,0 otherwise
Building characteristics	Decoration quality	The quality of the decoration is measured in three degree: good(score of 3),common(score of 2),poor(score of 1)
	Property fee	The property management fee paid by every month(yuan/m²)
	Greening rate	Percentage share of the site area that is covered by greenery
	Plot ratio	The ratio of total floor area of the entire development to its site area
Neighborhood characteristics	Education facilities	Is there any kindergarten, primary, secondary school in or 1000m surrounding the district, 3 scores, one for each
	Living facilities	Is there any shopping mall, hospital, post office, bank in or 1000m surrounding the district, 4 scores, one for each

4 HEDONIC PRICE MODEL VERIFICATION

4.1 *Descriptive statistics analysis for each variable*

The first step of statistical analysis is the descriptive statistics analysis, as described in Table 2.

According to the sample statistics, the mean of housing price is 9670.58, which is lower than that of the other metropolitans like Beijing and Shanghai. The means of education facilities and living facilities are respectively 2.49 and 3.61, which indicate that education and living facilities of Chengdu district are relatively complete

Table 2. Descriptive statistics of dependent variable and independent variables.

	N	Min	Max	Mean	Std.dev
Housing price	231	4907	18500	9670.58	2459.078
CBD distance	231	1	4	2.58	0.987
Bus route	231	1	30	9.99	5.398
Metro station	231	0	1	0.21	0.410
Decoration quality	231	1	3	1.22	0.611
Property fee	231	0.3	15.0	1.924	1.2932
Greening rate	231	15	50	31.71	6.405
Plot ratio	231	1.1	10.0	3.722	1.5362
Education facilities	231	1	3	2.49	0.703
Living facilities	231	1	4	3.61	0.663

4.2 *Significance test and analysis of variance*

It can be seen from Table 3, the coefficient of determination, R2 and the adjusted R^2 are respectively 0.355 and 0.329. That means the basic model can explain about 33% of the difference of dependent variables, indicating that the model has a relatively good fitting and explanation.

Table 3. Model summary.

R	R square	Adjusted R Square	Std. Error of the Estimate	Durbin-Watson
.596[a]	.355	.329	2014.063	2.040

From Table 4, the significance testing value of variance analysis is 0.000, which means that the equation is highly significant and rejects the assumption that all coefficients are 0. It also proves that log-linear relationship between the housing characteristics and listing price is tenable. The Durbin-Watson value of the model is 2.040, which is close to 2. It means that there is no statistically significant auto correlation.

Table 4. ANOVA.

Model	Sum of Squares	df	Mean Square	F	Sig.
regression	4.943E+08	9	5.493E+07	13.541	.000[a]
Residual	8.965E+08	221	4.056E+06		
total	1.391E+09	230			

4.3 *Collinearity diagnostics*

From Table 5, the maximum of VIF value of all variables is 1.566, which is far smaller than 10. As a result, the collinearity between the independent variables can be regarded as not serious.

Table 5. Analysis of collinearity.

	Co linearity Statistics	
Model	Tolerance	VIF
CBD distance	.713	1.402
bus route	.903	1.108
metro station	.867	1.153
decoration quality	.889	1.125
property fee	.881	1.135
greening rate	.831	1.203
plot ratio	.652	1.534
education facilities	.639	1.566
living facilities	.704	1.421

4.4 *Normality test of residuals*

It is can be seen from the residual histogram (Figure 1) and cumulative probability diagram (Figure 2), the residual distribution is similar to normal distribution.

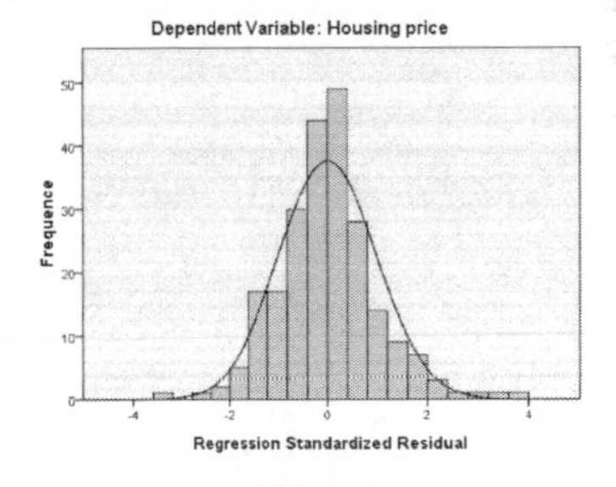

Figure 1. Residual histogram.

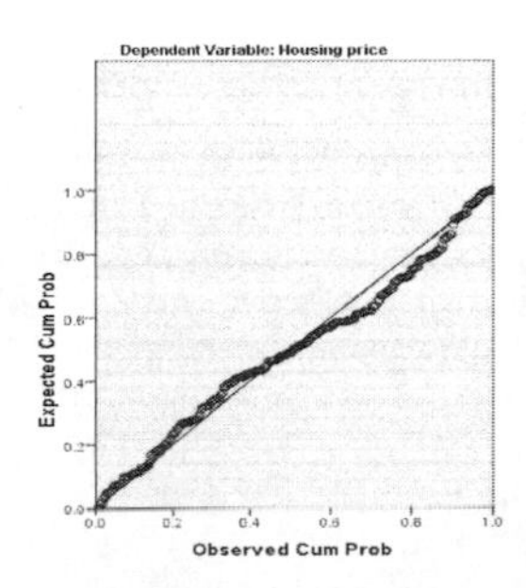

Figure 2. Residual cumulative probability graph.

As stated above, it can be concluded that the basic model has a relatively good explanatory ability and no heteroscedasticity or collinearity problems, and residual meets the normality assumption. Therefore, the model can use to analyze and explain the relationship between housing characteristics and housing prices.

4.5 *Analysis of regression coefficient*

In log-linear model, the independent variable and dependent variable all enter the model. From Table 6, the regression coefficient' significance levels of CBD distance, metro station, decoration quality, property fee and greening rate are lower than 5%. However, the rest variables are not significant.

Table 6. Analysis of regression coefficient.

Model	Unstandardized Coefficients B	Unstandardized Coefficients Std. Error	Standardized Coefficients Beta	t	Sig.
(Constant)	9018.720	1432.647		6.295	.000
CBD distance	-868.023	159.298	-.348	-5.449	.000
Bus route	19.552	25.895	.043	.755	.451
Metro station	712.558	348.096	.119	2.047	.042
Decoration quality	896.718	230.743	.223	3.886	.000
Property fee	394.133	109.409	.207	3.602	.000
Greening rate	46.383	22.745	.121	2.039	.043
Plot ratio	128.384	107.063	.080	1.199	.232
Education facilities	-191.609	236.344	-.055	-.811	.418
Living facilities	-215.144	238.718	-.058	-.901	.368

5 CONCLUSIONS

In this paper, we apply Hedonic Pricing Theory to the study of housing markets in Chengdu, analysis main characteristics that affect housing price. There are 5 significant variables in regression results: CBD distance, metro station, decoration quality, property fee, and greening rate. The coefficient of CBD distance is negative, which means that decreasing the distance to the CBD increases housing price. The coefficients of metro stations, decoration quality, property fee and greening rate are positive, which indicate that these variables can improve the housing price. Bus route, plot ratio, education facilities and living facilities are not significant, which mean that these variables don't have a significant effect on housing price. Taking the regression results above into consideration, it can be concluded that the buyers and investors in Chengdu are willing to pay an additional price for location and building characteristics.

ACKNOWLEDGMENT

It is a project supported by the Postgraduate Innovation Program of Southwest University for Nationalities (CX2014SZ10) and supported by Construction Foundation for the postgraduate degree site of Southwest University for Nationalities (2014-XWD-B0304, S1201). The corresponding author is M.S. ZHANG

REFERENCES

[1] Ozanne, L. & Malpezzi, S. 1985. The efficacy of hedonic estimation with the annual housing survey. *Journal of Economic and Social Measurement* 13(1): 153–172.

[2] Haurin D. R. & Brasington, D. 1996. School quality and real house prices: inter- and intrametropolitan effects. *Journal of housing economics* 5(4): 351–368.

[3] Adair, A., Mcgreal, S., Smyth, A., Cooper, J. & Ryley, T. 2000. House price and accessibility: The testing of relationships within the Belfast urban area. *Housing studies*. 15(5):699–716.

[4] Kestens, Y., Theriault, M. & Rosier, F.D. 2006. Heterogeneity in hedonic modelling of house prices: looking at buyers' household profiles. *Geograph Syst* 26(8):61–96.

[5] Jiang, Y.J & Gong, J.H. 1996. Real Estate Price Index and Hedonic Model. *Appraisal Journal of China* 19(6): 30~32.

[6] Wang, D. & Huang W.S. 2005. Hedonic House Pricing Method and Its Application in Urban Studies. *Journal of City Planning Review* 20 (3): 62~71

[7] He, D. & Jin, F.J. 2013. An Analysis of the Spatio-temporal Impactsof Major Infrastructure on Real Estate Prices———Take Beijing Metro Line 4 as an Example. *Journal of Beijing Union University*, 23(3):7-15+26.

[8] Lancaster, K. J. 1966. A new approach to consumer theory. *Journal of Political Economy* 74 (1):132–157.

[9] Rosen, S. 1974. Hedonic prices and implicit markets: Product differentiation in pure competition. *Journal of Political Economy* 82 (1):35–55.

Computing, Control, Information and Education Engineering – Liu, Sung & Yao (eds)
© 2015 Taylor & Francis Group, London, ISBN: 978-1-138-02800-5

Empirical research on influence of health qigong to the flexible quality of college students

Bao Han
Sports Department, Tianjin University of Traditional Chinese Medicine, China

ABSTRACT: Through teaching experiment, from the scores of body flexibility in sitting position and shoulder rotation of 120 students in Tianjin University, this study finds that health qigong is obviously higher than other projects to increase the muscle motion amplitude for college students' hip muscle, thigh posterior muscle and lumbar back muscle. For the shoulder joint flexibility, health qigong is also more obvious advantages, and it can enhance the students' ability of self control.

KEYWORDS: Health qigong; flexible quality; sports.

1 INTRODUCTION

The content of Health Qigong with Chinese characteristics, which brings together Confucianism, Legalism, Taoism and other various philosophies and theory of Chinese medicine, Such as "yin and yang," "Heaven one", "Heaven corresponding", "Harmony" holistic view of life, played a huge role in promoting and guiding to the emergence and development of Health Qigong. Health Qigong achieves the purpose of controlling the thought and act through body exercises is a kind of exercise through the use of consciousness. This paper discusses whether health qigong has an effect on the flexible quality of the college students through Teaching experiment.

2 RESEARCH OBJECTS AND METHODS

2.1 *Research objects*

The students from the PE class of 2013 grade enrolled on a voluntary basis, After initial testing, the selected experimental and control groups. All students committed not participate in any sports activity except the PE class during the experiment.

Experimental group: 30 boys and 30 girls from the PE class of Health Qigong Control group: 30 boys and 30 girls from the PE class (except the Health Qigong class)

2.2 *Research methods*

2.2.1 *Teaching experiment methods*

The experiment was operated by the physical education teachers of Tianjin University of Traditional Chinese Medicine in strict accordance with the "Tianjin University of Traditional Chinese Medicine Physical Education Teaching Syllabus", all tests were unified.

2.2.2 *Literature*

Compiling relevant literature, this study identified indicators, methodology and the use of writing papers provide a more adequate theoretical basis and foundation.

2.2.3 *Mathematical statistics*

The collected data, use SPSS16.0 statistical software for statistical analysis.

2.2.4 *Logical analysis*

Based on the reach of the information, drawn research results from the analysis of relevant statistics.

3 RESEARCH RESULTS AND ANALYSIS

Flexibility comes from the Latin word fleetere or flexibis. Relatively large differences in the definition of flexibility in physical education, sports medicine, health sciences and applications, probably the most simple definition of flexibility is "in a joint or a joint range of motion available in the (Rang of Motion, referred to as the ROM) " Others include: Flexibility also means freedom of movement, flexibility is a part or parts of the body at the speed required to engage in a substantial special ability to move; A body part of the ROM, it may reach the potential the entire range of movement (within the limits of pain); Normal joint and soft tissue of the ROM on the active and passive stretching of the response; Smooth movement of a joint through its entire ROM capacity; Smooth move of a single joint or a series of joint and easily through the ROM capacity non-binding and without pain, Gajdosik,who recommended flexibility is defined as the ratio of a change in muscle length or joint angle change or force change in the ratio of torque.

In summary, the precise flexibility should refer to the extension or stretching ability of the muscles, ligaments and other tissues. Range of motion (ROM) and muscle length, joint angle, force, torque is measured by the ratio of change in flexibility of the indicators, and ROM is the most commonly used indicators. Many researchers have used the maximum ROM or a limb musculoskeletal flexibility as an indicator. Therefore, this study use sit and reach and transfer test scores to reflect the subjects shoulder flexibility.

Table 1. Comparison of the flexible quality of subjects before experiment.

Gender	Indicators	Experimental group	The control group	P
Male	Sit and Reach	12.4±3.24	12.1±3.42	P>0.05
	Shoulder Turn	32.7±11.32	30.9±10.87	P>0.05
Female	Sit and Reach	15.6±2.27	15.3±2.32	P>0.05
	Shoulder Turn	29.4±9.31	28.9±9.32	P>0.05

The independent sample T test showed: Before the wexperiment, Observation group and control group boys Sit and Reach score was 12.4 ± 3.24cm, 12.1 ± 3.42cm; Shoulder turn test scores were 32.7 ± 11.32cm, 30.9 ± 10.87cm, There was no significant difference (P> 0.05), That means the boys of the observation group and control group have the same pliability. The Sit and Reach score of the observation group and control group girls was 15.6 ± 2.27cm, 15.3 ± 2.32cm; Shoulder turn test scores were 29.4 ± 9.31cm, 28.9 ± 9.32cm, there was no significant difference (P> 0.05), That means the girls of the observation group and control group have the same pliability. Therefore, we can think there is no significant between the observation group and control group before the experiment.

Table 2. Comparison of the male subjects pliability before and after the experiment.

Gender	Indicators	Before the experiment	After the experiment	P
Male	Sit and Reach	12.4±3.24	15.5±2.47	P<0.01
	Shoulder Turn	32.7±11.32	15.2±5.24	P<0.01

As can be seen from Table 2, After eighteen weeks of Health Qigong exercise, the experimental group boys to reflect hip, thigh muscles and lower back muscles range of motion of the Sit and Reach score on the front of the subjects 12.4 ± 3.24cm, raised to be post-test 15.5 ± 2.47cm, There are significant differences (P <0.01); Shoulder turn results from the 32.7 ± 11.32cm, increased to 15.2 ± 5.24cm, there are significant differences too (P <0.01). Shows that, after 18 weeks of Health Qigong exercise, the experimental group boys had a very significant improve on pliability, that Health Qigong has a very significant improvement for the flexibility of the boys.

As can be seen from Table 3, after eighteen weeks of Health Qigong exercise, the subjects of the girls on the Sit and Reach test scores have markedly improved, increased from 15.6 ± 2.27cm to 19.8 ± 3.19cm, there existed a significant difference (P<0.01), Test data in the shoulder turn, the girls increased from 29.4 ± 9.31cm 12.1 ± 4.13cm, there are also significant differences (P<0.01). Shows that, after 18 weeks of Qigong practice, flexible quality of the experimental group girls has very significantly improved, that means Health Qigong have a very significant improvement for the flexibility of the girls too.

Table 3. Comparison of the female subjects pliability before and after experiment.

Gender	Indicators	Before the experiment	After the experiment	P
Female	Sit and Reach	15.6±2.27	19.8±3.19	P<0.01
	Shoulder Turn	29.4±9.31	12.1±4.13	P<0.01

To sum up, after eighteen weeks of fitness Qigong system, the experimental group of boys and girls in the shoulder, hip, thigh muscles and lower back muscles have a very significant range of motion improved, significantly enhanced pliability, that proved Health

Table 4. Comparison of improving the range of flexible.

Gender	Indicators	Experimental group	The control group	P
Male	Sit and Reach	3.1±2.85	2.3±1.28	P<0.05
	Shoulder Turn	17.5±7.16	13.4±4.71	P<0.01
Female	Sit and Reach	4.2±2.57	2.6±1.46	P<0.05
	Shoulder Turn	17.3±5.34	11.4±4.21	P<0.01

Qigong can improve the flexibility of the quality of boys and girls of the experimental group.

As can be seen from the table, the boys of experimental group in Sit and Reach and shoulder turn test results increased by 3.1 ± 2.85cm and 17.5 ± 7.16cm,higher than results of the control group: 2.3 ± 1.28cm and 13.4 ± 4.71cm, The independent sample T test showed that Sit and Reach score of the experimental group boys improved significantly higher than control group($P < 0.05$), Shoulder turn results very obvious higher than control group($P < 0.01$), The shoulder turn and Sit and Reach test scores of female students in the experimental group increased by 4.2 ± 2.57cm and 17.3 ± 5.34cm, Compared with the control group, there were significant differences ($P < 0.05$) and very significant difference ($P < 0.01$).

In summary, the Health Qigong for college students hip, thigh muscles and lower back muscles to improve range of motion was significantly higher than other projects, and Health Qigong has very clear advantage for the flexibility of the shoulder joint.

Research shows that exercise the muscles and joints can improve blood circulation, access to adequate nutrition, thus enhancing the flexibility of the muscles, stretching and flexibility. Often stretch the ligaments and muscles of each joint, so that cartilage compression and decompression by alternately role, so that by the intra-articular synovial fluid into cartilage, improving their nutritional supply, thus ensuring its viscoelastic and improve joint mobility.

Improving the quality of participants flexibility, mainly because in Health Qigong exercise ,the upper and lower limbs, trunk movements should be more fully flexion and extension, development of collection and rotating motion, each joint as the body was multi-faceted, wide angle activities, thus "pulling bone" to " Stretching muscles " campaign Extend the size of the human body muscle groups, the fascia and the joints of the tendon, ligament, joint capsule and other connective tissue.

Flexibility of hip, thigh muscles and lumbar back muscles improved mainly by straight leg bent and bent leg movements acts . For example, in Baduanjin, some acts request straight legs to the waist axis, body bend forward, hands and feet climbing pause; these actions so that all parts of the body muscles are involved in sports, elongated muscle fibers, promoting bone stretch, strengthen bone blood supply, metabolism. Long-term practice of these actions will certainly enhance the hip, thigh muscles and lower back muscles and range of motion, thus improving flexibility.

Increace of the shoulder flexibility is due in a large number of health qigong development arm, rotating, stretching and other activities. For example: Seventh actions of Yi Jin Jing: Asked the left hand bend elbow back down and put behind him, fingers upward; right hand bent backward from the shoulder stretch, pulled left hand fingers, the right hand clinging to his neck, breathing, his hands tight, exhale to relax, Long-term practice of these actions, the muscles and ligaments of shoulder strength and scope of activities must be improved, thereby enhancing the flexibility of the shoulder joint.

4 SUMMARY

Health Qigong is a precious cultural heritage of China, which contains the philosophy of the Chinese nation, traditional Chinese medicine theory. Through long-term, regular exercise, fitness function effectively improves the air quality of students flexibility.

REFERENCES

[1] Hu Ping Jiao, Liu Hong Fu. Ba Duan Jin for College Students, psychological and mental health effects [J]. Mudanjiang Medical College, 2008 (4): 89–91.

[2] Ju Xiang Yang. Chinese Qigong efficacy and mechanism of psychological research [J] Hebei Institute of Physical Education, 2010 (6) :142–144.

[3] Zeng Yun Gui, Zhou Xiao Qing, Wang An li, Yang Bo Long, Wang Song-Tao. [J]. Beijing Sport University, 2005, (09): 139–141.

Structure analysis and mathematical modelling based optimization approach to robotics mechanical system: A case study

Jie Chang Ruan & Wen Kai Shao
YiBin Vocational and Technical College, Sichuan, China

ABSTRACT: In this paper, we propose a novel mathematical modelling and structure analysis based approach to optimize the mechanical system with case studies. We investigate the effect of turning delays on the behavior of groups of differential wheeled robots and show that the group-level behavior can be described by a transport equation with a suitably incorporated delay. Our mathematical analysis is based on the results of numerical simulation and experimental support E-Puck robot. The experiments we compare the correct model number of the average time to find the target area is robot in unknown environment. The transport equation with delay better predicts the mean time to find the target than the standard transport equation without delay. The experiment shows the effectiveness and robustness of the proposed approach, further optimization proposals are also calculated and discussed.

KEYWORDS: Mathematical Modelling; Mechanical System; Optimization Approaches.

1 INTRODUCTION

1.1 *Background analysis*

More theories have developed into a distributed coordination and control of autonomous agents, where the robot set performance in environment only short-range communication is possible [1]. By performing actions based on the presence or absence of signal, algorithms have made some county regiment on creating a task; Reconnaissance, for example, areas of interest, at the same time to collect data or keep formation. In this paper, we will investigate an implementation of searching algorithms, similar to those used by flagellated bacteria, in a robotic system [2]. Many flagellated bacteria such as E. coli, using the run-and-tumble search strategy of movement by more or less interrupted fall in the short run directly. When their motors rotate counter-clockwise the flagella form a bundle that propels the cell forward with a roughly constant speed; when one or more motors rotate clockwise the bundle flies apart and the cell 'tumbles'. Similar behavior can be observed in swarms of animals, avoiding predators and coordinating themselves within a group [3]. The behavior of E. coli is often modelled as a velocity jump process where the time spent tumbling is neglected as it is much smaller than the time spent running.

1.2 *Overview of our research*

In this paper, we will study an experimental system based on E-Puck robots. We model these differential wheeled robots to follow a run-and-tumble searching strategy in order to find a given target set. In the first set of experiments, we concentrate on the simplest possible scenario: an unbiased velocity jump process in two spatial dimensions with the fixed speed $s \in \mathbb{R}^+$, the constant mean run time $\lambda^{-1} \in \mathbb{R}^+$ and the turning kernel which is independent of u defined:

$$T(v,u) = \frac{\delta(\|v\| - s)}{2\pi s} \tag{1}$$

E-Puck robotic system is a special example. They can be executed on the spot to classical speed jump process described in [1]. In this article, we will study the extent to which [1] proposed a good robot system behavior description, we will develop an extended results of [1] in a better match experimental data and mathematical model. We then the extension speed jump theory applies to offset random walk through the experiment of signal Settings. The following sections will discuss the issue in detail.

2 VELOCITY JUMP PROCESS MODELLING

2.1 *Experiment setup and initiation*

To obtain the empirical data, an experimental system consisting of 16 E-Puck robots was used. E-Puck robots are small differential wheeled robots with a programmable microchip (Bonani and Mondada, 2004). The diameter of each robot is e =75mmwith a height of 50mmand weight of 200g. Throughout the experiments, the speed was chosen to be s = 5.8×10−2m/sec. The robots turn with an angular velocity w = 4.65/sec. The figure 1 indicated the set up and lay-out of the experimental initiation and setup procedure. As a consequence, we discuss the importance of robot-robot collisions on the experimental results in the next section.

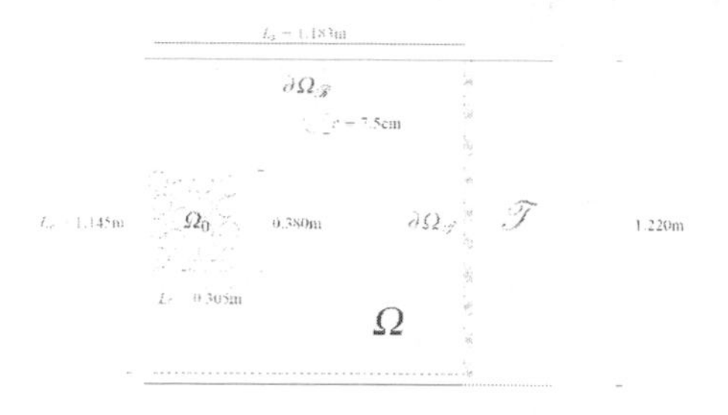

Figure 1. Schematic showing of the experimental set-up.

2.2 *Relevance of collision analysis*

For non-interacting particles which can change direction instantaneously, equation 1 accurately describes the mesoscopic density through time. However, in our experiments the robots undergo reflective collisions when they come into close contact, rather than passing through or over each other. For a low number of particles, we used Monte Carlo simulations to demonstrate that collisions are not the dominant behavior and have little effect on the distribution of particles. In panels (a) and (b) of Figure 2, we compare two Monte Carlo simulations: (a) in which particles are allowed to pass through one another and (b) in which collisions are modelled explicitly. In Figure 2(c) we present the solution of equation 1. This comparison demonstrates that the mean density of the underlying process converges to the solution of transport equation 1.

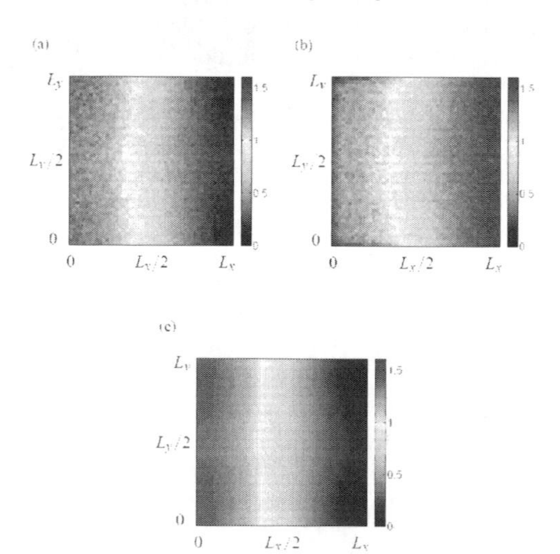

Figure 2. Comparison of individual-based simulations.

The parameters employed in this model comparison are taken directly from the equivalent robot experiment. In Figure 2(c), for the differential equation$(s,\lambda,\varepsilon)=\left(5.8\times10^{-2}m/\sec,0.250\sec^{-1},7.5\times10^{-2}m\right.$, we use a first-order numerical scheme with $\Delta\theta=\pi/20$. In the Monte Carlo simulations, we initialize particles in the effective pen for 20sec where they undergo hard-sphere collisions. They are then released into the larger arena where in one simulation they are point-particles and in the other they undergo reflective collisions as hard-spheres. Instead of removing particles at the target boundary as shown in Figure 1, we optimize the model as follows:

$$p(0,x,v)=\frac{\chi\Omega_0\delta\left(\|v\|-s\right)}{L_0^2 2\pi} \tag{2}$$

In the formula, we denote the $\chi\Omega_0$ to be the indicator function of the initial region Ω_0. The corresponding boundary condition is $p(\mathrm{t},x,v)=p(\mathrm{t},x,v')$ where the reflected velocity v' is defined as:

$$v'=v-2\left(v\cdot n_\Omega\right)n_\Omega \tag{3}$$

After 20 seconds, the density of our record in every scene, and now the results are shown in figure 2. Have a minimum of obvious differences in figure 2 gives a monte carlo simulation to our choice of parameter values. In order to compare three simulations given in figure 2 we also hired two Kolmogorov Smirnov has test.

2.3 *Comparison: theory and experiments*

We compare this result to the variation of the remaining mass with time from a numerical solution of equation 1 combined with the following boundary conditions:

$$\begin{aligned} &p(\mathrm{t},x,v)=0 \qquad x\in\partial\Omega t, v\cdot nt<0 \\ &p(\mathrm{t},x,v)=p(\mathrm{t},x,v') \qquad x\in\partial\Omega t \end{aligned} \tag{4}$$

The mass remaining in the domain is then defined as the following formula:

$$m(t)=\int_\Omega\int_V p(t,x,v)\,dx\,dv \tag{5}$$

Then it is plotted as a dotted (red) line in Figure 3(a). The initial mass is normalized to 1. An obvious observation from Figure 3(a) is that the transport equation description does not match the experimental data well, with the robots exiting the arena significantly slower than predicted. In this figure, we use a first-order finite volume method with the parameter: $\Delta\theta=\pi/20,\Delta x=1.183m/200,\Delta t=10^{-3}$.The average exit time of those 708 robots was 121.92sec. In order to be able to compare experimental exit times with the mean exit time problems, it is necessary to estimate the mean exit time of all 800 robots. Using the best exponential fit on the mass over time relation, we can estimate the mean exit time of the remaining 92 robots to be 424.69sec. The approximate mean exit time established in the experiments is therefore 156.74sec; this value is plotted as the solid (black) line in the b figure. In order to be able to compare this value to analytic results, one has to reformulate the transport equation into a mean exit time problem.

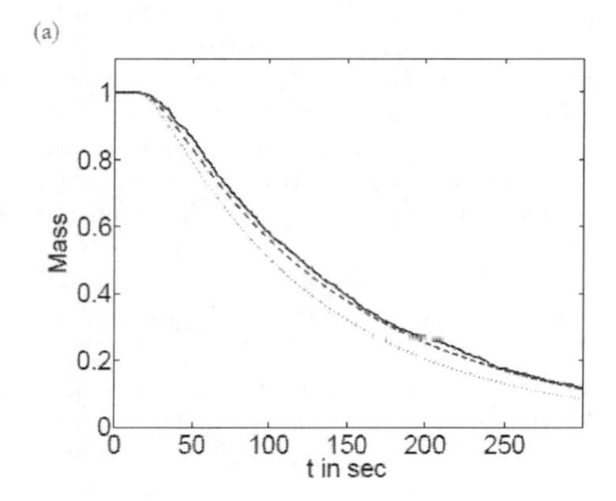

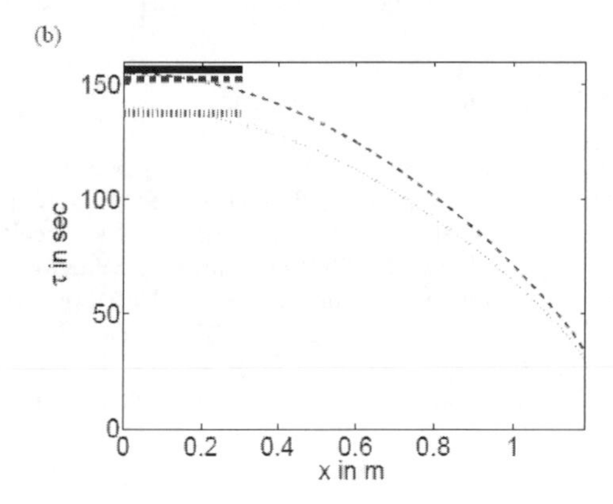

Figure 3. Comparison of the methods.

3 INCORPORATION OF THE SIGNAL GRADIENT

3.1 *General analysis*

In this section, we are aiming to formulate velocity jump models that incorporate changing turning frequencies λ. In particular, we are interested in turning frequencies that depend on the current velocity of the robot as well as its position in the domain. The general velocity jump model for this case can be formulated as:

$$\frac{\partial p}{\partial t}+v\cdot\nabla_x p=-\lambda(x,v)p+ \int_V \lambda(x,u)T(v,u)p(t,x,u)du \tag{6}$$

With the boundary conditions, we can formulate this system by incorporating the resting period:

$$\begin{aligned}&\frac{\partial p}{\partial t}+v\cdot\nabla_x p=-\lambda(x,v)p(t,\mathrm{x},\mathrm{v})+r\left(t,x,v,0^+\right)\\&\frac{\partial r}{\partial t}-\frac{\partial r}{\partial \eta}=\int_V \lambda(x,u)p(t,\mathrm{x},\mathrm{u})T(v,u^*)du\end{aligned} \tag{7}$$

The system can again be formulated in the form of a delay differential equation:

$$\frac{\partial p}{\partial t}+v\cdot\nabla_x p=-\lambda(x,v)p+\int_V \lambda(v,u)T(v,u)p\,du \tag{8}$$

3.2 *Experiment with signal gradient*

In order to compare these generalized velocity jump models to experimental results, we introduce an external signal into the robot experiments presented previously. The signal is incorporated in the form of a color gradient that can be measured by the light sensors on the bottom of the E-Puck robots. The color gradient is laid out in such a way that it changes along the x-axis in Figure 1 with the darker end closer to the target area. The reaction of the robots to this color gradient is implemented using the internal variable z and a changing turning frequency $\lambda(z)$ that are updated according to:

$$\frac{dz}{dt}=\frac{S-z}{t_a},\lambda=\lambda_0+\lambda_0\left(1-\alpha(S-z)\right) \tag{9}$$

In this formula, $S\in[0,1]$ represents the measured signal with increasing values of S indicating a darker color in the gradient. The way the turning frequency is changed is motivated by models of bacterial chemotaxis. According to results from Erban and Othmer (2005), a macroscopic density formulation for the robotic system is given through the hyperbolic chemotaxis equation:

$$\frac{1}{\lambda_0}\frac{\partial^2 n}{\partial t^2}+\frac{\partial n}{\partial t}=\frac{s^2}{d\lambda_0}\Delta n-\nabla\left(n\frac{\alpha\lambda_0 s^2 t_\alpha}{d\lambda_0\left(1+\lambda_0 t_\alpha\right)}\nabla S\right) \tag{10}$$

Where, $S:\Omega\to\mathbb{R}$ denotes the color gradient and $n(t,x)$ describes the concentration of robots in Ω. Equation 6 can be approximated by the velocity jump process 4 with the form for the turning frequency given by:

$$\lambda(x,v)=\lambda_0-\gamma v\cdot\nabla S(x),\qquad \gamma=\frac{\alpha t_\alpha\lambda_0}{1+t_\alpha\lambda_0} \tag{11}$$

Because the gradient of the color signal S was chosen to be parallel to the x-axis in the experimental setting, we can again simplify the formulation of the exit time problem 4 by averaging along the y-axis. The resulting equation takes the form the previous equation.

3.3 *Experiment result analysis*

We now want to compare the experimental data to the generalized velocity jump models presented prior. The numerical solutions were achieved using the exact same methods and parameters as in Section 2.3 and the results can be seen in figure 4.

(a)

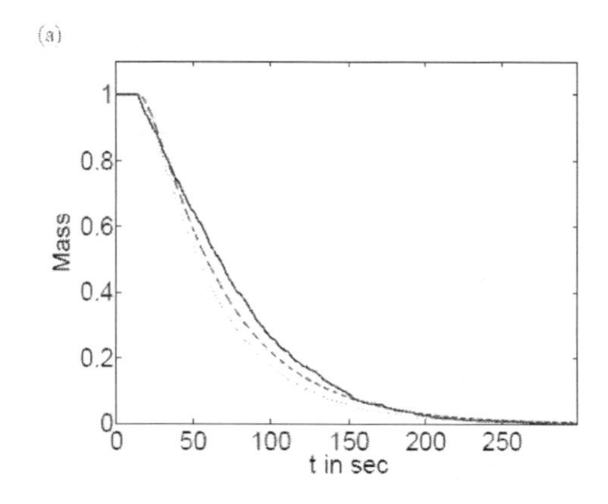

(b)

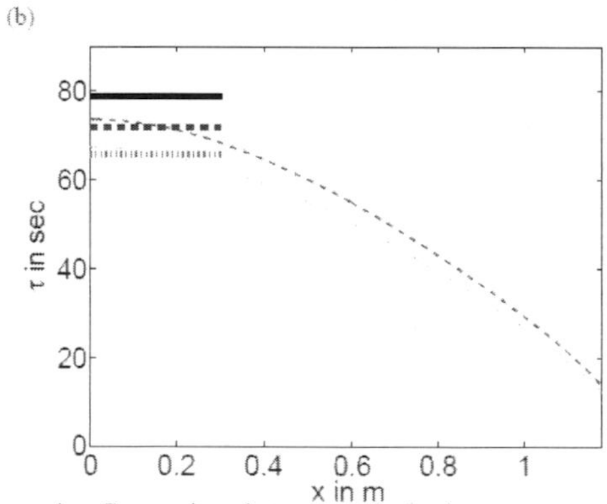

Figure 4. Comparison between velocity jump process and experimental data for experiment including color gradient.

The parameter values used for the robots is: $\lambda_0 = 0.25\,\mathrm{sec}^{-1}$. The experimental procedure was equivalent to the one presented in Section 2, i.e. we repeated the experiment 50 times with 16 robots, each time waiting until all of the 16 robots have left the arena. In figure 4 (a), we show the quality of our drawing system. Solid (black) line shows the proportion of the robot is still on the stage of time. (red) line is a speed jump equations and numerical solution of the corresponding boundary conditions. Dotted line (blue) is a numerical solution of speed jump system hibernation in 2 and boundary conditions in 7. In both plots in Figure 4, we see that the models including finite turning delays (represented through the dashed (blue) lines) give an improved match compared to the models without this delay. The remaining difference between the models and the experimental data can be explained by noisy measurement of the signal S(x) as well as the fact that we used linear approximation averaged over all robots to obtain the numerical results. We can conclude from this brief study of robot experiments, including a color gradient signal that this signal indeed improves the target finding capacity of the robots and that the models developed in Section 3 can be generalized to incorporate turning frequencies that change according to external signals.

4 SUMMARY AND CONCLUSION

In this paper, we have studied an implementation of a run-and-tumble searching strategy in a robotic system. The algorithm implemented by the robots is motivated by a biological system – behavior of the flagellated bacterium E. coli. Bio-inspired algorithms are relatively common in swarm robotics. Algorithms based on behavior of social insects have been implemented previously in the literature, see for example [4-7]. We have studied a relatively simple searching algorithm motivated by E. Coli behavior, but the transport equations and velocity jump processes naturally appear in modelling of other biological systems, such as modelling chemotaxis of amoeboid cells (Erban and Othmer, 2007) or swarming behavior as seen in various fish, birds and insects (Carrillo et al., 2009; Erban and Haskovec, 2012). We conclude that the same delay terms as in formula 9. In the future, we plan to do more research on the proposed method.

REFERENCES

[1] Berg, Howard C., and Robert A. Anderson. "Bacteria swim by rotating their flagellar filaments." (1973): 380–382.

[2] Petroff, Alexander, Xiao-lun Wu, and Albert Libchaber. "Rotating bacteria aggregate into active crystals." Bulletin of the American Physical Society 59 (2014).

[3] Srivastava, Disha. To stick or swim: Cyclic-di-GMP mediated inverse regulation of biofilms and motility in Vibrio cholerae. Diss. MICHIGAN STATE UNIVERSITY, 2014.

[4] Baabour, Magali, Carine Douarche, and Dominique Salin. "Response of a Motile/Non-Motile Escherichia

coli Front to Hydrodynamic excitations." Bulletin of the American Physical Society 59 (2014).
[5] Mahdavifar, Alireza, et al. "A Nitrocellulose-Based Microfluidic Device for Generation of Concentration Gradients and Study of Bacterial Chemotaxis." Journal of The Electrochemical Society 161.2 (2014): B3064-B3070.
[6] Lushi, Enkeleida, Hugo Wioland, and Raymond E. Goldstein. "Fluid flows created by swimming bacteria drive self-organization in confined suspensions." Proceedings of the National Academy of Sciences (2014): 201405698.
[7] Khalil, Islam SM, and Sarthak Misra. "Control characteristics of magnetotactic bacteria: Magnetospirillum magnetotacticum strain MS-1 and M. magneticum strain AMB-1." IEEE Transactions on Magnetics 50.4 (2014): 1–11.

Computing, Control, Information and Education Engineering – Liu, Sung & Yao (eds)
 ISBN: 978-1-138-02800-5

The application of laser technology in the information warfare

Z.H. Lan, M.Y. Hou, C.Y. Tian, Y.G. Ji & H.Y. Wang
Air Force Aviation University, Changchun, China

H. Mei
Changchun Yuheng Optics Co., Ltd., China

ABSTRACT: In the information age warfare, the both warfare parties must pay much more attention to fight to achieve the mastery of information and the right of the space and sky administration in order to win the battle. In this situation, all kinds of weaponry, especially the precision homing weapon, were put on the modern battlefield. The modern information technology was thoroughly changing the form and ways of the modem warfare and creating a new battle style. A great deal of usage of the laser weaponry in the information age warfare makes the technological weak side exist more and more difficult on the battlefield. The technologically weak part must fight and defeat the enemy by a surprising action. This paper tries to introduce the application of laser technology in the future information battle field about how to achieve the mastery of information, the right of the administration of space and sky and how to handle the precision strike technology.

KEYWORDS: Warfare; Laser radar; Information; Laser guided missile.

1 INTRODUCTION

In the information age warfare, the both warfare parties must pay much more attention to fight to achieve the mastery of information and the right of the space and sky administration in order to win the battle, In the modern war, the right of information is very important for the army like as the blood for the human body. The information, especially the information superiority will play a decisive role in the future war, so the key is to access and control information. If we obtain the control of space to gain the initiative of battle, we can detect and warn In real time on the globe, achieve high quality remote intercontinental communication and carry out the long-distance real-time command and control. The precise strike has become a main way in the information battle. The degree of long-range precision strike and accurate positioning of it can be improved by a space guidance and location. Without the right of space, it is impossible to grasp the information superiority and the air in battle, there would not be the sea and land power. Who controls the space, who would master the right of winning wars.

2 THE CHARACTERISTICS OF LASER

2.1 *Monochromaticity*

The laser has very good monochromaticity. The ordinary light source emits photons with different frequency, so it contains all sorts of color. But for laser, photons of the laser are the same in frequency, so the laser is one of the best monochromatic light source.

2.2 *Coherence*

The laser has good spatial coherence, because the phase of stimulated radiation photons is consistent and the effect of resonant that make the laser beam on the cross section has a fixed phase relation between points. The laser provides us with the best coherent light source. It was due to the advent of the laser that prompts coherence technology to advance in development by leaps and bounds and holographic technology to achieve.

2.3 *Direction*

The divergence angle of laser beam which is almost a parallel light is small. When a laser shines on the moon, the spot diameter on is only about 1km. But for an ordinary light source with light in all directions, the light spot diameter will expand to more than 1000 kilometers.

2.4 *High brightness*

The brightness of a laser is 1012-1019 times higher than an ordinary light source that it is the most right light source. The strong laser even can produce hundreds of millions of degrees of high temperature which can be used in laser machining industry and national defense, etc.

3 THE APPLICATION OF LASER TECHNOLOGY IN INFORMATION WARFARE

In the future information combat, the information and air superiority are key to win the war. As the laser technology is used widely in modern information warfare, it not only makes a command and control more flexible and accurate and operational cycle shorten, but also improves the strike precision of the weapons and promote weapons intelligent and unmanned.

3.1 *Laser radar has become the main means of access to information*

Laser radar is better than traditional radar in performance. For example, the laser radar can achieve very high detection sensitivity and resolution for small divergence angle of laser and energy concentration. Its very short wavelength can make antenna and the system size are very small which is incomparable for traditional radar. Compared with microwave radar, laser rangefinder is more suitable for carrying on the spacecraft and the measurement of space target distance. Laser ranging technology can meet the high accuracy, large measuring range distance space target requirements and won a wide application in the military field.

Laser radar is working in the light of the electromagnetic wave band which probe targets by laser that is the combination of a traditional radar technology and the modern. Laser radar system is composed by the receiving system, tracking the launch system, data processing, display and transmission system, etc. The launch system of a laser beam to the target which is reflected by corner reflectors, is back to measure station along the direction parallel to the beam and is received by the receiving system, detected and treated. The distance between goal and measuring station is calculated by the time interval between the laser signal. When a target is deviating from the optical axis, it will turn the laser beam tracking system to make a target accurately.

The laser radar's resolution is very high who can collect 3d data, such as azimuth, pitching angle, distance, speed and strength, and the data are shown in the form of images. They are the geometric distribution of radiation image, range gating image and speed. The laser radar has become a very important means of reconnaissance in the modern information warfare. The laser radar was installed on the P-3C Airborne Warning And Control System, Apache helicopter and B-2 bomber which played a huge power in previous campaigns.

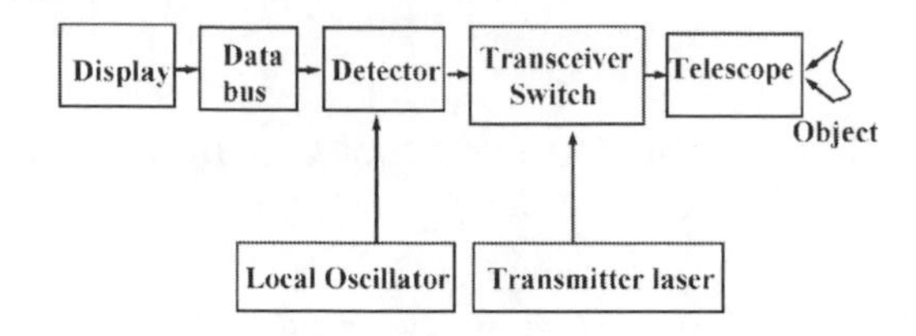

Figure 1. Sketch Map of Lidar System.

3.2 *The application of laser warning in warfare*

As a large number of laser weapons were used in combat that has made the battlefield threat increasingly serious. It has speeded up the development of the laser reconnaissance warning technology. The laser warning system can intercept, measure and identify the enemy laser threat signals and give an alarm real time. It is usually loaded on aircraft, ships, tanks and individual helmet or installed on the important ground targets. The laser warning econnaissance is a kind of special use. To laser threat sources on the battlefield, the laser system can detect timely the launch of the laser signal, determine the incident direction and report an emergency which can make carriers evade or countermove.

The laser warning usually has the best features. Its field of view is very large, so it can cover the entire alerts airspace; Because of frequency bandwidth, it can determine the enemy of all possible military laser wavelength; It has low false alarm, high detection probability and wide dynamic range; it's direction recognition ability is very good; the reaction time is short; It has small volume, light weight and low price.

The laser warning equipment is mainly composed of a laser optical receiving system, photoelectric sensors, signal processor, display and alarm device and other parts that can measure wavelength the enemy laser, pulse repetition frequency laser radiation source direction and other technical parameters. The laser optical receiving system is used to intercept the enemy after the laser beam, filter out most of the stray light will converge to the laser beam on the photoelectric sensor. Photoelectric sensor optical signals into electrical signals sent to the signal processor, after treating with signal processor sent to the display, which can display the target type, threat level and bearing and other relevant information, and an alarm signal.

3.3 *The application of laser weapon in warfare*

The laser weapon system is a kind of using a laser beam directly targets or disable the partial function of target, function of attack and defense in order to achieve the new concept of weapon system. Depending on the platform , laser weapons can be divided into a ground-based laser weapons, shipboard laser weapon, airborne laser weapon and space-based laser weapons. According to the target ability of laser weapons can be divided into strategic, operational laser weapons and tactical laser weapons. The strategic laser weapon is mainly used in the strategic missile attack, reconnaissance satellite, etc.; Main attack battle laser weapon fighters, anti-aircraft positions, etc.; The tactical laser weapon damage mainly photoelectric detection equipment, cockpit canopy of goals. The airborne laser installed on Boeing 747-400E is the battle laser weapon system. Now the ground-based laser weapon and shipboard laser weapons can be used to attack the satellite. For the airborne laser weapon, it is in the research and the experiment stage for countries. Tactical laser weapons (ATL) have become a research hotspot which will be applied to the four or five generation fighter in the future. It can improve the survival ability and combat efficiency of planes.

The characteristics of laser weapons are to attack target precisely and damage fast. In the situation of active attack, the course is just a few minutes from finding targets, tracking, aiming at target, to complete the operational mission. When the laser weapon is launched, it does not produce recoil, therefore it has no effect on aircraft. According to the operational requirements analysis, The multiple laser weapons can be set in the plane to achieve active attack and defense in different azimuth of space. When the photoelectric guided weapon attacks target, laser weapons will irradiation and damage the detection elements of photoelectric guided weapons to affect its killing effects. This completes the laser weapon of defense and improves the survivability of the aircraft.

3.4 *Laser weapon of the information warfare*

3.4.1 *Laser guided bomb*

The world's first laser guided bomb is the 1965 U.S. Raytheon Pave Way. Pave Way has developed to the fourth generation that can be used in the case of a low visibility and in a lower high and applicable to all fighters, attack aircraft in the United States, It is the production of one of the most precision bombs.

Us and British united army had thrown a total of 8716 guided bombs in the Iraq war. The guided bomb's operational effectiveness was higher than dozens of times of operational effectiveness. The GBU damage radius is 8 knots. The bomb was used to attack bridges, C^3I system and bunkers in "Desert storm". GBU-28/B can penetrate 30m thick land and 6 m thick reinforced concrete. To ensure the minimum collateral damage, the "scalpel" subminiature laser-guided bombs were researched in the US which is 250 pounds equivalent to 40 kg charge.

3.4.2 *Laser guided missile*

The laser guided missile appeared at the end of the 1970s. The American "Maverick" and the French AS-30L were its representative. Hellfire was a typical representative which was applied to attack tanks and equipped to the Apache helicopter gunships. In "Desert Storm" action, the apache AH-64 has launched Hellfire missile that destroyed accurately several kilometers of Iraqi radar station and opened the "electromagnetic gap". The laser-guided missile has developed into a model, multi-functional missile family.

In the case of AGM series, the semi-active laser guidance of AGM-114A is the basic type which is equipped with the U.S. army; AGM-114B has a semi-active laser, radio frequency/infrared and infrared imaging three kinds of seeker and low smoke rocket engine. It is equipped with fuzing insurance for blasting device, equipped with the United States Marine corps.

4 CONCLUSIONS

The laser has many excellent properties such as good monochromaticity, coherence, directionality and high brightness, so there are a lot of applications in the military field.

Compared with the traditional radar, laser radar has good anti-interference ability and high detecting precision. Laser radar has simple structure and small volume, which widely used to obtain information in combat.

The laser warning system can intercept information, measure and identify the enemy laser threat signals and give an alarm real time to improve the defense capability.

The laser weapon system is a kind of using a laser beam directly targets or disable the partial function of target, function of attack and defense in order to achieve the new concept of weapon system.

The laser precision guided weapon has become the main weapon of the information warfare. To sum up, the laser technology has already permeated various fields of combat.

REFERENCES

Meng.X.F. 2005. Infrared and LaserEngineering. *Application and protection of laser technology*.34:139.

Ma. C.J. 2005. Electro-Optic Technology Application. *Status and Trend of Laser Stealth technology*. 6:37–38.

Liu K.X. 2006. Measurement and Analysis of the Outfield Target's Laser Scattering Characteristics. *Chinese Journal of Lasers*. 33:207.

Yao. L.X. 1995. Optical Characteristic of the Object and Environment Beijing: *Astronavigation Press*.

Computing, Control, Information and Education Engineering – Liu, Sung & Yao (eds)
© 2015 Taylor & Francis Group, London, ISBN: 978-1-138-02800-5

Analysis on elasticity of substitution of migrant workers

Jian Dong Bu
College of Continuing Education, Shenyang University, Shenyang, China

Hui Pang
College of Economics, Shenyang University, Shenyang, China

Mi Zhou
College of Economics and Management, Shenyang Agricultural University, Shenyang, China

ABSTRACT: The growth of cities had scale effect and crowding effect, which determined migrant workers to become collaborators or competitors for local residents. We used CGSS2008 data and found that the elasticity of substitution between different levels of education groups are far less than the elasticity of substitution between level of education within the same group. This shows that in the urban labor market, different education degree leads to the labor market differentiation, and have stronger substitution effect between the laborers in the same education group but in different work experience.

KEYWORDS: Elasticity of substitution; Migrant workers; Wage.

1 INTRODUCTION

The previous literature analysis the difference of the immigrants' education, experience and other characteristics on the wages of city residents, laying the foundation for the analysis of this study. In the city scale perspective, the immigrants Education and experience of immigrates residents pay differentials in their wages and its mechanism, which is our contribution. We mainly test whether the entrance of migrant workers from different city sizes would reduce the local residents' wages. How do differences between foreign labor education and experience bring the impact on local public employment? It will affect their wage level? How much? For a long time, there is still a lack of empirical evidence that whether city expansion is conducive to the formation of the professional division of labor, improves the efficiency of city overall production and improve labor wages, nor do research for different level of education and work experience to improve Return of the city expands. Research hypotheses are presented in this paper for the mega city: industrial agglomeration ability is strong, but the city bears big pressure, and complementary effect and substitution effect exists at the same time. The big city (capital city) industrial agglomeration ability is big and bears less pressure, and migrant workers and local residents complementary effect are stronger than the substitution effect. Small and medium-sized city bears less pressure, but industrial agglomeration ability is weak and the substitution effect between migrant workers and local residents is stronger than the complementary effect.

2 THEORIES AND METHODS

According to Card (2001) and Borjas (2003), the theory about the relationship of substitution between capital and labor, we built a CES production function:

$$Q_t = [\lambda_{kt} K_t^v + \lambda_{Lt} L_t^v]^{\frac{1}{v}} \quad (1)$$

Says the total output of the city t , K means the Q_t capital used in the process of production of the city t, L says the workforce used in the process of production of the city t. Then, we can $v = 1 - \frac{1}{\sigma_{KL}}$ get

σ_{KL} means the elasticity of substitution between capital and labor, and
According to $\lambda_{kt} + \lambda_{Lt} = 1$ the differences in the levels of labor education, we establish the following equation

$$L_t = [\sum_i \theta_{it} L_{it}^{\rho}]^{\frac{1}{\rho}} \quad (2)$$

L_{it} represent the number of labor force under i education level in t city, $\rho=1-1/\sigma_E$, σ_E is as the same city size, different labor elasticity of substitution of different education level, θ_{it} says education returns differences under different urban scales , $\sum_i \theta_{it} = 1$ and

According to difference of different types labors'work experience between the local citizens and foreign residents , we establish the following equation:

$$L_{it} = [\sum_j \alpha_{jt} L_{ijt}^{\eta}]^{\frac{1}{\eta}} \tag{3}$$

L_{ijt}, the number of labor in i level of education and j type work experience in t city. $\eta=1-1/\sigma_X$,σ_X in the same level of education and the same city size, elasticity of substitution among different types of labors, α_{jt} says the technical efficiency of labor in the different city sizes and different work experiences, and $\sum_j \alpha_{jt} = 1$

If w_{ijt} is used as labor wages of i level of q_t education and j type work experience in the city t, says the products sales prices in the city t, according to the profit maximization of the first-order condition, we can get $\log w_{ijt} = \log \lambda_{Lt} + (1-\upsilon)\log Q_t + (\upsilon - \rho)\log L_t + \log \theta_{it}$ the following conclusion:

$$+(\rho\text{-}\eta)\log L_{it} + \log \alpha_{jt} + (\eta\text{-}1)\log L_{ijt} \tag{4}$$

Then,

$$\log w_{ijt} = \delta_t + \delta_{it} + \delta_{jt} - (1/\sigma_X)\log L_{ijt} \tag{5}$$

Among $\delta_t = \log \lambda_{Lt} + (1-\upsilon)\log Q_t + (\upsilon - \rho)\log L_t$ them, means the factors associated with urban size only, and $\delta_{it} = \log \theta_{it} + (\rho - \eta)\log L_{it}$ is the only factors associated with labor skill levels in the same city scale, $\delta_{jt} = \log \alpha_{jt}$,is only related to labor experience factor in the same city scale and level of education.

3 MODEL AND RESULTS

3.1 *Elasticity of substitution of different labor experience in the same level of education groups*

In order to calculate different elasticity of substitution between floating population's experience in the same level of education groups, we first need to estimate wage rate of return of different experience migrants in the same education level groups, and then through the formulas (5) to calculate the elasticity of substitution.We first establish a logarithmic wage equation, in which interpreted variable is logarithm of local citizens'wage income and explanatory variables is foreign population with same education level but different experiences.Estimation methods are the OLS method, estimation results are as shown in table 1, the report number are the weighted estimation coefficient and elasticity of substitution calculated the formulas (5). We found that the different experience of floating population has a significant substitution effect. In addition to the megacities, workers who have the same level of education and elasticity of substitution of different experience of labor between 3.79 to 4.13.

However, the results of OLS estimates may be biased. In the process of examining how much effect different experience of floating population have on local residents incomes, there will be a two-way causal relationship between the labor number and local citizens wages,which can lead to endogenous problems.In addition, although we have carried out regression every size cities in order to reduce possible omitted variable bias, but unobservable labor demand and supply factors in the urban labor market , may affect wages and bring the estimated errors.In view of this, we use the number of foreign labor force as a tool variable for different kinds of Labour quantity variables. Different types labors contain migrant workers labor force, and the structure of the urban labor have related to the number of foreign labor, such as megacities skilled Labour is more, the number of foreign labor is more too. In addition, foreign labor force is determined by sampling quantity of sample regions, it is an exogenous variable, not related to the urban characteristics.Actually, for the choice of the tool variable, the previous literatures have similar treatment methods, such as Borjas (2003), etc.IV estimation results are showed in Table 1.In addition to the megacities, under the same level of education, elasticity of substitution of different experience of labor between 3.94 to 4.42.

Table 1. Elasticity of substitution of different labor experience in the same level of education groups.

	city		City=1	
lwagen$_{it}$	OLS	IV	OLS	IV
Ls_{it}	-0.248***	-0.207***	-0.0915	0.0841
	(0.0473)	(0.0565)	(0.102)	(0.149)
σ_X	4.03	4.83	10.93	-11.89

	City=2		City=3	
lwagen$_{it}$	OLS	IV	OLS	IV
Ls_{it}	-0.264***	-0.226**	-0.242**	-0.254**
	(0.0867)	(0.0997)	(0.0937)	(0.104)
σ_X	3.79	4.42	4.13	3.94

Note:(1) those in parentheses are standard deviations, ***p<0.01, **p<0.05, *p<0.1;(2) the explanatory variables is lnSit, the number of labor force under the different level of education and experience in different cities , instrumental variable is lnMit ;(3) in the OLS and IV regression adopts weighted regression, the weight is sample size in each group, which is used to measure the influence the number of labor force has on average wage income of local people.

3.2 *Elasticity of substitution of labor in different level of education group*

In order to estimate elasticity of substitution of labors in different level of education group, we established the regression equation between the wages and the labor force in different education levels to measure wage returns of the degree of education, the interpreted variable is the logarithmic average wage of local citizens and explanatory variables are number of local labor force in different levels of education. Table 2 reports the weighted OLS estimation results and IV estimation results in the overall and each city scale.

Table 2. Elasticity of substitution of labor in different level of education group.

	city		City=1	
$lwagen_{ijt}$	OLS	IV	OLS	IV
ls_{ijt}	-0.295**	-0.379***	-0.304	-0.349
	(0.101)	(0.113)	(0.225)	(0.230)
σ_E	3.39	2.64	3.29	2.87

	City=2		City=3	
$lwagen_{ijt}$	OLS	IV	OLS	IV
ls_{ijt}	-0.940**	-0.971**	-0.569**	-0.565**
	(0.108)	(0.118)	(0.0989)	(0.101)
σ_E	1.06	1.03	1.76	1.77

Note: (1) those in parentheses are standard deviations, ***p<0.01, **p<0.05, *p<0.1;(2) the explanatory variables is lnSijt, the number of labor force under the different level of education and experience in different cities , instrumental variable is lnMijt;(3) in the OLS and IV regression adopts weighted regression, the weight is sample size in each group, which is used to measure the influence the number of labor force has on average wage income of local people.

We found that the degree of education groups exist substitution effect between the labor force, but IV estimation result shows that the substitution effect in a large city nearly 1.As a result, different education workers have larger complementary effect in large cities. In small and medium-sized cities, the substitution effect between different levels of education groups is less, OLS and IV estimation results are very similar between 1.76 to 1.77. Contrast Tables 1 and 2, we can be found that the elasticity of substitution between different levels of education groups is far less than the elasticity of substitution between level of education within the same group. This shows that in the urban labor market, different education degree leads to the labor market differentiation, and have stronger substitution effect between the laborers in the same education group but in different work experience.

ACKNOWLEDGEMENTS

Mi Zhou is a corresponding author. This work was supported by the National Science Foundation of China(NO.71203146;NO.71273179;NO.71373163; NO.71273177), the program for excellent talents in Liaoning province (NO.WJQ2014016); and agriculture youth science and technology innovation talent training plan in Liaoning province (NO.2014055).

REFERENCES

[1] De-wen wang, Yao-wu wu, Fang Cai. 2004. Migration, unemployment and urban labor market segmentation - why unemployment of rural migrants is low? *Journal of world economy* (1).

[2] Xiao-lu Wang , Xiao-lin Xia. 1999.Optimized city scale to promote economic growth. *Journal of economic research* (9).

[3] Yun-yan Yang, Ying-mei Xu , Shu-jian Xiang. 2003. Alternative employment and Labour mobility: a new analytical framework. *Journal of economic research* (8).

[4] Aydemir A.,Borjas G.J. 2007. Cross-country Variation in the Impact of International Migration: Canada, Mexico, and the United States. *Journal of the European Economic Association*5(4):663–708.

[5] Borjas G.J.2003. The Labor Demand Curve Is Downward Sloping: Reexamining the Impact of Immigration on the Labor Market. *The Quarterly Journal of Economics* 118(4):1335–1374.

[6] Card D. 2009. Immigration and Inequality. *National Bureau of Economic Research.*

[7] Card D. 2001. Estimating the Return to Schooling: Progress on Some Persistent Econometric Problems. *Econometrica* 69(5):1127–1160.

Computing, Control, Information and Education Engineering – Liu, Sung & Yao (eds)

 ISBN: 978-1-138-02800-5

Innovative research on construction of college student management team from perspective of applied talents training

Xin Guang Ren & Yong Min Cui
Harbin University of Science and Technology, Harbin, China

ABSTRACT: This paper focus on the social urgent training demand for the applied talents. With innovation in the construction of the college student management team as research background, by analyzing the current situation of student management team construction through research and strengthening the importance of student management team construction, it proposes specific measures to innovate construction of the college student management team. Only the continuous innovating construction of the college student management team and improving education and overall quality of college students can cultivate outstanding applied talents.

KEYWORDS: Applied Talents; Student Managers; Cultivate; Construction.

1 THE CURRENT SITUATION OF STUDENT MANAGEMENT TEAM CONSTRUCTION

1.1 *Large number of transactional works continued weakening student managers' education work*

Daily management and service of students can't be separated from the student managers who not only bear the duties of teaching and educating, but also bear the responsibility of management and service-based education. Currently, the college student managers are busy with the daily management and administrative affairs every day, serving as roles of "interpreter" and "nanny", which have a serious impact on effectively play of managers' educational function to college students , make student managers' education work standstill, make the overall quality of college student management team construction stay at a low level for a long term, and lead to lacks of proper guidance and cultivation in ideological education work for college students.

1.2 *Low degree of specialization and lack of reasonable professional knowledge structure in college student management team*

in recent years, the academic degree level of the recruited college student managers have been greatly improved, most of whom have a master degree, which fully proves that our country attaches more importance to college students management team, but most student managers are relatively scarce for the professional background and knowledge base for ideological education of college students. Therefore, facing with new requirements for college students, there are still varying degrees of "knowledge panic" and "skills crisis". From the current personnel structure of the student management team of colleges and universities cultivating applied talents, they still lack systematic, clear professional construction, with unitary professional skills.

1.3 *Operational capacity of student managers cannot fully meet the development needs of applied talents*

College students' living and learning are directly affected by college student managers, problems encountered in the growth of the university, such as emotional confusion, psychological problems, career choices, relationships and other issues, need the correct guidance from students' managers. So it puts forward higher requirements for college student managers, which needs student managers not only have the ideological and theoretical knowledge, but also have a basic knowledge of multiple disciplines, the requirements are higher than ordinary teachers.

At present, the distribution of student managers in many colleges and universities (department) is only a few ones, which means that, on average, one student manager has to manage more than 300 students, and some even should manage over 400 students. In this state, how to form a scientific and orderly management in college students, to make the formation of a unity and progressive culture between student groups, so as to form a positive energy in the classroom and faculty, and to advance the goal of applied talents training, which requires student managers to have a strong comprehensive ability. However, according the practical situation of our country, college student managers lack comprehensive theoretical knowledge, whose comprehensive, practical ability is weak, and cannot fully meet the level requirements of developing cultivation of applied talents.

1.4 *Inconspicuous development goal of college student managers, affecting the work enthusiasm*

Career aspiration that college student managers possess is the soul of giving birth to career progress and promoting career development of student managers. However, due to various reasons, generally student managers' positioning of their own work is not high, lacking of clear direction, which leads to shortage of confidence in career development. Therefore, most college student managers will not treat it as a long-term career development or to develop as a lifelong career. In the current severe employment situation, some student managers treat student management as a springboard, and their unstable thought leads to a poor sense of responsibility and thus they couldn't settle down in their job. Such student managers lack the competence to construct self-awareness and a sense of urgency, that they are satisfied with dealing with the repeated work of daily affairs, and they have no research, no summary and no innovation, which has a serious impact on the degree of specialization and career development.

2 THE IMPORTANCE OF STRENGTHENING CONSTRUCTION OF COLLEGE STUDENT MANAGEMENT TEAM TO CULTIVATE APPLIED TALENTS

College students are currently facing with a pressure of severe social, employment, college students have a more strong desire of success and thirst for knowledge, and more eager to get practical guidance and help. College student managers are not only the students' good teachers and helpful friends, but also the backbone of the ideological education work for college students. Thus promoting the construction of the college student management team plays a significant role in the healthy growth of college students.

2.1 *Strengthen professionalism and specialization of college student management team is the requirement of improving the overall quality of college students*

College students are the future backbones of the construction of Chinese characteristic socialism, and they shoulder historical responsibility to build a harmonious society and realize the Chinese dream. Only greatly improving the scientific and cultural qualities and health quality, and guiding college students in establishing the correct ideals and beliefs to promote healthy growth of students can we lay a solid foundation for college students employment and entrepreneurship in the future, and train more talents for society.

And the improvement of college students' comprehensive quality needs hard work from high-quality college student managers. Only the professionalism and specialization of the college student management team and strengthening and improving ideological education of college students can make more and more college students to be consciously devoted to constructing a harmonious society. So it is extremely important to promote professionalism and specialization of the college student management team.

2.2 *Strengthen professionalism and specialization of college student management team is the requirement of guaranteeing harmony and stability in colleges*

Practice tells us that student managers doing detailed, in-depth education work in colleges are one of the important effective methods to promote the development and maintenance of harmony and stability in colleges under the current new situation. At present, with the rapid development of society, college students are faced with the problems in learning, life, emotion, employment and other aspects. Due to the lack of proper understanding of college students and student manager's specialized guidance, current college students' mental health problems keep increasing, and more suicides happen frequently, which has the closest relationship with non-professional and non-specializing college student management team. Currently, the college student manager is low in the academic status of the university generally, which makes the student managed to treat their job as a springboard, resulting in disharmonious and unstable phenomenon within the teachers team. Therefore, only strengthening construction of the college student management team can guarantee harmony and stability in colleges more effectively.

3 SPECIFIC MEASURES ON STRENGTHENING CONSTRUCTION OF COLLEGE STUDENT MANAGEMENT TEAM FROM THE PERSPECTIVE OF APPLIED TALENTS TRAINING

3.1 *Educational function to enhance student management team for professional development*

For the current situation of college student managers in our country, classification and comb of the job functions of student managers are needed. Straighten out the responsibilities of college student managers and various functional departments of school, and gradually establish the responsibilities and working system matching with the identity of the student managers, establish students' "Psychological Consultation", "Career Guidance" and other services to make student managers free from the complex daily affairs, to allow them have fully plenty of time to actually put into students' ideological education guidance and research, so as to improve the overall educational level of college students.

3.2 *Strengthen the training of student managers, enhancing the educational level of student managers*

College student managers are ideological torchbearer and guiders for contemporary college students, so the student managers play an important role in the colleges, thus student managers are required to have strong theoretical and moral qualities. They should conduct education and management, psychology and other aspects of professional counseling and training for student managers, and carry out more scientific research activities related to student managers work. Support student managers to study the relevant professional degree based on doing well in undergraduate education work, encourage and support full-time student managers to become specialists in educational work, and take the road of professionalism and specialization. Establish training bases, broaden the work horizons of student managers, sum up experience and establish a typical model, intensify propaganda, so as to motivate the improvement of the quality of student management team; through a rigorous system and quantitative indicators of pragmatic work, it helps student managers to enter into the role as soon as possible, and through advanced management methods and means, it helps student managers to gradually become professionals for college student work.

3.3 *Enhance applied quality of student management team, improving cultivation of applied college students*

To cultivate applied talents, colleges should make students have applied quality, firstly, we should strengthen the construction of applied quality of the student management team, which is to keep a foothold before teaching others. During this period, as most colleges and universities focus on teaching in different degrees, they make light of college students' education from student managers. There are many cases cannot meet the actual work in the recruitment and training of the college student management team, in morality, analysis and solving problems, practical ability and so on, they fail to give timely and accurate education and guidance to students, which will directly affect and restrict the development of students' application-oriented ability.

In the daily management work, colleges and universities can choose to allow student managers to lead students to practice in enterprises, or they could send properly educated student managers to front-line business management for testing exercise, to strengthen student managers' learning and understanding of professional practical ability. Therefore, student managers can carry out all types of activities to cultivate students' applied quality in school's daily management.

3.4 *Guide the construction of college student management team with the goal of training the applied talents*

3.4.1 *Cultivate college student managers with the goal of training the applied talents*

Linked the success of the student manager work with colleges and universities' goal of applied talents training, it requires the daily management, education, and service work of student managers should be based on the goal of training successful and talent students. Educate and inspire student managers by training goal and success concept of college students, and make every student managers to establish a sense of responsibility and a positive attitude to cultivate successful college students, which requires "One should be strong to forge iron", in order to make college students succeed and become application-oriented talents of the new era, as the first-line student managers for college applied talents training, they should strengthen their own applied talents training firstly.

3.4.2 *Cultivate the college student managers with advanced working concept*

The goal of applied talents training is to make every student successful and become a useful person in society. In order to accomplish this goal, it requires all student managers to have the subject awareness, social awareness, a sense of cooperation, etc., they should guide students in the daily management. They should regard assists students in being successful as their own success, and realize their personal values in the process of helping students succeed. Student managers should guide students in learning to think from multiple perspectives and possessing the ability to solve problems, continuing to innovate and improving their overall quality. Student managers should timely master students' status, narrow the distance with students, make bosom friends with students, and constantly innovative working methods, so as to create a new situation of college student management work.

3.4.3 *Enhance specialization of the student management team, reinforcing task motivation of student managers*

High level of understanding is a deeper problem in student managers thought, which requires student managers to establish a correct world view, life view and values, and firmly establish lifelong struggle ideals and beliefs for socialists in education and student work. We need to attach great importance to the cultivation, selection and use of student managers, provide room for their development, and concern about the growth and progress of student managers, and then makes student managers have a high sense of honor and pride of doing their job well. Think about further development to achieve professionalism and specialization of student managers, establish and improve the professionalism of incentive mechanisms and treatment. Starting from the actual work, encourage and support college student managers to take their jobs seriously and work hard, strive to become experts in ideological education of college students, and effectively promote the college goal of applied talents training.

In conclusion, the construction of the student management team plays an important role in the process of training applied talents in college, work performance of student managers should not be overlooked, we can only cultivate outstanding applied talents by strengthening construction of college student management team continuously and enhancing college students' education and application-oriented quality.

REFERENCES

[1] Chunmei Huang, Yanggan Yao. Political Ideological Construction of Counselor Team in Independent Institutions under the Goal of Applied Talents Training [J]. Journal of Huaihai Institute of Technology, 2011, (8)

[2] Ying Wu. Discussion on the Role of College Counselors in Implementation of Applied Talents Training Program[J]. Northern Literature, 2013, (4)

[3] Yanxin Li.On How to Improve Student Administration Work in the Process of the Construction of Application-oriented Private Colleges [J]. Journal of Jilin Huaqiao Foreign Languages Institute, 2011, (1)

[4] Xiaolin Ji. Discussion on Development of Specialization and Professionalism in University Instructor Team[J]. Forward Position, 2012, (17)

[5] Mingna Shi.On Specialization and Professionalism in University Instructor Team [J]. Cultural Education, 2013, (5)

[6] liu Yang. Discussion on How Counselor to Work Around Applied Talents Training [J]. Journal of Jilin Huaqiao Foreign Languages Institute, 2012, (11)

[7] Yuanhong Wang, Deqin Jiang, Youwei Wang. Research on the Reform of Universities Applied Talents Ideological and Political Education[M]. Nanjing University Press, 2013.

Computing, Control, Information and Education Engineering – Liu, Sung & Yao (eds)
 ISBN: 978-1-138-02800-5

On the enrollment issues of private colleges and universities

Shui Ping Huang & Mei Ling Guo
Guangdong University of Science and Technology, Guangdong, PR China

ABSTRACT: Recently, the private colleges and universities are generally going through a hard time of recruiting students. How to survive from the draining of student pool is a crucial problem the private colleges and universities have to confront. Through analyzing the causes of why the recruitment of private higher education is so difficult, which would effectively guide the enrollment work of private higher education.

KEYWORDS: The private colleges and universities; Recruitment; Problem; Causes.

1 INTRODUCTION

As the implement of *Non-state Education Promotion Law*, the private higher education has been remarkably developed and become an essential part of higher education in China. However, since 2009, the number of students who attend the college entrance examination has decreased annually meanwhile the state-owned higher education colleges and university's enrollment expansion continues guided by the related national policies. Combined with an increasing number of overseas colleges and universities start to seize the market share of higher education in China, the recruitment for non-state run colleges and universities has to strike against them. In some particular undeveloped area, due to the uncompleted recruitment plan recently, the private colleges and universities there are threatened to close down. The golden time that they could make a profit through plans only has gone.

Table 1. Statistics of students who attended college entrance examination and the number of its admission over the years (from1999–2013).

	Students number who attends exam	Admission students number	Admission Rate
Year	10 thousand	10 thousand	%
1999	288	160	56
2000	375	221	59
2001	454	268	59
2002	510	320	63
2003	613	382	62
2004	729	447	61
2005	877	504	57
2006	950	546	57
2007	1010	566	56
2008	1050	599	57
2009	1020	629	62
2010	957	629	69
2011	933	675	72
2012	915	685	75
2013	912	700	78

Seen from Table 1, the student pool reached at the peak (10.50million) in 2008 and then declined since 2009. It has decreased 2 million in the recent two years in total and it has shown a downward trend. In contrast, the admission rate is substantially increased. All these put the privately-run colleges and universities in a severer competition by enrolling students as they are initially in the weak position of higher education system.

Take the year of 2003 for example, in Beijing, Shanghai, Guangdong, Shandong, Hu'nan, He'nan etc.; some of the privately-run colleges and universities cannot achieve their enrollment goals. Among those, the colleges and universities which start late and have relatively lower education quality have been through a serious shortage of student pool and confronted a severe survival issue.

With the increasing number of colleges and universities and the expansion of their capacity, the shortage of student pool will exist for a long time. As the decreasing of the potential student's number, the competition of enrollment is not only between the same level and same type institutions, but also between the private vocational colleges and public vocational colleges and they even have to compete with the public colleges and universities. Therefore, the student's enrollment of non-government colleges and universities has been a problem cannot be ignored and will endanger their survival and development.

To fully understand the issue of enrollment and ensure the private colleges and universities' sustainable development, it is necessary to analyze these factors restricting their enrollment. Through investigation and research, the fundamental reasons that why the private colleges and universities are going through enrollment issues are as follows.

2 THE MAIN CAUSES

2.1 *There are an increasing number of students who give up attending the college entrance examination or give up submitting college applications or give up registering. Meanwhile, more and more students choose to study aboard. Therefore, generally the student pool is shrunk*

In recent years, high school graduates giving up the college entrance examination or college application or registering is a nationwide phenomenon, and it is an upward trend. According to the statistics published by the Education Ministry, the number of students quitting college entrance examination is 840 thousand in 2009, in 2010 this figure has become about one million and in 2011, there are 1.2 million students giving up on the examination. These candidates could have been going to those private colleges or vocational and specialized colleges. Thus the shrunk of the student pool becomes a more serious issue for those colleges and universities.

Currently, a colossal amount of high school graduates from superior family background chooses to study aboard over the colleges and universities in China. Most of those students belong to the student pool of private colleges and universities and higher vocational colleges. Usually their college entrance examination scores are underperforming and they are not willing to go to 2B level colleges and universities or higher vocational colleges. Moreover, as they are financially affordable, they prefer to go aboard to further their study. All these intensify the shortage of the student pool for those colleges and universities. According to a survey about the overseas student market, in terms of the number of students who intend to study aboard, undergraduates rank first, followed by the high school students. The figure of high school graduates who want to study aboard shows a strong growing trend and it is predicted that after a few years, they will almost be equal with the number of undergraduates who want to go aboard to study, which unavoidably will drain the student pool further.

2.2 *The expanding of public colleges and universities takes more students from private colleges and universities' student pool*

The number of colleges and universities increased and their capacity expanded as well; meanwhile, the enrollment plan grows every year. Therefore, colleges and universities are competing in recruiting students let alone those privately-run colleges and universities. With the economic development of China, society requires more from higher education. Against this background, the government not only encourages the development of private colleges and universities, but also expands enrollment plan for public colleges and universities since 1999. This practice makes the student pool competition expands from the circle of private colleges and universities only to the competition among public and private colleges and universities. The existing public higher vocational colleges are expanding rapidly and new ones are established. In terms of school funding, government input more budget to help the public higher vocational colleges, according their nature, level, scale, etc. Those colleges have a large number of admissions quota and lower tuition fees, which attract more graduates and give the private one a hard time to enroll students.

2.3 *The private colleges and universities are not highly recognized by the society and criticize its education quality*

The private colleges and universities have a history of 20 years and have been developed with the state's recognition that they are an important part of higher education. However, the problems such as short time of development, the deficient of school funding; instability of teaching staff and loss of daily management do exist. During the development of private colleges and universities, there have been some behaviors violating the rules. A few operators are money-orientated and their attitude towards education is incorrect. Some of the schools are like family-run business and regarded the school as personal property and run it as enterprises and do not follow the basic rules of school-running. Some colleges and universities are obviously set for short-term without any school-running characteristics. Many private colleges and universities rush to start popular programs like computer science or laws, no school-running characteristics are shown. Other problems are lack of competence in terms of teaching staff and low level of education and teaching quality. Some colleges and universities emphasize that they have better dormitory than key universities when competitiveness is concerned. Although they have been equipped with telephones, televisions and sanitary equipment, they are not qualified in terms of teaching staff. Their teaching staffs are mainly consisted of retired teachers from public colleges and universities plus some fresh graduates and part-time teachers. With these issues, it is hard for private colleges and universities to ensure the quality of their education and gain a good reputation in society, which makes it is harder for them to enroll students. Take all these reasons into consideration, most of the graduates prefer to choose government-run colleges and universities over the private ones even those public colleges and universities actually are relatively lower ranked. No one would voluntarily choose to study in a private college or university unless there is no other option.

2.4 *The difficulty of employment among fresh graduates and the impact of the idea of "study is useless"*

In recent years, with the expansion of enrollment of colleges and universities, the number of graduates spurred annually; however, the new vacancies in the society are limited, which is hard to meet the need of employment of college students. Unavoidably, the college students' employment situation is more and more serious. In 2013, there are 7 million graduated which lead the phenomena "the most impossible year to get a job". As a colossal amount of graduates can not find an ideal job, "graduation means unemployment" is not an unbelievable thing any more. Under this circumstance, the graduates of state-run universities or even key universities are not able to obtain a good job, let alone those who graduate from private colleges and universities. The huge pressure of finding a good job causes the students as well as the parents lose their interest in private colleges and universities and they start to believe that study is pointless. They tend to argue that if finding a job is not guaranteed with a degree, why do not just save the money and time. All these put more difficulties into private colleges and university enrollment.

2.5 *The relatively high tuition fee forces the financially disabled candidates to give up*

At present, the tuition fee of private colleges and universities is obviously higher than public colleges and universities. Take Guangdong Province for instance, in terms of tuition fee, the public one charge within 5,500 RMB per year, whereas the tuition fee of private colleges and universities is usually over 10,000 RMB per year and some even charge almost 20,000 RMB per year. The potential students have to quit their study because of the expensive tuition fee, especially those who are from low income families. There are some private colleges and universities investing a lot and lacking financial support from government. Students' tuition fee is their main income to keep the schools running. Without students and the tuition fee, those colleges and universities are hard to operate and reducing tuition fee is not a solution, which makes it into a vicious circle.

2.6 *The talents cultivation and program construction lack characteristics and features*

A crucial factor that whether a private college and university can complete its enrollment plan or not, is its characteristics and features, whether they can cultivate talents that can meet its regional and industrial need or not. Due to the short time of the private colleges and universities' development, most of those schools do not have their unique characteristics and features yet.

Private colleges and universities have unreasonable major setup and layout, prefer popular majors, and have similar majors, and lack unique majors. In the mode of professional cultivation, the private one does not show any characteristics of higher vocational colleges or applied undergraduate education. They have no obvious advantages during the competition with public colleges and universities because they have similar major setup and talents cultivation mode.

On the one hand, the major setup is almost the same, lack special programs and emphasis on low cost majors.

Most of the programs in private colleges and universities are set according to students' career planning and the need of social and economic construction. This practice undoubtedly is a progress compared with the previous one which ignored the social need while setting a program; however, it has some disadvantages as well. Namely, the programs' setting is too similar without any consideration of its own characteristics and irrationally following the trend in society regardless of their conditions and teachers' qualification.

Table 2. The analysis of majors setup among 10 private colleges and universities in Guangdong.

Major	Number
Accounting	10
Logistic Management	10
Business Administration	10
Computer Networking Technology	9
Marketing	9
English	9
Hotel Management	8
Japanese Langusge	7
Financial Management	6
Software Technology	6
Administration	5
Mode Tooling Design	5
Mechanotronics	5
Project Cost	4
Architectural Engineering	4
CNC Technology	3

It is seen from Table 2, almost every private college and university is focusing on developing programs like Financing and Accounting, Computer Science, Foreign Languages. Although it is required by the law that private colleges and universities cannot be profit-oriented, these colleges and universities still give priorities to the majors are cheap to low the running cost and balance the input and output and give little consideration to the rationality of the program setting structure. For instance, the most common majors are about the Arts and the needless lab equipment science ones.

On the other hand, the major setting is short-term oriented and varying frequencies, which are vicious for its long-term development.

Stemming from the need of survival and the competition in students' pool, the private colleges and universities have paid more attention to the short-term need for social development and have considered less about the long term trend. This phenomenon has intensified their catering to the social focus and has led directly to the frequent change major setting. Constantly changing majors, textbooks; constructing training places and practice bases; purchasing new equipments and updating facilities led to the waste and destruction of educational resources and give no guarantee in education quality. Schools care only about students' enrollment and put students future employment aside. This short-term oriented behavior is passively following the market rather than meeting the need of the society. Therefore, it will be punished by the law of market and educational development. As soon as the supply exceeds the demand, parents and students are not the only victims so are the schools. On a long term view, it is detrimental for the sustainable development of private colleges and universities as well.

Viewing through the development history of private colleges and universities, in the recent years, these schools in China have confronted an unprecedented "cold current" though; it does not mean that private higher education has been shrinking. From a long time perspective, the private higher education has a promising future. The key to solve this problem, for the private colleges and universities, is locating their own characteristics and features precisely and enhancing their core competence. Moreover, they should improve their inner quality and enroll students with honesty and integrity and regulate the charging of fees to cultivate a healthy social image. Meanwhile, the government should give more support to private higher education and promote the academic reform within private colleges and universities to make sure their sustainable development.

REFERENCES

[1] Li, D.C. & Chen, H.F. 2014. On the Current Situation and Strategies to the Enrollment Issues of Private Colleges and Universities, *Intelligence*, 14(8):103.

[2] Lin, H.K. & Li, X.S. 2013. On the Problems and Strategies to the Enrollment Issues of Private Colleges and Universities in Guangdong Province, *Course Education Research*, 13(29):11.

Computing, Control, Information and Education Engineering – Liu, Sung & Yao (eds)
© 2015 Taylor & Francis Group, London, ISBN: 978-1-138-02800-5

The evaluation of Donghai Bridge offshore wind project

Ru Liu, Lu Ji Zhang, Qiong Wu & Jun Chao Zhao
Beijing Municipal Institute of Science and Technology Information, Beijing, China

ABSTRACT: In this article, we will try to make a Cost-Benefit Analysis (CBA) evaluation in the case study. A project has to be economically feasible to be implemented, especially offshore wind project with huge demand of initial investment cost. Due to the huge cost, investors are facing huge risk of losing money in the business. Therefore a CBA is necessary to be made in order to obtain a vision for the investors throughout the project lifetime. CBA will be used in the case study and in our case the CBA will be calculations on all the positive and negative impacts of the whole project period and see which is grater to determine if the project is advisable or not. Calculations will be in a broad sense just to acquire a general understanding of the economics embedded. This project was funded by the Beijing financial fund.

KEYWORDS: offshore wind power, CBA (Cost-Benefit Analysis), economic CBA, Donghai Bridge offshore wind farm.

1 INTRODUCTION

Compared to onshore wind power, offshore wind power has a relatively short history. The first offshore wind farm was constructed in Vindeby, Denmark in 1991, with 11 turbines of total capacity of 4.95MW. By the end of 2000, there were in total 30MW offshore wind farms worldwide. In 2002, a wind farm with 80 turbines was built in the North Sea, with a generation capacity of 160MW. By the end of 2005, the world offshore wind farm capacity reached 700MW. By the end of 2007, offshore wind capacity reached 1,080 MW, taking 1.5% share of total installed capacity.

It belongs to Shanghai East China Sea offshore Wind Project which will install 34 wind-driven generators along the Shanghai East China Sea Bridge, each with a capacity of 3 MW. The total installed capacity will be 102 MW.

2 DEFINITION OF COST-BENEFIT ANALYSIS (CBA)

As one kind of clean energy, offshore wind power development will be long-term strategy but it is not best investment way in the short term. CBA could confirm and compare the project cost and benefit that are valued through economic market as well as some social cost and benefit. As an investment analysis, CBA focuses on all benefits and all costs. As the outcome of the analysis, the Benefits ratio (Benefits/Costs) could determine whether the project is socially valuable and financially feasible, or if another project should be pursued. If the Benefits/Costs exceeds 1, that means investor could undertake this project.

CBA could be analyzed from different points of view. And it could be divided into two aspects: economic CBA and social CBA. From project stakeholders' point of view to consider this project's cost and benefit, economic CBA will use the market price to measure all benefit and cost of project. This kind of economic CBA shows if the project has the positive NPV(Net present value) under the market price or not. This NPV is the total value of project under the market price. NPV compares the present value of benefits with the present value of costs in the time t. As show as following formula, "t" is the time of the cash flow along with the lifetime of the project; "i" is the rate of return that could be earned on an investment; and X is the present value of cost (cash outflow) or the present value of benefit (cash inflow). So the NPV compares the cash outflow and cash inflow. It is sensitive to the reliability of future benefit that an investment will yield.

3 BENEFITS OF DONGHAI BRIDGE OFFSHORE WIND FARM

As a largely untapped energy resource, offshore wind power in China is one of the key projects to address the climate goals and energy demands with lower environmental impact and lower social risk. It will be one of important components in renewable energy resource structure. In the period up to 2020, however, there are some barriers to limit development.

Offshore wind power could help China deliver its recently agreed target of 15 percent (Vestas Annual report 2008) of all energy must come from renewable sources.

Donghai Bridge offshore wind farm in China has several benefits as follow:

Donghai Bridge offshore wind farm can help China contribute to CO2 reduction.

Offshore wind power produces less pollution than conventional energy sources. Most countries in the world are pursuing wind power as a zero-emission energy resource, because of current environmental challenges. Global warming is certainly one of the motivating factors. The energy sector is the largest source of CO2 emission. So reform of the current energy supply mix in China and more detailed and effective environmental policies at international level are required. Wind power causes zero GHG emissions during their life cycle and no environmental damage through resource extraction and waste management. Offshore wind power is a good way to reduce the future carbon and it is expected to play an important role in meeting the Kyoto Protocol.

Donghai Bridge offshore wind farm will contribute to a more supply of electricity beyond coastal area at east of China.

Offshore wind power could meet the demand for increased internal energy demand and stimulate upgrade of national grid capacity. The eastern area of China is developed and it is lacking conventional energy. The offshore wind power will supply more electricity at east of China.

Donghai Bridge offshore wind farm could increase more employment opportunities; it also could develop regional economy.

When compared to other fuel sources power, offshore wind power has the lowest social costs. Offshore wind turbines just have three possible environmental impacts: noise, wildlife and visual impact.

Donghai Bridge offshore wind farm in China does not only mean contribute to CO2 reduction, but it is also reduced fossil fuel imports dependence in the future and increases in export opportunities, sustainable local economic growth, high quality job and so on. These benefits put Donghai Bridge offshore wind farm on the forefront of availability and competitiveness. The investors needn't to pay primary energy (wind) and wind is never runs out. Wind farm has a stable life-cycle cost with low investment risk. And it has so many positive benefits for various stakeholders: not only effects on the investors but also other actors. From the Social CBA's standpoint, the Benefits ratio (Benefits/Costs) could determine this project is socially valuable. The Benefits/Costs exceeds 1 apparently, that means investor would undertake this project from social aspect.

4 ECONOMIC CBA

We will assess the costs and benefits of Donghai Bridge offshore wind farm from economic aspect in this part. This part makes a calculation of the true cost of the Donghai Bridage offshore wind power under the current condition and compares to the price of electricity on grid. In particular, this case can become a valuable component in the electricity supply of Shanghai, only if it has the positive NPV under the market price. Offshore wind power could influence the fuel price volatility.

In fact, cost of energy power is a very significant method to selecting the way of power generation. It is very important to analyze the local cost of offshore wind power in China and the related factors to the cost, so that investors could reduce the cost and promote development of offshore wind power in East China Sea in the future.

5 COST COMPONENTS OF WIND ENERGY

In order to do a cost-benefit analysis of the ongoing offshore wind project by the East China Sea Bridge in Shanghai, we shall first learn the factors that the cost is dependent on so we could see in which factors cost could be saved and in which factors cost is tend to increase. The basic cost of wind farm includes both upfront and O&M (operation and maintenance) cost. And moreover, there are some variable costs such as tax, salary of employees, depreciation of the offshore wind turbine, institutional arrangements, etc. So the cost of wind energy could fall into initial investment and the annual recurring costs. And there are in general three factors that affect the cost of wind power projects and they are:

Annual recurring costs: Mostly Operation and Maintenance costs (site works) and other variable costs such as tax, salary of employees, depreciation of the offshore wind turbine and institutional arrangements, which varies between different countries. Like other industrial equipment, offshore wind turbines require operation and maintenance (O&M), and it is a biggest share in the total annual recurring costs.

Wind farm lifetime: The lifetime of this case is assumed to be 25 years.

The offshore wind farm has a longer technical lifetime than onshore wind farm, because of lower turbulence. Figure 1 shows the cost sensitivity change with the different wind farm lifetime. In this graph, to compare with 25 years project lifetime and 20 years, it makes 9% cost lower.((0.355-0.325)/0.325DKK/kWh) (Danish Wind Industry Association web site, 2003)

Although the cost of electricity is different from Denmark to China and year to year, the relationship between cost and project lifetime is the same. Longer project lifetime produce the cheaper cost of electricity.

After 25 years, the offshore wind farm adds an overhaul, which costs 25% of the initial investment. So that this offshore wind farm has more than 40 years in all, we could get cheaper cost f electricity.

The capital cost is considerably high and is required to be available when the wind farm is going to be built. Therefore, capital access and good repayment becomes essential. Choosing of wind resource, namely rationally locate each wind turbine, also largely affects a wind project's profitability. And offshore wind power is substantially more expensive than onshore wind power because of several factors:

Foundations are considerably more expensive. For example, a conventional onshore wind turbine foundation takes 4-6% of the windmill's total cost, whereas in the two largest offshore wind farms in Denmark, the percentage taken by foundation is 21% (Maria Isable Blanco,2008). The cost basically depends on water depth and soil conditions.

The cost per unit of electricity generated have the relationship with the capital cost of wind farm, wind power annual energy output and cost of operating and maintaining the wind farm per unit of energy output. The calculation of European offshore wind power cost resulted in 1800-2500 €/kW, which yields a generation cost of 6-11.1 €cent/kWh. The calculations were based on an estimation of 25-30 years lifetime of offshore wind turbines (Maria Isable Blanco, 2008). The results vary between different study methods. We will calculate some costs in China so that we can make comparison between cost and price from economic aspect.

The known data:
Project life: 25years
Manufacturer: Sinovel Wind Co., Ltd
Model: Sinovel V90-3MW – 90m
Capacity per unit: 3MW
Number of units: 34
Capacity: 102MW (Power International Energy, 2008)
h is the number of hours in a year (365days×24hours=8760hours/year)
P is the rate power of each wind turbine in kilowatts (3000kw/each)
T is number of wind turbine (34)
F is the net annual capacity factor of the turbines at the site: 2600hours/8760hours =29.7% (The yearly operation hour of the turbines is 2600 hrs, and there is 8760 hrs per year)
Here, hF = 2600 hours
So we can calculate the offshore wind energy output per year (electricity exported to grid)
$E=(hPF)T=(hF)PT=2600\text{hours}\times 3000\text{KW/each}\times 34=265.2\text{ GWh/year}$
The operating and maintenance cost per year
$M'=KC/N=(0.025\times 1892\text{ million MB})/25\text{years}=1{,}892{,}000\text{ CNY}$

In our project, we use the RETScreen Clean Energy Project Analysis Software to evaluate the offshore wind power and savings, costs, emission reductions, financial viability. This software is one of decision support tools developed with the contribution of numerous experts from government, industry and academia (http://www.retscreen.net/ang/home.php). Here, it is used to help us estimate costs associated with the Donghai bridge offshore wind farm. We have calculated some data above, and then put them into the software form. These data are addressed from initial investment, annual O&M cost, price on grid and so on.

We now conclude by saying that the need of pursuing clean and secure energy as well as reform of energy sector from coal dependent to diversified pattern in China has selected wind power as one of the approaches. We have proven that this is a rational choice in respect of economic, social and environmental benefits. China is representing the rising force in the global wind power industry by climbing up to the world fourth place after doubling the installations three times in the recent years. With all the above considered and implemented, we believe that offshore wind in China will acquire competitiveness and lead China toward a higher level within the world wind industry.

REFERENCES

[1] Danish Wind Industry Association web site, 2003, http://www.windpower.org/en/tour/econ/offshore.htm
[2] Danish Wind Industry Association web site, 2003, http://www.windpower.org/en/tour/econ/offshore.htm
[3] InfoLib EXPRESS, Wind power development report in 2008, page 17, www.istis.sh.cn
[4] Godfrey Boyle, 1996, Renewable Energy-Power for a sustainable future, page 302

Computing, Control, Information and Education Engineering – Liu, Sung & Yao (eds)
 ISBN: 978-1-138-02800-5

A new method for evaluating priming effect in speech recognition

Jing Fei Yang & Wen Tao Ying
School of Information Science and Technology, Xiamen University, Fujian, P.R. China

Zhi Ling Hong
Software School, Xiamen University, Fujian, P.R. China

ABSTRACT: Automatic Speech Recognition still poses a problem that the introduction of acoustic background noise induced stress causes speech recognition algorithms to fail, and in everyday communication situation, listeners usually feel it difficult to attend to target speech when there was multi-people talking. Priming effect is very useful in speech recognition under noise environment. In this study, a new computational method for evaluating the priming effect in the target speech recognition under noise conditions was investigated. The proposed method can help to distinguish which paradigm is helpful to facilitate speech recognition in the noise environment.

KEYWORDS: Speech Masking, Noise Masking, Speech Recognition, Computation, Priming Effect.

1 INTRODUCTION

Speech Communication is very important in the everyday life. In every-day speech communication situations, the transmission of speech can be affected by the environment noises. Speech recognition algorithms, though increasingly successful in a naturally quiet environment, fail in the presence of background sound which does not trouble a human listener. Recent years much effort has been directed to reducing this deficit in auto speech processing [1]. In order to further understand the nature of speech processing, it is critical to find experimental paradigms to build an effective computational method for evaluating speech intelligibility in the noisy environment.

Humans can attend to target speech when there was multi-people was talking. Processing of masking speech interferes with processing of target speech leading to impaired processing of target speech. However, the amount of speech masking is highly dependent on the similarity of the target and masker voices. Previous study has found that priming effect can significantly release target speech from not only speech masking but also noise masking conditions [2].

Since features derived from speech have proven to be the most effective in automatic systems, in order to automatically extract information transmitted in speech signal, figuring out how the speech signal was recognized when masked by a background sound is much more crucial.

In this study, the priming effect in target speech recognition under the noisy environment was investigated, and two masking condition including noise masking and speech masking were introduced in the present study. This paper is organized as follows: firstly, we described the experimental condition and proposed the controlled procedure. Then, we analyzed the data and give a detailed description to the results. Finally, we give a discussion about data distribution and at the end we make a conclusion.

2 METHODS

2.1 *Speech materials*

The acoustic analog outputs were delivered to a loudspeaker (Dynaudio Acoustics, BM6 A, Dynaudio, Risskov, Denmark). Speech stimuli were syntactically correct but semantically-anomalous Chinese "nonsense" sentence. Since in Chinese, a large number of words are two-character compound words in which each of the two characters has its own semantic representation. In the present study, each of these experimental nonsense sentences including three key words component.

Both target speech and priming speech was speaking by young-female voices, and acoustic signals of both target speech and priming speech for each of the three target young-female voices were different in this study.

About 6,000 double-syllable verbs, which were rated as having high frequencies of occurrence, and 12,000 double-syllable nouns, which were also rated as having high frequencies of occurrence, were used. These words were combined randomly into 6000 syntactically correct sentences .To ensure that sentences used in experiments were not meaningful, the probability of co-occurrence of two nouns with a verb in a normal sentence was determined according to the utterance database. Only sentences in which probability of co-occurrence of key words in the database was zero were used as the nonsense sentences for the present study. Since Chinese is a tonal language, further selection was made to balance syllable tones across sentences. A double-syllable pronoun was then placed before a noun, and an auxiliary verb was placed before a verb, making a selected sentence more natural. Finally, all sentences were examined by the experimenters to ensure that selected sentences were nonsensical (see more details in [3]).

In the present study, speech signal was sampled at 16KHz, windowed by a 10-ms Blackman window with a 5-ms shift, and then mel-cepstral coefficients were obtained by the mel-cepstral analysis technique [4]. The speech masker was digitally-combined continuous recordings for Chinese nonsense sentences and the noise masker was a stream of steady-state speech-spectrum noise. Both priming and target speech sounds were presented at a level of 60 dBA. The sound pressure levels of maskers were adjusted to produce four signal-to-noise ratios (SNRs): -8, -4, 0, and 4 dB.

2.2 *Design and procedure*

There were three types of priming conditions: (1) 0-priming condition: no priming condition, baseline; (2) I-priming condition: early part of the target speech sentence (including first two key words); (3) II-priming condition: identical to the target sentence, except that the last key word was replaced by a noise burst.
Twelve Mandarin-speaking young university students participated in this study. All the participants had normal and bilaterally balanced pure-tone-hearing thresholds. A loudspeaker was in the central front of the participant.

For a testing session, participants were informed of both the masking condition and the priming type. Each trial was started with the priming phase, then a single target sentence and an either noise masker or speech masker was presented. Participants' task was to determine which stimulus was the target sentence and the performance for each participant was scored on the number of correctly identified last keyword of target sentences.

3 RESULTS

A logistic psychometric function, $P(y) = 1/(1+e^{-\sigma(x-\mu)})$ was fit to each of the 12 participants' data, using the Levenberg-Marquardt method [5], where y is the probability of correct identification of last keywords in target sentences, x is the SNR corresponding to y, μ is the SNR corresponding to 50% correct on the psychometric function, and σ determines the slope of the psychometric function.

Figure 1 illustrates group-mean percent-correct word identification as a function of SNR, along with the group-mean best-fitting psychometric functions (curves) under the syllable-correct scoring scheme when the masker was noise (left panel) or speech (right panel). As shown in Figure 1, under either noise-masking or speech-masking conditions, word identification performance was substantially improved as the SNR increased from -8 to 4 dB. Also, under noise-masking conditions, the word-identification performance was not affected by the priming-stimulus type. However, under speech-masking conditions, performance following the presentation of the II- prime condition was much better than that following the presentation of the 0-prime condition or that following the presentation of the I-prime condition.

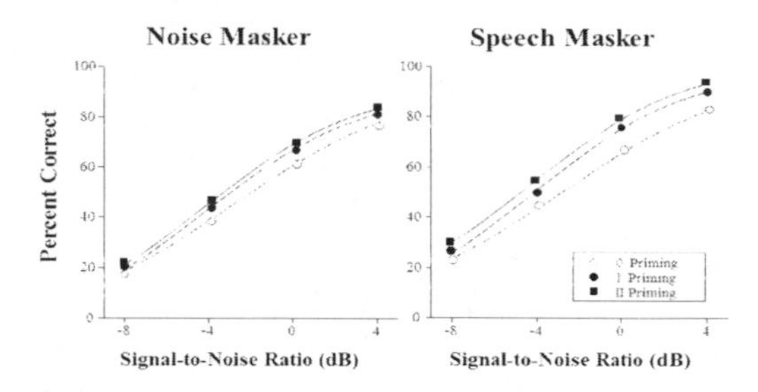

Figure 1. The mean percent-correct identification of the last target keyword across 12 listeners as a function of SNR for the three priming conditions when the masker was noise speech.

Figure 1 illustrates group-mean percent-correct word identification as a function of SNR, along with the group-mean best-fitting psychometric functions (curves) under the syllable-correct scoring scheme when the masker was noise (left panel) or speech (right panel), and the two panels also show the best-fitting psychometric functions under this three priming conditions.

As shown in Figure 1, under either noise-masking or speech-masking conditions, word identification performance was substantially improved as the SNR increased from -8 to 4 dB. Also, under noise-masking conditions, the word-identification performance was not affected by the priming-stimulus type. However, under speech-masking conditions, performance following the presentation of the I-prime was much better than that following the presentation of 0-prime or that following the presentation of the II-prime.

The differences in threshold m between conditions were examined. A one-way Analysis of variance (ANOVA) confirms that the effect of priming type was not significant ($F_{2, 22} = 1.049$, $p = 0.361$). However, when the masker was human-like speech, the performance under the 0-priming condition was much more poorer than those under other priming conditions. A one-way ANOVA shows that the effect of priming type was significant ($F_{2, 22} = 7.341$, $p = 0.003$). Post hoc analyses show that the threshold under the II-priming condition was significantly better than that under the 0-priming condition ($t_{11} = 3.671$, $p = 0.003$) and that under the I-priming condition ($t_{11} = 2.925$, $p = 0.011$). There was no significant difference in threshold between the 0-priming condition and the I-priming condition ($t_{11} = 0.678$, $p = 0.543$).

4 DISSCUSSION AND CONCLUSION

The results of this study show that priming effect can significantly improve recognition of the last target keyword when the masker was human-like speech. Under each of the stimulus conditions, percent-correct word scores increased monotonically with the increase of SNR from -8 dB to 4 dB, without displaying plateaus. The absence of non-monotonicity is in agreement with the results reported by [2] and [3].

Priming effect is useful in specifically reducing speech masking in the noise environments. Priming sentence can assist listener to attend to the trial-specific vocal features of the target voice, therefore facilitating them to follow the target stream.

Since the introduction of acoustic background noise causes speech recognition algorithms to fail in the auto speech recognition process, the proposed new approach can provide a useful subjective method for improving the automatic speech recognition systems.

ACKNOWLEDGEMENT

This work is supported by National Natural Science Foundation of China (31200769), the 2014 Program for New Century Excellent Talents in Fujian Province University. The corresponding author is Dr. Zhiling Hong, Xiamen University, email: hongzl@xmu.edu.cn.

REFERENCES

[1] Gong, Y. Speech recognition in noisy environments: A survey. Speech communication, 16(3), pp.261–291. (1995)

[2] Freyman, R. L., Balakrishnan, U., and Helfer, K. S. Effect of number of masking talkers and auditory priming on informational masking in speech recognition, J. Acoust. Soc. Am. 115, pp.2246–2256.(2004)

[3] Yang, Z. G., Chen, J., Wu, X. H., Wu, Y. H., Schneider, B. A., and Li, L. The effect of voice cuing on releasing Chinese speech from informational masking, Speech Communication, 49, pp. 892–904,2007.

[4] Tokuda, Keiichi, Heiga Zen, and Alan W. Black. An HMM-based speech synthesis system applied to English. Speech Synthesis, pp. 11–13.(2002)

[5] Wolfram, S. Mathematica: A System for Doing Mathematics by Computer. Addison-Welsey, New York .(1991)

[6] Yoshimura, T., Tokuda, K., Masuko, T., Kobayashi, T., and Kitamura, T. Simultaneous modeling of spectrum, pitch and duration in HMM-based speech synthesis, Proc. EUROSPEECH, 5, pp.2347–2350.(1999)

[7] Wu, M.-H., Li, H.-H.,Hong, Z.-L., Xian, X.-C., Li, J.-Y., Wu, X.-H., Li, L. Effects of aging on the ability to benefit from prior knowledge of message content in masked speech recognition. Speech Communication, 54,529–542. (2012)

[8] Wu, M.-H., Li, H.-H., Gao, Y.-Y., Lei, M., Teng, X.-B., Wu, X.-H., Li, L. Adding irrelevant information to the content prime reduces the prime-induced unmasking effect on speech recognition. Hearing Research, 283,136–143. (2012)

[9] Cao, S.-Y., Li, L., and Wu, X.-H.Improvement of intelligibility of ideal binary-masked noisy speech by adding background noise. Journal of the Acoustical Society of America, 129, pp.2227–2236.(2011)

[10] Masuko, T., Tokuda, K., Kobayashi, T., and Imai, S. Speech synthesis from HMMs using dynamic features, in Proceedings of the International Conference on Acoustics, Speech, and Signal Processing, ICASSP-96, Atlanta, GA, pp.389–392.(1996)

[11] Greenberg S., and Ainsworth W A. Speech processing in the auditory system: an overview. In: Greenberg S., Ainsworth W A., Popper A N., and Fay R R. (eds) Speech Processing in the Auditory System. Springer.(2004)

Computing, Control, Information and Education Engineering – Liu, Sung & Yao (eds)
© 2015 Taylor & Francis Group, London, ISBN: 978-1-138-02800-5

A novel method for nasal-oral airflow measurement

Long Ma, Jian Guo Wei & Yu Qing He
Tianjin University, Tianjin, China

Jian Wu Dang
Japan Advanced Institute of Science and Technology, Japan

ABSTRACT: This paper shows a novel nasal-oral airflow measurement system. Airflow is important to speech production field, particularly for the generation of voiced sounds and turbulent voiceless sounds. The traditional velocity-based instrument is fixed on a support and big and weight. In our nasal-oral airflow measurement system the subject wears a new mode mask with sensors that can measure the air flow output. The output signal is collected by a data collection module and synchronizes with the audio signal in the mean time. The computer can get the data via the data transportation port.

KEYWORDS: Phonetics, Air Pressure Transducer, Low-pass Filter and Amplify Circuit.

1 INTRODUCTION

Process of speech production includes many voice organs [1], such as the piriform fossa [2]. To generate complex voice signals, they produce a variety of physiological activities. These physiological activities of speech production process occur mutually coordinated. One set an appropriate width of the glottis by the throat, while adjust the airflow from lungs to match the movement of the larynx [3]. When we say, organs of tongue [4], lips, and chin and so on cooperate to adjust sound track [5], and interact with throat to adjust the tone. They constitute a completely physiological model [6], a division of labor to produce speech.

Here are some examples of articulation recording instruments: Electroglottography (EGG) is used to measure the contraction degree between the vocal folds. Photoelectroglottography (PGG) is used to measure the open degree of the glottis. Endoscope is for the direct observation of the larynx movement. Ultrasound allows us to clearly see the outline of the tongue. Dynamic or static palate digital photography equipment is to get the data of contact degree between the tongue and palate. Nasal microphone and vibration sensors can obtain nose vibration and nasal sound. Douglass Phonetics Laboratory also studies airflow velocity and pressure testing.

This study can indirectly advance the development of physiological speech, which is, through design and implement of a variety of detecting instruments on getting information of articulator movements previously, and through statistical analysis, applying the results in physiological speech filed, will provide the basis for development and expansion of phonetics.

2 THE NASAL-ORAL AIRFLOW MEASUREMENT SYSTEM

2.1 *The chosen of the mask*

Copy the template file B2ProcA4.dot (if you print on A4 size paper) or B2ProcLe.dot (for Letter size paper) to the template directory. This directory can be found by selecting the Tools menu, Options and then by tabbing the File Locations. When the Word programme has been started open the File menu and choose New. Now select the template B2ProcA4.dot or B2ProcLe.dot (see above). Start by renaming the document by clicking Save As in the menu Files. Name your file as follows: First three letters of the file name should be the first three letters of the last name of the first author, the second three letters should be the first letter of the first three words of the title of the paper (e.g. this paper: balpcc.doc). Now you can type your paper, or copy the old version of your paper onto this new formated file.

Airflow is important to speech production, particularly for the generation of voiced sounds and turbulent voiceless sounds. The airflow rate (liters/s) is often measured using a rigid facemask with a vent filled by a layer of stainless-steel fine mesh that gives small airflow resistance. The major inconvenience is that the rigid mask creates an acoustic cavity in front of the mouth and nostril openings. Its acoustic effect as a low-pass filter is so severe that the recorded audio signals cannot be used for speech analysis.

Our new pneumotachography uses a mask made of synthetic fibers, instead of conventional rigid types. This mask is almost acoustically transparent, and thus the radiated speech sound through the mask is almost free from acoustic distortion. In this preliminary stage of the development, we use shell-type protection masks commonly available in the market, which are designed to have relatively small airflow resistance in order to ensure the subject's comfort. The mask supplies a small resistance, which is necessary to measure airflow without perturbing sound propagation.

The mask itself is disposable and can be used in a hospital for clinical evaluation as well as in a laboratory with no risk of infection. A microphone is placed in front of the mask for sound recording. It is evident that only the weakened first and second formants are visible in the top part because of filtering, whereas the vowel formants are complete in the bottom. The airflow measurements using our shell-type protection mask give comparable results with those from the rigid mask.

2.1.1 *The design of circuit*

The circuit of our nasal-oral airflow measurement system includes two parts, they are the level shifter circuit, and the low-pass filter and amplify circuit.

2.1.2 *The level shifter circuit*

The level shifter circuit includes a pressure transducer and a voltage conversion circuit.

2.1.3 *The low-pass filter and amplify circuit*

The low-pass filter and the amplify circuit is used to process the output of the level shifter circuit. It can filter the noise and amplify the signal.

2.2 *Data recording*

We use USB data acquisition unit to record airflow pressure signal and voice signal at the same time. By setting different parameters under different conditions, we can record data at different frequencies. The data include the voltage from airflow pressure difference and speech signal. We can observe the data on PC and save the data in Excel file.

3 EVALUATION

To test the performance of our system, we use six plosive /p/, /b/, /t/, /d/, /k/, and /g/ in Chinese. We divide them into completely plosive /p/, /t/, /k/ and incomplete plosive /b/, /d/, /g/.

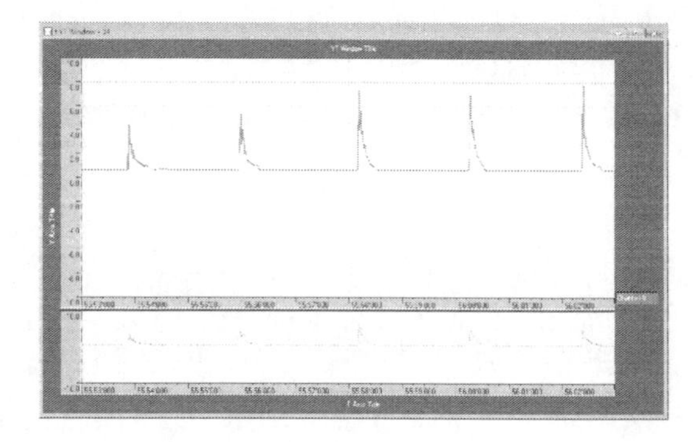

Figure 1. The data recorded by the device not with the low-pass filter circuit and amplify circuit (from /p/).

Form Figure 1, Figure 2 and table 1, we can see that there is an obvious difference in voltage amplitude. So these two parts of circuit work very well.

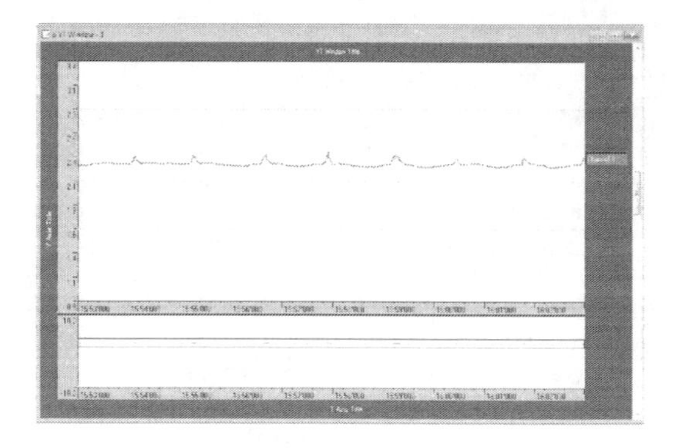

Figure 2. The data recorded by the device with the low-pass filter circuit and amplify circuit (form/p/).

Table 1. Comparison of the data.

	Not with amplify circuits	With amplify circuits
The voltage value when air pressure is 0	2.3[V]	0.6[V]
Peak Voltage Value	2.4[V]	9.0[V]
Voltage Amplitude	0.1[V]	8.4[V]

The following experiment is the contrast of completely plosive voltage value of /t/ and incomplete plosive /d/ recorded by the devices with amplify circuits.

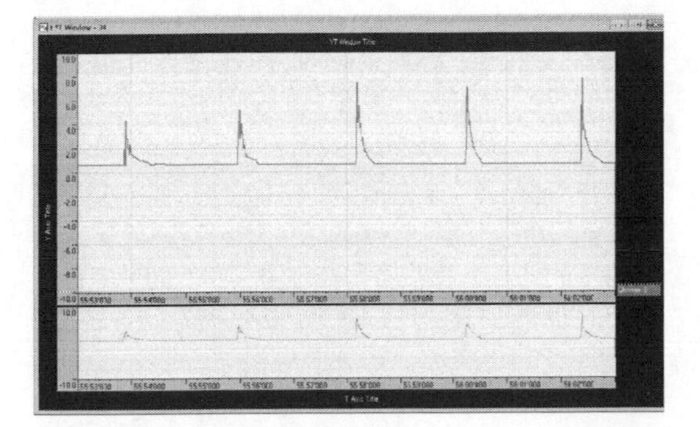

Figure 3. The data of /t/ recorded by the device.

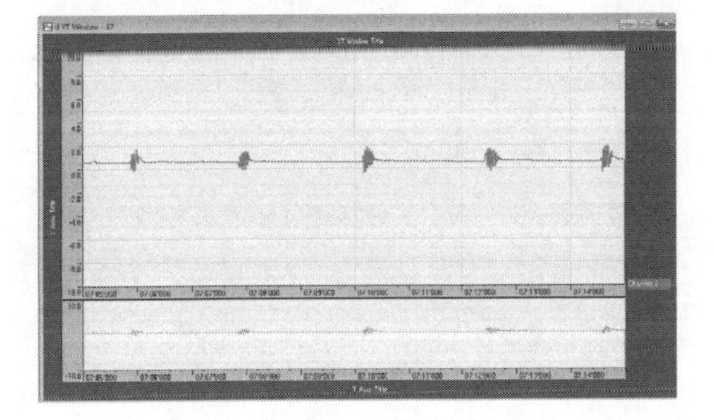

Figure 4. The data of /d/ recorded by the device.

From Figure 3and Figure 4 , the voltage amplitude of complete plosive /t/ is 5.9[V], and the voltage amplitude of incomplete plosive /d/ is 1.2[V]. All of the results are in accord with our general knowledge. And we can find that the air pressure of complete plosive is bigger than that of incomplete plosive.

4 DISCUSSION

Although this research has achieved the goal of making a nasal airflow pressure detection device, the device still needs to be improved in a lot of details. For example, when the pressure difference is zero, the output voltage conversion circuit is 100[mV]. An error occurs. You need to continue improving the circuit welding cases. You can even consider using a more accurate pressure sensor to reduce errors.

After an access to the amplification circuit, an error will also be amplified. In order to reduce the error, we need to improve the design.

The design of the device is for the measurements of oral and nasal air pressure simultaneously, not for a single measurement. Improvement program can be designed into a single measurement for oral and nasal air pressure.

5 CONCLUSIONS

This paper analyzes the data of the pronunciation of Chinese plosives. We can furthur analyze the data, and create the appropriate database which can help other researchers to study.

The study of speech production will become increasingly important, and have more and more broad prospects. Speech production is bound to be widely used in the field of artificial intelligence and multimedia to create a more colorful and richer sensory world of communication means for humanity.

ACKNOWLEDGEMENTS

This work is sponsored by the national natural science foundation of china with contract No.61175016 and No. 61304250. It is also supported by 973 Program (Contract No. 2013CB329305).

REFERENCES

[1] Dang J, Honda K, Suzuki H. Morphological and acoustical analysis of the nasal and the par nasal cavities [J]. The Journal of the Acoustical Society of America, 1994, 96: 2088.

[2] Dang J, Honda K. Acoustic characteristics of the perform fosse in models and humans [J]. The Journal of the Acoustical Society of America, 1997, 101: 456.

[3] Honda K, Hirai H, Masaki S, et al. Role of vertical larynx movement and cervical lordships in F0 control [J]. Language and Speech, 1999, 42(4): 401–411.

[4] Honda K. Organization of tongue articulation for vowels [J]. Journal of Phonetics, 1996, 24(1): 39–52.

[5] Takeout H, Honda K, Masaki S, et al. Measurement of temporal changes in vocal tract area function from 3D cine-MRI data [J]. The Journal of the Acoustical Society of America, 2006, 119: 1037.

[6] Dang J, Honda K. Construction and control of a physiological articulator model [J]. The Journal of the Acoustical Society of America, 2004, 115: 853.

Computing, Control, Information and Education Engineering – Liu, Sung & Yao (eds)
© 2015 Taylor & Francis Group, London, ISBN: 978-1-138-02800-5

Optimal orthogonal designs

Wei Yan Mu & Jian Jie Chen
Science School, Beijing University of Civil Engineering and Architecture, Beijing, China

Shi Feng Xiong
Academy of Mathematics and Systems Science, Chinese Academy of Sciences, Beijing, China

ABSTRACT: A class of optimal orthogonal designs are presented for computer experiments. This kind of designs optimize some optimal criterion such as the max-min distance criterion over all orthogonal designs. A block coordinate descent algorithm is proposed to compute such designs. Some two-dimensional examples are given.

KEYWORDS: Computer experiments; stochastic approximation.

1 INTRODUCTION

There are many designs introduced for computer experiments; see [1, 5, 6, 7] among others. Some authors proposed orthogonal space-filling designs and showed their good properties [2, 4, 8, 9]. In this paper we consider a class of orthogonal designs that optimize some optimal criterion but are not limited as Latin hypercube designs. A block coordinate descent algorithm is proposed to construct them.

2 CONSTRUCTION

We consider an orthogonal design $X = (\mathrm{x}_1,\ldots,\mathrm{x}_p)$, where $\mathrm{x}_j = (x_{1j},\ldots,x_{nj})' \in [-1,1]^n$, $\sum_{i=1}^{n} x_{ij} = 0$ for $j = 1,\ldots, p$, $\mathrm{x}_j'\mathrm{x}_{j'} = 0$ for $j \neq j'$, and $m = n - p > 1$. The set of such designs is denoted by D (n, p). It is desirable that X has some optimal property, such as space-filling property. Let f (X) be the criterion function. We need to solve

$$\min f(X) \text{ subject to } X \in D(n,p) \tag{1}$$

The block coordinate descent algorithm for solving (1) is stated as follows.

For given n and p , do the following:

1 Initialization: Start with $X^0 \in D(n, p)$.
2 Iteration: *Given* $X^K = (x_1^K,\ldots,x_P^K) \in D(n, p)$,

for j = 1,..., p, solve

$$\mathrm{x}_j^{k+1} = \arg\min_{\mathrm{x}_j} f\left(\mathrm{x}_1^{k+1},\ldots,\mathrm{x}_{j-1}^{k+1},\mathrm{x}_j,\mathrm{x}_{j+1}^{k},\ldots,\mathrm{x}_p^{k}\right), \tag{2}$$

$$\text{subject to } M_j'\mathrm{x}_j = 0 \text{ and } \mathrm{x}_j \in [-1,1]^n \tag{3}$$

Where $M_j = \left(1_p,\mathrm{x}_1^{k+1},\ldots,\mathrm{x}_{j-1}^{k+1},\mathrm{x}_{j+1}^{k},\ldots,\mathrm{x}_p^{k}\right)$.

Let $X^{k+1} = \left(\mathrm{x}_1^{k+1},\ldots,\mathrm{x}_p^{k+1}\right)$.

Constraint (3) guarantees the design $\left(\mathrm{x}_1^{k+1},\ldots,\mathrm{x}_{j-1}^{k+1},\mathrm{x}_j,\mathrm{x}_{j+1}^{k},\ldots,\mathrm{x}_p^{k}\right)$, to be an element of D (n, p).

Let f be the maxmin distance criterion [3] . There are some maxmin distance orthogonal designs (MMDODs).

$$D^*(4,2) = \begin{pmatrix} 1 & 1 \\ 1 & -1 \\ -1 & 1 \\ -1 & -1 \end{pmatrix}, f\left(D^*(4,2)\right) = 2,$$

$$D^*(5,2) = \begin{pmatrix} 0 & 0 \\ 1 & 1 \\ 1 & -1 \\ -1 & 1 \\ -1 & -1 \end{pmatrix}, f\left(D^*(4,2)\right) = \sqrt{2}$$

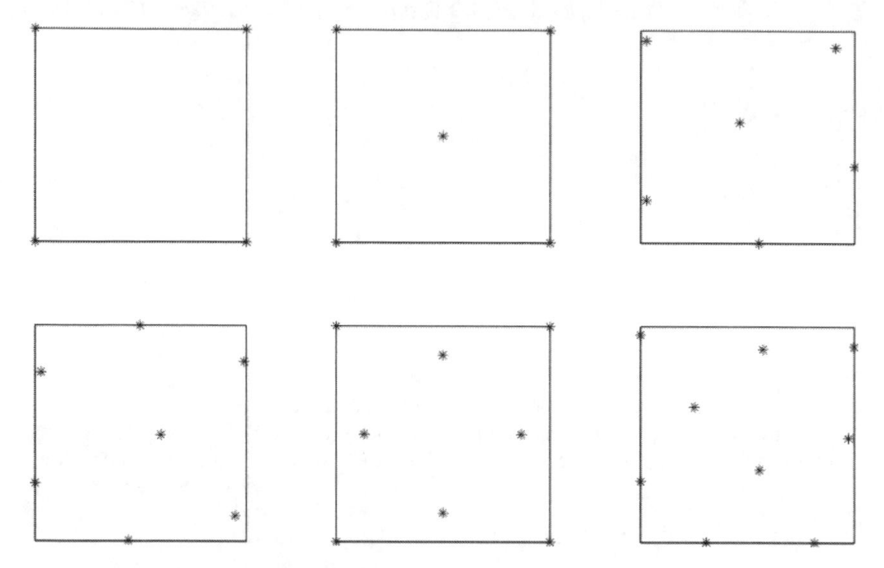

Figure 1. Some MMDODs with different n for p = 2.

From the above figure, the MMDODs have both orthogonality and good space-filling property.

ACKNOWLEDGEMENTS

Mu's work is supported by Funding Project for Talents of Beijing Municipal Party Committee Organization Department (Grant no. 2013D005017000016), Science and Technology Plan Project of BeijingMunicipal Education Commission (Grant no. KM201410016003), and Academic Innovation Team of Beijing University of Civil Engineering and Architecture (Grant no. 21221214111).

Chen's work is supported by BUCEA Urban Rural Construction and Management Industry Research Development Collabration Post Graduate Training Centre(Grant no. 31062014001).

Xiong's work is supported by the National Natural Science Foundation of China (Grant no. 11271355, 11471172).

REFERENCES

[1] Fang,K.T., Li, R. Z., and Sudjianto, A. (2005). Design and Modeling for Computer Experiments. Chapman Hall/CRC Press, New York.

[2] Lin,C.D.,Mukerjee,R.,and Tang, B. (2009), Construction of orthogonal and nearly orthogonal Latin hypercubes. Biometrika, 96, 243–247.

[3] Johnson, M., Moore, L., and Ylvisaker, D. (1990), Minimax and maximin distance design, Journal of Statistical Planning and Inference, 26, 131–148.

[4] Joseph, V. R. and Hung, Y. (2008), Orthogonal-maximin Latin hypercube designs, Statistica Sinica, 18, 171–186.

[5] McKay,M.D.,Beckman,R.J., and Conover,W. J.(1979),A comparison of three methods for selecting values of input variables in the analysis of output from a computer code, Technometrics, 21, 239–245.

[6] Qian, P. Z. G. (2009), Nested Latin hypercube designs, Biometrika, 96, 957–970.

[7] Tang,B.(1993), Orthogonal array-based Latin hypercubes, Journal of the American Statistical Association, 88, 1392–1397.

[8] Yang, J-F., Lin, C. D., Qian, P. Z. G., and Lin, D. K. J.. (2013), Construction of sliced orthogonal Latin hypercube designs, Statistica Sinica, 23, 1117–1130.

[9] Ye, K. Q. (1998), Orthogonal column Latin hypercubes and their application in computer experiments, Journal of the American Statistical Association, 88, 1392–1397.

Computing, Control, Information and Education Engineering – Liu, Sung & Yao (eds)
 ISBN: 978-1-138-02800-5

Improvement chain supplies a company of development of technological solutions in the commercial areas and area of consultancy

Amelec Jesus Viloria Silva
University Sergio Arboleda, Bogotá, Colombia

ABSTRACT: The company under study is dedicated to consulting and development of technological solutions using CRM software (Customer Relations Managing). Because of its low productivity, failure to demand and low efficiency in the use of technology for the generation of knowledge, it decides to carry out a project called "efficiency 3.0", which aims to develop proposals for improvement in the supply chain between commercial and consultancy area and propose a system of performance indicators for measuring results. Through the analysis of the current situation, improvement proposals were developed in three (3) key factors: people, processes and technology. A new working model based on Centers of Expertise to industrialize the process of implementing projects in the area of Consulting was designed. Turn a table Skills was established to measure the skills of resources and a system of performance indicators developed. Finally, he based Skills a linear programming model for optimal allocation of resources was created. Following the implementation of these improvements the company increase productivity by 75% within one year.

KEYWORDS: supply chain consulting, management indicators, centers of expertise.

1 INTRODUCTION

There are three (3) fundamental aspects to be continually revised in any business consulting: Processes, People and Technology [1]. The company under study is dedicated to consulting and development of technological solutions using CRM software (Managing Customer Relations). The group has offices in Spain, Venezuela and Colombia and executes projects in sixteen (16) countries.

Although the market has grown at a steady pace, current methodologies in the core business processes of study does not allow it to be possible to meet the demand. Additionally, current technology that supports much of the generation of knowledge in the business, is being used poorly. In this and due to the growing global economic crisis, the policy has been the need to overhaul the company to reduce costs.

The board intends to undertake a process reengineering project called "efficiency 3.0". This is carried out by an external consultant and its overall objective is to propose and implement improvements that reduce costs and increase business productivity. To achieve this, it is concluded that it is necessary to make proposals for improvement within three (3) fundamental aspects described above. This article aims to present design proposals for improvement in the supply chain to the company and creating a system of indicators to measure the performance of their resources.

2 METHODOLOGY

The methodology is divided into four (4) stages, all within the company under study [2]:

1 Analyze the supply chain between Commercial and Consulting.
2 Determine the key factors to improve the supply chain.
3 Propose improvements in the supply chain between Commercial and Consultancy and design tools for the allocation of human resources in the area of consulting.
4 Implement the proposed improvements in the supply chain between Commercial and Consulting.

3 RESULTS AND ANALYSIS

In this section the results obtained by using the methodology presented above are exposed.

Phase 1: Analysis of Supply Chain

Figure 1 shows the supply chain of the company under study.

Phase 2: Determination of Key Factors

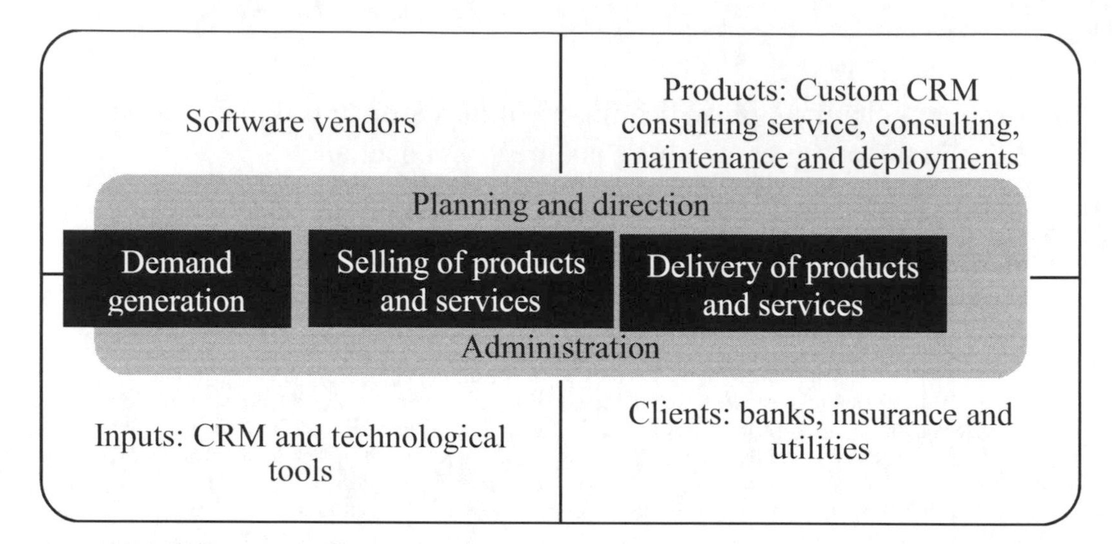

Figure 1. Supply Chain Company in studio.

To identify the strengths, weaknesses, opportunities and threats of the supply chain process internal and external agents were analyzed both. Figure 2 shows the SWOT matrix.

strengths	opportunities
Strategic alliance with Microsoft vendor Highly qualified human resources Horizontal communication flow segmented market	New products available for sale Potential for expansion in the North American market
weaknesses	**threats**
Individual and group no responsibility for the tasks Little reuse of knowledge and information generated in previous projects No measuring process	Marked economic crisis in Spain High dependence on a single customer Downsizing

Figure 2. SWOT Matrix for supply chain enterprise study.

Based on the analysis of key SWOT Matrix identifies three (3) components around which should focus improvement proposals: Processes, People and Technology. Which are described below:

Processes: improvements are needed in the current business model for the execution and delivery of projects in the area of consulting. Currently the company works with a Model of Multidisciplinary Teams, where an employee performs different roles. It should restructure or evolve the current model to one that increases quality, through group and individual responsibility not traditional projects. A turn is due to emphasize the Deliverable and not the activity that performs the action.

People: improvements in this component covering everything related to Employees and refines some of its entities and relationships. The concept of KPI (system management indicators) that will be used to measure the performance of resources and firm performance is introduced. On the other hand, the creation of a system to measure the skills of employees to track and control their experience as time progresses is necessary.

Technology: in this regard the creation of a single technology platform that facilitates the management of information throughout the supply chain is necessary. Due to the difficulty of locating previous papers on the current platform, create a technological space and a new methodology capable of improving the management and control of the information developed in the company.

Phase 3: Proposals for Improving and measuring tools

The following describes each of the proposals for improvement.

Design of a new working model based on Centres of Expertise

The Centers of Expertise (CE) are the answer to a search of quality, through the (not individual as is traditional) group responsibility and better use of resources by means of intermittent allocating them to reduce efforts required to run a project [3], [4] and [5].

With the implementation of the EC and the emphasis control and measurement Deliverable level, incentives for acceleration and consequent reduction of effort are created. The EC group related roles or one role, have their own portfolio of products or deliverables and their "clients" are the projects both customers and internal. Deliverables are any documents, programming code, record, etc. it is performed by a resource in the company [6] and [7].

The advantages of CE are: concentrated talent and knowledge to implementation and facilitate the transfer of knowledge, foster the creation of affinity groups with similar skills are the incubator of new skills and ideas, offering a capacity of more balanced service with less risk, reduce costs and improve quality through continuous improvement of the Deliverables for which they are responsible, provide Industrial property and linear processes of the company [8] and [9].

The EC should create are: Operations CE, CE Development & Management Functionality CE, CE and CE Project Control Knowledge & Quality.

Design a table Skills for evaluation of human resources

To assign employees to each role within the EC, it is necessary to measure their skills. Under the proposed model, employees are associated with a charge and one or more Roles. Therefore, based on the Role Skills to identify those measurable character needed to carry analysis is performed. It is proposed to divide based Skills thirteen (13) categories:

Database, Communication, CRM, Languages, Industries, Integration, Programming Languages, Processes, Reportadores, Bi, project management tool, Functional and (functional and technical) systems.

Designing a system of performance indicators

To give measurable character to current business processes, creating a system of performance indicators is proposed. The KPI (Key Performance Indicators) are a set of indicators that apply to all employees and will be a key process performance evaluation element. KPIs are designed based on four (4) different perspectives: Financial, Customer, Processes and Resources.

<u>Phase 4: Implementation and evaluation of improvements</u>

The improvements were implemented for one year in the company under study and its evolution monitored based on a productivity indicator defined as the ratio between successful and generated projects. At the beginning of implementing the indicator value is 30% a year later this was increased to 95%, representing 75% improvements in it.

4 CONCLUSIONS

Supply chain between the commercial and consultancy is analyzed. This allows to know the process from creating a business opportunity to sell the final product or service. As a result it is concluded that the company under study has some shortcomings in the current model based on multidisciplinary teams. Although effective in the implementation of projects, it is considered that the current methodology is of great complexity, which accommodates the development of proposals to improve the process.

A study of the key to be improved in the supply chain through the SWOT analysis of the same factors ago. Through various interviews and meetings with staff and directors of the company, it is concluded that should improve the chain from three (3) different components: people, processes and technology.

Aims to simplify the current model of consulting through the creation of Centres of Expertise and the introduction of the concept of Deliverable. This new methodology industrialization in some way the process for building projects. Each Expertise Center is considered a specialist in one area of activity and not as a multidisciplinary team equipment.

Using Skills table model resource allocation is designed through the use of Excel. This proposed model works as support for decision-making in project management. The model is able to assign the best possible combination for each of the roles required in a project resources. Therefore, the selection process and resource allocation will be easier.

The company productivity measured as the ratio between successful and generated projects increases after a year of implementation of the proposed 75%.

REFERENCES

[1] Alonso, B. Proyecto Lanka 3.0: Etapa de Diagnóstico. Informe preliminar para la consultoría. Madrid, 2012.

[2] Fuentes M. y Bulmaro A. La gestión de conocimiento en las relaciones académico- empresariales, un nuevo enfoque para analizar el impacto del conocimiento académico. Tesis Phd. Universidad Politécnica de Valencia, 2010.

[3] Goldratt, E. Cadena Crítica. Ediciones Diaz de Santos. Madrid, 2001.

[4] Grupo Lanka. (2012). 10 Razones Para Confiar en Grupo Lanka. Disponible en Internet: http://www.grupolanka.com/quienes-somos/por-que-somos-diferentes/ consultado 20 de agosto de 2012.

[5] Grupo Lanka. (2012). ¿Quiénes Somos? Disponible en Internet: http://www.grupolanka.com/quienes-somos/ consultado el 20 de agosto de 2012.

[6] Jiménez, G. Investigación Operativa I. Centro de publicaciones Universidad Nacional de Colombia. Primera Edición. Manizales, 1999.

[7] Microsoft. (2011). Noticias de Sharepoint. Disponible en Internet: http://sharepoint.microsoft.com/es-mx/Paginas/default.aspx consultado el 20 de agosto de 2012.

[8] Moreno, M. (2010). ¿Qué es la Cadena de Suministro? Disponible en Internet: http://www.elblogsalmon.com/conceptos-de-economia/que-es-la-cadena-de-suministro consultado el 02 de septiembre de 2012.

[9] Ortega, E. y Peñalosa, J. (2012). Claves de la Crisis Económica Española y Retos para Crecer en la UEM. Banco de España. Madrid, 2012.

Computing, Control, Information and Education Engineering – Liu, Sung & Yao (eds)
© 2015 Taylor & Francis Group, London, ISBN: 978-1-138-02800-5

Model parameter identification of vehicle frame using operation model method

Hai Teng Zheng, De Jian Zhou, Bing Li, Yu Tang & Jian Wei Jiang
Department of Mechanical Engineering, Guangxi University of Science and Technology, Guangxi Liuzhou, China
Guangxi Manufacturing Systems and Advanced Manufacturing Technology Laboratory, Guangxi Liuzhou, China
Key Laboratory of Sea Machinery and Equipment Design and Manufacturing and Control in Guangxi Province, Guangxi Liuzhou, China

ABSTRACT: This article is to get the vehicle frame model parameters. The method is based in PolyMax by LMS SCADAS Mobile SCM05 portable data acquisition front-end vehicle frame work take experimental knowledge model parameters, and with the theoretical simulation modes compares and calculate the relative error, and the error is less than 10% described method can PolyMax accurate knowledge of the actual mode parameters remove the vehicle frame.

KEYWORDS: PolyMax Method, Vehicle Frame, Operation Model.

1 INTRODUCTION

With the improvement of modern vehicle NVH technical level, the model parameters of the traditional method measured in the laboratory cannot meet the requirements. Because the model parameters of the actual environment of a large gap, and thus directly affect the accuracy of the frame design.

Operational Model Analysis, is the inverse of the theoretical model analysis, first by the experimentally measured excitation and response time history, the transfer function by FFT transform, and then get a non-parametric model of the system; finalize frame model frequencies, mode shapes and damping parameters, as well as to reflect the actual conditions of the frame dynamics [1].

2 POLYMAX OPERATION MODEL

PolyMax operation model, also known as multi-model method of least squares complex frequency-domain reference point method, the basic idea is as follows:

2.1 *Establish the frequency response function model*

Polyreference least squares complex frequency domain method (PRLSCF or PolyMAX) is available right matrix fraction (RMFD) to describe the expression of the right matrix fraction model. Expression model for

$$H_o(\omega)=U_o(\omega)D(\omega)^{-1} \quad (1)$$

Formula: $H_o(\omega)\in C^{l\times N_i}$—Theoretical FRF line 0, N_i Is input points, namely the number of incentive;

$U_o(\omega)\in C^{l\times N_i}$ — molecular polynomial performed vector;
$D_o(\omega)\in C^{N_i\times N_i}$ — Denominator polynomial matrix.

2.2 *Minus and Solving equations reduced standard*

Solving yield
natural frequency ω_r and model damping ratio ζ_r;
Relationship as follows:

$$\lambda_r,\ \lambda_r^*=\sigma_r\pm i\tilde{\omega}_r \text{or } \lambda_r,\ \lambda_r^*=-\zeta_r\omega_r\pm i\sqrt{1-\zeta_r^2}\,\omega_r \quad (2)$$

2.3 *Calculation of frequency and damping ratio points*

By the equation(2), Δ_i use $\mathrm{Re}(\Delta_i)+i\cdot\mathrm{Im}(\Delta_i)$ Description, then obtained frequency ω_i and model damping ratio ζ_i

$$\begin{cases} \omega_i=\sqrt{\tilde{\omega}_i^2+\sigma_i^2} \\ \zeta_i=-\dfrac{\sigma_i}{\omega_i} \end{cases} \quad (3)$$

Establishment of the stabilization diagram map is an effective method.

2.4 *Establishment of the stabilization diagram*

By gradually increasing the order N polynomial, and repeatability of analysis to calculate the corresponding stabilization diagram can be established[2].

3 OPERATIONAL MODEL EXPERIMENT

3.1 *Arrangement of the measuring points*

Measuring points should follow the following principles: determining the number of measuring points according to vehicle frame size and model orders need to be concerned; measuring points evenly distributed as possible vehicle frame surface; try to avoid the position of the node. As shown in Figure 2-1:

Figure 1. Vehicle frame node layout.

As can be seen from Figure 2-1 layout plan and location of the vehicle frame shape, and measuring point.

Table 1. Experimental system components.

No.	Device Name	Model	Quantity
1	Acceleration sensor	PCB 356A16	12
2	Acquisition equipment	LMS SCADAS Mobile SCM051	1
3	Data Analysis System	LMS Test.Lab Software	1

As can be seen from Table 1 Type and quantity of work required for the experimental model test device.

Experimental model testing system diagram is as follows:

3.2 *Experimental system components*

Experiment included test signal acquisition device and the data analysis software, as shown in Table 1

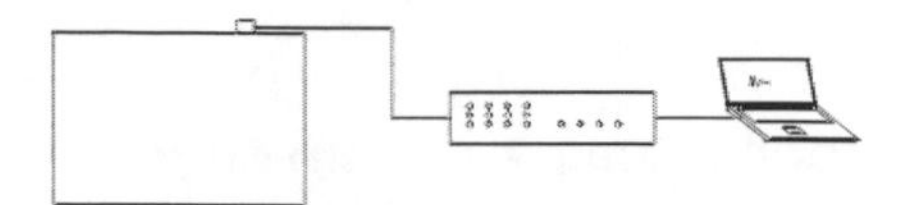

Figure 2. Model testing system schematic.

As can be seen from Figure 2-2 Principle of the model testing system, left to right is the frame, sensors, data acquisition front-end, consisting of interconnected computers and model testing system.

3.3 *Experimental conditions choice*

According to the factory to the customer to reflect that, the agricultural vehicle vibration at idle speed and driving conditions is relatively large, so this article, select idling, linear constant speed, wrong-way wave plate were three conditions to work model data acquisition. Three conditions shown in Fig.

Figure 2. Idle condition.

Figure 3. Straight uniform driving conditions.

Figure 4. Wiggler road driving condition.

Figure 2-3 which is idle condition, Figure 2-4 is a uniform linear driving conditions, Figure 2-5 is a wiggler road conditions, they are actually measuring the framework of the mode parameters in different conditions.

4 THE EXPERIMENTAL RESULTS

The experiment tested the system after LMS Test.Lab data processing. The acquisition rate test to 1024Hz. The experiment received a total of 50 modl, Table 2 lists the order idling compared to the previous four model frequencies and mode shapes with the theoretical frequency.

Table 2 Operational model comparison with theoretical model.

Order Number	Calculation of model frequenciesHz(A)	Operational model frequencies Hz(B)	Relative error $\frac{B-A}{B}$	Model characteristics
1	23.76	22.68	10%	Vehicle frame overall torsional
2	29.97	29.23	5%	Overall vehicle frame bending and torsion
3	32.31	33.91	5%	Torsional transverse plane within the vehicle frame
4	38.50	41.04	7%	Curved inner vehicle frame longitudinal plane

Table 2 shows comparative results, idling mode frequency model frequencies with the theoretical calculation results in a 10% overall error within the result was roughly the same. The frequency of the first four frames is mainly between 20Hz-50Hz, more reasonable, to avoid the car engine operating frequency. In conclusion, from the theoretical and experimental results verify the work comparing the model parameters identified results reflect the actual situation can be more accurate and reliable.

5 CONCLUSION

Based on the above analysis, it can be concluded as follows:

1 Operational Model method PolyMax model identification of the vehicle frame, can be more true, accurate and reflect the vehicle frame of the model parameters and modes provide the basis for improving and optimizing the vehicle frame Model parameters
2 PolyMax work model parameters obtained model law more in line with the actual situation the framework. Using a wide range of automobile vibration and ride comfort research has important reference value

ACKNOWLEDGEMENTS

This work is supported by a Guangxi Province Education Department of scientific research project (project name: study on vibration control of agricultural transport vehicle based on transfer path analysis method), and the open fund of Guangxi manufacturing systems and advanced manufacturing technology key laboratory (Project: No.1305109011K, name: study on vibration control of agricultural transport vehicle based on transfer path analysis method), and Guangxi universities seaside machinery and equipment design and manufacturing and control Key Laboratory of open funds (project name: Key Technology Research and Application of Hybrid Transfer Path Analysis Method)

REFERENCES

[1] Okuzum i, H Identification of the rigid Body Characteristics of a Powerplant by U sing Experimental Obtained Transfer Function. Central Engineering Laboratories, Nissan Motor Co, Ltd, Jun 1991.

[2] LMS Inernational Transfer path analysis:The qualification and quantification of vibro-acoustic transfer path[Z].LMS Internation,Application Notes,1995.

Computing, Control, Information and Education Engineering – Liu, Sung & Yao (eds)
© 2015 Taylor & Francis Group, London, ISBN: 978-1-138-02800-5

Optimization of multi-hole porthole die for hollow aluminum profile based on FEM analysis

Xing Rong Chu, Liang Chen, Shu Xia Lin & Jun Gao
School of Mechanical, Electrical and Information Engineering, Shandong University, Weihai, China
Key Laboratory for Liquid-Solid Structural Evolution and Processing of Materials (Ministry of Education), Shandong University, Jinan, Shandong, PR China

ABSTRACT: The multi-hole porthole die extrusion of hollow aluminum profiles was comprehensively studied. The influence of porthole areas, die orifice location and bearing length on the material flow behavior was investigated, and the designed extrusion die was optimized accordingly. After the optimization, the standard deviation of the velocity (SDV) of the flowing velocity was reduced from 2.43 mm/s to 0.42 mm/s. The purpose of the present study is to provide a logical and effective route for designing multi-hole porthole die.

KEYWORDS: Porthole die; aluminum profile; Optimization; FEM analysis.

1 INTRODUCTION

The porthole extrusion die has been widely applied for producing hollow aluminum profiles, since it has the advantages in producing profiles with complex cross-section and low manufacturing cost. In recent years, the multi-hole porthole die has been developed and becomes the interest of many researchers [1]. Several profiles could be extruded out during one extrusion stroke by using the multi-hole porthole die, and thus greatly enhance the productivity.

Hot extrusion process involves high temperature, high pressure and different fiction conditions, which makes it become a complex process. And there are many factors could affect the extrusion process, such as the material flow behavior, extrudate temperature, seam weld quality and die service life. Among these factors, the material flow behavior is one of the most important one, since the distortion of curve, bend and twist easily appear in the profiles without homogeneous material flow behavior, which directly affects the dimensional accuracy of the final product.

The porthole die as the main tool for conducting extrusion process plays an essential role for controlling material flow behavior. Zhao [2] studied the porthole die extrusion process for producing hollow profiles with irregular shape using FEM simulation, and optimized the porthole die using Pareto-based genetic algorithm. Jo [3] investigated the material flow and extrusion force with respect of different process parameters through performing non-steady state FEM analysis and experimental work. Sun [4] modified the shape and height of 2nd-step welding chamber of a porthole die using response surface method and genetic algorithm based on the steady-state simulation work. Liu [5] modified the porthole die for a large, multi-cavity aluminum profile by resizing portholes, adding bosses, chamfering mandrels, and adjusting the length of the bearings.

From the above open literature, it can be known that the numerical simulation is a powerful tool in studying extrusion process, and the die optimization is an effective method to well control the material flow behavior. However, rare attention was put on the multi-hole porthole die. Thus, in this paper, the multi-hole porthole extrusion of hollow aluminum profiles was studied by means of FEM simulation. The effects of some die structures on the material flow behavior were investigated. Accordingly, the multi-hole porthole die was optimized based on the simulated results.

2 SIMULATION MODELING

Figure 1 shows the dimensional geometry of the studied aluminum profiles. The thinnest part of the profile is 0.6 mm and the area of the cross section is around 62.44 mm^2. According to the characteristic of the profile, the multi-hole porthole die was designed, as shown in Fig. 2. Four potholes and two die mandrels were symmetrically designed in the upper die to distribute the material flow rationally. The width of port bridges in horizontal and vertical directions is 18 and 12 mm, respectively. The welding chamber has a height of 7mm to get a good quality of the seam weld. The bearing was set to have identical length of 5 mm in the original scheme.

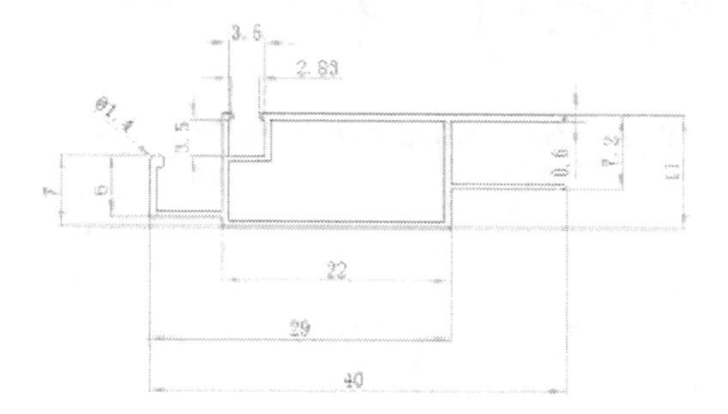

Figure 1. Aluminum profile shape. (Unit: mm).

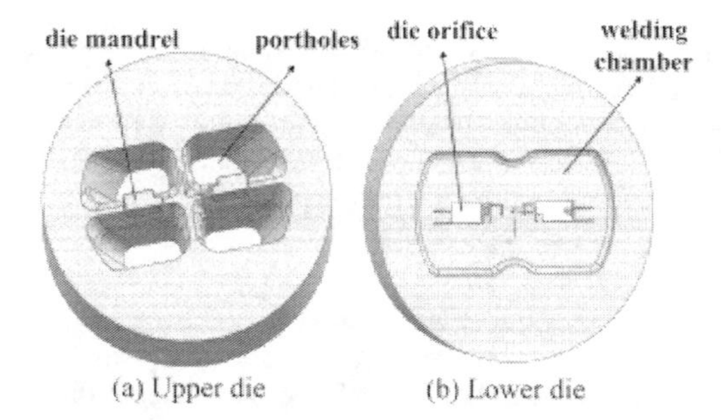

Figure 2. Original scheme of the die design.

The commercial code of HyperXtrude based on Arbitrary Lagrangian Eulerian algorithm was used to model the extrusion process. Since the designed multi-hole porthole die is symmetrical, one half of the material flow domains were chosen for simulation, as shown in Fig.3. In order to obtain accurate simulated results and to save the computational time, the elements with different sizes and types were set in the components of billet, portholes, welding chamber, profiles and bearing, respectively.

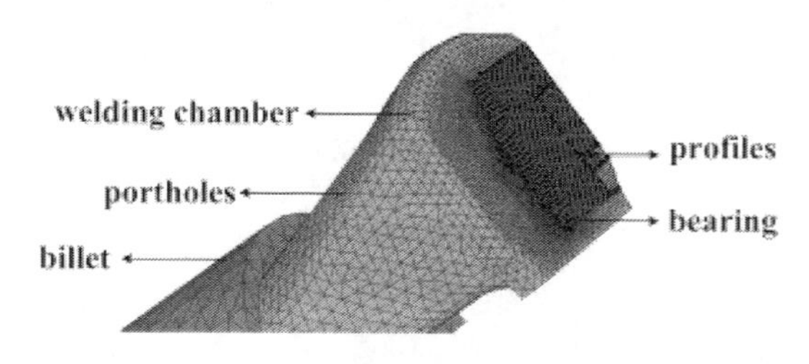

Figure 3. Mesh generation scheme in HyperXtrude.

Table 1. Process parameters used in the present study.

Billet diameter	Ram velocity	Tool Temperature	Billet temperature
94 [mm]	1.0 [mm/s]	430 [°C]	460 [°C]

The crucial process parameters were determined according to the actual condition, as listed in Table 1. The slop friction condition was assumed on the surfaces in contact with the bearing, while the sticking condition is applied for the other interfaces of billet/tools. The materials for extrusion billet and dies are Aluminum Alloy 6063 (AA6063) and H13 tool steels, respectively. The billet was assumed to be visco-plastic, and the flow stress was described using Sellars-Tegart model [6], which is expressed as:

$$\sigma = \frac{1}{\beta}\sinh^{-1}\left(\frac{\dot{\bar{\varepsilon}} e^{Q/RT}}{A}\right)^{1/n} \tag{1}$$

where s is the flow stress, $\dot{\bar{\varepsilon}}$ is the equivalent strain rate, R is the universal gas constant, Q is the activation energy of deformation, T is the absolute temperature, β, n and A are the material constants.

3 RESULTS AND DISCUSSION

Figure 4 shows the distribution of the flowing velocity in the cross-section of the profile using the original die design scheme. One can see that the maximum and minimum values of the flowing velocity are 58.17 and 50.49 mm/s, respectively. In order to quantitatively evaluate the distribution of flowing velocity, the standard deviation of the velocity (SDV) was introduced, which is described as,

$$\mathrm{SDV} = \sqrt{\frac{\sum_{i=1}^{n}\left(v_i - \bar{v}\right)^2}{n}} \tag{2}$$

where n is the total number of the selected nodes, v_i is the axial velocity at node i, and $\bar{v}$ is the average velocity for all selected nodes. All of the nodes in the cross-section of the profile were selected for the calculation, and the calculated SDV value is 2.43 mm/s.

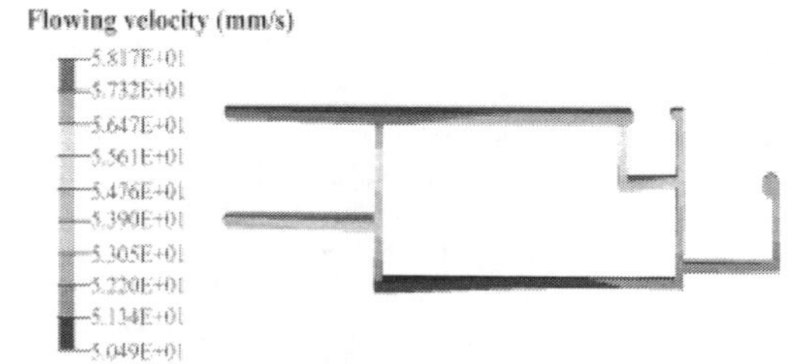

Figure 4. Flowing velocity distribution in the profile with original design scheme.

One can also see from Fig. 4 that some distortions like twist and curve emerged in the profiles, which indicates that the material flow behavior using the original scheme of die design cannot satisfy the requirements. Thus, some optimization work on the multi-hole porthole die should be carried out. The area of the portholes at the entrance side of the billet determines the amount of the material flowing inside the portholes, which can be used to control the material flow behavior. As shown in Fig. 4, the flowing velocity in the left side of the profile is higher than that in the right side. Thus, if the part of the portholes near die center is reduced and the other parts are enlarged, the uniformity of the material flow might be improved. According to this principle, several design schemes with various porthole areas were tried in this study, and the calculated SDV for each case is plotted in Fig. 5. When the area of the portholes was set to be 754 mm^2, the SDV was reduced to be 1.58 mm/s, which provides better velocity distribution.

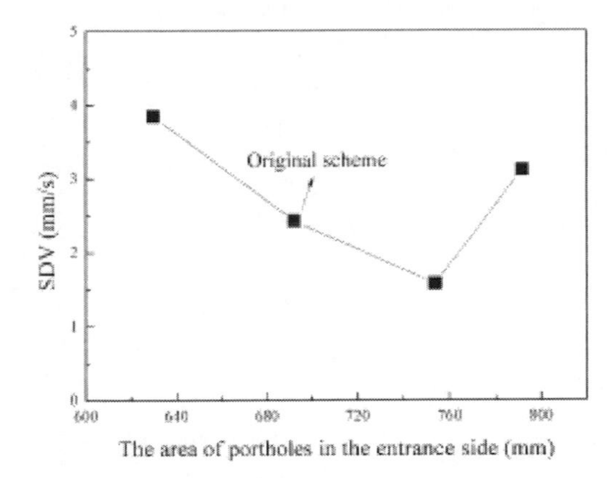

Figure 5. SDV *v.s.* porthole areas.

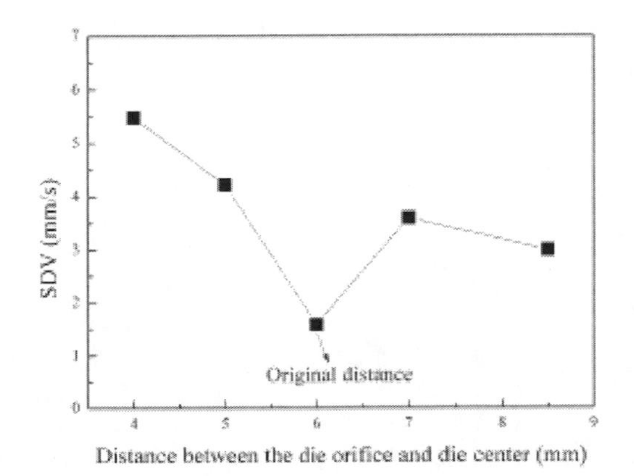

Figure 6. SDV *v.s.* distance between die orifice and die center.

In multi-hole extrusion, the location of die orifice is one of the major factors that should be well considered. Thus, as the second step, the distance between the die orifice and the die center was varied to study their effects, as shown in Fig. 6. It should be pointed out that the area of the portholes was set to be 754 mm^2 for all the cases in Fig. 6. It can be seen that the location of die orifice has significant influence on the SDV value. And the original distance of 6 mm was proved to be the optimal one, which should not be modified.

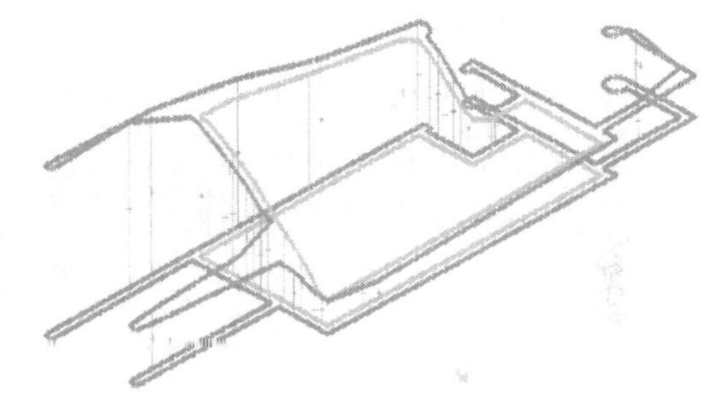

Figure 7. Modification on the bearing length.

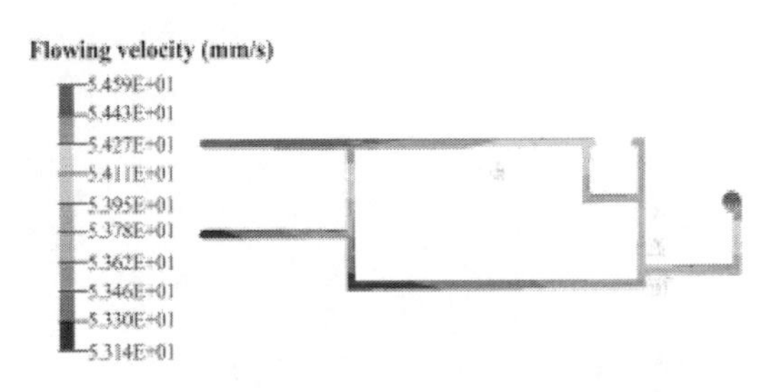

Figure 8. Velocity distribution using unequal bearing length.

In the above simulation, an identical bearing length of 5 mm was utilized for all cases. However, the bearing length is usually uneven in actual production. Longer bearing should be applied to provide additional friction resistance in the regions having higher velocity, while shorter bearing should be applied in other regions to reduce the fiction effect. In this study, the bearing length was further designed after the second modification. The proper design of uneven bearing length was drawn in Fig. 7, and the corresponding distribution of flowing velocity is shown in Fig. 8. It can be seen that, after die optimization by choosing the best porthole areas, die orifice location and bearing length, the material flow behavior has been greatly improved. The calculated SDV was reduced from 2.43 mm/s (Fig. 4) to 0.42 mm/s (Fig. 8), and the distortion of the profiles has been released.

4 CONCLUSIONS

In this study, the multi-hole porthole die extrusion for hollow aluminum profile was studied by means of FEM simulation. And it was found that the porthole areas at the entrance side of the billet, the die orifice location and the bearing length have significantly influence on the exit flowing velocity. More importantly, an effective route for multi-hole die optimization was proposed in this study.

ACKNOWLEDGEMENT

The authors would like to acknowledge the financial support from National Natural Science Foundation of China (Grant No. 51405268, 51405266). The corresponding author is Liang Chen.

REFERENCES

[1] M.K. Sinha, S. Deb, R. Das, and U.S. Dixit: Mater. Des. Vol. 30 (2009), p. 2386.
[2] G. Zhao, H. Chen, C. Zhang, and Y. Guan: Int. J. Adv. Manuf. Technol. Vol. 69 (2013), p. 1547.
[3] H.H. Jo, S.K. Lee, C.S. Jung, and B.M. Kim: J. Mater. Process. Technol. Vol. 173 (2006), p. 223.
[4] X. Sun, G. Zhao, C. Zhang, and Y. Guan: Mater. Manuf. Processes. Vol. 28 (2013), p. 823.
[5] P. Liu, S.S. Xie, and L. Cheng: Mater. Des. Vol. 36 (2012), p. 152.
[6] C.M. Sellars, and W.J. McTegart: Acta Metall. Vol. 14 (1996), p. 1136.

Computing, Control, Information and Education Engineering – Liu, Sung & Yao (eds)
 ISBN: 978-1-138-02800-5

Evaluation of the sheared edge of the metal blanking process with one ductile damage model combined the induced anisotropy

Zhen Ming Yue, Xin Cun Zhuang, Xing Rong Chu, Houssem Badreddine & Jun Gao
School of Mechanical, Electrical and Information Engineering, Shandong University, Weihai, China
Institute of Forming Technology and Equipment, Shanghai Jiao Tong University, Shanghai, China
University of Technology of Troyes, Troyes Cedex, France

ABSTRACT: The prediction of the sheared edge quality of the blanking process still needs more accurate numerical model, which can be of great value. In this paper, a recently proposed fully coupled ductile damage model, which takes into account of the initial and induced anisotropy, isotropic and kinematic hardening, isotropic ductile damage, are used to estimate the blanking process of sheet metal DP1000. The work is focused on the effect of the induced anisotropy on the sheared edge quality. Series tests have been conducted for the calibration of the model parameters. Through the comparison of the numerical responses with and without the induced anisotropy, the significance of induced anisotropy on the crack initiation and propagation especially under blanking process is given.

KEYWORDS: Blanking; Simulation; Damage; Induced anisotropy.

1 INTRODUCTION

Blanking process is one of the most common techniques in high volume production in industry. Some small product components are obtained using blanking technique. The operation consists of a cutting process with the shear stress, and the punch and the die get close and made the rupture of the sheet. Lots of factors, including the geometry of the punch and die, punch rate, lubrication, clearance between two parts (blanking clearance) will really affect the quality of the rupture surface. In long history, the evaluation of the industry blanking quality is mainly based on empirical knowledge, which can lead some error procedures, so a proper and accurate approach becomes more necessary, which can give a more reasonable physical explanation of the forming process.

Lots of approaches have been developed to study the blanking process. Some analytical approaches have been proposed to optimize the process control and tool design with the assumption of pure shear stress states, while some with tension and bending processes [1]. On the other hand, the physical background of the rupture for the blanking can be known as the initiation, growth and coalescence of voids, which can be presented with some ductile damage model, like McClintock model [2], Lemaitre model [3], Gurson model [4], etc. The Void can initiate around the inclusions, secondary phase particles and dislocation pile-ups, and grow and propagate together with the plastic deformation. These ideas can be implemented with different Finite element method (FEM) to simulate the blanking process, especially to estimate the sheared edge quality.

In this study, a recently proposed fully coupled damage model with considering the initial and induced anisotropy is chosen. Based on some experimental data about DP1000, the asymmetric blanking process for this material is simulated. In order to investigate the effect of the induced anisotropy, two sets of values are assigned to distortion parameters. Through the comparison of the simulation responses, the proof of the effect of induced anisotropy on the sheared edge quality can be obtained.

2 CONSTITUTIVE EQUATIONS

In the fully coupled damage model, the initial (Hill48) and induced anisotropy, isotropic and kinematic hardening, isotropic ductile damage are incorporated together. Their state variable and evolution equations are respectively shown in Table 1, detailed description can be found in previous studies [5].

Table 1. Constitutive equations for the fully coupled damage model.

	State variables	Evolution equations
Cauchy stress	$\underline{\sigma} = 2\mu_e\left[(1-d)\langle \underline{e}^e\rangle_+ + (1-hd)\langle \underline{e}^e\rangle_-\right] + k_e\left[(1-d)\langle tr(\underline{\varepsilon}^e)\rangle - (1-hd)\langle -tr(\underline{\varepsilon}^e)\rangle\right]\underline{1}$	$\underline{D}^p = \frac{\dot{\lambda}}{\sqrt{1-d}}\frac{\underline{\underline{H}}:(\underline{\sigma}-\underline{X})}{\lVert\underline{\sigma}-\underline{X}\rVert} = \frac{\dot{\lambda}}{\sqrt{1-d}}\tilde{\underline{n}}^p$ $\underline{n}^p = \frac{\underline{n}^d}{\sqrt{1-d}}:\left[\underline{\underline{1}}^{dev} + \frac{\underline{X}\otimes\underline{S}_0}{X_{l1}^p(R/\sqrt{1-d^\gamma}+\sigma_y)}\right]$
Kinematic hardening	$\underline{X} = 2/3(1-d)C\underline{\alpha}$	$\dot{\underline{\alpha}} = -\dot{\lambda}\frac{\partial F}{\partial \underline{X}} = \frac{\dot{\lambda}}{\sqrt{1-d}}(\underline{n}^x + a\cdot\underline{\alpha})$ $\underline{n}^x = \underline{n}^d - \frac{(\underline{S}_0:\underline{S}_0)\underline{n}^d}{2X_{l1}^p(R/\sqrt{1-d^\gamma}+\sigma_y)}$
Isotropic hardening	$R = (1-d^\gamma)Qr$	$\dot{r}_i = \dot{\lambda}(\frac{n^r}{\sqrt{1-d}} - b_i r_i)$ $n^i = \frac{(\underline{S}_0:\underline{S}_0)(\underline{n}^d:\underline{X})}{2X_{l1}^p(\sqrt{1-d^\gamma}R+(1-d)\sigma_y)^2}+1$
Ductile damage	$Y = Y^e + Y^\alpha + Y^r$ $Y^e = 2\mu_e\left[\langle\underline{e}^e\rangle_+ : \langle\underline{e}^e\rangle_+ + h\langle\underline{e}^e\rangle_- : \langle\underline{e}^e\rangle_-\right] + k_e\left[\langle tr(\underline{\varepsilon}^e)\rangle^2 + h\langle -tr(\underline{\varepsilon}^e)\rangle^2\right]$ $Y^\alpha = \frac{1}{3}C\underline{\alpha}:\underline{\alpha}; Y^r = \frac{1}{2}\gamma d^{\gamma-1}Qr^2$	$\dot{d} = \frac{\dot{\lambda}}{(1-d)^\beta}\left(\frac{Y-Y_0}{S}\right)^s$

The induced anisotropy is introduced through the Yield criterion f and potential equations F with the assumption of active and latent slip systems. The normal deviatoric stress is replaced by a new term which can be named as the distortion hardening $\underline{S}_d^i$ ($i = \{c, p\}$). This novel hardening includes the term parallel to kinematic hardening $\underline{S}_X$ and the term orthogonal to the kinematic hardening $\underline{S}_0$.

$$f = \frac{\lVert \underline{S}_d^c - \underline{X}\rVert_H}{\sqrt{1-d}} - \frac{R}{\sqrt{1-d^\gamma}} - \sigma_y \quad (1)$$

$$F = \frac{\lVert \underline{S}_d^p - \underline{X}\rVert_H}{\sqrt{1-d}} - \frac{R}{\sqrt{1-d^\gamma}} + \frac{3a}{4(1-d)C}\underline{X}:\underline{X} + \frac{b}{2(1-d)Q}R^2 + \frac{S}{s+1}\left\langle\frac{Y-Y_0}{S(\theta)}\right\rangle^{s+1}\frac{1}{(1-d)^\beta} \quad (2)$$

$$\underline{S}_d^c = \underline{S} + \frac{\underline{S}_0:\underline{S}_0}{2X_{l1}^c(R/\sqrt{1-d^\gamma}+\sigma_y)}\underline{X} - \frac{\underline{X}:\underline{X}}{2X_{l2}^c(R/\sqrt{1-d^\gamma}+\sigma_y)}\underline{S}_0 \quad (3)$$

$$\underline{S}_d^p = \underline{S} + \frac{\underline{S}_0:\underline{S}_0}{2X_{l1}^p(R/\sqrt{1-d^\gamma}+\sigma_y)}\underline{X} \quad (4)$$

$$\underline{S}_0 = \underline{S} - \underline{S}_x \quad and \quad \underline{S}_x = \frac{\underline{S}:\underline{X}}{\underline{X}:\underline{X}}\cdot\underline{X} \quad (5)$$

The stress norm $\lVert\;\rVert_H$ is the anisotropic Hill48 equivalent stress characterized by an anisotropic operator $\underline{\underline{H}}$, and σ_y is the initial yield stress. In Table 1, The parameters a and b represent the non-linearity of the kinematic and isotropic hardening respectively, while S, s, β and Y_0 govern the ductile damage evolution.

3 SIMULATION RESULTS

In order to estimate the effect of the induced anisotropy on the sheared edge quality in blanking process, the DP1000 are chosen as the test objective. The normal material parameters are shown in Table 2, and the distortion parameters will be assigned to zero or assumed value to test their influences (X_{l1}^{c}=0, X_{l1}^{p}=0 and X_{l2}^{c}=0; X_{l1}^{c}=300, X_{l1}^{p}=300 and X_{l2}^{c} =1000). The simulation will be conducted with the software ABAQUS/Explicit with User subroutine VUMAT.

Table 2. Determined material Parameters.

E (Gpa)	ν	σ_y(Mpa)	Q (Mpa)	b	C(Mpa)	a
208	0.3	809	4000	13.0	32000.0	150.0
F	G	H	L	M	N	S
0.525	0.546	0.454	1.5	1.5	1.67	12.5
s	β	Y_0	γ	D_c	h	
1.15	2.5	2.0	4.0	0.99	0.25	

The numerical geometry and the mesh condition for the blanking process are shown in Fig.1. The thickness of the sheet is 2.0 mm. 0.2 mm is chosen as the blanking clearance and constant mesh size 0.01 mm is chosen in the critical test zone. Loading velocity is set to be 10.0 mm/s, and constant mass scaling 1.0E 8 is chosen to reduce the calculation consuming time.

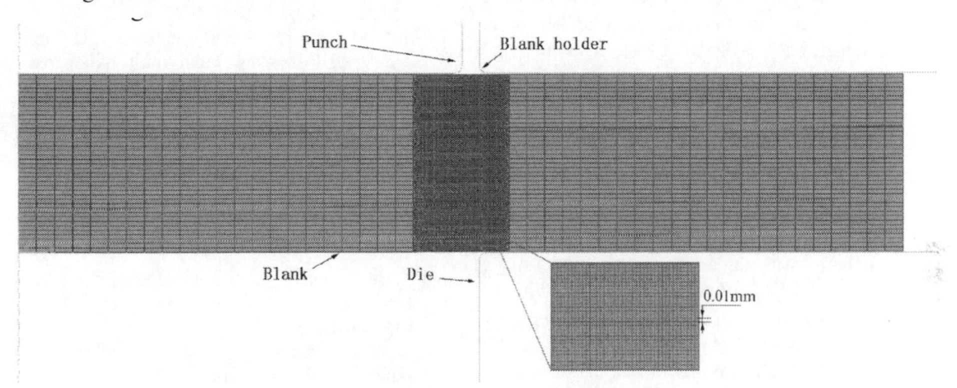

Fig.1. Numerical geometry of the asymmetric blanking process for DP1000.

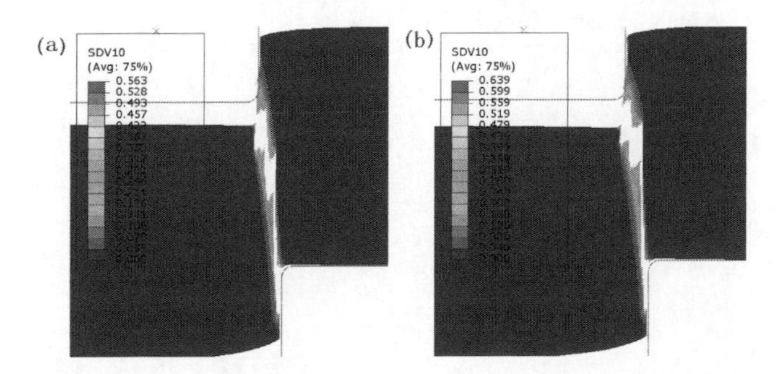

Figure 2. Ductile damage contour after total rupture. (a) response with X_{l1}^{c} =0, X_{l1}^{p} =0 and X_{l2}^{c} =0 and (b) response with X_{l1}^{c} =300, X_{l1}^{p} =300 and X_{l2}^{c} =1000.

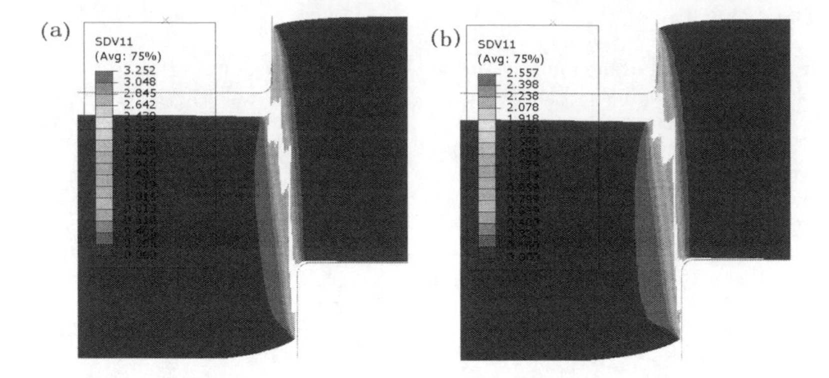

Figure 3. Equivalent plastic strain contour after total rupture. (a) response with X^c_{l1}=0, X^p_{l1}=0 and X^c_{l2}=0 and (b) Response with X^c_{l1}=300, X^p_{l1}=300 and X^c_{l2}=1000.

Through the comparison of the numerical responses about the ductile damage and equivalent plastic strain, separately shown in Fig.2 and Fig.3, it can be found that the crack profiles with and without inducing anisotropy give different profile, especially the fracture depth and burnish depth. The burnish depth in Fig.3a will be smaller than Fig.3b, while the fracture depth will be larger. Meanwhile the rollover and burr zone do not give much difference.

4 CONCLUSION

In this study, one symmetric blanking forming process about DP1000 was conducted with a recently proposed fully coupled ductile damage model, which takes into account the induced anisotropy due to plastic deformation. With the comparisons of the crack profile, ductile damage and equivalent plastic strain with and without inducing anisotropy effect, the influence of the induced anisotropy on sheared edge quality has been proved to be significant. Much deeper research can be conducted to optimize the fine blanking technique in the coming studies.

5 ACKNOWLEDGEMENT

The authors would like to acknowledge the financial support of the National Natural Science Foundation of China (Grant No. 51405266). The corresponding author is Jun Gao.

REFERENCES

[1] C. Husson, J.P.M. Correia, L. Daridon, S. Ahzi, Finite elements simulations of thin copper sheets blanking: Study of blanking parameters on sheared edge quality, J Mater Process Tech. 199, (2008), 74–83.

[2] F.A. McClintock, A criterion for ductile fracture by the growth of holes subjected to multi-axial stress states. J. Appl. Mech. 35 (1968), 363–371.

[3] J. Lemaitre, A continuum damage mechanics model for ductile fracture. J. Eng. Mater. Technol. 107 (1985), 83–91.

[4] A.L. Gurson, Continuum theory of ductile rupture by void nucleation and growth. Part I. Yield criteria and flow rules for porous ductile media, J. Eng. Mater. Technol. 99 (1977) 2–15.

[5] Z.M. Yue, H. Badreddine, K. Saanouni, E.S.Perdahcioglu, C. Soyarslan, A. E. Tekkaya, A.H. van den Boogaard. On the distortion of yield surface in sheet metal forming under complex loading paths. IDDRG2014, Paris, France.

Computing, Control, Information and Education Engineering – Liu, Sung & Yao (eds)
© 2015 Taylor & Francis Group, London, ISBN: 978-1-138-02800-5

The research of wear mechanism on spiral leaves with variable diameter and variable pitch

Jian Ming Wang, Jian Xun Li & Xu Ming Ye
Department of Mechanical Engineering, Shenyang University, Shenyang, China

ABSTRACT: The working principle of the spiral press filter is described and the theoretical basis of the spiral leaves with variable diameter and variable pitch analyzed. The parametric design model of the spiral shaft with variable diameter and variable pitch is established, which provides a theoretical basis for the structure design of spiral shaft with variable diameter and variable pitch. By three-coordinates measuring machine, it verifies the assumption that the theoretical analysis is consistent with the actual worn situation, and then including the wear law of spiral leaves. Namely, the closer to the outer edge, the thinner the thickness of the spiral leaves, the larger the actual amount of wear. At the inner edge of leaves rarely wear and tear. On the same shaft, the wear of spiral leaves is not the same in different positions. The amount of wear to the outer edge of leaves in outlet is larger than inlet.

KEYWORDS: Spiral press filter; Spiral leaves with variable diameter and variable pitch; Wear theory; Inlet; Outlet; The outer edge of leaves.

1 PREFACE

Under the effect of hydrostatic pressure, spiral press filter is the continuity of solid-liquid separation filtration equipment. Mainly used in petrochemical industry, biological engineering, chemical industry, etc[1], spiral press filter has simple structure, continuous working, extrusion pressure, and it is an ideal environmental equipment.

2 THE WORKING PRINCIPLE OF SPIRAL PRESS FILTER

Spiral press filter consists of the actuator, active helix, cylinder shell, rack, etc. In the working process, the volume between the helix and shell decreases, because the external diameter of spiral shaft along the conveying direction is increasing when the inner diameter of the shell is unchanged. The pressure of material is increasing, so as to realize the function of the solid-liquid separation.

3 THE THEORETICAL ANALYSIS OF WEAR ON SPIRAL LEAVES WITH CARIABLE DIAMETER AND VARIABLE PITCH

Taking the material particles M that the distance reaching the helix axis is r as the researching object[2], to analyze of the forced state of a particle M. Particle M force is shown in Fig.1.

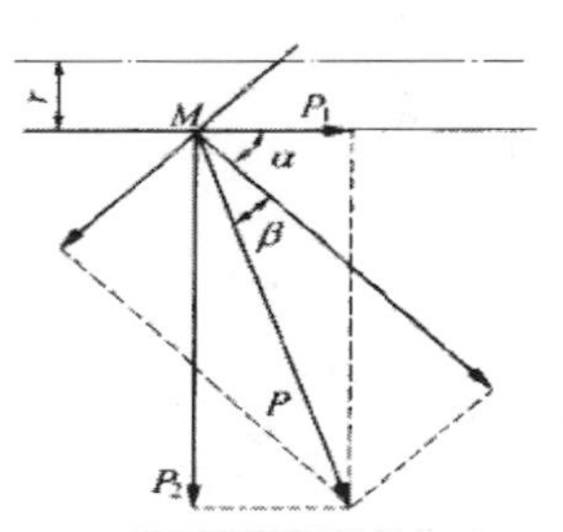

Figure 1. The analysis of force on particle M.

Assuming that the resultant force of the spiral surface acting on the particle M is P. But due to the existence of friction, which makes the direction of the resultant force P deviating from the normal angle of β , The angle of β depends on the friction angle which the slurry material to the rotation surface and the roughness of spiral surface, so taking $\beta = \theta$ (friction angle).

Taking resultant force P to decompose into P_1 of axial force and P_2 of radial force, knowing that:

$$\begin{cases} P_1 = P\cos(\alpha + \theta) \\ P_2 = P\sin(\alpha + \theta) \end{cases} \tag{1}$$

According to the needs of design, the friction angle θ and resultant force P can be regarded as constant

value. From (1), it can see that P_1 of axial force and P_2 of radial force are functions of spiral angle α.
Taking the approximate expansion method to expand the spiral surface[3], it can retain the relationship among spiral raising angle α, spiral distance s and the distance the point of spiral surface to the spiral axis. As showing in the Fig.2.

$$\alpha = \arctan \frac{s}{\pi d} \tag{2}$$

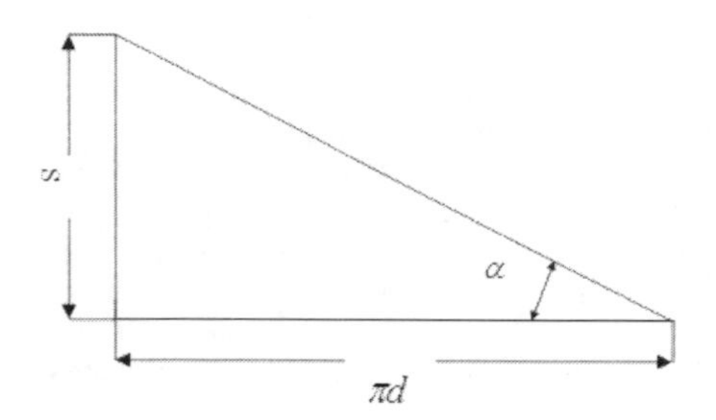

Figure 2. The approximate expansion of spiral surface.

As s and d are changing, In order to determine the relationship of three qualitatively, now the helix pitch is derived, now the distance of spiral line is derived. Assuming that the spiral line equation of spiral shaft is tapered spiral line of equal right angle, its parameter equation as following[4]:

$$\begin{cases} x = ne^{m\omega} \cos \omega \\ y = ne^{m\omega} \sin \omega \\ z = ce^{m\omega} \end{cases} \tag{3}$$

In equation, $n = a\sin\varphi; b = a\cos\varphi; m = \sin\varphi\cos\alpha;$; a is constant, φ is the half angle of the cone top; ω is a parameter, its value is $0, 2\pi, 4\pi, \ldots 2k\pi$.
It can make a conclusion from the equation:

$$\omega = 0,\ x = n;\ y = 0,\ z = c;$$

$$\omega = 2\pi,\ x = ne^{2\pi m};\ y = 0,\ z = ce^{2\pi m};$$

By analogy, we can know:

$$\omega = 2k\pi,\ x = ne^{2k\pi m};\ y = 0,\ z = ce^{2k\pi m}; k = 1, 2, 3 \ldots$$

The kst spiral distance of the spiral line is

$$s_k = z_k - z_{k-1} = c[e^{2k\pi m} - e^{2(k-1)\pi m}]; k = 1, 2, 3 \ldots \tag{4}$$

The changing curve of s_k along with k as following Fig.3:

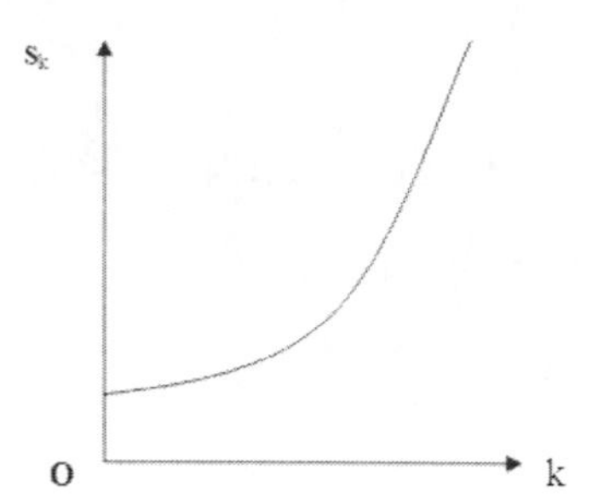

Figure 3. The changing curve of s_k.

It can be seen from the fig.3 that sk is exponential function. With the continuous growth of k, sk has increased gradually. Which show that the spiral distance is gradually increasing with the growth of k.

On same cross section of the helix, spiral distance s can be seen as unchanged. It can see from equation (2) that the spiral raising angle α is a function that the point on the surface of the spiral to the spiral axis distance d[5]. According to the stress analysis of material particle M above, It know that the axial force is smaller near the spiral shaft, however the radial force is larger.

Due to the diameter of the spiral axis gradually enlarging, spiral distance s is gradually reduced. As can be seen from equation (2), the spiral raising angle is decreased gradually on the spiral axis busbar. Therefore, on the helix, the axial force near the exit is larger than the axial force near the entrance.

4 THE ACTUAL SITUATION OF THE SPIRAL LEAVES WEAR

In order to get the actual situation of the spiral leaves wear accurately, different parts of the screw filter press worn are measured by three-coordinates measuring machine, and the result is shown in the Table 1.

Measuring results show that: the closer to the outer edge, the thinner the thickness of the spiral leaves, the larger the actual amount of wear. At the inner edge of leaves rarely wear and tear. On the same shaft, the wear of spiral leaves is not the same in different position. The theoretical analysis is consistent with the actual worn situation. Therefore, in the spiral filter press, the quantity of wear outside part of the leaves is bigger than inside, and the wear occurs mainly in the outer spiral leaves. When spiral filter press is working, the force of materials outlet is larger than inlet location, and the axial force of outer leaves is the largest near the outlet location, which indicate that the deformation of outer edge leaves is the biggest and the wear of leaves is the most serious.

Table 1. The thickness of leaves in different position of inlet/outlet.

Technical indicators	Measuring point							
	A	B	C	D	E	F	G	H
The distance from inner edge /mm	5	15	25	35	45	55	65	75
The thickness of leaves/mm	3.19/3.15	3.15/3.12	3.06/3.01	2.91/2.83	2.78/2.54	2.58/2.16	2.44/1.73	2.25/1.24

5 CONCLUSION

1 In the theoretical analysis processing, the parametric design model of the spiral shaft with variable diameter and variable pitch is induced, which provides a theoretical basis for the structure design of spiral shaft.
2 The amount of wear outer leaves is larger than inside on the same cross section.
3 In spiral filter press, the amount of wear is different on different working position, and the mount of outer place is the largest.

REFERENCES

[1] Rushton,A.S.Ward,R.G. Solid-Liquid Filtration and Separation Technology [M].Beijing: Chemical industry press, 2005.

[2] Chenanjiang, Lvchuanshan SP The structure and production application of single screw extrusion machine, 2009 06:0017—03.

[3] Licaiqin The expansion of spiral leaves with variable diameter variable pitch [J] Light industry machinery, 2001 03: 027—029.

[4] Xiadingli The forming principle and projection drawing method of isometric conical helix [J] Journal of Hefei industrial university (natural and science edition), 1990 13(03):099—105.

[5] Zhuwei The research on spiral filter press working [J] Science and technology of coal, 2006 05—0064—03.

Computing, Control, Information and Education Engineering – Liu, Sung & Yao (eds)
© 2015 Taylor & Francis Group, London, ISBN: 978-1-138-02800-5

A study of mechanical properties in 5083 aluminum alloy at different rolling states

Wei Bo Zhu, Xian Quan Jiang, Quan Li & Si Ya Wang
Advanced Materials Research Center, Chongqing Academy of Science and Technology, Chongqing, China

ABSTRACT: For excellent mechanical properties, rolling 5083 aluminum alloy sheets have very high value of application and wide prospects in coal car plates. Large flat ingots of 5083 aluminum alloy have been prepared by melting in furnace and casting in the large-size flat ingot mold. For the preparation of hot rough rolling sheets, large flat ingots of 5083 aluminum alloy were submitted to a hot rough rolling process, until the average thickness of the specimens was reduced by 95% respectively. Finally, in order to obtain hot finish rolling sheets in 6 mm thickness, hot rough rolling sheets were submitted to a hot finish rolling process, until the average thickness of the sheet was reduced by 81%. The mechanical behavior of 5083 aluminum alloy during roll forming process was investigated by tensile test at room temperature. From the experimental results obtained, the mechanical properties of ingots are almost the same from the edge to the central part. Mechanical properties of hot rolling sheets are almost the same in rolling direction and non-rolling direction, too. Hot finish rolling sheets showed surprisingly high yield stress values, tensile strength values and elongation values.

KEYWORDS: Mechanical properties; Aluminum alloy; Rolling state.

1 INTRODUCTION

Aluminum and aluminum alloys have become increasingly used in production of automobiles and trucks, packaging of food and beverages, construction of buildings, transmission of electricity, development of transportation infrastructures, production of defense and aerospace equipment, manufacture of machinery and tools and marine structures with its unique properties such as corrosion resistance, thermal conductivity, electrical conductivity, high strength with low density, fracture toughness and energy absorption capacity, cryogenic toughness, workability, ease of joining (welding (both solid state and fusion), brazing, soldering, riveting, bolting) and recyclability[1-3].

5083 aluminum–magnesium alloys are strain hardenable. They are widely use in structural applications, such as automotive parts, railway vehicles and aeronautics, due to the fact of presenting relatively high strength, good corrosion resistance and high toughness combined with good formability and weldability[4-6].

Generally, performance of 5083 aluminum alloy sheets in coal car plates depend mainly on the mechanical properties, including yield strength, tensile strength and elongation. Therefore, we studied the mechanical properties of 5083 aluminum alloy sheets at different rolling states, providing data as reference for improving rolling process.

2 EXPERIMENTAL PROCEDURES

Hot finish rolling sheets of 5083 aluminum alloy are prepared by the following process: proportioning → Melting→ Converter → furnace carbon black refining → standing →turning down → adding grain refiner online → degassing and filtering online → Casting(Large flat ingots) → sawing → milling surface → heating → hot rough rolling(hot rough rolling sheets) → hot finish rolling(hot finish rolling sheets) → straightening → crosscut → accurate sawing(product).

In the above process, large flat ingots have been prepared by melting in furnace and casting in the large-size flat ingot mold, and their thicknesses are almost 650mm. For the preparation of hot rough rolling sheets, large flat ingots were submitted to a hot rough rolling process, until the average thickness of the specimens was reduced from 650 mm to 32 mm, decreased by 95% respectively. Finally, in order to obtain hot finish rolling sheets in 6 mm thickness, hot rough rolling sheets were submitted to a hot finish rolling process, until the average thickness of the sheet was reduced from 32 mm to 6 mm, decreased by 81% respectively.

Select four parts from the edge to the center of a large flat ingot to make four groups tensile specimens, numbered ingot 1, ingot 2, ingot 3, ingot 4. Tensile specimens rolling direction and non-rolling direction were prepared by hot rough rolling sheets and hot finish rolling sheets. The tensile test was completed on RGM 6300 electronic universal testing machine. Reference GB/T 228 - 2002, the stretching rate is 1mm/min.

3 RESULTS AND DISCUSSION

The elemental composition of large flat ingots. The elemental composition of large flat ingots was shown in Table 1.

Table 1. The elemental composition of large flat ingots.

Elements	Si	Fe	Cu	Mn	Mg	Cr	Zn	Ti	Na
Contents	0.061	0.304	0.017	0.510	4.580	0.110	0.027	0.022	<0.0001

Mechanical properties of large flat ingots. The study of mechanical properties is of primary importance for determining the performance of materials, especially that of metal materials. Figure 1(a) shows tensile strength of large flat ingots from the edge to the center. Figure 1(b) shows yield strength of large flat ingots from the edge to the center. Figure 1(c) shows elongation of large flat ingots from the edge to the center. Figure 1(d) shows elastic modulus of large flat ingots from the edge to the center.

As can be seen from Fig.1. Tensile strength, yield strength and elastic modulus of ingots are almost the same from the edge to the central part, elongation of the edge is lower than other portions.

The average of tensile strength, yield strength, elongation and elastic modulus is about 218MPa, 124MPa, 12%, 2.9GPa respectively. Based on the above performance, ingots had good organizational uniformity and mechanical properties.

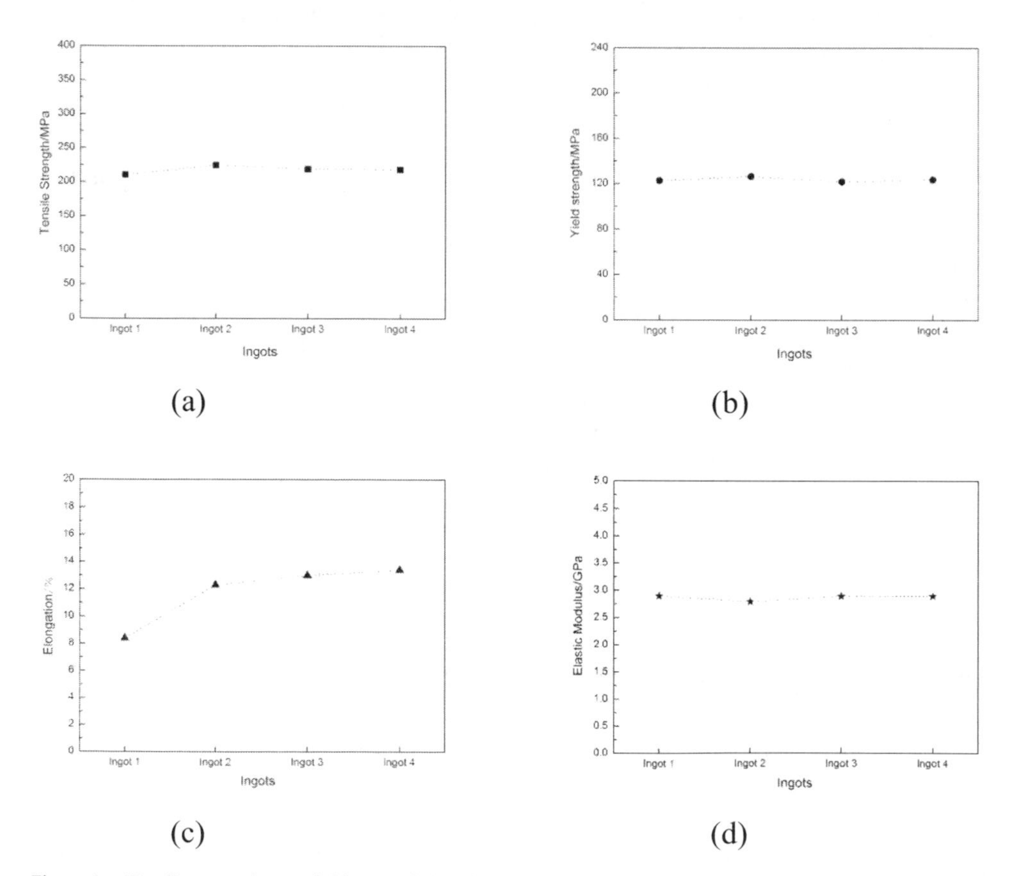

Figure 1. Tensile strength (a), yield strength(b), elongation(c) and elastic modulus(d) of large flat ingots from the edge to the center.

Mechanical properties of hot rough rolling sheets and hot finish rolling sheets. Mechanical properties of hot rough rolling sheets were shown in Table 2. Mechanical properties of hot finish rolling sheets were shown in Table 3.

Table 2. Mechanical properties of hot rough rolling sheets.

Hot rough rolling sheets	Tensile strength /MPa	Yield strength /MPa	Elongation / %	Elastic modulus /GPa
Non-rolling direction	316.9	197.4	18.7	4.3
Rolling direction	316.3	189.3	20.3	3.9

Table 3. Mechanical properties of hot finish rolling sheets.

Hot finish rolling sheets	Tensile strength /MPa	Yield strength /MPa	Elongation / %	Elastic modulus /GPa
Non-rolling direction	335.6	204.6	24.1	4.9
Rolling direction	331.6	196.9	18.4	4.8

As can be seen from Table 2 and Table 3. Tensile strength, yield strength, elongation and elastic modulus of rolling direction and the non-rolling direction of the hot rough rolling sheets. Hot finish rolling sheets, too. The average of tensile strength, yield strength, elongation and elastic modulus of the hot rough rolling sheets is about 317MPa, 193MPa, 20%, 4.1GPa respectively. The average of tensile strength, yield strength, elongation and elastic modulus of the hot finish rolling sheets is about 334MPa, 201MPa, 21%, 4.9GPa respectively. Based on the above performance, organization of hot rolling sheets is uniform in each direction.

Mechanical properties of 5083 aluminum alloy dependence of the rolling status. Figure 2(e) shows tensile strength of 5083 aluminum alloy dependence

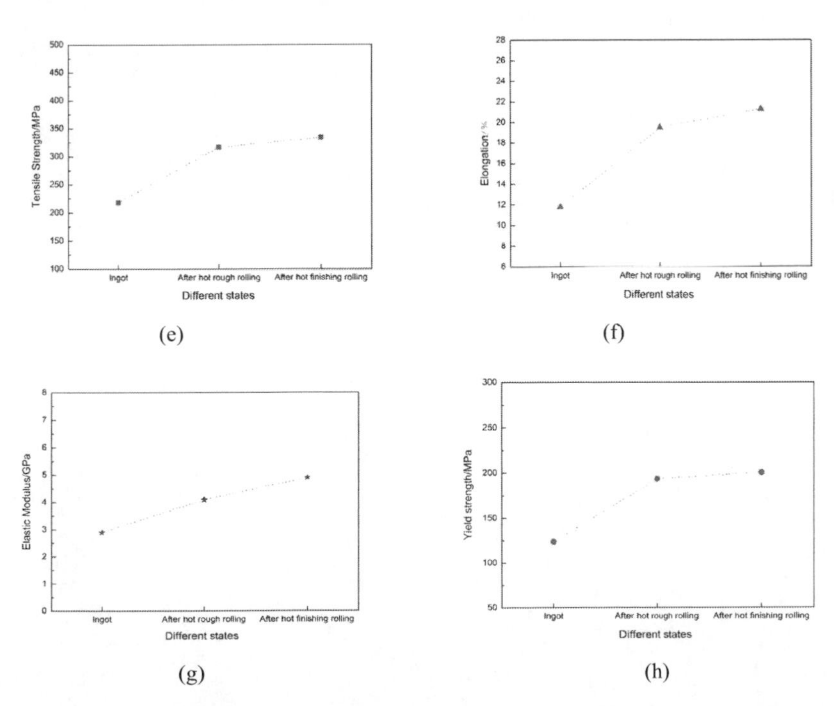

Figure 2. Tensile strength (e), yield strength(f), elongation(g) and elastic modulus(h) dependence of the rolling status.

of the different rolling status. Figure 2(f) shows yield strength of 5083 aluminum alloy dependence of the different rolling status. Figure 2(g) shows elongation of 5083 aluminum alloy dependence of the different rolling status. Figure 2(h) shows elastic modulus of 5083 aluminum alloy dependence of the different rolling status.

As can be seen from Fig.2, changing trend of tensile strength, yield strength, elongation and elastic modulus dependence of the rolling status are the same. After the ingot was hot rough rolled and hot finish rolled, mechanical properties gradually increased, it shows that hot-rolled can significantly improve the mechanical properties of the ingot.

4 SUMMARY

The mechanical properties of ingots are almost the same from the edge to the central part. Mechanical properties of hot rolling sheets are almost the same in rolling direction and non-rolling direction, too. Mechanical properties gradually increased, after the ingot was hot rough rolled and hot finish rolled. Hot finish rolling sheets showed surprisingly high yield stress values, tensile strength values and elongation values. Based on the above performance, ingots had good organizational uniformity and mechanical properties, organization of hot rolling sheets is uniform in each direction and hot-rolled can significantly improve the mechanical properties of the ingot.

ACKNOWLEDGEMENTS

This study is financially supported by National and International Special Scientific and Technological Cooperation Project (2011DFR50950), Joint Development of New Connection Technology and Equipment (2011GJHZ50001) and Chongqing Overseas High-level Entrepreneurial Talent Base Construction Projects. Corresponding author: Tel: 023-67301229; E-mail address: xwswxxzwb@163.com

REFERENCES

[1] Totten G E, Mackenzie S. Handbook of aluminum. Physical metallurgy and processes. Marcel Dekker Inc: USA; 2003.

[2] Jiang D M, Hong B D, Lei T C, et al. Influence of Aging Condition on Tensile and Fatigue Fracture Behavior of an Al-Mg-Si Alloy 6063[J]. Materials Science And Technology, 1991, 7(11):1010–1014.

[3] Field J E, Walley S M, Proud W G. Review of experimental techniques for high rate deformation and shock studies[J]. International Journal of Impact Engineering, 2004, 30: 725–775.

[4] Starink M J and Sinclair I. Development of New Damage Tolerant Alloys for Age-forming[A]. Aluminum Alloys 2002-their Physical and Mechanical Properties[C]. Cambridge, UK, 2002, 601–606.

[5] Kaufman J G. Introduction to aluminum alloys and tempers. ASM International; 2003.

Computing, Control, Information and Education Engineering – Liu, Sung & Yao (eds)
 ISBN: 978-1-138-02800-5

Improved design and implementation of position and pitch trim device based on transposition part

Li Qin Miao, Li Guo Lai, Hang Yu, Ji Chun Jiang, He Ping Jiang, Da Shun Zhang, Jing Wang, Ruo Ping Wang & Xiao Jiao Gao

Department of Numerical Design and Manufacture, Changchun Research Institute of Equipment and Technology, Changchun, P.R.China

Yong An Zhang

Henan Pinguang Optics Electronics CO., LTD, Jiaozuo, P.R. China

ABSTRACT: In view of the assembly requirements of transposition parts, to design azimuth and pitch axis fast balancing device, azimuth axis and pitching axis perpendicular to each other, azimuth shafting load rotates around the vertical axis. In previous designs, when the azimuth shafting was balancing and pitching axis was balancing, their balancing quantities influence each other and need to do repeatedly balancing, for the improved balancing device, the principle is based on the center-of-mass coordinates information, by the calculation formula of moment balance, it can be accurately calculated the azimuth axis load and pitching axis load balancing and balancing block placement, so as to guide the operator to quickly finish balancing work.

KEYWORDS: Azimuth axis; Pitch axis; Balancing; Moment.

1 INTRODUCTION

Transposition parts weight is heavy, assembly position often needs to be moved in the workshop, for balancing work before assembly, and the characteristics of the heavy weight and lifting inconvenience increased the difficulty of balancing, previous balancing devices adopts manual rotating parts in the form of dislocation motion, not only need many people to operate, and have certain risk in security, in this article based on the analysis on the existing alignment device, the structure of the balancing device is improved, the improved balancing device balancing can be convenient for operation, can be free to move in the workshop, reduces the number of operators, has obvious increase in balancing efficiency and safety.

2 EXISTING SCHEMES

Balancing device of the existing balance scheme as shown in Fig. 1, with the usual 4-5 people, needs to cooperate to complete the trim operation, the device itself cannot move, because it has not a mobile device, a transposition device and a dividing device, all actions need manual operation, because the match paperback device has a great weight, coupled with the weight of the translocation parts, so the difficulty of balancing operation is big and the balancing efficiency is low.

Figure 1. Existing balancing device.

3 THE IMPROVED SCHEME

For transposition parts with the adjustable system azimuth and elevation of biaxial loading, to develop a special device, the device has the azimuth and pitching accurately adjustment and dynamic balancing function, balancing software is used to calculate the data, we can quickly achieve the equilibrium of shafting. Static unbalance torque is caused by the structure of the center of gravity and the center of mass is not coincidence. Pitch and azimuth shafting static balancing torque calculation principle is shown in Fig. 2. As shown in Fig. 3 for static balance coordinate system.

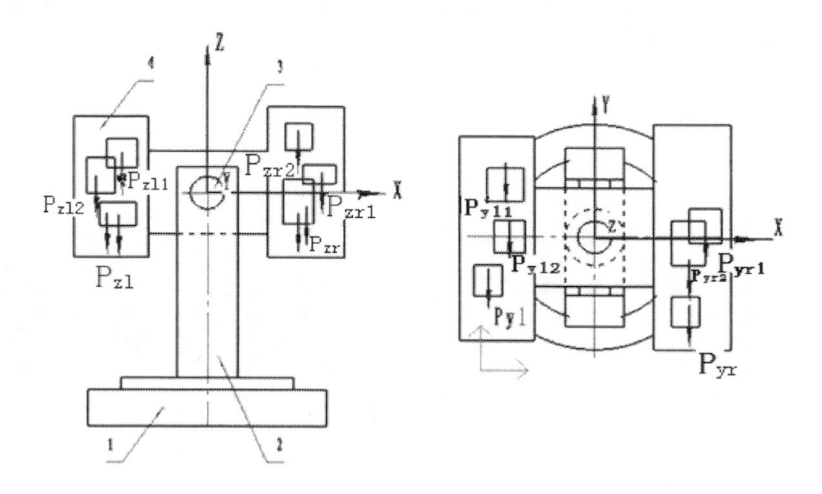

(a) Main view (b) Top view

1 as the base, 2 for the azimuth shafting, 3 as the pitch axis, 4 for the pitch axis double structure parts

Figure 2. Pitch and azimuth shafting static balancing torque calculation principle diagram.

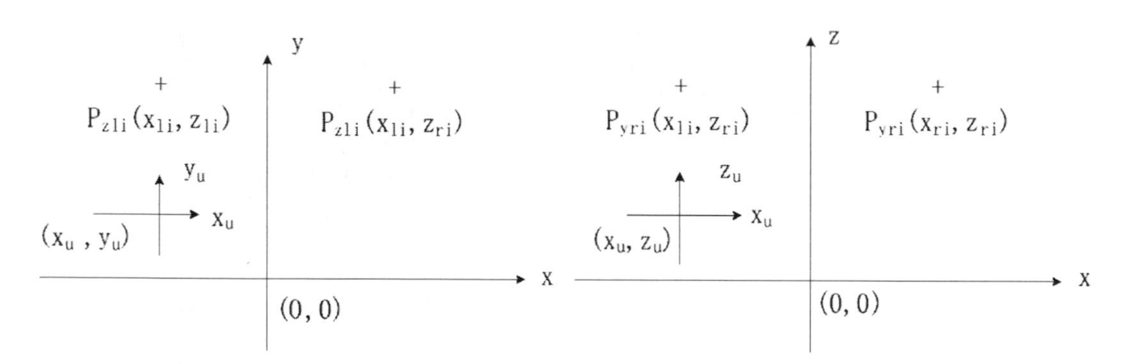

Figure 3. Static balance coordinate system.

Pitch shafting balance calculation is as follows: By applying the principles of static equilibrium, pitch shafting puts pitch axis as the balance shaft (y direction) and the left and right parts produce moment balance.

$$\sum_{i=1}^{n} p_{zli}|(x_{li}-x_u)| = \sum_{i=1}^{m} p_{zri}|(x_{ri}-x_u)| + p_{zx}l_{x1} + p_{yx}l_{x2} \quad (1)$$

Type: x_u is the bearing axis, y_u is pitching axis.

$p_{zli}(x_{li}, y_{li})$—To y_u is on the left side of the balance shaft parts weight and center of mass coordinates

$p_{zri}|(x_{ri}, y_{ri})|$—To y_u is on the right side of the balance shaft parts weight and center of mass coordinates

p_{zx} is pitching axis balancing, l_{x1} is balancing moment

p_{yx} is pitching axis balancing, l_{x2} is balancing moment

Through a calculation of the amount of unbalance, it can be calculated according to the set of particles need to add balance weight.

The transfer parts flip 90º, and pitch axis static balancing torque calculation by the same token, the structural parts by azimuth axis to pivot around static moment balance. It can be calculated by azimuth axis to balance axis (z direction) of the static unbalance.

$$\sum_{i=1}^{n} p_{yli}\left|(x_{li}-x_u)\right| = \sum_{i=1}^{m} p_{yri}\left|(x_{ri}-x_u)\right| + p_{yx}l_{x1} + p_{zx}l_{x2} \quad (2)$$

$p_{yli}(x_{li}, z_{li})$——The z_u as a weight for the parts on the left side of the balance shaft and center of mass coordinates

$p_{yri}(x_{ri}, z_{ri})$——Taking z_u as a weight for the parts on the right side of the balance shaft and center of mass coordinates when balancing, we needs the transfer parts to be moved from assembly workshop to another workshop for balancing, because we do not balance every day, when there is no need to balance and the device is put in assembling workshop, balancing device will affect assembly line to work well with usual work, balancing device is heavy, so in a fixed position, is not convenient to use at ordinary times, after the balancing will have to transfer parts to assembly workshop. For this situation, the article design the mobile car, the balancing device installed in the car, increase the movement the flexibility of the balancing device. As shown in Fig. 4 for the improved balancing device.

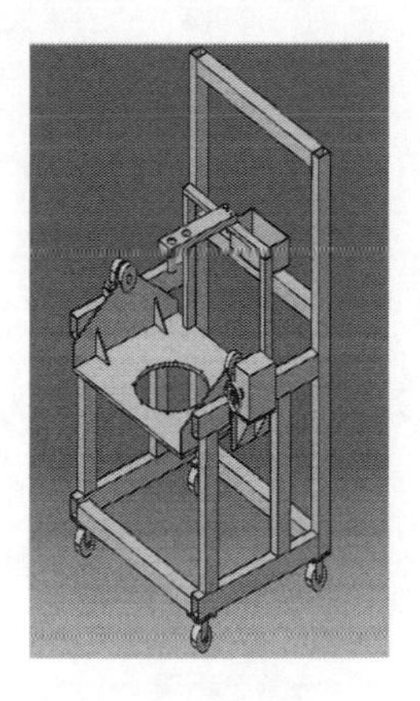

(a) Balancing device

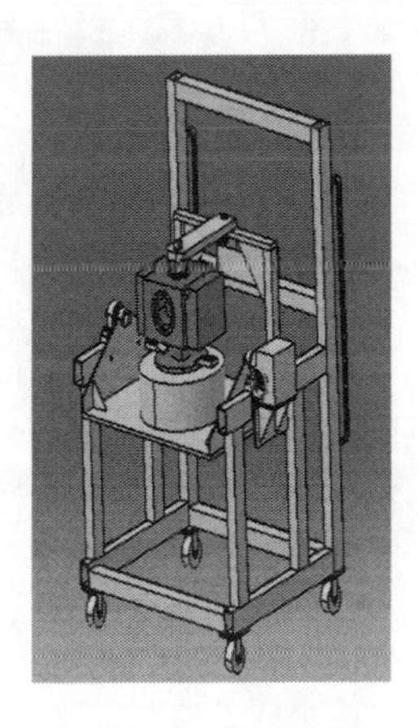

(b) 90° balancing position

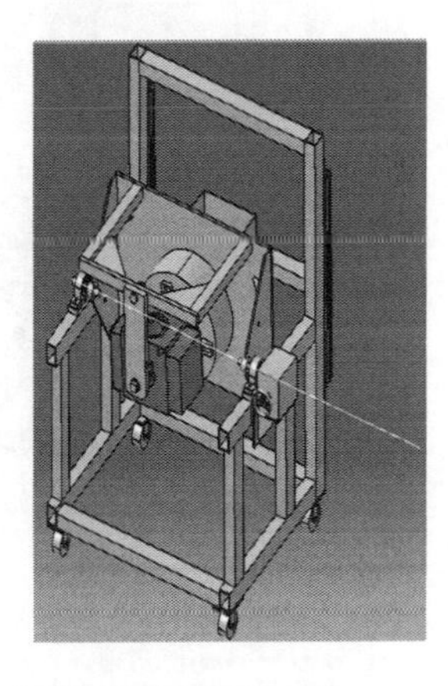

(c) 180° balancing position

Figure 4. 180° Arbitrary angle flip.

To meet the photovoltaic system turret azimuth and elevation shafting trim needs, the article designs a special device. The device consists of artificial carts, flip agencies and computer management system. Photoelectric turret 180° at any Angle, azimuth axis and pitching axis with equal function at a time.

After the improvement, just 1 or 2 people can operate to complete balancing. The balancing device is composed of a rotating device, trolley, reducer, etc, the system is automatically controlled by single chip microcomputer, rotary motion achieved transmission by gear. Due to the large overall weight, the balancing device after the design is simulated by the finite element analysis, the analysis results are shown in Fig. 5.

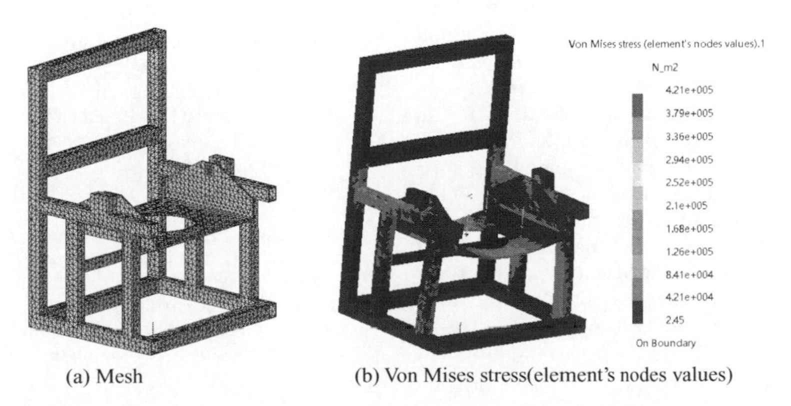

(a) Mesh

(b) Von Mises stress(element's nodes values)

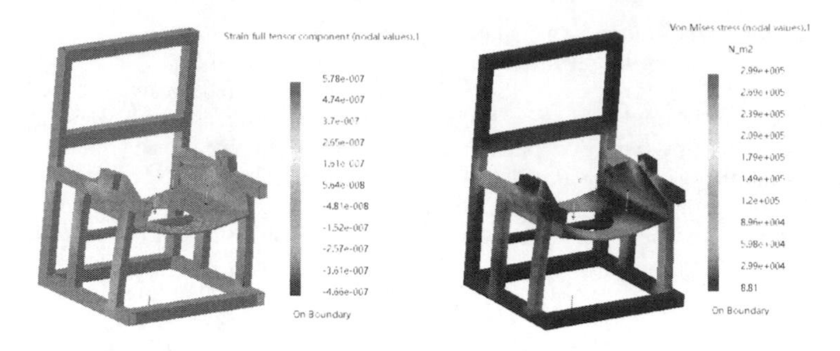

(c) Strain full tensor component(nodal values) (d) Von Mises stress(nodal values)

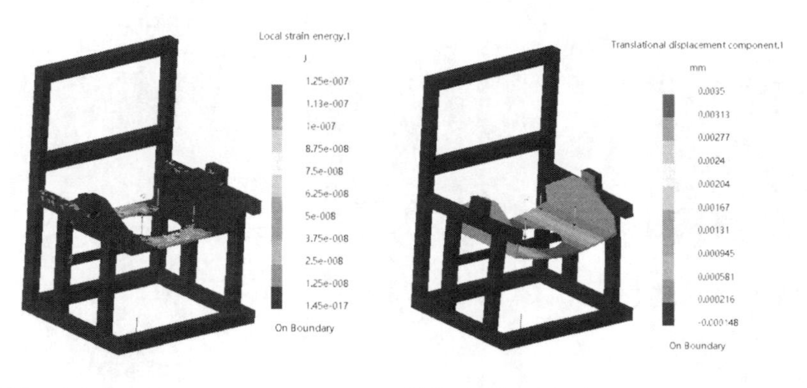

(e) Local strain energy (f) Translational displacement

Figure 5. Balancing device by finite element analysis results.

The finite element analysis results show that the improved balancing device can satisfy the use requirement, the amount of deformation of the whole after the stress switches to the balancing permissible error range.

4 CONCLUSION

Through analyzing the device on the translocation of part balancing of the original scheme, the article designs a new improved balancing, the original device needs 4-5 people to complete the balancing work, and the improved device requires only 1-2 people to finish balancing. The device can realize the precise adjustment of azimuth, pitch and dynamic distribution equality, balancing software is used to calculate the data, we can quickly achieve the equilibrium of shafting, establish the static balance calculation formula of moment of pitch and azimuth shafting and static balance coordinate system, the device was analyzed by finite element analysis, the analysis results show that the balancing device can satisfy the use requirement, balancing efficiency and balancing precision are improved.

REFERENCES

[1] CHEN Ye. Design of Automatic Leveling Control Device. New Technology & New Process, 2013(3):67–68.

[2] WANG Xuan, LI Zhe. Research on the Pitch Automatic Trim System for a Certain Commuter. Journal of System Simulation, 2008(2):260–262.

[3] CUI Zhao, LI Jianbo, WANG Junchao, et al. Trim and stability characteristics of an auto gyro equipped with Gurney flaps. FLIGHT DYNAMICS, 2013,31(5):385–388.

[4] LI Zixing, LI Gaofeng, HUANG Ruiling. A Vehicle Rolling-Guidance Law Based on Fixed Trimmed Angle of Attack[J]. Aerospace Control and Application, 2012,38(6):23–26.

[5] ZHANG Zijun, WANG Lei, WANG Lixin, et al. Three-axis stability characteristics of flying wing with high aspect ratio. Systems Engineering-Theory & Practice,2012,32(5):1129–1135.

[6] CHEN Shenglai,ZHU Changchun, DENG Zhigang. Dynamic balance analysis of suspended basket in centrifugal test of gasoline tank. Equipment Environmental Engineering, 2010,7(6):243–246.

Computing, Control, Information and Education Engineering – Liu, Sung & Yao (eds)
 ISBN: 978-1-138-02800-5

Quality of reinforcing steel of JSC «Arselor Mittal Temirtau»

O. Krivtsova
Karaganda State Industrial University, Temirtau, Kazakhstan

T. Seisembinov
JSC «ArselorMittal Temirtau», Temirtau, Kazakhstan

A. Naizabekov
Rudny Industrial Institute, Rudny, Kazakhstan

A. Nassonov
Novosibirsk State Technical University, Novosibirsk, Russia

ABSTRACT: The results of the statistical analysis of defects reinforcing steel, rolled in the rolling mill of JSC "ArcelorMittal Temirtau", and described the analysis of its geometry are presented in this paper. Built a computer model of the rolling process of reinforcing steel, in which there is the possibility of obtaining of billet, corresponding to the standards is also presented.

KEYWORDS: reinforcing steel, defects, model, simulation.

1 INTRODUCTION

Continuous shape mill installed in JSC "ArcelorMittal Temirtau", despite the constant improvement of the technology of rolling in the production of long-rolled steel products, there is still the output of the metal in the marriage.

The major share of the total production is reinforced steel, intended for the manufacture of an unstressed longitudinal tension rod operating the valve in the crocheted frames and grids. This type of rolling products meet the strictest demands. Therefore, research aimed at improving the quality of reinforcing steel, are relevant.

The aim of this work is the analysis of defects of reinforcing steel that occur during rolling in JSC "AMT". The objectives of this work include statistical analysis of the defects of reinforcing steel, the analysis of its geometry and the proposal of measures to reduce occurring defects.

2 ANALYSIS OF DEFECTS

For the static analysis were taken two samples from product rolling plant JSC "ArcelorMittal Temirtau" for 2012 and 2013. Fig. 1 shows a rejection for different types of defects identified in 2012 and 2013.

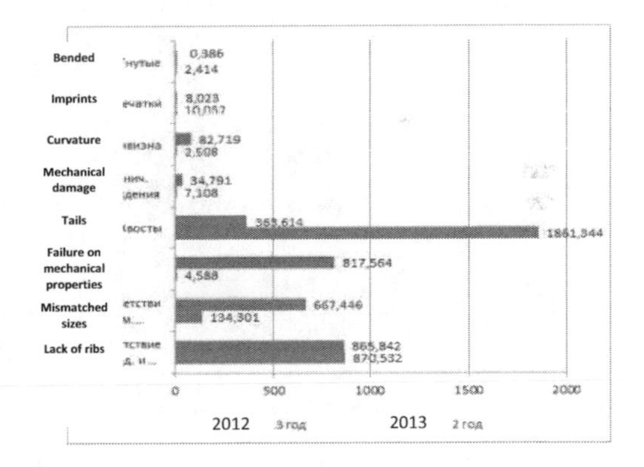

Figure 1. Identified defects calibrated rolled metal.

As can be seen from Fig. 1, the most frequently occurring defects in JSC "AMT" are: "lack of longitudinal and transverse ribs" - 866 and 871 tons respectively by year; "mismatched sizes" - 667 and 134 tons; "failure on mechanical properties" - 818 and 5 tons; "tails" – 364 and 1861 tons.

Other kinds of defects, due to small numbers, it was decided not to consider, then the sample takes the final form presented in Fig. 2.

Histogram in Fig. 2 shows that most often occurs defect "tails". Moreover, in comparison with 2012, the following year saw a sharp decrease of the defect 5 times.

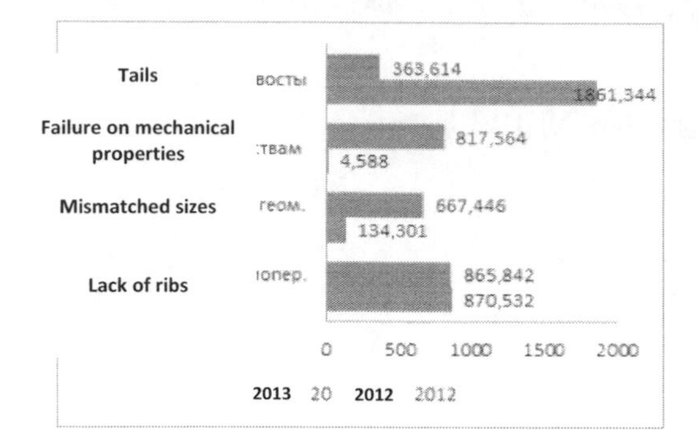

Figure 2. Main types of reinforcing steel defects.

Defects "failure on mechanical properties" and "mismatch size" on the contrary have increased in 2013 in 164 and 5 times respectively. Most often there is a defect in the absence of longitudinal and transverse ribs.

According to technological instructions [1], all defects in rolled sections are divided into recoverable and unrecoverable. In accordance with this, it held the distinction for the specified parameter (Fig. 3, 4).

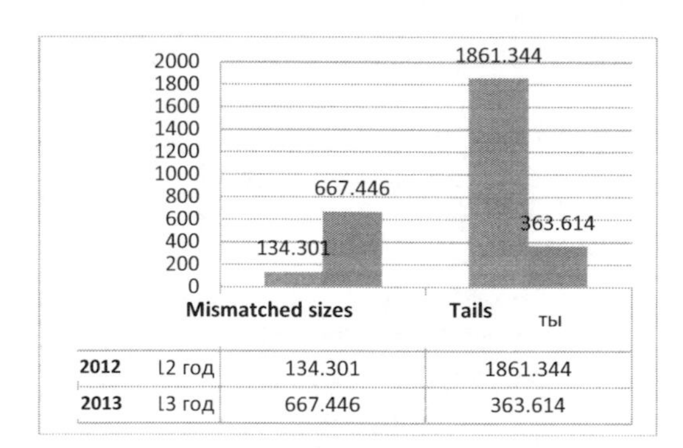

Figure 3. Recoverable defects calibrated rolled metal.

The above defects are most often found during rolling of reinforcing steel in the shape rolling mill of JSC "AMT". Despite the fact that this mill is "youngest" at the plant and equipment plant is a modern, completely get rid of many defects still does not work. Part of the reason is that some defects (e.g., "tail") is "natural" that accompany the process as plate and section rolling, and completely get rid of them is not possible. However, by observing the technology and application of new technical solutions it is possible to minimize the proportion of occurrence and, thereby, reduce scrap and increase the volume of good metal.

All four of the above-described defect is a consequence of non-compliance with various process procedures. By fine tuning of stripping plates and gauge rolls along the line of rolling, the correct settings scissors, hot cutting, observance of speed limits can reduce the occurrence of these defects is approximately 8-10%. However, as world practice shows, these measures are not enough.

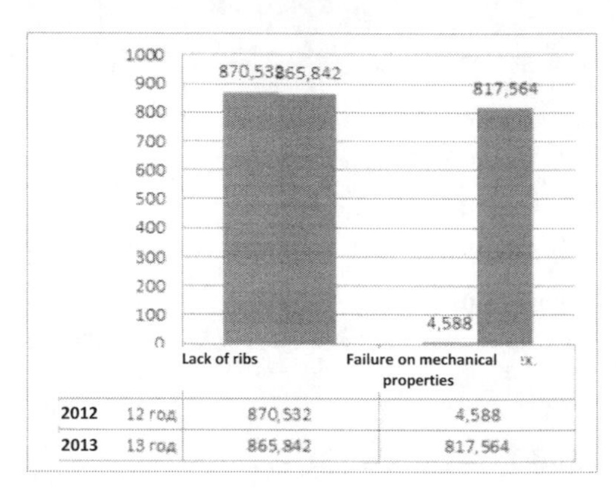

Figure 4. Unrecoverable defects calibrated rolled metal.

One of the most specific defects of reinforcing steel is the lack of longitudinal and transverse ribs (Fig. 5-6), the correct execution of which provides a desired profile. The main reason for the appearance of this defect is incomplete filling of the screw insets fine caliber due to uneven deformation in the cross section of the workpiece.

Figure 5. Lack of longitudinal ribs.

Figure 6. Lack of transverse ribs.

According to the results of studies [2, 3] it was revealed that one of the key factors to full fill the fine caliber during rolling of reinforcing steel is the rational form of pre caliber. In work[4] was proposed calibration of pre-finishing stand in the form of a flat oval with double-sided concavity. This form of roll when hit in finishing the caliber has more scope for broadening and the shape of the side edges as close to a circle, allowing a greater extent to enhance the stiffness of the metal insets fine caliber.

However, authors of work [5] proposed as a pre caliber to use rollers with a smooth barrel. Therefore, to determine the most optimal form pre calibers, as well as optimization of the shop rolling parameters, it was decided to build a computer model of the rolling process of reinforcing steel. Because the mill in the shape rolling shop is designed for rolling reinforcing steel № 8-32, it was decided to simulate the rolling average profile №20.

Model geometry of workpiece for light gauge is constructed in the form of flat oval two-way concavity. Dimensions of billets selected according to the recommended data (Fig. 7) [2]. On the model of the billet with a length of 150 mm suffered a grid of 200 000 finite elements with ratio 4. Chosen elastic-plastic model material, material - steel AISI 1015, initial temperature of the workpiece is assumed to 950° C, as the optimum temperature of the metal before rolling in last stand, according to the technological instruction of workshop; the temperature of the rolls is 60° C.

Figure 7. Model of workpiece.

The values of heat transfer coefficients are the following: between metal and rolls - 3 (kWh/m^2)°C; between metal and air - 0,1 (kWh/m^2)°C. The friction coefficient is 0.35, ambient temperature of 20 °C.

3 RESULTS AND DISCUSSION

In the result of the model calculation produced the following results (Fig. 8-9).

After rolling in finishing stand metal is completely filled the contour of caliber, and screw pull-quotes transverse ribs. Longitudinal ribs are also formed completely.

Figure 8. Common view of deformed workpiece.

The main requirements of the dimensions according to requirements [1], are under vertical and horizontal diameter $\mathbf{d_1}$ and $\mathbf{d_2}$, which characterize the height of the longitudinal and transverse ribs. Values and deviation of diameter $\mathbf{d_2}$ correspond to data of requirements for diameter d_1. The ovality (difference between $\mathbf{d_1}$ and $\mathbf{d_2}$ in one section) shall not exceed the sum of the positive and negative extreme deviations in size $\mathbf{d_1}$. In addition, an important feature of the geometry of reinforcing steel is the value of **t**, which determines the distance between two adjacent transverse ribs.

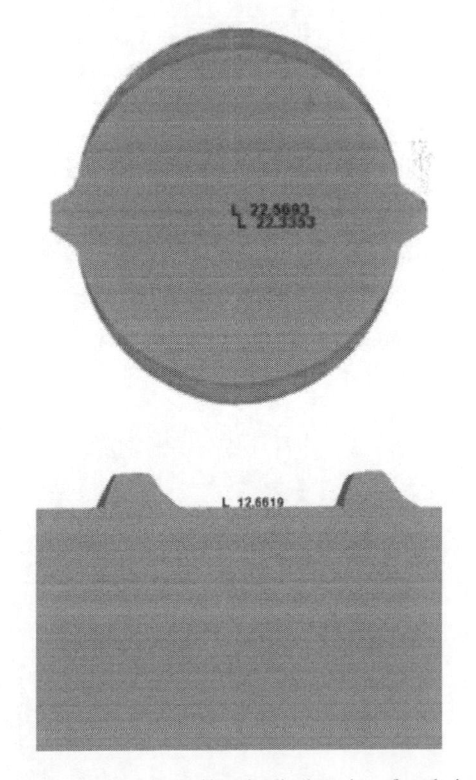

Figure 9. Transverse and longitudinal section of workpiece.

As shown in Fig. 9, the value $\mathbf{d_1}$ in the model is 22,569 mm, the value $\mathbf{d_2}$ is 22,335 mm. Deviation for $\mathbf{d_1}$ is 0,269 mm; for $\mathbf{d_2}$ - 0,035 mm, that fully corresponds to the specified values extreme deviations. Ovality of the rod is 0,234 mm, which also meets the requirements. The value of **t** is 12,66 mm, which fully corresponds to the table value with the permissible limit deviation ±15%, (in our case it is equal to 5%) [6].

4 CONCLUSION

The resulting model is a standard, which if carefully selected settings, there is a possibility of providing of workpiece, fully corresponded the requirements [1] by sizes. In the future, this model will be used for optimization of workshop rolling parameters of the reinforcing steel, the ultimate goal is to minimize the geometric defects.

In order to reduce the percentage of occurrence of these defects, it is proposed to further investigate in the model three variants pre calibers: "traditional oval-round" and "flat oval-round" and "smooth barrel-round". The most effective option will be the form of pre caliber, which will provide the most uniform distribution of accumulated strain in the cross section.

REFERENCES

[1] Technological instruction. TI SR-01–2010. Temirtau, 2010. – 63 p.

[2] Diomidov, B.B. Calibration of rolls. Textbook for high schools / B.B. Diomidov, N.V. Litovchenko. – Moscow: Metallurgy, 1970. – 312 p.

[3] Smirnov, V.K. Calibration of rolls. Textbook for high schools / V.K. Smirnov, V.A. Shilov, Y.V. Inatovich. – 3rd edition., rev. and add. – M.: Heat-technic, 2010. – 490 p.

[4] Levchenko, L.N. Production of reinforcing steel / L.N. Levchenko, A.S. Natapov, L.F. Mashkin, Y.T. Hudik, S.L. Baskin. - Moscow: Metallurgy, 1984. – 136 p.

[5] Asanov, V.N. Improving the calibration of rolls for rolling round and reinforcing profiles [Text] / V.N. Asanov, A.B. Steblov, O.N. Tulupov, D.V. Lenartovich // Steel. 2008. №11. p. 90–91.

[6] Naizabekov A., Lezhnev S., Panin E., Volokitina I. Computer modeling of the rolling process of reinforcing steel. Advanced Materials Research. Vol. 1030–1032 (2014), pp 1286–1291.

Computing, Control, Information and Education Engineering – Liu, Sung & Yao (eds)
© 2015 Taylor & Francis Group, London, ISBN: 978-1-138-02800-5

Qualimetric assessment of the quality of forgings obtained by hammering tool with elastic elements that implements the transverse and longitudinal shift

S.N. Lezhnev, E.A. Panin & A.O. Tolkushkin
Karaganda State Industrial University, Temirtau, Kazakhstan

A.B. Naizabekov
Rudny Industrial University, Rudny, Kazakhstan

ABSTRACT: Paper describes forging technology of billets in dies with elastic elements and the comparison of results of forging of billets with existing technology in the flat dies by the differential and integrated quality indicators. Also the dependence of the complex index of forging quality of the deformation degree described.

KEYWORDS: dies with elastic elements, qualimetric assessment, forging.

1 INTRODUCTION

The flourishing of market economy and the related huge variety of products, over-saturation of the market with high quality products have led to increased competition in the markets. The most important condition for the success and prosperity of high quality steel products is satisfying the ever-growing demands for it. Quality improvement has become a leitmotif of industrial and scientific activities of enterprises and associations, even companies with a worldwide reputation. The basis for quality assurance has been and remains the level of production technology. That level of engineering technology today has become the most important factor, not only the economic power of thc country, but also its national prestige, authority in the world.

Forging and stamping production is an essential part of most manufacturing companies. It refers to the number of advanced production facilities, as it provides improved mechanical and operational characteristics of parts and products in general, high productivity, reducing waste of materials, energy and labor costs.

The current technology of forging based on the use of traditional Blacksmithing tools and modes of deformation, characterized by a low level of mechanical properties and their uneven distribution over the cross section of the metal forgings. Use of the new forging tool that implements the alternating deformation can improve the quality of metal forgings.

One technology that realizes alternating deformation is the technology of forging ingots and billets in dies with elastic elements (Fig. 1) [1]. Technology of forging ingots and billets in these dies allows to realize significant additional shear deformation in the transverse and longitudinal direction.

Earlier in work [2] has been described technology of deformation of blanks in this instrument, which included transverse shear and straightening of the billet in the same dies and it was proved the advantage of this technology of billet deformation compared to the existing technology of deformation of blanks in flat dies. In addition to the possibility of deformation of billets of any size in these dies in work [2] also presented a formula for calculating the stiffness of the elastic elements (springs in compression).

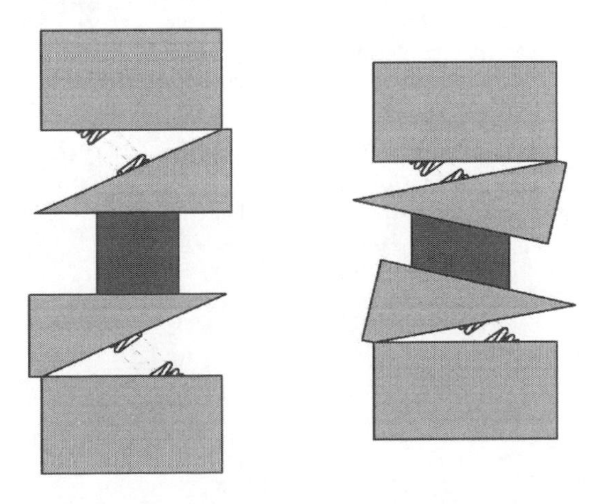

Figure 1. Dies with elastic elements.

2 RESEARCH METHODOLOGY

To study the effect of the new deformation scheme of ingots and billets in dies with elastic elements, allowing to implement in the volume of metal transverse and longitudinal shear, on the quality of forgings forging of billets made of steel W 108 with sizes b×h×l = 200×200×300 mm was carried out in Forging-pressing site at machine shop of JSC "ArcelorMittal" on a hydraulic press with a force of 12.5 MN. As the quality of forgings characterized by a multitude of quality indicators, each of which can serve as a criterion for the selection of process parameters, the evaluation of the quality of forgings, deformed by the current and proposed technology, made with parameter (K_0 - complex quality index) characterizing qualimetric assessment of the level of mechanical properties of the metal.

Forging of billets was produced by the proposed technology in the dies with elastic elements and the existing technology in the flat dies. Billet before deforming was heated in reheating furnaces prior to the start of the forging temperature of 1200°C.

Deformation of billets was carried by following way:

1 Deformation of the 1st billet in dies with elastic elements carried by following scheme:

[compression (transverse shear)→the rotation around its axis by 180°→straightening→ the rotation around its axis by 90°→compression (longitudinal shear)→the rotation around its axis by 180°→ straightening] – 1st cycle.

2 Deformation of the 2nd billet carried by following scheme:

After 1st cycle billets rotates around its axis by 90° and deform according to 1st cycle (2nd cycle).

3 Deformation of the 3rd billet carried by following scheme:

After 1st cycle billets rotates around its axis by 270° and deform according to 1st cycle.

4 Deforming of 4th, 5th and 6th billets carried in flat dies with compression and deformation schemes corresponding to deformation in the dies with elastic elements.

Thus, the 1st billet with sizes 200×200×300 mm was forged in dies with elastic elements to size b×h×l = 168×215×332 mm, deformation degree in height was 16.8%. Second billet was forged in dies with elastic elements to size b×h×l = 157×221×346 mm, deformation degree in height in that case was 21.5%. Third billet deformed in dies with elastic elements with relative reduction of 35% allowing to receive equal forging sizes 130×251×368 mm. Forging reduction thus amounted to 1.19, 1.27 and 1.54 accordingly.

From the second batch of billets with sizes b×h×l = 200×200×300 mm, in a flat tool were forged three forgings so that their sizes correspond to the size of forgings obtained in dies with elastic elements. The deformation of the billet heated to the forging beginning temperature of 1200°C was produced in the flat dies on a similar scheme and with the same degree of deformation in height, as in the forging in the dies with elastic elements.

All six forgings were marked. Then they were cut into the templates in the longitudinal and transverse directions for the production of specimens for mechanical testing, microstructure analysis.

After cutting of templates from all forgings, forged on existing and developed technologies standard samples for tensile and toughness tests were made. Testing of the samples in tension was performed on a universal torsional testing machine MI-40KU, used in conjunction with an IBM-compatible computer that provides plotting the dependence of the load and moment of deformation on the display of a computer. Test specimens in tension carried by the method of "Checking in accordance with RD 50-482-84". Impact tests conducted on a pendulum impact-testing machine MK-30A.

3 RESULTS AND DISCUSSION

During mechanical testing identified the following mechanical properties of steel a: tensile strength R_m, yield stress $R_{0.2}$, elongation δ, relative narrowing ψ, impact strength KCU. On the basis of the obtained values of the mechanical properties of steel W 108 using a computer program written by the method given in [3] was calculated complex quality indicator. The calculated results are shown in Tables 1 and 2 and in Figures 2 and 3.

Table 1. Differential and integrated indicators of quality of steel W 108 (samples with longitudinal section).

Cut place (area)	Deformation method	Deformation degree ε	Differential indicators					
			$\hat{E}_{R_{0.2}}$	$\hat{E}_{R_m}$	K_δ	K_ψ	K_{KCU}	K_0
								(Continued)
surface	flat dies	0,168	0,689	0,741	0,690	0,701	0,704	0,699
	new dies	0,168	0,691	0,716	0,701	0,712	0,709	0,706
	flat dies	0,215	0,735	0,748	0,733	0,723	0,732	0,734
	new dies	0,215	0,757	0,755	0,753	0,749	0,749	0,761
	flat dies	0,35	0,783	0,773	0,772	0,748	0,775	0,770
	new dies	0,35	0,788	0,783	0,806	0,900	0,794	0,814
axial	flat dies	0,168	0,662	0,676	0,663	0,679	0,692	0,674
	new dies	0,168	0,683	0,645	0,684	0,695	0,702	0,682
	flat dies	0,215	0,702	0,709	0,703	0,701	0,726	0,708
	new dies	0,215	0,738	0,731	0,740	0,744	0,738	0,778
	flat dies	0,35	0,745	0,761	0,736	0,729	0,761	0,746
	new dies	0,35	0,785	0,773	0,790	0,788	0,785	0,784

Table 2. Differential and integrated indicators of quality of steel W 108 (sample with cross-section).

Cut place (area)	Deformation method	Deformation degree ε	Differential indicators					
			$\hat{E}_{R_{0.2}}$	$\hat{E}_{R_m}$	K_δ	K_ψ	K_{KCU}	K_0
surface	flat dies	0,168	0,669	0,684	0,676	0,679	0,664	0,677
	new dies	0,168	0,685	0,694	0,692	0,697	0,697	0,693
	flat dies	0,215	0,717	0,726	0,716	0,707	0,699	0,713
	new dies	0,215	0,746	0,746	0,732	0,728	0,737	0,738
	flat dies	0,35	0,744	0,753	0,747	0,731	0,725	0,740
	new dies	0,35	0,783	0,771	0,780	0,758	0,785	0,781
axial	flat dies	0,168	0,648	0,672	0,661	0,654	0,651	0,637
	new dies	0,168	0,677	0,657	0,72	0,674	0,664	0,669
	flat dies	0,215	0,685	0,665	0,680	0,676	0,673	0,676
	new dies	0,215	0,732	0,714	0,728	0,728	0,726	0,726
	flat dies	0,35	0,734	0,750	0,728	0,731	0,685	0,725
	new dies	0,35	0,769	0,761	0,775	0,796	0,775	0,775

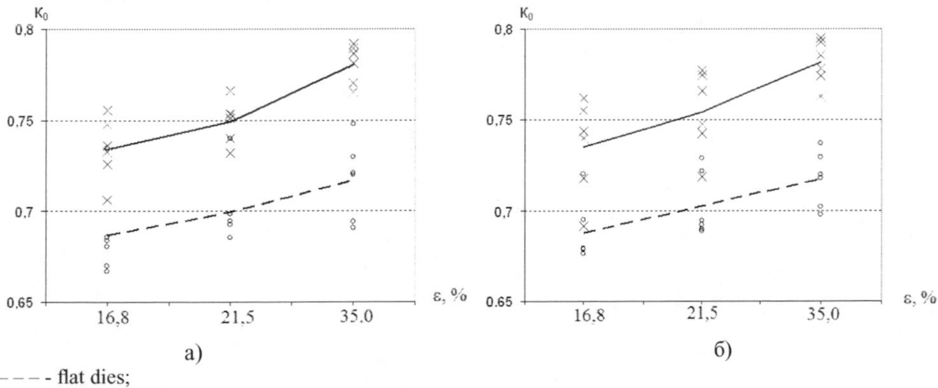

o, – – – – – - flat dies;

×, ——— – proposed tool;

a) surface area;

б) axial area.

Figure 2. The dependence of the complex index of forging quality made of steel W 108 from the degree of deformation (longitudinal samples).

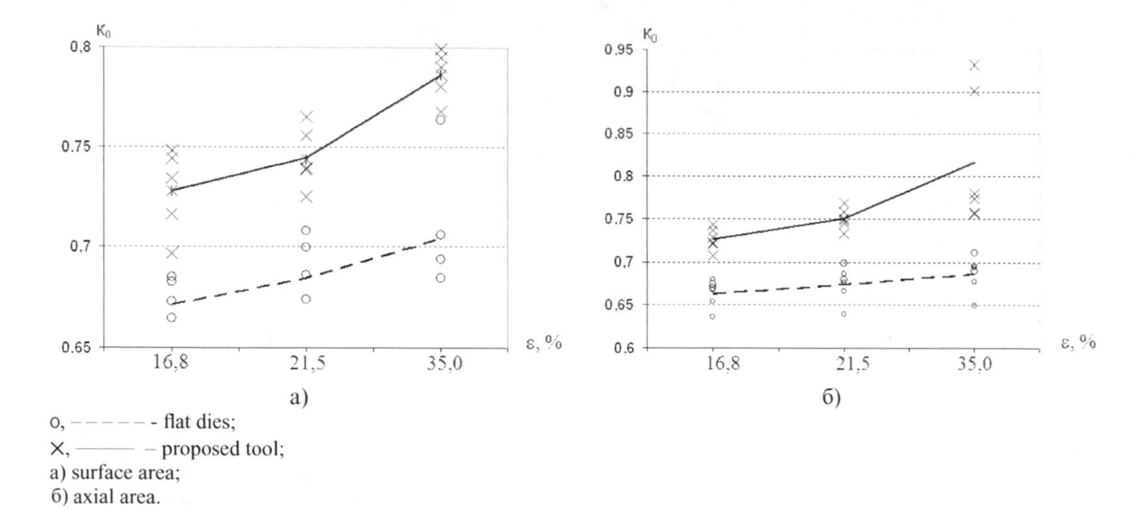

o, – – – – – - flat dies;
×, ——— – proposed tool;
a) surface area;
б) axial area.

Figure 3. The dependence of the complex index of forging quality made of steel W 108 from the degree of deformation (transverse samples).

Comparative analysis of the graphs shown in Figures 2 and 3, based on the results of calculation of the integrated quality indicator, shows that the values of the complex index of forging quality deformed on the proposed technology in dies with elastic elements to implement the transverse and longitudinal shift, on average 8-12% greater than the values in forgings deformed on existing technology in a flat dies.

4 SUMMARY

The results of the qualitative assessment of the quality of forgings, forged under the current and proposed technologies, once again proving the advantage of forging ingots and billets in dies with elastic elements to implement the transverse and longitudinal shift, compared to their forging in flat dies, because of the calculations shows that the proposed technology of forging provides the best quality of metal compared to the current technology.

REFERENCES

[1] Naizabekov A.B., Lezhnev S.N. Investigation of the billet deformation process in the dies with elastic elements realizing a cross and a longitudinal shear. Materials Science Forum, Stafa-Zurich, Switzerland, 2008, Vol. 575–578.

[2] Naizabekov A.B., Lezhnev S.N., Golumbovskaya S.Yu. Study of the process of deformation of the workpieces in dies with elastic elements/ Procuring production in mechanical engineering. 2006 №7.

[3] Naizabekov A.B., Talmazan V.A., Shmidt N.Yu. Qualimetry in metal forming. Almaty: Publ. RIK for educational and methodical literature, 2002.–142 c.

Computing, Control, Information and Education Engineering – Liu, Sung & Yao (eds)
 ISBN: 978-1-138-02800-5

Amorphous-nanocrystalline composite based on nanocarbon, obtained by sintering under high pressure

D.V. Kuis, G.P. Okatova, N.A. Svidunovich, P.V. Rudak & I.L. Tobolich
Belarusian State Technological University, Minsk, Belarus

S.N. Lezhnev & A.O. Tolkushkin
Karaganda State Industrial University, Temirtau, Kazakhstan

V.S. Urbanovich
Scientific-Production Center for Materials Science of NAS of Belarus, Minsk, Belarus

ABSTRACT: The paper investigated the possibility of preparing a solid C-Fe composite with a predominance of superhard carbon phase and a reflux ratio of the components of iron and carbon. Conducted a study of the fine structure of the nanocomposite, refinement of the phase composition and the degree of disorder of the crystal structure.

KEYWORDS: nanocrystalline composite, nanocarbon, microstructure.

1 INTRODUCTION

Sintering of composites based on nano-dispersed components of the system Fe-C at high pressures is of considerable interest because of the possibility to obtain improved physical and mechanical properties, in particular hardness and wear resistance [1]. Previously it was shown that during sintering under high pressure (4-5 GPa) and temperature (950-1200°C), based on the nanocomposite with addition of Fe 3-10 wt. % of nanocarbon formation of superhard carbon phase occurs not only because of the fullerenes, but also from other, cheaper nanodisperse carbon materials - soot containing fullerenes, multi-walled nanotubes, fullerene black [2]. As a result, it assumed a leading influence on the formation of "superhard carbon phase" in the Fe-C composites dispersion of nanocarbon component and sintering technology of material [3]. Therefore, it seemed appropriate to explore the possibility of preparing a solid C-Fe composite with prevalence of superhard carbon phase and a reflux ratio of the components of iron and carbon.

2 RESEARCH METHODOLOGY

As the source was used nanopowder extracted fullerene soot (C_{efs}) micropowder and carbonyl iron with a particle size of 5-100 microns in a ratio of 90 wt. % C_{efs}: 10 wt. % Fe. Extracted fullerene soot was used (C_{efs}) after nearly exhaustive extraction of fullerenes from the product of the electric arc evaporation of graphite. Composite sample preparation procedure described in [3]. Sintering is performed at a pressure of 4 GPa. Modes of sintering properties of the samples shown in Table 1.

To study the fine structure of the nanocomposite, refinement of the phase composition and the degree of disorder of the crystal structure were used methods of transmission electron microscopy (TEM), electron diffraction in the high resolution TEM JEM2100, company JEOL, Japan and Raman spectroscopy and combination scattering spectrometer RAMANOR U-1000 firms Jobyn Yvon Instruments SA Inc., France.

3 RESULTS AND DISCUSSION

By light and scanning electron microscopy, X-ray diffraction and electron microprobe analysis determined that the obtained composite material is ~ 90% is a continuous amorphous carbon phase constituent and nanocrystallites different morphology and degree of dispersion (1.5 ... 14.5 nm) also contains less than 7 ... 10% of the dispersed phase particles of different sizes on the basis of iron carbides distributed fairly evenly in volume (Fig. 1). It was established that the samples contain different superhard carbon modification gray phase (Fig. 1), of which a dominant microhardness up to 78 GPa plays the role of the binder being sintered at a quasi-liquid state.

Table 1. Sintering conditions of the composite 90 wt. % C_{efs}-10 wt. % Fe.

Sample number	Sintering temperature, °C	Sintering time, s	Density, g/sm³	Microhardness (Load), GPa Type of a gray phase
EFS-15	1200	120	2,14	31,5 (1 H), «zigzag»
EFS-16	1500	120	2,18	81,1 (0,5 H), «zigzag»
EFS-17	1500	43	2,18	107 (5 H), «speckled»
EFS-18	1500	30	2,15	26,8 (0,25 H) «zigzag»

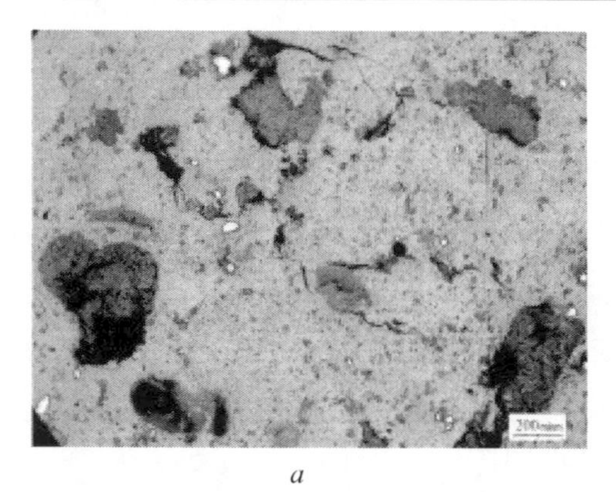

a

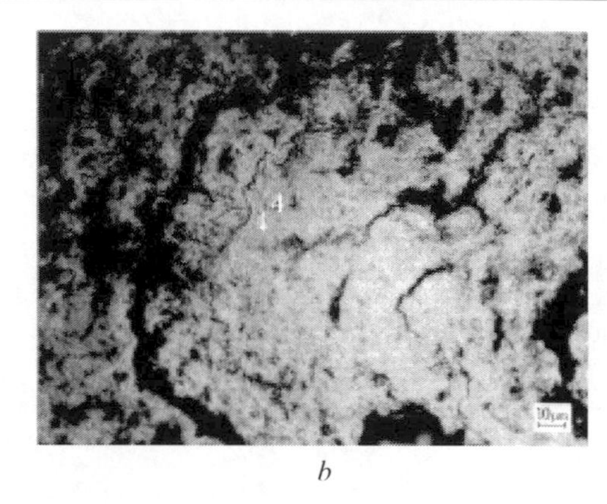

b

a – general view, *b* – superhard phase particles with a relief " speckled zigzag ", Hμ~107 GPA; *a* – ×50, *b* – ×1000

Figure 1. Microstructure of the nanocomposite composition 90% C – extracted fullerene soot + 10 % Fe.

Microhardness inclusions of superhard phase reaches 107 GPA, phases on the basis of Fe – 9,2…10,8 GPA. Specific weight of the highly rigid carbon composite 2,14…2,18 g/sm³. It has specific weight 2.14…2.18 g/sm³ and typical vitreous-like fracture [3].

Type of communication gray phase "basis" in the fracture (SEM) (Fig. 2, *a*) almost smooth, typical for fully vitreous-like amorphous, non-crystallinity carbon X-ray diffraction, which shows only "amorphous halo".

The surface of the particles with the gray "globular" phase relief with increased resolution is in turn composed of smaller "globules", soldered together (Fig. 3, *b*); microprobe elemental analysis showed that the ultrahard particles with a "globular" relief composed of carbon – C (Fig. 3, *c*).

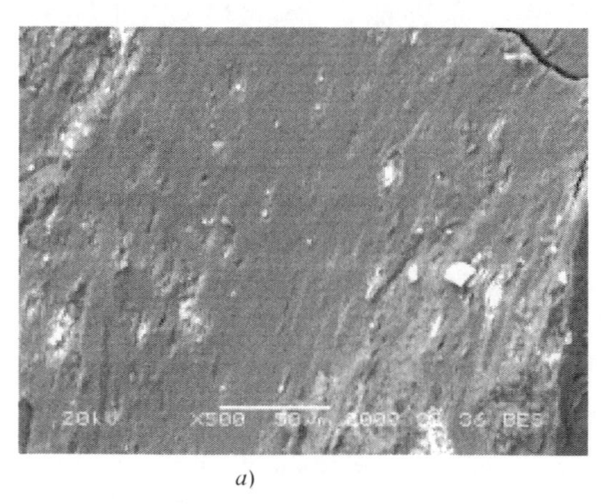

a)

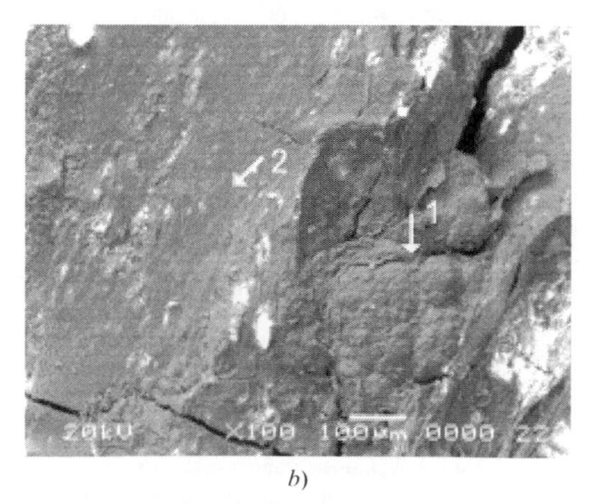

b)

a, b (arrow 2) – gray phase "basis", smooth vitreous-like, *b* – superhard phase particles with the "globular" relief (arrow 1)

Figure 2. Surface of the composite sample C - 10% Fe in fracture (SEM).

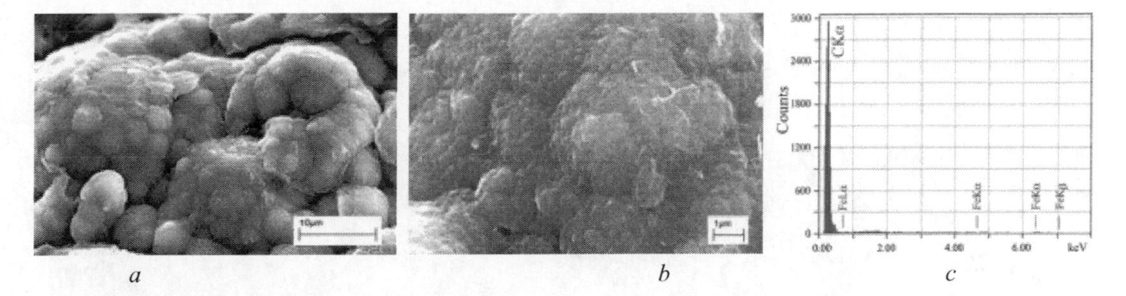

a *b* *c*

Figure 3. Surface of the particles with a super-hard "globular" relief in the fracture of the sample C-10% Fe, SEM (*a*, *b*), results of EDX analysis (*c*) from image square on fig. *a*.

Gray phase "basis" – consists of carbon with Fe inclusions from 1.8 to 7-10 wt. % (Fig. 4, *a*), which close to laid in the batch number – 10% Fe, distributed Fe quite evenly in the form of dispersed particles of different size (Fig. 4, *b*).

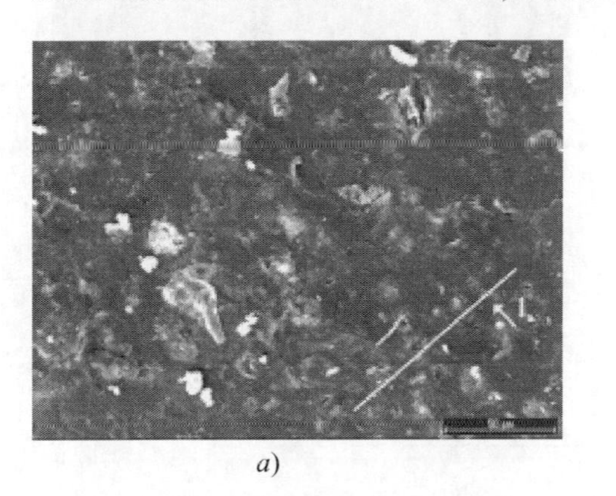

a)

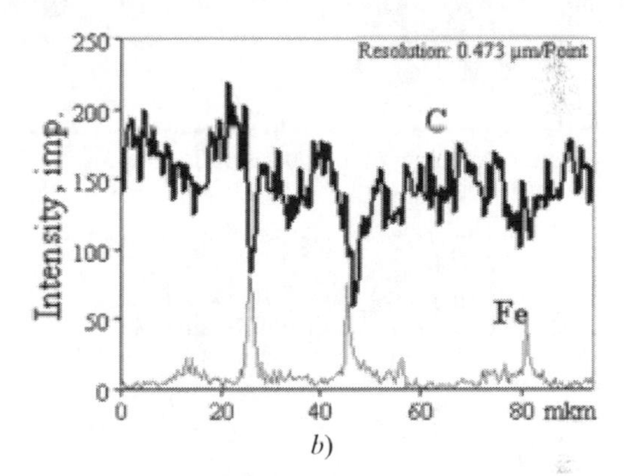

b)

Figure 4. Surface of the gray phase "basis" in the fracture of the sample C - 10% Fe (*a*), results of EDX analysis (*b*) during scanning of line image (on fig. *a*, arrow 1).

Complex diffraction profile of nanocomposite C+10%Fe at angles interval 2θ≈19…31° contains several superimposed X-ray lines with broad, diffuse peaks - "amorphous halo" (Fig. 5 lines 2, 3); peak of line 1 corresponds to the nanocrystalline state; thus, the structure of carbon nanocomposite C+10%Fe is X-ray amorphous nanocrystalline.

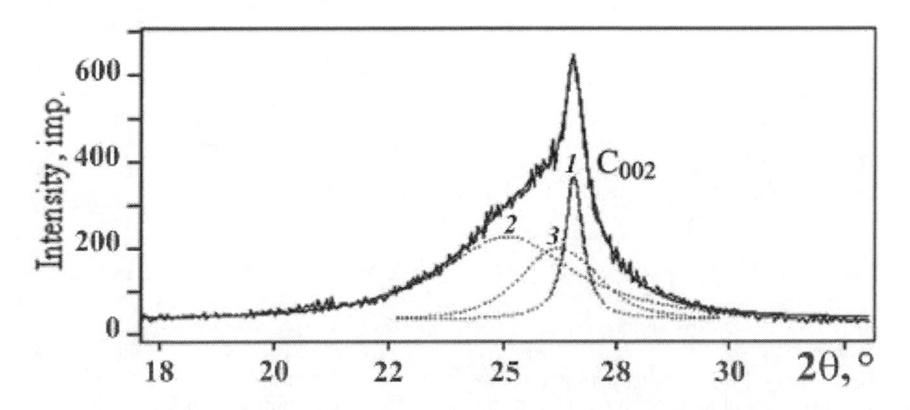

Figure 5. Diffractogram of nanocomposite C – 10 % Fe with decomposition profile singlets at 1, 2, 3.

When through the transillumination in the TEM in the nanocomposite observed the field of nanocrystalline structure and unstructured areas (Fig. 6, *A*); diffraction pattern with structureless area (Fig. 6, b) is a two diffuse Laue rings corresponding to first and second fields of carbon, indicating the complete disordering, i.e. amorphous state.

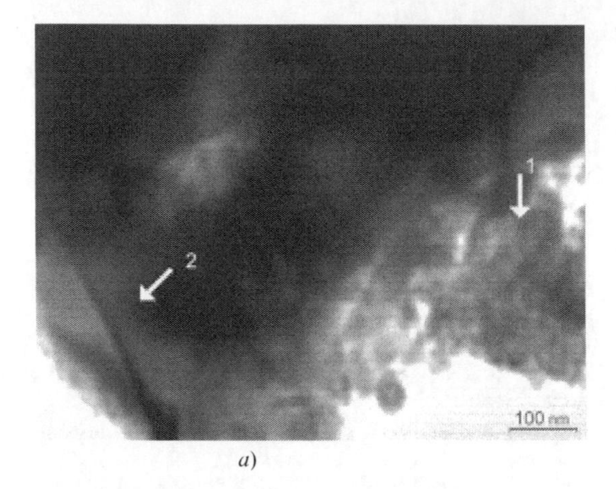

a)

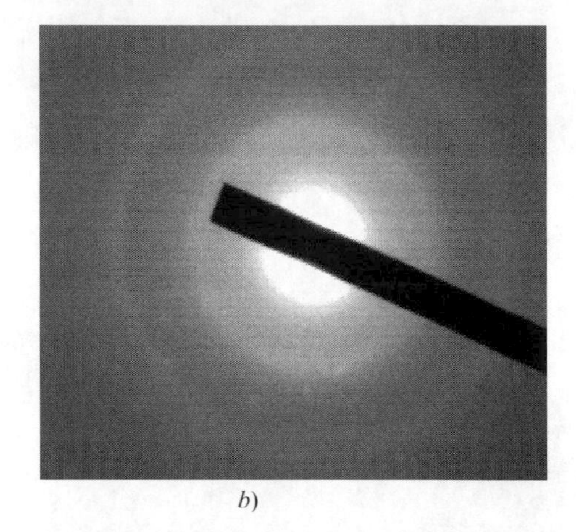

b)

a –fine structure, arrow 1 - nanocrystalline area, arrow 2 – amorphous; b – diffraction pattern from the amorphous area (fig. A, arrow 2)

Figure 6. Results of research in TEM of nanocomposite C_{efs}–10 wt.% Fe.

The results of Raman spectroscopy (Fig. 7) confirm the TEM data of disordering - amorphization nanocomposite structure. Position of D-line v_D=1350 sm^{-1} and high intensity ratio I_D/I_G typical for amorphous carbon [4].

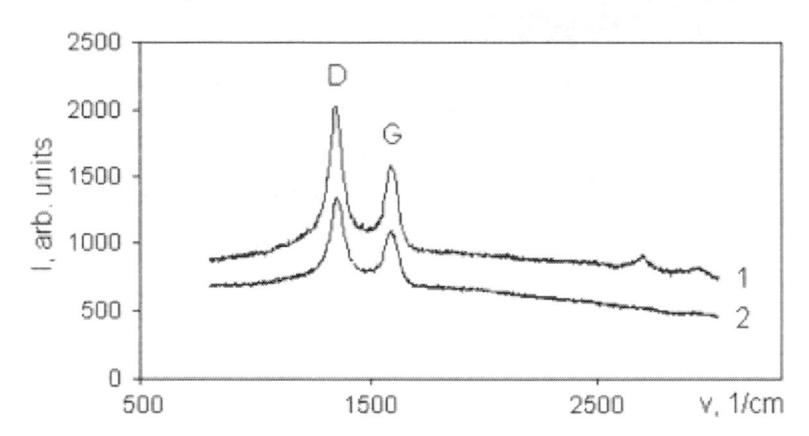

1 –highly rigid phase with "globular relief", 2 – gray phase "basis"

Figure 7. Raman spectra of the nanocomposite C_{efs}-10 wt. % Fe, typical for amorphous carbon.

4 SUMMARY

Thus obtained carbon-based nanocomposite Cefs-10% Fe, is amorphous vitreous-like carbon, comprising ultrahard particles. Moreover, its hardness is isotropic, i.e. equally high in all directions. The results are of interest in connection with the communication of scientists from the Geophysical Laboratory of the Carnegie (US) about the new super-hard carbon allotropes - amorphous diamond [5], which has a potential advantage over conventional diamond - its hardness is also isotropic, that is the same in all directions, Unlike a conventional diamond for which its value depends on the direction in the crystal lattice.

REFERENCES

[1] Chernogorova O.P., Drozdova E.I., Blinov V.M., Bulenkov N.A. // Russian Nanotechnologies, 3 (5–6) 150 (2008).

[2] Okatova G.P., Svidunovich N.A., Kuis D.V., Urbanovich V.S., Oychenko V.M., Korzhenevsky A.P. // Izvestiya VUZ, Ser. "Chemistry and chemical technology", 53 (10) 90 (2010).

[3] Urbanovich V.S., Kuis D.V., Okatova G.P., Svidunovich N.A., Oychenko V.M., Baran, L.V. // Carbon: Fundamental Problems of science, material science, Technology: proceedings of the 8th Intern. Conf. Mosk. reg., Troitsk. Sept. 25–28, 2012. – Troitsk: Trovant. – 2012. – P. 500.

[4] Compan M.E. Raman scattering in self-assembled nanoporous carbon based on silicon carbide / M.E. Compan, D.S. Krylov, V.V. Sokolov // Semiconductor Physics and Technology. – 2011. – V. 45. – Iss. 3. – P. 316–321.

[5] Carnegie Institution for Science. News. New form of superhard carbon observed. [Electronic resource]. USA, Washington. 11.10.2011 (http://www.carnegiescience.edu).

Computing, Control, Information and Education Engineering – Liu, Sung & Yao (eds)
© 2015 Taylor & Francis Group, London, ISBN: 978-1-138-02800-5

Mini-language for efficient and comprehensive definition of time intervals with the possibility of recurrence

Petr Voborník
University of Hradec Králové, Hradec Králové, Czech Republic

ABSTRACT: The article presents the new mini-language for efficient and comprehensive definition of time intervals with the possibility of recurrence. Through this code can be defined arbitrarily distributed periodic time intervals that covers all possible exceptions. The language can work with values related to the week parity, dates derived from the irregular end of the month, etc. Implementation of the presented mini-language could extend the capabilities of the software working with time intervals.

KEYWORDS: mini-language; time; date; repeat; periods; interval; XML.

1 INTRODUCTION

The project Universal testing environment, introduced in [1], allows you to limit the availability of tests to predefined time intervals. This definition had to take into account for all possible cases that may arise.

A time interval can be defined by only two dates (and times), i.e. by values from–to, in the simplest case. For investigational repeating tests are usually needed to limit their availability not only for a fixed period, but also on the periodic time, e.g. weekly or fortnightly intervals (for school exercises). However, any other repetition period is not excluded. Also need to be able to exclude from the defined time and certain periods, e.g. holidays, vacation, and other long-term forward unplanned outages. Certain periods, e.g. holidays, vacation, and other long-term forward unplanned outages also need to be able to exclude from the defined time.

The new mini-language, the structure of which will be described here, was created for the purpose the definition which enables to cover all these possible cases.

2 SPECIFICATION OF THE MINI-LANGUAGE

The simple XML-based language (as e.g. in [2]), that combines with the attribute value text mini-languages for writing time ranges, was created for easy and clear writing conditions for limiting the time available. It specifies four elements: **add**, **remove**, **only** and **repeat**.

The <add> element adds a certain time range to the list of allowed period, conversely the <remove> element adds forbidden intervals during the allowed period. The other two elements may only appear as a sub element of the previous two. The `<only>` element limits the validity of the parent element (<add> or <remove>) for more specified periods, the `<repeat>` element allows setting the periodic repetition for the parent's time interval.

```
<add date="2014-09-01/2015-06-30">
<!-- Entire school year 2014/15 -->
  <only week="o1" time="08:35~20m" />         <!--
only odd Mondays 08:35-08:55 -->
  <only week="e2" time="09:40~20m" />         <!--
or even Tuesdays 09:40-10:00 -->
  <remove date="2014-12-22/2015-01-02" />  <!--
but without at Christmas holidays. -->
</add>
```

Code 1 Example of a combination of elements <add>, <only> and <rcmove>

All of these elements except `<repeat>` may contain the following three attributes: **date**, **time** and **week**. Attribute **date** can contain several types of values:

- specific date (e.g. "2014-12-18") – valid only for this day
- a date range
- determined by two dates (e.g. "2014-09-01/2015-06-30") – if the time is not specified, whole days start and end date are included (e.g. "2014-09-01 00:00/2015-06-30 23:59:59")
- determined by two dates with the time (e.g. "2014-12-18 13:05/2014-12-18 14:50")
- determined by the start date with time and by the duration (e.g. "2014-12-18 13:05~45m")
- open range determined only by the start or by the end (e.g. "2014-09-01/" or "/2015-06-30")

- enumeration of dates (e.g. "2014-10-28;2014-11-17;2015-05-01;2015-05-08")
- combination of enumeration and ranges (e.g. "2015-01-06;2015-01-21 13:05~45m")
- only by the duration (e.g. "~1h") – it is necessary to add the sub element `<repeat>`, which determines the start of this interval (see Code 5)

```
<add date="2015-02-01~1M" />
```

Code 2 Interval which includes the whole of February 2015 (from 2/1/2015 00:00:00 to 3/1/2015 00:00:00)

In these entries, the date is in format "yyyy-MM-dd", the date with time is in format "yyyy-MM-dd hh:mm:ss" and seconds is not necessary to state. A semicolon (";") separates the entries in the enumeration, a forward slash ("/") is used as a separator for ranges with the start time and the end time and a tilde ("~") in the range defined by the time of start and by the duration. This is attributed to the time of start exactly at the seconds (e.g. if "10:00~1h" indicates the interval X, then it is true that 10:00:00 ≤ X < 11:00:00). Duration of interval is specified by the value (decimal separator is a dot) and by the units that are: **s** – seconds, **m** – minutes, **h** – hours, **d** – days, **w** – weeks, **M** – months (e.g. see Code 2) and **y** – years.

The procedure is similar for **time** attribute, but values are concerned about the time only. Time is written in the format "hh:mm:ss", where the hours are in 24 hour format (00-23) and seconds is not required. The time limit is determined only by ranges (e.g. "13:00/14:00" or "13:00~1h") or as a list of ranges (e.g. "13:00/14:00;16:00~1h").

If attributes **date** and **time** are in the element at the same time, both definitions must be valid, to the range has been allowed (add) or rejected (remove). This can limit the range at only certain hours of every day (e.g. see Code 3).

```
<add date="2015-01-05~1w"
     time="13:00~1h;16:00~1h" />
```

Code 3 Definition of two hourly intervals, each day for one week

Attribute **week** allows define an enumeration or range of days of the week in numeric format (i.e. 1 = Monday, 7 = Sunday) and thus further define the validity period in relation to the weekly cycle (e.g., "1;3;5" or "1/5"). If it is necessary to distinguish the odd and even weeks, it is possible to write a character for determining parity of the week before the number of the day of the week, namely "**o**" for odd and "**e**" for even (e.g. "o1;e2;3" – Monday of odd weeks, Tuesday of even weeks and every Wednesday, also see Code 4). The week number of the year is determined by ISO 8601, i.e. the first week of the year is the one whose Thursday is in that year [3].

```
<add date="2015-01-01~1y" week="1/3;o4/5;e6" />
```

Code 4 All year 2015, Monday to Wednesday, Thursday and Friday in odd weeks and Saturday in even weeks

3 REPETITION

The <repeat> element allows set to repeat for allowed or restricted periods. Its **type** attribute specifies the type of repetition derived from the standard units of the time (seconds, minutes, hours, days, weeks, months and years). All types of repetition have common attributes the **start** (sets the start date or even the time of repetition), the **period** (specifies the numeric value of the repetition period, e.g. 7 for the day type means repeating every 7 days) and the **end**. The end can be determined in several ways: by the number of repetitions (**count** attribute containing an integer, e.g. see Code 5) or by the end date (attribute **end**). If there are no any possible attributes determining the end date, it is an endless repetition.

```
<add date="~5m">
  <repeat type="hour" period="12"
start="2015-01-05 08:00" count="14" />
</add>
```

Code 5 Five minutes from 8 A.M. and P.M. (20:00) from the date 1/5/2015 with fourteen repetitions (i.e. 7 days)

Types of repeating **week**, **month** and **year** also allows more specific settings. Days of the week are for weekly recurrence, its enumeration or range can be specified in **weekDay** attribute (e.g. see Code 6). Odd and even weeks are not distinguished against of definition of weekdays in the main element (**add**, **remove** or **only**), because there is the concrete start date and the repetition period, and days are therefore always only as numbers (e.g. "1;3/5").

```
<add time="18:00~2h">
  <repeat type="week" weekDay="1;4" period="4"
start="2015-01-01" count="10" />
</add>
```

Code 6 Every fourth week on Monday and Thursday from 18:00 for 2 hours for 10 weeks starting at 1/1/2015

Monthly repetition allows to relate it to a specific day or week of the month. For a day or days is sufficient to indicate their list or range (e.g. "1/3;12/15") to the **day** attribute. In the enumeration can be also used the character "L" (last), which meaning the last day of the month. The number that indicates how many days before the end of the month is it, can be written after the "L" character (e.g. "L1" is the penultimate day of the month) if is needed to count from the month end. Day of the week and the number of the week in the month can be determined (e.g. the second Sunday of the month). The combination of these two values is entered in the form of an enumeration to the attribute **weekDay** as two-digit numbers so that the first digit indicates the number of the week in the month and the second digit is day of the week (e.g. the third Friday = "35"). The first number can be replaced by the letter "L" which means the last day of the week in the month (e.g. the last Wednesday = "L3", also see Code 7).

```
<add date="~1d">
  <repeat type="month" weekDay="L7" period="2"
start="2015-01-01" end="2015-05-31" />
</add>
```

Code 7 All day every last Sunday of every second month between January and May 2015

Annual repetition can be applied to days and to months. Days can be defined again by the enumeration or ranges in the attribute **day**, but these numbers mean the day of the year, i.e. 1-365 (in leap year then 1-366). The day can also be determined by the date (or by enumeration or by data ranges) in the attribute **date** in the format "MM-dd" (e.g. "15-24" or "07-01/08-31"). Annual repetition in certain months and its days has a similar format as in the case of monthly repetition in the attribute **weekDay**. These days are defined in the attribute **monthDay** and they are complemented by numbers of months and separated by commas (e.g. "LD,3;27,5" – the last day of March and the second Sunday in May; also see Code 8).

```
<add date="~1d">
  <repeat type="year" monthDay="LD,2"
period="4" start="0000-01-01" />
</add>
```

Code 8 All last days of every February in a leap year

4 COLLATION OF RULES

Individual elements with the rules are written into the root element, either below the other (at the same level of the XML hierarchy) or nested in themselves. The rules at the same level are processed sequentially and the later rule (element) has a higher force ("overrides") than previous rules. E.g. this can allow certain range, disable part of it and from this disabled part again allow some subsection (e.g. see Fig. 1 and Code 9).

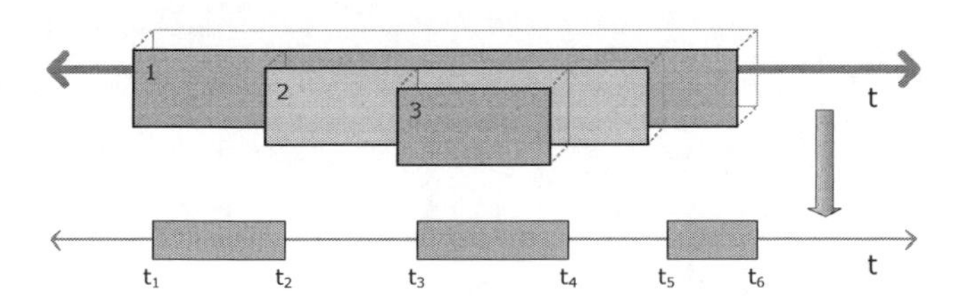

Figure 1. Illustrative scheme of folding of rules (see Code 9) [1].

```
<add    date="t1/t6" />
<remove date="t2/t5" />
<add    date="t3/t4" />
```

Code 9 Code for Fig. 1

Nested elements supplement and clarify the parent element at the same level of validity (e.g. see Code 10). Nesting depth has no limits. More of sub elements may be at the same level, then the collating of validity of rules is the same as in the root element.

```
<add date="t1/t6"> <remove date="t2/t3" />
<remove date="t4/t5" /> </add>
```

Code 10 Code with an equivalent result as Code 9, using two-level elements

5 CONCLUSION

Created a mini-language described in this article allows very efficient and understandable definitions of time intervals including extensive options of recurrence. This language, readable for a man [4], can be applicable not only in the project for which it was created (determine the availability of tests), but it can be useful in many other cases (e.g. [5]). For example, it can limit the user's group membership, telecast or its record, repeating of calendar events, etc.

A caching method can be used for fast database processing. It evaluates the XML code when changing and it stores current status (enabled / disabled) and the following earliest timestamp of the change of this state of a special cache fields. Parsing of the code will be necessary to only one time on the first use after the change of the state.

REFERENCES

[1] Petr Voborník: *Universal Testing Environment*, Ph.D. thesis, University of Hradec Králové, (2012), online <http://download.petrvobornik.cz/docs/disertace.pdf> [2014-11-03].

[2] Roman Borkovec, Josef Šedivý, Štěpán Hubálovský: *Effective use of the UML-Language in Small Companies*, Applied Mechanics and Materials, Vols. 336–338, (2013), pp. 2111–2114. ISSN 1660–9336.

[3] Markus Kuhn: *A summary of the international standard date and time notation*, online, University of Cambridge, Computer Laboratory, Faculty of Computer Science and Technology, (2004), <http://www.cl.cam.ac.uk/~mgk25/iso-time.html> [2014-10-21].

[4] Věra Strnadová: *Interpersonální komunikace*, Hradec Králové, Gaudeamus (2011), ISBN 978-80-7435-157-0.

[5] Philipp M. Hund, John Dowell, Karsten Mueller: *Representation of time in digital calendars: An argument for a unified, continuous and multi-granular calendar view*, International Journal of Human-Computer Studies, Vol. 72, Issue 1, January (2014), pp. 1–11, ISSN 1071–5819.

Computing, Control, Information and Education Engineering – Liu, Sung & Yao (eds)
© 2015 Taylor & Francis Group, London, ISBN: 978-1-138-02800-5

Fabrication of polystyrene microneedles for transdermal drug delivery

W. Luangweera, S. Jiruedee, A. Pimpin, C. Rattanasumawong, T. Palaga & W. Srituravanich
Department of Mechanical Engineering, Chulalongkorn University, Pathumwan, Bangkok, Thailand

K. Patoomvasna & B. Sookyu
International School of Engineering, Chulalongkorn University, Pathumwan, Bangkok, Thailand
Department of Microbiology, Chulalongkorn University, Pathumwan, Bangkok, Thailand

ABSTRACT: Currently, microneedles are attracting a lot of attention because microneedles can deliver drugs in patients without pain as opposed to conventional hypodermic needles. Furthermore, microneedles can be safely used by untrained people. Generally, microneedles consist of tens of micrometer-sized needles with the needle length ranging from 150 μm to 2 mm. Several materials such as poly (lactic-co-glycolic acid) (PLGA), polyglycolic acid (PGA), poly(l-lactic acid) (PLA) [2], maltose [3], stainless [4, 5], titanium [6] and ceramic have been used as materials for fabricating microneedles. This work aims to develop a fabrication process of plastic microneedles using polystyrene in a toluene solution. According to the experimental result, the fabrication process of polystyrene microneedles offers several advantages such as low cost, simple fabrication, and short processing time, whereas the properties of the fabricated polystyrene microneedles are suitable for practical applications. Furthermore, this proposed method can be used to fabricate replicas of microneedle master.

KEYWORDS: microneedles, polystyrene, replica.

1 INTRODUCTION

Recently, microneedles have emerged as an alternative to hypodermic needles. Microneedles combine the advantages from both hypodermic needles and transdermal patches since microneedles can deliver big molecule drugs through the epidermis without causing pain from the injection. Microneedles can be divided into four types, [1] solid microneedles, coated microneedles, dissolving microneedles, and hollow microneedles. Solid microneedles are designed to open holes on the skin, so drugs can be delivered into the body through the holes, while coated microneedles can release drugs that are coated on the microneedles after injection. Dissolving microneedles are made of dissolvable materials where the microneedles and the drugs are dissolved after injection. Hollow microneedles can release liquid drugs via the holes in them. Till now, several materials were used to produce microneedles, for example, poly (lactic-co-glycolic acid) (PLGA), polyglycolic acid (PGA), poly (l-lactic acid) (PLA) [2], maltose [3], stainless [4, 5], titanium [6] and ceramic [7]. However, dissolvable polymers such as PLGA, PGA and PLA are very expensive. Maltose microneedles are brittle and easy to dissolve in humid environments [3]. Moreover, the fabrication process for stainless, titanium and ceramic microneedles are complicated and expensive. Therefore, there is a strong need to develop low-cost micro needles that are easy to fabricate within short processing time. This work aims to develop a fabrication process of plastic microneedles using polystyrene solution which could offer several advantages such as low cost, simple fabrication, and short processing time. Furthermore, the fabricated polystyrene microneedles can be modified to produce replicas of a microneedle master.

2 FABRICATION OF METALLIC MICRONEEDLE MASTER

Fabrication of polystyrene microneedles started with microneedle master fabrication followed by PDMS mold fabrication and polystyrene microneedle casting. The fabrication of aluminum (Al) microneedle master was described in reference [8]. First, an Al workpiece was machined in a CNC machine to get a 7x7 microneedle array template with a microneedle height of 1.8 mm, a base width of 0.4 mm and a tip width of 0.2 mm (Fig. 1(a)). After etching the microneedle template in Al etchant Type A for 4 hrs, the Al microneedle master was obtained. The final base width and height were approximately 0.3 mm and 1.5 mm, respectively as shown in Fig. 1(b).

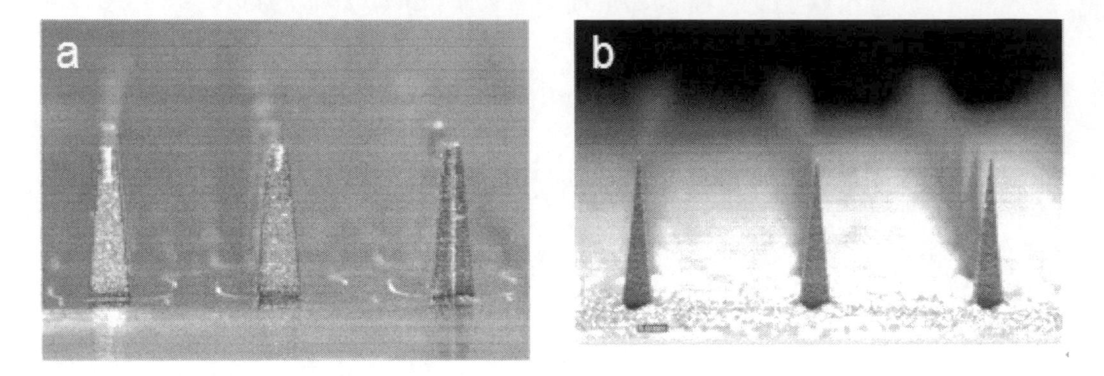

Figure 1. Al template was machined using a CNC (a), and followed by wet etching in Al Etchant Type A for 4 hrs to obtain microneedle master (b).

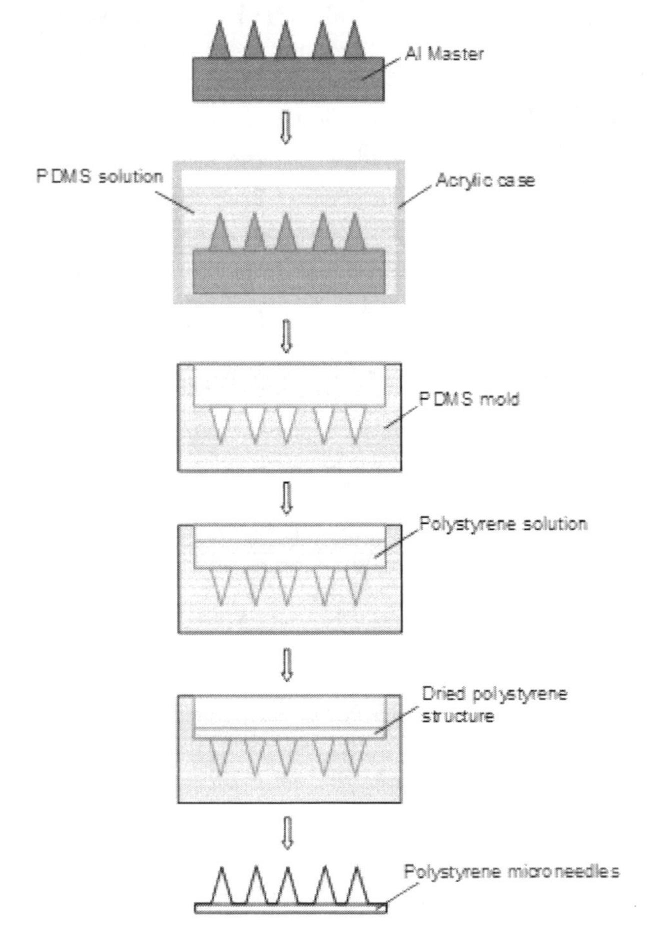

Figure 2. Schematic diagram of polystyrene microneedle fabrication: it starts with the fabrication of Al microneedle master and subsequent PDMS mold casting. Then, a solution of polystyrene in toluene is poured into PDMS mold and left until it hardens. Finally, polystyrene microneedles are removed from PDMS mold.

3 FABRICATION OF PDMS MOLD

In this work, the mold of microneedles was made of polydimethylsiloxane (PDMS) due to its good heat and temperature resistances, high transparency and flexibility. The fabrication process of PDMS mold is schematically shown in Fig.2. First, an acrylic case was prepared by cutting acrylic pieces using a laser cutting machine and glued with dichloromethane (CH_2Cl_2). After that silicone elastomer base and silicone elastomer curing agent (Sylgard 184, Dow Corning) were mixed by a ratio of 10:1 by weight and poured into the acrylic case. Then, the prepared sample was subjected to vacuum to remove air bubbles and left on a hotplate at 70°C for 1 hour. Finally, the PDMS mold was removed from the acrylic case.

4 FABRICATION OF POLYSTYRENE MICRONEEDLES

The fabrication process of polystyrene microneedle is as follows. First, PDMS mold was subjected to oxygen plasma treatment in a plasma cleaner to modify the surface property of PDMS from hydrophobic to hydrophilic. Secondly, 5% polystyrene in toluene was poured into PDMS mold and left at room temperature for 4 hours to let the sample dry. After it solidifies, polystyrene microneedles were removed from the PDMS mold. According to the fabrication results, polystyrene microneedles were easily fabricated in a short processing time. It should be noted that the processing time might be shortened, but the minimum processing time was not determined. It was found that the fabricated polystyrene microneedles replicate the PDMS mold very well as shown in Fig. 3. The fabricated microneedles have good mechanical properties while the microneedle patch provides sufficient flexibility. Such properties are ideal for practical applications. These microneedles could be used as solid microneedles to open holes for pre-treatment or could be coated with drugs or vaccines and used as coated microneedles to deliver drugs or vaccines.

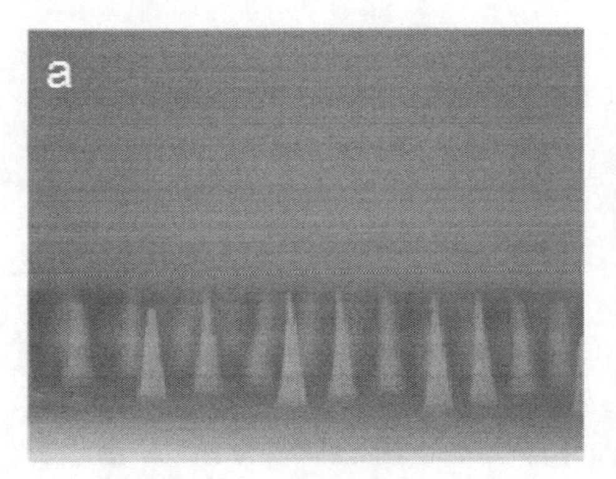

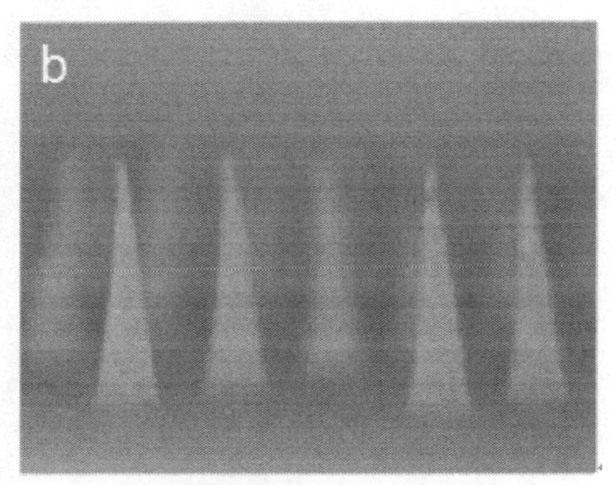

Figure 3. Polystyrene microneedls were made of 5% polystyrene in toluene (a), and a zoom-in image (b).

5 FABRICATION OF MICRONEEDLE MASTER REPLICAS

The fabrication process of conventional microneedle masters is complicated, expensive and time consuming. Thus, we modified the fabricated polystyrene microneedles to produce replicas of the microneedle master. The fabricated polystyrene microneedle patch was attached on top of an epoxy base made of epoxy glue. It should be noted that the fabricated replicas could reproduce the exact shape and dimension of the master without any distortion while other methods using polymerization of polymers suffer from distortion or shrinkage due to residual stress.

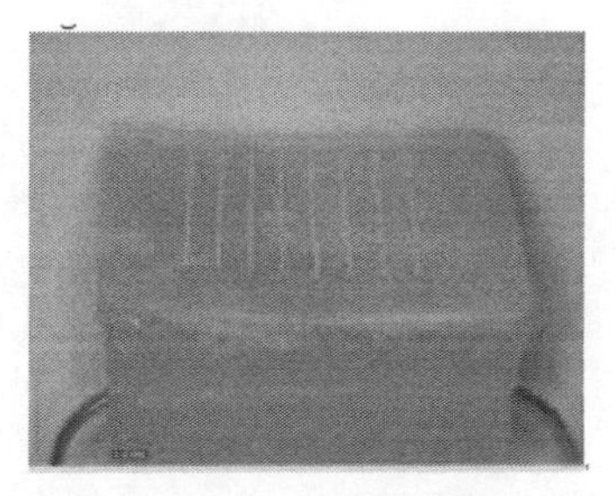

Figure 4. A replica of microneedle master was fabricated from polystyrene microneedle attached on an epoxy base.

6 SUMMARY

In this work, novel polystyrene microneedles were fabricated using polystyrene in a toluene solution. These polystyrene microneedles could be used as a pre-treatment microneedles or as coated microneedles. Such microneedles offer several advantages, including good mechanical properties, high flexibility, low production cost, simple fabrication and short processing time. Furthermore, this method can be used to fabricate replicas of microneedle master without any distortion.

ACKNOWLEDGEMENT

This research has been supported by the Ratchadaphiseksomphot Endowment Fund of Chulalongkorn University (CU-57-004-HR) and Special Task Force for Activating Research (STAR) of Chulalongkorn University, through Micro-Nano Fabrication Technology Research Group (GSTAR 56-005-21-002).

REFERENCES

[1] Y.C. Kim, J. H. Park, M.R. Prausnitz, "Microneedles for drug and vaccine delivery," Advanced drug delivery reviews, 64 (2012) 1547–1568.

[2] J.H. Park, M.G. Allen, M.R. Prausnitz, "Biodegradable polymer microneedles: fabrication, mechanics and transdermal drug delivery," J. Control. Release 104 (2005) 51–66.

[3] G.H. Li, A. Badkar, S. Nema, C.S. Kolli, A.K. Banga, "In vitro transdermal delivery of therapeutic antibodies using maltose microneedles," Int. J. Pharm. 368 (2009) 109–115.

[4] H.S. Gill, M.R. Prausnitz, "Coated microneedles for transdermal delivery," J. Control. Release 117 (2007) 227–237.

[5] W. Martanto, S.P. Davis, N.R. Holiday, J. Wang, H.S. Gill, M.R. Prausnitz, "Transdermal delivery of insulin using microneedles in vivo," Pharm. Res. 21 (2004) 947–952.

[6] J.A. Matriano, M. Cormier, J. Johnson, W.A. Young, M. Buttery, K. Nyam, P.E. Daddona, "Macroflux (R) microprojection array patch technology: a new and efficient approach for intracutaneous immunization," Pharm. Res. 19 (2002) 63–70.

[7] S. Bystrova, R. Luttge, "Micromolding for ceramic microneedle arrays," Microelectron. Eng. 88 (2011) 1681–1684.

[8] K. Tsioris, W.K. Raja, E.M. Pritchard, B. Panilaitis, D.L. Kaplan and F.G. Omenetto, "Fabrication of Silk Microneedles for Controlled-Release Drug Delivery," Adv. Funct. Mater. 22 (2012) 330–335.

Computing, Control, Information and Education Engineering – Liu, Sung & Yao (eds)
 ISBN: 978-1-138-02800-5

The research on the peer-to-peer trust model under the internet financial environment

Qi Wang & Yao Liu
Beijing University of Posts and Telecommunications, Haidian District, Beijing, China

ABSTRACT: P2P (peer to peer) internet finance is developing very rapidly, but fraud and bad faith phenomenon appeared at the same time. It's necessary to contribute the P2P internet financial trust environment. This paper introduced the background of IOF (Internet of Finance) and development of P2P lending, compared some typical trust models and analyzed their own strength and weakness, explained some relevant definitions and connotations, at last proposed a trust model based on lender's own experience, others' recommendation and environmental factor.

KEYWORDS: internet of finance; P2P lending; trust model.

1 INTRODUCTION

IOF (internet of finance) is a new financial model that combines traditional finance and rising internet [1,2]. As one of the basic forms of IOF, P2P internet lending refers to personal lending on network platform [3]. It's developing rapidly around the world, especially in China [4,5].

Among the business community, some P2P platforms have constructed credit-rating system to ensure users' property security. For example, to assist lenders in judging whether to trust a borrower, Paipai Lending provides real-name, telephone-number, personal-photo, driving-license and resident-certification authentication to lower the risks. In academic circles, there are many reputation-based trust models with different mathematical methods. EigenTrust [6] was proposed by Kamvar (2003), PeerTrust [7] was proposed by Li Xiong (2004), they are the overall trust model in P2P system. Beth (1994)proposed an experienced-and-probability-statistics-based trust model [8], which introduced the concept of experience to describe and evaluate credit rating. There're also some calculating methods that are based on evidence space and opinion space [9] (Jsang A, 1997), reinforcement learning method in machine learning that was put forward by Claudiu (2005) and Bayesian-based trust model [10] (Yao Wang, 2004). They all provide the way to calculate reputation scoring and are decentralized but couldn't adapt completely to the changing environment.

This paper introduced some of classical and new trust model and compared their strengths and weaknesses, put forward some advice and proposed a trust model based on experience, recommendation and environment. In the end, we illustrated our model's advantages and practicability.

2 BACKGROUND

Trust management is to adopt a unified approach to describe and explain security policies, security credentials and trust relationships used for directly authorizing critical safety operation [11]. In the open, accessible public service system, the lack of centralized management system provides a breeding ground for malicious nodes' attacks[12]. This problem has drawn widespread attention of experts and scholars. These trust models can be classified into 2 types according to their arith-method. One is linear literation, the other is a probability distribution.

2.1 *Linear-literation models*

This kind of models has two representatives, EigenTrust and PeerTrust. EigenTrust is a kind of global reputation model. In the model, global reputation is based on local reputation. The followings are their formulas.

$$T_k = \sum C_{ij} * T_{jk} \quad (1)$$

$$C_{ij} = (Sat_{ij} - UnSat_{ij}) / \sum(Sat_{ij} - UnSat_{ij}). \quad (2)$$

T stands for the global trust, C is local trust. Sat is the times of satisfying trades and UnSat is unsatisfying times in the historical record.

EigenTrust is better in expansibility and shrinkage and is convergent to the whole situation. But there exists the problem of feasible solutions. It could just protect the node from simple attack.

PeerTrust is a peer-to-peer trust model based on a reputation that was given by Li Xiong and Ling Liu. It contains five parameters in a coherent scheme. The algorithm is:

$$T(u) = \alpha * \sum I(u)S(u,i)*Cr(p(u,i))*TF(u,i)+\beta *CF(u) \quad (3)$$

The model is composed of two parts. One is a weighted average of the amount of transaction a peer receives for each transaction. The other is the trust value adjusted to the different communities. PeerTrust introduces TF(u,i) as the parameter to evaluate trust value, which reflects the actual contribution of peer. However, there is the problem of convergence in the model and the algorithm is complex. These two disadvantages brings difficulty to the calculation of the trust value.

2.2 *Probability-distribution models*

This kind of models introduces some concepts to describe and measure the trust relationship. They all use the method of calculating probability. Two representatives are Beth's and Jsang's models.

Beth proposed a trust model based on experience and probability-statistics. He suggested that trust classified into two kinds, direct trust and recommendation-trust. And he considered the differences among recommendation from different ways. His model is relatively considerate, but his definition of direct trust is extremely rigid, which couldn't describe the trust relationship in real deals.

Jsang introduced evidence space and opinion space to measure the trust relationship. And provided a suit of subjective logic operators for the trust's calculating. It's convenient and efficient to some degree. But it doesn't distinguish direct trust and recommendation and couldn't eliminate the influence from bad nodes.

Considering the advantages and disadvantages of the models, we propose a relatively comprehensive method to estimate a node's reputation and decide whether to trust the node in one trade.

3 THE TRUST MODEL BASED ON EXPERIENCE, RECOMMENDATION AND ENVIRONMENT

The comprehensive trust from this trust model is composed of three parts, that is the direct trust based on lenders' own experience, indirect trust coming from other nodes' recommendation and environmental factor changing with different conditions. Let Tij denote the comprehensive trust, its formula is:

$$T_{ij}=\alpha *DT_{ij}+\beta *\sum \rho *DT_{kj}+\gamma *EF, \alpha+\beta+\gamma=1. \quad (4)$$

It means that the comprehensive trust form i to j = the direct trust from i to j + the indirect trust form i to j + the environmental factor, α, β, γ are used to balance the relationship in proportion among the three.

Next, I'll introduce and explain how to attain the values.

3.1 *The direct trust*

There are always new completion of trades and new nodes joining the credit rating system, so it's necessary to update the trust of nodes constantly. This determines the dynamics and the real time of the trust model. The direct trust means one node calculates the trust value of another according to the two's history of direct trades. Let it denote DTij. When i is going to trade with j, firstly i will search for its own direct transaction record with j. If i didn't trade with j before, the direct trust will be a fixed number. If i have transaction record with j, the historical direct trust value will be a part of the comprehensive trust. The way to acquire historical direct trust value is as follows:

$$DT^{(k)}_{ij} = \begin{cases} d^{k}_{ij}*TF^{k}, & k=1. \\ \varphi *DT_{ij}^{(k-1)} + (1-\varphi)*TF^{k}*d^{k}_{ij}, & k>1. \end{cases} \quad (5)$$

In this formula, k denotes the number of trades. TF denotes the context information of the service. φ denotes the coefficient decaying with time. k is a positive integer. $0 \le \varphi \le 1$.

The DT (k)ij in this formula denotes the historical direct trust value of i to j resulting from the k services in the past.dkij is the direct trust value of i to j in the k-th trade, that is the credit rating of i to j in the k-th service, divided into six levels (0,0.2,0.4,0.6,0.8,1). The larger the number is,the higher the integrity degree is. TFk denotes the context information of the relating service in the k-th service, including the principal amount, interest rate and duration. The introduction of TF not only improves the practicality and authenticity of credit, but also curb the malicious deception activities effectively. φ is the time-decaying coefficient of the direct trust value ($0 \le \varphi \le 1$).

When the k-th trade between i and j is over, i will calculate dkij, and the system will multiply φ with the direct trust value resulting from k-1 services before in the storage, and multiply 1-φ with the direct trust and context information this time. In the end, i will get the direct trust value of all the k services by combining the two multiplication results, and record it into the system to renew the history record.

3.2 *The indirect trust.*

After acquiring the direct trust toward j,i will make a request to other nodes in the system for their direct trust of j, and work out the proportion of every recommendation in accordance with I's familiarity and trust degree of these nodes.

To be specific, the indirect trust value of i to j is acquired by multiplying the trust degree of k's recommendation with a k's direct trust value toward j, usually k has traded with j before. The trust degree of recommendation reflects the degree that i trust k, its formula is:

$$\sum \rho * DT_{kj} , \tag{6}$$

ρ is the coefficient of indirect trust. It stands for the proportion of k's recommendation in all the recommendations as for i. Different familiarity means different formulas. If i is strange with k, in other words, i didn't trade with k before, the formula is $\rho=1/n$. Or else, the formula is:

$$\rho = DT_{ik} / \sum_k DT_{ik}. \tag{7}$$

This formula means the value of ρ will be acquired by dividing the direct trust of i to k with the sum of direct trust of i to all its familiar nodes. Because the nodes that make recommendation have their own trust degree, the introduction of ρ distinguishes these recommendations well. The DTkj in the third formula stands for the direct trust of k to j. Multiply every trust degree of recommendation with direct trust value given by the corresponding node and sum up the numbers, we can finally acquire the indirect trust value. The indirect trust value makes the node combine its own experience with others'. Its introduction makes the trust degree more objective and reduce the occasionality.

3.3 *The environmental factor*

Let EF denote the environmental factor. EF stands for the influence of environment on the credit rating. The value of it changes with the trade's condition. It's determined by many environmental factors during the transaction, such as the market conditions, inflation rates, operating level of industry. Usually speaking, if the environment is stable and the market is safe, the value of EF will be large.

By means of the formulas above, we can attain the direct trust, indirect trust and environmental factor. After summing them up, we can get the value of comprehensive trust. According to the size of the value, we could have knowledge of the node's credit and decide whether to trade with it. When listing facts use either the style tag List signs or the style tag List numbers.

4 CONCLUSION

This paper introduces the development of IOF and discuss the progress of research for the trust environment at the background of P2P network lending. After analyzing other models, I propose a new trust model based on experience, recommendation and the environment, according to my own understanding and thoughts. Its advantages could be concluded as follows:

1 Introduce the context information of sevice, which increase the pertinence.
2 The time coefficient keeps balance between historical trust and new trust, which contributes to a real-time model.
3 The coefficient of indirect trust distinguishes familiar nodes and strange nodes. It makes their recommendation have a different impact.
4 The indirect trust and environmental factor to improve the objectivity and reliability of the model.

P2P internet of finance is developing rapidly, the prospect of its market is very broad. The research of its trust environment will be deeper and deeper. Next, I'll further improve this model, perfect the computing methods and test its practicability through experiments.

ACKNOWLEDGEMENT

It is a project supported by Beijing University of Posts and Telecommunications. The corresponding author is Qi Wang,nino940607@163.com.

REFERENCES

Bruce Mitchel Kogut: The global Internet Economy(MIT Press,2003).

Xie Qinghe:The Research on the Development of IOF in China,Economic Research(2013).

Du Xiaoshan:The Development of P2P Network Lending Institutions at Home and Abroad and the Reflection on Related Policies,West China Finance(2013).

Song Pengcheng, Wu Zhiguo, Melissa Guzy:Way of Survival: the Research on Business Mode of P2P Lending Platform, New Finance(2013).
Zhang Zhiqiang.:The Situation and Advice on Development of P2P Lending in China,Huabei Finance(2013).
Kamvar SD,Schlosser M T,Garcia-Molina H:The Eigentrust AlgorithmforReputationManagementinP2PNetworks// Proceedings of the 12th International World Wide WebConfefence,Budapest,Hun49ary:ACM Press, 2003:640–651.
Xiong L,Liu L:PeerTrust:Supporting reputation-based trust in peer-to-peer communities,IEEE Transactions on Data and Knowledge Engineering,Special Issue on Peer-to-Peer Based Data Management,2004,16(7):843~857.
Beth T,Borcherding M,Klein B:Valuation of Trust in Open Network//Proceedings of the European Symposium on Research in Security(1994).
Jsang A:A model for trust in security systems//Proceedings of the 2nd Nordic Workshop on secure Computer Systems(1997).
Wang Y,VassilevaJ:Bayesian network trust model in peer-to-peer networks//Moro G,ed.Proc,of the 2nd Int'l Workshop on Agents and Peer-to-Peer Computing. Berlin:SpringerVerlag,2004:23–34.
Hu Jian-li Wu Quan-yuan Zhou Bin.:Research on Reputation Based Trust Model for P2P Environment,COMPUTER SCIENCE(2009).
Kinateder M,Baschny E,Rothermel K:Towards a generic trust model-Comparison of various trust update algorithms// Proc.of the iTrust 2005.LNCS 3477.2005:177~192.

Computing, Control, Information and Education Engineering – Liu, Sung & Yao (eds)
© 2015 Taylor & Francis Group, London, ISBN: 978-1-138-02800-5

The numerical simulation method of diesel fuel atomization field characteristics of diesel engine

Zhe Peng Liu
School of Energy and Power Engineering, Wuhan University of Technology, China

ABSTRACT: In this paper, based on previous experience and a large amount of literature, summarizes the methods and some conditions which must be satisfied by the numerical simulation of diesel fuel spray, established the calculation basis by numerical calculation method of the simulation of, and to simulate fuel atomization based on an experimental device.

KEYWORDS: summary; numerical simulation of diesel fuel atomization; calculation basis.

1 FLOW CHART

The first to carry on the theoretical analysis on the atomization field characteristics of diesel engine, and numerical simulation. In this paper, This paper uses the FLUENT software to simulate the process of atomization of fuel. The detail flow chart as shown in figure 1.

2 MODELING AND MESH

The first to carry out the theoretical analysis and numerical simulation on the atomization field characteristics of diesel engines. Based on the experiment by Yin Zijia et al., the establishment of the atomization field model corresponding to the experimental device, the experiment was completed in the

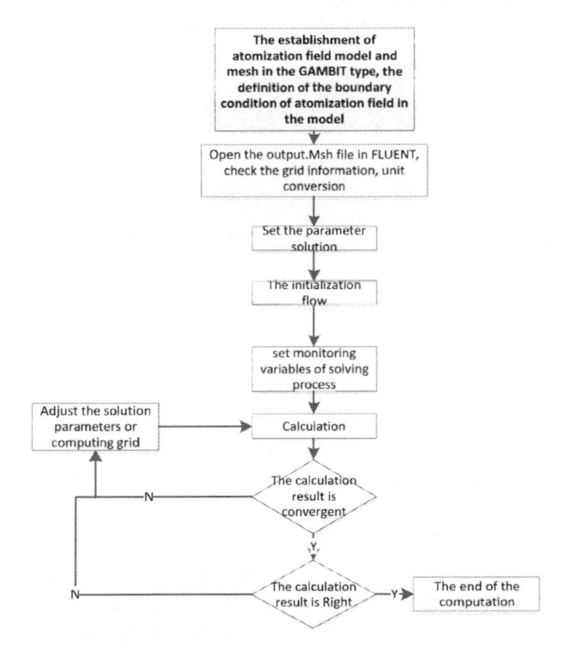

Figure 1. Diesel engine fuel spray field numerical simulation flow chart.

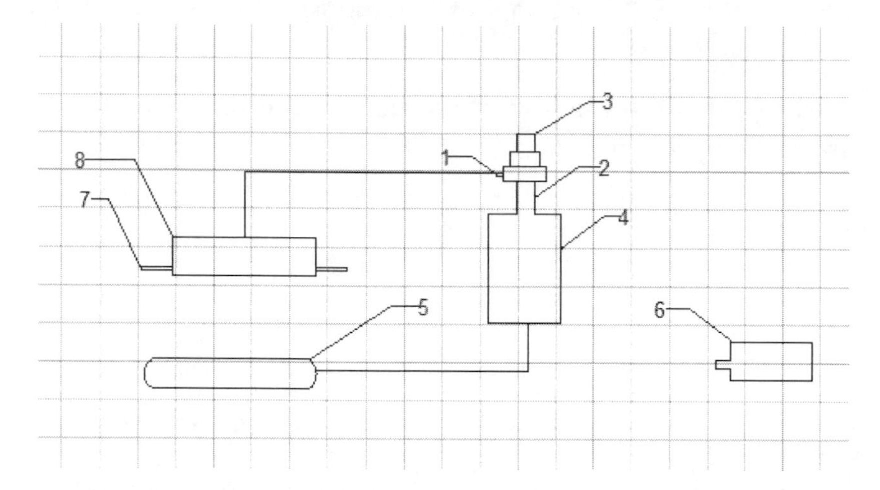

1-a pressure sensor; 2 - needle lift sensor; 3-Fuel injector; 4-pressure vessel; 5 - high pressure cylinders; 6-high speed camera; 7 - speed sensor 8-High pressure oil pump;

Figure 2. Schematic diagram of the experimental apparatus.

simulation experiment device as shown in Figure 2. Fuel injector injection to the container in the experiment, high-speed photography in the right side of the container the same time.

State of the air in the container is no trouble flow and normal temperature. The center of the top container installation of a single hole nozzle, By the plunger to supply oil, when jet inches, shooting spray process with high speed camera. Because the vessel in the experiment is the axial symmetry structure, this paper builds a two-dimensional axisymmetric model, the combustion chamber model size is 090X200 (MNI). Grid, Atomization field calculation area as shown in figure 3.

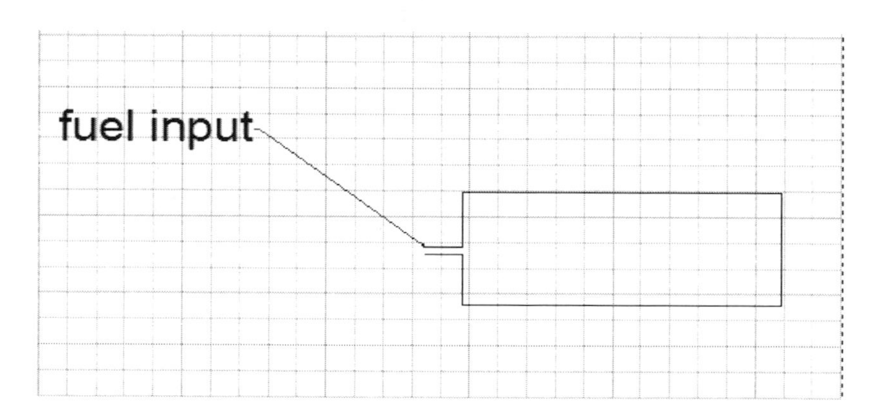

Figure 4. Calculation of regional spray field.

3 CALCULATION OF INITIAL AND BOUNDARY CONDITION

In the atomization process, the flow process of fuel is the unsteady flow, therefore, in the numerical calculation of the problem, at the same time to set the initial conditions and boundary conditions.

3.1 *Initial conditions*

The continuous phase of initial conditions: setting the air as the continuous phase, Calculation of gas phase initial conditions is the initial time of air turbulence intensity and hydraulic diameter. When the initial conditions for the calculation is the oil phase at the initial time of fuel and hydraulic diameter turbulence intensity.

3.2 *Boundary conditions*

The continuous phase boundary conditions: gas phase boundary conditions using the pressure boundary conditions on the outlet, inlet boundary as pressure inlet, outlet boundaries is set to the operating pressure on the environment, i.e.

$$p_{in} = p_0;\ p_{out} = p_e$$

Phase boundary conditions of discretization: the velocity inlet boundary of discretization using the discrete phase boundary conditions, and the outlet boundary is set to the operating pressure on the environment, i.e.

$$\upsilon_{in} = \upsilon_0;\ p_{out} = p_e$$

And for the discrete phase, using the method of surface injection to make the fuel ejected from a nozzle and oil droplets by spherical considered, the calculation of deformation is ignored during the process of oil droplets.

4 CONCLUSION

1 A brief description of the research on the importance of improving the combustion efficiency of fuel atomization of diesel engines; introduced the research method of diesel fuel spray field.
2 Describes the physical model and mathematical model of diesel fuel spray field, and the discrete and continuous phase control equation in the mathematical model.
3 Three kinds of combustion chamber model was established according to the zurna mouth different orifice diameter, and respectively on two kinds of model grid was divided; introduces respectively boundary conditions of the discrete and the continuous phase in the initial ; introduces the numerical simulation used in the process of calculation model and calculation method.

REFERENCES

[1] Trolinger J. D., Bentley H. T.,Lennert A. Application of Electronical Technique in Diesel Engine Research[J], SAE paper 740125.
[2] Reitz R. D. Mechanisms of Atomization Processes in High-Pressure Vaporizing Sprays[J], Atomization and Spray Technology, 1987(3): 309–337.
[3] Reitz R. D.,Bracco F. V. Mechanism of Atomization of a Liquid jet[J]. Physics of Fluids, 1982,25(10): 1730–1742.

Computing, Control, Information and Education Engineering – Liu, Sung & Yao (eds)
 ISBN: 978-1-138-02800-5

The experimental research of horizontal soil washing machine on power and torque characteristics

Xu Ming Ye & De Xing Liu
Department of Mechanical Engineering, Shenyang University, Shenyang, China

ABSTRACT: The selection of power supplied power and load torque calculation has a particularly important practical significance and economic benefits in the design of mechanical equipment. This paper, on the basis of brief introduction on the structure and working principle of horizontal soil washing machine, attains the curve of P-n and T-n through experimental analysis of horizontal soil washing machine on power and torque under different rotational speed, which provides a reliable basis for the selection of power and torque in the process of design on equipment.

KEYWORDS: Washing machine; The experimental research; Power; Torque.

1 INTRODUCTION

In the process of the design of the mechanical equipment, determining that the power of the motor and load torque exactly is the basis of guaranteeing the functions that stable operation and higher mechanical efficiency [1]. But the working environment for irregular movement of material mixing or centrifugal equipment design is mostly based on the ideal mechanical model to estimate the load torque and power, leading to the design results failing to accurately determine the actual power and torque.

According to the above situation, this study takes the horizontal soil washing machine designed autonomously as an experiment platform. Through the data acquisition system recording and analyzing the date, the study finally gets the actual value of load torque[2] and power in different conditions. At the same time, the experimental results provide reference for the design research of domestic soil pollution equipments.

2 THE STRUCTURE AND WORKING PRINCIPLE OF THE HORIZONTAL SOIL WASHING MACHINE

2.1 *The main structure of the horizontal soil washing machine*

The horizontal soil leaching machine has an inner and outer cylinder double layer structure. The machine is mainly composed of the rotating inner cylinder, fixed outer tube, spray system, transmission system, frame, etc. Its structure is shown in Fig. 1.

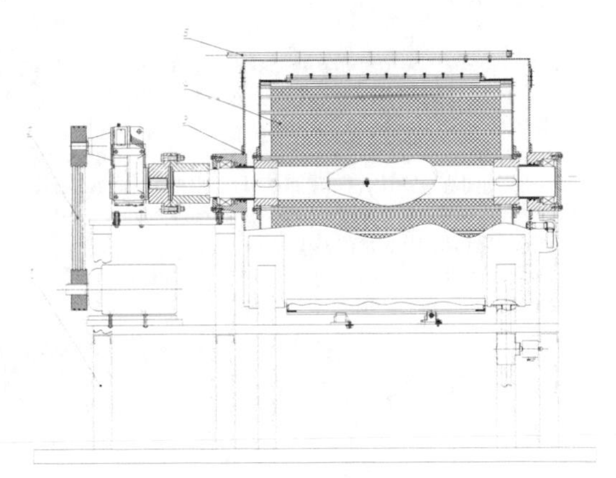

1-Frame 2- Transmission system 3-Outer cylinder 4- Inner cylinder 5- Spray system.

Figure 1. The structure of leaching machine.

The transmission system concludes: Motor connects reducer through pulleys; Reducer connects to the main shaft through the coupling; The main shaft drives the inner cylinder rotating.

Inner cylinder and outer cylinder take an industrial filter cloth as separation net, squirrel-cage inner cylinder, realizing the isolation between the inner cylinder and outer cylinder filling soil. The filter fills in fixed outer cylinder containing solvents. Due to the permeability of filter cloth, solvent can freely pass through filter cloth into the soil of the inner cylinder. Inner cylinder rotating can fully mix soil and solvent, realizing the leaching of soil. However, most of the soil particles are much larger than the pores of filter cloth, so that the soil is still retained in the filter cloth interior.

2.2 *The working principle of the horizontal soil washing machine*

Soil washing machine can achieve both washing and solid-liquid separation working process. Washing process works by circulating movement of a changing speed of the inner cylinder to achieve the full effect of washing. The solid-liquid separation process is to separate the water in the soil by high speed centrifugal way. The motor speed in the two working process is changed by adjusting the inverter input frequency to achieve changes, then changing the speed of inner cylinder through the transmission system.

3 EXPERIMENTAL CONDITIONS AND METHODS

In this paper, the experimental sample is a mixture of soil 100 kg and water 200 kg. Getting the data acquisition system by writing the data collection configuration software[3], experiment can collect the inner cylinder at different speeds within the inverter real time current, output power, output torque value, L.e., To obtain the actual torque and power washing machine required under working condition. Data acquisition route in Fig. 2.

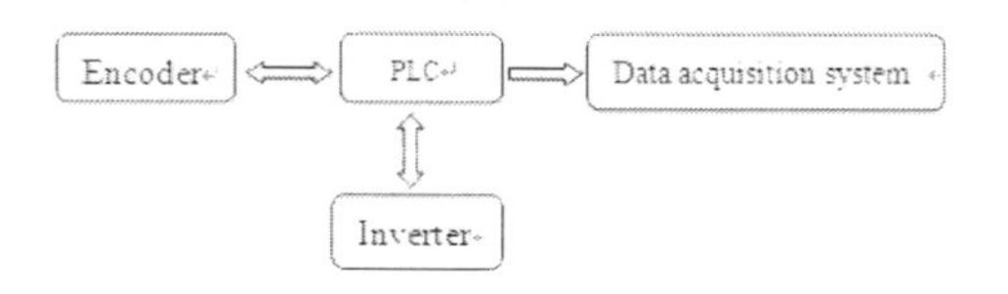

Figure 2. Data acquisition route.

When the inner cylinder began to turn, the speed of it is set respectively in 20r/min 、30 r/min 、40 r/min, 50 r/min, finally, Get torque, power curves at different speeds.

4 THE EXPERIMENTAL RESULTS AND ANALYSIS

Within the inner cylinder speed under the condition of 20 r/min, the washing machine began to run until smooth operation for some time. Power and torque curves in Fig. 3.

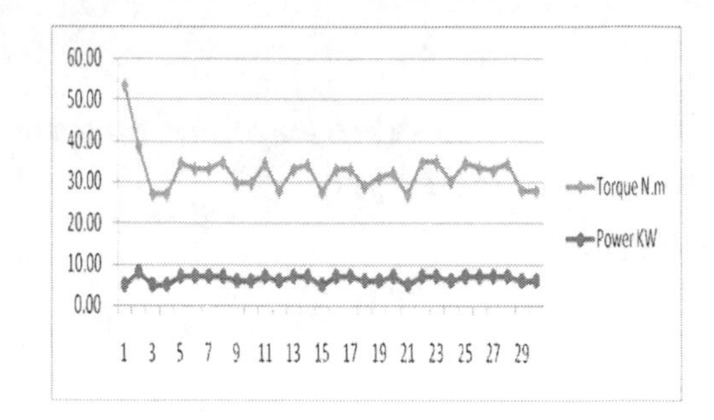

Figure 3. Power and torque curves.

Within the inner cylinder speed under the condition of 30 r/min, the washing machine began to run until smooth operation for some time. Power and torque curves in Fig. 4

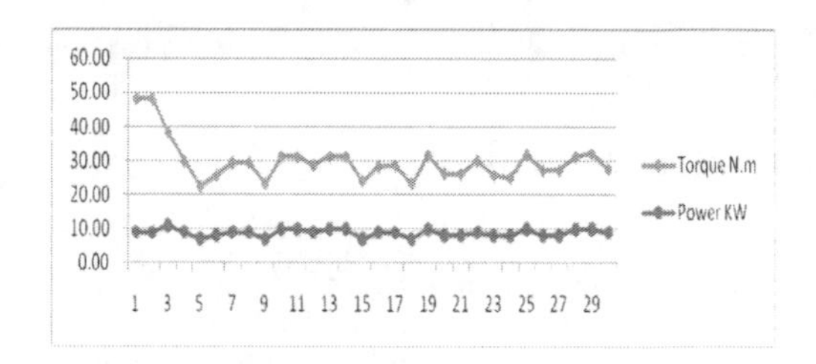

Figure 4. Power and torque curves.

Within the inner cylinder speed under the condition of 40 r/min, the washing machine began to run until smooth operation for some time. Power and torque curves in Fig. 5

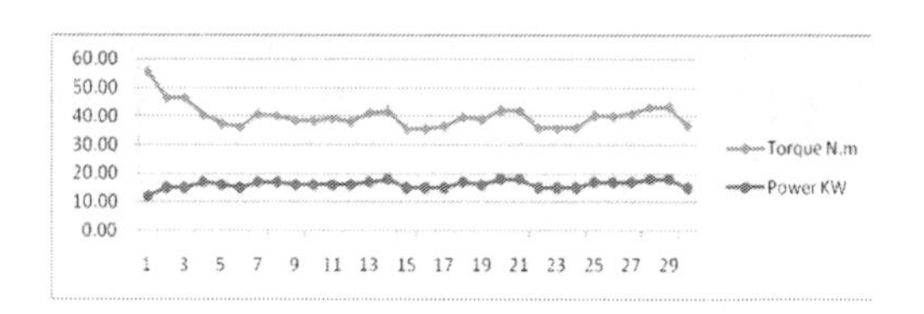

Figure 5. Power and torque curves.

Within the inner cylinder speed under the condition of 50 r/min, the washing machine began to run until smooth operation for some time. Power and torque curves in Fig. 6

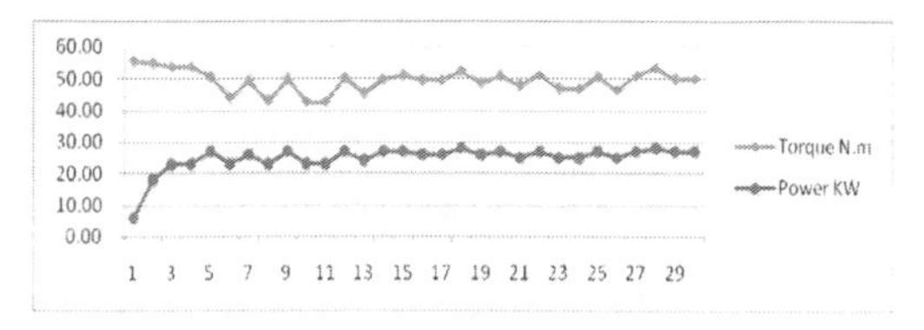

Figure 6. Power and torque curves.

This paper obtains the changeable curve of torque and power in working condition for irregular movement through experimental study. By the experimental data can be obtained: When washing machine starts to sun smoothly under the inner cylinder speed of 20r/min, starting torque is53.43N.m and starting power 5KW. After starting, the maximum of torque and power is 34.52 N.m and 8 KW respectively, the minimum 28.63N.m and 5KW. Under the inner cylinder speed of 30r/min, starting torque is 48.12 Nm and starting power 9KW. After starting, the maximum of torque and power is 31.8 N.m and 11 KW respectively, the minimum 22.4N.m and 7KW. Under the inner cylinder speed of 40r/min, starting torque is 55.51N.m and starting power 12KW. After starting, the maximum of torque and power is 43.25 N.m and 18 KW respectively, the minimum 32.87N.m and 15KW. Under the inner cylinder speed of 50r/min, starting torque is 55.50 Nm and starting power 12KW. After starting, the maximum of torque and power is 55.52 Nm and 18 KW respectively, the minimum 46.35 Nm and 23KW.

The results found that: Under the inner cylinder speed of 20r/min, starting torque and average torque is higher than the speed of 30r/m. The reason is that industrial filter cloth requires solvent to have the characteristic of freely passing, and there must be smaller soil particles pass filter cloth into the solvent, leading to the quality of soil in the inner cylinder, reducing relatively (reduction of working load), so as to reduce the torque to a certain value.

5 CONCLUSION

1 The starting torque is greater than the torque after starting.
2 After material mixed, whatever the speed is when starting, the needed starting torque is 52.52 Nm and starting power is 12 KW.
3 Under the inner cylinder speed of 50r/min, the needed power is 18 KW to make the washing machine run.

REFERENCES

[1] Guo Wei, Hu Jing-tao Solid-Liquid Motor selection and Energy Analysis Model [J]. Microcomputer Information, 2008(04)

[2] Wang Chongren Han li, Electromagnetic torque calculation method of variable frequency variable speed asynchronous motor [J]. Small & Special Electrical Machines,2012(12)

[3] WU Shunpen, Configuration Software PC monitoring system design and development [J]. Science & Technology Information, 2008(21)

Computing, Control, Information and Education Engineering – Liu, Sung & Yao (eds)
© 2015 Taylor & Francis Group, London, ISBN: 978-1-138-02800-5

The affecting factors and countermeasures on college students group competition

Yong Peng & Yong Li
Wuhan University of Technology, Wuhan,China

ABSTRACT: With the current situation that students are involved in declining physical fitness and low enthusiasm in exercise, a questionnaire survey of college student satisfaction with group activities was conducted in Hubei province, it turns out that tissue management, publicity guidance, students' engagement and teacher instruction are those key factors that influence the satisfaction of college student group activities. According to this, that strategies like optimizing organization scheme of group activities in college, strengthening the advocacy and outreach of group activities, strengthening awareness and motivation of group activity engagement, enhancing the professional ability of physical education teachers and both maintaining and improving sports facilities are proposed to improve the satisfaction of the student population activities.

KEY WORDS: Colleges and universities; Group competition; Affecting factors; Countermeasures.

1 INTRODUCTION

Since the 1980s, with the continuous development of college physical education reform, the number and types of student group competition keep increasing, the impact of which on students has been further strengthened. Being the carrier of that, university group competitions has a positive effect on increasing the practical application ability, teamwork skills and social responsibility. Furthermore, it's irreplaceable for group competitions help students raise their sports participation, awareness and develop good exercise habits, as a result of which, school sports work is easier to carry out. This paper aims to find out the key factors that influence students to participate in group competitions by questionnaire survey about Hubei University Students' satisfaction with it. And then propose appropriate countermeasures to assist the development of college students in group activities.

2 SUBJECTS AND METHODS

Undergraduates of Hubei province are chosen as the main research object to research the satisfaction of group activities. The questionnaire College Students Group Activities Satisfaction Questionnaire, which is consisted with basic information survey and satisfaction survey, 33 questions included, is designed owing to a lot of literature and interviews with experts.1000 questionnaires were sent out and 907 of which have been recovered, the recovery rate was 90.7%. In the recovered questionnaires, 824 were valid questionnaires, the valid rate was 90.85%.

3 RESEARCH RESULTS

After doing a principal component analysis and the biggest factor analysis of variance model with the results of the questionnaires and being examined by

Table 1. Test of KMO and Bartlett.

Measure of sufficient sample Kaiser-Meyer-Olkin	.951
Test of Bartlett sphericity approximated chi-square	12267.675
df	435
Sig.	.000

Table 2. Varimax rotated factor loadings matrix.

Factor	Principal Component				
	1	2	3	4	5
Contest organized by the school	.767				
Contest organized by the Institute Organization	.719				
The individual associations and community activities	.709				
Tournament nature	.649				
Single sports contest	.629				
Organizational form	.625				
Publicity		.690			
Propaganda		.674			
Organization times		.616			
Venues and facilities		.626			
Students' participation attitude			.713		
Students' participation motivation			.679		
Student engagement			.653		
Teachers' organizational capacity				.847	
Teachers' work attitude				.837	
Teachers' operational capacity				.805	
Tournament place					.786
ExtractionMethod: Principal Component Analysis					
Rotation method: a Kaiser standardizedorthogonal rotation method					

KMO and Bartlett's test of sphericity, it turns out that the KMO value is 0.951 which close to 1, and the significant probability (Sig =.000) is less than 0.001 which means highly significant. Rejecting the null hypothesis, factor analysis supposed to be suitable.

Table 2 is the result of the Principal Component Analysis Method, which used to acquire factors whose eigenvalues are greater than 1. Finally, 5 factors whose cumulative contribution rate is 57.502% are chosen to be the key factors. And it is named when further analysis of classification with the 5 class main factors and 17 high load index is done, as shown in Table 3.

According to the result of factor analysis, the first principal component, variable has a higher load factor, which is related organizations factors that affect the

Table 3. High load indicators and factor named.

Ingredient factor named	High load index and their values	
Contest organized by the school	.767	Tissue Factor
Contest organized by the Institute Organization	.719	
The individual associations and community activities	.709	
Tournament nature	.649	
Single sports contest	.629	
Organizational form	.625	
Publicity	.690	Publicity Factor
Propaganda	.674	
Organization times	.616	
Venues and facilities	.626	
Students' participation attitude	.713	Participation Factor
Students' participation motivation	.679	
Student engagement	.653	
Teachers' organizational capacity	.847	Teacher Factor
Teachers' work attitude	.837	
Teachers' operational capacity	.805	
Tournament place	.786	Tournament Factor
ExtractionMethod: Principal Component Analysis		
Rotation method: a Kaiser standardizedorthogonal rotation method		

activities of the groups is named as tissue factor; the second, third and fourth component are respectively designated as publicity factor, participation factor and teacher factor for similar reasons; the last main ingredient that with higher load 0.745 only about venue facilities is defined as the tournament place vector.

4 COUNTERMEASURE

4.1 *Improve organization and management capacity of school*

Since the result shows that the capacity to organize and manage group competition influence the satisfaction of students, schools should take the combination of school, Institute Organization, the individual associations and class into account when they designing college group activities program schedule .P.E teachers are encouraged to direct the organization of group competition, and cooperation between different departments needs to be strengthen, so that the organizational capacity in school can be improved comprehensively, and students will enjoy the goodness bringing by sport and improve physical fitness awareness and develop scientific, healthy and civilized lifestyle.

4.2 *Strengthen publicity of group competition*

Advocacy attaches great influence to group activity engagement, regardless of publicity or propaganda. The school should regard group competition as a platform to enhance culture construction of P.E. With the modern means of information dissemination, vinculum between teachers and students can be strengthened. It should become a tradition that there's sports activities every week and contest each month. That providing opportunities for students to watch high-level sporting events and attracting them to involve in it should be a good idea[1].

4.3 *Raise engagement awareness of students*

It is believed that university students generally lack specific education is one of the most direct cause that leads to lower engagement in group competition[2]. To this end, teaching and various publicity measures should be taken to raise further engagement awareness of students. On the other hand, both family and society should pay attention to the physical health of college students either.

4.4 *Strengthen faculty team construction and teachers' professional capacity*

Teachers playing an important role in effecting the enthusiasm of the college students to group activities, their cognition to physical and teaching skills have direct influence on the formation of good habit for students in physical training[3]. Moreover, they are the primary way to affect the formation of exercise habits of the students systematically, all these have made teachers an irreplaceable role in this matter. Therefore it is imperative to strengthen the faculty team by enhancing teachers' professional ability, and they have to meet three requirements as followed; First, gym teachers have to firmly establish the concept that health goes first as the guiding ideology; when setting goals and choosing the content of teaching and making course evaluation they must have that promoting students' health conditions is the purpose in mind.Second, it is essential for teachers to learn advanced knowledge of physical education, and set a sound physical health evaluation criterion for students. Third, in promoting the physical conditions of the students, they should be an instructor instead of a supervisor, a promoter instead of a teacher. So that they can improve students' initiative and enthusiasm of learning and involve in sports by students' autonomous study as well as deep communication and cooperation between students and teachers. Thus students will get themselves engaged in various sports activities using their imagination and creativity[4].

4.5 *Provide comprehensive opportunities of group competition for college students*

The conditions of sports facilities in college for students to conduct group activities being one of the crucial factors in effecting the initiative and satisfaction of engaging in competitions, measures have to be taken to increase the maintenance and input in those facilities [5]. We need to increase the proportion of group competition in organizing sports activities, create more opportunities for students, and provide more convenient conditions for students by setting up more divisions according to the number of campus and distance between students and activity site.

5 CONCLUSION

In summary, the conduction of group competition for universities at present has to take the comprehensive development of the students as a core, and take improvement of students' physical conditions as purpose, and to promote initiative participation of students finally. As to satisfaction of engaging in group activities, it can be improved by improving organization and management capacity of the school, strengthening publicity of group competition, raising the engagement awareness of students, strengthening faculty team construction and teachers' professional capacity, and providing comprehensive opportunities of group competition for college students.

REFERENCES

[1] Lu Chun. Reform Theories and Practice of College Group Activities [J] Wuhan Institute of Physical Education, 2009,7 (23): 97–100.

[2] Zhang Luogeng, He Xiaozhi. The Situation and Countermeasures of After-Class P.E Contest in Universities, Taking Hunan as an Example [J] Journal of Guangzhou Sport University, 2009, 1(29): 68–72.

[3] Zhang Junying. Research into and Analysis of the Factors That Influence College students To Take Physical Exercise After School [J]. Shandong Sports Science & Technology, 2002, 24 (1): 47–48.

[4] Pan Xiugang, Chen Shanping. Effects of Sports Community on Physical Exercise Motives of the Undergraduates [J]. Beijing Sport University, 2010,7 (33): 71–74.

[5] Xu Liyun. Study on the Current Situation and Operational Mechanism of College Group Contests in Anhui Province [D]. Fujian Normal University, 2009.

Computing, Control, Information and Education Engineering – Liu, Sung & Yao (eds)
 ISBN: 978-1-138-02800-5

A research of art design education and market economy

Xin Yue Zhang
Department of Art, Lijiang College, Guangxi Normal University, Guilin, China

ABSTRACT: The art education in colleges and universities, especially art design education, develops quickly in recent years and accumulates a lot of successful experiences in many aspects. But at the same time, some problems appear. Because of the lagging research on the theories and the different recognitions of the art design education, the successful experiences accumulated in practice are always not summarized well, and theories cannot be raised to a higher level. And the existing problems are not paying attention to and solved in time. With time going by, these problems must affect the development of art design education. Now the art design education is lack of the interaction with market economy. In colleges and universities, a lot of students major in the art design courses. But when they graduate, they find it difficult to get a job. In society, the demand exceeds the supply, but in fact, the supply cannot satisfy the demand. Focusing on the main problems existing in the interactive teaching of art design education and market economy, this paper studies and discusses the problems in the art designing education.

KEYWORDS: Art design;Market Economy; Interaction;Teaching.

Chinese art design began to develop since China began to reform and open up to the outside world. During the past more than 20 years, design promotes the economical development in many fields, and appear thousands of new architectural design offices, indoor design companies, advertisement companies, industrial design offices, dress designing companies, enterprise image plan companies. At the same time, lots of design departments and schools are set up in many colleges and universities; this makes art design education develop quickly. With the help of economy market, both design industry and design education have made great achievements. But the short history of the real art design education results in some shortcomings and unbalanced situation: education cannot meet the demands of the market economy at all, resulting in many college graduates' failure to find a job at their graduation and students and parents' beginning to doubt the education. On the one hand, in the past years, many students are enrolled in art design departments and schools, which make people feel great. On the other hand, the graduates cannot find a job and companies cannot find competent designers. So what is the problem? We need to consider the relation between art design education and market economy.

1 THE RELATION BETWEEN THE ART DESIGN EDUCATION AND MARKET ECONOMY

The socialist market economy is actually an economy in which market plays a role to allocate the resources. After the Chinese economy turned into a market economy from the former planning economy, art design education derived from craft art education. The new born art design education ever lost itself in the market economy.

To avoid this, we need to recognize the real natures of market economy and art design education; that is, we need to have a clear idea of the negative and positive effects that market economy has made on the art design education.

1.1 *The negative effects of market economy*

The pursuit of interest results in money worship. Market develops with pursuit of interest and reasonable profits are acceptable. But the pursuit goes too far, money worship becomes people's highest principle and aim. When it goes to the extreme, people do every illegal way to get the biggest and illegal profits. They see money as the only goal in their life and think money can talk. Some people work and design for others' demand. The philosophy of "design for money" results in worldly-minded and disgusting works.

"Demand" results in unchecked materialistic pursuit. The demand is one of the tools to market to operate. It is reasonable for people meet their normal demand, but we need to keep a suitable balance between demand and the principle. The principle cannot press the demand too much, or the market economy can become the dull planning economy and market cannot be active. When demand surpasses the principle, the market will be out of order and full of unchecked materialistic pursuit and people will be the slave of demand and unchecked materialistic pursuit. Pleasure-seeking makes people act like an animal and the spiritual world becomes wasteland. This is just like what Fromm said, "The problem with 19th century is that God has died and the one with 20th is that man has died." "What art expresses is the attitudes of artists." So during the interaction between the art design education and market economy, we must prevent students from materializing the market and man.

The extreme pursuit of personal principle results in extreme individualism and selfishness, making individuals and society lose the balance between them and harm the society's interest as a whole. The relation between individual and society is very important for a society to keep balance and develop. Though art stresses its characteristics, it just belongs to art aesthetic, not personal behavior. The aim to educate team work spirit in art design education is to avoid the negative effect when personal principle goes to the extreme in a market economy.

1.2 *The positive effects reflected in a series of valuable spiritual qualities.*

What comes to the market economy is the spirit and idea of freedom, democracy and equality. And the freedom and democracy of markets reflect the highest cultural spirit of design. The design work that is culturally, meaningful and popular with clients is more valuable in the economy.

What comes with the market is the open mind and spirit. Since the principle of market competition is a free-competition one, free competition must be open. Free competition activates the market and the ways to design, providing the scientific direction to educate excellent talents in design.

What comes with the market is the spirit of recognizing and advocating pluralism. This is because that market that takes free competition as its basic principle operates only in domestic or pluralistic situations. The development of art design education should take pluralistic education as its character, enriching its teaching contents and teaching ways.

What comes with the market is the spirit of innovation. Only in the market economy, can the creativity and talent of people can be freed to its full application.

What comes with the market is the scientific spirit. To operate the market economy and produce competitive commodities, we cannot do it without science. So, since science and technology are producing power, the art design is too.

At the very beginning, the effect of market shows us the tempt of the exterior "interest" and "demand", and just like the loss it brings to us, "the theory of design crisis" appears. But as time goes by with the choice by history and human wisdom, the market effect develops according to the development of human society and "design crisis" turns out to be "design chance". The art design that meets the human development and is valuable in the economy is the trend of design, is the hope of design, is the result of the interaction between the market economy and art design education.

2 THE INTERACTION BETWEEN ART DESIGN EDUCATION AND MARKET ECONOMY

The final aim of the interaction of art design education and market economy is to improve the students competition ability.

"Interaction" doesn't mean that teaching should be organized and structured at will. This "interaction" demands both depth and width in teaching. It at least reflects the following three relations of interaction: the interaction between "teaching" and "learning" in the market situation, the interaction between "teaching" and market economy, and the interaction between "learning" and market economy.

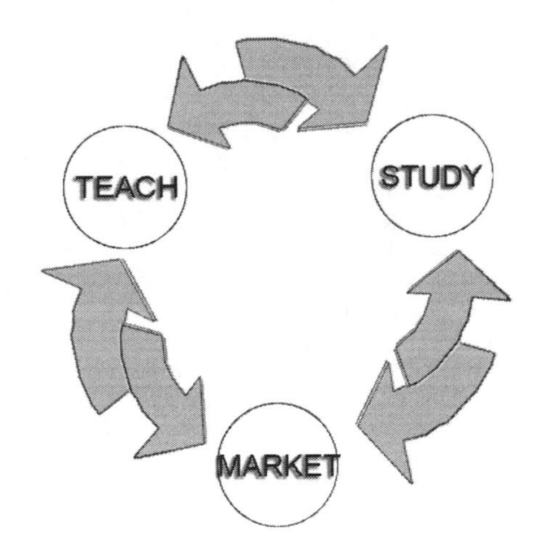

(Illustration 2-1 Interactions among teaching, learning and market).

2.1 *The interaction between teaching and learning*

The interaction in teaching and learning is one of the contents of new curriculum reform. The new curriculum advocates the equal dialogue between the teacher and the students and the interaction between teaching and learning. And the traditional relation between teachers and students have changed basically; the role of teachers also has an essential change too—they change from the role teaching knowledge to the one supporting enthusiastically and cooperating with students. Students also change from the passive role to the one studying positively and supporting teaching.

2.1.1 *The interaction in teaching*

"Teaching innovation and learning to cooperate" is the real meaning of the interaction in teaching and learning. "Teaching innovation" is based on teaching students how to learn. A teacher should play the role as a guide, a revealer and a consultant. A teacher should change from "teaching only knowledge and ethics" to "setting up good example with personality", from "feeding existing knowledge" to "constructing knowledge together", from "providing standard answers" to "pursuing new knowledge together".

The real cooperation is the reflection of the interaction. One's energy is limited and cannot do everything in person; he or she must learn to share various educational technology and curriculum resources with others, and this is the interaction in teaching, or the cooperation among the educators. "In the three accompanies, there must be one I learn from."(Confucius) and "He who is able is teacher" are actually the cooperation and interaction among teachers and students. At present, a concept of "course leader" is making an impact on traditional "course management" pattern, the real cooperation is a kind of equal interaction relation, is the accompany relation to new curriculum construction. This is because art cannot be taught and an excellent design work needs inspiration. The design works done by teachers are not necessarily better than the ones by students and the fact is that teachers just know the design techniques earlier than students. So when we discover an excellent design work by some student, we should let the student organize the relevant lecture in order to inspire more students' thoughts on design.

2.1.2 *The interaction in learning*

Interaction learning is the reflection of the team work spirit. Interaction learning can develop students' ability to learn from the people around them, to communicate and cooperate, to set up their confidence, to understand that there is no so called status in design dialogue and everyone communicates on a platform. The traditional design learning pattern stresses too much on students' reception and command of knowledge, keeping students in a "being fed" situation and making learning a procedure of pure passive reception and memorization. Changing the learning patterns is changing situation, activating the discovery, exploration, communication and consensus among the learning process, making the process of design become one of discovering, analyzing, and solving problems. Emphasis on discovering learning, exploration learning, research learning and communication learning is the important character of interaction curriculum.

2.2 *The interaction between learning and market economy*

In the process of "teaching", we need to emphasize the educating and developing professional talents needed in the market with the guide of market and according to teaching contents. Firstly, teachers should pay great attention to the research and development of new projects and new products which is greatly demanded in the market, adaptable in market and market-oriented. These new projects and products include industrial design, advertising design, packing design and environment art design; they should be taken as "market cases" in the teaching lectures, thus making students be able to act as real designers and at the same time this can reflect the teachers' research and teaching ability in a comprehensive way. Secondly, achievements in science and technology should be paid great attention to be turned into productive power. Take the course of industrial design, for example, when a teacher produces teaching research achievement that is high-tech and meets market demand, he or she should contact some enterprise to produce; and this can make school have a close relation with market. Thirdly, we should conduct the "inviting" open teaching pattern; that is the "Theory and Practice Lecture Room". First, we should invite famous home or foreign experts or scholars to give lectures on the real problems in the design education and market practice; second, we should invite entrepreneurs, market economists and professionals to give lectures, making it possible that students can learn market knowledge in their classes.

2.3 *The interaction between learning and market economy*

Learning does mean confining students for their study in their classroom; the school should create more chances for students to go out and learn in society and the market. "Going out" means conducting on-the-spot teaching. By "going out and learning", students can learn a lot in aperiodic "expert forum" and professional "design expert conference", taking advices from the experts who work in enterprises and adjusting their learning goal and future job direction after graduation.

3 CONCLUSION

As is known to all, the history and accumulation of Chinese art design are far from letting us rest on the achievement we have made. On the contrary, because of its shortcomings, we need to make every effort in many ways. The development of market economy inevitably leads to the conducting of teaching pattern with interaction between art design education and market economy. And making the design education meets market and international situation is just to follow the principle of "Production relation meets the development of productive power". As society develops, the art design education changes accordingly. By following the pattern "survival-development-market-education (art design education)", we should try our best to educate and develop high-quality, all-around art design talents who can advance with the times and meet demand in society.

REFERENCES

[1] Li Dechao. Art Is Me, Design Is Him. Suzhou University Journal (Science Edition), 03, 2003.

[2] Luan Changda. Market Economy and Art. Jilin Fine Arts Press. 12, 2000.

[3] Zong Mingming. Modern German Design Education Concept and Practice. Shengyang: Liaoning Fine Arts Press, 2001.

[4] Xu Hengchun. Design Aesthetics. Beijing: Qinghua University Press, 07, 2006.

Computing, Control, Information and Education Engineering – Liu, Sung & Yao (eds)
 ISBN: 978-1-138-02800-5

Approaches to knowledge reduction based on interval and set-valued decision information systems

H. Wang & R. Gao
College of Science, Zhongyuan University of Technology, Zhengzhou, Henan, P. R. China

X.Q. Wang
XiJing University, Shaanxi, Xian, P. R. China

ABSTRACT: Interval and set-valued information systems are the set-valued information systems of a kind of promotion model.This paper deals with knowledge reduction of interval and set-valued decision information systems. The concepts of distribution reduct based on interval and set-valued decision information systems is introduced. The judgement theorems and discernibility matrices associated with tolerance relation were examined, from which we can obtain the specific operation methods to interval and set-valued decision information systems.

KEYWORDS: Rough set; Interval and set-valued decision information systems; Knowledge reduction; α Tolerance relation.

1 INTRODUCTION

Rough set theory, proposed by Pawlak [1], is an extension of set theory for the study of intelligent systems characterized by insufficient and incomplete information. It has conceived as a tool to conceptualize and analyze various types of data. With more than twenty years development, rough set theory has important applications to intelligence decision, cognitive sciences, machine learning, pattern recognition, and so on.

In Pawlak's original rough set theory, partition or equivalence relation is an important and primitive concept. But partition or equivalence relation is still restrictive for many applications. It is unsuitable for handing incomplete information systems or incomplete decision systems. To address this issue, several interesting and meaningful extensions to equivalence relation have been proposed in the past. Such as tolerance relations, dominance relations and others. Knowledge reduction is performed in information systems by means of the notion of a reduct based on a specialization of the general notion of independence. By such one subsets the information for classification purposes provided is the same as the condition attribute set done. Such subsets are called reducts. To acquire brief decision rules from decision systems, knowledge reduction is needed. In recent years, more attention has been paid to knowledge reduction in inconsistent systems in rough set research. Many types of knowledge reduction have been proposed in the area of rough sets [2-7].

In this paper, we are concerned with the approaches to knowledge reduction of interval and set-valued decision information systems. In the next section, we give some basic notions of interval and set-valued information systems. In Section 3, we examine the judgement theorems and discernibility matrices associated with α-level reduction, from which an approach to knowledge reduction of interval and set-valued information systems based on α-level reduction, which are significant both in the theoretic and applied perspectives. A computative example is also given to illustrate our approaches. We then conclude the paper with a summary and outlook for further research in Section 4.

2 BASIC NOTIONS RELATED TO INTERVAL AND SET-VALUED DECISION INFORMATION SYSTEMS

Definition 2.1. Let P and Q are ordinary sets, if the range of variable R takes set P as the lower limit, set Q as the upper limit, then the variable R is called interval and set-valued variable.

Definition 2.2. An information system is a quadruple $S=(U,A,V,f)$, where the universe $U$is a non-empty finite set of objects, A is a non-empty finite set of attributes, V is the union of attribute domains ($V = \bigcup_{a \in A} V_a$), V_a is the set of all possible values for attribute a and V_a is an interval and set-valued variable, $f : U \times A \to V$ is a function that assigns particular values from attribute domains to objects, then the information system is called an interval and set-valued information system.

In this paper, we mainly focus on following type .

$\forall x \in U,\ a \in A$, the value of attribute a for object x is denoted by $V_{f_a^-(x)}^{f_a^+(x)}$, where $f_a^-(x), f_a^+(x)$ are the finite sets, and satisfy condition :

$$\min(f_a^+(x)) \ge \max(f_a^-(x)) \min(f_a^-(x)) \le V_{f_a^-(x)}^{f_a^+(x)} \le \max(f_a^+(x))$$

the minimum of set $f_a^+(x)$ at least equal to the minimum of set $f_a^-(x)$, at most equal to the maximum of set $f_a^+(x)$. The value of attribute a for object x is interpreted disjuctively. For example, if a denotes the environmental risk assessment index of enterprise investment, then $V_{f_a^-(x)}^{f_a^+(x)} = V_{\{1\}}^{\{2,3\}}$ can be interpreted as: the lowest grade of the environmental risk assessment index is 1, the highest grade may be 2 or 3.

Definition 2.3. Let $S = (U, A, f, d, g)$ be an interval and set-valued decision information system and $B \subseteq A$. we denote

$$C_{ij}^k = \frac{|[f_{a_k}^-(x_i), f_{a_k}^+(x_i)]| \cap |[f_{a_k}^-(x_j), f_{a_k}^+(x_j)]|}{|[f_{a_k}^-(x_i), f_{a_k}^+(x_i)]| \cup |[f_{a_k}^-(x_j), f_{a_k}^+(x_j)]|}$$

$$R_B^\alpha = \{(x_i, x_j) \in U \times U \mid \forall a_k \in B, c_{ij}^k \ge \alpha\}$$

$$R_{\{d\}} = \{(x_i, x_j) \in U \times U \mid g(x_i) = g(x_j)\}$$

R_B^α is α tolerance relation, $R_{\{d\}}$ equivalence relation.

Definition 2.4. Let $S = (U, A, F, d, g)$ be an interval and set-valued decision information system, If $R_A^\alpha \subseteq R_{\{d\}}$, then S is α consistent interval and set-valued decision information systems, otherwise, it is called inconsistent.

Example 2.1. Given a consistent set-valued decision information system (Table 1).

For $\alpha = 0.5$, then

$$[x_1]_d = [x_3]_d = [x_4]_d = \{x_1, x_3, x_4\}$$

$$[x_2]_d = [x_5]_d = \{x_2, x_5\}, [x_1]_A^\alpha = \{x_1\} [x_2]_A^\alpha = \{x_2, x_5\}$$

$$[x_3]_A^\alpha = \{x_3\}, [x_4]_A^\alpha = \{x_4\}, [x_5]_A^\alpha = \{x_2, x_5\}$$

Obviously, $[x_1]_A^\alpha \subseteq [x_1]_d, [x_2]_A^\alpha \subseteq [x_2]_d$

$[x_3]_A^\alpha \subseteq [x_1]_d, [x_4]_A^\alpha \subseteq [x_2]_d, [x_5]_A^\alpha \subseteq [x_2]_d$,

by the above, we have $R_A^d \subseteq R_d$, so the system in table 1 is consistent.

Table 1. A consistent DT.

U	a_1	a_2	a_3	a_4	a_5	a_6	d
x_1	$V_{\{3\}}^{\{4\}}$	$V_{\{4\}}^{\{5\}}$	$V_{\{3\}}^{\{4\}}$	$V_{\{3\}}^{\{4\}}$	$V_{\{2\}}^{\{3\}}$	$V_{\{4\}}^{\{5\}}$	1
x_2	$V_{\{1\}}^{\{2\}}$	$V_{\{1\}}^{\{2\}}$	$V_{\{1\}}^{\{2,3\}}$	$V_{\{1\}}^{\{2,3\}}$	$V_{\{2\}}^{\{3\}}$	$V_{\{1\}}^{\{2,3\}}$	2
x_3	$V_{\{3\}}^{\{4\}}$	$V_{\{4\}}^{\{5\}}$	$V_{\{3,4\}}^{\{5\}}$	$V_{\{3\}}^{\{4\}}$	$V_{\{3\}}^{\{4,5\}}$	$V_{\{4\}}^{\{5\}}$	1
x_4	$V_{\{2\}}^{\{3\}}$	$V_{\{4\}}^{\{5\}}$	$V_{\{2\}}^{\{3\}}$	$V_{\{2\}}^{\{3,4\}}$	$V_{\{2\}}^{\{3\}}$	$V_{\{3\}}^{\{4,5\}}$	1
x_5	$V_{\{1\}}^{\{2\}}$	$V_{\{1\}}^{\{2\}}$	$V_{\{1\}}^{\{2,3\}}$	$V_{\{1\}}^{\{2\}}$	$V_{\{2\}}^{\{3\}}$	$V_{\{1,2\}}^{\{3\}}$	2

The following we deal with attribute reduction of an consistent interval and set-valued decision information systems.

3 APPROACHES TO KNOWLEDGE REDUCTION OF INTERVAL AND SET-VALUED INFORMATION SYSTEMS BASED ON TOLERANCE RELATION

This section provides approaches to interval and set-valued information systems based on $\alpha - level$ tolerance relation. Let us first give the following notions.

Definition 3.1. Let $S = (U, A, F, d, g)$ be a consistent interval and set-valued decision information systems based on α tolerance relation. Define

$$D_{ij}^\alpha = \begin{cases} \{a_l \in A : C_{ij}^k < \alpha\} & g(x_i) \ne g(x_j) \\ \varnothing & g(x_i) = g(x_j) \end{cases}$$

Then D_{ij}^α are referred to as x_i and x_j α discernibility attribute sets. And $D^\alpha = \{D_{ij}^\alpha, x_i, x_j \in U\}$ is referred to as αdistribution discernibility matrices.

Theorem 3.1. Let $S = (U, A, F, d, g)$ be a consistent interval and set-valued decision information systems based on $\alpha-level$ tolerance relation, $B \subseteq A$, then B is α consistent set of Sif and only if $B \cap D_{ij}^{\alpha} \neq \varnothing (x_i, x_j \in U)$.

Proof. Since B is $\alpha-level$ consistent set, therefore, $R_B^d \subseteq R_d$. If $D_{ij}^{\alpha} \in D^{\alpha}$, then $D_{ij}^{\alpha} \neq \varnothing$. Thus $g(x_i) \neq g(x_j)$. Therefore, $(x_i, x_j) \notin R_B^{\alpha}$, there exists $a_k \in B$ such that $C_{ij}^{\alpha} \prec \alpha$. Thus $a_k \in B \cap D_{ij}^{\alpha}$, $B \cap D_{ij}^{\alpha} \neq \varnothing$.

Conversely, if for any $x_i \in U$ such that $[x_i]_B^{\alpha} \subseteq [x_i]_d$. For any $x_j \in [x_i]_B^{\alpha}$, we assume that $g(x_i) \neq g(x_j)$, then $[x_i] \cap [x_j] = \varnothing$, and S is a consistent interval and set-valued decision information systems, thus $[x_i]_A^{\alpha} \subseteq [x_i]_d$ $[x_j]_A^{\alpha} \subseteq [x_j]_d$, we have $[x_i]_A^{\alpha} \cap [x_j]_A^{\alpha} = \varnothing$, $x_j \notin [x_i]_A^{\alpha}$, thus means there exist $a_k \in A$, such that $C_{ij}^k \prec \alpha$, so $D_{ij}^k \notin D^{\alpha}$. Since $B \cap D_{ij}^{\alpha} \neq \varnothing$, then exist an element $a_m \in B$, such that $C_{ij}^m \prec \alpha$, this is a contradiction.

Definition 3.2. Let $S = (U, A, F, d, g)$ be a consistent interval and set-valued decision information systems, $D^{\alpha} = \{D_{ij}^{\alpha}, x_i, x_j \in U\}$ is $\alpha-level$ discernibility matrix of S. Denote by

$$M = \wedge\{\vee\{a_k : a_k \in D_{ij}^{\alpha}\}, x_i, x_j \in U\}$$
$$= \wedge\{\vee\{a_k ; a_k \in D_{ij}^{\alpha}\}, (x_i, x_j) \in D^*\}$$

Then $M$are referred to $\alpha-level$ discernibility functions.

Theorem 3.2. Let$S = (U, A, F, d, g)$ be a consistent interval and set-valued decision information systems based on $\alpha-level$. The minimal disjunctive normal form of discernibility formula is $M = \vee_{k=1}^{p}(\wedge_{s=1}^{q_k} a_s)$

Denote $B_k = \{a_s : s = 1, 2, \cdots q_k\}$, then $\{B_k : k = 1, 2, \cdots p\}$ is just a set of all reductions of S.

Proof. It follows directly from Theorem 4.1 and the definition of minimal disjunctive normal of the discernibility formula.

Example 3.1. (continued from Example 2.1) a consistent set-valued decision information system is given in Table 1.

Table 2. α (=0.5) level discernibility matrix.

D_{ij}	x_1	x_2	x_3	x_4	x_5
x_1	$\varnothing$	$\{a_1a_2a_3a_4a_6\}$	$\{a_3a_5\}$	$\{a_1a_3a_4\}$	$\{a_1a_2a_3a_4a_6\}$
x_2	$\varnothing$	A	$\{a_1a_2a_3a_6\}$	$\varnothing$	
x_3		$\varnothing$	$\{a_1a_2a_5\}$	$\{a_1a_2a_3a_4a_6\}$	
x_4	$\varnothing$	A			
x_5	$\varnothing$				

By Theorem 3.2 we obtain$F = a_1 \wedge a_3$. Therefore, $\{a_1, a_3\}$ is all reductions of S in Table 1.

4 CONCLUSIONS

Many types of attribute reduction in inconsistent decision tables have been proposed based on rough set theory. In this paper, we have studied knowledge reduction of set-valued decision information systems based on tolerance relation, The distribution reduction is introduced. The judgement theorem and discernibility matrix are obtained, from which We have obtained new approaches to knowledge reduction of set-valued decision information systems based on $\alpha-level$ tolerance relation by providing the discernibility matrices. In further research, we will develop the proposed approaches to more generalized and more complicated information systems such as incomplete information systems and fuzzy information systems.

ACKNOWLEDGEMENT

This work was supported by the grants of the National Nature Science Foundation of China(No. 61473237).

REFERENCES

Pawlak, Z. 1982. Rough sets. International Journal of Computer and Information Science 11(5): 341—356.

Kryszkiewicz, M. 1998. Rough set approach to incomplete information systems. Information Sciences 112: 39–49.

Kryszkiewicz, M. 2001. Comparative study of alternative types of knowledge reduction in inconsistent systems. International Journal of Interligent Systems 16:105–120.

Kumar ,A. 1998. New techniques for data reduction in a database system for knowledge discovery applications. Journal of Intelligent Information Systems 10: 31–48.

Zhang, W. X. & Qiu, G. F. 2005. Uncertain Decision Making Based on Rough Sets. Qinghua Publishing Company, Beijing.

Leung, Y., Wu, W.Z. & Zhang ,W.X. 2006. Knowledge acquisition in incomplete information system: a rough set approach. European Journal of Operational Research 168 : 164–180.

Meng, Z.Q. & Shi Z.Z. 2012. Extended rough set-based attribute reduction in inconsistent incomplete decision systems.Information Sciences 204: 44–69.

Computing, Control, Information and Education Engineering – Liu, Sung & Yao (eds)
 ISBN: 978-1-138-02800-5

Reference and influence of Korean furniture on Chinese national furniture

Zhen Gao
Tibet University, lasa, P.R. China

Rui Bo Hu, Jun Fang An & Xin Yu Suo
Guizhou Normal University, Guiyang, P.R. China
School of Vocational Technical Education, Yunnan University of Nationalities, Kunming, P.R. China

ABSTRACT: South Korea has been historically called the "Little China" and has a big influence in Chinese cultural circle. In the past few decades, under the inspiration of "cultural renaissance", Korea has strived to develop its national culture and closely combined it with modern industry. Korea constantly tries to integrate art with science and uses modern business concepts and forms its national furniture which learns from the classic and also has modern features. As Korea is greatly influenced by Chinese culture, the varieties of the furniture are similar. The success of Korean furniture provides a good example for the still exploring Chinese furniture industry. Researches on modern Chinese furniture include theoretical study, technological study and production practice.

KEYWORDS: National furniture, Classics, Innovation, Differences and similarities, Reference points, Education.

1 INTRODUCTION

Korean culture, i.e. culture of the Korean Peninsula, has been influenced by Chinese culture since ancient times. Since Tang Dynasty, Korea has been in active connection with China in areas such as politics, economy, culture and technology and has been integrated with Chinese culture. Korean art includes painting, calligraphy, print, craft and decoration, not only inheriting traditions but also absorbing foreign art strengths. Even its national flag-taegeukgi is influenced by Chinese Taoism and Zhouyi. Confucian culture has a deep influence on Korea, for example, Gyeongguk Daejeon was issued in Jeseon Dynasty in Korea according to the Law of Ming Dynasty and the Korean social ethics and ideas are also greatly affected. In some way, Chinese culture is effectively preserved and promoted in Korea. The development of Korean furniture is also under great influence of China. Back in the Qin and Han dynasty, most of Korean land belonged to China. Therefore, the low furniture of Han Dynasty was completely introduced to Korea. However, due to its regional remoteness, the workmanship was very coarse but the furniture was generally low and stout. Silla unified Korea and learned from systems in Tang Dynasty. After Song Dynasty, China started to form the tradition of sitting on chairs through national integration while Korea which was not invaded kept sitting on the ground. During this period of time, Korea constantly integrated and exchanged with its neighbors and by combining with its own mountainous geographical features and conventions, formed Korean national furniture which had the same shape but different soul of the Chinese furniture. Due to the lack of rosewood, its domestic wood was adopted and decorated with red paint and bronze.

2 THE COURT STYLE, FOLK STYLE AND MODERN STYLE OF KOREAN NATIONAL FURNITURE

2.1 *The court style*

Court Style: The Korean court style is different from that of China. It inherits traditions and has a wider collocation of colors and besides dragon and phoenix and other auspicious decorative signs, it also adds fashionable images such as butterflies. Pure colors are often used to match with each other, for example, the match of black with red and that of black with white. It is very colorful but not excessive, showing a feeling of classics, elegance, comfort and fashion. Its decorative patterns have a tendency to be simple, but are becoming richer and more gorgeous. Reference points: The Chinese court style is vigorous and powerful but not so approachable. A variety of fashionable decorations can be adopted and the patterns can be simplified. The traditional methods of engraving, embedding and buckling do harm to the structures and decrease the intensity. These should be replaced by methods such as gold-painting, flower-drawing and bronze-painting.

2.2 *The folk style*

Folk Style: The Korean folk style has generally remained unchanged, and modern adhesives only partly replace traditional copper joints, making the structures more robust. Decorations on the surface become simpler and the complexity of the workmanship has been decreased, making it easier for mass production.

2.3 *The modern style*

Reference points: The Chinese folk furniture mainly takes example from the court style but the workmanship is coarser and there is a variety of materials. These materials are mainly from local places and firewood is the main form of material. Compared to the Korean furniture, the Chinese furniture are more fragile and less durable. This is mainly because of its complicated structures. Its designs and structures should be simplified on the basis of keeping its soul. Modern style: The Korean style is aesthetic, warm, simple and elegant, having a clean and warm family atmosphere. The fact that Korean furniture is suitable for modern living concept based on post-modern architecture is the interpretation of Korean traditional culture against current grounds and is also the fusion of multiculture. This provides choices for people pursuing cultural integration and fashion mix-match.

3 CULTURAL CONNOTATION

The culture, values and conceptions and conventions of China and Korea are similar to each other and both are influenced by Confucianism . The connotations of furniture not only insist on traditions but also integrate with the West, for example, the Ruyi cloud pattern and passion follower and so on. The Korean national furniture also integrates methods in the meticulous painting such as portray with itself, keeping the ancient feeling while making the work less hard at the same time. This is worth learning from. The Taoists think that "iridescent colors cause blindness" and "more colors are less beautiful". The small- jasper- styled yet also graceful and dignified Korean furniture is more acceptable by the public than the high-end and powerful Chinese furniture. If the Qing furniture is considered having the domineering atmosphere of the coming of an emperor, then the Korean style is with the easiness of ruling on its own. Since the Jeseon Dynasty, Korea has started to value folk aesthetics and its furniture had a sense of beauty and the sensibility of common people. Perfection is not advocated and instead, the beauty of honesty and hard work are promoted. Decoration is not emphasized and yet natural beauty is pursued.

4 STRUCTURAL INNOVATION

The modeling of Korean furniture is affected by the Japanese furniture. The structure of Korean furniture is not complicated and is simple and robust. Squares and rectangles are mainly adopted. The surface is properly decorated and the materials are very light, having a primitive and honest affinity. Even for the most complex chairs, wood is connected in a flat way, and the chair legs are inserted. This method not only saves materials but also makes it easy to assemble and also uses small and delicate materials, catering for the modern people's aesthetic pursuit of simplicity. The Chinese furniture, especially those in the Qing Dynasty, is bigger in size and rougher in materials. It gives people a feeling of fear. Even the most delicate Ming styles are too exquisite and would be too complicated for practical use as modern products.

5 DECORATIVE PATTERNS

The decoration of the surface of the Korean national furniture tends to be simple. For example, the hinges and the bronze decorations change from Ruyi cloud patterns to ellipses or squares or circles. In regard of the important decorations, modern elements (such as pear-shape and grass shape) are used, not only integrating the relaxed life style of Daoism but also stressing simple production elements. The Korean furniture favors detail portraying and carves carefully and uses small decorations and only engraves simple images and is full of picturesque feelings. All these have something to do with Korean females' preference to pink collocation. (8) The images of Korean furniture are various, including Ruyi cloud patter, calabash, lotus and butterflies and so on. It usually uses red to match yellow, black to match yellow and red to match black. It not only inherits the classics but also represents fashion and could meet requirements of clients with different age. Reference Points: Structures of the surface should be simplified and the important decorations should be more diversified. Therefore, it is easy for mass production and is also attractive for different classes.

6 MATERIAL CHOOSING

Ma Weidu says "Chinese people have always been peculiar about materials and think that only the heavy materials are good." The Chinese style is intended for the high end users and uses expensive material to show the status of the users and thus believes that the more expensive the material is the better. The designers dare not to innovate on materials and this makes it difficult for the Chinese furniture to meet the requirements of the public. The choice of the Korean materials can be learnt by the Chinese style. The ancient Korea was too poor to import rosewood and had to use their national soft wood. The surface is decorated with clear varnish and together with the texture expresses a special beauty. The Chinese style emphasizes too much on the polish of the surface, but if we use hard materials and thin the paint and highlight the decorative function of the texture, then we can decrease the cost while preserving the ancient beauty. In today's society where rosewood is very rare, this is worth a try.Reference points: Use a thin layer of paint and no polish and emphasize the texture. The texture goes with the paint, making it not only primitive and elegant but also modern and fashionable.

7 AN ANALYSIS OF THE MARKET

The economical situation of the development of China and Korea are similar and the buying ability of its consumers is about the same. The Korean products have taken the international and national road. Patriotism has become a function and an aim of Korean products. Buying Korean furniture means being influenced naturally by Korean culture. However, the Chinese furniture has not done this yet. The Korean customers have a difference of being high, middle and low. Different furniture has different production lines which have serial services of design, material choosing, production, package, promotion, after-sales service and market analysis. The Industrial Wood Law also greatly supports the production and wipes out false frauds in order to ensure the safety of the furniture. While looking at the Chinese furniture, the domestic law construction is not complete. This is mainly because of the unbalanced regional development and the weakness and unbalanced distribution of the buying ability. All these make it difficult to form a robust and flexible law protection system. If the Chinese furniture wants to realize the industrial upgrading, it must rely on the developmental advantages of the creative industry and let innovative design play its role.

8 FAMILY EDUCATION

In this aspect, the Chinese government, educational institutes and enterprises should improve the awareness. When there is a lack of fund and allowance, co-operations between schools and enterprises should be called upon. Sustainable courses with flexible agenda should be opened and pedagogical teaching should be strengthened. The width and depth of the students' knowledge should be improved. It should be people-oriented and provides an unlimited source of energy to the rejuvenation of national furniture. National cultural education: Most of the Chinese students lack cultural cultivation. While looking at the Korean students, they are not only talents in engineering designs but also advocators of national culture. "Patriotism" is a key standard for talents cultivation in Korea. China needs to strengthen the construction in this respect especially the construction of higher education. Reference points: Patriotic education and cultural education should be learned from. Support from the government should be improved and international exchanges should be strengthened. The positive direction of co-operation between schools and factories should be taken.

REFERENCES

[1] Ruibo Hu, Renping Xu, Kunqian Wang and Jingyuan Li. The Research of Virtual Tactile Design Application in Dongba Sculpture[C] .Machinery, Materials Science and Engineering Applications. November 228–229,2011,wuhan China, 185–190.

[2] Ruibo Hu, Renping Xu, Kunqian Wang and XingRan Mao . The Research of Developing and Constructing Tourist Attractions in Dong Jiahe[C].Advanced Building Materials 250–253,2011,haikou, China, 3885–3888.

[3] Ruibo Hu, Jingyuan Li ,Liyan Chen, Renping Xu , Kunqian Wang and Xingran Mao. Application study of conceptual design in the product design[C]. Advanced Science Letters. 2011,taiwan, China, 460–464.

Computing, Control, Information and Education Engineering – Liu, Sung & Yao (eds)
© 2015 Taylor & Francis Group, London, ISBN: 978-1-138-02800-5

The development tendency of information design and the transformation of design discipline

Da Wei Li
Wuhan University of Technology, Wuhan, China

ABSTRACT: Information design which has significant influences on human life and work is not only the symbolic product of the information age, but also gave birth to the interdisciplinary: Information Art & Information Design. In information age, the forms of information transmission change along with the progress of science and technology. Information and art design as a branch of art design discipline, how to adapt to the rapid development of the information age puts forward a serious proposition for our design education. This paper attempts to analyse the development tendency of information design and put forward the direction of the development of art design education in this environment.

KEYWORDS: Information design; Tendency; Design discipline; Transformation.

1 INTRODUCTION

The information age or information society is defined as the nonmaterial society by foreign scholars. In the information society, knowledge and information which are gradually replacing the substantive material becomes the most important resources for social development. After industry revolutions such as industrialization, electrification, telematics, human dependence on the dissemination of knowledge and information is growing. Today's society is certainly in this special historical node, with the help of the third industrial revolution and the development of information technology, human society is experiencing a hitherto unknown "information explosion" era in which the importance of information design is particularly prominent. Similarly, how should the design discipline based on human oriented change to adapt to this revolution is worth considering. In addition, the design object and category of art design also have been changed accordingly: Design objects changing from the objective material to nonmaterial is a significant characteristic of the contemporary art design. Experience, service, system, structure, interaction and so on have become the main objects of information and art design.

This paper puts forward some suggestions on the development and transformation of art design discipline based on the analysis of the process of the development of information design and the new trends in current combined with the current situation of design education.

2 DEFINITION AND TANSFORMATION OF INFORMATION DESIGN

Information design is generally defined as: a universal design field to create harmonious interaction and experience for consumers with digital content as research object, digitalization as the means and concise and beautiful information interface, product or environment as the media. The original information design arose in graphic design in 1970s takes the responsibility of "effective transmission of information". Human beings have been committed to transmit information through different forms for thousands of years, and information design is playing an important role in people's life and all research fields. Information is transmitted more accurate and fast through the approaches of the mining, screening, art treatment and so on.

Traditional information design mainly depends on texts, graphics, symbols, colors and other visual elements for information transmission, however, in the information era, the design objectives change from traditional tangible visual communication design in intangible service based on user experience because of the tremendous change of the definition and concept of information design under the circumstance of giant progress in the fields of computer and information technology. It is necessary for the designers to learn more, even beyond their major, to adapt to this alteration.

3 DEVELOPING TENDENCY OF INFORMATION DESIGN

New technologies and inventions indeed enrich the forms of information design and make it more diverse, also make the generation, transmission and the feedback process of information more rapidly. The same is true in the field of computer technology, the birth of Internet makes the earth more smaller, the relationship between people more closer, and communication more convenient; The combination of personal mobile communication terminals and geospatial technologies allow people to travel more convenient; Smarter Planet and Smarter City service human life ubiquitously; The integration of e-commerce and large data provide convenience for people to buy hot selling products recommended from the Internet and so on. The facts show that people enjoy the convenience brought by information technology.

Technology is always in progress, so new methods of information design will arise inevitably. If we talk about the form of presentation, the paper media are dominant in the industrial era. On the contrary, it should be network and multimedia in today's information age. Especially in the background of " Big Data", the popular of personal communication terminal makes the presentation forms of information design more diversified. Although many scholars worry about " information overload " or " information explosion" if things go on like this, but the convenience of information transmission indeed optimize people's life.

Because of the constraint of technology, information design is confronted with the changing alteration and a challenge. With the developing computer technology, information design trend in future will be in the following direction:

3.1 *Interaction design of human-machine interface*

Interaction design of human-machine interface affects the user experience directly. Nowadays, information design pays more attention to the study of user experience, and user centered humanistic design has become the focus of interaction of information design. The endless desire for efficiency and the reasonableness and convenience of operation, so people-oriented information design has become the main driving force in the new information environment.

Interface design is a graphical interactive form that largely needs professional knowledge of design. As a dialogue bridge connecting human and machine, interface plays a very important role. A good interactive interface design has high requirements for designer, besides the basic user interface design, other aspects of knowledge, such as human behavior, psychology and perception are necessary. The correct transmission of information, as the first level of usability of information design, is far from enough, interactive interface design need to manifest more emotional level. No matter which carrier to obtain information, the reflection of emotion should move the user's heart. Therefore, interactive interface design of information is a new goal and challenge to art design of information.

3.2 *Virtual reality*

The forms of new information solutions must be determined by their functions and existing technical process, and all these technological inventions as a wealth provide powerful support for modern design.

Virtual Reality (VR for short) which is formed in the process of exploring nature is a scientific method or technique used to understand, to simulate, to adapt to and to use nature. The development of computer technique helps to create 3D simulation and Augmented Reality Technique, which is a kind of dynamic simulation system based on multi-source information and 3D simulating the real scene generation and give the sense true feeling by operating various kinds of professional virtual equipments. With the increase of productivity, the demand for VR is growing in all walks of life and VR has made tremendous progress.

VR technology has become a very important research direction in information design, and been widely used in science, art, medicine, military and other fields. Its a function that is indispensable to information design is shown in the fields of earth, information, navigation simulation, game design, space simulation, etc.

3.3 *Cloud computing and internet of things*

Ballmer defines cloud computing as a revolutionary computing model. It combines computing devices near at hand and the abilities of processing and storage of ultra - large - scale data centers, and changes calculation of limited resources to an almost unlimited platform. Cloud computing and Internet of things that have broad prospects for development are undoubtedly another wave following computer, Internet and mobile Internet. Cloud computing is an inevitable trend in the information age, provides a good processing platform for "Big data" which generated in the background of information explosion and virtual, cross-regional service model, brings large-scale data flow as well.

The realization of Cloud computing makes Internet of things more perfect. Information interaction is not limited between man and machine, Internet of things also extends the interaction to objects and objects through cloud technology. Those collected mass information is sent to internet for cloud computing, processing and analysis, then the feedbacks manipulate objects intelligently. In fact, these wonderful life vision need the support of powerful technology, however, there is a long way to go to achieve this technology.

In addition to the above-described directions, a new application technique such as Augmented Reality Technique, Holographic Technique and Artificial Intelligence are peculiar to information era as well. Although these technologies are still undeveloped, some results have been achieved more or less and some of them there are many related technologies have been put into practice. All in all, science and technology are developing rapidly; it is difficult to anticipate what kind of a new era will be confronted with after the information age.

4 REFORMATION AND TRANSFORMATION—NEW REQUIREMENTS FOR DESIGN DISCIPLINE

4.1 *Promotion of information art design*

The need for the development of creative and cultural industries art design of information is a special field based on information technology and culture, a new development of artistic design in the information age, meets the developmental need of creative and cultural industries, covering the frontier of design, art and culture, and reflects the major growth point of art design.

Features of the art design of information: aesthetic-centered "morphology expression", processing and reconstruction of information with knowledge and skills in other disciplines, combination between Design activities and social cognitive science, pursuing rationalization of information, people and the whole environment. How to utilize information efficiently is the prime aim of the art design of information. In the new era, it is necessary for a healthy discipline to explore new directions. The discipline of design should develop the art design of information in this context. The ultimate purpose of design is to solve problems by various approaches, to create value for humanity through the integration of rationality of technology and sensibility of art.

4.2 *Crossover and integration of disciplines*

Information design is a cross discipline integrating science, technology and art. To define the information design standing in an isolated point is not scientific. We must clear the definition, then to carry out design education based on their own professional characteristics with consideration of other disciplines. In addition to the existing research on design theory, design application and design thinking of a design education system, we should also search for the new direction of design education, set up an interdisciplinary research team, pay more attention to the cultivation of compound talents and the promotion of scientific research competitiveness. Building interdisciplinary is a strong promotion of the integration of advantage academic resources. The growth points of new subject maybe arise through the common research from a wide range of disciplines, especially for art design discipline in comprehensive university that should make uses of the platform advantage to implement the goal. In a word, we should construct an integral design education system for the information age that beyond the model based on operation and performance.

4.3 *Mastery of emerging technology*

Today, technology is a part of life just like metabolism. Art, science and technology are merely part of the surrounding environment for modern industrial designers, there are some other influential factors such as sociology, biology and psychology. The change of the external environment and conditions mean a new method chance for designers. Putting a high value on new technical means is important to the development of art design discipline. With the progress of science and technology, the application of information technology is very extensive in many fields, also in design discipline: Generative Design is used in the architectural design, Virtual Reality in industrial design, and Interaction design in visual communication design, etc. Therefore, it is easy to see that technology determines and impact the forms of artistic expression to some extent. Traditional knowledge systems of design education cannot meet the requirements of the information age, to master knowledge of different disciplines and techniques is essential to adapt to this background.

5 CONCLUSION

It goes without saying that the rapid development of the information age is vital to the development of design education and design industry, design discipline in such environment inevitably face more opportunities and challenges. The eternal basic principle of design is to design for life and human, so all the problems of design can be summarized as "Design for life", furthermore, design activities should be launched based on the fundamental principle. Besides, design thinking and method should be combined with the development characteristics of the times to follow the footsteps of technical innovation and apply new technologies with the times for a better future.

REFERENCES

[1] Lu Xiaobo, Jiang Shen, “Mission & Transformation : Information Design in the Cloud Computing Era”, DESIGN RESEARCH, Vol.5 No.5 2011.

[2] Lu Xiaobo, “New Development Trends of Design Subject in Information Society--Infomation Design”, DECORATION, Vol.50.2008.

[3] Zhao Qinping, “The Overview of Virtual Reality”, Science in China(Series F:Information Sciences), Vol.1.2009.

Computing, Control, Information and Education Engineering – Liu, Sung & Yao (eds)
 ISBN: 978-1-138-02800-5

Heterogeneous data integration system based on ontology

Yong Liang Jiang
Network Center, Hainan Normal University, Haikou, China

Ya Min Zhang
Department of Computer, Luohe Medical College, Luohe, China

ABSTRACT: To effectively solve the problem of data integration between different departments in colleges, universities, data integration technology and ontology technology was studied. Considering the characteristics of all kinds of data integration technology and combining with the actual situation of each data source we used the intermediate technology for a university to build a set of heterogeneous data integration system. The ontology technology was introduced to the system development, which effectively solved the semantic heterogeneity problem of different data sources. Global ontology and local ontology hybrid ontology approach were used to create the ontology. A web service for each data source was created. That improved the flexibility of the system and could effectively cope with changes in the data sources.

KEYWORDS: Ontology; Heterogeneous data; Data integration; Semantic integration.

1 INTRODUCTION

With the development of university informatization construction, various departments of universities have established the information management system suited to their own needs. These management information systems have played an important role in the information construction of colleges and universities. However, due to the lack of unified planning and different development technology made the data between those systems exist great differences. These differences made the data of different departments, independent of each other and not be shared. Existing data integration technology, such as a federal database, middleware, data warehouse, etc., just takes the information systems' data to carry on the simple extraction and restructuring, and cannot solve the problem of semantic heterogeneity. So, realizing the semantic integration among different department data is an urgent problem in the process of informatization construction in current universities.

2 ONTOLOGY

Ontology was originally a philosophical concept which was used to describe the nature of things. It is proposed by the philosopher R. Goclenins. Along with the development of the computer technology, ontology was introduced into the informatics and artificial intelligence. In the field of computer, ontology is a clear formal specification of the shared conceptual model [1]. The emergence of the semantic web made a further application of ontology. Now the ontology has become a core of semantic integration technology. In addition, the data integration based on ontology technology can effectively solve the data integration of knowledge acquisition, knowledge renewal and knowledge sharing. If the ontology technology is applied to the heterogeneous data integration system in universities, it can effectively solve the problem of semantic integration of data. That is, Web Services and XML technology unable to complete.

3 SYSTEM DESIGN

Ontology with its strong ability of semantic expression, has been applied in terms of semantic integration. Different departments created their own information systems in the information construction. Those systems played a huge role in the information construction of universities. Different creation time, development tools and development vendors made the same objection, synonymous names and other semantic heterogeneity problems exist in different information. That has become an important factor restricting the development of information construction of universities. Considering the characteristics of distributed heterogeneous data sources in universities and the practical application requirement of heterogeneous data integration system we developed a heterogeneous data integration system based on ontology. It effectively solved the semantic integration between different department information. The system is designed based on middleware technology. It uses three layer structure[2]: data layer, middle layer and application layer. The system structure is shown in the figure 1.

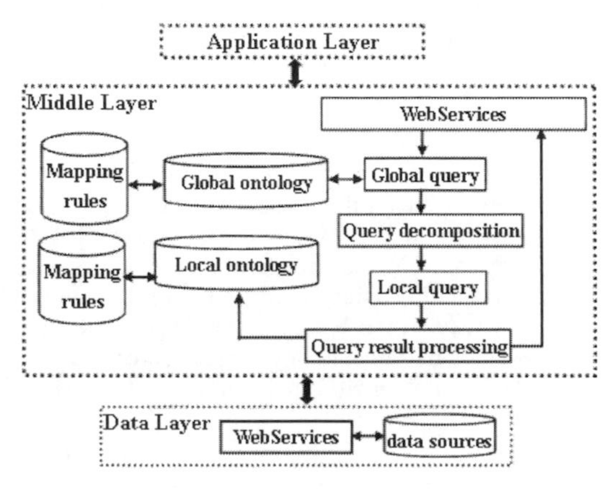

Figure 1. System architecture.

The data layer is composed of relational databases which are distributed in different departments of information system. The system creates a wrapper for each data source. Each wrapper is encapsulated into a web service. Each wrapper is used to receive queries from the middle layer. After receiving query requests of the mediation layer, the wrapper can convert them into ontology database support query. In addition, the wrapper is used to extract the data source of metadata, and carried on the package in the form of web service, then sent to the middle layer.

The middle layer is the core of the system. It is mainly composed of ontology library, mapping rules library, query processing module. An ontology library is composed of global ontology and local ontology. Global ontology is a normalized description of data model, which is used to define the data source object model and semantics. Global ontology can also provide a common semantic model for the application layer users. The local ontology is a semantic description of each independent and autonomous data source. The mapping rule base is made up of mapping rules which are made up of global ontology and local ontology mapping rules and mapping rules between the data source and the local ontology. It is the key to the data sharing among heterogeneous data sources. The query processing module is the query system of execution part. Receiving query request from the application layer, firstly it will handle the user's input based on the global ontology and generate a global query. Secondly, it will put the global query into a local query according to the generated global query and mapping rules in the library rules. Thirdly, it will execute the local query for each data source. Finally, it will process the query results and send them to the application layer.

4 SYSTEM IMPLEMENTATION

4.1 *Ontology design*

This system aims to solve the semantic integration of heterogeneous data sources. Hybrid ontology approach was used to design the ontology for the system.Drawing lessons from the theory of knowledge management technology, refering the domestic teaching resources management technology, and according to the school of digital campus construction requirements for data integration we design the following seven types of ontology as global ontology: teaching resource ontology, course ontology ,learning object ontology, teaching management of ontology ,department of ontology, management of object ontology and user ontology. Teaching resource ontology, course ontology, learning object ontology and user ontology are used to solve the problem of heterogeneous data integration of teaching resources. Teaching management of ontology, department of ontology,management of object ontology and user ontology are used to solve the problem of teaching management of heterogeneous data integration.

After studying the construction methods, construction principle, describing language of ontology we proposed a Teaching System Design algorithm to design the ontology for the system. The working principle of algorithm is shown in Figure 2.

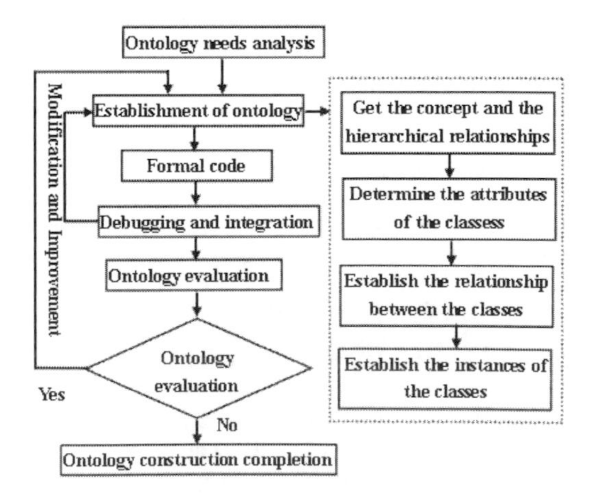

Figure 2. Teaching system design algorithm.

A relational database is composed of metadata and tuples. Metadata is mainly used to describe the relationship name, attribute name, data type, primary key, foreign key and integrity constraint information. A tuple is the real data. An ontology is usually made up of concepts and its relationships. In this system, the ontology is described by the OWL language. Based on the OWL ontology, the concept is known as the class of the object. The relationship between the concepts is characterized by how to rule the class hierarchy, attribute, value of the constraint, disjoint description and the logical relationship between objects. To convert a relational database to an ontology two work needs to be done as follows: the first one is that extracting useful information from a database metadata and converting it to the concept of ontology and the relationship between the concepts. The second one is that extracting useful information from tuples and converting it to instances of the ontology. In the process of various data sources to ontology, for the metadata of transformation rules, here no longer detail, you can reference [4]. The information of tuples in a relational database is mainly used to describe entities of a database. We can convert the tuples to instances of the class. If an entity's primary key has only one attribute we can use the attribute value for the instance name which is transformed from the entity. If the entity's primary key has multiple attributes we can first transform the attribute value into characters and connect them up. After that the instance can be named after the connected characters.

Relational database tuples usually imply a large number of semantic information. In order to ensure the ontology semantic richness we can extract semantic information in the process of from tuple to the instance. The extraction process is: first, by analyzing the tables' primary and foreign key relationships determine a link between entities, and then by further analyzing the tables' primary and foreign key relationships obtain the specific relationship between them.

4.2 *Ontology mapping*

The process of ontology mapping is the discovery of the semantic connection between ontologies[5]. The ontology of this system is based on the hybrid ontology approach to build, so it mainly involves the global ontology mapping to local ontology and the local ontology mapping to the data source. Local ontology mapping to data source has been done in the process of a relational database of ontology conversion. We will not mention it. Here we only introduce the global ontology to local ontology mapping. The global ontology to local ontology mapping is completed in accordance with the following rules[6]:i If two elements have the same URI, these two elements are regarded as the same elements;ii If the two concepts have the same parent, or have the same concept, or have the same brothers,these two concepts may be the same;iii If the two concepts have the same instance, or the same properties, these two concepts may be the same;iv If the two attributes have the same domain and range, these two properties may be the same;v If the two attributes have the same father, or the same properties, these two properties may be the same.

4.3 *Query processing*

Receiving the user's query reques the query processing module immediately generate global query and determine which data source will be queried. Then the global query are converted into queries of data source. After the queries execution of each data source, the query results are integrated and sent to the application layer. SPARQL language is used to query the ontology in this system. Data query is completed by using SQL query language. Query transformation algorithm is used the algorithm of reference[7].

5 CONCLUSION

In this paper, the ontology technology was applied to the heterogeneous data integration system in universities,which effectively solved the problem of semantic integration that was the traditional data integration technology can not complete. The system is based on the intermediate technology development that ensures the real-time and the flexibility of the system for the integrated data. The implementation of this system for information construction of universities is of great significance.

ACKNOWLEDGMENT

I would like to express my gratitude to all those who helped me during the writing of this thesis. This work was supported by Science Foundation of Hainan Province, China (612136) and Scientific Research Foundation of the Higher Education Institution of Haina Province,China (Hjkj2012-59).

REFERENCES

[1] Borst , W. Construction of Engineering Ontologies for Knowledge Sharing and Reuse. Ph.D. Dissertation, University of Twente(1997).

[2] Yongliang Jiang,Chuanyi Fu,Jie-qing Xing,in: Modern Computer,vol.205,pp.164–166(2009) In Chinese.

[3] Hainaut J.L.. Research in database engineering at the University of Namur. SIGMOD Recor(2003),pp:124–128.

[4] Guiqing Jiang,Jiao Lu, in:Application Research of Computers,vol 28,pp.3018–3021(2011),In Chinese.

[5] Xiaomeng Su, Technical Report, Norway,Department of Computer and Information Science, Norwegian University of Science and Technology (2002).
[6] Lili Yang, Research and Implementation of Ontology-based Heterogeneous Data Integration System(2011), Northwest Agriculture and Forestry university, In Chinese.
[7] Xin Guo, Research and Implementation of Ontology based Heterogenous data Integration Technology, The Second Academy of China Aerospace(2008) In Chinese.

Computing, Control, Information and Education Engineering – Liu, Sung & Yao (eds)
© 2015 Taylor & Francis Group, London, ISBN: 978-1-138-02800-5

Adaptive compressive sensing using optimized projection matrix

Ya Peng & Xiao Qin Song
Nanjing University of Aeronautics and Astronautics, Nanjing, China

Yong Gang Zhu
Nanjing Telecommunication Technology Institute, Nanjing, China

ABSTRACT: A two-stage adaptive compressive sensing procedure was proposed in this paper. In the first stage, the initial matrix is optimized via decreasing its mutual coherence. In the second stage, i.e. adaptive stage, each ensuing row of the projection matrix is sequentially designed towards minimizing the Cramer-Rao bound of recovery errors. The proposed approach can be regarded as the combination of the projection matrix optimization with adaptive process, improving the SNR of the measurements. Simulation results show that the proposed adaptive compressive sensing procedure performs better than many traditional compressive sensing methods in the aspect of anti-noise ability and recovery accuracy.

KEYWORDS: compressive sensing; adaptive sensing; optimized projection; subspace pursuit.

1 INTRODUCTION

Compressive Sensing (CS) has a variety of potential applications, and is currently at the forefront of signal processing research domain. It states that the signals that have a sparse or compressible representation on a dictionary can be recovered from far less linear measurements than those required by the classical Shannon-Nyquist theorem [1]. The measurements used in CS correspond to linear projections obtained by a projection matrix. Existing research has shown that nonadaptive random sensing matrices with Gaussian or Bernoulli distributions satisfy the fundamental theoretical requirements of CS [2]. On the other hand, a projection matrix that is optimally or adaptively designed based on prior measurements can further improve the reconstruction accuracy or take fewer samples. The idea behind adaptive compressive sensing is that one should focus sensing power toward the non-zero components of the spare signal in order to increase the signal-to-noise ratio (SNR). The potential advantages of an adaptive projection scheme are demonstrated in [3]-[5].

In this paper we proposed a two-stage adaptive compressive sensing procedure, stated in Algorithm 1, that delivers such improvements. In the first stage of our approach, the initial projection matrix is optimized with less coherent columns which effectively improves the reconstruction quality. Then the second stage uses previous recovery results to guide the design of the next sensing vector, which makes it possible to reconstruct the signal exactly at lower SNRs. Simulation results (see Fig. 1, 2 and 3) show that the developed procedure consistently outperforms traditional (non-optimized, non-adaptive) CS by 2-8 dB. Indeed, as more information is obtained from the measurements, subsequent samples could be focused more directly onto the relevant signal elements, providing a significant improvement of the estimation performance.

The rest of this paper is organized as follows. In Section 2 we formalize the signal model that will be employed throughout. We propose our adaptive sensing procedure in Section 3, and present experimental results demonstrating the developed approach in Section 4. Finally, conclusions of this work are discussed in Section 5.

2 SIGNAL AND OBSERVATION MODELS

Let $x = [x_1 \ldots x_N]^T$ be an unknown sparse vector with $K << N$ non-zero entries. We assume that we can learn about x by taking M noisy linear measurements of the form

$$y_m = \phi_m^T x + e_m, \quad m = 1, \ldots, M. \tag{1}$$

where $\phi_m \in R^N$ are sensing vectors, and e_m are assumed to be white Gaussian with variance σ^2, i.e., $e_m \sim N(0, \sigma^2)$. Collect all the measurements and equation (1) is equivalent to the linear model

$$y = \phi x + e. \quad (2)$$

where $y = [y_1, \dots, y_M]^T$, Φ is the $M \times N$ matrix whose rows are the sensing vectors $\{\phi_m\}_{m=1}^M$ and $e = [e_1, \dots, e_M]^T$ whose entries are independent to each other. In addition, we assume that the rows of Φ are normalized, essentially limiting the amount of sensing energy available for each projection observation.

The goal of the problem we consider is to recover the sparse signal x from the measurement y. Generally speaking, the approaches to this problem can be divided into three categories: optimization algorithms, greedy iterative algorithms, and iterative thresholding algorithms. Subspace pursuit (SP) [6], one popular greedy iterative method capable of achieving stable sparse signal reconstruction against noise, is selected as recovery algorithm in this paper.

3 ADAPTIVE COMPRESSIVE SENSING APPROACH

As indicated in the introduction, an outline of the proposed two-step adaptive sensing approach is described in Algorithm 1. With a budget of M measurements of the form of equation (1) or (2), the algorithm proceeds as follows. Here, we divide the projection matrix Φ into two parts which are designed in two stages respectively. The aim of the first stage is to optimize the initial matrix $\Phi_{1:M_0}$ (see Section 3.1) via decreasing its mutual coherence, which will lead to higher recovery quality. Then in the sequential sensing stage, x is reconstructed sequentially, and each row of $\Phi_{M_0+1:M}$ (derived in Section 3.2) is designed adaptively based on prior estimate of x. The key advantage of this adaptive algorithm is that the signal can be recovered at much lower mean-squared error (MSE) compared to traditional (non-optimized, non-adaptive) CS methods.

Algorithm 1: Two-stage adaptive compressive sensing

Input: total measurement budget M, coherence threshold t, shrinkage scaling factor Υ, number of iterations L

Initialize: Set $M_0 = \lfloor M/2 \rfloor$ (number of rows of projection matrix in Stage 1).

Set $\Phi_{1:M_0} \in R^{M_0 \times N}$ to be a random matrix.

Stage 1: (Optimize the Initial Projection)

for $l = 0, \dots, L$

1 Normalize: Normalize the columns in the matrix $\Phi_{1:M_0}^{(l)}$ and get $\hat{\Phi}_{1:M_0}^{(l)}$.
2 Compute Gram Matrix: $\hat{G}^{(l)} = \left(\Phi_{1:M_0}^{(l)}\right)^T \Phi_{1:M_0}^{(l)}$.
3 Shrink: Update $G^{(l)}$ and obtain $\hat{G}^{(l)}$ via (4).
4 Reduce Rank: Apply SVD and force the rank of $\hat{G}^{(l)}$ to be equal to M_0.
5 Update $\Phi_{1:M_0}$: Compute the squared-root $\left(\Phi_{1:M_0}^{(l+1)}\right)^T \Phi_{1:M_0}^{(l+1)} = \hat{G}^{(l)}$ and get $\Phi_{1:M_0}^{(l+1)}$ of size $M_0 \times N$.

end

Stage 2: (Sequential Sensing)

for $m = M_0, \dots, M-1$

1 Collect Measurements: $y_{1:m} = \Phi_{1:m} x + e_{1:m}$.
2 Estimate: Reconstruct x using SP algorithm and obtain an approximation $\hat{x}^{(m)}$.
3 Update Support: Find the support set $\Phi^{(m+1)}$ of $\hat{x}^{(m)}$.
4 Compute eigenvalue decomposition: $\Phi_{1:m,\Phi^{(m+1)}}^T \Phi_{1:m,\Phi^{(m+1)}} = V^{(m+1)} \Phi^{(m+1)} U^{(m+1)}$.
5 Generate Sensing Vector: Find the minimum eigenvalue in $\Phi^{(m+1)}$ and get the next sensing vector $\phi_{m+1,\Gamma^{(m+1)}}$ which has unit norm.

end

Output: the estimated signal $\hat{x}$.

3.1 *Stage 1: Optimization of initial projection*

In CS theory, it is realized that in order to be able to attain a reasonably small M and yet have a stable recovery, columns of Φ must have small mutual coherence. In general, the smaller the mutual coherence, the better the reconstruction quality. Elad [7] proposed a shrinkage technique to iteratively decrease the t-average mutual coherence of the effective dictionary. As our work follows in part Elad's idea, its mutual coherence defined as the average of all absolute and normalized inner products between different columns in Φ (denoted as g_{ij}) that are above the predefined threshold t. Formally

$$\mu_t\{\Phi\} = \frac{\sum_{1 \le i,j \le k \text{ and } i \ne j} (|g_{ij}| \ge t) \cdot |g_{ij}|}{\sum_{1 \le i,j \le k \text{ and } i \ne j} (|g_{ij}| \ge t)}. \quad (3)$$

For $t = 0$, we obtain an average of all the absolute off-diagonal entries of Gram matrix $G = \Phi^T \Phi$. As the value of t grows, we see that $\mu_t\{\Phi\}$ grows accordingly, and $\mu_t\{\Phi\} \ge t$.

In the optimization stage, described in Algorithm 1, we start with a random matrix $\Phi_{M_0 \times N}$. As the main objective is to iteratively minimize $\mu_t\{\Phi\}$, Grassmanian frame is adopted to reduce the inner products that are above t in absolute value. We denote

$\hat{g}_{ij}$ as the shrunken element of given g_{ij} which is obtained by applying the following shrinking operation [8]:

$$\hat{g}_{ij} = \begin{cases} \gamma g_{ij}, & |g_{ij}| \geq t \\ \gamma t \cdot sign(g_{ij}), & t > |g_{ij}| \geq \gamma t \\ g_{ij}, & \gamma t > |g_{ij}| \end{cases} . \quad (4)$$

where $0 < \gamma < 1$ is the down-scaling factor used to shrink the values of the Gram matrix.

In general, the aforementioned shrinking operation causes $\hat{G}$ to become full rank. Thus, in the next two steps, force the rank of $\hat{G}$ to M_0 by applying singular value decomposition (SVD) and obtain the emerged projection matrix $\Phi_{1:M0}$ which is given by the M_0 eigenvectors corresponding to the M_0 largest eigenvalues of the new Gram matrix. By iterating many times, $\Phi_{1:M_0}$ is optimized to a less coherent matrix which leads to better CS performance.

3.2 *Stage 2: Sequential sensing*

Considering the problem of estimating a sparse vector x from the measurements y, we measure the quality of an estimator using the MSE, defined as

$$E\{\|\hat{x} - x\|_2^2\} \quad (5)$$

where $\hat{x}$ is the recovery result. Our goal in the current subsection is to minimize the MSE. The sequential sensing stage we proposed is related to other recent efforts in adaptive CS. As shown in [5], minimizing the Cramer-Rao bound (CRB) related to minimizing the MSE for reconstruction purposes. When the bound is tight, decreasing the CRB means that MSE can be decreased accordingly. A result of the CRB, given in [9], on recovery errors is derived as

$$E\{\|\hat{x} - x\|_2^2\} \geq \sigma^2 \mathrm{Tr}((\Phi_\Gamma^T \Phi_\Gamma)^{-1}), \qquad \|x\|_0 = K. \quad (6)$$

where $\mathrm{Tr}(\cdot)$ denotes the trace of a matrix and Φ_Γ represents columns of Φ indexed in Φ.

In sequential sensing stage (see Algorithm 1), the sensing vectors $\phi_{M_0+1}, \dots, \phi_M$ are sequentially optimized via minimizing the CRB, $\sigma^2 \mathrm{Tr}((\Phi_\Gamma^T \Phi_\Gamma)^{-1})$. It was shown in [5] that the sampling vector ϕ_{m+1} is the eigenvector of $\Phi^T_{1:m,\,(m)} \Phi_{1:m,\,(m)}$ corresponding to the minimum eigenvalue. In the initial measurement step, obtain a collection of M_0 measurements using sensing matrix $\Phi_{1:M_0}$, which is optimized in the previous stage. Next, estimate x from the measurements $y_{1:m}$ and update the corresponding support set of $\hat{x}^{(m)}$. In the following step, the eigenvector corresponding to the minimum eigenvalue of the Gram matrix can be used to generate the next sensing vector ϕ_{m+1}, and then the $(m+1)th$ measurement will be obtained in the ensuing iteration. This process is iterated several times, resulting in much lower recovery errors, which finally can estimate the sparse signal exactly.

4 EXPERIMENTAL RESULTS

To illustrate the effectiveness of the proposed algorithm, we perform some numerical experiments. Specifically, we compare the performances of four methods as follows:

- The standard SP algorithm (denoted as StdSP, [6]): We set the sensing matrix Φ of size $M \times N$ to be a Gaussian matrix with i.i.d zero mean and $1/N$ variance Gaussian entries.
- Optimized SP algorithm (denoted as OptSP, [6] [7]): The initialized sensing matrix $\Phi_{M \times N}$ is the same as the counterpart in StdSP and then we use the algorithm as described in [7] to optimize it. The parameters are selected as $t = 0.7$, $\gamma = 0.95$ and number of iterations $L = 100$.
- Adaptive SP algorithm (denoted as AdaSP, [5]): We set $M_0 = 64$ and $\Phi_{1:M_0}$ is the same as the counterpart in StdSP.
- The proposed algorithm (denoted as ProASP): The two-stage adaptive sensing approach was implemented according to Fig. 1 with inputs as above specified. And each row of the sensing matrix is normalized.

Four methods are compared under different measurements M, sparsity order K, and SNRs. Our results represent an average of 500 independent trials. Throughout our simulations, the test signals $x \in R^N$ $(N = 256)$ are constructed to have K nonzero entries (at randomly selected locations) and generated from identical distribution $N(1, 0.1)$ in each trail. We use SP algorithm in reconstruction part and iterate 50 times. In addition, the signal to noise ratio is defined as

$$\mathrm{SNR\ (dB)} = 10 \log_{10}\left(\frac{\|x\|_2^2}{N\sigma^2}\right). \quad (7)$$

In all experiments, Gaussian noise with zero mean and variance σ^2 is added to each measurement, as specified in (1).

Fig. 1 shows the MSEs of four methods versus observation length M. In this experiment we set $K = 5$ and SNR = 10 dB. At the origin $M = M_0 = 64$, ProASP achieved much lower MSE than AdaSP, but the same as OptSP. It is mainly because the effect of the adaptive stage in ProASP has not worked enough. As M increases, we can see that the MSEs achieved by ProASP quickly degrades and outperformed the other discussed methods by > 2dB. It suggest that our proposed method often yields lower average reconstruction errors than other methods.

Fig. 2 depicts the performance comparison of four methods at different sparsity order K, where SNR = 10 dB and the total number of measurements $M = 128$, and MSEs are calculated when all samples are obtained. Again, the MSEs achieved by ProASP were significantly lower than those achieved by the other methods. At $K = 10$, ProASP outperformed AdaSP by approximately 2 dB and other methods by > 8 dB, and at $K = 28$, ProASP outperformed AdaSP by approximately 3 dB and the other methods by even more.

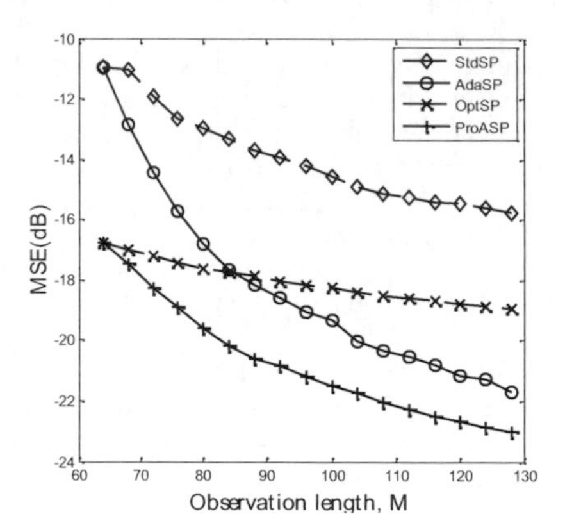

Figure 1. MSEs versus Observation length M.

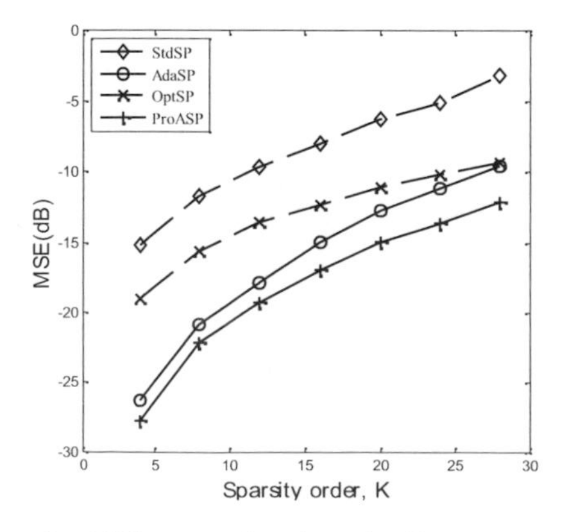

Figure 2. MSEs versus Sparsity order K.

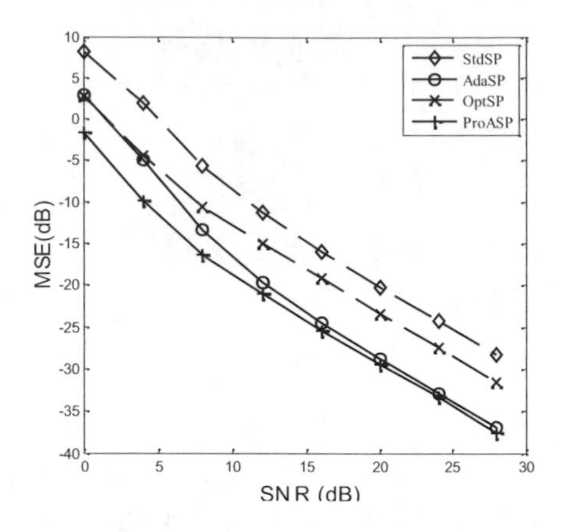

Figure 3. MSEs versus SNR.

Fig. 3 compares the performance of four methods at several SNRs. We consider $K = 10$ and $M = 128$. MSEs are calculated after all M measurements are collected. There we see that PorASP achieved lower MSE than the other algorithms. In particular, it outperformed AdaSP by approximately 4 dB at low SNR.

5 CONCLUSIONS

In this paper, we have developed a novel two-stage adaptive CS procedure for recovery of sparse signals in additive Gaussian noise. The procedure combines the projection matrix optimization with adaptive process, which makes our algorithm be able to reconstruct the signal exactly at lower SNRs. We have demonstrated that this joint method outperforms both the use of optimized sensing matrices and those matrices that are adaptively designed.

ACKNOWLEDGMENT

This work was supported by the National Natural Science Foundation of China (No.61301103, 61401505), Jiangsu Province Natural Science Foundation (No.BK20130069), Foundation of Science and Technology on Communication Information Security Control Laboratory (No. 9140C130306130C13060) and China Postdoctoral Science Foundation (No.2012M521853, 2013T60914).

REFERENCES

D. L. Donoho, "Compressed sensing," IEEE Trans. Inf. Theory, vol.52, no. 4, pp. 1289–1306, Apr. 2006.

E. J. Candes, J. K. Romberg, T. Tao, "Stable signal recovery from incomplete and inaccurate measurements," Communications on Pure and Applied Mathematics, vol. 59, no. 8, pp. 1207–1223, 2006.

S. Ji, Y. Xue, and L. Carin,"Bayesian compressive sensing," IEEE Trans. Signal Process., vol. 56, no. 6, pp. 2346–2356, Jun. 2008.

J. Haupt, R. Castro, and R. Nowak, "Distilled sensing: Adaptive sampling for sparse detection and estimation," IEEE Trans. Inf. Theory, vol. 57, no. 9, pp. 6222–6235, Sept. 2011.

T. Huang, H. Meng, and X. Wang, "Adaptive Compressed Sensing via Minimizing Cramer-Rao Bound," IEEE Signal Process. Lett., vol. 21, no. 3, pp. 270–274, Mar. 2014.

W. Dai and O. Milenkovic, "Subspace pursuit for compressive sensing signal reconstruction," IEEE Trans. Inf. Theory, vol. 55, no. 5, pp. 2230–2249, May. 2009.

M. Elad, "Optimized projections for compressed sensing," IEEE Trans. Signal Process., vol. 55, no. 12, pp. 5695–5702, Dec. 2007.

I. S. Dhillon, R. W. Heath, Jr, and T. Strohmer, "Designing structured tight frames via alternating projection," IEEE Trans. Inf. Theory, vol. 51, no. 1, pp. 188–209, Jan. 2005.

Z. Ben-Haim and Y. Eldar, "The Cramer–Rao bound for estimating a sparse parameter vector," IEEE Trans. Signal Process., vol. 58, no. 6, pp. 3384–3389, Jun. 2010.

Computing, Control, Information and Education Engineering – Liu, Sung & Yao (eds)
© 2015 Taylor & Francis Group, London, ISBN: 978-1-138-02800-5

Heterogeneous ensemble of Chinese message sentiment classification on automatic animation generation

J.L. Zhang
Beijing University of Technology, Beijing, China

ABSTRACT: As an important part of mobile phone auto-generated animation technology, message sentiment classifications is to analysis and classify the emotional tendency of the message, provide emotional information for subsequent animation plot planning and animation scene. Sentiment classification is divided into two steps: subjective and objective classification of the message and sentiment tendency classification of subjective message. Due to single classifiers used directly for these two perform low accuracy in Chinese message sentiment classification. The aim of this paper is to use a heterogeneous ensemble of classifiers to improve the accuracy, which combines some different classifiers by ensemble techniques instead of focusing on ensemble techniques within a classifier. This ensemble method is applied to a Chinese message corpus from manual collection. The results show that the proposed heterogeneous ensemble approach yields higher accuracy when compared with using only a single classifier, better meets the needs of mobile phone auto-generated animation technology.

KEYWORDS: Auto-generated animation technology; Chinese message sentiment classification; Heterogeneous ensemble.

1 INTRODUCTION

Auto-generated animation technology on a mobile phone is an application of 3D auto-generated technology on phone messages, which aim to send back 3D auto-generated animation according to the sender's message. The whole system process is divided into four parts: information extraction, plot planning, scene calculation and animation rendering. Information extraction is extracting key information from the Chinese message content. As a part of it, the Chinese message sentiment classification is the analysis and classification of the emotional tendency of the message, provide emotional information for subsequent animation scene, it is important to the mobile phone auto-generated animation technology.

The paper is organized as follows. In Section 2, we review related work. Section 3 discusses some classification algorithms and heterogeneous ensemble. Experiments and results are discussed in Section 4 followed by the concl lusions and prospects in Section 5.

2 RELATED WORK

At present, the existing research about Chinese message sentiment classification in our laboratory is using the Naive Bayes classification algorithm on subjective and objective classification of the message in the first stage, and using multi-label K nearest neighbor algorithm for sentiment tendency classification of subjective messages in the second stage, which concerns about classifying four big granularity sentiment, that is happy, anger, sadness and fear. But accuracies of these two parts are not high. It is often unable to extract the correct results in the practical application, which directly affects the subsequent animation effects. The error results lead that scene animation expresses wrong emotion, which is not consistent with the message content. So it is necessary to improve accuracy of Chinese message sentiment classification.

In Chinese message sentiment classification, both subjective and objective classification and sentiment tendency classification are using the single classifier, so we consider combination of different multiple classifiers to improve accuracy. That is an ensemble learning method. Ensemble learning is the most important hot research topic in machine learning (Dietterich 1997). It is mainly training multiple classifiers to solve a problem (Polikar 2006). In recent years, ensemble techniques are becoming increasingly important as they have repeatedly demonstrated the ability to improve upon the accuracy of single classifiers (Tahir et al. 2007). Ensembles can be homogeneous, in which every base classifier is constructed using the same algorithm (such as bagging, boosting), or heterogeneous in which base classifiers are constructed using different algorithms and the outputs of these classifiers are combined by some way (such as an average value method, the majority voting method, Bias method of voting) (Yu 2003). The aim of this paper is to use heterogeneous ensembles of base classifiers to improve accuracy.

3 CLASSIFICATIONS ALGORITHMS AND HETEROGENEOUS ENSEMBLE

The core problem of text classification is constructing a classification function or a classification model with a large number of labeled training samples according to some strategies and using this model to maps unlabeled testing samples to the specified class space. Chinese message sentiment classification includes the following two subtasks:

1 Subjective and objective classification: Chinese message can express either subjective emotion or objective emotion. This is the most common binary classification, which belongs to the single-label classification.
2 Sentiment tendency classification: a sample can belong to more than one category in practical problems. For example, in document classification problems (Kazawa et al. 2003), each document may belong to more than one predefined themes; in image classification (Boutell et al. 2004), each image may contain different semantics. That is, the text content can be divided into a plurality of predefined classes. Likewise, a Chinese message may also convey a variety of emotions. For sentiment tendency classification, this paper focuses on four granularity emotions: happy, anger, sadness and fear for multi-label classification.

3.1 *Single-label classification algorithm*

There are many single-label classification algorithms, such as Naive Bayes, K-Nearest Neighbor, Support Vector Machine and so on (Mitchell 1997). Among them, NB is based on Bayes theorem and characteristic conditions independence assumption. KNN is a classification algorithm of supervised learning. SVM solves machine learning problems with optimization methods. Decision Tree represents a mapping relationship between the object properties and the object values. The representative algorithm is C4.5.

3.2 *Multi-label classification algorithm*

There are many multi-label classification algorithms, such as Random K-Labelsets, Multi-Label KNN, Classifier Chain and so on.

1 RandomK-Labelsets: Multi-label classification can be reduced to the conventional classification problem by considering each unique set of labels as one of the classes. This approach is referred to as label Powerset (LP) in the literature. However, this approach leads to a large number of label subsets with the majority of them with a very few examples and it is also computationally expensive. Random K-Labelsets (Tsoumakas & Vlahavas 2007) (RAKEL) constructs an ensemble of LP classifiers where each LP classifier is trained using a different small random subset of the set of labels. RAKEL not only consider the correlations between labels, but also avoid a large increase in the number of labels with only a few instances.
2 Multi-Label KNN: As a lazy learning algorithm, Multi-Label KNN (Zhang & Zhou 2005; MLKNN) has been proposed in 2007.This method is derived from the KNN algorithm. It is mainly divided into two steps, in the first step, for each test sample, its k nearest neighbors in the training sample are determined; in the second step, according to the statistical information acquired from these neighbor instances, a set of labels of the test sample is determined with maximum a posteriori probability principle The maximum a posteriori probability is the product of prior probability and posterior probabilities for each tag in KNN. This method was shown to perform well in some domains, e.g. in predicting the functional classes of genes in the Yeast Saccharomyces cerevisiae (Cheng & Hullermeier 2009).
3 Classifier Chain: This approach is referred to as Binary Relevance (BR) learning in the literature. In BR learning, the original multi-label dataset Dis divided into $|Y|$ data sets d_j (j=1...q)where Y is the finite set of labels. Each dataset d_j contains all samples of original data set D, and each instance in d_j is marked as either Positive or Negative. When classifying a test instance X, BR learns one binary classifier h_a: $X \rightarrow \{\neg a, a\}$ for each label a will output aY.BR learning is theoretically simple and has a linear complexity with respect to the number of labels. But the limitations of the BR algorithm are not considering correlations among the labels. The

Classifier Chain (Read et al. 2009) (CC) approach uses a chain to make the mark information transfer in the classifiers based on BR. It considers the correlation between labels.

4 Binary Relevance K-Nearest Neighbor: Binary Relevance K-Nearest Neighbor (Spyromitros et al. 2008; BRKNN)is the adaptation algorithm of KNN. This method combines KNN with Binary Relevance (BR). The specific process consists of two main steps: in first stage, transform the multi-label dataset D to q single-label datasets $\{d_1,d_2,...d_q\}$(q is the total number of class labels of samples), dataset D_i is the corresponding category for label L_i. Then for a test instance X, predict the class label of it on every single-label datasets D_i by KNN algorithm. At last, the combination of the q results is the final forecasting result.

3.3 *Heterogeneous ensemble*

Ensembles can be homogeneous, in which every base classifier is constructed using the same algorithm (such as bagging, boosting), or heterogeneous in which base classifiers are constructed using different algorithms and the outputs of these classifiers are combined by some way (such as an average value method, the majority voting method, Bias method of voting). In this paper, both subjective and objective classification and emotional tendency classification adopt heterogeneous in order to get higher accuracy. Figure 1 represents the basic ideas of heterogeneous ensemble.

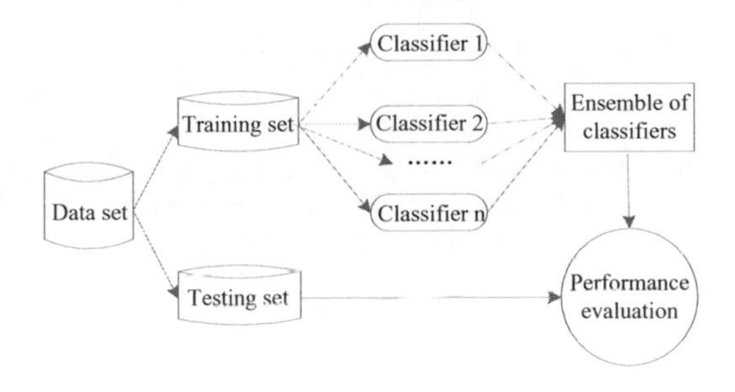

Figure 1. Heterogeneous ensemble.

4 EXPERIMENTS AND RESULTS

4.1 *Message sentiment classification process*

Sentiment classification is divided into two steps: subjective and objective classification of the message and sentiment tendency classification of subjective message. The process of Chinese message is illustrated in Figure 2.

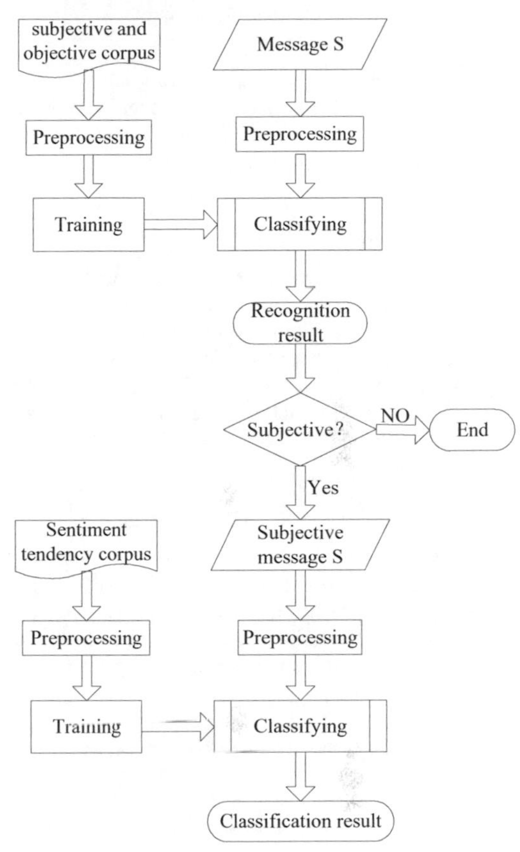

Figure 2. Message sentiment classification process.

4.2 *Data sets*

Due to the particularity of the message, there is not the Chinese message corpus used in the study of Chinese text emotion classification. So the experimental data sets are all from manual collection of real message set, with a total of 9600 Chinese messages.

4.3 *Method of heterogeneous ensemble*

There are many ways of combining the outputs of the base classifiers. Among them, the majority voting decision rule is in accordance with the principle of the minority is subordinate to the majority to obtain the final result. It is simple and classical. We use this rule to combine the outputs of these base classifiers.

4.4 *Evaluation measures*

In the mobile phone animation auto-generation project, accuracy will directly influence the classification effect. Therefore accuracy is the indicator of the experimental results of the experiment.

4.5 *Benchmark methods*

In this paper, the experiment classifiers are all from Weka and Mulan open source libraries. Among them, Naïve Bayes, KNN, support vector machine and decision tree algorithm these single-label classifiers are from Weka open source library. RAKEL, MLKNN, Classifier Chain and BRKNN these multi-label classifiers are all from the Mulan open source library.

4.6 *Subjective and objective classification*

In this part, the goal of the experiment is using the combination of different base classifiers to obtain the optimal ensemble. These base classifiers contain NB, SVM, KNN and C4.5. The heterogeneous ensemble of these base classifiers is illustrated in figure 3.

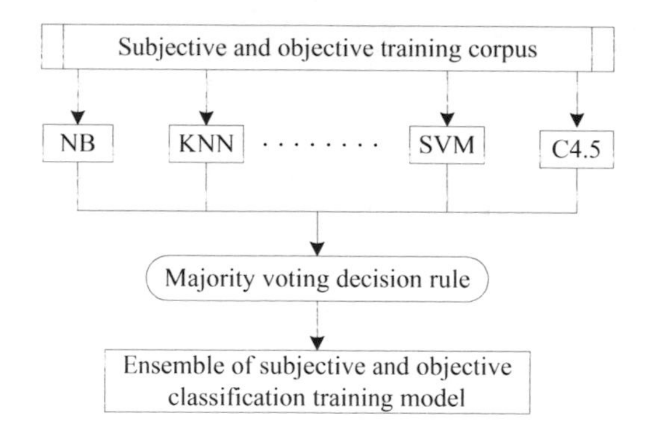

Figure 3. Heterogeneous ensemble of subjective and objective classification.

In the experiment, training corpus contains 3600 Chinese text messages totally, including 1800 subjective messages and 1800 objective messages. In experimental stage, training data are divided into N equal parts. The first equal part is regarded as the testing dataset and the remaining N - 1 equal parts are regarded as the training dataset. The experimental results are shown in Table 1:

Table 1. Accuracy of different heterogeneous ensemble.

Base	Accuracy			
classifiers	Three	Four	Five	Six
NB SVM	0.8	0.814	0.817	0.819
NB KNN	0.739	0.765	0.768	0.758
NB C4.5	0.761	0.764	0.765	0.771
SVM KNN	0.772	0.82	0.823	0.809
SVM C4.5	0.802	0.815	0.816	0.82
KNNC4.5	0.731	0.766	0.764	0.758
NB SVM KNN	0.831	0.856	0.864	0.863
NB SVM C4.5	0.79	0.801	0.806	0.806
NB KNN C4.5	0.78	0.786	0.795	0.797
SVM KNN C4.5	0.833	0.856	0.866	0.864
NB SVM KNN C4.5	0.806	0.821	0.829	0.863

Seen from Table 1, the results of an ensemble of NB, SVM and KNN and an ensemble of SVM, KNN and C4.5 are superior to accuracies of other ensembles. It is observed from Table 2 that training time of C4.5 is longer than the other classifiers, so in this paper, ensemble of NB, SVM and KNN is the final scheme in the part of subjective and objective classification.

Table 2. Training time of base classifiers.

Classifiers	KNN	SVM	NB	C4.5
Training time(s)	1	18	20	558

4.7 *Sentiment tendency classification*

Similarly, in sentiment multi-label classification experiment, there are MLKNN, RAKEL, CC and BRKNN as a base classifier to constitute different combinations to obtain the optimal ensemble. The heterogeneous ensemble of these base classifiers is illustrated in figure 4.

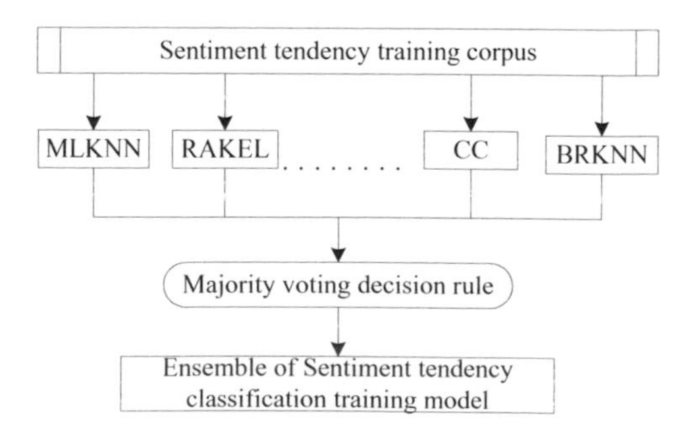

Figure 4. Heterogeneous ensemble of sentiment tendency classification.

In experimental stage, training data are divided into N equal parts. The first equal part is regarded as the testing dataset and the remaining N - 1 equal parts are regarded as the training dataset. The experimental results are shown in Table 3:

Table 3. Accuracy of different heterogeneous ensemble.

Base classifiers	Accuracy			
	Three	Four	Five	Six
MLKNN RAKEL CC	0.605	0.615	0.607	0.619
MLKNN BRKNN CC	0.609	0.611	0.602	0.623
MLKNN BRKNN RAKEL	0.59	0.664	0.595	0.601
BRKNN RAKEL CC	0.597	0.61	0.59	0.612
MLKNN RAKEL CC BRKNN	0.607	0.6.7	0.597	0.619

Table 3 shows that ensemble of BRKNN, RAKEL and MLKNN deliver a litter better performance than other ensemble. Ensemble of BRKNN, RAKEL and MLKNN is the final scheme in the part of subjective message sentiment tendency classification. In addition, Table 4 shows accuracies of these multi-label classifiers.

Table 4. Accuracy of base classifiers.

	Accuracy			
Classifiers	Three	Four	Five	Six
MLKNN	0.41	0.426	0.408	0.412
RAKEL	0.288	0.28	0.286	0.283
BRKNN	0.344	0.374	0.36	0.399
CC	0.406	0.406	0.396	0.425

Table 4 shows that the accuracy of MLKNN, RAKEL, BRKNN and CC is about 40%, 28%, 37% and 40% respectively. Heterogeneous ensemble of multi-label learners can provide a better performance when compared with the single multi-label classifier.

5 CONCLUSION AND PROSPECT

This paper adopts heterogeneous ensemble of classifiers on Chinese message sentimental classification. The experiment results show that the heterogeneous ensemble of classifiers can yield higher accuracy than single classifier. It is a kind of effective method of emotional information extraction. This paper applies a method of heterogeneous ensemble to the mobile phone animation auto-generation project and improves the accuracy of Chinese message sentimental classification significantly. It is of great significance to mobile phone animation automatic generation technology. But there are still several problems to be solved. So the further work as follows:

1. In current Chinese message multi-label sentiment corpus, message with single-label make up the most part. The latter experiments will consider adding more multi-label corpus to enhance the label correlation of classifications.
2. This paper adopts majority voting decision rule to combine the output of different base classifiers. There are many other ways of combining the outputs of classifiers, such as MAX, MIN, MEAN and so on. The latter experiments should consider using other ways of heterogeneous ensemble.

REFERENCES

Boutell, M. R., Luo, J. & Shen, X. etal. 2004. Learning multi-label scene classification. Pattern Recognition 37(9):1757–1771.

Cheng, W. & Hullermeier, E. 2009. Combining instance-based learning and logistic regression for multilabel classification, *Mach. Learn.* 76 (2–3):211–225.

Dietterich, T.G. 1997. Learning Research: Four Current Directions, AI Magazine 18(4):97–136.

Kazawa, H., Izumitani, T., Taira, H. et al. 2003. Maximal margin labeling for multi-topic text categorization, Proceedings of Advances in Neural Information Processing Systems, Vancouver, Canada.

Polikar, R. 2006. Ensemble based systems in decision making, IEEE Circuits and Systems Magazine 6(3):21–45.

Read, J., Pfahringer, B., Holmes, G. & Frank, E. 2009. Classifier chains for multi-label classification, *ECML/PKDD* 5782: 254–269.

Spyromitros, E., Tsoumakas, G. & Vlahavas, I. An empirical study of lazy multilabel classification algorithms: *Proc.5th Hellenic Conference on Artificial Intelligence, Syros, Greece, 2–4 October 2008.*

Tahir Atif Muhammad, Kittler Josef & Bouridane Ahmed. 2012. Multilabel classification using heterogeneous ensemble of multi-label classifiers, Pattern Recognition Letters 33:513–523.

Tom M. Mitchell. 1997. Machine Learning. New York: McGraw Hill.

Tsoumakas, G. & Vlahavas, I. 2007. Random k-labelsets: An ensemble method for multilabel classification: Proc. Of the 18th European Conference on Machine Learning, Warsaw, Poland, 17–21 September 2007.

Yu Shi Xin. 2003. Feature Selection and Classifier Ensembles: A Study on Hyperspectral Remote Sensing Data, Antwerp: Univ. of Antwerp.

Zhang, M.L. & Zhou, Z.H. 2005. A K-nearest neighbor based algorithm for multi-label classification: *Proceedings of the 1st IEEE International Conference on Granular Computing, Beijing, China, 25–27 May October 2005.*

Computing, Control, Information and Education Engineering – Liu, Sung & Yao (eds)
 ISBN: 978-1-138-02800-5

Adaptive compressive wideband spectrum sensing with adjustable measurement length

Yong Gang Zhu
Nanjing Telecommunication Technology Institute, Nanjing, China

Ya Peng
Nanjing University of Aeronautics and Astronautics, Nanjing, China

ABSTRACT: The sparisty of the received signal is unknown or even variable in wideband wireless spectrum sensing application. The required number of measurements of compressive sampling will change according the sparsity level of the received signal. A novel compressive spectrum sensing algorithm with adjustable measurement length is proposed. Adding the measurement length into the traditional object function of compressive sensing, the problem is formulated as one of the optimal solution involving a measurement length penalty, and its solution is obtained using an online adaptive manner. The simulation results illustrate that the new algorithm can estimate and track the slow time-varying sparsity of the wideband received signal validly, and it can reduce the compressive sampling rate remarkably.

KEYWORDS: adaptive compressive sampling; dynamic sparse signal; spectrum sensing; tracking.

1 INTRODUCTION

Compressive sensing or compressive sampling is a technique that can effectively acquire a sparse signal using relatively fewer measurements than Nyquist sampling does. As the wideband spectrum is inherently sparse due to its low spectrum utilization, compressive sampling becomes a promising candidate to realize wideband spectrum sensing by using sub-Nyquist sampling rates in recent years [1][2][3]. However, in practice, the sparisty level of the received signal of second users (SUs) is unknown or even variable, because of either the dynamic activities of primary users (PUs) or the time-varying fading channels between PUs and SUs. The required number of measurements of compressive sampling should be chosen according to the sparsity level of the received signal, in order to fully exploit the advantages of compressive sampling technology.

To relax the sparsity level requirement, a sequential compressive sensing approach was proposed in [4]. In this approach, the measurements are got in sequence, and the reconstruction error between them is computed, which is used to decide whether enough samples have been obtained. An adaptive compressive sensing algorithm based on cross-validation is proposed in [5]. In this algorithm, the measurements are divided into two complementary subsets, i.e., the training subset for reconstructing the original signal, and the testing subset for validating the signal recovery. Recently, a two-step CS algorithm has been proposed in [6] and [7], in which the sparsity order is estimated in the first step and the total number of collected samples is adjusted in the second step based on the estimated sparsity order. By combing the measurements of multiple local CRs, [8] proposed a dynamic sampling rate adjusting algorithm. An adaptive method, based on Cramer-Rao bound minimization, is proposed to design the sensing matrix in [9]. In this method, the elements of the sensing matrix are updated sequentially, and the reconstruction error is improved dramatically. But all the aforementioned methods can only be used effectively when the sparisty order is time-invariable, and will be invalid when the sparisty level is time-varying.

Involving the measurement length, a new object function is constructed, which can reflect not only the reconstruction error, but also the effect of measurement length. Then, an online adaptive solving method is proposed for the object function. The simulation results illustrate that the new algorithm can estimate and track the slow time-varying sparsity of the wideband received signal validly, and it can reduce the compressive sampling rate remarkably.

2 RELATED WORK

The goal of compressive sampling (CS) is to reconstruct a sparse signal from much fewer samples. In particular, suppose $\mathbf{x} \in R^N$ is a K-sparse signals, i.e., the number of nonzero elements of X is K, where $K = N$. The task of CS is to recover X from a reduced set of M, i.e., $M = N$, samples $\mathbf{y} \in R^M$ that are acquired by sub-sampling X with a compressive measurement matrix $\Phi \in R^{M \times N}$:

$$\mathbf{y} = \Phi \mathbf{x} \tag{1}$$

Classical linear algebra indicates that the linear system (1) is underdetermined and that it is impossible to find the unique solution without additional information. In the help of sparsity of X, theoretical work indicates that X can be reconstructed exactly when the measurement matrix Φ satisfies the so-called restricted isometry property (RIP) [10]. An estimation X of original signal X can be computed by solving convex optimization problem, as follows:

$$\hat{\mathbf{x}} = \arg\min_{\mathbf{x} \in R^N} \|\mathbf{y} - \Phi \mathbf{x}\|_2^2 + \lambda \|\mathbf{x}\|_1 \tag{2}$$

where λ is a regulation factor. The RIP offers a lower bound on the minimum number of samples needed for sparse signal reconstruction. For example, for Gaussian and Bernoulli random matrices, a log-like expression $M \geq CK \log(N/K)$ holds for some constant C. As a result, the minimum measurement length for signal reconstruction is decided by K, given N. If the sparsity level K is determined and known *a priori*, the measurement matrix can be designed effectively. However, in practice, the sparisty level is usually unknown a priori or even dynamically changing. In order to choose the appropriate number of measurements, a family of adaptive compressive sampling algorithms has been proposed [4] - [8]. The framework of basic structure of adaptive compressive sampling method is shown in Fig. 1, in which the measurement length is regulated sequentially, and limited to the best measurement length M_{BEST} in some sense. The sampling and the reconstruction formulations of the original signal should be modified, as follows:

$$\mathbf{y}(i) = \Phi(i)\mathbf{x} \tag{3}$$

$$\hat{\mathbf{x}} = \arg\min_{\mathbf{x} \in R^N} \|\mathbf{y}(i) - \Phi(i)\mathbf{x}\|_2^2 + \lambda \|\mathbf{x}\|_1 \tag{4}$$

where i is the index of the sampling frame. It can be seen from the up formulations, that the original signal X is invariant among different sampling frame.

The difference between these algorithms is the regulation rule of measurement length. In [4], the sequential reconstruction error is compared with the preset gate, and the algorithm stop increasing measurement length when the reconstruction error is lower than the gate. But this algorithm can not reduce the measurement length effectively. There is the same problem about the algorithm in [5]. The algorithms of [6] and [7] use a part of samples to estimate the sparsity level all the time. It is a waste of computing and sampling resource. The algorithm of [8] can only be used in networking.

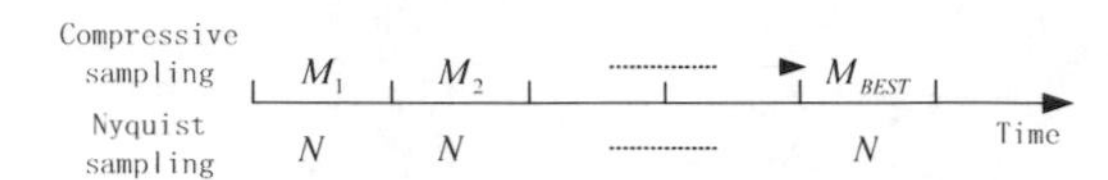

Figure 1. Basic structure of adaptive compressive sampling.

It can be concluded that the algorithms proposed in aforementioned references [4] - [8] can increase measurement length effectively if it is not enough to reconstruct the original signal. But these algorithms can not reduce measurement length when the number of the samples is overused. The reason is that the reconstruction error is decreasing according to the increasing of measurement length, when the measurement length is not enough to reconstruct the original. But the reconstruction error hardly decreases when the measurement length is enough. Fig. 2 depicts an example of the relationship between reconstruction error and measurement length.

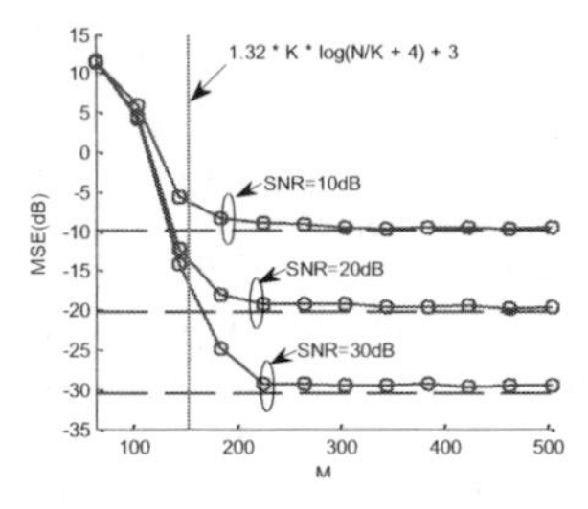

Figure 2. Relationship between reconstruction error and measurement length.

3 DYNAMIC ADJUSTING MEASUREMENT LENGTH

Different from the objective function, Equation (2), of traditional compressive sensing, a new objective function is proposed in this section, as follows:

$$\left(\hat{\mathbf{x}}, M\right) = \arg \min_{\mathbf{x} \in R^N, M \in [1, N]} \|\mathbf{y} - \Phi \mathbf{x}\|_2^2 + \lambda \|\mathbf{x}\|_1 + \gamma M \tag{5}$$

It can be seen from the comparing Equation (2) and (5) that the only difference between them is that there is an additional item, γM, at the right hand of Equation (5). γ is a regulation factor, Intuitively, when the measurement length is small, the decreasing amount of the first two parts of right hand of

Equation (5) is larger than the increasing amount of the third part. So Equation (5) decreases along the increasing of the measurement length, M, when M is small. When the measurement length is so large that the reconstruction signal is limiting to the original signal, the decreasing amount of the first two parts is smaller than the increasing amount of the third part. So Equation (5) increases along the increasing of M. As a result, there must be an "optimal" measurement length with the objective function of Equation (5). And this "optimal" measurement length is limiting to the "best" measurement length of compressive sensing.

The optimal problem of Equation (5) can be solved by enumerating all possible M, when the original signal X is time-invariant. Alternatively, a sequential method is proposed in this paper. As a counterpart of Equation (3) and (4), the problem of dynamic adjusting measurement length can be displayed as following:

$$\mathbf{y}(i) = \Phi(i)\mathbf{x} \tag{6}$$

$$\left(\hat{\mathbf{x}}, M\right) = \arg \min_{\mathbf{x}\in R^N, M\in[1,N]} \left\|\mathbf{y}(i) - \Phi(i)\mathbf{x}\right\|_2^2 + \lambda\left\|\mathbf{x}\right\|_1 + \gamma M(i) \tag{7}$$

To solve the optimal problem effectively, an online adaptive solving system is proposed, displayed in Fig. 3. It can be seen from Fig. 3 that the last recovered original signal $\hat{x}(i-1)$ is used as the objective signal. The cross-validation method can also be used, i.e. divide the compressive signal into the training and the test signal, but the training signal is wasted in this method.

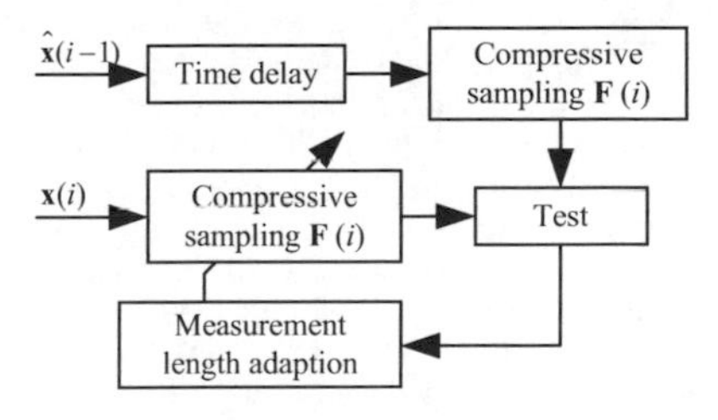

Figure 3. Block diagram of an adaptive adjusting measurement length system.

The adaptive algorithm can be displayed as following steps:

Initialize: Given the initial compressive sampling length M_0, the Nyquist sampling length N, the step size μ and the preset gate δ.

First Step: Compressive samples the received Nyquist sampling signals, $\mathbf{x}(0)$ and $\mathbf{x}(1)$, with compressive sampling length M_0 and $2M_0$, respectively.

Second Step: Solve the optimal problem of Equation (7), and compare the objective function $f(0)$ and $f(1)$: if $f(0) > f(1)$, then $2M_0 + \mu$ compressive sampling signals should be taken at the next sampling frame, else if $f(0) \le f(1)$, then the compressive sampling length of the next frame should be set as $M_0 - \mu$.

Third Step: Repeat the step (1) and (2), until the difference between $f(i)$ and $f(i+1)$ smaller than the preset gate δ.

It can be seen from the algorithm that there is no need of the noise power which is time-varying in the application of wireless communication system all the time. So the proposed algorithm is unrespective about the power noise. Secondly, the algorithm is solved in the adaptive manner, so the proposed algorithm can be tracking the changing of the signal sparsity.

4 NUMERICAL EXAMPLE

In this section, computer simulation results are provided to show the validation of the proposed algorithm in this paper. As an example of the compare of the proposed objective function, Equation (5), and the traditional objective function, Equation (4), Fig. 4 depicts the objective function, Equation (5), with the same simulation parameters as used in Fig. 2. It can be seen that, when the compressive sampling length is small the objective function decrease as the increasing of the compressive sampling length. This is the same as the traditional objective function behavior, as depicted in Fig. 2. But different from Fig. 2, the objective function increases al the increasing of the compressive sampling length, when the compressive is enough to reconstruct the original signal. And the minimum of the objective function is near the "optimal" compressive sampling length proposed in [6]. The simulation results with different SNR and regulation factor, r, are displayed in this figure. It can be seen that the result of the objective function is hardly affected by SNR. And the regulation factor hardly affects the "best" measurement length.

The following computer simulation is used to show the tracking performance of the proposed algorithm. Suppose there are at most 30 primary users over the frequency range $F_{max} = 50\text{MHz}$. The Nyquist sampling length of every sampling frame is $N = 512$. So the sampling time for each frame is $5.12*10^{-6}$ second. To simply the simulation, each primary signal is supposed to be a single-tone signal with random frequency lower than F_{max}. The amplitude of every user signal is random, and the whole power of the primary users is unitary. The sparsity of the spectrum is changed at time $i = 50$ and 100, respectively. The sparsity is 10 when $i \le 50$, the sparsity is 30 when $50 < i \le 100$ and the sparsity is 15 when $100 < i \le 150$.

The simulation parameters are set as follows: the initial compressive measurement length is $M_0 = 20$, the step size is $\mu = 10$, the regulation factor is $\gamma = 0.001$, the gate is $\delta = 0.01$. Gaussian measurement matrix is used in the simulation. The experimental results are obtained by ensemble averaging over 100 independent trials. Figure 5 shows the tracking curves of measurement length with different SNR. The result of [6] is alas displayed as a compare. It can be seen from the simulation results that the proposed algorithm can change the measurement length along with the changing of sparsity, and converge to the "optimal" measurement length approximately. And the performance of the proposed algorithm is independent with the variance of the noise.

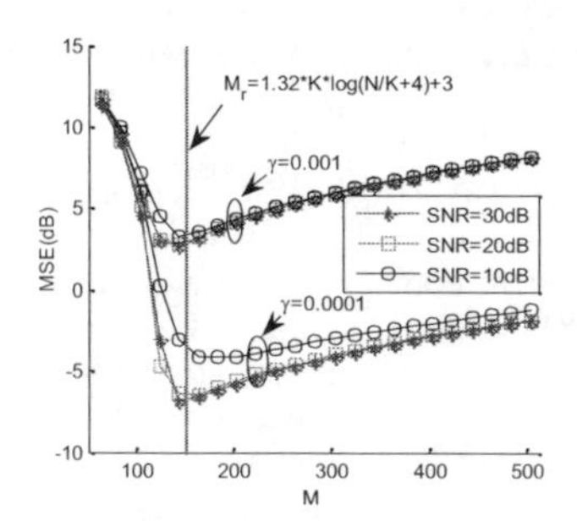

Figure 4. Relationship between the proposed objective function and the measurement length.

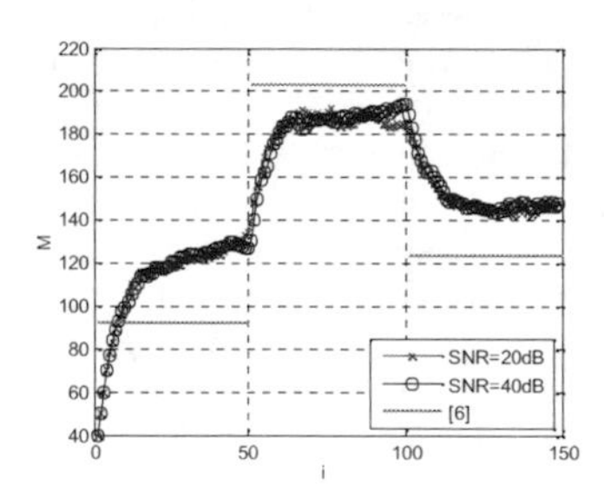

Figure 5. The adaptive tracking curves of the measurement length.

5 CONCLUSION

A novel compressive spectrum sensing algorithm with adjustable measurement length is proposed in this paper. Adding the measurement length into the traditional object function of compressive sensing, the problem is formulated as one of the optimal solution involving a measurement length penalty, and its solution is obtained using an online adaptive manner. The simulation results illustrate that the new algorithm can estimate and track the slow time-varying sparsity of the wideband received signal validly, and it can reduce the compressive sampling rate remarkably.

ACKNOWLEDGEMENT

This work is supported by the National Natural Science Foundation of China under Grant No. 61401505, Jiangsu Province Natural Science Foundation under Grant No. BK20130069, Foundation of Science and Technology on Communication Information Security Control Laboratory under Grant No. 9140C130306130C13060 and China Postdoctoral Science Foundation under Grant No. 2012M521853, 2013T60914.

REFERENCES

[1] Z. Tian and G. Giannakis, "Compressive Sensing for Wideband Cognitive Radios," *Proc. IEEE Int'l. Conf. Acoustics, Speech, and Sig. Proc.*, Honolulu, HI, April 2007, pp. 1357–60.

[2] Z. Tian, Y. Tafesse, and B. M. Sadler, "Cyclic Feature Detection with Sub-Nyquist Sampling for Wideband Spectrum Sensing," *IEEE J. Sel. Topics Sig. Proc.*, vol. 6, no. 1, Feb. 2012, pp. 58–69.

[3] H. Sun, A. Nallanathan, C. Wang, Y. Chen. Wideband Spectrum Sensing for Cognitive Radio Networks: A Survey [J]. IEEE Wireless Communications, 2013, 6(4): 74–81.

[4] Dmitry M. Malioutov, Sujay R. Sanghavi, Alan S. Willsky. Sequential compressed sensing [J]. IEEE Journal of Selected Topics in Signal Processing, 2010, 4(2): 435–444.

[5] H. Sun, W. –Y. Chiu, A. Nallanathan. Adaptive Compressive Spectrum Sensing for Wideband Cognitive Radios [J]. IEEE Communications Letters, 2012, 16(11): 1812–1815.

[6] Y. Wang, Z. Tian, C. Feng. Sparsity Order Estimation and its Application in Compressive Spectrum Sensing for Cognitive Radios [J]. IEEE Transactions on Wireless Communications, 2012, 11(6): 2116–2125.

[7] S. K. Sharma, S. Chatzinotas, B. Ottersten. Compressive Sparsity Order Estimation for Wideband Cognitive Radio Receiver [J]. IEEE Transactions on Signal Processing, 2014, 62(19): 4984–4996.

[8] C. Huang, L. Wang. Dynamic Sampling Rate Adjustment for Compressive Spectrum Sensing over Cognitive Radio Network [J]. IEEE Wireless Communications Letters, 2012, 1(2): 57–60.

[9] T. Huang, Y. Liu, H. Meng, X. Wang. Adaptive Compressed Sensing via Minimizing Cramer–Rao Bound [J]. IEEE Signal Processing Letters, 2014, 21(3): 270–274.

[10] E. Candès and T. Tao, "Decoding by linear programming," *IEEE Trans. Inf. Theory*, vol. 51, no. 12, pp. 4203–4215, Dec. 2005.

[11] R. Baraniuk, M. Davenport, R. DeVore, and M. Wakin, "A simple proof of the restricted isometry property for random matrices," *Constructive Approximation*, vol. 28, no. 3, pp. 253–263, Dec. 2008.

Computing, Control, Information and Education Engineering – Liu, Sung & Yao (eds)
© 2015 Taylor & Francis Group, London, ISBN: 978-1-138-02800-5

Study on the detection of genetic algorithm optimization based on OBB tree collision

Bing Heng Lai & Chang Hua Li
School of Information and Control Engineering, Xi'an University of Architecture and Technology, China

ABSTRACT: According to the covariance matrix based on the traditional (CV) model for the generation of the bounding box, it is difficult to find the optimal OBB bounding box, in order to solve this problem, this paper proposes a genetic algorithm to optimize the generation of OBB bounding boxes method improved (OBB), hierarchical bounding box collision generated for detection between objects. The genetic optimization algorithm to adjust geometry oriented bounding box model, and volume of the surface OBB bounding box, which can be real-time collision detection of objects. Finally, through experiment, using the genetic algorithm to optimize the OBB tree collision detection, collision detection effect is better than the traditional CV method, at the same time, OBB bounding box model is rigid, not easily deformed.

KEYWORDS: collision detection; genetic algorithm; OBB box; covariance matrix.

1 INTRODUCTION

Collision detection technology is an important research measure used in virtual reality of computer graphics, especially in the fields such as video games, architectural design and path planning system, it is used very widely [1]. In general, the implementation of collision detections mainly includes two methods-space decomposition and bounding box[2]. The first method is to divide the virtual space into some small cells with different volumes, and conduct intersection test of the same cell or the object occupying adjacent cell [3]. Bounding box method is to use the bounding box with a volume slightly larger than the geometric object and with a simple geometric feature to approximately simulate an object with geometric features more complex, the most commonly-used bounding box - OBB bounding box [4,5] is a specific-direction oriented bounding box [6], rather than the axis-aligned bounding box AABB [7, 8].

Generally speaking, the bounding box is used to conduct collision detection between objects through fast and accurate calculation and statistics [9, 10]. The purpose of this study is to put forward a precise calculation method of bounding box to as much as possible reduce the calculation time of bounding box collision detection. This paper puts forward an improved genetic algorithm to optimize the generated method of oriented bounding box (OBB), and the generated level bounding box is used to detect the collision between objects. The optimized genetic algorithm can be used to adjust the geometric direction of the bounding box model, and can stand for the volume of the OBB bounding box size, so as to conduct real-time collision detection between objects. Finally, it can be seen that through experimental verification, the OBB tree collision detection methods optimized by genetic algorithm has a better effect than the traditional CV collision detection method, at the same time, OBB bounding box model is rigid, it is not easy to become deformed.

2 ANALYSIS OF BASIC PRINCIPLE

2.1 *Collision detection*

To detect a collision, it is not to detect the collision between the triangles of all the interested objects, but to check whether the collision between the bounding boxes of the possible detection objects has happened. If no bounding box collision happened, so it is clear that there would be no intersection between the triangles. The OBB tree leaf nodes form a rectangle to the single triangle. So, the triangle intersection test is calculated through the similar tests defining their rectangles.

There are two cases - intersection collision and disjoint collision. The collision process will continue until:

1 Because no bounding box collision, so the detection stopped.
2 Because the leaves cannot be divided, so the triangles in the bounding box must be tested one by one, the dimensions of the level bounding box is small, so only a few triangles are mutually contains.

2.2 *OBB tree*

Oriented bounding box OBB can be defined as a rectangle bounding box in an arbitrary direction in a 3D space, essentially, it is a rectangle relatively close to the object, which can randomly revolve according to the object's shape, OBB can be used to describe the bounding box information of node sets of objects, and the intersection test between 2 OBB bounding boxes is the most key link in the whole collision detection system. It will give specific confirm to whether the two wanted node sets intersect.

OBB tree and AABB tree are two very effective and simple hierarchical structures applied to the collision detection between geometric models. Both kinds of model trees belong to the category of binary tree, and each OBB tree will undergo a subdivision along the moment direction of the object's three reaction shape distribution, and each node set is a cuboids the most close to the object. OBB bounding box test can get a structure more precise, drawing less objects.

2.3 *OBB bounding box tree construction method*

Assuming the vector of the triangle's *i*th vertex is p^i, q^i, r^i, the number of triangle surface in the OBB bounding box is n, so:
The triangle surface vertex average of bounding box is expressed as:

$$C = \frac{1}{3n}\sum_{i=1}^{n} p^i + q^i + r^i$$

Covariance elements can be represented as:

$$1 \leq j,k \leq 3$$

$$D_{jk} = \frac{1}{3n}\sum_{i=1}^{n}(\overline{p}_j^i \overline{p}_k^i + \overline{q}_j^i \overline{q}_k^i + \overline{r}_j^i \overline{r}_k^i)$$

Where, n represents the amount of triangle surfaces, $\overline{p}^i = p^i - u$, $\overline{q}^i = q^i - u$, $\overline{r}^i = r^i - u$ and each is a matrix.

Getting the three eigenvectors of covariance matrix based on numerical calculation, we can unitize the eigenvectors into a base. Limiting the size of the OBB tree according to the extreme vertices on each axis, we take the base axial direction as the direction of the OBB tree.

Because we adopt the above mentioned methods but do not take into account the existence of convex hull, the object's triangle surfaces show some uniform sizes. This paper adopts the optimized genetic algorithm to adjust the geometric direction of bounding box model, and represents the volumes of OBB bounding boxes, so we can conduct real-time collision detection of objects.

3 OPTIMIZING THE OBB TREE WITH GENETIC ALGORITHM

3.1 *Genetic algorithm*

Genetic algorithm (GA) is a random search method simulating natural biology's evolution behavior. Genetic algorithm operation adopts the principle of survival of the fittest, successively produces an approximately optimal solution among the potential solution populations, and makes individual selection in every generation of genetic algorithm according to the fitness values of the individuals in the problem domain and the reconstruction method profited from nature genetics, producing a new approximate solution. This process leads to the evolution of the individuals in the population, and the produced new individuals can adapt to the environment better than the original individuals, just like the renovation in nature.

Genetic algorithm is a gradually evolutionary algorithm (EA), mainly including three operations - Selection, Crossover and Mutation. The definitions are as follows:

$$EA = (\mathrm{C},\mathrm{E},\mathrm{P}_0,\mathrm{M},\zeta,\psi,\vartheta,\mathrm{T})$$

1 Selection The purpose of selection is to choose excellent individuals from the current population, make them have the opportunity to breed the next generation of descendants as the parent. According to each individual's fitness value, certain rules or methods, some excellent individuals are selected from the previous generation of populations and hand down them to the next generation of populations. Genetic algorithm reflects this thought through selecting operation, the principle of selection is to select some individuals with stronger adaptiveness and greater probability of contributing one or more descendants. Thus it embodies the Darwinian principle of the survival of the fittest.
2 Crossover: Crossover operation is the main genetic operation in genetic algorithm. Through the crossover operation we can get a new generation of individuals, which combine the characteristics of the parent individuals. The individuals in the population are randomly matched in pairs, for every pair of individuals, part of chromosomes are exchanged between them at a certain probability (Crossover Rate). Crossover reflects the idea of information exchange.
3 Mutation: The first step of mutation operation is to randomly select a individual in the population, and randomly change the value of a string in the string-structure data of the selected individual at a certain probability, namely, to change the values of genes on one or a few locus of each individual in the population to other alleles at a certain rate

(Mutation Rate). Like the biosphere, in the genetic algorithm, the probability of variation is very low. Mutation provides the opportunity for the production of new individuals.

3.2 *Algorithm implementation*

In this chapter, we mainly analyze using the genetic algorithm to implement the OBB tree mentioned in this paper, which mainly includes the chromosomes, the genetic algorithm parameters and other implementation methods.

3.2.1 *Chromosome*

Oriented bounding boxes are expressed in accordance with the characteristics orientation. Once its direction is determined, the objects surrounded by the minimum bounding box can be easily calculated. This paper adopts quadruple method to build the bounding box, and the quaternion structure is as follows, q = (X, Y, Z, W), meaning that the box can revolve around any axis. To show a quaternion, we need to set up four floating-point values. The chromosome is made up of four alleles, and as shown in figure 1, axes of X, Y, Z form a vector and revolve with W in the vector.

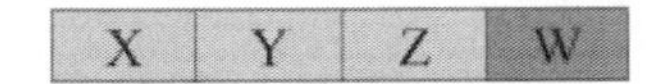

Figure 1. Chromosome.

The multiplicative group of non-vanishing quaternion has a conjugate action in the part whose R3 real part is zero, and the conjugate action can realize the rotation. If the real part of a unit quaternion (quaternion with the absolute value of 1) is cos (t), its conjugate action is a rotation with an angle 2t, and the axis is in the direction of the imaginary part.

3.2.2 *Genetic algorithm parameters*

The genetic algorithm put forward in this paper has the following main characteristics:

1 Generation gap: in a non-overlapping model, maternal line never compete with own children, so the whole maternal line are always replaced by offspring groups, while in an overlapping system, parents always compete with their children for survival.
2 Fitness function: the fitness function is used to calculate which bounding box is more suitable for the tested object, the scheme adopted in this paper can surround the objects more compactly, which means the best oriented bounding box can be more close to the object with the smallest volume. Assuming $f = \frac{1}{v}$,

Wherein, v represents the volume of the bounding box.

3.2.3 *Oriented bounding box*

As shown in the above section, like quaternion, if the quaternion of a chromosome are represented as (X, Y, Z, W), then the rotation axis of the quaternion is represented as (1, 0, 0), (0, 1, 0), (0, 0, 1), in the event of the rotation, you will get three new axes, and the three new coordinates will form a OBB bounding box.

Use these values to calculate the center of the bounding box and the length of each side, and in order to make the OBB tree make cycle recursion, the longest axis is used to divide the bounding box into two parts.

4 EXPERIMENT AND RESULT ANALYSIS

In order to verify the effectiveness of the proposed method in this paper, collision detection experiment was carried out in this paper. All of the tests were run on a computer with a quad core 2.4 GHz processor and a 4GB RAM. Four models were presented in this paper, as shown in figure 2, the geometric features of the model are shown in table 1. The experiment mainly compared the OBB tree with genetic algorithm adopting the covariance (CV).

Table 1. Test model.

model	The number of nodes	The number of triangle
Tibia	7735	22662
Jacky	3304	6604
Sphere	12	20
Prosthesis	7252	9874

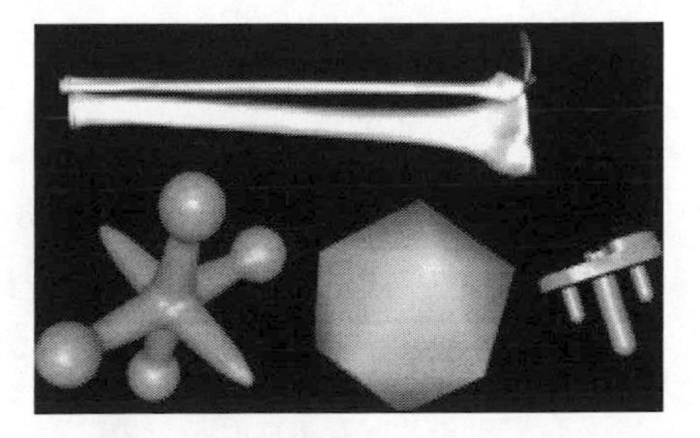
Figure 2. Experimental geometry.

Experimental results are shown in figure 3, figure 3 shows the fitness function of a generation of genetic algorithm, figure 3a show a pair model (tibia) OBB and figure 3b shows a Jacky (toy) OBB. It can be seen from the figure that the adaptive function always increases with the increase of generation.

The adaptive function can almost continue for several generations. As the adaptive function in figure 7b continuously increases, if the amount of generations is higher than the function, it will continue to grow to a high point, namely will become constant.

As previously explained, the main advantage of OBB tree is to reduce the number of triangles in each leaf. The following experimental purpose is to weigh the merits of OBB tree, as shown in figure 4, four kinds of models all adopted GA method to optimize the OBB tree. It can be seen that the triangles got by GA method is always less than the average of triangles on each leaf.

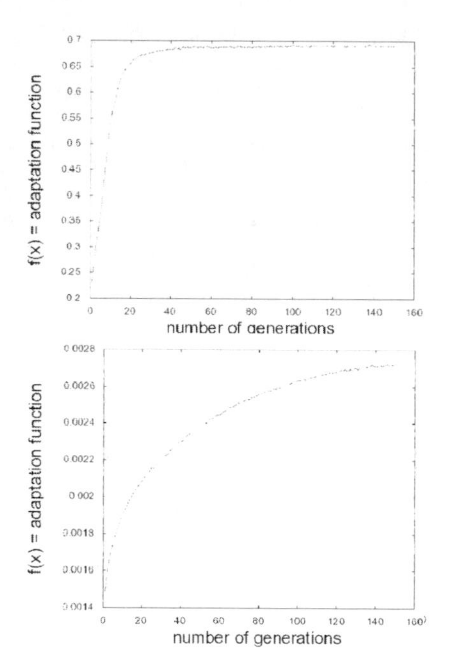

Figure 3. Fitness function.

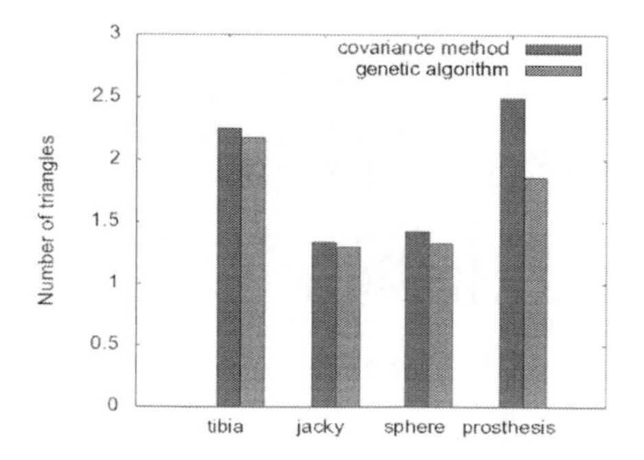

Figure 4. The number of triangles on each leaf.

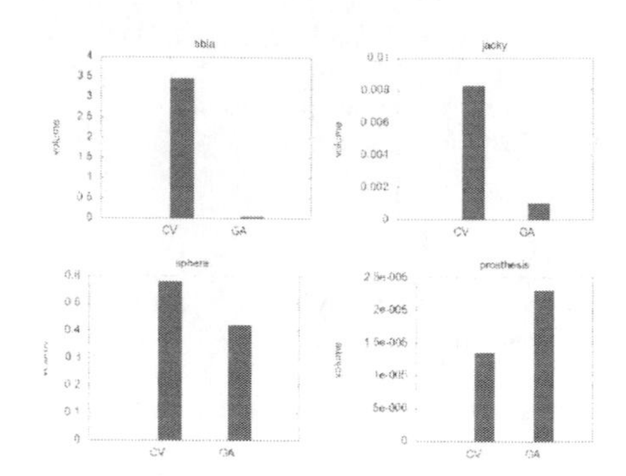

Figure 5. GA and CV performance comparison results.

Then we compared the traditional covariance-theory (CV) OBB tree with GA-optimized OBB tree in this paper, the experimental results are shown in figure 5, which can clearly show the advantages of the algorithm adopted in this paper, the OBB tree collision detection method optimized with genetic algorithm can get a collision detection effect better than traditional CV method, especially, the former showed a stronger rigidity, not easy to become deformed.

5 CONCLUSION

This paper mainly studied the collision detection method of OBB tree optimized using genetic algorithm, and used the optimized genetic algorithm to adjust the geometric direction of bounding box model, and changed the volume of OBB bounding box through fitness function, and compared the algorithm with the traditional covariance (CV) OBB tree collision detection method, and verified the advantages of this algorithm through the experiments.

ACKNOWLEDGEMENTS

The corresponding author of this paper is Bingheng Lai.This paper is supported by Natural Science Foundation of China NO.50878176 and NO.61373112; Xi'an University of Architecture and Technology Youth Fund No.1123

REFERENCES

[1] Carmona R, Navarro H. An Image-Space Approach for Collision Detection Between Multiple Volumes and a Surface[J]. International Journal of Creative Interfaces and Computer Graphics (IJCICG), 2012, 3(1): 16–27.

[2] Yuanfeng Zhu,Jun Meng,Guanghua Xie,ect.Research on Real-Time Collision Detection Based on Hybrid

Hierarchical Bounding Volume [J].Journal of System Simulation, 2008,20:372–377.
[3] Zhaowei Fan,Huagen Wan,Shuming Gao.Streaming Real Time Collision Detection Using Programmable Graphics Hardware[J].Journal of Software, 2004,15:1505–1514.
[4] Lauterbach C, Mo Q, Manocha D. gProximity: Hierarchical GPU-based Operations for Collision and Distance Queries[C]//Computer Graphics Forum. Blackwell Publishing Ltd, 2010, 29(2): 419–428.
[5] Qianru Xie,Guohua Geng.Fast Collision detection method in virtual surgery[J].Journal of Computer Applications, 2012,32:719–721
[6] Chang J W, Wang W, Kim M S. Efficient collision detection using a dual OBB-sphere bounding volume hierarchy[J]. Computer-Aided Design,2010,42(1):50–57.
[7] Zhenhua Zhang,Wenli Zhou,Fujun Quan,ect.Collision detection algorithm research based on spatial domain in virtual scene[J].Journal of Computer Applications, 2012,32:51–54.
[8] Tu C, Yu L. Research on collision detection algorithm based on AABB-OBB bounding volume[C]//Education Technology and Computer Science, 2009. ETCS'09. First International Workshop on. IEEE,2009,1: 331–333.
[9] Yanfei Zhou,Ziniu Wang.Research on NC Collision Interference Algorithm Based on the OBB and Octree[J].Journal of Guizhou University(Natural Sciences),2012,29:68–71.

Computing, Control, Information and Education Engineering – Liu, Sung & Yao (eds)
© 2015 Taylor & Francis Group, London, ISBN: 978-1-138-02800-5

Use of web-assisted technology in "2+2" college English teaching model

Ling Yi Huang
College of Foreign Languages and Cultures, Xiamen University, China

ABSTRACT: Recent years have witnessed great changes in College English Teaching models in Xiamen University. The expanding enrollment and changing needs of the students propel the Department of College English in Xiamen University to implement the new "2+2" teaching paradigm from 2012. In face of the reduced meeting time and increased after-class tasks, students and teachers rely largely on web-assisted technologies to achieve the teaching goal. A survey has been conducted to find out the students' view of the current model, and the effect of the use of these technologies in the daily learning process, whose result will give an insight into the model and technology itself and the possible problems it may still incur. Reflections will be made and adjustment can be made in the following reforms.

KEYWORDS: Web-assisted; College English; "2+2" Teaching Model; Online Writing System.

1 INTRODUCTION

The expanding enrollment in universities has been an increasingly common phenomenon in recent years. Te mounting number of students has brought huge impact on the teaching quality of the universities. In the field of College English teaching, besides the rising insufficient number of qualified teachers, how to balance the different levels of students' English proficiency and how to maintain the teaching quality have become a big task for most managing sectors.

From Year 2012, the Department of College English of Xiamen University has implemented a new model called "2+2" teaching model. It is assisted by the web-based technology and has already attained preliminary confirmation by most teachers and students.

2 "2+2" TEACHING MODEL

2.1 *"2+2" Teaching model*

The teaching reform in College English has undergone several changes in Xiamen University in recent years. From year 2005, the level-placement exam has been set up and all freshmen have to take the test right after they enter the university. Based on the test result, students will be placed in different levels and take different courses.

Therefore the 5000 students are divided into 5 levels and have different requirements in their learning syllabus.(See Figure 1)[i]

Table 1. Level-based college English teaching model (2013).

	First Year		Second Year	
Level	First Term	Second Term	Third Term	Fourth Term
Preliminary level	Preliminary Course	Integrated English One	Integrated English Two	Integrated English Three
Level One	Integrated English One	Integrated English Two	Integrated English Three	Improved Course
Level Two	Integrated English Two	Integrated English Three	Improved Course	
Level Three	Integrated English Three	Improved Course	Selective Course	
Level Four	Improved Course	Selective Course		

* Jiang, College English syllabus of Xiamen University,2013

This level-based teaching reform enables the teachers to teach the students in accordance to their English abilities. And this ensures the possibility of further reform.

In year 2012, the new "2+2" teaching model was implemented in view of bringing the initiatives of teachers and students into full play. This is a model that combined both face-to-face instructions and students' self-learning.(See Figure 1)

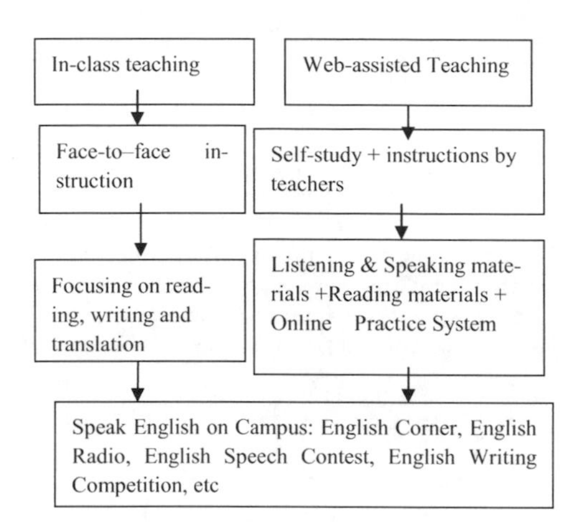

Figure 1. "2+2" College English teaching paradigm of Xiamen University.

This teaching model has a 2-hour in-class instruction time, in which teachers will focus on the learning methods and skills, and the major ways used are mostly instructions, interactions, communications, etc. The key feature of this first "2" is fact-to-face communications. The second "2"means, besides the 2-hour in-class time, the students will spend 2 hours learning the materials assigned by the teachers by themselves. Teachers will mark the homework and ensure two "online office hours" every week to provide advice or instructions online. The key feature of this 2-hour pattern is the web-assisted technologies that aim to reduce the burdens of teachers and give more advanced technological help to students' language learning process.

2.2 *Web-assisted technology in college English teaching model*

The success of the current 2+2 teaching model relies a lot on the advanced web-based technology, represented here by Online Writing System—pigai.org, Online Practice System developed by the Department of College English of Xiamen University and the instant communication tool—QQ Group.

2.2.1 *Online writing system–Pigai.org*

The Online Writing System used in Xiamen University is pigai.org, the largest Online English Writing Practice Platform in China. Now over 1000 colleges, universities and related institutions used their product. The system has over 10 million users, and 5 million checking daily.

Xiamen University started the trial use of the platform in 2012, and officially launch the system in school in year 2013. From the first term's report (from September 1st 2013 to January 9th 2014), 65 teachers and 9518 students have registered for the use of the system. Students have submitted 40739 compositions, in which 36348 compositions are assigned by the teachers and 4391 articles are written purely on students' willingness. 65 teachers had assigned 444 topics and on average, each one assigned 6.83 topics in one term. The teacher who received the largest number of papers is Associate Professor Li Suying, who assigned 22 topics and received 3511 articles.[ii]

The online writing system provides spelling, grammatical and lexical, cohesive and other linguistic checks to a submitted article. The major linguistic mistakes that students often make in their writing process can be checked by the system, and once the student submits the article, the score and comments can be seen at one. Students can rewrite the article based on the advice given by the system, which is as clear and detailed as on a sentence basis. Therefore, most students would get the incentive to revise their articles and try to get a higher score until they feel satisfied. The most diligent student in my class has revised his article for 212 times. On average students revised each article for 6.322 times.[iii] Students could also get help from the linguistic database and tools to improve the details of their writing. The feedback is instant and inexhaustibly, and would not involve constrains that may be posed by the limited number of teachers. On the other hand, the system greatly reduce the trivial things the paper work would present. Teachers only have to give the assignments on website and tell students to finish the task before the deadline. They can set the marking standard and control several detailed settings. The scores and revision frequency will be automatically produced in forms and be downloaded in excel, easy for teachers' references. Most importantly, the linguistic problems in all the articles would be divided into several categories (such as grammar, spelling, cohesion, etc) and the system would give a report on examples and frequency of each key mistake. This helps teachers to get a general idea of the key problems, and leaves the trivial tasks of checking spells and simple grammatical mistakes to the system. Thus the teachers can focus on the main questions in the writing process and give personal check afterwards.

Another key function of this system is that it also provides originality check, which can effectively prevent students from plagiarism. All copied sentences or paragraph will be marked in red and give the percentage of plagiarizing. And this is not an easy task for teachers if it is done on paper.

Now all the undergraduate students in Xiamen University have used this online writing system to write papers every two to three weeks and all the key writing competition on campus has used this system to hold the preliminary contest.

2.2.2 *Online practice system*

from 2012, in order to meet the requirement of self-learning under the framework of "2+2" teaching paradigm, the Department of College English in Xiamen University developed an Online Practice System, which provides different forms of English learning exercises, such as linguistic quizzes, exercises, reading comprehension, listening materials of different levels, vocabulary building up, Band Four or Six Model Test, IELTS or TOFEL GRE Model Test, etc. All these exercises are offered free and students can choose different stages and levels according to their levels. All the practice scores will be recorded automatically and can be provided as a reference as students' after-class performance .

To fully utilize the textbooks used in the instruction class, the System develops some exercises based on the texts in integrated course, such as designing the listening dictation practice on the basis of listening & speaking textbook, revising the reading course textbook into exercises like cloze, gap filling, summary.

All these exercises can offer score and checking right after the students submit the task and give students freedom of choice in the language learning process on the basis of their ability.

2.2.3 *Interactive communicative tool—QQ group*

The most popular instant communicative tool in China today is the QQ, and its group function QQ Group has become a good way for teachers and students to get connected.

The "2+2" Teaching Paradigm requires teachers to have a 2-hour online office hour every week. So most teachers use this tool to communicate with other teachers and students. Problems encountered in the learning process, exercises and assignment that need to be further emphasized, all these can be discussed on the group talk.

The online communicative tool is instant, efficient and convenient and free. Students can leave the message via internet without worrying the limited meeting time. Students also use this tool to communicate with other classmates to discuss their group work assigned by the teachers. Since students come from different departments, effective contact means becomes a necessary in the big campus.

3 SURVEY OF THE WEB-ASSISTED TECHNOLOGIES IN COLLEGE ENGLISH TEACHING REFORM

After two and a half years reform, the "2+2" College English Teaching model has been accepted by most students and teachers. It is time to analyze the preliminary feedbacks from the students to find out their reactions to the new teaching model and whether the web-assisted technology has fully come to play in the learning process.

A survey titled "The Use of Web-assisted Technology in '2+2' Teaching Model" has been conducted on the website www.sojump.com in January,2015.Questions were designed focusing on the issues in the current teaching model and the use of web-assisted technologies in the learning process.

3.1 *Findings*

The survey link was posted on the QQ group and was spread by teachers from the Department of College English Department. 355 students answered the survey on the website. 41.41% of the students surveyed are from Grade 2013, and 58.59% from Grade 2014. Students of these two grades are the major parts that have College English course now.

3.1.1 *The use of online writing system*

when asked about the topic "Do you like to use the Online Writing System to write composition ?", 37.46% say they like this practice, 36.9% express their dislike and 25.63% say it makes no difference to them.

Of the reasons given by students on why they favor the online system, 86.47% say the system can give instant feedback, 89.47% say they can revise the article unlimited times, 45.86% say they can get technological support by the website.

On the other hand, of the top 3 reasons why students dislike the writing system, 81.68% say they don't think the machine has done a better work than the teacher, 51.15% say they dislike writing articles on internet, 9.16% say they don't know how to use the technological helper.

3.1.2 *The use of online practice system*

as for "the use of Online Practice System", the attitude don't seem to diverse. The majority of students (65.35%) don't like to use this online system to do exercises. They attribute this to the following three reasons: they don't like to do exercises online (79.13%), the system sometimes crashes down (37.07%), the exercises are too old, not updated enough (14.22%).

For the rest who love to use this online system (34.65%), the main three reasons why they like the system are: they can do the exercises anytime, anywhere (86.99%), the exercises are abundant and rich in forms (65.85%), these exercises can improve their ability (60.16%).

The result is in great contrast with that of the online writing system, which are favored by most students. The reason may be lie in the system itself. The system is developed by a professor in the Department of College English and without a strong team behind, the technical problems do occur sometimes. And since the system is still new, the exercises database is not large enough to provide all the practice forms students need.

But we have to admit that the practices offered by the system give a good complement to the textbook used in class. It offers course-related exercises and various levels of model tests, which to some extent, reduce the unfavorable impact of a large class that make personal needs unheard. Students can choose the exercises according to their personal plans and learning styles.

3.1.3 *The use of QQ group*

The way to communicate with teachers in the age of internet—QQ group has been favored by a large majority of students. 81.69% of the students surveyed express their likeness of this contact tool. For them, the reasons why they like to use this means to contact their teachers are: they feel at ease in communication (89.31%), they can express their opinion anonymously (26.9%) and some feel fearful to talk to teachers directly (13.1%).

For all the rest (18.31% of the all students surveyed) those who dislike this means say that they think it is not easy to explain the things clearly (73.85%), they feel remote and cold in this form of communication (33.85%) or they just dislike the way to communicate online (20%).

3.2 *Possible improvements*

The system of online writing is largely accepted by the students. The problem of technical support can be solved by giving a detailed instruction course before the students use the system for the first time. Relevant paper and digital version of instructions can be provided beforehand.

Manual marking and revision can be done after the paper is submitted online. Teachers can give more options to students who really cannot adjust to the way of writing online.

On the Online Practice System, technical support could be sought from engineers from other colleges or companies and more exercises could be enriched through various means, either by collected by teachers or bought from specialized companies. Teacher should also pay enough attention to the negative reactions to the online practice way and balance between paper work and online tasks.

And for the communication means between teachers and students, effective way should still put first and teachers could offer more face-to-face communicative time after class if possible. Anyway, language learning is still a process that involves human interactions.

4 CONCLUSIONS

To meet the requirement of the internet era, and in a university that aims to become an internationalized one, College Teaching reform would not be an easy task. After years of experiment, the web-assisted College English Teaching Model has been gradually accepted by both students and teachers.

Although problems of the web-based technologies still need to be improved and new challenges may come up in the future use, the new practices have given us surprises and really brought unexpected advantages. Adjustments could be made and suggestion could be given on those shortcomings. Admittedly, the current new model is in line with the changing era and developing society, and gets support from students, which can be manifested in the survey when 74.93% of the students surveyed feel satisfied with the "2+2" teaching model.

REFERENCES

[1] Jiang,G.Y.2013.*College English Syllabus of Xiamen University*. Xiamen: Xiamen University Press.

[2] & [3] Report on The Use of Pigai.org in Xiamen University (2013.1—2014.1.9), www.pigai.org.

Computing, Control, Information and Education Engineering – Liu, Sung & Yao (eds)
 ISBN: 978-1-138-02800-5

The effect of ERP on accounting and financial management of enterprises in China

J.Q. Xiu & N. Zhang
Langfang Polytechnic Institute, Hebei, China

ABSTRACT: With the rapid development of market economy, the concept of modern management is frequently introduced to the enterprise's financial management. ERP (Enterprise Resource Planning) bases on information technology and fully integrates the enterprise's information resources to form an all-round, systematic platform in order to strengthen the enterprise's decision-making, controlling, planning, and performance evaluation. This paper expounds basic content of ERP, its great significance on accounting and financial management in China, analyzes the problems of ERP existing in the application of the enterprise's accounting and financial management and points for attention in the implementation process, and discusses the influence of ERP on the financial and accounting management of the enterprise.

KEYWORDS: ERP; the enterprise's accounting and financial management; overview; significance; problems; points for attention; influence.

In order to stand out in the fierce market competition, the enterprise must introduce new management patterns and technical methods, optimize the internal information management, improve the production efficiency and economic benefit, and enhance the comprehensive competitiveness of the enterprise. ERP is a kind of advanced financial and accounting management mode; the main function of it is to have financial management and accounting. This new management idea will regulate the enterprise's internal financial management and maintain the consistency of the internal information data.

1 AN OVERVIEW OF ERP

1.1 *The definition of ERP*

ERP is not only a means of modern management but also a kind of financial management software. The core concept of ERP is the integration of the enterprise's information resources based on a series of effective analysis, the standardization and informationization of the whole supply chain of the information resources, and the implementation of planning in advance, process control and solving methods afterwards. The main function of ERP in the financial and accounting management in the enterprise is the financial planning, financial analysis, and financial decision-making to provide supports for the managers. The enterprise of our country uses financial and accounting subsystem and module of the system of the ERP software to implement ERP management. In spite of different functions between the various subsystems and modules, they exist integrally. The application of ERP software improves the enterprise's accounting computerization level and effectively realizes the sharing of the accounting information resources which inject new vitality into the development of the enterprise.

1.2 *The significance of ERP on the enterprise's financial and accounting management in China*

ERP technology, based on advanced network information technology, is a scientific management mode of resource management and planning management and is a necessary prerequisite for large scale enterprise's survival and development. It can reasonably allocate the enterprise's human resources, material resources and money and also can play the enterprise's role of operation and management to fully integrate basic information of production, logistics, and customer service to improve the utilization of the enterprise's resources and to optimize the enterprise's internal management by improving the management system of the resource utilization. If the enterprise can better use the ERP technology, it helps the systematization of the internal management of the enterprise, improves the production efficiency and economic benefits and makes up the defects of the enterprise's management to build a good internal environment, reduce the enterprise's financial risk, improve the comprehensive competitiveness of the enterprise, and promote its sound and rapid development.

2 THE PROBLEMS OF ERP EXISTING IN THE ENTERPRISE'S ACCOUNTING AND FINANCIAL MANAGEMENT OF OUR COUNTRY

2.1 *The lack of correct understanding of ERP in chinese enterprises*

Chinese enterprise managers do not grasp the great significance of ERP. Many companies which apply ERP systems don't give full play to the main function of ERP and don't have a deep understanding of the value of software. Although having applied ERP technology in the enterprise's financial and accounting management, it still can't reduce the risk due to the lack of financial control and can't achieve to standardize the enterprise's internal information and fulfill the goal of improving the internal management system. Some enterprises lack basic knowledge of the function of ERP system software, they only treat the application of ERP as a purely technical issue, partially put emphasis on the function of its information flow and neglect its application in terms of logistics. Some enterprises ignore their actual needs and job requirements in the configuration of the accounting software, purchase the most advanced software system, partially emphasize the best quality of software, and ignore the enterprise's staff quality and the existing management means to see whether they can match with the advanced software. On the contrary, some enterprises only consider the price factor of the software configuration, blindly control the cost and ignore the quality of the ERP software and the ability of the software supplier for technical maintenance service which may cause the configuration of the accounting software not to up to the standard. Some enterprises do not pay enough attention to the initialization of software modules when applying the ERP software. Basic information prepared according to the module division before the initialization can't be input into a computer which hinders the system initialization, causes the accounting informatization hard to continue, and increases the difficulties for customer management and customer classification work someday. Some companies do not fully understand special functions of ERP. They only establish the electronic version of manual accounts with ERP, don't realize the advantages of the accounting information processing compared with the traditional accounting work, and give up miscellaneous functions of the ERP system such as department accounting, project accounting, customer contacts and special functions such as cashing management, bank reconciliation. They don't make full use of resource sharing between various modules of the system which may hinder the integration process of the enterprise's financial business. The enterprises are lack of knowledge of the ERP system, and the application of it is still in the exploratory stage. Thus they need to deepen their understanding of ERP and give full play to this advanced management system in order to improve the enterprise's financial and accounting management.

2.2 *A lot of risks due to low quality of the accounting and financial management staff*

The comprehensive quality of the enterprise's accounting and financial management staff directly determines the ERP system's function level, and deeply affects the effectiveness of ERP in financial management. At present, the overall quality of accounting and financial management staff in our country is low, which is the hidden trouble to hinder the smooth implementation of the ERP system, may even cause distortion of the enterprise's financial information and affects the integrity and the timeliness of information. The continuous promotion of the ERP system has brought new changes to the enterprise's operation mode and management mode. The practical application of accounting and financial management in Chinese enterprises doesn't give full play to the ERP control of the enterprise's capital expenditure, the reduction of financial risks of the enterprise, the improvement of the enterprise's economic efficiency and productivity, which results in a large number of financial risks in the financial management and doesn't achieve an effective risk evasion.

2.3 *The lack of funds and many factors affecting the ERP financial system*

The application of the ERP software requires corresponding hardware and software configuration, which needs the enterprise to invest a lot of money. To popularize the ERP system will also generate a large number of personnel training costs and system maintenance costs. The shortage of funds will restrict the application of ERP system in the enterprise's accounting and financial management, which causes the enterprise to only buy report forms, general ledger module in addition to other financial accounting module or even to abandon the purchase of ERP software. The application of the ERP system in enterprises can make up for the traditional financial and accounting management deficiencies in a large extent. At present, the enterprise lacks perfect accounting management mode and internal system mechanism. There are many other factors affecting the ERP financial system. These factors affect the performance of the ERP system, which are not conducive to standardize the enterprise's internal information, to reasonably allocate its resources, and thus can not guarantee the normalization of the system operation and can't make full use of favorable conditions brought by ERP.

3 POINTS FOR ATTENTION IN THE IMPLEMENTATION OF ERP IN THE ENTERPRISE'S ACCOUNTING AND FINANCIAL MANAGEMENT

3.1 *The application of ERP financial management mode based on the enterprise's own needs*

Before implementing ERP accounting and financial management mode, the enterprise needs to use the system to survey its own conditions and makes a comprehensive judgment of applying the ERP mode to see if it can meet the enterprise's own needs for development and the enterprise has certain requirements for implementing the ERP mode. The managers of the enterprise should not only need to make a macro analysis of the current operation and management problems by taking the enterprise's development characteristics into consideration, but also need to make a microscopic analysis of the enterprise's accounting and financial management work content and working procedures. The enterprise's operation and management characteristics should integrate with that of the ERP mode. The application of ERP financial management mode is based on the enterprise's own needs.

3.2 *To pay attention to ERP software selection by using the enterprise's accounting and financial management*

The ERP software selection by using the enterprise's accounting and financial management will directly determine whether the implementation of ERP management is successful or not. The current technology of ERP software has fully developed, such as a short cycle of implementation, low risks, easy to use, easy to operate, reasonable costs. The enterprise can purchase right ERP software according to its own needs for development and the nature of different kinds of software. Choosing a suitable ERP software and timely preparing data can ensure the smooth and stable operation of ERP system and guarantee the authenticity and timeliness of the accounting and financial management information to improve the enterprise's level of financial management.

3.3 *To deepen the enterprise managers' understanding of the ERP system about its effectiveness in the accounting and financial management*

During the process of implementing the ERP system by the accounting and financial management staff, they need to have a comprehensive understanding of the characteristics of the system and to standardize detailed rules for the implementation of the ERP system. They should also strengthen the test of each module after constructing the ERP system. Documentation system should be established in the implementation of the ERP system to ensure the stability of the system, give full play to its advantages in the accounting and financial management, and improve the enterprise's level of financial management.

4 THE EFFECT OF ERP ON THE ENTERPRISE'S ACCOUNTING AND FINANCIAL MANAGEMENT IN CHINA

4.1 *To improve the enterprise's quality of accounting and financial management, ability of financial budget and financial control*

The application of ERP system in the accounting and financial system is mainly through the integration mode to process all kinds of information. Through the system service module, the information needed can be directly produced. The information integration mode can combine financial accounting and the accounting balances information, which will be of great help to the refinement of commodities and contracts. The application of ERP system is conducive to the establishment of the profit mechanism to improve the enterprise's accounting and financial management quality and enhance its economic efficiency. The data of financial budget and financial control are transferred with a high integrated level in the implementation of the ERP system which will help practice the enterprise's financial budget and financial control ability. The enterprise can scientifically and reasonably calculate its future operating results and financial conditions based on its historical data and its own operational conditions. The financial prediction of the ERP system applies methods of structure analysis and comparative analysis and analyses the enterprise's financial, profit and loss, and core index based on an integrated summary and budget in order to improve the enterprise's ability of financial prediction, enhance the scientificalness of the managers' decision-making, and increase the control ability of the enterprise.

4.2 *To be of great help for the enterprise to collect market information and integrate its internal information*

The enterprise needs to pay close attention to market information. If the enterprise can accurately grasp first-hand market information, it can timely adjust its management strategy and make right decisions according to the change of market information. The traditional way of information collection has been unable to adapt to the environment of market economy which causes increasingly fierce competition and increasingly advanced information technology. The ERP system has functions of efficient data collection and information sharing, which greatly enhance the enterprise's efficiency of the market information collection , improve its production efficiency and economic benefits. Timely and accurate market information is helpful for the enterprise to reduce financial risks and improve its ability of risk evasion and market competitiveness. If the enterprise gives full play to the ERP system's market information collection ability in the process of its financial management, it can not only enhance the timeliness and accuracy of market information, but also can improve the authenticity and integrity of the accounting and financial management information. It can also improve the enterprise's structure of the accounting information, enhance its accounting and financial management working efficiency, and greatly improve the level of financial management.

4.3 *To be helpful for the enterprise's long-term development*

The application of ERP system in the accounting and financial management is helpful for the enterprise to make decisions, and to adjust its management strategy from long-term interests which is favorable to the long-term development of the enterprise. It can timely learn its own production and operation situations to achieve cost control and make judgment on its development direction. It can also better allocate resources in links of purchasing, production, and sales which will enhance the enterprise's ability to control the entire organization structure and optimize its supply chain management. The ERP system's information integration and collection functions are conducive to the enterprise's financial prediction and strengthen financial control which provide information support for the enterprise's decision-making, reduce financial risks, improve the production efficiency and management level, thus are conducive to enhance the core competitiveness of the enterprise and promote its long-term development.

5 SUMMARY

ERP is a kind of advanced management concept; the implementation of it is helpful to improve the enterprise's accounting and financial management quality, and enhance the ability of financial prediction and financial control. It is also of great help to timely collect market information, integrate the internal management information, optimize the organization structure of accounting information, improve the production efficiency and operation efficiency, increase the comprehensive competitiveness, and promote the long-term development. At present, there are some problems existing in the ERP system of the application of financial management in our country, the enterprise lacks correct understanding of ERP. The overall quality of financial staff is low. There are a large number of financial risks in the business finance. The enterprise is lack of funds. Many factors affect the system and therefore the enterprise needs to select ERP financial management mode according to the development needs, pays attention to the selection of ERP software types, and deepens the managers' knowledge on ERP.

REFERENCES

[1] Liu Ying. The analysis of the influence of ERP on the enterprise's accounting and financial management in China [J]. Value Engineering, 2014, (10): 168–168,169.

[2] Li Linghui. The study on the effect of ERP on the enterprise's accounting and financial management in China [J]. Financial Circles, 2014, (32): 194–194.

[3] Hu Zhihua. The impact of ERP on the enterprise's accounting and financial management in China [J]. Management Experts, 2014, (8): 62–62.

Computing, Control, Information and Education Engineering – Liu, Sung & Yao (eds)
 ISBN: 978-1-138-02800-5

Problems and strategies of the enterprise's cost accounting

N. Zhang & J.Q. Xiu
Langfang Polytechnic Institute, Hebei, China

ABSTRACT: In recent years, the rapid development of China's economy has caused increasingly fierce competition among all professions and trades. Cost accounting is very important because the enterprise's goal is to increase its economic benefits. So it has a tremendous influence on enhancing the enterprise's competitiveness and on its cost input and control. This paper briefly analyzes the enterprise's cost accounting problems and puts forward corresponding strategies.

KEYWORDS: accounting; cost accounting; problems; strategies.

With the constant development of the economy, the accelerating process of economic globalization, and the increasingly intense competition in the market, if the enterprises want to enhance their competitiveness and improve their economic benefits, they should control the cost. So the enterprise should pay more attention to cost accounting and solve some existing problems to improve the level of accounting and effectively reduce certain costs to increase economic benefits.

1 EXISTING PROBLEMS OF THE ENTERPRISE'S COST ACCOUNTING

1.1 *The definition and functions of cost accounting*

The definition of the enterprise's cost accounting is that it mainly relies on accounting as its basis and uses currency as the unit of account. It is essential to conduct cost accounting since it is a key part of cost management. Cost accounting should cover the costs of the entire production management process which need to be distributed and collected according to the categories of objects, and should work out unit cost and total cost.

The functions of cost accounting are as follows: whether the enterprise can carry out cost accounting well not only involves many control work, such as cost plans, evaluation, prediction, analysis and improvement, but also determines whether the decisions like cost decisions and operational decisions can play a decisive role. The consumption and cost of the enterprise produced in production management process can be clearly reflected by cost accounting. Cost accounting is also a feedback process. Cost information becomes more clear through cost management. It enables the enterprise make targeted adjustments to the decisions according to different situations. Therefore, cost accounting is very important for an enterprise to make cost plans and control costs. It can make the enterprise smoothly achieve its development goals.

1.2 *Existing problems of the enterprise's cost accounting*

Cost accounting plays a very important role in the development of enterprises. It is often affected by some factors, which hinder its smooth progress. The following is a brief introduction to some problems in cost accounting.

Having a less control of cost accounting. Cost accounting should run through the whole process of the enterprise's development. It should not only involve the control and check of the cost, but also the production management costs. If accountants have a less control of the consumption costs, there will be certain costs which don't match the production and operating plans or even exceed the budget which may cause waste. These additional costs and expenses are produces because cost accounting doesn't work so well which will bring considerable economic losses for the enterprise.

Cost management has been dragging behind. Nowadays, with the rapid development of social economy, management requirements are also not the same as before. The traditional mode of cost accounting in the past has been unable to fully meet the needs of modern enterprises. Most enterprise managers often view the economic benefits as the most important issue and the primary objective and pay less attention to cost accounting. There is no specialized personnel to carry out the management of cost accounting. That's the reason why most of the enterprises still use very traditional cost accounting methods. The enterprise with traditional methods can not keep up with the pace of social development which is not good to enhance the enterprise's productivity and competitiveness as well as it economic benefits.

Not having comprehensive knowledge of cost accounting. The problem exists in most enterprises which is helpful to the normal operation and development of the enterprises. With the constant development of economy, the progress of science and technology, and the rapid change of knowledge, knowledge economy is also very important. So the enterprise should add some relevant information about knowledge economy to cost accounting. But there are always some enterprises which only pay attention to the management of tangible assets and ignore the importance of intangible assets, for example, knowledge economy is not added to cost accounting. So there is incomprehensive knowledge about cost accounting, which is no good to the enterprise's normal development, to the increase of its competitiveness, level of cost accounting and its economic benefits, and to the smooth development of cost management.

Financial personnel's professional quality is not very good. In order to be qualified for the work of accounting, accounting practitioners should be equipped with rich professional knowledge, professional qualities, moral qualities and individual qualities. However, when some enterprises allocate and arrange the personnel in the financial department, the staff's knowledge level and professional ability can reach the basic requirements, but they lack enthusiasm in learning and don't have innovative consciousness. So they can't adjust and improve cost accounting methods. Some of the accountants and financial staff don't have accounting qualifications, professional quality, and strong innovative consciousness. They can not timely grasp the concepts of accounting. Some accounting personnel not only have low professional qualities, but also lack certain occupational morality, which makes the cost accounting results unreliable and improper. Enterprises can not use these statements to learn the enterprise's production and operating conditions.

2 CORRESPONDING SOLUTIONS TO THE ENTERPRISE'S COST ACCOUNTING PROBLEMS

2.1 *To make the enterprise's cost accounting content more comprehensive*

To make the enterprise's cost accounting content more comprehensive. Some problems arise because business managers or accountants do not attach importance to cost accounting. So the enterprise should pay more attention to it. When the enterprises carry out cost accounting, especially the accounting and management of their intangible assets, they need to make effective classifications. According to different categories of intangible assets, financial and accounting personnel should use different cost accounting methods. The calculation of intangible assets should be counted in the cost of the product. In addition, when accounting knowledge assets, accountants should realize that this kind of cost will be affected by refresh rate as well as its own features. Appropriate amortization of knowledge resources can be used to deal with this account and a flexible amortization method is expected. In recent years, our country has paid more and more attention to environment, so when the accountants conduct the calculation of the environmental costs, they need to be attentive and calculate the cost for beautifying the environment. It can enhance the enterprise's environmental protection and beautification consciousness, helps to develop the spiritual civilization of the enterprise, and enables the enterprise to have a harmonious development with the society and environment.

2.2 *To make the enterprise's cost accounting methods more scientific*

To make the enterprise's cost accounting methods more scientific. There are many cost accounting methods, such as quota method, batch method and category method. These are the more important cost accounting methods. Each method has its own merits and defects if used on different accounting objects. Therefore, enterprises have to base on their own conditions to select an appropriate method for cost accounting in order to display its advantages and increase the enterprise's economic benefits. Besides, enterprises also have to consider their development directions and goals to choose a suitable cost accounting method, which makes the development directions and goals combine with cost accounting methods.

2.3 *To make the enterprise's cost accounting more sustainable*

To make the enterprise's cost accounting more sustainable. Cost accounting is an inseparable part of the enterprise's development. It also affects the enterprise's cost management. Therefore, accountants should give prominence to the work of cost accounting based on the enterprise's economic goals, its production management conditions and the development of society and economy, and should establish a more perfect cost accounting system.

2.4 *To enhance financial personnel's professional level*

The staff's training needs to be conducted in order to constantly improve the financial management and accounting standards. Accounting practitioners should enrich their knowledge by learning which is the foundation of the accounting construction. They continue learning new knowledge which can not only increase their own professional quality, but also can increase the transparency of supervision between various departments. Special departments and personnel have to be organized to supervise accounting practitioners, and to have regular inspection and unscheduled random inspection. The staff who have made serious mistakes in their positions can't be forgiven. Of course, incentive mechanism needs to be set up as well to motivate the accounting personnel to constantly improve themselves by increasing their professional knowledge, and improving their professional quality and moral quality. To establish effective management system controls the financial expenditure. Certain financial funds need to be distributed to each department to achieve the enterprise's maximum benefits which is also the objective of financial management. Financial work is very complex and needs to be conducted carefully. Financial personnel must enhance their professional quality and comprehensive quality, and take a responsible and meticulous attitude towards their work.

2.5 *The innovation of the enterprise's cost accounting methods*

Enterprises should actively innovate cost accounting methods and shouldn't always stick to the traditional one. Changing traditional ideas of development can not only maximize the enterprise's interests, but also can make it enhance its economic benefits by means of cost accounting. Good methods and experience of cost accounting both at home and abroad can be used by absorbing the best and discarding the worst. Good preparations have to be made for the innovation of cost accounting methods. Of course, the innovation of methods can not exceed legal restrictions. The enterprise can use these cost accounting methods, such as accounting on the cash basis, accrual basis, quota method, output method. Each accounting method has its advantages and disadvantages. Enterprises should choose the best scheme according to their actual situations and the characteristics of production management. When choosing the accounting method, they must take those factors like their products, product features and service into consideration. After making an overall assessment and analysis of the enterprises, they can select the most suitable and effective cost accounting method.

3 SUMMARY

In short, if the enterprise wants to have further development and improve its competitiveness, the level of cost accounting needs to be enhanced and existing problems have to be solved. The solution to the problems can not only improve the enterprise's economic benefits, but also can promote its development. Enterprise managers are expected to pay more attention to cost accounting problems in order to improve the economic benefits of the enterprise.

REFERENCES

[1] Liu Yide. Problems and strategies of the enterprise's cost accounting [J]. Foreign Investment in China, 2013,08:86–89.

[2] Zhang Yuping. Problems and strategies of the enterprise's cost accounting [J]. Chinese market, 2013,33:140–141–144.

[3] Zhang Lili. Problems and strategies of the enterprise's cost accounting [J]. Modern Business, 2013,23:214–215.

[4] Yan Zhuo. An analysis of the problems and methods of the enterprise's cost accounting [J]. The Fortune Time, 2014,02:37–39.

[5] Wang Kong. An analysis of the problems and methods of the enterprise's cost accounting [J]. Foreign Investment in China, 2014,03:88.

Computing, Control, Information and Education Engineering – Liu, Sung & Yao (eds)
© 2015 Taylor & Francis Group, London, ISBN: 978-1-138-02800-5

Analysis of the innovative construction of entrepreneurship courses in higher vocational education

H.Y. Wang & H.Y. Liu
Langfang Polytechnic Institute, Hebei, China

ABSTRACT: Nowadays, with the constant changing in the requirements for talents in the society, higher vocational colleges begin the reform of talent cultivation mode. In the process of the reform, they pay more attention to the entrepreneurship education, and hope to improve the university graduates' employment competitiveness. But entrepreneurship education in higher vocational colleges still exists some problems which have hindered its development. This paper mainly describes present situations of the innovative entrepreneurship education in higher vocational colleges, and also analyzes ways of constructing innovative entrepreneurship courses in higher vocational colleges.

KEYWORDS: higher vocational education; innovative entrepreneurship courses; present situations; ways.

In recent years, the enterprise's demand for talents tends to be saturated and the continuous expansion of enrollment in colleges has made many graduates can not find their satisfactory jobs and even a part of college students face the risk of not finding jobs after graduation. Under such circumstances, our country begins to call on college students to start a business on their own which will increase employment. Therefore, higher vocational colleges begin to establish innovative entrepreneurship courses which will help change college students ideas and let them have a more comprehensive understanding of entrepreneurship.

1 CURRENT SITUATIONS OF INNOVATIVE ENTREPRENEURSHIP EDUCATION IN HIGHER VOCATIONAL COLLEGES

1.1 *The weakness of innovative entrepreneurship education practice in higher vocational colleges*

At present, many higher vocational colleges still focus innovative entrepreneurship education on textbooks which leads to the weakness of its practice. They often encourage students to start their own business to improve their innovation ability. However, these methods have drawbacks which can't really develop the students' awareness of innovation.

1.2 *The lack of perfect educational assessment system in higher vocational colleges*

Most higher vocational colleges lack a sound assessment system of the innovative entrepreneurship education. They focus the assessment only on the theory, but ignore teaching research of innovative entrepreneurship education. Higher vocational colleges carry out the innovative entrepreneurship education just in response to the country's call, but can't really play its role.

1.3 *The lack of a reasonable system of entrepreneurship education in higher vocational colleges*

There are various entrepreneurship competitions in some of the higher vocational colleges. The enterprise competition is not only helpful to cultivate the students' innovation consciousness, but also helps to improve their innovation ability. But this activity just plays a quite limited role and can't really improve the students' innovation ability. The project of entrepreneurship education in higher vocational colleges is systematic and complex, so it is impossible for them to achieve the purpose of entrepreneurship education if they just rely on entrepreneurship competitions. In addition, innovative entrepreneurship courses are different from other courses, thus it is difficult to form a complete scientific system.

1.4 *The lack of accurate knowledge of entrepreneurship education in higher vocational colleges*

At present, college students' employment situation is more severe, so higher vocational colleges have carried out the entrepreneurship education. But most of colleges graduates choose to start their own businesses because they could not find suitable jobs instead of realizing their own dreams. If graduates can find jobs, then they will give up starting businesses. There are a lot of students after graduation who can not find suitable jobs, so they're forced to choose to do pioneering work. Besides higher vocational colleges have set up Employment Guidance Center, but it just propagates some successful stories of entrepreneurship, which leads to the lagging concepts of the innovative entrepreneurship education in higher vocational colleges. In addition, many students have a rather weak entrepreneurship consciousness and are lack of enough understanding of entrepreneurship, so they don't really put entrepreneurship activities into practical actions.

1.5 *The lack of professional teachers in higher vocational colleges*

The teachers' professional level in higher vocational colleges is low, and they lack rich business knowledge, which will affect the quality of teaching activities. As an entrepreneurship education instructor, he or she should not only has rich business experience, but also has marketing, management knowledge. But in fact, many teachers lack entrepreneurial experience and they can just cite some successful business people's examples, but they are not able to help the students solve problems that they will encounter in their actual businesses.

2 WAYS OF CONSTRUCTING INNOVATIVE ENTREPRENEURSHIP COURSES IN HIGHER VOCATIONAL COLLEGES

2.1 *To determine teaching concepts and improve course standards*

Course standards refer to a programmatic document for the completion of the teaching task. Some vocational colleges opened the innovative entrepreneurship course in the freshman year. At that time, the students are full of curiosity about this course thus they have a strong enthusiasm in the class. But due to the lack of a unified course standard and the lagging teaching methods, the students gradually lose interest in learning. In order to make some changes, vocational colleges should improve course standards, reposition the innovative entrepreneurship courses, and view improving the students' occupational ability as the teaching goal. When having a clear positioning of the course, vocational colleges should design some appropriate practical activities to enhance the students' innovation ability.

2.2 *To select typical cases and cultivate the students' innovation consciousness*

In the background of the new course reform, case teaching method can enhance the students' learning interest and also can improve the efficiency of classroom teaching. In the teaching process, the use of case teaching method enriches teaching content, and students can acquire some relevant knowledge from the case. In traditional teaching, teachers often use the "spoon feeding" teaching method to carry out teaching activities, which always makes students in a passive learning state and is not helpful to cultivate their creative thinking. Therefore, in the process of teaching, teachers should introduce cases having been heard many times to the students. Introducing these characters can stimulate the students' pioneering consciousness. At the same time, teachers should use quotations from successful people in the classroom and encourage each student by them, which can achieve a better result of teaching.

2.3 *To play games in class and stimulate the students' innovative thinking*

Innovative thinking is an advanced form of thinking. In the process of teaching, teachers should pay more attention to cultivate the students' divergent thinking, reverse thinking, which can stimulate their creative potential. Teachers should also grasp some strategies to stimulate the students' creative thinking . Under normal circumstances, teachers often introduce some creative games, such as the left limb sports game, combined imagination games to achieve the goal of teaching. In order to cultivate the students' divergent thinking, teachers will carry out a game like this: teachers present a pencil in front of the students, and then ask them what purposes it can be used. The teacher needs to give them ten minutes to think about this question, and asks them to list as many purposes as they can. When time is up, the teacher should let every student give their answers. After all of them have given their answers, the teacher has to summarize which one of the uses is a more creative imagination and which student has given original answers. The teacher should encourage the students who have creative ideas. Students' creative thinking can be motivated by playing such a simple game.

2.4 *To encourage the students to design innovative works and show their personality*

When students begin to slowly develop their own innovative thinking, the teacher should ask them to gather material after class and to observe various things in life. At the same time, the teacher needs to ask the students to creatively design their own works and requires them to pay attention to the following tips: first, making their own innovation objective clear; second, briefly expressing their innovative ideas; third, generally outlining a sketch. When the students finish the design of their new works, teachers have to look at every student's design carefully and write down comments. The teacher asks each student to show their work in class. When presenting the works, he or she should express his or her innovative ideas to the rest of the class. After that other students can put forward the defects of the work, so as to help him or her better improve it. After every student finishes showing the work, the teacher will ask the students to select eight works with the most creative themes. This teaching method can arouse the students' enthusiasm towards designing and also can develop their creative thinking which will slowly improve their innovative ability.

2.5 *To build a practice platform and encourage students to start businesses on their own*

Relevant data have revealed that most students want to do pioneering work after graduation, and they hope that college courses can cover entrepreneurship preparation and methods and higher vocational colleges can provide a practice platform for them. Therefore, teachers should impart some basic knowledge of business to the students and also tell them the relationship between innovation and entrepreneurship, which can make the students master rich innovation knowledge. At the same time, teachers should tell students if they want to start their businesses, they must have certain innovative thinking; if they lack innovative thinking, and then it is difficult for them to succeed. In entrepreneurial activities, teachers should use innovative ideas to guide each student in order to improve their independent innovation ability.

Higher vocational colleges should create "innovation and entrepreneurship club" where students can learn the development direction of the current society and also can know what the current business projects are. Under the guidance of the club, students can grasp some relevant business knowledge and also can have a clue about their entrepreneurial direction. At the same time, the club should strive to seek cooperation partners and provide a wider platform for students to carry out practical activities.

2.6 *To change assessment methods and inspire students an innovative entrepreneurship*

Students' study habits and learning interest will be affected by the way of assessment. If the teacher has chosen an inappropriate assessment method, it will dampen the students' enthusiasm. So the teacher has to apply reasonable assessment methods to motivate the students' entrepreneurship. At present, if each student is evaluated only by high and low test scores, it can't train talents who are needed by society. Therefore, higher vocational colleges should explore new ways of assessment in order to inspire each student's potential. They should adopt an assessment mode which combines creativity and achievements to stimulate the students' innovation enthusiasm. Changing the way of assessment not only can stimulate students' interest in learning, but also can improve their learning efficiency. For example, when commenting on the students' works, teachers can ask several students to make their comments. The process of commenting will also improve the students' innovation ability and their creative thinking.

3 SUMMARY

In the process of educational reform, higher vocational colleges should attach great importance to the innovative construction of entrepreneurship courses. At the same time, teachers in vocational colleges should continue to improve their own teaching methods and change their ways of assessment in order to stimulate the students' enthusiasm of innovation.

REFERENCES

[1] Zhu Xiaofeng. Practice and thinking about the development of innovative entrepreneurship courses in higher vocational colleges [J]. Vocational Education Forum, 2012, (18): 76–78.

[2] Tao Yan. The innovative entrepreneurship education and the construction of course system of the accounting major in higher vocational education [J]. Exam Weekly, 2012, (54): 160–161.

[3] Yan Qi. The construction and application of training platform in computer assembly and maintenance under the background of of the innovative entrepreneurship [J]. Journal of Chifeng University: Natural Science Edition, 2012, (5): 201–202.

Computing, Control, Information and Education Engineering – Liu, Sung & Yao (eds)
© 2015 Taylor & Francis Group, London, ISBN: 978-1-138-02800-5

A study of the significance of tax planning for the enterprise's financial management

L.D. Su & J. Zheng
Langfang Polytechnic Institute, Hebei, China

ABSTRACT: Scientific and reasonable tax planning can reduce certain tax burden for the enterprise, and save costs to increase its economic profits. So tax management is an important part of the enterprise's financial management. This paper expounds the definition of tax planning and its necessity. It analyzes the present situation of tax planning and specific points for attention during the process of its implementation. It also discusses the significance of tax planning for the enterprise's financial management.

KEYWORDS: financial management; tax planning; definition; necessity; present situation; points for attention; significance.

Tax planning is a privilege enjoyed by the enterprise to achieve maximum interests through legal means. The emphasis of tax planning for the enterprise's financial management can reduce its tax burden and also can improve its external competitiveness. During the process of financial management, some agents can be entrusted to formulate points for attention covering all aspects from design to tax under the premise of abiding by the tax law in order to improve the enterprise's economic efficiency. The scientific and systematic management of tax planning is favorable to a long-term development of the enterprise.

1 AN OVERVIEW OF THE ENTERPRISE'S TAX PLANNING

1.1 *The definition of tax planning*

The theoretical definition of tax planning is that natural person and the legal person of the enterprise as taxpayers comply with relevant laws and regulations; they need to make scientific and systematic tax plannings in accordance with the provisions of the tax law, such as the enterprise's financing activities, management control activities, investment activities, in order to reduce operational costs and tax and increase the enterprise's profits. While the actual meaning of tax planning is that under the premise of protecting the interests of the country, the enterprise's tax consciousness needs to be enhanced; and the management of tax planning will save the enterprise's operational costs and improve its market competitiveness. Tax planning will ensure that the enterprise's financial management can be orderly conducted. It runs through the entire process of the enterprise's earnings management, strategic investment, capital circulation. To strengthen tax planning management is helpful to improve the enterprise's financial management level.

1.2 *The necessity of implementing the enterprise's tax planning*

The enterprise's tax planning is not just simple financial accounting. As an important part of financial management, it helps the enterprise obtain maximum economic benefits with the minimum input in order to realize the maximization of the economic efficiency. The enterprise can make full use of preferential tax policies in compliance with corresponding tax laws and regulations to reduce payable taxes, tax expenditure and management control costs and improve the economic benefits. Paying taxes is the citizens' obligation in our country, so the tax is charged for free. The enterprise's tax expenditure is a net outflow of funds. Some enterprises partially pursue the maximization of the economic interests and take illegal means to reduce payable taxes which is unfavorable for the enterprise to build a good social image, damage its reputation, and also makes it get severe punishment by law affecting its healthy development. The systematic and reasonable tax planning helps the enterprise perform the taxpayer's obligations. It is helpful to increase the enterprise's social benefits and enable it to have healthy and stable development. Tax planning will reduce the payable tax and it will also reduce the country's fiscal revenue. But in a long run, to strengthen tax planning under the government's macro-control policies can reduce the enterprise's tax bearing, and promote its long-term development.

The enterprise's development can promote the development of the socialist market economy. In fact, tax planning will stimulate tax revenue and tax policies of the country and promote the economic development.

2 AN ANALYSIS OF THE CURRENT SITUATIONS OF THE ENTERPRISE'S TAX PLANNING IN OUR COUNTRY

2.1 *Less attention payed by the enterprise to tax planning*

At present, tax authorities and some related departments in our country haven't made tax planning knowledge popularized due to the lack of effective publicity, which causes most enterprise's managers lack a correct understanding of tax planning. Many of the managers do not pay enough attention to tax planning and don't realize its value for financial management. They aren't aware of its significance for the enterprise's development. Some enterprises during the management control process neglect to conduct tax planning; some hold an one-sided view that the improvement of the comprehensive competitiveness only relies on improving the production efficiency and production technology, and to obtain the share of the market will achieve the maximum interests. Wrong management ideas won't help enterprises fulfill their taxpayers' obligations, and even make them punished by laws which will increase their tax burden.

2.2 *Without hiring professional tax planning staff*

There are higher technical requirements for tax planning. Tax planning staff should be equipped with higher professional and legal accomplishments. Related working personnel should not only have a solid professional tax knowledge, but also should be familiar with the basic knowledge of law and business management knowledge. They can flexibly deal with different situations when facing market changes and promptly take appropriate measures to reduce tax burdens and tax payments. Tax planning currently isn't conducted very well in most enterprises of our country. The overall professional ability of financial personnel is low. Most of them can only do a simple job like financial accounting and are lack of tax planning experience. They can't have a correct and comprehensive understanding of the government's preferential tax policies, so the enterprise can not fully enjoy relevant tax preferential policies. Besides, financial personnel lack a correct understanding of the significance of tax planning and neglect basic management of it which also increases the difficulties of tax planning. At present, many enterprises are lack of professional talents with higher comprehensive quality in tax planning. Some tax professionals have low professional accomplishments. It is difficult to fundamentally change tax planning situations.

2.3 *The excessive pursuit of tax relief in enterprises*

Many enterprises only focus the attention on the advantages of tax planning for it can reduce the enterprise's tax burden. They only attach importance to the immediate interests regardless of their long-term development. The excessive pursuit of tax minimization is not conducive to the smooth development of tax planning, is not good to the enterprise's sustainable development, and will even cause a loss. Enterprises can not only consider tax minimization when they invest. Tax reduction does not mean an overall growth of the enterprise's interests. When choosing an investment scheme, they shouldn't only consider the scheme having the lightest tax burden, but should consider the enterprise's long-term interests to select the best plan for its sustainable development and guarantee its interests.

3 POINTS FOR ATTENTION WHEN THE ENTERPRISE IMPLEMENTS TAX PLANNING

3.1 *A systematic consideration of tax planning*

A reasonable tax planning is to achieve the goal of financial management. In order to increase the capital recovery rate, the enterprise has take some measures, such as reducing the tax burden, controlling tax costs. It hopes that tax planning can help to gain maximum profits after paying taxes by means of tax deferment to save tax expenditure. Tax planning, as one of the subsystems of the financial management, needs to be conducted according to the financial management. Enterprises need to have a comprehensive analysis of tax planning schemes and take all influencing factors into consideration including tax and non tax factors. The maximization of economic benefits can be achieved not just by the minimization of non tax burdens. To reduce tax burdens by tax planning can't guarantee the increase of the enterprise's benefits. Hence tax planning should be conducted on an overall basis.

3.2 *Tax planning should make a cost - benefit analysis*

The enterprise's management control has dual characters. By planning a scheme, the enterprise can obtain certain tax benefits, but the implementation of the scheme will pay certain costs and the selection of one scheme also means to give up others which may make more benefits. When choosing the tax planning scheme, the enterprise has a variety of choices. It has to weigh and balance costs and benefits and select the best one to improve economic benefits.

3.3 *Tax planning should serve for financial decision-making*

The enterprise's tax planning will have a significant impact on its operational and financing activities. Tax planning can be well conducted by the enterprise if it can make better business arrangements. Tax planning must be subordinated to financial decision-making. If the former deviates from the latter, it will have a bad influence on the feasibility of financial decisions which is of no help for managers to make correct decisions.

4 THE SIGNIFICANCE OF TAX PLANNING FOR THE ENTERPRISE'S FINANCIAL MANAGEMENT

4.1 *The importance of tax planning in the enterprise's operational activities*

Enterprises can choose inventory valuation methods by tax planning. When materials prices continue to drop, they can choose first-in first-out (FIFO) method; when the material price isn't stable, they should choose moving weighted average or weighted average method which can avoid the profit being affected by costs. Enterprises' sales revenue will be affected by different time of tax payment and changes in incomes due to different means of settlement. If money is collected after the goods are sold, the confirmed time is the day when the seller receives the payment for goods and offers the buyer bills of lading; if the goods are sold by collection payment, the confirmed time is the day when the seller sends out goods and finishes procedures for collection; if the goods are sold on credit and installment, the confirmed time is the appointed time in the contract. To choose a better sales settlement method and timely adjust the confirmed time enable the enterprise to delay paying taxes. In the process of tax planning, enterprises need to list period charges and adopt a withholding method to keep accounts. Losses that may arise can be expected beforehand which can shorten the amortizing period, delay the time of paying taxes, and pay less income taxes to reduce the tax burden and achieve the maximum benefits. The costs can be increased by choosing the right fixed assets to accelerate depreciation. It is helpful to delay paying taxes. Tax planning can help the enterprises improve their management control level. There is a close relationship between the period cost and period profit. In the process of tax planning, enterprises' managers need to weigh and consider various factors to improve the management control level and ensure a sustainable development of the enterprises. The enterprise's tax planning can choose the way of increasing inventory costs which will lead to the rise of operational costs and a relative fall of the profit to decrease the enterprise's tax burden.

4.2 *The role of tax planning played in financing activities*

There are two main ways of financing in the enterprise, that is, equity financing and indebted financing. Enterprises need to adopt different tax planning methods according to different forms of financing. As to equity financing, the distribution of profits mainly includes those after paying taxes. It can't reduce taxes by way of decreasing the profit. Equity investment doesn't need to pay fixed interest and can't play the role of paying debts. While the indebted financing requires the enterprises to pay fixed interest before charging taxes and the enterprise's borrowing costs can be deducted. So debt investment has the function of paying debts. It enables the enterprises to lower their profits so as to reduce the tax burden. But indebted financing needs a regular repayment of the principal. Excessive indebted financing will cause financial risks. Enterprises should try to reduce equity financing and properly increase indebted financing, which will effectively reduce the enterprise's tax burden. Equity investment needs to pay more money, but there is no need to repay the funds, so the risk of smaller. Enterprises need a scientific balance of the proportion of the indebted financing and equity financing to make them have the best financing and achieve the maximum economic benefits.

4.3 *The significance of tax financing in investment activities*

When investing money in projects, enterprises should make a scientific and rational analysis of the investee, organizational forms, and investment businesses and should weigh all factors to select the optimal investment plan. They have to make full use of preferential tax policies to reduce their tax burden. Different organizational forms apply to different tax policies. Incorporated enterprises are required to pay personal income taxes and they can enjoy a variety of preferential tax policies. The enterprise's legal person can deduct value-added tax (VAT). Partnership enterprises cannot enjoy the preferential tax policies, but they can pay the individual income tax based on their incomes. Generally, they are a small number of taxpaying bodies and VAT can't be deducted. When investing, the enterprise should consider the actual operating situations to ensure that tax financing can go smoothly in order to reduce the tax burden and improve the economic benefits.

5 SUMMARY

Tax planning is of great help to reduce the enterprise's tax burden, increase its economic benefits and promote the sustainable development of the economy. But at present, it still has some problems, such as paying less attention to it, without hiring professional tax personnel to conduct financial management, the excessive pursuit of tax relief. Enterprises need to emphasize cost-benefit analysis and consider tax planning on an overall basis. Moreover, tax planning must be subject to financial decision-making and it should play a role in financial management in terms of operational, financing and investment activities. Tax planning can strengthen the enterprise's financial management level and guarantee its maximum interests.

REFERENCES

[1] Liu Yan. A study of the significance of tax planning for the enterprise's financial management [J]. Business Horizon, 2014, (13): 192–192.

[2] Dong Hongyun. A study of the significance of tax planning for the enterprise's financial management [J]. Financial World, 2014, (21): 264–264,269.

[3] Sui Yajuan. A brief analysis of the influence of tax planning on the enterprise's financial management [J]. Foreign Investment in China (second half of the month), 2013, (11): 84–85.

Computing, Control, Information and Education Engineering – Liu, Sung & Yao (eds)
© 2015 Taylor & Francis Group, London, ISBN: 978-1-138-02800-5

Incorporating of lean manufacturing techniques in supply chain of automotive company

Amelec Jesus Viloria Silva
University Sergio Arboleda, Bogotá, Colombia

ABSTRACT: This article describes the development stages for the incorporation of lean manufacturing techniques in the supply chain of a company of the automotive sector is presented. The methodology is based on a continuous improvement cycle of Deming. After implementation of the manual setting was achieved lean techniques adapt the delivery time of products according to customer type, work according to the Takt Time, achieve continuous flow in 90% of transactions and increase the reliability of the supply chain 95%.

KEYWORDS: Automotive industry, Logistics, Supply Chain, Lean Manufacturing.

1 INTRODUCTION

From a strategic standpoint, lean manufacturing is to identify the value chain, the sequence of business activities that contribute to the generation of value to the product and / or service, and thus gain competitive advantage in the long-sought market [1]. This value can only be achieved when customer needs at the right cost and at a precise time are met, justifying the importance of creating value in the assembly processes and activities responsible for transforming raw materials and deliver products and / or services to end customers [2].

The company study sample is an automobile manufacturer with a global presence. This under constant quest for continuous improvement, focuses its efforts on optimizing each area of the Directorate for Planning and Logistics Operations and capitalize on their knowledge, thus creating the "Excellence System PSR" designed to be implemented in all areas of the company: both vertical (Research and Development, Production, Sales, etc.) and cross-sectional areas (Supply Chain, Quality, Human Resources, Finance, etc.).

The term "System of Excellence" encompassing all the policies that frame the operation of the company based on their strategic priorities. According to [3] and [4] is the set of management policies, behaviors and tools to meet the objectives of the vision of the company.

Composed of a series of working methods and management supported the principles of lean manufacturing and continuous improvement culture, the PSR offers [5]:

- Operation and simplest methods of work
- Best individual and collective efficiency
- Better customer satisfaction
- Better partnership with suppliers
- Competitive advantage for the company
- Culture of continuous improvement

This article aims to show the incorporation of lean manufacturing techniques to supply chain an automotive company based on the excellence PSR system.

2 METHODOLOGY

The project was divided into four (4) distinct phases. These continue the iterative process of planning, implementation, revision and correction, according to the Deming Cycle (PDCA) [6].

- Phase 1. Diagnosis: Diagnose the current state of implementation of the system PSR Excellence in the supply chain of the company.
- Phase 2. Characterization: To characterize the supply chain of the company.
- Phase 3. Integration: Determine Lean Manufacturing principles established in the PSR Excellence System applicable to the supply chain.
- Phase 4. Development: Develop manual for the implementation of Excellence System PSR in the supply chain of the company.

3 RESULTS AND ANALYSIS

The results obtained are shown in each of the steps outlined in the methodology.

3.1 *Phase 1. Diagnosis of the current situation*

Table 1 presents a summary version of the diagnosis made Supply Chain Company by structuring the generated observations and corresponding corrective actions.

Table 1. Summary of the analysis of the current situation Supply Chain Company

Observation	Plan of action
Lack of documents within the company clearly stating the concept of Supply Chain PSR	Generate a concept of Supply Chain PSR and include it in the manual.
The existence of various documents stating different objectives PSR Supply Chain	Consolidate the objectives of PSR Supply Chain and include it in the manual.
The existence of the general idea within the company that supplies PSR Chain has a single client: the final customer or retailer.	Define the different customer supply chain and include it in the manual.
Different directions establish the existence of different key performance indicators (KPI) for Supply Chain	Define what the KPI will be used to measure the performance of the supply chain and include it in the manual
Little detail in the existing graphical representation of perimeter Supply Chain PSR	Create a new graphical representation of the perimeter of the company in detail and include it in the manual.
Lack of clarity regarding who is responsible for what in the flow of material and information along the supply chain	Define the responsibilities of the flow of material and information along PSR Supply Chain and include it in the manual
Personal goals based employees as the Deming cycle	When integrating the principles of Lean Manufacturing in the Supply Chain PSR, structure them in a way that is based on the culture of continuous improvement of PDCA and include it in the manual.
Lack of knowledge about the business side of the company and its importance.	Develop the topic of demand planning to give a general idea of how it affects the industrial part and include it in the manual.
General knowledge of the principles of Lean Manufacturing and the existence of Excellence System PSR	When developing the manual: - Do not stop to explain in depth the concepts, but rather how they all work together to achieve the desired improvement.
The existence of a process map planning outdated Supply Chain PSR	Create SIPOC diagrams of the major planning processes generating a detailed map of a section of the current state.
Lack of information about the future state of the planning of the supply chain PSR	Create documents necessary to establish the basis of the desired future state. When developing the manual: - Generate a diagram that integrates three main level planning processes. Strategic, tactical and operational

3.2 *Phase 2. Characterization chain PSR supplies*

The second phase began with corrective action as to what diagnosed in the previous phase to map planning processes: work with updating and documentation of processes, further comprising further the current planning structure that presents the company. Twelve (12) SIPOC diagrams current map of PSR planning processes were created, which are listed below [7]: “Setting the resources for handling materials (Weekly)”, “Create and send orders to the parts”, “Break into pieces the volume per period”, “Break into pieces calculating net requirements of parts (annual)”, “Breakdown parts calculating net requirements of parts (weekly)”,”Break into pieces the calculation of gross requirements of parts (weekly)”, “Prepare material handling (annual)”, “Prepare material handling (per period)”, “Prepare the packing process (per period)”, “Prepare the orders provisioning (per period),” “Prepare provisioning orders (weekly)” and “Sequence DMB production”.

3.3 *Phase 3. Integration responsible for the implementation of PSR team*

Different Lean Manufacturing principles established in the PSR Excellence System applicable to Supply Chain Company are listed below:

Analyze and segment the customer demand: adapt the diversity of products and delivery time for different types of customers to better respond to your request.

Increase the reliability of the supply chain: increase the reliability of commercial and industrial forecasts to meet the needs of customers at the lowest cost and inventory level possible.

Flexible and adapt the supply chain: eliminate or reduce the limitations of plants and suppliers in order to evolve at the same pace with demand.

Level production: reduce variations in customer demand in order to optimize investments and labor.

Establish a pull system: only produce what the customer uses to reduce inventory and WIP.

Working under Takt Time: producing customer pace optimizing investments and labor in addition to reducing inventory and WIP.

Ensure the continued flow: a process parts continuously in order to optimize investments, labor and reduce inventory and WIP.

These seven (7) principles were integrated into a wheel for improvement which, following the methodology of the Deming Cycle, you find that the Chain of PSR Supplies company understands the needs of enterprise customers, improve responsiveness to these needs and increase profitability of the company under a culture of continuous improvement. This wheel of continuous improvement is presented in Figure 1 [8], [9] and [10].

3.4 *Phase 4. Development manual for the implementation of PSR excellence system*

The project culminated in the development and implementation of the Manual for the implementation of the PSA Excellence System in the supply chain of the company. Fulfilling formats and standards set by the Department of Communications, it explains the concepts primarily through diagrams and figures made after long periods of analysis. The index presents the final version of the manual is as follows [10]:

1 Customer Supply Chain.
2 Key Success Factors of Supply Chain.
3 Definition, structure and scope of the Supply Chain.
4 Position Supply Chain within the company.
5 Excellence System Principles PSR applied to the Supply Chain.
6 Macro-planning processes Supply Chain.

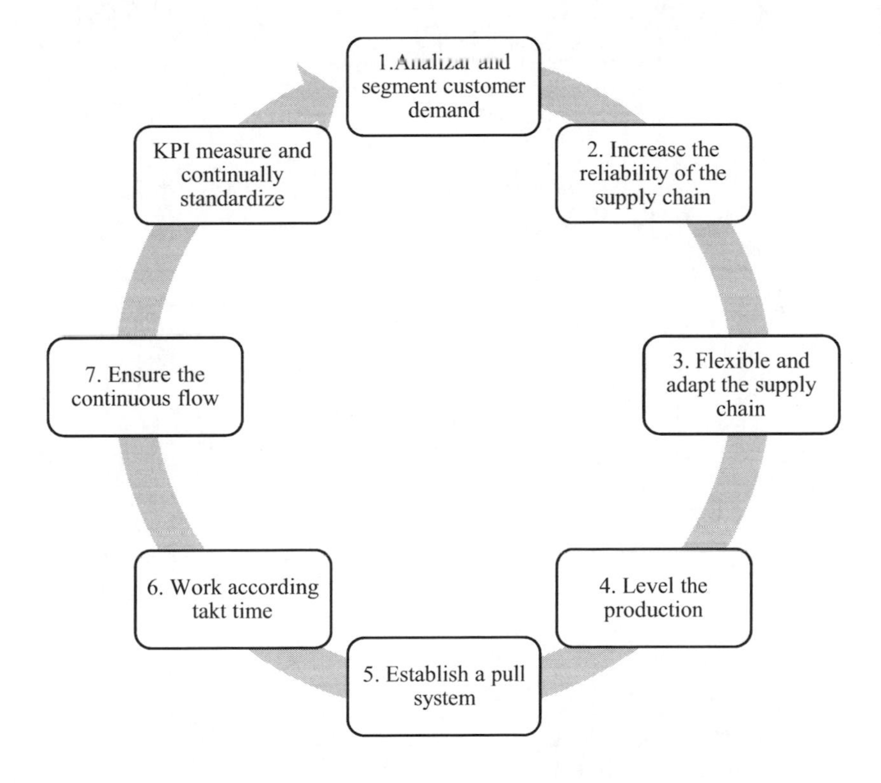

Figure 1. Wheel continuous improvement of the Supply Chain PSR.

4 CONCLUSIONS

After a year of incorporating lean manufacturing techniques within the supply chain of the company the following results were achieved: adapt the time netrega product by type of client, work according to the Takt Time, achieve continuous flow in a 90% of operations and increase the reliability of the supply chain by 95%.

REFERENCES

[1] Blackstone, J. APICS Dictionary. 13ava edición. Illinois: APICS The Association for Operations Management, 2010.

[2] Casanovas, A. y Cuatrecasas Ll. Logística Integral. Lean Supply Chain Management. 1ra edición. Barcelona: Profit editorial, 2011.

[3] Cortés, J. (s/f). "La Navaja de Ockham". Disponible en Internet: http://www.economiadigital.es/es/notices/2012/11/la_navaja_de_ockham_35123.php.

[4] Arnold, T. Chapman, S. y Clive, Ll. Introduction to Materials Management. 7ma edición. Nueva York: Pearson Prentice Hall, 2010.

[5] Belohlavek, P. OEE: Overall Equipment Effectiveness. 1ra edición. Buenos Aires: Blue Eagle Group, 2006.

[6] Evans, J. Lindsay, W. Administración y control de la calidad. 7ma edición. México. 2008.

[7] Gerstein, M. Encuentro con la tecnología. Estrategias y cambios en la era de la información. 1ra edición. México: Addison Wesley Iberoamericana, 1988.

[8] Kalakota, R y Robinson, M. E-Business: Roadmap for Success. 1ra edición. Nueva York: Addison-Wesley, 1999.

[9] Sobek, D. y Smalley, A. Understanding A3 thinking. A Critical Component of Toyota's PDCA Management System. 1ra edición. Illinois: Productivity Press, 2008.

[10] Porter, M. Ventaja Competitiva. 1ra edición. México: Editorial C.E.C.S.A, 1986.

[11] Russell, J. y Cohn, R. Kiss Principle. 1ra edición. Nueva York: Book on Demand Ltd, 2012.

Computing, Control, Information and Education Engineering – Liu, Sung & Yao (eds)
© 2015 Taylor & Francis Group, London, ISBN: 978-1-138-02800-5

Research on the sports industry competitiveness index system construction based on Analytic Network Process (ANP)

S. Li
Department of Sports, Guizhou University of Finance and Economics, China

ABSTRACT: On the foundation of analyzing the sports industry competitiveness' composition requirements and influencing factors, follow scientific, practical, and the principle of combining dynamic analysis with static analysis, hard targets with soft targets, this paper designs sports industry competitiveness evaluation index system, uses analytic network process, takes full account of the logical relationship among the various indicators, combines the ANP evaluation results with sports industry development strategy, then provides a decision basis for enhancing China's sports industry competitiveness.

KEYWORDS: sports industry competitiveness, index system, analytic network process.

1 INTRODUCTION

The sports industry is a concept of collection; it is the sum of the various sports industry's sports goods, sports services, and the chain of related industries [1]. With the reform and opening up, the continued development of socioeconomic, China's sports industry has developed rapidly, and plays an important role in promoting economic growth. However, China's sports industry only sharing 0.7 percent of GDP compared to developed countries sports industry such as United States sharing more than 7 percent of GDP, there is a big gap [2]. From the perspective of the industrial economy, China's sports industry has experienced 30 years development after reform and opening up, is still at a relatively low level; mainly due to China's sports industry is still a lack of competition. In the development of the sports industry, the major factor deciding sports industry's competitive advantage is its own development status, to enhance the competitiveness of China's sports industry, we must first properly evaluated it, find the existing problems, to facilitate take appropriate countermeasures. However, the sports industry has a diversity of content, the complexity of the level, to evaluate its competitiveness must combined with its actual characteristics, taking a comprehensive evaluation and reasonable assessment, can be consistent with the actual situation, has a guiding role and real value for the sports industry competitiveness evaluation. Thus, this paper combines with China's sports industry own actual, objectively analyzes the competitive strengths and weaknesses, and through establishing sports industry competitiveness evaluation index system, to provide scientific methods for China's sports, industrial competitiveness evaluation, to explore how to quickly improve the international competitiveness of China's sports industry.

2 BASIS AND PRINCIPLES OF ESTABLISHING THE SPORTS INDUSTRY COMPETITIVENESS EVALUATION INDEX SYSTEM

2.1 *Basis of establishing sports industry competitiveness evaluation index system*

Present the literatures about sports industry competitiveness evaluation are variations in the perspective of the index system establishment, which mainly establish the index system from the perspective of content composition. However, there are many impact factors of the sports industry competitive, both major factors, also non-essential factors, for the factors with a different influence degree, how to scientifically sort, empower, are questions worthy of serious consideration. The solution to these problems must have a reasonable basis, to take a scientific process of evaluation and calculation, ensure that the selected indicators and indicator system established are scientific, reasonable and effective. Due to in the sports industry factors, some factors are more obvious and clear, also are easy to quantify, but some other factors, although have a great role, but are not easy to determine, or to separate, which to some extent increase the difficulty for the index system establishment. However, in order to improve the effectiveness of sports industry competitiveness evaluation index, we must carefully analyze the sports industry and its influencing factors' key features; fully understand that "To be competitive must have a difficult imitability, value-added, scalability, integration, prospective etc."[3], through a full range of analysis, to clear the core competitiveness of the sports industry. The evaluation from this perspective is intuitive, clear, and this method in the process of the index system establishment naturally makes a distinction between primary and secondary influencing factors, also can make the index more streamlined, simple, more convenient to operate.

2.2 *Principles of establishing sports industry competitiveness evaluation index system*

To make a reasonable evaluation of China's sports, industrial competitiveness, index system design is the most important; the established index system not only needs to reflect China's sports industry competitiveness well, but also must be easy to apply in practice. For the establishment of the sports industry competitiveness evaluation index is also the same, we need to be able to objectively evaluate the competitiveness of China's sports industry, but also has a certain role to enhance the competitiveness of China's sports industry. Therefore, when designing the competitiveness evaluation index system of China's sports industry, should follow these principles:①Scientific principle. Scientific refers to that when the design competitiveness index system, the definition of the selected indicators, description, calculation methods, problems designation and solutions, evaluation criteria and other procedures should be scientific and accurate, only then can ensure that the assessment results are accurate enough. Thus China's sports industry competitiveness index system establishment must comply research direction, and based on the properties of the sports industry, take the relevant influencing factors and the impact extent of China's sports industry competitiveness fully into account, reasonably analyze linkages interaction and joint action between these factors, so that make each designed index with exact concepts, clear meaning, explicit calculation range.②Practicality principle. Practicality, simply to say, refers to the direct application value of the established sports industry competitiveness evaluation index system, whether easy to operate, whether simple and feasible, namely require the established sports industry competitiveness evaluation index system is facilitate to use. For example, the sports industry competitiveness index evaluation system should not contain too many indexes; too many are very complicated and not easy to statistics, not easy to calculate. Some indicators are numerous, seemingly comprehensive, but is difficult to operate, very inconvenient to use, make them difficult to put into practical application, thereby reduce the practical application value of research results [4]. Because the evaluation index system takes a comprehensive assessment of China's sports industrial competitiveness, so the design of this competitiveness index system should fully take the actual situation of the sports industry development into account. ③Principle of combining dynamic analysis with static analysis. The factors affecting the competitiveness of China's sports industry, there are both macro aspects influencing factors, also micro aspects influencing factors. Because these two aspects influencing factors are constantly changing, so when we establish sports industry competitiveness evaluation index system, should take full account of the dynamic change characteristics, which requires the sports industry competitiveness evaluation indexes must also have a dynamic evaluation function. Therefore, the establishment of sports industry competitiveness evaluation index system should based on a comprehensive analysis, mutual integrate the dynamic evaluation indexes and static evaluation indexes, research potential development of the sports industry and potential competitiveness, predict future competitiveness, thus pointedly to foster and enhance the competitiveness of China's sports industry. ④Principle of combining hard targets and soft targets. The distinction standard of hard targets and soft targets lies in that hard target can be quantified, and soft targets only can be qualitative. When we establish sports industry competitiveness evaluation index system, must take an objective and comprehensive analysis, consider the various key constraints in the process of the sports industry development, and then improve the accuracy by taking the future. Therefore, when designing the evaluation index system, should take the quantitative hard target operation under the guidance of soft targets these qualitative indicators, fully understand the complexity of factor analysis, insist on meticulous in the work process, and strive to achieve that the extracted hard targets can truly reflect external notable features, the extracted soft targets can truly reflect the inherent personality characteristics.

3 CONSTRUCTION OF THE SPORTS INDUSTRY COMPETITIVENESS EVALUATION INDEX SYSTEM

According to the related definitions of industry, industry is understood to an ecosystem formed by a group of companies compete directly through production or labor services [5]. In modern economic development, competitiveness is a diverse and complex system, its basic constituent elements, also can be said the essential elements should include human resources, material resources, market competitiveness, production capacity, technological innovation, management innovation capability, comprehensive management capabilities, industry and culture, etc.

Table 1. China's sports industry competitiveness.

Target layer A	Criteria layer B	Index layer C	Serial number
Sports industry competitiveness evaluation index system	Human resources B1	Quality of employees	1
		Reasonable degree of staff team	2
		Leaders' ability	3
		Effectiveness of performance management	4
		Effectiveness of incentive mechanism	5
		Sports human resource development	6
		Employees' satisfaction degree	7
		Employees' loyalty degree	8
	Material resources B2	Sports equipment advanced degree	9
		The scale of fixed assets	10
		New proportion rate of fixed assets	11
	Market competitiveness B3	International sports market share	12
		Sports marketing business growth rate	13
		Ability to open up the international sports market	14
		Response capability to sports market	15
		Sports industry service quality	16
		Customers' satisfaction degree	17
	Production capacities B4	Production running punctuality rate	18
		Production capacity effective utilization rate	19
		Security productivity	20
		Labor productivity	21
	Technological innovation capabilities B5	The proportion of R & D investment	22
		The proportion of adopting new equipment, new technology	23
		The occupancy of core technology, equipment, etc.	24
		Technological progress contribution rate	25
	Management innovation capabilities B6	Organizational innovation	26
		Mechanism innovation	27
		Tactical and strategic	28
	Comprehensive management capabilities B7	Profitability ability	29
		Business decisions ability	30
		Operational Capability	31
	Industrial culture B8	Employees mental outlook	32
		Leaders' style and charismatic	33
		Industrial values	34
		Teamwork spirit	35
		Sports industry's image	36

[6]. In the process of constructing sports industry competitiveness evaluation index system, this paper follows the principles and basis of establishing index system, based on the constituent elements of competitiveness, closely according to the connotation of sports industry competitiveness, combines the key features of sports industrial competitiveness, designs 8 evaluation contents and 36 evaluation indexes, establishes a simple and clear sports industry competitiveness evaluation index system, respectively from human resources, material resources, market competitiveness, production capacity, technological innovation capability, management innovation capability, comprehensive management capabilities, industrial cultural eight aspects to take a relative comprehensive evaluation, each sub-index under evaluation contents is independent, through data passing layer by layer, eventually form a comprehensive sports industry competitiveness evaluation index system (Table 1).

As shown in Table 1, the indicators of index system, some are quantitative indicators, also some are qualitative indicators. Consider that the acquisition of some index data may not be easy, in order to make the whole index system simple and easy to operate, thus when design some indicators have been adjusted, finally form the practical indicator system with better operability. Due to use the analytic hierarchy process to select indicators, also limited by the article length, and each index is clear and concise, so we don't make a further elaboration for each index.

4 SUMMARY

The sports industry evaluation is a complex project, due to the statistical indicators of sports industry is not unified, the existing statistical data cannot be directly used to reflect the actual situation of sports industry competitive, this paper still cannot obtain international comparison evaluation results of sports industry competition. However, apply the ANP method in sports industry this complex social system, fully consider the various feedback and constraint relationships among it, has an important practical significance for the scientific evaluation of the sports industry competitiveness. The ANP as a decision-making tool combining qualitative with quantitative, subjective initiative with scientific method, is conducive to clearer objectively understand their current situation and development direction, according to the evaluation results, find out the factors impact greater on their own, avoid the subjective, arbitrary and blindness in the process of development. They can effectively combine the industrial competitiveness evaluate results with the sports industry development strategy, according to the situation of their own and competitors, to develop a scientific and rational development strategy.

REFERENCES

[1] LIU Yanwu, Explore the meaning of sports industry [J], Sports Culture Guide, 2006(4): 6–9.

[2] LI Liang, Comparative analysis of China's sports industry brand development [D], Capital University of Economics Master Degree Thesis, 2009.

[3] LI Guang, Research on the 21st century enterprise competitiveness evaluation index system [J], Operation and Management, 2000(6): 137–138.

[4] BI Jinjie, LIANG Jin, YE Jiabao, Theoretical study on sport goods industry international competitiveness evaluation index system [J], Tianjin Institute of Physical Education, 2005,20 (5): 16–18.

[5] CHEN Ming, Initial idea for setting China's sports industry index system [J], Sports Science, 2006,26 (2): 86–90.

[6] LV Ping, LI Zhengzhong, Research on standards competition strategic based on core competence [J], Technology Progress and Policy, 2004 (1): 79.

Computing, Control, Information and Education Engineering – Liu, Sung & Yao (eds)
© 2015 Taylor & Francis Group, London, ISBN: 978-1-138-02800-5

Study on model innovation of internet finance

Y. Shen
Guizhou University of Finance and Economics, China

ABSTRACT: In the context of the rapid development of internet technology, the internet finance comes into being. The internet finance is an emerging financial model relying on the payment, cloud computing, social networking and search engines and other internet tools, mainly includes third-party payment platform model, P2P network microfinance model, model-based big data finance services platform, crowd funding model, information financial institutions, internet finance portal and other model innovation.

KEYWORDS: internet finance, model innovation, the emerging finance model.

1 INTRODUCTION

With the development of the internet and big data, the rise of the internet finance sector forms a shock to plurality of traditional finance sector areas and develops in the core areas of the finance sector. Internet finance refers to an emerging finance relying on payments, cloud computing, social networking, search engine and other internet tools to achieve financing, payment and information intermediary services[1]. Internet finance is not a simple combination of internet and financial sector but in achieving security, mobile and other network technology level, familiar and accepting by the user, (especially accepting by e-commerce), and then naturally emerge the new model and new business to adapt to new needs. The difference between the internet finance and traditional finance lies not only on the finance business using different media, more importantly, on the finance participant is well versed in the internet essence of "open, equality, cooperation, sharing", through the internet and mobile internet tools, make the traditional finance services with greater transparency, higher participation, better collaboration, lower intermediate costs, more convenient operation and a series of features. Theoretically any related to the internet applications of generalized finance, should be internet finance, including but not limited to third-party payment, online sales of finance products, credit evaluation audit, finance intermediaries, and finance e-commerce model. Internet finance development has undergone a multi-stage: online banking, third party payments, personal loans, corporate finance, and increasingly develop in finance intermediation, matching finance supply and demand and other core aspects of the traditional finance services[2].

2 THE BASIC FEATURES OF INTERNET FINANCE

Overall, the basic characteristics of the Internet finance include: (1) Low cost. Under internet finance model, both sides of the funds' supply and demand through the network platform to complete information screening, matching, pricing and trading, without the traditional intermediaries, no transaction costs, no monopoly profits. On the one hand, financial institutions can avoid opening outlets' capital investment and operating costs; on the other hand, consumers can quickly find finance products suitable to them in an open and transparent platform, weaken the degree of information asymmetry, more effective. (2) High efficiency. Internet financial services mainly process by computer, operational process fully standardized, customers do not need to wait in line, and business is processed faster, user experience is better. (3) Wide covering. Under Internet finance model, customers can break through time and geographical constraints, to find financial resources needed on the internet, finance services more direct, customer base more extensive. In addition, the internet finance clients are mainly small and micro enterprises, covering the part services blind of the traditional finance sector, are conducive to enhancing the efficiency of resource allocation, promote economic development entity. (4) Rapid development. Relying on the development of big data and e-commerce, the internet finance has been growing rapidly. Take the Yu Ebao for example, after 18 days on the line, the cumulative number of users reached over 250 million, totaling transferred funds reached 6.6 billion yuan. (5) Management weak. First, risk control is weak. Internet finance has not access to the People's Bank credit system, does not exist credit information sharing mechanism, does not have a similar bank's risk control, compliance and collection mechanisms, prone to various types of risks, there are the credit net, the net to win the world and other P2P network platform to declare bankruptcy or stop the service. Second, the regulator is weak. Internet finance in China is at the initial stage, there is no regulatory and legal constraints, lack of barriers to entry and industry standards, the industry is facing many policies and legal risks. (6) High risk. First, there is high credit risk. At this stage, China's credit system is not perfect, the internet finance-related laws have yet to be developed, the cost of internet finance defaults is low, likely to cause malicious defraud loan, escaped on foot and other risks. In particular, a P2P net loan platform due to low barriers to entry and lack of supervision, become a hotbed for criminals to engage in illegal fund-raising, fraud and other criminal activities. Second, there is large network security risk. Chinese internet security issues are outstanding, network finance crime cannot be ignored. Once hit by a hacker's attack, will affect the proper functioning of the internet finance and endanger the security of consumer financial and personal information.

3 THE MAIN MODE OF INTERNET FINANCE INNOVATION

3.1 *The third-party payment*

The narrow sense of the third-party payment refers to the non-bank institutions having a certain strength and credibility of security, by means of communications, computer and information security technology, through contracting with major banks to establish connect electronic payment mode in payment clearing system between user and bank; Broadly speaking third-party payment refers to the network payment, prepaid cards, bank cards acquiring and other payment services identified by People's Bank of China, offered by non-finance institution as the beneficiary's and the payer's payment intermediaries. The third payment is not limited to the initial Internet payment, but has become a comprehensive payment tool with coverage of online, offline and richer scenarios. From the development path and the way accumulated by the user, the current operating mode of the third-party payment companies on the market can be classified into two categories: One is an independent third-party payment mode refers to the third-party payment platform completely independent of the e-commerce sites, and don't live a secured function, only provide users with the pay products and payment systems solutions, typical representative by quick money, EPRO payments, remittance world, Kara and so on. The other is relying on its own B2C, C2C e-commerce website's security function to provide third-party payment, led by Paypal, Tenpay. Payment temporarily hosted by the platform, when the payment reach, the platform notify the seller to ship.

3.2 *P2P Network loan platform*

P2P is namely point to point credit. P2P network loan refers to match both funds sides of borrow and loan by the third-party Internet platform, people need loans can look through the web platform to find the crowd has lent able and willing to lend based on certain conditions, help lenders to share a loan credits with other lenders together to spread the risk, and also help borrowers to choose an attractive interest rate conditions in sufficient comparable information[3]. The earnings of P2P platform is mainly one-time fee from the borrower as well as the assessment and management fees charged to investors. The loan interest rate is determined by the lender bid or based on the reference interest rate provided by borrower's reputation condition and the bank's interest rate. Due to no barriers to entry, no industry standard, non-regulatory agency, there has not been defined in the strict sense of the P2P net loan concept, its business model has not been fully finalized. At present, there have been several operational modes, one is pure online mode, pat loans, loan together, everybody loan (part of the business), etc. are such patterns' typical platform, its' characterize is that money lending activities are online, not combined with audit offline. Another one is a combination of online and offline mode, such model is represented by pterosaurs loan. After the borrower submits loan applications online, the platform notifies city agents to review the borrower's credit, repayment ability and so on by the way of household survey.

3.3 *Big data finance*

Big data finance refers to collection of massive unstructured data, through real-time analysis, can provide a full range information of customers for the internet finance institutions, grasp customer's spending habits through analysis and mining customer transactions and consumer information, and accurately predict customer behavior, so that finance institutions and finance service platform have definite object in view of marketing and risk control areas[4]. Finance services platform based on big data mainly refers to finance services carried out by e-commerce companies have huge amounts of data. The key of big data is the ability to quickly obtain useful information from large amounts of data, or to quickly realize from large data assets, therefore, the information processing of big data tends to cloud-based. Currently, the big data services platform business model can be divided into microfinance, represented by Ali platform mode and Jingdong, Suning, represented by supply chain finance model. Big data through massive data verification and assessment, to increase risk controllable and management efforts, to discover and solve possible risk points, there is an accurate grasp for the risk's regularity, will promote finance institutions more in-depth and thorough data analysis needs. Big data will drive finance institutions to innovate brands and services, fine service, personalized customization, use data to develop new predictive and analysis models, achieve the client's consumption pattern analysis in order to improve customer conversion rate.

3.4 *Crowd funding model*

The general meaning of crowd funding model is public financing, refers to through the form of buy + pre-order model to raise project funding from users. The original meaning of crowd funding model is to use of the spread feature of internet and SNS, so start-ups, artists or individuals can show their creativity and projects to the public striving for everyone's concern and support, and then get the needed finance assistance [4].The operation mode of crowd funding platform is similar --- individual or team needing funds gives project planning to the platform, after relevant audits, they can create their own pages on the website platform for informing the public about the project situation. There are three rules for crowd funding: first, each project must set funding targets and days; second, if within the setting days reaching the target amount means successful, the sponsor can obtain financing; and when project fails funding has been funded will be entirely refund to supporters; third, crowd funding is not donation, all supporters must have appropriate returns. The crowd funding platform will take a certain percentage of service fees from the fundraising successful project. At present, the domestic provisions about public fundraising and particularly vulnerable to stepped on the red line of illegal fund-raising make the stake system of crowd funding in domestic developing slowly, it is difficult to become bigger and stronger in domestic, short-term the impact on the finance sector and corporate finance is very limited. This requires the crowd funding website's operation reflecting their differences, highlighting its vertical oriented features.

3.5 *Information finance institution*

The information finance institution is through the use of information technology to transform or reconstruct traditional operational processes, realize operating, managing fully electronic banking, securities, insurance and other finance institutions [5]. Finance information is one development trend of the finance sector, and information finance institution is a product of financial innovation. From the point of view entire finance industry, the bank's information technology has been in the industry-leading level, not only has the world's leading finance, IT platform, and builds a three-dimensional electronic banking service system consists of self-service banking, telephone banking, mobile banking and online banking and dominates in the industry with information - data centralization project. At present, some banks are self-building electronic business platform, from the perspective of the banks, the core values of the electricity supplier are to increase user stickiness, accumulate authentic user data, and then the bank can rely on their own data to explore the needs of users. From the business model, the traditional bank loan is a process-oriented, immobilization, from the perspective of cost savings and risk control, banks tend to be service for large organizations, through IT can ease or even solve the problem of asymmetric information, build a platform for SMEs and bank direct cooperation and enhance the finance institution service for the real economy. But more importantly, the bank through constructing electronic business platform, actively get through the various bank departments within data silos, format a trinity internet platform of "website bank + finance supermarket + electricity supplier", respond to the challenge of the internet finance wave.

3.6 *Internet finance portal*

Internet finance portal refers to using the internet to sell financial products as well as providing a third-party services platform for finance product sale. Its core is the "search + compare price" model, through the approach of vertical comparing finance products' price, put the various finance institutions' products onto the platform, then the users select suitable finance products after comparing[5].Internet finance portal diversifies innovation and development, formats the third-party finance institution providing high-end wealth management investment services and wealth management products, such as insurance portal provides the services of consulting insurance products, comparing price, purchasing. This model does not exist too much policy risk, because the platform is neither responsible for the actual sale of finance products, nor assumes any adverse risk, while funds are totally through the middle of the platform. Currently for credit, finance management insurance, P2P and other industry segments in the field of internet finance portal, there are finance 360,91 finance supermarket, haodai network, Bank rate network, the grid banking, datong network, net loan home, etc. Internet finance portal's greatest value lies in its channel value. Internet finance diverse the customers of bank, trust, insurance, and intensify competition in these industries. With the advent of internet finance time, for the demand side of the capital, as long as within a certain period of time, within an acceptable range of cost, specifically where the money comes from, ICBC, CCB, P2P platform, small loan company, trust fund, private placement bond, etc., is not so important.

4 SUMMARY

Overall, the emergence of internet finance not only makes up the service gap of the traditional finance institutions represented by the bank, but also improves the efficiency of social capital, more critical is through the internet make finance popularity, not only significantly reducing the cost of financing, but also closer to the people and people-oriented. Its impact on the financial sector is not just grafted the information technology onto finance services, and promote changes in the finance business structure and service concept, more important is to improve the entire community finance functions. The development and growth of internet finance will bring some impact to the banking sector, but also bring new opportunities to the fund companies, securities companies, insurance companies and trust companies.

REFERENCES

[1] Huang Lingchao, Network Finance Analysis of Business Model Innovation [J], Finance Sector, 2013(18):9.

[2] Li Yan, Reflections on the Internet Finance Models and its Development [J], Business, 2013 (19):144–145.

[3] Qian Jinye, Yang Fei, Status and Prospects of the Development of Chinese P2P Networks [J], Finance Forum, 2012(01):45–51.

[4] Gong Xiaoling, Internet Finance Model and its Impact on Traditional Bank Sector [J], Southern Finance, 2013(05):86–88.

[5] Song Guoliang, Zhang Yi, Inquiry on Business Models and Risk Management of Internet Finance [J], Modern Management Science, 2014(05):82–84.

Computing, Control, Information and Education Engineering – Liu, Sung & Yao (eds)
© 2015 Taylor & Francis Group, London, ISBN: 978-1-138-02800-5

Applied research on the internet of things in logistics management

T. Peng
School of Business Administration, Guizhou University of Finance and Economics, China

ABSTRACT: The internet of things is a polymeric applications and technology upgrading appeared after modern information technology to develop a certain stage, integrates a variety of sensing technologies, modern network technology, artificial intelligence, automation technology and application aggregation. The internet of things technology will have a significant impact in modern logistics management. Based on this, the paper analyzes the connotation and the characteristics of the internet of things, insufficient existing in modern logistics management, and the internet of things' impaction on the logistics management, then focus on t expounding the internet of things applications in logistics management.

KEYWORDS: internet, the internet of things, logistics management.

1 INTRODUCTION

Logistics management refers to in the process of social reproduction, according to the regularity of material entities flow, apply the basic principles and scientific methods of management, through planning, organizing, directing, coordinating, controlling and supervision for logistics activities, so as to make all logistics activities to achieve optimal coordination and cooperation, then reduce logistics costs, improve logistics efficiency and economic benefits [1]. Modern logistics is the advanced stage of traditional logistics development, based on the advanced information technology, focus on a comprehensive integration of service, personnel, technology, information and management, and is the manifestation of modern production methods modern management methods modern information technology combination in the field of logistics [2]. It emphasizes standardization and efficient of logistics, with relatively low cost to provide a higher level of customer service. Rapid, real-time, accurate information collection and processing is an important foundation to achieve standardization and efficiency of logistics. The internet of things in the modern logistics management application will make a significant impact.

2 THE INTERNET OF THINGS' CONNOTATION AND CHARACTERISTICS

The internet of things is based on the internet, extend users to the field of materials, objects according to the agreed-upon protocol in advance, through radio frequency identification devices, infrared sensors, and GPS systems to connect to the internet, to form a the new intelligent network system, between things no need outside help can exchange information, managers can through phone or computer terminal, intelligently identify, locate, monitor and manage the object [3]. Namely a network through radio frequency identification (the internet of things), infrared sensors, global positioning systems, laser scanners and other information sensing device, according to the agreed protocol, connect any items with the internet, exchange information and communicate so as to achieve intelligent identification, locate, track, monitor and manage.

The internet of things concept is proposed in 1999, the internet of things is "the internet with things connect to things". There are two meanings: first, the core and foundation of the internet of things is still the internet is the extension and expansion of the network based on the internet; Second, the extension and expansion of the internet users to any goods and items, exchange information and communicate. The internet of things is a polymeric applications and technology upgrading appeared after modern information technology to develop a certain stage, integrates a variety of sensing technologies, modern network technology, artificial intelligence, automation technology and application aggregation, make the wisdom dialogue between people and things, create a wisdom world. The internet of a thing's essence summed up is mainly reflected in three aspects: First, the internet features, namely the internet ensures the objects needing for networked to be able to achieve interoperability. Second, identification and communication features, which the things included in the internet of things, must have the function of automatic identification and material to material communication (M2M). Third, intelligent features, the network system should have the features of automate, self-feedback control and intelligent [4].

The Internet of Things systems requires three steps to achieve the management for objects: Firstly, identify objects, storage after classifying; Secondly, use the smart identification apparatus to read objects' attribute and convert information; Finally, upload the objects' information to the network, through the Internet, transmit the information to the control center, then centralize management of the objects. (1) Information exchange between things and things. Through intelligent recognition technology to scan, organize, upload the things' information, store the things' information to management center, so that they understand each other's needs between things and things, and then make the appropriate response. (2) "Cloud computing" deal. Due to the wide range of the internet of things applications, need to deal with the huge amount of data. This needs the help of cloud computing. Cloud computing concentrates information, resources to the network through utilizing the network, users can use readily available. The internet of things through cloud computing platform uses pattern recognition, M2M and other computing technologies to compute and process larger data information.

3 DEFICIENCIES EXISTING IN LOGISTICS MANAGEMENT

3.1 *Logistics management tools backward*

Many enterprises in logistics management is still stuck in the era of pen and paper, although some companies are equipped with computers, but has not established management information systems, and not formatted network management, while also lack applications of EDI, PC, artificial intelligence, expert systems, communications, bar code scanning and other advanced information technologies in logistics operation [4].

3.2 *Lack the philosophy of "the third profit source"*

Logistics enterprises in China Generally are lack of the concept of modern logistics as "the third profit source", do not see logistics as the key to optimize the production process, strengthen market management, but put the logistics activities into a subordinate position, most companies make warehousing, transportation, handling, procurement, packaging, distribution and other logistics activities scattered in different sectors, are not incorporated in a functional department to make system planning, unify management for logistics activities.

3.3 *Lack of the concept of cooperative competition*

In increasingly competitive market, customer needs are constantly changing environment, the speed that individual enterprises rely on their own resources for self-adjustment is difficult to catch up with the speed of market change, so companies have to focus limited resources on its core business, strengthen their own core competencies, and they do not have their own core competencies in the form of contracts.

4 THE INTERNET OF THINGS' IMPACT ON LOGISTICS MANAGEMENT

4.1 *Increase the visibility of supply chain, improve the adaptability capacity of supply chain*

By using the internet of things technology in the whole process of supply chain, from the production of goods to the supplier and then to the end-user, the goods distribution in the entire supply chain and the information of product itself can completely real time and accurately reflect in the enterprise information system, greatly increase the visibility of enterprise supply chain, make the enterprise's entire supply chain and logistics management process become a fully transparent system. Rapid, real-time, accurate information allows businesses and even the whole supply chain can in the shortest possible time respond quickly to the complex and volatile markets, improve the supply chain's ability to adapt to market changes.

4.2 *Reduce inventory levels, improve inventory management capabilities*

Modern logistics management takes reducing costs and improving service levels as the main purpose. Inventory cost is an important part of logistics costs, therefore, reduce inventory levels become a core element of modern logistics management. Apply the internet of things technology into inventory management, enterprise can real time master the inventory information about goods, learn about patterns of demand of each commodity, timely Complement goods, combine with automatic replenishment systems and vendor managed inventory solutions to improve inventory management capabilities, reduce inventory levels.

4.3 *Help companies to realize assets management visualization*

Use the internet of things technology in enterprise asset management, to real-time track the production operational processes of transport vehicles and other equipment through the way of label, can real-time monitor usage of these devices, to realize the visualization of enterprise assets management, help enterprise to make rational planning applications for its overall assets.

4.4 *Accelerate the process of enterprise information, improve customer service levels*

Informatization is the main characteristics and development trends of modern logistics. Using the internet of things technology, greatly accelerate the process of enterprise informatization, facilitate information sharing between various departments within the enterprise, allow companies to integrate their business processes more effectively, improve the ability of quickly responding to market changes [5]. At the same time, companies can provide customers with accurate, real-time logistics information, and reduce operating costs, provide customers with personalized service, greatly improve the company's customer service levels.

5 THE INTERNET OF THINGS' APPLICATION IN LOGISTICS MANAGEMENT

Currently in logistics process, there exists logistics information asymmetry, the information not timely and other defects, it is difficult to achieve timely regulation and collaboration. With the advance of economic globalization, scheduling, management and balance resource between the supply chains has become increasingly urgent. With the electronic product code (EPC) and the internet of things at the core, constructing "Internet of Things" on the internet will in worldwide change the management level of goods flow monitoring and dynamic coordination in production, transportation, warehousing, sales, all the aspects of the fundamental. The internet of things can achieve non-contact automatic identification of multi-objective, moving targets, the internet of things emphasizes the interaction between material and information, applying the internet of things technology into information collection and logistics tracking in the logistics industry, can greatly improve the industry service level. Specific performance:

5.1 *Use the internet of things to improve the logistics enterprises' ability to obtain information*

The internet of things combines the EPC technology and internet technology, can achieve automatic, rapid, parallel, real-time, non-contact processing for individual items' information, through the network to realize information sharing, so as to achieve efficient management in the supply chain. Logistics companies can use this platform to expand value-added information services, mainly reflected in providing a unique service to customers through access to accurate, comprehensive and timely information. Therefore, we must improve the information access capabilities of logistics enterprise, and the emergence of the internet of things, is just meeting the needs of logistics enterprises in this regard.

5.2 *Use the internet of things for logistics information value-added services*

Logistics information value-added services are based on the traditional basic logistics services, but also is different from the traditional logistics services, which is a modern logistics management tool built on the basic traditional logistics services and promote basic logistics services for further development [6]. The internet of things taking EPC as the core technology in, collects coding technology, grid technology, RF technology, breaks the bottlenecks of previous mode to obtain information, innovate in standardization, automation, networking and other aspects, thus the logistics company can accurately , comprehensive and timely access to logistics information, and according to different information levels respectively provide enterprise-level industry-level and supply-chain level information value-added services based on these.

5.3 *Actively establish cooperative competition mode of circulation enterprises*

In the internet of things era, especially when the internet of things technology has not yet matured, need the enterprise to focus resources and cultivate its core competencies, develop core business, make the main business bigger, stronger, doing fine, and seeking logistics outsourcing in worldwide or make a strategic alliance to the global external enterprises, take the overall advantages of the entire supply chain to participate in domestic and international competition.

5.4 *Realize information collection, information processing automation*

Can be applied to aspects of supply chain management, equipment preservation, traffic transportation, factory production, provide users with real-time accurate delivery status information, vehicle tracking positioning, transport path selection, logistics network design and optimization services, greatly enhance the comprehensive competitiveness ability of logistics enterprises [7].

5.5 *Automate the movement of goods in kind and other operational aspects*

Such as sorting, handling, loading and unloading, storage, etc., depending on the circumstances to distinguish goods in kind, reduce labor intensity, optimize operating procedures, save manpower and material resources for the user.

6 SUMMARY

The logistics industry is not only one of national top ten industrial revitalization plan, also an important area in information and physical networking applications. Its information technology and integrated logistics management, process monitoring can not only bring efficiency to logistics, cost control and other logistics benefits for enterprises, but also improve the enterprise and related areas the level of information as a whole, so as to achieve the purpose of promoting the whole industry. Although the logistics field application the internet of things is not very extensive, but the application in the field of logistics better reflects the value the internet of things, when the internet of things technology matures, the logistics industry is perhaps the best field embodying the values the internet of things. In the era of the internet of things, logistics should actively utilize the internet of things to reduce costs, improve efficiency, promote the development, be integrated into the internet of things economy, with the development and expansion of the internet of things, the logistics industry will be able to obtain a broader space for development.

REFERENCES

[1] Fang Hong, Logistics Enterprise Management [M], Beijing: Higher Education Press, 2005.

[2] Li Genzhu, Du Xiaoxiao, Discussion the logistics entity throughout real-time tracking system based on modern information technology [J], Logistics Technology, 2003(11).

[3] Zhou Hongbo, The Internet of Things: Technology, Application, Standards and Business Models [M], Beijing: Electronic Industry Press, 2010.

[4] Zheng Chunping, Modern logistics application research based on RFID technology [J], Logistics Technology, 2007(05).

[5] Wang Xiaojing, Zhang jing, Review of the internet of things [J], Liaoning University Journal (Natural Science), 2010(01).

[6] Wu Hequan, Summary of the internet of things' applications and challenges [J], Chongqing University of Posts and Telecommunications Journal (Natural Science), 2010(05).

[7] Dai Dingyi, The internet of things and smart logistics [J], Logistics and Purchasing, 2010(08).

Computing, Control, Information and Education Engineering – Liu, Sung & Yao (eds)
© 2015 Taylor & Francis Group, London, ISBN: 978-1-138-02800-5

Approximation of Fejèr means for vector-valued Dirichlet function

Y.W. Chen & Z.F. Li
College of Mathematics and Statistics, Hebei University of Economics and Business, Shijiazhuang, China

Y.J. Liu
College of Career Technology, Hebei Normal University, Shijiazhuang, China

ABSTRACT: For vector-valued Dirichlet class on the unit disc D in the complex plane, we get the inclusion relation between vector-valued Dirichlet class and vector-valued Hardy space, and then obtain an exact rate of best polynomial approximation and of upper bounds for the deviations of Fejèr means in the metric of vector-valued Hardy space.

KEYWORDS: Dirichlet class; Vector-valued function; Hardy space; Fejèr means.

1 INTRODUCTION

In many problems arising in engineering and control one requires approximation methods to reproduce physical reality as well as possible. Very schematically, if the input data represent a complicated discrete/continuous quantity of information, then one desires to represent it with the less-complicated output information. Function approximation also plays an important role in the application of information engineering, such as identification system and optimal control theory. Function approximation provides a technical method and learning method. This paper is concretely desired to research vector-valued function approximation by the Fejèr mean operator.

Let D denote the unit disc$\{z:|z|<1\}$in the complex plane and $H(D)$ the collection of all holomorphic functions on D. The Dirichlet class $D_p(D)$ is defined by $1\le p<\infty$

$$D_p := \{f\in H(D): \| f\|_{D_p}^p = \int_D |f|^p\, dv \le 1,$$

where dv is the normalized Lebesgue measure on D.

The Hardy space $H^p(D)$ consists of all holomorphic functions f on D for which

$$\| f\|_{H^p} := \sup_{0<r<1} \left(\int_0^{2\pi} |f(re^{i\theta})|^p\, d\theta\right)^{\frac{1}{p}} < \infty,$$

It is well known that (see [5,11]) $D_p \subset H^p$. For any $f\in H(D)$ there exists its Taylor expansion

$$f(z)=\sum_{k=0}^{\infty} a_k z^k, \qquad a_k = \frac{f^{(k)}(0)}{k!} \quad ,z\in D.$$

The sequence of Fejèr means is given by

$$\sigma_0(f)(z)=0, \sigma_n(f)(z)=\sum_{k=0}^{n-1}(1-\frac{k}{n})a_k z^k, n\in N.$$

Suppose that U is a subset of $H^p(D)$ and denote that

$$F_n(U,H^p) := \sup_{f\in U}\{\| f-\sigma_n(f)\|_{H^p},$$

$$E_n(U,H^p) := \sup_{f\in U}\{\inf_P\{\| f-P\|_{H^p} : P\in P_{n-1}\}\},$$

where$P_{n-1}$is the collection of algebraic polynomials of degree at most $n-1$.

The approximation of periodic functions by the Fejèr means of their Fourier series has a long history (see, for example, [14]). There are many results in the unit disk with regard to the approximation of holomorphic functions (see [10,12,13,14]).

Recently, Savchuk has obtained the exact approximation of a Dirichlet class by Fejèr means. In the case of the vector-valued holomorphic function in the unit disc, the situation is complex and analogous. The purpose of this article is to obtain the corresponding results for vector-valued Dirichlet class $D_p(D)$ on the unit disc D through Fejèr means.

Let $X(\cdot,\|\cdot\|)$ denote a complex Banach space and $H(D,X)$ be the collection of all vector-valued holomorphic functions on the unit disc D. It is well known that [8,9]: If $f\in H(D,X)$, then $f(z)$ has the Taylor expansion

$$f(z)=\sum_{k=0}^{\infty} x_k z^k, z\in D, x_k\in X.$$

The vector-valued Hardy space H^p consists of all vector-valued holomorphic functions f in D for which

$$\|f\|_{H^p}^p := \sup_{0<r<1} \frac{1}{2\pi}\int_0^{2\pi} \|f(re^{i\theta})\|^p \, d\theta < \infty.$$

Let $dm(z)$ be the normalized Lebesgue measure on D. We define the Dirichlet class by the derivative as $1 \le p < \infty$,

$$D_p := \{f \in H(D) : \|f\|_{D_p}^p := \int_D \|f'(z)\|^p dm(z) \le 1\}$$

and also observe the fact [1]:

$$M_p(g,r) \le \|g(0)\| + \int_0^r M_p(g',s)\,ds, 1 \le p < \infty,$$

where $M_p^p(g,r) = \frac{1}{2\pi}\int_0^{2\pi} \|g(re^{i\theta})\|^p \, d\theta$. Then we are easy to know that for $1 \le p < \infty, D_p(D) \subset H^p(D)$.

Therefore, on the boundary of D, each function in the Dirichlet class D_p has angular boundary values generating a function from the Banach space $L_p(T,X)$ in which the norm:

$$\|\cdot\|_{L_p(T)}^p := \frac{1}{2\pi}\int_{-\pi}^{\pi} \|\cdot\|^p \, d\theta.$$

It is well known [2,9] that $\|f\|_{H^p(D)} = \|f\|_{L_p(T)}$ for any $f \in H^p$.

Along with D_p, we also consider a class of holomorphic functions

$$H_p^1 := \{f : f \in H(D). \|f'\|_{H^p} \le 1\}, 1 \le p < \infty,$$

which, obviously, is a subclass of D_p.

Our aim is to study the approximation properties of the Fejèr means σ_n for the vector-valued Dirichlet class D_p in order to compare the effectiveness of approximation by these sums of functions from D_p with that of polynomials of best approximation in the metric of vector-valued Hardy space H^p.

In this paper, we extend to that of vector-valued function class and obtain that when X=C, the result is the same as it was before.

Theorem 1.1 Assume $1 \le p < \infty$ and $\frac{1}{p} + \frac{1}{q} = 1$. Then

1 for any $n \in N$,

$$\frac{\frac{1}{2}\min\{p,q\}}{((\frac{q}{2})(n-1)+1)^{1/q}} \le E_n(D_p, H^p) \le F_n(D_p, H^p) \le \frac{1}{((\frac{q}{2})(n-1)+1)^{1/q}}.$$

2 for any $f \in D_p$,

$$\|f - \sigma_n(f)\|_{H^p} = o(n^{-1/q}), \qquad n \to +\infty \; .$$

A class of holomorphic functions is defined by

$$H_p^1 := \{f : \|f\|_{H^p} \le 1\}, \qquad 1 \le p < \infty.$$

Theorem 1.2 Assume $1 \le p < \infty$ and $\frac{1}{p} + \frac{1}{q} = 1$. Then, for any $n \in N$,

$$n^{-1} \le E_n(H_p^1, H^p) \le F_n(H_p^1, H^p) \le 2n^{-1}.$$

2 SOME LEMMAS

We give some notations:

$$Q_{n,\rho}(f)(z) := \sum_{k=0}^{n-1} (1 - \frac{k}{n}\rho^{2(n-k)}) x_k z^k, x_k \in X.$$

Poisson kernel $P(re^{it}, e^{i\theta}) = \frac{1-r^2}{|1 - re^{i(\theta - t)}|^2}$.

Lemma 2.1 Suppose $f \in H(D)$. Then for any $\theta \in [0, 2\pi]$, $\rho \in [0,1)$ and $n \in N$

$$f(\rho^2 e^{i\theta}) - f(0) = \frac{e^{i\theta}}{\pi}\int_0^{\rho}\int_0^{2\pi} \frac{f'(re^{it})}{1 - re^{i(\theta - t)}} dt \cdot r\, dr,$$

$$Q_{n,\rho}(f)(\rho^2 e^{i\theta}) - f(0) = \frac{e^{i\theta}}{\pi}\int_0^{\rho}\int_0^{2\pi} f'(re^{it}) \sum_{k=0}^{n-2} (1 - r^{2(n-k-1)}) r^k e^{ik(\theta - t)} dt \cdot r dr.$$

Proof: We only need to prove the second equation, because the proof of the other is similar and even more easier.

First, $f'(re^{it}) = \sum_{k=1}^{\infty} k x_k r^{k-1} e^{it(k-1)}$. By the property of integral orthogonality of trigonometric functions, we obtain

$$\frac{e^{i\theta}}{\pi}\int_0^{\rho}\int_0^{2\pi} f'(re^{it}) \sum_{k=0}^{n-2} (1 - r^{2(n-k-1)}) r^k e^{ik(\theta - t)} dt \cdot r dr = \sum_{k=1}^{n-1} (1 - \frac{k}{n}\rho^{2(n-k)}) x_k \rho^{2k} e^{ik\theta} = Q_{n,\rho}(f)(\rho^2 e^{i\theta}) - f(0).$$

Lemma 2.2 Suppose $f \in H(D)$. Then for any $\theta \in [0, 2\pi]$, $\rho \in [0,1)$ and $n \in N$

$$f(\rho^2 e^{i\theta}) - Q_{n,\rho}(f)(\rho^2 e^{i\theta}) = \frac{e^{in\theta}}{\pi}\int_0^{\rho} r^n \int_0^{2\pi} f'(re^{it}) e^{-i(n-1)t} P(re^{it}, e^{i\theta})\, dt\, dr.$$

Proof: Transforming the kernel in Lemma 2.1,

$$\sum_{k=0}^{n-2}(1-r^{2(n-k-1)})r^k e^{ik(\theta-t)}$$
$$=\frac{1-r^{n-1}e^{i(n-1)(\theta-t)}}{1-re^{i(\theta-t)}}-r^n e^{i(n-1)(\theta-t)}\frac{1-r^{n-1}e^{-i(n-1)(\theta-t)}}{1-re^{-i(\theta-t)}}$$
$$=\frac{1}{1-re^{i(\theta-t)}}-\frac{r^{n-1}e^{i(n-1)(\theta-t)}}{1-re^{i(\theta-t)}}-\frac{r^n e^{i(n-2)(\theta-t)}}{1-re^{-i(\theta-t)}}+\frac{r^{2n-1}e^{-i(\theta-t)}}{1-re^{-i(\theta-t)}}$$
$$=\frac{1}{1-re^{i(\theta-t)}}-r^{n-1}e^{i(n-1)(\theta-t)}P(re^{it},e^{i\theta})+\frac{r^{2n-1}e^{-i(\theta-t)}}{1-re^{-i(\theta-t)}}.$$

Substituting the obtained expression for the kernel, using the relation for $0<r\le\rho<1$

$$\int_0^{2\pi}f'(re^{it})\frac{e^{-i(\theta-t)}}{1-re^{-i(\theta-t)}}dt=0.$$

We can obtain the desired result.

We also need the dual relation in the theory of approximation.

Lemma 2.3 [4] Let X be a normed linear space with norm $\|\cdot\|_X$, Y be a closed subspace of X, and $f\in X\setminus Y$. Then

$$E(f,Y):=\inf_{g\in Y}\|f-g\|_X=\sup_{\lambda\in Y^\perp,\|\lambda\|=1}\lambda(f),$$

where $Y^\perp:=\{\lambda\in X^*:\lambda(g)=0,\forall g\in Y\}$.

3 APPROXIMATION BY FEJÈR MEANS

Take $h(z)=\dfrac{\min(p,q)xz^n}{2(\frac{q}{2}(n-1)+1)^{1/q}},x\in X,\|x\|=1$, it is easy to get $\|h\|_{D_p}\le 1$. We will use $h(z)$ to find a lower bound for $E_n(D_p,H^p)$.

Proof of Theorem 1.1: We first to prove (1).

It is easy to know the duality (see [3]):

$$H_p(D,X)^*=H_q(D,X^*).$$

From the dual relation (see Lemma 2.3) we have

$$E_n(h,H^p)=\sup\{|\langle h,g\rangle|:g\in L_{q,n}(T,X^*),\|g\|_{L_q(X^*,T)}=1\},$$

where

$$\langle\langle g(e^{i\theta}),h(e^{i\theta})\rangle\rangle=\frac{1}{2\pi}\int_{-\pi}^{\pi}\langle g(e^{i\theta}),h(e^{-i\theta})\rangle d\theta,$$

$$L_{q,n}(T,X^*):=\{g\in L_q(T,X^*):\langle\langle g,xz^m\rangle\rangle=0,$$
$$\|x\|=1,m=0,1,\cdots,n-1\}.$$

So we can take $g(e^{i\theta})=x^*e^{in\theta}$, where $x^*\in X^*$, $\|x^*\|_{X^*}=1$ and observe the equation (see [14])

$$\int_0^{2\pi}e^{im\theta}e^{-il\theta}\,d\theta=\begin{cases}0,n\ne l;\\1,n=l.\end{cases}$$

It is easy to know that $g\in L_{q,n}(T,X^*)$ and $\|g\|_{L_q}=1$. Therefore,

$$E_n(D_p,H^p)\ge E_n(h,H^p)\ge\left|\langle\langle h,g\rangle\rangle\right|=\frac{\frac{1}{2}\min\{p,q\}}{((\frac{q}{2})(n-1)+1)^{1/q}}.$$

Now let us obtain an upper bound for $F_n(D_p,H^p)$.

For any $f\in D_p$, by Lemma 2.3 and Hölder's inequality, we have

$$\|f(\rho^2e^{i\theta})-Q_{n,q}(f)(\rho^2e^{i\theta})\|^p$$
$$\le\left(\frac{1}{\pi}\int_0^\rho\int_0^{2\pi}\|f'(re^{it})\|r^nP(re^{it},e^{i\theta})dtdr\right)^p$$
$$\le\frac{1}{\pi}\int_0^\rho\int_0^{2\pi}\|f'(re^{it})\|^pP(re^{it},e^{i\theta})dt\cdot r^{1+p}\,dr\left(\frac{q}{2}(n-2)+1\right)^{-p/q}.$$

In the last step we use the assumption $n>2/q$, so $nq-2q+2>0$.

Hence integrating over θ on $[0,2\pi]$ and applying Fubini's theorem, we obtain the estimate

$$\frac{1}{2\pi}\int_0^{2\pi}\|f(\rho^2e^{i\theta})-Q_{n,p}(f)(\rho^2e^{i\theta})\|^p\,d\theta$$
$$\le\frac{1}{\pi}\int_0^1\int_0^{2\pi}\|f'(re^{it})\|^p\,dt\cdot rdr\times\left(\frac{q}{2}(n-2)+1\right)^{-p/q}$$
$$=\|f\|_{D_p}^p\cdot\left(\frac{q}{2}(n-2)+1\right)^{-p/q}.$$

Then for $f\in D_p,\forall\rho\in[0,1)$

$$\|f(\rho\cdot)-Q_{n,p}(f)(\rho\cdot)\|_{L_p(T)}\le\left(\frac{q}{2}(n-2)+1\right)^{-1/q}.$$

Passing to the limit as $\rho\to1^-$ above inequality, taking into account the limiting relation

$$Q_{n,\rho}(f)(\rho\cdot)\to\sigma_n(f)(\cdot)$$

and applying the so-called Riesz theorem (see [2,8])

$$\|f(\cdot)-f(\rho\cdot)\|_{L_p(T)}\to0,\rho\to1^-,$$

we obtain the relation

$$\| f-\sigma_n(f)\|_{H^p} \le \left(\frac{q}{2}(n-2)+1\right)^{-1/q}.$$

Because f is an arbitrary function in D_p,

$$F_n(D_p, H^p) \le \left(\frac{q}{2}(n-2)+1\right)^{-1/q} \sim n^{-1/q}.$$

(2) Suppose that, just as above, f is an arbitrary function from D_p. From Lemma 2.3, using the same method as above, we get

$$\frac{1}{2\pi}\int_0^{2\pi} \| f(\rho^2 e^{i\theta}) - Q_{n,p}(f)(\rho^2 e^{i\theta})\|^p \, d\theta$$

$$\le 2\int_0^1 r^{n-1}\int_0^{2\pi} \| f'(re^{it})\|^p \, dt\, dr \cdot \left(\frac{2}{n}\right)^{p/q},$$

in which, the last inequality is obtained mainly through Hölder inequality.

Hence, using the same argument as (1) above, we obtain the estimate

$$n^{1/q}\| f-\sigma_n(f)\|_{H^p}$$

$$\le 2\left(\int_0^1 r^{n-1}\frac{1}{2\pi}\int_0^{2\pi}\| f'(re^{it})\|^p \, dt\, dr\right)^{1/p}.$$

It remains to note that, for any fixed $\varepsilon > 0$ and all $n \ge 2$, there exists a fixed $R, 0 < R < 1$, such that

$$\int_R^1 r^{n-1}\frac{1}{2\pi}\int_0^{2\pi}\| f'(re^{it})\|^p \, dt\, dr \le \varepsilon.$$

We notice that

$$\int_0^R r^{n-1}\frac{1}{2\pi}\int_0^{2\pi}\| f'(re^{it})\|^p \, dt\, dr \le R^{n-2}\| f\|_{D_p} \le R^{n-2}.$$

So

$$n^{1/q}\| f-\sigma_n(f)\|_{H^p} \le 2(\varepsilon + R^{n-2})^{1/p} \to 2\varepsilon^{1/p}, n\to\infty.$$

The theorem is proved.

Proof of Theorem 1.2: Take $h(z) = \frac{xz^n}{n}, \| x\| = 1$ and $g(e^{i\theta}) = xe^{in\theta}$. It is easy to calculate that $h(z) \in H_p^1$, for $\| h(z)\|_{H_p^1} = 1$. The same as the proof of Theorem 1.1 (1), we obtain

$$E_n(D_p, H^p) \ge E_n(h, H^p) \ge \left|\langle h, g\rangle\right| = \frac{\Gamma(n)}{\Gamma(n+1)} = n^{-1}.$$

Suppose $f \in H_p^1$. Then, since

$$\sup_{0<r<1}\frac{1}{2\pi}\int_0^{2\pi}\| f'(re^{i\theta})\|^p \, d\theta \le 1,$$

we can obtain that: $\forall \rho \in [0,1)$,

$$\frac{1}{2\pi}\int_0^{2\pi}\| f(\rho^2 e^{i\theta}) - Q_{n,p}(f)(\rho^2 e^{i\theta})\|^p \, d\theta$$

$$\le 2\int_0^1 r^{n-1}\frac{1}{2\pi}\int_0^{2\pi}\| f'(re^{i\theta})\|^p \, d\theta\, dr \cdot \left(\frac{2}{n}\right)^{p/q} \le \left(\frac{2}{n}\right)^p.$$

Just like the proof of Theorem 1.1 (1), passing to the limit as $\rho \to 1^-$, we obtain the required assertion.

ACKNOWLEDGEMENTS

This research was supported by the NNSF of Hebei (Grant No. A2015207007) and the Youth Project of Science and Research of Hebei Education Department (Grant No. QN20131027) and the Youth Project of Science and Research of Hebei University of Economics and Business (Grant No. 2013KYQ07).

REFERENCES

[1] J. Arregui, and O. Blasco: *Bergman and Bloch spaces of vector-valued functions*, Math. Nachr. vol. 261–262 (2003) , p. 3–22.

[2] O. Blasco: *Boundary values of functions in vector-valued Hardy spaces and geometry on Banach spaces*, J. Funct. Anal. vol. 78 (1988), p. 346–364.

[3] O. Blasco: *Hardy spaces of vector-valued functions: Duality*, T. Am. Math. Soc. vol. 308 (1988) , p. 495–507.

[4] R. A. DeVore, and G. G. Lorentz: *Constructive Approximation* (Springer-Verlag, Berlin 1993).

[5] J. B. Garnett: *Bounded Analytic Functions* (Academic Press, New York 1981).

[6] M. Jevtic: *Bounded projections and duality in mixed-norm spaces of analytic functions*, Complex Variables, vol. 8 (1987), p. 293–301.

[7] N. P. Korneichuk: *Extremal problems of approximation theory* (Nauka, Moscow 1976).

[8] J. Latila: *Vector-valued BMOA and Composition Operators* (University of Helsinki, 2004).

[9] J. Mujica: *Complex Analysis in Banach Spaces* (North-Holland, Amsterdam 1986).

[10] A. Pinkus: *n-Widths in Approximation Theory* (Springer-Verlag, Berlin 1985).

[11] V. V. Savchuk: *Approximation of functions of Dirichlet class by Fejèr means*, Math. Notes, vol. 81 (2007), p. 665–670.

[12] S. B. Stechkin: *An estimate of the remainder of the Taylor series for some classes of analytic functions*, Izv. Akad. Nauk SSSR Ser. Mat. vol. 17 (1953), p. 461–472.

[13] È. A. Storoženko: *Approximation of functions of class H_p, $0 < p \le 1$*, Math. USSR Sb. vol. 34 (1978), p. 527–546.

[14] A. Zygmund: *On the degree of approximation of functions by Fejèr means*, Bull. Amer. Math. Soc. vol. 51 (1945), p. 274–278.

Computing, Control, Information and Education Engineering – Liu, Sung & Yao (eds)
© 2015 Taylor & Francis Group, London, ISBN: 978-1-138-02800-5

On knowledge migration and knowledge transfer

L. Wang
School of Business, Wuchang Institute of Technology, Wuhan, Hubei Province, China

ABSTRACT: Knowledge migration theory emphasizes the method of individual acquiring knowledge and the ability to apply. Knowledge transfer theory regards knowledge as a flow of information resources and it researches knowledge sharing in different subjects. But, people are apt to confuse the two issues. In this paper, the author analyzed themes of more than 3700 articles where come from *CNKI*. Through the comparative study on the connotation and the extension of both, it illustrates their differences and connections. At last, it points out that the use of knowledge migration theory can be more effective in promoting knowledge transfer.

KEYWORDS: Learning transfer; Knowledge transfer; Internalization; Educational psychology; Knowledge management.

1 INTRODUCTION

Knowledge acts as a very important role in the knowledge economy, today's society and which has been given a huge promotion to the progress of humanity. It not only can arm individual brain, making the organization more competitive, but also create wealth to the community. Knowledge exists objectively and is actively used to create wealth for individuals or organizations. Who a body of knowledge is, he must go through the initiative to acquire knowledge, to absorb knowledge and use of knowledge in order to create wealth. The process made of the above four steps is the same as knowledge management one. It just shows that effective knowledge management is inseparable from the process of learning and knowledge-sharing. In the research field of knowledge management, knowledge migration theory refers to knowledge acquisition, absorption and utilization while knowledge transfer focus on the research of transmission and using between different subjects. For the difference and connection between knowledge migration and knowledge transfer, the author tries to get results through access to relevant reference. So, in December 16, 2014, the author took subject word as the search term, with *knowledge transfer* as the retrieval words having a search in *CNKI* database, resulted that there were more than 3700 articles. According to a certain point view of Fei-cheng Ma and Xiao-guang Wang in the article *social network model of knowledge transfer research*, researchers were not from the perspective of cognitive psychology to study how individuals learn, absorb, memory and apply, they preferred to take the knowledge and information as the flow of social resources, and studied its rules and influence factors in social organizations.[1] What are the differences and connections of knowledge migration and knowledge transfer, the paper will compare with the intension and extension.

2 THE CONNOTION OF BOTH

2.1 *The connotation of knowledge migration*

The term *knowledge migration* comes from psychology. Cognitive psychology calls influence that the gained knowledge and the obtained movement skill, emotion and attitude on subsequent learning activities and follow-up one each other as learning transfer, and it also calls knowledge migration.[2] In fact, knowledge gained either promote or hinder learning. Positive transfer results in correct performance while negative transfer results in mistake.[3] Xiao-xia Zhang in the article *self and knowledge transfer* pointed out that learning transfer is the ability to apply the knowledge to new situations and to solve new problems embodied, including the new context awareness, the link of old knowledge and the new one as well as recognition and solution.[4] Positive transfer enhances the body's learning abilities. In this paper, the author takes the positive transfer as the main study object. Knowledge Migration is defined as a process in that the body of knowledge gains something, including recognition, experience and skills in order to deal with new problems. The promotion of transfer helps the body of knowledge to have an efficient learning. According to the common elements of knowledge transfer theory, generalization theory, situational theory and cognitive structure theory, the author conclude the way to improve knowledge migration mainly in the following four aspects. First, master the basic concepts and principles in the learning process then construct meta-cognitive knowledge from internal and find the same elements in objective knowledge, avoid rote. Second, accurately grasp the similarities and differences by comparison and analysis. Third, create the same problem situation. The theory of situated cognition holds that learning cannot be separated from the situation, so the creation of the problem situation is conducive to the realization of the positive transfer. [5] Fourth, construct knowledge based on the existing knowledge and experience, psychological structure and beliefs, promote interaction between old and new knowledge. Above these, it can help us to perfect our cognitive structure and be systematic, comprehensive and holistic.

2.2 *The connotation of knowledge transfer*

Knowledge transfer is a stage of knowledge management. Owing to the body of knowledge can flexibly make use of knowledge which is an important corporate resource to solve practical problems, it is valuable. Stage of knowledge management can be divided into acquisition, coding, transference, communication and sharing. 1977, American technology and innovation management expert *Teece* first proposed the concept of knowledge transfer, he believed the international transfer of technology enterprises can help companies to accumulate valuable knowledge and promote technology diffusion, thus narrow the technology gap between regions.[6] So we can see that knowledge transfer is a foreign word. Currently, there is not an accepted concept of knowledge transfer. After analyzing the scholars' understanding about the topic, it can be summed up several things in common. First, knowledge is a resource which is different from the traditional factors of production such as the resource of human, money and material. If it is used effectively, can create a tremendous wealth with a minimum output, the ratio of input and output is higher than the preceding three resources. Second, as a kind of energy, knowledge is spread and be used in certain circumstances. Third, although there isn't unified standard, the successful knowledge transfer depends on whether it can be internalized and be created benefits of output as well core competencies for individuals and organizations. Thus, the concept of knowledge transfer can be summarized as that it is a process in which knowledge is disseminated from the sender to the receiver for narrowing the knowledge gap between human individuals or organizations in a controlled environment. [6]

3 THE RELATIONSHIP BETWEEN KNOWLEDGE MIGRATION AND KNOWLEDGE TRANSFER

3.1 *The distinction between knowledge migration and knowledge transfer*

3.1.1 *Subjects are different*

Knowledge migration generally occurs between teachers and students, or individual students in independent study.[7] In the former, teachers teach students without reservation, not related to economic interests. As a recipient of knowledge, learning ability of students that above mentioned positive transfer is a great impact on the efficiency. Students are the subject of knowledge migration. Now, the study about knowledge migration focuses on teaching and learning.

Both sender and recipient are the body of knowledge transfer. From this perspective, the sender's knowledge storage, the will, the awareness for protection of knowledge, the motivation, the expression skills as well as all about receiver will affect the efficiency of knowledge transfer. In the field of economics and management, the sender and the recipient can be an individual or organization.

3.1.2 *Objects are distinct*

The object here refers to the knowledge itself—knowledge entities. In knowledge migration, objects are generally recognized as explicit knowledge which includes axioms, theorems, laws and common sense etc. Students master them to apply to solving practical problems. While in knowledge transfer, objects includes explicit and tacit knowledge. The latter is very difficult to be transmitted efficiently. But when successful, the payoff is huge. Currently, all the research topics about knowledge, tacit one is the mainstream issue.

3.1.3 *Assessment methods are different*

The results of knowledge migration can be described as hidden and difficult to measure. The purpose of knowledge-migrated is that the body internalizes knowledge into quality, focusing on the cultivation of individual capacity. It is difficult to be successful in short time. On the contrary, the general effect of knowledge transfer is tangible things, such as technological or management results and easily be quantified.

3.1.4 *Knowledge migration is a social unpaid activity, but knowledge transfer is an economic one*

Knowledge migration is generally free, as it emphasizes how to obtain, absorb and make use of knowledge, or the communication between teachers and students. Knowledge transfer is generally paid, which involves the following economic indicators, the opportunity cost, the risk, social capital and so on. The person in charge of the enterprise may be able to pay huge economic costs to import a certain technology that can increase their core competitiveness. Here, the technology can be seen as tacit knowledge.

3.2 *The relation of knowledge migration and knowledge transfer*

In the above-mentioned over 3700 articles published in accordance with the order of time, the first article was in April 1980, *Zhu Bo*, published in *South China Normal University (Philosophy and Social Science Edition)* entitled *our experience with the reform of teacher education teacher education abroad*, it belongs to Education Papers. The second was published in August 1982, belongs to economics and management; the third one also was economically management papers. When looking carefully at the number of papers published, we found that from 1980 to 1993 the annual output was less than three, after 1994 an increase arose, to 2000 it reached 15. Research papers before 2000 are each half of education and economic management. Since 2000, there were a lot of economics and management articles. Take the year of 2000 for example, a total of 15 papers was retrieved, of which 12 papers for Economics and Management, 3 for educational papers. From then on, the number of published papers about knowledge transfer massively grew, reached 463 in 2010. Among them, economic and management thesis accounts for the proportion was on the increase. During 2009-2012, knowledge transfer paper annual was close to 500, has become a hot research field of economic management, as shown in Figure 1.

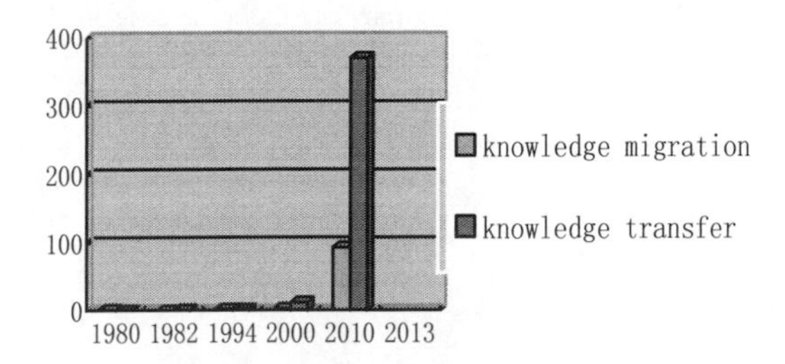

Figure 1. The number of annual papers in knowledge migration and knowledge transfer.

The author analyzed above-mentioned papers in recent years found that some authors confused the concept of knowledge migration and knowledge transfer. In fact, from the literal meaning, both are great similarities but not identical. The paper will elaborate the similarities between knowledge migration and knowledge transfer below.

3.2.1 *The same transmission channels and means*

Whether knowledge migration or transmission, the body of them can't go into without communication. There are many ways to obtain knowledge, such as searching the knowledge base, carrying out by way of Internet, face-to-face communication, or participating in the meeting and on-site education so on.

3.2.2 *The same process*

Knowledge migration and transmission can be divided into preparation, transmission, integration three processes. In the preparation process, the lack of knowledge made the recipient have the motivation to migrate knowledge. In the process of transmission, knowledge itself, the media and knowledge body influences each other to achieve the sending, transmission, reception. During the integration process, the recipient constructs their own knowledge base, updates his personal knowledge as well as absorption and utilization to improve knowledge innovation.

3.2.3 *The purpose of knowledge migration with the same knowledge transfer*

Based on the above, they have the ultimate goal which is to promote the exchange of knowledge, sharing and utilization, and use the knowledge to create value, promoting the progress of society.

4 CONCLUSIONS

4.1 *Learning transfer is the basis of knowledge transfer*

From the chronological point of view, Knowledge migration occurs earlier than knowledge transfer. So, learning transfer is the basis of knowledge transfer. The sender and recipient must have a stock of knowledge, and the knowledge has been internalized by a body of knowledge. Then knowledge maybe creates value in the future. For example, a student learns to write programs by means of recursive method in C language. They carry out knowledge migration within a lot of practice of a *recursive programming method.* When the student encounters a similar problem again, he can program by a recursive method in C language to solve it, thus this process is called knowledge transfer. In order to better solve the problem, knowledge sender and recipient must fully express their views in the process of communication, encode and decode their existing knowledge associated with the problems. The meaning of encode and decode is to express as any form of knowledge which is acceptable by the body of knowledge.[8] Chronological order and process about knowledge migration and knowledge transfer are shown in Figure 2.

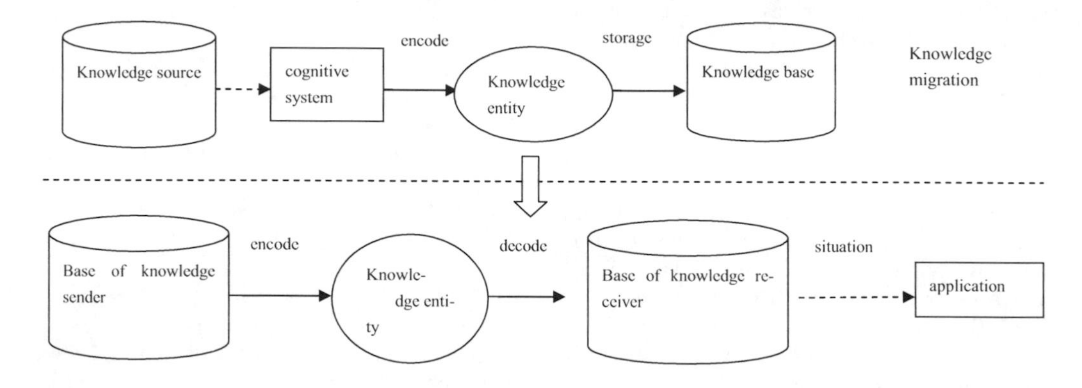

Figure 2. A chronological order and process about knowledge migration and knowledge transfer.

From the above chart, we can see that knowledge migration achieves, the knowledge stored in the knowledge base can be applied to guarantee the success of knowledge transfer in exceptional situations.

4.2 *The ability of learning transfer influence the effectiveness of knowledge transfer*

From the efficiency and success rate, learning transfer plays a very important role in successful shift. In the process of knowledge transfer, the recipient receives new knowledge, and associates with the existing in his knowledge network, to find the same elements of them, and constructs knowledge structure based on the context. Only in this way can we use it to have a solution. Therefore, when evaluating the effect of knowledge transfer, the receiver's ability of learning transfer is a very important factor. It can improve the efficiency and success rate.

REFERENCES

[1] Fei-cheng Ma, Xiao-guang Wang 2006. The research of social network model of knowledge transfer. *Jiangxi social science* 7: 38–44.
[2] Mei-lin Yao 1999. *The rules of learning*. Wuhan: Hubei Education Press.
[3] Ru-de Liu 2010. *The psychology of learning*. Beijing: Higher Education Press.
[4] Xiao-xia Zhang 2005. Self-study and knowledge transfer. *Inner Mongolia Science & Technology and Economy* 18: 113+129.
[5] Da-jun Zhang 1997. *Eeducational psychology*. Beijing: People's education press.
[6] Da-peng Tan, Guo-qing Huo et al 2005. Knowledge transfer and its related concepts. *Analysis of library and information service* 49 (2): 7–10.
[7] Linda Argote, Paul Ingram, John M. Levine and Richard L. Moreland 2000.Knowledge Transfer in Organizations: Learning from the Experience of Others. *Organizational Behavior and Human Decision Processes* 82 (5):1–8.
[8] Garavelli, A.C. Gorgoglione, M., and Scozzi, B 2000. Managing knowledge transfer by knowledge technologies. *Technovation Journal* 22:269–279.

Computing, Control, Information and Education Engineering – Liu, Sung & Yao (eds)
© 2015 Taylor & Francis Group, London, ISBN: 978-1-138-02800-5

An evaluation method for node importance based on pagerank in complex undirected weighted networks

F. Li, W.T. Zhao, Z.F. Sun, B. Dong & Y.J. Wang
National University of Defense Technology, Changsha, China

ABSTRACT: The existing evaluation methods for the research on the node importance in complex network mainly focus on undirected and unweighted complex network or directed and weighted complex network. In this paper, based on undirected and weighted complex network model, and inspired by the PageRank ranking algorithm, we propose a new evaluation method for node importance in undirected weighted complex networks on the basis of PageRank (referred UW_PageRank), by simulating the edges and nodes of complex network as the pages and hyperlinks in the web respectively. Experiment of transfer matrix decomposition shows that the proposed method is efficient and suitable for engineering applications.

KEYWORDS: Undirected and weighed; Complex network; Topology; PageRank.

1 INTRODUCTION

The important nodes of a complex network is defined as some special nodes that can affect the structure and function of the network to a greater extent compared to other network notes, which is a network structure constructed with a huge number of nodes and the complex relationships between them . In mathematical description, it is a diagram with enough complex topology features [1]. Put it into the real world, when nodes are used to represent all sorts of individuals while edges represent the links between individuals, many things can be analyzed with the network graph theory. Numerous studies have shown that complex network possesses the small-world property [2], scale-free property [3], robustness against random attacks, vulnerable to malicious attacks and so on. To the study of complex network getting deep, discussions of many basic issues become more important. It is of high practical value to explore the important nodes in the network and assess their importance. This paper firstly introduces several research methods on the importance of the important node in complex network in recent years, and on the basis of the previous study, we proposed an evaluation method on the importance of important nodes in the weighted complex network based on PageRank. This method not only reveals the characteristics of the network topology structure, but also accurately reflect the relative importance of the nodes. Result of the experiment shows that this evaluation method based on PageRank performs effective with low overhead, and its operation speed is fast. It can get a perfect calculation capability in large-scale complex network

2 RELATED WORKS

There are lots of methods of evaluating the importance of nodes in complex networks, most of which are based on graph theory and data mining. Initial research originated in the field of sociology, then researchers from other scopes begun the study of such problems, too. To sum up, the main research methods are as follows:

1. *Sorting method based on neighbor node.* A class of the most simple and intuitive method, with degree centrality [4] investigating the number of direct neighbors of nodes, which believes that the larger number of neighbors a node owns, the bigger influence it will have; Chen et al. [5] considered the information of neighbor nodes in 4 layers with semi-local centrality; Kitsak et al [6] proposed to determine the location of nodes in the network using k- shell decomposition, which can be seen as an extension of degree centrality. It defines the importance of nodes according to their positions in the network with the view that the more a node is close to the core, the more importance it has.
2. *Sorting method based on path.* This class of methods assumes that the network information flow travels only through the shortest path, but in the real communication network, in addition to the path length, the number of intermediate nodes on the path also have significant impact on the spread. The most significant indicator is betweenness, which reflects the influence of a node.

- The betweenness centrality that frequently mentioned generally refers to the shortest path betweenness centrality (shortest path BC), it considers that among all the shortest paths of all node pairs in the network (usually there are more than one shortest path between a pair of nodes), the more the shortest paths pass through a node, the more important the node is;
- Flow betweenness centrality [7] considers that among all distinct paths in the network, the higher proportion of paths pass through a node, the more important this node is;
- Routing betweenness centrality [8] considers that during all packet transmission, the expected value of packet number through a node could reflect the importance of the node in the network.

3 *Sorting method based on feature vectors.* The method based on the feature vector considers not only the number of nodes in the neighbor, but also the impact of its quality to the importance of the node. Eigenvector centrality [9] considers that the importance of a node depends on the number (i.e., the degree of the node) of its neighbor nodes, as well as the importance of each neighbor node, but the scoring of each node is completely determined by the neighbors, causing the convergence process to be slow. In order to make the convergence work well and work fast, the cumulative nomination [10] method simultaneously take the value of a node and its neighboring nodes into consideration in each iteration, and this value changes with every update; PageRank algorithm [11] and LeaderRank algorithm [12] are usually used in a directed network, increasing the scores of nodes along the access path by simulating the process of users browsing the web, to identify the importance of pages.

4 *Sorting method based on node removal and contraction.* Node (set) removal and contraction method coincides with the idea in system science of determining a system's core [13], its most notable feature is that the structure of the network will dynamic change in the process of sorting the important nodes, at the same time, the importance of nodes is often reflected in the destruction of the network after the node is removed. The destruction reflects the importance. Node deletion shortest distance method [13] suggests that the destruction is related to the distance changes caused by the removal of a node: removing a node (set) will cause the network differentiate, and form a number of connected components, the greater of the short distance between the network nodes change, the more important the removed nodes are; Node deletion spanning tree method [14] suggests that the less of the number of network spanning tree after a node deleted, the more important the node is. Node contraction is to contract a node and its neighbor nodes into a new node [15]. When a node is a very important core node, the entire network will concentrate better after the contraction.

The above methods on assessing node importance are all based on the node degree and the network structure, but in the real world, almost all of the networks are weighted, where each edge in the network not only represents whether there is a connection between two nodes, but also show some characteristics, such as the different intimacy of relationships between different members in a social network, and the flow among nodes in the communication network. If each edge of the unweighted network is given an appropriate weight, the network becomes a weighted network, which could be able to describe the tightness of the links between network nodes, and provide a more realistic and detailed expression of the structure of complex networks. This paper starts with the network topological structure, meanwhile takes the strength and the amount of the connections between nodes into account to quantificational evaluate the importance of a node, and thus proposes an evaluation method based on weighted complex networks, which is more accurate and efficient compared to some other methods.

3 ALGORITHM

PageRank algorithm is a very classic Web page ranking algorithm currently widely used in many fields. In this paper, inspired by PageRank ranking algorithm, by means of simulating the edges and nodes of complex network as pages and hyperlinks respectively in the Web, we proposed a new evaluation method for node importance in complex undirected weighed networks based on PageRank.

3.1 *The basic idea of PageRank algorithm*

PageRank algorithm was proposed and published by Larry Page and Sergey Brin in 1998. The main idea of PageRank algorithm is based on linkage analysis of network structure. It was constructed on the idea that the more a website is accessed, the higher quality the website owns, when browsing the web, the users jump between pages mainly through the links, so you can analyze the hyperlink topology to calculate the access frequency of each page. When a user stays on a page, if we assume that the possibility of jumping to each linked page is the same, then for the page, the PageRank value is defined as:

$$Pagerank(p_i) = \frac{1-q}{N} + q\sum_{p_i} \frac{Pagerank(p_j)}{L(p_j)} \tag{1}$$

Where the q is the damping coefficient, generally is 0.85; $p_1, p_2, p_3,\ldots,p_N$ is the analyzed page; $L(p_j)$ is the linked page number of p_j; N is the number of all the pages.

Formula (1) shows that the more source links p_i has, the larger number of links pointing to a page P_i, namely N, and also the more important p_i is; Meanwhile, the higher the level of pages linking to the source page p_i, in other words PageRank(p_j) has the greater values, the most important page p_i is, and the greater the PageRank value of p_i is. Therefore, learning from the PageRank algorithm, we take the concept of link value as a critical factor of importance ranking, and introduce it to the access of node importance in a complex network, finally propose an indicator to assess the node importance of the undirected and weighed network.

3.2 *Background definition*

Abstract the network as graph $G(V, E)$ where, V is the set of all vertexes in the network, E is the set of all edges of the network, N is the number of vertexes in the network, $w(v_i, v_j)$ is the weighted value of the edge from vertex v_i to vertex v_j in the graph. Since we only take the undirected and weighted network into consideration, so as $w(v_i, v_j)= w(v_j, v_i)$.

Definition 1 Adjacency matrix M. M is used to describe the connection in an undirected and weighted network with N vertexes, when there are connections between two vertexes, the matrix element represents the weight (or connection strength), otherwise the matrix elements is 0:

$$M = (m_{ij})_{n\times n} = \begin{cases} w(v_i, v_j), (v_i, v_j) \in E \\ 0, otherwise \end{cases} \tag{2}$$

Where $w(v_i, v_j)$ is the weight from vertex i to vertex j.

Definition 2 For a network node v_i, the greater weight the connected vertex v_i has, the greater contributes the edge due to the whole network, than the greater probability of the vertex to be jumped to. Thus, the probability of jumping from node v_i to every connected vertex v_j in the network is:

$$P(v_i, v_j) = \frac{w(v_i, v_j)}{\sum_{k=1}^{N} w(v_i, v_k)} \tag{3}$$

Where, $p(v_i, v_j)$ is the probability of jumping to vertex v_i. $w(v_i, v_j)$ is weight from vertex v_i to vertex v_j, $\sum_{K=1}^{N} w(v_i, v_k)$ is the sum of weights connected to the vertex v_i. From formula (2) and formulas (3)we can conclude

$$Pagerank(p_i) = \frac{1-q}{N} + q \sum_{j=1\cdots N} \frac{w(v_i, v_j)}{\sum_{k=1}^{N} w(v_i, v_k)} Pagerank(p_j) \tag{4}$$

Definition 3 probability transition matrix Q. Divide each line of the adjacency matrix M by the sum of non-zero elements in this row, thus normalize the matrix to obtain the probability transition matrix Q, and the matrix elements record the probability from one node to another.

$$Q = \begin{cases} P(v_i, v_j) \; (v_i, v_j) \in E \\ 0, Otherwise \end{cases} \tag{5}$$

Definition 4 According to the formula (4) and formula (5) a new matrix P is generated

$$P = q \times Q^T + (1-q) \times \frac{1}{N} \times e \times e^T \tag{6}$$

Where, $e \times e^T$ is a matrix in which each element is 1, elements of Pin formula (6) are all nonzero, which means P is a full rank matrix, so the Markov process that the transition matrix representation is homogeneous, ergodic, irreducible, and aperiodic, Therefore, there is a unique solution.

Step1 input the adjacency matrix M of the current network.
Step2 calculate the probability transition matrix Q of matrix M according to the formula (5).
Step3 calculate the new matrix P according to the formula (6).
Step4 calculate PageRank[i] of each node by means of iterative of the decomposition transfer matrix.
Step5 If PageRank [i] -PageRank [i-1]> delt, then go to Step4.
Step6 normalize PageRank.
Step7 sort and output the result of node importance.

3.3 *Algorithm and analysis*

Based on the above definition, the proposed evaluation algorithm based on the node importance of PageRank is as follows:

The input of the algorithm is the graph $G(V, E)$, and the output is PageRank of each node, namely the node importance. According to formula (3), the algorithm eliminates the irrationality of the average distribution of the traditional PageRank algorithm. By means of giving different weights to different nodes, the algorithm improves the PageRank values of the important nodes, at the same time reduces the PageRank value of unimportant ones.

The algorithm uses a decomposition transfer matrix iterative method during the iterative process, with specific reference to [17]. In each iteration, complexity O(n) and space complexity O(n) can be calculated for each iteration time, then the time complexity of the whole algorithm is O(mn), and space complexity is O(mn) where m is the time of iterations. For large, complex network, m is much smaller than the number of nodes N. Therefore, transfer matrix decomposition iterative method can greatly reduce the complexity of the calculation, which is suitable for large complex networks.

4 EXPERIMENTAL VERIFICATION

Experiments use network topology as shown in Figure 1 [16], the network contains a total of 10 vertexes, 10 edges, on which the value represents the corresponding weight. We use dissimilarity weights in this paper, which means the greater the weight is, the longer distance is indicated, and the smaller contribution to the network it dose [16].

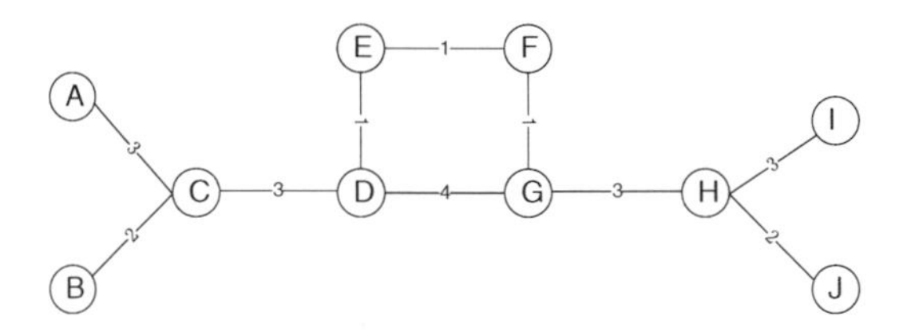

Figure 1. Topology of Network G.

We applied degree centrality, betweenness centrality, node contraction method, PageRank method and UW_PageRank method to assess the importance of each node in network G. To facilitate the comparison, the data in the tables are all normalized, the specific evaluation results are shown in Table 1.

The results in Table 1 show that the nodes E and F are most important, which means that they do the largest contribution to the entire network in reality, this fits the intuitive judgment; The pages linked from a number of high quality pages, must be of high quality as well, therefore, for node C and node D, the importance of node C is less than node D; for node A and Node B, the contribution of C to A is 3/8, and the contribution to B is 2/8, so the importance of node A is less than that of node B.

Table 1. Node importance evaluation results.

node	degree centrality	between-ness centrality	node contrac-tion	Page Rank	UW_ PageRank
A	0.05	0.0580	0.0340	0.0558	0.0418
B	0.05	0.0580	0.0474	0.0558	0.0577
C	0.15	0.1065	0.1290	0.1527	0.1236
D	0.15	0.1226	0.1340	0.1408	0.1272
E	0.10	0.1548	0.1558	0.0950	0.1497
F	0.10	0.1548	0.1558	0.0950	0.1497
G	0.15	0.1226	0.1340	0.1408	0.1272
H	0.15	0.1065	0.1290	0.1527	0.1236
I	0.05	0.0580	0.0474	0.0558	0.0418
J	0.05	0.0580	0.0340	0.0558	0.0577

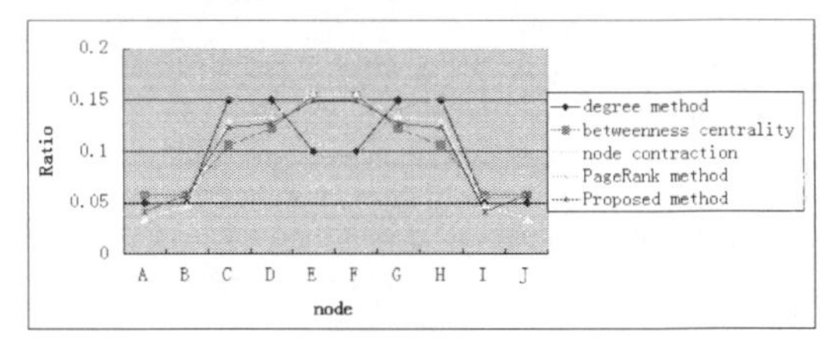

Figure 2. Node importance evaluation results.

It can be seen from Figure 2 that the importance of UW_PageRank algorithm, betweenness centrality and node contraction ranks substantially the same , but the complexity of betweenness centrality is O(n3), the time complexity of the node contraction is O(n3), which means that they do not suitable for large-scale network. While the complexity of UW_PageRank algorithm is O(cn), where c is the times of iterations, it approaches to be linear when n is large enough, which is slightly different from the degree centrality and PageRank method. But we can see from the results that the nodes E and F are more important than C, D, G and H, which is consistent with an intuitive sense. Therefore, by considering the edge weights, the algorithm further characterizes the differences in the importance of nodes, which makes the results more accurate and efficient

5 CONCLUSIONS

In this paper, inspired by the use of the regression relationship based on The pages linked from a number of high-quality pages, must still be of high quality of the PageRank algorithm to determine the importance of all the pages, we proposed an evaluation method for node importance based on PageRank on complex undirected weighed networks. Experimental results show that the UW_PageRank algorithm achieved good results in the evaluation of node importance in the complex undirected weighed networks, and its operation speed is fast. It can get a perfect calculation capability in the large-sized complex network.

ACKNOWLEDGEMENT

This work is funded by the National Science Foundation (NSF) of China (61271252) and key laboratory for information system security technology.

REFERENCES

[1] Zhou Tao, Bai wen-jie, Wang Bing-Hong, et al. A brief review of complex networks [J].Physical.2005.34 (1):31–36.

[2] Watts D J, Strogatz S H. Collective Dynamics of Small-world Networks [J].Nature, 1998:440–442.

[3] Albert R, Barabási a L. Emergence of Scaling in Random Net-works [J].Science, 1999, 286:509–512.

[4] Bonacich P. Factoring and weighting approaches to status scores and clique identification. J Math Sociology, 1972, 2: 113–120.

[5] Chen D B, Lü L, Shang M S, et al. Identifying influential nodes in complex networks. Physical A, 2012, 391: 1777–1787.

[6] Kitsak M, Gallos L K, Havlin S, et al. Identification of influential spreaders in complex networks. NatPhys, 2010, 6: 888–893.

[7] Freeman L C, Borgatti S P, White D R. Centrality in valued graphs: A measure of betweenness based on network flow. Social Networks, 1991, 13: 141–154.

[8] Dolev S, Elovici Y, Puzis R. Routing betweenness centrality. J ACM, 2010, 57: 25.

[9] Bonacich P. Factoring and weighting approaches to status scores and clique identification. J Math social, 1972, 2: 113–120.

[10] Poulin R, Boily M C, Mâsse B. Dynamical systems to define centrality in social networks. Social Networks, 2000, 22: 187–220.

[11] Brin S, Page L. The anatomy of a large-scale hyper textual Web search engine. Computer Networks ISDN Sys, 1998, 30: 107–117.

[12] Lü L, Zhang Y C, Yeung C H, et al. Leaders in social networks, the delicious case. PLoS One, 2011, 6: e21202.

[13] Li Peng-Xiang, Ren Yu-Qing, Xi Qiu-Ming. A metric for node (set) importance of network. Systems Engineering, 2004, 22.13–20.

[14] Chen Yong, Hu Ai-Qun, Hu Xiao. Evaluation method for node importance in communication networks. Journal of China Institute of Communications, 2004, 25:129–134.

[15] Tan Yue-Jin, Wu Jun, Deng Hong-Zhong. Evaluation method for node importance based on node contraction in complex networks. Systems Engineering Theory and Practice, 2006, 26:79–83.

[16] Wang Jia-Sheng, Wu Xiao-Ping, Liao Wei, et al. Improved method of node importance evaluation in weighted complex networks. Computer Engineering, 2012, 38:74–75.

[17] Liu Song-Bin, Du Yun-Cheng, Shi Shui-Cai. A method of computing PageRank based on transition matrix decomposition. Journal of Chinese Information Processing.2007, 21(5):41–45.

Computing, Control, Information and Education Engineering – Liu, Sung & Yao (eds)
 ISBN: 978-1-138-02800-5

Analysis of the cultivation of college students' environmental literacy in the age of internet

Zhe Zhang
Publicizing Department of Sichuan Agricultural University, Sichuan Ya'an, China

ABSTRACT: Environmental literacy is an important part of environmental education. It focuses on environmental awareness, cognitions, skills and values. For college students, new ways can be adopted to foster their environmental literacy at the age of internet. To emphasize news reports on the internet, organize activities by social networking site and diffuse knowledge and information about environment may be effective ways.

KEYWORDS: Environmental Literacy; the Age of Internet; Cultivation.

1 INTRODUCTION

In 1975, The *Belgrade Charter* was adopted by the United Nations conference on the environment. It states: "The goal of environmental education is to develop a world population that is aware of, and concerned about, the environment and its associated problems, and which has the knowledge, skills, attitudes, motivations, and commitment to work individually and collectively toward solutions of current problems and the prevention of new ones."[1] Mostly environmental education focus on environmental literacy. As an important part of it, college students' environmental literacy needs to pay more attention. At the age of internet, new ways can be adopted to foster college students' environmental literacy to make it more efficiently.

2 THE CONTENTS OF COLLEGE STUDENTS' ENVIRONMENTAL LITERACY

"A common definition of environmental literacy which satisfies everyone may never be written. However, what most professionals in the field of environmental education do agree on are the concepts or strands which should be included: knowledge, skills, affect, and behavior. These are the foundations upon which the assessment of environmental literacy should be based."[2] According to it, we can conclude that the contents of college students' environmental literacy mainly focus on four dimensions:

2.1 *On environmental awareness*

As the environmental problems are complex at home and abroad, environmental education ought to make the students recognize the importance of environmental quality and the existence of environmental problems and issues and have appreciative and caring attitude toward the environment. As a result, the students will have the willingness to work toward the preservation and remediation of environmental problems and issues.

2.2 *On environmental cognitions*

In order to improve cognitive ability to the environment, basic knowledge of ecological and socio-political foundations is needed. The youth should know that natural environment, along with society, culture, politics and economy are interrelated and mutually constrained. In addition, the relationship between human and nature is also an important part of the course, which is dynamic and stable.

2.3 *On environmental skills*

Environmental skills refer to the abilities that dealing with regional or global environment problems, including identifying, analyzing, investigating and evaluating environmental problems and issues. On the basis of identifying it accurately, the problem will be analyzed scientifically and rational solutions can be put forward.

2.4 *On environmental values*

The education should make the students fully realize the values of environment and its importance to mankind. When people really know the intense connections between human and the environment and serious consequences of that, the willingness to solve environmental problems will be inspired strongly.

The values also include aesthetic appreciation of environment, the respect and love for the nature and the willingness to join the work of nature conservation actively.

3 THE CHARACTERISTICS OF THE AGE OF INTERNET

At present, the internet has become an inalienable part of our daily life. People's life has changed a lot in the age of internet. The characteristics of the age of internet mentioned below may result in the changes.

3.1 *A tremendous amount of information*

The internet connects all the computers that form huge database around the world. Every event occurred at any place and any time in the world will be widespread by the internet. It breaks time and regional limit. In the past days, we record the information on papers or tapes. If the papers and tapes lost, the information would never be found again. Nowadays the digital library can solve the problem. It is convenient for people to search and save what they are interested in. You may just press the button "Save", instead of finding a storeroom.

3.2 *Multimedia information*

Generally the newspaper spread the information by text and graphics, while the broadcast spread it by sound and the television transmit it by sound and image. In the age of internet, people save, express and deliver the information in multiple ways in text, graphics, sound and image simultaneously. The information turns more vivid for us to get in that situation. So people are willing to choose it, rather than that from traditional media.

3.3 *Cyberized social relations*

According to *the 27th Report of China Internet Network's Development*, the number of China's social networking site users was 235 million by the end of 2010, an increase of 59.18 million over last year. The netizen rate was 51.4%, which rose 5.6 percent from 2009.[3]We can conclude that social networking site attract more and more people's attention. The most important reason of that is the cyberized social relations it provides. People can write diary, post photos and comment after registration. They'll share their happiness and sorrows with friends by the internet in spite of far distance, enjoying the pleasure of communicate with friends in the reality. The real interpersonal relationship network will increase the users' viscosity of social networking site.

4 THE WAYS TO CULTIVATE COLLEGE STUDENTS' ENVIRONMENTAL LITERACY BY THE INTERNET

Nowadays, more and more college students spend longer and longer time in surfing the internet. They receive scientific and cultural knowledge and make friends here. It's a more efficient way to cultivate college students' environmental literacy. We suggest that it can be strengthen in these three ways:

4.1 *To emphasize news reports on the internet*

There are many memorial days relate to environment issues we can celebrate, such as World Environment Day, World Earth Day, World Water Day, World Car Free Day, Arbor Day and so on. On these days, our campus media especially news website should spend a large space of pages on the reports of relevant activities and knowledge to cooperate with the environmental education. It is more effective to raise college students' environmental awareness by the media's agenda-setting function.

After the publication of regulated daily behaviors by mass media, the powerful public opinion environment is brought to the society. It's a kind of significant external pressure, which forces people to obey. Besides the mass media play an important role in lead us "how to do". To choose reusable bags and public transportation, say no to disposable goods and excessive packaging. These conceptions are spread by mass media to the public, which make them shift from "have the consciousness" to "behavior change", reaching the purpose of environmental education.

Additionally many news events are very good opportunities for environmental education. For instance, at the time of the APEC Economic Leaders' Summit held in Beijing November 2014, a new term "APEC Blue" became popular. That means PM2.5 fall down to the lowest level, so the blue sky and white clouds return in Beijing. On our internet media, we read a lot of editorials on this term. These news articles will evoke college students to think about relevant problems, such as "will the temporary measures be extend to secular one?", "at the turning point of the recovery of haze, a series of deep changes may emerge", "the key point of solving the problem is our determination of slowing down the speed of economy growth" and so on. Deep thinking about these issues will help college students' to enhance environmental consciousness and social responsibility.

4.2 *To organize activities by social networking site*

As was said in the former part of the article, the social networking sites attract more and more people's attention, especially the young people. The sites can organize various kinds of activities relevant to environmental issues on memorial days. Because of the cyberized social relations of the site, the activities held by social networking sites will attract more and more college students to take part in. When people see all their friends interested in the same thing, they are willing to participate in it too.

The social networking sites regularly arrange and plan some activities to raise public concern about a certain issue and lead the public to attend. On the basis of the public traditional psychology, it's an effective way to arose people participate enthusiastically by massive publication under the influence of strong public opinion. According to these environmental activities, the public can not only get plentiful knowledge and message, but also raise the ability to participate in solving the local environmental problems.

Nowadays environmental protection and low carbon economy has become more and more popular topic in our daily life. In 2011, *Global People* organized "Green China" Environmental Protection Summit in Beijing. The experts from Chinese Academy of Social Sciences and Beijing University and famous environmentalists were involved. The topic for discussion was the age of low carbon, the challenges and chances brought by new energy industry, imagination of low carbon lifestyle. If you were the followers of *Global People's* official page, you would have the chance to attend the summit after registration. Many followers came to get the chance, most of them influenced by their friends, and made the decision to attend the same event. The impact of the summit became enormous after the friends' viral communication on the internet.

4.3 *To diffuse knowledge and information about environment*

One of mass media's important functions is to hand knowledge, values and regulations down from generation to generation in the society, which makes all the people get close together. It helps people to integrate into the society constantly before or after school education. Environmental education is a kind of all-round lifelong education, which will help people respond to the world changes properly. There is no definite start or end of the education. At any age, all of the people need new information and knowledge about environment and get the ability to solve related problems. So mass media, especially the internet, is the most convenient and constant way to get that.

The contents what mass media spread today are diverse and comprehensive, including science and technology, domestic and world news, literature and art, life encyclopedia etc. All the contents may be in connected with environmental education and supply with related knowledge.

In China most of authoritative internet media have their own green page, which provide with environmental information and knowledge, Sina.com, for example. The contents of the green page are plentiful, including news information, observation and commentary, dialogues, micro-interview, aural and visual images and activities. The youngsters may just hit on the page to get to know recent news about environmental problems, read the experts' opinions on these issues, exchange ideas about it with other people, enjoy beautiful pictures and interesting videos and take part in the games. The media information is much more vivid and easier to receive. It breaks the limit of time and space of school education. It plays an important role in raising the level of people's environmental literacy.

5 CONCLUSIONS

A great deal of examples has proved that fostering college students' environmental literacy at the age of internet is much more effectively. It is an easier way for students to receive and accept. Fully taking the advantage of it is worth further and careful analyzing and exploring.

REFERENCES

[1] UNESCO-UNEP.(1976). The Belgrade Charter. *Connect*: UNESCO-UNEP *Environmental Education Newsletter*, 1(1): 1–2.

[2] McBeth, William C.(1997). *An Historical Description of the Development of an Instrument to Assess the Environmental Literacy of Middle School Students. Dissertation.*:Unpublished doctoral dissertation, Southern Illinois University Carbondale, Carbondale, IL.

[3] CNNIC. *The 27th Report of China Internet Network's Development* (2012), p. 40.

Computing, Control, Information and Education Engineering – Liu, Sung & Yao (eds)
© 2015 Taylor & Francis Group, London, ISBN: 978-1-138-02800-5

Exploration and practice of materials chemistry graduate training mode

Liang Hu & Shu Wang Duo
Jiangxi Science and Technology Normal University, Nanchang, Jiangxi, P.R. China

ABSTRACT: Pros and cons of postgraduate training mode is directly related to the level of postgraduate training, is a direct manifestation of the University of comprehensive strength. This paper analyzes the current situation of Chinese universities graduate training mode and problems, combined with the concrete reality of Materials Chemistry, Materials Chemistry four elements proposed graduate training mode, training mode and materials chemistry graduate with a general build.

KEYWORDS: Graduate training mode; high-level personnel; Materials Chemistry; exploration; Practice.

1 INTRODUCTION

Energy, information and materials are the three pillar industries of the modern economy, and the material is the material basis and ensure energy and information technology industries. Rational use of natural materials, new energy development, information engineering information collection, processing and execution of various functional materials are required. Therefore, preparation, application and development of new materials is often a milestone in the development process of human civilization, but also the fields of energy, information, environment and manufacturing development of the cornerstone. As an important part of materials chemistry materials science is the basis of materials science and materials engineering development, materials science and chemical cross-disciplinary formation. With the rapid development of technology, materials and chemical cross-integration, more and more modern energy materials, information materials, functional materials and ecological materials, often have to involve two aspects of materials and chemicals. Therefore, training materials chemistry professional high-level personnel, both for research and development of new materials to meet the application needs of the community are still very important. And graduate education is an important way and effective way to train high-level professionals in materials chemistry. The pros and cons of the postgraduate training model is one of the important factors that determine the level of training of graduate students, is directly reflected in the comprehensive strength of a university and school levels.

Better training for graduate students in materials chemistry, our training model materials chemistry graduate students conducted in-depth research.

2 CULTIVATION MODE OF MATERIALS CHEMISTRY GRADUATE PROBLEMS FACING OUR COUNTRY

Under the guidance of the State Council Academic Degrees Committee and fully learns the other postgraduate training mode, training mode of graduate professional degree in Materials Chemistry has been established, but because of the development time is not long, each sponsored training mode institutions is still some problems, each hospital school is constantly adjusted and improved. Our group of related institutions for materials chemistry graduate student training mode analysis, that the main problems are as follows:

First, the material chemistry graduate training objectives have not been well represented. Existing materials chemistry graduate teaching experimental system has been difficult to meet, "reflects the development of cutting-edge disciplines, graduate training ability and innovative thinking" of graduate training in the material necessary to achieve and many other requirements; although the material culture of graduate students application of high-level talents to meet the materials science and engineering, but from understanding materials chemistry graduates degree view, many schools just put a professional degree in education as a high level of continuing education in order to expand the knowledge of the main objective, rather than members work in order to meet actual needs. From a practical point of view, the existing curriculum materials postgraduate education system does not reflect the needs of industry professionals a knowledge structure and ability.

Second, Materials Chemistry Graduate Degree form to be further adjustment, the choice is mainly limited to students on school students. In fact, in addition to full-time graduate enrollment of full-time outside, may have practical work experience enrollment in engineering and technical personnel, the use of full-time enrollment is not the way training materials chemistry graduate students, help provide quality training.

Third, the material Postgraduate Education manner to be further improved. First, using a single form of full-time study, due to the obvious contradiction between engineering, leading to the teaching task can be completed in accordance with the teaching plan though, leading to weakening of practical work and research skills. Second, although the curriculum emphasizes the complex structure of knowledge, but ignores the expertise and practical knowledge of teaching and training, but does not reflect the characteristics of the culture by area. Third, there is no dual mentoring, or though a dual mentoring, but the second did not play its due role mentors.

Fourth, the quality of materials chemistry graduate student specializing in control there are still insufficient. First Course Assessment excessive use of paper in the form of courses, and graduate students cannot effectively mastered the basics of professional conduct a comprehensive evaluation. The second is still not out of the traditional academic degree postgraduate dissertations weight training, focusing on scientific research, theses are not closely linked with the actual work.

In addition, materials chemistry graduate student in the teaching process, there are also courses system is imperfect, curriculum unreasonable, outdated teaching content simple, keep up with new functional materials, construction materials, and the pace of development of nano-materials, trained professional graduate their knowledge base narrow, low level of professional experimental skills, innovation and pioneering consciousness is weak, and many other problems.

3 ENLIGHTENMENT OF AMERICAN GRADUATE TRAINING MODE

3.1 *Reasonable position postgraduate training mode*

Cultivation Model of American major professional degree graduate who is studying for a degree in vocational preparatory nature of professional training or to improve education, transport of high-level expertise in the field of professional practice or development work to the community, more focus on "pragmatism." Graduate training process has failed to shake off the shackles of Chinese academic degrees, errors on the application of this concept, complex, high-level talents in our society shortage of very negative impact of full-time graduate education reform fundamentally and development. Therefore, only the changing concepts of education, postgraduate training mode reasonable position to make graduate education in our country get better development.

3.2 *Outstanding graduate training mode features*

According to a different academic degree or professional degree, in the training mode prominent distinction, highlighting their characteristics.

First professional degree and academic degree, although both belong to a degree level, but after all, not the same type, so professional degree education itself must also have academic degrees and different qualities, or if in training mode to academic degree requirements of professional graduate degree, it will also lead to an academic professional degree. The postgraduate Degree purpose of education is to cultivate community senior special talent, and thus its training model should focus on the standards applied talents to develop, such as the requirements of students, learning form, curriculum and teaching methods and other aspects. Should be to broaden the students choose to enroll some admissions that have the equivalent surface and corresponding professional experience of the candidates; the learning organization is not limited to full-time, depending on the actual situation to adopt a flexible learning time system, you can use the more modern and advanced teaching organization teaching methods of teaching high-end technology, the introduction of case teaching, field teaching and online teaching, the formation of a more diversified teaching methods.

Secondly, in the training methods of each ring can be cultured in a follow certain principles based on the combination of the characteristics of a professional degree in professional training mode characteristics of a professional. Degree culture can engage in different professional expertise in different occupations, in order to achieve the professional graduates different requirements, and truly reflects the professional, occupational characteristics, it is necessary to train professional degree in general based on the principles established with professional features with a corresponding disciplines, professional training mode features.

In short, China's professional degree training model should be based on the degree to pursue those who have to make the application more practical ability to improve their professional and creative, so obviously more emphasis on the ability to distinguish academic postgraduate training as the American Professional Master Mode and academic degree open.

3.3 *Quality assessment mechanism to improve the diversification of the market as the main*

Develop training methods, no matter how, if not a corresponding quality evaluation system, its ultimate effect is still not guaranteed, professional degree graduate education must also follow this rule. What set of quality assessment mechanism, depending on the fundamental concept of quality choices and identified and evaluated to determine the appropriate subject and evaluation criteria, which are based on the training objectives to be achieved based on the results. US graduate training objectives are diverse, application-oriented talents of its evaluation of the subject is the social and market evaluation criteria mainly by the market rather than government regulations, personnel training and quality are directly affected by the results of evaluation of the market and society, in order to meet the market can demand for standards and quality assessment mechanism to reflect the diversity and refine specific training methods, but also shows a wide range of features.

US professional degree development experience tells us that the social forces involved in graduate education in a fair and objective evaluation of graduate education is one of the hallmarks of healthy development, but also to achieve the scientific management of education, an important way to democratization.

4 CONSTRUCTION OF A MODE OF MATERIALS CHEMISTRY GRADUATE

4.1 *Training objectives of the materials chemistry graduate*

The overall objective is to develop postgraduate training all-round development of vocational teachers and application-oriented talents with innovative spirit and practical ability, adhere to the principles of the comprehensive development of both ability and integrity, moral, intellectual, physical, and require graduate should achieve:

Love the socialist motherland, support the leadership of the Communist Party of China; law-abiding, have good moral character, professionalism and responsibility.

Firm grasp of the basic theories and systems in the field of materials chemistry and related disciplines of professional expertise and skills to master modern experimental methods of the discipline, to understand the status and trends of the development of the subject, has engaged in the practical work and research work skills, management ability, innovation and the ability to analyze and solve problems.

Proficiency in a foreign language and basic computer knowledge, with a strong foreign language skills and computer skills.

4.2 *Full-time master's training methods degree*

4.2.1 *Learning theory and practice of combining learning*

Postgraduate training methods mainly theoretical study and practical learning to take a combination of approaches. Master both a deep understanding of the basic theory and expertise to grasp the meaning of the professional knowledge of vocational and technical education, with the ability to engage in professional and technical work. Requirements to participate in postgraduate academic activities, social practice and exam-related qualifications. For lack of practical experience and expertise shortage graduate, try to create the conditions for them to make up for deficiencies. Students with a solid grasp of the theory and techniques of professional basis, understanding of the professional development of cutting-edge trends, have the ability to conduct independent research and professional work in the field, become expertise in this specialized field.

4.2.2 *Tutor system*

All postgraduate training process, the main instructor is responsible for the implementation of the system, has hired engineering experience within the mining industry or engineering department instructors and high service levels recommended by the unit, if necessary, responsible person with senior technical titles of the joint guidance.

According to regulations instructor graduate and postgraduate training programs, to develop a practical guide for each graduate training plan, give full play to the leading role of the instructor, and graduate students to strengthen ties, teaching and learning. Excellent graduate training instructor team is the basic guarantee, you can graduate to the guidance to help them choose the right research direction; the professional background of good research can provide a good learning, the experimental conditions for graduate students; in addition, other academic exchange and scientific research units contribute to graduate to broaden their horizons.

4.3 *Research*

4.3.1 *Polymer materials chemistry*

This research direction in composition, structure, properties, preparation and application of basic theory and application of chemical methods of research materials mainly between chemical synthesis, structure (chemical structure, physical structure) and properties of the material of the structure-activity on the basis of the relationship, mainly engaged in organic polymer materials, polymer matrix composites, chemical synthesis of micro and nano materials and various functional materials, surface modification, structural characterization, and application of research and preparation. In active controlled radical polymerization, step polymerization, plasma polymerization chemical polymerization method as the basis for all kinds of functional macromers, functional polymer membrane materials, design and synthesis of biologically active materials, and for its functionality and characteristics of applied research, developed a light, electricity, magnetism, heat, adsorption, separation and other special features of functional materials.

4.3.2 *Materials and interface chemistry*

The research tracked metal corrosion protection and frontier areas around the metal temperature oxidation theory, and chemical stability of the material to carry out related research work surface modification of materials technology and corrosion and control of the production process and so on. The main contents include: the relationship between the nano-metal materials, high temperature chemical stability and high temperature corrosion (oxidation) acts as the focus, selectivity of nanomaterials in different oxidation conditions and environmental media (02, C12 and mixed atmosphere, etc.) in the early an oxide film (nucleation) growth intrinsically linked a grain size of a temperature of the question to reveal the impact of nanomaterials chemical stability under different conditions of material corrosion (oxidation) properties forthedevelopmentofspecialpurposehightemperature corrosion resistance nano-coating to provide theoretical and experimental evidence. Using PvD, CvD technical design and preparation of high performance protective coatings to improve the material life of special corrosive environment.

4.3.3 *Nanomaterials chemistry*

The direction is mainly engaged in the chemical synthesis of inorganic nano-materials, surface modification Sui, structure, characterization and application of research and various functional materials. Chemical andtablestudytableinterfacialboundarylayerstructure of the material is based on the one hand, hydrothermal, solvothermal, sol-gel method for the synthesis of new nanoparticles and single-molecule crystals, and structural and functional characterization, on the other hand by surface chemical modification, solubilization graft material and other technical means to achieve chemical, physical and complex change l Health, functional devices and the synthesis and application of novel nano-scale structure of the material and more.

4.4 *Assessment methods*

Course Assessment exams and test is divided into two kinds. Closed book exam can be written in different forms of open-book examination, oral, coursework, practical assessment (e.g. test), etc; degree examination should be carried out, according to percentile scores assessed 70 into conformity; non-degree courses can take the exam or test mode 60 divided qualified; using test compulsory part of the way, etc., according to the results "excellent", "good", "qualified" and "unqualified" four-tier system assessment.

5 CONCLUSION

After comparative study and practice, training work my school material chemistry graduate students have been greatly improved compared to the past. This article is supported by Jiangxi Science and Technology Normal University Graduate Teaching research project: research and practice materials chemistry postgraduate training mode (KSDYJG-2014-05) research. The corresponding author is Shuwang.Duo(Email: 85904917@qq.com).

REFERENCES

[1] Reform Liangrui Jie, Wangyue Hui, Sun Yan a, Qu Xiaoyue. Materials Chemistry Practice Teaching System and optimization [J]. Science Week 2014 (07).

[2] Hu Yang sword. Exploration and Practice of Applied Materials Chemistry Professional Practice Teaching System [J]. Technology Vision 2012 (20).

[3] Wangyue Hui, a Sun Yan, Qu Xiaoyue, Zhong Jianjun, Ma now. Materials Chemistry Application Personnel Training mode [J]. Jilin College of Education (mid) 2012 (01).

Computing, Control, Information and Education Engineering – Liu, Sung & Yao (eds)
 ISBN: 978-1-138-02800-5

Analysis of creative cooperation of stage lighting and sound

Jun Chen
Jiangxi Science and Technology Normal University, Nanchang, Jiangxi, P.R. China

ABSTRACT: Through a combination of theory and examples, this paper elaborates the usage of stage lighting and sound on the stage arts. On the stage arts, the cooperation of lighting and sound and the contrast of the plots make the portrayal of characters more vividly and the performance of stagecraft more fascinating.

KEYWORDS: Lighting and music control system; wireless RF; Stage Lighting; Sound; Drama.

1 INTRODUCTION

With a richer entertainment and cultural life of people, they pursuits more on stage arts. Lighting and stage, as two important elements of stagecraft, play an irreplaceable role in the theatrical arts performances. Some performances of stage arts need stage lighting; others need sound; but most of those performances need the mutual cooperation of stage lighting and sound.

2 THE EFFECTS OF STAGE LIGHTING AND SOUND APPLIED

2.1 *Performing time*

Performing time is the effect of the application of stage lighting. Stage lighting highlights the features of a particular era through rendering, contrasting, close-up by lighting colors.

2.2 *Performing environment*

In different environments, the performances of sound will not the same. Different audios give different light color scenes; on different stages we can use different features of sound to render atmospheres, so that scenes and roles on the stage will be more vivid.

2.3 *Depict the characters' inner feelings*

Shading, the collocations between various colors and different sounds have effects on the rhythm and atmosphere of the stage, which make the figures" emotional performance clearer and reveal the change of each character's inner emotions [2].

2.4 *Replace sets and props*

The arena of art often takes a lot of props and sets, but we can use reasonable lighting and sound to replace them. In stage art performances, lighting can cut the stage space clearly and separate the scene and space for each actor, as well as allocate the stage to the distant shot, medium shot and close shot in order to give audiences the full visual effect.

For example, in the use of colored lights, people often classify lights into cold light and warm light, while in general, cold light mainly means dark blue and warm light generally yellow or red. Two lights will bring different feelings, which contrast the performances of the roles and make theirs portray more deeply.

However, we can not deny the significance of stage lighting and sound on stage arts, but the usage of lighting and sound must be determined by the discussion among the director, makeup artist and scene layout staff. In addition, in the rehearsal process, each relatively person can not absent and each performer's gestures, eye contacts and facial expressions should match with lighting and sound, so that some unrealistic design is not appeared [3].

3 THE ANALYSIS OF STAGE LIGHTING AND SOUND APPLIED

3.1 *The application of stage lighting*

The standards of rationality, economy and advanced should be upheld in the use of stage lighting. 1, using the stage lighting scientifically and rationally; 2, allocating the stage lighting economically in order to achieve the best value for money ; 3, ensuring the stage lighting system advanced in their field and fully considering its upgrades and expansion in the future.

The following aspects should be paid attention to in the use of stage lighting: Firstly, the angle and strength of the light. In general, the lights not only have a certain artistic effect, but also for the front lights can be illuminated from the sides and top of the object; the light source in the horizontal direction and vertical direction should be controlled between 34.0 to 45.0 degrees [4].If the lights clarity permits, we can exposure the object from the frontage, thereby ensuring the performer's appearance does not change when the performer's position changes; secondly, LUX degree should be controlled to ensure that its degree moves smoothly and naturally in the hundreds between thousands; then, stage lighting applications need both the hardware (lighting, equipment) and the appropriate software (computer programming and memory) to meet the conditions of work lighting designer ; finally, the stage lighting should able to lighting , color, contrast and other functions, ensuring the effectiveness in practical applications.

3.2 *The application of sound*

The following aspects should be considered in stage sound applications: 1、microphone selection; stage performer now mostly use the microphone condenser microphone which can be supplemented with sound compensation, so the actor in the microphone selections should choose suitable microphone to achieve the best performance results; 2、stereo reverb; reverb switch and adjustment should determine according to different interpretations; for example, when the host speech, reverberation must be turned off, so as not to affect the host announcer; reverberation should be adjusted according to the performers and theirs songs' features when the performer playing popular songs; symphony performance will need to completely shut down the reverb, avoiding destroying the clarity of sound; 3,、coordinated sound. By coordinating sound, to make the music sound emitted pleasant euphemism, lively, beautiful melodies, so there is a sense of immersive and the stage is more expressive.

3.3 *The cooperation of stage lighting and sound applied*

If we want to ensure the integrity of the theatrical arts, so the cooperation of stage lighting and sound is inseparable. Especially in some special sound effects, stage lighting is more inseparable. The operation of lighting should be timely right; the mistake on the cooperation between the stage lighting and sound will directly affect the performance results and performance level and atmosphere will be greatly reduced. Therefore, the only way to meet the demands of the creative stage arts is to ensure the cooperation between the stage lighting and sound coordinately and effectively in the stage performances.

4 EXAMPLES OF THE USE OF STAGE LIGHTING AND SOUND EFFECTS ON STAGE ART

4.1 *Use of stage lighting and sound effects in the drama "Teahouse"*

Stage lighting and sound effects can be used to express the character's emotional changes, such as used in the drama "Teahouse" in the first act (see below).This is a morning at the end of the Qing Dynasty, smoke curl, the crashing sound rang between spoon and pan in the kitchen of the teahouse, three or five guests chatting, orange-red ray of morning light incident to the house through the windows, the bright color contrast of indoor and outdoor, Cooperate with each other to form a pair of busy scene which is booming across the screen light and sound processing.

Figure 1. Teahouse Act I.

4.2 *Use of stage lighting and sound effects in the drama "Thunderstorm"*

"Thunderstorm" as Cao Yu's works, was later repeated choreography, but no matter who the director is, who the actors are, what will never change is the use of lighting and sound in choreography. Sound effect is especially important in the drama "Thunderstorm ", you can say its role in the drama may be more important than the actor. The sound of wind and rain heighten the atmosphere and bring the audience into the scene throughout the entire drama. Lighting and the sound of muffled thunder came from air push the the drama atmosphere and rhythm to a new climax. Here are a few examples to illustrate specific application about stage lighting and sound in the drama "Thunderstorm ".

The location of Act III is Lu house (As Figure of Lu room). This is a night under a heavy rain, the dark clouds outside the window, lightning constantly flickering in the sky, sound of thunder appearing loudly frequently. A faint sound of noisy could be heard, coming from a drunk beating his wife. You can also hear a woman's voice coaxing the child to sleep. All these sounds are intertwined together forming the environment of night of LU house. When the curtain is rising , the lights on the stage not only show the shiny of chamber but also use "far flash" and "near flash" to represent lightning in the sky , showing the prevailing weather.

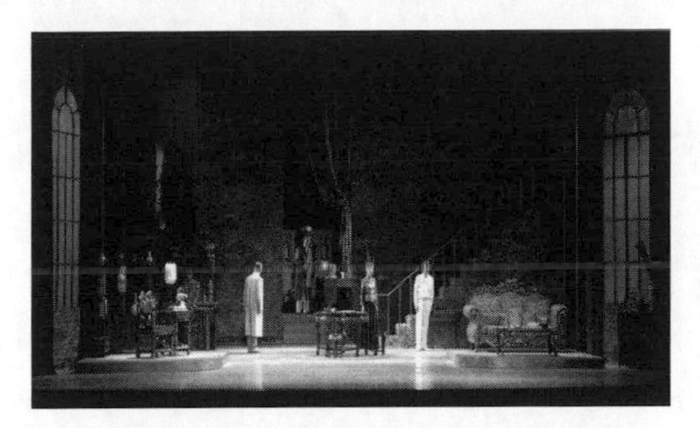

Figure 2. Figure of LU house indoor.

At the fragment of Lu Sifeng swearing, her mother was worried about her, so she made her swear not to see Zhou family any more.

Lu ShiPing: fall to your knees and say something now.

Lu SiFeng: Mom, I promise you (the sky flashes bright lightning) that I will never see Zhou family (thunder rolling).

Lu ShiPing: In the name of heaven thunder, if you forget my advice and then meet with Zhou's people?

Lu SiFeng: Mom, I do not, I will not.

Lu ShiPing: My child, you must say, you have to say. (Sky thunder and lightning) if you forget what I say...... (Sky thunder rolling).

Lu SiFeng: Oh!

Lu SiFeng: That ... (a light injection room) the thunder hack me. (Huge thunder is rolling over).

This is a very tense drama with high needs for cooperating of lighting and sound effects. Lightning controlled by Lighting, thunder controlled by acoustics must follow the character's changes in mood. Besides, the drama must manage the plot development very accurately.

5 CONCLUSION

Through the application of stage lighting and sound analysis, this paper carried out a detailed analysis on the cooperating of both. At the same time know the importance of mutual cooperation between the stage lighting and sound through the use of stagecraft instance, and learn the practical matters that need attention.

This paper is one of the research results of Jiangxi Arts and Sciences Project Cultural Industry Development Strategy of Poyang Lake Ecological Economic Zone (Issue number: YG2010041).

REFERENCES

[1] Guo XiuQuan. A simple analysis of the application of stage lighting [J]. Lighting technology.2010 (30).

[2] Liu Wei. A simple analysis of the stage sound [J]. Home theater.2010 (11).

[3] Liu JianJie. Cooperation of stage lighting and sound in the creation [J]. Entertainment equipment and technology. 2010(6).

[4] Huang RongDong. Meta-analysis on equipment resources [J]. Big stage.2010(5).

Computing, Control, Information and Education Engineering – Liu, Sung & Yao (eds)
© 2015 Taylor & Francis Group, London, ISBN: 978-1-138-02800-5

Preparation of nano hydroxyapatite and β-cyclodextrin composite microspheres used in release of theophylline

Yu Ping Guo
School of Pharmacy, Jiangxi Science and Technology Normal University, Nanchang, P.R. China

Liang Hu, Yu Qiong Chen, Xiao Feng Xing & Wei Min Gao
Jiangxi Key Laboratory of Surface Engineering, Jiangxi Science and Technology Normal University, Nanchang, P.R. China

ABSTRACT: Nano hydroxyapatite (HA) and β-cyclodextrin (β-CD) polymer composite microspheres and their theophylline inclusion carriers were prepared. The morphology of nano HA was tailored by Gly and theophylline release from the composite microspheres was evaluated in vitro. Nano HA and the composite microspheres were characterized by TEM. The average diameter of products was 350μm and the drug loading and drug encapsulation efficiency were 1.79% and 89.50% respectively. The drug release profile could be described by first-order release equation and Korsmeyer-Peppas equation. The composite microspheres of nano HA and β-CD polymer have a special absorption effect of drugs with special group, which might highlight the research for drug delivery system.

KEYWORDS: Nano hydroxyapatite; β-Cyclodextrin; Microsphere; Theophylline.

1 INTRODUCTION

Theophylline is a kind of methyl purine drug extracted from tea with the effect of smooth muscle relaxation, coronary artery dilatation and central nervous system excitement. It is used in clinical mainly as a bronchodilator [1]. The blood concentration of theophylline in vivo showed peak valley phenomenon obviously, which results in a series of adverse reactions. In order to avoid the side effect, the sustained drug release is often needed in clinical [2].

HA [Ca10(PO4)6(OH)2], the main component of bones and teeth, was widely used as a medical material because of its avirulence, biocompatibility and bioactivity [3]. Besides, it could be made into nanoparticle and used as the sustained drug release materials [4].

β-CD polymer microspheres have been developed in recent years as a new kind of drug carrier with the non-toxic character and biocompatibility [5]. Because β-CD polymer not only retains the characteristics of the molecular structure of β-CD itself, but also obtains the three-dimensional network structure, its microspheres in the performance not only keep the controlled release ability used in catalysis and recognition, but also have good chemical stability and mechanical strength polymer[6]. As one of the important functional materials, β-CD polymer microspheres can be effectively used as drug controlled release in modern pharmaceuticals.

In this paper, the nano HA and β-CD composite microspheres intends to be prepared as theophylline sustained-release carriers, in order to provide a new platform for the development of theophylline sustained release and other needs accurate control of drug release.

2 EXPERIMENTAL

HA Synthesis. HA was prepared using the method reported by Suzuki [7]. 25 mL 0.05 mol/L

Ca(NO3)2 solution(pH=11) was dropped into 25 ml 0.03 mol/L (NH4)3PO4 (pH=11) using a peristaltic pump with vigorous stirring. In both solutions, Gly could be added at a concentration of 0.1 mol/L. The reaction product was poured into an autoclave with Teflon inner liner and aged for 20h at 155oC. HA powders were recovered by high-speed centrifuge, washed with deionized water for several times and suspended in 50mL deionized water.

Blank Composite Microspheres Preparation. 10mL sodium solution with mass concentration 40%, 10mL HA suspension and 6g β-CD was completely dispersed in 30oC at constant speed stirring (800 r/min). 7.34g ECH was dripped slowly for cross-linking and emulsion (Span80-Tween203: 1) in kerosene 0.8g was added. Heating to the reaction temperature for 2h, the products were obtained after washing by sodium hydrogen carbonate solution, ethanol, distilled water and acetone respectively and vacuum drying.

Drug Loading in Coprecipitation Method. Theophylline was dissolved in a certain amount of water, then the blank composite microspheres of nano HA and β-CD polymer 1g were added. After remaining at 70oC for 2h, the system was cooled to room temperature in static state, and stored in the refrigerator for 24h. Theophylline composite microspheres were obtained after filtering, fully washing and vacuum drying to constant weight at 60oC.

Characterization. Transmission electron microscope (TEM, JEOL JEM 2100 at 200kV) was used to observe the morphology of composite microspheres and HA powers in and without the present of Gly. The drug loading and drug encapsulation efficiency were tested after dissolving theophylline composite microspheres by dilute hydrochloric acid and amylase. The drug release in vitro was evaluated in phosphate buffer solution.

2.1 *Results and discussion*

Fig.1 showed TEM images of HA in the present of Gly and without. Gly not only inhibited the growth of HA in nature's way, but also improved the nanoparticles dispersion obviously. All of these made the specific surface area increasing and HA suiting for drug carrier.

During the vivo biomineralizing process of HA, an amino acid is proven to be a contributable regulator besides the cellular modulation and the substrate function [8]. In this regard, an amino acid is an ideal candidate as a possible crystalline inhibitor of bioinorganic HA due to its ability to interact with HA surface. Compared to other compounds used in the literatures, amino acid has other advantages, mainly because amino acid is a kind of physiological substance and no toxicity. The remaining amino acid can be controlled by physiological mechanisms such as an introduction into the cell and further metabolism.

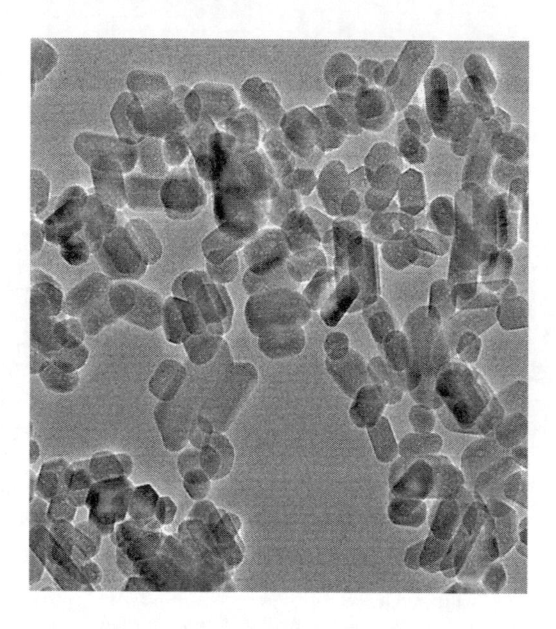

a

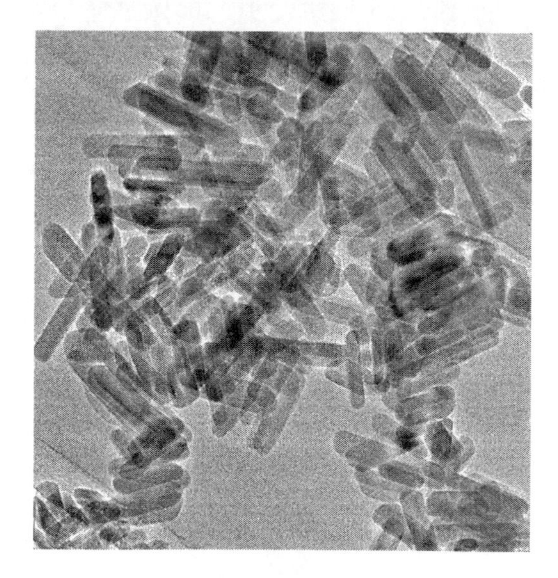

b

Figure 1. TEM images of HA.
a HA without the present of Gly
b HA in the present of Gly

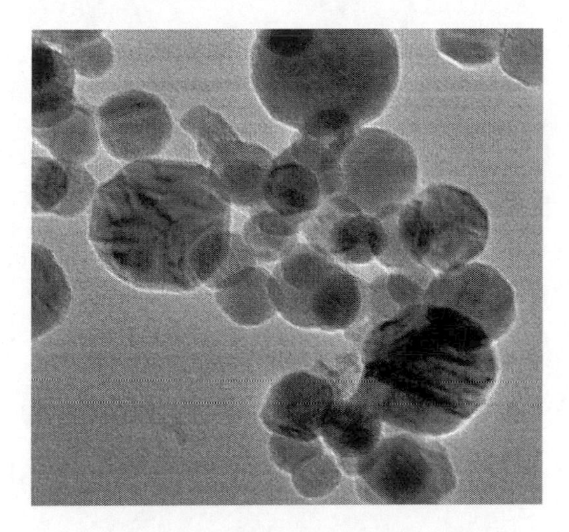

Figure 2. TEM image of composite microspheres.

Fig. 2 shows a TEM image of composite microspheres of nano HA and β-CD. The average diameter of composite microspheres was about 350μm.

Theophylline content in composite microspheres was measured by ultraviolet-visible light detector after dissolving microspheres by dilute hydrochloric acid and amylase. The drug loading and drug encapsulation efficiency were 1.79% and 89.50% respectively.

Theophylline concentration in phosphate buffer solution was measured by ultraviolet-visible light detector at 2, 4, 6, 8, 12, 18, 24, 30 and 42h respectively. The drug release curve was showed in Fig. 3. Theophylline sustained release slowly. The drug release profile could be described by first-order release equation and Korsmeyer-Peppas equation. It was showed that theophylline release from microspheres' network structure was simple diffusion primarily, supplemented by frame erosion.

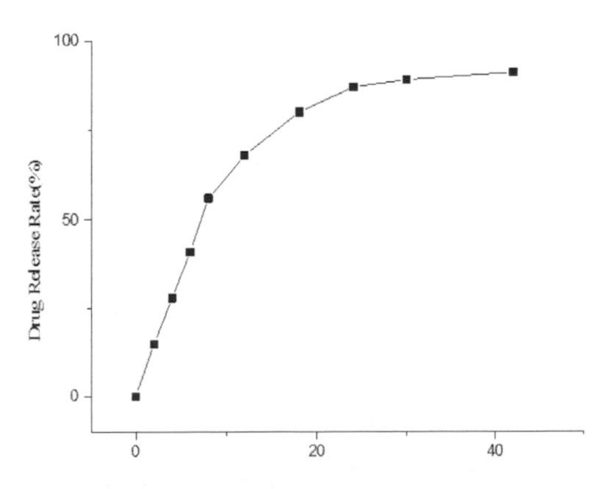

Figure 3. Theophylline release curve in vitro.

3 CONCLUSIONS

Nano HA and β-CD polymer composite microspheres and their theophylline inclusion carriers were prepared. The present of Gly improved the HA dispersion. The average diameter of microspheres was 350μm and the drug loading and drug encapsulation efficiency were 1.79% and 89.50% respectively. The drug release profile could be described by first-order release equation and Korsmeyer-Peppas equation.

ACKNOWLEDGEMENTS

It is a project supported by the National Natural Science Foundation of China (51162010), the Natural Science Foundation of Jiangxi Province (2009GZC0031), the S&T plan projects of Jiangxi Provincial Education Department (GJJ10590) and the science research projects of Jiangxi Science and Technology Normal University (KY2011ZY12). The corresponding author is Weimin Gao (Email: wmgao@163.com).

REFERENCES

[1] H. S. Samanta and S. K. Ray: Carbohydr. Polym. Vol. 106 (2014), p. 109

[2] M. Hirose, H. Yokoyama and K. Iinuma: Brain Dev. Vol. 26 (2004), p. 448

[3] I. S. Neira, Y. V. Kolenko, O. I. Lebedev, G. Van Tendeloo, H. S. Gupta, F. Guitián and M. Yoshimura: Cryst. Growth Des. Vol. 9 (2008), p. 466

[4] A. Jillavenkatesa and J. E. Kelly: J. Nanopart. Res. Vol. 4 (2002), p. 463

[5] A. Yotaro, K. Shigeru and Y. Fumiyoshi: Bio. Pharm. Bull. Vol. 28 (2005), p. 1679

[6] A. Ruebne: Macromol. Chem. Phys. Vol. 201 (2000), p. 1185

[7] S. Suzuki, M. Ohgaki, M. Ichiyanagi and M. Ozawa: J. Mater. Sci. Lett. Vol. 17 (2002), p. 381

[8] E. Dalas, P. V. Ioannou and P. G. Koutsoukos: Langmuir Vol. 6 (1990), p. 535

Computing, Control, Information and Education Engineering – Liu, Sung & Yao (eds)
© 2015 Taylor & Francis Group, London, ISBN: 978-1-138-02800-5

Comparative analysis on risk assessment methods of small and medium-sized enterprises

X.F. Lei & Z. Wang
Department of Computer Science and Technology, Jilin University Changchun, China

Y.T. Li
WHY-E Science and Technology Co., Ltd, Changchun, China

ABSTRACT: The development of small and medium-sized enterprise of Science and Technology (SMES) is restricted by the financial difficulties. As a result, constructing a targeted credit evaluation model to provide data support for the financing become an emerging research area. The key challenge of this problem is to establish a set of special models of credit rating with high precision for SMES. In this paper, several classical classification algorithms are adopted to set up different risk assessment models. The experimental comparison is illustrated using R (A famous programming language, which can achieve the best evaluation model more clearly). In addition, our experimental comparison can reduce the impact of asymmetric information between enterprises and financial institutions, and to break the bottleneck of financing difficulties for SMES fundamentally.

KEYWORDS: Financing difficulties, Credit evaluation model.

1 INTRODUCTION

The credit rating of research in China started late, and based on own limitations of SMES, such as less credit records, caused the financing difficulties which severely curb the development pace of independent innovation. The basic way to solve this problem is to establish a set of high accuracy and specifically evaluation model for SMES. At present, the domestic development has also been gradually, Originally, the neural network technology was introduced to the commercial bank credit risk assessment in [1], and then gradually the decision tree, genetic algorithm and support vector machine (SVM) method were also applied to risk assessment, now many scholars made many improvements on them, but nobody compare these methods .In this paper, we utilize four classical methods of machine learning, including Support Vector Machine (SVM), Decision Tree, Random Forest and Boosting, to model the risk assessment for SMES and make a comparison to select the most appropriate model. The contribution of the paper is listed as follow:

1 Many scholars introduced different methods of machine learning into credit evaluation, however, no person make a comparison to these ways and select the most appropriate method to model credit assessment.
2 Selecting the best model by this way has a win-win situation. On the one hand, evaluation institutions may according to this model to get a more accurate assessment, so as to minimize risk. On the other hand, enterprises can summarize its own weakness through the unified model, and in order to get more money, it will pay more attention to its credit standing, which promotes the development of the whole financial market to a certain degree.

The paper structure is as follows: In Section 2, we optimized the predecessor's indicators and selected some representative ones as our credit indicators of the model. In Section 3, several classical classification algorithms in machine learning are adopted to set up different risk assessment models, such as Decision Tree, SVM, etc. Analysis of the experimental comparison is shown by the R language in Section 4, which summarizes the characteristics of the various methods, and excavate the highest accuracy of risk assessment methods, to minimize the risk of a financial institution. And give a conclusion in Section 5.

2 THE CREDIT RATING INDEX SYSTEM OF SMES

Selection of the index must be representative, feasibility and comprehensiveness, especially for SMES. It must integrate into the enterprise's own characteristic, more dynamic and complicated. This paper adopted the evaluation method, taking the quantitative index as the main and a qualitative index as supplemented. This paper optimized the predecessor's summary and picked out four aspects as qualitative factors, a total of 12. Enterprise Quality aspect which includes management quality, staff quality, and management system three indexes; Technology of Enterprise aspect contains the advance of technology, life cycle of technology, and update rate of equipment; Historical Records includes two indexes, repayment willingness and fulfillment situation of contract, which are the effective indicators reflecting the credit; As well as the prospects for development, which is the proper index for SMES, including market prospects, industry prospects and supplying chain of finance[2]. At present all risk measurement research institutions are using expert evaluation method to quantify the qualitative indexes, this paper use the scoring criteria as follow: if the score is between 0.7 and 1, it means good performance, while between 0.4 and 0.7, means not bad. However, if the score is less than 0.4, it refer to terrible.

For quantitative indicators, combining previous summarizes, we select five aspects, a total of 19. Profit ability, which includes net interest rate of sales, return on equity and total return on assets; Solvent, including current ratio, quick ratio, asset-liability ratio, cash flowing ratio, and total assets; Management Ability, which contains total assets turnover, accounts receivable turnover, inventory turnover rate, and current assets turnover [3]; Development Capacity, including operating profit margin, net rate of profit growth, growth rate of net assets, and equity ratio, and we infused the Innovation Capacity which including R&D investment, researcher proportion, and intellectual property rights three indexes.

So in this paper, we pick out 31 indicators in total to set up the credit rating index system. In the next section we will establish a different credit assessment model using the unified index system.

3 THE METHODS TO MODEL RISK ASSESSMENT OF SMES

In the section, we select four methods to model the credit rating for SMES as follows:

- Support vector machine(SVM)

Support vector machine(SVM) was proposed by Vapink in 1974. The base principle of SVM classifier is mapping a linear inseparable space to another higher dimensional, linear and divisible space through a non-linear Transformation. And then establish a classifier, which has a pole Small VC dimension [4]. The classifier is determined by a few number of the vector in samples, and with the largest boundary width. The advantage of the algorithm (SVM) is simplify the calculation greatly.

- Decision Tree

The decision tree method is first developed in the 1960 s, ID3 algorithm is put forward by Quinlan by the end of the 70s[5]. It is a kind of typical classification method. The decision tree is essentially the process of classifying data through a series of rules. C4.5 algorithm is improved on the basis of the ID3 algorithm, which had greater improvement in the missing value in predicting variables processing, pruning technology, derivation rules and so on [6], it is suitable for classification problems, and is suitable for regression problems.

- Boosting

Boosting is a kind of method to improve accuracy of the weak learning algorithm. In 1990, Schapire proved the equivalence between the weak learning and strong learning by constructing a polynomial algorithm, which is the first Boosting algorithm [7]. The algorithm mainly obtains sample subset through the operation of the sample set, and to generate a series of base classifiers by using weak classification algorithms on training sample subset [8]. Boosting algorithm fused the n base classifier with weight, for good training results, give a low weight, opposite to the bad training result, which makes the poor training effect can get more attention in the next round of training, and so on to produce the final result of a classifier. In addition, it can avoid overfitting.

- Random Forests

Random Forests (RF) are a kind of statistical learning theory, it is the combination of a lot of decision tree classification models, and the parameter set is a random vector that is independent identically distributed. The basic idea of RF: first of all, use the bootstrap sampling to extract five samples from the original training set, and each sample size is the same as the original training set; Second, for k samples, a decision tree model is set up, respectively, and get k kinds of classification results; Finally, according to the classification results voting, to determine the final classification [9]. A lot of theoretical and empirical studies have proved that the RF has a higher prediction accuracy, and have a very good tolerance for outliers and noise.

The four the classic classification methods all can be introduced into evaluation, but different ways have own characteristics, we make a comparison using the R language in the next section and sum up the most suitable model.

4 EMPIRICAL ANALYSIS

1 Experiment platform and tools

This experiment platform is configured to:

Processor: Intel (R) Core (TM) 2 Duo CPU E7400 @2. 80 GHz to 2.79 GHz

Install the memory (RAM) : 2.00 GB

System type: a 64 - bit operating system

This experimental tool for the R programming language, version for 3.1.0.

2 Data acquisition

This study used data from Sina Finance Network 2014 Top 50, the financial data of 44 companies of the listed in 2010-2014 was selected, distribution is shown in Table I, which meets the requirements of the model.

Table 1. The Proportion of "Good" and "Bad".

	"Good"	"Bad"	Proportion
Sample set(753)	521	232	2.25:1
Train set(551)	354	197	1.8:1
Test set (202	144	58	2.48:1

3 The establishment of the model

We adopted four classic classification methods in machine learning, including Decision tree, SVM, Random forest, Boosting in this experiment, as follow:

- Modeling by SVM Algorithm

This experiment used R package "e1071" and the function"svm"in the training data. Function: svm(x, y = NULL, scale = TRUE, type = NULL, kernel="radial",degree=3, gamma = if (is.vector(x)) 1 else 1 / ncol(x),coef0 = 0, cost = 1), details are shown in Table II.

Table 2. The statistics of kernel function.

Kernel function	Number of SVM	Related Parameters	Accuracy
Linear	167	Gamma: 0.052 63158	0.8354229
Sigmoid	182	Gamma:0.05263158 Coef0: 0	0.8751219
Radial	92	Gamma: 0.05263158	0.8925373
Polynomial	339	Gamma0.05263158 Degree: 3, Coef0: 0	0.8656716

From the table, we know the choice of kernel function is crucial. For different kernel functions, the accuracy is different. Radial's accuracy can reach 89.25%, while Linear is 83.54%. But for the Radial, there is only one parameter "Gamma" that can influence the accuracy, so we can optimize the result by choosing suitable "Gamma".

- Modeling by Random Forest Algorithm

The experiment used R package "randomForest" and the function "randomForest". Function model: randomForest (x, y=NULL, xtest=NULL, ytest=NULL, ntree=500), the accuracy varies with the parameter "ntree", which means the trees to grow in the model. Details are shown in TABLE III.

Table 3. Impact analysis of ntree.

Ntree	20	50	100	500
Accuracy	0.8756	0.89054	0.89552	0.89552

From the table, the accuracy goes up with the parameter "ntree" increased, but the result is same when ntree is 100 or 500, considering the complexity,we choose ntree=100. Details are shown in Fig. 1.

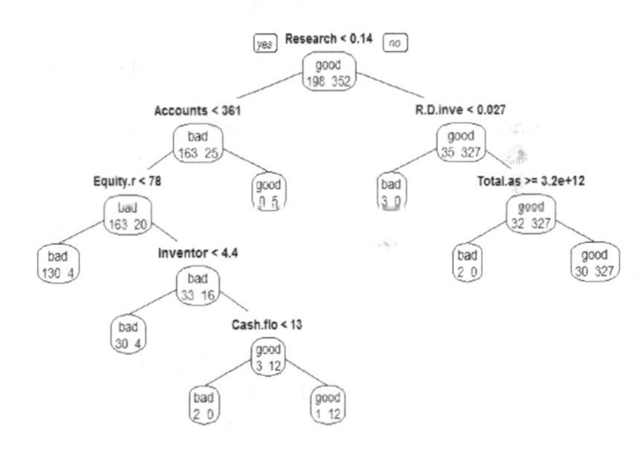

Figure 1. The Result of DT (ntree=100).

- Modeling by Decision Tree Algorithm

This experiment used function "rpart" in R package "rpart" and return to a model to predict.

Function model: rpart(formula, data, weights, subset, na.action = na. rpart , method, model = FALSE, x = FALSE, y = TRUE, cost, ...). Details are shown in Table IV(In the table, cp is complexity, nsplit means partial node, xerror is variance, xstd means standard deviation.)

Table 4. The analysis of rpart.

CP	nsplit	Rel error	xerror	xstd	accrucy
0.0254	1	0.2994924	0.3045685	0.3711300	0.79342
0.0152	4	0.2284264	0.3045685	0.3711300	0.85467
0.0100	7	0.1928934	0.3197970	0.3791292	0.87055

- Modeling by Boosting Algorithm

This experiment used function "boosting" in R package"adabag". Function model: boosting (formula, data, boos = TRUE, mfinal = 100, coeflearn ='Breiman', control), where mfinal refer to the number of iterations. Accuracy and the importance of the evaluation index ranking were changed with mfinal as Table V (the importance of the top 5 only were listed in the table).

Table 5. The analysis of boosting model (X%).

mfinal	20	50	70	100
NO.1	Proportion of researcher 18.3	Proportion of researchers 18.3	Proportion of researchers 18.3	Proportion of research ers18.3
NO.2	Net rate Of profit growth 8.7	Growth rate of total assets 7.5	Net rate Of profit growth 9.5	Accounts receivable turnover 7.9
NO.3	Turnover rate of inventory 6.7	Proportion of R&D 7.1	Accounts receivable turnover 7.3	Accounts receivable turnover 7.3
NO.4	Growth rate of total assets 6.5	Turnover rate of inventory 6.9	Turnover rate of inventory 6.9	Cash Ratio 6.25
NO.5	Cash Ratio 6.3	Cash Ratio 6.6	Cash Ratio 6.62	Proportion of R&D 5.4
Error	12.9353	11.9403	10.4478	8.45771

The result, namely the proportion of innovation and R&D (Research and Development) are most important node, indicate that the credit rating index system of SMES was optimized by infusing the element of innovation,which is the characteristic of the SMES.We also can see when the precision floats with mfinal. We choose mfinal = 100 as the default, whose error is 0.0845771, that is the accuracy is 91.6%, and won't appear over fitting phenomenon, which is also an outstanding characteristic of Boosting method. Details are shown in Fig. 2.

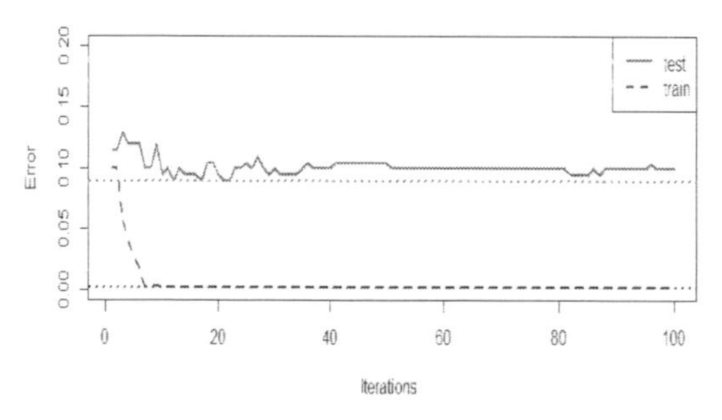

Figure 2. Final influence of error (accuracy).

This experiment compared the four methods (DT, SVM, RF, Boosting) using the unified data set of precision, and found that although the accuracy of these methods were almost on the same level, while Boosting is higher than the other three methods, as high as 91.65%. Details as shown in Fig. 3

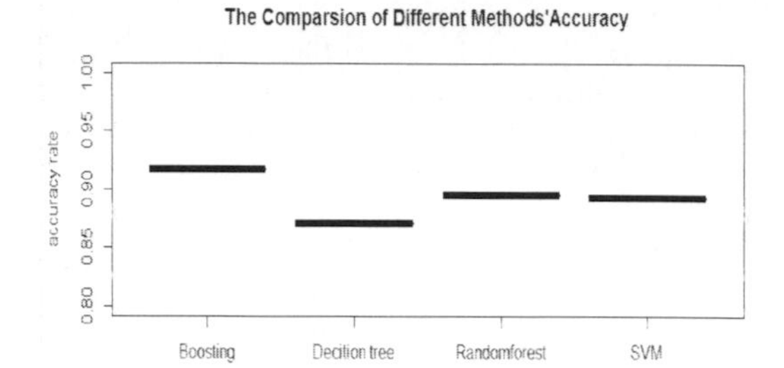

Figure 3. The comparison of four methods

5 CONCLUSION

According to the experimental analysis, we can get the following conclusion:

1 We summarized and optimized the indexes of ancestors credit rating index system of SMES, put the element of innovation into the index, and outstanding the characteristic of SMES, which make the evaluation more targeted.
2 The accuracy of different models is changed with the metabolic parameters in the model, so the accuracy of the method can be improved by the optimization of parameters.
3 We choose four kinds of classic methods, but from the point of accuracy, Boosting is the highest, and can avoid over fitting phenomenon, which is the characteristic of it.

So Boosting is better to be selected for modeling risk assessment for SMES. The next work is to model risk assessment for SMES based on Boosting, and make a further improvement on it.

ACKNOWLEDGMENT

This work was financially supported by the National Science and Technology Support Projects - Jilin Financial Institutions Application Demonstration Project (Project No.2013BAH07F05).

REFERENCES

[1] SuLin Pang, Rongzhou Li, Jianmin Xu. The application of BP algorithm On credit risk analysis [J]. Control theory and application. 2005, (1) : P139–144.

[2] HuoHaiTao. Credit Risk Index System and Evaluation Method of High-tech Small and Medium-sized Enterprise[J]. Journal of Beijing institute of technology (social science edition), 2012 (01) : P60–65.

[3] Limei Wang.Credit Evaluation Model of Small and Medium-sized Enterprise Bbased on Ensemble Learning [D].Changchun,Jilin University, 2014 .

[4] Wei Chen,Yeqiu Wang.The Optimizaton Research of the Small Business Credit Rating Based on Support Vector Machine (SVM) method[J]. Journal of Yunnan Finance & Economics University,Vo1.26, NO.6.

[5] Yuanyuan Li.The Decision Tree Algorithm and the Application on Credit Risk Control[D].Shandong University, 2013.

[6] Sulin Pang,Jizhang Gong.C5.0 Classification Algorithm and the Application on Personal Credit Rating of Bank [J].Systems Engineering — Theory & P ractice,2009,Vo l.29 , NO. 12.

[7] Xuehua Shen, zhihua zhou, Jianxin Wu and zhao-qian Chen.Boosting and Bagging review[J].Computer Engineering and Application. 2000, (12): P31–33.

[8] Xia Li, Liyun He, Chao Liu, Boosting algorithm and its application in traditional Chinese medicine and health data classification [J]. China health statistics. 2008, (02) : 158–162.

[9] Kuangnan Fang, Jianbin Wu,Jianping Zhu.The Reviewe of Random Forests Method[J].Statistics&Information Forum,2011,Vol.26,NO.3.

Computing, Control, Information and Education Engineering – Liu, Sung & Yao (eds)
© 2015 Taylor & Francis Group, London, ISBN: 978-1-138-02800-5

Design of remote monitoring system of embedded smart home based on ARM and ZigBee

Z. Gao
Hubei Communications Technical College, Wuhan, China

J.F. Hu
School of Mechanical Engineering, Hubei University of Technology, Wuhan, China

ABSTRACT: With the development of the Internet of things technology and embedded technology, smart Home Furnishing ushered in new opportunities. Aiming at the smart home system wiring trouble, high maintenance costs, poor mobility shortcomings, and this paper presents the design and implementation of smart home remote monitoring system based on ARM chips and ZigBee technology. Design is based on ZigBee technology to build a home network, using S3C6410 as the hardware platform, and porting with the Linux operating system to design a home gateway, using the GSM digital mobile communication system as a family outside the network. The experimental results show that the system has the advantages of simple wiring and good mobility, and can realize the local and remote control of smart home system.

KEYWORDS: Smart home; ARM; ZigBee; S3C6410; remote monitoring.

1 INTRODUCTION

Along with rapid development of information technology, sensor technology and network communication technique, and growth in the living standard, people request increasingly higher of the quality of life and attach greater importance to home life that is closely connected related to daily life. Thus, such concepts as smart home, digital family and building atomization emerge at the right moment. Smart home will provide people with great convenience and rapidness in work and life and make home life comfortable, safe, humane and intelligent [1]. It means to organically integrate such subsystems related to home life through advanced computer network communication technique, embedded technique and sensor technology and facilitate home life to be more comfortable and safer through planning management.

Smart home system is generally divided into home intranet, home gateway and outer network. Presently, home intranet is mostly designed in the form of wired connection based on bus and power line with the disadvantages of complex wiring, corrosive wire, inconvenient maintenance and unaesthetic interior [2-4]. Additionally, home gateway is designed to be controlled by the singlechip; so it is difficult to upgrade, maintain and debug the system due to limited software and hardware resources.

ZigBee is a kind of new wireless network technique with close range, low complexity, low power consumption, low data rate and low cost which has extensive application prospects in a great many fields. S3C6410 is an ARM11 processor produced by the Korea Samsung Electronics based on ARM1176JZF-S core; it adopts 32-bit RISC processor and the master frequency reaches up to 667MHz. Besides, it has a lot of peripheral interfaces such as 4-channel UART, 32-channel DMA and I2C bus port.

In view of features of ZigBee technique and S3C6410, the author builds up a home intranet based on ZigBee technique, uses S3C6410 as a hardware platform and GSM digital mobile communication system as the outer network designs a home gateway by Linux operating system and realizes local and remote control of the smart home system finally.

2 MAIN FUNCTIONS AND OVERALL STRUCTURAL DESIGN OF THE SYSTEM

The smart home remote monitoring system mainly consists of a smart home controller and several domestic ZigBee monitoring modules as shown in Fig.1, which can realize information interconnection and interflow through GSM network and ZigBee personal area network[5]. Users can regulate and control home appliances and lighting devices by the mobile phone and receive environmental information such as temperature and luminance and alarm information such as burglary prevention, fire prevention, gas leakage prevention out of doors. The smart home controller can realize intelligent control of home appliances and lighting devices, according to environmental information collected by the sensor.

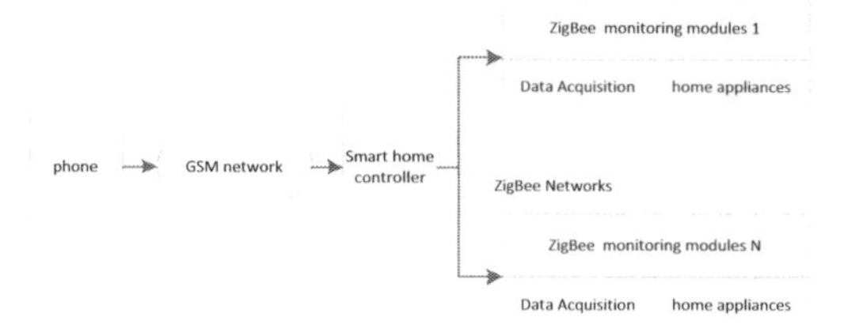

Figure 1. System framework.

3 SYSTEM HARDWARE DESIGN

3.1 *Smart home monitoring module and hardware design*

The smart home monitoring module is used to acquire detection information and send to the smart home controller, to receive control commands from the smart home controller and realize intelligent control of home appliances. It mainly consists of master module, environmental detection module, security detection module, home appliance control module and power module and its hardware framework is shown in fig.2.

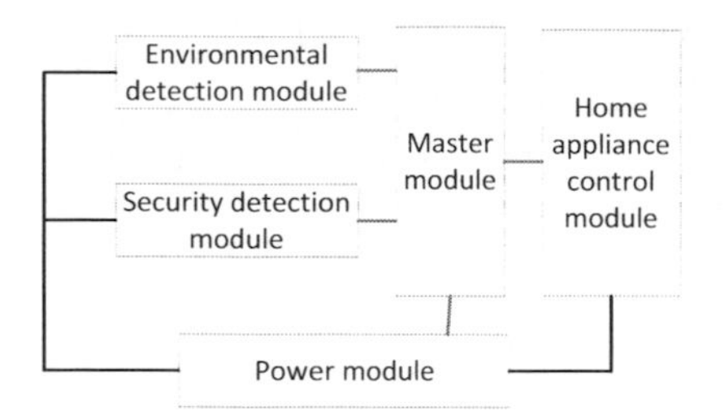

Figure 2. Smart home monitoring module hardware framework.

CC2530 is the core of the master module and a true solution for the system on chip (SoC) used by 2.4-GHz IEEE 802.15.4, ZigBee and RF4CE application which can build up a powerful network node by an extremely low material cost. Moreover, CC2530 combines superior properties of leading RF transceiver, standard enhanced 8051 CPU, intrasystem programmable flash memory, 8-KB RAM and many other powerful functions. Its hardware supports carrier sensing of multi-access/ conflict detection (CSMA/CA) and the working voltage of 2.0-3.6 V makes for realizing low consumption of the system. A wireless star like ZigBee network is built up indoors through connecting the ZigBee coordinator module of the smart home controller to realize wireless ZigBee network control of indoor security and home appliances.

Environmental detection module consists of a temperature detection module and humidity detection module. The temperature and humidity sensor used in this paper is DHT11 which carries out communication with the processor in the form of a single bus. Communication data are divided into two parts: integer and decimals. This paper takes a complete 40bit data transmission as an example to illustrate the data transmission form of DHT11 as below: sent an 8bit integer of humidity and 8bit decimals, then an 8bit integer and 8bit decimals of temperature, and an 8bit checksum. In case of correct data transmission, the checksum equals to the last 8 figures of "humidity data+ temperature data". In case the system is running, the master control will issue an original signal, DHT11 will switch from low power consumption mode into working mode after detecting information and wait for the host engine to transmit information. Then, DHT11 will issue a response signal and transmit 40bit temperature and humidity data after receiving the information from the control site; and the user can know about the indoor temperature and humidity timely through the human-computer interaction interface. Meanwhile, DHT11 will determine whether to acquire data of temperature and humidity according to the information received. Data of temperature and humidity acquired by the terminal node through DHT11 will be instantly transmitted to the coordinator through ZigBee network, and then to the master control through the serial port. Finally, the master control displays current temperature and humidity in the form of local LCD and serial port printing so as to realize intelligent control of home appliances.

The security detection module is used to detect indoor combustible gas leakage and rubber invasion and consists of combustible gas detection module and pyroelectric infrared detection modules. The combustible gas detection module is composed of MQ-2 gas sensor, buzzer and GPRS module. This paper adopts MQ-2 gas sensor to acquire the concentration of sensitive gases. In case a certain concentration of such combustible gases as liquefied gas and oxygen is detected, the MQ-2 gas sensor triggers the buzzer to alarm and transmit the alarm to the user's mobile phone through GPRS module. In case combustible gases are detected, the conductivity of MQ-2 gas sensor will rise with the increase of concentration and pass through A/D (analog-digital conversion) conversion channel; namely, to convert changes of the conductivity into the output level signal corresponding to the gas concentration. Then,

CC2530 connected through I/O port can detect the level change and transmit the acquired information to the coordinator through ZigBee network. And ARM control analyzes the acquired data, judges fire risks and determines whether to transmit the alarm information.

Home appliance control module realizes control of equipment based on equipment functions. An electric fan is taken as an example in this paper. The smart home controller transmits the control command to the corresponding ZigBee monitoring node through the ZigBee network to control the electric fan. While home appliances have different identification codes; for example, the identification code of the electric fan specified by the system is 088 and that of the lighting switch is 089, so the smart home controller is able to identify different home appliances. With respect to the same command code, different home appliances execute different functions.

3.2 *Smart home controller hardware design*

The smart home controller is the core of the smart home system and its core component is the embedded processor that consists of ZigBee processor, GPRS module, touch screen and warning device. The hardware structure of the smart home controller is shown in fig. 3.

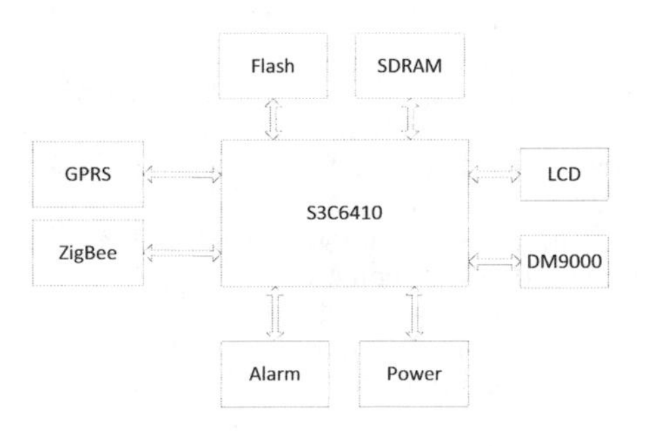

Figure 3. Hardware structure chart of smart home controller.

The smart home controller is the control center of the smart home remote monitoring system and its core hardware is the central processing unit (CPU) that is specific to all equipment control, task scheduling, communication protocol conversion, data acquisition and transmission, data management. In consideration of function, expandability, operating system and power consumption, this paper selects the powerful 32-bit ARM11 microprocessor S3C6410 as the main processor.

In view of man-machine interface, the touch screen is used to display status information of home appliances and information forwarded by the user's mobile phone so that the user can input the command and control home appliances conveniently and easily.

The controller contains 2 communication modules: GPRS module and ZigBee coordinator module. GPRS module consists of host module, SIM card interface, voice frequency and radio frequency and realize data transmission with ARM through SIM300 chip (GPRS module) produced by Simcom Company; besides, it adopts standard RS-232 interface, receives and sends information through operating the module by AT command. ZigBee coordinator works at building up ZigBee network and adding all monitoring modules into the network. After ZigBee network is built up, it acquires environmental detection information and warning information from all monitoring modules and carries out data exchange with ARM11 processor through RS-232 interface.

4 SYSTEM SOFTWARE DESIGN

System software contains transplantation of Linux operating system, edition of serial driver, program design of ZigBee coordinator and ZigBee terminal node.

4.1 *Porting linux operating system*

The system selects and uses 2.6.39 edition of Linux kernel. Download the kernel source from the official website ftp://ftp.kernel.org/pub/linux/kernlel/ first, modify document Makefile in the root directory prior to compiling and set up the target platform ARCH=arm and the cross compiler CROSS_COMPILE=arm-linux-, then, operate command make menuconfig to configure necessary options in the menu, save and exit; execute the kernel compile command make uImage and the mirror image document uImage guided by u-boot is generated; finally, program the kernel compile command into NAND Flash by u-boot [6].

4.2 *Serial driver program*

Serial deriver program is designed by the third party of Qt concerned with posix_qextserialport.cpp, posix_qextserialport.h, qextserialbase.cpp and qextserialbase.h. This system adopts the serial category defined in these documents, establishes the serial object in the original function and realizes serial port communication. Then, define serial object, appoint the serial port name and the query mode and set up Baud rate, data bits, odd-even check, stop bit, data flow control and delay, and so on.

4.3 *ZigBee coordinator program*

ZigBee coordinator of the system is mainly used to build up the wireless network, allocate the website address of ZigBee monitoring node, transmit control commands to the monitoring node, receive environmental information and security information from the monitoring node, upload the received data to ARM11 microprocessor S3C6410 and forward to the remote user side in form of short messages through mobile network. The coordinator initializes the application layer first, then I/O port and the opened global interrupt. Then, the coordinator initializes an information channel and establishes a ZigBee network; ZigBee monitoring node is identified and added into the network. The flow chart of ZigBee coordinator program is shown in fig.4.

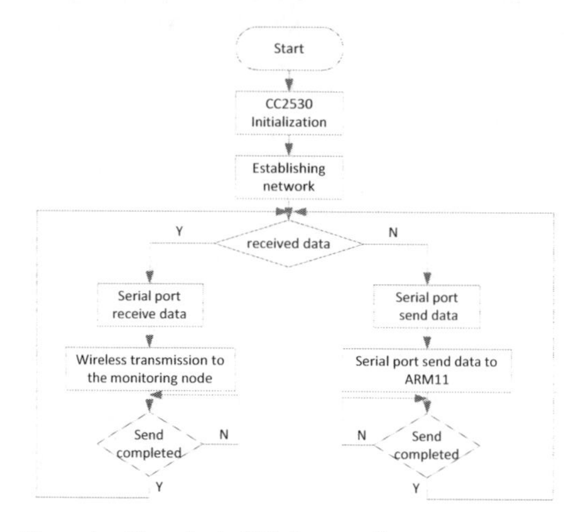

Figure 4. Flow chart of ZigBee coordinator program.

ZigBee terminal node refers to wireless ZigBee node controlled by ZigBee coordinator which means monitoring node in the system. Initialization of ZigBee terminal node also contains initialization of the application layer, initialization of I/O and initialization of opened global interrupt. Then, it tries to join in ZigBee network. It needs to be stressed that only the terminal node consistent with the ZigBee coordinator can join in the network (such as setting of the information channel). After joining in the network, it forwards log-on message to ZigBee coordinator and ZigBee coordinator fulfills the registration.

ZigBee terminal node acquires environmental information at regular intervals and uploads the data to ZigBee coordinator. In case security detection module detects combustible gas leakage or rubber invasion, ZigBee terminal node enters into interrupt processing immediately and uploads warning information to the smart home controller so as to inform the user and start the warning device. Receiving control commands from the smart home controller, it controls home appliances directly during the interrupt processing of wireless data receiving. The flow chart of ZigBee terminal node is shown in fig.5.

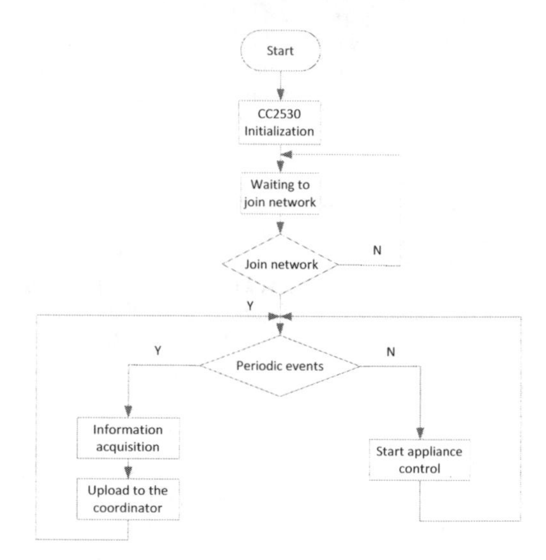

Figure 5. Flow chart of ZigBee terminal node.

5 EXPERIMENTAL RESULT

This paper carries out comprehensive test and debugging in the laboratory after system software and hardware are designed. The system is a starlike network topology consisted of a smart home controller and three ZigBee monitoring modules. The distance between the smart home controller and the ZigBee monitoring module is 10m and that between adjacent ZigBee monitoring modules is 15m. It is tested that the system runs stably and acquires information timely and accurately, actuators respond rapidly, which meets the requirements of family information networking and realizes the advantages of simple wiring and superior mobility.

6 CONCLUSION

The smart home remote monitoring system designed in this paper adopts ZigBee technology to build up a family wireless network, designs a home gateway through transplanting the embedded Linux operating system on the hardware platform with S3C6410 as the control cure, and realizes remote control and warning through connecting GPRS module. Its powerful function, low power consumption and simple operation realize intelligent management and remote monitoring of household facilities and have vast application prospects.

ACKNOWLEDGMENT

In this paper, the research was sponsored by the Department of Transportation of Hubei Province.

REFERENCES

[1] Ricquebourg Vincent,Menqa David etc. The smart home concept: Our immediate future[C], 2006 1st IEEE International Conference on E-Learning in Industrial Electronics,2006:23–28.

[2] MA Yanzheng, WANG Meng, XU Xiaohui, SONG Tao,WANG Zhixue. Design of Home Intelligent Control Network Based on HBS Protocol [J] .Modern Electronic Technique,2009,32(13):189–192.

[3] Xu Feng, Liu Xin, Fang Jia-bao. Design of Smart Home Remote Control System[J]. Low Voltage Apparatus, 2009 (4): 21–24.

[4] Peng Xiaojun, Li Rong. Research on Embedded Smart Home Control System Based on ARM [J]. Low Voltage Apparatus, 2009 (18): 42–45.

[5] Hoffman Christian, Weigand Christian,Bernhard. WirelessmedicalsensornetworkwithZigBee[J],WSEAS Transaction Communications,2005,5(10):1991–1994. Hoffman Christian, Weigand Christian, Bernhard. Wireless medical sensor network with ZigBee [J], WSEAS Transaction Communications, 2005,5 (10): 1991–1994.

[6] Lan Ping Wei, Wang Tao. Building embedded Linux application development environment [J] Net Security Technologies and Application, 2008 (2): 57–58 .

Computing, Control, Information and Education Engineering – Liu, Sung & Yao (eds)
© 2015 Taylor & Francis Group, London, ISBN: 978-1-138-02800-5

Architecture and implementation of new medical information system

X. Jia & M.J. Chang
Department of Electronic Information Engineering, Handan Polytechnic College, China

ABSTRACT: This paper mainly introduces the premise of environmental, technical means of development and design of system description, the core functions of the system and the system operation maintenance plan implementation strategics for the implementation of new medical information system, method at the same time on the system architecture, development and design are used in detail etc.

KEYWORDS: PowerBuilder, Sybase, The new type of medical information system, C/S.

1 INTRODUCTION

A full range of solutions to new medical information system for hospital information management, the use of hospital existing computer network equipment and database technology, the hospital each department, auxiliary acquisition mechanism of hospital pharmacy drug information, hospital pharmacy store all department information, hospital medical students information, diagnosis and treatment of tests the project information and all other details of hospitalized patients with charge entry the basic data, using the network environment, realize the hospital information and data in the model of medical information system platform for paperless and humanized management.

Hospital installation and the operation model of medical information system, greatly improve the management level of the hospital all kinds of information, not only reduce the cost, improve the quality of medical service, reduce man-made errors, but also bring obvious economic and social benefits to the hospital.

The system adopts the software architecture of traditional client / server (C/S) mode, with strong database front-end application development tools PowerBuilder 11 platform, back-end database use Sybase the United States large relational database management system as the server database architecture development environment.

2 THE FUNCTION OF SYSTEM OVERVIEW

Based on the main business logic model of medical information system of C/S mode are: large two outpatient management and hospital management. The core function module of new medical information systems include: medical information basic data maintenance (hospital, doctors, medicine, treatment, laboratory examination, etc.), dynamic user author ization, outpatient registration fee management, in hospital management, management, management of inpatient pharmacy of outpatient pharmacy, pharmacy store management, Dean ward nursing station management, query management, financial management, system maintenance management module.

3 TECHNICAL POINTS

C/S (Client/Server) architecture is the traditional application development framework for the response and request of the business logic on the basis of the network environment by using the client and server to interact in a way. The mode of C/S environment to run the client application needs to bear great pressure because system based on all of the business logic and transaction processing request, display all need implemented on the client, and the server only needs to accept requests to respond to system based on the corresponding rules, so bear the load of the server is light.

Sybase database is a relational database management system currently popular, application is very wide, such as finance and insurance, industrial and mining enterprises and enterprises manufacturing, health care, government agencies, securities investment, catering services and other industries, in view of the Feixiang County Chinese Medicine Hospital Department of orthopedics new medical information system development environment is PowerBuilder, in order to realize seamless link application the development of the system and the backstage database, so the hospital medical information system using Sybase's

Sybase 12.5.1, using special interface in connection with the application procedure and the database on the butt joint, realize the real meaning of the seamless connection.

The life cycle method, also known as a structured system development method, is one of the information system development and design method of the current more popular at home and abroad, is widely used and concerned with the application system development process, especially in the application system of the business logic in the perplexing, highlighting the advantages of this method. So far, this method is still the most common method of application development in a mature.

Object oriented method is a kind of object oriented concept is applied to the software development process, method to guide system development activities, referred to as OO (Object-Oriented) method. Attribute oriented object method including are: inheritance, encapsulation, polymorphism and encapsulation. The package refers to the object oriented method, the object is the set of data and allow for the integration, corresponds directly to the objective entity, defines an object class can contain a set of objects with similar properties. Each inheritance is a style attribute and operation of hierarchical relationships of sharing the. Polymorphism refers to generating the parameter effects similar to different results of different methods. While the coverage refers to the superclass and subclass generated according to different situation results will directly cover by the original results.

By using the method of design and development of a new medical information system in this paper is the comprehensive application of the life cycle method and object oriented method for the whole.

4 SYSTEM ANALYSIS AND DESIGN

Organization of Handan city Feixiang County Hospital Department of orthopedics has two main departments, one is the outpatient department and inpatient department is two, all the daily business work in hospital is all around the two major departments, the hospital organization such as: Outpatient registration、The outpatient service charge、Outpatient Dispensary、Surgery、Outpatient department、Financial affairs、Equipment department、Drug Storage、Inpatient department、Out/in-patient Department、Hospital pharmacy、Medical Center、Logistics Office.

Comprehensive analysis according to the specific requirements, hospital departments and agencies business logic that the main function of the hospital information system includes: basic data management and maintenance fees, outpatient registration, outpatient pharmacy, in hospital management, hospital pharmacy, ward nursing station, financial supervision, Dean integrated query, pharmacy store management, system management, user dynamic authorization the core function.

4.1 *Outpatient registration fees*

Outpatient service registration fee is mainly responsible for outpatient treatment of patients registered fee. According to the prescription of documents doctors were provided to patients, provided by the use of outpatient charging subsystem prescriptions for related information entry, fees, outpatient invoice printing work .

4.2 *Outpatient pharmacy*

Mainly of outpatient pharmacy management method and is responsible for outpatient drugs. Patient related expenses paid in the outpatient cashier department, the system will print related clinic invoice for patients, patients after they can rely on the invoice to receive the corresponding drugs in outpatient pharmacy.

4.3 *In hospital management*

Access hospital management mainly is responsible for the registration, admission of patients with hospital deposit payment, hospital billing, and clearing invoice printing work.

4.4 *Nursing station*

In patients hospitalized for the relevant procedures, can go to the corresponding departments of hospital treatment, hospitalization patients all the inspection, diagnosis, drug etc..

According to the admission and discharge management of nursing station and functional description, design of admission and discharge management data flow diagram as shown in Fig.1.

4.5 *Financial supervision*

Financial supervision is mainly responsible for the total invoice and settlement of the hospitalization invoice management; work in hospital management department and outpatient cashier department expense report and check.

4.6 *Dean integrated query*

The Dean query: the core functions of query the dean is responsible for all the subsystems in the whole medical information system query, browse, in order to better grasp the medical information system operations in the hospital

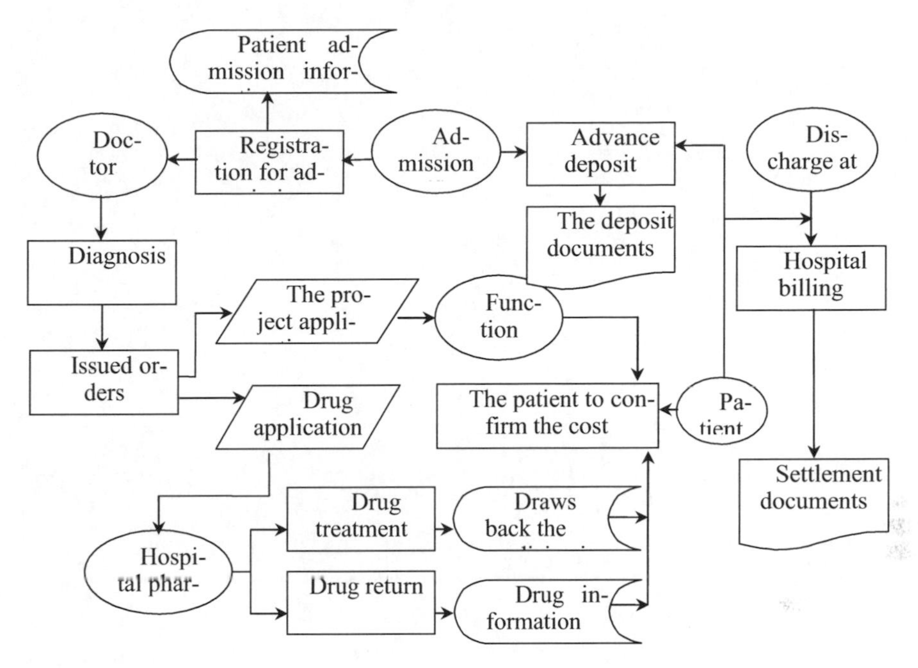

Figure 2. The discharge data flow chart.

4.7 *The management of pharmacy store pharmacy store management*

The main object is the outpatient pharmacy and hospital pharmacy. Including the pharmacy claim, payment and check, and a variety of statistical statements of the printing.

4.8 *Dynamic user authorization*

The highest level is the only system user login system, users can function module of system dynamics for existing systems for dynamic authorization, user login system authorized after the operating system only administrator authorization function.

4.9 *Systems management*

Responsible for all the function modules, system management in hospital information system user management, the cost of design and so on in the whole medical information system all management and setting in the module, so the subsystem is the medical information system command.

5 CONCLUSION

In this paper, the development of hospital information system design process to make the detailed introduction and description according to the design principle and method of life cycle of software engineering method and object oriented will. The new medical information system after the official operation of Feixiang County Chinese Medicine Hospital Department of orthopedics, system stability, good results. The practice has proved that the application of the system in advancing the construction of hospital information at the same time, but also creates a good economic and social benefits for the hospital, improve the medical environment and the quality of services for the majority of patients.

REFERENCES

[1] Xu Renzuo. [M]. software reliability engineering. Beijing: Tsinghua University press (2007).
[2] Li Dong. The management information system theory and its application in [M]. Peking University press, 06(2004).
[3] Chen Ming, Yang Jinsong.PowerBuilder programming database connection method of [J](2006) May micro computing applications.
[4] Bao Yonggang, Wang Degao. The core technology of PowerBuilder8.0 and [M]. Examples of the development of the electronic industry press.06 (2002).

Computing, Control, Information and Education Engineering – Liu, Sung & Yao (eds)
© 2015 Taylor & Francis Group, London, ISBN: 978-1-138-02800-5

Parallel differential reduction algorithm for multi-domain modelling

Hong Jun Wang
School of Computer Science and Technology, Chongqing University of Posts and Telecommunications, Chongqing, China

Wen Yuan Wu
Chongqing Key Laboratory of Automated Reasoning and Cognition, CIGIT, CAS, Chongqing, China

ABSTRACT: Reduction of large scale differential algebraic systems, which are automatically generated by computer from unified multi-domain models, is the key issue in simulation and optimization. From the previous work by others, we found there are many parallel coupling blocks in the whole symbol matrix of a system. This article presents a parallel algorithm using this structure and the implementation based on multi-thread pool. The analyses of time-efficiency and speed-up are given. The speedup from our experiments is consistent with the theoretical result.

KEYWORDS: Differential Reduction, parallel computing, Kuhn-Munkres, structural analysis.

1 INTRODUCTION

With the fast development of design of integrated circuit, large-scale control systems and other various applications, the mathematical models in the form of differential-algebraic equations (DAEs) become more complicated. People use computers to generate such models automatically by unified multi-domain langrage e.g. Modelica [11]. To solve these large DAE models, one of the key steps is the differential reduction of the DAE to ODE which can be handled by traditional numerical approaches. The most efficient reduction technology is "the structural analysis method" proposed by Pryce [1]. It is a fixed-point iteration algorithm aiming to get equation differential offset and variable differential offsets. A rough cost estimation for the fixed point algorithm can be found in [3] and it was extended to the fixed point algorithm with parameters for the systems with upper-triangular block structure.

In the fixed-point algorithm with parameters, there are many parallelizable sub-blocks which are independent with similar sizes and procedures. Naturally the parallel scheme based on CUDA/GPU [8, 9], is suitable for this situation. Although OpenMP [10] is very simple to convert the serial program into parallel algorithm, the speedup of fixed-point algorithm with parameters is quite low in practice. On the other hand, the design of the thread pool to handle concurrent tasks has become a basic model and standard technique [5]. So we design a new multi-threaded parallel version of the fixed-point algorithm to solve the large-scale, sparse blocked systems.

2 PROBLEM BACKGROUND

Through [1], we can represent any DAE model in the form of $F=(f_1,f_2,...,f_n)=0$ $f_i=f_i(t,x_j(t), \text{derivatives of } x_j(t))$ $1\le i\le n\, x_i(t)(i=1,2,...,n)$. And the *symbol matrix* can be defined as $\Sigma=(\sigma_{ij})_{n\times n}, i,j=1,2,...,n$ where the element σ_{ij} represents the highest differential times of the variable x_j about t, when $f_i=0$. *"Differential reduction"* of DAEs is used to convert a DAE problem to an ODE problem which can be solved by classical numerical algorithms [7]. It can be formulated as the following optimization problem.

$$\begin{cases} \min \sum_j d_j - \sum_i c_i \\ s.t.\ \ d_j - c_i \ge \sigma_{ij} \\ c_i \ge 0, i=1,2,...,n \end{cases} \quad (1)$$

Because of block design in unified multi-domain modelling, the symbol matrix Σ is sparse with blocked structure. Naturally it can be converted to an upper triangular matrix, using the sub-block approach [3], as below.

$$\begin{bmatrix} A_{11} & A_{12} & \cdots & A_{1p} \\ & A_{22} & \cdots & A_{2p} \\ & & \ddots & \vdots \\ & & & A_{p,p} \end{bmatrix} \quad (2)$$

For each block, we define *sparsity* as the ratio of the number of non-negative elements and the number of the entries of the whole matrix. In the following description, the diagonal matrices are labelled as $A_{ii}, i = 1, 2...., p$, and the coupled sub-matrices are labelled as $A_{i,j}, i < j$.

By Kuhn-Munkres algorithm [13], we can get the maximum cross-section [1] of the optimization problem which is the input of the New-PFP algorithm below. In our algorithm, it plays a role as the bridge of updating equations differential offset and the coupled sub-matrix.

One of the key problems is the dependency between A_{ii} and $A_{i,j}$. The dealing of A_{ii} depends on the dealing of $A_{1,i}, A_{2,i}, ..., A_{i-1,i}$. The dealing of $A_{i,j}, i < j$ depends on the dealing of A_{ii}. The other is that the time of dealing with $A_{i,j}$ is much less than the time of dealing with A_{ii}.

3 THE BASIC ALGORITHMS FOR BLOCKS

This section mainly describes the different procedures for diagonal blocks $A_{ii}, i = 1, 2...., p$ and the coupled sub-blocks $A_{i,j}, i < j$ in (2). The algorithm in 3.1 is based on [3] with some modifications so that the problem $A_{ii}, i = 1, 2...., p$ can be solved more efficiently. The algorithm in 3.2 originates from [3].

3.1 *The improvement of Pryce fixed point iterative (New-PFP)*

Input parameters: the diagonal sub-block A_{ii}, d_i

Output parameters: d_i and c_i

Step1. Initiate the equations differential offset $c = 0$ *Array* is assigned 1, ..., n;

Step2. Calculate variable differential offset d as $d_j = \max_i \{c_{Array[i]} + \sigma_{Array[i],j}\}$;

Step3. Update differential equations offset c as $c_i = d_j - \sigma_{i,j}, (i, j) \in M$;

Step4. If the value of every c's part is constant, the algorithm is over. Otherwise, turn to step6;

Step5. Clear the value of *Array*, put the subscript into *Array*, which does not keep constant in step5, and turn to steps3.

Compared with original Pryce fixed point iterative algorithm, we only store the indexes of the changed components of c, in our algorithm New-PFP. These indexes indicate the columns which will be updated in the next iteration. It leads to less computation. Here, we will use an example to explain the efficiency of the new algorithm.

$$\begin{bmatrix} 7 & 0 & 0 & 0 & 13 & 2 \\ 0 & 6 & 9 & 8 & 0 & 1 \\ 4 & 0 & 8 & 3 & 6 & 10 \\ 3 & 0 & 9 & 7 & 0 & 0 \\ 0 & 0 & 10 & 0 & 12 & 0 \\ 4 & 0 & 0 & 7 & 0 & 11 \end{bmatrix}$$

For the above example, we know that Pryce original fixed point iteration algorithm requires a total of 5 iterations, involving a total of 180 basic operations (e.g. summation operation and comparison operation). Although the New-PFP algorithm also requires five iterations, the basic operations in each iteration will decrease until the end of the algorithm. Totally it only needs 78 basic operations. Obviously, compared with the original fixed point iterative algorithm, the efficiency of the new algorithm has been improved significantly.

3.2 *Coupled Block Processing Algorithm (CBPA)*

Input parameters: Coupled matrix $A_{i,j}$, c_i

Process: $k = 1 \rightarrow n$

$d_{j,k} = \max_i (\sigma_{i,k} + c_{i,j})$;

End loop

Among them, $d_{j,k}$ represents the jth sub-block in the main diagonal, corresponding to the kth component of the variable differential offset. Among them, $\sigma_{i,k}$ represents the element of the $i-th$ row of $k-th$ columns in the sub-matrix $A_{i,j}$; $c_{i,j}$ represents the jth component of the ith sub-block in the main diagonal, corresponding to the differential equation offset.

4 THE DESIGN AND IMPLEMENTATION OF PARALLEL PROGRAM

Usually the symbol matrix of a DAE model is sparse and the elements of symbol matrix just need to be visited orderly, so we use data structure triple to store each element of symbol matrix and the element of the largest cross-section. Every triple consists of row, data and column. Every block matrix uses a one-dimensional array to store.

Because of the dependency between A_{ii} and $A_{i,j}$ mentioned in Section 2, we introduce two single-lists [6] for maintaining the thread pool. One is used to store the blocks which can be processed now, named as *task list*, the other is used to store the blocks which cannot be processed now, named as *limited_task list*. In the task list, each node is sorted by the column index, but the node in *limited_task list* is sorted by the row index. Each node of two lists consists of sub-matrix blocks and the corresponding processing function. There are two mutexes, which are used to keep the data of the lists consistency. One is *task_mutex* for the task list, the other is *limited_task_mutex* for the *limited-task list*. They are used to keep data consistency. Otherwise, because the equation differential offset and variable differential offset of the block matrices are divided into p components, 2p mutexes are created to make sure execution sequences, when we update the equation differential offsets and the variable differential offsets right. The dealing process follows figure 1. In figure 1, the updating source of *task list* comes from the *limited_task list*. The executing order is labelled by the number in the line.

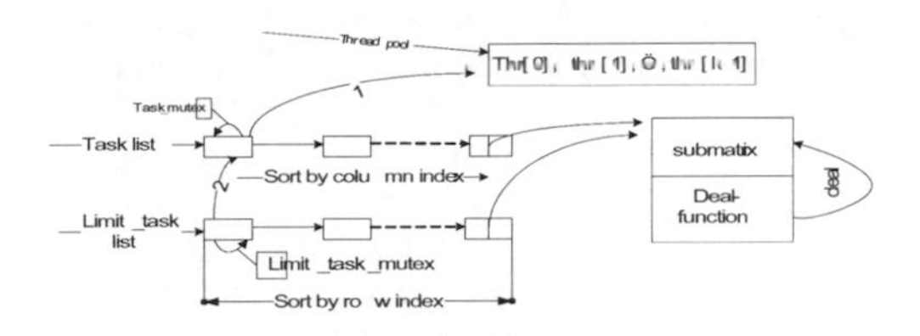

Figure 1.

Large-scale sparse matrix can be converted into an upper triangular block matrix in nature [2, 3]. Clearly, in the form of (2), the left lower position of the matrix is empty. At this point, we can see that if $c_1, c_2, ..., c_{ii}$ of $A_{11}, A_{22}, ..., A_{ii}$ is got, we can compute the corresponding coupled block and the diagonal matrix blocks $A_{i+1,i+1}, A_{i+2,i+2}, ..., A_{p,p}$, to get the value of $c_{i+1}, c_{i+2}, ..., c_p$. The elements of lower triangular matrix are negative infinity, so $c_1, c_2, ..., c_{ii}$ will keep unchanged. Thence, it is reasonable to compute $c_1 \to c_p$ as the order of dealing with $A_{11} \to A_{pp}$. Finally $\vec{c} = (\vec{c}_1^T, \vec{c}_2^T, ..., \vec{c}_p^T)^T$consisting of $\vec{c}_i, i = 1, ..., p$, is the optimal solution of problem in (2). The coupled algorithm separates the dealing of $A_{i,j}$ from the whole algorithm [3] and updates the value of d_j, by c_i. And overall the result is same with the fixed-point algorithm with parameters. From the order of adding task submatrix $A_{ii} \to A_{i,j+1} \to A_{i+1i+1}$ $j = 1, 2, ..., p$, we know that the logical order does not break during the process of parallel algorithm. So the whole parallel program is reasonable and its termination can be ensured. The next sections will introduce the base dealing function and the updating rules about the single-lists.

4.1 *Task scheduling algorithm*

4.1.1 *Thread handler*

The main tasks are to get sub-blocks from task list and to call handler to deal with sub-blocks. And moreover it is responsible for releasing the memory block and determines whether the thread can be terminated. The key code appears as follows:

```
While (1)
{
Pthread_mutex_lock (&task_mutex);
If (! limit_work)
   Pthread_cond_broadcast (&task_ready);
While (! task && limit_work)
   Pthread_cond_wait (task_ready, task_mutex);
If (! limit_work &&! task)
  {
    Pthread_mutex_unlock (&task_mutex);
    Pthread_exit (NULL);
  }
Work = task;
Task = work->next;
Pthread_mutex_unlock (&task_mutex);
Work->routine (work->mymatrix);
Free(work->mymatrix->sub_data);
Free (work->mymatrix);
Free (work);
}
```

4.1.2 *Main diagonal block handler*

It mainly consists of two parts. One deals with matrix blocks, the other is to complete the updating operations about the *task list* and *limit-task list*.

Input parameters: mymatrix, which is submatrix block to be processed

Process:

Pthread_mutex_lock (&DiffD_mutex [mymatrix->col_pos]);

Pthread_mutex_lock (&DiffC_mutex [mymatrix->row_pos]);

Call handler New-PFP to deal with mymatrix;

Pthread_mutex_unlock (&DiffC_mutex [mymatrix->row_pos]);

aPthread_mutex_unlock (&DiffD_mutex [mymatrix->col_pos]);

Add the data blocks into the *task list*, which have the same row index with mymatrix. The elements in the task list, which have the same column index, is put together. This operation will be over till there is no data block in the *limit_task list*, which has the same row index with mymatrix.

4.1.3 *Coupled block handler*

Input parameters: mymatrix is submatrix block to be processed

Process:

Pthread_mutex_lock (&DiffD_mutex [mymatrix->col_pos]);

Call function CBPA to deal with mymatrix;

Pthread_mutex_unlock (&DiffD_mutex [mymatrix->col_pos]);

If the row_pos of the *limit_task list*'s first node, task_head, is equal to task_head's col_pos, then we should make a judge about the value of the *task list*'s first node. If it is equal to task_head->mymatrix->row_pose-1and limit_task_head->mymatrix->col_pos < task_head->mymatrix->col_pos, then we add the first node of *limit_task list* into the task list. Meanwhile, it involves the use of the appropriate lock to maintain data synchronization.

5 NUMERICAL EXPERIMENTS AND EFFICIENCY ANALYSIS

The program is achieved in linux/g++. To let the numerical experiments do easily, we set the number of blocks in the main diagonal as q. In the main diagonal, the size of sub-matrices is n and n is a random integer. Its distributions are 40~80 and 80~120. Numerical experiments come from the two main distributions. The time of dealing with $A_{i,j}$ is much less than the time of dealing with A_{ii} in (2), so we set the sparsity of main diagonal submatrixes as 0.05 while setting the sparsity of the coupling block as 0.2. Every element of the sub-blocks is generated by random function, ranging between 0~n. And the computer memory is 128GB and the CPU is Inter® Xeon CPU E5-2620 @2.00GHz. The experiments is done in two above situations. And every time data is the average of 100 groups under the same blocks. The number of kernels, which participate in the test, is 1~6 and the number of diagonal sub-blocks is from 100~900, increasing by 100.

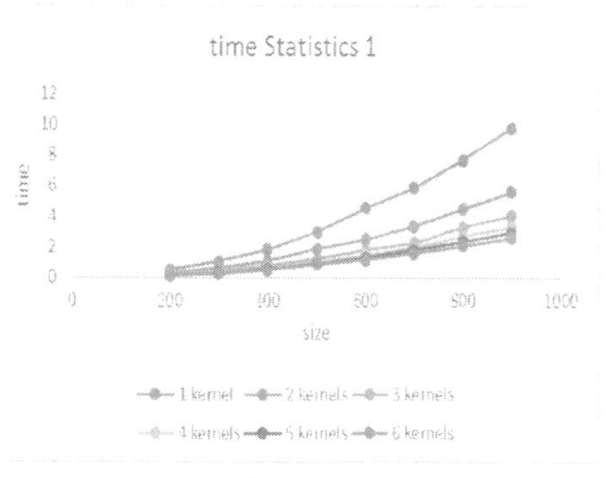

Figure 2.

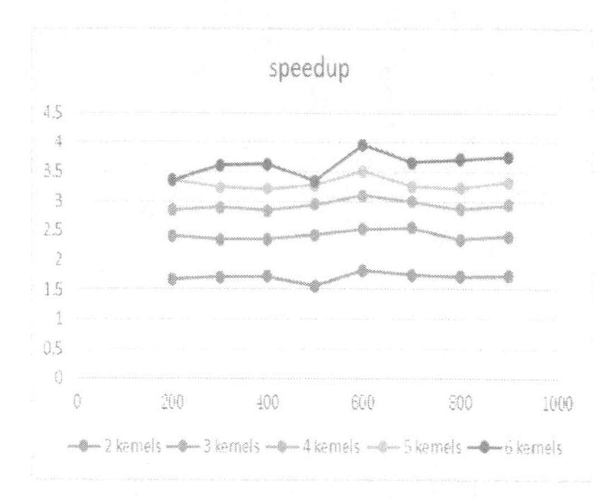

Figure 3.

In Figure 2, 3, 4, 5, i kernels means that parallel algorithm uses $i$threads to produce the data in the curve. In Figure 2 and 3, the size of the submatrix in the main diagonal is 40~80. In Figure 4 and 5, the size of the submatrix in the main diagonal is 80~120. In Figure 2 or 4, although the time of completing computation decreases, the time efficiency becomes smaller, with the number of available threads increasing. After reaching a certain level, the effective used by multi-thread program to enhance the time efficiency, will continue to reduce as the kernel numbers increase. Comparing the results from figure 3 and figure 5, we can know that when the size of sub-matrixes in the main diagonal increases, the speedup is improved. So we can get that the speedup gets an apparently improving, along with the size increasing.

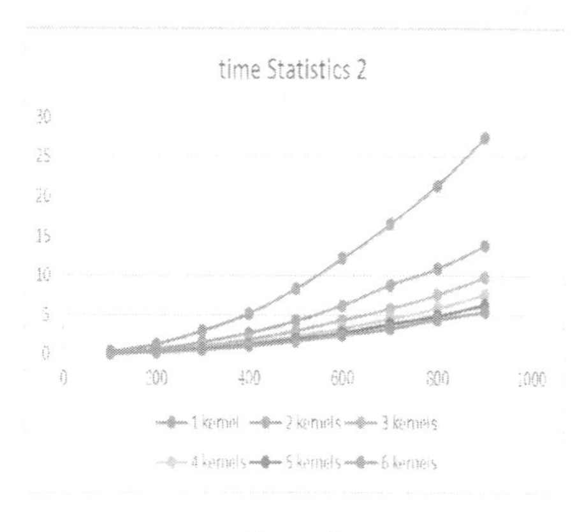

Figure 4.

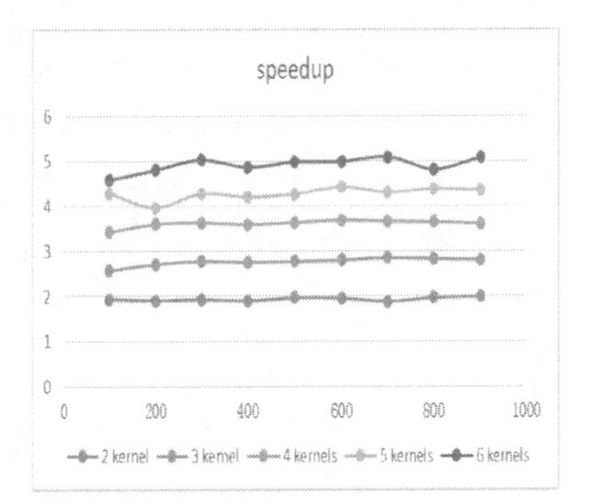

Figure 5.

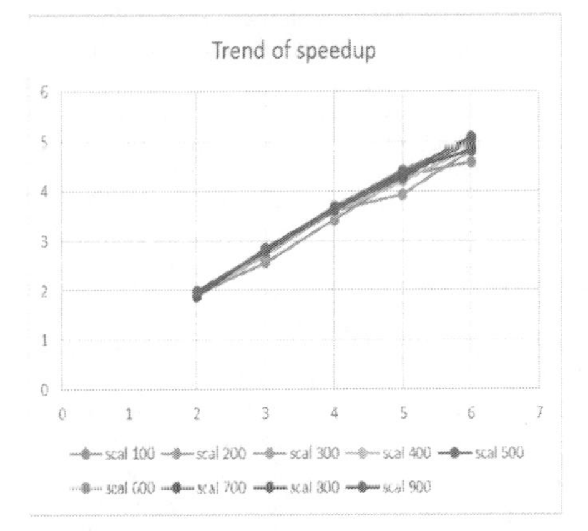

Figure 6.

5.1 *Speedup analysis*

Assuming the symbol matrix has been processed by the method which is mentioned in [3, 12]. For each main diagonal block, the time complexity is $O(n^2 * M)$ by using Pryce fixed point algorithm, where M represents the number of iterations in the algorithm(one iteration means New-PFP algorithm visits all the elements of the block once). Because this article uses New-PFP algorithm to process the main diagonal block and the matrix is sparse, the basic operation involved in each iteration is much less than n^2. In experiments, the number of processor kernels is k. Assuming that the number of blocks on the diagonal of the symbol matrix is P and the size of each block is less than 200*200. By the task scheduling algorithm, we can see that if we want to let the speedup of the whole algorithm to achieve an optimal value, we should let the $k$threads be in full load condition. At this time, because the number of the blocks that can be handled is not more than the total of coupled blocks, we can obtain the inequality $\{M*(k-1)+k\}*(p-2)+k \le (p^2-p)/2$, it can be simplified to$H_0*(p-2)+k \le (p^2-p)/2$, where

$$H_0 = (k-1)*M+k.$$

The result is $p_0 \approx \left\lceil H_0 + (1+\sqrt{H_0^2 - 3H_0 + 2k})/2 \right\rceil$. When the number of the sub-matrixes in main diagonal reachesp_0, we will get an ideal speedup. Here, the costing time of keeping synchronization is not considered. In this case, the time cost by the parallel algorithm is$(p_0 * M + p_0 - 1)*n^2$.

The time cost by corresponding serial algorithm is

$$\{p_0 * M + (p_0^2 - p_0)/2\}*n^2$$

The speedup can be calculated as

$$(2p_0 * M + p_0^2 - p_0)/(2p_0 * M + 2p_0 - 2).$$

When$p \le p_0$, the speedup is

$$(2p*M + p^2 - p)/(2p*M + 2p - 2).$$

When $p > p_0$, the speedup is $(M + (p_0 - 1)/2)/(M + L_0/p_0 + 1 - 1/p_0)$, where $L_0 = \{(p^2-p)/2 - (p-1)\{M*(k-1)+k\} - k\}/k$.

From the figure 3 and figure 5, we can see that the speedup of the parallel algorithm just changes in a very small scope under the same number of threads. This phenomenon shows that when the number of the sub-blocks exceeds the critical point ($p > p_0$), the speedup in experiments is consistent with the one in theory. In Fig.6, we find that at the same scale, the speedup increases approximately linearly, along with the number of threads increasing, when$p \le p_0$.

5.2 *Time efficiency analysis*

With same assumptions in the speedup analysis, we mainly consider the time complexity of the parallel algorithm from two cases. The first one is that when $p \le p_0$, the time complexity is

$$t_1 = \sum_{i=1}^{p} n_i^2 + \sum_{i=1}^{p-1} n_i n_{i+1} + \sum_{i=1}^{p-2} (p-i-1)$$

Where n_i represents the size of the i-th submatrix in main diagonal. Because of $n = \sum_{i=1}^{p} n_i$, then $n^2 \gg t_1$. When $p > p_0$, the time cost by the parallel algorithm is

$$t_2 = t_1 + (p^2 - p - 2k)/2k * \max\{n_i^2 \mid i = 1,..., p\}.$$

6 CONCLUSION

The results from experiments match the theoretical analysis very well. But we need to point out that the difference between experiments and theory comes from two reasons. One is keeping information synchronization, which is not considered in theoretical analysis. Another is that the sub-matrix is too small so that the communication time between threads cannot be ignored.

Experimentally, our parallel scheme solves the problem (2) successively. In general, our parallel algorithm is suitable for the situation when the size of sub-blocks is large. Especially, when the size of sub-blocks is less than 80 or the number of sub-blocks is less than 100, we recommend to use 2~3 threads.

ACKNOWLEDGEMENT

This article is partially supported by the West Light Foundation of Chinese Academy of Sciences and NSFC (Grant No. 11471307).

REFERENCES

[1] Pryce J D.A SIMPLE STRUCTURAL ANALYSIS METHOD FOR DAES [J]. BIT. 2001: 41(2):364~394.

[2] Pryce J D,Nediakov N S, Tan Guangning. DAESA – a Matlab Tool for Structural Analysis of Differential-Algebraic Equations: Theory [EB/OL]. (2012) [2013-11-12].

[3] Tang Juan, YANG Wen-Qiang, WU WEN-Yuan, FENG-Yong. A heuristic Algorithm for Block Fast Index Reduction in Differential Algebraic Systems [J]. JOURNAL OF SICHUAN UNIVERSITY (ENGINEERING SCIENCE EDITION). 2014:Vol.46 No.4.

[4] Peter.S, Pacheco. An introduction to Parallel Programming [M]. Beijing: Machinery Industry Press. 2011:209–258.

[5] David R. Butenhof. Programming with POSIX Thread [M]. Addison-Wesley.1997: 28–111.

[6] Thomas H.Cormen, Charles E. Leiserson, Ronald L.Rivest, Clifford Stein. Introduction to Algorithms (third Edition) [M]. Beijing: China Machine Press.2013:31–34.

[7] Lamour R, März R. Detecting structures in differential algebraic equations: Computational aspects [J]. Journal of Computational and Applied Mathematics, 2012, 236:4055–4066.

[8] F. De Angelis, F. Gentile, F. Mecarini, G. Das, M. Moretti and etc. Breaking the diffusion limit with super-hydrophobic delivery of molecules to plasmonic nanofocusing SERS structures [J]. 2011, Nature Photonics 5. 2011: 682–687(2011).

[9] Duhu Man, Kenji Uda, Hironobu Ueyama, Yasuaki Ito, Koji Nakano. Implementations of a Parallel Algorithm for Computing Euclidean Distance Map in Multicore Processors and GPUs.2011, International Journal of Networking and Computing. 2011: Volume 1, Number 2,260–276.

[10] YUNHENG WANG, YOUNGSUN JUNG. A Hybrid MPI-OpenMP Parallel Algorithm and Performance Analysis for an Ensemble Square Root Filter Designed for Multi-scale Observations.2013, JOURNAL OF ATMOSPHERIC AND OCEANIC TECHNOLOGY. 2013: Volume 30, 1382–1397.

[11] Ding Jianwan, Zhang Chenjian, Chen Liping. Index analysis of discontinuous differential-algebraic models for performance simulation [J]. Journal of computer-aided design & computer graphics. 2008: 20(5):585–590.

[12] ALEX POTHEN, CHIN_JU FAN. Computing the block triangular form of a sparse matrix. ACM Transactions on Mathematical Software.1990: Vol. 16, No. 4:303–324.

[13] Douglas B. West. Introduction to Graph Theory. Pearson Education, Inc. 2001:108–118.

Computing, Control, Information and Education Engineering – Liu, Sung & Yao (eds)
 ISBN: 978-1-138-02800-5

Information engineering curriculum system reform of local undergraduates universities

X.M. Yang, W. Jun, X. Ling & T.Q. Li
School of Electrical and Electronic Information Engineering, Xihua University, Chengdu, China

ABSTRACT: According to talent cultivation practice experiences of information engineering for years, a curriculum system is proposed to improve the quality of education. The curriculum system has distinctive local characteristic. It conforms to the national policy and local enterprises' demand for talents. Furthermore, the importance of practice is embodied adequately in the curriculum system. So it can be better applied to talent cultivation due to consideration of local undergraduate universities' own characteristics. The employment rate of graduates and admissions to Chinese universities for postgraduate education are reached more than 95% and 30% respectively in our university.

KEYWORDS: Curriculum system; Information engineering; Talent cultivation; Local undergraduate university.

1 INTRODUCTION

The information industry has become one of the pillar industries of the national economy. Cultivating a large number of outstanding industry professionals and enhancing the level of development of the information industry are two key subjects of educators' research [1-3]. The reform of curriculum system plays a very important role in cultivating talents. According to the teaching reform and Practice for years, the characteristics of local universities and the precise positioning of talent training goal, a curriculum system of talent cultivation is proposed, which is well adapted for local undergraduate universities.

2 TALENT DEMAND AND CULTIVATION

The information industry is rapidly developing and new technologies emerge continually, which puts forward higher requirements for the talents. The characteristics of demand for talents are that: supply and demand of talent has been basically balance; more talents are demanded in operation, maintenance, system integration, technical support and other positions; the demand for talent of experience of skill has risen steadily; Development, high-end management personnel should have high professional quality; the most popular talents are engineers with comprehensive and innovative ability.

From the above characteristics, we can be seen that one curriculum system cannot meet all these requirements. So the aim of talent cultivation must be emphasized. In national key universities, the aim is the cultivation of innovative and high-end professional talents. In higher vocational education, the training of personnel of low-end professional skilled technology should be taken care of. But it is difficult to formulate the aim of talent cultivation for many local undergraduate universities. How to accurate positioning of personnel training objectives is a key problem about the survival and development of them.

3 OBJECTIVE OF TALENT CULTIVATION

In the "national medium and long-term educational reform and development plan (2010-2020)", "higher education should enhance the social service ability" and "optimize the structure of running characteristics" are the important functions of universities and colleges [4]. It is an orientation of the local undergraduate universities cultivation. The servicing for social is the guidance on setting objectives in talent cultivation. Meanwhile, the needs of local economic development, analysis of the development of dynamic and hot industry must be fully considered. This goal includes at least the following three points:

1 Students are cultivated mainly for obtaining application and compound ability in local

undergraduate universities and the cultivation of innovative talents is also given attention to.
2 The talents who are trained in local undergraduate universities must have good professional basics and professional direction of skilled.
3 Specialized quality should be paid adequate attention to in talents cultivation.

4 THE CURRICULUM SYSTEM

Colleges and universities are the main body to implement the personnel training and perfect the curriculum system is the key to realize the goal of cultivating talents. The curriculum system should fully reflect the advantages of universities in the aspects of professional training and has the following characteristics: enterprises of professional basis; consideration talent demand and strengthen the practices, enterprising the cultivation of students' innovative ability, training the professional quality in every teaching link. Curriculum system can be divided into three courses group and an engineering practice, as shown in figures 1.

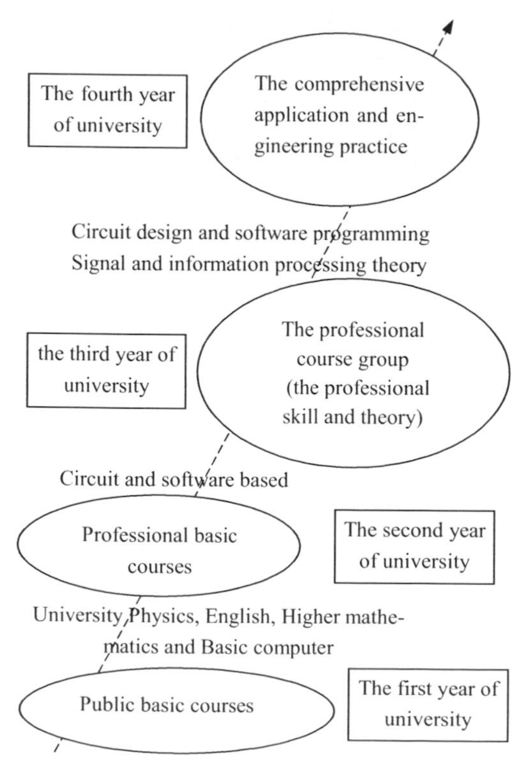

Figure 1. The professional curriculum system of information engineering.

4.1 *Public basic course group*

The public basic course group consists of College English, computer application basis, college physics, and higher mathematics, as shown in figures 2. Some courses must be optimized the teaching content according to the information engineering, such as strengthening of teaching the electromagnetic in physics of the University and the calculus, Fourier transform in higher mathematics content. The course of computer application basis is very important to learn later professional courses.

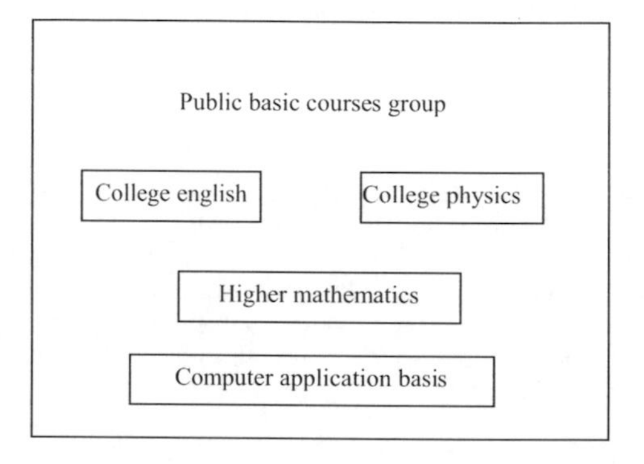

Figure 2. The public basic course group.

4.2 *Professional basic course group*

Professional basic course group consists of all kinds of circuit, signal, information and communication basic course, as shown in figure 3. First, students should study the professional introduction course. They will know what the information engineering is and work out a plan of study for four years. This course also helps them to become real university students. After finishing these courses, they can understand the basic knowledge and skills of information engineering.

4.3 *Specialists course group*

Professional knowledge and skills are acquired from Professional courses. In order to meet the needs of the enterprise, university student should master a few Professional courses instead of learning more as they can. These courses must be divided into a few modes according to characteristic of information industry, such as wireless communication, network communication, signal and information processing. Students can choose one of these modes by their own situations. In addition, some courses in the course group can be replaced by others when employment is changing.

These replaceable courses enhance the flexibility of curriculum system and adapt to the demands for specialists. Figure 4 gives an example of course modes division.

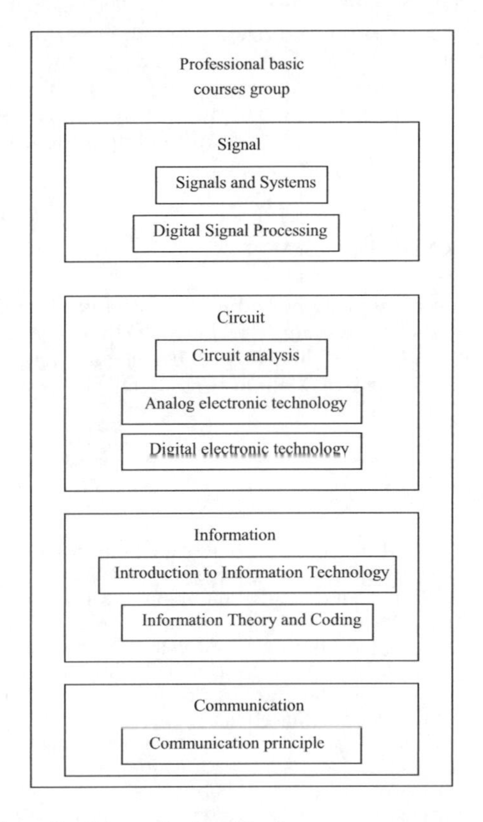

Figure 3. The professional basic course group.

4.4 *Comprehensive application and engineering practice*

Compared with science talent, engineering talent lays a lot of emphasis on the application of knowledge [4]. To improve the quality of teaching, theory teaching hours should be reduced for more practice teaching which is a variety of experiments, curriculum design and practice. The application knowledge and practical ability of university student is improved in the practice teaching. The structure of comprehensive application and engineering practice is shown in figure 5, which is a combination of .

The professional, comprehensive application is composed of open experiment design and integrated curriculum design. It is different with experiments and curriculum design in a professional basic course group and Specialists course group. Students must apply comprehensive knowledge in engineering practices. They design and finish themselves experiments with a few professional courses knowledge in open experiment design requirements. Teachers only give these students some advice and grade their work. In integrated curriculum design, students are required as a group to design a product, by using soft and hardware knowledge. And then create and test it. After being trained in two experiences, students' application knowledge, ability and team cooperation, consciousness will be greatly improved.

That students practice in the enterprise is engineering practices and it is an opportunity for them to improve skills and knowledge. The engineering practice includes consisted of enterprise practice, graduation practice, graduation design and project practice. The visit of enterprises is assigned in second year at the university. Students will know enterprise

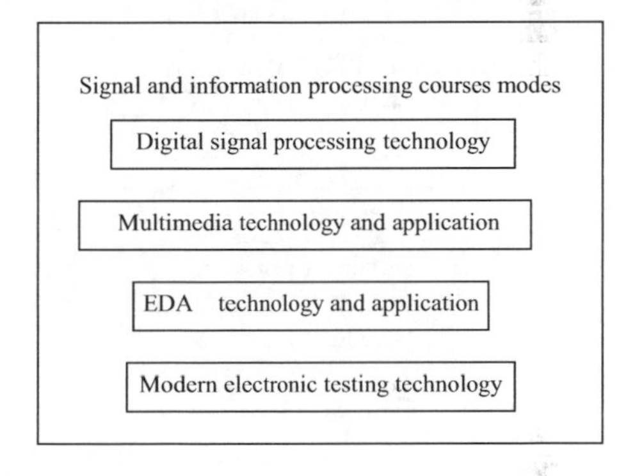

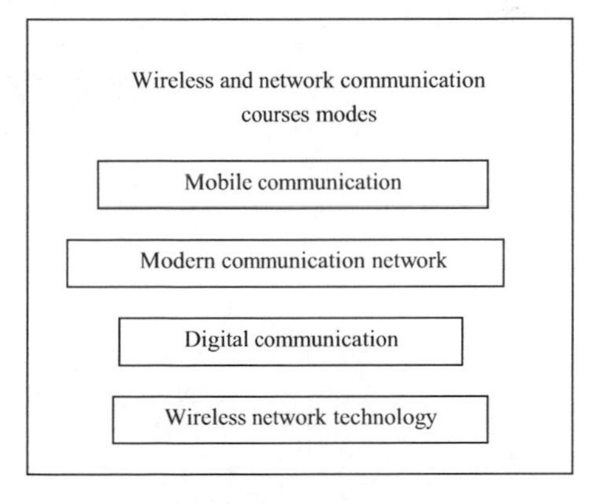

Figure 4. An example of course modes division.

activities and their learning purpose are made clear. The other engineering practices are assigned in third and fourth year at the university. In graduation practice, Students receive training in the different work sectors such as production, sales, development and services. The graduation practice is geared to preparing students for further work. In project practice, students are assigned different enterprises and choose a job. Under strict management, they must carefully finish their task. The knowledge and practical ability of the students professional is tested well in graduation design. It is very important to choose an appropriate design subject. Only students with good engineering practice can complete graduation design successfully. Students often are encouraged to complete their work in enterprise for a better combination of practice and knowledge.

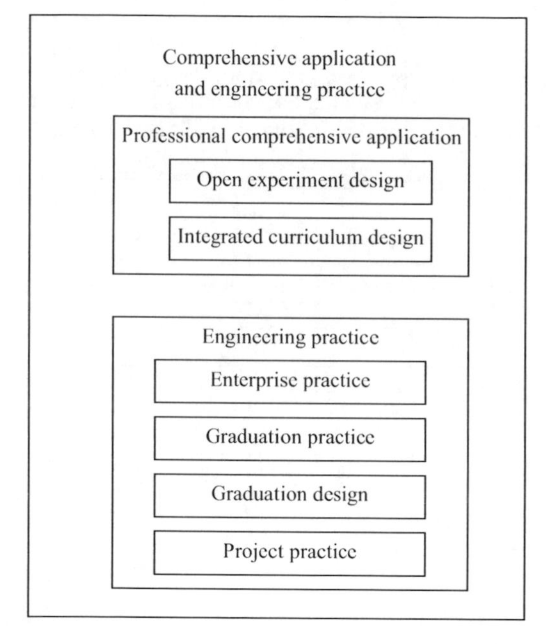

Figure 5. The structure of comprehensive application and engineering practice.

5 CONCLUSIONS

The reform of curriculum system has been researched since 2003 and completed in the main so far. Notable achievements has been achieved by the adoption of a new curriculum system. The employment rate of graduates reached more than 95% at a time and admissions to Chinese universities for postgraduate education increase to above 30%. So the reform of a curriculum system has proved successful.

ACKNOWLEDGMENTS

The work was supported by the Teaching Reform Project of Sichuan Provincial Department of Education (No.700337) and Excellent Engineers Training Plan of Sichuan Provincial Department of Education.

REFERENCES

[1] Gu, T.L. et al. 2012. A Research on the Training Mode Reform of College Engineering Practical Talents. China Higher Education Research.

[2] Liu, J. et al. 2010. Construction of Talents Cultivation System of Mobile Communications Specialty of Higher Vocational Education Based on Working and Learning Combination. Vocational and Technical Education.

[3] Huang, X.B. 2010. Countermeasures to cultivate innovative talents of information and electronic field. China Adult Education.

[4] Ministry of Education of PRC. 2012. Undergraduate professional directory and professional presentation of ordinary university and college. Higher educational department of the Ministry of Education.

Computing, Control, Information and Education Engineering – Liu, Sung & Yao (eds)
© 2015 Taylor & Francis Group, London, ISBN: 978-1-138-02800-5

Educational exploration based on computational thinking capacity cultivating

Xia Ling Zeng
Jiangxi Science and Technology, Normal University, Nanchang, China

ABSTRACT: How to cultivate the students' capacity of computational thinking during the educational process of computer programming is one of the basic goals of computer basic education. Considering the generally reflected problems in education of programming courses such as boring, not enthusiastic about learning, grammatical structures and teaching cases are dispersed, etc; this paper introduces the gamified educational mode for the purpose to cultivate the students' capacity of computational thinking. Educational model based on the gamified education was constructed, and the cultivation of gamified education to students' computational thinking ability was analyzed through specific case analysis. Gamified educational mode can effectively stimulates students' motivation and interest in learning, and then cultivate students' self-learning ability. It better tap the abstract algorithm thinking which cultivate students' computational thinking capacity.

KEYWORDS: Computational Thinking; Programming; Gamified Education; Case; Cultivation.

1 INTRODUCTION

Programming is an important and required course in computer basic curriculum system, and it is the only way for people to enter the computer world. In the "Computer basic education curriculum System of China Colleges 2008", it puts forward three requirements to college students' computing application capability: operating using capacity, application development capacity, and research innovation capacity [1]. Wherein application development capacity and research innovation capacity are mainly in the computer programming that is the second basic course. Through the learning of programming course, one hand enables students to master how to make computer for us to solve a variety of practical problems. The other hand, cultivates students' abstract thinking ability, logic judgment ability and problem-solving ability. But from the practical educational effects, students generally reflect that the programming courses are boring and difficult, and the education less than ideal results.

In "the sixth forum of university computer courses in 2010", academician Guoliang Chen pointed out that computational thinking ability is an important training objective in the education process of university computer basic courses[2]. He also pointed out that the three pillars of the human science development are theoretical science, experimental science and computational science. Corresponding the three thinking ways that people to understand and transform the world are theoretical thinking, experimental thinking and computational thinking [2]. The computational thinking is an important ability that students must master.

How to train and develop students' computational thinking ability and how to enable students to think and solve problems using computer's thinking way in the educational process of programming courses, is a very worthwhile exploring the problem.

2 COMPUTATIONAL THINKING

In March 2006, Yizhen Zhou, who was Professor of computer science department of the America Carnegie·Mellon University first at Communications of the ACM which is the authoritative journal of American proposed the concept of "computational thinking". In his definition computational thinking is the use of the basic concepts of computer science of problem solving, system designing and thinking activity understand human behavior [3]. Professor Zhou pointed out that computational thinking is everyone's basic skills that is not only belongs to computer scientists. We should make every child not only master Reading, wRiting and aRithmetic (3R), but also learn computational thinking.

Computational thinking exudes a powerful magic once available, leading the domestic and international computer science community, especially the development direction of the computer education sector. MIT, Stanford, Carnegie·Mellon University, Purdue University and other famous universities have set up "computational thinking" general education courses at the core of problem solving for the whole school [4-7]. They have achieved remarkable results in the computer basic education that is at the guidance of computational thinking. In 2010, nine universities of china published "Nine School Alliance (C9) computer basic education strategy joint Statement", it takes the cultivation of computational thinking ability as the core task of computer basic education [8].

The computational thinking proposes higher requirements to people's abilities of using computer analysis and solves the problem, but also puts forward higher requirements for computer basic education for non-computer professional colleges. Computer basic education needs training students' computational thinking skills, and improving their information literacy. Then when dealing issues students can better use the computer as an indispensable tool for thought and expression, and can better apply grasped expertise to scientific research and production

3 EDUCATIONAL SITUATION ANALYSIS OF PROGRAMMING COURSE

For most non-computer professional colleges, the purpose of learning programming is not to be a programmer, but to learn the methods, processes, and ideas of analyzing and solving problems from the perspective of a computer, which is an important method and content of culturing computational thinking. However, from the current educational practice of most colleges, the cultivation of computational thinking to students has not achieved inadequate attention. The following problems are most existed:

1 When learning programming students need to learn a computer language. Thus, it is easy to fall into the misunderstanding of explaining and teaching of the language basics in class teaching. Then emphasis on language and ignore programming. The education is boring and the interest in learning of students is not high.
2 Students' understanding of computational thinking is not enough. Most think that programming course is to learn programming, and generally considered that it is not related to his profession, and will never engage in related work. So their learning motivation is not high.
3 The lack of better educational methods and means. Most educational cases are designed to explain the use of grammatical structure, and they are scattered, short, and little correlation. After learning students just master the grammar, do not know what the real problem can be solved, and even do not know how to solve problem.

Therefore, in the educational design and implementation of programming course, we must clear the ultimate goal to cultivate and improve students' computational thinking abilities.

4 EDUCATIONAL EXPLORATION AND CULTIVATION OF COMPUTATIONAL THINKING ABILITY

The principle of the Constructivism Education Theory states that the complex field of study should be directed learners' previous experience and interests. It is the only way to stimulate learners' learning motivation and only by this way learning may be active [9]. The author introduces the gamified educational case for the purpose to cultivate the students' capacity of computational thinking. Gamified educational mode can effectively stimulate students' motivation and interest in learning, and then cultivate students' self-learning ability. It better tap the abstract algorithm thinking which cultivates students' computational thinking capacity.

4.1 *Educational model*

For the abstraction of computational thinking, we introduce the gamified educational case into class teaching. The way through the game and entertaining way to cultivate students' computational thinking ability can play a multiplier effect. According to the educational characteristics of programming course and features of gamified education, we construct following educational model that is shown in figure 1.

In the implementation of gamified educational activities, educator can take advantage of games to the new lesson into, creating scenarios and stimulating interest in learning. Then game-assisted education and knowledge, explain make the learning process change from passive to active. Educator also can use the game to learner for consolidation exercises. For example educator determines the goal of the game and learner complete by team cooperation and is encouraged to complete by a variety of algorithms. Then to improve learner's practical skills and to cultivate a learner's computational thinking ability.

In education, the educator is the key. So strengthening the cultivation of learner's computational thinking firstly must raise the awareness of educators of computational thinking and the importance of the cultivation of computational thinking. It requires educators designing educational content and gamefied cases on the basis of understanding computational thinking. And it requires educators trying different educational methods in the educational process, then to innovative educational methods. It also requires educators refine and demonstrate the computational thinking way that hidden behind knowledge, so that

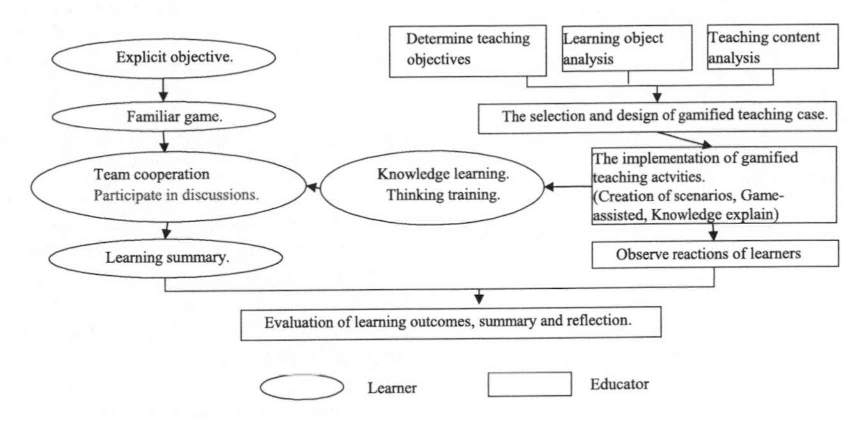

Figure 1. The educational model based on gamified education.

learners can feel the charm and importance of computational thinking, and can inspire learner's desire for knowledge and psychological resonance.

4.2 *The design of gamified educational case*

The gamefied cases can be educational games that are question exploring approach, task-driven, hands-on exercises, analog training style, thinking guided style and so on. The purpose of games is enhancing students' interest and understanding of abstract content of programming, and establishing a suitable logical thinking and way of thinking in the process of the games, then to master the ideas and methods to solve problems from the perspective of a computer. Gamefied learning can strengthen the training to students' programming mode, algorithmic thinking ability and program practical ability. It can help learners internalizing computational thinking way in imperceptible method, and then cultivate the students' computational thinking ability.

The guessing game which integrates three basic control structures of programming, the scheduling problem and the Hanoi Tower game which is representative of the algorithm field are all belong to the typical gameified cases. Following through a Lucky 52 guessing game that is gradual and layering for example, we describe gemified education in the programming course for students' cultivation of computational thinking ability.

We choose a fun guessing game, and make the game from easy to difficult and step by step through setting different game conditions. It integrates the main knowledge points together such as selection, circulation, function and array. Then students can complete the learning and use of relevant knowledge points in a gradual and fun process of solving the case. Simple guessing game is to guess a number and its different game conditions and corresponding knowledge points are shown in figure 2.

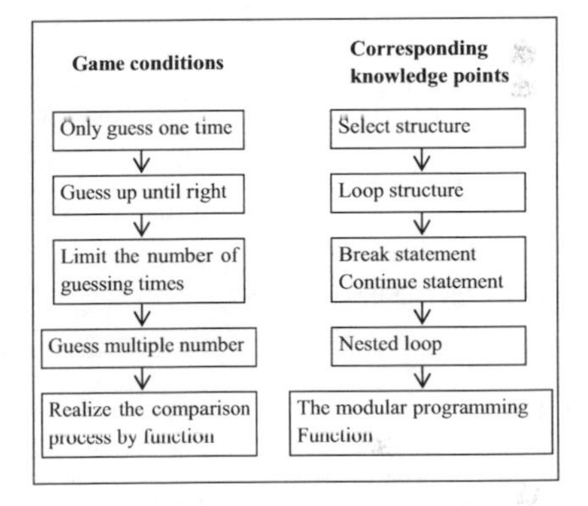

Figure 2. The different game conditions and corresponding knowledge points of the simple guessing game.

Combined with the form visual interface of Visual Basic programming, we can design more interesting guessing game, for example Lucky 52 guessing game. It is to guess the price of a commodity and the name and price of commodity are assigned when the program is initialized. The game interface when the user only guesses one time to master the use of assignment statements and selection statement is shown in figure 3.

Further escalation of the game is achieved by increasing a button of out a commodity randomly. The name and price of commodity are randomly generated in diverse commodities to master the use of random function. The game interface is shown in figure 4. In the next design of the game, we can give the user several times to guess the price, in order to master the use of loop statement. Finally, many commodities can be designed in a library to wait for a random selected. Here can introduce the concept of array and master the use of arrays.

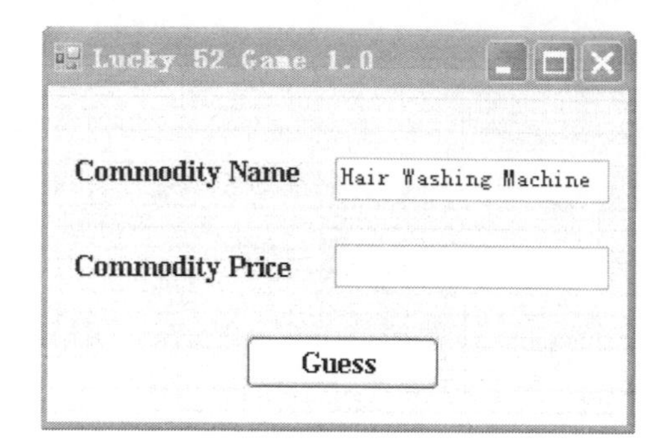

Figure 3. Lucky 52 guessing game 1.0.

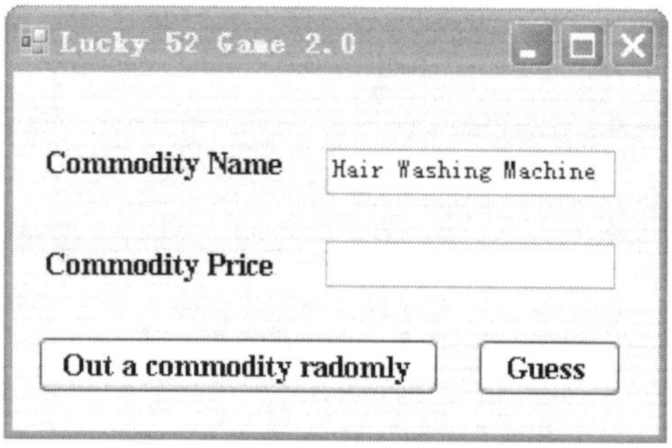

Figure 4. Lucky 52 guessing game 2.0.

4.3 *Education effectiveness analysis*

In order to cultivate students' computational thinking ability, author design tiered game cases which are progressive approach. It changes the traditional educational way that focus on introducing grammatical structure and creates an autonomous exploration space for students. It inspires students' learning interest and sense of innovation.

1. Interesting gamified cases can stimulate and sustain students' interest and motivation to learn. In the completing process of gamified cases, students can master the learning of related knowledges, and then cultivating students' self-learning ability.
2. The hierarchical design of cases achieves a comprehensive study of knowledge. Teachers can take care of all students' learning progress by hierarchical teaching. Also, teachers can guide students that have spare capacity to develop deep-step improvements in current cases, to cultivate students' ability of self-exploration and innovation.
3. Progressive evolution of cases can stimulate students' learning psychology of self-exploration, and guide students step by step to the learning of in-depth knowledge and skills. In the process, students can master the methods and ideas of analyzing problems, modeling problems and solving problems and their computational thinking can be cultivated.

5 CONCLUSIONS

This paper introduces the gamified educational case into classroom teaching for the purpose to cultivate the students' capacity of computational thinking. An educational model based on the gamified education was constructed, and the cultivation of gamified education of students' computer thinking ability was analyzed through specific case analysis. Gamified educational mode can effectively stimulate students' motivation and interest in learning, and then cultivate students' self-learning ability. It better tap the abstract algorithm thinking which cultivates students' computational thinking capacity. It puts forward new ideas and exploration of education of programming courses and cultivation of computational thinking.

ACKNOWLEDGEMENTS

In this paper, the research was sponsored by the Teaching reform project of Jiangxi Science & Technology Normal University (JGYB-13-40-35).

REFERENCES

[1] Computer basic education reform research group of Chinese institutions of higher learning. 2008. *2008 computer basic education curriculum system of Chinese institutions of higher learning*. Beijing: Tsinghua University Press.
[2] Guoliang Chen. 2010. *Computational thinking and the university computer basic education*. Jinan: Sixth university computer course forum.
[3] Jeannette M. Wing. 2006. Computational Thinking. *Communication of the ACM*, 49(3): 33–35.
[4] MIT. 2008. Introduction to Computer Science and Programming. In: http://ocw.mit.edu/courses/electrical-engineering-and-computer-science/6-00-introduction-to-computer-science-and-programming-fall-2008.
[5] Stanford.IntroductiontoComputerScience|Programming Methodology. In: http://see.stanford.edu/see/courseinfo.aspx?coll=824a47e1-135f-4508-a5aa-866adcae1111.
[6] CMU. Principles of Computation. In: http://www.cs.cmu.edu/~tcortina/15-105sp09.
[7] Purdue. Introduction to Computational Thinking. In: http://secant.cs.purdue.edu/cs190c:start-s09.
[8] Nine school alliance (C9). 2010. Joint Statement of computer basic teaching development strategy from nine school alliance. *Chinese University Teaching*, 9.
[9] Jing Wang. 2010. The application of PBL learning model in "computer application foundation" course. *Journal of Hunan Science and Technology College*, 31(4): 123–125.

Computing, Control, Information and Education Engineering – Liu, Sung & Yao (eds)
© 2015 Taylor & Francis Group, London, ISBN: 978-1-138-02800-5

A novel skin detection method based on local information consistency

Jiang Xue & Zhi Yan Wang
School of Computer Science and Engineering, South China University of Technology, Guangzhou, Guangdong, China

ABSTRACT: The traditional skin detection algorithm only use prior knowledge to identify the skin pixel. The recognition result is often unsatisfactory because the threshold is difficult to select. A novel skin color detection method with spread seed in sequence of frames based on local information consistency is proposed. The sequence of frames is obtained by tapered threshold. The algorithm consists of two steps. One is to select the initial seed pixels. And the other is to detect skin pixel according to spread method based on the local information connectivity and consistency. Experimental results show that after combining with our proposed method the traditional skin segmentation algorithm are significantly improved.

KEYWORDS: skin detection; Sequence of frames; local information; seed pixels; spread.

1 INTRODUCTION

Skin detection is an important part of digital image processing, and it is widely used in the face recognition, content-based image retrieval, pornographic image filtering and so on. There are many factors affecting the skin detection results, such as uneven illumination, different camera, different ethnic, and the information similar to skin.

Skin detection has been extensively studied over the years, and many scholars are committed to further improve skin detection results. Skin detection can be divided into four categories: explicitly defined skin region method, parametric skin distribution modeling method, nonparametric skin distribution modeling method and dynamic distribution modeling method. Explicitly defined skin region method is very simple, it is based on that the human skin color cluster in a small region in color space[1]. Basilio et al. [2] used fixed threshold to segment skin in YCbCr color space, Wang and Yuan defined skin color in rg and HSV color space, the skin color is fall in the range Rr=[0.36,0.465], Rg=[0.28,0.363], RH=[0,50], RS=[0.2,0.68], RV=[0.35,1.0]. Parametric skin distribution modeling method include Gaussian model and elliptical boundary model. Prem et al[3] using a single Gaussian model (SGM) to calculate the probability of pixels are skin, Jones and Regh [4] were trained skin color model and non-skin color models, each uses a 16 Gaussian kernel. This method is called Gaussian mixture model (GMM). Nonparametric skin distribution modeling method mainly refers to the statistical model. Jedynak[5] built three skin models based on the maximum entropy principle, which are Baseline Model, Hidden Markov Model and First Order Model. Dynamic distribution modeling method, including Dynamic histogram model[6] and adaptive skin model[7], is mainly used to solve the uneven illumination, brightness change and other issues.

Explicitly defined skin region method is simple, but the effect is often unsatisfactory. The modeling method include parametric, nonparametric and dynamic distribution is a hot research. Although many scholars have continued to improve the optimal model, there are still two problems they ignored. The first problem is that the threshold used to distinguish skin and non-skin in the criteria is hard to determine. For one thing, if images contain no skin-like color, the accurate threshold of different image is different. For another, if images contain skin-like color, there is not an accurate threshold which can perfectly differentiate skin from non-skin. The second problem is the information used to detect skin is insufficient. Human detect skin from an image based on not only the priori knowledge, but also the association information, but the traditional skin segmentation algorithm ignore this issue.

A novel skin color detection method with spread seed in sequence of frames based on local information consistency is proposed to solve the above two problems. Firstly, sequence of frames are produced based on the gradual change threshold. And then select the initial seed points and detect skin color using seeds spread method according to local information connectivity and consistency. The result of the traditional color segmentation algorithm are significantly improved after being combined with the proposed method.

2 PROPOSED METHOD

2.1 *Selection of the initial seed pixels*

Suppose E is set of the skin color, A={A_1,A_2, ..., A_m} is the set of the skin area of the input image, which contains M disjoint skin region, R={(x_1,y_1),... (x_n,y_n)} is the initial set of seed points, the ideal R is described as follow:

1 $R \subseteq \boldsymbol{E}$, that means the initial seed points must be skin, which will ensure the correctness of skin recognition in the spread algorithm.
2 $\forall A_i \in A, \exists r(x_j, y_j) \in R$, meet $r \in A_i$, that means at least one skin pixel of each skin area is to be selected in the initial set of seed points.

The actual selection of the initial set of seed points is often difficult to strictly meet these two constraints. There are many reasons, for example, some skin area is too small, or the color of the skin area is distortion because of overexposed. So the initial set of seed points can only approximate perfect.
Search algorithm is adopted to select the initial seed pixels. Firstly, let the initial ca larger value (such as 100), in this case there are no skin pixel is detected according to formula (1). And then, reduce θ by a certain percentage, until a few skin pixels are detected. These skin pixels are the initial seed pixels.

2.2 *Local consistency computation based on improved LBP operator.*

There are two key points in spread algorithm: connectivity and local consistency. Connectivity refers to the pixels are whether adjacent, and local consistency means the local information of the skin pixels are similar in the same image, meanwhile there are large difference between the skin pixels and the non-skin pixels. The computation of the local information consistency is adopted improved LBP operator which is proposed in this paper.

1 Improved LBP operator
LBP(Local Binary Pattern) refers to the local information of a pixel, which is proposed by Ojala[8], and then It is improved to become uniform LBP operator[9]. LBP operator is simple and effective, however, as the represents of the region information, LBP only considers the relationship between each center pixel and its neighborhood, while ignoring the center itself and the relationship between its neighborhood pixels. All of them are very important, so this paper proposed the improved LBP operator which consists of three parts: the original LBP information, center pixel information and neighborhood pixel information. The improved LBP operator is described as follow:

$$LBP = \sum_{p=0}^{N-1} s(y_p - y_c)2^p + s(y_c - m)2^N + \sum_{p=0}^{N-1} s(y_p - y_{(p+1)\%N})2^p$$
$$m = \frac{1}{N}\left(\sum_{p=0}^{N-1} y_p + y_c\right), s(x) = \begin{cases} 1, & x \ge 0 \\ 0, & x < 0 \end{cases} \quad (2)$$

Where the first part is the original LBP information, it ranges [0,255]. The second part is the center pixel information, it ranges {0,256}. The last part is the neighborhood pixel information, it ranges [0,255].

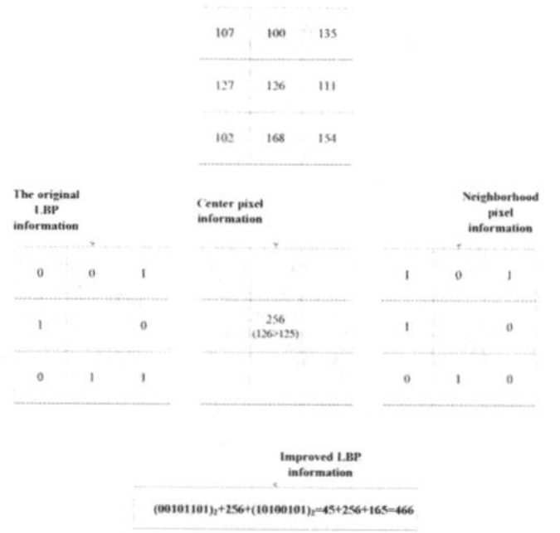

Figure 1. Improved LBP operator.

2 The local information consistency
Suppose R={(x_1,y_1),...(x_n,y_n)} is the set of the initial seed pixels, T is the input pixel. C={c_1,...,c_n} and c_t are the local information of the initial seed pixels and the input pixel T, which is calculated by formula (3). The local information of the input pixel T and the initial seed pixels are consistent when they satisfy the following conditions:

$$\exists c_i \in C, dist(c_i, c_t) \le \lambda$$

where $dist(a,b) = \sum_{i=1}^{k} a(i) \oplus b(i)$, $a(i)$ and $b(i)$ respectively represent the i-th bit of a and b.

2.3 *Skin color detection with spread seed in sequence of frames based on local information consistency*

the sequence of the frames are produced from the input image, the current seed image is generated by the past seed image and the current frame image. The final skin segmentation results are obtained after the seed spread algorithm are carried out in all the sequence of the frames. The detail of the process is described as follow:

1 Generate the sequence of the frames based on the gradual change threshold θ in formula (1) of the input image, the probability of the pixel is skin P(skin/F_G) and the probability of the pixel is non-skin P(non-skin/F_G) are obtained according to

the trained skin and non-skin models. Suppose I(n),n=2…N are the sequence of the frames.
2 Select the initial seed pixels, suppose the image of the initial seed pixels is S(1). The detail is described in section 1.2
3 Deal with the sequence of the frames I(k),k=2…N successively, generate the corresponding seed image S(k), ,k=2…N. the first k seed images S(k) is obtained based on the seed image S(k-1) and the current frame image I(k), the detail of the spread algorithm is described as follow: If the foreground pixel of the frame image I(k) and the seed pixels of the seed image S(k-1) satisfy connectivity and local consistency, let the pixel retain in the seed image S(k), otherwise discard it. The pixel of the seed image S(k) is consisted of the retain pixel of the I(k) and the foreground pixels of the S(k-1). This process can be completed by two scanning process, that is a forward scanning process and a reverse scanning process.

Forward scanning process is described as follow. starting from the top left corner pixel of the image S(k-1),according to the order from left to right , top to bottom, visit each pixel of the image S(k-1). Suppose G(x,y) is the pixel in image S(k-1) which is visited right now, its right neighbor is G(x,y+1), bottom neighbor is G(x+1,y), and its right bottom neighbor is G(x+1,y+1), judge the three corresponding pixel of the three neighbor pixel in image I(K) respectively. If it is foreground pixel and satisfy local consistency with the seed pixel in image S(k-1), the neighbor pixel is marked as seed pixel(skin pixel).Reverse scanning process is similar to the Forward scanning process
4 (4)the finial S(N) which is the skin detection result.

3 EXPERIMENTAL VALIDATION

Confirm the truth of the effectiveness of the skin detection algorithm requires a large number of test samples, so the famous Compaq skin database[4] is adopted. The Compaq skin database is the largest skin database, including more than 14,000 images. More than 2 billion pixels are labeled by Jones and Regh, the pixel is either belong to skin or non-skin. All the images are collected from the Internet randomly by crawlers to ensure its broad applicability. The database was split into two equinumerous subsets, one is training dataset, the other is test dataset. Training dataset has 5000 images, 2000 of them are skin images and the other 3000 images are non-skin images, including 25,465,467 skin pixels and 421,713,655 non-skin pixels. Test dataset has 5849 images, 2662 of them are skin images and the other 3187 images are non-skin images, including 54,844,888 skin pixels and 650,979,755 non-skin pixels.

In order to verify the validity of the proposed algorithm, FSS(frames spread segmentation) is called for short, four typical skin detection method is adopted. The first is the histogram statistical model[4] which is proposed by Jones and Regh. This method use naive Bayesian as the criteria, and its color space is RGB. The second one is histogram statistical method based on CbCr color space without the luminance component[10]. The third one is Gaussian Mixture Model(GMM) [11] which is proposed by Yu et al. It is assumed that the skin subject to Gaussian distribution. The last one is Max Entropy Model(MEM) [5] which is proposed by Jedynak et al. Its basic idea is the entropy will be get largest after the input pixel is classified.

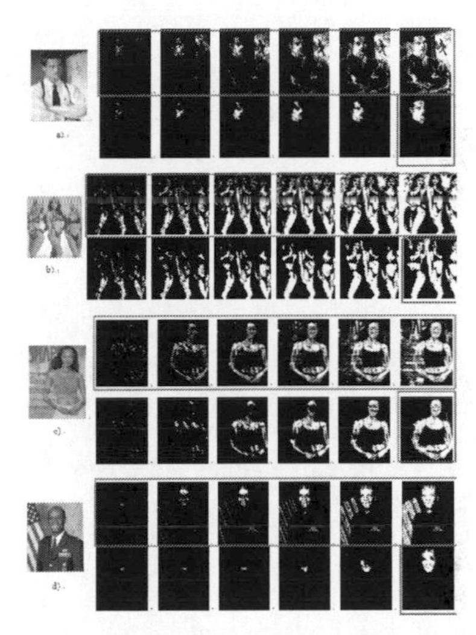

Figure 2. Experimental results.

As shown in Fig. 2, the images in the blue box are the sequence of the frames generated by GMM algorithm(GMM is widely used and has a good effect in skin detection), in which different thresholds corresponding different segmentation results, but all the segmentation results are obvious mistake segmentation. The images in the red box are the skin detection results after combined with our proposed method, the effect has been significantly improved. A key of the proposed algorithm is the choice of the initial seed pixels. Perfect initial seed pixels should not only contains no non-skin pixel, but also cover all the skin area. As shown in Fig. 2(a), the initial seed pixels miss the finger area, so the detection result miss this area too. In Fig. 2(b), the initial seed pixels contains non-skin pixels, the segmentation result has mistakenly segmented regions in the upper left corner. On the other hand, even if the initial seed pixels are perfect, the segmentation result may have some mistaken area. As shown in Fig 2(c), the initial seed pixels are correct, but there have been some mistake segmentation in the edge section of the detected result because the background of the edge portion is similar between skin area and non-skin area. In Fig 2(d), the above problems do not exist, so the skin detection result are very accurate.

Table 1. Experimental results.

	TPR	*FPR*	*Precision*
RGB	0.8367	0.1551	0.8436
RGB+FSS	0.8904	0.1432	0.8615
CbCr	0.8252	0.1440	0.8514
CbCr+FSS	0.8813	0.1401	0.8628
GMM	0.8479	0.1469	0.8523
GMM+FSS	0.9012	0.1413	0.8645
MEM	0.8226	0.1429	0.8520
MEM+FSS	0.8798	0.1397	0.8630

Table 1 shows the results of comparison of the four original skin detection method and improved method(combined with the proposed FSS algorithm). It can be seen that after combined with the proposed FSS algorithm, the skin detection algorithm can be significantly improved. This is due to there is no exact threshold in the original skin detection, FPR increase with TPR, this will lead to the detection result either lost a lot of skin pixels, or contains a large number of non-skin pixels. After combined with our proposed method, the skin detection algorithm can not only retains the skin pixels, but also exclude the non-skin pixels. This is due to two guarantee. First is the proposed method is starting from the initial seed image in which the pixels belong to skin. And the second is in the spread process, during which the skin pixels will be detected and the non-skin pixel will be abandoned.

4 CONCLUSION

A novel skin color detection method with spread seed in sequence of frames based on local information consistency is proposed. Firstly, sequence of frames are produced based on the gradual change threshold. And then select the initial seed points and detect skin color using seeds spread method according to local information connectivity and consistency. The result of the traditional color segmentation algorithm are significantly improved combined with the proposed method. However, there are two problems in the proposed algorithm, one is the incorrect initial seed pixels will lead the incorrect segmentation, and the other is over spread may appear in the border section, how to solve the two problems is the study emphasis in future research.

REFERENCES

[1] Yang, J., W. Lu, and A. Waibel, Skin-color modeling and adaptation1997: Springer.

[2] Basilio J A M, Torres G A, Pérez G S, Medina L T, Meana H M. Explicit image detection using YCbCr space color model as skin detection [C]. Applications of Mathematics and Computer Engineering, 2011: 123–128.

[3] Prem K, Prasad G, Subbanna B P, Sumam D S. Human face detection and tracking using skin color modeling and connected component operators [J]. IETE Journal of Research , 2002, 48(3–4): 289–293.

[4] Jones M, Regh J. Statistical Color Models with Application to Skin Detection [J]. International Journal of Computer Vision, 2002, 46(1): 81–96.

[5] Jedynak B, Zheng H C, Daoudi M. Maximum Entropy Models for Skin Detection [J].Energy Minimization Methods in Computer Vision and Pattern Recognition, 2003, 2683: 180–193.

[6] Soriano M, Martinkauppi B, Huovinen S, Laaksonen M. Skin detection in video under changing illumination conditions [C]. 15th International Conference on Pattern Recognition, 2000:839–842.

[7] Yang G L, Li H, Zhang L, Cao Y. Research on a Skin Color Detection Algorithm Based on Self-adaptive Skin Color Model [c]. 2010 International Conference on Communications and Intelligence Information Security,2010:266–270.

[8] Ojala, T., M. Pietikainen, and D. Harwood. Performance evaluation of texture measures with classification based on Kullback discrimination of distributions. in Pattern Recognition, 1994. Vol. 1 - Conference A: Computer Vision & Image Processing., Proceedings of the 12th IAPR International Conference on. 1994.

[9] Ojala T,Pietikainen M,Maenpaa T. Multiresolution gray-scale and rotation invariant texture classification with local binary patterns [J]. IEEE Transactions on Pattern Analysis and Machine Intelligence, 2002, 24(7): 971–987.

[10] Xue J, Wang Z Y. An Improved Statistic Model in Skin Detection [J]. Journal of Information and Computational Science, 2014, 11(11): 3643–3651.

[11] Yu C C, Cheng H Y, Lee C C. Mixture models with skin and shadow probabilities for fingertip input applications [J]. Journal of Visual Communication and Image Representation, 2013, 24(7): 819–828.

Computing, Control, Information and Education Engineering – Liu, Sung & Yao (eds)
 ISBN: 978-1-138-02800-5

Research on constructing integrated information service platform of "Three Rural Issues" based on agricultural cloud computing technology

Li Wei Geng & Juan Zhou
Agricultural University of Hebei, Hebei, China

ABSTRACT: Cloud computing is a new compute mode, whose computing power is strongest, economical and flexible, therefore it is very meaningful to apply this technology to resources sharing of agricultural information. It is very feasible to use cloud computing in resources sharing of agricultural information in terms of the theoretical foundation, technical advancement, practical operability, operating costs, etc. The author explores a new mode to deal with the current "Three Rural Issues" in the paper—agricultural informatization planning, analyzing the current situation of agricultural informatization construction with the help of unique research and academic strength of Shenyang Agricultural University. On the other hand, the paper studies the information client and the specific features of the softwares of the "Three Rural Issues" based on the Android System to build agricultural information service platform under the condition of cloud computing, and further provides a theoretical and technical support for the operation and upgrade of the platform.

KEYWORDS: Three Rural Issues; cloud computing; information platform; construction.

In 2012, a report released by CNNIC showed that the number of rural Internet users had reached 136 million at the end of 2011, which accounted for 27% of China's total number of Internet users and 21% of the total rural population; in addition, the number of internet users on mobile phones was 360 million, and internet users of mobile phones in rural areas was close to 100 million, which was 71% of the total number of internet users in rural areas. The popularity of smart phones with lowest price made in China makes it easy for farmers search the Internet, and there are a large number of new users per day from rural areas, especially mobile phone-based Internet searching model is rapidly spreading to the vast countryside. In addition the Ministry of Agriculture showed that they had signed a series of agreements with telecom operators in 2012, at the same time they brought 3G technology, the Internet of things, cloud computing technology, etc. to "Three Rural Issues" category, which laid a solid foundation for the construction of agricultural information technology. China Telecom shows high enthusiasm for the joining to "Three Rural Issues" informatization construction, which effectively promotes the construction process of of rural e-government platform, joint-defense information sharing, agricultural products express, the development of modern agriculture, etc.

From the perspective of farmers, the "Three Rural Issues" information they need includes: the latest national agricultural policy, local agricultural support information and new government requirements on agricultural construction; agricultural technologies in the production process of sowing, cultivation, irrigation, deinsectization, etc.; market of harvested agricultural products are naturally the issue farmers concern, so they need to understand and master the information of the latest business information and market dynamics information of sales section before they can better arrange specific planting plan. Because the object of "Three Rural Issues" information is for farmers, so there is a huge user group and at the same time information services should also be a high priority. The high-speed development of information technology generates network technology and cloud computing technology. The B/S operating platform is used more widely, and cloud computing technology platform based on SaaS (Software as a Service) has become the mainstream of the industry with its unique advantages of quickness, large capacity, zero cost, etc. Many IT service firms with certain size combine 3G technology with cloud computing technology to create cloud mobile phones to meet the needs for information searching, reading, storing and applying. The authors discuss how to build "Three Rural Issues" information platform based on cloud computing technology by analyzing and exploring "Three Rural Issues" information platform and cloud computing technology.

1 AN INTRODUCTION TO CLOUD COMPUTING TECHNOLOGY AND "THREE RURAL ISSUES"

1.1 *An introduction to cloud computing technology*

This technology is a computing mode opening up to qualified users to provide them with computing services through internet technology and collecting and making information processing capacity and data into batching and multi-layered mode, and cloud computing technology allows all types of operating systems have some computing power, storage capacity and information service capability. Under the condition of cloud computing technology, the application program's user operates will not run on the electronic device used by them, but run on the specific Internet server. And the data reading and processing will not exist on the electronic device used by the user, but stored in a specific database on the Internet. Once the user operates it, he can use any device anywhere to read the past data or do other treatments. Specifically, cloud computing technology has the following advantages:

a. Economical property. It can save costs effectively, changing spending into business expense with a rate of return. Because the equipments cloud computing technology needs are generally provided by third parties, so users can enjoy the services of cloud computing technology without having to pay any extra costs to buy equipments.
b. Flexibility. Service can be used at any time and anywhere. The service is fast and convenient, and users will be able to get the desired service without having to understand the specific operating principle of cloud computing.
c. Sharing property. Information resources can be shared among users, and single user does not need to pay the high cost.
d. Security. Since it implements a centralized management mode, the security of cloud computing technology is very good, and suppliers have already done the information security and protection work before the data are given to users.

1.2 *An introduction to "Three Rural Issues"*

We may hear the term of "Three Rural Issues" every day, in fact, it is a general term for agriculture, rural areas and farmers and also a comprehensive problem of industry, residence and identity in nature, but each has its own focus areas and comprehensive consideration and exercise due diligence should be noticed. China is an agriculture-based country, "Three Rural Issues" not only affects the overall quality and economic progress of the country, but also affects the social development and other the states' prosperity. Understanding "Three Rural Issues" from its literal meaning, then it is farmers who make a living by planting crops, breeding animals the vast rural areas, and identity is a social problem of the vast majority of the improvement of national living conditions and agricultural structure adjustment and upgrade. Looking at it from the combination of the problem evolution, then it means the apparent current dual economy of China, and the cities develop rapidly and the living standards of citizens improve significantly; due to the poor infrastructure in rural areas and the lack of information, the improved speed of farmers' living standards is far lagging behind. China's economy contains infinite potential for development, if we only consider the quantity of economic growth without concerning the quality, then the economic problems troubling the government are closely related to rural areas, agriculture and farmers. In essence, the "Three Rural Issues" is actually an important manifestation of the imbalance between urban and rural development and structural proportion.

1.3 *Construction situation of agricultural informatization*

Currently, many of the rural areas in China are still in the isolated and information-blocking state, and cloud computing technology has almost not been applied. The lagging information construction is mainly manifested in the following aspects: a. information resources do not have the integrity, low degree of integration and sharing of resources, there is no effective incentive mechanism to promote the further use and development of information resources, low development level, database construction situations are diversified into different areas, delayed related database maintenance, data update, market research and other works, resulting in a serious lack of effective information resources; b. Information contents and site layout and operation are almost the same, and even the function keys and window of the websites are very similar; c. monotonous information form, information needs of farmers include a wide range of items, such as: planting techniques, breeding techniques, etc., which must rely on pictures and videos to show its functions, but agricultural sites now are almost all static pages, and there is no dynamic presentation and information is conveyed in text form, so the standardized level is not high, the websites are lifeless without any personality and characteristic.

2 FRAMEWORK AND FUNCTIONAL ANALYSIS OF "THREE RURAL ISSUES" INFORMATION PLATFORM

"Cloud" platform of agricultural information users can be divided into internal and external styles, so as to achieve the purpose of extending the platform resources effectively. The operation mode of internal user provides data access and data query services for external users through connecting the subsystem to database by the experts tean of Shenyang Agricultral University with the help of heterogeneous integration and subsystem based on cloud computing. The system is based on design specifications that supports the use under different operating systems environments, so it can be installed to various operating systems. In addition, the system also sets different levels of administration authorities, applying administration + Category + managing logs, which strictly regulates administrator's authority and operation; in order to facilitate the replacement of management positions, the administrative privilege is set up by administrator uniformly, supporting multi-threaded information uploading, information searching, information retrieval, etc.

2.1 *Construction mode of the platform framework*

The technical framework platform can be divided into five levels: system hardware and software layer for basic operation, retrieval, storage layer of the underlying database, real-time sharing layer of data and specific service layer of the user. The construction of agricultural information platform can be divided into information resources construction and sharing of information resources. Information resource construction technologies are mainly: the establishment and setup of the database, the processing and summary of heterogeneous data; sharing of information resources are: search navigation, self-searching, expert diagnosis, asking advice, analysis and forecasting, extraction and sharing of data, etc.

2.2 *Platform's specific functions*

Platform integrates network and agriculture-related information effectively, and it collects information distributed in heterogeneous and put them into platform to construt a comprehensive agricultral information platform to help users find the information or data they need in a fast and convinient way through consistent data element interface, information description element, information-sharing agreements, etc. The platform's design uses a organic system with the combination of centralization and decentralization, and the service network allows users to enter the system easily from any service network platform to obtain any information in the service network through multi-layers design. The specific informations include: agricultural policy issued by the government, the government work dynamics, etc.; all kinds of market demand, labor demand, advanced agricultural technology, rural market conditions and so on; current employment situation, farmer entrepreneurship policy, state support conditions, work disputes, legal aid, etc.; the latest prices and supply and demand for agricultural products throughout the country, etc.

3 DESIGN OF AGRICULTURE INFORMATION PLATFORM

For "Three Rural Issues" information platforms in all areas, the relevant staff plays a crucial service role. For the current system IT resources, to build "Three Rural Issues" information platforms based on cloud computing technology and develop clients and relevant softwares based on the Android system enable farmers and the associated staff enjoy efficient cloud computing services at any time and anywhere.

The Information Interaction Mode applied by "Three Rural Issues" information platform is seen in Figure 1. Farmer raises the problems he wants to resolve with cell phone cloud platform, and agricultural technician analyzes and processes the information he received from the farmer with professional knowledge to answer questions raised by the farmer, providing "Three Rural Issues" technical support for the farmer and send the information back to the farmer's client quickly. To interpret the latest agricultural policies issued by government sectors, according to the local specific circumstances, and then sends the information to the farmer's phone platform through information platform. Workers of supply and marketing cooperative gather agricultural marketing sales information, and to summarize and organize them, enabling farmers to be more clear about the business information.

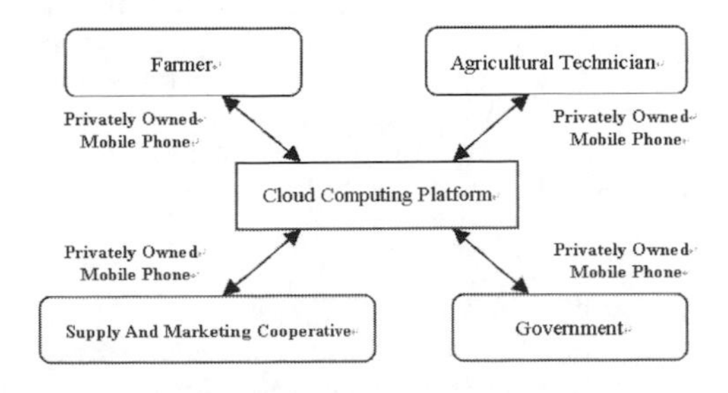

Figure 1. Information interaction mode.

3.1 *Cloud computing platform*

in fact, when farmers exchange information with "Three Rural Issues" staff, it relates to the access to the central database, Web information and data. The basic structure of private cloud storage system WFS (Web File System) includes: data storage layer, data management layer, software butt layer and user access layer. Data storage layer can be effectively compatible with all kinds of storage devices, and thus to achieve rapid arrangement of the server in the supply and marketing cooperative system, which can significantly reduce the cost of inputs; data management layer store data by means of the virtual path, so as to enhance the safety of the system and reduce redundant data effectively; software butt layer has multi-access interfaces of many types, so users can achieve connectivity with storage systems in various ways; user access layer can achieve multi-path access to the user, thereby enhancing service efficiency of the data.

Data storage layer is the most basic part inWFS, WFS can store firewall into the storage device in the shortest time with the help of basic storage management function. Data management layer is the most critical part, it can achieve the collaborative work of the different memory devices with the cluster and distributed processing technology, allowing different memory devices, combined with each other to provide the same services for external users and provide a more convenient and efficient access to data function, especially for those service platforms with large user base and a huge amount of real-time access. In the software butt layer, WFS provides various types of services interfaces, providing exclusive services in the nature of each interface, so users can not only browse and store files through a browser, but also can share files with others. Any authorized users can get access to the system directly, enjoying data query and processing services provided by the system. WFS make authorization in accordance with the identity of farmers, agricultural technicians, relevant government departments, administrators, etc., providing the appropriate level of services contents. Users can use different interfaces to upload files onto WFS, and the system will process these data and information appropriately; when it stores data, the independent technology of data and metadata will be used, and all of the metadata information is automatically stored in the database, thus achieving multi-user, simultaneous access, real-time sharing, self-search, etc. For data information of "Three Rural Issues".

3.2 *Terminal system of mobile phone*

On the Android system of cloud phones, make the relevant services of "Three Rural Issues" information as the basic theme to design system interface according to farmers' preference of easy acceptance and operation. For agricultural services, providing agricultural services online and design the related softwares can be convinient for farmers to upload the situation they encounter onto system in the form of photographs, so that technician can understand the situation in various aspects; in addition, the system should set up a special "Three Rural Issues" technical information database, so that farmers can easily find their own solutions to problems they encounter and achieve online communication with "Three Rural Issues" experts through network and the. Cloud system can automatically store all interaction informations, and classify and store them. microblog and information push cloud module of mobile phone can help farmers keep abreast of weather information, policy trends, product market, etc. Program that the cloud mobile phone has will automatically be upgraded, so no additional downloading and installation are needed. For individual users, batch storage capability can meet user's information storage needs.

4 CONCLUSION

Open eyes to the whole world, the application of cloud computing technology in high-tech industry is becoming more and more widely. American government has improved the total budget funding for the development and application of cloud computing technology; Japanese has invested huge cloud infrastructure(Kasumigaseki Cloud); India and some other countries regard the emerging cloud computing technology as a new development direction of the information industry. In the government's strong support, many countries do in-depth research on the specific application of cloud computing technology in agriculture, for example: the recent rising of agriculture cloud operators Farmeron can provide corresponding data record for farm, so that scientific analysis on animal breeding on the farms, costs and benefits, the farm operations, etc., providing guidance and other arrangements of the next season's production for farmers and helping them to save costs and improve earnings, and the company has set up agricultural management services platforms in 15 countries. Compared with other countries, theory and practice in this area has lagged behind a lot in China, and it is basically still in its infancy. under this environment, promoting agricultural development and progress with the help of cloud computing technology can not be successful without the support and guidance of government departments. Of course, technology research requires a certain process, it is impossible to achieve the desired effect in a short time, but we believe that as the most respected calculation method in the field of IT, cloud computing technology will give its full play and show its unique functions in the development process of agriculture.

ACKNOWLEDGEMENTS

1 2014 Planning Research of Baoding Science and Technology Bureau, code 2014ZG019;
2 2012 Planning Research of Baoding Science and Technology Bureau, code2012ZN004;
3 2012Agricultural University of Hebei fund, code LG20120203.

REFERENCES

[1] GUO Meirong, LI Jin, QIN Xiangyang. *Discussion on the Cloud Service Platform Architecture of Agricultural Information Services*[J]. Agriculture Network Information, 2012, 2, 10(12): 104–105.

[2] Cui Wenshun. *Application and Developing Prospect of Cloud Computation in the Agricultural Informationization*[J]. Agricultural Engineering, 2012(1): 41–42.

[3] LUO Jun-zhou, JIN Jia-hui, SONG Ai-bo, DONG Fang. *Cloud computing: architecture and key technologies*[J]. Journal on Communications, 2011, 5, 20(7): 5–6.

[4] Li Qiao ZHENG Xiao. *Research Survey of Cloud Computing*[J]. Computer Science, 2011, 38(4): 34–35.

[5] Zhang Xingwang, Li Chenhui, Qin Xiaozhu. *Research on Digital Information Resources Construction Model Based on Cloud Computing*[J]. Information Studies: Theory & Application, 2011, 8: 101–102.

About the Author

Geng Li Wei (1982-), female, Baoding City, Hebei Province, Master Degree, Lecturer, research direction: Measurement and Control Technology and Intelligent Monitoring.

Computing, Control, Information and Education Engineering – Liu, Sung & Yao (eds)
© 2015 Taylor & Francis Group, London, ISBN: 978-1-138-02800-5

An application of the intelligent controller on the pulp concentration adjustment

Xiao Xi Zheng & Chao Hua Ma
Zhengzhou Technical College, Zhengzhou, Henan, China

ABSTRACT: In the paper production process, it is essential to adjust the pulp concentration with high stabilization. However, the controlling system of the pulp concentration suffers from the algorithm's long lags and the model's indetermination. The neural network technology, which is characterized by its self-learning and arbitrarily nonlinear expressing ability, is introduced to optimize the controlling system. An intelligent controlling method of auto-tuning the PID parameters with the neural network is proposed to adjusting the pulp concentration. The experimental results indicate that the system is of high adaptiveness, high stabilization and high robustness with the proposed method.

KEYWORDS: Neural network; Pulp concentration; PID; Parameter self-turning.

1 INTRODUCTION

Pulp concentration with high stabilization is an important part in the paper production process, It is also one of the difficult problems. The controller designed by traditional method is difficult to satisfy the requirement of practical production, Which Can't keep the optimal operation and even appear the stability problems.

PID control is a commonly used method In the process of industrial control. However, in the production of paper pulp concentration adjusting process, the mechanism is more complex, time-varying uncertainties exist, using a conventional PID controller is difficult to obtain the satisfactory control effect. The neural network has a good approximation ability of nonlinear mapping and adaptive learning, parallel distributed processing, etc.. In order to have better adaptability of PID controller, Combining the advantages of both, using a PID controller based on neural network to control the concentration, the system has a self-adaptive, can automatically adjust the control parameters, To adapt to the change of the controlled process and improve the control performance and reliability.

Automatic control system of pulp concentration as shown in Figure 1, which was composed of sensor, controller and control valve, etc. Slurry concentration is converted into a current signal of 4-20mA by the transmitter, and it will be sent to the controller of concentration. Current signal is converted into numerical signals by the A/D to display, and compare the concentration value and the preset value, the controller adjusts the opening of electric valve according to the difference, adjust the dilution water into the slurry pump automatically, slurry concentration change at this moment, the concentration of transmitter will detect a new concentration value. Repeat the above process, makes the slurry concentration gradually close to the level set by the user, the resulting stable slurry concentration.

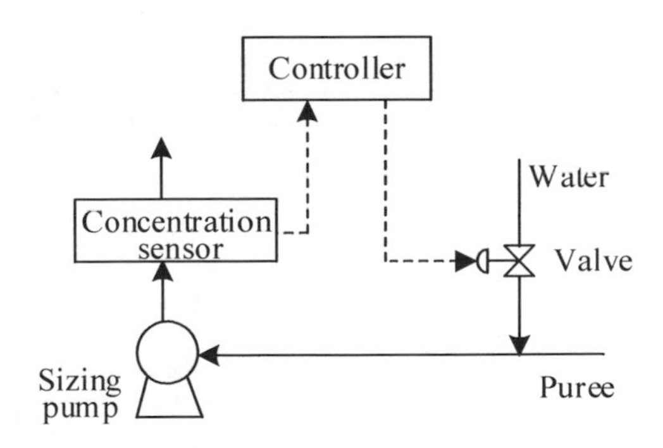

Figure 1. Pulp concentration adjusting process.

The simplified transfer function of the pulp concentration system is

$$\frac{Y(s)}{U(s)} = \frac{K}{(Ts+1)} e^{-\tau s} \tag{1}$$

Determine the transfer function of the parameters T,τ and K using the step response method.

2 NEURAL NETWORK BASED PID CONTROLLER

The system structure of PID controller based on BP neural network as shown in Figure 2.

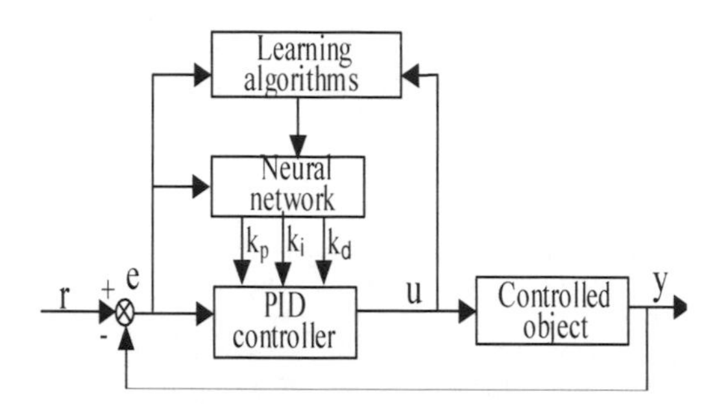

Figure 2. System structure of PID controller based on BP neural network.

PID controller based on BP neural network is composed of the traditional PID controller and neural network.

Traditional PID controllers: The controlled object is controlled directly using Closed-loop, on-line tuning parameters k_p,k_i, k_d.

Neural network: In order to achieve certain performance optimization, Adjust the parameters of the PID controller according to the system's operating status, Three adjustable parameters of PID controller k_p, k_i, k_d, respectively corresponding to the output of the output layer neurons, by neural network to the weighted coefficient of adjusting, self-learning, make the neural network output corresponding to some optimal control rule of PID controller parameters. Using the neural network to adjust the weighting coefficient, self-learning, make the neural network output corresponding to some optimal control rule of PID controller parameters.

Based on self-learning neural network controller, PID self-learning controller constructed using BP neural network by its inputs and outputs to achieve incremental digital PID controller correspondence.

3 IMPLEMENTATION OF NEURAL NETWORK BASED PID CONTROLLER

The discrete form of incremental PID control algorithm is

$$u(k)=u(k-1)+k_p(e(k)-e(k-1)) \\ +k_i e(k)+k_d(e(k)-2e(k-1)+e(k-2)) \tag{2}$$

Where $u(k)$ is controlled variable of the current sampling time, $e(k)$ is the error between the desired value and the actual output value.

According to the multi-layer feed-forward network of approximation theory, Under the condition of a sufficient number of hidden neurons, Three layer network with sigmoid nonlinear characteristics can be any precision, the realization of arbitrary nonlinear mapping from the input space to the output space. BP neural network in this paper adopts three layers of structure, the structure as shown in Figure 3.

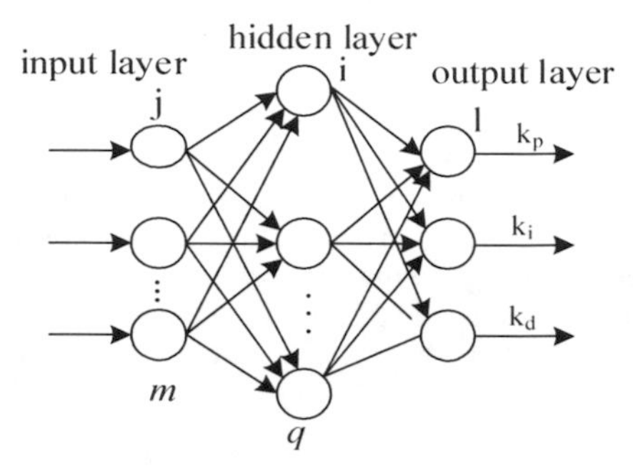

Figure 3. structure of BP neural network.

The input layer of the network structure with m nodes, q nodes of the hidden layer, the output layer has 3 nodes. The output nodes correspond to the three parameters of PID controller: k_p, k_i, k_d, because of k_p, k_i, k_d can't be negative, take the nonnegative sigmoid function as the output layer neuron activation function, take the symmetrical sigmoid as the excitation function of hidden layer neurons. By the figure 3 can be seen, the input function of the BP neural network as follows:

$$o_j^1 = x(j) \quad j=1, 2,..., m \tag{3}$$

The number of input variables m, depending on the complexity of the controlled system. The input function of hidden layer as follows:

$$net_i^2(k) = \sum_{i=0}^{m} w_{ij}^2 o_j^1 \tag{4}$$

The output function of hidden layer as follows:

$$o_i^2(\mathrm{k}) = f(net_i^2(k)) \quad i=1, 2,\ldots,q \tag{5}$$

Where$\left\{w_{ij}^2\right\}$ is a Hidden layer weighting coefficient, Superscript 1,2,3 represent, respectively the input layer, hidden layer and output layer. Excitation function f(x) of hidden layer neuron is the hyperbolic tangent function, namely

$$f(\mathrm{x})=\frac{e^x - e^{-x}}{e^x + e^{-x}} \tag{6}$$

Three nodes function of input and output in network output layer for each are

$$net_l^3(k)=\sum_{i=0}^{q} w_{li}^3 o_i^2(k) \quad l=1, 2, 3\ldots \tag{7}$$

$$o_l^3(k)=g(net_l^3(k)) \quad l=1, 2, 3\ldots \tag{8}$$

Where w_{li}^3 is the weighting coefficient of output layer, the neuron activation function of the output layer is:

$$g(x)=\frac{e^x}{(e^x+e^{-x})} \tag{9}$$

The output of the BP neural network output layer is three control parameters of PID controller, I.e.

$$o_1^3(k)=k_p \tag{10}$$

$$o_2^3(k)=k_i \tag{11}$$

$$o_3^3(k)=k_d \tag{12}$$

Take performance index function for

$$E(k)=\frac{1}{2}(r(k)-y(k))^2 \tag{13}$$

Fixed network weights according to the gradient descent method, the weighted coefficient according to $E(k)$ of the negative gradient direction search, with a fast convergence of global minimum inertia item search.

$$\Delta w_{li}^3(k)=-\eta\frac{\partial E(k)}{\partial w_{li}^3}+\alpha\Delta w_{li}^3(k-1) \tag{14}$$

In the formula, η is the learning rate;α is inertia coefficient.

$$\frac{\partial E(k)}{\partial w_{li}^3}=\frac{\partial E(k)}{\partial y(k)}\cdot\frac{\partial y(k)}{\partial u(k)}\cdot\frac{\partial u(k)}{\partial o_l^{(3)}(k)}\cdot\frac{\partial o_l^{(3)}(k)}{\partial net_l^{(3)}(k)}\cdot\frac{\partial net_l^{(3)}(k)}{\partial w_{li}^{(3)}(k)} \tag{15}$$

$$\frac{\partial net_l^3(k)}{\partial w_{li}^3}=o_i^{(2)}(k) \tag{16}$$

$$\frac{\partial u(k)}{\partial o_1^{(3)}(k)}=e(k)-e(k-1) \tag{17}$$

$$\frac{\partial u(k)}{\partial o_2^{(3)}(k)}=e(k) \tag{18}$$

$$\frac{\partial u(k)}{\partial o_3^{(3)}(k)}=e(k)-2e(k-1)+e(k-2) \tag{19}$$

The learning algorithm of network output layer is as follows:

$$\delta_1^3=e(k)\operatorname{sgn}\left(\frac{\partial y(k)}{\partial u(k)}\frac{\partial u(k)}{\partial o_l^3(k)}\right)g'(net_l^3(k))$$

$$(l=1, 2, 3) \tag{20}$$

$$\Delta w_{li}^3(k)=\eta\delta_l^3 o_i^2(k)+\alpha\Delta w_{li}^3(k-1) \tag{21}$$

The learning algorithm of hidden layer weighting coefficient are as follows:

$$\delta_i^2=f'(net_i^2(k))\sum_{i=1}^{3}\delta_l^3 w_{li}^3(k-1) \quad i=1,2,\ldots,q \tag{22}$$

$$\Delta w_{ij}^2(k)=\eta\delta_i^2 o_i^1(k)+\alpha\Delta w_{ij}^2(k-1) \tag{23}$$

Where

$$g'[.]=g(x)[1-g(x)];f'[.]=[1-f^2(x)]/2 \tag{24}$$

The PID controller based on BP neural network, the control algorithm is summarized as follows:

1 Determine the number of nodes in the input layer and hidden layer nodes, to determine the structure of the BP neural network, gives the layer weighted coefficient of the initial value and selected, and the learning rate η and inertia coefficient α, make k=1;
2 According to the sampling to get the output values of $r(k)$ and the output value $y(k)$, to calculate the current moment of error $e(k)=r(k)-y(k)$;
3 Calculate the BP neural network input, output, the output of the output layer which is a three control parameters of PID controller k_p, k_i, k_d;
4 Calculate the output of PID controller;

$$u(k)=u(k-1)+\Delta u(k)$$
$$\Delta u(k)=k_p\left[e(k)-e(k-1)\right]+k_i e(k)+k_d\left[e(k)-2e(k-1)+e(k-2)\right]$$

5 Through learning, the BP neural network adjust the weighting coefficient online, in order to realize the online adjustment of PID control parameters;
6 For k=k+1, return to step (1).

4 PULP CONCENTRATION SIMULATION

The process parameters of the pulp concentration control system change with the requirements on the difference of the paper. In this paper, Taking a test example for producing 60g/m^2 paper, the sampling time is 20s, the BP network structure is 4-5-3, the network has four input layer neurons, BP network input vector for $x_1=[r(k), y(k), e(k), 1]$; output vector corresponding to the three parameters of PID controller, take the BP network learning rate $\eta = 0.15$, inertia coefficient $\alpha = 0.1$; network weights taken within [-0.5, 0.5] in the range of random numbers, input $r(k) = 1$. Traditional PID controller and BP neural network PID controller were used to simulate the pulp concentration system with the software M ATLAB .The simulation curves shown in Figure 4.

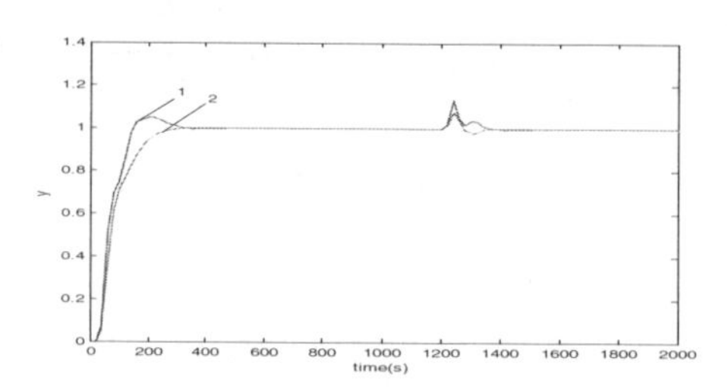

Figure 4. Response on the interferential model.

Simulation curves as shown in figure 4 when the system added to the interference of 0.3 magnitude, where the curve 1 for the simulation results with a conventional PID controller, the curve 2 for the simulation results with BP neural network PID controller.

The control effect of PID controller is obviously superior to the traditional PID neural network BP controller in Figure 4. From the perspective of the anti-interference ability of the system, there is interference in the system, PID controller based on neural network can adapt to changing conditions, and with the deepening of the learning process, has the very good regulator, can run more smooth and fast return to a stable state, anti-jamming is better than traditional PID control. From the overshoot of system, PID controller based on BP neural network can make the system without overshoot reached set value, the traditional PID control has a larger overshoot. The PID control from either the adjustment time, the overshoot of the system or anti-interference does better than the traditional PID control.

5 CONCLUSION

This paper combines the advantages of neural network and PID control, using a PID controller based on BP neural network to control pulp concentration, online tuning and optimization of PID parameters, avoid the tedious work in Artificial tuning of PID parameters and defects of control accuracy is not high, By reasonable selection of hidden layer and output layer activation function, the algorithm has the very strong generalization ability, and the pulp concentration gets effective control. The simulation results show that the effect of PID controller based BP neural network in the pulp concentration control system is better than the traditional PID controller, and obtain a better control effect.

REFERENCES

Bing Han & Min Han. 2010. Nonlincar time delay systems identification based on dynamic BP algorithm. Journal of Dalian University of Technology(05):777–779.

Dongqing Feng & Xiaoxi Zheng. 2009. Application of Neural Network Based on PID Controller in Crystallizing Process Temperature System of the Sorbitol Production. Computer Applications and Software (12):21–23.

Hai-jun Zhou & Mengxiao Wang.2011.The Application of Expert PID Control Algorithm in Pulp Consistency Control Loop.Paper Science & Technology(2):57–60.

Lu Cao, Zhixin Xiong & Muyi Hu. 2012. Simulation and Research on Pulp Consistency Control System. Computer Applications and Software (6):176–179.

Wei Zou, Yu Sun & Haijun Zhou.2005. Simulated Human Intelligence PID Control for Pulp Consistency. China Pulp &Paper(8):45–46.

Xiaoying Zhou, Genbao Zhang & Huihai Ma. 2006. The Application of Improved Fuzzy-Smith Prediction Control in the Pulp Concentration Control System. Hunan Papermaking(4):32–34.

Xing Zhang, Ying Dai & Zheng LI.2006. Journal of Hefei University of Technology (Natural Science) (11):1375–1376.

Yifei Chen.2011.Study and Simulation on PID Control Based on RBF Neural Network. Computer Simulation (04):212–215.

Computing, Control, Information and Education Engineering – Liu, Sung & Yao (eds)
© 2015 Taylor & Francis Group, London, ISBN: 978-1-138-02800-5

Statistical adjustment at module advanced planner and optimizer generated of SAP to the case of a food production company

Amelec Jesus Viloria Silva
University Sergio Arboleda, Bogotá, Colombia

ABSTRACT: An analysis of forecast system SAP-APO from a producer of food demand is made. It is detected that the process does not generate accurate forecasts compared to sales of products in previous months due to the mismatch of parameters and forecast periods models. Therefore the following suggestions for improvement are generated: redefining the parameters of statistical models of demand module SAP-APO, change the statistical model of adaptive average seasonality and trend historical data and categorize the products according to an analysis segmentation to assign a range of different tolerance to each when comparing their demand forecasts with sales in recent months. After implementing the improvements to the product percentage decreased manually adjust 12.58%, compared to the level of previous attention. This decrease of attention, resulting in an increase in the utility system more than 85%.

KEYWORDS: SAP Advanced Planner & Optimizer Module, information systems, management of inventory.

1 INTRODUCTION

Currently organizations require an information infrastructure that allows them to make accurate decisions based on real time data and market intelligence elements. New developments in technology improvements in planning the supply chain are increasingly sophisticated and have become gradually into key performance differentiators.

One of the most efficient and important today, technological developments is generated by SAP APO module, which has many features specifically designed to support companies in the process of Planning Supply Chain. The Demand Planning is one of its features, which is responsible for generating accurate forecasts and production plans by analyzing the historical planning, and business intelligence [1], [2] and [3].

The company under study is one of the leading producers and distributors of food in Latin America, this has offices in Venezuela and Colombia. Currently, statistical forecasts generated by their system optimization and advanced planning (SAP-APO, for its acronym in English); are not adjusted to the volumes needed by the Business. The statistical model that chooses the system is not the most appropriate for each product because forecasts show very high deviations or equal proposals every month, so it must be adjusted manually.

The research is to propose improvements to the SAP-APO system by analyzing and adjusting their statistical models and parameters to the behavior of sales for the following categories of products that makes up the company's sales portfolio: ABA (Balanced Food for animals), oils, Rice, Tuna, corn Flakes, Detergents, corn Flour, Cream, Soap, Ketchup, Margarines, Pets, Mayonnaise, Modifiers dairy and pasta. It is expected that the above increase the efficiency of it.

2 METHODOLOGY

The methodology in which the steps involved in each of the phases [4] and [5] are pointed occurs.

Step 1. Define: Business induction interaction with tools and system generation process statistical forecasting.

Phase 2. Measure: choice of sample size studied, obtaining previous forecasts, SKU classification, story time and life cycle; obtaining average sales last month, description of statistical models and parameters.

Step 3. Analyze and diagnose: analysis of the current situation through the signal indicator tracking, identification and causes of restrictions, SWOT analysis forecast generation system.

Phase 4. Plan and execute: setting and defining the parameters of the models proposed change of statistical models, categorizing SKU for review of forecasts, results and analysis implemented and establishment of improvements obtained in the process proposals.

5. Phase Control: generation of indicators.

3 RESULTS AND DISCUSSION

The results of each phase are described below.

3.1 *Phase I: Definition*

Four (4) basic steps for generation of statistical forecasting of demand establishes the company's study, using historical values and statistical forecasting methods, which are described below:

- Collection of historical data: this activity is loaded into the information system each product portfolio. We review and adjust the horizon of history to use (number of months in the past) for the calculation of statistical forecasts. This is done on the first day of each month for the following month forecasts on and takes one and a half to complete.
- Validation of history: historical settings are enabled to use (number of months) for the calculation of statistical forecasts. If inconsistencies in history (months sales figures to zero) are observed is channeled to the IT department. This activity takes place at the end of the data collection should be completed no later than the second day of the current month.
- Calculation Forecast: calculating corrected orders for previous months, i.e., orders of the month unless unmet demand the previous month running. Then, the product automatically segmented to the forecast in the corresponding string is run in three, six or twelve months (time to make history for each product). To finally run the chain statistical forecast for twelve (12) months, which includes statistical models and leads to the final results by choosing the model that yields less error. This activity is carried out in half a day of work after the validation of the story is finished.
- Forecast Revision: cases where the automatic statistical model selection is not the best are analyzed, based on the comparison of the forecast generated for the following month course with average sales of the last three (3) months for each product by establishing a tolerance of 5%.

3.2 *Phase II: Information collection*

Foods that distributes the company under study are divided into fifteen (15) categories and a total of three hundred (300) products. In this project we study just a bunch of these, establishing their choice in calculating the sample size for a finite population of 108 products taking into account the SKU for the categories of flour, detergents and soaps.

3.3 *Phase III: Analysis and diagnosis*

After selecting the products to be studied, these forecasts were collected, thrown in the month of July 2012 and containing forecasts for the next twelve (12) months. By observing the data generated by the system, the products are divided into three (3) large groups, these are:

Forecasts Constants: Products bearing the same proposals for the next twelve (12) months.

Forecasts outside the tolerance range: products presenting proposals that, when compared with average sales of the last three (3) months, differ by more than 5%.

Forecasts that need to be adjusted manually: products having constant forecasts, outside the range of tolerance or both.

One hundred and eight (108) were taken from products shown, fifty-nine (59) of these forecasts yielded constants, representing 54.63% of the total. Moreover, eighty-two (82) forecasts that products present are outside the set tolerance range, that is, 75.93% of the sample. This leads to the 87.96% of the products need to be adjusted manually, that is, ninety-five (95).

The models used by the SAP-APO system are:

Model 1: constant adaptation alpha

Model 2: moving average

Model 3: weighted moving average

Model 4: test for seasonality

Model 5: adaptive forecast

Generic model: linear regression

3.4 *Phase IV: Proposed improvements and implementation results*

As to model 1 variation was performed on the alpha parameter, leaving its forecast period of twenty-four (24) months as it was before, but failing to take into account only six (6) for the last month history (offset).

In model 2, the forecast period remained the same, but the offset was modified in order to take closest to the current date to describe the behavior coming out the products of the sample values as a feature of the model is which is not easy to determine a forecast period according to the data, therefore remained interim periods if the model is not readily adapted to changing trends or seasonality.

For model 3, the offset is modified to not varied considerably from the current date exponentially decreases as the alpha parameter, whose values were left in the same interval as they were configured (0.05 to 0.90).

Model 4 acts with exponential smoothing, so as model 3 assigns more weight to recent data history, according alpha range (0.05 to 0.090), in this case the forecast period was extended, but the offset is decreased because if is much fluctuation with respect to recent data will shed many mistakes. Model 5 for adjustment as the model 3 because they have the same forecasting strategy is the weighted moving average was performed.

For the generic model, which represents a linear regression model the alpha value was considered appropriate since it was quite high, but the beta and gamma values increased to 0.5 because this way could soften the lines of trend and seasonality of those products that required it. Overall the sigma parameter is increased for all models 1.5 as by default the system assigned the value of 1.25 and how it corrects historical values that fall short of the forecast, while higher their value, the prognosis resemblance to actual history.

After defining the new parameters of each model, and the model adaptive replaced by the test average trend and seasonality, runs to obtain forecasts for the next twelve (12) months, were made from the month of August 2012, where improvements were observed, generating more precise and reliable forecasts. Also forecasts were evaluated as performed previously, noting that percentage of these were constant and which went beyond the margin of error established from the outset (5%).

By obtaining these statistics the percentage decrease from each group were noted with respect to the run of the previous forecast month of July 2012, yielding a rate of 5.56% constant forecasts, which represent only 6 products from the sample. For those forecasts outside the range of tolerance of 5% the figure fell to 23 products in total, which involves 21.30% of the sample to reach a total of 27 products that need adjustment, meaning the 25.01% .

3.5 *Phase V: Control*

Finally, two (2) monitoring indicators and analysis in order to take direct control and continuous planning process demand is generated, and the proposals of forecasts generated by the system.

Tracking signal indicator: corresponds to an accuracy indicator for evaluating the accuracy of the forecasts generated by the SAP-APO system. The main objective of this indicator is to assess whether the process is carried out is under control, a once considered normal parameters defined. Since the profile of demand forecasting has different parameters influencing the data produced, review each case why after obtaining the demand forecast. Furthermore, this indicator can also be assessed globally, in which the level of care needed by statistical models that are in the system will be reviewed.

Alert Indicator: with the premise that forecasts thrown by the system are the final proposals that will define the dynamic plan of Business Polar Foods, the second indicator is framed in the percentage of attention needs to be paid to the forecasts based on the usual calculations from the actual sales. In this sense, the main objective is to evaluate forecasts and much attention requires the system, contributing to profitability. The desired level of this indicator is 0%, since if the system is able to generate effective proposals, Analysts should not have the need to change these forecasts generated by the system, in this sense, the goal is to have the system generate distribution proposals adjusted to the realities and requirements Planning Analyst.

4 CONCLUSION

Improvement proposals were reviewed by the “Management of demand planning and master data” and then approved, began the implementation process. Once implemented, ongoing monitoring of the results of the proposed demand generated by the system and the percentage of predictions that did not meet previously established requirements is performed. It is observed that due to changes in the DP module “Demand Planning”, the percentage of products to be manually adjusted decreased by 12.58% compared to the previous level of care. This decrease of attention, resulting in an increase in the utility system more than 85%.

REFERENCES

[1] Stang, D.; Arcuri, G. (2002). SAP mySAP advanced planner & otpimizer supply chain planning (SCP) applications, Gartner Product Report.

[2] Peterson, K. (2003). Multienterprise SCM solutions will come of age by 2012, Gartner Research Note.

[3] Walravens, P.; Shu, M. (2001). Understanding supply chain management software, Lehman [1].

[4] Brothers. Stadtler, H.; Kilger, C. (2002). Supply Management adn Advanced Planning, Ed Springer.

[5] Davies et al. (2002). How to get the most out of your supply chain?, Deloitte Consulting Research.

Computing, Control, Information and Education Engineering – Liu, Sung & Yao (eds)
 ISBN: 978-1-138-02800-5

Research on legal case data management system based on information fusion

Hui Liu
Weifang University of Science and Technology, Shandong, China

ABSTRACT: In this work, we conduct research on legal case data management system based on information fusion. To address the privacy problems of an M&A process, we have proposed the SIMS for securely accessing and disclosing the sensitive personal information. Our system manages the private HR data based on privacy policies using XACML, and it supports evidence or a legal proof complies with the security laws. Our proposed system will play an adequate role of facilitating the protection of personal information in an M&A transaction. From the experimental simulation, we could conclude that our proposed management system is able to safely manage the legal cases and analyze the data with classified result. Further research is also discussed in final.

KEYWORDS: Legal Case; Information Fusion; Data Management System.

1 INTRODUCTION

1.1 *Background research*

At present, more and more enterprises need BI software to support their statistics, analysis, forecast and decision-making. Some big companies, such as Oracle, SAS, BO, Cognos, MS, SAS and SPSS, have developed their own BI software. But those BI softwares are developed for some special application and only used in a limited range. Currently, the existing BI software is deficient in three points [1-5]. Firstly, the current BI software can only provide solutions for specified situation. If the situation is beyond its range, it can't recognize it and respond to it. Secondly, the current BI software can't deal with the dynamical requirement of the enterprise. If the requirement changes with time, the responding capability of the BI software is weakened. Lastly, the update speed of current BI software is slow. The source code of BI software must be always rewritten when a new requirement is added. So the updating cost is high. In the current market, there is a need for a universal BI software that can meet the dynamical and various requirement of heterogeneous enterprise and can fuse all kinds of heterogeneous intelligence technologies together on one BI software [6-8].

In order to develop such BI software, proposes a fusion method of intelligence resource named based on intelligent agent network and complex network. This approach will BI as a complex network composed of proxy node. An agent is a unit of information resources on behalf of the calculation model or algorithm. Each BI technology is by a group of agents. A complex network of large-scale accumulation ability can help keep aggregate all useful new agents to the system and make the system update more seamless and cheap.

1.2 *Related work*

Multi-agent systems as a standard communication platform can interchange data and tasks. The characteristics of the multi-agent system are autonomy, reasoning, reactivity, social ability, initiative, availability and adaptability. Recently, the multi-agent technology is more and more used in BI system modeling and research. Using multi agent technology can make the BI system more adaptable, renewable, flexible and extensible. Goals can be achieved by multi-agent intelligent fusion. It seems that it's feasible to use multi-agent technology in BI system. But currently, all the existing researches are put forward to dealing with specified business tasks for one or such a kind of enterprise. And there is no such a BI software that can meet the requirement of all kinds of heterogeneous enterprises. So in this paper, we propose a method that tries to solve this problem.

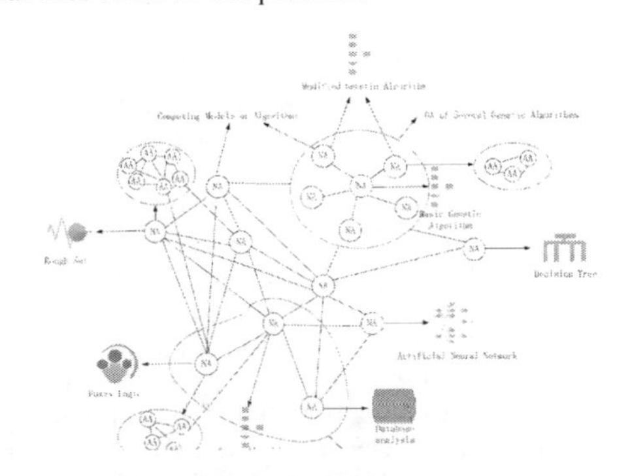

Figure 1. The weighted edge of agents.

2 OUR LEGAL CASE MIS MODEL

2.1 *Pre-discussion*

Our proposed SIMS aims to manage the private HR data based on privacy policies in M&A transaction. It could make the privacy policies using the extensible access control markup language (XACML) to specify role-based access control (RBAC) policies, and helps to reasonably disclose to the private information of the target company's employees. The Figure 2 illustrates the system structure.

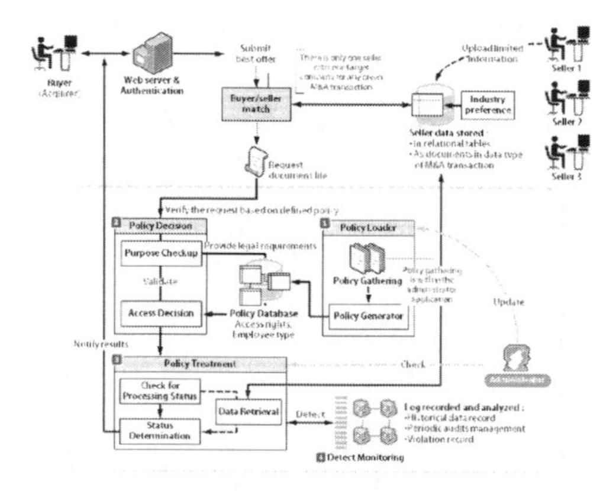

Figure 2. The legal data MIS model.

In this system, the authentication is a personal identification number recognition method based on the Internet. This may be related to verify whether the user authentication method such as the buyer for mergers and acquisitions. We can have an effective inspection request document authenticated and matching. Based on the XACML request document, this is the technology, and the disclosure of the privacy information of the target companies and access to private information required for merger and acquisition process. Support XACML privacy policy, this mechanism can automatically generate a machine-readable XML format.

2.2 *The policy decision mechanism*

It can automatically analyze the request document with checking their authenticity or suitability. Each objective classification processing of personal data of the four stages of the lifecycle (that is, from the collection, retention, use and destruction), as shown in Figure 3. The access decision function allows the applicability of the established privacy policy, and then it supports privacy policy complies with security laws, including South Korea's privacy laws, regulations and the OECD guidelines. It is also important to check one or more conditions. The condition is defined as a basic requirement for information such as the owner agreed, must meet or meet the privacy policy. This mechanism is check whether the condition is satisfied, then decide whether to or reject the request should be allowed to operate. Including the applicable legal requirements must inform those involved in the data processing. For this reason, in this paper, the mechanism for employees of the target company has the right to informed consent before the processing of personal information.

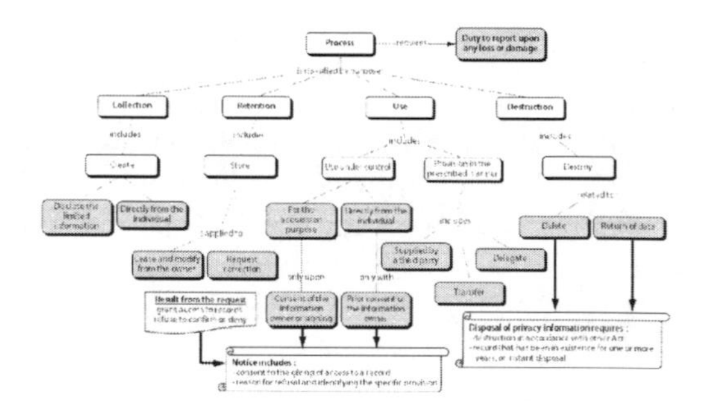

Figure 3. The type of process and relationship.

2.3 *The treatment and monitoring*

The retrieved information is then provided to the acquirer according to that of the result of processing, and the information owner will be notified about the occurrence of such a decision. The result in this mechanism is built on XACML. This document indicates the status of the process and specifies the protection policies. Measuring the safety of the system administrator, it also records the history data. In particular, the mechanisms found suspicious activity, such as trying to get a large amount of unauthorized access or access to personal data. System administrators can limit for personal data of similar incidents occur, unauthorized access. Access should also be limited to a specific IP address and the name of the identifier.

3 PROTOTYPE AND IMPLEMENTATION

3.1 *The prototype description*

To demonstrate the feasibility of our system, SIMS, which protects the personal information based on privacy policies, was developed using JSP and JAVA technologies. We use MySQL for our database server and Apache as the interface for the web server system. We designed a policy database, which manages privacy policies, and it is developed based on the laws related to privacy.

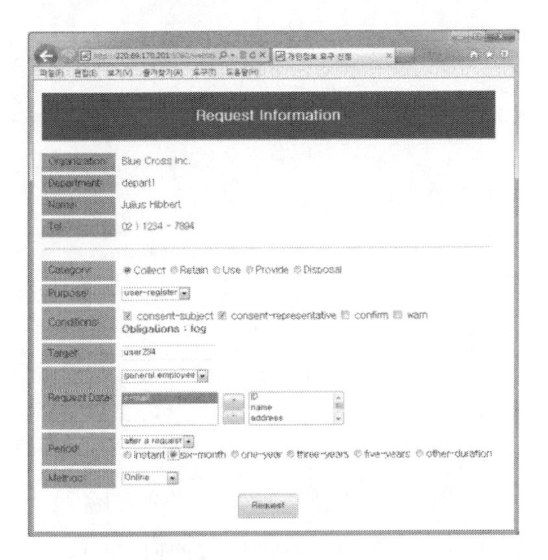

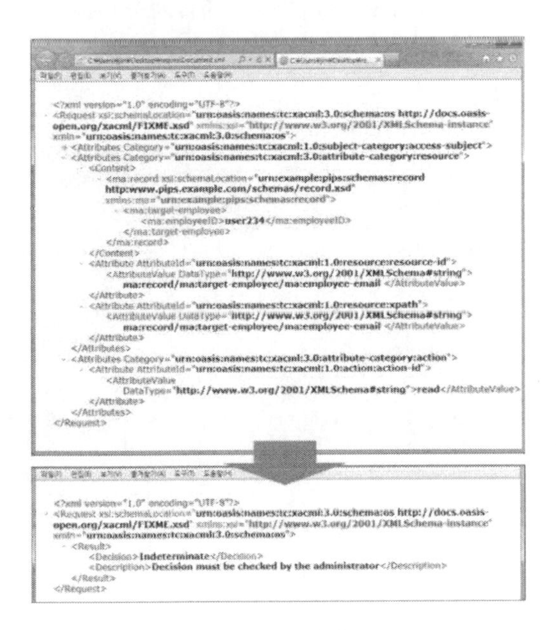

```
<?xml version="1.0" encoding="UTF-8"?>
<Request xsi:schemaLocation="urn:oasis:names:tc:xacml:3.0:schema:os http://docs.oasis-open.org/xacml/FIXME.xsd" xmlns:xsi="http://www.w3.org/2001/XMLSchema-instance" xmln="urn:oasis:names:tc:xacml:3.0:schema:os">
  <Attributes Category="urn:oasis:names:tc:xacml:1.0:subject-category:access-subject">
  <Attributes Category="urn:oasis:names:tc:xacml:3.0:attribute-category:resource">
    <Content>
      <ma:record xsi:schemaLocation="urn:example:pips:schemas:record http:www.pips.example.com/schemas/record.xsd" xmlns:ma="urn:example:pips:schemas:record">
        <ma:target-employee>
          <ma:employeeID>user234</ma:employeeID>
        </ma:target-employee>
      </ma:record>
    </Content>
    <Attribute AttributeId="urn:oasis:names:tc:xacml:1.0:resource:resource-id">
      <AttributeValue DataType="http://www.w3.org/2001/XMLSchema#string">
        ma:record/ma:target-employee/ma:employee-email </AttributeValue>
    </Attribute>
    <Attribute AttributeId="urn:oasis:names:tc:xacml:1.0:resource:xpath">
      <AttributeValue DataType="http://www.w3.org/2001/XMLSchema#string">
        ma:record/ma:target-employee/ma:employee-email </AttributeValue>
    </Attribute>
  </Attributes>
  <Attributes Category="urn:oasis:names:tc:xacml:3.0:attribute-category:action">
    <Attribute AttributeId="urn:oasis:names:tc:xacml:1.0:action:action-id">
      <AttributeValue
        DataType="http://www.w3.org/2001/XMLSchema#string">read</AttributeValue>
    </Attribute>
  </Attributes>
</Request>

<?xml version="1.0" encoding="UTF-8"?>
<Request xsi:schemaLocation="urn:oasis:names:tc:xacml:3.0:schema:os http://docs.oasis-open.org/xacml/FIXME.xsd" xmlns:xsi="http://www.w3.org/2001/XMLSchema-instance" xmln="urn:oasis:names:tc:xacml:3.0:schema:os">
  <Result>
    <Decision>Indeterminate</Decision>
    <Description>Decision must be checked by the administrator</Description>
  </Result>
</Request>
```


Figure 4. The prototype system design.

In its prototype, figure 4 below said sample request. In figure 4(a) shows that the user can request personal information, including name, upon others, financial status and other specific entity. This request can convert a privacy policy language, known as the XACML. (b) refers to the XACML policy specifies a request information, and is suitable for processing the validation decision in the system when the transfer or disclose any sensitive personal information. In such cases, the decision must be checked again by the administrator.

3.2 *Architecture and implementation*

There are two functions in the user interface layer: request acceptation and service result return. Request admitted to provide registration, booking and cancellation of the user. At the same time, the service task is generated, and then sent to the service layer. Service results returned to provide users with the return results show and a satisfaction survey. Each responding of the user's request is a new cooperation of a group of agents in the agent - network. The dispatching mechanism is to determine which agent can participate in cooperation with others according to the dynamical property of task. Service reaction layer is an intelligent scheduling and control center. All kinds of BI service integration system application in response to the service layer after the completion of tasks. Task analysis is to identify the structured degree, complexity, workload and select task decomposability, and provide the basis for the task decomposition strategy and response strategies. Task shunt is policy set by the mission decomposition, suitable conditions set and select rule sets. This task is the task decomposition plan according to the proper decomposition strategy and distributor decomposed granularity. The optimal cooperation that can realize fusion of all kinds of BI on system application is ensured by selecting the most suitable cooperative strategy of current tasks and completing the final agent selection according to the corresponding process and algorithms.

The cooperative plan executor is to send command to the selected agents and receive task execution result according to cooperative plan. At the same time, it's responsible for communicating and coordinating between agents. A fusion of cooperative result is to gain the final result by dealing with the return result of cooperative agents. The service managers is to load, upload, import, export, and maintain agents' record which is including list maintenance, warehousing registration, cooperation registration and performance records. The service manager is also to assist network management layer to maintain agent list in the warehouse consisting with node list of networks. The agent list in the warehouse must contain a node list of networks, otherwise the no-existing agent would be assigned to respond to service requests. A network analyzer is to calculate and analyze the network statistical characteristics, such as all nodes, all edges, average degree, average path length and cluster coefficient.

4 EXPERIMENT AND SIMULATION

We designed a program to simulate its performance: Privacy policies using XACML affect the response time perceived by the user. Therefore, the response time is the performance of the prototype system. By randomly processing the request, we compare the response time of two processing mechanisms, as shown in Figure 5. The lines on the graph represent the existing and proposed system response times. Compared to the existing system, the policy processing with XACML shows better performance, since the XML based output reduces the time considerably during the process.

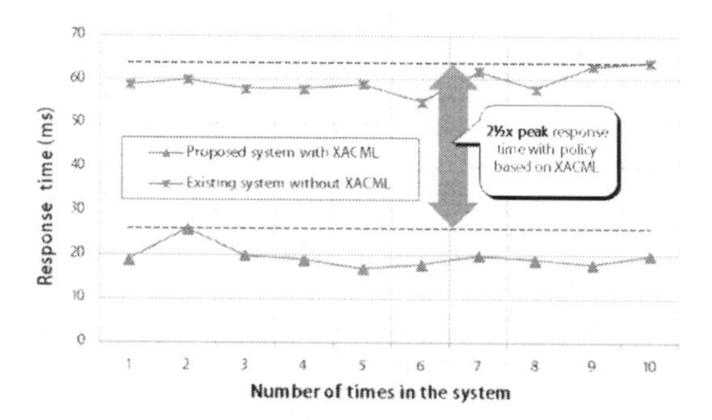

Figure 5. The performance evaluation.

5 SUMMARY AND CONCLUSION

In this work we conduct research on legal case data management system based on information fusion. To address the privacy problems of an M&A process, we have proposed the SIMS for securely accessing and disclosing the sensitive personal information. Our system manages the private HR data based on privacy policies using XACML, and it supports evidence or a legal proof complies with the security laws. Our proposed system will play an adequate role of facilitating the protection of personal information in an M&A transaction. The simulation result illustrates the effectiveness of our method. In the future, we will focus our research on compliance issues in M&A activity, and continue to study the automated policy management for private data.

REFERENCES

[1] Maxwell, John M. "Case management for a personal injury plaintiff's law office using a relational database." U.S. Patent No. 6,098,070. 1 Aug. 2000.

[2] Geldermann, Martin, Viktor Kister, and Michael Rohbeck. "Apparatus and Method for Acquiring a Data Record, Data Record Distribution System, and Mobile Device." U.S. Patent No. 20,150,018,047. 15 Jan. 2015.

[3] Barbosa, Frank A., and Luis M. Ortiz. "SYSTEM AND METHODS FOR MANAGEMENT OF MOBILE FIELD ASSETS VIA WIRELESS HANDHELD DEVICES." U.S. Patent No. 20,150,018,007. 15 Jan. 2015.

[4] Gianni, Maria, and Katerina Gotzamani. "Management systems integration: lessons from an abandonment case." Journal of Cleaner Production 86 (2015): 265-276.

[5] Ibbotson, Craig A., Shalabh Kakkar, and Eirene A. Shipkowitz-smith. "APPARATUS AND METHOD FOR DETERMINING CONTEXT-AWARE AND ADAPTIVE THRESHOLDS IN A COMMUNICATIONS SYSTEM." U.S. Patent No. 20,150,024,735. 22 Jan. 2015.

[6] Lamers, Nathan John, Caleb Robert Krisher, and Michael Anthony Gill. "WELDING SYSTEM DATA MANAGEMENT SYSTEM AND METHOD." U.S. Patent No. 20,150,019,594. 15 Jan. 2015.

[7] Fischer, Wolfgang, Stefan Mahn, and Dennis-indrawan Soebagio. "CABIN MANAGEMENT SYSTEM HAVING A THREE-DIMENSIONAL OPERATING PANEL." U.S. Patent No. 20,150,007,082. 1 Jan. 2015.

[8] Smith, Martin Gregory, et al. "DATA MANAGEMENT PROCESS UTILIZING A FIRST-PARTY TECHNIQUE." U.S. Patent No. 20,150,006,302. 1 Jan. 2015.

Computing, Control, Information and Education Engineering – Liu, Sung & Yao (eds)
© 2015 Taylor & Francis Group, London, ISBN: 978-1-138-02800-5

The application and exploration of the informatization of art design teaching

Xin Yue Zhang
Department of Art, Lijiang College of Guangxi Normal University, Guilin, China

ABSTRACT: This paper aims to propose the appropriate balance among media, technology and design of the informatization of art design teaching through the analysis of the shortage of education mode and the advantage of teaching informatization in art design, and elaborates the application of educational informatization in art design teaching through three aspects as application environment, software support and transformation in order to establish the teaching philosophy of people oriented in technology and design.

KEYWORDS: art design; informatization; education; teaching.

With the arrival of "digital living"[1] times, the technique of art design teaching is facing huge challenges, digital education has become a new trend, which is in response to the educational informatization under the International Education Quality Certification System: the cultivation of new "digital" talents is an inevitable process of educational reform. The informatization of art design teaching includes various kinds of art information processing technologies such as applied computing technology, applied software and network technology and multi-media technology. It represents the advanced educational productivity of current art design and the innovative ability of high efficient configuration and optimization in art design. Meanwhile, it plays an important role in promoting the construction of the whole educational informatization.

1 THE TRADITIONAL ART DESIGN (AD) TEACHING AND ITS SHORTAGE

The traditional teaching mode of artistic design in our country is based on such teaching mode as "one textbook, one chalk, one blackboard; teacher speaks, students listen; the teacher writes, students copy; teacher tests, students recite; teacher asks, students answer", which is short of the initiative, discussionality and creativity of the class. With the features of a single textbook, limited information, obsolete media devices and outdated teaching methods, this kind of teaching mode was often used in art design teaching and once made theory rather preferable than practice. The practical course, even if there is any, may only stay at the making of design proposal, or dictation of artistic cases in order to state different viewpoints of their own. It either illustrates the working process by showing designs have been seen or done, or fulfills the practical teaching by assigning homework. Students cannot enter the "on-site" state of "seeing is believing" under such educational environment for lack of practical teaching, along with the lack of information sharing, interaction and collaboration. Therefore, almost every talent cultivated by traditional art design education is "solo artist" who can hardly adapt to competitive, modern market due to the lack of team spirit.

2 THE APPLICATION OF INFORMATION TECHNOLOGY IN (AD) AND ITS ADVANTAGE

With the development of media and network technology, the traditional art design, teaching has been badly impacted and pushed to a new high-tech level of modernization and technicalization, which guides the development of art design education.

First of all, digital informatization makes art design and media technology complement each other. Digital informatization is the nature of media network technology based on digital network technique, an interactive system integrating voice, graphic images, audio and video information in order, connected with the artistic integration of graphics, image, modeling, text, audio, video, two dimensional (2D) technique and three dimensional (3D) technique. Art design has the feature of marginality and comprehensiveness, which generalize the media network technology into all walks of life and popularize its recognition. On the contrary, as an important means of artistic design, teaching technicalization, media network technology deduces art design excellently. Usually network media are needed to provide convenient, clear, editable, pioneering and sharing modern technique support when art design working and teaching are in process. The combination of art design working and teaching can complement each other so well.

Secondly, digital informatization enhances work efficiency. Information technology deviates teaching from a single form of blackboard-writing and realize the diversification of artistic teaching resources. The cultivation of information technology is beneficial to improve student's capability of collecting design information, design analyzing as well as choosing a design orientation. It is beneficial to diversify abundant teaching methods: by offering students the study environment of "running classroom" through an excellent course network, video-on-demand (VOD) system and shared teaching resource library; by raising student's self-awareness and capability of information technology through electronic teaching plan, electronic courseware, video recording, etc., thus to adapt themselves to the requirements of educational development at new times.

Thirdly, digital informatization help with the liberation of teaching load and realize the intelligent management of teaching. With the expanded enrollment of modern artistic education, the work of the teacher is overloaded. The liberation of the teacher's workload depends on the intelligent management of informative teaching. The information intelligentization of network media technology can share responsibility for general teaching and management during the art design teaching process. General teaching refers to the consistency of educating each student with knowledge and technology, and the uniformity of knowledge, exercise, tutorship and answer. The general teaching of teaching information intelligentization sets teacher free with more time guiding students in personalized artistic creation through interactive functions set by application software and courseware of media, such as intelligent information management of study, review and homework correction.

3 THE APPLICATION AND EXPLORATION OF (AD) TEACHING INFORMATIZATION

3.1 *Application environment*

Technicalization and modernization in art design teaching cannot be realized without the support of media technology. However, not all art design courses require media technology support, some don't fit in a media environment while some call for the combination of media technology and traditional teaching technology. Information-based teaching should be customized according to course environment.

Generally speaking, the courses of art design can be divided into public basic course, professional basic course and specialized course. For basic course, there are design sketching, color, freehand sketching, sketching and three "compositions" (plane composition, color composition and 3D composition). These courses aim to train student's basic painting skill, which is the same as hand skill taught by traditional art education, media technology support is not necessarily needed. But for all kinds of software design courses of professional basic course, especially for the advanced professional course, multi-media technology support is definitely required. For courses as <Photoshop Image Processing>, <Auto CAD Interior Design of Architecture>, <3Ds Max> and so on, multi-media technology is applied during the whole process, whereas for specialized courses as <Font Design>, <VI Design>, <Display Design>, <Commercial Space Design>, the teaching method combining multi-media technology and traditional teaching method can be used. The general teaching method of these courses is to make a crossover with all teaching information on the course first, then display or project it on a certain media interface to let students understand and grasp the knowledge and working process. The amount of information on teaching can be more visualized and introspective through media display, which is good for student's comprehension of abstract conception.

3.2 *Software support*

Information-based teaching in art is a complex and creative task, which requires various kinds of multi-media technology and software technology support. The five major forms of media technology in art design are text (format as: DOC, WPS, TXT, RTF); image (format as: BMP, GIF, JPEG, etc.); animation (format as: FLA, SWF, GIF, etc.); Audio (format as: WAV, MIDI, MP3, etc.); Video (format as: AVI, MOV, DAT). The text conveys messages by words and all kinds of specialized symbols, mainly used in description of knowledge, such as concept and definition. The image is composed by color and graphics, which is one of the most important information forms of multi-media technology to show visual effects. The animation is to take advantage of human's visual persistence, quickly display a series of moving and changing photo images, which could visualize abstraction and make boredom vivid. The audio is to deliver messages, exchange emotions with three main expression forms as explanation, music and effect. The video is a sounding and colorful medium used to show how things develop as an evolutionary process, for example, the film and TV we are familiar with.

For information-based teaching in art design, software application resources are abundant: Word,

WPS are softwares on text; Photoshop, Cereldraw, Freehand are softwares on graphics and image; Video for Windows, Adobe Premiere, Quicktime for Windows are softwares on Video; Soundsystem WoveEdit are softwares on Audio; Animator, Maya, 3Dmax are softwares on 3D animation. There are many kinds of plug-in technologies in 3Dmax to help with making 3Dmax effect images, among which the Virtual Reality Technology (VRT) is a highly life-like human-computer interaction technology simulating human behaviors as seeing, listening and touching. The VRT is a new area in information-based teaching of art design as well as an important area for future informatization development.

3.3 *Transformation and application*

Art teaching itself is very technically difficult, with information technology teaching added, which makes it a challenging creative task. The application software system is extensive and complex, the format conversion of different softwares is the key factor in teaching. Namely, powerful Word, WPS can be converted into PPT or JPG format; 3Dmax, AutoCAD can be saved or output as BMP or JPG, and converted as a supportive format for the application of multiple softwares and production of various coursewares. For example, the production of PowerPoint (PPT) is to put each static photo on continuous display, or alternate with some built-in animation effect. But it will be replaced by FLASH due to its staid style. Flash is a network-based animation producing software with the features of good interaction and powerful function: free control of actions technique, easy playback; the function of fade-in and fade-out can reach the unique prelude effect with film slide show and vivid animation; setting up background music control button to meet the on-site playing demand. The flash courseware not only displays the overall comprehension on Aesthetic aspects with details considered, but also pays attention to harmonious matching of colors with free creation, which exploits a new world for teaching informatization.

4 CONCLUSION

The core of art design education at digital times is the fusion of art and technology, art design teaching cannot rely on media technology too much. There are some problems in media technology as insufficient hardware and software support, short lifetime, obsolete version lack of large-scale application. As the "nuclear power" (human driving factors) of design, media technology should be developed and applied at maximum with technical coordination. Handling the relations between media, technology and design reasonably[3] should focus on the design itself rather than show off the technology. It should be people-oriented and service-oriented and applying the most advanced media technology to accomplish the teaching process of art design, in order to make information technology reach the ideal level of technical teaching serving for students and technical design serving for all humans.

REFERENCES

[1] Nicholas.Negroponte. *Digital Life*.Hainan:Hainan Press,1997.

[2] Li Zhang.Try to Analysis Deficiency and Improvement Countermeasures of The Traditional Teaching Method. *Chinese technical information*.pp192.2005(19).

[3] Ganqin.Yao.Art Design Education in the information era.*Art Education*.pp119–120.2013.

Computing, Control, Information and Education Engineering – Liu, Sung & Yao (eds)
© 2015 Taylor & Francis Group, London, ISBN: 978-1-138-02800-5

Research on reliable transmission of VANET's safety messages based on motorcade

Li An, Jun Zhang, Ming Fu Tuo, Yong Mei Zhao & Hong Mei Zhang
College of Science Air Force Engineering University, Xi'An, China

ABSTRACT: Based on the background of the motorcade's safety running, this paper puts forwards a heart-beat message transmission power and frequency joint algorithm which is distributed and network situation adaptive. In the case of no network congestion, the transmission power and frequency of each node should be used as large as possible to improve the transmission range and the accuracy of the state message updating. It is proved by the simulation experiment that this method can significantly improve the acceptance rate news at close range, effectively reduce the packet collisions and meet the safety requirements of applications.

KEYWORDS: Vehicular Ad Hoc Networks; Motorcade; Heart-beat Safety Message.

1 INTRODUCTION

In recent years, With the rapid development of wireless communication technology, Vehicular Ad Hoc Networks (VANET) is an intelligent transportation system in urgent need of transportation information field. The nodes in the VANET are vehicles as shown in Figure 1-1.It has many characteristics such as autonomy, no center control, no fixed structure, multi hop, dynamic network topology, limited network capacity, complex and changing network environment, high real-time requirement on information transmission, large network scale as shown in Figure1.

Figure 1. VANET.

The vehicle team is everywhere in our daily life. For example, self driving tour is pretty hot right now. In the condition of many vehicles, More and more people drive out in a convoy of caravans. Some companies establish their own goods transport convoy based on their needs and so on. The vehicle team is often up to dozens of vehicles, stretching several kilometers, quite inspiring. However, this situation is very easy to cause behind or rear end collision accident. In order to indicate the travel direction, traffic conditions, urgent situation report, and grasp the convoy in dynamic, people often use the intercom to communicate. In some remote location, the wireless signal is poor. The vehicle team cannot communicate effectively. In the process of the running team, if a self organizing inter vehicle communication network(VANET)is constructed rapidly. Each driver can known well about the information on other vehicles and real-time traffic information in the beyond-visual range by VANET. So the survival ability of the vehicle team will be improved consequentially.

2 THE SAFETY INFORMATION TRANSMISSION OF THE VEHICLE TEAM

2.1 *The safety information*

The VANET distributes Safety Related Information(SRI) and Non-Safety Related Information (NSRI) to provide various application services for drivers through the running vehicles. Mac Torrent proposed two kinds of security message in 2005 : periodic secure message and the message triggered by emergency accident. The former is used for traffic accident early warning, called Heart-beat Safety Message (Emergency Alarm Safety Message, EASM). The safety of the vehicle team is guaranteed by using the two kinds of safety messages. The running convoy periodically broadcasts their state information in a single hop (such as vehicle position, speed, direction), referred to as HSM. When an emergency (such as traffic accidents) occurs, the Message is broadcasted in a way of multi-hop, usually including information about time, location, degree of urgency, called EASM.

The every vehicle of the marching convoy informs the other vehicles their running state, sending HSM (including vehicle identification, position,velocity, acceleration, also some information of the transmission parameters, etc.) constantly, receives from the other vehicle HSM at the same time, obtains the environmental information of the other vehicle, in case for taking measures to avoid accidents prior to dangerous situations.

If the marching convoy encounters an emergency situation (such as traffic accidents, the ice road collapse, etc.), the accident information should be informed to the other vehicles as soon as possible. It is in the case that the EASM is needed. When the vehicles around receive the EASM, quickly take measures (such as an emergency brake, etc.) to avoid secondary accidents, as shown in figure2.

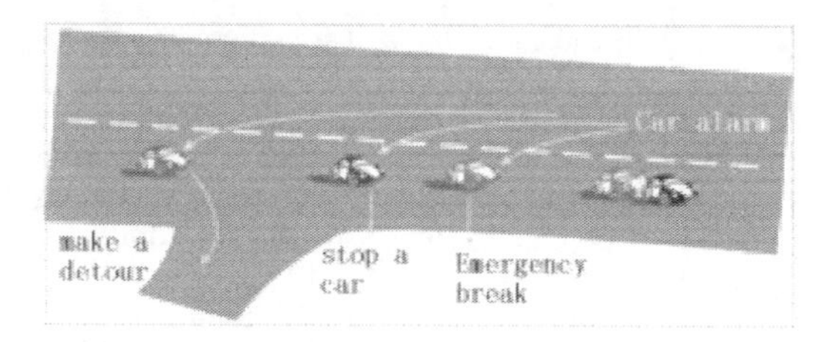

Figure 2. EASM.

2.2 *Transmission channel*

In VANET, security information interaction between vehicles depend on the wireless channel, which is called Safety-Related Information Transmission Channel (SITC). The channel bandwidth standard size of Dedicated Short-range communication is also used in the SITC. The two kinds of safe messages which is defined as before are transmitted in the SITC. The SITC is a public channel for nodes of the VANET, which is transparent to the vehicles involved. Therefore, each vehicle can send and receive safe information directly.

With each node sending HSM periodically, the VANET could be overloaded and its performance decline by degrees, leading to the heartbeat message congestion in the highly dynamic changing network, as shown in Figure 3. Because of the multi-hop transmitting of various EASM, if every node transmits message without restraint, the load of the VANET will be raised enormously, which badly affects the transmission of two kinds of the safety message.

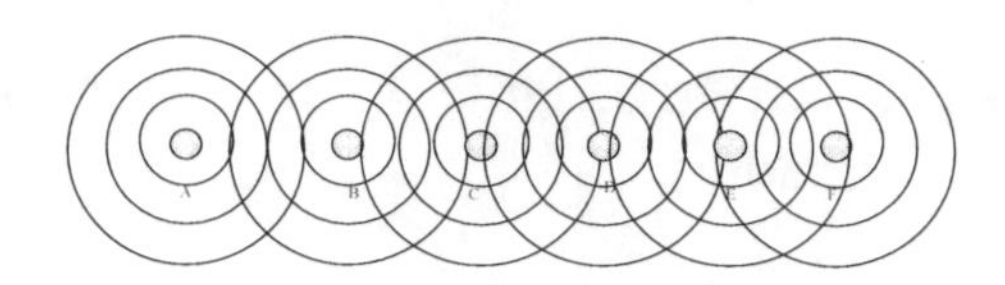

Figure 3. Network congestion caused by HSM sent by vehicles.

3 POWER CONTROL AND FREQUENCY CONTROL ALGORITHM

The Distributed Network Condition Adaptive Power Control and Frequency Control algorithm(PCFC) proposed in this paper is on the basis of the two judging criteria of network capability relied on distance and feedback: the local density of vehicles ρ_{Est} and the estimated rate of heartbeat message transmissionP_{Est}. According to current network condition, the PCFC adjusts constantly transmitting power and frequency of the heartbeat messages, whose core idea is that every node in the VANET increases the transmission power and frequency as large as possible, without causing network congestion, in order to improve the safety message transmission range and the accuracy of message status updating. Once finding the decline of the network capability, the PCFC reduces the transmission power and frequency, in order to decrease network congestion due to excessive channel competition caused by network nodes.

An scene could be supposed: the heartbeat message of the all vehicles has a maximum value and a minimum value of the same, upper frequency and lower frequency of the same .At the end of each cycle, the vehicle estimates the vehicle density and the network data transmission rate estimation, according to the PCFC presented as before.

The parameters used in the PCFC are defined as follows:

Definition 1 transmitting power distribution, transmitting frequency distribution

The running vehicle collection is $N=\{V_1,V_2,\cdots V_n\}$. The Power distribution strategy is used for distributing transmission power for the heartbeat messages of each node $V_i(i=1,2,\cdots,n)$ in the network . And the Power distribution strategy works with a proportion. (Power Setting) $PS(i)\in[0,1]$. The power of each node V_i in the network is $P_i\leftarrow P_{\min}+PS_{(i)}\times(P_{\max}-P_{\min})$. In the whole process of adaptation adjustment, the Power distribution strategy influences the final transmitting power by changing the value of the $PS_{(i)}$. And also the transmitting frequency is set in the manner of proportional division(Frequency Setting) $FS(i)\in[0,1]$, whose final value is set by $PS_{(i)}$.

Due to the dynamic nature of the network, vehicles joining at any time, the VANET sends heartbeat messages to the new member and assign them the initial power. The initialization of power is $PS_{(i)} = 0.5$, $P_i = \frac{P_{\min} + P_{\max}}{2}$. At the same time, the initialization of frequency is $f_i = \frac{f_{\min} + f_{\max}}{2}$, $FS_{(i)} = 0.5$

Definition 3 the threshold of vehicle density

The local density of vehicles distinguishes the boundary of dense and sparse ($\rho_{\_UpBoundary}$ $\rho_{\_DownBoundary}$).When the vehicle density is higher than $\rho_{\_UpBoundary}$, the period is called the dense phase. Conversely when the vehicle density is lower than $\rho_{\_DownBoundary}$, the period is called the sparse phase, and the middle period is called the transition phase. In the dense phase, a gradual growth and the rapid decline are used to power adjustment. In the sparse phase, a rapid growth and slow decline of power is used. In the transition phase, the fixed step adjustment works. The power sent by PCFC defined as previous chapter is associated with the vehicle density, which is different corresponding to different vehicle density.

The data transmitting rate is $p_{\_Standard}$. When the data transmission rate is higher than, the network capability shows well; when the data transmission rate is less than $p_{\_Standard}$, the network capability declines. High node density shows that there is a network congestion, which need to adjust the transmission parameters.

Definition 5 local vehicle speed estimation The local vehicle speed estimation, considering the current vehicle and the mobile situation of vehicles around , is used to judge the dense degree of vehicle. The speed of current vehicle and the vehicles around are v_i, v_j.

$$V_{Est}(i) = \alpha v_i + (1-\alpha)\frac{\sum_{j=1}^{j=n} v_j}{n}, \alpha = 0.75$$

4 THE SIMULATION VERIFICATION

This paper uses the network simulation software ns-2.33 to verify the network capability. Fixed Scheme First, the fixed transmitting strategy and the PCFC should be compared. In Fixed Scheme, the transmitting power and the transmitting frequency are fixed in the heartbeat message transmission process. Compared with the Fixed Scheme, PCFC can adjust the transmitting power and the transmitting frequency in real time on the basis of the network capability. A measure of the standard is the receiving rate of the HSM.

As shown in Figure 4, When nodes sent no EASM in the network, with the fixed transmitting power and transmitting frequency as Fixed Scheme, a severe data packet collision occurred due to higher network load. In the 200 meters, the reception rate of the data packet was lower than 50%. With the distance increasing, the receiving rate showed a downward trend. On the other hand, the PCFC near the sender can obtain higher receiving rate. In 200 meters, data receiving rate was close to 75%, increased by 25%, which could satisfy the requirements of safety applications. Compared with Fixed Scheme, this conclusion indicated that the PCFC could get higher heartbeat receiving rate in close distance to the adjustment of the power and frequency.

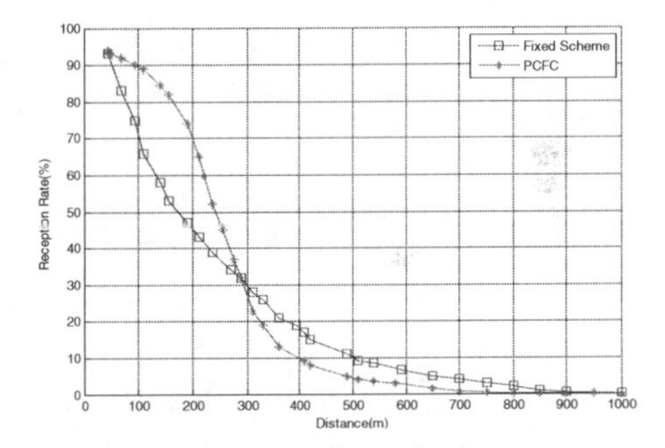

Figure 4. The receiving rate comparison of the heartbeat message under the PCFC and Fixed Scheme.

In the network transmitting, the data packet collision caused by hidden terminal or inappropriate channel scheduling is one of the two main reasons leading to message transmitting failure. The other is the loss of data packets caused by signal attenuation.

As shown in Figure 5, the data packet collision ratio under the Fixed Scheme was significantly larger than the PCFC. At a distance of 600 meters from the sender, the data packet collision ratio under the Fixed Scheme reached to 90%, and the data packet was nearly all conflicted. Under the PCFC, the data packet collision ratio has only been 58%, in the 300 meters. The simulation showed that under the adaptive strategy of the PCFC, the data packet collision obviously reduced.

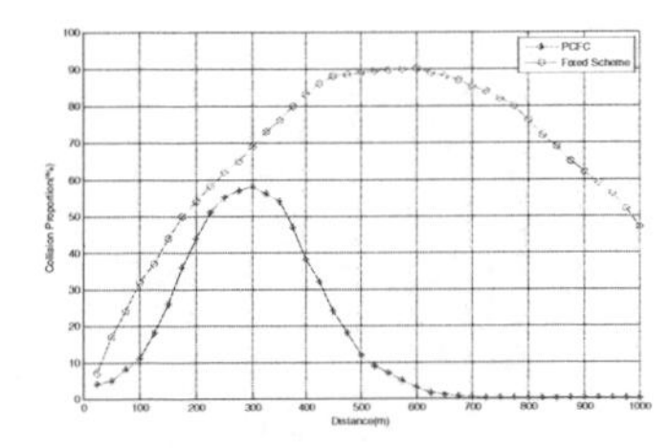

Figure 5. The data packet collision ratio comparison under PCFC and Fixed Scheme.

Security applications are sensitive to delay. So another important indicator is the message transmitting delay which measures the network capability. From the simulation results of Figure 6, the combined control manner could effectively reduce the transmitting in different vehicles density.

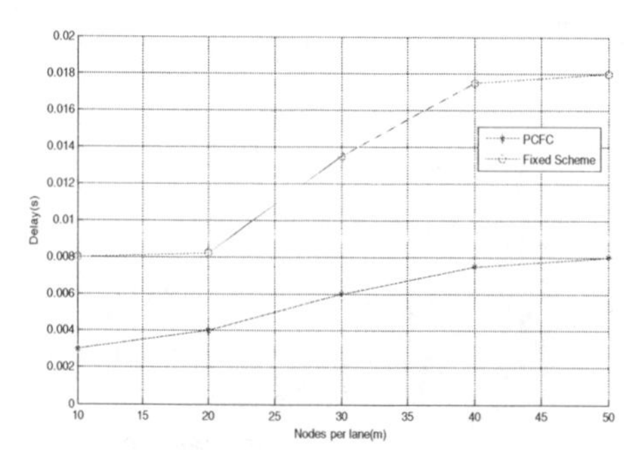

Figure 6. The relationship between the transmitting delay and the number of nodes under different strategies.

From the analysis of simulation, we can conclude that the PCFC could effectively reduce the number of nodes in the collision zone, packet collision, improving the delay of end-to-end and the correct receiving rate of safety message of nodes surround by.

From the simulation results we can draw the conclusion: At the distance near the transmitting node, low transmitting power is a good choice to improve the receiving rate of HSM. From a security point of view, the zone near the sender is more related to the safety degree. PCFC is used to adjust the transmitting power and transmitting frequency of HSM, and then control the load in the transmitting channel, improve the reliability on HSM receiving of the sender node near by.

5 SUMMARY

This paper analyzes the shortcomings of secure message transmitting in VANET, propose a combined control strategy on the power and frequency of heartbeat message. The corresponding control strategy algorithm is proposed. And the simulation of the NS2 network simulation software work out a clear result.

The simulation result shows that the method proposed can well improve the receiving rate of the heartbeat message in the near distance, effectively reduces the data packet collision, satisfies the requirements of safety application.

REFERENCES

[1] D. Reichardt, M. Miglietta, L. Moretti, et al. CarTALK 2000: safe and comfortable driving based upon inter-vehicle-communication. IEEE Intelligent Vehicle Symposium. 2002. IEEE Press, 545–550.

[2] Jiang, D., Taliwal, V., Meier, A., Holfelder, W. & Herrtwich, R., "Design of 5.9 GHz DSRC-based vehicular safety communication," IEEE Wireless Communications, voU3, no.5, pp.36–43, October 2006.

[3] Jiang, D., Taliwal, V., Meier, A., Holfelder, W. & Herrtwich, R., "Design of 5.9 GHz DSRC-based vehicular safety communication," IEEE Wireless Communications, voU3, no.5, pp.36–43, October 2006.

Computing, Control, Information and Education Engineering – Liu, Sung & Yao (eds)
© 2015 Taylor & Francis Group, London, ISBN: 978-1-138-02800-5

Half-automatic welding for low position operation using CO_2 and spot welding on bus body

Hui Gan
Department of Automotive Engineering, Hunan Communication Polytechnic, Changsha City, Hunan, P.R. China

ABSTRACT: Based on the characteristic of automatic welding for low position operation on bus roof panel and top skeleton, through analyzing its welding equipment requirement and welding technological characteristic when combining using CO_2 welding and single-face double-spot resistance welding, this research adopts proper welding technological process and process specification, to find a solution improving welding quality by eliminating influence factors on the welding quality, so as to give full play to its advantage of this kind of welding of small deformation, high strength, energy saving, and high efficiency.

KEYWORDS: CO_2 welding; spot welding; low-position operation; welding equipment; welding technological process; process specification; welding quality; influence factor.

1 INTRODUCTION

As we all know, CO_2 welding is a welding method with the advantage of highly efficient and energy-saving. Although this welding method is not the latest in science and technology, there are still a lot of researches to explore its application field for its wide adaptation.

Currently, handwork (manual) CO_2 gas shielded arc welding is employed in the production line by the most bus factory, to weld the roof panel together on large and medium-sized passenger car in the sites. Spot welding is used at the junctions between the panel and either of top inner sides among the stringers of the frame after simply pulled back with tension, fully-length welding is used between the panels and then welding joints are cemented with sealant. The main advantage of this welding method is that the operation is simple and flexible, investment in equipment is less, power consumption is low, and adaptability to different bus models is strong, while its defects are that welding quality is poor, panel tension is not enough, the labor intensity to welder is large, the workload of grinding after spot welding is large, working conditions is bad. For these reasons, a few bus manufacturers adopt low position welding operation, and use a resistance welding. Though it is a trend to use the resistance welding for the bus envelop panel, to single-point /spot resistance welding on single or double sides, the major problems are the large energy consumption and large cooling system being required for most bus plants. A new generation of automatic welding equipment used for the bus body roof panel and top skeleton is developed based on previous work for low position operation, which strives to overcome the above drawbacks in conjunction with the use of CO_2 gas shielded arc welding and low-power single-face double-spot resistance welding.

2 REQUIREMENTS AND FEATURES OF WELDING EQUIPMENT

Welding the roof panel in the lower position could be incomplete off line of production, which may reduce the workload on the assembly line, meet the requirements of mass production; and manipulation convenient at the low position can ensure the production process and product quality more easier.

Hydraulic tensioning and automatic welding can improve the degree of tension of roof panels.

With reliable welding quality and well weld appearance, CO_2 automatic welding is employed in long seams between the panels at the top center and the top side, which relieves the workers from the hard work of hand soldering for long-time squatting.

If the automatic resistance welding is used to weld the top skin and laptop shelf between the two sides within the stringers, the grinding workload for original manual CO_2 gas shielded arc welding is cut down, the labor intensity is greatly reduced, and the workers can be freed from full noise and dust environment.

The position of a resistance welding machine, CO_2 welding torch, locating and clamping mechanism for the skeleton, skin tensioning mechanism can be adjusted to adapt to the requirements of different models roof cover.

Power consumption is low. The power consumption of single-sided double-point resistance welding machine is only 1/5~1/6 of the single point resistance welding machine, and 1/8 of double-sided single-point resistance welding machine. As well as the cooling system is small and simple, and easy to maintain.

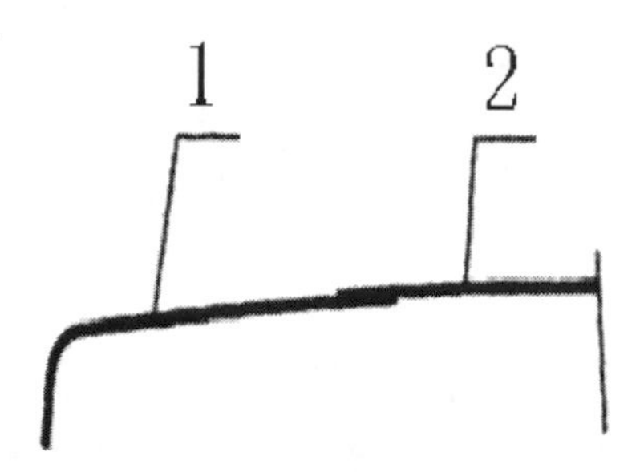

Figure 1. Roof panel Composition diagram
1-Top side roof panel, 2-Top-middle roof panel.

3 WELDING EQUIPMENT

The equipment mainly consists a rackack, gantry, welding molding bed, panel tensioning mechanism, CO_2 protection welder, single-sided double-point resistance welding machine and some auxiliary tooling components. A stepping motor and MCU control, improved the precision of gantry running, thus nsuring the accuracy of welding. As an example, to weld the roof with three panels of two sides and one top, shown in Figure 1,

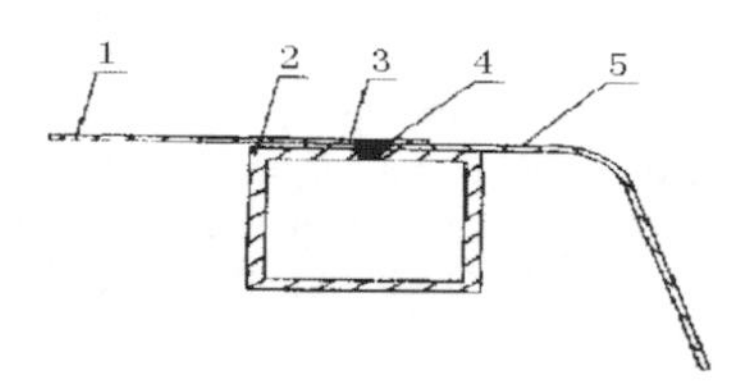

Figure 2. Gluing spot welding diagram.
1- Top-middle of the roof panel, 2- sectional material, 3-sealant, 4-spot welding point, 5- the top side of the roof panel.

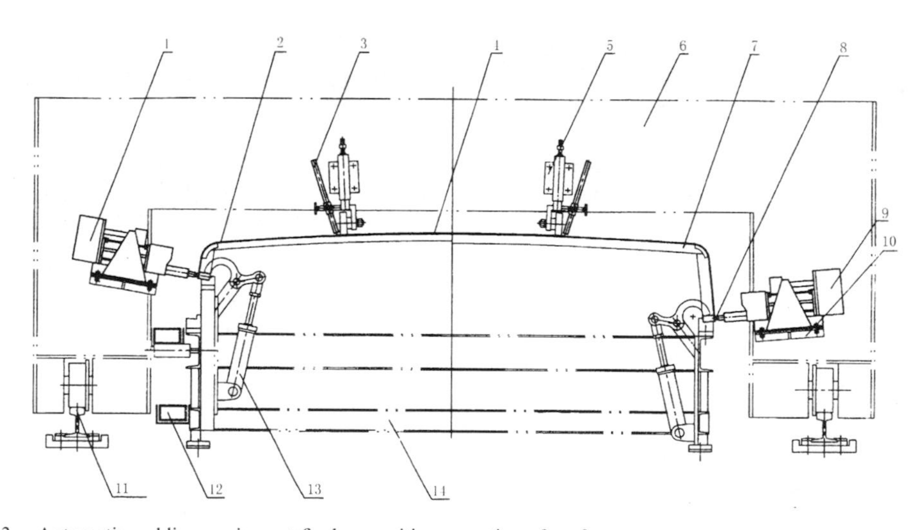

Figure 3. Automatic welding equipment for low position operation of roof.
1 - spot welder (position 2), 2 - the top side of the roof panel, 3 - CO2 welding torch, 4 - Top-middle of the roof panel, 5 - welding gun bracket, 6- portal frame, 7- roof frame, 8- welding mechanical and electrical pole, 9-spot welder (position 1), 10 - spot welder machine bracket, 11- pathway, 12-tanks panzer chain, 13- clamping mechanism, 14 - mould.

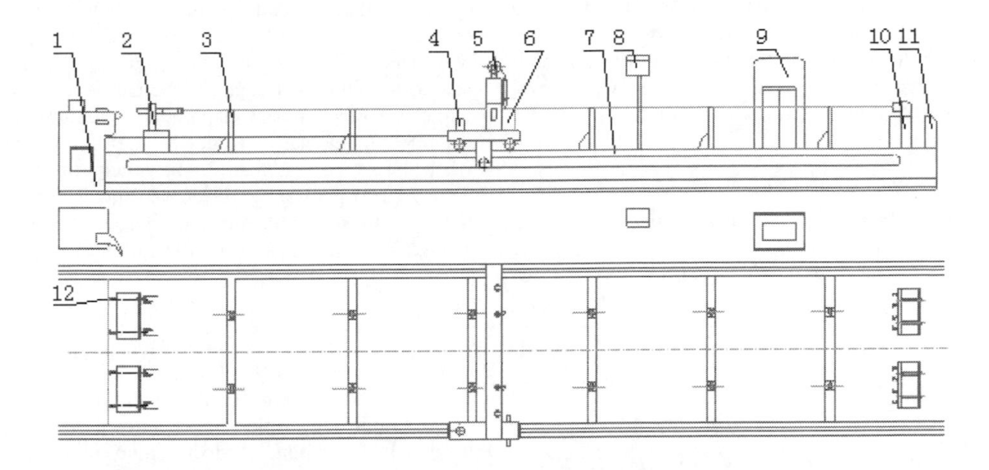

Figure 4. The schematic of automatic spot welder used for roof skin manufactur.
1 - hydraulic cylinder station; 2-tretch tension system; 3 - fixed positioning; 4- stepper motor; 5-welding machine; 6-gantry frame; 7-pathway(rack); 8-bench board; 9- console; 10 - fixed clamping system; 11- limited block; 12 – tension cylinder.

uses two units CO_2 gas shielded welders of NBC-200G type for the top and top side panels, two units single-sided double-point resistance welders of the Great Wall DN3-25Q type of soldering the junctions between the sides of the roofs and top skeleton stringers, as shown in Figs. 2 through 4. Figure 2 shows two models roof welding, so the left and right are asymmetry.

4 MAIN TECHNOLOGICAL PROCESS

4.1 *Assemble material*

The skeleton of the roof is hung onto the roof mold, positioning, clamping. In turn, the top side and the top center of the panel are put on, and then positioned and fixed with clamps. One side of the roof is connected with the first beam roof frame using manual CO_2 intermittent welding.

4.2 *Tension*

Hydraulic tensioning mechanism clamping device clamps the other side of the roof panel, stretching. The roof panel and the first beam of the roof frame are connected to fix using CO_2 plug welding.

4.3 *Welding*

Intermittent mobile gantry front to back, while resistance welding machine start, spot start. After completion of the spot welding gantry return, with continuous movement returns, while CO_2 Welding starts. Gantry welding work on a round trip, due to single-sided double-point resistance welding machine can be both time points, so the higher the efficiency, a 12m-long gantry cover a round trip of about 40min. Taking into account the conditions of use of resistance welding, welding work should be carried out before treatment to rust, etc.

4.4 *Repair welding*

Check welding quality, hand-repaired welding. In additional, selectively drill 9 to 12 plug weld holes φ6 from front to back on the panel, and to prevent skin agitation uses CO_2 plug weld is used to fix the panel to the top frame stringers.

4.5 *Closeout*

Cut unnecessary roof panels; seal all the CO_2 welding seam sealant with sealant. Low job is finished.

5 WELDING PROCEDURE SPECIFICATION

5.1 *Material*

The material of the roof panel is ordinary low-carbon steel, thick 1.0mm, galvanized or painted conductive primer inside surface. And that of rectangular tube is ordinary low-carbon steel, thick 1.75mm, galvanized or coated conductive primer on ektexine.

5.2 *CO_2 welding*

The current is 80-100A, voltage is 20-22V, and gas flow of CO_2 is 15L/min. The speed of torch travel is 600 ~ 700mm / min. The extension length of wire from current contact nozzle is 10mm, the distance from the nozzle to the panel is 10mm, the diameter of wire is 1.0mm, the material ingredient is HO8Mn2SiA, and in theory, it is required generally that the wire extension length is about 10 times the diameter of the wire.

5.3 *Resistance welding*

The adjustable heat reading of the controller of spot welding is 65 to 85 units, which is equivalent of the secondary current about 5500 ~ 6500A. The readings of pressing, welding, maintain, ceasing are 35 ~ 45, 40 ~ 50, 35 ~ 50, 30, the unit is 0.02S. The pressure of compressed air ≥0.4mPa, water pressure of being cooled reaches normal water pressure or as so.

6 WELDING QUALITY AND INFLUENCING FACTORS

Although the combination of CO_2 welding and resistance welding is relatively advanced, due to restrictions by the objective conditions and there is non-uniform standards in domestic, the welding quality has not reached the ideal state, we must continue to improve. Because the bus skeleton panel is mainly made of special shaped steel tube sheets and a variety of connectors welded together, the basic form of welding are two types of butt and fillet welds, these materials are belonged to light gauge welding, and the main factors affecting the quality of welding panel skeleton including welding specifications, assembly space, CO_2 gas flow and purity, air and water pressure, and the operation level of the welding, and the objective response to quality problems are mainly poor shaped and producing blowholes and so on.

Assembling and welding process is to install the special shaped steel tube (galvanized) and a variety of fittings onto welding mould to assemble and weld, and then using CO_2 Welding to weld skeleton into one unit. Because the precision of welding mould and unit mass caused weld gap unequal, which have a greater impact on the quality of welding. Usually the thickness of the work piece of welding is 1~ 2mm, the gap should be less than 1.5mm or no gap. If the gap is too much, it will cause difficulties in shaping the weld line, and solder lutetium increases. Under normal circumstances, when the gap is less than 2mm, which can be remedied by the operator's level of welding. To overcome the problems above, the need to improve the quality of a single piece can ensure assembly and welding gap.

7 CONCLUSION

The result of any kind of new technology methods is continuing to explore and innovation, but also the level of personnel and the environment are two factors that cannot be ignored. For the impact of welding equipment, you can use the advanced unified CO_2 welding machine, for other aspects, we can strengthen management, continuous improvement, improve quality to encourage the process advantages for the production.

REFERENCES

Dengshi Zhen, Guan Dong, Welded Structures and Production Processes [M], Tianjin; *Tianjin University Press, 2010, in Chinese.*

Cao Hongliang. Welding Technology and Equipment Used for the Bus Outer Roof-panels [J], *Bus Technology and Research, 2013(02), in Chinese.*

Gan Hui, Design and debugging of semi-self-automatic welding equipment for bus roof-hood skin, *Applied Mechanics and Materials, v 184–185, pt.1, p 121-4, 2012.*

Changsha bus body welding process information of the original bus manufacturing plant.

Jiang Huan, Arc Welding and Electro slag Welding [M], *Shanghai: Shanghai Jiao tong University Press*, 2010, in Chinese.

Computing, Control, Information and Education Engineering – Liu, Sung & Yao (eds)
© 2015 Taylor & Francis Group, London, ISBN: 978-1-138-02800-5

Research on storage optimization and encryption algorithms for CD disc management system

Wei Gang Feng & Wei Ting Zhang
Xianyang Vocational Technical College, Shaanxi, China

ABSTRACT: With the bursting and fast development of computer science and CD-ROM technology, the storage optimization and related encryption algorithms are studied popularly. Due to the emergence of heterogeneous cloud storage systems, we generalize the concept of the FR code and propose the IFR code. A key property of the FR code is its un-coded repair process. To determine the repair pattern, which we call the repair overlay, and the storage allocation, we formulate the whole problem based on a new irregular model, with the aim of minimizing the system repair cost by properly designing the MDS-IFR code and the retrieval sets. The simulation result prove the effectiveness of our method.

KEYWORDS: Storage Optimization; Encryption Algorithms; CD Management System.

1 INTRODUCTION

1.1 *Background*

In the last few decades, the use of digital content increased dramatically. Many forms of digital products in the form of CDs, DVDs, TV broadcasts, data over the Internet, entered our life. Classical cryptography, where encryption is done for only one recipient, was not able to handle this change, since its direct use leads to intolerably expensive transmissions. Moreover, new concerns regarding the commercial aspect a rised. Since digital commercial contents are sold to various customers, unauthorized copying by malicious actors became a major concern and it needed to be prevented carefully. Therefore, a new research area called digital rights management (DRM) has emerged. Within the scope of DRM, new cryptographic primitives are proposed [1–3].

1.2 *Overview of our research*

In this research paper, we consider three of these: broadcast encryption (BE), traitor tracing (TT), and trace and revoke (T&R) schemes and propose methods to improve the performances and capabilities of these primitives. In particular, we first analyze the recipient set up in order to improve the transmission size in one of the most popular programs. Then we study and solve the problem of the optimal distribution of hitchhiking so far one of the most effective solution. Next, we try to close the nontrivial gap and transceiver plan put forward a general method to add the traitor tracing ability to plan, thereby gaining a transceiver. Finally, we study the problem of a neglected: the privacy of the recipient to send and receive. Right now, most schemes [4–6] do not keep the recipient set anonymous, and everybody can see who received a particular content. As a generic solution to this problem, we propose a method for obtaining anonymous T&R scheme by using anonymous BE schemes as a primitive. In the following sections we will discuss the problems more.

2 GENERAL ENCRYPTION MODEL

2.1 *The basic structure*

Since we will explain these algorithms formally when needed in the following sections, we briefly explain the general structure here. In different settings, some concerns such as user domain size, security, bandwidth, or hardware may be more important than others. However, usually, two concerns are inherent in almost all BE systems. First, the amount of key storage must be adequate because the long-term secure storage size at the receiver side is very limited since it has to be tamper resistant. Second, the amount of additional data sent along with the content through the communication channel, called the transmission overhead, must be adequate because of the limited nature of the bandwidth of communication channels. A broadcast encryption scheme can be defined in terms of three algorithms. (1) Key distribution algorithm is used to create a key and assign it to the receiver is the construction time system.

(2) An encryption algorithm to encrypt the message, only the user can decrypt. (3) Decryption algorithm, by receiver and a successful only if the receiver. Otherwise have no valuable information to the message. Safety plan is confidential data. As a classical cryptography, it is a game indistinguishability and form definition. In the section we will detail for the game. Basically, revoke users must be indistinguishable from an arbitrary messages sent from the message space. We leave detailed explanation of security to the relevant sections.

2.2 *Evaluation parameters and standards*

As in all computational systems, the speed of the key distribution, encryption and decryption algorithms are important parameters by default. Speed is the most sensitive of key distribution algorithm, however, because it usually runs only once when the system is ready in advance. The speed of the encryption algorithm is usually more important equipment, will run since the decryption algorithm will no longer so strong. Broadcasting center equipment, for instance, is often a more powerful than the set-top box in a house. However, there are two more specific broadcast encryption scheme evaluation parameters: the key store and transport costs. The size of the key storage is basically key/receiver. This storage must be minimized as key must security to prevent malicious violations actor like stealing keys and forging a pirate decoder. Transmission overhead is broadcast encryption additional costs and not safe. Both cases are infeasible in almost all broadcast encryption systems since the number of users is typically huge and we simply have neither that much key storage space in the user devices to carry out the first idea nor that much bandwidth capacity to carry out the second.

3 DATA STORAGE OPTIMIZATION MODEL

3.1 *Data storage model: overview*

From the figure 1 to figure 3, we could have an overview of the current data storage architecture.

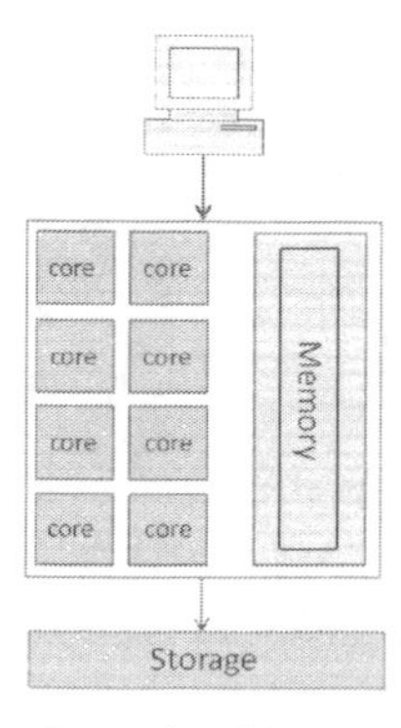

Figure 1. Single, advanced multi-core machine.

The figure 1 illustrates the structure of traditional single core based structure for data storage. We can sense the low effectiveness of this kind of single channel structure.

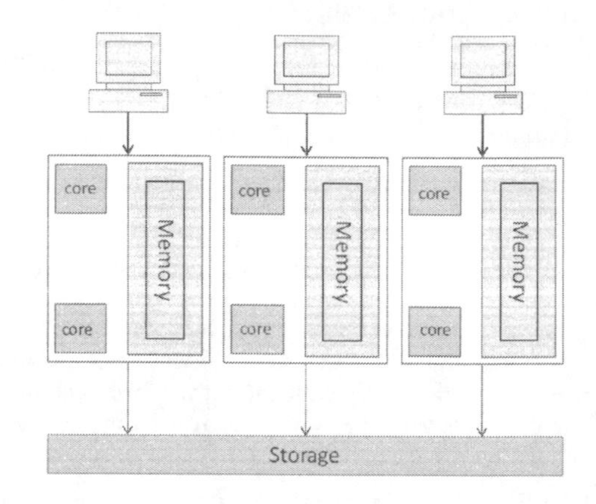

Figure 2. Cluster, consisting of multiple weaker machines.

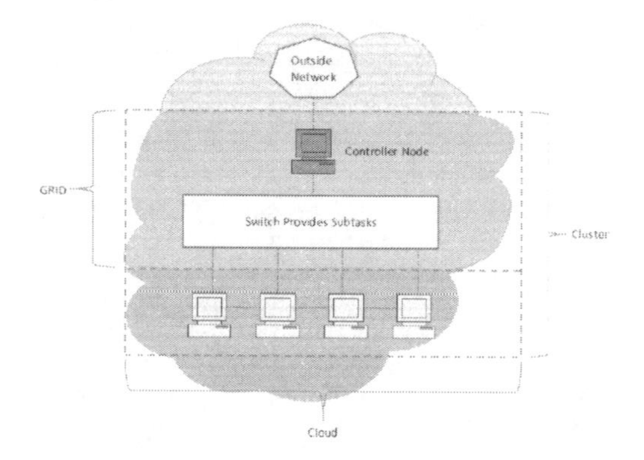

Figure 3. A computer cluster is built up out of several connected computers that cooperate as a single powerful unit.

3.2 *Core algorithms*

In this section, we discuss the advanced MDS-IFR code optimization algorithm. Our objective is to design the MDS-IFR code so as to minimize the system repair cost, if a non-repairable failure patternγ occurs with possibility, download and decode the data object will be data from a set of storage retrieval then re-encoded. We assume that the system is properly designed and the probability of non-repairable failure mode is very small. Therefore, we focused on minimizing the expected maintenance cost per unit data, the expected all repair mode. The repair cost can be denoted as the following formula one:

$$c_r(\tau,b) \triangleq \sum_{\gamma:repairable} p(\gamma)\tilde{c}_r(\tau,b,\gamma) \tag{1}$$

The formula 2–6 present the optimization operation procedures.

$$x_i \triangleq \begin{cases} 0 & if\,\widetilde{E}_i \notin H^\tau \\ 1 & if\,\widetilde{E}_i \in H^\tau \end{cases} \tag{2}$$

$$\sum_{i:v\in\widetilde{E}_i} x_i \le d, \forall v \in V \tag{3}$$

$$0 \le \beta_i \le Bx_i, i = 1, 2, \ldots, \binom{n}{\rho+1} \tag{4}$$

$$\alpha_v = \sum_{i:v\in\widetilde{E}_i} \beta_i, \forall v \in V \tag{5}$$

$$c_s(\tau(x), b) \triangleq \frac{1}{R}\sum_{v=1}^{n} s_v \alpha_v \le C_s \tag{6}$$

Part or all of these sets may be predetermined based on considerations other than storage and repair costs. For example, if a data object according to user's requirements is mainly in a specific geographical area, it will be more convenient, if one or more sets of searching are the storage nodes in the region, so that users can shorten the response time to download the object.

4 PROPOSED APPROACH

4.1 *Traitor tracing method*

The majority of the black-box traitor tracing schemes shares the same tracing strategy that is called 'hybrid coloring' in [7] or 'linear tracing' in [8] and is inherent almost in all black-box traitor tracing mechanisms. This strategy can be summarized as the following way: a pirate decoder query sequence of special trace ciphertexts, randomized receiver decryption way gradually. In this sequence, and the first special cipher is decryptable receiver, the last is a decryptable didn't. The technique above yields a trivial traitor tracing system with each user having a unique decryption key. In the case of tracking and cancel, found that the same technology is useful, but potential multi-user encryption scheme is the property of the need to design a more demanding. Boneh meet this by introducing an enhanced broadcast encryption scheme support to cancel any subset of the program can use pure broadcast encryption scheme, and further allow enable any subset in the process of walking. This will eventually lead to the application of basic linear tracking strategy, thus to identify the traitor. Traitors and can easily cancel plans to further support. The next section will conduct in-depth discussion.

4.2 *Encryption in KEM structure*

As we know from the previous sections, a broadcast encryption (BE) scheme is a method for encrypting messages in a way that only an intended recipient set, which we call the privileged users, will be able to decrypt it, and even if all other users, which we call the revoked users, collude, they cannot get any information about the message. In a content distribution setting with revocation, the actual data is usually encrypted with a standard, symmetric message encryption scheme while the symmetric key used in that encryption is transmitted with the BE scheme. The reason is that the data that needs to be encrypted is typically too long and encrypting the whole data directly with the BE scheme is too expensive. Tracking and cancel (transceiver) system is a system, tracking and undo function. Track users in this case, the leak of their decryption key, and revoke have failed key users such as leakage, the key is neutral. Tracking and revocation system by adding a tracking algorithm is adopted. Tracking algorithm ensures the detection of a traitor receiver button may be used to create the pirates decryption box be traced. Later, they may only be added to the blacklist, whenever a new radio, the system can ensure that the receiver in the cancellation. The flowchart of the algorithm is shown in the figure 4.

Input: $\check{G} = (\mathcal{V}, \check{\mathcal{E}}), d, \rho$
Output: $\tau = (\mathcal{V}, \mathcal{H}^\tau)$

1) Compute the metric closure G of $\check{G}$.
2) Initialize τ with $\mathcal{H}^\tau \leftarrow \emptyset$, and $N_v^\tau \leftarrow 0$ for all $v \in \mathcal{V}$. (Note that N_v^τ represents the degree of vertex v in τ.)
3) Sort all the $(\rho+1)$-subsets of $\mathcal{V}$ in ascending order of MST weight and get the sequence $\check{E}_1, \check{E}_2, \ldots, \check{E}_{\binom{n}{\rho+1}}$.
4) **for** $i - 1$ to $\binom{n}{\rho+1}$ **do**
   if $N_v^\tau < d$ for all $v \in \check{E}_i$ **then**
   $\mathcal{H}^\tau \leftarrow \mathcal{H}^\tau \cup \{\check{E}_i\}$
   $N_v^\tau \leftarrow N_v^\tau + 1$ for all $v \in \check{E}_i$
   end
   end
5) Return $\tau = (\mathcal{V}, \mathcal{H}^\tau)$.

Figure 4. The flowchart and description of our proposed methodology and algorithm.

In order to prove the traceability of our generic construction, we have to prove that no polynomial time attacker A that forges a perfect decoder can win the tracing game with some non-negligible probability. More specifically, we will bound the winning probability of such an attacker A by a function of the security bounds of the underlying primitives. In the next section, we will conduct experimental analysis to simulation the scene.

5 EXPERIMENT AND SIMULATION

In this section, we consider heterogeneous storage systems. We compare the optimal tradeoff between system storage cost and system repair cost that can be achieved by the MDS-IFR code with that achieved by the regenerating code. Moreover, we compare the minimum system repair cost that can be achieved by the MDS-IFR code with that achieved by the regenerating code for different network size. Here, we use the term "regenerating code" to refer to any code that achieves points on the tradeoff curve under the regular model. The curve connecting all Pareto-optimal Bachievable cost pairs is the optimal tradeoff between system storage cost and system repair cost that can be achieved by the MDS-IFR code. The following figure 5-7 shows the simulation result for our approach.

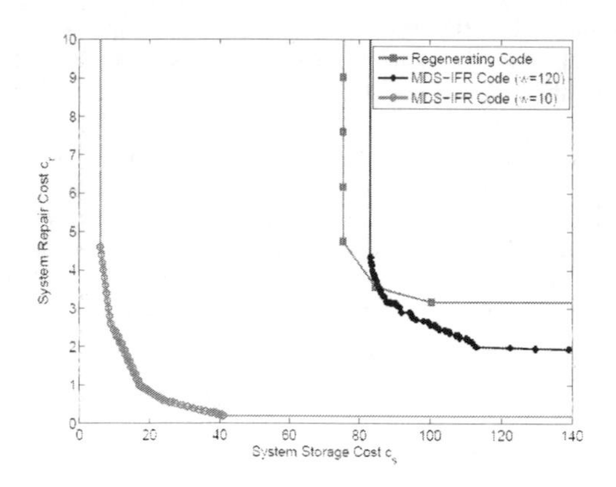

Figure 5. Optimal tradeoff between system storage cost.

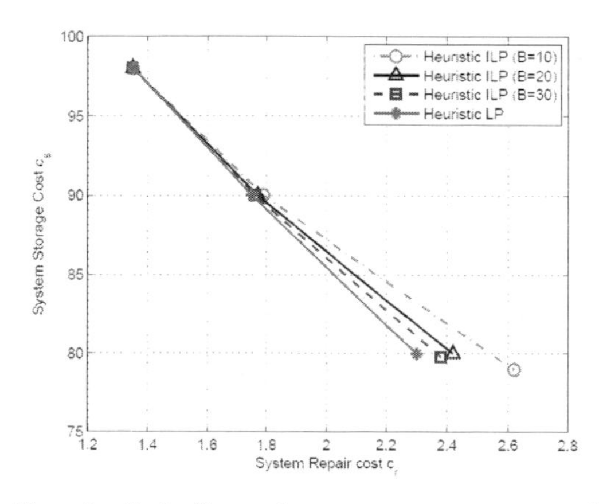

Figure 6. Tradeoff curves between system storage cost and system repair cost.

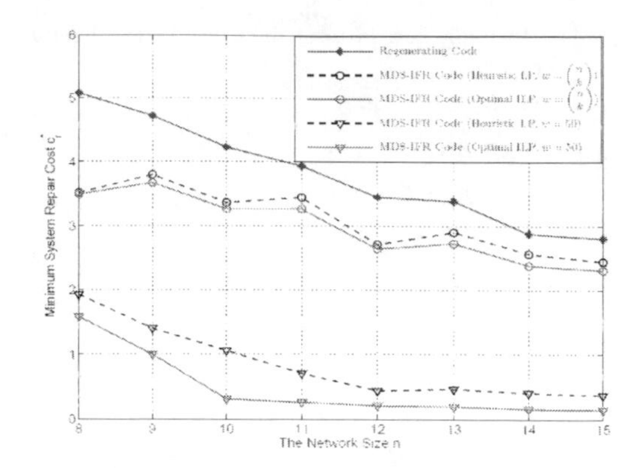

Figure 7. The minimum system repair cost for different network size demonstration.

ACKNOWLEDGEMENT

This paper is supported by the Xianyang vocational technical college scientific research fund project (NO. 2012KYA07).

6 SUMMARY

Due to the emergence of heterogeneous cloud storage systems, we generalize the concept of the FR code and propose the IFR code. A key property of the FR code is its un-coded repair process. To determine the repair pattern, which we call the repair overlay, and the storage allocation, we formulate the whole problem based on a new irregular model, with the aim of minimizing the system repair cost by properly designing the MDS-IFR code and the retrieval sets. While the optimization framework established in this paper concerns mainly on system repair cost, it can be modified to include other system objectives and extended by incorporating more resource constraints. On the other hand, as it is based on the MDS-IFR code, it provides very low repair cost at the expense of higher storage overhead. The simulation result demonstrates the effectiveness and feasibility of our proposed methodology.

REFERENCES

[1] Saravanan, P., and K. K. Thyagarajan. "A Novel Block-Based Selective Embedding Type Video Data Hiding Using Encryption Algorithms." Power Electronics and Renewable Energy Systems. Springer India, 2015. 1087–1097.

[2] Zhang, Xin Zheng, and Ya Juan Zhang. "On Data Security and Encryption Algorithms in Cloud Environment." Applied Mechanics and Materials. Vol. 701. 2015.
[3] Zhu, Bo. Analysis and Design of Authentication and Encryption Algorithms for Secure Cloud Systems. Diss. University of Waterloo, 2015.
[4] Liao, Xuefeng. "Improved Tent Map and Its Applications in Image Encryption." (2015).
[5] Ali, Sammoud, and Adnen Cherif. "Performances analysis of image encryption for medical applications." Journal of Information Sciences and Computing Technologies 1.1 (2015): 78–87.
[6] Ahmad, Musheer, and Faiyaz Ahmad. "Cryptanalysis of Image Encryption Based on Permutation-Substitution Using Chaotic Map and Latin Square Image Cipher." Proceedings of the 3rd International Conference on Frontiers of Intelligent Computing: Theory and Applications (FICTA) 2014. Springer International Publishing, 2015.
[7] Ullagaddi, Vishwanath, Firas Hassan, and Vijay Devabhaktuni. "Symmetric synchronous stream encryption using images." Signal, Image and Video Processing 9.1 (2015): 1–8.
[8] Zhou, Yicong, et al. "Fast Fourier transform using matrix decomposition." Information Sciences 291 (2015): 172–183.

Computing, Control, Information and Education Engineering – Liu, Sung & Yao (eds)
 ISBN: 978-1-138-02800-5

A stroke intersection points extraction method for Chinese calligraphy of stone rubbings

Pei Zhi Wen

Guangxi Colleges and Universities Key Laboratory of Intelligent Processing of Computer Images and Graphics, Guilin University of Electronic Technology, Guilin, China

Chao Bi, Meng Zhao, Long Cheng

College of Electronic Engineering and Automation, Guilin University of Electronic and Technology, Guilin, China

ABSTRACT: In the study of automatic extraction and recognition of Chinese calligraphy strokes on the stone rubbing using a computer, the key lies in confirming the stroke intersection points. In this paper, we present an algorithm to extract intersection points on the strokes' contours. First of all, we preprocess the stone rubbing image and extract the contours; Secondly, Douglas-Peucker algorithm is used to extract the feature points which indicate the calligraphy contour information; Finally, a feature point selection method based on maximum local contour curvature method is used to remove the feature points which are not the intersection points, and get the real intersection points. The experimental results show that this algorithm is effective in extracting stroke intersection points of stone rubbing calligraphy character.

KEYWORDS: Calligraphy; Contour; Douglas-Peucker algorithm; Stroke intersection points.

1 INTRODUCTION

Stone rubbing is the practice of creating an image of surface features of a stone on paper and always been used to transfer the calligraphy from stones by Chinese scholars. As a culture carrier and an expression of the ancient Chinese art, the words on the stones are illegible or missing under natural conditions due to corrosion, weathering, oxidation and other causes. So it is meaningful to research the calligraphy on the stone rubbing and even regenerate the missing words with computer technology. Computer calligraphy is also promising, for example, the generation of a particular style of calligraphy, building a digital calligraphy library, calligraphy tutoring systems and so on. Because rubbings calligraphy is a unique art form in China, so the study in western countries is few, but a certain amount of research have been done in many universities and institutes in China such as the China Academic Digital Associative Library. At present, the study of ancient Chinese calligraphy rubbings focused on two main aspects, one is the calligraphy strokes extraction[1][2][3], and the other is the generation of a specific style of calligraphy[4][5][6].

Strokes are the building blocks of characters; different calligraphers have different writing styles. Therefore, stroke extraction is a research focus in computer calligraphy. Contour information method[1] detects feature points on the contour of a character, and then matches contour feature points or contour segments to extract strokes. An intersection point lies on two or three strokes' outlines, and it is also a separation point of respective strokes. As a key step to extract strokes in contour information method, the extraction of intersection points hasn't been solved effectively yet. In this paper, we present a new algorithm for extracting stroke intersection points. Firstly, we preprocess the calligraphy image and extract the contour of the character. Then the Douglas-Peucker algorithm is used to extract the feature points which describe the contour. Finally, the characteristics of the contour curvature at the feature points and the distance between them are used to get the primary result of intersection point selection; After calculate the forward direction and the opposite of the afterward direction of selected feature points, if the two directions of a selected feature point both go into the internal of the contour of the character, this point is an intersection point; otherwise it isn't an intersection point. The test result shows that the algorithm is effective in the extraction of stroke intersection points.

2 CALLIGRAPHY CHARACTER CONTOUR ACQUISITION

2.1 *Image denoising*

Due to the weathering of the stone or interference during the photographing process, there is a lot of noise in rubbing images (see Figure 1 (a)). In this paper, the adaptive median filter is used to remove most of the noise in the image. Adaptive median filter solves the contradiction between denoising and protection of the image details. It can eliminate the noise, protect the edge information of strokes, highlight the main part of the image and suppress noise and high frequency interference (see Figure 1(b)).

2.2 *The binarization of image*

Binarization can make rubbing image becomes simple, reduce the amount of data, and especially highlight the contour of calligraphy character. But the threshold has a tremendous impact on the results. If the threshold is too large, the dark portion of the character will be classified into the background; If the threshold is too small, the dark background can't be completely removed, and it is hard to distinguish the character and the background. The Otsu method is an adaptive threshold determination method, which minimizes the intraclass variance and maximizes the interclass variance of the object and the background to get the threshold. This method also minimizes the probability of misclassification of object and background (see Figure 1 (c)).

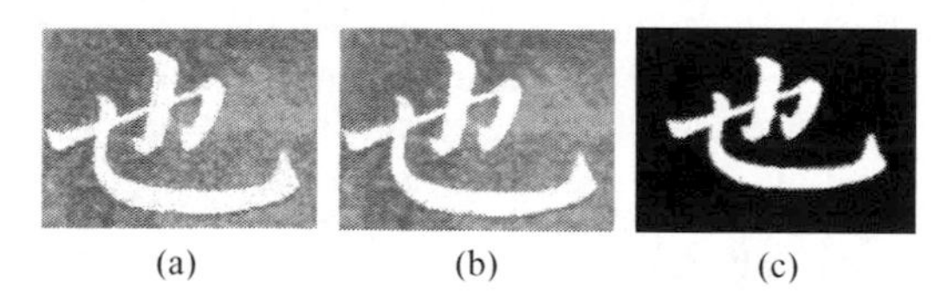

Figure 1. The result of preprocessing. (a) The original rubbing image. (b) The result of using an adaptive median filter. (c) The result of binarization using Otsu method.

2.3 *Contour acquisition*

A contour acquisition that using image edge detection has many methods, such as Sobel operator, Prewitt operator, Roberts operator and Canny operator. These algorithms are used to extract the contour of the image (see Figure 2). Comparative experiment shows that the Sobel operator can smooth the noise, but there exist some false edges and the edge positioning ability is a little lower; Prewitt operator method is easy to miss the corner points; the Roberts operator has a high positioning accuracy, but particularly sensitive to noise; Canny operator method seeks the brightness gradient after the removal of image noise, and finally tracks edge to get the contour. This algorithm has strong noise suppression and edge continuity, which is suitable for noisy images like stone rubbings.

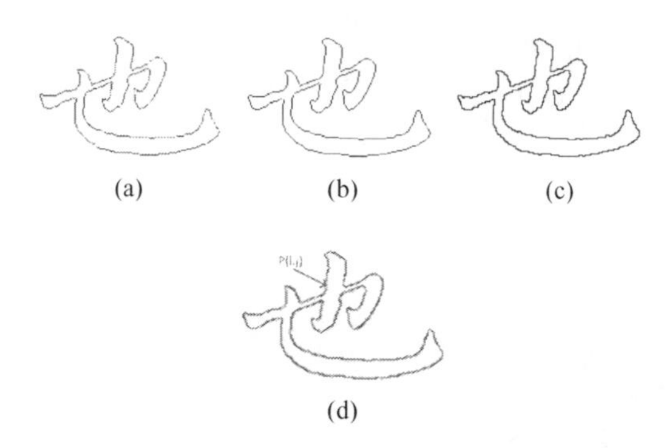

Figure 2. The comparison of edge detection methods. (a) Sobel operator. (b) Prewitt operator. (c) Roberts operator. (d) Canny operator.

In order to facilitate the detection and selection of feature points, the points need to be arranged clockwise along the contour, which is the contour points sequence. The j-the pixel on the I-the contour is denoted as P(i,j). The coordinates of P(i,j) is [x(i,j), y(i,j)] if the top left of the image is the coordinate origin.

3 FEATURE POINTS DETECTION AND SELECTION ALGORITHM

The feature points of a contour characterize the stroke information and the topology of a character. There are mainly three types of stroke feature points: the stroke endpoints, the stroke turning points and the stroke intersection points (see Figure 3).

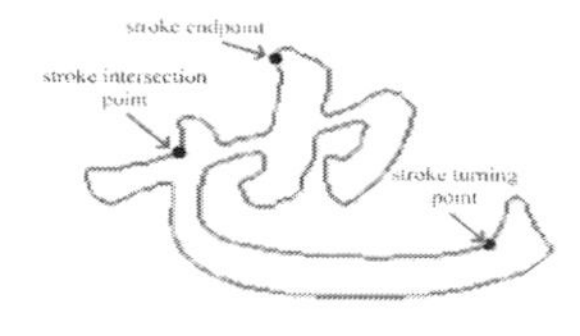

Figure 3. Types of stroke feature points.

3.1 *Feature points detection*

At present, the main stroke feature point detection algorithms are: feature point detection based on morphology [7], maximum local contour curvature method[8], feature point detection based on Douglas-Peucker algorithm(D-P algorithm)[9]. Feature points detection based on morphology has a higher complexity, while the maximum local curvature method is sensitive to the noise so that it is bad for subsequent processing.

D-P algorithm is a curve approximation method, which represents the curve by a series of points. The following are the basic ideas of the D-P algorithm. Even a straight line between the first and last point of the curve and calculate the distances from every point on the curve to the line. After finding the maximum distance Dmax, compare Dmax with the distance threshold Th. If Dmax > Th, set the corresponding point as a new feature point and demarcation point, which divides the curve into two parts. Then process these two parts with the same method like before. In figure 4, even a straight line between A and B, and then calculate the distances from points on curve AB to the straight line AB. Find point C corresponding to the maximum distance Dmax, and if Dmax > Th, set C as the new feature point. C divides AB into two parts, then use this algorithm in AC and BC until no new feature point generated. D-P algorithm can control the degree of density of the feature points by adjusting the distance threshold.

Feature points that detected using D-P algorithm are arranged clockwise as KP(i,j), j=1,2,3…N, and N is the number of feature points on the i-th contour (see Figure 5).

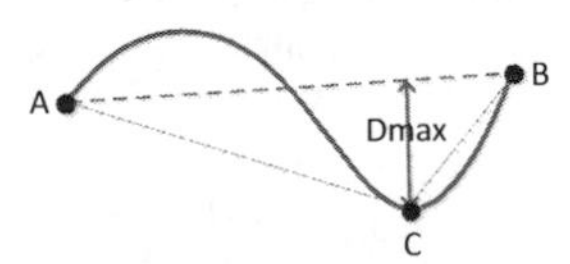

Figure 4. Basic idea of D-P algorithm.

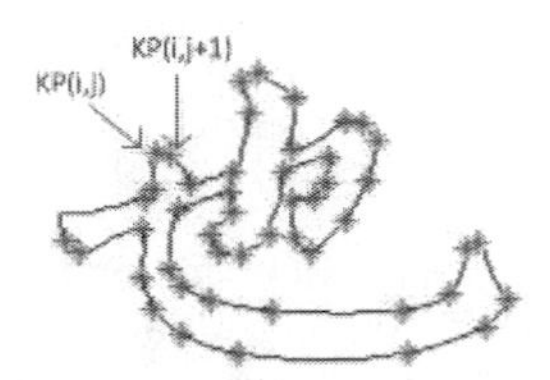

Figure 5. Result of D-P algorithm.

3.2 *Feature points selection*

The result of feature point detection includes not only stroke intersection points, but also other types of feature points. The feature point selection algorithm that we proposed based on maximum local contour curvature method includes two steps (see Figure 6).

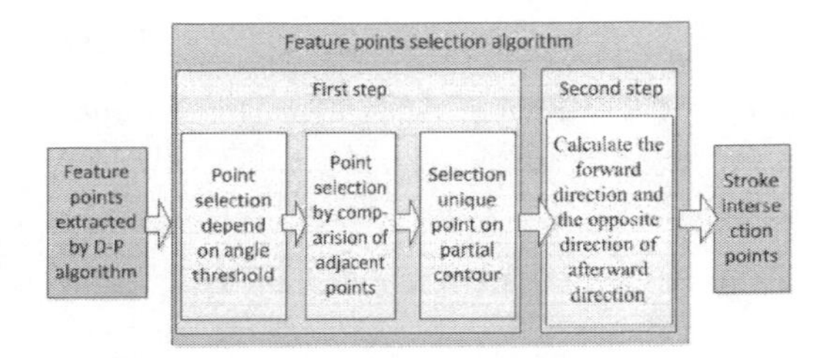

Figure 6. Flow diagram of feature point selection algorithm.

After a lot of analysis and observation of the results of feature detection, we discovered that the curvature of the intersection area is larger than other places, where the feature points usually gather together. The curvature of the contour can be measured by the contour angle at the feature point. So, the stroke intersection points can be selected using the contour angle of the feature points.

The first step of the algorithm, compare the contour angles of feature points with the angle threshold and remove the points whose contour angle is bigger than the threshold. Then compare the contour angle of the adjacent feature points and selected the points whose contour angle is smaller. After comparative selection, there may exist a few feature points in a small region, so select the point with a smallest contour angle in a local contour region and remove the other points.

According to Chinese writing habits, the forward direction and the opposite of the afterward direction of stroke intersection points go into the internal of the contour, but the other types of feature points are not like this. So in the second step, calculate the forward direction and the opposite of the afterward direction of feature points, then distinguish the stroke intersection points from the others.

In order to automatically implement the algorithm on a computer, we define the following key parameters.

Definition 1. The angle of the direction from point P(i,j) to P(s,t) in the coordinates is D[P(i,j), P(s,t)], D[P(i,j),P(s,t)] ϵ [0,2π) (see Figure 7)[10]. That is,

$$\theta = \arctan\frac{|y(i,j) - y(s,t)|}{|x(i,j) - x(s,t)|}$$

$$D[P(i,j),P(s,t)] = \begin{cases} 0 & y(i,j) = y(s,t)且\ x(i,j) < x(s,t) \\ \frac{\pi}{2} & x(i,j) = x(s,t)且\ y(i,j) > y(s,t) \\ \pi & y(i,j) = y(s,t)且\ x(i,j) > x(s,t) \\ \frac{3\pi}{2} & x(i,j) = x(s,t)且\ y(i,j) < y(s,t) \\ \theta & x(i,j) < x(s,t)且\ y(i,j) > y(s,t) \\ \pi - \theta & x(i,j) > x(s,t)且\ y(i,j) > y(s,t) \\ \pi + \theta & x(i,j) > x(s,t)且\ y(i,j) < y(s,t) \\ 2\pi - \theta & x(i,j) < x(s,t)且\ y(i,j) < y(s,t) \end{cases} \quad (1)$$

Definition 2. The forward direction FDir(i,j) of point P(i,j) is a group of points in front of P(i,j) move toward it. The afterward direction ADir(i,j) of point P(i,j) is the direction P(i,j) moves to its successive points (see Figure 8)[10]. That is,

$$\mathrm{FDir}(i,j) = \frac{1}{T}\sum_{n=1}^{T} D[P(i,j-n), P(i,j)] \quad (2)$$

$$\mathrm{ADir}(i,j) = \frac{1}{T}\sum_{n=1}^{T} D[P(i,j), P(i,j+n)] \quad (3)$$

Where, T is the number of adjacent points, which is generally the width of the strokes.

Definition 3. The contour angle Ang[P(i,j)] at a feature point P(i,j) is the angle between the forward direction and the afterward direction (see Figure 8), $\mathrm{Ang}[P(i,j)] \in [0, \pi]$.

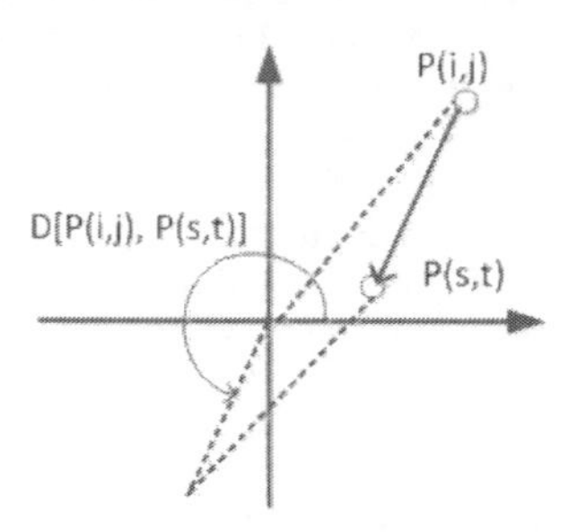

Figure 7. The direction and afterward direction.

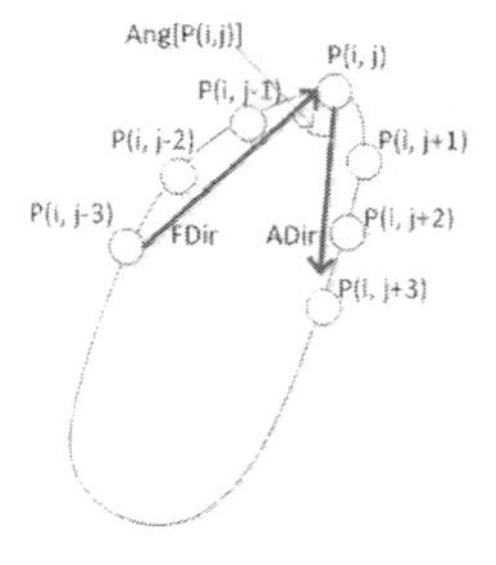

Figure 8. Forward direction from P(i,j) to P(s,t).

According to the definition above, calculate the contour angle of every feature point, then using the feature point selection algorithm to get stroke intersection points.

The implement of the algorithm is as follows.

Step 1: Enter the feature point sequence P (i, j) extracted by the DP algorithm.

Step 2: For any P(i,j)$\in \mathrm{KP}(i,j)$, calculate the Ang [P (i, j)], if the angle is greater than the threshold (generally 140°), remove the feature point P (i, j) from KP (i, j) .

Step 3: For any P(i,j)$\in \mathrm{KP}(i,j)$, if the Euclidean distance between P (i, j) and P(i,j+1) is less than the distance threshold Th(generally the stroke width), and Ang [P (i, j)] is greater than $\mathrm{Ang}[P(i,j+1)]$, remove the feature point P (i, j) from KP (i, j).

Step 4: For the all candidate feature points P(i,j) $\in \mathrm{KP}(i,j)$, in a certain contour segment Th, choose the feature point with a smallest contour angle and remove the remaining feature points.

Step 5: For any P(i,j)$\in \mathrm{KP}(i,j)$, if the forward direction FDir(i,j) and the opposite of afterward direction ADir(i,j) of the point go into the internal of the contour, this point is a stroke intersection point, otherwise it is not a stroke intersection point.

The stroke intersection points of calligraphy character 'ye' are shown in figure 9.

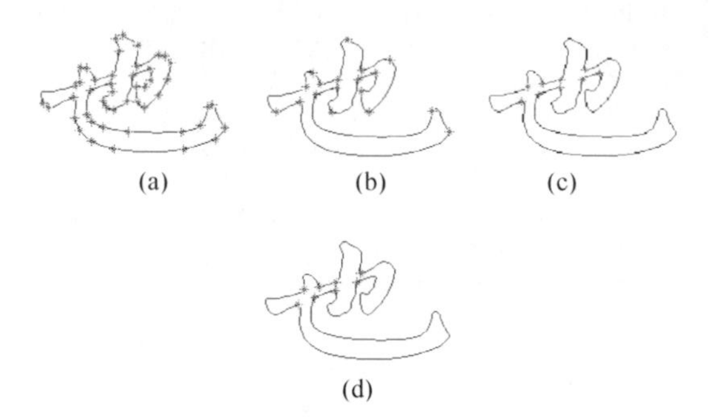

Figure 9. The result of stroke intersection points of character 'ye'. (a) Original feature points. (b) The result of the first step. (c) The result of the second step. (d) The final result.

The experimental results show that this algorithm is effective in extracting stroke intersection points of stone rubbing calligraphy character, and lays a necessary technical foundation for automatic analysis and repair of calligraphy.

4 THE EXPERIMENTAL ANALYSIS

In order to verify the validity of the algorithm in this paper, we selected 150 digital images of the authentic stone rubbings of the ancient calligraphy masters (such as Zhenqing Yan, Xizhi Wang, Ouyang Xun, etc.). Some of the experimental results are shown in figure 10. The experimental results show that the algorithm can accurately extract the stroke intersection points of stone rubbings calligraphy character, applicable to the stroke intersection points of calligraphy character of the running script, official script and other fonts.

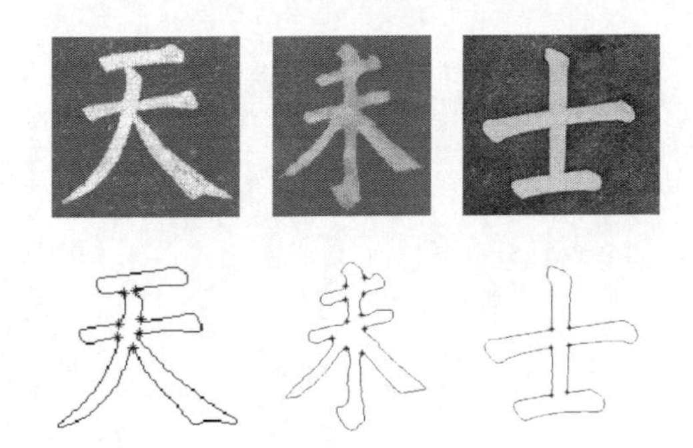

Figure 10. Some of the experimental results.

5 CONCLUSION

In this paper, stroke intersection point extraction algorithm is proposed. Firstly, the algorithm does pre-processing and contour extraction for the calligraphy character image, then uses D-P algorithm to get the contour feature points, and then use the feature point selection algorithm proposed to get stroke intersection points. Experiments show that the algorithm has higher accuracy of the stroke intersection point extraction of more standardized fonts. But the accuracy is unsatisfactory for the stroke intersection point extraction of calligraphy character with complex structure, writing more untidy and stroke conglutination, and the angle calculation and selection algorithm of feature points need further improvement and optimization to adapt to the more scrawled calligraphy font.

6 ACKNOWLEDGEMENTS

The research work reported in this paper is supported by Guangxi Colleges and Universities Key Laboratory of Intelligent Processing of Computer Images and Graphics (NO. LD14138), "Science and Technology Plan of Guangxi Province, China" (NO. 14122007-48, 20140108-4).

REFERENCES

[1] Cheng, L., Wang J.Q., 2013. Algorithm on Strokes Separation for Chinese Character Based on Edge. Computer Science 40(7), 307–311.

[2] Yuan, Y., Liu, W., 2010. A Method for Extracting the Handwritten Chinese Strokes Based on Shape Decomposition. Computer engineering & science 32(12), 57–60.

[3] Chung, L., Bo, H.W., Wen, C. H., 1997. Integration of Multiple Levels of Contour Information for Chinese-Character Stroke Extraction. Proceedings of the Fourth International Conference on Document Analysis and Recognition 2, 584–587.

[4] Song, H.X., Hao, J., Francis, C. M., 2007. An Intelligent System for Chinese Calligraphy. Proceedings of the 22nd AAAI Conference on Artificial Intelligence, 1578–1583.

[5] Song, H.X., Francis, C.M. Lau, Williams, K Cheung, Yun, H.P., 2005. Automatic Generation of Artistic Chinese Calligraphy. Intelligent System, 20(3), 32–39.

[6] Wei, L, Chang, L.Z., 2013. Automatic Creation of Artistic Chinese Calligraphy. Journal of Software, 8(12), 3048–3054.

[7] Li, H., Liu, W.Y., Zhu, Y.T., Zhu, G.X., 2002. A Uniform Model of Corner Detection Based on Morphology. Journal of Image and Graphics, 7(6), 543–547.

[8] Kai, Y, 2010. Researches on Some Key Technologies of Computer Calligraphy. Zhejiang University.

[9] Cheng, L., Wang, J.Q., Tian, W., 2012. Application of Improved D-P Algorithm in Image Edge Smoothing. Computer Engineering, 38(17), 232–234.

[10] Rong, H., Hong, Y., 2000. Stroke extraction as per-processing step to improve thinning results of Chinese characters. Pattern Recognition Letters 21, 817–825.

Computing, Control, Information and Education Engineering – Liu, Sung & Yao (eds)
© 2015 Taylor & Francis Group, London, ISBN: 978-1-138-02800-5

Measuring analysis of magnetization characteristic curve and exciting current of single-phase transformer under DC bias

Dong Xia Wang
Dezhou Vocational and Technical College, Dezhou, China

ABSTRACT: Under DC bias, the magnetization characteristic curve and exciting current of the single-phase three limb transformer under no-load condition are measured. The results show that the magnetization characteristic curves do not shift up, but they are no longer symmetrical under DC bias, they get saturated at one side of X axis and hysteresis loops become narrow when DC increases. The corresponding peak values of exciting current get sharp. Their waveforms shift up and are no longer symmetrical too. The harmonics content become more and more abundant. The 1st and 3rd order harmonics increase steadily with increases of DC. Especially the 2nd order harmonic increases rapidly with increases of DC. At last, the paper simulates the magnetic flux density distribution in the three legs. The simulating results indicate that the magnetic flux density increases when the DC current grows into the transformer. And the magnetic flux density gets saturated in the central limb.

KEYWORDS: Transformer; DC bias; Magnetization characteristic curve; Exciting current.

1 INTRODUCTION

As an important element in AC network, power transformers are often subject to some external disturbances. DC bias is one of those external disturbances and will take effect when 1) there is geomagnetically induced current[1-2] (with a frequency less than 0.01HZ and resembling the DC); or 2) direct current transmission system is operated by means of unipolar ground return. Under this circumstance, it will generate a DC potential distribution[3-5] on the ground and disturb all the neutral directly grounding power transformers within that range.

When DC flows into the transformer neutral point, the transformer core will reach magnetic saturation and there is an increase in winding magnetic leakage, which results in local overheating, intensification of vibration and noises of the core and winding, richness in harmonic wave constituent. This phenomenon is called DC bias, which is harmful to the transformer and power grid[6-8].

Up to now, there have been some calculations and experimental studies on DC bias and their research contents mainly include: 1) simulated analysis of exciting current[9-10] ; 2) experimental study on the vibration, noise and local overheating of transformers[11]; 3) study on winding magnetic leakage[12]; 4) Study on measures of restraining DC current[13-15]. In those aspects mentioned above, research scholars home and aboard have obtained certain achievements and progress.

In this paper, the magnetization characteristic curve of the transformer under DC bias is measured and the result will be used to analyze the features of the magnetization characteristic curve of the transformer under DC bias as well as the waveform character of no-load exciting current.

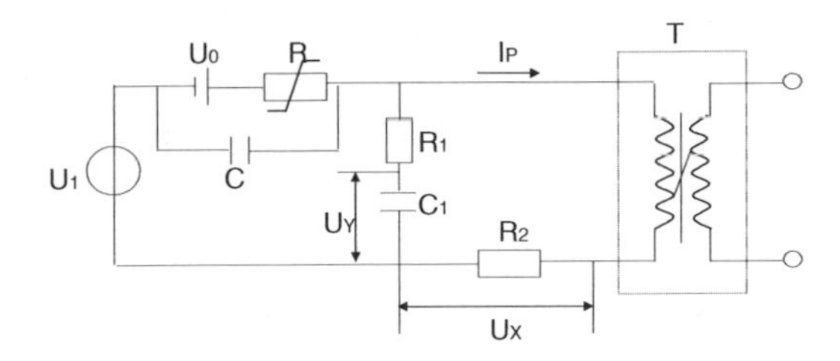

Figure 1. The circuit model of measuring magnetization characteristic curve of single-phase transformer under DC bias.

2 MODEL OF THE MEASURING CIRCUIT

Figure 1 shows the circuit model of measuring magnetization characteristic curve of single-phase transformer under DC bias, in which T stands for the single-phase transformer to be measured, U_1 and U_0 stand for AC power supply and DC power supply respectively. The function of capacitance (C) is to restrict AC flows into DC power supply and variable resistance (R) is used for obtain diverse DC. R_1 and C_1 is the resistance and capacitance in integral circuit while $R_1 \gg \frac{1}{\omega C_1}$;$U_Y$ formulates the voltage in each end of the capacitance and is in direct proportion to magnetic induction intensity(B).The derivation process of the formulas is described in the following:

$$u = R_1 i \approx R_1 C \frac{dU_Y}{dt} \tag{1}$$

$$u = ns \frac{dB}{dt} \tag{2}$$

subsequently deduces:

$$B = \frac{R_1 C_1}{nS} U_Y \tag{3}$$

In the above formulas, n and S stand for the number of primary winding turns and the sectional area of transformer core limb while u is the total voltage of R_1 and R_2.

Resistance R_2 is connected in series with the primary winding of transformer and meets the condition of $R_2 \ll R_1$.U_X is the voltage drop between the two ends of resistance R_2 and is in direct proportion to magnetic field intensity (H). The derivation process of the formulas is described in the following:

$$H = \frac{n}{L} I \tag{4}$$

$$I = \frac{U_X}{R_2} \tag{5}$$

subsequently deduces:

$$H = \frac{n}{LR_2} U_X \tag{6}$$

In the above formulas, L is the length of transformer core limb and I stands for exciting current.

In formula (3) and formula (6), R_1, C_1, n, S, L, R_2 are all known constants, therefore B and H can be determined after U_X, U_Y are measured.

3 ANALYSIS OF THE EXPERIMENT RESULT

The transformer model undergoes a series of experiments under no-load condition and the results of its magnetization characteristic curve are illustrated in fig. 2~3. According to the results, the positive axis and negative axis of magnetization characteristic curves are symmetrical when there is no DC flowing into the transformer. When there is DC flowing in, they become obviously unsymmetrical although the magnetization characteristic curves do not shift up. The positive axis tends to become saturated (if DC flows from the opposite direction, the negative axis tends to become saturated) and its saturation state will be intensified with the increase of DC current. Examples

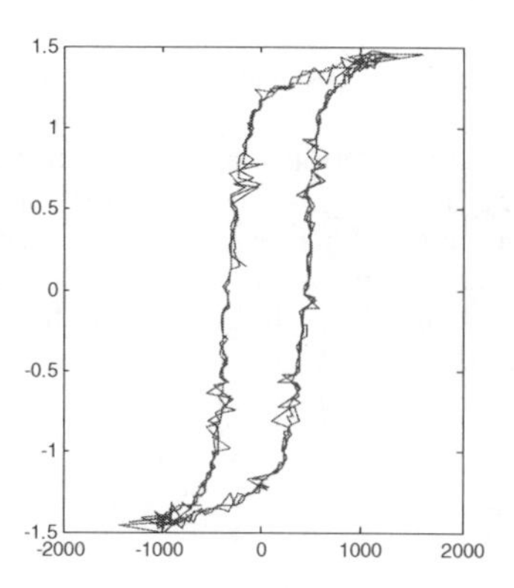

Figure 2. Experiment result of B-H curve without DC current.

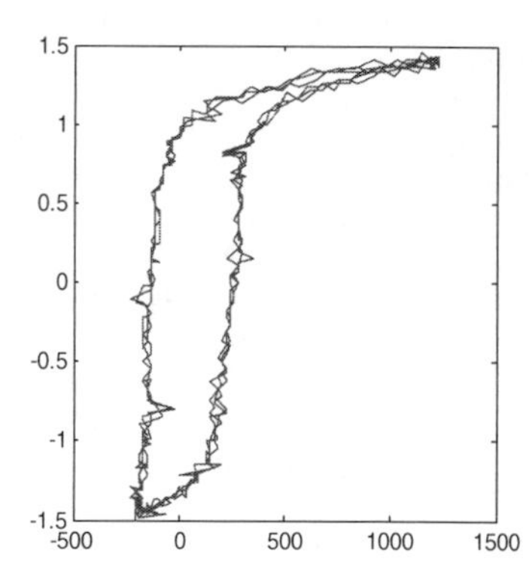

Figure 3. Experiment result of B-H curve under 0.22A DC current.

The experiment results of the exciting current are illustrated in fig. 4~5. According to the figures, the waveform of exciting current will become distorted under DC bias and its distortion will intensify with the increase of DC current. As a result, the positive peak will become more and more pointed and the positive and negative axes are no longer symmetrical, which proves that there appeared plentiful even harmonic.

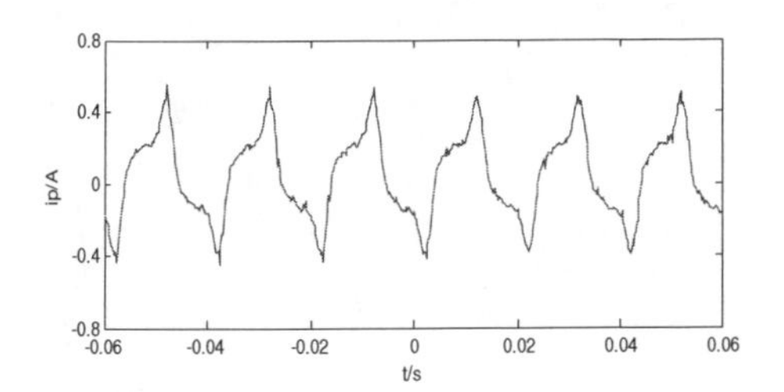

Figure 4. Experiment result of exciting current without DC bias.

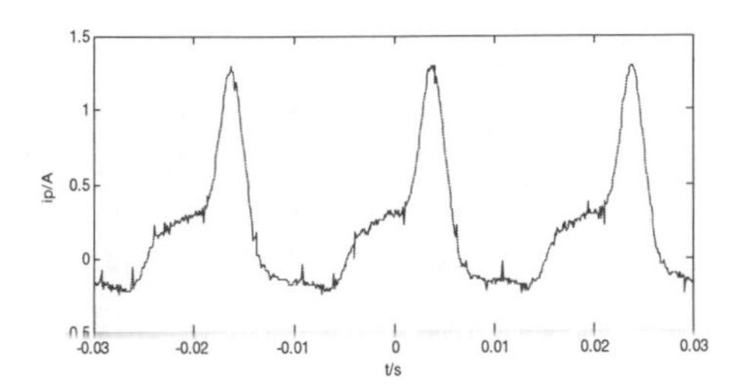

Figure 5. Experiment result of exciting current with 0.22A DC current.

4 CONCLUSION

a. When there is DC flowing into the transformer, the magnetization characteristic curves do not shift up but their positive and negative axes become obviously unsymmetrical. The positive axis tends to become saturated and its saturation state will be intensified with the constant increase of DC current. Furthermore, the two hysteresis loops become closer and closer, nearly resembling a non-linear single-valued curve.

b. The waveform of exciting current will become distorted under DC bias and its distortion will intensify with the increase of DC current. As a result, the positive peak will become more and more pointed and the positive and negative axes are no longer symmetrical.

REFERENCES

[1] Molinski T S, Feero W E, Damsky B L. Shielding grids from solar storms[J]. IEEE Spectrum, 2000, 37(11): 55–60.

[2] Price P R. Geomagnetically induced current effect on transformers[J]. IEEE Trans. on Power Delivery, 2002, 7(4): 1002–1008.

[3] Zhang Wenliang, Yu Yongqing, Li Guangfan, et al. Researches on UHVDC technology[J]. Proceedings of the CSEE, 2007,27(22):1–7

[4] Wang Mingxin, Zhang Qiang. Analysis on influence of ground electrode current in HVDC on AC power network[J]. Power System Technology, 2005, 29(3):9–14

[5] Shu Yinbiao, Zhang Wenliang. Research of key technologies for UHV transmission[J]. Proceeding of the CSEE, 2007, 27(3):3–8

[6] EPRI EL-1949. Investigation of geomagnetically induced currents in the proposed Winnipeg-Duluth-Twin cities 500KV transmission line[R]. California: Electric Power Research Institute, 1981.

[7] IEEE Working Group on Geomagnetic Disturbances and Power system Effect. Geomagnetic disturbance effects on power systems[R]. New Jersey: Institute of Electrical and Electronics Engineers, 1993.

[8] Price P R. Geomagnetically induced current effect on transformers[J]. IEEE Trans. on Power Delivery, 2002,7(4) : 1002–1008.

[9] Li Xiaoping, Wen Xishan, Lan Lei, et al. Test and simulation for single-phase transformer under DC bias[J]. Proceedings of the CSEE,2007, 27(9):33–40

[10] Takasu N, Oshi T, Miyawaki F, et al. An experimental analysis of DC excitation of transformers by geomagnetically induced currents [J]. IEEE Trans. on Power Delivery, 1994, 9(2): 1173–1182.

[11] Picher P, Bolduc L, Dutil A, et al. Study of the acceptable DC current limit in core-form power transformers[J]. IEEE Trans. on Power Delivery, 1997, 12(1): 257–265.

[12] Cao Lin, Zhao Jie, He Jinliang, et al. Research on the withstand performance of three-phase three-limb power transformer under DC current biasing. High Voltage Engineering, 2007, 33(3):71–74

[13] Li Xingyuan, Chen Lingyun, Yan Quan, et al. Design of nonlinear complementary controllers for multi-infeed HVDC transmission systems[J]. Proceedings of the CSEE, 2005, 25(15):16–19

[14] Zhu Yiying, Jiang Weiping, Zeng Zhaohua, et al. Studying on measures of restraining DC current through transformer neutrals[J]. Proceeding of the CSEE, 2005, 25(13):1–7

[15] Ma Yulong, Xiao Xiangning, Jiang Xu, et al. Optimized grounding resistance configuration against DC magnetic bias of large capacity power transformer[J]. Power System Technology, 2006, 30(3):62–65

Computing, Control, Information and Education Engineering – Liu, Sung & Yao (eds)
© 2015 Taylor & Francis Group, London, ISBN: 978-1-138-02800-5

Discussion on China's automobile cover mold development

Yu Ming Zhou
Liaoning Jidian Polytechnic, Department of Material Engineering, Liaoning, Dandong, China

ABSTRACT: In recent years, more and more development of the mold has been valued, it embodies all kinds of high-tech, rapid precision directly to the material molding, welding, assembly of parts, components or products, its efficiency, accuracy, streamline, Ultra Miniature, energy saving, environmental protection, and product performance, appearance, all is the traditional process of reach. The mold is an important modern manufacturing equipment technology industry, it is a symbol of manufacturing level and production capacity is a measure of a country or enterprise. Due to the rapid development of the automobile industry makes the panel die design and development is becoming more and more important in the development of automobile die. This paper mainly introduces the automotive panel die design significance and some basic knowledge, and the development prospects of the car cover mold.

KEYWORDS: Automotive panel die design; development prospects.

1 A SUMMARY OF AUTOMOBILE COVER MOLD DESIGN

1.1 *Important significance of the automobile cover mold*

Mold design is one of the important parts of automobile die of automobile cover, its production and manufacturing quality directly affects the subsequent production of automobiles, and it is also an important guarantee vehicle personalization and upgrading. Therefore, how to effectively improve the die of automobile cover production efficiency has become an important work.

In the process of manufacturing car mold, mold design is not only a time-consuming work, and is the important foundation for the practical processing, according to the relevant data display.

1 Trucks drive car company 70% manufacturing cost depends on the stage of design.
2 A Ford automobile company in the design, labor, raw materials, and the enterprise general management expenses four manufacturing factors, design innovations have brought 70% save for its production.
3 Product design usually only accounts for about 5%~15% of the total product cost, manufacturing cost and 80% of the product quality and performance, but he decided to 75%.
4 We can see that the design work is very important, but in the actual work is often a lot of time is spent in the design of the mold structure, spent time on the corresponding surface design is compressed, which coincided with the opposite of what we need, so how to effectively reduce the design time short structure, make the staff can be more time with the automobile design is particularly important.

1.2 *The present situation of automobile covering parts mold design*

1 In the present design, often happen designers design work and technology, manufacturing, assembly personnel in conflict with each other, thus causing rework and revision of extra work, which not only increases the production cost, but also in the serious delay, delay the products on the market timing, resulting in serious consequences loss of market competitiveness.
2 And this one the most direct impact should be conflicting design and mold design to mold design personnel on the other often happen designer cannot achieve, resulting in both rework and negotiation constant revision.

1.3 *The general process of mold design of automobile covering*

The general design process of automobile panel die is as follows:

1 Analysis of mold concept designs according to the manufacturer's demand.
2 On the die working conditions, technical analysis to draw the drawing process.

3 Modeling design of die structure process according to drawings.
4 CAD/CAM to NC code, to prepare for the NC machining.
5 EPC modelling, if not the proper place, and then transferred to the CAD/CAM design coordination.
6 Casting and surface processing.

A mold design are summarized in the whole design process of automobile panels, CAD/CAM modeling is the key of design and it played a role, and the important place of die casting is that it determines whether the designer's idea into reality, so in actual design, this two part time accounts for the proportion is the largest. Die casting and other requirements of workers is relatively high, the intention of the design should not only can read the designer, each part and the design of using solid expressed, especially the status quo at home mainly uses a 2D design based, is particularly difficult, the same job must than foreign already use 3D design that spend more time.

2 DEVELOPMENT OF AUTOMOBILE COVER MOLD DESIGN

EPC outline shape design EPC is EPC (Taiwan called packet Lilong casting) the process has been widely used in foreign automobile mould castings, making model is based on plastic (polystyrene) instead of wood, real production casting economic model for cost saving, casting, machining allowance small, high precision. Because this process through the entity model casting, simplifies the complicated process of previous modeling, casting without using distribution box, mold repair type, falling core complex process, thus casting blank without parting surface, no burr, no cavity backstop, hole smooth. This technology is suitable for single, complex mold body use, so it is the best choice for automobile die manufacturing.

2.1 *Concurrent engineering concept overview*

1 Concurrent engineering concurrent engineering is relative to the previous serial production technology and provides a new mode of product design and manufacture, America defense analysis research as early as in 1988 December put forward the definition of concurrent engineering: "concurrent engineering is an integrated method of system, which uses parallel processing method of product design and related process, including the manufacturing and support processes, this method tries to make the product development staff from the beginning to consider all factors in product design from concept to product scrapped the entire product life cycle, including cost, quality, job scheduling and the needs of users". From this definition, it can be seen that the theory of concurrent engineering is the requirement of the product design personnel should as far as possible considering the design thought of the whole life cycle of products in the early stage of product design.
2 The concept of concurrent engineering into the auto panel die design reasons in the traditional automotive panel die design, often need to design engineer with manufacturers to design their own one's ability and cleverness, play creatively to die, but this kind of "creativity" to the lack of follow-up processing considerations, this often led to design modifications and rework, so if in the design process can take into account the EPC habit and experience of manufacturing engineer, it can effectively avoid unnecessary rework, and as part of the standardization of mold structure, stylist can put more energy in the car shape design, which is a key consideration some of the needs of the design.

In order to improve the efficiency of manufacturing of automobile panel, reduce the manufacturing cycle, in the design of automotive panel die, should follow the idea of concurrent engineering, namely to consider the follow-up processing and future design needs, also to the involved in the design of product information as part of the consideration of the standardization and digitization. After such once 3D modeling, design of mold standard parts to complete, all concerned in the design process can be expressed by statistical chart form, the difficulty of the work so as to simplify the EPC of stylist, can effectively save the picture and the modelling work time.

Based on these considerations, we passed on some automobile panel die is analyzed, can see, except for covering parts of complex surface, the other parts of the mould in general have symmetries, very similar, even repetitive, such as plate, stiffened plate, lug, boss and so on, these characteristics for in the design of die structure characteristics of standardization and digitization has brought the possibility and necessity of great.

Concurrent engineering idea is introduced into the automotive panel die design basis if we design covering on the mold to cavityless casting workshop model as the main basis of words, so this is not only for the mold design to bring the convenience, but also in the future design of the die is completed, you can get the information data of the above standard parts used in relevant design immediately. Such as, in the design uses several cylindrical, several bosses, and their respective characteristics of size, so that you will get the data information tab, directly in the database and real casting workshop to realize the information sharing, full mold casting workshop immediately according to the data information to produce models needed parts, and no longer need to analyze the drawing as before, it can not only reduce the EPC workload and the difficulty of the work of the workshop, but also can further reduce the automobile covering parts of the production cycle, reduce real casting material (polystyrene plastics) consumption and waste, so as to save energy to reduce the pollution and environmental protection purposes, but also reflect the thought of concurrent engineering.

3 CONCLUSION

Through the discussion of the introduction of the concurrent engineering in the design process of automobile covering parts of thought, has opened up a new way for the design ideas of automobile panel, caught two work steps in the design of the most time consuming (modeling design CAD/CAM design of die and mold casting), put forward how to more effectively using the idea of concurrent engineering to help enterprises to save time, improve work efficiency, to win the precious time in the market competition.

ACKNOWLEDGMENT

This project was supported by The Scientific Research Foundation of Liaoning Jidian Polytechnic2014008.

REFERENCES

[1] Cui Lingjiang. The Automobile Panels Stamping Forming Technology. The mechanical industry press.2005.

[2] Zhou Tianrui. Technology of automobile panel stamping forming machinery industry press..2001.

[3] Xie Yongdong. Basic auto manufacturing process. Mechanical Industry Press.2008.

[4] Li Tibin. The stamping process. The Chemical Industry Press.2008.

[5] Xue Qixiang. Analysis of stamping process and die design examples. Machinery Industry Press.2008.

Computing, Control, Information and Education Engineering – Liu, Sung & Yao (eds)
© 2015 Taylor & Francis Group, London, ISBN: 978-1-138-02800-5

Research on manage measures of college accounting information system based on network database

Chun Ming An & Yin Xing Li
School of Economic Management, Beihua University, Jilin, China

ABSTRACT: In the network database, along with the rapid development of the construction of college accounting information system, how to establish the internal control system is safe and effective to avoid the risk of computer network security is particularly important. In this paper, according to the main problems existing in the development process of information construction, put forward to strengthen the awareness of network security, concrete measures to strengthen the internal control.

KEYWORDS: College accounting information system; Network security; Information construction.

1 INTRODUCTION

With the wide application of information technology in the field of accounting, the accounting system more and more to the network environment from the single user environment, accounting information system based on the network, more and more attention of the accounting circles. The implementation of accounting information system, accounting methods and data processing technology has undergone a qualitative leap, also present new challenges for data security problem of accounting information system. High school is an implementation of the accounting information system unit, basically completed by the transformation of single user of accounting to the network accounting information system, and realizes the automation of accounting bookkeeping, accounts and reimbursement and achieves resource sharing. However, according to the extensive investigation and the author of the recent years, the accounting information system exists many problems and risks in the process of construction and development, and did not attach great importance to the financial department, the phenomenon of paralysis data is destroyed, lost and the system has occurred, causing serious consequences to the school.

2 PRESENT SITUATION OF THE ACCOUNTING INFORMATION SYSTEM CONSTRUCTION IN COLLEGES AND UNIVERSITIES

Accounting information system is the information system with the aim of providing accounting information, is a system for accounting information processing and accounting management activities, is the use of modern information processing technology, acquisition, storage, processing and transmission of the accounting information system, accounting reflects, complete control functions. Its establishment and development, and basically realized to replace the traditional manual accounting information system goal.

At present, the following college accounting information system construction and development status: Accounting information system is composed of single user accounting system the past development becomes multi user network accounting system, the implementation of parts of the system to change the accounting contents, ways and methods, and promote the development of its own accounting. However, we must also see that the current establishment of accounting information system has not realized the innovation of accounting information system and revolutionary change, distance requirements of modern university management of it is still a big gap.

The system structure of the accounting application system of the past the client/server (C/S) structure, development of the browser/server (B/S) structure, the B/S structure data are focused on the school headquarters database service system, therefore, the higher requirements for hardware performance, at the same time on the network safety are also put forward higher requirements.

The basic function of accounting is to reflect and control, the accounting information system under network environment has had a significant impact on the two major functions of accounting. The school of economic business, the accounting information system can automatically collect the relevant accounting data from the school inside and outside, real-time reflect, therefore, reflect the function of accounting becomes more and more desalination. From the point of view of the control function of accounting, accounting information system for real-time automatic processing, especially for the multi campus network accounting has been introduced in schools, remote and real-time monitoring of economic activity in the network of schools, play to accounting supervision and participation in decision making functions will become more and more important. At the same time, the data security, timeliness and accuracy requirements are also higher.

With the accounting information system from accounting to management, the transition from micro management of macro management, the existing accounting system of post responsibility, personnel management system, the accounting procedures, records management system has been unable to meet the needs of the construction and development of accounting information system, and manual accounting system integrated accounting information management system will be fully formed.

3 THE MAIN RISKS EXISTING IN THE DEVELOPMENT PROCESS OF ACCOUNTING INFORMATION SYSTEM

Extensive application of computer technology in accounting, but also makes the computer data processing under the environment of accounting risk increase the difficulty, the accounting information system security should be attached to. To strengthen the risk of accounting information system control to prevent information leakage, to protect the integrity of the financial data, has a very important role to protect the user's legitimate rights and interests of users. The network accounting information system in Colleges and universities is a double-edged sword, the implementation of financial management at the same time in the use of the network, will also be their exposure to risk; financial network has many hidden troubles of safety. According to the analysis, the risk existing in current college accounting information system is mainly manifested in the following aspects: Run the risk of accounting information system mainly has the human factors and natural factors, two. Human factors refer to the user of illegal occupation of cyber source, steal or deliberately blocking network communication, through the computer virus to reduce the network performance due to paralysis of computer network, etc. Non-human factors mainly refers to the influence of natural disasters, such as storm of thunder and lightning, earthquake, fire and other system operation caused by faults.

From a technical point of view, any data in the network are likely to be "steal". Illegal users not only can steal the accounting information content, can also steal flow and flow, information without knowing the information content under the condition of the communication frequency and length, thus to obtain the useful information. Transmission risks include operators using improper means to obtain, tampering with data, misappropriation of resources (computing resources, communication resources, storage resources) etc.

The integrity of the risk of accounting information system data can be divided into the human factors and natural factors, two. Human factors refer to the operator the illegal invasion of accounting information system, unauthorized user processing of accounting data and other accounting data of destruction; non-human factors of data destruction means transferring interference, the data of computer hardware failure, software running error. The above risk could tamper with the accounting information content, form and direction, will cause the entire financial system data loss, unable to run even paralysis, cause huge economic losses to the school.

Computer fraud includes when data enters the system operator to falsify the data, delete, modify or increase the accounting matters. The most common means are: access to the database from a remote address, browse to find the copy of useful data in the file, in order to achieve a variety of personal purposes. The destruction program is logic for computer virus. Such as a Trojan horse is a commonly used method of calculation of failure in a computer program. A computer virus has. A computer virus has spread, latent, is becoming one of the most commonly used means of computer fraud for damage to computer network. The program can directly access data in a database file, change or delete or change the accounting records, accounting matters.

The success or failure of accounting information, talent is a key. College accounting information talents in China lack is mainly manifested in two aspects: the lack of accord with the requirement of an accounting information management leader. The realization of accounting information is needed to mobilize all resources, the formation of an efficient team, to play every employee enthusiasm and expertise, consistent with the goal of school information. These need financial leaders have a strong coordination ability, innovation ability and insight, but such financial leadership talent is the lack of current university. The implementation of college accounting information polarization status after personnel of technology of knowledge presentation, know computer people, lack of knowledge of financial management; financial management experience, computer technology and not strong enough. Want to carry on the deep level of accounting information, to cultivate compound talents not only to understand the computer and understand accounting.

4 STRENGTHEN THE CONSTRUCTION OF ACCOUNTING INFORMATION SYSTEM SAFETY CONTROL MEASURES

With the awareness of network security is an important premise to guarantee the network security, many network security events and lack of awareness of security related.

1 Physical measures. The protection of key network equipment (such as switches, a large computer network security, etc.), formulate strict rules and regulations, take radiation proof, fireproof and install an uninterruptible power supply (UPS) and other measures.
2 Access control. Authentication and strict control are user access to network resources. For example, user authentication, the password encryption, update and identification, set user access directory and file permissions, control network equipment configuration of the authority, etc.
3 The network isolation. Network isolation in two ways: one is the use of isolation card implementation, one kind is to use the network security gap to achieve.
4 Technical measures. In recent years, around the network security problem presents many technical solutions, such as data encryption, firewall, vulnerability scanning, intrusion detection, data processing, data backup protection function. Data encryption is to encrypt the data transmitted in the network, after arriving at the destination and then decrypted revert to the original data, the purpose is to prevent the illegal user intercepted after theft information security information intercepted after couldn't understand its meaning. The firewall technology is through isolation and restricted access method on the network to control network access.

The existing problems of the accounting information system are largely due to the internal control deficiencies, therefore, we must establish and perfect the internal control system of electronic data processing, establish the general and application controls measures to adapt to them. General risk control is applicable to a wide range, the key is to control program; application of control the main risk to a specific system, the focus is to control the input process.

The main method of internal control has taken:

1 The analysis of the weak link in the accounting information system and the threat may have.
2 Security measures in the formulation and implementation of the system operation.
3 Establish a multilayer control system, physical isolation of potential intruders, such as site into the control, system access control and document into control etc.
4 To strengthen the construction of internal control system. After the implementation of accounting information system, with characteristics of accounting information system to make an internal control system of a series, such as network maintenance and management system, equipment management system, the operation of access control system, business process and accounting system, to ensure the normal operation of accounting information system.
5 To strengthen the system of management personnel and operator safety knowledge education and vocational moral education, through safety education to reach a consensus on the issue of data security, improve the basic quality of personnel.

5 THE EXISTING PROBLEMS IN THE IMPLEMENTATION OF ACCOUNTING RISK MANAGEMENT

According to the comprehensive risk management system requirements, the four targets of the enterprise strategic target, target, management should report the target and supervision target. According to the strategic objective design of each business risk business objectives, clear reporting objectives to make reporting accurate and effective, made a clear regulatory objectives and ensure compliance. The enterprises of our country although there are strategic targets, a clear business objective and reporting objectives and regulatory objectives, but the coordination of strategic objectives and the lack of overall business objectives.

Risk identification of enterprises is the combination of both past and future recognition technology, at the same time to the corresponding support as auxiliary tool. In order to improve the response speed of risk management and initiative, improve the economic capital allocation efficiency, must establish the mechanism of the risk identification system and systematic. The risk identification mechanism of Chinese enterprises lacks systematic risk identification activities, mainly for lack of integrity, lack of standard risk recognition; risk identification responsibility is not clear, front, middle and back of the division of labor and cooperation is not enough; risk identification activities limited coverage, identification activities have blind spots.

The perfect corporate governance structure is on the internal control mechanism to be effective. The internal control of enterprises in our country has many problems, mainly has four aspects: first, understanding of internal control of uncertain, incomplete, senior management self discipline is not enough; second, our country enterprise cognition deviation in internal control and development, the relationship between internal control and risk, focusing on the immediate benefits, catch grasping the scale, relax the prevention and management of risk, illegal phenomenon is serious; third, belongs to the branch control ineffective, task arrangement, light inspection management, lack of the implementation of internal control regulations, serious formalism; fourth, internal control rules and regulations are not perfect, assessment indicators, assessment content and method and modern enterprise business development and risk control, do not meet.

Risk management of enterprises in China the focus of the work still focuses on the risk of post management, loss control work in advance, against the risk of. At present, still lack the risk monitoring and early warning system of effective most enterprises, for the existence of blank on the early risk prevention.

6 THE COUNTERMEASURES TO STRENGTHEN THE ACCOUNTING RISK MANAGEMENT

To carry out the outlook on life, values education. Under the market economy, strengthen the accounting personnel of the ideal and life outlook education are very necessary. Enterprise accounting work leadership should pay close attention to the accounting personnel, especially the concept of dynamic, important position of staff behavior change and job status, patiently do ideological work to change the backward financial personnel, timely for resolving the difficulties in life and work, so that each of the accounting staff to maintain a good attitude. On the important position of accounting personnel should strictly periodic evaluation, continuous rotation, timely will not suitable for accounting personnel transferred from the accounting post. The two is to carry out the risk awareness education. To reverse the accounting department has been emphasizing accounting, light regulation of thought will prevent the accounting risks referred to an important position to catch up, the daily accounting business in the non normal circumstances, should be vigilant and careful investigation, to prevent the occurrence of accounting risk.

Perfecting the accounting system is to strengthen the risk prevention, basic links to the accounting work. The enterprise internal accounting control, according to the effective, comprehensive, timely and effective principle, standardized business operation, solve the accounting system lags behind relatively caused the problem, enterprises should regularly revise and improve the accounting rules, a new account management approach promulgated, it is necessary to timely carry out investigations and studies, timely supplement and improvement to the enterprise accounting system. Due to the strict implementation of the accounting system is the enterprise internal control in the center, established a relatively perfect accounting system, the key is to implement the plan. For the illegal operation of the staff, regardless of whether they cause losses, they should be held responsible for serious.

7 CONCLUSION

The implementation of accounting information system is the inevitable trend of modernization of accounting development, is an important means to improve the modern management level of colleges and universities, only to discover and solve the relevant problems in the information construction, establish a sound internal control system of network accounting information system, to ensure safe and stable operation of accounting information system under network environment, in order to give full play to advantages and the role of financial information network, to better promote the healthy and orderly development of higher education.

REFERENCES

The Ministry of finance. The guiding opinions on promoting our country's accounting information work (Accounting No. [2009]6) [Z].2009.

Chen Xu, Mao Huayang. The accounting information system analysis and design [M]. Beijing: Tsinghua University press, 2009.

Wang Xiaoman: "optimization" of accounting environment. Pioneering with science and technology, 2007,8.

Zhu Yi: "the causes and prevention of accounting risk", "Journal of Southwest Institute for Nationalities (PHILOSOPHY AND SOCIAL SCIENCES EDITION)" in 2003 Fourth Period.

Wang Jiping: "on the development of accounting internal and external environment". "Shanxi tax", in 2003 08 period, 20.

Li Lianjun: the accounting environment and its impact in China [J]. Chinese agricultural accounting, 2001.5.

Xie Qin: on the accounting risk "group" [J]. economic research, 2005,18:33–37

Dou Shan: financial accounting risk management analysis of [J]. "accounting communications", 2005,8:39-43.

Wang Yan: Research on the risk of [J]. Master Thesis of accounting, 2009,5: 22–43.

Zhao Qiang: on the enterprise risk management and discussion of [J]. research, 2007,3:26–29.

Computing, Control, Information and Education Engineering – Liu, Sung & Yao (eds)
© 2015 Taylor & Francis Group, London, ISBN: 978-1-138-02800-5

Analysis of art teaching effect based on computer vision techniques

Jian Guang Gong
Teacher's College, Beihua University, Jilin, China

ABSTRACT: According to the special feature of CVT (Computer Vision Techniques), the significance and function are analyzed in art education, then the application of multimedia teaching to art education is discussed from three aspects. Prospects for art teaching are given at last.

KEYWORDS: Computer vision Techniques; Art education; Application.

1 INTRODUCTION

With the continuous deepening of quality education in the regular teaching, multimedia technology as a new means of teaching, has appeared in the high school art teaching. In art class of multimedia teaching in the full reasonable use, the best state close to the level of psychological development of students, to improve the overall quality of students, and better able to make and vivid education technology combination. How multimedia as an advanced means used for art education will have become a big problem in contemporary teaching.

As a teacher should not only have this kind of idea, simultaneously also should as soon as possible use and design a variety of means to use provides security just as this teaching, in the art teaching in the use of multimedia can make students better understand art, performance art to create art, the art of teaching is more be lively and vivid, the best state close to the level of psychological development of students to improve the overall quality of students, and better, make and vivid education technology combination.
In traditional art teaching in the students to listen to the teacher, a single performance (singing) or tape recording feel boring, monotonous, so often makes the students in the appreciation of art is very passive, student's easily distracted, so teacher's teaching is not reach good results, full of beauty MIDI music, text, pictures, animation, video and other multimedia integrated information, can appreciate the enthusiasm, the initiative art natural mobilize students, students will take the initiative to feel the art strength, speed, mood under the guidance of teachers, students in the appreciation of the beautiful picture and the MIDI music at the same time, a variety of vivid imagination generated the mind, like be personally on the scene, and stimulate the artistic expression of strong desire, the desire of creation.

2 THE ROLE OF NEW MULTIMEDIA TECHNOLOGY IN TEACHING ART

At the end of last year the sixth grade final exam, art out of 60 points, this is not a small proportion, higher than that of science, society and information technology, second only to the Chinese, mathematics, English in fourth. It relates to the music part to the instrumental music appreciation, appreciation and folk art appreciation and so. In art, oil painting, traditional Art education, but also relates to the drawing, animation photos; not only to cultivate students aesthetic ability, but also through the game production experience friendship precious, for emotional education to students. How to get the best teaching effect? The only means is to show the teaching video tape. Through video, students can put the band size lineup, command style gesture, the emotion of art image, musical instrument timbre characteristic as well as the actor's costume props, singing action expressions look as clear as noonday, hear clearly. As in the appreciation of Modern Peking Opera "tiger" fifth field in a section of Yang Zirong's "human" ushered in the spring for singing and other combined fragments in the process, I followed the picture describes the Peking Opera students, Dan, net, ugly four businesses, to enable students to understand.

According to the psychological and the student's thinking characteristics, appropriate use of multimedia teaching, teaching situation, create a good atmosphere, not only can stimulate students' interest in learning, to mobilize the enthusiasm of the students to appreciate art, but also with interesting to stimulate thinking, improve teaching effect.

I'm in guides the student to appreciate the music "moonlit night" at the beginning of the end of three, the climax, a fragment of art appreciation teaching, before I guide the students to the combination of art and pictures to feel every piece of music, music, speed strength of sound and music emotion. While the students appreciate the music at the beginning part of art, shows the students face is hazy night, Sundowners sunrise Dongshan, beautiful scenery, hear the art began to part by a free beat intro, played by the clarinet and bassoon, clarinet playing art, flute solo a decorative phrases, this part of the art the intensity of freedom, speed growing fast, beautiful lyrical mood, people will come into the beautiful night mood; students in time to enjoy music as part of the climax of art, displayed in front of students is a vivid picture of Jiang Tao surged, tourists, wild duck swimming joy, hear is the art of using a variety of orchestration, various methods of woodwind ensemble the music to a climax, this part of the art of strength, speed fast, crescendo gradually warm and cheerful mood, now students in high spirits, can be excited.

Students in the appreciation of an end part, show in front of students is the boat gone wild duck, the rest of the beautiful picture, this part of the art of the clarinet in the bass playing theme melody, gentle and soothing, again make the students feel the quiet moonlight, immersed in the beautiful reverie...... Students in the appreciation of the time, very devoted immersed in the beautiful art and artistic conception, the mind Thoughts thronged one's mind will own and picture, art as one solution. Therefore, for students to appreciate the music, multimedia creating vivid artistic conception, played a strong role in infection. After each period of art appreciation is finished, the student is very easy to distinguish each piece of strength, speed, and the instrument timbre musical artistic conception of the legal emotion. Students in the appreciation of music "full moonlit night" when, can be combined with a vivid picture, beautiful scenery to feel the music speed changes, efforts to change the tone. Natural produce beautiful reverie.

3 NEW MULTIMEDIA TECHNOLOGY IN TEACHING ART

The students have a better appreciation of audiovisual experience in time, can be vivid, graceful language piece said expression of artistic expression was most incisive, vivid image. Many students described the language art; the screen itself is more beautiful than! For example: students in the end to enjoy...... Students in the appreciation of each period of art when, can easily distinguish each piece of strength, speed, and the performance of the instrument timbre music mood and emotion. Thus, played his part in house using multimedia, fully mobilize students' a variety of sensory organs, from the visual, listening, think aspects to appreciate art, to analyze the art from the narrative point of view, greatly improves the students' appreciation, feeling, imagination, understanding ability and expression. If the "belly" in teaching, students only a single teacher to tell the story of Chaplin, the student's attention is easily scattered, only by a single auditory, imagination to feel art, far less than the use of multimedia art effect is obvious; and the use of multimedia in teaching is to arouse students' appreciation of the appreciation of the positive, great to play his role, from the aspects of cultivating the students' comprehensive ability.

The diversity of art lies not only in singing his form, method, the creation technique, and it also has a rich cultural background, take Chinese folk songs, each nation will have different customs, architecture, language, clothing, dance and so on, many culture infiltration in the art. Teachers in the teaching of language, pictures and other ways to impart to students is far from enough, up for the purpose, and the use of multimedia, teachers can put the information into the courseware, let the students to experience, so that both the real and natural, students not only learned a song, but the understanding of this family name. Knowledge of students has been extended, but also the real art is as a kind of culture, let the students accept. For example, in my junior high school a year teaching material "home" this first song, the Russian church in West China, song form, customs, clothing, environment make courseware, let the students fully understand the national culture, through these students to the song content and how the performance of the song also to understand nature.

In short, the multimedia art teaching for the inevitable trend of modern education, make full use of multimedia, make classroom teaching more lively and interesting art teachers will be necessary qualities.

4 IMPACT OF CONTEMPORARY ART EDUCATION DEVELOPMENT

Art education as a symbol of Chinese art, has developed a number of years, has experienced a long history. Its development and development of traditional art education is basically synchronized. A traditional art education model for studying the development of self-contained Art education provides a strong guarantee, but the crisis has brought to the development of Art education, Art education may lead to the development of stagnation and rigid.

With the introduction of Western culture, contemporary art education has been given a new content and features, traditional art education has been seriously challenged, and Art education is no exception. Chinese and Western cultures fierce collision, given the contemporary art education conform to the trend of the times, and the Chinese have a tendency to weaken the function of painting, we can see the impact on Art education contemporary art education development is enormous. Therefore, Art education to have sustainable development, we must look for a path.

Given the cultural heritage and cultural traditions there are significant differences among different ethnic, fusion of contemporary Art education art education and the presence of significant challenges, mainly as follows: Western contemporary art education curriculum is based on the theory, although trying to achieve Western reconcile, western system in order to mark the traditional Art education art education is based on the poetry of learning-based, it is trying to achieve a comprehensive innovation in the original basis. Since the starting point of contemporary art education and traditional art education, the focus is different. Should contrary to objective laws, not according to the actual situation of contemporary art education and the development of Art education, blindly impose two very different systems together, will generate a lot of negative impact, but we should make an objective assessment, can not ignore the contemporary the positive role of art education for the development of Art education.

The diversity of art lies not only in singing his form, method, the creation technique, and it itself also has a rich cultural background, take China folk songs, each nation will have different customs, architecture, language, clothing, dance and so on, many culture infiltration in the art. Teachers in the teaching of language, pictures and other ways to impart to students is far from enough, up to the purpose, and the use of multimedia, teachers can put the information into the courseware, let the students to experience, so that both the real and natural, students not only learned a song, but the understanding of this name family. Knowledge of students has been extended, but also the real art is as a kind of culture, let the students accept. For example, in my junior high school a year teaching material "home" this first song, the Russian church in West China, song form, customs, clothing, environment make courseware, let the students to fully understand the national culture, through these students to the song content and how the performance of the song also to understand nature.

Contradictions are opposites, we study the impact of Chinese contemporary art education in the process of painting effects, specifically requires an objective analysis of contemporary Art education art education teaching, writing produced. This article from the aspect of two dimensions, the use of comparative approach, in teaching, through the teaching objectives, teaching methods, teaching materials, teaching content, teaching evaluation analysis of differences in traditional and contemporary art education art education; in the creative aspects of , will be an inspiration on Art education, technique, emotion comparative study, summed up the pros and cons of contemporary art education for the development of Art education, but overall, pros and cons, more harm than good.

Because of our long-term in a feudal society, and the ancient social, political, cultural and systems are designed to meet the needs of the ruling class. The prevalence of the traditional hierarchy is art education, just to satisfy the interests of the ruling class and the service. In this case, the role of art education is major with a strong "into enlightenment, helping Fallon," the. In this historical context, the traditional art education does not have a complete and standardized education system. In particular the teaching process, the main idea is to master personal education based, then painting learners main objective things through observation and copying works of the classical way to achieve the purpose of drawing creation. Contemporary Art education is committed to the community to cultivate creative talents, and the fine arts education in the country gaining in popularity, becoming a national art education. Up to now, we have been gradually achieving the transformation of quality education by exam-oriented education to the "moral, intellectual, physical, aesthetic, labor" comprehensive development.

Students in the appreciation of an end part, show in front of students is the boat gone wild duck, the rest of the beautiful picture, this part of the art from the clarinet in the bass playing theme melody, gentle and soothing, again make the students feel the quiet moonlight, immersed in the beautiful reverie...... Students in the appreciation of the time, very devoted immersed in the beautiful art and artistic conception, the mind Thoughts thronged one's mind. Will own and picture, art as one solution. Therefore, for students to appreciate the music, multimedia creating vivid artistic conception, played a strong role in infection. After each period of art appreciation is finished, the student is very easy to distinguish each piece of strength, speed, and the instrument timbre music artistic conception of the legal emotion. Students in the appreciation of music "full moonlit night" when, can be combined with a vivid picture, beautiful scenery to feel the music speed changes, efforts to change the tone. Natural produce beautiful reverie.

Visible, education has become the most basic rights of every person, and not a few aristocratic privileges, which will benefit the construction of a socialist harmonious society. Contemporary art education teaching objectives conducive to the development of our comprehensive quality education, improve the overall level of China's national art knowledge. For the construction of a beautiful home that has everything to gain but no harm. However, we also see the shortcomings cannot be ignored him an objective reality. Universality corresponds to the peculiarities of contemporary art education is universal knowledge for all students in the art of Art education teaching goal setting, and not as a traditional art education that, according to each student's specific conditions, can amount tailor the most appropriate teaching painting learning objectives and teaching goals at any time to adjust according to the needs of learning. Contemporary Art by differences in education and Art education teaching objectives analysis, summed up contemporary art education exists on the pros and cons of teaching objectives. We can see that they are closely related, the latter is dependent on the former, while drawing on Western art education.

5 CONCLUSION

New art multimedia technology in teaching, not only to speed up the progress of the art of teaching, improve learning outcomes, more important is that it can give students a more novel stimulus, resulting in the best area of the cerebral cortex related development, so as to stimulate their creative evolving thinking. Two innovative works are art majors. The information age, New multimedia teaching the art of teaching is the direction of development, however, stressed the role of modern teaching media is not to deny the traditional teaching, but rather calls attention to a combination of both in teaching practice, flexible use. Contemporary art teaching focus on the integrated is use of multimedia in order to continuously improve the quality and efficiency of the art of teaching. The main trends in the art of teaching is the traditional teaching media development and integration of modern teaching media, and gradually form a complete, current information society to adapt to the development of art education system.

REFERENCES

Li Si Hui. On Myth and Art Education of College Students Comprehensive [J]. "Heilongjiang Science and Technology Information." 2008.07

Liangtai Sheng. Reform of university teaching of Art education Rethinking [J]. "Zong Tai'an College of Education Science" .2003 01

Han Jing. Confront contemporary art education principal institutions of higher art China Forum on --2010 [J]. "Art Watch." 2010.05.

Zhang Yao-guang. Lilley. Chinese modern art education in the "Western painting" Complex [J]. "Grand Art". 2011 09

Ge Xintong. Reflections on Contemporary Art Education [J]. .2010. O5 "business culture."

Zhang Jun to stay. On Art education teaching traditional culture [J]. "Arts education research" .2011.10

Lee slip down modern art education model and the concept of cultural and ecological harmony - of Our universities Normal Art Education in the 21st Century [J]. "Inner Mongolia Normal University (Educational Science Edition)." 2006.11

Reflections by Lisa Shuai. Contemporary art trends in China [J]. "Industry and Technology Forum." 2012.

Zhang Bing. Explore the aesthetic perspective of Art Education [J]. "China-school education" .2010.05

Computing, Control, Information and Education Engineering – Liu, Sung & Yao (eds)
© 2015 Taylor & Francis Group, London, ISBN: 978-1-138-02800-5

Research on foreign linguistics corpus on teaching Chinese based on multimedia environment

Yu Chun Chen
Teacher's College, Beihua University, Jilin City, China

ABSTRACT: The application of modern educational technology in the teaching of Chinese as a foreign language has provided a wider scope and prospect for this area. Compared to traditional teaching means, it shows irreplaceable advantages in many aspects. It has become a tradition of information technology for us to teach by means of a computer network, but what bothers us most is how to organize the present teaching resources and make it run effectively online. To solve this problem, it is indispensable for us to construct a systematic and standardized multi-media database of Chinese as a foreign language Teaching.

KEYWORDS: Foreign linguistics; Foreign linguistics corpus; Multimedia environment.

1 INTRODUCTION

National Philosophy and Social Science Foundation (hereinafter referred to as the National Social Science Fund) represents the highest level of research in the humanities and social sciences our country, because of its authority, guidance and representation, macro reflects the philosophy of social science research in various disciplines Status and trends. This study was based on the National Social Science Fund Corpus Linguistics research topic information on recent linguistic research and development trend of China's foreign hotspot quantitative statistics and qualitative analysis.

National Social Science Fund Project is led by the state, most scientific and authoritative Social Science Fund Project, funded projects number of widely-disciplinary coverage; macro reflects the history, current situation and development trend of the various disciplines of social sciences. Therefore, the study of the National Social Science Foundation-funded was researched project to help understand the dynamics of various disciplines, to provide guidance for future research topics.

From the date of the establishment of the National Social Science Fund in 1993, "linguistics" was into them, it has been 18 years. It is worth mentioning that, 1993-1996, National Social Science Fund of foreign linguistics, language teaching and translation studies classified under "Foreign Literature" category, beginning in 1997, before they are classified as "linguistic" category.

2 STUDY DESIGN

National Social Science Fund subject of "linguistics" discipline includes Chinese linguistics, including foreign languages and linguistic minorities in China. According to this discipline is set, this study, we will, "Linguistics" is defined as the National Social Science Fund "Linguistics" class of non-Chinese and non-linguistic minority languages of our country, as well as outside the Han and minority and foreign languages the comparative study.

As used herein, the calendar year to the National Social Science Fund project information data are from the "National Planning Office of Philosophy and Social Sciences" homepage "historical data", "calendar year funded project" section under Information. Research Methods for text analysis and probability and statistics-based corpus research methods, research tools used are commonly used in research corpus indexing software AntConc and Chinese word segmentation tool Yacsi.

1. Corpus screening: Screening of "foreign linguistics" class research topic from "Linguistics" project task information. Screening Project name only encountered the problem difficult to determine ownership of a subject discipline, the further reference to the subject moderator work unit, background and other factors.
2. Corpus Pretreatment: screening out "Linguistics" research topic how to save each year for the index word software support UTF-8 plain text files.

And applying segmentation tool Yscis batch word corpus.

3 Corpus proofreading: Taking into account the expected Corpus Linguistics research topic particularity of the text, in order to ensure the accuracy of segmentation, automatic segmentation before we import foreign linguistics and applied linguistics common terms as Yacsi segmentation auxiliary dictionary, and the results of automatic segmentation proofread manually one by one, in order to ensure the accuracy of the final statistics.
4 Retrieval analysis: the complete word and proofreading corpus import AntConc, Run "word clusters / N-Gram" tool, word cluster size is set to 1-3 words appear frequency is set to 1, click the "Start" menu, AntConc the statistics of the number of frequency corpus all words composed by 1-3 blocks and words appear in descending order. Finally, the results are the import to Excel spreadsheet software for processing and statistical analysis next.
5 Results statistics: statistics on AntConc filter and merge. First, remove the high frequency filtering rules but appear unrelated research function words or content words, such as "the," "and" "and" and "research", the second is a high frequency, but not the full meaning of the word block, such as "contrast" "corpus", retaining only the high-frequency words appear more than three times the frequency and with research-related; high frequency words merge rule is to block some words or words with similar meaning or significance to merge with each other included as a word (group) process. After screening and merger, we have come over the past decade IF Corpus Linguistics research topic than three times the number of 50 high-frequency words (group), and classify statistical analysis.

English Vocabulary Teaching in senior middle schools, collocation plays a very important role. But there are a lot of words or phrases in the dictionary is a source of knowledge, or based on the learner derived from the relevant knowledge on the basis of the collocation form, which is syntactically correct, but whether it is an authentic American expression but can make nothing of it. Corpus intervention can make us to understand more comprehensive collocation model, typical collocations using the high-frequency paradigm and common real-life lexical collocation.

Linguistics NSSF's distribution (2001-2010) is the National Social Science Fund Total linguistics research project 853, the total annual project was increasing year by year trend. Initiating a project is divided into four major categories: key projects, general projects, youth projects and self-funded projects. Among them, the general projects and youth programs are basically a yearly basis, key projects and self-funded project to maintain a small amount at a fixed level. To 2010, the number of youth projects increased by 6.83 times compared with 2001, far more than the general items 2.32 and 2.73 of the total annual growth of multiple projects. This indicates that the National Social Science Fund focus on training young academic team, earnestly implement the "Note foster youth social science researchers," the general guiding ideology, also shows that more and more young scholars have the ability to undertake large-scale research projects.

The corpus vocabulary related to true statements can be translated into Chinese, let students do the translation exercises, vocabulary usage; can be collected a total of the current context, the related lexical meaning and usage of hollowing out, let the students guess the missing words; you can also put a word collocation vocabulary collected, with practice so that the corresponding fill in the blanks sentence. In the "Oxford high school English" module eight units 2 reading in the word cast as an example. Cast is a word frequency in BNC is 4154. Present 50 lines of context by random co-occurrence, we can find that as the verb cast is not on the books said "select...... Play a role (drama, role); for the casting of actors "meaning. Teachers can according to the context co-occurrence of collocation in selected some typical sentences, made in the blanks for students to practice, and on this basis, let the students analyze typical usage summary of the verb cast.

3 STATISTICAL ANALYSIS OF DATA AND DISCUSSION

After observation, the 50 high-frequency words (group) into the relevant language, type of research related to the study and related research areas related to four categories.

1 Language-related high-frequency word analysis. Before 50 high-frequency words, the language associated with the high-frequency words have 6 (see Table 1), in which high-frequency words on behalf of English study appeared in 154 words, the first 50 high-frequency words that appear more frequently, suggesting that an important feature of our foreign based linguistic research in English linguistics research-based. And with the frequency followed by Chinese and Chinese-English bilingual high frequency corresponding to the word "Chinese/Chinese", "English/Chinese" high-frequency appears, showing the linguistic study of foreign Another major feature: My linguistic study of foreign, especially linguistic study of English and Chinese are often closely related, the researchers hope to discover the inherent law of comparative study of Chinese and English languages. The number of representatives from the

high-frequency word frequency point of view Russian and Japanese, Russian, and Japanese still is an important second and third foreign language, but its influence and project and issue a far cry from the number and English.

Foreign language project schedule management is the whole process of shipbuilding to management as a project, the use of project management thinking, in the project implementation process, each stage of the degree of progress and the project was finally completed delivery cycle management. This requires a reasonable and effective schedule development, optimize the allocation of resources, dynamic tracking and control and implement the plan for the manufacturing process.

In a meaningful context in vocabulary teaching is the foundation of high school English vocabulary learning. English Vocabulary Teaching in senior high school can the real corpus based compaction based on. English teachers can combine with the related corpus in senior high school English vocabulary analysis, frequency and frequency index analysis and match analysis, and on this basis, according to the students' actual level of English vocabulary exercises, teaching design, according to the demand of various forms, strengthen the students' vocabulary understanding and use, improve the students' English communicative competence.

Improve the teaching means can promote teaching reform, improve teaching quality. Modern teaching today in order to implement multimedia oriented has become a symbol of the reform of English teaching methods. In recent years, with the rapid development of modern education technology, Computer technology and communication technology have gradually entered the classroom. Especially widely used CAI many educational technologies, put forward a severe challenge to the traditional classroom teaching model, but also accelerate the reform of teaching methods and means. Application of CAI technology in education, not only can freely dispose of voice, text, picture and video information, can also use the vivid ways will present historical events, tasks, location to the students, and easy to impress the students deeply, learning is easy to remember and forget, no feeling of boredom. At the same time, to provide a lot of material for the use of multimedia technology, so that we achieve the sharing of resources. Conveniently mutual exchange is needed products, to improve the quality of teaching. At the same time, the computer simulation technology cans effectively virtual real scene, and the use of interactive means to give students the most intuitive language training.

In view of the advantages and disadvantages of and other scholars have put forward one step forming method, this chapter presents the consideration of stamping forming linear elastic one step forming the initial solution algorithm. The basic idea of the algorithm is based on the assumption that the stamping parts of the final configuration is through the linear elastic reverse deformation to the initial configuration, the premise to effect instead of die and external load, according to the equilibrium equation to establish the initial sheet linear elastic finite element method, according to the characteristics of triangular element and quadrilateral element, the convergence conditions set in advance as the constraint, to obtain the initial flange on one step forming solution by using iterative method.

4 SENIOR HIGH SCHOOL ENGLISH VOCABULARY EXERCISES DESIGN BASED ON CORPUS

Corpus can enrich the content of high school English vocabulary, enable it to have the authenticity and authentic. At the same time, teachers can also according to the actual level of students learning with corpus design their own English vocabulary exercises. Senior high school English vocabulary exercises design forms based on corpus is varied. The corpus vocabulary related to true statements can be translated into Chinese, let students do the translation exercises, vocabulary usage; can be collected a total of the current context, the related lexical meaning and usage of hollowing out, let the students guess the missing words; you can also put a word collocation vocabulary collected, with practice so that the corresponding fill in the blanks sentence. In the "Oxford high school English" module eight units 2 reading in the word cast as an example. Cast is a word frequency in BNC is 4154. Present 50 lines of context by random co-occurrence, we can find that as the verb cast is not on the books said "select...... Play a role (drama, role); for the casting of actors "meaning. Teachers can according to the context co-occurrence of collocation in selected some typical sentences, made in the blanks for students to practice, and on this basis, let the students analyze typical usage summary of the verb cast.

In recent years, with the rapid development of modern education technology, Computer technology and communication technology has gradually entered the classroom. Especially widely used CAI many educational technologies, put forward a severe challenge to the traditional classroom teaching model, but also accelerate the reform of teaching methods and means. Application of CAI technology in education, not only can freely dispose of voice, text, picture and video information, can also use the vivid ways will present historical events, tasks, location to the students, and easy to impress the students deeply, learning is easy to remember and forget, no feeling of boredom. At the same time, to provide a lot of material for the use of multimedia technology, so that we achieve the sharing of resources. Conveniently mutual exchange is needed products, to improve the quality of teaching.

In meaningful context in vocabulary teaching is the foundation of high school English is vocabulary learning. English Vocabulary Teaching in senior high school can the real corpus based compaction based on. English teachers can combine with the related corpus in senior high school English vocabulary analysis, frequency and frequency index analysis and match analysis, and on this basis, according to the students' actual level of English vocabulary exercises teaching design, according to the demand of various forms, strengthen the students' vocabulary understanding and use, improve the students' English communicative competence.

The index is also known as the "key word in context" (key word in context, referred to as KWIC). It refers to a word or phrase in the corpus query shown by actual examples of use of indexing tools. In general, the search word, center, periphery appeared for the retrieval of word context. We can understand the specific usage of the word through the observation of search terms and the surrounding words.

Of course, English Vocabulary Teaching in senior high school with the corpus also exist some problems. For example, the time and effort required to teacher before class to spend more using the corpus preparation, a lot of time need to allow students to observe, analyze and summarize data in class. A better approach is to use a combination of teacher corpus dictionary and corpus based retrieval. The dictionary provides typical usage, corpus provides a wealth of examples, the teacher can select suitable for teaching students from actual level of.

5 CONCLUSION

In short, to senior high school English vocabulary teaching corpus assisted put them in a real language environment; they were exposed to typical usage and examples, various exercises suited to their actual level of practice. This model will inject new vitality into the vocabulary teaching currently single, boring.

REFERENCES

Qu Peimin. Chinese curriculum and teaching theory [M]. Zhejiang Education Publishing House, 2003.

Xiao Ying. Aesthetic strategies of Contemporary Aesthetic Culture [J]. "Academic Monthly", 1995, 2.

Zhou Xin, Xu, Liu. Address translation technology in firewall system using [J]. Technology Plaza, 2012,10:79–81.

Jin Yaohong. A mixed strategy of patent Machine Translation system [J]. computer engineering and applications, 2012,04:29–32.

Dong Xiaofang, Cao Hui, Jiang Tao. The Tibetan and Chinese statistics Machine Translation phrase [J]. technology based on 2012,17:60–61.

Chen Chi-yu. Chinese modern aesthetics and Chemistry theory [J]. Central China Normal University, 2000,2.

NiGuti Najimy, Xi Xiaogang, Ma Bin, Maikomti Maimaiti. Research and implementation of [J]. computer knowledge and technology support multilingual translation system, 2012,02:345–350.

Guo Zhigang. Chinese language and Literature Teaching Analysis of [J]. new curriculum teaching, 2009 (5).

Zhou Xin, Xu 白龙, Liu. Address translation technology in firewall system using [J]. Technology Plaza, 2012,10:79–81.

Chen Meimin. Aesthetic under Contemporary Aesthetic Culture [J]. Hainan Radio and Television University, 2005,3.
Lu Chang. On the relationship between Chemistry s education and aesthetic education [J]. Theory and practice of education, 2002,2.

Jin Yaohong. A mixed strategy of patent Machine Translation system [J]. computer engineering and applications, 2012,04:29–32.

[NiGuti Najimy, Xi Xiaogang, Ma Bin, Maikomti Maimaiti. Research and implementation of [J]. computer knowledge and technology support multilingual translation system, 2012,02:345–350.

Qu Peimin. Chinese curriculum and teaching theory [M]. Zhejiang Education Publishing House, 2003.

Guo Zhigang. Chinese language and Literature Teaching Analysis of [J]. new curriculum teaching, 2009 (5).

Computing, Control, Information and Education Engineering – Liu, Sung & Yao (eds)
© 2015 Taylor & Francis Group, London, ISBN: 978-1-138-02800-5

Study on employability training strategy of college students based on computer start incubators platform

Bing Bai & Xiao Wei Wang
Beihua University, Jilin, China

ABSTRACT: Modern universities are faced with the dual pressures of university reform and international competition, how to overcome difficulties, to play the role of modern employability training strategy in colleges and universities to enhance the core competitiveness of universities, colleges and university's modern employability training strategy problems to be solved. Based on the current situation of modern universities in the premise of the Internet platform fully demonstrated the characteristics and viability of the Internet employability training strategy, the necessity of modern employability training strategy of college Internet. Discusses the Internet targeting employability training strategy and important role is in the employability training strategy to guide the development of modern universities Internet.

KEYWORDS: Computer Start Incubators Platform; Employability training strategy; College students.

1 INTRODUCTION

The rise and fall of colleges and universities, a direct impact on livelihood issues of national energy security, economic development and the people, plays an important role in the process of building socialism with Chinese characteristics. Therefore, how to deposit is the university sustained, rapid and efficient development of great economic significance and far-reaching strategic implications. Employability training strategy in colleges and universities in the development process has been to protect the university sustained, healthy and rapid development has made a significant contribution. As the market economy continues to develop, in-depth reform of state-owned university, college party mobility, independence, increasing dispersion, coupled with the Internet age young party members active thinking, resulting in an employability training strategy faced unprecedented difficulties and challenges. Therefore, to explore the use of the Internet platform to carry out the effective employability training strategy, and promote the healthy and orderly development of universities, colleges and universities to strengthen cultural cohesion, enhance the international competitiveness of modern universities, colleges and universities has become a new topic of modern employability training strategy research needs.

2 THE IMPORTANCE OF MODERN EMPLOYABILITY TRAINING STRATEGY

As a large state university college of energy is the basis for the development of the national economy, is to enhance China's comprehensive national strength, reflecting the superiority of the socialist market economy, the window, I was an important foundation for a stable ruling party, but also realize the important position of our Party's mass line. The modern employability training strategy of college will be directly reflected in the central role of political party organization in colleges and universities are an important aspect of management and operation mechanism. With the industry's market economic internationalization, universities have from the previous administration into a national monopoly to participate in international competition in the market economy subject. Therefore, improving the modern employability training strategy of college based work, for speeding up the economic development of the country's construction and reform of decisive significance, is a modern development of university important organizational guarantee. Employability training strategy of the guiding ideology of "economic catch around employability training strategy, a good job of employability training strategy to promote economy," which indicates that the promotion of economic development is the central task and the goal of employability training strategy work, and employability training strategy but also for the economy spiritual foundation for development, promote market is economically healthy, orderly and rapid development.

Dialectical relationship between employability training strategy and the economy shows the employability training strategy an indispensable position and role, only in this dialectical relationship guidance to determine the exact ideas and methods of modern employability training strategy in order from a strategic height strengthen the party's base, and promote sustained universities, stability and development. In the full international competition in modern society, we should take the new situation and problems, "eighteen" spirit study employability training strategy is facing, so the modern college party organizations maintain the advanced nature forever, purity, and full combat effectiveness and creative and cohesion, to enhance the core competitiveness of universities to contribute to a new era.

3 PRACTICAL NECESSITY OF THE EMPLOYABILITY TRAINING STRATEGY

Employability training strategy reform after the existence of "serious economic, employability training strategy of light", staff mobility and strong employability training strategy less attractive forms and other issues, needs to carry out new forms, content diversity and effective employability training strategy form, solving complex problems of modern employability training strategy colleges face. The rise of the Internet as a twenty-first century employability training strategy of modern college gives new vitality, by the group continues to explore the network platform for network platform party activities to promote party exchange in the form of new programs to enhance the Party's theoretical study, exercise of party members party, provide for the development of the core competitiveness of universities. The necessity of the employability training strategy in the following areas:

3.1 *Emerging "network politics" is the leading employability training strategy of new ideas*

The development and use of Internet technology are changing the way the work of employability training strategy. January 5, 2010, Xi Jinping, employability training strategy through national grassroots work phone information system, to the national grass-roots party secretary, students "Village" issued a greeting message. This regards access to a vast grassroots ideological and political educator's enthusiastic response, become my party grassroots advantage of emerging Internet media an important manifestation of the employability training strategy. February 22, 2010, People named "Hu" microblogging Home Information bar shows: the CPC Central Committee General Secretary and State President and CMC Chairman. Users quickly add attention, fans in one day, "Hu microblogging" on the super-people, and the number of fans still rising. Party and state leaders convey decree through SMS, microblogging and other network form, send messages, listen to suggestions, as a new form, try new ideas and new Chinese Communist Party's employability training strategy.

3.2 *A large number of "network citizen" is the Internet employability training strategy of the new object*

Since 2009, the cumulative incidence of "Network mass incidents" show, "the network platform in the field" has been formed, "network citizen" has grown, matured. "Statistical Report of the 32nd China Internet Development" China Internet Network Information Center (CNNIC) 2013 released in July showed that as of the third quarter of 2013, Chinese Internet users reached 608 million, the size of mobile phone users reached 464 million people . Above that, the employability training strategy can be the party's theory with practice combine, through a modern Internet platform, party and government activities continue to explore new forms of ideological education, improve grassroots party organizations. Promote the adoption of the employability training strategy platform party activities to strengthen party building theory, improving their party level.

4 FEATURES INTERNET PLATFORM

Development of Internet technology and the user, the network communication into public acceptance of information exchange. The internet is increasingly becoming an important means of business, education, cultural exchanges, news media, and other areas of ideological propaganda. Internet efficiency is to achieve real-time characteristics of the information capacity, transmission speed, time-sensitive functional requirements, and direct or indirect impact on people's thinking through different forms, values and spiritual pursuits. The internet platform has the following features:

1. Openness and sharing of information the Internet platform for knowledge acquisition and exchange of information to facilitate the condition, a new way to acquire new knowledge and communicate with each other.
2. An internet platform to ensure the transmission of information regardless of time, geographical restrictions, information convenient, fast and widely disseminated.

3 Internet platform with strong real-time interaction, the user can place unlimited, you can anytime, anywhere to communicate, interact, express ideas and communicate accordingly.

5 THE MAIN PURPOSE OF LOCATING EMPLOYABILITY TRAINING STRATEGY

The main purpose of employability training strategy is the use of the Internet platform, according to the status of employability training strategy to upgrade to the traditional forms of employability training strategy, so as to realize the party organization at the university to join the international competitive environment to play their own public authority in charge of the construction and management of social functioning the role and effectiveness of the positive energy generated by other aspects. Employability training strategy is the new era of the Chinese communists to actively adapt to changes in the social environment and working methods based on the current situation of innovation and network environment, the basic objectives include:

5.1 *Close liaison party unity*

Internet employability training strategy can be achieved "were scattered parties together," the purpose of using the Internet to break the geographical limits of the original platform for employability training strategy, and extends the work of the geographical space. By between regions, between departments, the lateral interaction between party members, the advantages of the network open to get the effective development, college party anywhere in the world, you can get relevant information and assistance through the network. Communication and interaction platform are Internet employability training strategy, employability training strategy to achieve unity of party members and liaison purposes.

5.2 *Strengthening democratic life*

Internet employability training strategy platform is an easy, fast, real-time, efficient and understanding, supervise the work of the party platform. An internet platform for employability training strategy should focus on the protection and implementation of the party and the people's right to know, the majority of party members and cadres to strengthen the implementation of the right of expression, the right to effectively implement and safeguard the democratic participation of party members and cadres to strengthen the implementation and protection of the majority of party members and cadres of supervision. Through the platform to enhance people's awareness of the work the party, strengthen the supervision of the work of the party sensing capabilities, the real purpose of the network guarantee their democratic political power.

5.3 *Strengthen the cohesion of the party organization, serving the people*

New forms of organization are covering the full features of the Internet and ultra-strong space, developing a network of grassroots party organizations, online organizational life and online services platform. Entirely new employability training strategy of life, create new operational and organizational models and innovative party service system, develop and maintain the Party's advanced nature, thus ensuring the vitality and energy of party organizations, strengthen party cohesion. By adapting to the new situation and changes in the modern college party members, were the party organization and network service system, the establishment of a full range of employability training strategy service platform, forming a real network platform to serve the masses.

6 ROLE OF THE INTERNET IN MODERN EMPLOYABILITY TRAINING STRATEGY

Dissemination of information on the Internet as a way of being permeated all areas of social life, quickly changing the way people work, live and ways of thinking. Advantages of using the Internet more and more to the employability training strategy of college administration get the practice and promotion in the employability training strategy in different forms, to play its role to protect the core competitiveness of universities.

The Internet has become the new carrier spread advanced ideas. With the rapid development of the Internet, the university can build the unit's website. By employability training strategy of the comprehensive section of the site, such as "employability training strategy Expo", "Friends of party members" and other party knowledge platform, the party's political theory classics, the Party's basic line, principles and policies, the latest of the party and state leaders speech and the employability training strategy and other dynamic data sharing. Provide for an all party workers and members abreast of the latest information on the employability training strategy, the exchange of ideas and political education work experience, explore the carrier of employability training strategy theory, easy to learn about the latest current political party, for the majority of party members to provide a new theoretical ideas of the carrier.

The Internet has become a new model of Party work. College party management through the establishment of an integrated electronic system for party work can improve efficiency and reduce the cost of party work. By using an employability training strategy of the electronic accounting system can achieve a paperless office, party workers can be recorded via the network party work, access to documents on previous work subordinates affairs and transport, which saves the cost of an office, but also improves the work efficiency. Internet-based digital party management and daily work, improve the efficiency of party work, to achieve a comprehensive new model of party building and run party affairs management services.

The Internet has become the new position of party members in management education. You can use the Internet features across time and space, construction of employability training strategy information dissemination, open Internet Party, online ideological and political work, honest government, mass organizations and other online employability training strategy positions. You can also take advantage of the openness of the characteristics of the Internet, the employability training strategy platform links to other units, and learn from the best practices and advanced experiences of other units' employability training strategy, expanding the learning dimension of party members, improve the demonstration effect of advanced employability training strategy. By constantly enrich the content of the employability training strategy network, enhance the depth and breadth of the party of education, employability training strategy create a good atmosphere of culture, strengthen the network of employability training strategy of the sound. Make the Internet platform to become the party's knowledge mastered college party, the party's theory, become party to enhance the sense of responsibility, improving their quality of new political positions.

The Internet has become a new platform for interaction between the parties. Open interactive use of the Internet, the majority of party members through the "employability training strategy exchange," "Hot Topics"and other Internet platforms, ideological report, comments, ideas and other work of the party. A university Party platform through the Internet can grasp the ideological trend of the masses, for the masses of party members or share this problem-solving.

7 CONCLUSION

New media technology to break the boundaries between the real world and the virtual world, changing the way people exchanges fundamentally. In the new media, interpersonal communication between the conditions, everyone can hide their true situation; you can speak freely with impunity, to express their views. Thus, it is conducive to the education of college students who learn about the true thoughts, so that their employability training strategy targeted; also conducive to a more in-depth discussion of the relevant issues, employability training strategy should produce some of its practical effect.

In the employability training strategy is the degree of trust between teachers and students how to have been an important factor in the impact and effectiveness of education and quality of education. In traditional teacher-student relationship, both teachers and students are always in a state of inequality, which makes student teachers are often reluctant to tell the truth, it prevents enhance the effect of employability training strategy. And with the help of SMS, blog, forums and other new media, the exchange between education and the educated have some hidden, and thus bring a sense of equality of the two sides in the personality, rights and status, there is conducive to the formation of a relaxed and harmonious atmosphere, thereby eliminating the gap between teachers and students, thereby enhancing the degree of trust between teachers and students, employability training strategy to have good teaching.

ACKNOWLEDGEMENTS:

1 Based Employability Training Strategy Students Volunteer Service - Beihua University Project
2 Status and Strategy of Graduate Employment Training Capability - Jilin Province Educational Science Planning Project

REFERENCES

Liu Hongjun. Several colleges and universities scientific development thinking [J]. Learning and exploration, 2009 (5).

Cold dives. Adapt to the new situation and to explore new ways-on the petroleum and petrochemical Employability training strategy [J]. Theoretical front, 2006.

Chen Li. Ways to enhance the effectiveness of grassroots employability training strategy in colleges and universities [J]. PARTY SCHOOL, 2006.

Mr. Zhao. Universities in New employability training strategy of ideological and political work study [M]. Sichuan People's Publishing House, 2009.

Jiang Tiezhu. Famous Employability training strategy [M]. Wenhui Press, 2008.

Sun Chuanming. Employability training strategy Problems and Solutions [J]. Coastal universities and science and technology, 2005.

Xiao Xuebin, Julie. The impact of new media on the employability training strategy and its response [J]. Ideological Education Research, 2009 (7).

Wang Huancheng. New trends in the new media environment of employability training strategy of college students [J]. Contemporary Education Forum (Management), 2010 (8).

Lai Yong. Employability training strategy in the new media environment exploration [J]. Net wealth, 2010 (7).

Computing, Control, Information and Education Engineering – Liu, Sung & Yao (eds)
© 2015 Taylor & Francis Group, London, ISBN: 978-1-138-02800-5

Analysis of mechanics principle for the technology in tug-of-war sport

Zhi Xiang Jin & Qian Zhang
Huazhong Agricultural University, Wuhan, China

ABSTRACT: Tug-of-war as a popular sport is now catching on all around the world. The primary aim of the paper is to reveal the technology, including in tug-of-war and extend some relating knowledge. First, the paper introduces the significance of research on the technology in tug-of-war. Then, the basic mechanics in tug-of-war on the basis of the simplified rope model is analyzed. Based on this, the mechanical principle of two main sport forms of translation motion and rotational motion in tug-of-war is discussed. At last, the conclusions are summed up. The paper finds that friction and gravity are two influencing factors of great importance in tug-of-war. Therefore, increasing friction and gravity contributes to the victory in tug-of-war.

KEYWORDS: mechanics principle, force analysis, tug-of-war, friction.

1 INTRODUCTION

Tug-of-war sport is a very popular sport all over the world. Since the year of 1900, tug of war world championship has been held every two years [1]. Unfortunately, very few fans of tug-of-war know more about the technology in tug of war. In general, many people frequently take it for granted that a greater strength will create more change to acquire success. Indeed, a part of people persistently consider that there is no any technology in tug-of-war except more power. In fact, Tug-of-war is not only a sport; but also it is a common physical phenomenon. A large amount of physical and mechanical knowledge is included in tug-of-war.

In addition, these issues, including in tug-of-war are able to provide a good point of view to understand the physics and mechanics principle. In order to correct the inaccurate point of view and extend the knowledge of tug-of-war, the paper takes the technology in tug-of-war as the research object and attempts to provide some correct line with respect to the technology of tug-of-war.

The structure of the paper is as follows: the paper starts with the introduction of comprehending the significance of technology in tug-of-war; then, it puts forward the basic mechanics analysis in tug-of-war on the basis of rope mechanics in section 1; mechanics principle analysis of two motion forms of translation and rotation in tug-of-war is conducted in section 2 to present important influencing factors for tug-of-war. The last section sums up the research conclusions

2 BASIC MECHANICS ANALYSIS IN TUG-OF-WAR

Tug-of-war is a match carried out by a rope. In this sense, the analysis of tug-in-war is to pay attention to mechanics of rope [2]. Therefore, a rope of mass m withstands two tensions F_1 and F_2 in its both sides respectively. We attempt to analyze the tension in the rope. For simplification of the research, provided that the rope is averagely divided into n sections. In addition, the mass of each section is concentrated to one point in the section, which is shown in Fig. 1 as follows. In Fig. 1, a is the acceleration produced by the two tensions.

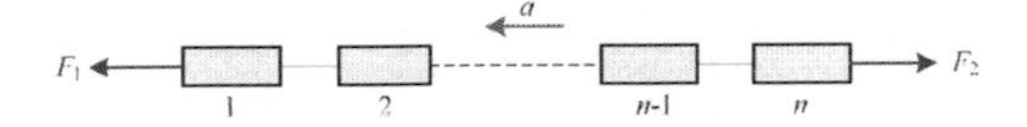

Figure 1. Simplified rope model.

The rope is considered as a whole, the kinetic equation is constructed as the following through whole method:

$$F_1 - F_2 = n\left(\frac{1}{n}m\right)a = ma \tag{1}$$

Arrangement of Eq. (1), we can obtain:

$$a = \frac{F_1 - F_2}{m} \tag{2}$$

The tension between every two sections can be got as follows by means of isolating method:

$$\begin{cases} T_1 = \dfrac{(n-1)F_1 + F_2}{n} = F_1 - \dfrac{F_1 - F_2}{n} \\ T_2 = \dfrac{(n-2)F_1 + 2F_2}{n} = F_1 - \dfrac{2(F_1 - F_2)}{n} \\ \vdots \\ T_{n-1} = \dfrac{F_1 + (n-1)F_2}{n} = F_1 - \dfrac{(n-1)(F_1 - F_2)}{n} \end{cases} \quad (3)$$

where T_i expresses the tension in the ith section rope.

Eq. (3) illustrates that if $F_1 > F_2\ (a > 0)$, we can get:

$$F_1 > T_1 > T_2 > \cdots > T_{n-1} \quad (4)$$

which describes that the rope will move along the direction of F_1.

Besides, if $F_1 = F_2\ (a = 0)$, we can obtain:

$$F_1 = T_1 = T_2 = \cdots = T_{n-1} = F_2 \quad (5)$$

Eq. (5) indicates the tensions in every point are equal with each other.

3 MECHANICS PRINCIPLE ANALYSIS IN TUG-OF-WAR

According to the theory proposed above, two common sport forms in tug-of-war, which are translation motion and rotational motion [3], are analyzed by means of mechanics principle.

3.1 *Mechanics principle analysis of translational motion*

Section 1 analyzes the mechanics of rope. In a tug-of-war, the both teams and rope can be considered as a whole junction. For convenience of research, the gross mass of both teams M_1 and M_2 are assumed to be equal and the game starts from a standstill.

Hence, the junction's move is owing to its resultant force is not equal to zero. In terms of the analysis process in section 1, if the external friction forces acting on these two teams are F_1 and F_2, and $F_1 > F_2$, the junction will move towards with F_1, shown in Fig. 2, where G_1 and G_2 describe the force of gravity for two teams respectively, N_1 and N_2 describe the counter-acting forces of G_1 and G_2 respectively.

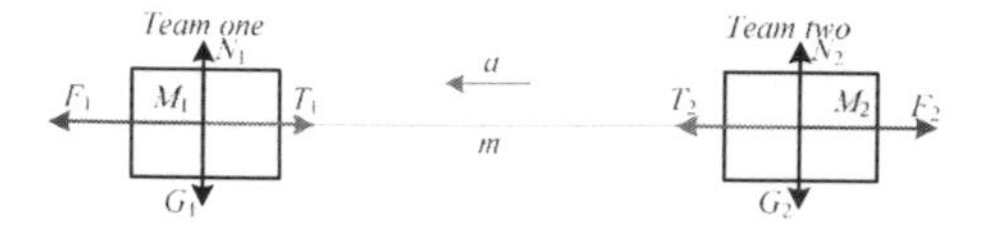

Figure 2. Simplified analysis model in tug-of-war.

Similarly, the isolating method is utilized to construct the kinetic equation of junction and indicated as follows:

a. For team one:

$$F_1 - T_1 = M_1 a \quad (6)$$

b. For the rope:

$$T_1 - T_2 = ma \quad (7)$$

c. For team two:

$$T_2 - F_2 = M_2 a \quad (8)$$

The whole method is used to model the whole system, we can get:

$$F_1 - F_2 = (M_1 + M_2 + m)a \quad (9)$$

Simultaneously combining and solving Eqs. (6), (7), (8) and (9), we can get:

$$T_1 - T_2 = (F_1 - F_2)\left(1 - \frac{M_1 + M_2}{M_1 + M_2 + m}\right) \quad (10)$$

In actual tug-of war, the friction forces F_1 and F_2 have only a relatively small distinction. Additionally, the comparison between $M_1 + M_2$ and m will be $(M_1 + M_2) \gg m$, we can get:

$$\begin{cases} 1 - \dfrac{M_1 + M_2}{M_1 + M_2 + m} \approx 0 \\ F_1 - F_2 \approx 0 \end{cases} \quad (11)$$

Substitute Eq. (11) into Eq. (10), we can obtain:

$$T_1 - T_2 \approx 0 \quad (12)$$

Eq. (12) indicates that if the rope mass is much less than the total mass of athletes, the tensions in a rope's two ending can be considered to be equal. This conclusion not only holds in stationary or uniform motion, but also in accelerated movement. In other words, for a light rope, its tension in every place will be equal.

Hence, under the condition of equal tensions, a larger friction force can weaken the other side's pulling force, which can increase the chances of winning. In general, due to the friction force is equal to friction coefficient μ multiplying by positive pressure N, hence there are two solutions to increase the friction

a. One is to enhance the friction coefficient μ, for instance, the footwear of rubber and flat bottom shoe with larger friction coefficient should be put on.
b. Another is to increase the positive pressure, the directly effective solution is to choose bigger mass athletes in the same number of players.

3.2 *Mechanics principle analysis of rotational motion*

Consider the person in tug-of-war as a rigid body, the rigid body in tug-of-war will rotate along the fulcrum of the feet, shown in Fig. 3.

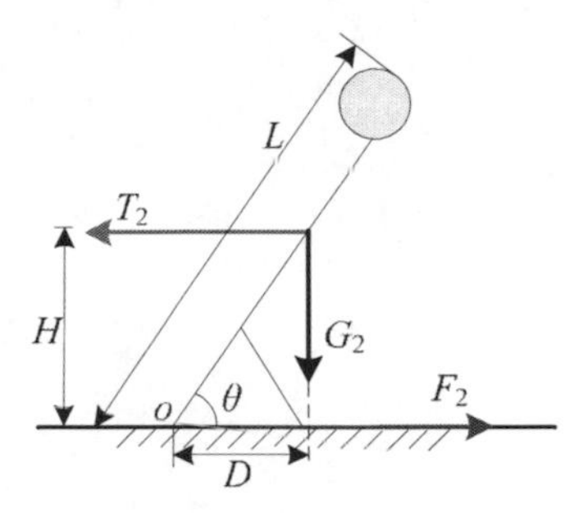

Figure 3. Mechanistic analysis of rotational motion in tug-of-war.

Fig. 3 indicates the positive moment in rotational motion is gravitational moment. In order to keep the balance of the body, the slope angle between the person and the ground must be less than 90°. The person is assumed to be L tall, the slope angle between the person and the ground is θ. The heights of the points of action of T_2 and G_2 are L_1 and L_2 respectively. The distances between T_2 and G_2 and fulcrum o are H and D, respectively. If the tension torque is equal to gravity torque, we can obtain:

$$T_2 \cdot H = G_2 \cdot D \tag{13}$$

thus,

$$\begin{cases} H = L_1 \sin\theta \\ D = L_2 \cos\theta \end{cases} \tag{14}$$

Hence,

$$T_2 \cdot L_1 \sin\theta = G_2 \cdot L_2 \cos\theta \tag{15}$$

Eq. (15) indicates under the situation of a certain tension T_2, the torque primarily depends on G_2, distances L_1, L_2, and the slope angle θ. It should be noted that the range of θ is [0, 90°]. In general, for a person, the distances L_1 and L_2 are frequently certain. In other words, the ration of L_1 to L_2 is a certain value, we assume it to be r. Hence, Eq. (15) can be rewritten as the following:

$$r \cdot T_2 \cdot \sin\theta = G_2 \cdot \cos\theta \tag{16}$$

Fig. 4 indicates the changes of sinθ and cosθ, we can observe that when θ belongs to [0, 90°], sinθ is an increasing function, and cosθ is a decreasing function. In addition, the point of 45° is the point where sinθ = cosθ.

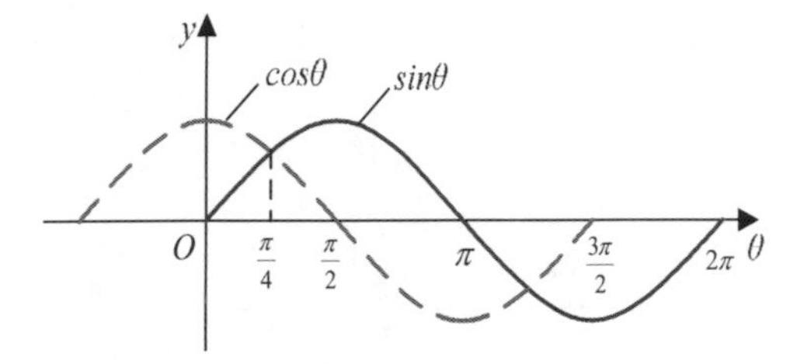

Figure 4. Functions of sinθ and cosθ.

Fig. 4 demonstrates that when θ belongs to [0, 45°], sinθ < cosθ. In this interval, gravity torque as a positive torque will play a greater effectiveness. Hence, the slope angle θ should be restricted in [0, 45°]. Obviously, the value of θ should not be too small, 30° is a reasonable value.

In addition, in terms of Eq. (16), we can derive the relationship between θ and friction coefficient μ as follows:

$$\theta = \tan^{-1}\frac{G_2}{r \cdot T_2} = \tan^{-1}\frac{f_2}{r\mu T_2} \tag{17}$$

Eq. (17) depicts that if friction coefficient μ is greater, the slope angle θ will be smaller. That is to say, θ will be more reasonable in a more rough ground.

Furthermore, Eq. (16) also illustrates that when tension T_2 and slope angle θ are determined, adding weight G_2 can enhance gravity torque as well. In other words, a greater weight can not only increase friction, but also improves the gravity torque.

4 CONCLUSIONS

This paper studies on the mechanics principle of the technology for tug-of-war. The conclusions are summarized as follows:

a. The forces of friction and gravity are two mechanics factors of comparative importance in tug-of-war.

b. A larger friction force can weaken the other side's tension force.
c. The reasonable slope angle value is about 30°, and a greater weight is beneficial to acquire positive gravity torque in both translation motion and rotational motion.

REFERENCES

Stalder G., Tug-of-war, in: *Magglingen*, 1991, p. 6–9.

Hexner D, Kafri Y., Tug of war in motility assay experiments, in: *Physical biology*, Vol. 6, 2009, p. 036016.

Müller M J I, Klumpp S, Lipowsky R., Bidirectional transport by molecular motors: enhanced processivity and response to external forces, in: *Biophysical journal*, Vol. 98, 2010, p. 2610–2618.

Computing, Control, Information and Education Engineering – Liu, Sung & Yao (eds)
 ISBN: 978-1-138-02800-5

Change detection of ship target in port based on Candidate region

Yan Li Sun, Wei Zhou, Jian Hai Li & Jia Qi Chen
Department of Basic Experiment, Naval Aeronautical and Astronautical University, Yantai, China
Department of Electronic and Informrtion Engineering, Naval Aeronautical and Astronautical University, Yantai, China

ABSTRACT: The basic idea is to use of the electronic map, rapid extraction of ship possible docking area and rule it. Then operate the image segmentation at a different phase, and project along the horizontal and vertical direction, find the maximum interval exceeds the average projected energy, and determine the change region of the ship. The algorithm can avoid the large range of useless search, improve the efficiency of detection.

KEYWORDS: Change detection; Candidate region; Projection energy.

Usually changes to harbor ship target detection mainly rely on manual interpretation. The port range is large, and the target is relatively dispersed, so the change and development of an automatic detection algorithm can reduce the burden of technicians in a certain extent. Considering the port is a fixed target, its geographical location and the main ship in the parking area are relatively fixed for a long time. Therefore, we only need to carry on the contrast analysis of specific regions. So we propose a change detection method of harbor ship target based on Candidate region on map. Change the complex detection problem into several small, relatively simple test problems. Set reasonable rules, then the analysis and automatic change can be realized.

1 THE CANDIDATE REGION EXTRACTION BASED ON ELECTRONIC MAP

We first extract the coast contours from the map, then operate equidistant sampling, extracting the key point of the corner, identify potential ship docking area.

1.1 *The coastline inflexion*

First extracting complete waters, electronic map change into a two value images, operate edge detection by two value images, use classic Canny operator can achieve very good results. Then find an endpoint of the coastline, a traverse counterclockwise, get a lengthy sequence of $\{l_{0i}=(x_{0i}, y_{0i}), i=1, \cdots, n\}$, if get multiple sequences, choose the longest one. In order to reduce the burden of subsequent inflection point analysis, can operate equal interval interpolation sampling of the sequence, get the new sequence of $k\{l_i=(x_i, y_i), i=1, \cdots, k\}$. Using (1) to calculate the new sequence points,

$$\theta_i = \tan^{-1}\left[\frac{y_{i-1}-y_{i+1}}{x_{i+1}-x_{i-1}}\right] + \pi \cdot \mathbb{T}(x_{i+1}-x_{i-1}<0) \qquad (1)$$

Among them, $\mathbb{T}(t)$ refers to the logic operation, when t is true, the output is 1. Inflection point defined shoreline as local changes in the direction of the larger point, if {lj} is the inflection point, there are

$$|\theta_{j+1}-\theta_j| > \Delta\theta \qquad (2)$$

Often think that if the direction difference of the two points exceed $\Delta\theta$, that has changed the direction partly. The Pearl Harbor, which the image shown in Figure 1, the length of the sequence from n=12647, k=400 sampling points, the results obtained in Figure 3 with a red "O" mark. On the basis of (2)defined rules, take$\Delta\theta = \pi/18$, extract 97 inflection points, illustrated by the Yellow marked "*"in Figure1.

1.2 *Extraction of ship Candidate region*

The ship has a certain length, apparently, not all two inflection points are suitable for ships to dock. Thus, according to the electronic map on the scale or resolution with small and medium-sized ship size setting pixel distance threshold appropriated, followed by calculation of the distance between two adjacent inflection points. If exceeded a threshold, for possible ships to dock coast section. In Figure 1 the Pearl Harbor regional electronic map, the resolution is 0.6m, we set the minimum ship docked interval length is 60m, namely the pixel distance threshold is 100, extract 26 possible ship docked coast section, the results in Figure 1 shown in the blue line segment.

For any potential ship docked coastal section, inflection point from both ends of the $P_{t1}(x_{t1}, y_t1)$ and $P_{t2}(x_{t2}, y_{t2})$, the direction can be recorded as θ, its length is denoted as d. Obviously, the sea is always in the left of the $\overline{P_{t1}P_{t2}}$. Move $\overline{P_{t1}P_{t2}}$ respectively, along the $\theta+\pi/2$ of d_2, the area is the ship anchored region of the coastline by the sea side. Move $d1$ along the $\pi/2-\theta$ direction. The area of the coastline is on the land side of the narrow area. Two sections constitute the ship candidate the coastal section, then the region along the rotation angle clockwise rule. This ship candidate rooms are rules into the height of d_1+d_2 width of d rectangular region[1]. Using the above method, the extraction of the 26 ship candidate region, where d_1 and d_2 are 30 and 60 pixels, as shown in Figure 2.

Figure 1. Inflection point extraction results.

(a) QuickBird image phase1 (b) QuickBird image phase2

Figure 2. In the different time phase of remote sensing image are labeled on the 26 ship candidate.

2 THE CANDIDATE OF CHANGE DETECTION

Because the difference of local gray level differences between image is large, and methods of direct comparison are difficult to extract changing target. We note that, in the local ship candidate has three main targets: the land, the ship and the sea. In spite of the spectral difference, the local also has serious aliasing phenomenon. Therefore, we first to different candidate areas are adaptive segmentation, extraction of ROI area, land and ship target is obtained through morphological processing, then compare the segmentation results, get the change information[2]. Combining with the ship size and other information to carry on the comprehensive analysis, gives the final results of the target change.

2.1 *An area of land and ship target segmentation*

Because the land area is relatively stable in the different time phase image, the difference operation can be well eliminated, therefore before the change analysis, we first to different candidate image were land adaptive threshold segmentation and extraction potential target and as completely as possible using morphology [3].

Although usually brightness, ship land than sea water is high, can be a color image into a gray image and then the threshold segmentation, but for submarines, due to its close with seawater gray, the segmentation effect is not ideal. To this end, we will first image from RGB space to the HSV space transform, three components were denoted as I_H, I_S and I_V, with (3) of three component fusion can enhance the target.

$$I = \frac{\min(I_H, I_V)}{I_S} \tag{3}$$

Figure3 shows the grayscale image fusion of twelfth candidate regions at different image enhancement and channel averaged. Compared with the gray image, land and ship targets have been significantly enhanced, applying a global threshold is influenced by the background is relatively large. We design a simple way to set up an adaptive threshold.

(a) Fusion enhancement results in HSV of phase1 image

(b) Grayscale image of phase1

(c) Fusion enhancement results in HSV of phase2

(d) Grayscale image of phase 2

Figure 3. Results compared with gray and enhancement in same candidate regions at different image.

Let $T(i, j)$ for the enhanced image in $(i, j)$of the pixel. In $1\le i\le M$, $1\le j\le N$, were calculated for each column and row μ_i,μ_j and the mean value of the whole image mean μ.

$$\mu_i = \frac{1}{N}\sum_{j=1}^{N} I(i,j) \tag{4}$$

$$\mu_j = \frac{1}{M}\sum_{i=1}^{M} I(i,j) \tag{5}$$

$$\mu = \frac{1}{M}\sum_{i=1}^{M} \mu_i = \frac{1}{N}\sum_{j=1}^{N} \mu_j \tag{6}$$

In the(i, j) threshold at $T(i, j)$,choose μ_i,$\mu_j$and the median μ, i.e.

$$T(i,j) = median(\mu_i, \mu_j, \mu) \tag{7}$$

The image was segmented by the threshold, ROI area, preliminary next, considering the constraints of ship size and location of the target, corresponding to the larger regional connectivity, therefore, using morphological method to set the area threshold can effectively eliminate isolated independent point and broken small area, filling and internal hole of large area the final, complete land and ship target. The results obtained as shown in figure 4 by fusing, enhancing an adaptive segmentation the candidate region two phase image In Figure 3.

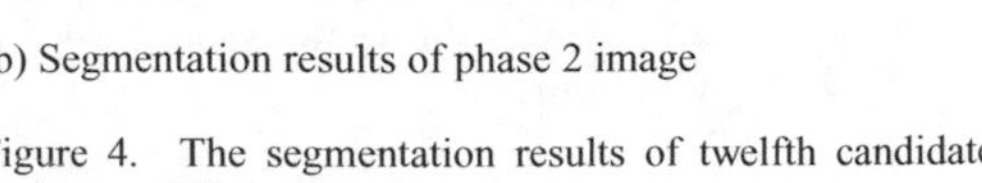

(a) Segmentation results of phase 1 image

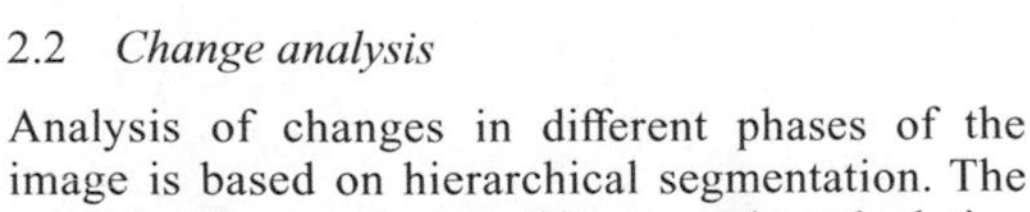

(b) Segmentation results of phase 2 image

Figure 4. The segmentation results of twelfth candidate regions at different image.

2.2 *Change analysis*

Analysis of changes in different phases of the image is based on hierarchical segmentation. The segmentation results are difference, the calculation of the total area change region, if less than a certain threshold value, can from the overall decision did not change significantly in the region, or to further determine the changes in regional position, if the change of main body part of land along the coast side, no change is determined to ship, or ship, according to the size of the ratio of length to width geometric characteristics of the ship targets to be recognized, and then determine the target disappear or appear[4].

Total change area threshold setting, usually depends on the resolution of the image and the size of the ship, the main purpose of this step is to quickly remove the no obvious target for changing scene. To determine the position change region, we of the differential image respectively to the projection along the horizontal direction and the vertical direction to realize, because the image change analysis have been the rule, the ship bow and stern axis and the coastline are generally in a horizontal direction, so the difference results projection along the horizontal direction, exact location change of ship two side, further along the vertical direction projection, precise position can be changed the fore-and-aft ends.

The result of image segmentation at different phase difference operation, get the results as shown in Figure 5 (a) below, and then along the horizontal direction projection, find the maximum interval exceeds the average projection energy, if its width exceeds a set threshold (usually corresponding to the width of the ship), the preliminary determination for the ship might fluctuate, but if the interval most coastline reference position right, that change the interval is not from the ship. As shown in Figure 5(b), a reference position left between [23,55] on the coastline, there is a large energy projection interval, next to the projection ship might fluctuate along the vertical direction, the results as shown in Figure5 (c) shows, from which we can find 4 energy projection larger interval, in which, the most the left interval width is small, can be ruled out, the other three interval A:[87,240], B:[414,53] and C:[482,507] in A can be directly recognized as the target, and the distance between B and C and their respective width close to, can not be directly determined as target, the area labeled target change area possible, need to go back to the image on the use of other pattern recognition.

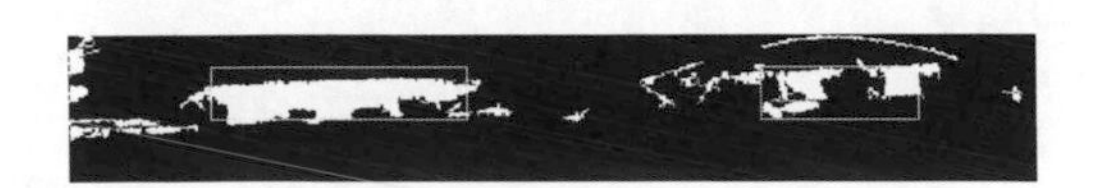

(a) The segmentation difference results and final confirmation of the two targets

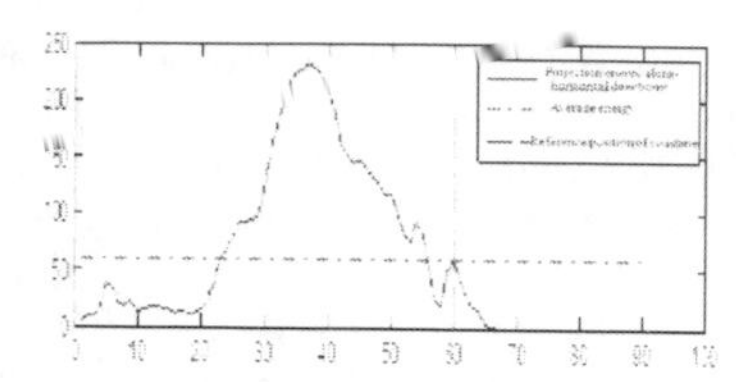

(b) Determine the fore-and-aft position along the horizontal dir ection projection

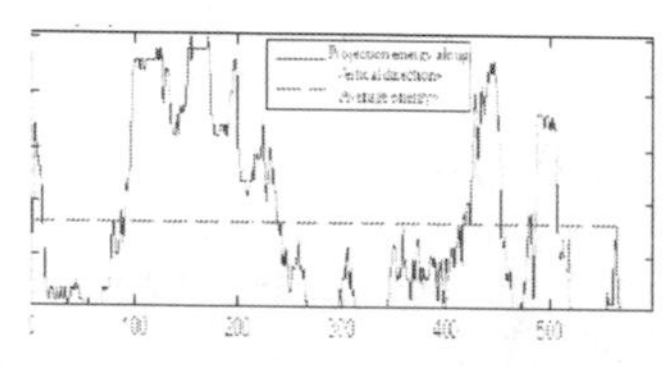

(C) Determine the fore-and-aft position along the vertical direction projection

Figure 5. Example of ship targets change analysis.

To find the target after change of the region, according to the region in the different time phase image on the average energy, can be divided into two kinds of situation, the image segmentation is different, for the same target change region, phase 1 less energy than if a phase 2, then determine the ship appear, otherwise, determine the ship disappeared.

At this point, we can change the analysis of each ship candidate region using the above method, the port of the final change detection results. Because of the limited space, Figure 6 gives only 9, 22 and 25 three ship candidate results. The results indicated in the image is 2, the "red," mark ship target emerging, "ship yellow–" signs disappeared.

(a) No. ninth candidate regions (only 35% of the original size)

(b) Twenty-second, the candidate region

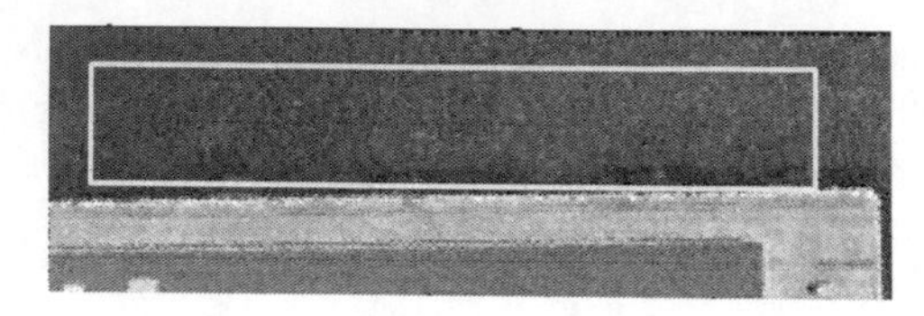

(c) Twenty-fifth, the candidate region

Figure 6. Part ship candidate change analysis results.

3 CONCLUSIONS

This paper discusses the detection method based on map changes. The basic point of departure is the use of electronic map information to find the candidate quickly and accurately and rule it. Extract the target and analysis of changes in the standard operating environment can realize automatic processing. It has strong application potential. For single image harbor ship detection and recognition, we can also make registration with the electronic map at first. Extract regional sections of ship candidate rules, which can avoid the large range of useless search, significantly reduce the computational cost, improve the efficiency of detection and recognition algorithm.

REFERENCES

[1] Y.L.Sun, W.Zhou, Y.F. LING, Candidate region detection of ship target in port based on electronic map[C]. Frontiers of Manufacturing Science and Measuring Technology IV. (2014)1420–1424.

[2] Durucan E, Ebrahimi T, Change detection and background extraction by linear algebra[J].Proceeding of the IEEE,2001,89(10):1368–1381.

[3] Charalampidis D, Stein G W. Target detection based on multiresolution fractal analysis[C]. In: Proceedings of SPIE-The International Society for Optical Engineering Signal Processing, Sensor Fusion, and Target Recognition XVI. Orlando, 2007.

[4] Kwon Oh-Kyu , Sim Dong2Gyu , Park Rae2Hong. Robust Hausdorff Distance Matching Algorithm Using Pyramidal Structure [J] . Pattern Recognition, 2001,34 (7) :2005–2013.

Computing, Control, Information and Education Engineering – Liu, Sung & Yao (eds)
© 2015 Taylor & Francis Group, London, ISBN: 978-1-138-02800-5

The design of fatigue life test bed for certain air type shock absorber

Kun Yang, Lan Tang, Ping Wu
Key Research Laboratory for Vehicle, Xihua University, Sichuan, China

ABSTRACT: The fatigue life test bed for certain air type shock absorber is designed on the basis of eccentric motion mechanism, and the system composition and the basic structure of test bed are detailed in this paper. SolidWorks is used to establish a virtual prototype model of test bed, and ANSYS Workbench is used to analyze the structure reliability of test bed on the key position. The results show that the structure of test bed meets the test requirements, at the same time the stroke adjustment of test bed is broader than before.

KEYWORDS: air type shock absorber; fatigue life test bed; SolidWorks; ANSYS Workbench.

1 INTRODUCTION

The fatigue life test bed is the key equipment for the design of shock absorber[1], and the current domestic test bed is relatively backward, which restricts the domestic improvement of shock absorber performance[2,3]. With this situation, the author has designed the fatigue life test bed based on certain air type shock absorber. The strength of each parts of shock absorber can be reflected by it, and then the performance of shock absorber will be better.

2 THE OVERALL DESIGN OF TEST BED

Bench motion model. The diagram of motion mechanism [4] is shown blow in Figure 1. The motion model of test-bed is eccentric motion mechanism. The sine excitation wave is used to simulate the car passing in the face of a quarter suspension model of sine wave excitation, which is outputted by the model at point D.

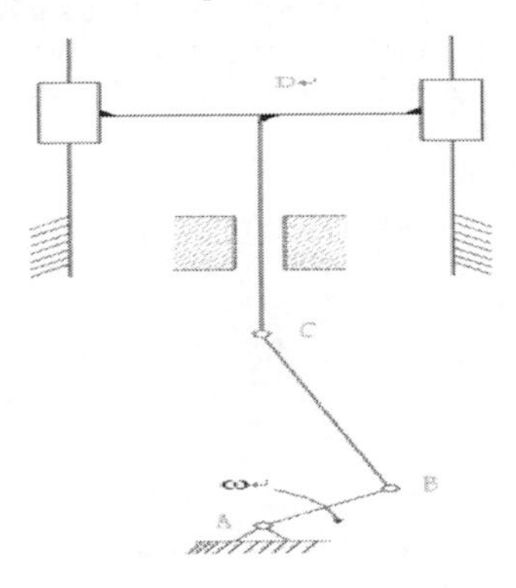

Figure 1. Diagram of motion mechanism.

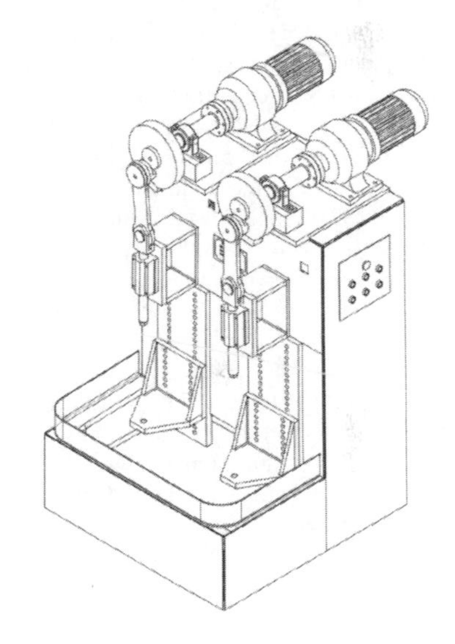

Figure 2. Structure diagram of test bed.

The system composition and the basic mechanism of test bed. The Structure diagram of test bed is shown in the above Figure 2[5]. The system is mainly composed of mechanical transmission system, control system, cooling system and air pressure detection system.

Mechanical transmission system. It is mainly composed of motor, reducer, coupling, bearing chock, slider, mounting rack, mounting plate, base, etc. This system is a major component of the overall system, in which the motor and reducer are connected together, and reducer is connected with axle by a coupling. The other end of the axle is connected with baffle by a bearing chock, and the other end of baffle is connected with connecting rod. The lower end of the connecting rod is connected with the slider, and the lower end of the slider is connected with the upper end of the shock absorber. The conversion process of the form of the movement is that electrical power is passed to the eccentric wheel mechanism by the reducer, to drive it to do rotation. The slider converts rotary motion to linear motion with the constraints of the guide, thereby to achieve the output of the sine excitation.

Eccentric motion mechanism is composed of connecting rod, nut baffle, axle baffle, swing link baffle, balance wheel, silk pole, etc. The adjustment of the silk pole ranges from 100 to 200 mm, which meets the needs of a variety of different strokes in the shock absorber testing work.

Control system. Control system consists of inverter, power control switch, counter display, magnetic proximity switch, etc. In order to ensure the test results, the test frequency should be convenient and adjustable. The test system including a motor and MM440 inverter, and it is easy and convenient to continuously adjust the vibration frequency by the panel operation mode.

According to the "Regulations of QC/T545-1999 automobile shock absorber bench test method", the shock absorber cycles of experimental work of fatigue life testing are at least one hundred thousand times. The number of cycles is about 1.1 million times when we increase testing efforts in the actual testing process. ZN48JR single reversible counter display that the counts range from 0 to 99999999 meets the experimental requirements.

Cooling system. The cooling system is mainly composed of pump, water pipe, outlet connector, water tank, etc. The water pipe is a common plastic pipe, and pipe outlet is mixed by two ways to increase the contact area between water and shock absorber. The relative position of the water outlet for shock absorber is adjustable, and the outer wall of shock absorber can be cooled in 360 °. The water tank is located at the bottom of the fatigue life test bed, so that the cooling water can directly fall into the water tank, and then the water is pumped by the electric pump to form a water circulation system.

Air pressure detection system. Due to internal working medium for gas, it is essential for shock absorber to detect internal gas pressure in the test. The measuring range of the barometer is 0~3MPa. Barometer is connected with shock absorber by copper pipe. Both ends of the pipe are welded by copper rotary joints, and the copper pipe is connected with shock absorber in a rotary mode.

3 TEST THE STRENGTH OF THE KEY POSITIONS OF THE TEST BED.

The strength of the output axle. The strength of the axle is calculated by the torque, and torsional strength condition for solid axle is [6].

$$\tau_T = \frac{T}{W_T} \approx \frac{9550000\frac{P}{n}}{0.2d^3} \le [\tau_T] \tag{1}$$

$$d \geqslant A_0 3\sqrt{\frac{P}{n}} \tag{2}$$

Where: τ_T—Torsion shear stress, MPa. T—Axle torque, $N \cdot mm$. W_T—Axle torsional cross section coefficient, mm^3. N—Axle speed, $r/\min$. P—Axle transmission power, KW. d—Axle diameter, mm. $[\tau_T]$—Allowable torsional shear stress, MPa.

The A value of common materials is shown in table 1.

Table 1. A value of common materials.

material	$[\tau]/N \cdot mm$	A
Q235,20	12-20	160-135
35	20-30	135-118
45	30-40	118-107
1Cr18Ni9Ti	15-25	148-125
40Cr,35SiMn,2Cr13,42SiMn	40-52	100.7-98

Axle material is No.45 steel, A value is between 107 and 118, P= 2.2KW, n = 1450$r/\min$ (in theory), reduction ratio is 11:1, therefore n_1=131.8 r/min, the power factor is 0.82.The value of d is greater than or equal to 30mm by the calculation. The minimum axle diameter that is designed by the author is 40mm in the actual design process, which meets the requirement.

Finite element analysis of the key positions. *The finite element analysis of bottom mounting plate.* Physical attribute parameters of plate material is No.45 steel, and the yield strength is 355MPa. The model of bottom mounting plate is imported into ANSYS software and loaded force. The mesh diagram and the mesh quality diagram are shown blow in Figure.3 after meshing. The mesh distortion degrees are mostly less than 0.7, which meets the engineering requirements.

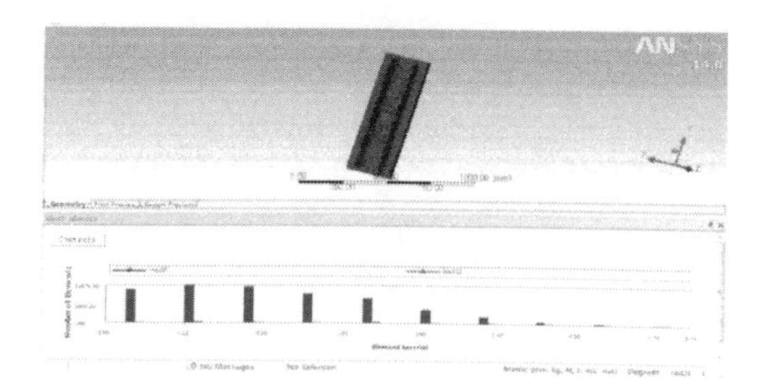

Figure 3. Mesh diagram and mesh quality diagram.

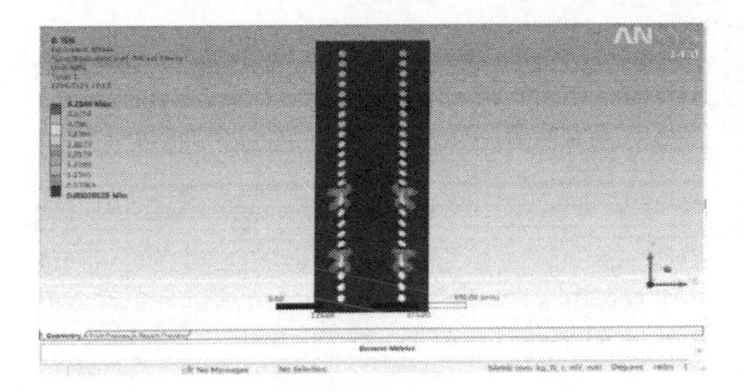

Figure 4. Maximum stress nephogram.

The maximum stress nephogram of bottom mounting plate is shown in the above Figure.4. The maximum stress of bottom mounting plate is 5.2148MPa, which is less than the yield strength of materials. To sum up, the mounting plate meets the requirement of the strength.

The finite element analysis of mounting plate. Physical attribute parameters of plate material is No.45 steel, and the yield strength is 355MPa.The mesh diagram and the mesh quality diagram are shown blow in figure 5 after meshing. The Figure 6 is the total deformation nephogram of mounting plate, and the maximum equivalent stress nephogram of mounting plate is shown in Figure 7.

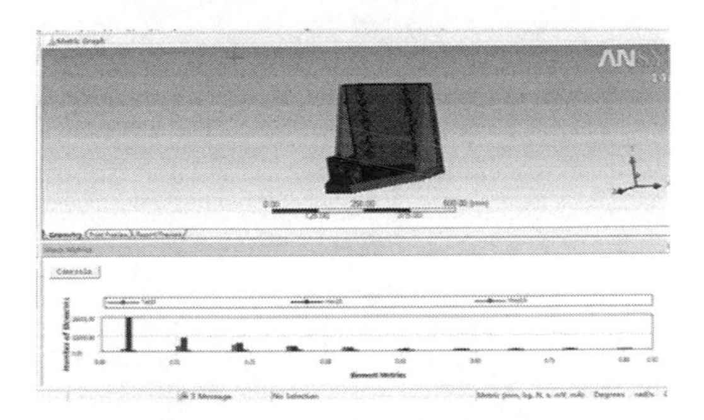

Figure 5. Mesh diagram and mesh quality diagram.

Figure 6. Total deformation nephogram.

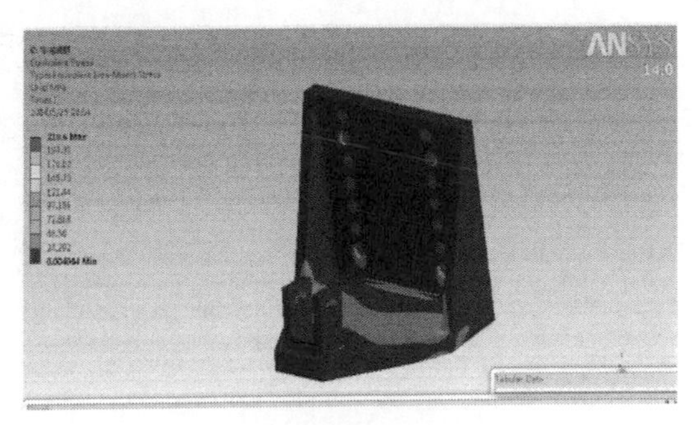

Figure 7. Maximum equivalent stress nephogram.

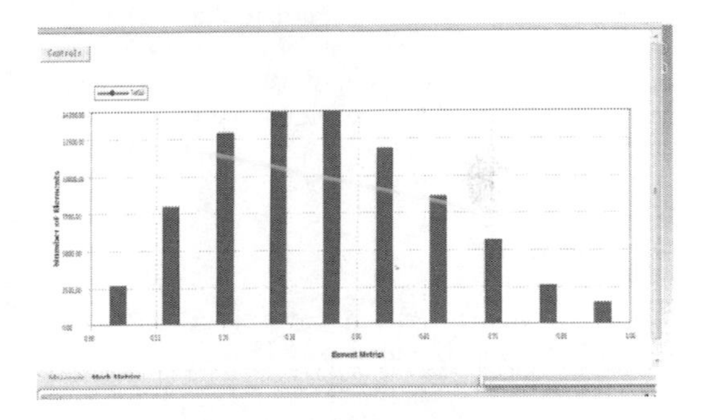

Figure 8. Mesh quality diagram.

Visible by the above figure, the maximum deformation of plate is only 0.37073mm under the maximum force. And the maximum stress is 218.6MPa, which is less than the yield strength of 355MPa. To sum up, the plate meets the requirement of strength and rigidity.

The finite element analysis of eccentric mechanism. The mesh quality diagram of eccentric mechanism is shown in the above Figure 8 after meshing. The Figure 9 is the total deformation nephogram of eccentric mechanism, and the maximum equivalent stress nephogram of eccentric mechanism is shown in Figure 10. The mesh distortion degrees are mostly less than 0.7, which meets the engineering requirements.

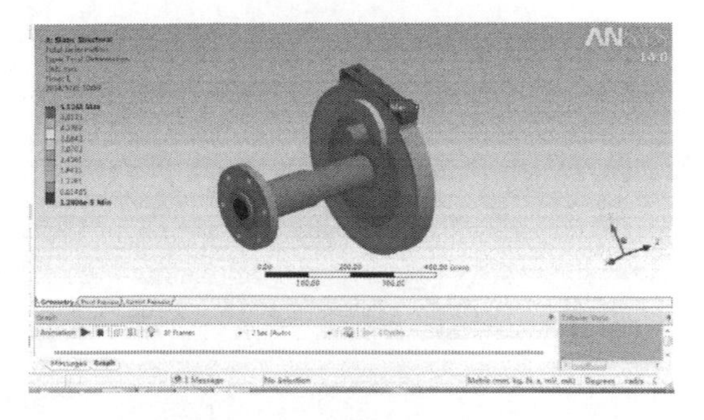

Figure 9. Total deformation nephogram.

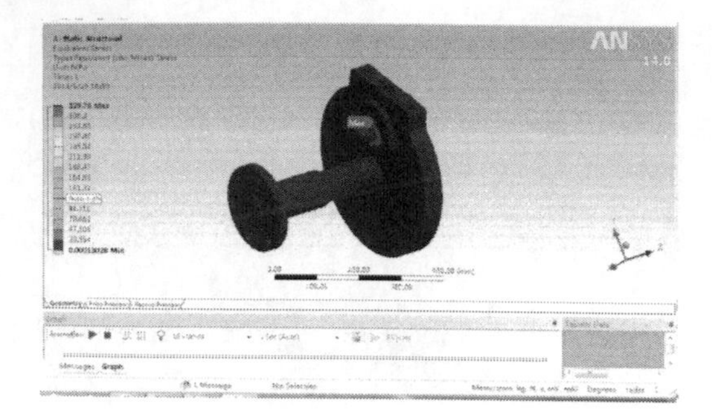

Figure 10. Maximum equivalent stress nephogram.

Visible from the above diagram, the largest stress is 329.76MPPa under the maximum stress, which is less than the yield strength of 355Mpa. In summary, the mechanism meets the requirements of strength and rigidity.

4 SUMMARY

Both the national standards of shock absorber test-bed and additional equipments added to the test bench can meet the requirements. At the same time the structure of test bed is simple and reliable, and so it is convenient to be repaired. The stroke adjustment of the eccentric mechanism is very broad, thus test force can be increased in the process of actual test. This is a breakthrough design. Finite element analysis is used to test the key positions of test bed, which proves that they can fully meet the test requirements. The design of the adjusting mechanism (eccentric) schedule is a process of constant exploration progress, and it can accumulate certain experience for future design.

ACKNOWLEDGEMENTS

The support of the key subject project for vehicle engineering of Sichuan (SZD0410),Automobile Engineering Laboratory of Sichuan (SGXZD9902-10-1) are acknowledged.

REFERENCES

[1] Sanhuai Wang, Wensi Ding. Design of Railway Damper Fatigue Test Bench[J]. Mechanical and electrical products development and innovation. 2004, 1; 23–24 In Chinese.

[2] Xuejun Liu, Cunxiang Liu,Zhikang Wei. Development of Electro Hydraulic Servo Damper of the Test Platform Based on LabVIEW[J]. Measurement and control technology. 2009, 36 (7); 88–89 In Chinese.

[3] Jinghua Shang, Zhaohua Yu. Durability Test Bed of Oil Damper[J]. Railway Technical Supervision. 2014, 42(7);45–46 In Chinese.

[4] Wei Jiang, Hongli Gao, Jianjun Yin. Research on Automobile Shock Absorber Performance Detection System[J]. Machine design and research. 2003,19(3);48–50 In Chinese.

[5] Chengtong Qu, Tao Jiang. The Design of Automobile Shock Absorber Double Dynamic Durability Test Rig[D]. Changchun: Changchun University of Science and Technology. 2013:12–20 In Chinese.

[6] Yunfei Chen, Yumin Lu. Machine Design Foundation[M]. The fifth edition. Higher Education Press, 2008.

Computing, Control, Information and Education Engineering – Liu, Sung & Yao (eds)
 ISBN: 978-1-138-02800-5

A study of developing innovative concepts of schooling in higher vocational education

H.Y. Liu & H.Y. Wang
Langfang Polytechnic Institute, Hebei, China

ABSTRACT: At present, with China's rapid economic development, higher vocational education has also developed well. In recent years, higher vocational colleges in our country have trained a large number of professional talents for the society, which has played a positive role in promoting social and economic development. However, there are also a series of problems and contradictions in the course of development which have seriously hindered the further development of higher vocational education. Therefore, this paper discusses and analyzes measures of innovative concepts of schooling in higher vocational education based on years of working experience.

KEYWORDS: higher vocational education; concepts of schooling; innovation.

Deepening education reform of higher vocational education requires to change teaching ideas and cultivate innovative modes. Despite the booming development of higher vocational education, there are some contradictions and problems, such as low innovation ability, unscientific majors offered in the college, unclear talent training goals, imperfect management system. Facing new challenges, new situations and new requirements, higher vocational colleges should persist in the reform and innovation and deepen the thought of schooling to promote healthy and sustainable development of the higher vocational education.

1 FUNCTIONS AND EXISTING PROBLEMS IN HIGHER VOCATIONAL EDUCATION

Higher vocational education in China plays a positive role in promoting talent training, family harmony, social and economic development. It can be manifested in the following aspects: firstly, having cultivated a large number of high quality talents with professional skills. Higher vocational education focuses the talent training on humanistic quality and sustainable development of humans. It not only pays attention to improve comprehensive ability to make the talent have special technical quality, also focuses on promoting professional ethics to develop the students' technical operation ability and innovation spirit and make them become responsible and thoughtful social talents; secondly, higher vocational education can make full use of the scientific research to retrain the employees. By means of a cooperative mode —"production, education, research", a large number of technical talents have flowed into society which not only reflects the characteristics of higher vocational education, but also promotes the development of social economy; thirdly, it has relieved the pressure of public education and promoted social harmony and family stability. As a member of the family, the embodiment of the student's value and rewards will affect the quality of life. Besides the harmony of the family is also conducive to social stability and development.

Higher vocational education is of great significance and value, but it also has some problems. First of all, the lagging management mechanism of higher vocational education is difficult to keep pace with its own development. Many managers have focused their attention on administrative work and neglected the innovation of theory and practice. Secondly, the curriculum system of higher vocational education is unreasonable. Basic knowledge and skills haven't formed a system, which is not helpful to educate people; thirdly, there is a loose link among different higher vocational colleges which pay attention to the competition and lack mutual understanding.

2 METHODS OF DEVELOPING INNOVATIVE CONCEPTS OF SCHOOLING IN HIGHER VOCATIONAL EDUCATION

The development of higher vocational education can not be ignored. How to solve the problems existing in running schools, how to improve the quality of teaching, how to enhance the development level of higher vocational education, and how to make higher vocational education better serve the society have become primary concerns in the educational circle. According to China's national education plan and years of working experience, this paper puts forward the following suggestions in order to develop innovative concepts of schooling in higher vocational education and promote its better development.

2.1 *To exploit features and have an innovative positioning in the major construction*

To improve the quality of major construction is not only the task and direction of developing higher education, but also is the foundation to achieve a powerful country. Higher vocational education should exploit major features and optimize talent training modes.

Firstly, the optimization of major structure. Major structure should be established according to the social development and the needs of the market. Therefore, higher vocational colleges should communicate and contact all kinds of enterprises and timely grasp the social trends and market demands. They should learn the current situation, change ideas, carry out research in various industries, and actively listen to senior professionals' opinions to truly understand the demand of the society and market and make the professional talent training goal specific. According to the needs of the market and the goals of cultivating professional ability, higher vocational colleges should carry out scientific and reasonable adjustments and reform on the course arrangement and innovate and optimize teaching modes to realize the unity of theory and practice and improve the teaching quality. According to the teaching characteristics of higher vocational education, research and discussions on construction planning of majors will promote specialty construction and realize a better development of majors.

Secondly, the timely adjustment of professional training plan. In order to better meet the needs of the students in terms of employment, job selection and starting a business, higher vocational colleges establish a course management mechanism which enables the students to choose their own direction of development, offer more practical elective courses, and increase minor and second majors in order to help students have better improvements in their professional field. Higher vocational colleges should pay attention to shaking off traditional ideas. According to their actual situations, they should divide majors and have grading teaching and improve credit conversion mechanism. The professional competence requirements need to combine with national standards and professional needs. The obtaining of professional qualification certificates will be closely connected with the credit system to improve the students' professional ability and enthusiasm for learning. Higher vocational colleges should also do a good job in employment guidance to provide career planning training and professional role education to students and let them have a better understanding of their majors and future employment situations. They should create internship training opportunities for students to increase their entrepreneurship awareness and ability. Besides they need to provide financial and technical supports, so that students can fully understand the tendency and demand and constantly enrich their professional skills which will be of great help for their employment.

Thirdly, the innovation of the talent training mode. Higher vocational colleges should improve the teaching of practical courses, and strengthen the monitoring and management. The training base needs to be fully utilized to enrich the students' practical activities, such as organizing professional skills competition, professional innovation and invention competition. A standard competition and incentive mechanism should be formed in the process of teaching and practice to enhance the students' level of practice. Higher vocational colleges can strengthen cooperation with enterprises. The cooperation between colleges and enterprises enables them to gain mutual benefit, that is, colleges provide talents and technology for the enterprise, and the enterprise provides funds and practice places for the college. It helps the students integrate theory with practice and enables them to constantly enhance their research ability. To experience the working atmosphere in practice makes it easier for the students to adapt to their future work and the social development.

2.2 *The innovative construction of the teaching team*

The teachers' quality and practical ability have an important influence on improving the students' learning and comprehensive ability. To develop innovative concepts of schooling in higher vocational education should emphasize the construction of innovative teaching team to improve the quality of teachers.

The construction of teachers should be strengthened, and the structure of teachers should be optimized. When recruiting teachers, schools should pay attention to teachers who have rich experience in business. For mid-career teachers, schools have to take different measures accordingly. Teachers having practical experience and ability should be encouraged to actively participate in professional qualification exam, obtain relevant professional qualification certificate, and enrich their theoretical level. For some young teachers, they can be organized to train in the production or management spot where they can gain knowledge and experience in practice. All teachers are encouraged to do scientific research in order to improve their professional ability and enrich the teaching content. Thus the optimization of the teaching team's overall and individual quality can realize the construction of teachers.

A new management mechanism needs to be introduced for part-time teachers. Although part-time teachers are lack of stability, they can also improve the quality of teaching. Therefore, the introduction of part-time teachers should emphasize the following points: first, when choosing part-time teachers, they need to have solid professional knowledge, rich experience and operational skills. Moreover, they also need to have a certain understanding of the local area and companies; second, part-time teachers and full-time teachers can exchange experience and help each other to form a good working mechanism and effectively improve the quality of teaching.

2.3 *The innovation of the management system in vocational education*

Developing innovative concepts of schooling in higher vocational colleges requires constant innovation of the management system in higher vocational education and to form a distinctive developing pattern.

The evaluation of professional titles needs to adopt a double track system. At present, the standards of evaluating the teachers' professional titles in higher vocational colleges are implemented in accordance with the standards of regular colleges, which is not scientific and reasonable for teachers in higher vocational colleges. Teachers in regular colleges are mainly engaged in theoretical teaching and research, and therefore the evaluation of their professional titles focuses on their theoretical innovation. While teachers in higher vocational colleges should not only pay attention to theoretical research, but more importantly to the study of skills. There is a certain gap between higher vocational colleges and regular colleges. Therefore, the former has to set practical standards of evaluating professional titles according to the teaching characteristics. The evaluation of skills should also be added in the standard of evaluating professional titles and a double track system needs to be adopted at the same time, which requires teachers to have teachers' titles and also have professional qualification certificates. So teachers' ability level will be better promoted and the educational management characteristics of the higher vocational colleges will be better presented.

Research center needs to be established. At present, higher vocational colleges in China have not formed unified scientific research institutions and there are less research results in the aspect of teaching management, academic research and technical practice. They also lack academic and skills exchange activities. So it is difficult to realize a long term development in the professional field. Mind is the guide of actions. Higher vocational colleges conduct relevant research which can better explore the source of vocational education, realize to develop innovative concepts of schooling, and broaden educators' horizon to improve their teaching ability. And the establishment of the center for education and research can promote the development of education, strengthen the colleges' overall strength, and make it possible to achieve undergraduate education, postgraduate education and doctoral education. To establish a specialized research center relies on teachers who have rich experience and theories. Their jobs mainly involve the following: firstly, they need to carry out localized research projects and put research results into practice; secondly, they need to release advanced research results both at home and abroad timely and broaden their horizons, so that educators can renew their teaching concepts and master the latest technology; thirdly, academic exchange and research need to be carried out among different schools to expand educational space; fourthly, scientific educational concepts need to be developed in order to promote the major construction, course construction and assessment system construction and improve the teaching quality.

3 SUMMARY

By means of constructing professional features, positioning innovation, improving the quality of teachers and teaching, creating an effective educational management system, and establishing teaching centered and human-oriented educational concepts, higher vocational colleges in China can better develop innovative concepts of schooling, realize healthy and harmonious development of higher vocational education and cultivate more professional talents for the society to provide better service for social development.

REFERENCES

[1] Cui Zhongliang. A study of the combination of learning with working in higher vocational education in China – Based on the concepts of schooling in NanYang Polytechnic [J]. Guangxi Education.2012.19 (05): 64–66.

[2] Li Jin. Philosophical thinking about the sustainable development of higher vocational education [J]. Chinese Higher Education Research.2010.02 (01): 6–10.

[3] Jiang Jun, Cui Junshan. Research on the concepts of schooling and practice mode in higher vocational education [J]. Liaoning Vocational College Journal.2012.06(06): 1–4,12.

Computing, Control, Information and Education Engineering – Liu, Sung & Yao (eds)
© 2015 Taylor & Francis Group, London, ISBN: 978-1-138-02800-5

Method of reliability computer simulation assessment for microelectronic device under multi-failure mechanism

B. Wan & G.C. Fu
Reliability and Systems Engineering School, Beihang University, Beijing, China

ABSTRACT: The reliability computer simulation assessment of microelectronic device is based on the PoF (Physics of Failure) model, which contains the parameters such as material, structure, technology and stress. The failure mechanism models and stress damage model are utilized to carry out stress analysis and failure time calculation. A great amount of TTF (Time To failure) samples under every failure point and every failure mechanism can be achieved by modeling and computing. Nevertheless, there is no assessment method for estimating the reliability parameters like failure rate and MTTF (Mean Time to Failures) of the whole microelectronic device. Aiming at this problem, the data processing method in reliability simulation assessment of microelectronic device, by which failure rate and MTTF can be obtained is presented, in which the step of failure distribution fitting and multi-point distributions fusing are given. A case for the method is also presented, which proved suitable for engineering application.

KEYWORD: Computer Simulation; Reliability; Physics of failure.

1 INTRODUCTION

Reliability simulation assessment of microelectronic device based on physic of failure (Pecht, M. A. et al. 2007) is different from traditional method of reliability analysis and assessment. Performance parameters such as material, structure, technology and stress are used to build the digital model of the product, and FMMEA (Failure Mode, Mechanism and Effect Analysis) is proceeded (Ganesan, S. 2004, Ganesan, S. et al. 2005), to obtain the PoF mode and all the corresponding potential failure points. Then the TTF(Time To failure) estimation of every failure mechanism based on PoF model and stress damage model is utilized to carry out for every unit of microelectronic device, which is got a great amount of TTF samples under every failure mechanism simulation. Then “the earliest failure time” is utilized to calculate the TTF of the microelectronic device. Nevertheless, in microelectronic devices application, we hope to get the reliability index such as failure rate and MTTF (Mean Time To failure) to provide basis for later overall unit reliability assessment and evaluation. So regarding TTF as the final result of reliability assessment can't meet the demand for microelectronic devices application development. A method to compute the reliability index such as failure rate and MTTF through single point failure distribution fitting and multipoint distribution fusion and other steps is established, and the implementation process of the algorithm is presented. A study case is presented by certain microelectronic device to prove that the method is suitable for engineering application. And this paper solved the practical problem that present reliability simulation for microelectronic devices application.

2 RELIABILITY MODE OF MICROELECTRONIC DEVICE

A microelectronic device generally consists of package, pin, die and bonding wire. Its function and performance mainly depend on the chip. All these basic parts are collectively known as units.

In reliability simulation, the reliability hierarchical model is as Figure 1. The difference from common reliability model is that the partition of units and failure mechanisms are more detailed. According to the theory of reliability simulation assessment, it is the basic mechanical, electronic, thermal, and chemical stresses that lead to the products' failure. So, under every unit, there are more one possible failure points caused by different failure mechanism.

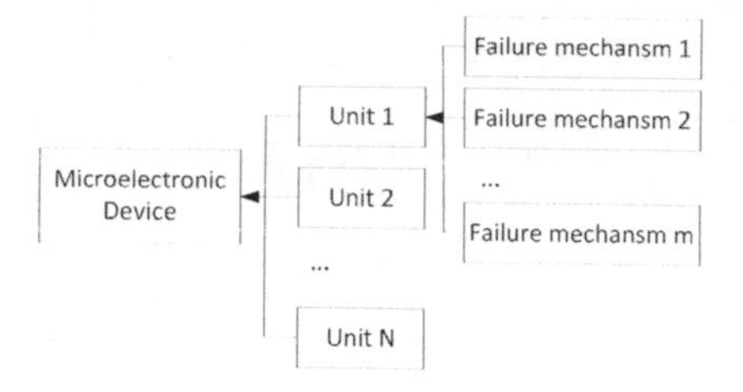

Figure 1. The reliability model of microelectronic device.

3 DATA PROCESSING

According to the reliability model, to calculate the reliability index, the failure data (failure time matrix) of every failure point is utilized and analysed in the order of every layers. And the process is somewhat different from the traditional reliability assessment which is from units to system.

3.1 *Failure distribution fitting*

According to the obtained failure time, $T=\{T_1,T_2,\ldots,T_n\}$(column vector of failure). Then calculate the single point failure mechanism distribution of every point preparing for the following Failure clustering and Multi-point distributions fusing. When doing failure distribution, we need to choose the best distribution type and calculate the distribution parameters by using goodness-of fit test. Common failure distribution types include Exponential distribution, Weibull distribution, Normal distribution, and Lognormal distribution.

3.2 *Multi-point distributions fusing*

The purpose is to obtain the failure distribution of every failure mechanism data. In fact, the failure of a unit may be the result of more than one failure mechanism simultaneously. In this paper, it is supposed that every failure mechanism is independent of each other, and the failure time of every unit can be got according to the competitive mechanism of the failure mechanism. Then, by means of Monte-Carlo simulation reliability assessment method from unit to system (Gong, Q. X. 2007), we can get the failure distribution and reliability index of the whole device.

4 MULTI-POINT DISTRIBUTIONS FUSING

The purpose is to get the failure distribution and reliability index of the microelectronic device under multi-failure mechanisms. The multi-point distribution fusing process is as Figure 2.

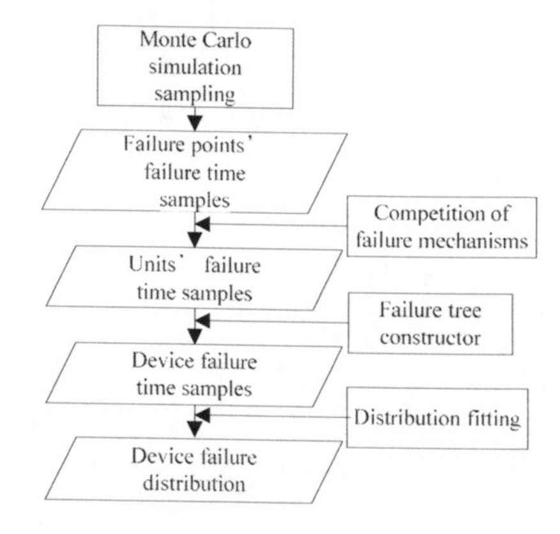

Figure 2. Multi-point distributions fusing.

Define the total number of times of simulation N, and the simulation sequence number *n*, $n=1,2,\ldots,N$.

For the jth failure point of the ith unit, the failure distribution $F_{i,j}(t)$ is obtained by single point failure distribution fitting.

In the nth simulation, the failure time of *jth* failure point of the *ith* unit is generated by Monte Carlo simulation randomly, $t_{i,j,n}=F_{i,j}^{-1}(\xi_n)$, where ξ_n is the random reliability of nth random selection. So the failure point sample values are $t_{i,1,n},t_{i,2,n},\ldots,t_{i,l_i,n}$, and l_i is the failure number of the ith failure point.

According to the competitive mechanism of failure mechanism, the earliest failure point failure time is the failure time of the ith unit.

$$TTFd_i=\min(t_{i,jn}),\ j=1,2,\cdots,l_i \tag{1}$$

Create the failure tree of every unit, and obtain the failure tree structure function $\Phi(\{t_k\})$

Utilizing the structure function, traverse the failure tree and find failure time of the whole device in the nth simulation.

$$TTm_n=\Phi(\{TTFd_i\}) \tag{2}$$

According to the sample $\{TTFm_n\}$, $n=1,2,\cdots,N$, obtain the whole device MTBF.

$$MTBF_m=\frac{1}{N}\sum_{n=1}^{N}TTFm_n \tag{3}$$

And then fitting with exponential distribution, and obtain the distribution parameters.

$$\lambda_m=\frac{1}{MTBF_m} \tag{4}$$

5 CASE

In this paper, we select a MOS device with plastic dual inline-pin (DIP) package which provide the system designer with direct implementation of the NOR function. We will depict the process of reliability evaluation by simulation in detail with this device.

The MOS device is divided into three general classes firstly: package, bonding wire and die, 3 units. From FMMEA, the potential failure mechanisms of the specific MOS device include: Solder Joint Thermal Fatigue(SJTF), Corrosion, Gate-oxide Time Dependent Dielectric Breakdown (TDDB), Electro–Migration(EM) and Hot Carrier Injection (HCI). The relibility hierarchical model is constructed as Figure 3.

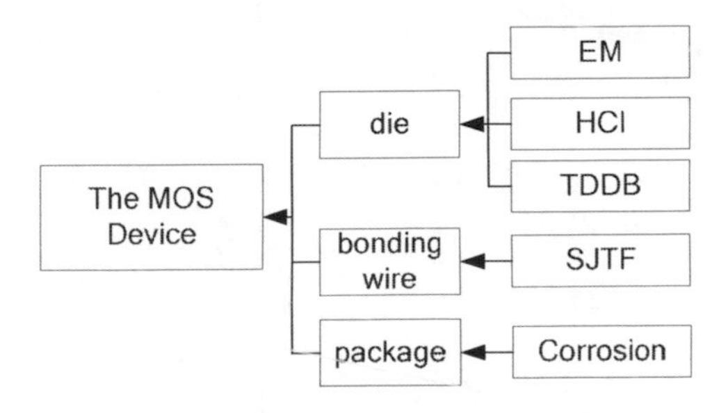

Figure 3. The reliability model of the MOS device.

Then according to PoF model and Monte-carlo sampling, vector of *TTF* samples are calculated, such as *TTF(hrs)* for EM failure mechanism.

$$TTF = \frac{WdT^{m}}{Cj^{n}}\exp(\frac{Ea}{kT}) = [3442317, 3448932, \cdots, 3568532] \quad (5)$$

Proceeded the Multi-point distributions fusing and Monte-carlo sampling to failure distributions for every failure mechanism, calculate the *TTF(hrs)* samples of every unit according to the competitive mechanism mentioned in section 4.

$$TTFd = \begin{bmatrix} 2348741, 2486393, \cdots, 2496321 \\ 2753612, 2893437, \cdots, 3087412 \\ 1934322, 1957329, \cdots, 2043226 \end{bmatrix} \quad (6)$$

Calculate the vector of TTF samples of the whole MOS devices using he competitive mechanism mentioned in section 4 and the result is as follows.

$$TTFm = [1323832, 1394423, \cdots, 1823392] \quad (7)$$

Finally, we obtained the failure probability density function curve as shown in Figure 4, failure rate and MTTF of the MOS device.

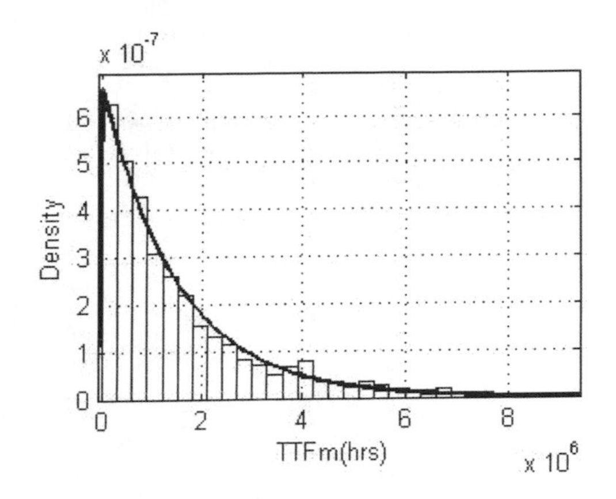

Figure 4. Failure density function curve of the device.

$$f(t) = \lambda e^{-\lambda t} = 6.6375891\times10^{-7}\times e^{-6.637589\times10^{-7}t} \quad (8)$$

Failure rate :

$$\lambda = 6.6375891\times10^{-7} \quad (9)$$

MTTF :

$$MTTF = \frac{1}{\lambda} = 1506,571 \text{ hrs} \quad (10)$$

6 SUMMARY

Reliability computer simulation analysis and assessment method has been expanded in the development of the microelectronic devices application, but still relies on the reliability assessment method. Relevant research and application aimed at reliability assessment is not enough. According to development requirement of microelectronic devices application, this paper comes up with methods of computing MTTF and failure rate based on reliability assessment simulation data, and provides reliability index assessment for a kind of MOS device as an example. This paper's method and process is innovative. It is significant that generalizing and applying the method in the field of microelectronic devices reliability analysis and electronic product reliability design.

REFERENCES

Pecht, M. A. Dasgupta, D. Barker, C. & Leonard, T. 2007. The Reliability Physics Approach to Failure Assessment Modeling. *Quality and Reliability Engineering International* 6(4): 267–273.

Ganesan, S. 2004. System Level Approach for Life Consumption Monitoring of Electronics. *Master of Science.*
Ganesan, S. Eveloy, V. D. D. & Pecht, M. 2005. Identification and Utilization of Failure Mechanisms to Enhance FMEA and FMECA. *Proceedings of the IEEE Workshop on Accelerated Stress Testing & Reliability (ASTR)*.
Gong, Q. X. 2007. *Model Reliability Engineering Handbook*. Beijing: Defense Industry Press.

Computing, Control, Information and Education Engineering – Liu, Sung & Yao (eds)
 ISBN: 978-1-138-02800-5

Inventory reduction in chain supplies finished products for multinational enterprises

Amelec Jesus Viloria Silva
University Sergio Arboleda, Bogotá, Colombia

ABSTRACT: In times of recession, there are fewer consumers are willing to pay for items that do not add value to the final product, such as high inventories, or unnecessary costs. This article aims to propose strategies for the reduction of inventories in the supply chain of multinational companies finished product. For this one of the largest producers of consumer goods in the world with presence in Europe, the Middle East and Africa was studied. Through analysis and business reporting interviews diagnosed and identifies high inventory levels and inventory policies inadequate as key factors that improve supply chain study sample. Also formulating and implementing redistribution processes and forecasting to combat Productive Inventory No impairments, as well as designing processes to be executed to deepen them. The end result is achieved reduce inventory by more than half of Currency Units (UM) million and, in addition to identifying future risks that could jeopardize service levels of the company.

KEYWORDS: supply chain product finished, inventory management, inventory not productivity.

1 INTRODUCTION

The inventory takes up space, extends the timeout production generates transport and storage needs, and absorbs financial assets. Materials, work in process and finished products that occupy space in the factory or warehouse non-value added, quite the contrary, they are worsening and may even become obsolete quickly. Therefore, one of the biggest concerns that companies have for their operations is the reduction of inventory. However, it often happens that when trying to reduce inventory other issues that directly impact the level of customer service as product availability are neglected. That is why high performers are working on maintaining a balance between reducing inventory and customer service provided by precision, communication and teamwork [1], [2].

Through analysis and business reporting interviews, this article aims to propose strategies for reducing inventory supply chain finished product (PUC) of multinational companies. For this sample is studied as one of the largest producers of consumer goods in the world whose supply chain extends throughout Europe, the Middle East and Africa (EMEA).

2 METHODOLOGY

The methodology in four (4) phases designed to achieve specific objectives in the project. Distribution channels selected in the company of study are shavers, razors and products for skin care. The phases are as follows [3]:

1 The supply chain and information systems that are used in the department of the category under study were diagnosed.
2 Activities are developed to identify factors and opportunities in the supply chain to improve key performance indicators (ICD).
3 The process of developing proposals for improvements that would meet the overall objective of this project starts.
4 The tools and processes to implement each of the improvement proposals are developed.

3 RESULTS AND DISCUSSION

Phase I: Diagnosis the PUC in EMEA

After the diagnosis of Gillette supply chains in EMEA, it is determined that all possess Strengths, Weaknesses, Opportunities and Threats very similar to each other. This is because shared by all processes, Markets, Key Performance Indicators, and even demand planning staff. Because of this a unique SWOT matrix, which is presented in Table 1 is developed.

Table 1. SWOT Matrix of the PUC in EMEA.

Strengths	Opportunities
Effective monitoring processes and inventory control have	Possibility of reduction policies bolder inventory.
Have remained products and Service Level	There is support from senior management to promote improvements.
There is close relationship between the Department of Planning Supply Category (DPCSC) and plants	The DPCSC is looking for new processes to reduce inventory.
Weaknesses	**Threats**
Currently there are budgetary constraints that limit contingency plans	There is the risk of major downsizing in case the global economic crisis is maintained.
High levels of inventory Low Income (UBI) lowering the performance of the PUC Gillette	Risk of matching employee vacation periods Plant, ODM and DPCSC.
Difficulty levels decrease INP and Excess	Occasionally failures have occurred in production due to quality problems of products.

Phase II: Identifying key factors to improve the PUCT in EMEA.

Based on the analysis in the previous stages the following factors that allow improved inventory levels and service [4] were identified:

Excess Inventory Reduction: improvement in this factor will improve IBU levels, as well as reduce overall inventory levels without risk of affecting the indicator Percent Complete Orders (POC).

Forecast Inactive Inventory: the ability to foresee the appearance of Inactive Inventory lets you take measures to eliminate them before they appear. This reduces INP indicators, IBU and reduce inventory in general.

Recalculation of inventory adjustments Safety and inventory policies: improvement of this factor would make better use of safety stock while improving service indicators as the mathematical model used properly comply with the parameters and supply chain requirements. In turn, allow process control and continuous monitoring to handle contingencies are reduced, since it would provide a safety stock (IS) that is mathematically designed to manage the volatility correctly and so eventually release staff DPCSC and plant to focus on other activities.

Setting Cyclical inventory levels for MDGs: the setting of this factor would reduce total inventory level and prevent the creation of large amounts of Idle Inventory improving indicators INP and IBU.

Phase III: Improvement proposals for the PUC in EMEA

This stage is divided into four (4) parts. Each part focuses on improving one of the factors identified in the previous phase.

Redistricting process excess:

The distribution process requires a computational tool to identify when a Stock Keeping Unit (SKU) is in excess in this lawsuit MDGs and other MDGs.

The process is based on the following steps:

1 SKU candidates for redistribution are determined: from the DPCSC computational tool is used to determine which SKUs could be distributed as there are levels of excess on one side and demand in the Central Supply Depot.
2 The MDGs is selected to redistribute: these MDGs must possess more than 67,000 UM in merchandise to be distributed and prioritized according to the value of redistribution. This is due to the parameters given by the Finance Department in charge of paying redistribution.
3 Planners production plants and planners demand in selected MDGs: the SKU candidates to be redistributed for approval is sent.
4 Quantities are determined to be redistributed: planners agree on the amount to be redistributed and Authorization Redistribution is created from the DPCSC to start running.

5 The ODM organizes the shipment of the products: after receiving authorization Redistribution, the MDGs should make the shipment of the products within ten (10) business days.
6 This process is repeated monthly.

Removal process inventory preventive orange:

This proposal seeks to specifically attack the Orange inventory Inactive Inventory.

For this proposed development is needed a computational tool to identify when new inventory orange to verify the date the demand for the product stops appears.

The process consists of the following:

1 The tool runs and SKUs are identified that represent potential risks of INP Orange.
2 The causes of risk identified planners demand ODM is verified.
3 The MDGs must formulate a plan for the removal of that future INP and present it to DPCSC a period not exceeding two (2) weeks.
4 The plan developed by the ODM and monitored weekly progress will be executed.

Reprogramming cyclic inventory:

This proposed improvement consists in reprogramming Cyclical inventory levels in information systems. The ultimate goal is that the maximum IC in ODM corresponds to the Days Prior To Next Replenishment of inventory, or what is the same, the frequency of shipments to the MDGs. However, the process must be gradual so as not to affect the ICD once.

The action plan would be:

1 Maximum parameter Inventory (IMax) is reduced by two (2) days of demand.
2 ICD, specifically the INP and the IBU are reset.
3 Once adjusted indicators in two (2) weeks, the MDGs must develop a plan for the elimination of INP that could generate parameter change.
4 It runs the plan developed
5 The process is repeated until the IMax parameter is equal to the send to the MDGs.

Recalculation and correction policy inventory:

This proposed improvement is recalculated inventory levels Security in the ODM and also proposes changing the managed inventory policy in the Central Supply Depot Reading (ACSR) so that it can maintain levels of IS to enable protect the warehouse market volatility. However, modification of the IS may involve significant risks in service levels. Because of that, the transition should be gradual and follow the following procedure:

1 Using a system capable of modeling a supply chain multi-step to establish appropriate levels of inventory in the Central Supply Depot.
2 Add the levels of IS in the Central Supply Depot without changing the levels in the MDGs.
3 After three (3) months, IS levels are reduced in ODM to a midpoint between the recommendations of the tool and current levels of IS.
4 Verified that in the next three (3) months are not risks in the service attributable to changes, and if this condition is met proceed to step 5.
5 IS levels in ODM are reduced to the recommendation of the tool SKU except where risks have been presented at the level of service.

Phase IV: Implementing improvement proposals

This process applies to Supply Chain Shavers and Paprika flakes as pilot phase. The first tests were run on the MDGs in Italy, Europe, South East, UK and France,

4 CONCLUSIONS

During the pilot testing Redistribution Process The following results were obtained:

Redistricting Process Excess: 341 pallets are achieved redistribute merchandise related to Supply Chain Shavers and Paprika flakes valued at 984,000 Currency Units (UM), while reducing the cost of transporting additional 23 pallets that had to be re-packaged. These results are achieved only in a period of 14 weeks.

Preventive Elimination Process Inventory Orange: 529.500UM preventing future Inactive Inventory is achieved over a period of eight weeks, preventing representing an increase of 16% over the boundaries of the target.

REFERENCES

[1] G. Pundoor. Supply Chain Simulation Models for Evaluating the Impact of Rescheduling Frequencies. Master Thesis. Institute for Systems Research. University of Maryland, 2002. pp. 23–41. 40.

[2] R. Teunter. E. van der Laan, D. Vlachos. "Inventory strategies for systems with fast remanufacturing". Journal of the Operational Research Society. Vol. 55. 2004. pp.475–484. 41.

[3] C. A. Soman. D. Pieter van Donk, G. Gaalman. Combined make-to-order make-to-stock in a food production system. International Journal of Production Economics. Vol. 90. 2004. pp. 223–235. 42.

[4] H. S. Abhyankar. S. Graves. "Creating an Inventory Hedge for Markov Modulated Poisson Demand: An Application and Model". Manufacturing and Service Operations Management. Vol. 3. 2001. pp. 306–320.

Computing, Control, Information and Education Engineering – Liu, Sung & Yao (eds)

ISBN: 978-1-138-02800-5

Study on risk of internet finance

Y. Shen
Guizhou University of Finance and Economics, China

ABSTRACT: The inherent development need of the internet industry and the financial sector, especially deep-seated need of financial reform, drives the rapid development of internet finance. Internet finance is convenient for the masses in daily life, takes tremendous impetus to country's economic activity, but also exacerbates the instability of financial markets. Internet finance addition to having the risk categories in traditional financial areas, but also brings a new kind of risk, such as the specific risk of laws and regulations, the model innovation risk, technology security risk, operational risk, etc.

KEYWORDS: Internet finance, risk, regulation.

1 INTRODUCTION

The modern information technology represented by the internet, especially the emergence of social network, search engine, mobile payment, cloud computing, network, big data and other information technology, will have a substantial impact on modern financial model [1]. Over the past 10 years, a similar disruptive effect has occurred in multiple industries: book, music, merchandise and retail [2]. The inherent development need of the internet industry and the financial sector, especially deep-seated need of financial reform, drive the rapid development of internet finance. Financial sector can expand the breadth and depth of the internet service functions, also help finance innovative products and services, expand with low-cost, make a useful supplement to the traditional financial system, and meet the growing financial needs of heterogeneous. In internet finance model framework, labor division and specialization of modern financial industry are bound to greatly weaken, and will be replaced by the internet and its related technologies; Financial market participants will be more civilianize, causing through the exchange in internet finance market not only enable investors to be benefit, also even ordinary consumers will be very profitable; Entrepreneurs, investors and the general public are all capable to take a wide range of financial transactions through the internet platform, and risk pricing, duration matching and other more complex financial transactions will be greatly simplified, easy to operate [3]. However, when the internet finance makes convenience to the masses daily lives, tremendous impetuses to national economic activity, also exacerbates the instability of financial markets. Internet finance has the same risk categories of traditional finance sector, such as market risk, operational risk, the risk of asymmetric information, reputational risk, policy risk, but these risks due to incorporate the internet finance characteristics, are different in evoking reasons, performance forms, the degree of harm and other aspects. In addition, the internet finance has also brought new kinds of risk, such as technical risk, the unique legal and institutional risk, information security risk, etc.

2 THE ANALYSIS OF RISKS FACED BY INTERNET FINANCE DEVELOPMENT

2.1 *Legal and institutional risk*

Legal risk is mainly reflected in a lack of internet finance legal documents. The regulation object of China's relevant financial laws and regulations is primarily aimed at the traditional financial sector. Because cannot cover many aspects of internet finance, neither cannot fit the unique characteristics of internet finance, there will be bound to cause some legal conflicts. Understand about professional legal document is blank, standardize of internet finance company's workplace, the legitimacy of the operation mode, the trader's identity authentication, electronic contracts and electronic signatures valid confirmation, etc., there are no clear detail legal norms[4]. Internet users in the process of through the internet to provide or enjoy financial services, they will face the risks of legal lack and legal conflict, are easy to fall into legal disputes, not only increase the transaction costs, but also affect the healthy development of the internet finance. In the management system, the supervisory authorities have not issued clear regulatory documents of internet finance. China's internet finance firms walk in the legal blind spots and regulatory loopholes, conduct illegal business, or even illegal deposit from the public, illegal fund-raising and other phenomena, accumulate a lot of risks.

2.2 *Model risk*

In recent years, domestic and foreign internet finance model emerges one after another, mostly because of does not fit with financial market environment and the specific needs of the customers, and then fails. When some domestic internet finance enterprises imitate foreign internet finance services model, due to the subjective or objective reasons, emerges distortion and alienation, cannot obtain the same commercial success as foreign enterprises [5]. Internet finance model risks mainly include model innovation risk and pattern distortion risk two categories. One is model innovation risk. Model innovation risk refers to some development models of original internet finance innovate overly or inadequate, in the reality of socio-economic status, due to the development bottleneck ultimately lead to the risk of failure. Second is the pattern distortion risk. Pattern distortion risk refers to the replica of internet finance model distorts, departs from the true meaning of internet and practical foundation, derives many variants, the security boundary of the original model is broken, the risks continue to accumulate and eventually lead to a crisis.

2.3 *Technical security risk*

Technology security risks mainly include the technology apply risk and the technology ability risk. First is the technology applications risk. Technology application risk refers to in the technology application stage, due to the virtualization of the internet finance services, blurring of the business boundaries, opening business environment, the technical security risks present characteristics of openness, diffusion, dynamic, etc.; or due to the design idea one-sidedness or compatibility inadequate cause internet finance technology systems existing congenital defects[6]. Computer viruses can rapidly spread and infect through the internet, poor key management and poor compatibility of certain internet finance institution's technology cause the risk outbreak. Second is the technical capability risk. Technical capability risk refers that due to internet finance platform's lack of technology, in some special moment need a timely response to sudden large-scale trading in time, then emerge the risk of adverse consequences[6].The risk is mainly present in the "Tanabata" "double eleven", "Christmas" and other traditional electricity providers discount promotion day. Due to massive online transactions in a day or even a certain point, the amount of data far exceeds the daily basis number; the system is prone to be unstable, server failure and other problems. Taobao, tmall, poly cost-effective as well as Jingdong, Dangdang, etc. are directly main electricity suppliers involving in the day's promotions, in the large promotional activities of the past few years appear pages crashes, single system cannot open, bank payment system congestion and so on in varying degrees.

2.4 *Data security risk*

Data security risks mainly include the data management risk, data transmission risk and data processing risk. First is the data management risk. With the explosive growth of data, massive data storage centralization facilitates data analysis, processing, but if improper safety management, can easily lead to information leakage, loss, damage. Second is the data transmission risk. In internet finance time, the high-speed flow of data between transaction parties and trading platform, brings convenience to finance economic activity, also brings data secure transmission risk. Including cloud computing systems, social media and high-performance mobile devices, a number of new technical capabilities provide convenient to enterprise personal financial data flow, but the financial security of data transmission is still unable to get adequate protection. The following are the main reasons. First, the internet finance data exchange dependents on FTP transfer services or database data sharing. Second, a large number of network protocol and cross-application software affect the speed of the file transfer process in a certain extent. Third, the amount of encryption products including data security encryption software, secure transmission software are insufficient. The big data analysis ability of China's internet finance institution is weak, the technology in database, data warehousing, data mining and cloud computing and other areas are lagging behind the international advanced level. There are few internet finance institutions equipped with advanced big data processing capabilities. Data processing focuses on data preparation and raw data search stage, lacks the data factory and data product, data mining and analysis are inadequate.

2.5 *Market risk*

Market risk is the risk inherent in the traditional financial system. As the combination of internet technology and financial sector, the internet finance market risk has its own unique side. First is the interest rate risk. Changes in interest rates will affect the pricing of financial products. Because of the convenience and preferential, internet finance can absorb more deposits, more loans, with more customers to trade, face greater interest rate risk. Second is the liquidity risk. Liquidity risk refers to the risk of internet finance institutions at some point they do not have sufficient amount of funds to meet the needs of customer withdrawals. Internet finance institutions often play a role in cash flow, sedimentation funds may remain in third-party intermediaries ranging from two days to a few weeks, due to a lack of effective security and monitoring, is likely to cause misappropriation of funds, if lacking liquidity management, once the capital chain ruptures, will trigger payments crisis. Third is the credit risk. Credit risk is the risk that the internet finance transactions, fulfill its obligations incompletely in the contract expiration date.

2.6 *Operational risk*

Operational risk refers to the possibility of potential losses caused by major defects in the system reliability, stability and security, mainly due to errors of staff[7].For example, the internet finance changes the traditional bank's settlement mean stamp as payment instruction, and uses digital signatures to confirm the validity of the payment instruction. Because the "virtualization "of the internet, the reliability of digital signature depends entirely on the control system rigorous of the internet finance platform. If trade peoples do not understand the operating specifications and requirements of internet financial services, may occur improper operation, even in the transaction process appear illicit, payment and settlement interruption and other issues, cause unnecessary financial losses. Because the virtual of internet finance services way, internet finance business activities break the network limitations of traditional financial services, with obvious geographical openness. During internet financial services, the transaction process operational errors may be passed quickly in all areas, accumulate risk to the entire internet finance system; even from a purely operational risk, evolve into a systemic risk, impact to the overall national internet finance normal operation and payment settlement.

2.7 *Reputational risk*

Reputation risk refers to the negative public opinion environment does damage to reputation of internet finance institutions, and then cause a serious loss of customers; result in the risk of adverse effects on their earnings and capital. Internet finance is established on the basis of the internet, highly virtualization and dissemination, once some errors happen, the damage suffered by internet finance institutions will be amplified exponentially in the internet effect. When appear reputational risk, lending platform will not be able to maintain a normal client resource, cannot continue to provide products and services to existing and potential customers, when serious even unable to engage in financial lending business. Reputational risk will bring litigation, financial loss, customer loss and other unfavorable situation of internet financial institution; easy to produce the same degree of influence on other unrelated organizations in the industry, and in extreme cases can lead to paralysis of the internet finance system. In China's internet finance companies, senior managers are generally weak in reputational risk awareness, creating a good public image consciousness is not strong, emphasis not enough, cause enterprise internal lack resources and mechanisms to match reputation risk management. Internal staff lack related knowledge about public relations management and the awareness and ability of reputational risk management, not properly deal with the customer's complaint, not timely treat the negative case reflected by news media, resulting in reputation damaging incidents continue to occur, eventually evolve into risk events, and even induce a wider range of risk occurrence.

2.8 *Policy risk*

Policy risk is the risk brought to internet finance due to fluctuations in the country's macroeconomic policy, industry policy, regulatory policy, tax policy and other policy environment. There is no clear norm about internet finance policy regulations and behavior definite; there is a big uncertainty in future policy efforts and regulatory depth. In China, internet finance is a new thing, it is sensitive to policy changes, and policy risk has a direct role in guiding other risks. Internet financial institutions are generally homogeneous, a policy fluctuation may cause various agencies select operating in the same direction, the "resonance effect" may impact internet finance system. Therefore, stable and lasting policy expectation is the key factor of internet finance institutions' successful operation and financial industry's orderly development.

2.9 *The risk of new crime*

Internet finance crime mainly has the characteristics: mean diversify; way intelligent, professional; highly interactivity, hidden; crime diversification, younger; low cost of crime, criminal cases involving internet finance crime are the following several types: One is to establish a false financial institution website, lure customers to login, require customers to fill in personal account, password, bank detail information, and then wait to carry out illegal activities; Second is to fake high visibility online shopping mall, cheat customer information, or in the specification online mall spread false information, entice customers to payment, but afterwards do not shipment accordance with the provisions; Third is to use Trojans, viruses and other computer hacker programs implant bank customers' computer, attack online information system, automatically obtain user's information to take customer's funds; Fourth is the criminal activities of directly using hacking techniques to attack internet sites; Fifth is to engage in money laundering.

3 SUMMARY

Since 2012, China's internet finance is developing rapidly, its innovations about ideas and technologies are emerging endlessly, and plays a positive role in the service for the real economy, promoting inclusive financial development, its industry scale has been expanded[8]. Internet finance fills the gaps and weaknesses of traditional financial services, but also exacerbates the instability in financial markets. Internet finance brings new risks in addition to the traditional financial risks. Through this article's analysis, we believe that the internet finance has brought at least nine risks, such as legal and institutional risk, model risk, technical security risk, data security risk, market risk, operational risk, reputational risk, policy risk and the risk of new crimes. At present, China's internet finance is still in the preliminary stage of exploration, but the development is in full swing. The rapid rise of internet finance brings new issues and opportunities for wealth management, but also brings new risks, proposes a deeper, higher standards and requirements for risk management.

REFERENCES

[1] Xie Ping, The opportunities and challenges of meeting internet finance model [N], 21st Century Economic Report, 2012-09-03.

[2] Xie Ping, Zou ChuanWei, Research on internet finance models [J], Finance Research, 2012(12):11–22.

[3] Chen Qian, The development trend of internet finance [J], Modern Business, 2014(09):132.

[4] Yan ZhenYu, Reflection on the current internet finance risks [J], Zhejiang Finance, 2013(12):40–42.

[5] Rui XiaoWu, Liu LieHong, The report of Chinese internet finance development [M], Beijing: Social Sciences Academic Press, 2014.

[6] Zhang Ming, Alert to the potential risks of internet finance sector [J], Economic Herald, 2013(10):10–12.

[7] Luo MingXiong, Internet finance [M], Beijing: China Financial and Economic Publishing House, 2013.

[8] Xu Shen, Research of China's new rural financial organizations development [D], Beijing: Central Party School PhD Thesis, 2012.

Computing, Control, Information and Education Engineering – Liu, Sung & Yao (eds)
 ISBN: 978-1-138-02800-5

Several methods of generating teachers' practical knowledge in vocational colleges

Cai Xia Wang
Information and Technology College, Tianjin University of Technology and Education, Tianjin, China

ABSTRACT: Practical knowledge was very important for professional development of teachers in vocational colleges. In vocational colleges, constitutes of teachers' practical knowledge included not only self-knowledge, knowledge of the subjects, students' knowledge, and knowledge of the scenario , but also vocational skills and teaching experience in the professional training of teaching scene. In accordance with the special law of vocational education and the principle of knowledge conversion, teachers' practical knowledge in vocational colleges was constructed by watching demonstration lessons, practicing in enterprise, writing educational narrative based on reflection and action research.

KEYWORDS: vocational colleges; teachers' practical knowledge;explicit kowledge;knowledge conversion.

1 INTRODUCTION

Many scholars believed that teachers' practical knowledge was an important part of teachers' knowledge. According to professor Yelan's opinion, the wealth of practical knowledge was important in the growth and development of teachers, and the profession of teachers relied on practical knowledge, which teacher used the comprehensive insight to expand the awareness of the problem and problem solving maturity. [1] This paper mainly discussed the structure of teachers 'practical knowledge in vocational colleges, then put forward the ways of constructing teachers' practical knowledge.

2 TEACHERS' KNOWLEDGE

About definition of teachers' knowledge, many scholars had given different definition, according to Manabu Sato's opinion, who was Japanese scholar, teachers' personal knowledge was practical knowledge, which depended on certain background (concepts, beliefs, and values), and was summed up from experiential knowledge in practice. Compared with conceptual knowledge which was used by other researchers, practical knowledge was multi-righteous and the lack of rigor and universality, but it was functional knowledge which had a certain performance. According to zhang lichang's opinion, teachers' personal knowledge was effectiveness knowledge which was summed up by teachers in specific teaching practice situational through their own experience, meditation, feelings and understands. Practical knowledge was obtained and updated in the self-deconstruction and constantly reconstructed by individual as subject, practice as a way and reflection as strategy. According to the extensive literature, many scholars at home and abroad thought that teachers' practical knowledge was an important part of teachers' knowledge.

3 THE STRUCTURE AND CHARACTERISTICS OF TEACHERS' PRACTICAL KNOWLEDGE

Teachers' practical knowledge was formed and refined by teachers reflecting on their own teaching experience, and teachers' practical knowledge was applied and practiced through teaching action. [2]Teachers'practical knowledge mainly included self-knowledge, knowledge of the subjects, students' knowledge, and knowledge of the scenario. [2]The four aspects of teachers' practical knowledge were influenced by the teachers' educational beliefs. Characteristics of teachers' practical knowledge included practicality, situationality and tacit, individuality and interactivity.

3.1 *Practicality*

Teachers' practical knowledge came from practicing, and pointing in practicing. The specific problems in the practice were a direct incentive for teachers to construct practical knowledge, and practice was the starting point and end point of the practical knowledge of teachers. Knowledge which teachers accumulated through the practice continued to be updated and deepened, in result it was turned into education philosophy of teacher.

3.2 *Situationality*

Practical knowledge was to explore the specific teaching problem in the particular context, such the specific time, place, and person. At the same time, practical knowledge provided cases and suggestion to the similar educational context.

3.3 *Tacit*

Teachers' practical knowledge was a special kind of tacit knowledge, which existed in the daily lives of teachers and teaching experience. Practical knowledge was formed for teachers in unaware of the way, which was difficult to use language and writing to express.

3.4 *Individuality*

Teacher was the main body of the practice, and their practical knowledge in specific teaching scenarios was affected by teachers' personal experiences, subjective beliefs, values and experience, so the cognitive results of teacher was not the same, and teacher formed practical knowledge of the personality.

4 TEACHERS' PRACTICAL KNOWLEDGE IN VOCATIONAL COLLEGES

The fundamental task of the vocational education was to train highly skilled talented person who serviced the needs of the society in production, construction, management, and met the needs of society as a fundamental objective, so it required students to have a strong practical ability. So, teachers of vocational colleges had not only theoretical knowledge, but also professional skill, who was called "Double-professionally-titled Teachers". They must have certain professional practice, and the specific performance were the practical ability and guiding practice teaching ability.So,in vocational colleges, constitutes of teachers' practical knowledge had self-knowledge, knowledge of the subjects, students' knowledge, and knowledge of the scenario , but also vocational skills and teaching experience in the professional training of teaching scène. Practical knowledge was the professional value for teachers in vocational colleges. The teachers should think that practical knowledge was the basis, and realize their own wisdom and strength, but also find a space of their own professional development, in order to promote the professional development of teachers.

5 CONSTRUCTION OF TEACHERS' PRACTICAL KNOWLEDGE IN VOCATIONAL COLLEGES

In vocational colleges, sources of teachers' practical knowledge mainly had three aspects:(1)conversion of the knowledge of educational theory and professional theory;(2)accumulation of teachers' long-term practice of classroom teaching experience, and(3) practice of teachers in specific training site or off-campus enterprise. Teachers' practical knowledge was automatically generated in a specific cultural context, which implicated in the words and deeds of the people and could be felt but unspeakable. The practical knowledge of teachers must be converted to let knowledge flow, share, communicate and store.Nonaka who was scholars in Japan proposed the model of knowledge conversion , On this basis, this paper presented the model of teachers' knowledge conversion shown in figure1.

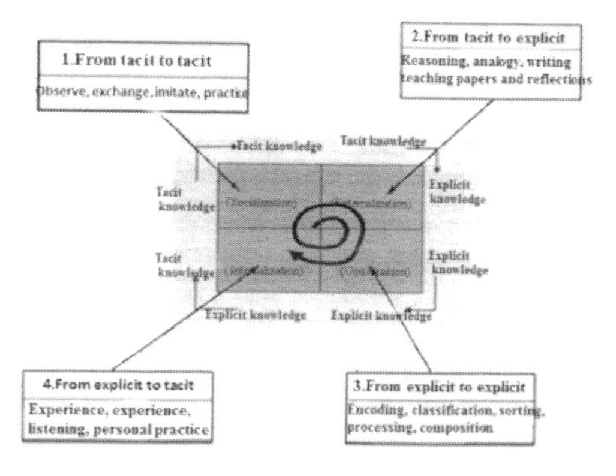

Figure 1. The model of teachers' knowledge conversion.

From above figure, teacher' practical knowledge was transformed through socialization internalization and externalization. Therefore, teachers' practical knowledge in vocational colleges was constructed by the following ways.

5.1 *Watching demonstration classes*

The demonstration classes as boutique which were formed by testing at every point and embodied teaching experience and wit of excellent teachers, were complexity of explicit knowledge and tacit knowledge of teachers. Watching demonstration lessons realized internalization of teachers' practical knowledge, knowledge sharing and exchange, and the flow of knowledge among teachers. Watching demonstration lessons, we could analyze the characteristics of teaching behavior, organizational forms of the classroom and teaching ideas, so as to absorb reasonable information and experience in teaching, and mine practical knowledge dependent on specific educational events. After this knowledge was integrated into the cognitive structure of teachers, they could be internalized into practical knowledge of individual teachers. Demonstration lessons were excellent off-the-shelf video teaching cases or teaching scene. Video teaching cases generally came from educational resource library in college or the Internet, which could be paused, reminded and repeatedly observed, in addition, it also came from observing the teaching scene, where you could feel the atmosphere in the classroom, learn teaching skills and expertise of the instructor. Training teaching was the most important form of teaching in vocational colleges. Through observing demonstration lessons, teachers could not only imitate the operation of professional skills, but also enhance the practical knowledge of guiding training teaching.

5.2 *Writing educational narrative based on reflection*

Teachers' practical knowledge was showed through specific teaching scene, with the characteristics of complicated, unstable, and fleeting. If the practical knowledge of teachers was not conscious to pay attention and intervene, a part of the experience would be lost because you forget to save, and a part of the experience in mind existed in the wrong form because of the lack of timely review and clarify. The other part of the experience which could not been coded and summarized, was not easy to be found at any time. Writing educational narrative on the base of reflection, practical knowledge could be externalization and preserved in long-term. When writing educational narrative, it was very important to fully embody the concept of reflection, which was not just a simple description for teaching classroom and stories in education, but "narrative and discussion", on the base of full original course of events presented, by inserting personal feelings, attitudes, motives and perspectives. In the teaching narrative "Committed, said, think, feel" should be embodied. The following has been noted.

1. Present form: by narrative and discussion, the content was long or short, according to the specific circumstances of the incident, which could be performed by text, pictures and video etc.
2. Content: the main object described was some of the real events occurring in teaching, and the content included the characters, plot, and problems of the event which reflected the theme of teaching and educational ideas.
3. Tools: there were many online tools which were available for teachers to write the narrative of education, such as the popular blog which was personal information release system, and teachers could apply for their own blog, and then write educational narrative.

5.3 *Practicing in enterprises*

Due to the ongoing changes in the vocational areas, and close contact between vocational education and the economy of society, teachers' professional knowledge and skills must evolve with the development of technology and the changes of corporate organizational model. Teachers in vocational colleges should pay close attention to the technical knowledge of professional frontier and state-of-the-art production equipment, in order to apply them in teaching. [3] Teachers in vocational colleges through the company practice and training , could enrich work experience in corporate , enhance vocational skills and practical teaching ability, so, company practice had become an important way of training "Double-professionally-titled" Teachers. Vocational colleges had actively explored and established the system of enterprise practice for teachers, and try hard to explore some ways of enterprise practice which were diversity, innovation, effectiveness. There were generally three forms for teachers to practice in company :(1) teachers were organized by vocational college to visit company during the holidays, in order to understand the production process and workshop process. (2) In-service teachers practiced in enterprises about one or two months every two years, and the new teachers practiced in the enterprise for a period of time. And (3) teachers and students went together enterprises to practice, in this way; we could reduce the burden for businesses, but also teachers and students to learn together, both teaching and learning.

The teachers in vocational colleges should cherish the opportunity of professional practice in the enterprises, during the practice, Theoretical knowledge of teachers was transformed into professional skills, and teacher could understand the social needs of profession. Through practice in enterprises, the theoretical knowledge of teachers would be deepened and teachers' practical knowledge would be promotion.

5.4 *Participating in subject study, and carrying out action research*

When participating in research team, teachers in vocational college could internalize explicit knowledge into their own wisdom, to quickly construct their own knowledge. By the project, teachers would build up connection between educational theory and educational practice, and update their knowledge. In the process of this, explicit knowledge would be internalized. Meanwhile, knowledge of teachers would be applied and innovated by curriculum development and innovative teaching. Under normal circumstances, the research team was composed of a number of specialized researchers, such as professors of university, experts in other educational institutions and first-line teachers. In the research community, the experts and professors had extensive knowledge about educational theory, and teachers had the practice knowledge in teaching. In the process of collaborative research, knowledge among team members continued to convert and upward spiral. In the process of dynamic transformation of knowledge, it enabled teachers to constantly enrich and improve individual's knowledge, and in the exchange and sharing, knowledge would be updated and created to improve professional standards of teachers. In the research community, knowledge conversion was shown in figure2.

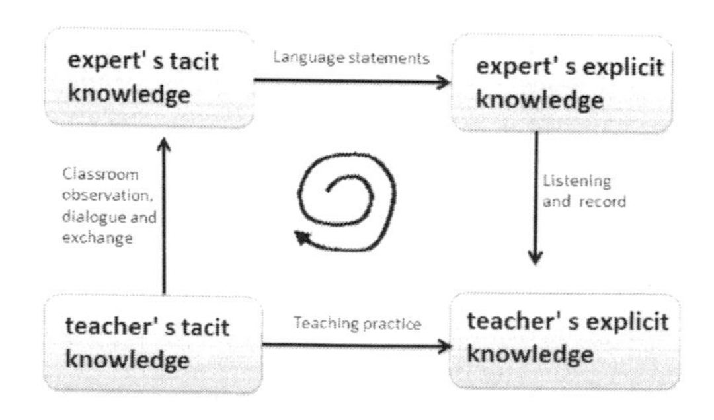

Figure 2. Process of knowledge conversion.

It could be seen from the figure, when the front-line teachers joined the research team of project and carried out action research, teacher could internalize explicit knowledge into their own practical tacit knowledge. As a result that knowledge would be applied and innovated to enhance comprehensive professional ability.

6 CONCLUSION

In accordance with the composition and characteristics of teachers' practical knowledge in vocational colleges, some ways were put forward about constructing practical knowledge. In addition, some systems should be built for the construction of teachers' practical knowledge.

Teachers should establish correct concept of education, focus on its own development, and use some network tools to facilitate the conversion and storage of practical knowledge.

ACKNOWLEDGMENT

This study was funded by "The Research on Methods of Generating and Getting for Teachers' Practical Knowledge in Vocational Colleges "which was the Young Subject of the Ministry of Education with National Plan of Education (project number: EJA140376). Thank you very much for this study to provide funds, and also thank the other members of the research group. In the course of the study, some literature was referred, thanking the authors.

REFERENCES

[1] Ye Lan, Exploration of the role of teacher and teacher development, Education and Science Press:Beijing,2001,pp. 215–216.
[2] Chen Xiangming, Build a bridge of practice and theory – research of teachers' practical knowledge , Education and Science Press:Beijing,2001,pp.77–78.
[3] Liuming, Vocational colleges - capacity building and management of teacher, China Science and Technology Press:Hefei,2012,pp16–17.
[4] Connelly, F. M, Clandinin, D. J. Teachers Personal Practical knowledge on the Professional Knowledge Landscape,Teaching aml Teacher Education, July,1997.
[5] F. Michael Connelly, D. Jean Clandinin. Teachers' Persanal Practical Knowledge on the Profes-sional Knowledge Landscape, Teaching and Teacher Education,July,1997.
[6] Nico Verloop, Jan Van Driel and Paulien Meijer,Teacher knowledge and the knowledge base of teaching, InternationalJournal of Educational Research,May, 2001.
[7] Polany Miehael, Personal Knowledge, Chieago: The University of Chieago Press, 1958.
[8] Sehon. DA. the Refraetional Praetioner: How Professionals think in Aetion. . NewYOrk.Basie book, 1983.
[9] ElbazE, The teaeher's Praetieal knowledge, RePort of case study.CurrieulumInquiry, November, 1981.

Computing, Control, Information and Education Engineering – Liu, Sung & Yao (eds)
© 2015 Taylor & Francis Group, London, ISBN: 978-1-138-02800-5

Optimization design of intermittent rotary mechanism of large rotor

Xin Yu Liu & Jian Gang Yi
Hubei Key Laboratory of Industrial Fume and Dust Pollution Control, Jianghan University, Wuhan, China
School of Electromechanical and Architectural Engineering, Jianghan University, Wuhan, China

ABSTRACT: Large rotor is the main part of rotating machinery. In order to provide convenient conditions for large rotor maintenance, a novel intermittent rotary mechanism based on grooved gear mechanism is proposed. The design scheme is provided and the whole grooved gear mechanism is designed. By using computer simulation, the stress is analyzed and calculated. The established grooved gear mechanism improves the reliability and the security of the large rotor maintenance system.

KEYWORDS: Optimization design; Intermittent rotary mechanism; Computer simulation; Finite element analysis.

1 INTRODUCTION

Large rotor is likely to exist some faults in application, which need maintenance with intermittent rotary mechanism [1, 2]. The work method of the intermittent rotary is converting mechanical power from high speed, low torque to high torque, low speed, in which motor provides power and is reduced by reducer. Generally, the force is transferred to the runner to achieve low rotate speed with tire, which has good elasticity and strength. However, when starting or stopping such mechanism, it exists dangers because of the failure of elastic component [3]. Therefore, it is necessary to design a novel intermittent rotary mechanism for large rotor maintenance.

2 DESIGN SCHEME

In order to realize the intermittent rotary motion, the groove wheel mechanism is used to improve traditional rotor maintenance device. As shown in Figure 1, a set of groove wheel is added to the reducer, so that the continuous rotary motion is converted to the intermittent rotary motion and the device has enough time to adjust at starting and stopping steps.

From Figure 1, it can be seen that the set of mate gears in the two stage reducer is instead of groove gear mechanism. The movable groove wheel is coaxial with the small gear, and the power source is connected in the active turntable. Groove wheel mechanism includes outer meshing grooved wheel, inner meshing grooved wheel and spherical grooved wheel [4]. The grooved sheave and the rotating arm turn in the opposite direction in the outer groove wheel. On the contrary, the grooved sheave and the rotating arm turn in the same direction in inner groove wheel [5, 6]. The spherical grooved wheel can realize intermittent transmission between two intersecting axis. The typical structure of groove gear mechanism is shown in Figure 2, which consists of an active turntable, a driven grooved gear and a frame.

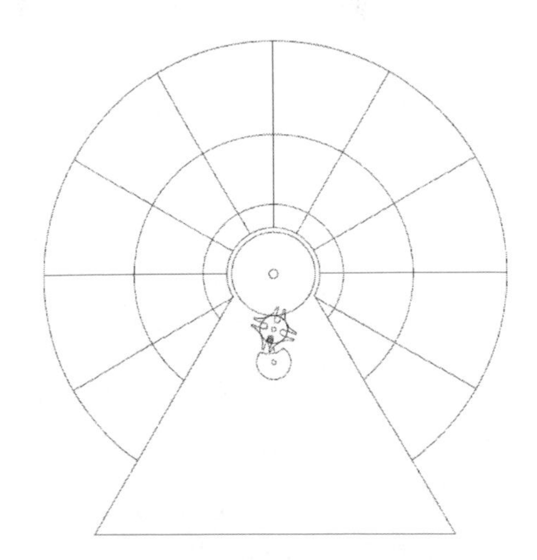

Figure 1. Intermittent rotary motion mechanism.

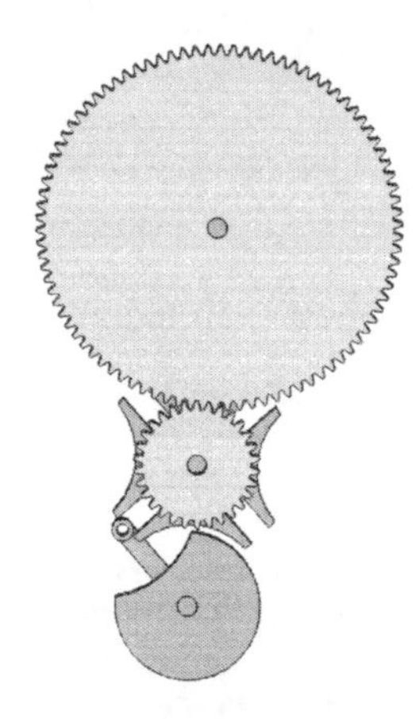

Figure 2. Groove gear mechanism.

3 STRUCTURE DESIGN

According to the working parameters, it is assumed the rotor diameter is 100m and the angular velocity of rotation is 1o/s. In this case, the grooved gear mechanism is designed and calculated as follows.

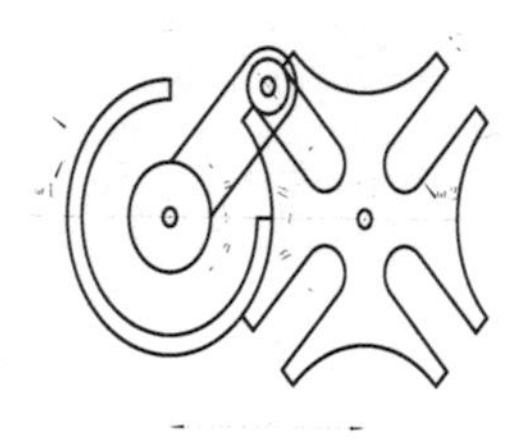

1- active turntable, 2- driven grooved gear, 3-frame
Figure 3. Diagram of mechanism.

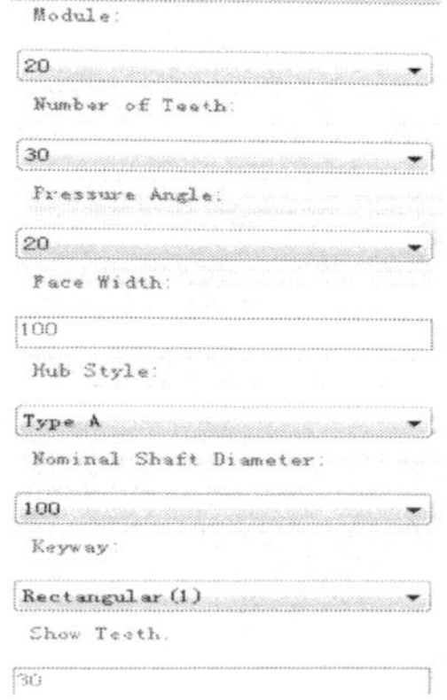

Properties
Module:
20
Number of Teeth:
30
Pressure Angle:
20
Face Width:
100
Hub Style:
Type A
Nominal Shaft Diameter:
100
Keyway:
Rectangular (1)
Show Teeth:
30

Figure 4. The simulated interface.

From Figure 3, the number of grooves is 4. Considering the friction wheel size, the center distance L between the grooved wheel and the gear is 4220mm. Define α is the driving wheel movement angle and β the driven wheel movement angle, and then the driving crank length can be calculated as 5105mm. According to the related formulas of mechanics theory, the teeth number is 90 and the pressure angle is 20 in the driven gear, while the teeth number is 30 and the pressure angle is 20 in the driving gear. The transmission ratio is 3. The computer simulated interface and the simulated structure of the driving gear are shown in Figure 4 and Figure 5.

Figure 5. The structure of the driving gear.

4 FINITE ELEMENT ANALYSIS

The basic idea of the finite element method is considered a continuous elastic body as an equivalent assembly composed of a finite number of units, which are connected by a finite number of nodes called grid unit. Because there are no restrictions on the use of unit approximation function of the global field function equation, and there are no limitations that the corresponding equation of each unit be the same, the finite element method can be used to solve all kinds of problems in physics.

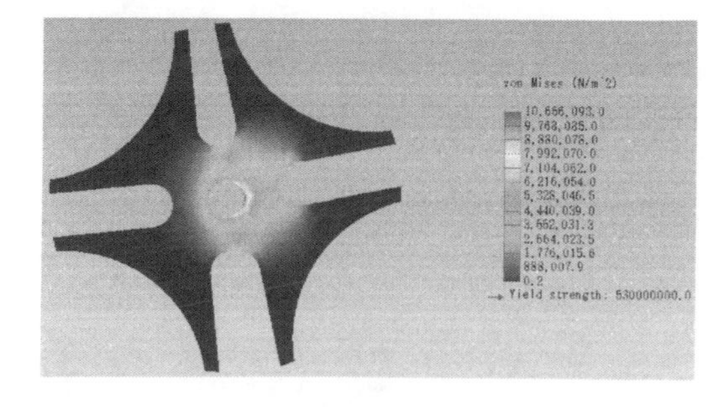

Figure 6. The simulation result.

With the 3D design software SOLIDWORKS, the model of the proposed grooved gear mechanism is built. The procedure is as follows. Firstly, the part drawings of the components are finished, including the groove wheel, the shaft, the motor, the reducer, the base, etc. Secondly, the 3D model is established through rotation, stretching and resection. Finally, the components are assembled to become the whole mechanism. With the computer simulation, the analysis result by using the Slidworks CosmosXpress toolbox in SOLIDWORKS is shown in Figure 6. The analysis shows the maximum stress of the entire device is in the cylindrical pin and slot wheel contact points, which is 530MPa. The material is chrome plated stainless steel. Therefore, the maximum stress is less than the allowable stress.

5 CONCLUSIONS

The main advantage of the improved grooved gear mechanism is that the failure rate is reduced and the large rotor device is more safe and reliable. Moreover, compared with traditional structure, the designed intermittent rotary mechanism has a more simple structure and easy for assembly, which provides convenient conditions for the whole equipment repairment. The designed intermittent rotary mechanism can not only be used in the large rotor device, but also be applied in other occasion where intermittent rotary motion is required.

ACKNOWLEDGEMENTS

This work is supported by the Industry-university-research project of the Wuhan Education Bureau (Granted No: cxy02) and the open project in Hubei Key Laboratory of Industrial Fume & Dust Pollution Control (Granted No: HBIK2014-05), China. The authors also gratefully acknowledge the helpful comments and suggestions of the reviewers, which have improved the presentation.

REFERENCES

[1] Khlaief Amor, Boussak Mohamed, Gossa Moncef. 2013. Model reference adaptive system based adaptive speed estimation for sensorless vector control with initial rotor position estimation for interior permanent magnet synchronous motor drive. *Electric Power Components and Systems*, 41(1): 47–74

[2] Yi Jiangang. 2015. Modelling and Analysis of Step Response Test for Hydraulic Automatic Gauge Control. *Strojniški vestnik - Journal of Mechanical Engineering*, 61(2): 115–122

[3] Ulrika Nyström K G. 2009. Numerical studies of the combined effects of blast and fragment loading. *International Journal of Impact Engineering*, 36(8): 995–1005

[4] Zhou Linren, Ou Jinping. 2010. Study of parameters selection in finite element model updat-ing based on parameter correction. *ICIC Express Letters*, 4(5): 1831–1837

[5] Key Nicole L. 2014. Compressor vane clocking effects on embedded rotor performance. *Journal of Propulsion and Power*, 30(1): 246–248,

[6] Dalili N., Edrisy A., Carriveau R. 2009. A review of surface engineering issues critical to wind turbine performance. *Renewable and Sustainable Energy Reviews*, 13(2):428–438.

Computing, Control, Information and Education Engineering – Liu, Sung & Yao (eds)
© 2015 Taylor & Francis Group, London, ISBN; 978-1-138-02800-5

Optimization approach of faults samples for heavy automobile engine

Mei Ye
Hubei Key Laboratory of Industrial Fume and Dust Pollution Control, Jianghan University, Wuhan, China
Wuhan Dongxihu Vocational-Technical School, Wuhan, China

Jian Gang Yi
Hubei Key Laboratory of Industrial Fume and Dust Pollution Control, Jianghan University, Wuhan, China

ABSTRACT: It is verified that neural networks can be used in engine fault diagnosis. However, the diagnosis effect depends on the samples, especially for heavy automobile engine. Aiming at the determination of the best hidden layer neural number and the best training function, many maintenance samples are summed up. The sample data tables are established. Based on it, different training functions are compared and the BP neural network structure is optimized, which is greatly useful for improving diagnosis accuracy rate.

KEYWORD: Fault Sample; Optimization Approach; Heavy Automobile Engine; Neural Network.

1 INTRODUCTION

Nowadays, the faults diagnosis approach based on artificial neural networks has achieved great progress [1, 2]. However, in the maintenance work, it is found the accuracy rate of faults diagnosis of heavy automobile engine is low [3, 4]. To eliminate the work faults, it is necessary for technical workers to check and compare faults symptoms carefully. Therefore, the diagnosis effect greatly depends on the experience and professional knowledge of technical workers.

Figure 1. Heavy automobiles.

In many factories, heavy automobiles are commonly used for products transportation. Because of the bad industrial environment such as heavy load, high temperature and high concentrate dust, the heavy automobile engine is more likely to be damaged, as shown in Figure 1. To solve this problem, the optimization approach based on artificial neural networks is proposed in this paper. With the common fault symptoms of heavy automobile engine as input samples and the relative causes as output samples, the BP artificial neural network model in the MATLAB toolbox is used to establish the optimized faults diagnosis frame, which lays a foundation of diagnosis expert system for more accurate fault diagnosis of heavy automobile engine.

2 FAULT SAMPLES ANALYSIS

The common faults of heavy automobile include cannot start, difficult starting, idling instability etc. Assume some faults correspond with some causes, the relative code is 1, or the code is 0. From many maintenance cases, 13 faults and their symptoms are summed up, in which every symptom is related with 21 faults causes.

1 Engine is unable to start. Causes: the battery is consumed; starter does not work; the gasoline pump does not work; idle speed control valve failure; the ignition timing is wrong; lubrication system failure; air filter clogging; air flow meter failure; pressure sensor fault; ECU fault;
2 Engine starting difficulty. Causes: fuel pressure is insufficient; the idle speed control valve failure; lubrication system failure; blocked air filter;

air flow meter fault; pressure sensor fault; cooling system failure; engine internal sinter;

3 Engine idling instability. Causes: idle speed control valve fault; intake system leakage fault; lubrication system failure; the air filter is clogged; the throttle fail, air flow meter fault; pressure sensor fault; ECU fault;

4 Bad acceleration. Causes: spark plug aging; idle speed control valve failure; solar term door fault; wrong ignition timing; intake system leakage fault; air flow meter fault; pressure sensor fault; ECU fault;

5 Lack of motivation. Causes: insufficient fuel pressure; three element catalytic plug; wrong ignition timing; air filter stuck; air flow meter fault; pressure sensor failure; cooling system fault;

6 Bad speed reduction. Causes: air inlet system fault; idle speed control valve failure; ignition system fault; lubricating system fault; cooling system fault;

7 High oil consumption. Causes: wrong ignition timing; solar term door fault;

8 Abnormal ignition. Causes: spark plug aging; ignition coil failure; idle speed control valve failure; the intake system leakage; air flow meter fault; pressure sensor fault, ECU fault, cooling system fault;

9 Tempering in intake pipe. Causes: ignition timing wrong; lubrication system failure; the air filter is clogged; cooling system failure; the fuel supply system fault;

10 Exhaust pipe blasting. Causes: idle speed control valve failure; ignition system failure: ignition timing wrong; lack of power; air flow meter fault; pressure sensor fault; ECU fault;

11 Engine shaking. Causes: air intake system leakage; air filter jam; solar term door fault; air flow meter fault; pressure sensor fault; ECU fault; internal sinter;

12 Sometimes stalling. Causes: Ignition fault; lubrication system fault; air flow meter fault; pressure sensor failure; cooling system fault;

13 Intermittent misfire. Causes: intake air system leakage; solar term door fault; air flow meter fault; pressure sensor fault; ECU fault; fuel supply system fault;

3 FAULT SAMPLES ESTABLISHMENT

Assume F is learning sample function, then F=(A, Y), A=(a1,a2, a3 ...... a13); Y=(y1,y2, y3 ...... y21), where A is input samples, Y is output goal function, ai is fault symptoms. ai∈A (i=1,2,3......13), yi is output value of output layer, yi∈Y(i=1,2,3......21). When ai=0, it means no relative fault symptom. When ai=1, it means the relative fault exists.

When Y={1,0,0,0,0,0,0,0,0,0,0,0,0,0,0,0,0,0,0,0,0}, it indicates the battery is consumed. The input and output samples is shown in Table 1, where x1 means engine is unable to start, x2 means engine starting difficulty, x3 means Engine idling instability, x4 means bad acceleration, x5 means lack of motivation, x6 means bad speed reduction, x7 means high oil consumption, x8 means abnormal ignition, x9 means tempering in intake pipe, x10 means exhaust pipe blasting, x11 means engine shaking, x12 means sometimes stalling, x13 means intermittent misfire. Corresponding to the 13 fault symptoms, 21 typical fault causes are selected as output values, in which y1 means the battery is consumed, y2 means starter does not work, y3 means the gasoline pump does not work, y4 means fuel pressure is insufficient, y5 means spark plug is aging, y6 means solar term door fault, y7 means three element catalytic plug, y8 means intake system fault, y9 means Ignition coil failure, y10 means idle speed control valve failure, y11 means the ignition system fault, y12 means the ignition timing is wrong, y13 means intake air system leakage, y14 means lubrication system failure, y15 means air filter clogging, y16 means air flow meter failure or pressure sensor fault, y18 means ECU fault, y19 means cooling system fault, y20 means fuel supply system fault, y21 means internal sinter.

Table 1. The input and output values.

Input signals	Output signals
x1~ x13	y1 ~ y21
1 0 0 0 0 0 0 0 0 0 0 0 0	1 1 1 0 0 0 0 0 0 1 0 1 0 1 1 0 1 1 0 0 0
0 1 0 0 0 0 0 0 0 0 0 0 0	0 0 0 1 0 0 0 0 0 1 0 0 0 1 1 0 1 0 1 0 1
0 0 1 0 0 0 0 0 0 0 0 0 0	0 0 0 0 0 0 0 0 0 1 0 0 1 1 1 1 1 1 0 0 0
0 0 0 1 0 0 0 0 0 0 0 0 0	0 0 0 0 1 1 0 0 0 1 0 1 1 0 0 0 1 1 0 0 0
0 0 0 0 1 0 0 0 0 0 0 0 0	0 0 0 1 0 0 1 0 0 0 0 1 0 0 1 0 1 0 1 0 0
0 0 0 0 0 1 0 0 0 0 0 0 0	0 0 0 0 0 0 0 1 0 1 1 0 0 1 0 0 0 0 1 0 0
0 0 0 0 0 0 1 0 0 0 0 0 0	0 0 0 0 0 0 0 0 0 0 0 1 0 0 0 1 1 0 0 1 0
0 0 0 0 0 0 0 1 0 0 0 0 0	0 0 0 0 1 0 0 0 1 0 1 0 1 0 0 0 1 1 1 0 0
0 0 0 0 0 0 0 01 0 0 0 0	0 0 0 0 0 0 0 0 0 0 0 1 0 1 1 0 0 0 1 1 0
0 0 0 0 0 0 0 0 0 1 0 0 0	0 0 0 0 0 0 0 0 0 1 1 1 0 1 0 0 1 1 0 0 0
0 0 0 0 0 0 0 0 0 0 1 0 0	0 0 0 0 0 0 0 0 0 0 0 0 1 0 1 1 1 1 0 0 1
0 0 0 0 0 0 0 0 0 0 0 1 0	0 0 0 0 0 0 0 0 0 0 1 0 0 1 0 0 1 0 1 0 0
0 0 0 0 0 0 0 0 0 0 0 0 1	0 0 0 0 0 0 0 0 0 0 0 0 1 0 0 1 1 1 0 1 0

4 MODELING OF BP NEURAL NETWORK

The input nodes of BP neural network of heavy automobile indicate the fault symptoms, and the output nodes indicate the fault causes. The fault samples are used to train the BP neural networks to determine the network structure. The transfer function and the number of neurons can be selected in MATLAB neural network toolbox. After training, the faults classification is the nonlinear mapping process based on the given set of fault symptoms to the fault set [5]. The BP neural network structure contains three layers. The input layer nodes number is 12, which means 12 fault symptoms. The output nodes number is 21, which means 21 faults causes. The hidden layer nodes number is selected from 12 to 17 to determine the best structure. The learning rate is chose as 0. 05, and the model is shown in Figure 2. The Sigmoid transfer function is used in the hidden layer and the output layer, and the traingda and traingdx fuctions are used to train the model. The training epoch is selected as 500 and the goal MSE is 0.001. The training results and the working parameters are recorded after every 100 epochs, as shown in Table 2 and Table 3. According to the working parameters, the best hidden layer nodes number and the best training function can be selected.

Table 2. The running results with traingda function.

Nodes	Epoch	Time	MSE	Gradient	Learning Rate	R
11	456	2s	0.0159	9.71×10^{-6}	12726.0453	0.996119
12	454	2s	2.72×10^{-5}	8.74×10^{-6}	20349.3662	0.99998
13	482	2s	0.0159	9.87×10^{-6}	19445.5526	0.9612
14	578	2s	0.00792	9.86×10^{-6}	6967.9165	0.98068
15	553	2s	0.023841	9.82×10^{-6}	8039.1871	0.94145
16	514	2s	0.0159	9.89×10^{-6}	11620.6993	0.96121
17	420	192s	0.007975	9.83×10^{-6}	18676.2834	0.98056

Table 3. The running results with traingdx function.

Nodes	Epoch	Time	MSE	Gradient	Learning Rate	R
11	334	4s	0.0333	5.19×10^{-6}	2766.8804	0.92765
12	264	3s	0.0421	6.68×10^{-6}	3076.2984	0.89256
13	482	3s	0.0645	8×10^{-6}	7465.6886	0.83586
14	300	3s	0.0464	9.74×10^{-6}	3351.8833	0.88095
15	300	2s	0.0464	4.69×10^{-6}	2586.8159	0.88778
16	282	129s	0.0245	4.07×10^{-6}	1989.6811	0.93984
17	389	7s	0.0145	7.67×10^{-6}	8686.3743	0.96444

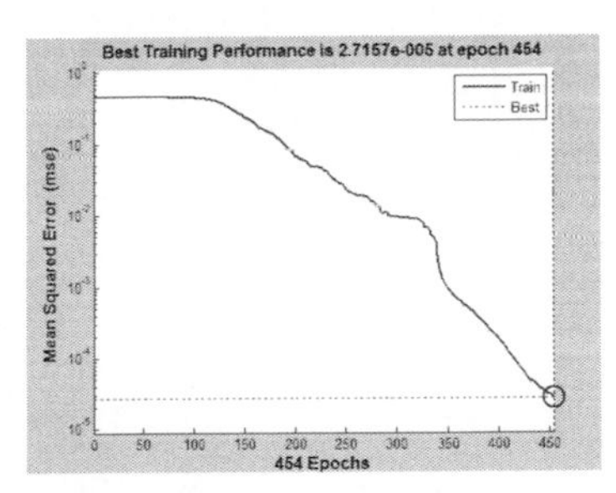

Figure 3. MSE curve.

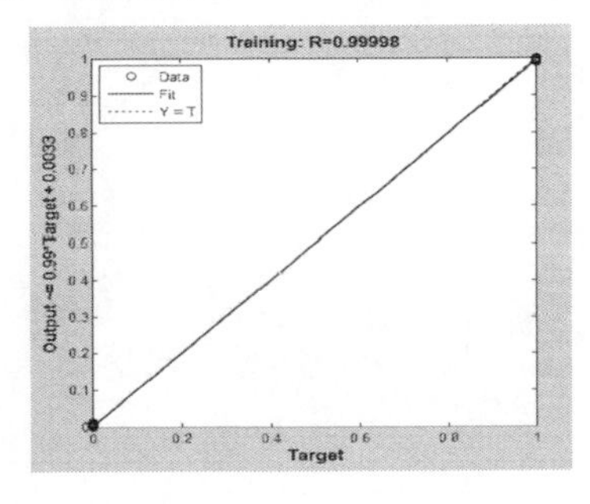

Figure 4. Correlation curve.

5 CONCLUSIONS

In terms of the comparing approach mentioned above, the optimized BP neural network model can be determined, which has high fault diagnosis accuracy rate. The shortage is that the general computational model cannot be established because different sample structure means different model structure [6]. However, if the neural network model is determined, the diagnosis software with human-computer interaction can be developed with computer platform, which is greatly useful for technician to maintain heavy automobile engine without much experience and professional knowledge.

ACKNOWLEDGEMENTS

This work is supported by the Industry-university-research project of Wuhan Education Bureau (Granted No: cxy02) and the open project of Hubei Key Laboratory of Industrial Fume & Dust Pollution Control (Granted No: HBIK2014-05), China. The authors also gratefully acknowledge the helpful comments and suggestions of the reviewers, which have improved the presentation.

REFERENCES

[1] Fu Zhouyu, Robles-Kelly Antonio, Zhou Jun. 2010. Mixing linear SVMs for nonlinear classification. *IEEE Transactions on Neural Networks*, 21(12):1963–1975.

[2] Khlaief Amor, Boussak Mohamed, Gossa Mon-cef. 2013. Model reference adaptive system based adaptive speed estimation for sensorless vector control with initial rotor position estimation for interior permanent magnet synchronous motor drive. *Electric Power Components and Systems*, 41(1): 47–74.

[3] Orkisz M., Ottewill J. 2012. Detecting mechanical problems by examining variable speed drive signals. *2012t International Symposium on Power Electronics, Electrical Drives, Automation and Motion:* 1366–1371.

[4] Neethu S., Shinoy K.S., Shajilal A.S. 2011. FEA-aided design, optimization and development of an axial flux motor for implantable ventricular assist device. *International Journal of Electrical & Electronics Engineering*, 5(1): 58–64.

[5] Yi Jiangang. 2015. Modelling and Analysis of Step Response Test for Hydraulic Automatic Gauge Control. *Strojniški vestnik - Journal of Mechanical Engineering*, 61(2): 115–122.

[6] Key Nicole L. 2014. Compressor vane clocking effects on embedded rotor performance. *Journal of Propulsion and Power*, 30(1): 246–248.

Computing, Control, Information and Education Engineering – Liu, Sung & Yao (eds)
 ISBN: 978-1-138-02800-5

Students' information literacy in the business colleges for AACSB international accreditation

Xia Xu
Personnel Division Business College, Beijing Union University Beijing, China

Jing Hua Bi
Office of Academic Affairs, Beijing Union University Beijing, China

ABSTRACT: The study is mainly about students' information literacy at Business College of Beijing Union University for AACSB international accreditation. By tracking and analyzing students' on-line behaviors at BCBUU for one semester, the characteristics of the students' on-line behaviors at business colleges have been obtained, which can typically reflect the contemporary college students' information literacy in study, living, entertainment, consumption, etc.. These data and behavioral characteristics provide a realistic basis for college teachers to educate students in many aspects.

KEYWORDS: business; college students; information literacy; AACSB international accreditation.

1 INTRODUCTION

On-line behavior, also known as network behavior from a psychological point of view, is defined as, in the cyberspace, individuals show interactive or non-interactive behavior in order to satisfy a certain need or obtain some experience with text (multimedia) as an intermediary [1].

On-line behavior as an integral part of human behavior, like other human behavior, is the purposeful activity generated by the stimulation of external factors via the refraction of internal experience, and closely related to human emotion, consciousness, cognition and environment.

John Suler explained the on-line behaviors from the perspective of Need Satisfaction and he emphasized the influence of potential demand on individual on-line behavior. He believed some functions of cyberspace can meet the potential needs of the individual and different on-line behaviors among individuals are the refraction of these potential needs. Therefore, different motivations and needs will inspire individuals to use different network functions and show different network behavior preferences [2].

Individuals' different online behaviors are caused by a variety of motivations and incentives. Here, we take the students from Business College of Beijing Union University as a research object, concern about the group's motives under the interaction of a variety of factors, relatively stable attitudes and behavioral tendencies manifested by using internet as an intermediary.

The group we study is a miniature of contemporary college students whose majors are financial management. Through the study, we can further understand how they use internet in their daily life and study, provide the basis for the further development of the day-to-day education and on-line behavior control, and guide them to form the concepts and behaviors of the active use of network.

Our approach was: SINFOR's internet behavior management device accessed BCBUU's core switch to collect the group data and we analyzed the data collected. This method can ensure a true and objective collection of student on-line data, and avoid such drawbacks as unreliable previous questionnaire, online surveys and other forms of data, low recovery rate.

2 STUDENTS' ON-LINE BEHAVIOR DATA

Students' on-line behavior data records in one semester were collected at the following three stages, namely at the beginning of the semester, in the middle of the semester, and at the end of the semester. We would analyze the characteristics of their online behaviors from the following several angles.

2.1 *The types of on-line applications*

After collecting the statistics of top 20 specific on-line application behaviors at the three stages, we sorted them into five categories to form ranking table. It can be seen from the ranking table that students' overall use of the network was relatively stable, the difference was small and students used the internet less frequently at the beginning of the semester compared to the other two stages. Students' main application at the three stages focused on website visits, file downloads, IM, e-mail and entertainment.

Visiting websites was the main application for students. The websites that students visit were relatively concentrated, and they were mainly about search engines, MLM advertising, online shopping, online payment, news portal, BBS, blog, WebBBS, among which the applications of search engine and MLM advertising ranked first and second. According to the number of behavior shown in the table, it can be seen that file download constituted a very large proportion, mainly including P2P behavior and the use of other download tools such as BT, FlashGet, emule. The third largest category of applications was mainly students' real-time information exchange through a variety of tools, such as the most frequently used tools QQ, mobile fetion, MSN, and Yahoo messenger etc.. Ranked fourth was Web mail. Using the internet to send and receive e-mail has become a common online behavior in students' day-to-day school life. Network Entertainment, was the fifth largest category of major applications. Playing online games, watching on-line movies, and listening to on-line music have become options for students to relax and entertain themselves after school.

2.2 *The areas of on-line applications*

Table 1. The areas of on-line applications.

the beginning of the semester		the middle of the semester		the end of the semester	
areas	times	areas	times	areas	times
finance	32499	finance	24407	finance	22568
IT	16436	IT	15663	IT	18068
education	15513	education	7851	education	5459
life-related	5245	life-related	4564	life-related	5848

Network applications have gone deep into every aspect of people's study, work, life, and entertainment. There is no exception into students'. According to the features of the websites the students visited, finance, IT, education and life-related applications were the four areas that our students concerned most.

2.3 *The website visit URL ranking*

Table 2. The website visit URL ranking.

URL	type	times	total
www.baidu.com	search engines	58559	82075
www.google.com.hk		11842	
www.Sogou.com		5265	
dict.youdao.com		3392	
hao123.com		2272	
www.soku.com		745	
login.taobao.com	Online shopping	3116	3116
mail.qq.com	Web-mail	1172	1930
mail.bcbuu.edu.cn		758	
alipay.com	Online payment	1622	1622
www.renren.com	Dating chat	1366	1366
t.sina.com.cn	Forum blog	757	757
www.sina.com.cn	News portal	630	630

Visiting websites was the first application type for students. Getting access to the website URL ranking made us have a better understanding of what websites students have visited. From the ranking it can be seen that search engine was the most widely used at the stage of mid-semester. Baidu, Google and Sogou were the widely used search tools, YOUDAO dictionary was the widely used online dictionary, hao123 was the most popular navigation website, and SOKU was the widely used search video website. Modern college students have adapted to online shopping, most of them have their own online banking and online payment. Taobao was the most popular on-line shopping website. The most widely used e-mail systems were the qq-mail and college's mailbox. The most popular SNS was RenRen. Sina microblogging held first place in websites of the forum Blog, and Sina was the most popular news portal.

2.4 *Online duration and time*

Judging from the entire semester, students' on-line behaviors were not significantly different at the three stages. In the number of on-line applications, the number at the beginning of the semester was less than the numbers in the middle of the semester and at the end of the semester. In the specific application types, there was no difference at each stage.

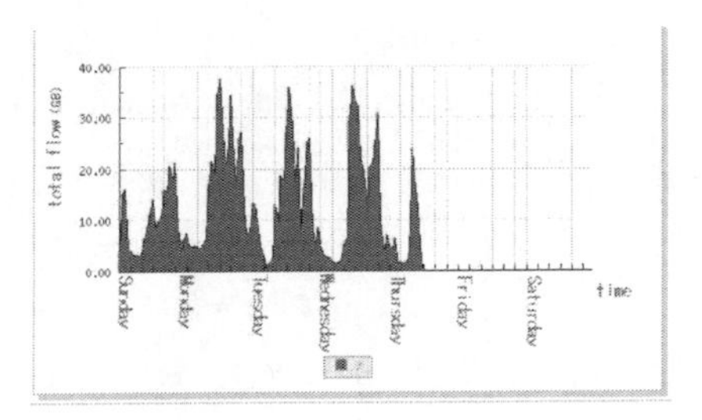

Figure 1. The students' on-line time distribution trends in a week.

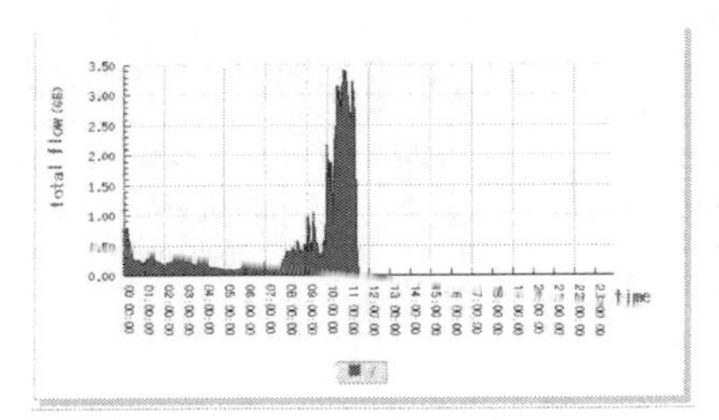

Figure 2. The students' on-line time distribution trends in a day.

According to the statistics of students' daily on-line time and weekly on-line time in one semester, the distribution diagrams for students' daily on-line time and weekly on-line time can be obtained. It can be seen from Figure 1 and Figure 2, our students' online time mainly focused on the period from Sunday to Thursday. Monday, Tuesday and Wednesday were more focused. The online time in a day was mainly in the morning.

2.5 *Keywords*

On the other hand, keywords which the students retrieved through a search engine can also reflect their concerns. Beijing Union University, Beijing, network, download, Postgraduate examinations, Business, Renren.com, and PPT were long-term focuses for the group. This fully showed that the content related to the college was their focus, Renren was the students' favorite SNS, and PPT was an indispensable skill in daily learning.

3 CHARACTERISTICS OF STUDENTS' ONLINE BEHAVIORS

3.1 *Network application deep into vaious aspects and fields*

Network has been deep into students' study and life. For the object of this study selected, the students focused more on the areas of finance, IT, education, and life.

3.2 *Interactive behaviors and non-interactive behaviors*

Students' online behaviors were both interactive and non-interactive. From the number of on-line behaviors, there was no big difference between interactive and non-interactive. Interactive behavior was mainly reflected in the application of real-time communication tools, such as QQ, mobile fetion, MSN and Yahoo Messenger. The applications of File download and search engines were major non-interactive behaviors.

3.3 *The tendency of specific applications*

The students tended to use search engines to find interesting topics, such as Baidu, Google, and to visit Sina news portal and Sina microblogging more often. They preferred to use qq-mail and college's mailbox, chat in the RenRen, use Youdao Dictionary and Google Translate tools, and visit the shopping site Taobao.

3.4 *The main contents concerned*

The theme the students searched was related to their college, Postgraduate examinations, business and operations, and daily assignments and coursework. WEBBBS posting contents highlighted the importance of the completion of the day-to-day school learning tasks.

3.5 *Online duration and time*

Students' on-line time at college was closely related to the time of the course. The more courses they had, the longer the time they spent on-line. They had a longer time online in the morning.

3.6 *Other applications*

The application of MLM advertising among students was significant.

4 ANALYSIS OF THESE FEATURES

4.1 *Network teaching requirements*

Network teaching has spread to ordinary colleges and universities and built an interoperable platform for the communication between teachers and students. Our college required all the courses except the elective ones use the network to assist teaching and students use the network to complete their learning tasks, which was an important reason to promote students' network application. In addition, the college compressed the face-to-face time for the curriculum in experiment district, strengthened the proportion of network learning, which also prompted the students to use the network to complete the various disciplines of the learning task.

4.2 *The impact of the learning environment*

Our college doesn't have campus accommodation for our students. Most students are from Beijing and go back home on weekends, so students surf the internet at home or off-campus in their spare time. The first-year students and the second-year students go to night classes on campus. The rate of Internet access is low, thus a unique distribution of students' on-line time has been created.

4.3 *Development and limitations of network technology*

There were a large number of applications in MLM advertising for the students. We could not determine whether the types of students' discipline and the application had a significant relationship, but it could be sure that there was a close relationship between these characteristics and the development of network technology. For example, online advertising push has direct impact on students' click and view of the ads. This, to some extent, explained those students had limited ability to avoid network garbage interference.

Strategies on improving students' on-line behavior

In summary, as the representative of the college business students, the students at BCBUU had positive online behaviors and their primary need was to meet the day-to-day learning tasks.. At the same time, the network has gone deep into students' study, life, entertainment, making friends etc. On the premise of strengthening the current good momentum, the college should further focus on how to make students' on-line behaviors more efficient, healthier, enhance the instrumental role of the network and use the virtual space created on the network to influence students' thinking and behavior. The following strategies can be used to guide such students to improve their on-line behavior.

4.4 *Use technical means*

By using the core switch, and the on-line behavior management software, the unhealthy websites will be blocked so that students can browse more active and effective information in the limited time [3].

4.5 *Focus on curriculum*

Through the curriculum, we can further strengthen to cultivate students' information literacy to enable them to improve the ability to retrieve and identify valid information.

4.6 *Offer good service*

We can regularly collect business related information, process and classify the resources according to a certain theme or topic, provide students with a quick search to save their time online and guide them to surf the net in a healthy way.

4.7 *Strengthen education and guidance*

Guide students to handle the relationship among surfing internet, study and life, and create a positive atmosphere for network behavior in which everyone can supervise and help each other [4]. For example, we can guide students who are addicted to online games to recognize that after learning statically indoor for a long time, more dynamic ways such as outdoor sports should be taken to relax or release pressure, which are healthier and more effective than sitting in front of the computer in the room, and immersing in the virtual space.

ACKNOWLEDGMENT

This study was funded by "The Research on the Support of the Students in Business colleges for AACSB International Accreditation" which was the Beijing Union University of education reform project. Thank you very much for this study to provide funds, and also thank the other members of the research group. In the course of the study, some literature was referred, thanking the authors.

REFERENCES

[1] Zhou Lin, "The Research on the Internet-Behaviors Preferences of Undergraduates," Shanghai: Shanghai Normal University, 2005, May, pp.3–4.

[2] John Suler. "Healthy and pathological Internet use Cyberpsychology and Behavior," To get what you need, 1999, February, pp.385–394.

[3] Lei Fang, "Our students Internet survey analysis and guidance countermeasures," Library World, 2008, April, pp.30–31.

[4] Zhou Bo, "Present Situation Survey on University Students' Internet Using——Colleges and Universities in the Kunming Area as the Case, " Journal of Kunming University,2009, vol.31, pp75–77.

[5] Li Xiu-min and YIN Guo-en, "Consideration on Relation Between Online and Personality characteristics for College Students," Psychological Development and Education, 2004, January, pp.34–37.

[6] Liang Duo-hong, Shi Xin-zhu and Wang Feng-zhi, "The analysis of on-line behaviors and their relative factors among medical college students," Modern Preventive Medicine, 2005, Vol.32, pp.1644–1645.

Computing, Control, Information and Education Engineering – Liu, Sung & Yao (eds)
© 2015 Taylor & Francis Group, London, ISBN: 978-1-138-02800-5

Exploration and analysis on the reform of NCO English teaching in military academies under the condition of informationization

Guang Hong Pei, Yi Ding, Yao Yao &Yu Bin Wang
Engineering Academy, PLA, Xuzhou, Jiangsu, China

ABSTRACT: The application of computer technology in English teaching in military academies has the function of stimulating the students' studying interest. As a kind of teaching method, computer technology is the supplementation and development of English teaching method when it is applied in English classes, which will improve the teaching efficiency of English classes, and have significant meaning in cultivating the students' comprehensive ability in English academies.

KEYWORDS: Promotion of information technology, English Teaching, The education of Military Academies.

1 THE CONNOTATION OF THE INFORMATIONIZATION OF EDUCATION IN MILITARY ACADEMIES

The informationization of education in military academies is an important part of the informational reform of our army, which means fully applying the integrated information and communication technologies to the military education system. It is a series of processes that realize the development and utilization of military academies' informational resources as well as the unique and significant theoretical organizing and managing methods in educational field. It has the following four contents:

First, informational resources are the core of the informationization of education in military academies;

Second, the purpose of the informationization of education in military academies is the widespread application of informational resources and informational technology;

Third, the network is the basis for effectively delivering information in a wide range;

Fourth, as a part of the educational reform of military academies, informationization will surely be influenced and restricted by people's concepts, ideals, wills, skills, group interests, social organizations and structures as well as many other factors. Therefore, there should be corresponding safeguard mechanism for educational informationization.

2 TRADITIONAL MODES OF ENGLISH TEACHING IN MILITARY ACADEMIES FACES REFORM

In traditional modes of English teaching in military academies, the English teaching concept and method are relatively old-fashioned, and the teaching of English pronunciation, vocabulary and grammar is conducted without specific language environment, the teachers explain in "broadcast" style, and the students accept passively, so the lack of two-way communication makes it difficult to stimulate the students' interest and bring their subjective initiatives into play.

In recent years, our country especially emphasizes that the guiding ideology of English teaching should be turned from teachers-centered mode into students-centered mode, and the important point of teaching should be changed from "how to teach" to "how to learn". But the target will be hard to achieve if we still follow the old mode of teaching, especially under the circumstance that to a large extent, the purpose of teaching is to pass the exams. While the introduction of computer-assisted language learning (CALL) and web-assisted language learning (WALL) has solved this problem. Through taking advantage of information technology, during the teaching processes, besides teacher-students study, there will be students-computer mutual communication, which provides the guarantee for transforming from teachers-centered mode to students-centered mode. In the developing process of the informationization of English teaching, CALL and WALL are bound to bring major changes to English teaching.

3 THE APPLICATIONS OF INFORMATION TECHNOLOGY IN ENGLISH TEACHING

There are many advantages of computer-assisted language learning, for example, it has a large amount of information, it is interactive, vivid and interesting, it is practical, it provides the latest knowledge, and hence it has great influences on listening, speaking, reading, and writing in English teaching. Specific manifestations are as follows:

3.1 *Applications in teaching English listening*

1 Using magnetic storage media to store listening materials. External hard drives, USB flash drives and other storage media have large capacity, are portable, easy to save, and quick to copy. Moreover, there are abundant software and online listening resources, which provide more choices for students than before.
2 Using multimedia computers to play the students the listening materials. By mixing the texts, images and sound together, this method is vivid and lively, and stimulates the students' interest of learning and hence can solve the problems in listening for them.
3 Selecting authentic English listening software with pure pronunciation to clear away the accent obstacles for students' English learning.

3.2 *Applications in teaching oral English*

1 Man-computer interaction. Students can choose some software to practice their own pronunciation, intonation and ways of expression independently. They can imitate the contents played by the computers in front of the microphones, and the computers can conduct feedbacks to motivate the students.
2 Online chatting. With the teachers' guidance, students can communicate in English with English speakers or even foreigners through the internet according to their own language levels and preferences, they can also choose different chatting contents and chatting partners.

4 THE COMBINATION OF INFORMATION TECHNOLOGY AND ENGLISH COURSES

4.1 *The combination of the information technology and English courses based on computer courseware*

1 Courseware for classroom demonstration. There are some principles for making this kind of courseware: the courseware should be designed to reflect teaching purposes and be helpful for solving the important and difficult points in the teaching contents, it should have rigorous structure, friendly and unified interface, and should be controlled easily. Besides, the emphasis of the technology should be laid on stressing the teaching contents and improving the teaching effect.
2 Interactive courseware. The ideology of making this kind of courseware is to provide the students with learning materials of different levels according to different students' mastery of knowledge, and the students can select the learning materials according to their own needs, or evaluate that whether they have the ability to carry out certain kind of learning through interactive exercises.
3 Electric books. They are suitable not only for teachers' teaching but also for students' independent learning.

4.2 *The combination of information technology and English courses based on the internet*

1 Online classroom teaching model that centers on students' learning with the procedure of creating situations→ proposing subjects→ exploring the subjects independently→ cooperating online→ briefly summing up for the subjects→ testing online→ extending the subjects. In this mode, teachers are the designers and instructors during the teaching process, they develop and constantly perfect the teaching resources, and they are also the evaluators and managers of the teaching resources. The students are in the center of this mode, and they can choose the modules and ways of learning that are suitable for themselves. The advantage of this mode is that it is beneficial for shaping the students' personalities and protecting their self-esteem; while the disadvantages is that it is limited to classroom teaching so teachers and students seldom have emotional communication.
2 Web quest mode with the procedure of teachers putting forward a question→ dividing the students into several groups and every group plays a role→ students carrying out their own tasks (looking for

materials that they need, discussing, analyzing, organizing materials and writing down reports) → making a speech to report→ teachers' assessment→ next question. This mode attaches importance to the training on methods and skills, the cultivation of students' abilities of using knowledge and their learning enthusiasm.

All in all, the development of information technology has promoted the improvement of modern educating methods in military academies, and through exploring the modernization of education and the reform of teaching, the cognition of students of military academies on English curriculum will be significantly enhanced. As teachers in military academies, we still have to grope our way forward on modernization reform of teaching in military academies and make our efforts to improve the classroom environment through informational education. This is also the responsibility and mission for every educator in military academies.

Computing, Control, Information and Education Engineering – Liu, Sung & Yao (eds)
© 2015 Taylor & Francis Group, London, ISBN: 978-1-138-02800-5

Author index